CONCISE ENCYCLOPEDIA OF

MATERIALS PROCESSING

Elsevier Science and Technology Homepage - http://www.elsevierdirect.com
Consult the Elsevier homepage for full catalogue information on all books, major reference works, journals, electronic products and services.

Elsevier Titles of Related Interest

K.H.J. BUSCHOW
The Encyclopedia of Materials: Science and Technology (2001)
Major Reference Work
ISBN-13: 978-0-08-043152-9

R. CAHN
Concise Encyclopedia of Materials Characterization (2004)
ISBN-13: 978-0-08-044547-2

K.H.J. BUSCHOW
Concise Encyclopedia of Magnetic and Superconducting Materials (2005)
ISBN-13: 978-0-08-044586-1

A. MORTENSON
Concise Encyclopedia of Composite Materials (2006)
ISBN-13: 978-0-08-045126-8

J. MARTIN
Concise Encyclopedia of Structure of Materials (2006)
ISBN-13: 978-0-08-045127-5

J. MARTIN
Concise Encyclopedia of Mechanical Properties of Materials (2006)
ISBN-13: 978-0-08-046525-8

Concise Encyclopedia of Chemical Properties of Materials (2007)
ISBN-13: 978-0-08-046526-5

Concise Encyclopedia of Properties of Materials Surfaces and Interfaces (2008)
ISBN-13: 978-0-08-054811-1

Related Journals:
Elsevier publishes a wide-ranging portfolio of high quality research journals, encompassing the materials science field. A sample journal issue is available online by visiting the Elsevier web site. Leading titles include:

Materials Letters
Materials Science and Engineering: A
Materials Science and Engineering: B
Materials Today
Mechanics of Materials
Metal Finishing
Metal Powder Report
Progress in Material Science
Reinforced Plastics

All journals are available online via ScienceDirect: www.sciencedirect.com

To contact the Publisher
Elsevier welcomes enquiries concerning publishing proposals: books, journal special issues, conferences proceedings, and all other media. Should you have a publishing proposal you wish to discuss, please contact the Publisher responsible at http://www.elsevierdirect.com.

General enquiries, including placing orders, should be directed to the relevant Regional Sales Offices at the Elsevier website.

CONCISE ENCYCLOPEDIA OF

MATERIALS PROCESSING

Editor

JOHN MARTIN

University of Oxford, United Kingdom

ELSEVIER

Amsterdam—Boston—Heidelberg—London—New York—Oxford
Paris—San Diego—San Francisco—Singapore—Sydney—Tokyo

ELSEVIER Ltd
The Boulevard, Langford Lane
Kidlington, Oxford OX5 1GB
UK

2009

Library of Congress Catalog Number: 2009934947

ISBN: 978-0-08-096492-8

Printed and bound in the United Kingdom

09 10 11 12 11 10 9 8 7 6 5 4 3 2 1

CONTENTS

U

V

W

Z

EDITOR'S PREFACE

The Concise Encyclopedia of Materials Processing draws together in a single volume a range of articles which appear in the *Encyclopedia of Materials Science and Technology (EMSAT)*. This very substantial work first appeared in 2001, the editors-in-chief being Jürgen Buschow, Robert Cahn, Mert Flemings, Bernhard Ilschner, Ed Kramer and Subbash Mahajan; it has been kept up-to-date by its editors-in-chief in the form of periodic updates loaded on to the internet version of EMSAT.

In the present volume we have deliberately excluded articles devoted to the processing of semiconducting materials which appear in EMSAT; and we have also excluded a consideration of the primary production of engineering materials (these appear in *The Concise Encyclopedia of The Chemical Properties of Materials*). We have concentrated on those processes aimed at producing final useful products, so we have included crystal growth, various casting techniques, deformation processing, heat-treatment, machining and joining techniques, as well as manufacturing by the production and consolidation of powders.

The processing of metallic materials, ceramics and glasses, as well as the forming of polymeric materials have been considered. An article on the chemical treatment of wood has also been included.

137 articles have been selected which discuss the processing of these groups of materials, with probably the greatest emphasis being placed upon metals and alloys. The whole assembly provides the reader with an up-to-date understanding of the processing of the entire range of materials of engineering importance.

John W. Martin
Editor

ALPHABETICAL LIST OF ARTICLES

A

Aluminum Alloys, Heat Treatment of

Controlled heat treatment of aluminum alloys can significantly influence properties such as strength, ductility, fracture toughness, thermal stability, residual stresses, dimensional stability, and resistance to corrosion and stress corrosion cracking. The main heat treatment procedures are homogenization, annealing, and precipitation hardening involving solution treatment, quenching, and aging.

1. Homogenization

The as-cast microstructure of direct chill (DC) cast ingots normally displays significant segregation and supersaturation. Microsegregation in the cast ingots, i.e., segregation on a microstructural scale, is manifested by coring in the dendrites and also nonequilibrium eutectics and intermetallic phases at the grain boundaries. Coring describes the gradual change in composition from the center to the edge of dendrites and grains. Elements such as iron, chromium, zirconium, and manganese have a low solid-state diffusion rate and, therefore, exist in supersaturated solid solution in the aluminum dendrites owing to the high cooling rates during solidification.

The main aim of homogenization is to improve workability and mechanical properties by dissolving the nonequilibrium brittle interdendritic constituents and by providing a more homogeneous structure throughout the ingot. Some of the intermetallic phases are only spheroidized or transformed into more stable phases. In addition, the controlled precipitation from the supersaturated dendrites during homogenization is of major importance for the recrystallization and grain growth behavior of the alloys during subsequent processing. In particular, chromium, zirconium, and manganese precipitate small particles, dispersoids, of 10–100 nm, which are uniformly distributed in the microstructure.

Homogenization is normally performed on both heat-treatable and non-heat-treatable wrought aluminum alloys. It is carried out at temperatures between the solvus and the solidus, ~450–600 °C, for extended periods of time, 2–10 h. The heating rate of the product is critical to avoid local melting of the precipitates which may cause brittle constituents, voids, microporosity, blistering, cracking, or heavy external oxidation. The homogenization temperature and time influences the extent of the diffusion processes. A low cooling rate after homogenization ensures workability by re-precipitation and coalescence of solute as relatively coarse particles. Homogenization treatment in continuous furnaces where the temperature history of each individual billet can be controlled and monitored are now common.

2. Heat Treatment and Temper Designation

A heat treatment and temper designation system, Table 1, has been developed by the Aluminum Association to describe the processing of wrought and cast aluminum alloys. This does not apply to remelt ingots.

3. Annealing

Different types of annealing treatments with different objectives exist depending on the alloy system and prior processing. Full annealing (O temper) produces the softest, most ductile, and workable conditions for both heat-treatable and non-heat-treatable wrought alloys. Reduction or elimination of the strengthening from cold working is accomplished by heat treating at a temperature between ~250 °C and 450 °C for times ranging from a few seconds to several hours. The exact time and temperature depends on the amount of prior cold work and solute concentration. To remove precipitation hardening effects, the annealing treatment is designed to dissolve the small hardening precipitates. The annealing temperature should not exceed 415 °C to avoid oxidation and grain growth. Heating and cooling rates must be controlled to avoid precipitation hardening/softening in heat-treatable alloys. Relatively slow cooling is recommended for all alloys to minimize distortion. However, for heat-treatable alloys, a slow cooling rate will result in the formation of coarse precipitates.

Annealing of cold-worked, non-heat-treatable, wrought alloys to obtain intermediate mechanical properties (H2 tempers) is known as partial annealing or back annealing. During this process, the material undergoes recovery, partial recrystallization, or full recrystallization. Partial annealing results in improved bendability and formability compared to the alloy at a similar strength level in a H1 temper. It is important to have close control of temperature to achieve uniform and consistent mechanical properties.

Annealing of castings is not common. However, it may be used to provide maximum dimensional stability for elevated temperature service and a high ductility. When used, the treatment involves 2–4 h at

Table 1
Heat treatment and temper designation system for aluminum alloys.

Suffix	Treatment and/or condition		
F	As-fabricated		
O	Annealed		
H	Strain hardened by cold work		
	H1 Cold worked	2nd digit	Hardness
	H2 Cold worked and partially annealed	2	1/4 hard
	H3 Cold worked and stabilized	4	1/2 hard
		6	3/4 hard
		8	hard
		9	extra hard
T	Heat treated to a stable condition, excluding annealing (O)		
	T1 Cooled from an elevated temperature forming process (partial solution) followed by natural aging		
	T2 Cooled from an elevated temperature forming process, cold worked and naturally aged		
	T3 Solution heat treated, quenched, cold worked and naturally aged		
	T4 Solution heat treated, quenched and naturally aged		
	T5 Rapidly cooled from elevated forming temperature and then artificially aged		
	T6 Solution heat treated, quenched and then artificially aged		
	T7 Solution heat treated, quenched and overaged		
	T8 Solution heat treated, quenched, cold worked and then artificially aged (amount of cold work in % indicated by subsequent digit)		
	T9 Solution heat treated, quenched, artificially aged and then cold worked		
	T10 Cooled from an elevated temperature forming process, cold worked and then artificially aged		
W	Unstable temper applied for alloys which age spontaneously at room temperature after solution treatment. (Only specific if followed by the time of natural aging.)		

315–345°C to provide the optimum relief of residual stresses and precipitation of phases formed by any excess solute retained in solid solution in the as-cast state.

4. *Solution Treatment*

Solution treatment to achieve the maximum concentration of solute in solid solution is carried out at a sufficiently high temperature for sufficiently long times to produce a nearly homogeneous solid solution. The temperature is determined based on the composition and maximum solid solubility, and is normally achieved by heat treating near, but below, the eutectic temperature. It is important to avoid local melting both during heating, i.e., heating sufficiently slowly to redissolve nonequilibrium precipitates, and at the solution treatment temperature. Normally, temperature variations within ± 6°C are allowed, but may be even stricter for some high-strength aluminum alloys. Underheating results in incomplete solution and reduces the strengthening potential of the alloy. Solution treatment time depends on microstructure, section thickness, and furnace loading and can vary from less than a minute to 20 h. Generally, the soak times for castings are longer than for wrought products due to coarser microstructures. Aluminum–silicon casting alloys in which the eutectic silicon is modified by the addition of elements such as strontium, sodium, or antimony, undergo rapid spheroidization of the silicon particles, while complete spheroidization is never achieved in unmodified alloys.

Some products, particularly those which have been cold worked, may be susceptible to recrystallization and grain growth during solution treatment which may, in some cases, add restrictions on the allowable solution treatment temperature and time.

Solution treatment is normally performed in air, but molten salt baths or fluidized beds can be used to obtain more rapid heating. Distortion due to creep must be considered in designing the furnace loading equipment. High-temperature oxidation causing surface blistering and rounded pores within the material can result if the solution treatment is performed in an atmosphere with high moisture content and is aggravated by sulfur containing gases.

5. *Quenching*

Subsequent to solution treatment the material must be cooled at a sufficient rate to retain the solute in solid solution and a high number of vacancies at room temperature. Precipitation during cooling could result in coarse particles that are detrimental to mechanical and corrosion properties. It also reduces the level of supersaturation and, therefore, the response to subsequent age hardening. Generally the highest attainable strength levels, the best combination of strength and toughness, and best corrosion resistance and resistance to stress corrosion cracking are, therefore, associated with the most rapid quenching rates. The sensitivity of alloys to quench rate and the allowable delay between solution treatment and quenching is determined from quench factor analysis which is based on avoiding the tip of the nose of the continuous cooling transformation (CCT) diagram. Precipitation kinetics is a compromise between degree of supersaturation and diffusion rate which depend on temperature in opposite ways, resulting in maximum nucleation and growth occurring over a critical temperature range which is between 290°C and 400°C for most alloys. Quench sensitivity is generally higher for higher solute levels and in alloys containing dispersoids which can act as nucleants for coarse precipitates.

Water is normally used as the quenching agent. Components with complex shapes and with different section thicknesses, such as castings and forgings, are often cooled at slower rates to avoid distortion and excessive residual stresses. This can be achieved by using water at 65–80°C, boiling water, polyalkylene glycol, or some other fluid medium such as forced air or mist.

Forming, straightening, or stress relief directly after quenching is advantageous as the material is almost as soft as in the annealed condition.

6. Aging

Aging is the controlled decomposition of the supersaturated solid solution to form finely dispersed precipitates in heat-treatable alloys, usually by soaking for convenient times at one or sometimes two temperatures. The decomposition is normally complicated and occurs through several precipitate stages. These finely dispersed precipitates have a dominant effect in raising yield and tensile strengths, and may also affect other properties such as corrosion resistance and dimensional stability. The age hardening response of some alloys is significantly improved by introducing a controlled amount of cold work prior to aging. Some alloys retain a supersaturated solid solution and high vacancy concentration, without quenching, by rapidly cooling the material after mechanical working at elevated temperatures.

6.1 Natural Aging

Natural aging refers to the decomposition of the solid solution that occurs at room temperature. Most alloys age harden at room temperature and some alloys decompose over a few days to yield stable and adequate properties for many applications. Natural aging is practiced industrially with alloys that display the most significant changes in microstructure and properties over a period of a few days (<4 days), while the subsequent changes are minor. Other alloys can continue to naturally age for many years. During natural aging the high level of supersaturation and the high vacancy concentration causes rapid formation of clusters or coherent stable Guinier–Preston (GP) zones. In some alloys, notably the 2xxx series alloys, the aging response is greatly improved by cold working the recently quenched material, i.e., T3 and T4 treatments are common for these alloys.

6.2 Artificial Aging

Artificial aging involves decomposition at elevated temperature, normally in the range of 100–200°C, for times between 2 h to 48 h. Multiple aging treatments can also be used. The artificial aging treatment is designed to produce optimum size, distribution, constitution and morphology of precipitates and amount of solute in solid solution. Hardening is often obtained from partially coherent precipitates. Temperature control during artificial aging is important while there are often large tolerances on treatment time. Paint-baking operations of many products are often performed in the artificial aging temperature range and can result in significant hardening. The T6 temper normally yields the highest practical strength combined with useful engineering performance with respect to other specific requirements.

Castings have the highest combination of strength, ductility, and toughness in the T6 temper. The T5 and T7 (overaging) treatments are more commonly used for castings than for wrought products. Overaging often involves heat treating at higher temperatures to carry the alloy beyond the maximum strength level. It results in lower strength and hardness than the T6 temper but yields improved dimensional stability during elevated temperature service. The level of residual stress is reduced during artificial aging, ranging from 10% to 35% for T6, while T7 provides even more substantial reductions. Although the alloys are suitable for age hardening, thermal treatments are normally not beneficial to the mechanical properties of high pressure die-castings, owing to problems such as blistering from entrapped air.

Bibliography

Aluminum Standards and Data 1997 The Aluminum Association. Washington, PA

Aluminum and Aluminum Alloys 1993 ASM Specialty Handbook. ASM International, Materials Park, OH

Chandler H (ed.) 1996 *Heat Treater's Guide, Practices and Procedures for Nonferrous Alloys.* ASM International, Materials, Park, OH

Hatch J E (ed.) 1984 *Aluminum.* American Society for Metals, Metals Park, OH

Polmear I J 1995 Light Alloys, 3rd edn. Arnold, London

A. K. Dahle

Aluminum Alloys: Thermomechanical Processing

Mechanical deformation and heat treatment of aluminum alloys are normally considered separately in both theory and practice. Traditionally, mechanical working operations are designed to change the shape efficiently and economically, and heat treatments are used for structure control to achieve desired physical, mechanical, and chemical properties. However, since one operation can greatly influence the other, manufacturing procedures have been developed that combine mechanical working and heat treatments for microstructure and property control. These procedures, called thermomechanical treatments (TMTs), have been used by the aluminum industry since 1944 when it was found that the aging response of aluminum–copper alloys could be enhanced by mechanical working after solution heat treatment and prior to aging.

TMTs have been developed not only for strength improvements, but also for optimizing grain size and shape for subsequent forming operations, for improved fracture toughness, and for stress-corrosion and fatigue-crack-initiation resistance. For convenience, TMTs are divided into three classifications: mechanical thermal treatments (MTTs), intermediate thermomechanical treatments (ITMTs), and final thermomechanical treatments (FTMTs). MTTs are applied to single-phase or non-heat-treatable alloys to develop a stabilized subgrain structure or a desired texture. ITMTs are used to develop a fine grain structure and are applied to the ingot as part of the rolling operation, prior to the final heat treatment. FTMTs are used to modify the dislocation structure and the hardening precipitates of heat-treatable alloys and are applied to the wrought product after the final hot-working operations.

1. *Conventional Processing*

Beginning with the as-cast ingot, conventional processing includes a homogenization treatment followed by hot working for ingot breakdown and shape change to the appropriate product form. Since aluminum and its alloys have high stacking-fault energy, sufficient dynamic recovery normally occurs during hot deformation (i.e., at temperatures above 0.5 T_{m}) to give rise to a stable polygonized substructure. The grains elongate in the direction of flow and do not recrystallize. Most alloying additions are usually in solution during hot working but the dispersoid-forming elements have already precipitated. The dispersoids, and any constituents and primary phases that may be present, are strung out in the working direction; their spacing increases in the working direction and decreases normal to the working direction. In some cases large constituent phases, if present, are broken up during the working operation. The fine distribution of dispersoids delays or prevents static recrystallization and aids in retaining the elongated or "pancake" shaped grains during subsequent processing. The wrought product is heat treated to develop the desired microstructure—that is, annealed for stress relief or solutionized, quenched, and aged. The elongated grains developed during hot working persist after this final treatment, although some static recrystallization may occur during solutionizing if the material is highly strained. The pancake structure is considered responsible for poor fracture resistance in the short transverse direction of the plate material, but is beneficial for stress-corrosion resistance of thin sheet products.

2. *Mechanical Thermal Treatments*

Since engineering materials are exposed to mechanical treatments as part of their fabrication, work hardening is an important method of producing high strengths in single-phase or non-heat-treatable alloys. Conventionally, work-hardening aluminum alloys are designated by the H temperature nomenclature. However, processes developed especially to produce a stabilized subgrain or cell structure for strength improvement are frequently referred to as MTTs. This terminology differentiates them from TMTs that are used to modify grain structure and/or precipitate distribution. MTTs involve the introduction of a high dislocation density during cold- or warm-working operations and a modification of the dislocation distribution by some form of post-deformation recovery heat treatment.

The final thermal treatment rearranges the dislocations into more stable configurations, normally called low-energy dislocation structures (LEDS), without excessive reduction in the overall dislocation density, and allows solute or dispersoid pinning of the substructure. The subgrain and cell size decreases with increasing dislocation density which increases with the amount of deformation. For dislocation subgrain and cell structures, similitude leads to the widely observed empirical relationship

$$\tau - \tau_o = \frac{KGb}{D}$$

where τ is the resolved shear stress, τ_o is the frictional stress, K is ~ 10, G is the shear modulus, b the Burgers vector of the dislocation, and D is the average subgrain or cell diameter.

EC aluminum (99.6 wt.% Al) that is used for conductor wire is given a MTT, i.e., strain hardened to the H19 temper, to increase its strength. Other aluminum conductors contain small amounts of alloying additions and are given MTTs to develop a fine subgrain structure that is stabilized by dispersoids. This results in a good balance of electrical conductivity (⩾60% of the conductivity of copper) and good ductility, strength, flexibility, creep resistance, and thermal stability. Aluminum alloys used for the packaging industry must have both high strength and excellent formability. In many cases, this combination is obtained by MTTs that utilize the baking process, required during coating, to form well-defined subgrains. The subgrains lead to more favorable work hardening and homogeneous deformation during subsequent forming operations than are obtainable in unrecovered structures.

3. *Intermediate Thermomechanical Treatments*

ITMTs involve a series of heat treatment and deformation steps directed toward producing a fine grain size by discontinuous precipitation. Depending on the mechanical-working schedule, random textures may also be produced. The processes have been applied to high-strength aluminum alloys prepared by both standard ingot methods and powder metallurgy consolidation. Significant increases in elongation, stress-corrosion resistance, and fatigue-crack-initiation resistance is obtained in ITMT-processed materials when compared with conventionally processed counterparts. However, a reduction in crack-propagation resistance has been associated with the fine-grained structure.

A representative ITMT is shown in Fig. 1. After initial homogenization and hot working, an appropriate overaging treatment is used to precipitate a high density of particles larger than ~0.75 μm in diameter. Cold work can be used prior to the overaging step to improve the distribution of overaged precipitates, by providing a uniform network of dislocations to serve as heterogeneous nucleation sites. The overaging step is followed by mechanical

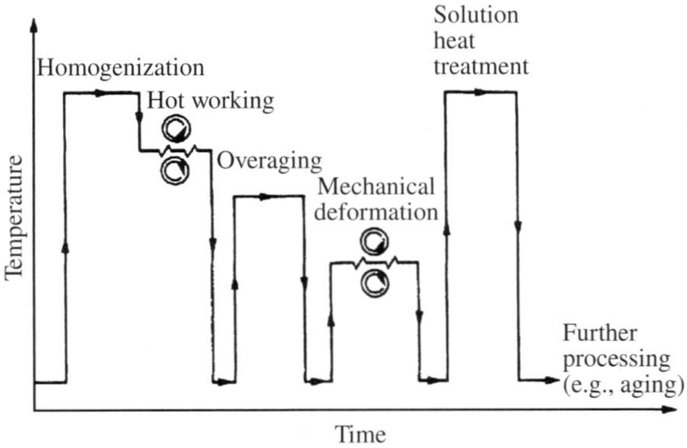

Figure 1
Grain refinement by ITMT processing

deformation at lower than conventional hot-working temperatures to introduce a relatively high degree of work-hardening and deformation zones around the large nondeformable particles. A recrystallization treatment follows, during which the highly strained regions of the deformation zones serve as nucleation sites for recrystallized grains, a process known as particle-stimulated nucleation. The recrystallized grains grow to consume the deformation zones until they impinge or until the drag force exerted on them by submicron dispersoids balances the driving force for grain growth. The submicron particles can also retard or prevent grain growth during subsequent high-temperature working operations by the same mechanism. The recrystallized grain size can be correlated with the density of particles and the amount of deformation, decreasing with increasing values of these parameters to some minimum value. In many cases, the grains developed by ITMT may be sufficiently small to allow superplastic forming.

Superplastic forming normally requires temperatures of at least 0.4 T_m and low strain rates, in the vicinity of $10^{-4}s^{-1}$. Faster strain rates, on the order of $10^{-1}s^{-1}$ have been obtained in some aluminum metal matrix composites. It is generally believed that grain-boundary sliding is the primary deformation mechanism during superplastic flow. However, other processes must also occur to permit accommodation of grain shape changes and contiguity of sliding grains. In any event, the basic microstructural requirement is for small, in the order of 10 μm diameter, equiaxed and stable grains. ITMT processing produces small equiaxed grains in many aluminum alloys, and grain size stability is maintained by dispersoid-forming alloying additions.

4. *Final Thermomechanical Treatments*

FTMTs were initially developed for aluminum–copper alloys and consisted of room-temperature deformation after solution heat treatment and before age hardening. The mechanical-working step accelerates decomposition of the supersaturated solid solution by increasing the vacancy concentration, thereby aiding the diffusion necessary for precipitation, and also by increasing the dislocation density and thereby the number of nucleation sites for heterogeneous precipitation. Precipitation pins the dislocations, producing higher strengths than obtained for undeformed alloys. The strengthening-precipitates in aluminum–copper alloys are responsible for large interfacial strains that are minimized by heterogeneous precipitation on dislocations. The effect of dislocations on the nucleation of precipitation is different from that of precipitation on grain boundaries. In grain boundary precipitation, the destruction of portions of the grain boundary surface by the critical nucleus reduces the interfacial energy associated with the nucleation process. On the other hand, the interaction of the critical nucleus with a dislocation makes some of the strain energy associated with the defect available for the reduction of the total free energy change due to the formation of the nucleus. The transformation strain energy associated with the nucleation event is also reduced by nucleation of the precipitate within the strain field of the dislocation.

The combination of cold-work and low-temperature aging to produce a T8 temper often results in an optimum balance of strength and fracture toughness when compared with the peak strength T6 temper for aluminum alloys that have strengthening precipitates with large interfacial strains. This is associated with the dislocations increasing the aging kinetics of the matrix precipitates that produce the high strength. However, the deformation has no effect on nucleation and growth of grain boundary precipitates. Consequently, the shorter aging times required for optimum matrix precipitation also reduce the amount of grain boundary particles that lower fracture toughness and ductility, resulting in an optimum balance of strength and fracture toughness. Attempts to strengthen aluminum–zinc–magnesium alloys in a similar manner, i.e., by cold-working prior to aging, has been unsuccessful since the strengthening precipitates in this class of alloys do not create large interfacial strains, and heterogeneous nucleation on dislocations is infrequent.

FTMTs have been applied successfully to many aluminum alloy systems. Optimum processing requires that the quench temperature, degree of deformation, temperature of deformation, and aging times and temperatures be controlled so as to obtain both a uniform deformation structure and a uniform precipitate distribution.

5. *Age Forming*

Often the complex shapes of aluminum parts made by forging, extrusion, or from machining of plate, prevent stretching or cold-working operations prior to aging. If subsequent forming of these complex shapes is required, distortion often occurs due to the difficulty of achieving uniform forming forces, especially when parts have abrupt changes in thickness. Age forming is a process that offers a solution to these problems since it utilizes the concept of stress relaxation that occurs during artificial aging of precipitation-hardenable alloys by the conversion of elastic strain to a plastic state. The forming operation is usually done in an autoclave that applies fluidized deformation forces uniformly about the entire surface of the part; thereby overcoming the forming limitations normally associated with complex shaped parts. Deformation is applied at the aging temperature and aids in the nucleation of strengthening precipitates. The final aged structure also has lower residual stresses as compared to structures formed by conventional means. This improves the long-term performance of the part since it improves the resistance to both fatigue and stress-corrosion-cracking.

6. *Powder Metallurgy Processing*

Rapid solidification and powder metallurgy consolidation may be considered to be a form of TMT since it combines thermal and mechanical processes for microstructure and property control. During conventional casting of alloys that form two or more phases in the solid state, large primary crystallites of intermetallics may form which are deleterious to mechanical properties. During rapid solidification, extended ranges of solid solution are obtained. In addition, any primary crystallites that form are very small in size, 0.5 μm or less in diameter, and are generally beneficial to mechanical properties. Impurities, which give rise to large undesirable inclusions or constituent particles, may be tolerated in rapid solidification processing (RSP) alloys because, again, the resulting particle size is small. The process permits the manufacture of high-strength and high-temperature aluminum alloys. Fabrication typically consists of prealloying, atomizing, cold compaction, hot pressing, extruding into billets, and further processing similar to that used for ingot metal. The atomizing allows solidification rates in the range 10^2–$10^6\,\mathrm{K\,s^{-1}}$ and is used to reduce or eliminate segregation, reduce the grain size, extend solid solubility, and refine the dispersoid particle size. The small grain sizes give increased yield strength as well as improvements in other properties, e.g., stress-corrosion resistance, ductility, fracture toughness, and fatigue-crack-initiation resistance.

7. *Limitations*

TMTs have frequently resulted in significant improvements in such properties as strength, stress-corrosion resistance, fracture toughness, and fatigue-crack-initiation resistance. Simultaneous improvements in several of these properties are sometimes observed, but they often are obtained at the expense of other properties, such as fatigue-crack-propagation resistance. It is interesting to note that in some cases the desirable properties obtained by complex TMTs have been duplicated by simple modifications in chemical composition and heat treatment. Even so, some structures and properties are not economically feasible by manufacturing methods other than TMT.

TMTs have some limitations associated with product size, shape, and fabrication. Large sections cannot be cold-worked, and it is difficult to achieve uniform deformation in complex shapes. Moreover, welding and machining of hard TMT materials may also limit applications. Manufacturing problems arise from the logistics involved in these complex processing schedules. However, the normal processing of some alloy systems routinely incorporates the same controls required by TMT processes. Material properties and cost will make the final determination.

See also: Aluminum Alloys, Heat Treatment of

Bibliography

Arieli A, Mukherjee A K 1982 The rate-controlling deformation mechanisms in superplasticity—a critical assessment. *Metall. Trans. A* **13**, 717–32

Brewer H M Jr, Holman M C 1990 *World Aerospace Technology 1990*. The Sterling Publications International Ltd., London, p. 41

Chia E H, Starke E A Jr 1977 Application of subgrain control to aluminum wire products. *Metall. Trans. A* **8**, 825–32

Fine M E, Starke E A Jr (eds.) 1986 *Rapidly Solidified Powder Aluminum Alloys*, ASTM STP 890. American Society for Testing and Materials, Philadelphia, PA

Humphreys F J, Hatherly M 2004 *Recrystallization and Related Annealing Phenomena*. 2nd edn., Elsevier, London

McElroy R J, Szkopiak Z C 1972 Dislocation-substructure-strengthening and mechanical thermal treatment of metals. *Int. Metall. Rev* **17**, 175–202

Williams J C, Starke E A Jr. 1982 The role of thermomechanical processing in tailoring the properties of aluminum and titanium alloys. *Deformation, Processing, and Structure*. American Society for Metals, Metals Park, OH

E. A. Starke, Jr.

Atomization

The starting material for powder metallurgy consists of small particles (i.e., powder) characterized by a size <1 mm. Elemental metallic and alloy powders can be fabricated by liquid metal atomization, mechanical comminution, chemical reactions, and electrolytic deposition (German 1994). Of these approaches, liquid metal atomization is now the dominant method for the production of a broad spectrum of ferrous and nonferrous metal powders on a commercial scale. Capacity worldwide is estimated to be $>1 \times 10^6$ $t\,yr^{-1}$. The growth in atomization technology, at the expense of the other methods of powder production, reflects the intrinsic benefit of a high production rate, the ability to alloy in the molten metal, and the control that can be exercised in terms of powder characteristics, in particular size, size distribution, and shape.

Atomization is the process by which a liquid is disintegrated into droplets which then solidify. With few exceptions, any metallic element or alloy that can be melted is amenable to atomization. The fundamental principles of atomization are governed primarily by fluid dynamics, heat and mass transfer, and physical metallurgy.

1. Methods of Atomization

There are three modes of atomization in use commercially (Lavernia and Wu 1996, Lawley 1992, Dunkley 1998, Yue and Dunkley 1994): two fluid atomization in which break-up of the molten metal occurs by impingement with water or gas; atomization utilizing centrifugal force for melt disintegration; and vacuum (soluble gas) atomization. In the latter, a melt supersaturated with gas under pressure is exposed to vacuum, resulting in melt

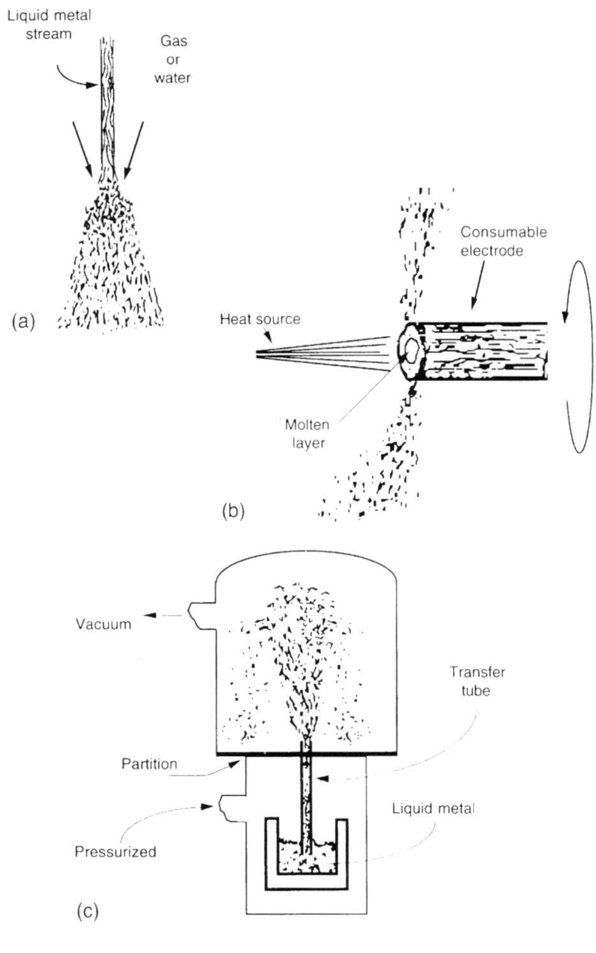

Figure 1
Commercial modes of atomization: (a) water/gas; (b) centrifugal; (c) vacuum (soluble gas)—schematic (reproduced by permission of Metal Powder Industries Federation from 'Atomization: The Production of Metal Powders', 1992).

disintegration. Configurations representative of these three modes of atomization are shown schematically in Fig. 1.

1.1 Water Atomization

In the commercial production of metal powders water atomization is the pre-eminent mode; attendant capacity worldwide is at least 700000 $t\,yr^{-1}$ with production rates up to 500 $kg\ min^{-1}$. Water atomization is used primarily for ferrous compositions but can be used to produce a range of nonferrous alloys. It is lower in cost than other modes of atomization but limitations exist in relation to powder purity, particularly with reactive metals and alloys.

Nozzles used in water atomization are either in the form of discrete multiple nozzles or an annular slit concentric with the metal stream. Annular cone and V-shaped jet geometries are used widely. The "free-fall" configuration is common in which the liquid metal exits from the base of the tundish and falls under gravity before impingement by the jets of water.

Representative operating conditions in water atomization are summarized in Table 1 (Lawley 1992). Commercial water atomized powders are normally irregular in shape with a median particle size of about 100 μm (Fig. 2(a)). Particle size of distributions are relatively broad (10–300 μm). Cooling rate is a function of particle size and is estimated to be $>1\times10^{3}\,°C\,s^{-1}$.

1.2 Gas Atomization

In gas atomization, the stream of liquid metal is disrupted by a high velocity gas (air, nitrogen, argon, or helium). Worldwide annual tonnage of gas atomized powders is much less than that of water atomized powders: about 300,000 $t\,yr^{-1}$ for air atomization and 50,000 $t\,yr^{-1}$ for inert gas atomization of nonferrous alloys. Melt size and melt feed rate ($<120\,kg\,min^{-1}$) are lower than in water atomization.

The nozzles used in gas atomization are either of the "free-fall" configuration or "confined." The latter, used almost exclusively with annular nozzles, enhances the efficiency of the process since there is a rapid decay in gas velocity as the gas moves away from the jet. A schematic of a confined gas atomization nozzle is shown in Fig. 3.

Representative operating conditions used in the gas atomization of a nickel base superalloy are summarized in Table 2 (Lawley 1992). Gas pressure is significantly lower than the water pressure in water atomization. Commercial gas atomized powders are normally spherical in shape with small satellites attached (Fig. 2(b)). Median particle size is in the range of 50 μm to 300 μm. For a given particle size, cooling rate is about one order of magnitude lower than in water atomization.

Table 1
Representative operating conditions in water atomization.

Metal flow rate (single nozzle)	4.5–90 $kg\ min^{-1}$
Water flow rate	110–380 $l\,min^{-1}$
Water velocity (at nozzle exit)	70–230 $m\,s^{-1}$
Water pressure (at nozzle exit)	5.5–21 MPa
Metal superheat	75–150 °C

1.3 Centrifugal Atomization

The primary commercial adaptation of centrifugal atomization is the rotating electrode process (Fig. 1(b)). The metal or alloy to be atomized, in the form of a consumable cylindrical electrode (65 mm diameter × 1524 mm), is rotated at speeds of about 15000 rpm while the end of the electrode (anode) is melted by an arc. The molten metal is ejected by centrifugal force in the form of droplets from the periphery of the bar. A transferred arc helium plasma torch (plasma rotating electrode process) is used in order to avoid contamination by tungsten, characteristic of earlier designs. Helium gas in the working chamber enhances arc stability and convective cooling efficiency.

Powders atomized by this technique are of high purity since there is no containment of the molten metal. The powders are spherical and satellite free (Fig. 2(c)) and in the size range 50 μm to 400 μm with a median particle size of about 200 μm. Droplet cooling rates ($<1\times10^{2}\,°C\,s^{-1}$) are lower than in water or gas atomization.

1.4 Vacuum Atomization

When a molten metal, supersaturated with gas under pressure, is suddenly exposed to vacuum, the gas expands and comes out of solution with attendant atomization of the molten melt. This principle is used in vacuum atomization, a commercial batch process. It is also termed *soluble gas atomization* or *melt explosion atomization*. The method was developed primarily for the production of high purity nickel and cobalt base superalloy powders, but has also been used to atomize alloy powders of aluminum, copper and iron.

The major features of vacuum atomization are shown schematically in Fig. 1(d). The alloy is melted under vacuum and the melt is then pressurized (1–3 MPa) in the lower chamber with a mixture of inert and soluble gas (hydrogen). The molten alloy is then atomized by exhausting the liquid stream into the upper vacuum chamber via a syphon tube. Both the

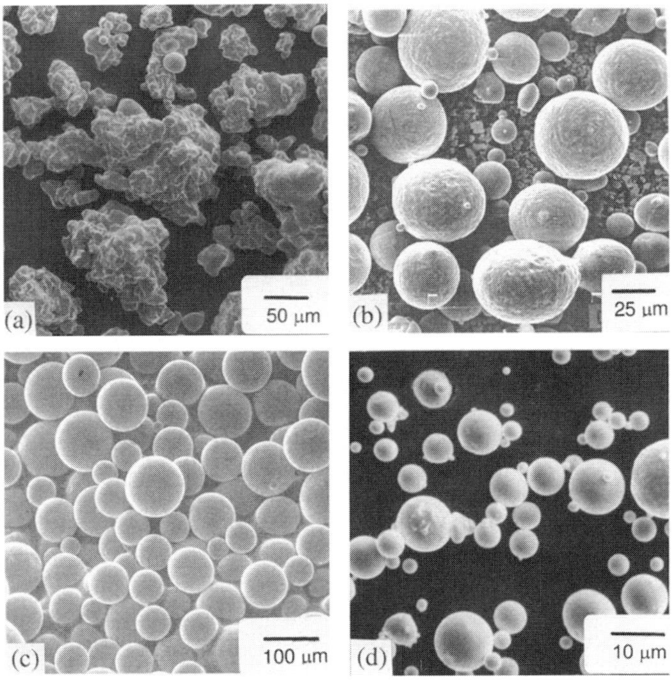

Figure 2
Scanning electron micrographs of atomized powders. (a) water atomized iron powder; (b) gas atomized high carbon steel; (c) centrifugally atomized nickel-base superalloy (plasma rotating electrode process); (d) vacuum atomized nickel-base superalloy (reproduced by permission of Metal Powder Industries Federation from 'Atomization: The Production of Metal Powders', 1992).

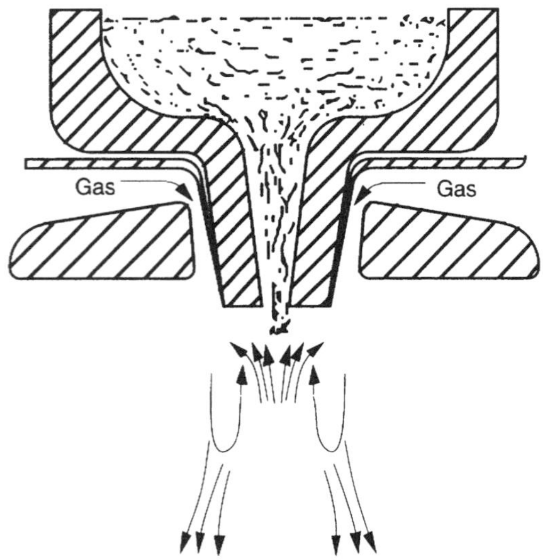

Figure 3
"Confined" nozzle for gas atomization—schematic (reproduced by permission of Metal Powder Industries Federation from 'Atomization: The Production of Metal Powders', 1992).

Table 2
Representative operating conditions in gas atomization[a].

Metal flow rate	20 kg min^{-1}
Gas pressure	2 MPa
Gas velocity	100 ms^{-1}
Gas flow rate	8 $m^3 min^{-1}$
Metal superheat	150°C

[a]Argon atomization of nickel base superalloy; angle between gas jets 80°.

high velocity and gas desaturation provide the energy for disintegration of the molten alloy. Hydrogen concentrations in solution in the molten alloy are in the range 0.0001 to 0.001 w/o.

Vacuum atomized powders are smooth and spherical with very few satellite particles in evidence. Powder size range is broad (1–500 μm) with a median particle size of 40–70 μm. Because the particles cool primarily by radiation, cooling rates are low ($\sim 1 \times 10^{2}$°Cs^{-1}), depending on particle size. A representative micrograph of a vacuum atomized nickel-base superalloy powder is shown in Fig. 2(d).

2. *Particle Size in Atomization*

Size is an important property of the metal powder since it dictates its utility in subsequent powder metallurgy processes. In commercial methods of atomization, empirical relations have been identified which relate the mean particle size to atomizing conditions and the properties of the molten metal.

In water atomization, the magnitude of the median particle size is dictated by the pressure of the water jets. Water pressure influences water velocity and changes the size of the water droplets, and in turn the size of the metal droplets. Quantitatively, the relation between median particle size (d_m) and the water pressure (P), shown in Fig. 4 for a range of metals and alloys, is (Dunkley 1998)

$$d_m = KP^{-m} \tag{1}$$

where K and m are constants. The magnitude of K is a function of the molten metal or alloy and is a measure of the resistance of the metal stream to disintegration. Examination of Fig. 4 shows that molten copper, copper alloys, and the iron base alloy disintegrate at a lower pressure than aluminum or zinc. For a given level of water pressure, median particle size can vary by a factor of about six, depending on the metal or alloy being atomized.

There is an extensive literature base for gas atomization in terms of the effect of atomizing conditions on median particle size. The dependence of median particle size on gas pressure is described by a relation similar to that for water atomization. In gas atomization, the ratio of the mass flow rate of metal to the mass flow of gas is important; median particle size decreases as this ratio decreases.

For a given nozzle design, specific gas consumption (measured as the volume of gas per unit weight of powder produced) is an important parameter. Data relating median particle size (d_m) to specific gas consumption for several metals and alloys are shown in Fig. 5; the line conforms to an equation of the form (Dunkley 1998)

$$d_m = KF^{-1/2} \tag{2}$$

where K is a constant and F is the specific gas consumption in $m^3 kg^{-1}$.

Metal powders produced by the rotating electrode process exhibit narrower size distributions than gas atomized powders. Median particle size decreases with increasing rotational speed of the consumable electrode.

Based on the mechanism(s) of disintegration of the molten metal stream, atomization models have been developed which predict the median particle size.

In water atomization, Grandzol and Tallmadge (1973) derived a relation between median particle size (d_m) and water velocity (V_W) using a simple two fluid atomization model:

$$d_m = A/V_W \tag{3}$$

where A is a constant. The same relation was established experimentally. The model reflects the dispersed nature of the water and an inverse proportionality between d_m and the component of water droplet velocity normal to the liquid metal stream. Conceptually, this is consistent with droplet formation by impact, rather than shear as illustrated

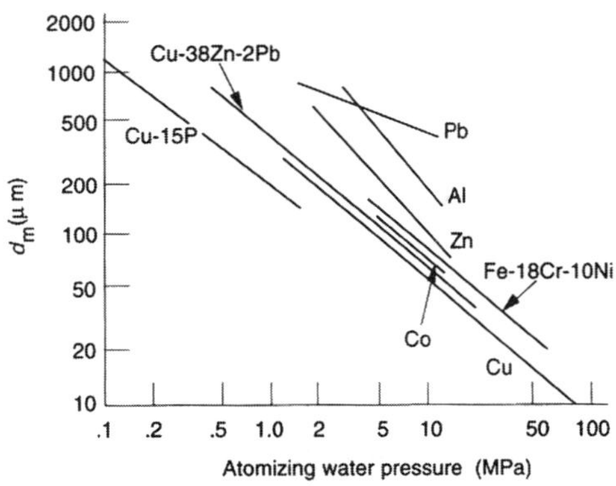

Figure 4
Median particle size (d_m) as a function of water pressure (reproduced by permission of Metal Powder Industries Federation from 'Atomization: The Production of Metal Powders', 1992).

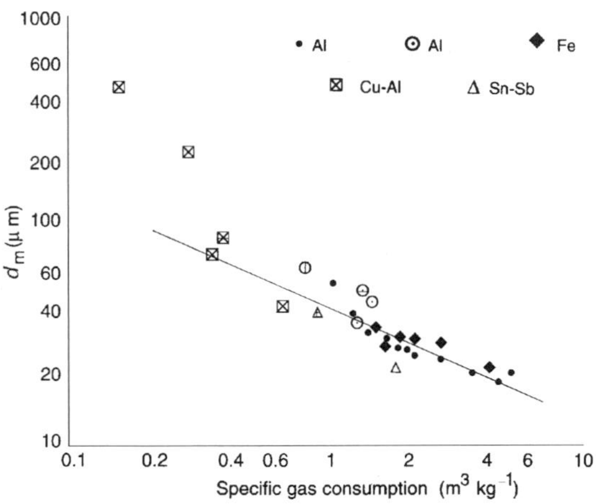

Figure 5
Median particle size (d_m) as a function of gas consumption (reproduced by permission of Metal Powder Industries Federation from 'Atomization: The Production of Metal Powders', 1992).

in Fig. 6. Without altering the elements of the model, it is possible to vary the number of metal droplets produced by each water droplet.

In gas atomization, the conceptual mechanism of droplet formation involving the disintegration of a liquid sheet by a high velocity jet of gas is shown in Fig. 7. Three stages are involved: the initiation of sinuous waves that increase rapidly in amplitude, fragmentation of a ligament from the bulk liquid, and breakdown of the ligament into droplets. The first two stages have been modeled mathematically (Bradley 1973) and involve a determination of the wave number ($k_{\max}$) associated with the fastest growing amplitude.

The third stage is governed by Rayleigh instability. The model leads to a relation between droplet diameter d and $k_{\max}$ of the form:

$$d = 11.86\varepsilon / k_{\max} \qquad (4)$$

where ε is a parameter determined independently from experiments on the air atomization of water. The value of $k_{\max}$ is dictated by the values of the gas velocity, the sonic velocity of the gas, the liquid/gas interfacial energy, and the gas density.

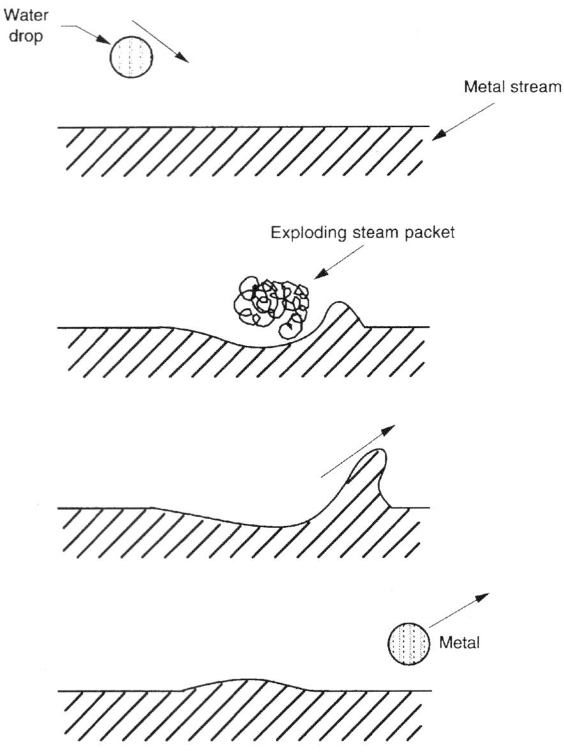

Figure 6
Mechanism of droplet formation in water atomization (reproduced by permission of Metal Powder Industries Federation from 'Atomization: The Production of Metal Powders', 1992).

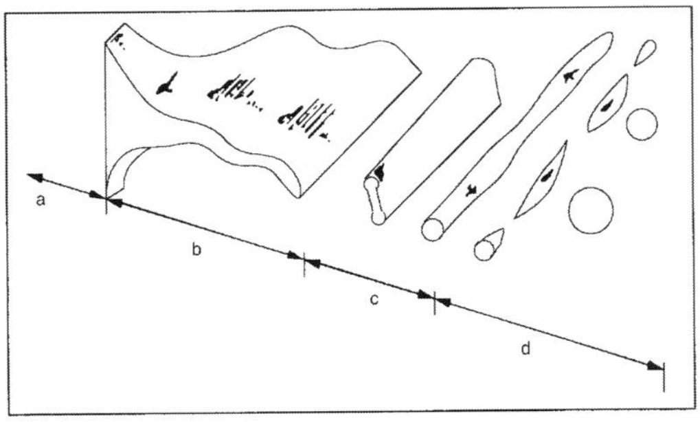

Figure 7
Stages in the disintegration of a sheet of liquid: (a) stable sheet; (b) growth of waves in sheet; (c) ligament formation; (d) ligament breakdown (reproduced by permission of Metal Powder Industries Federation from 'Atomization: The Production of Metal Powders', 1992).

It is also possible to predict the droplet diameter (d) in gas atomization by considering the balance between impact force and work. From the model (Tornberg 1992):

$$d = \sqrt{\frac{A\sigma}{\rho_{\mathrm{m}} \dfrac{M_{\mathrm{g}}}{M_{\mathrm{m}}} V_{\mathrm{g}} (1 - B\eta D)}} \qquad (5)$$

where σ is surface tension, ρ_{m} is the alloy density, η is the dynamic viscosity of the metal, V_{g} is the velocity of the gas leaving the nozzles, D is the diameter of the metal stream, and M_{g} and M_{m} are the mass flow rates of the gas and metal, respectively. A and B are atomizer-specific constants.

In the rotating electrode process, three modes of atomization have been identified: direct droplet formation, ligament disintegration, and liquid film disintegration (Fig. 8). The transitions between these modes are well defined and dictated by specific values of the atomization parameters, in particular the rate of supply of liquid metal. As the melting rate is increased, the atomization mode changes from direct droplet formation to ligament disintegration and then to liquid film disintegration.

In the direct droplet formation mode, the droplet diameter (d) can be predicted from a simple model based on the force balance between surface tension and centrifugal force, if account is taken of melting rate (Champagne and Angers 1980). The regression relation for d is:

$$d = A\omega^{-0.98} D^{-0.64} (\gamma/\rho)^{0.43} Q^{0.12} \qquad (6)$$

where γ is surface tension, ρ is the melt density, D is the diameter of the consumable anode, ω is the

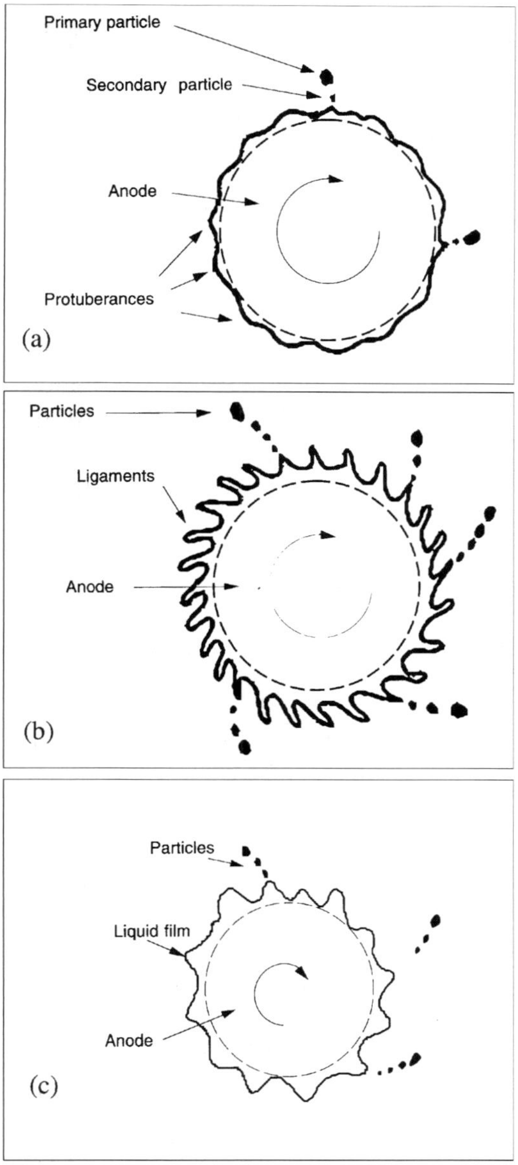

Figure 8
Mechanisms of atomization in the rotating electrode process: (a) direct droplet formation; (b) ligament disintegration; (c) liquid film disintegration (reproduced by permission of Metal Powder Industries Federation from 'Atomization: The Production of Metal Powders', 1992).

angular velocity of the anode, Q is the melting rate, and A is a constant.

The mechanism of disintegration of the molten metal in vacuum atomization is complex. To date, no conceptual mechanisms or models have been developed for this mode of atomization.

3. Particle Shape in Atomization

In addition to size, the shape of a metal powder particle exerts a major effect on its utility in subsequent powder metallurgy processes. Since the metal powder particles are produced by solidification of the atomized liquid droplets, particle shape is controlled by the relative magnitudes of the time for a liquid droplet to spheroidize under the action of surface tension forces and the time for the droplet to solidify.

In commercial atomization, spheroidization times are at least two orders of magnitude shorter than solidification times. Thus, atomized metal powder particles are expected to be spherical in shape. In general, gas atomization results in spherical powders, whereas water atomized powders are typically irregular in shape. This morphology is attributed primarily to the presence of an oxide film on the droplet surface prior to solidification. The oxide film opposes the forces of surface tension, thereby inhibiting spheroidization.

Bibliography

Bradley D 1973 On the atomization of a liquid by high velocity gases. *J. Phys. D. Appl. Phys.* **6**, 1724–36
Champagne A, Angers R 1980 Fabrication of powders by the rotating electrode process. *Int. J. Powd. Metall. Powd. Technol.* **16**, 359–67
Dunkley J J 1998 *ASM Handbook*, ASM International, Metals Park, OH, Vol. 7, pp. 35–52
German R M 1994 *Powder Metallurgy Science*, 2nd edn. Metal Powder Industries Federation, Princeton, NJ
Grandzol R J, Tallmadge J A 1973 Water jet atomization of molten steel. *AIChE J.* **19**, 1149–58
Lavernia E J, Wu Y 1996 *Spray Atomization and Deposition*. Wiley, New York
Lawley A 1992 *Atomization: The Production of Metal Powders*. Metal Powder Industries Federation, Princeton, NJ
Tornberg C 1992 Particle size prediction in an atomization system. *Adv. Powder Metall.* **1**, 137–50
Yue A J, Dunkley J J 1994 *Atomization of Melts for Powder Production and Spray Deposition*. Oxford University Press, Oxford

A. Lawley

B

Beryllium: Alloying, Thermomechanical Processing, Properties and Applications

Beryllium has long been known for its high elastic modulus (3.03×10^5 MPa), low density ($1.848\,g\,cm^{-3}$), high specific heat ($1.88\,J\,g^{-1}\,K^{-1}$ at room temperature), high melting point (1289°C), and x-ray transparency. The density of beryllium is approximately 25% that of steel- or nickel-based alloys, and only 66% that of aluminum, yet the elastic modulus is 50% greater than that of steel or nickel alloys. Therefore, the specific modulus of beryllium (E/ρ) is 6.8 times that of conventional materials, which renders it very efficient in weight-sensitive structures. Although such a combination of attractive properties would be expected to result in many structural applications for beryllium and its alloys in the aerospace industry, its use has been limited to the most critical applications.

Until the 1970s commercial grades of beryllium exhibited marginal ductility for structural use. Competitive materials had higher strength and lower cost in addition to higher ductility. Therefore, alternatives were generally specified in all but the most weight- and stiffness-critical components, such as space optics, space capsule heat shields, and satellite antenna booms. Significant changes in the manufacturing process for beryllium have provided at least a threefold improvement in ductility, and generate the potential for higher strengths than could previously be achieved reliably. Modern structural grades of beryllium are produced which have specification minimum elongations of 3%, but typical values exceed this. S-200FH, a common structural grade, typically exhibits 0.2% yield and ultimate strengths of 348 and 450 MPa, respectively, with elongation of 5%. Mechanical properties of available beryllium grades are presented in Table 1. It should be noted that instrument grades, those intended for use as inertial guidance or optical components, are designated by the prefix "I", and structural grades are designated by the prefix "S."

Beryllium is a powder metallurgy-derived metal. The majority of the property improvements cited above are a result of improved powder manufacturing and consolidation processes. Beryllium powder comminution and consolidation processes are compared in Table 2. It may be seen that attrition milling has been largely replaced by impact grinding and inert gas atomization. Attrition milling results in flat powder particles, the flat surfaces of which are the easily fractured basal planes. Due to its high speed and low temperature, impact grinding gives rise to a blocky particle whose shape is less closely associated with the basal planes, and which therefore yields a less anisotropic consolidated billet. Inert gas atomization produces spherical particles, and results in an isotropic billet. In anisotropic billets, high ductilities can exist in one direction at the expense of the ductility in other directions.

Table 1
Characteristics of commercial beryllium grades.[a]

Grade[b]	Powder type/ Consolidation process	Fe (max ppm)	Al (max ppm)	%BeO (max)	Si (ppm)	0.2% Y.S. (MPa)	U.T.S. (MPa)	% elongation
S-200F	IG/VHP	1300	1000	1.5	600	263	380	3.4
S-65	IG/VHP	800	600	1.0	600	270	420	3
S-200FH	IG/HIP	1300	1000	1.5	600	348	450	5
S-65H	IG/HIP	800	600	1.0	600	325	482	5.3
I-70	IG/VHP	1000	700	0.7	700	270	420	4
I-220	IG/VHP	1500	1000	2.2	800	338	439	2.6
I-70H	IG/HIP	1000	700	0.7	700	323	482	5
I-220H	IG/HIP	1500	1000	2.2	800	441	530	3.2
I-250	IG/HIP	1500	1000	2.5	800	517	551	4
O-30	Atomized/HIP	1200	700	0.5	700	345	207	2

[a] Chemistries are specification requirements, not typical values; mechanical properties reflect typical values. [b] IG = impact ground, S = structural grade, VHP = vacuum-hot-pressed, I = instrument grade, HIP = hot-isostatic-pressed.

Table 2
Comparison of beryllium manufacturing processes.

A. Superseded process
1. Vacuum melt pebble, vacuum cast ingot
2. Machine ingot into chips
3. Attrition (disk) mill, or ball mill into powder
4. Vacuum hot press into billet
B. Current process—Ground powder
1. Vacuum melt pebble, vacuum cast ingot
2. Machine ingot into chips
3. Impact grind (similar to airstream grinding) into powder
4. Vacuum hot press or hot isostatic press into billet
C. Current process—Atomized powder
1. Vacuum melt pebble, vacuum cast ingot
2. Inert gas atomize ingot to powder

Some sheet products require attrition milled powder billets for rolling.

Specification minima must be based upon the lowest property value that can be encountered in the final product, resulting in low specification minimum values. Powder consolidation by hot isostatic pressing (HIP), introduced in the mid-1980s, also permits greater isotropy. In addition, HIP allows greater latitude in the consolidation temperature, which in turn allows for greater control of microstructure and properties. This can readily be seen in Table 1. HIP'd grades are denoted by an "H" after the grade number (e.g., S-200FH). The improvement in strength and ductility produced by HIP can be seen by comparing the properties of S-200F (263 MPa 0.2% yield tensile strength, 380 MPa ultimate tensile strength, 3.4% elongation) with those of S-200FH (348 MPa, 450 MPa, and 5%, respectively).

There is only one integrated manufacturer of beryllium: Brush Wellman Inc. of Cleveland, Ohio. Brush Wellman operates a bertrandite ore mine at Spor Mountain in Utah, and a mill in Delta, Utah to convert the ore to refined beryllium hydroxide. Final extraction is performed at a production facility in Elmore, Ohio. The product of the extraction process is "pebble," which consists of porous beryllium spheres, of the order of several millimeters in diameter, contaminated with fluorine and several other high vapor pressure elements. The pebbles are vacuum-induction melted to produce an ingot of high-purity beryllium. The beryllium ingot is porous and has such a large grain size that it must be converted into powder and then consolidated for it to have useful properties. Beryllium is available primarily as vacuum-hot-pressed (VHP) or HIP'd block, and rolled sheet. Extruded bar is available to a limited extent. High-purity foil for use in x-ray windows and detectors is also available. Other forms which have been made are cold-pressed and sintered tiles which are used in aircraft brakes, and as a first wall material in the Joint European Torus (JET) Fusion Reactor.

A principal barrier to the extensive use of beryllium in structural, optical, and thermal management systems is its cost. The cost derives largely from the difficulty of winning the metal from its ores, bertrandite and beryl. Bertrandite is the most common ore of beryllium; beryl ore (nonprecious stone quality emerald), which was the only commercial ore of beryllium until 1969, is a by-product of emerald mining. Beryl ore is still used to a small extent as a means of maintaining a secondary source of supply. The high cost of beryllium relative to other, less capable materials narrows its use primarily to aerospace and other "exotic" applications wherein the high cost can be justified by the weight savings or by the lack of a viable substitute.

The most important physical properties of beryllium are its high thermal conductivity, high heat capacity, and low density combined with a thermal expansion coefficient that is compatible with many other metals. The physical properties of beryllium are presented in Table 3. The nuclear properties of beryllium should also be noted. It has a low thermal neutron capture cross-section and a high scattering cross-section. It also acts as a neutron multiplier, with each neutron incident upon the material giving rise to two neutrons. The most important optical property is the near-infrared reflectivity. At wavelengths below about 3 microns the reflectivity is 50%, which increases to 99% between 4 and 10.6 microns. This characteristic has dictated its use in infrared optical systems, such as space surveillance mirrors, and infrared targeting systems. Its low mass absorption coefficient suits it to x-ray windows for medical and industrial applications. The high specific modulus coupled with low density has made it a primary material for use in precision inertial navigation systems.

1. *Impurity Effects*

The beryllium atom is very small, and the Hume–Rothery rules accurately predict limited solubility of most elements in the beryllium matrix. The elements which are important to beryllium properties and processing are discussed below.

1.1 Oxygen

The high surface area presented by metal powder coupled with beryllium's high affinity for oxygen results in beryllium oxide being an unavoidable constituent. Evidence suggests that the oxide can act to pin grain boundaries, and thereby limit grain growth and creep deformation. This effect is particularly noticeable when the oxides are fine and

Table 3
Physical properties of beryllium.

Elastic modulus (GPa)	280–300
Density (g cm^{-3})	1.848
Specific heat (J g^{-1} K^{-1})	1.88 at room temperature
Melting point (°C)	1289
Specific modulus (GPa $cm^3 g^{-1}$)	157
Crystal structure	h.c.p. to 1254°C (c/a = 1.568) b.c.c. 1254–1283°C
Lattice parameters	h.c.p.: a = 0.22854 nm; c = 0.35829 nm b.c.c.: a = 0.255
Thermal conductivity (W m^{-1} K^{-1})	
room temperature	220
200°C	140
400°C	110
Thermal expansion coefficient (10^{-6} K^{-1})	
2–38°C	10.66
100°C	14.55
200°C	15.5
Poisson ratio	
elastic region	0.024—0.030
plastic region	0.5
Heat of fusion (J g^{-1})	1080
Reflectivity	
white light	50–55%
10.6 μm	98–99%
Sound velocity (m s^{-1})	12 600

well distributed in the grain boundaries. The high strengths and ductilities of I-220 and I-250 are evidence of the strengthening effect of oxides resulting from grain refinement. A very high oxide content, however, can limit ductility. An outmoded grade, I-400, had a minimum of 4.3% BeO which gave high strength, but no ductility requirement was placed on this grade. Low oxide levels are thought to be required for high-temperature formability.

1.2 Iron and Aluminum

Iron and aluminum occur in commercial beryllium in the several hundred to several thousand parts per million range. Even at these low levels, their effects upon properties are very important. The heat treatment of beryllium is directed primarily at the control of iron–beryllium and iron–aluminum–beryllium precipitates. In order to obtain the best levels of ductility, $FeAlBe_4$ is allowed to precipitate in preference to grain boundary aluminum or $FeBe_{11}$. However, through heat treatments ductility and strength can be achieved by controlling $FeBe_{11}$ precipitate morphology. Typically, a 2.5:1 ratio of iron to aluminum (by weight) is maintained so as to permit the control of properties by heat treatment.

1.3 Silicon

Silicon acts as an aid to consolidation in vacuum-hot-pressed and vacuum-sintered material. However, too high a silicon content can produce low melting point, embrittling grain boundary phases. Typical silicon levels are kept below 500 ppm.

2. *Heat Treatment*

The heat treatment of beryllium is related directly to the presence of the alloying species iron and aluminum. These two elements are in solution at temperatures at which the consolidation of powder takes place. Therefore, control of these precipitates conventionally takes the form of a controlled cooling cycle after powder consolidation is complete. An extended hold in the 150–690°C range for up to three days before it is cooled to room temperature is used to prevent the segregation of aluminum to grain boundaries where it acts to embrittle the material. The extended hold allows the formation of $FeAlBe_4$ precipitates. The elimination of aluminum at grain boundaries is particularly important in elevated temperature applications, such as aircraft brakes. In addition, the formation of $FeBe_{11}$ is also an important precipitation reaction. Iron which is in solution results in yield points, and various aging treatments can be applied to achieve a range of strength and ductility. $FeBe_{11}$ dissolves at about 700°C, and precipitation progresses at discernible rates to temperatures as low as 300°C. Overaging results in ductility and strength reduction.

Stress relief of beryllium after machining or mechanical working is generally carried out at temperatures in the range of 700–800°C followed by cooling at 100°C per hour or less. Mechanical machining leaves a thin layer of disturbed material which is generally removed by etching, or can be healed by heat treatment. A stress relief at 800°C for two hours is effective in restoring mechanical properties, particularly ductility, to machined parts. Precipitation effects after stress relief must be taken into account because of the overlap between solution treatment and stress relief temperatures.

3. *Major Alloying Elements*

Aluminum is the most important alloying element used with beryllium. It does not form any intermetallic phases with beryllium, and the minor amount of mutual solid solubility of aluminum and beryllium causes this material to behave as though it is a composite of the two. The most common commercially available alloy of aluminum and beryllium contains 62 wt.% beryllium and 38 wt.% aluminum. It is designated Lockalloy, referring to its invention by Lockheed Aircraft, or AlBeMet™ 162. Compositions that contain more aluminum that AlBeMet 162 are available, but are not as well characterized.

AlBeMet is manufactured by inert gas atomization of prealloyed powder. The powder is consolidated by HIP, or by cold isostatic pressing followed by extrusion. Extrusions may be used as mill shapes, or rolled to produce plate. The physical and mechanical properties of AlBeMet are presented in Table 4. It can be seen that its physical characteristics are closely approximated by the rule of mixtures (ROM). The density, thermal expansion coefficient, and elastic modulus are volume-weighted averages of the properties of aluminum and beryllium. The physical properties of AlBeMet are somewhat inferior to those of beryllium, but are far superior to those of aluminum. It can readily be electron beam- or TIG-welded, which is an important advantage over beryllium in terms of fabrication cost. A major impetus to the development of AlBeMet was an expected improvement in fracture toughness. Beryllium fracture toughness values are typically 8–11 ksi $in^{1/2}$, while AlBeMet values can exceed 20 ksi $in^{1/2}$ depending on the thermomechanical history of the material.

AlBeMet162 utilizes commercially pure aluminum as the Al-rich phase. Alloying of the aluminum with magnesium and other elements to produce higher strengths and improved elevated temperature properties are being developed but are not commercially available.

The limited availability of suitable elements for alloying with beryllium itself results from the limited solubility of candidate solute elements. The small size of the beryllium atom limits the choices. Many elements have been intentionally added to beryllium without any significant benefit being demonstrated.

4. *Mechanical Working*

It is possible to roll, extrude, form, and forge beryllium at elevated temperatures. The procedures used do, of course, introduce crystallographic texture in the final product and this enhances anisotropic behavior.

Rolling of bare ingot sheet from thicknesses of 5–8 mm down to 1 mm is accomplished by preheating

Table 4
Properties of AlBeMet 162 (Lockalloy).

A. Physical properties			
Density (g cm^{-3})			2.1
Elastic modulus (GPa)			193
Poisson ratio (elastic)			0.16
Thermal conductivity at 25°C (W m^{-1} K^{-1})			210
Thermal expansion coefficient (10^{-6} K^{-1})			
20°C to 80°C			14.6
Specific heat (J g^{-1} K^{-1})			1.500
Electrical resistivity (μΩ cm)			2.3
B. Mechanical properties (typical values)			
Form	0.2% Y.S. (MPa)	U.T.S. (MPa)	% elongation
Extrusions	320	430	<3–10%
Sheet	309	412	10
HIP'd billet	221	288	2

for 60 min at 790 ±14°C in an inert atmosphere and then rolling at 730 ±14°C at a 10–16% reduction in thickness per pass with reheats after each pass. At thicknesses below 2 mm the frequency of edge cracking is increased and special handling is required to prevent heat losses at the edges. Room temperature properties that can be expected are an ultimate tensile strength of 230 MPa and a yield tensile strength of 175 MPa with 3–5% elongation in the plane of the sheet.

Although beryllium processed by ingot metallurgy is not available commercially, investigations have shown that this material can be competitive in properties with its powder metallurgy counterparts. By applying special thermomechanical treatment to ingot-processed beryllium, low-oxide, equiaxed grains of < 30 μm are produced. A combination of upset forging at 760°C, rolling at 650–760°C to reduce the thickness by 90%, and partial recrystallization of the heavily deformed material at 705°C is used. For the low-oxide ingot-processed material with grain size < 30 μm, grain-boundary sliding is observed at room temperature, thus adding another deformational mode to beryllium. Ductilities as high as 7% with strengths in the 250–300 MPa range are possible. Precaution must be taken to minimize texture formation, otherwise degradation of properties in the short transverse direction may result. Charpy impact values of 4–6 J have been reported, which are very nearly equal to those for 7075-T6 aluminum alloys. In comparison, hot-pressed billets made from powder only exhibit a Charpy impact energy of 0.7 J.

Similar thermomechanical treatment of the higher oxide powder-processed material (upset forging, rolling, and partial recrystallization) can also reduce the grain size, but higher recrystallization temperatures are needed. The finer the size of the BeO, the more effective it is in stabilizing the grain boundaries and the higher is the recrystallization temperature. However, the grain size must be reduced to < 3 μm before appreciable grain boundary sliding is observed and impact resistance is improved.

Extrusions are typically used for satellites and other aerospace applications. To prevent galling of, or sticking to, the extrusion tools beryllium is placed in a carbon steel container. This solves two other problems: it limits oxidation and protects the worker from exposure to beryllium dust. The temperature range for extrusion is 850–950°C with a preheat of 60 min, although lower temperatures are used if subsequent forgings are to be made.

In extruded tubing, the yield strength decreases and the ductility increases with increasing texture. Properties in the bar are, of course, directional, particularly ductility, which ranges from 5% in the extrusion direction to 0.8% in the transverse direction. The ultimate and yield tensile strengths are about 310 MPa and 207 MPa, respectively.

Forming is usually carried out at 600–750°C but has been successful only when the grain structure is equiaxed, the grain size is < 50 μm, the oxide content is < 1%, and the texture of the basal planes is minimized. For very severe forming operations like deep drawing, suitable lubrication is required to avoid galling. Graphite-impregnated asbestos paper has been used successfully, although caution is advised because of the use of the asbestos paper and the formation of smoke from the burning of colloidal material during rolling.

Forgings of special shapes can be produced but again a suitable starting material is required. Generally the requirements are the same as for forming. The material should be heated in the 760–820°C range and the dies should be preheated to 425°C. If a right circular cylinder of test material with rounded edges (height–diameter ratio of unity) can be forged to a height reduction of 60% without formation of surface cracks, the material can be forged successfully into special shapes. Generally the properties are improved in one direction at the expense of another, owing to texturing effects.

The new technology of isothermal forging has not been tried at the time of writing on the new grades of beryllium, but because this technique has been successful with other difficult-to-forge materials, there may be advantages for beryllium. The use of dies heated to the forging temperature should prevent heat loss at the edge of the part being forged, which normally is responsible for edge cracking. To what extent this will improve the impact resistance of beryllium is not known.

5. *Machining*

It has been known for some time that beryllium is susceptible to surface damage during machining and that its mechanical properties are likely to be degraded unless the damaged surface region is removed. The depth of the damage is dependent upon the severity of the machining operation and the grain size of the material. Care in the control of feeds, speeds, cutter geometry and cutter sharpening, or changing intervals must be exercised to minimize the severity of the deformation incurred during machining. Material of smaller grain size appears to be less susceptible to this surface damage, but etching to remove 0.1 mm from each surface is recommended, regardless of grain size. In some cases requiring close dimensional tolerances, heat treatment at 800°C for 2 h has proved adequate where machine damage consists only of twins and not microcracks.

The heat treatment removes damage from any depth, providing that there are no microcracks. On the other hand, etching removes only 0.1 mm but also eliminates the microcracks. Selection of the technique for machining damage removal depends upon the particular application.

To avoid the formation of airborne beryllium dust and at the same time minimize the amount of surface damage during machining, electrochemical machining or electrodischarge machining may be used. Surface roughness is controlled by the machining rate to give surface finishes in the range of 10–200 μin (25–500 μm) RMS, but for tensile load-bearing or critical components etching and/or a stress-relief heat treatment is recommended.

6. *Corrosion*

Corrosion of beryllium is generally not a serious problem. In dry air, the development of a protective oxide film, similar to those found on aluminum and titanium, protects the surface. However, in moist environments local corrosion at inclusions of beryllium carbide can occur. A chromate passivation coating applied by a simple dip treatment can provide suitable protection. A 25% saturated solution of KCr_2O_3 in 25% H_3PO_4 and 50% H_2O is suitable for a dip treatment of 30 min.

7. *Applications*

Beryllium has applications in instruments, optics, heat dissipation, nuclear energy, audio equipment, and aerospace structural components. Its dimensional stability and low inertia make it uniquely suitable in some cases. In instruments, beryllium is used in capstans for high-speed tape drives, precision shafts, proportional counters, inertia wheels, and optical benches. It is also used in inertial navigation systems, especially for military aircraft as well as commercial jet airplanes that fly over water on long-distance routes. The good dimensional stability and precise machinability of beryllium make it a natural choice.

In optics, mirror substrates and optical support assemblies are made of atomized powder, or impact-ground powder consolidated by HIP to minimize anisotropy. Beryllium substrates may be coated with electroless nickel or gold for use in visual wavelengths, or left bare for use in the near-infrared. The low density of beryllium has resulted in its selection for high-speed rotating mirrors where inertial effects must be minimized.

Because of the high specific heat of beryllium, heat-sink applications are numerous. Heat shields, aircraft-brake heat sinks, rocket engine nozzles, switch-gear components, and nose cones of re-entry vehicles all use beryllium.

Because of its superior specific stiffness (E/p), beryllium is the best choice for critical aerospace structures liable to buckling. By designing the structures so that the beryllium parts are in compression, buckling and excessive deflection are minimized or eliminated. Since the specific strength (ratio of ultimate tensile strength to density) is about the same as that of other materials, the lower ductility and high cost of beryllium have restricted structural applications in favor of less costly alternatives. However, no suitable alternative exists for certain primary missile structures. Complex forged brackets, longerons for low-weight stiffeners, and various satellite structures and Viking landing camera housings have made good use of beryllium. Complex satellite structures are made by brazing, soldering, or adhesive bonding of the parts in very rigid structures.

The low thermal neutron capture cross-section and high neutron scattering cross-section of beryllium have led to its use as a neutron reflector and/or generator. The transparency of beryllium to x-rays allows nearly all x-ray tube windows to be made vacuum tight without impeding transmission of the radiation.

For the future, secondary aircraft structures are potential uses of beryllium, but no such applications exist today. In hypersonic aircraft, especially high-maneuverability military aircraft, low density and high tangent elastic modulus at elevated temperature are required. Only beryllium can provide this combination. If such aircraft are built, significant portions may be made of beryllium. As a heat sink in brakes of the C-5A military aircraft, beryllium has performed well and commercial airliners may soon use it to reduce weight and improve brake performance. Lightweight bicycles and automobile engine pistons have also been proposed as possible uses. There might be additional applications if beryllium were not so expensive, or if the health and safety issues, which are generally associated with fabrication rather than use, were properly addressed.

Bibliography

Floyd D R, Lowe J N (eds.) 1979 *Beryllium Science and Technology*, Plenum, New York, Vol. 2

Hashiguchi D, Marder J M 1992 *Structure and Properties of Aluminum–Beryllium Alloys*. Presented at European Space Agencies Conference, Brussels, 1992

Institute of Metals 1963 Metallurgy of Beryllium. . Conference Proceedings, Chapman and Hall, London

London G J, Keith G H, Pinto N P 1976 Grain size and oxide content affect beryllium's properties. *Met. Eng. Q* **16**, 45–57

Marder J M 1984 Beryllium Powder Processing. *Metals Handbook–Powder Metallurgy*. American Society for Metals, Metals Park, OH, Vol. 7

Marder J M 1986 Beryllium in stress critical environments. *J. Mater. Energy Systems* **8**,

Martin A J, Webster D (eds.) 1977 *Beryllium 77*. Metals Society, London

Myers S M, Smugeresky J E 1978 Phase equilibria in the Be–Al–Fe system using high-energy ion beams: II. *Met. Trans. A* **9**, 1789–94

Schetky L M, Johnson H A (eds.) 1966 Beryllium technology. *Proc. 33rd Conf. AIME Metallurgical Society*. AIME Metallurgical Society Conference, Gordon and Breach, New York, Vols. 1 and 2

Webster D, Greene R L, Lawley R W, London G J 1976 Factors affecting the tensile strength, elongation and impact resistance of low oxide, hot isostatically pressed beryllium block. *Met. Trans. A* **7**, 851–6

Webster D, London G J (eds.) 1979 *Beryllium Science and Technology*. Plenum, New York, Vol. 1

1990 Beryllium. Metals Handbook, 10th edn. ASM International, Metals Park, OH, Vol. 2

J. M. Marder

Biomimetic Materials: Properties and Processing

The concept of biomimesis has long been used in chemistry in the context of compounds with enzyme-like catalytic action. Since the mid-1980s it has been applied to materials, particularly with a view to producing ceramic and composite materials with improved toughness, analogous to shell, tooth, and bone. There was some discomfort in the materials community with the idea of mimicking biological materials in the sense of producing an indistinguishable copy.

However, mimesis more generally refers to copying some essential aspects of a thing rather than duplicating or faking it. The phrase "bioinspired materials" is also used to express the idea in more familiar terms, but is grammatically less desirable as it is a Greek/Latin hybrid. The field has now spread to include a group of loosely linked goals in new materials and processes. These are surveyed below.

It should be kept in mind that materials development is considerably upstream from the development of new products. It is quite typical for new materials to find their way into commercial products about 20 years after their discovery (Calvert 1994). Examples include Kevlar, high-T_c ceramic superconductors, piezoelectric polymers, and gallium arsenide. It is also typical for the first applications to be quite different from those originally proposed, and for their impact to be modest compared to that suggested during the initial excitement. A personal view of the status and prospects for biomimetic materials is given at the end of this article.

1. Inorganic Particle Formation

Coccolithophores—single-celled marine algae—assemble an external skeleton from single crystals of calcite with very complex shapes, such as that shown in Fig. 1. The growing crystal is surrounded by a lipid membrane that controls the growth orientation in the crystal, but it is not known exactly how this is achieved. As shown in Fig. 1, a series of overgrowths on the initial ring lead to the whole calcite assembly. Sponges and diatoms show similar close control of the shape of silica particles on the micron scale. Magnetotactic bacteria form single-domain iron oxide (magnetite) crystals with a very closely controlled size of a few nanometers, which then aggregate into magnetic chains (Fig. 2). These biological

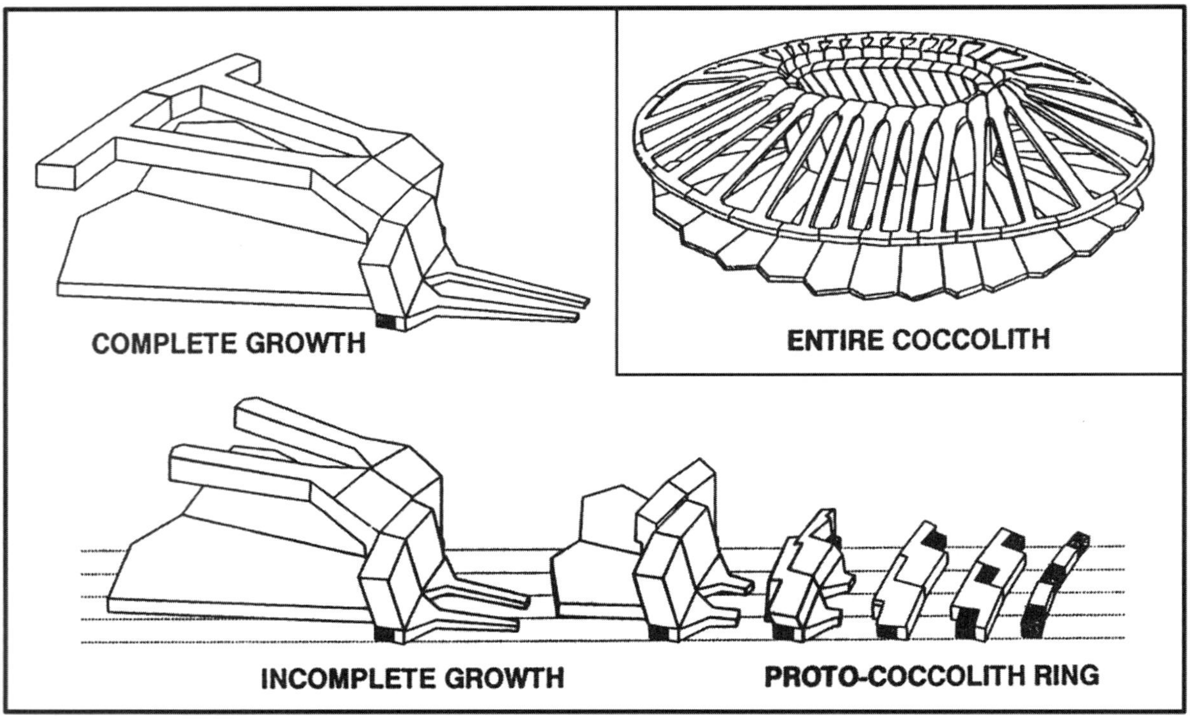

Figure 1
Assembly of a coccolith from calcite crystals (reproduced by permission of Macmillan from *Nature*, 1992, **356**, 516–8).

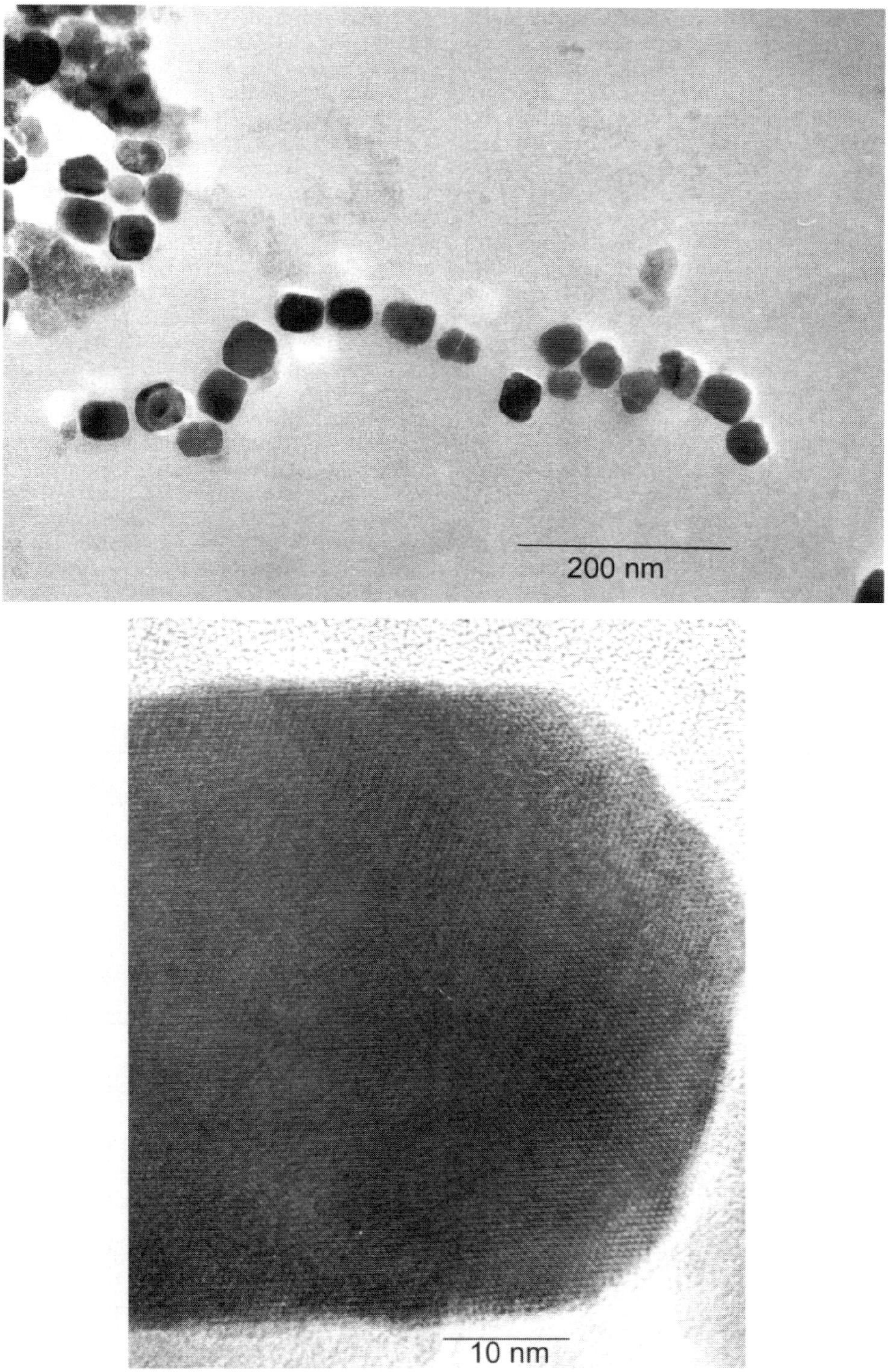

Figure 2
Magnetic particles from a magnetotactic bacterium, showing chain formation (courtesy of Professor S. Seraphin, University of Arizona).

examples all involve growth within a compartment surrounded by a membrane. For sponge spicules there is an organic template on which the mineral grows. In other cases there may be specific nucleation sites on the membrane surface. This suggests synthetic approaches where inorganic particles are grown within a micromold—a predefined space.

There have been many efforts to grow particles in liposomes, i.e., spherical shells with a lipid bilayer wall (Mann *et al.* 1997). One solution is trapped when the liposome is formed, and precipitation occurs when a second reagent, often a base, diffuses in through the wall. Generally the trapped solution must be quite dilute in order to avoid destabilizing the liposome. As a result, the precipitate therefore only occupies a small part of the internal volume. A method is needed to introduce a continuous feed of both reagents through the membrane.

There have also been many efforts to grow particles in multiphase polymer systems, such as block copolymers (Kane *et al.* 1996). For instance, a two-phase polymer can be soaked in one reagent, which selectively absorbs into one phase. A cadmium salt could be taken up by a polyether phase. Subsequent treatment with hydrogen sulfide results in precipitation of cadmium sulfide within the polyether. While the volume fraction of sulfide formed is quite small, repeated cycles can give rise to higher fractions of the mineral.

Work on precipitation in lyotropic liquid crystalline amphiphile solutions has led to composite structures (see Sect. 4).

One characteristic of biological minerals is their elaborate shapes. For mechanical reinforcement of soft matrices, a filler should consist of high-aspect-ratio rod or plate particles. These may grow as a natural outcome of differing crystal growth rates along different crystal axes. However, most simple minerals are not sufficiently anisotropic to form such elongated particles. The shape of biological crystals is probably controlled by selective inhibition of growth on specific crystal faces. Very elongated silica particles are also formed in lipid vesicles, but it is not known how important these are for controlling shape in crystals. While synthetic methods offer many ways of controlling particle size, good methods for controlling shape are lacking. One route is to form long micromolds by phase separation in a two-phase polymer, which is cold-drawn to elongate the included phase. The included phase is then swollen with a metal alkoxide, which is hydrolyzed to oxide, Fig. 3 (Calvert 1996).

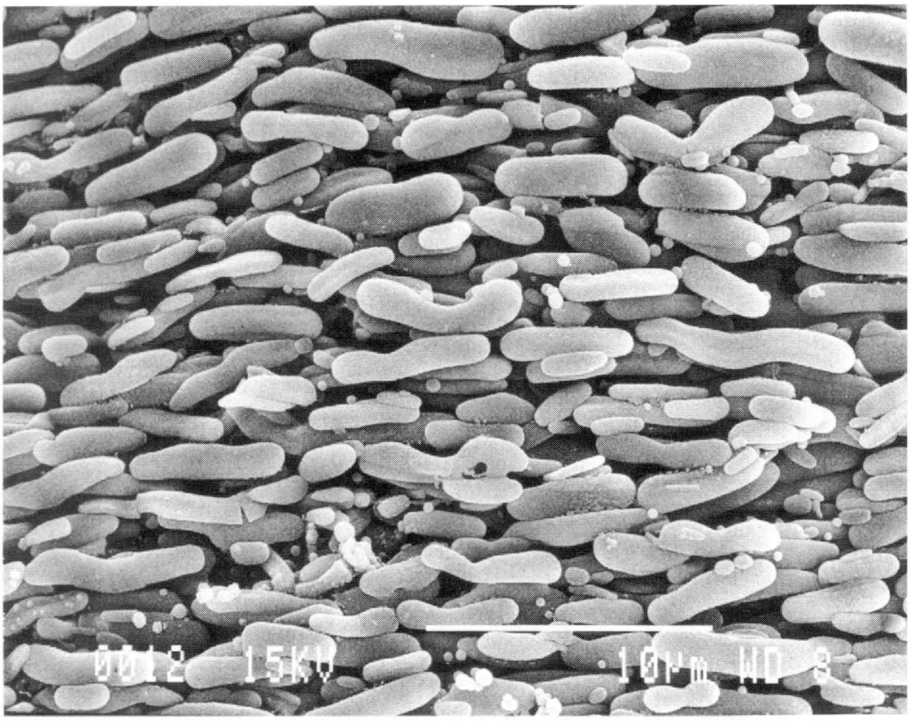

Figure 3
Elongated titania particles formed by impregnation of a stretched two-phase polymer with titanium alkoxide, followed by hydrolysis.

2. *Toughness and Layered Structures*

Large organisms operate in an environment that subjects them to fluctuating forces—from the action of wind and water on plants, and from the locomotion of animals. These fluctuating forces and collisions will often result in local damage that should not lead to catastrophic failure. As a result, many biological tissues contain structural features that add toughness without severely compromising stiffness or strength.

In mollusk shells and teeth, toughness may arise from added polymeric layers or from very fibrous structures. The addition of polymer could be especially effective if the polymer structure is capable of a large extension to break after yield, such as occurs in folded β-sheet proteins, where the unfolding of the chains leads to a large energy absorption, as has been discussed by Smith *et al.* (1999). However, polymer layers will also lead to a significant loss of elastic modulus when compared to a wholly inorganic material.

Much effort has gone into increasing the serviceability of ceramic materials by improving their toughness. Much work was initially devoted to fibrous materials (see *Fibrous Ceramics*), particularly silicon nitride, but the need for reliable, cheap processing methods was not met. Attention then switched to ceramic composites with fibrous reinforcements that would add toughness. These mostly proved to be unstable or reactive at the temperatures needed for turbine engines, while still being hard to process. Since then there have been many studies of ceramic/metal-layered structures, where the metal layers take the role of the polymers in mollusk shell. Such materials are of interest for use in engines and for armor. There is also interest in incorporating structured porosity into materials to improve fracture properties, or to modify elastic properties such as Poisson's ratio or piezoelectric response (Sigmund *et al.* 1998).

In the case of metal/ceramic-layered materials, the key problem is to absorb as much energy as possible within the metal layers. Hwu and Derby (1999) showed that the major energy absorption is due to metal drawing across the gap after the crack has passed. However, the extent of deformation occurring during the drawing was limited by the thinness of the metal layers. As a result, substantial increases in toughness were only seen when the volume fraction of metal was quite large. In contrast, mollusk shells show significant toughness with only a small percentage of polymer. It is likely that some combination of embedded metal, porosity, and interfacial debonding can give much enhanced toughness or impact resistance to ceramics. However, current methods for modeling fracture do not provide enough guidance for these designs and the biological models are only just being interpreted.

Recent studies on the influence of polymers on the crystallization of calcium carbonates *in vitro* has shown that metastable liquid complexes with anionic polymers can be formed which subsequently transform to calcite. Studies of carbonate biomineralization have shown that amorphous calcium carbonate does form either transiently or as a stable phase. It has also been shown that absorbed proteins do modify the fracture properties of carbonates. In particular, fracture surfaces show smooth conchoidal fracture, like glass, rather than the faceted fracture characteristic of normal crystals. This suggests that we have much to learn about the modification of the properties of crystals and amorphous solids by entrained polymer. In the example shown in Fig. 4, the polymer coating may serve to protect the surface from weakening cracks, and the entrained polymer may increase toughness. There are parallels with the Lanxide process for toughening alumina ceramic by entrained aluminum metal particles.

3. *Polymers*

The core difference between proteins and synthetic polymers is that protein synthesis provides total control of the sequence of units along a chain while the best-controlled polymerizations can only provide several blocks of functional units on a chain (see *Biosynthesized Materials: Properties and Processing*). One obvious goal is the formation of synthetic polymers with enzyme-like catalytic activity. Many enzyme active sites have an array of active groups held in close proximity so as to interact with the substrate (target molecule), and to reduce the activation energy for reaction by a precise spatial array of ionic or hydrogen-bonding interactions. To achieve such a precise spacing of active groups on a synthetic polymer would require a rigid structure, which would in turn normally render the material insoluble and so inactive. Studies of dendrimer molecules (see *Dendrimers: Polymerization and Properties*) with highly branched structures may lead to the required combination of flexible and soluble outer structures combined with a highly structured core. Many proteins also go through large changes in shape in response to the binding of substrates or other energy inputs. This seems to require a structure where two well-defined conformations are closely balanced in energy, so that the molecule can flip from one to another. Such a change is not likely in a wholly flexible system, and again some subtle combination of flexible segments and rigid units is required. There is still much to learn about these aspects of macromolecular design, but both the synthetic tools and the understanding are being acquired.

Structural proteins include regular and random regions in the chain structure that would be expected to give rise to crystalline and amorphous material

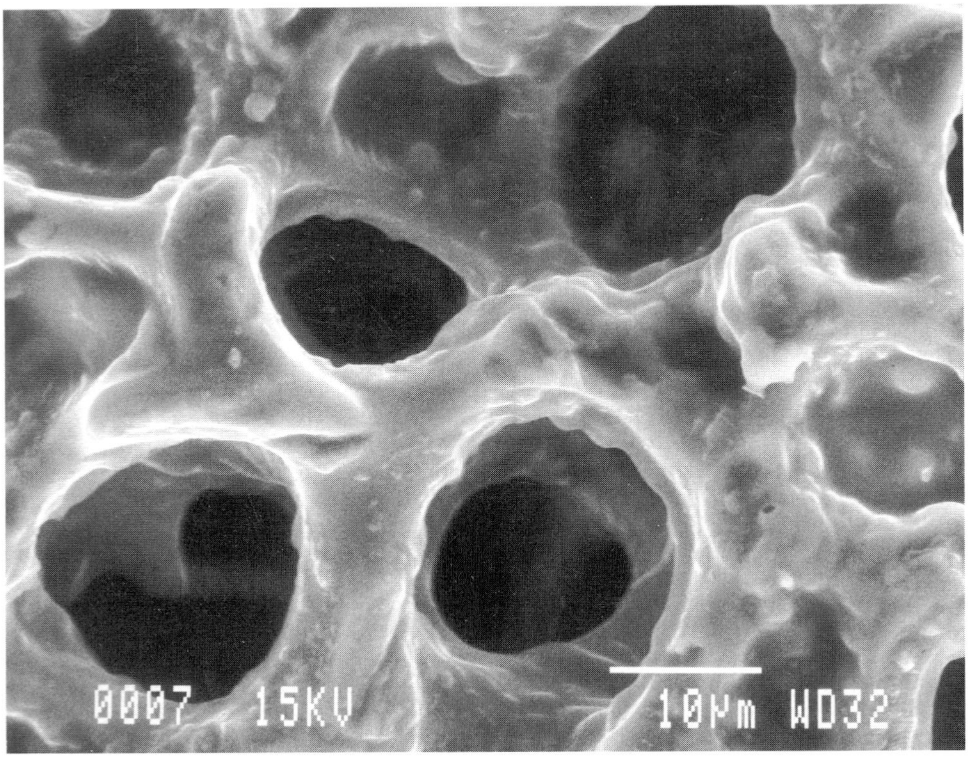

Figure 4
Porous interior of a sea urchin spine (calcite single crystal).

(Calvert 1998). In contrast, the amorphous component of synthetic polymers arises from entanglements of the coiled chains that cannot be resolved during the crystallization process. New polymer properties may be achievable once polymer chains can be designed with such controlled sequences.

There has been a report of enzyme-like activity in a block co-polypeptide which enhances the rate of hydrolysis of tetraethoxysilane (TEOS; a standard reagent in sol-gel chemistry) as a suspension in water (Cha *et al.* 2000). If this block structure is made up of a sequence of units of a hydrophilic amino acid followed by units of a hydrophobic amino acid, it would be expected to be active at a water-solvent interface. The morphology of the silica that forms is dependent on the structure of the copolymer. This system is biomimetic both in the sense of employing a polypeptide catalyst and in the sense of it functioning in a multiphase system, since biological processes rarely occur in homogeneous solutions.

Many tissues, such as cartilage, arterial wall, and the walls of soft marine organisms, are swollen polymer structures. Swollen polymers, such as plasticized polyvinylchloride (PVC), do occur in artificial structures but they are usually avoided because loss of plasticizer leads to shrinkage and cracking. Even the swelling of wood with changes in humidity is a major impediment to its use in structures—though years of experience have taught us how to design around this problem. Skin does change in volume and properties as it takes up or loses water. The structure of amphibian skin keratin is apparently different from mammalian keratin for this reason.

There have been many suggestions that designers should make more use of soft structures. These could take the form of composites of hard fibers with rubbers. In which case, tires (National Materials Advisory Board 1994) and reinforced plastic tubing could be considered as examples. Liquid-swollen soft structures could also be used more in the synthetic world. For purely mechanical systems this may not make much sense, but in active systems such as batteries or muscle-like actuators, a liquid component is necessary and should probably be viewed as a soft material rather than simply as a liquid to be contained.

4. *Surfactants and Self-assembly*

Self-assembly is a hallmark of biological systems, including assembly of protein subunits into holoenzymes, of proteins and nucleic acids into virus particles, and of tropocollagen into collagen fibers. There has been increasing interest in synthetic self-assembly. In addition to the assembly of molecules with complementary hydrogen bonding to form supramolecular clusters (Lehn 1990), there are many papers on the assembly of charged polymers and particles to form multilayers (Bertrand *et al.* 2000), and on the assembly of particles or particles and films coated with complementary biological recognition molecules, such as the biotin-streptavidin system.

Work at the Mobil Corporation showed that mesoporous silica could be formed by hydrolysis of TEOS entrained in a highly concentrated water/silane/surfactant system. In this regime, the three-component mixture forms ordered structures with a range of symmetries. In one hexagonal phase, rods of water are surrounded by surfactant and embedded in a hydrophobic silane matrix. Hydrolysis of the silane under suitable conditions followed by drying and sintering, results in a porous silica with aligned pores of a few nanometers in diameter.

Growth of the lyotropic liquid crystal precursor is very sensitive to the environment. Ozin and co-workers have shown that complex particle morphologies can result from growth of these mesoporous structures in quiescent solutions as diffusion fields and surface forces interact (Yang *et al.* 1999). Several workers have shown how the direction of the rods or plates of silica can be controlled. Polymers can be introduced to form composite structures very reminiscent of some biological composites (Sellinger *et al.* 1998). This does seem to parallel the proposed importance of liquid crystals in the growth of many biological structures.

While one would expect that this approach could be extended to many other material combinations, the rules are not understood. Efforts to form similar structures, other oxides such as titania or various crystalline materials, have been only partly successful. Any rapid or localized conversion process possibly also disrupts the liquid crystalline organization.

Stupp and co-workers have produced a range of amphiphiles, which assemble into various ribbon and wedge structures, and have explored their catalytic activity (Stupp *et al.* 2000).

5. *Muscles*

There is a clear need for actuators for robot arms with properties comparable to human muscle. Unlike human muscle, an artificial muscle ought to respond to an electrical impulse by a change in length, since any chemical energy source must involve a system of pipes. There are many candidate materials, but all have considerable drawbacks at the moment. Piezoelectric actuators have a rapid response but the total strain is small, less than 0.1% and, as a result, complicated linkages are needed to obtain significant motion. Electrically conducting polymers show volume changes of a few percent, but this is still much below that needed for a muscle. Shape memory alloys are promising but the response is slow and the cycle life is limited. Motorized or hydraulic arms tend to be heavy and clumsy. A soft polymer actuator is the obvious analog for a muscle for fast but gentle manipulations.

Ionizable gels that respond to electrical fields and demonstration devices are discussed in *Synthetic Muscle*. There is also much interest in composites of metal and ionic polymer as synthetic muscles (Shahinpoor *et al.* 1998).

6. *Processing Methods*

Freeform fabrication methods allow objects to be built as a series of layers directly from a three-dimensional computer representation. Current methods are described in *Freeform Fabrication* (00031). These allow only one material, or a material plus a soluble support structure, to be used at a time. It is clearly feasible to combine several different materials into a single object, which would allow the building of something much closer to an organism. A recent report on simple robots, which were allowed to evolve in a virtual environment and then were built by freeform fabrication, shows how evolutionary methods might be applied to manufacturing (Lipson and Pollack 2000).

To build a crude organism would require resolution at the scale of about 10 microns. Most current freeforming methods allow resolution down to about 100 microns and a limited materials set. Microcontact printing and related methods allow much higher resolution—down to about 1 micron but effectively are restricted to one layer (Xia and Whitesides 1998). There is much current interest in ink-jet printing methods which could provide the required 10 micron resolution while allowing many layers to be deposited (Fan *et al.* 2000).

In both synthetic and biological structures, it is useful to keep in mind the distinction between design and patterning. Phase separation, crystallization, and aggregation processes can give rise to patterns in two and three dimensions on a scale, from millimeters to nanometers, which reflects the kinetics of the separation and diffusion processes. To form working devices or organisms, we need to build to a non-recurring design, which may include patterned elements. If we seek to adopt biomimetic processes, we will need to exploit self-assembling structures and patterns but within an overall design.

In current silicon technology, photolithographic methods can form two-dimensional designs down to less than 1 micron. Much finer resolutions are achievable in laboratory methods. Three-dimensional designs can be formed to below 1 micron using two-photon methods (Cumpston *et al.* 1999), but commercial freeforming methods are limited to about

100 microns resolution. In biology there are many examples of patterns forming at the nanometer level, but most designs are on the scale of individual cells, i.e., a few tens of microns. There are structures, such as sensing hairs, which are much finer but the spacing between them is still on the 10 micron scale.

7. Cell Adhesion and Tissue Engineering

Biomedical engineering is becoming more concerned with the problems of the long-term biocompatibility of synthetic implants. Material wear and degradation, and tissue loss or chronic inflammation due to changing mechanical loads will limit the lifetime of most implants to about 10 years. In addition, biocompatibility is not a material property but is very dependent on the specific environment of the implant—in terms of implant site in the body, animal species, and animal age. Strategies to eliminate these problems include the use of biodegradable materials that will eventually be replaced by natural tissue, and tissue-engineered implants where as much of the device as possible is formed from tissue grown *in vitro* on a synthetic support before implantation.

Tissue engineering imposes a need to understand the interaction between neighboring cells, and between cells and synthetic surfaces. Cell binding to a biological surface proceeds through a series of stages including physical adsorption, interaction between surface macromolecules and specific binding sites on the surface, reorganization of the macromolecules in the cell membrane to bring more binding sites into contact with the surface, and then specific changes in the cell induced by the surface.

At the physical adsorption level, studies of cell attachment to self-assembled monolayers on silicon or glass have shown that attached polyethylene oxide chains form a structureless hydrophilic barrier layer and will thus prevent adsorption. Different polar end groups allow cells to absorb, and printed surface patterns with binding and non-binding areas can be used to control cell shape (Zhang *et al.* 1999). Cells recognize and bind to simple short sequences of amino acids in a protein exposed on a surface. Arginine-glycine-aspartic acid (RGD) and arginine-alanine-aspartic acid-serine (RADS) are sequences that can be used to induce strong cell attachment. Thus suitable polymers can be produced which will promote attachment of particular cell types on the surfaces for tissue engineering.

In the case of bone implants, there has been much interest in the use of bone morphogenic proteins (now produced from cloned bacteria). These constitute one group of a family of cell-signaling proteins known to induce the development of bone or other tissues when released locally in the body. A similar set of signals induces the growth of small blood vessels and is an important factor in the development of many cancers. By allowing implants to release such signaling proteins, it should be possible to speed up the integration of the implant. However, much remains to be learnt about the appropriate concentrations, gradients, and the timing of these signals. All these efforts take us in the direction of making synthetic organs that look more like a natural organ transplanted from an identical twin.

Belcher and co-workers (Whaley *et al.* 2000) have demonstrated that a phage library, displaying 10^9 different peptide sequences at the surface, can be used to identify short peptide sequences that selectively bind to inorganic semiconductor surfaces. Sarikaya and co-workers (Brown et al. 2000) have developed a similar method for the control of gold precipitation using *Escherichia coli* genetics.

8. Applications

Vogel has discussed the importance of biomimesis in engineering, asking whether it is possible to unequivocally attribute any engineering advance to a biological inspiration (Vogel 1998). There are no unambiguous examples and there are cases, such as attempts to imitate flapping flight, where a biological analogy may have inhibited progress. At the same time, many engineering advances clearly drew some inspiration from biology. In chemistry and materials, studies of the structure and properties of biological materials do suggest alternative approaches to particular problems and illustrate new properties that should be achievable in synthetic materials.

Biomimesis clearly will have an important role in new biomedical devices, and in new devices that try to combine biological structures or organisms with electronics in sensors and actuators. In structural materials, the obvious place for advance is the introduction of greater toughness into synthetic composites and ceramics. In electronics, there are not yet many signs of a conjunction between the hard, high-resolution, two-dimensional world of silicon and the soft, larger-scale, three-dimensional design of the brain and nervous system.

Bibliography

Bertrand P, Jonas A, Laschewsky A, Legras R 2000 Ultrathin polymer coatings by complexation of polyelectrolytes at interfaces: suitable materials, structure and properties. *Macromol. Rapid Commun.* **21**, 319–48

Brown S, Sarikaya M, Johnson E 2000 A genetic analysis of crystal growth. *J. Mol. Biol.* **299**, 725–35

Calvert P D 1994 Polymers for new materials. *Polymer* **35**, 4484–8

Calvert P D 1996 Biomimetic processing. In: Brook R J (ed.) *Materials Science and Technology*, VCH, Weinheim, Germany, Vol. 17B, pp. 51–82

Calvert P D 1998 Silk and sequence. *Nature* **393**, 309–10

Cha J, Stucky G, Morse D, Deming T 2000 Biomimetic synthesis of ordered silica structures mediated by block copolypeptides. *Nature* **403**, 289–92

Cumpston B, Ananthavel S, Barlow S, Dyer D, Ehrlich J, Erskine L, Heikal A, Kuebler S, Lee I, McCord-Maughon D, Qin J, Rockel H, Rumi M, Wu X, Marder S, Perry J 1999 Two-photon polymerization initiators for three-dimensional optical data storage and microfabrication. *Nature* **398**, 51–4

Fan H, Lu Y, Stump A, Reed S, Baer T, Schunk R, Perez-Luna V, Lopez G, Brinker C 2000 Rapid prototyping of patterned functional nanostructures. *Nature* **405**, 56–60

Hwu K L, Derby B 1999 Fracture of metal/ceramic laminates: I Transition from single to multiple cracking. *Acta Mater.* **47**, 529–43

Kane R S, Cohen R E, Silbey R 1996 Synthesis of PbS nanoclusters within block copolymer nanoreactors. *Chem. Mater.* **8**, 1919–24

Lehn J M 1990 Perspectives in supramolecular chemistry—from molecular recognition towards molecular information-processing and self-organization. *Angew. Chem. Int. Ed. Engl.* **29**, 1304–19

Lipson H, Pollack J B 2000 Automatic design and manufacture of robotic lifeforms. *Nature* **406**, 974–8

Mann S, Burkett S, Davis S, Fowler C, Mendelson N, Sims S, Walsh D, Whilton N 1997 Sol-gel synthesis of organized matter. *Chem. Mater.* **9**, 2300–10

National Materials Advisory Board 1994 *Hierarchical Structures in Biology as a Guide for New Materials Technology*. National Academy Press, Washington, DC

Sellinger A, Weiss P, Nguyen A, Lu Y, Assink R, Gong W, Brinker C 1998 Continuous self-assembly of organic-inorganic nanocomposite coatings that mimic nacre. *Nature* **394**, 256–60

Shahinpoor M, Bar-Cohen Y, Simpson J, Smith J 1998 Ionic polymer-metal composites (IPMCs) as biomimetic sensors, actuators and artificial muscles—a review. *Smart Mater. Struct.* **7**, R15–30

Sigmund O, Torquato S, Aksay I A 1998 On the design of 1-3 piezocomposites using topology optimization. *J. Mater. Res.* **13**, 1038–48

Smith B, Schaffer T, Viani M, Thompson J, Frederick N, Kindt J, Belcher A, Stucky G, Morse D, Hansma P 1999 Molecular mechanistic origin of the toughness of natural adhesives, fibres and composites. *Nature* **399**, 761–3

Stupp S, Pralle M, Tew G, Li L, Sayar M, Zubarev E 2000 Self-assembly of organic nano-objects into functional materials. *Mater. Res. Soc. Bull.* **25**(4): 42–8

Vogel 1998 *Cats' Paws and Catapults: Mechanical Worlds of Nature and People*. Norton, New York

Whaley S R, English D S, Hu E L, Barbara P F, Belcher A M 2000 Selection of peptides with semiconductor binding specificity for directed nanocrystal assembly. *Nature* **405**, 665–8

Xia Y N, Whitesides G M 1998 Soft lithography. *Annu. Rev. Mater. Sci.* **28**, 153–84

Yang S, Yang H, Coombs N, Sokolov I, Kresge C, Ozin G 1999 Morphokinetics: Growth of mesoporous silica curved shapes. *Adv. Mater.* **11**, 52–5

Young J, Didymus J, Bown P, Prins B, Mann S 1992 Crystal assembly and phylogenetic evolution in heterococcoliths. *Nature* **356**, 516–8

Zhang S, Yan L, Altman M, Lassle M, Nugent H, Frankel F, Lauffenburger D, Whitesides G, Rich A 1999 Biological surface engineering: a simple system for cell pattern formation. *Biomaterials* **20**, 1213–20

P. Calvert

Brazing

1. *Introduction*

The use of a lower-melting-point molten metal to join metallic components is perhaps the earliest example of processing which employs metallurgical bonding principles, having roots as far back as 4200 BC (Peaslee 2003). More than 6000 years later, brazing occupies a prominent position in our suite of joining processes where it offers cost and/or performance advantages in the fabrication of many structures.

More precisely, brazing can be described as the use of a molten filler metal to wet the closely fitting faying surfaces of a joint, leading to formation of metallurgical bonds between the filler metal and substrates. Historically, brazing processes employ filler metals whose solidus temperature exceeds 723 K (as opposed to soldering processes which use lower-melting-point temperature filler materials.

In the past several decades, technological advances have facilitated a broadening of applications for brazing while simultaneously contradicting some of the traditional perceptions of the process. However, many of those tenets remain appropriate for the majority of brazing processes and products. Accordingly, this article provides a brief description of traditional brazing and some important factors to be considered when designing and producing brazed structures. An additional section describes the technical advances in the field.

2. *Overview of Conventional Brazing Processes*

The brazing process begins with preparation of pieces to be joined, including machining and cleaning of joint details, fixturing of the pieces to maintain orientation and desired joint clearance, and placement of braze filler material adjacent to the joint. The joint area is then heated, either through the application of a localized heating source or by heating the entire assembly (e.g., in a furnace). Upon reaching a temperature sufficient to melt the filler metal, capillary forces draw the liquid metal into the gap; subsequent cooling results in solidification of the filler metal and formation of the brazed structure.

The evolution of braze process technology has advanced to a state where it is now frequently employed in applications requiring a high degree of mechanical performance in challenging service environments. Accordingly, a number of important factors need to be considered in the design of successful brazements.

2.1 *Wetting*

Although virtually all braze joints employ a capillary design, examination of the simplified example of a liquid metal droplet on a solid surface is useful for understanding the brazing process. The theory of wetting of liquids on solid surfaces (Hawkins 1952) maintains that a liquid will spread over a solid surface until surface energies are balanced. Accordingly,

$$\gamma_{SL} = \gamma_{SV} - \gamma_{LV} \cos \Theta$$

where γ_{SL} is the solid–liquid surface energy, γ_{SV} is the solid–vapor surface energy, γ_{LV} is the liquid–vapor surface energy, and Θ refers to the contact angle between the edge of a liquid droplet and a solid surface (Fig. 1). Wetting conditions are generally defined by contact angles (Θ) of $<90°$. Upon melting of filler metal, an imbalance in these forces leads to spreading (wetting) of the liquid across the solid surface, provided appropriate conditions exist. Improvements in wetting (which lead to reductions in Θ) are accomplished by a reduction in γ_{SL} or γ_{LV}, or an increase in γ_{SV}.

The presence of surface contamination ("dirt," oxide films, adsorbed water molecules, etc.) is perhaps the biggest impediment to the formation of sound braze joints. Removal of surface contaminants to provide a clean metallic surface is the best way to maximize γ_{SV} and improve wetting. Both γ_{SL} and γ_{LV} are constants at fixed temperature and pressure for particular material combinations. However, γ_{SL} is very temperature sensitive, in most cases decreasing rapidly with increasing temperature (Schwartz 1987), enabling improvements in filler metal wetting. Similarly, γ_{LV} is most rapidly reduced by decreasing the pressure of the brazing atmosphere, leading to the increased popularity of vacuum brazing processes.

2.2 *Filler Metal Selection*

While hundreds of commercial compositions are available, the choice of an appropriate filler metal is

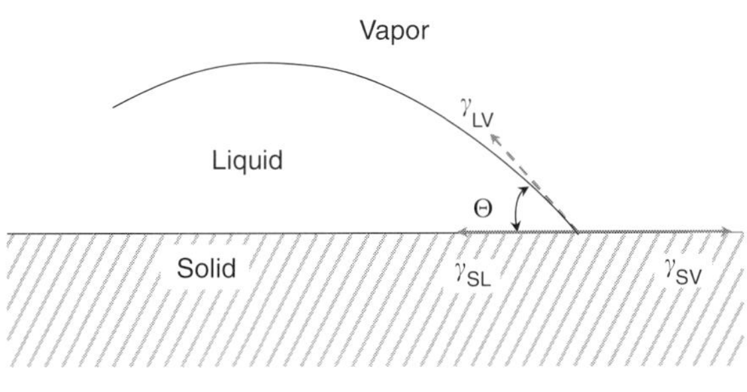

Figure 1
Surface-tension forces determine the contact angle Θ for a liquid droplet on a solid surface (See colour plate section).

determined by both processing and service considerations. Joints subjected to relatively low stresses at ambient temperatures can generally be made with a variety of filler metals; cost and processing issues are important factors. Pure metals or simple binary (e.g., eutectic) compositions can frequently provide adequate service in these applications. More severe service conditions (stress, temperature, chemical environment) limit filler metal choices as control of final joint microstructure becomes more important. In particularly demanding applications, where high-temperature strength is required in conjunction with oxidizing (or other aggressive) environments, highly alloyed filler metals are often required in order to provide satisfactory performance.

2.3 Filler Metal Flow

The molten filler metal, in addition to wetting the substrate, must also possess the capability to flow into and fill the braze joint. Thus, viscosity of the molten metal is an important property. In cases where there are significant chemical interactions between the substrate and molten filler metal, it is important that resulting compositional changes of the filler metal do not lead to unacceptable increases in viscosity. Typical interactions include dissolution of the base metal, diffusion of filler metal species into the base metal (particularly the rapid diffusion of interstitial elements along grain boundaries), and formation of intermetallic compounds.

2.4 Joint Gap

Control of joint clearance, both at the brazing temperature and in the finished joint, is important for two reasons. First, capillary forces that determine filler metal flow into the joint are strongly dependent upon the gap dimensions. Reduction in joint clearance enhances capillary forces and promotes filler metal flow, although minimum limits exist. Second, mechanical strength of the finished joint is also a strong function of joint width. Filler metals often possess inferior mechanical strength (but superior ductility) compared to structural metals. However, reduced clearances result in highly constrained joints; behavior of the filler metal is influenced by proximity to higher-strength substrate material and the resulting joint can have tensile strength far superior to that of the bulk filler metal (Orowan 1918–19). Conversely, the mechanical performance of wide joints is limited by the properties of unconstrained filler metal as shown in Fig. 2 (Saxton 1971).

2.5 Substrate Interactions

In addition to concerns regarding potential increases in filler metal viscosity, reactions with the substrate

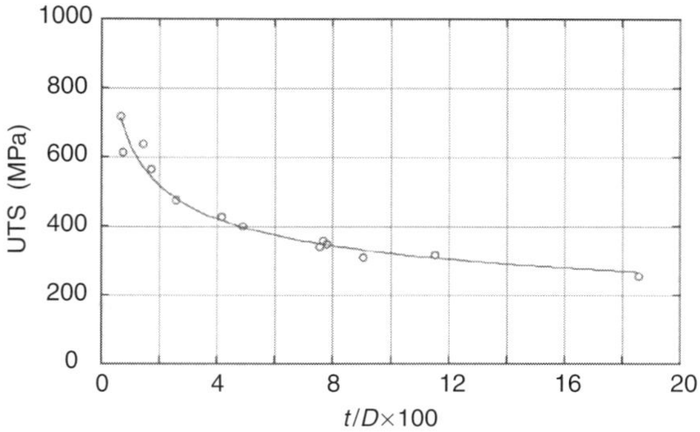

Figure 2
The effect of interlayer thickness on the ultimate tensile strength of maraging steel specimens brazed with Ag–4% Pd filler metal. t = interlayer thickness, D = specimen diameter. Reproduced from Saxton H J, West A J, Barrett C R Deformation and failure of brazed joints – macroscopic considerations, from *Metall. Trans.* volume **2**, 1971, page 999, with permission from The Minerals, Metals & Materials Society (See colour plate section).

materials need to be considered for other reasons, as these interactions can affect physical and mechanical properties of both the joint and substrate materials (Lugscheider *et al.* 1982). It can be anticipated that filler metals containing interstitial atoms (boron, silicon, and carbon) will exhibit rapid diffusion of these elements into base metals, particularly along grain boundaries, with mixed effects. Erosion of base metal, particularly in the brazing of thin sections, can cause concerns about mechanical performance. On the other hand, diffusion of these elements out of the joint area can also prevent formation of brittle intermetallic compounds, while increasing the melting point of the solidified filler metal.

Reactions between molten filler metal and base metal are enhanced by increases in brazing temperature. However, the equilibrium concentration of parent metal elements in the molten filler metal is usually reached in a matter of seconds (Humpston and Jacobson 1993).

2.6 Thermal Cycle

There are no rigid rules guiding the selection of a brazing temperature that will allow melting and flow of filler metal while minimizing undesirable filler-metal interactions with either the brazing atmosphere or substrate materials. Generally, a peak temperature between 30°C and 100°C above the filler metal liquidus temperature is employed.

In order to minimize undesirable metallurgical reactions at intermediate temperatures, it is generally preferred to employ a rapid heating rate. However, in many cases the difficulty in maintaining temperature uniformity limits the heating rate used for processing. Large assemblies are typically brazed using a thermal cycle that employs a short hold at a temperature just below the filler metal liquidus (to facilitate a reduction in thermal gradients) followed by a short, rapid climb to the peak temperature. The peak temperature is held long enough to ensure adequate time for complete melting of filler metal. Rapid cooling rates are also desirable, although thermal mass often limits achievable rates and excessively fast rates can cause part distortion.

2.7 *Fluxes and Atmospheres*

Most metals react with air at elevated temperatures to form surface films that interfere with wetting and flow of filler metals. One solution is to use brazing fluxes that dissolve oxide films on both metal surfaces and the surface of the molten filler metal, and protect those surfaces during the brazing cycle. Flux compositions are tailored for specific base metal and filler metal combinations, not only enabling the proper chemical reactions to occur throughout the brazing cycle, but also maintaining desirable physical properties (viscosity, surface tension) to ensure displacement of the flux, out of the joint, by the liquid filler metal. Oxide reduction, the principal function for fluxes, is usually accomplished through the reactions produced by combinations of fluoride, chloride, or borate ingredients. Fluxes are commonly used in torch- and induction-brazing processes, and dip-brazing of aluminum parts.

An alternative approach to the management of surface oxides is the utilization of controlled brazing atmospheres. Ellingham diagrams can be used to determine the thermodynamic stability of surface oxides, although these diagrams provide no insight regarding reaction kinetics. It is sometimes desirable for the brazing atmosphere to have oxidizing properties, at least temporarily, during the brazing cycle. Oxidizing atmospheres are utilized to help burn off the organic compounds used as binders in filler metal pastes. High-purity, low-dewpoint nitrogen is used as a base atmosphere in conjunction with small amounts of reducing gas. The additive gas contains a balanced mixture of hydrogen and water vapor (or other oxygen-containing species). A continuous flow of the nitrogen carrier gas allows tailoring of the atmosphere to maintain variable control of oxidizing or reducing conditions throughout the brazing cycle. Hydrogen is the most effective gas for reducing metal oxides; a high-purity hydrogen can effectively reduce surface oxides from almost all metallic alloys. However, many high-temperature superalloys contain alloying additions of aluminum and/or titanium, the oxides of which cannot be realistically reduced in hydrogen at practical brazing temperatures. Vacuum environments can prevent oxidation and, for some metals, serve to reduce surface oxides. However, vacuum-brazing equipment is expensive and not well suited for volume production; hence, it is used predominantly in the fabrication of high-value components including many superalloy applications.

3. *Brazing of Ceramics*

The nature of atomic bonding in ceramic materials (i.e., covalent or ionic) precludes the presence of delocalized electrons found in metallic structures. Consequently, metal–ceramic interfaces are electrical discontinuities (Pauling 1939), resulting in the nonwetting of most ceramic surfaces by liquid metals.

3.1 *Metallization-based Processing*

The traditional approach to brazing of ceramic structures consists of an initial application of a metallized layer to ceramic surfaces followed by joining using a conventional brazing process (Kohl 1995).

A slurry consisting of refractory metal particles (W or Mo) and glass is "painted" onto the ceramic surface using a variety of application processes, from hand-brushing to automated screen painting. Commercially prepared coating mixtures are available, although many users prefer proprietary formulations. The coating is then fired, usually in a reducing environment containing controlled moisture content, to facilitate wetting of the ceramic surface by the glass phase and prevent oxidation of the refractory metal. The fired coating consists of a continuous glass phase binding the metallic particles to the ceramic substrate. Most braze alloys will not wet the discontinuous metallic structure; consequently, a continuous metal layer of nickel, copper, or gold is applied over the fired coating, usually by electroplating. A properly designed metallization not only provides a surface that is easily wet by braze filler metals, but also serves to isolate the ceramic substrate from large stresses that can be induced during brazing.

Modern coating technologies (e.g., PVD) have also been used to directly deposit thin metallic layers on ceramic surfaces permitting subsequent braze processing with conventional filler metals.

3.2 *Active Metal Brazing*

Recently, the development of a new family of "active" filler metal compositions has enabled the direct brazing of ceramics, thereby eliminating the need for metallization steps. While the use of active filler metals simplifies the joining of ceramics from a manufacturing standpoint, control of processing steps to produce desired microstructural development and successful joint formation is no less important.

Many of the active filler metals are based on conventional compositions but contain an active element as an alloying addition. Titanium is the most common active agent, although compositions containing vanadium, zirconium, chromium, and hafnium are also employed. During the brazing process, the active element, in proper concentrations, reacts with the ceramic structure to form a thin interfacial reaction product that is strongly adherent to both the ceramic substrate and the solidified braze filler metal. In many material systems, the reaction product(s) is readily characterized (Fig. 3); in other systems, it is barely observable using high-magnification electron microscopy (Hosking *et al.* 2000).

Unlike metallization-based processes where a comparatively compliant structure is formed at the ceramic interface, direct brazing processes result in solidified filler metal separated from the ceramic interface by only a thin reaction layer. Thus, shrinkage stresses resulting from the solidification of filler metal and differential thermal contraction during cooling from the brazing temperature are more likely to be "seen" by the inherently brittle ceramic substrate. Accordingly, some active filler metal compositions are intentionally constructed to produce a two-phase microstructure during solidification, providing improved ductility at the expense of filler metal strength (e.g., Mo additions to Au–16.5Ni–2V).

Figure 3
TEM image showing two reaction layers formed at interface between sapphire and Ag–2Zr braze alloy. Layer 1 consists of epitaxially oriented columnar ZrO_2 grains and commingled alpha-Ag particles. Layer 2 consists of continuous ZrO_2 grains.

Additionally, when active filler metals are used to join ceramics to metals, or when a metallic shim is used in conjunction with ceramic–ceramic joints, there exists the potential for loss of the active ingredient through reaction with the metallic component. This "scavenging" of the active element can cause its concentration in the molten filler metal to fall below a level necessary to promote the desired reaction at the ceramic interface, leading to poor joint strength (Vianco *et al.* 2003).

3.3 Special Considerations for Ceramic Brazing

The unique properties of structural ceramics force a different approach to braze joint design compared to that used for metallic components. Most ceramics have a relatively low thermal expansion coefficient. More significantly, ceramics generally have lower thermal conductivity, shear strength, and fracture toughness. These attributes are important factors for both joint design and thermal cycle selection.

Braze joints should be designed with consideration of post-braze thermal contraction; tensile stresses in the ceramic substrate should be minimized. For example, cracks in the ceramic substrate are often found propagating beneath large fillets in a lap joint. Many creative solutions have been used to reduce residual stresses in ceramic braze joints, including the use of low-expansion metallic shims and composite filler metal systems incorporating low-expansion particles or fibers.

Reduced on-heating ramp rates are frequently selected for ceramic parts, enabling better temperature uniformity to be obtained. The on-cooling rates are also lower; in addition to maintaining more uniform temperature distribution in the ceramic, creep of the solidified braze-filler metal is permitted resulting in reduced residual stress upon completion of the cooling cycle.

4. Recent Developments in the Science and Technology of Brazing

The demand for brazing of new materials, new applications for existing materials, and the continued effort to improve current production processes has led to significant research and development activity in the field of brazing. Following is a brief review of recent research activities.

4.1 Processing

Perhaps the two most significant achievements in the processing area are the development of an improved technique for brazing of aluminum structures for automotive applications and wide-gap brazing, a process used to repair damaged gas turbine engine components. While fairly new, both of these technologies have been integrated into their respective industries although continued research is adding additional technical insight.

The Nokolok™ process (also referred to as controlled atmosphere brazing (CAB)) employs a noncorrosive flux coating on clad aluminum parts that are passed through a furnace in conjunction with a protective nitrogen atmosphere (Cooke *et al.* 1978) to produce brazed components. The flux leaves a passivated surface that has improved corrosion resistance.

After removing damaged material in a turbine component, the wide-gap process is used to perform a repair by placing a mixture of superalloy and braze alloy powders (Demo and Ferrigno 1992) in a machined groove followed by exposure of the entire part to an elevated temperature thermal cycle. Proper adjustment of material and processing parameters can produce a structure possessing performance characteristics comparable to that found in the original component.

Two promising new processes offer the potential for obtaining superior braze-joint properties through different approaches. Self-propagating reactions in synthesized multilayer foils have been used as a very localized heat source to melt filler metals in braze joints (Weihs and Knio 2003), an approach that can be used to join structures without significant heating of the bulk material, thus reducing residual stresses and enabling the brazing of large ceramic cross-sections. In addition to structural ceramics, this technology has applications in semiconductor thermal management designs. It has also been reported that an induction heating process coupled with preplaced filler-metal powders can produce localized heating and melting of those powders with minimal base-metal melting (Cossu *et al.* 2000). Type 304L stainless steel cylinders were joined using alloy 310 stainless steel powder as filler metal; sound joints were produced having transverse (to bond) tensile strengths approaching 75% of parent metal values with comparable ductility.

4.2 Materials Science

Filler-metal development continues to be a fertile research area as investigators strive to find optimum properties, particularly in braze joints involving specialized high-performance metals, intermetallics or ceramics. As our understanding of materials science improves, we will see increasing application of knowledge-based approaches to design of filler-metal compositions (Nishimoto and Saida 2000). The braze joints of many technically advanced materials, notably those operating in high-temperature environments, exhibit reduced performance (strength, creep, and environmental resistance) compared to base-material properties. Increased alloy content in filler metals can aid in reducing this performance penalty; however, conventional ingot-based processing is not possible with many compositions of interest due to workability issues. The range of manufacturable filler metal compositions (in foil form) has been expanded by the introduction of rapidly solidified foils (Datta *et al.* 1984); titanium and nickel-based alloys are seeing increased application in the aerospace industry. The amorphous structures produced by this technology have a very narrow melting range that is also advantageous in braze processing.

The understanding of reaction processes that occur at the interface between molten filler metal and surfaces of either ceramic or intermetallic substrates is a necessity for controlling reaction-product formation and subsequent mechanical performance of brazed assemblies. While numerous investigators have concentrated on post-braze microstructural analysis and characterization, researchers (Derby *et al.* 2000) have developed a technique that employs neutron reflection spectroscopy in conjunction with a high-temperature stage to examine, *in situ*, structures at liquid metal–ceramic interfaces. Their procedure does not provide chemical information directly; interfacial structures must be inferred by comparing experimental reflectivity data with predictions based upon possible reaction products. Nonetheless, this tool is unique in its ability to examine buried interfaces at high temperatures.

Finally, research to further advance our knowledge of filler metal wetting and flow are particularly useful for ceramic brazing, where wetting by many active filler alloys is marginal. Experimental techniques are being developed to calibrate, validate, and verify physics-based computer models that simulate braze-flow kinetics, including fillet formation and arrest (Hosking *et al.* 2003). The development of a high-temperature wetting balance (Solomon *et al.* 2001) enables the direct measurement of filler metal wetting behavior. This experimental capability also captures time-dependent changes in wetting force related to chemical reactions occurring at the solid–liquid interface.

4.3 Modeling

Efforts to model brazing processes have expanded with the widespread availability of inexpensive, high-power computing resources. The majority of activity has been focused on the calculation of residual stresses in brazed components incorporating brittle materials—predominantly structural ceramics, carbon composites, and semiconductor materials and means to mitigate undesirable stress levels

(Kovalev *et al.* 1996, Neilsen and Stephens 2000). A slightly different analytical approach has shown (Park *et al.* 2002) that in predicting performance of metal–ceramic joints, the amount of strain energy in the ceramic (from CTE mismatch) is a good strength metric.

Simulations have also been performed (Dempsey *et al.*, 2003) in order to establish operating windows for furnace brazing of production components. The authors used physics-based models to capture the thermal response of parts and associated fixturing in order to identify practical processing limits.

Efforts to predict both microstructural evolution and transient mechanical properties of solidified filler metal during the cooling cycle have also been reported. Neilsen *et al.* (2003) used uniaxial compression and creep-compression data combined with ultrasonically determined Young's moduli and Poisson's ratios to develop a sophisticated viscoplastic material model for Ag–28Cu. It is important to note that the accuracy of many-model predictions, particularly those addressing the mechanical response of brazed structures, is largely determined by the fidelity of complex constitutive material models used in the calculations (Fig. 4).

Diffusion-brazing processes have been specified for several decades in the fabrication of components, predominantly nickel-base superalloys for aerospace applications where the solidified joint is required to have both, high strength and a re-melt temperature in excess of the filler metal solidus temperature. Diffusion brazing is a preferred joining technique for the next generation of materials currently of interest for high-temperature service, notably intermetallics and silicides. However, generally the filler metals used to join these materials differ significantly in composition from the substrate materials; consequently, it is more difficult to predict isothermal solidification times and joint composition. Progress in both experimental and predictive capabilities has been reported. MacDonald and Eagar (1998) used a moving boundary analysis to successfully model isothermal solidification of Ag–28Cu filler metal with pure copper substrate, predicting concentration profiles and solidification times. An improved experimental technique was developed to maintain the mass balance necessary for measurement accuracy.

$$\gamma = 1.07\times10^{6}\left[\sinh\left(\frac{\sigma}{31.58}\right)\right]^{2.019}\exp\left(\frac{-43424}{R\theta}\right)$$

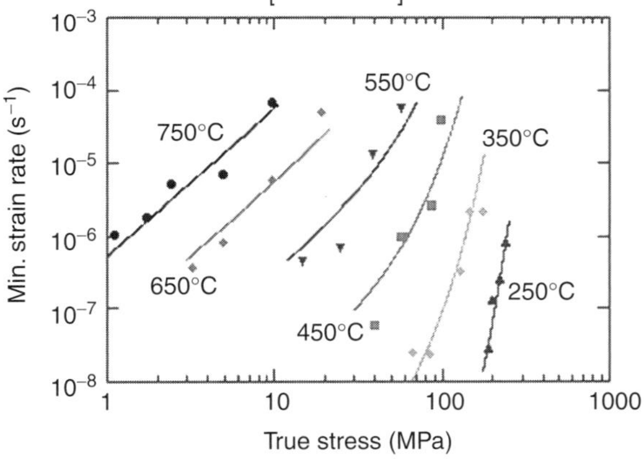

Figure 4
Steady-state creep vs. applied true stress for monolithic Ag–28Cu specimens tested at various temperatures. The data can be described by the expression presented above the graph, where γ = steady-state creep rate, σ = applied true stress, and Θ = temperature (See colour plate section).

5. *Concluding Remarks*

The use of brazing continues to grow as component designs increasingly incorporate high-performance structural materials that are not readily joined by fusion welding processes. Simultaneously, there is a demand for more efficient designs that employ traditional materials. These combined needs are driving ongoing research efforts to improve our understanding of both, the technology and science of brazing. Modeling and simulation efforts are helping to add insight to both materials science and processing knowledge. Materials-research activities continue to provide valuable data for validation of modeling tools as well as enhancing our physical understanding of the brazing process. The results of these efforts will allow the deployment of mechanically efficient braze joints for a wide range of applications and materials.

Bibliography

Cooke W E, Wright T E, Hirschfield J A 1978 Furnace brazing of aluminum with a non-corrosive flux. *SAE Technical Paper Series No. 780300*. Society of Automotive Engineers

Cossu C, Barberi D, Laille A, Harcouet P 2000 A new process for joining metallic pieces by induction heating. In: Vianco P T, Singh M (eds.) *International Brazing and Soldering Conference Proceedings*. American Welding Society and ASM International

Datta A, Rabinkin A, Bose D 1984 Rapidly solidified copper–phosphorus base brazing foils. *Weld. J.* **63**(10), 14–21

Demo W A, Ferrigno S J 1992 Brazing method helps repair aircraft gas turbine nozzles. *Adv. Matls Processes* **141**(3), 43–5

Dempsey J F, Dykhuizen R C, Crowder S V, Dudley E C, Hosking F M 2003 Braze process characterization using thermal analysis modeling and statistical design of experiments. In: Hosking F M (ed.) *International Brazing and Soldering Conference Proceedings*. American Welding Society and ASM International

Derby B, Edwards R, Webster J R P 2000 Reactions at active braze–ceramic interfaces studied by neutron reflection spectroscopy. In: Vianco P T, Singh M (eds.) *International*

Brazing and Soldering Conference Proceedings. American Welding Society and ASM International
Hawkins W D 1952 *Physical Chemistry of Surface Films*. Van Nostrand-Reinhold, New York
Hosking F M, Cadden C H, Yang N YC, Glass S J, Stephens J J, Vianco P T, Walker C A 2000 Microstructural and mechanical characterization of actively brazed alumina tensile specimens. *Weld. J.* **79**(8), 222–30
Hosking F M, Hall A C, Givler R C 2003 *In situ* visualization of braze flow through a parallel gap. In: Hosking F M (ed.) *International Brazing and Soldering Conference Proceedings*. American Welding Society and ASM International
Humpston G, Jacobson D M 1993 *Principles of Soldering and Brazing. ASM International, Metals Park, OH*
Kohl W H 1995 *Handbook of Materials and Techniques for Vacuum Devices*. American Institute of Physics, New York
Kovalev S P, Miranzo P, Osendi M I 1996 Finite element simulation of thermal residual stresses in joining ceramics with thin metal interlayers. *J. Am. Ceram. Soc.* **81**(9), 2342–48
Lugscheider E, Partz K D, Lison R 1982 Thermal and metallurgical influences on AISI 316 and Inconel 625 by high temperature brazing with nickel base filler metals. *Weld. J.* **61**(10), 329–33
MacDonald W D, Eagar T W 1998 Isothermal solidification kinetics of diffusion brazing. *Metall. Trans.* **25A**, 315–25
Neilsen M K, Stephens J J 2000 Residual stress in metal-to-ceramic braze joints advanced braze alloy constitutive models. In: Vianco P T, Singh M (eds.) *International Brazing and Soldering Conference Proceedings*. American Welding Society and ASM International
Neilsen M K, Stephens J J, Gieske J H 2003 A viscoplastic model for the eutectic silver–copper braze alloy. In: Hosking F M (ed.) *International Brazing and Soldering Conference Proceedings*. American Welding Society and ASM International
Nishimoto K, Saida K 2000 Computer-aided alloy design of insert metal for transient liquid phase bonding. In: Vianco P T, Singh M (eds.) *International Brazing and Soldering Conference Proceedings*. American Welding Society and ASM International
Orowan E 1918–19 Fracture and strength of solids, reports on progress in physics. *Phys. Soc.* **12**, 185–231
Park J W, Mendez P F, Eagar T W 2002 Strain energy distribution in ceramic-to-metal joints. *Acta. Mat.* **50**, 883–99
Pauling L 1939 *The Nature of the Chemical Bond*. Cornell University Press, Ithaca, New York
Peaslee R 2003 A short history of brazing. In: Hosking F M (ed.) *International Brazing and Soldering Conference Proceedings*. American Welding Society and ASM International
Saxton H J, West A J, Barrett C R 1971 Deformation and failure of brazed joints—macroscopic considerations. *Metall. Trans.* **2**(4), 997–1007
Schwartz M M 1987 *Brazing*. ASM International, Metals Park, OH
Solomon H D, DeLair R E, String L H 2001 *U.S. Patent No.* **6**, 286–368
Stephens J J, Hosking F M, Headley T J, Yost F G, Hlava P F 2004 A microstructural study of the reactions between a 98Ag–2Zr braze alloy and aluminum oxide. In: *Proc. 7th Intl Conf. Brazing, High Temp. Brazing and Diff. Bonding* Aachen, Germany
Vianco P T, Stephens J J, Hlava P F, Walker C A 2003 A barrier layer approach to limit Ti scavenging in FeNiCo/Ag–Cu–Ti/Al_2O_3 active braze joints. *Weld. J.* **82**(9), 252–62
Weihs T, Knio O 2003 A novel method for soldering and brazing materials: reactive multilayer foils as a local heat source. In: Hosking F M (ed.) *International Brazing and Soldering Conference Proceedings*. American Welding Society and ASM International

C. H. Cadden

Brazing Filler Metals

1. The Essence of the Brazing Process

Metal joining by brazing is an ancient practice that has become a sophisticated modern technology owing to the contribution of modern materials science. The ultimate goal is to join parts into an assembly through metallurgical bonding (ASM Handbook 1993). This is achieved by placing an alloy with a relatively low melting temperature, or "filler metal" (FM), in the clearance (or gap) between the pieces of base materials (BMs) to be joined, then heating the assembly until the FM has melted and spread throughout the gap. The molten metal fills the gap and reacts with the parts to be brazed, forming after solidification an integral solid whole. Assembly heating can be carried out using various means, including electromagnetic induction, Joule heating, oven, flame, etc. While joining temperatures above 723 K are arbitrarily associated with brazing rather than with soldering, these processes are essentially similar. In order to distinguish joining materials for the two processes, those used at temperatures below 723 K are called solders, whereas those used at temperatures above 723 K are called brazing filler metals.

As Fig. 1 illustrates, there are three major stages in any brazing or soldering process. The first stage occurs during heating of an assembled workpiece. At this stage, the FM melts and flows, filling completely

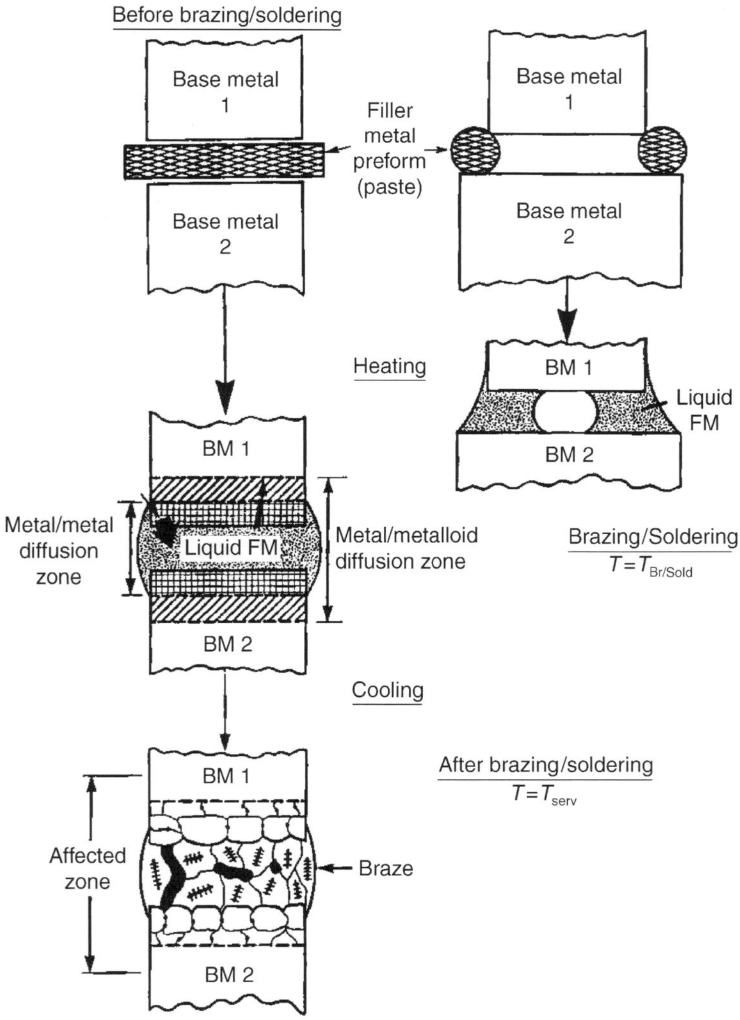

Figure 1
The essence of the brazing/soldering processes.

the gap normally existing between the parts. This gap is usually rather small, on the order of 10 to a maximum of a few hundred micrometers, or even close to zero. Thus, the degree of penetration of a molten alloy into the gap is determined largely by capillary effects and, of course, by wetting. Because physical parameters such as part dimensions and the surface tension forces play the major role, this stage may be conditionally called a physical one.

The second stage, which normally sets in at a given joining temperature, is characterized by an intensive solid–liquid interaction accompanied by a substantial mass transfer through the interface at strongly uneven rates. Indeed, some BM immediately adjoining the liquid filler metal dissolves at this stage. At the same time, a small amount of elements from the liquid phase penetrate into the solid BM. Such mass transfer unbalance is due to dramatically different diffusion rates in the solid and liquid phases. This redistribution of components in the joint area leads to changes of phase composition and sometimes to the onset of crystallization or, in other words, joint solidification. All the processes of this stage in joining have a distinctly metallurgical character and therefore the second stage may be properly called a metallurgical one. During soldering, in most cases, all these processes proceed to a much lesser degree because of a substantial difference in the melting temperature of a solder and a BM. However, often the BM surface still needs to be protected from erosion by molten solder.

The final stage of the brazing (soldering) process overlaps with the second and is characterized by the formation of the final joint microstructure, progressing vigorously during assembly cooling while the liquid phase is still present. Here, the crystallization process takes place. Subsequent cooling results in only minor changes such as annealing of the joint and partial relief of thermal stresses induced on cooling. This stage has very similar features in both brazing and soldering processes.

In any brazing/soldering process, a molten alloy comes in contact with a surface of solid, which may be an alloy, a ceramic, or a composite material. For a molten alloy to advance over the solid surface, a special relationship has to exist between surface energies of the liquid–gas, solid–gas, and liquid–solid interfaces. The same relationships should, in principle, hold in joining processes where a molten alloy has to fill the gaps existing between surfaces of the parts to be joined. In general, this means that the molten alloy should have lower surface tension than the base metal.

All transition metals, particularly iron, have surface tension higher than $1500\,\mathrm{mN\ m^{-1}}$, whereas nontransition metals, such as silver, tin, lead, and particularly metalloids such as silicon, have surface tension substantially lower than that of major groups of steels and other BM alloys. The addition of phosphorus, silicon, boron, etc. to aluminum, nickel, and copper results in a precipitous drop in the surface tension of the resultant materials. Therefore, for low-temperature joining, tin- and lead-based alloys have been used as solders, whereas aluminum-, copper-, and nickel-based alloys with major additions of P, Si, and B have been used as brazing filler metals. Both groups of alloys in the molten state have surface tension values favorable for strong wetting (see Eustathopoulos *et al.* 1999).

Because the surface tension of oxides such as Fe_xO_y, Al_2O_3, Cr_2O_3, etc. are much lower than those of the majority of metals and alloys, the presence of an oxide film on a BM surface impedes wetting. It is of paramount importance to keep the BM surface clean from so-called virgin oxides and oxides that may have formed during the brazing operation if the heat-treating atmosphere is oxidizing. This is particularly important during the brazing of stainless steels and high-temperature alloys containing active metals, such as chromium, titanium, and aluminum, that form corresponding protective oxide films. The wettability of base metals is strongly affected by cleanliness, not only with respect to oxides, but also by the presence of traces of grease, oil, etc. on part surfaces. Therefore, cleaning of the BM pieces to be joined and protecting them from oxidation during heating are essential steps in joining operations.

Chemically active substances (fluxes) are used to accomplish cleaning and to provide protective shielding when a vacuum or protective atmosphere is not available. However, the use of nonmetallic agents in fluxes is not always desirable, owing to the increased propensity for their entrapment and the subsequent formation of voids in the brazed joint. Alternatively, if the brazing environment is not oxidizing, then boron, silicon, and phosphorus, which are used as inherent constituents of FM alloys, can play the role of a flux, i.e., reducing original oxide films and making FM self-fluxing.

2. *Major Groups of Base Materials and Brazing Filler Metals*

2.1 *Base Materials*

Due to the universality of metallurgical mechanisms that work during brazing and soldering, practically all types of metallic alloys, steels, ceramics, and composite materials can be joined by either of these technologies. Detailed descriptions of optimal brazing filler metals and associated processing for a large number of metallic alloys, ceramics, and composite materials can be found in Brazing Handbook (1991), Hupston and Jacobson (1993), and ASM Handbook (1993). The range of examples is endless; it includes applications involving gold- and silver-based alloys, all types of carbon and stainless steels, and heat-resistant, hard, and refractory alloys, including those based on titanium, zirconium, and beryllium. Other materials such as high-temperature ceramics, graphite and graphite composites, polycrystalline diamonds, etc. are also widely engaged.

2.2 *Brazing Filler Metals*

The choice of both specific FM alloy composition and optimal brazing conditions are mostly determined by joint performance under specific service conditions. Ideally, the composition of an FM alloy must be such that the following four functions are achieved: (i) the FM melting temperature must be lower than that of the corresponding BM; (ii) the FM surface tension for both solid and liquid states must be lower than that of the BM to provide a driving force for wetting; (iii) the FM must be compatible with the BM in order to form good metallic bonding, i.e., FM and BM structure, composition, and properties should be similar; and (iv) the FM-containing elements should be able to bring about chemical reduction/decomposition or physical removal of any BM oxide film.

Accordingly, five families of brazing FM alloy compositions have emerged with respect to their metallurgical nature or type. Of these, the first four families of conventional brazing FMs have been classified by the American Welding Society (AWS) into eight well-defined classes, as indicated in Table 1. Table 1 also contains the preferred base metals with which each specific family is best compatible as well as the major areas of FM applications.

The first and largest Family, family I, contains eutectic-type alloys having aluminum, nickel, cobalt, and copper as a base to which silicon and/or boron (aluminum- and nickel-base alloys) and phosphorus (copper- and nickel-base alloys) are added. Family I encompasses AWS classes 1, 4, 6, and 7, as well as nonclassified nickel- and palladium-based alloys. The presence of one or more of these metalloid elements in alloys tends to impart the required characteristics to the FM. Silicon combined with a small amount of magnesium is successfully used in aluminum-based FM alloys, the surface tension and melting temperature of which are depressed, and good bonding is promoted. The magnesium addition serves as a fluxing agent. Boron and silicon are used in high-temperature FMs in which the presence of another potential melting temperature depressant and surfactant element, i.e., phosphorus, causes unacceptable joint brittleness.

Of the alloy additions, which promote self-fluxing of the FM during brazing, boron has the greatest penetrating power. Alternatively, phosphorus is a beneficial fluxing element for use in copper (low-temperature) brazing, whereas another element, e.g., silicon, would cause unacceptable copper joint brittleness. Most of these alloys are brittle because various intermetallic phases precipitate when processed by conventional technology. This brittle character has limited the available forms of conventionally produced brazing alloys to powder. The presence, however, of silicon, phosphorus, and boron in many conventional FM alloys having near-eutectic compositions facilitates the conversion of such alloys into ductile, thin amorphous alloy foil form by using rapid solidification (RS) technology (Sexton and DeCristofaro 1981, see also *Rapid Solidification Processing*).

The second family of brazing FMs, Family II, consists of solid solution alloys based on copper (class 2) and gold–nickel (class 8). These alloys are used mainly in vacuum-brazing applications and, therefore, require no alloying elements playing the role of fluxing agents. Family III consists of alloys having a phase diagram where a peritectic reaction exists, such as copper–zinc alloys (class 3) and non-classified Cu–Sn alloys. The fourth family of alloys, Family IV (class 5), is probably the most widely used. These alloys are based on the copper–silver binary eutectic system which is modified by substantial additions of zinc and cadmium (both providing fluxing activity), and minor additions of tin and nickel. A small but important FM group associated with Cu–Ag-based Family IV contains Cu–Ag–Ti/Zr compositions used for brazing ceramics. Here, oxygen-active Ti and Zr promote wetting of oxides such as Al_2O_3 due to their reducing reaction at joint interfaces (Moorhead *et al.* 1993). Family V of brazing FM alloys, although so far nonclassified, consists of purely metallic eutectic/peritectic titanium/zirconium-based alloys to which copper and/or nickel are added.

3. *Basic Forms of Filler Metals*

Filler metal forms include solid preforms and powders, the only form in which a wide variety of FMs, which are brittle, can be produced using conventional metallurgical means. The powders are mostly used in the compound form of paste and plastic-bonded tape. Conventional paste forms of joining alloys, e.g., FM powder, plus fluxing agent, plus binder/solvent, are applied at externally accessible locations of the clearance between BM pieces (Fig. 1). Such practice requires substantial FM liquid flow during joining to achieve the fusion of constituent FM powder particles and binder/solvent extraction. In addition, organic binders decompose when compound FM forms are used in the high-vacuum brazing of parts intended for critical high-temperature service. Such decomposition of binders can result in the formation of soot in the joint, which can act to degrade the performance of expensive vacuum equipment.

Table 1
Major AWS classes of brazing filler metals.

Class	Alloy family (I–V) and type	AWS designation	Forms	Base materials joined	Major applications
1	Al–Si, eutectic (I)	BAlSi	Preforms, wire, rods, foil, powder, RS foil[a]	Aluminum and aluminum alloys, steel to aluminum and aluminum to beryllium	Car radiators, heat exchangers, honeycomb aircraft structures, structural parts
2	Cu–X, solid solution (II)	BCu	Preforms, wire, rods, foil, powder, RS foil	Copper and copper alloys, copper to mild steel, copper to stainless steel	Heat exchangers, structural parts, automotive parts
3	Cu–Zn, peritectic (III)	RBCuZn			
-	Cu–Sn, peritectic (III)	None			
4	Cu–P, eutectic, (I)	BCuP	Preforms, wire, rods, foil, powder, RS foil	Copper to copper, copper to silver/oxide powdered metal composites	Electrical contacts, bus bars, heat exchangers
5	Cu–Ag, eutectic (IV)	BAg	Preforms, foil, powder	Most ferrous and nonferrous metals, except aluminum and magnesium	Most widely used utility filler metals
6	TM–Si–B[b], eutectic (I) (Ni/Fe + Cr) Si–B	BNi	Powder, tape[c], RS foil	AISI 300 and 400 series steels and nickel- and cobalt-base superalloys; carbon steels; low-alloy steels; and copper	Aircraft turbine components, automotive parts, heat exchangers, honeycomb structure
7	(Co, Cr)–Si–B	BCo	Powder, tape, RS foil	Cobalt-based heat-resistant alloys, steel	Aircraft engines, honeycomb marine structures
-	(Ni, Pd)–Si–B	None	Powder, tape, RS foil	AISI 300 series stainless steels, cemented carbide, superalloys	Honeycomb structures, cemented carbide/polycrystalline diamond tools, orthodontics, catalytic converters
8	Au–Ni, solid solution (II)	BAu	Preforms, wire, rods, foil, tape	Nickel-based heat-resistant alloys, steels	Honeycomb structures, structural turbine parts
-	Cu–(Ti, Zr)–Ni eutectic and peritectic (V)	None	Cladded strip, RS foil	Titanium/zirconium-based alloys	Titanium tubing, aircraft engines, honeycomb aircraft structures, aircraft structural parts, chemical reactors

[a]May be produced as rapidly solidified, ductile, amorphous/microcrystalline foil. [b]This family includes alloys based on transition metals, such as nickel, iron, cobalt, and palladium. [c]Brazing filler metal is carried on a plastic-bonded tape.

4. *Joint Requirements*

A basic requirement for any joint is that its strength and ductility be equal or, at least close to that of the BM. In general, the strength of a material increases with decreasing grain size, whereas ductility is affected by the presence of brittle phases. Therefore, in addition to limiting porosity in joints, it is important to limit grain size and the amount of brittle phases. Ideally, the maximum size that the grains in a joint may achieve is equal to half the clearance between the BM pieces. Thus, the use of a smaller clearance during BM joining necessarily limits maximum grain size, promotes higher cooling rates of the FM alloy, and thereby results in a refined joint microstructure. A smaller clearance in brazing also promotes improved retention of BM properties because of curtailed BM erosion by the use of a smaller volume of FM. For these reasons, a preplaced self-fluxing thin FM foil used as a preform is superior to the use of powder-containing paste. The latter requires larger clearances for filling joint cross-sections and results in deleterious effects on properties owing to a coarser joint grain size, more fully developed intermetallic compounds, and the presence of substantial amounts of contaminants.

The grain size and brittle intermetallic size and extent in the parent FM have a direct effect on the formed joint strength. This consideration is particularly important under transient heating/cooling conditions such as torch or belt oven brazing. The rapid dissolution of constituent FM alloy phases allows the use of higher joining throughput speeds and/or lower joining temperatures.

By this reasoning, newer rapidly solidified microcrystalline and amorphous brazing filler metals and solders are the epitome of FM microstructural possibility: they possess ultimate uniformity of their elemental components, which is beneficial for the formation of a fine microcrystalline joint microstructure in both soldering and brazing.

5. *Amorphous Alloy Brazing Foil*

Since the 1980s, the majority of all FM classes, even noneutectic alloys, have been processed successfully by rapid solidification (RS) technology (Sexton and DeCristofaro 1981, Rabinkin and Liebermann 1993, Rabinkin 1998, see also *Rapid Solidification Processing*). Indeed, the processing of FMs was one of the first practical, industrial applications of this technology. Their availability on the market was well timed; it directly advanced the brazing of high-temperature alloys used in the aerospace industry in the form of a flexible ductile foil which could replace inherently brittle alloys, such as the above mentioned (Ni/Co/Pd)–B–Si-based compositions. Since the discovery of RS potential in the manufacturing of flexible brazing foil (Sexton and DeCristofaro 1981), practically all developmental work on RS brazing foil has been carried out by a group associated with Amorphous Metals. This amorphous filler metal (AFM) is now known under the trade name METGLAS® foil; this has become close to a generic one and is designated as MBF-XX-, MBF-10X-, and MBF-100X-series.

The most important advantage of RS amorphous and microcrystalline AFM alloys is their flexibility and ductility. This opens an opportunity to use them as preplaced preforms with no need for large brazement gaps, such as those used with pastes, in order to achieve a complete filling of the braze cross-section. Here, AFM, for example, has a particular advantage over powder and polymer-bonded strip forms because of its superior flow characteristics and low oxygen concentration. Indeed, gas-atomized powder has a very large total surface area with subsequently large amounts of surface oxides. These oxides to a certain degree prevent fusion of individual metallic droplets into a uniform liquid pool and cause joint porosity. The AFM flows more freely upon melting than any powder form (DeCristofaro and Bose 1986). A smaller clearance also promotes improved retention of BM properties because of curtailed BM erosion by the use of a smaller volume of FM in the AFM form. The paste requires larger clearances for filling joint cross-sections and results in deleterious effects on properties owing to a coarser joint grain size, more fully developed intermetallic compounds, and the presence of substantial amounts of contaminants.

The modern manufacture of RS products is characterized by high cooling rates (approaching 10^6 K s^{-1}) which enable the stabilization of certain alloys into an amorphous solid state, having a spatial distribution of atoms similar to that of liquids with a random, spatially uniform arrangement of the constituent atoms. Because RS amorphous materials are compositionally much more uniform even after crystallization, their melting under transient heating occurs over a narrow temperature range. This is a consequence of the shorter distances over which atoms of different elements have to diffuse in order to form a uniform liquid phase. The resulting "instant melting" of RS materials is only one of their important features. This is particularly important when brazing fine-gauge honeycomb cores, for example, which have to be protected from erosion by molten filler metals during joining. A shorter brazing time is also beneficial in cases where BM parts, such as cold-deformed stainless steels and precipitation-hardened superalloys, may lose their inherent strength due to annealing during the brazing operation.

The absence of contaminating organic solvent bases, which are present in powder pastes/tapes, correspondingly eliminates soot formation and furnace fouling. The low level of gaseous impurities in

AFM, due to the specific characteristics of its production technology, is an attractive feature for vacuum furnace brazing. Because of these unique properties, AFM occupies an important and increasing role as the most advanced filler metal used in applications related to the aerospace industry, precise machinery and tools, and modern medical equipment, among others. Most AFM classifications may also be grouped according to their metallurgical type.

5.1 Metal–Metalloid Eutectic Group: Ni-, Co-, and Ni–Pd-based Alloys

(a) Nickel-based amorphous alloys (MBF-XX-series). A wide variety of ribbons with compositions similar to many conventional FMs are, at the start of the 2000s, in extensive industrial use. Compositionally these alloys originate from ternary Ni–B–Si alloys, the phase diagram of which has a number of ternary eutectics, and from a binary $Ni_{81}P_{19}$ eutectic alloy. These Ni–B–Si eutectic compositions have a rather wide range of values of their (B + Si) concentrations (12–28 at.%), whereas their solidus temperatures, T_S, are located within a range of about 960–1000 °C. The $Ni_{81}P_{19}$ binary eutectic has a temperature of 870 °C and the T_S of alloys containing phosphorus is close to this value (883 °C). The majority of all these alloys also contain a small amount of iron (3–4.5 wt.%) and chromium (3–19 wt.%). The former is added to improve brazability, whereas the latter improves the corrosion resistance of joints formed by these alloys.

Changes in (B + Si) concentration and in the B/Si concentration ratio give the customer a choice of FM alloys with melting temperatures and performance tailored to their specific needs. Still, there is a limit in the allowable range of B and Si concentrations, which is determined by the potential loss of alloy ductility in the as-cast state. Addition of metalloids to nickel may vary: for P it is 10–11 wt.%, for B it is 3.5–4.0 wt.% and can be as low as 1.25 wt.% if about 7 wt.% Si is added. Silicon can be used only together with B and may vary within a 0–7.5 wt.% range. All Ni-based AFM can be produced as a ribbon having 300 mm width and thickness varying within an 18–75 μm range.

At first, the prime application area of Ni-based AFM was confined to the aerospace industry where for years AFM was used in the brazing of aircraft structural parts, acoustic tail pipes (J-pipes), thrust reversers, turbine blades, and seals. Since the late 1990s, owing to growing attention to energy conservation and pollution elimination, there has been a substantial growth of AFM application in production of plate heat exchangers, fuel cells, heat pumps, and catalytic converters for internal combustion engines. The improvement in cutting very intricate shapes (Fig. 2) from these very hard materials using stamping, photo etching, and inexpensive high pressure water jet methods has been a significant economic factor in application expansion of Ni-based and other AFM. Strong, ductile, and corrosion-resistant stainless steel and superalloy joints, having a tensile/shear strength close to that of these BM in the virgin state, are produced as a result of the fine-tuned optimization of the processing parameters (Rabinkin *et al.* 1998, Heikinheimo *et al.* 2001).

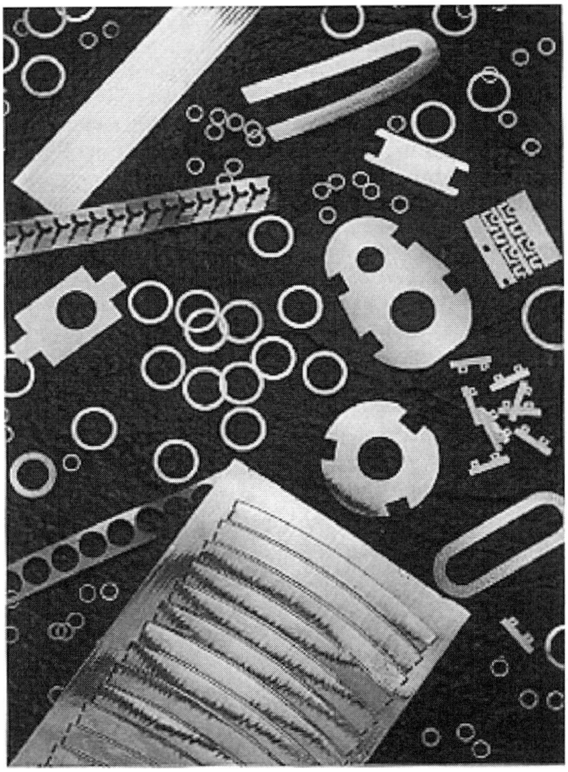

Figure 2
Various ribbons and preforms stamped and photo etched from METGLAS® amorphous brazing foil.

(b) Co-based Amorphous alloys (MBF-10X-series). The necessity to increase the working temperature of jet and stationary turbine engines to 1200 °C led to the development of Co-based AFM which can work better than their Ni-based counterparts. The BMs to be joined in this case are highly alloyed poly- and single-crystalline heat-resistant materials which have a substantial amount of aluminum, titanium, and refractory elements. These elements form a strong 0.3–0.8 μm thick complex oxide layer on the surface of parts in a virgin state that protects these parts from oxidation. Application of MBF-10X-series alloys overcomes wetting

problems associated with the presence of the oxide layer via the self-fluxing action of boron and silicon as AFM components. A long brazing time combined with long-term postbrazing annealing yields a very favorable joint microstructure free from intermetallic phases and excessive chromium borides at grain boundaries, resulting in very high mechanical properties (Heikinheimo *et al.* 2001)

(c) Ni/Pd-based alloys (MBF-100X-series). These materials have been used for applications where exceptional corrosion resistance and high operating temperatures are needed (Datta *et al.* 1985, Rabinkin 1990). To retain amorphous state upon RS they need a moderate amount of metalloids added to the Ni/Pd base resulting in a high joint ductility. They melt at very low temperatures—their liquidus T_L may be $\sim 820°C$—and also possess higher strength than that of substantially more expensive gold-based alloys. Those factors make them a superior replacement for the latter in joining jet engine parts and polycrystalline diamond blanks for oil drilling furnished with cemented carbides. Specifically, their low melting temperature guarantees that diamond will not be transformed into graphite upon brazing. Some compositions contain a sufficient amount of cobalt and tungsten. This makes them particularly compatible with cemented carbides while preventing the leaching of cobalt from the cemented carbides matrix and retaining a strong and nonporous joint interface (Rabinkin and Pounds 1989).

5.2 Metal–Metal Group: Ti/Zr–Cu–Ni Amorphous Alloys

This group contains Zr–Ni binary, Ti/Zr–Ni and Ti–Cu–Ni ternary, and Ti–Zr–Cu–Ni quaternary alloys, all of which are good glass formers upon RS and have a rather low melting temperature (Rabinkin *et al.* 1991). Copper and nickel are added to binary Ti–Zr compositions to decrease melting temperature and, most importantly, to facilitate conversion of these alloys into the amorphous state. The amorphous Ti/Zr–Cu–Ni alloys have excellent melting characteristics (narrow T_S–T_L range and T_L lower or close to β-transus of the majority of Ti-based alloys), ultimate uniformity of microstructure, and are compatible with Ti/Zr-based materials because they contain these elements. Conventional FM is made only as a clad strip, which melts in a high and wide temperature range. The existing RS vacuum-casting technology is refined now to such a degree that it is possible to produce a plurality of compositions in a good quality, flexible foil form, up to 50 mm wide. The Ti/Zr-based BMs, when brazed, form joints with a variety of complex multiphase microstructures that may be brittle.

It was shown that to achieve high joint ultimate tensile strength ($\sigma_{UTS} > 1000$ MPa and elongation $\varepsilon = 10\%$), the brazing time should be short and cooling should be carried out at a sufficient rate to obtain a very fine lamellar or cellular α-Ti + γ-(Ti,Zr)$_2$Cu eutectoidal microstructure throughout the joint (Botstein and Rabinkin 1994, Botstein *et al.* 1995). In spite of their superior performance, the industrial application of these materials is still limited because of their high cost. Still, some of the alloys have been successfully tested by NASA in brazing laminate composites made of SCS-6/β21S (Ti-15Mo-2.7Nb-3Al-0.25Si) titanium matrix reinforced by silicon carbide fiber (Hoffman *et al.* 1994) and titanium aluminides. Joints made of 70T-15Cu-15Ni AFM are 70 and 80% stronger than samples made using conventional 70Ti/(15Cu-15Ni) laminated strip at room temperature and 816°C, respectively.

6. Importance of Brazing Filler Metals in Modern Manufacturing

After four millennia of uninterrupted history, modern filler metals continue to play an extremely important, irreplaceable, and critical role in manufacturing. The wide scope of FM applications in the early 2000s is demonstrated by the following examples, all reliant on brazing filler metals for their manufacture: bundles of thousand-strong feeding/cooling pipes in the hearts of Saturn and shuttle rockets; polycrystalline diamond/cemented carbide bits brazed to cemented carbide crowns, used for drilling more than 40% of all oil wells in the world; all jet engines of civil and military airplanes with a total value of hundreds of billions of dollars; the vast majority of car and track radiators; titanium parts of the body skin of modern jet fighters; orthodontic braces for children; all crowns of diamond engagement rings and innumerable items of gold and silver jewelry.

Bibliography

1993 *Welding, Brazing and Soldering. ASM Handbook.* Metals Park, OH Volume 6, pp. 328–72, 617–40, 903–1002

1991 *Brazing Handbook.* AWS, Miami, FL, pp. 43–65

Botstein O, Rabinkin A 1994 Brazing of titanium-based alloys with amorphous 25Ti-25Zr-50Cu filler metal. *J. Mat. Sci. and Eng., A* **188**, 305–15

Botstein O, Schwarzman A, Rabinkin A 1995 Induction brazing of Ti-6Al-4V alloy with amorphous 25Ti-25Zr-50Cu brazing. *J. Mat. Sci. and Eng., A* **206**, 14–23

Datta A, Bose D, DeCristofaro N, Rabinkin A 1985 High strength nickel-palladium-chromium brazing alloys. *Welding J.* **64**, 23–9

DeCristofaro N, Bose D 1986 Brazing and soldering with rapidly solidified filler metals. In: In: Lee P, Moll J (eds.) *Rapidly Solidified Materials: Properties and Processing.* ASM, Metals Park, OH, pp. 415–21

DeCristofaro N, Datta A 1985 Rapidly solidified filler metals in brazing and soldering. In: Das S K, Kear B J, Adam C M (eds.) *Rapidly Solidified Crystalline Alloys*. TMS-AIME, Warrendale, PA, pp. 263–8

Eustathopoulos N, Nicholas M G, Drevet B 1999 *Wettability at High Temperatures*. Pergamon, Oxford

Heikinheimo L, Miglietti W, Kipnis J, Leone E, Rabinkin A 2001 Brazing of Co–Ni-based superalloys using new amorphous brazing filler metals. In: *Proc. LOT01, 212*, DVS, Aachen, Germany, pp. 30–4

Hoffman E, Bird R, Dicus D 1994 Brazed joint properties and microstructure of SCS-6/β21S titanium matrix composites. *NASA Technical Paper* **3396**, 1–39

Hupston G, Jacobson D 1993 *Principles of Soldering and Brazing*. ASM, Metals Park, OH, pp. 31–70

Moorhead A J, Elliot W H, Kim H E 1993 Brazing of ceramic and ceramic-to-metal joints. In: *Welding, Brazing and Soldering. ASM Handbook*, Volume 6, pp. 948–60

Rabinkin A 1990 Melting characteristics, glass-forming-ability and stability of Ni-Pd-Si and Ni-Pd-TM-Si alloys in the concentration range 10 to 22 at.% Si. *Mat. Sci. Eng., A* **124**, 251–8

Rabinkin A 1998 After two decades of history-new frontiers of amorphous brazing foil application In: *Brazing, High Temperature Brazing and Diffusion Welding, 5th International Conference*. DVS, Aachen, Germany, pp. 44–8

Rabinkin A, Liebermann H H 1993 Brazing and soldering with rapidly solidified alloys. In: Liebermann H H (ed.) *Rapidly Solidified Alloys*. Dekker, New York, Chap. 21, pp. 691–736

Rabinkin A, Liebermann H H, Pounds S, Taylor T, Reidinger F, Siu-Ching Lui 1991 Amorphous Ti-Zr-based METGLAS® brazing filler metals. *Scr. Met. et. Mat.* **25**, 399–404

Rabinkin A, Pounds S 1989 Brazing cemented carbides: specifics, braze optimization, and custom-designed METGLAS® brazing filler metals. *Int. J. Refr. Hard Met.* **8, N4**, 224–37

Rabinkin A, Wenski E, Ribaudo A 1998 Brazing stainless steel using a new MBF-series of Ni–Cr–B–Si amorphous foil. *Welding J.* **77**(2), 66–75

Sexton P, DeCristofaro N 1981 *US Pat.* 4 148 973

A. Rabinkin

C

Casting of Intermetallics

Several intermetallic alloys offer potential as structural materials in demanding applications. However, many of the unique properties of intermetallic alloys that make them attractive for high temperature structural applications, such as high melting and disordering temperatures, high stiffness, low diffusivity, etc., also make these alloys a challenge to process into useful products. The cast intermetallic alloys which are closest to commercial application are titanium- and nickel aluminide alloys, namely: TiAl ($L1_0$, tP4); Ni_3Al ($L1_2$, cP4); and NiAl (B2, cP2). These three alloys have relatively high melting temperatures, and retain their ordered structures to or nearly to their respective melting points. The alloys also have relatively high ductile-to-brittle transition temperatures, often making processing, machining, and handling difficult. Additionally, the low melting temperature and the high volatility of aluminum can pose challenges to traditional melting and casting technology.

In spite of these challenges, some success has been achieved in the casting of some intermetallic alloys utilizing existing or modified cast-process technology, when exercised with care. This chapter focuses on three classes of cast structural metal-aluminide intermetallic alloys that have achieved at least limited commercial success: near-gamma titanium aluminide alloys; and two types of nickel aluminide alloys.

1. Cast Gamma TiAl Alloys

The development of casting technology for gamma TiAl alloys has been aided by the ability to use conventional Ti melting and casting technology, with minor modifications. Many barriers to the successful cast processing of gamma alloys have been largely overcome, such as component fill, cracking and hot tearing, and surface-connected porosity (Ram and Barrett 1996, McQuay *et al.* 1999). Some significant remaining barriers include the cost of ingot, the ability to control dimensions, and the establishment of a vendor base. The vendor base issue extends beyond casting vendors, to include capable machining sources for gamma alloys.

1.1 Melt and Casting Technology

(*a*) *Melting technology.* Conventional titanium alloy melting processes are typically used for the manufacture of gamma alloy ingots and castings. Because of the reactivity of molten TiAl alloys with ceramic crucibles, cold-wall crucible melting is required for producing aerospace-grade ingots and castings. The most common ingot casting methods are vacuum-arc remelting (VAR), induction-skull melting (ISM), and plasma arc melting (PAM). To take advantage of economy of size, conventional titanium casting ingots are generally converted from large-diameter ingots (360–860 mm) via thermal-mechanical processing. These processes are currently not cost-effective for gamma ingots because of their poor workability, so the ingot is usually melted and cast directly to the appropriate size. Although still under development, the PAM process is gaining in popularity due to advantages in improved flexibility in ingot size and shape, more control over thermal stresses (which can cause cracking of the ingot), the ability to revert scrap easily, and better compositional control of volatile elements such as Al and Mn (Dowson *et al.* 1995). Ingot-production processes have matured to where the typical alloy ingot cost is in the range from \$35–60 per kg (Austin and Kelley, 1995, McQuay and Larsen 1997). Additional alloy ingot cost reductions should be realized with higher volume production.

(*b*) *Investment Casting Technology.* VAR-pour is perhaps the most common investment casting melt practice used for gamma alloys, again due to its availability, and it remains the only practical technique for castings with pour weights in excess of 60 kg. ISM is gaining in popularity in Ti alloy castings, including gamma TiAl alloys, because of its increased flexibility and higher productivity in casting compared with VAR-pour (Larsen and Govern 1995). However, care must be taken in order to control the amounts of volatile elements in the melt, such as Al and Mn (Rishel *et al.* 1999).

Cold-wall crucible melting limits the superheat of the melt compared with VIM, generally in the range of 10°C to 60°C, with the amount of available superheat being somewhat a function of the crucible and coil design, and the frequency and available strength of the power supply. The limited superheat can restrict the ability to fill thin-wall sections (<2.5 mm) in conventional gravity casting. Two pressure-assist processes have been developed which improve fill of thin sections: centrifugal casting and counter-gravity casting. Centrifugal casting, as practiced for Ti and TiAl alloys, usually involves spinning one or more molds on a turntable. The molten alloy is usually delivered from the crucible to the molds via a tundish or piping system, usually made of steel.

Another method of pressure-assisted investment casting, that is gaining in popularity, is the counter-gravity low-pressure inert-atmosphere (CLI) investment casting process developed by the Hitchiner Corporation (Chandley 1991). This process utilizes a pressure differential between the melt and mold chambers to drive the molten alloy up through a fill tube and into the mold cavity. The rate of rise of the metal is specified by a computer-controlled vacuum system, which allows for bottom feeding with a greater degree of control over metal velocity. Properly practiced, this technology can both improve fill of difficult-to-cast parts and reduce defects. Daido Precision Casting, a licensee of the Hitchiner counter-gravity casting technology, has developed a version of the Hitchiner technology which combines counter-gravity casting with an induction skull-melt system, named Levi-Cast, to produce gamma TiAl alloy turbocharger wheels (Isobe and Noda 1997, Noda 1998). This system has the benefit of clean, controlled induction-skull melting, with the superior fill control of counter-gravity casting.

The successful casting of gamma-alloy components requires careful attention to casting parameters, such as mold materials, mold preheat temperatures, and gating design (Larsen and Govern 1995, Ram and Barrett 1996, McQuay *et al.* 1999). In particular, conventional investment casting mold systems utilized in the casting of titanium and nickel-based alloys may be too strong for the lower ductility and strength of gamma TiAl alloys, which can lead to component hot tearing and cracking. Hence, modifications are often made to these mold systems to make them more suitable for gamma TiAl alloy casting.

Mold preheat temperature is one of the most important casting variables at the disposal of the caster. Higher preheat temperatures not only improve fill and feeding, but also reduce the thermal gradients and the cooling rate in a given casting. The reduction of thermal gradients and cooling rates can be critical in producing tear- and crack-free components. However, higher preheats can also lead to severe metal/mold reactions, and can increase the propensity for surface-connected porosity (Larsen and Govern 1995, Ram and Barrett 1996, Rishel *et al.* 1999). Slower cooling rates can also lead to coarser microstructures and inferior mechanical properties. Therefore, a balance must be found in order to achieve a component which meets the customers design requirements.

(*c*) *Metal mold casting technology.* While the majority of casting processes employed for gamma TiAl have employed investment casting, several companies have developed novel versions of metallic-mold casting processes to produce simple components, such as automotive exhaust valves, at costs potentially lower than that which can be achieved by investment casting. Two metallic mold casting processes have been developed which offer potential cost savings for simple parts, relative to investment casting. The Howmet Corporation has developed a gravity metal mold (GMM) process (Colvin 1995), while a German government and industry consortium, led by ALD Vacuum Technologies, has developed a centrifugal permanent mold casting process which utilizes heated refractory metal molds (Blum *et al.* 1999). The primary target market for both of these processes are automotive exhaust valves.

1.2 Cast Microstructure

In the as-cast form, most two-phase gamma alloys exhibit coarse, columnar, primary alpha colonies, with some amount of interdendritic gamma (Huang and Chesnutt 1995). The alloys also exhibit interdendritic-connected porosity and centerline porosity; so nearly all applications require hot isostatic pressing (HIP) to ensure soundness, just as with conventional Ti-alloy castings. Common HIP temperatures are from 1165°C to 1260°C, at pressures from 125 MPa to 175 MPa for up to four hours. The columnar grains generally form perpendicular to the mold surface, growing inward toward the center of the casting, leading to a cast texture (De Graef *et al.* 1999). In addition to closing porosity, HIP processing promotes the formation of equiaxed gamma grains, and heat treatments can be used to further modify the microstructure. The addition of B transforms the as-cast microstructure to predominately equiaxed lamellar colonies and gamma grains, and reduces the sensitivity of the microstructure and properties to casting conditions and component geometry (Larsen 1996). A more comprehensive review of cast gamma alloys and microstructures are given by Dimiduk *et al.* (2000).

1.3 Technology and Production Readiness

Over the past decade, significant government and corporate resources have been expended to bring gamma TiAl alloys to a state of production readiness for a variety of commercial and aerospace components. These efforts have made progress towards demonstrating technology and production readiness for cast gamma TiAl alloy applications. However, some significant work remains for some of the most challenging applications, such as turbine blades; while other applications, such as turbocharger turbine wheels, are finally emerging. Although cast processing has advanced considerably for gamma alloys, many important component-processing technologies require additional effort: joining and repair; surface treatments and coatings; and machining and finishing technologies.

The first reported commercial application for a gamma TiAl alloy is a turbocharger turbine wheel for the Mitsubishi Motor Lancer Evolution VI produced by Daido Precision Casting via the Levi-Cast process (Tetsui 1999). Although this is a small volume production launch, the importance of a successful commercial production launch for the development and design community should not be understated.

Nearly every major turbocharger and diesel engine manufacturer has successfully tested cast gamma TiAl alloy turbocharger turbine wheels, some of which have been reported at conferences and in the open literature: ABB (M. Nazmy, oral presentation at Aeromat '98, Washington DC, June 15–18, 1998), Honeywell Garrett (C. Baker, oral presentation at the High Temperature Materials Seminar at AES, Torrance, CA, February 26, 1998). MHI (Tetsui 1999), and Toyota (Nishino *et al.* 2000). These wheels range in size from 50 mm diameter for the Mitsubishi gasoline engine turbocharger to 250 mm diameter for an industrial diesel turbocharger turbine for ABB.

Nazmy (oral presentation at Aeromat '98, Washington DC, June 15–18, 1998), Baker (oral presentation at the High Temperature Materials Seminar at AES, Torrance, CA, February 26, 1998) and Tetsui (1999) have reported a significant reduction in "turbo lag," which is the delay time required to spin the turbocharger rotor up to speed where it can effectively provide more complete combustion for the engine. The reduction in turbo lag leads to a reduction in particulate emissions and an improvement in engine responsiveness. Nishino *et al.* (2000) reported that the application of a gamma turbine wheel in combination with an improved stiffness shaft can reduce the moment of inertia for the turbocharger wheel assembly by 42%, and increase the resonance frequency of the system 44%.

Many automotive and racing engine manufacturers have tested cast-gamma engine intake and exhaust valves because they can provide benefits such as increased horsepower, efficiency, and possibly lower emissions, due to their lower density versus steel alloys. Test results have been published by Ford (Dowling *et al.* 1993), General Motors (Hartfield-Wünsch *et al.* 1995), BMW (Blum *et al.* 1999) and others. Mass production of automotive engine valves requires prices dramatically lower than those available at the beginning of the 21st century that utilize prototype metal-mold and investment-casting technology. However, racing valves offer an opportunity where a premium will be paid for improved performance, and where reliability is crucial.

Within turbine engine applications, a number of successful engine tests have been reported. Perhaps the most publicized engine test was the GE CFC-80C low-pressure, turbine-blade test (Austin and Kelly 1995). Additional tests include: the Volvo high-pressure turbine-blade dampers (Larsen 1996); GE90 transition duct beams (Austin and Kelly 1995); GE F414 carbon seal housing (Austin *et al.* 1997); and a Pratt & Whitney high pressure turbine blade retainer (Stewart 1994).

One of the largest and broadest-based technology demonstration programs for gamma alloys was the USAF F119 derivative CAESER test program (Austin *et al.* 1997, Davidson 1996). The gamma alloy CAESER engine test components included ninth-stage compressor blades, compressor inner diameter shroud segments, and exhaust nozzle tiles. The blade and shroud segment test components were tested in an F119 derivative engine core beginning in 1998, followed by full engine testing for all three gamma components.

1.4 *Remaining Issues*

Although progress has been made in nearly every important step in the transition of the gamma alloys to production, a number of issues remain which require attention. This discussion is not a comprehensive list, but rather points out some of the most serious barriers to commercial production.

Alloy cost: although cost has steadily decreased over the last few years, continued improvements are required in the development of low-cost aerospace and automotive-grade melt-stock.

Low-cost machining technology and vendor base: progress has been made in nearly every type of machining, but additional improvements are required in low-cost conventional machining, especially in turning, grinding and drilling; additionally a competent vendor base needs to be established that can handle low-cost production.

Welding and joining: joining technology will continue to be required in nearly all gamma applications on the horizon, and weld-repair technology will be required for medium and large structural castings.

Production readiness: product development cycles continue to shorten, making it even more difficult to develop processes and to demonstrate production readiness within the given schedules; additionally, although much activity is currently underway to produce engine test components, these activities do not always lead to the development and maturation of low-cost production processes.

2. *Cast NiAl Alloy Technology*

The interest in cast, single-crystal NiAl alloys lies in the combination of high thermal conductivity and the potential for improved operating temperature versus single crystal superalloys. However, state-of-the-art single-crystal NiAl alloys lack the balance of properties available in conventional superalloys. In particular, no single alloy has both high-temperature creep resistance, and adequate ductility and toughness. Additionally, due to the higher melting temperature of the NiAl alloys, only an expensive fabricated blade and vane process has been developed, the cost of which might preclude or reduce its use. Because of these key barriers, development activity for single-crystal NiAl alloys for turbine engine use has been dramatically reduced (Williams 1997).

Noebe and Walston (1997) reported on the manufacture, rig testing, and successful engine testing of NiAl HPT vanes. These vanes were produced by the fabricated airfoil method from cast, oversized SX slabs (Darolia *et al.* 1996) A floating-vane design alleviated thermally induced stresses on the vane attachments. Although technical feasibility was demonstrated with the successful fabrication, installation, and engine test of the NiAl HPT vanes, several significant breakthroughs are required for the full-scale production of cast NiAl alloy components.

In spite of these implementation barriers, valuable insight and experience has been gained in the processing and engine testing of single crystal NiAl high pressure turbine vanes, which may be applicable to other high temperature alloys in the future (Williams 1997). These advances include the modification of ceramic mold systems for the higher melt temperature and lower strength and ductility of the NiAl alloys, and fabrication and machining technologies (Darolia *et al.* 1996).

3. *Cast Ni_3Al Alloy Technology*

There are two classes of cast Ni_3Al-based alloys currently under development or in production. The first are a family of equiaxed alloys, developed principally by the US Department of Energy at the Oak Ridge National Laboratory (ORNL), which are finding use in the metals processing industry. The second class is a DS alloy developed at the Beijing Institute of Aerospace Materials (BIAM), which is being applied to aeroengine turbine vanes and blades in China.

The development of casting technology for equiaxed Ni_3Al-based alloys has been aided by the development of an air-melting process that allows minimum pick-up of impurities such as oxygen and nitrogen. The air-melting process, known as Exo-Melt™ (Sikka and Deevi 1995, Deevi and Sikka 1997), uses air-induction-melting (AIM) furnaces, available at most iron- and nickel-based alloy foundries. Other barriers to successful casting of Ni_3Al-based alloys have been overcome through the development of data on physical properties and casting solidification modeling (Viswanathan *et al.* 1997) and the study of parameters controlling porosity (Ho *et al.* 1991, Cheng 1992, Sekhar *et al.* 1991). Also, for successful transition to commercial applications, identification of alloy compositions is important with the appropriate combinations of ductility, creep strength, and weldability (Aoki and Izumi 1979, Liu *et al.* 1988, Liu and Sikka 1986).

Further advances to successful transition to commercial applications have occurred through cooperative casting and in-service testing programs. The strong efforts made in transferring the technology for the Exo-Melt™ process, alloy compositions, property database, and operating experience to casting foundries (both static and centrifugal) and to potential users have further aided in achieving the current commercialization status (Sikka 1997) for Ni_3Al-based alloys. Some remaining barriers to broad commercialization include: the cost of Ni_3Al-based alloy, as compared to currently used alloys; a broader vendor base (at the beginning of the 21st century there were only three casting houses); lack of acceptable machining processes; lack of welding experience using commonly used weld methods; and limited industrial operating experience.

3.1 *Exo-Melt™ Process Technology*

For a long period, the lack of commercialization of Ni_3Al-based alloys was limited by lack of safe and economical melting practice, due to the strong exothermic reaction of Ni and Al and to the difference in melt temperatures between them. The Exo-Melt™ process takes advantage of this exotherm, and by so doing has decreased the power consumption for melting, improved safety and crucible life, and improved the quality of the melt. Since the successful demonstration of the Exo-Melt™ process in the laboratory, its details have been transferred to three commercial foundries, which have successfully melted large numbers of heats using the Exo-Melt™ process.

3.2 *Equiax Casting Technology*

Sand casting is the most common method used for the production of Ni_3Al-based alloy (IC-221M) components (Sikka *et al.* 1995, Viswanathan *et al.* 1997, Deevi and Sikka 1997). The finished sand-cast components so far produced have weighed from less than 50 kg to 200 kg, and have a varied in range of complexity. Typical foundry sands and molding procedures are acceptable for casting of nickel aluminides. The flushing of argon through the sand molds minimizes the formation of oxide skin on the cast surfaces. A computer modeling of the gating system and mold-filling orientations for nickel aluminide has been developed and experimentally verified (Viswanathan *et al.* 1997).

The inherent difficulty in machining nickel aluminides, lack of suitable machining tool materials, and lack of experience at the foundries, continue to make the cutting of runners and risers, grinding, machining, and weld repair of castings to be of concern for all foundries. Additional research effort is needed in the areas of cutting, grinding, drilling, and single-point machining.

Because they contain thin sections, centrifugal casting is the most economical method for the production of nickel aluminide for applications such as thin-wall (6 mm) radiant burner tubes for efficient heat transfer, steel plate austenization rolls (350–425-mm OD × 25 mm wall × 4 mlength), steel continuous casting guide rolls (75–150-mm OD × 25 mm wall), and steel strip hydrogen annealing furnace seals and rolls (150–200 mmOD × 12–25-mmwall). Centrifugal castings of nickel-aluminide alloys of all the size ranges listed above have been commercially cast by using conventional mold washes and parameters similar to those for commonly used alloys for similar applications. The surface quality of nickel-aluminide castings is similar to that of conventional alloys.

However, the inner diameter tubes of Ni_3Al-based alloys tend to contain oxide particles of ZrO_2 and Al_2O_3. The appearance of such oxides in Ni_3Al-based alloy tubes results from the presence of stable oxide-forming elements such as zirconium and aluminum, which are not present in the conventional alloys. The presence of ZrO_2 and Al_2O_3 particles on the inner diameter has no influence on the performance of centrifugal tubes of Ni_3Al-based alloys for furnace roll applications. However, the presence of ZrO_2 and Al_2O_3 particles does cause some difficulty in machining of inner surfaces, as required for assembly of components. This problem with oxide particles has also been manifested in investment cast turbocharger wheels, and has limited the application of these alloys (Sikka 1996).

Other than difficulties in cutting and machining, the centrifugal casting process for nickel-aluminide alloys is well established for commercial production. The primary manufacturers of centrifugally cast tubes and pipes in the United States are AEC and Sandusky.

3.3 DS Casting Technology

Han and Xing (1997) have reported the successful development of a DS alloy, named Alloy IC6, which offers a 10% density reduction versus competitive nickel superalloys and has exceptional strength and creep resistance without the use of expensive alloying elements such as Ta, Hf and Re. The aim alloy composition is Ni-7.8Al-14Mo-0.03B (in weight percent). Three major phases are present in this alloy: gamma; gamma prime; and borides.

The ingot production and DS casting processes used for the IC6 alloy are very similar to the processes used for conventional DS superalloys. The master alloy ingot is prepared by VIM with a refining temperature from 1570°C to 1600°C for 15–20 minutes. DS components are produced using the power reduction method in a VIM furnace. Alumina molds are used, connected to a water cooled copper chill plate, utilizing a mold withdraw rate of 6–8 mm per minute. Han and Xing (1997) also report that a solutioning heat treatment followed by rapid cooling is applied to components in order to obtain optimum mechanical properties.

The focus of recent developments has been on DS turbine vanes, with engine and flight tests being conducted successfully on multiple sets of vanes. Han and Xing (1997) report that the alloy has been selected for production for advanced aeroengines, including vanes with complex cooling passages.

Bibliography

Aoki K, Izumi O 1979 Cleavage fracture of the intermetallic compound Ni_3Ge single crystals. *Nippon Kinzoku Gakkaishi* **43**, 1190–250

Austin C M, Kelly T J 1995 Progress in the implementation of cast gamma titanium aluminide. In: Kim Y W, Wagner R, Yamaguchi M (eds.) *Gamma Titanium Aluminides*. TMS, Warrendale, PA, p. 413

Austin C M, Kelly T J, McAllister K G, Chesnutt J C 1997 Aircraft engine applications for gamma titanium aluminide. In: Nathal M V, Darolia R, Liu C T, Martin P L, Miracle D B, Wagner R, Yamaguchi M (eds.) *Structural Intermetallics 1997*. Seven Springs, Champion, PA, p. 413

Blum M, Choudhury A, Scholz H, Jarczyk G, Pleier S, Busse P, Frommeyer G, Knipscheer S 1999 Properties of low cost TiAl automotive valves produced by cold wall induction melting and permanent mold centrifugal casting. In: Kim Y W, Dimiduk D M, Loretto M H (eds.) *Gamma Titanium Aluminides 1999*. TMS, Warrendale, PA, p. 35

Chandley G D 1991 *US Pat.* 5 042 561

Cheng C T 1992 Microstructure and porosity evolution in equiaxed solidified multicomponent nickel aluminide. Ph.D. thesis, University of Cincinnati, OH

Colvin G 1995 Permanent mold casting of titanium aerospace and automotive hardware. In: Blenkinsop P A, Evans W J, Flower H M (eds.) *Titanium '95*, Institute of Metals, Birmingham, UK, Vol. 1, p. 691

Darolia R, Waltson W S, Nathal M V 1996 NiAl alloys for turbine airfoils. In: Kissinger R D, Deye D J, Anton D L, Cetel A D, Nathal M V, Pollock T M, Woodford D A (eds.) *Superalloys 1996*, TMS, Warrendale, PA, p. 561

Davidson D E 1996 Designing with gamma titanium-CAESAR program titanium aluminide component applications. In: Kissinger R D, Deye D J, Anton D L, Cetel A D, Nathal M V, Pollock T M, Woodford D A (eds.) *Processing of Superalloys 1996*. (eds. R. D. Kissenger *et al.*) TMS, Warrendale, PA, p. 539

De Graef M, Biery N, Rishel L, Pollock T M, Cramb A 1999 On the relation between cooling rate and solidification microstructure in as-cast titanium aluminides. In: Kim Y W,

Dimiduk D M, Loretto M H (eds.) *Gamma Titanium Aluminides 1999*. TMS, Warrendale, PA, p. 247
Deevi S C, Sikka V K 1997 Exo-melt® process for melting and casting intermetallics. *Intermetallics* **5**, 17–27
Dimiduk D M, McQuay P A, Kim Y W 2000 Gamma alloy technology 1999. In: Titanium '99. CRISM Promety, St. Petersburg, Russia, in press
Dowling Jr. W E, Allison J E, Swank L R, Sherman A M 1993 *SAE Technical Paper Series, no. 930620*, p. 31
Dowson A L, Johnson T P, Young J M, Jacobs M H 1995 Microstructure and chemical homogeneity of plasma arc cold hearth melted Ti-48Al-2Mn-2Nb gamma titanium aluminide. In: Kim Y W, Wagner R, Yamaguchi M (eds.) *Gamma Titanium Aluminides*. TMS, Warrendale, PA, p. 467
Han Y F, Xing Z P 1997 Development and engineering applications of a DS cast Ni_3Al alloy IC6. In: Nathal M V, Darolia R, Liu C T, Martin P L, Miracle D B, Wagner R, Yamaguchi M (eds.) *Structural Intermetallics 1997*. Seven Springs, Champion, PA, p. 713
Hartfield-Wunsch S E, Sperling A, Morrison R S, Dowling W E Jr., Allison J E 1995 Titanium aluminide automotive engine valves. In: Kim Y W, Wagner R, Yamaguchi M (eds.) *Gamma Titanium Aluminides*. TMS, Warrendale, PA, p. 41
Ho C T, Cheng C T, Sekhar J A 1991 Solidification microporosity in directionally solidified multicomponent nickel aluminide. *Metall. Trans* **22A**, 225–34
Huang S C, Chesnutt J C 1995 Gamma TiAl and its alloys. In: Westbrook J H, Fleischer R L (eds.) *Intermetallic Compounds: Practice,*. Wiley, Vol. 2, p. 80
Isobe S, Noda T 1997 Automotive applications of TiAl intermetallics. In: Nathal M V, Darolia R, Liu C T, Martin P L, Miracle D B, Wagner R, Yamaguchi M (eds.) *Structural Intermetallics 1997*, Seven Springs, Champion, PA, p. 427
Larsen D E 1996 Status of investment casting of gamma titanium aluminides in the USA. *Materials Science and Engineering A*, **A213**, 128
Larsen D, Govern C 1995 Advanced investment cast processing for gamma titanium aluminide alloys. In: Kim Y W, Wagner R, Yamaguchi M (eds.) *Gamma Titanium Aluminides*, TMS, Warrendale, PA, p. 405
Liu C T, Sikka V K 1986 Nickel alumides for structural use. *J. Met*. **38**, 19–21
Liu C T, White C L, Lee E H 1988 Effect of test environment on ductility and fracture behavior of doped Ni_3Al at 600°C. *Scr. Metall* **19**, 1247–50
McQuay P A, Larsen D 1997 Manufacturing-performance-cost relationships in gamma TiAl castings. In: Nathal M V, Darolia R, Liu C T, Martin P L, Miracle D B, Wagner R, Yamaguchi M (eds.) *Structural Intermetallics 1997*. Seven Springs, Champion, PA, p. 523
McQuay P A, Simpkins R, Seo D Y, Bieler T R 1999 Alloy and process improvements for cast gamma TiAl alloy applications. In: Kim Y W, Dimiduk D M, Loretto M H (eds.) *Gamma Titanium Aluminides 1999*. TMS, Warrendale, PA, p. 349
Nishino K, Kawaura H, Tanaka K, Horie T, Saito T, Uchida H 2000 High performance alloys developed for turbocharger. *Journal of the Gas Turbine Society of Japan* **28**(1), 58
Noda T 1998 Application of cast gamma TiAl for automobiles. *Intermetallics* **6**, 709
Noebe R D, Walston W S 1997 Prospects for the development of structural NiAl alloys. In: Nathal M V, Darolia R, Liu C T, Martin P L, Miracle D B, Wagner R, Yamaguchi M (eds.) *Structural Intermetallics 1997*. Seven Springs, Champion, PA, p. 573
Ram S V, Barrett J R 1996 In: Blenkinsop P A, Evans W J, Flower H M (eds.) *Titanium '95*, Vol. II, Institute of Materials, Birmingham, UK, p. 88
Rishel L L, Pollock T M, Cramb A W 1999 Induction skull melting of titanium aluminides. In: Mitchell A, Ridgway L, Baldwin M (eds.) *Proc. 1999 Int. Symp. Liquid Metal Processing and Casting*. American Vacuum Society, Santa Fe, NM, p. 287
Sekhar J A, Lin C S, Cheng C J 1991 Castability maps: how to design the processing of a multi-component alloy from knowledge of the semi-solid morphology and properties. In: Sekhar J A, Dantzig J (eds.) *Nature and Properties of Semi-solids Materials*. The Minerals, Metals and Materials Society, Warrendale, PA, pp. 267–90
Sikka V K 1996 Processing of aluminides. In: Stoloff N S, Sikka V K (eds.) *Physical Metallurgy and Processing of Intermetallics Compounds*. Chapman and Hall, New York, pp. 561–604
Sikka V K 1997 Commercialization of nickel and iron aluminides. In: Deevi S C, Sikka U K, Maziasz P J, Cahn R W (eds.) *Int. Symp. Nickel and Iron Aluminides: Processing, Properties, and Applications*. ASM International, Materials Park, OH, p. 361
Sikka V K, Deevi S C 1995 Safer and more cost-effective process for melting and casting intermetallic alloys enhances their commercial potential. *Mater. Technol* **10**(5/6), 97–100
Sikka V K, Deevi S C, Vought J D 1995 Exo-melt: a commercially viable process. *Advanced Materials and Processes* **147**(6), 29–31
Stewart D C 1994 Designing with gamma TiAl-CAESAR program titanium aluminide component applications, (oral presentation at Aero Mat '94, Anaheim, CA 6–9 June, 1994) ASM International, Materials Park, OH.
Tetsui T 1999 The effect of composition on the endurance of TiAl alloys in turbocharger applications. In: Kim Y W, Dimiduk D M, Loretto M H (eds.) *Gamma Titanium Aluminides 1999*. TMS, Warrendale, PA, p. 15
Viswanathan S, Duncan A J, Porter W D, Webb D S 1997 Modeling of nickel and iron aluminide castings. In: Deevi S C, Sikka U K, Maziasz P J, Cahn R W (eds.) *Int. Symp. Nickel and Iron Aluminides: Processing, Properties, and Applications*. ASM International, Materials Park, OH, p. 243
Williams J C 1997 Intermetallics for structural applications: potential, reality and the road ahead. In: Nathal M V, Darolia R, Liu C T, Martin P L, Miracle D B, Wagner R, Yamaguchi M (eds.) *Structural Intermetallics 1997*, Seven Springs, Champion, PA, p. 3

P. A. McQuay and V. K. Sikka

Casting of Semisolid Metals: Engineering Applications

Semisolid metal (SSM) casting, while known since the 1970s, largely emerged as a viable manufacturing process for engineered components only during the 1990s. The process is proving capable of as-formed part detail, complexity and section *thin*ness akin to high-pressure die casting, while using high-purity primary alloys, employing T-5, T-6, and T-7 heat treatments, and developing the high strength and ductility more typically associated with premium-engineered, high-integrity sand, permanent mold and squeeze castings. Tool life is long, the onset of heat checking is delayed and fine details remain intact, thus the SSM process is suitable for products having critical dimensional and surface requirements.

1. SSM Consistency

SSM cast components are made from billets, pre-cast using the same melt-control and semi-continuous casting methods that are applied to forging and rolling stock. The quality and consistency of SSM castings are thus positively influenced by those billet characteristics, rather than the typical shot-to-shot melt and pouring variables experienced when making parts from fully liquid metal.

Beyond the benefits of billet, SSM processing parameters and shot profiles must also be properly selected and controlled. The process is relatively new and many practitioners are still fine-tuning their methods, so available consistency is not always being achieved. Still, the properties and integrity of the more mature SSM cast components have proven exceptionally reliable and consistent.

2. SSM Alloys and Heat Treatments

Semisolid metal processing has been successfully applied to ferrous, magnesium, and copper alloys also, but the vast majority of commercially manufactured SSM parts are made from aluminum, thus this article will primarily address aluminum alloys. The commonly available billet materials are A356, 357, a 319-type, and 390 alloy. Limited processing of six thousand series (6xxx) normally wrought alloys and various metal matrix composites (MMCs) has also been accomplished and shows promise. In fact, SSM will probably prove to be a preferred method for making components from other alloys that have proven difficult to cast from liquid—normally wrought 2xxx, 5xxx, and 7xxx compositions, cast alloys of the 5xx and 7xx families, and other hyper-eutectic alloys.

Table 1 provides the chemistries of the more popular SSM billet feedstocks currently available.

Table 1
Aluminum SSM billet stock.

Alloy	Si (%)	Fe (%)	Cu (%)	Mg (%)
A356	7.0	<0.2	<0.05	0.3
357	7.0	<0.15	<0.05	0.55
319	6.0	<0.2	3.0	0.3
390	17.0	<0.5	4.5	0.45
206	<0.1	<0.15	4.6	0.25
6262	0.6	<0.7	0.3	1.0

Shown are nominal compositions of major alloying ingredients. Impurities and minor elements for *SSM billet* are the same as those found in premium *foundry ingot*-based castings, as specified by the Aluminum Association.

Typical heat-treating cycles for SSM cast aluminum parts are shown in Table 2. Note that solution heat treating (SHT) times can be much shorter than are typical for other casting methods. Allowed solidification dwell times for SSM are extremely short (mere seconds), which reflects also in shorter times being required to accomplish complete solid solution at customary SHT temperatures. SHT temperatures can also be somewhat reduced without sacrificing strength or ductility.

Note also the range of aging temperatures. The T-5 temper applied to castings poured from liquid typically reduces ductility to only half of that of either F (as cast) or T-6, so is normally used only for the purpose of achieving elevated temperature stability and is accomplished at the upper end of the range shown. SSM parts, on the other hand, can be quenched from the die and given a T-5 treatment to increase strength and hardness without significant loss of ductility, and that is accomplished at the lower end of the temperature range shown.

3. Mechanical Properties of SSM

The mechanical properties being realized from SSM cast components are typically equal to or better than those achieved using other high-integrity casting methods such as sand, permanent mold, and squeeze casting.

Typical tensile properties of test coupons cut from SSM cast components are shown in Table 3. As with any casting process, the level of properties to be realized in a given example will depend on the specific SSM cast section size and overall component configuration as well as the precise parameters selected for processing by the manufacturer, and could vary above or below the "typical" properties in Table 3.

Properties of some other metals as SSM formed are shown in Table 4.

Table 2
Heat treating cycles.

Alloy/temper	SHT (°C)	SHT (h)	Quench	Age (°C)	Age (h)
A356 T-5				160–225	2–5
A356 T-6	530–540	0.5–4	hot water	155–170	2–5
357 T-5				160–225	2–5
357 T-6	530–540	0.5–4	hot water	155–170	2–5
319 T-6	500–510	0.5–4	hot water	170	8–10
390 T-5				170–235	8–10
390 T-6	495–500	0.5–4	hot water	170–180	8
390 T-7	495–500	0.5–4	hot water	225–235	8–10
206 T-7	500–530	2–8	hot water	195–205	4
6262 T-6	540	0.5–4	hot water	175	12

SHT, solution heat treating.

Table 3
SSM mechanical properties.

Alloy/temper	UTS (MPa)	YS (MPa)	% El
A356 T-5	265	185	10+
A356 T-6	320	240	12+
357 T-5	290	200	7+
357 T-6	330	265	8+
319 T-6	405	310	5
390 T-5	205	=UTS	<1
390 T-6	350	=UTS	<1
390 T-7	300	=UTS	<1
206 T-7	390	320	6
6262 T-6	365	330	10

UTS, ultimate tensile strength; YS, yield strength.

Table 4
Properties of other SSM metals.

Material	UTS (MPa)	YS (MPa)	% El
C377 Brass	360	140	45
C642 Bronze	515	260	30
C142 BeCu	415–1520	140–1380	2–60
AZ91D Mg	235	160	3
AZ80 Mg	275	195	6

UTS, ultimate tensile strength; YS, yield strength; El.

4. SSM Engineered Applications

In spite of the fact that semi-solid metal casting of engineered high-integrity components has only been practiced on a significant scale since the early 1990s, some important applications have emerged. The following examples are intended to provide a representative cross-section of the types of parts where SSM proved to be both functionally and economically suitable.

4.1 Automotive

One of the most important potential uses for engineered SSM components is in the chassis and suspensions systems of automobiles. Already, there are several important examples of such components. Because auto makers are sensitive about sharing their designs and specifics of their applications, most of the examples that follow are *not* described in detail; only general information is given about the type of components, the alloys and heat treatments, a range of weights and section thickness. Those that are already in series production and have been previously noted in the trade literature are described in greater detail.

(*a*) *Multi-link subframe.* Figure 1 shows the Alfa Romeo multi-link. These 357 alloy link arm ends have been described by Chiarmetta (1998). Each casting weighs 7.4 kg and is used in the T-5 temper. A pair of SSM castings is welded to 6xxx alloy extruded members to form a completed sub-frame.

(*b*) *Body node.* Figure 2 shows an example of a body node. This type of component has been described by Midson (1999). Such parts are typically very thin-walled (1.5–4 mm) and must have very high ductility for crashworthiness (>12% elongation). A356 alloy has proven suitable, although a search continues to find alloys having even greater ductility. Magnesium-free alloys such as A444 (same as A356 but with no magnesium) in a T-4 temper (SHT only, no quench and no age), can achieve >20% elongation and alloys of the 5xx Al–Mg family achieve moderately high strength and ductility without any heat treatment (thus, less chance of quench-stress distortion).

(*c*) *Steering knuckles and wheel carriers.* These are also viable SSM components, described by

Figure 1
Alfa Romeo multi-link end.

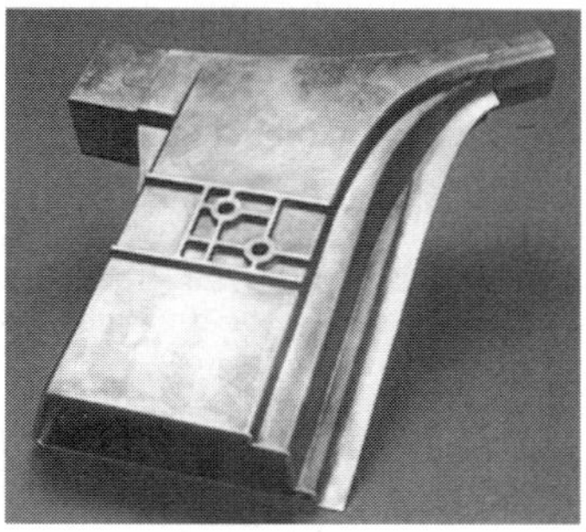

Figure 2
Example of body node.

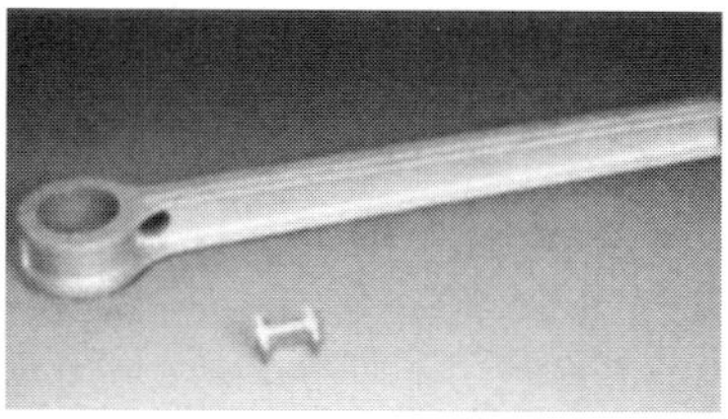

Figure 3
Porsche suspension control arm.

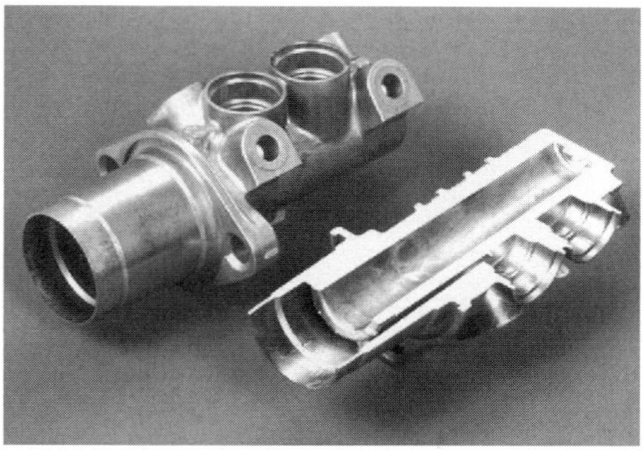

Figure 4
Master brake cylinder. Note sound appearance of cut section.

Chiarmetta (1999), Hirt (1999) and Jorstad (1999). Suitable alloys are A356 and 357. The T-6 temper would be used to maximize strength and ductility in front wheel knuckles, while the T-5 temper might prove sufficient for rear wheel carriers. Typical part weights range from 1.4 kg to 3 kg and sections can be relatively thick (up to 25 mm or more).

(*d*) *Control arms.* Eisen (1998) and Jorstad (1999) have reported examples of these components (Fig. 3). Again, A356 has been the preferred feedstock because it provides greater ductility in the T-5 or T-6 tempers than 357. The illustrated part has uniform thin walls (2–3 mm) and achieved ultimate and yield strengths in T-6 of 320 and 245 MPa, respectively, and elongation of 12%. Other control arms may be thicker (to 15 mm or 20 mm) and reach slightly lower T-6 strength and ductility.

(*e*) *Fuel rails.* Fuel rails have been described by Vinarcik *et al.* (1998) and Moschini (1996). Fuel rails were one of the earliest and still a continuing major application of the SSM process. They were first made in Europe and recently in North America. These are generally made from 357 alloy and typically weigh from 0.5 kg to 1 kg. A major requirement is soundness and freedom from leaks; equally important, of course, is cost-effectiveness.

(*f*) *Brake cylinders.* An example of brake cylinders (Fig. 4) has been given by Midson (1999), and illustrates freedom from porosity in the cut cross-section.

(*g*) *Compressor scroll.* Idegomori *et al.* (1998) and Jorstad (1999) have described compressor scroll parts (Fig. 5), made either from a 319-type alloy or from hypereutectic 390 alloy. The requirements are high integrity, high strength at operating temperature and good wear resistance.

4.2 *Non-automotive*

SSM has proven a preferred method for casting components such as *wave-guides* (very tight dimensional tolerances, fine detail and surface finish), *heat sinks* (very thin sections and close dimensional control), engine and compressor *pistons* (the wear properties, superior stiffness, elevated temperature strength and hardness of hypereutectic Al–Si alloys and MMCs) and other parts too numerous to mention. There are two notables, however.

Figure 5
Compressor scroll.

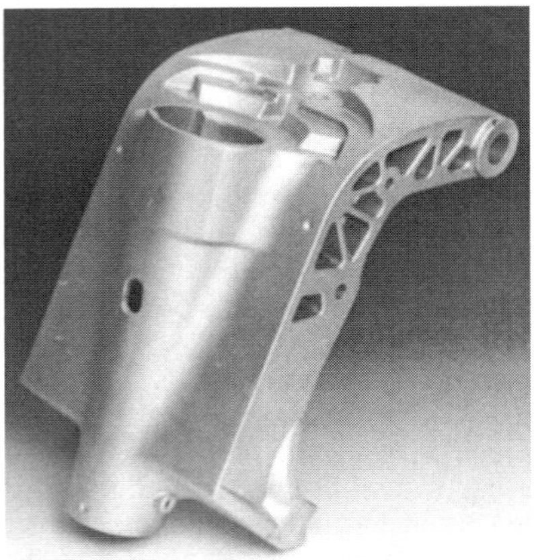

Figure 6
Outboard marine engine swivel bracket.

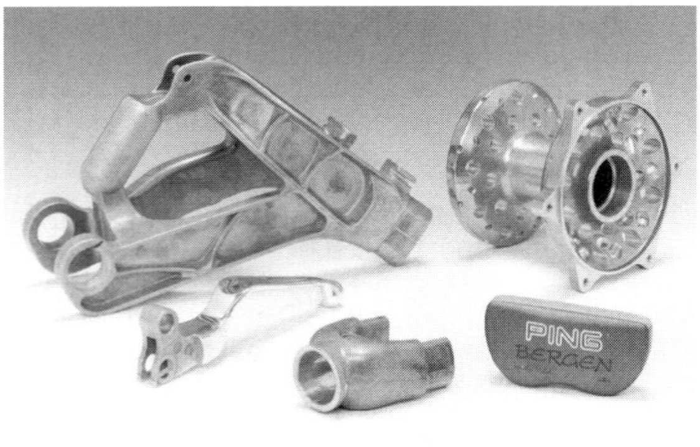

Figure 7
A variety of recreational products.

(*a*) *Marine engine swivel bracket.* A conversion from die casting and low pressure casting based on superior fatigue performance and competitive cost (Fig. 6). It is an A356 alloy and weighs 3.2 kg.

(*b*) *Recreational industry components.* Midson (1999) has illustrated a variety of components in this category (Fig. 7). The requirements, of course, vary but high integrity, strength, and ductility are necessary, as are thin sections and minimum weight. The most common alloys are A356 and 357 in the T-5 or T-6 tempers. Products such as golf clubs also require hardness and wear resistance and are good candidates for application of MMCs.

5. Concluding Remarks

SSM is now a proven process for a broad variety of engineered components. First and foremost, SSM provides high part integrity. It accommodates equally well thick and/or thin cast sections. It provides close dimensional control and excellent surface detail and quality. The process uses high-purity primary alloys, enables full heat treatment and provides high strength and ductility. Economics are achieved through material thrifting (thin sections and close dimensional control), minimized secondary processing and machining, short process cycle times and long tool life.

Bibliography

Chiarmetta G 1998 Thixoforming and weight reduction – industrial application of SeSoF. In: *Proc. 5th Int. Conf. Semi-solid Processing of Alloys and Composites.* Colorado School of Mines, pp. 87–98

Chiarmetta G 1999 Current and future designs utilizing SSM. In: *Proc. Semi-solid Metal World Users Conference.* Ormet: #3

Eisen P 1998 Introduction of SSM cast safety critical components for automotive applications. In: *Semi-solid Processing of Alloys and Composites.* Colorado School of Mines, pp. ix–xvi

Hirt G 1999 Recent thixoforming R & D at EFU. In: *Proc. Semi-solid Metal World Users Conference*, Ormet: #6

Idegomori T, Hirono H, Ito O, Kimishima S, Mizoui K 1998 The manufacturing of automobile parts using semi-solid metal processing. In: *Semi-solid Processing of Alloys and Composites.* Colorado School of Mines, pp. 71–8

Jorstad J 1999 Selecting parts and materials for conversion to SSM. In: *Proc. Semi-solid Metal World Users Conference*, Ormet: #4

Midson S 1999 SSM production costs and prototyping. In: Proc. *Semi-solid Metal World Users Conference*, Ormet: #7

Moschini R 1996 Mass production of fuel rails by pressure die casting in the semi-liquid state. *Metall. Sci. Technol.* **12**(2): 55–9

Vinarcik E J, Taylor J D, Farris K 1998 Automotive fuel system component design for manufacture using the semi-solid process. In: *Semi-solid Processing of Alloys and Composites.* Colorado School of Mines, pp. 105–14

J. L. Jorstad

Ceramics and Glasses, Tempering of: Residual Stresses

An important technique for increasing the strength and resistance to contact damage of ceramics and glasses is to place their surfaces in a state of residual compression. The term "tempering" was used initially in the glass industry as an analogy to the tempering of steel and other metals, in which the material is heated to a critical temperature and then rapidly quenched. In this latter process, the microstructure is modified, often leading to changes in hardness and toughness. For the case of glass, however, the aim of the thermal tempering is to introduce residual stresses into the material. The process is controlled such that the surfaces are "prestressed" into a state of residual compression. Glasses usually fail from surface flaws under the action of stresses applied in tension. Thus, by introducing the residual surface compression, the applied stresses must "overcome" the residual stresses before the net (tensile) stress can reach the fracture stress. It is important to note that when residual stresses are present in a body there needs to be a force balance. Thus, if the surfaces of a glass article are in residual compression, the interior must be in residual tension. The introduction ofresidual surface compression can be accomplished by means other than thermal quenching. Indeed any process that increases the volume of the surface with respect to the interior will lead to residual compression, provided no major stress relaxation processes are available. For example, various chemical processes can used in this manner and this is sometimes referred to as chemical tempering, although some prefer the term "chemical strengthening" to emphasize that it does not involve a quenching process. Thermal tempering and chemical strengthening have been commercially exploited successfully in the glass industry. Similar processes have also been applied to polycrystalline ceramic materials but with less commercial success.

1. *Thermal Tempering*

There has been an interest in tempered glasses since the seventeenth century when Prince Rupert of Bavaria introduced thermally tempered glass drops that are now known as Prince Rupert's drops. The tear-shaped drops were formed when a hot fluid glass was allowed to drip into water. The drops proved to be a scientific curiosity as they showed great strength but would explode into powdery fragments if their tails were broken. The process was developed into one of significance in the early part of the twentieth century and is used extensively in automobile and architectural glazing and as other types of "safety glass."

The residual stresses produced by rapidly cooling glass plates are two-dimensional plane stresses that are constant except near the edges but vary with depth. Figure 1 schematically compares the residual stress profile for thermally tempered and chemically strengthened glass. The stress profile for thermally tempered glass is approximately parabolic and the surface stress of a commercial "fully tempered" glass is typically about -100 MPa (the negative sign indicates a compressive stress). If it is assumed that the surface stress is uniform over the depth of the surface cracks, the material will be strengthened by an amount that is the negative of the surface stress (e.g., ~ 100 MPa). An important attribute of a fully tempered glass is that the failure process leads to a dicing or fragmentation process when the material fails. When an annealed glass fractures, the glass shards are often sharp and dagger-like, whereas the fragments formed by fracture of a fully tempered glass are usually small, harmless, and approximately cubic in shape with rounded edges. For lower levels of tempering, some strengthening is retained but the fragments become significantly larger. The degree of temper is sometimes related to the magnitude of the midplane tension. The stresses produced by tempering depend strongly on the thickness of the glass, the temperature prior to quenching (typically 100 K above the glass transition temperature), and the cooling rate. To obtain the higher tempering stresses, cooling using forced convection and impinging air jets is often used. Clearly, the process must be controlled such that fracture does not occur during the quenching process. The main drawbacks of thermal tempering is that the strengthening is moderate, surface distortion can occur, and the minimum thickness for successful thermal tempering is ~ 3 mm.

The formation of residual stresses by the thermal quenching of glass is based on the unique behavior of

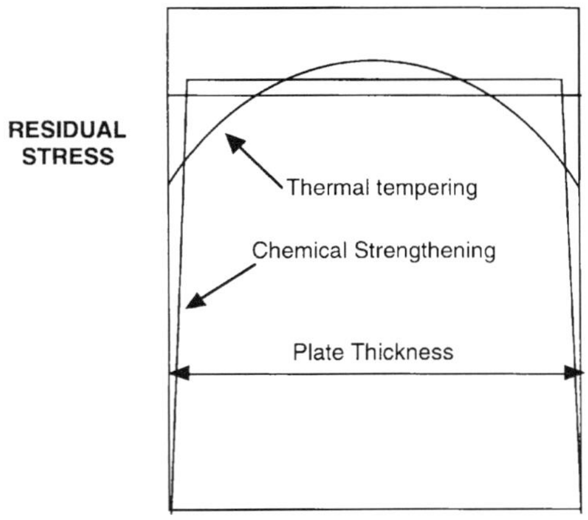

Figure 1
Schematic comparison of the residual stress profiles across the thickness of thermally tempered and chemically strengthened glass.

glass in the transformation temperature range during cooling, in which there is a gradual transition from a viscous liquid to an elastic body. If both surfaces of a glass plate are rapidly cooled from above the glass transition temperature, T_g, the thermal contraction stresses are expected to be tensile at the surface and compressive in the interior because the surface has cooled (and contracted) more than the interior. For the portion of the glass that is above T_g, however, these stresses can be relaxed. If it is assumed that relaxation is rapid above T_g then once the midplane temperature falls below this temperature, residual stresses will arise because the interior has to cool (and contract) more than the surfaces. The above ideas represent a simple approach for understanding the source of the residual stresses but clearly the process is more complex. For example, the thermal history of the glass will differ from the surface to the interior and the transformation into an elastic solid does not occur at a single temperature but rather over a temperature range in which the glass is viscoelastic. It is also known that the structure of the glass varies from the surface to the interior in a tempered glass: an effect that can also lead to residual stresses. These various sources of stress have been used to develop theories of thermal tempering, which have been reviewed by Gardon (1980).

2. *Chemical Strengthening*

As with thermal tempering, the aim of chemical strengthening is to place the surface of the glass into a state of residual compression. As shown in Fig. 1, the surface compression, though more shallow, has a much higher magnitude than that produced by thermal tempering. Thus, chemical tempering usually leads to a much higher degree of strengthening and glass strengths up to ~1 GPa are possible. There are two main approaches used to produce chemical strengthening. In the first approach, the composition of the surface layer is modified such that it has a lower thermal expansion coefficient. The processing usually occurs above room temperature so that during cooling the surface contracts less than the interior, and thus the surface is placed into residual compression. For example, one glass can be coated with a second (lower expansion) glass or ion exchange can be used to alter the surface composition. As an alternative, surface crystallization has been used to produce similar effects.

In the second approach, ion exchange is used and ions in the surface of the glass are replaced by larger ions. Provided this process can be accomplished without significant stress relaxation, the "crowding" effect or the "stuffing" of ions into a small space leads to the residual compression. This latter process is usually performed below T_g to ensure the stresses are not easily relaxed. It is this second technique that has received most of the attention the scientific literature. For example, sodium ions in the glass are often exchanged by larger potassium ions. The major difficulty with ion exchange is that exchange depths are often small (<100 μm). This is similar to the size of flaws found on glass surfaces and, thus, flaws are not necessarily subjected to the maximum surface compression. Of particular concern is the possibility that the tips of large flaws may reside in the region of interior tension. The presence of the steep stress gradient is a problem because it can lead to lower than expected strengths and a dispersion of the strength values. For these reasons, the strength of ion-exchanged glasses is often performed on abraded surfaces. There has been a strong research effort to increase ion exchange depths and this can be accomplished by adjusting the glass composition or by the application of an electric field. It is, for example, difficult to chemically strengthen soda lime silica (window) glass and produce acceptable abraded strengths.

The ion exchange process for ion stuffing is performed at elevated temperature, typically 75–150 K below T_g. The glass is immersed in a molten salt, such as potassium nitrate. The process is controlled by the diffusion of ions in the glass. The net diffusion is usually assumed to be unidirectional into a semi-infinite body. In binary ion exchange, the diffusing species are charged and tend to diffuse at different rates. This leads to an electrical potential gradient that acts to equalize the fluxes of the two ions. Clearly, the exchange depth will increase with increasing time and temperature. In the absence of stress relaxation, the linear strain associated with the exchange will be $\varepsilon = (\Delta V/V)/3$, where ΔV is the molar volume change produced by the exchange process and V is the molar volume of the unexchanged glass.

For the K^+–Na^+ exchange, ε is expected to be ~5% from the relative ion sizes. The stresses calculated from such strains appear to overestimate those found in practice and this has led to substantial discussion in the scientific literature. For example, it is now understood that stress relaxation occurs during the ion exchange process. Indeed, it is often found that strengths pass through a maximum as the time or temperature of the exchange is increased. There have been various attempts to further increase strength using multistep processes and it has been shown that certain residual stress profiles can lead to crack arrest. Many of the above ideas have also been applied to glass-ceramics.

3. *Compressive Surface Strengthening of Ceramics*

There have been various attempts to introduce residual surface compression into polycrystalline ceramics (see Kirchner 1979). Similar processes to those described above have been attempted and these have met with moderate success. There have also been attempts to use chemical reactions and phase transformations that involve molar volume changes as a means of introducing the residual stresses. The difficulty is that these materials often contain internal flaws that can initiate failure and their usage temperature is often high enough to induce stress relaxation. The use of residual stresses in ceramic hybrid laminates and surface coatings may, however, prove to be a useful future direction for exploiting compressive strengthening of these brittle materials.

Bibliography

Bartholomew R F, Garfinkel H M 1980 Chemical tempering of glass. In: Uhlmann D R, Kreidl N J (eds.) *Glass Science and Technology: Vol. 5. Elasticity and Strength in Glasses*. Academic Press, New York

Gardon R 1980 Thermal tempering of glass. In: Uhlmann D R, Kreidl N J (eds.) *Glass Science and Technology: Vol. 5. Elasticity and Strength in Glasses*. Academic Press, New York

Green D J 1999 Residual stresses in glass: an old curiosity. *Z. Metallkd.* **90**(12), 1059–63

Kirchner H P 1979 *Strengthening of Ceramics*. Marcel Dekker, New York

D. J. Green

Ceramic Matrix Composites: Matrices and Processing

Ceramic matrix composites (CMCs) generally consist of ceramic fibers or whiskers in a ceramic matrix. CMCs are designed to overcome the main drawback of monolithic ceramics, namely their brittleness. They are referred to as inverse composites, which is to say that the failure strain of the matrix is lower than the failure strain of the fibers, whereas it is the reverse in most polymer or metal matrix composites. Hence, under load it is the matrix which fails first. In order to prevent an early failure of the brittle fibers when the matrix starts to microcrack, the fiber/matrix (FM) bonding should be controlled during processing. CMCs are tough materials and display a high failure stress when the FM bonding is not too strong or too weak, which is usually achieved through the use of a fiber coating referred to as the interphase (Chawla 1993, Warren 1992, Mazdiyasni 1990). The fabrication of CMCs requires specific processing techniques. Gas-or liquid-phase routes (or a combination of both), in which the interphase and the matrix are formed around the fibers from gaseous or liquid precursors, are usually preferred.

CMCs are used as thermostructural materials under severe service conditions, e.g., high temperatures under load and in corrosive atmospheres, such as combustion gases. The most commonly used CMCs are nonoxide CMCs, namely carbon/carbon (C/C), carbon/silicon carbide (C/SiC), and silicon carbide/silicon carbide (SiC/SiC), the fibers being specified first.

1. Structure

There are three constituents in a CMC: a fiber (or whisker reinforcement), a matrix, and generally an interphase. The nature, volume fraction, and arrangement of the constituents depend on a number of considerations including the function of the part to be fabricated, the service conditions, and the cost.

Both oxide and nonoxide fibers are used in practice. Refractory oxide fibers, such as alumina-based fibers, display a high resistance to oxidizing atmospheres and are insulating materials. However, their strength retention and creep resistance at high temperatures are poor due to grain growth. The most attractive oxide fibers are those made of a mixture of alumina and mullite ($3Al_2O_3$, $2SiO_2$), their service temperature limit being 1000–1100 °C. Nonoxide fibers include carbon and SiC-based fibers. With respect to oxide fibers, they exhibit a lower density and much better strength retention and creep resistance at high temperatures. The failure strength of carbon fibers increases with increasing temperature up to 1500–2000 °C and that of the best SiC fibers is still significant at 1400–1500 °C. Depending on their composition and microstructure, they are either conducting or insulating. Unfortunately all of them react with oxygen, above 500 °C for carbon fibers and above 800 °C for SiC fibers. The fiber arrangement depends on the load pattern. It is rarely unidirectional (1D) but rather multidirectional (*n*D, with $n \geqslant 2$). In this latter case, which is very common in CMCs, the starting material is a fiber preform, i.e., a self-standing porous body made of fibers (continuous or short) and having roughly the shape and size of the part to be fabricated (Bunsell and Berger 1999).

The interphase has a key role in a CMC although its volume fraction is extremely low. It is actually a thin coating (less than 1 μm) deposited on the fibers or formed *in situ* during composite processing. It has three main functions: load transfer, crack deflection (the so-called mechanical fuse function), and diffusion barrier (in reactive FM systems). The best interphase materials might be those with a layered crystal structure (such as pyrocarbon (PyC) or hexagonal boron nitride (BN)) or a layered microstructure (such as $(PyC\text{–}SiC)_n$ multilayers, with $n = 5\text{–}20$) (Fig. 1(a)), the layers being oriented parallel to the fiber surface to promote crack deflection (Naslain 1998).

A variety of ceramic matrices are used. Silica-based glass–ceramics, i.e., ceramics made of a mixture of a crystalline refractory phase and a glass phase have been extensively studied (Mazdiyasni 1990). The most common are LAS, MAS, MLAS, CAS, and BMAS, with $L = Li_2O$, $A = Al_2O_3$, $S = SiO_2$, $M = MgO$, $C = CaO$, and $B = BaO$. All of them have two important advantages. First, before being ceramed, they are glasses which soften upon heating and display relatively low viscosities at high temperatures. This property is actually used to embed the fibers by hot pressing. Second, they exhibit a high resistance to corrosive media. However, the occurrence of residual glassy phase in the ceramed state lowers their creep resistance. Alumina and mullite, the latter having a better creep resistance than the former, are also used. However, their lack of plasticity at high temperatures is an important disadvantage in terms of CMC processing.

Nonoxide matrices comprise mainly carbon, silicon carbide, and silicon nitride. Carbon matrices display a variety of properties depending on their microtexture. Pyrocarbon can be isotropic, nongraphitizing, and insulating on the one hand, or strongly anisotropic (laminar), graphitizable through heat treatment at very high temperatures (2500–3000 °C), and highly conducting (such as the so-called rough laminar used for C/C brake systems) on the other hand. Carbon is extremely refractory ($\approx$3600 °C), light (1.8–2.1 $g\,cm^{-3}$), and biocompatible (C/C composites are used in prosthetic devices) (Savage 1993). Silicon carbide is less refractory ($\approx$2500 °C) and its density slightly higher (3.2 $g\,cm^{-3}$).

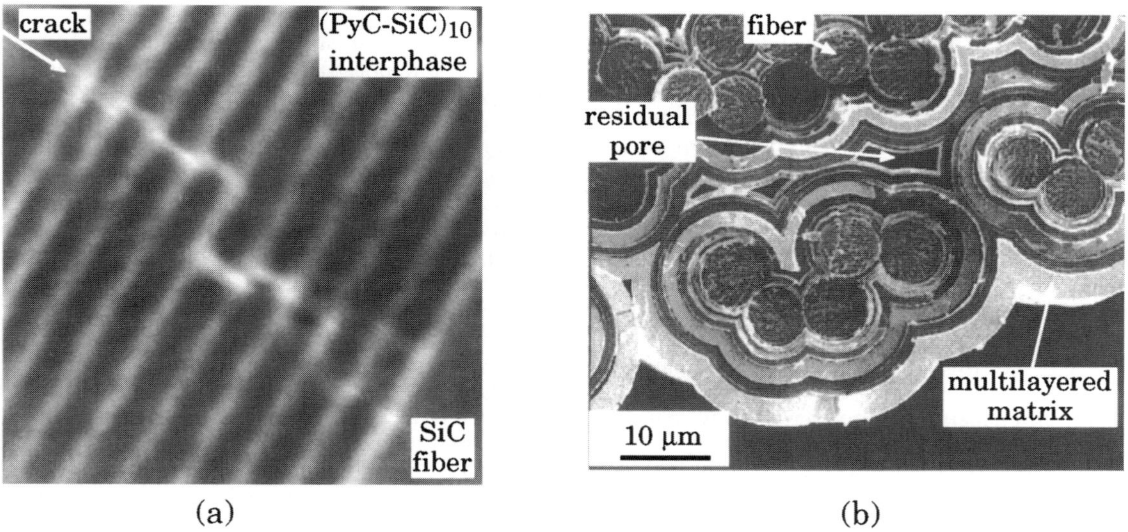

Figure 1
(a) Multilayered interphase and (b) self-healing matrix infiltrated in a fiber preform by P-CVI (courtesy of S. Bertrand and F. Lamouroux *et al.*).

It is extremely hard and, more interestingly, its reaction with oxygen results in the formation of a protective layer of silica explaining why SiC ceramics can be used up to about 1500°C in oxidizing atmospheres. Beyond this limit and/or at low oxygen partial pressure, the oxidation of SiC yields gaseous silicon monoxide (SiO) with no longer any protective effect. SiC matrices can actually be pure SiC or mixtures of SiC with free silicon or free carbon. When free silicon is present, the service temperature limit is lowered to ~1410°C. The occurrence of free carbon does not alter the refractoriness of SiC matrices. Silicon nitride matrices have properties close to those of SiC matrices; however, they are less thermally stable and exhibit lower conductivities.

When selecting a given fiber–interphase–matrix system, several considerations should be taken into account. First, the constituents should be compatible. The coefficients of thermal expansion (CTEs) of the fiber, α_f, and the matrix, α_m, should be roughly the same. If $\alpha_m > \alpha_f^R$ (the radial CTE of the fibers), the matrix clamps the fibers after cooling from processing to ambient temperature, resulting in strong FM bonding and a brittle character. Conversely, if $\alpha_m < \alpha_f^R$, the FM interface may spontaneously debond during cooling. The matrix should also be chemically compatible with the fibers. FM reactions may alter the strength of the pristine fibers and/or change the FM bonding. As an example, SiC fibers react with silica-based glass–ceramic matrices and a suitable interphase, such as a BN–SiC dual interphase, has to be used. To avoid these problems, the fibers and the matrix often have relatively close compositions in actual CMCs.

CMCs should also be compatible with their service atmospheres. Mostly, these service atmospheres are oxidizing (air or combustion gases) whereas the best CMC constituents are nonoxides, a somewhat paradoxical situation. To solve this key problem, two routes have been suggested: (a) the use of oxidation inhibitors and protective coatings and (b) the design of self-healing interphases and matrices. The addition of boron or phosphorous compounds (inhibitors) to carbon matrices decreases significantly their oxidation rate. Multilayered coatings containing glass formers (such as silicon or boron carbides) are also used to slow down the in-depth diffusion of oxygen in C/C composites. The concept of multilayered self-healing interphases (Fig. 1(a)) and matrices (Fig. 1(b)) has been developed for advanced SiC matrix composites, with a dramatic increase in their lifetime in oxidizing atmosphere at high temperatures and under cyclic loading (Lamouroux *et al.* 1999). Here, the interphase and/or the matrix, deposited according to a gas-phase route, are multilayered, involving crack arrester layers (such as PyC or BN) and glass former layers (such as B–C, Si–C, or Si–B–C compounds). When a microcrack is formed, oxygen diffusing along that crack reacts with the boron- or silicon-bearing species, yielding a low-viscosity glass which fills the crack and slows down oxygen diffusion. Under such conditions, the composite behaves somewhat as a smart material. It displays high mechanical performance with a lifetime long enough for applications under oxidizing atmospheres (e.g., in engines) (Naslain 1999).

2. *Processing Routes*

Most ceramic matrices being hard, refractory, and displaying no (or little) plasticity within a broad temperature range, the brittle fibers are preferentially embedded in a liquid or gaseous precursor of the matrix which is transformed into a ceramic matrix via chemical reactions or annealing treatments (Mazdiyasni 1990, Naslain 1999, Warren 1992).

2.1 Liquid-phase Routes

(*a*) *Reactive melt infiltration* (*RMI*). RMI processes can be used when one of the elements constituting the ceramic matrix displays a low (or relatively low) melting point and readily wets the fibers. Such is the case for aluminum alloys (m.p. $\approx 650^\circ C$) and even silicon (m.p. $= 1410^\circ C$). The former can yield an alumina matrix by reaction with an oxidizing atmosphere and the latter a silicon carbide matrix through a chemical reaction with a carbonaceous fiber preform.

In the so-called DIMOX process (from a company called Lanxide, DE, USA), the molten element, aluminum, reacts with the atmosphere, e.g., air, as it climbs by capillary forces in the preform pore network, yielding an alumina matrix. The alumina matrix grows according to a complex mechanism involving the dissolution of oxygen and the precipitation of alumina, in which magnesium (present in the alloy) plays a key role. A layer of magnesia is formed at the metal surface which actually prevents the formation of a dense protective scale of alumina that would otherwise stop the in-depth diffusion of oxygen. The process being run at about 1200°C, a variety of fibers, including Si–C–O Nicalon fibers (from Nippon Carbon), can be used. The process can be extended to nitride matrices (such as aluminum nitride and even silicon nitride) if it is conducted in a nitrogen atmosphere (Mazdiyasni 1990, Urquhart 1991).

In the so-called SILCOMP (from General Electric) and related processes, a ceramic fiber preform is first consolidated with carbon, e.g., according to the PIP process (see below). Then it is impregnated with liquid silicon at about 1450°C (Fig. 2). Liquid silicon spontaneously wets carbon and hence climbs by capillary forces in the preform pore network, reacting with the pore wall, i.e., carbon, to give the SiC matrix (Mazdiyasni 1990, Luthra *et al.* 1993). Since the process is run at high temperatures, only fibers with a high thermal stability, namely carbon or oxygen-free SiC fibers, can be used.

Both processes are relatively rapid and yield near-net-shape composites. Conversely, liquid aluminum and silicon being corrosive, the fibers have to be precoated. Dual BN–SiC coatings deposited from gaseous precursors (see Sect. *2.3*) have been successfully used in Nicalon/alumina and Hi-Nicalon/SiC composites. The BN sublayer acts as mechanical fuse (crack deflection function) and the SiC sublayer as diffusion barrier. Further, some residual free aluminum or silicon is usually present in the matrix as the result of an incomplete chemical reaction, which lowers the mechanical properties at high temperatures. Residual aluminum can actually be removed by

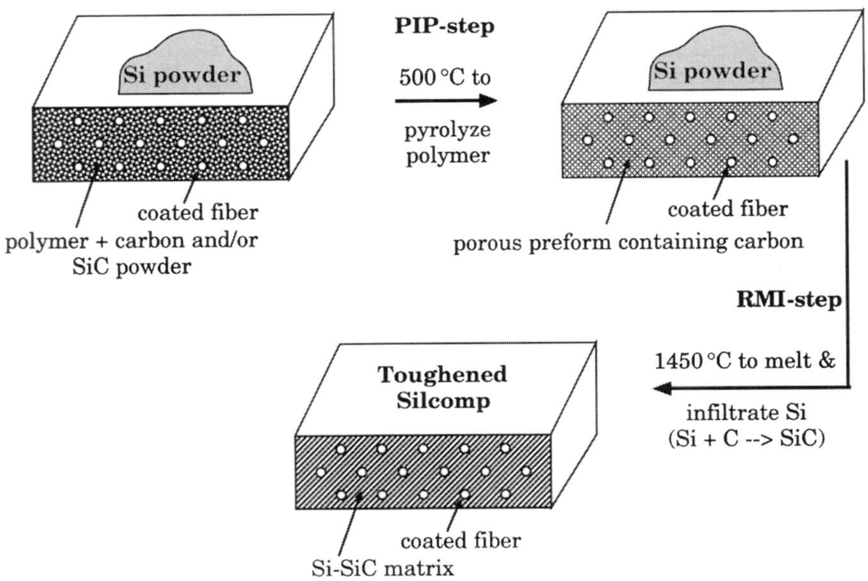

Figure 2
Principle of RMI processing of Si–SiC matrix composites (courtesy of K. Luthra).

wicking (thus yielding some open porosity) whereas residual silicon can be entrapped in the SiC matrix as a refractory silicide when necessary.

(*b*) *Polymer impregnation and pyrolysis* (*PIP*). In the PIP process, a porous fiber preform is first impregnated with a liquid organic (or organometallic) precursor of the matrix. After curing by heating, irradiation, or chemical reaction, the material is pyrolyzed at medium temperatures (700–1000°C). When necessary, the composite is further submitted to a heat treatment (2500–3000°C for carbon) to stabilize its microstructure and optimize its properties (Fitzer and Manocha 1998, Mazdiyasni 1990, Savage 1993). The liquid precursor is used either in the molten state or in solution. Examples of commonly used precursors are thermosetting polymers (e.g., polyphenolics) or mesophase pitch for carbon, polycarbosilanes for SiC, and polycarbosilazanes for Si–C–N matrices. The liquid phase should display a low enough viscosity and wet the fibers (in the impregnation step).

Further, the precursor should have a high ceramic yield, typically 70–85 wt.% (in the pyrolysis step). Even under such conditions a weight loss and shrinkage of the matrix occur during the pyrolysis due to the formation of gaseous species. This drawback can be overcome by loading the liquid phase with an inert filler (such as a fine powder of the ceramic matrix) and/or by repeating several times ($n=4$–10) the impregnation–curing–pyrolysis sequence, hence increasing processing time and labor cost at the plant level. Further, organometallic liquid precursors are often sensitive to moisture. The PIP process relies on relatively simple technologies.

(*c*) *Slurry impregnation and hot pressing* (*SIHP*). The SIHP process can be used to fabricate a variety of oxide and nonoxide matrix composites. In a first step, a fiber tow (or a fiber tape) is impregnated with a slurry, i.e., a stable suspension of matrix powder in a liquid containing a fugitive organic binder and various additives. The wet continuous material is taken up on a drum, yielding after drying a green composite prepreg somewhat similar to those employed for producing polymer matrix composites. After cutting and stacking, the prepreg layer assembly is heated to eliminate the fugitive binder and hot pressed yielding a dense matrix composite (Mazdiyasni 1990).

Two requirements should be fulfilled. First, the fibers should be stable under the hot-pressing conditions; hence, the fiber choice is limited to the most stable ceramic fibers, such as the oxygen-free SiC fibers. Second, the matrix should soften enough and flow during hot pressing to embed the fibers, supposing that some liquid phase is formed. This is typically the case for glass and glass–ceramic silica-based matrices, as already mentioned, the hot-pressing temperature being 1200–1400°C. For such matrices, a heat treatment (the ceramization step) is usually performed after hot pressing, to transform partly or totally the glassy matrix into the related stable crystalline phase(s). Specific sintering additives, which form a small amount of liquid phase at the sintering temperature, have to be added to the slurry, for refractory covalent nonoxide matrices (such as SiC and Si_3N_4), hot pressing being performed at much higher temperatures (up to ~1800°C), with some risk of fiber degradation. Some organic or organometallic liquid precursors (e.g., PCS for SiC) can also be added to the slurry. An important advantage of CMCs fabricated according to the SIHP process lies in their low residual porosity.

2.2 *Vapor-phase Routes*

In vapor-phase routes, also referred to as chemical vapor infiltration (CVI), the interphase and the matrix are successively deposited within the pore network of a fiber preform from gaseous precursors. The chemical reactions involved in the deposition process have to be activated. This is achieved by heating the fiber preform to moderate temperatures (typically 900–1100°C) either in cold- or hot-wall reactors, the latter being usually preferred. CVI is closely related to chemical vapor deposition (CVD), a process commonly used in material engineering to form thin films, thick coatings, or filaments. Both processes have the same chemical basis but are run under different temperature (T), pressure (P), and gas flow rate (Q) conditions. In CVD, deposition mainly occurs on the external surface of the substrate, whereas in CVI it takes place within the pores of the substrate, the deposition process being used to embed the fibers in the ceramic matrix, i.e., to densify the fiber preforms (Warren 1992).

Two different phenomena are involved in CVI: (a) chemical reactions occurring either in the gas phase or on the fiber surface with formation of the deposit and (b) mass transfer of the gaseous reactants and products outside the preforms and in their pore networks. As an example, carbon, silicon carbide, boron nitride, and alumina can be deposited according to the following overall equations (Naslain and Langlais 1986):

$$CH_4(g) \rightarrow C(s) + 2H_2(g) \quad (1)$$

$$C_3H_8(g) \rightarrow 3C(s) + 4H_2(g) \quad (1a)$$

$$CH_3SiCl_3(g) \overset{H_2}{\rightarrow} SiC(s) + 3HCl(g) \quad (2)$$

$$BX_3(g) + NH_3(g) \rightarrow BN(s) + 3HX(g) \text{ with } X = F, Cl \quad (3)$$

$$2AlCl_3(g) + 3CO_2(g) + 3H_2(g) \rightarrow Al_2O_3(s) + 3CO(g) + 6HCl(g) \quad (4)$$

Actually, the chemical mechanisms are much more complex and often involve intermediate gaseous species, the gas phase undergoing a maturation process as it diffuses in the pores. Mass transfer of the gaseous species occurs by convection and/or diffusion depending on *T–P–Q* conditions. Generally speaking, a key requirement in CVI is that the pores should remain open up to the end of the densification process.

(*a*) *Isothermal/isobaric CVI (I-CVI).* In I-CVI, the fiber preforms are set in the isothermal zone of a hot-wall deposition chamber (Fig. 3). The gaseous precursor flows outside the preforms by convection and inside the preforms by diffusion. In order to avoid an early sealing of the pores entrance by the deposit, deposition is performed at relatively low temperatures (to lower reaction rates) and under reduced pressures (to favor the diffusion of the gaseous species in the pores). Hence, the deposition yield and deposition rate are low. When necessary, the densification process is stopped and the preform surface machined in order to reopen the porosity or to enlarge the pore entrance (deposition occurring preferentially near the pore entrance) (Naslain and Langlais 1986, Savage 1993, Warren 1992).

(*b*) *Temperature/pressure gradient CVI (F-CVI).* The F-CVI process (the "F" signifying "forced") has been designed to increase the deposition rate and to shorten the densification time via the use of temperature and pressure gradients applied to the fiber preform in opposite directions. The reactant species are injected in the fiber preform through its cold face (where almost no deposition occurs), under pressure, whereas the porosity is rapidly filled near the opposite hot face (where the deposition rate is high). As deposition proceeds, a densification front moves from the hot face to the cold face (whose temperature progressively increases if the deposit is conducting, this being the case for both pyrocarbon and SiC), while the unreacted precursor and reaction products are extracted by pumping first through the hot face of the sample then through its lateral face(s). Under such conditions, the densification time is significantly shortened (Mazdiyasni 1990).

(*c*) *Pressure-pulsed CVI (P-CVI).* In P-CVI, the furnace and the fiber preform pore network are first evacuated by pumping. Then the reactants are injected in a very short time in the reactor and inside the preforms. Deposition occurs during a given time, referred to as the residence time. This sequence of events is repeated in time, the opening and closing of the gas flow valves being operated automatically. One of the advantages of P-CVI lies in the fact that the fiber preform pore network is periodically fed with fresh reactants whereas the reaction products (which often act as reaction inhibitors) are periodically evacuated. Second, when the residence time is low enough (typically a few seconds), the deposit thickness per pulse can become extremely

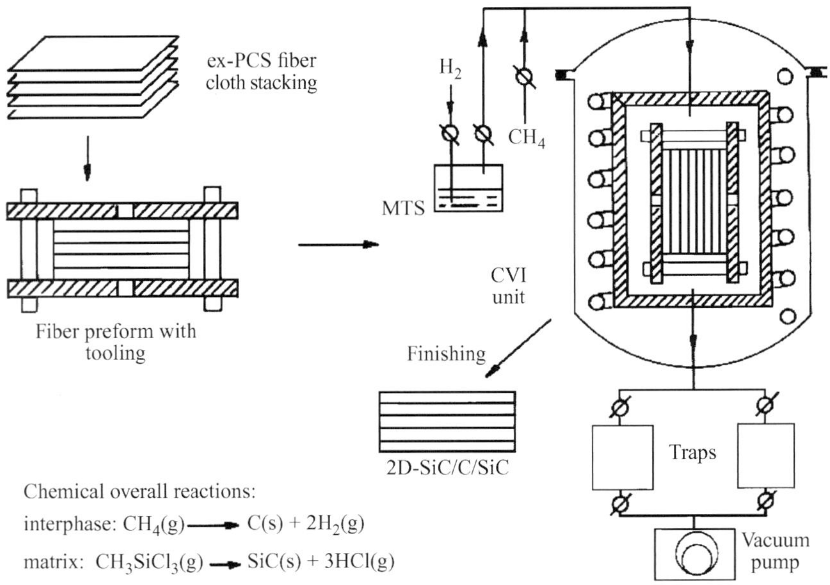

Figure 3
Principle of I-CVI processing of 2D fiber/SiC matrix composites (courtesy of Woodhead Publishers).

small (of the order of 1nm or less). Hence, if the composition of the precursor is periodically changed, P-CVI yields multilayered ceramic deposits. As an example, if one switches periodically from propane (Eqn. (1)) to methyltrichlorosilane/hydrogen (Eqn. (2)), a multilayered $(PyC–SiC)_n$ interphase can be deposited *in situ* on the fibers surface (Fig. 1(a)) (Naslain 1998).

(*d*) *Film boiling.* In the film boiling (or calefaction) process, a self-standing fiber preform, inductively heated with a susceptor, is directly immersed in a boiling liquid precursor, e.g., cyclohexane for pyrocarbon matrices. Assuming an axisymmetrical configuration, i.e., a cylindrical preform with a hole at the center in which the susceptor is set, a steep temperature gradient is generated radially in the preform. Further, precursor mass transfer in the preform, diphasic in character, is extremely efficient. As in the F-CVI process, a densification front moves vs. time from the preform hot surface (that near the susceptor) to the surface in contact with the boiling precursor. Densification times of the order of one day have been reported (David *et al.* 1999).

(*e*) *Advantages and drawbacks.* CVI processes are run at relatively low temperatures and hence are less damaging to the brittle fibers. They also yield interphases and matrices with high microstructural quality and excellent mechanical properties. Further, some of these processes are extremely flexible and all yield near-net-shape materials. However, CVI processes require important investments for volume production and the resulting CMC parts exhibit some residual open porosity. This residual porosity is not a problem from a mechanical standpoint since CMCs are tough materials; however, the occurrence of open residual porosity (that does not exist in composites fabricated according to some of the liquid-phase routes, e.g., the RMI process for SiC–Si matrix composites) favors the in-depth diffusion of oxygen in service conditions, which can be a drawback in nonoxide CMCs unless the materials are self-healing.

I-CVI has been already transferred to the plant level for the fabrication of C/C brake disks or C/SiC jet engine parts. Its main advantage is its high flexibility, a large number of preforms (which may have different sizes and shapes) able to be treated simultaneously. Even when intermediate surface machining of the partly densified preforms is performed, there is still much less handling than in, say, the PIP process. The flexibility of the process mitigates the fact that the densification times are relatively long. I-CVI is the preferred process for the volume production of CMC parts.

Both F-CVI and film boiling are attractive processes since densification times can be reduced by almost one order of magnitude with respect to I-CVI; however, their flexibility is much lower. In F-CVI, a specific fixturing has to be used for each fiber preform to establish the temperature/pressure gradients. As a result, this process is not suitable for the densification of preforms of complex shapes. In a similar manner, the use of a susceptor per preform (or pile of preforms) somewhat limits the flexibility of the film boiling process. To summarize, both processes may be advantageous when limited series of parts, with a relatively simple shape, have to be fabricated and delivered in a short time.

P-CVI is still an experimental process. It has a flexibility similar to that of conventional I-CVI but it requires more complex facilities. Although its use may result in shorter densification times, its main advantage lies in the fact that it is extremely suitable for the infiltration of highly engineered multilayered ceramics (interphases and matrices) in porous fiber preforms (Fig. 1).

2.3 Solid-state Routes

Solid-state routes, which are in principle closely related to the powder processing of multiphase ceramics, are limited to CMCs with short fiber reinforcements, such as whiskers (Mazdiyasni 1990). The starting material is a homogeneous powdered mixture of the fibers and the matrix, with appropriate additives (e.g., fugitive binder, sintering aids, stabilizers, etc.). It is often prepared by ball-milling in a protective liquid medium. After drying, the powder mixture is poured into the die of a press and sintered by uniaxial hot pressing. It can also be cold pressed, encapsulated in a suitable material (a metal or a glass), and sintered isostatically. During the ball-milling and hot-pressing steps some fragmentation of the brittle fibers usually occurs, leading to a decrease of the fiber aspect ratio and reinforcing capability. Further, the presence of the short fibers often renders the sintering of the powder matrix still more difficult (sintering of covalent carbides and nitrides, such as SiC and Si_3N_4, being known to be intrinsically difficult and to necessitate sintering additives), requiring high temperatures and pressures. Hence, only very stable fibers can be used (short carbon fibers or SiC whiskers). Finally, composites fabricated with short fibers according to solid-state routes display mechanical properties which are much lower than those of the related composites processed from continuous fibers.

SiC whiskers have been used in alumina-based or silicon nitride matrices, e.g., for producing cutting-tool inserts and other wear-resistant items, with an increase in toughness, hardness, and lifetime. It is important to note that the use of whiskers of small size raises health problems somewhat similar to those reported for asbestos fibers, which actually limit the development of these composites.

3. Concluding Remarks

The development of low-cost, high-performance CMCs relies on available fibers and it supposes an optimization of both the FM interfacial zone and the matrix itself. Some progress has been achieved in the field of ceramic fibers. As an example, oxygen-free SiC fibers and quasistoichiometric SiC fibers display a higher thermal stability than their Si–C–O counterparts. Hence, they are compatible with a variety of processing techniques including the high-temperature processes. However, available oxide fibers are not stable enough from a microstructural standpoint.

A key point in terms of material design is to improve the durability of nonoxide CMCs in oxidizing atmospheres. Taking the example of SiC/SiC composites, a promising approach is to replace the pyrocarbon interphase by boron nitride or multilayered $(PyC–SiC)_n$ or $(BN\text{-}SiC)_n$ interphases and the SiC matrix by self-healing multilayered matrices. In terms of processing, rapid densification techniques (such as RMI or film boiling) are becoming available beside well-established routes (such as CVI) which could still be improved. Even if these processes display less flexibility, they may be appropriate for specific matrix compositions and/or specific application areas.

Bibliography

Bunsell A R, Berger M H 1999 *Ceramic Fibers*. Dekker, New York

Chawla K K 1993 *Ceramic Matrix Composites*. Chapman & Hall, London

David P, Benazet J D, Ravel F 1999 Rapid deposition of carbon and ceramic matrices by film boiling technique. In: Vincenzini P (ed.) *Advanced Structural Fiber Composites*. Techna. Faenza, Italy, pp. 135–40

Fitzer E, Manocha L M 1998 *Carbon Reinforcements and Carbon/Carbon Composites*. Springer, Berlin

Lamouroux F, Bertrand S, Pailler R, Naslain R, Cataldi M 1999 Oxidation-resistant carbon-fiber-reinforced ceramic-matrix composites. *Compos. Sci. Technol.* **59**, 1073–85

Luthra K L, Singh R N, Brun M K 1993 Toughened Silcomp composites: process and preliminary properties. *Am. Ceram. Soc. Bull.* **72**(7), 79–85

Mazdiyasni K S 1990 *Fiber Reinforced Ceramic Composites*. Noyes, Park Ridge, NJ

Naslain R R 1998 The design of the fibre–matrix interfacial zone in ceramic matrix composites. *Composites*. **29A**, 1145–55

Naslain R R 1999 Materials design and processing of high temperature ceramic matrix composites: state of the art and future trends. *Adv. Compos. Mater.* **8**(1), 3–16

Naslain R R, Langlais F 1986 CVD processing of ceramic–ceramic composite materials. *Mater. Sci. Res.* **20**, 145–64

Savage G 1993 *Carbon–Carbon Composites*. Chapman & Hall, London

Urquhart A W 1991 Novel reinforced ceramics and metals: a review of Lanxide composite technologies. *Mater. Sci. Eng. A* **144**, 75–81

Warren R 1992 *Ceramic Matrix Composites*. Blackie, Glasgow, UK

R. R. Naslain

Cobalt Alloys: Alloying and Thermomechanical Processing

Cobalt alloys have been in use since 1907 when Elwood Haynes obtained the first patents on cobalt–chromium compositions. This article describes the nature of cobalt alloys and provides the basis for the selection of elements that are used in the alloys of commercial importance. The article also provides a brief description of the manufacturing methods that are employed.

1. *Alloying*

Like its neighbors in the periodic table (nickel and iron), cobalt can be alloyed with significant quantities of elements such as chromium, molybdenum, and tungsten to create very useful engineering materials. Indeed, cobalt imparts to its alloys some extremely important properties, such as resistance to several forms of wear, creep and fatigue strength at high temperatures, and resistance to sulfidation.

The resistance to wear of cobalt alloys derives partly from the fact that they undergo a structural transformation during mechanical deformation. Cobalt alloys also benefit from mechanical twinning and planar slip propensities.

Cobalt itself exhibits a h.c.p. structure at temperatures up to 417°C. At higher temperatures (up to the melting point) the atomic structure is f.c.c. Alloying elements such as chromium, molybdenum, and tungsten increase the transformation temperature, i.e., they stabilize the h.c.p. structure, whereas elements such as nickel and iron suppress the transformation temperature and thus stabilize the f.c.c. form.

Importantly, the transformation of cobalt and its alloys from f.c.c. to h.c.p. upon cooling from the molten state or some high temperature in excess of the transformation point is extremely sluggish, and, in most cases, cobalt and its alloys exist in a metastable f.c.c. form at room temperature, unless they have been subjected to mechanical deformation. It is possible to generate the h.c.p. phase in alloys with sufficiently high transformation temperatures by holding the alloys close to this temperature, but this requires considerable time. As already inferred, the easiest way to generate the h.c.p. structure (at least partially) is by cold working. The formation of so-called "h.c.p. platelets" is believed to progress through stacking fault coalescence. The stacking fault energy of the metastable f.c.c. cobalt alloys is exceptionally low, therefore wide stacking faults (which are atomically thin h.c.p. plates) are abundant in cold worked materials.

There are several cobalt alloy types. The first cobalt alloys were designed primarily for wear resistance and contained high carbon contents to encourage the formation of carbides in the microstructure during alloy solidification. Since these carbides tend to limit hot ductility, hence restrict hot forging and hot rolling, the high-carbon alloys are generally used in cast or weld overlay form, rather than in the form of wrought products (sheets, plates, bars, etc.).

Next to be developed were the cast cobalt alloys for dental and biomedical use, i.e., to resist wear and corrosion from body fluids. These alloys contained less carbon than the original materials. Experimentation with these alloys under high-temperature conditions led to the evolution of the cast and wrought high-temperature alloys, which provide high-temperature strength, oxidation resistance, and sulfidation resistance.

A few cobalt alloys have been designed with general resistance to aqueous corrosion in mind. These alloys have been optimized for corrosion resistance, and take advantage of the inherent wear resistance and/or the strengths that can be attained by cold working.

Cobalt alloys, therefore, may be grouped according to their primary uses, as follows:

(i) Wear
(ii) High temperature
(iii) Biomedical
(iv) Aqueous corrosion.

It is difficult to generalize about product forms, however, since some groups involve several forms (wrought, cast, weld overlay, and powder metallurgy). Also, even if optimized for a particular end use, all cobalt alloys share the characteristics of good wear resistance, good high-temperature strength, and reasonable resistance to aggressive environments (liquid and gaseous).

Although many of the same alloying elements are used to optimize cobalt alloys for their different end uses, their roles are not necessarily the same. Chromium, for example, is used to impart resistance to oxidation and sulfidation to the high-temperature alloys and passivity to the aqueous corrosion alloys, whereas its primary function in the wear alloys is as a carbide former. The chromium that remains in solution provides environmental resistance, some solid solution strengthening, and increases the f.c.c. to h.c.p. transformation temperature, thus having an impact on the transformation tendencies under the action of mechanical deformation.

The carbon contents of the wear alloys vary from about 0.5 wt.% to 3.3 wt.%. This much carbon results in significant carbide weight percentages in the microstructures; for example, approximately 13 wt.% carbide (predominantly chromium-rich M_7C_3) at 1.1 wt.% carbon, and 29 wt.% carbide (predominantly chromium-rich M_7C_3 and tungsten-rich M_6C)

at 2.5 wt.% carbon. As will be discussed, these microstructural carbides are of great benefit under low stress abrasion conditions.

In addition to chromium, the wear alloys typically contain tungsten. The primary function of this element is to provide solid solution strengthening and to promote transformation tendencies (being a stabilizer of the h.c.p. form); it also partitions to the carbides in the higher carbon alloys.

A small, but commercially significant, subset of cobalt wear alloys is alloyed with chromium, molybdenum, and silicon, rather than chromium, tungsten, and carbon. Instead of relying on carbides to optimize low stress abrasion resistance, these alloys rely on the formation of Laves phase (an intermetallic compound, promoted by the presence of silicon). In these alloys, the primary function of molybdenum is Laves phase formation. Unfortunately, their ductility is limited to such an extent that large castings and weld overlays are impractical. Indeed, these alloys have been most successful in powder form, as consumables for plasma spraying.

With regard to high-temperature alloys, it has already been stated that the primary function of chromium is to provide resistance to the environment. In oxidizing, high-temperature environments, protection is afforded by Cr_2O_3 films which form on the surfaces. In high-temperature, sulfur-bearing environments in which the oxygen partial pressure is not very low, Cr_2O_3 films are again important, since they provide protection in the early stages of the process. Chromium is also an effective alloying addition because its sulfides are more stable than those of cobalt and nickel and they have much higher melting points. Cobalt alloys are intrinsically more resistant to sulfidation than nickel alloys, possibly because of the lower diffusivity of sulfur and the higher melting point of cobalt sulfide. In corrosive, aqueous media, chromium is beneficial in that it forms protective (oxide or hydroxide) films, providing oxidizing species are present.

Tungsten is used to strengthen high-temperature cobalt alloys. It is a very effective solid solution strengthener because of its large atomic size as well as its tendency to lower the stacking fault energy, thereby making dislocation cross-slip more difficult. Molybdenum is also an effective strengthener for the same reasons, although it is used mostly in the aqueous corrosion alloys to enhance their resistance to reducing acids such as hydrochloric and dilute sulfuric.

The carbon contents of the high-temperature and aqueous corrosion alloys are much lower than those of the cobalt wear alloys. Most of the high-temperature alloys, for example, have carbon contents of about 0.1 wt.%, which produces a sparse dispersion of primary carbides, which effectively pin the grain boundaries during high-temperature exposure. In most alloys secondary $M_{23}C_6$ carbides precipitate on glide dislocations during high-temperature creep, thereby enhancing resistance to further deformation. The aqueous corrosion alloys have even lower carbon contents, within the soluble range, to minimize the formation of carbides which can reduce aqueous corrosion resistance, especially when they precipitate in the grain boundaries of heat-affected zones during welding.

Apart from chromium, tungsten, and molybdenum, the most important elements in the high-temperature and aqueous corrosion alloys are nickel and iron. As previously mentioned, these are f.c.c. stabilizers and are used to counteract the effects of chromium, tungsten, and molybdenum. In other words, they are used to stabilize the f.c.c. structure and indirectly to make the alloys more amenable to processing, particularly at room temperature (by reducing the transformation tendencies under the action of mechanical deformation). The only problem with nickel and iron is that, in reducing the transformation tendencies, they reduce resistance to wear. For the high-temperature materials this is not an important issue, but for the aqueous corrosion materials (which are typically only used when a combination of corrosion resistance and wear resistance is needed) it is important. In the aqueous corrosion alloys, therefore, the h.c.p. and f.c.c. stabilizers are carefully balanced, and other elements, such as nitrogen, are used to enhance properties.

In the past, attempts were made to develop age-hardenable cobalt-based alloys for high-temperature applications using additions of titanium or tantalum to form the compounds Co_3Ti and Co_3Ta, analogous to $Ni_3(Al,Ti)$ in nickel-based alloys. However, due to problems with the thermal stability of the precipitates because of their higher lattice mismatch and their much lower solvus temperatures, they were easily surpassed by the nickel-based superalloys, and none of these alloys were commercially successful. In cast alloys, additions of elements such as tantalum, niobium, and zirconium are made for solid solution strengthening and to form MC-type carbides which enhance high-temperature strength. In both cast and wrought alloys, small additions of boron are also used to promote creep and stress-rupture strength.

As mentioned previously, chromium is the principal alloying element used to provide oxidation resistance. Attempts have been made to use aluminum additions in order to promote the formation of protective Al_2O_3 surface films, but such additions result in the formation of β-CoAl which significantly reduces ductility. Minor additions ($<1\%$) of elements such as manganese and silicon are used to promote the formation of very protective spinel oxides. Small additions of rare-earth elements such as lanthanum have also proved to be very effective in reducing oxide scale spallation.

2. *Processing*

The melting and processing of wrought cobalt alloys is similar to that of wrought nickel alloys. Typically, they are melted in arc furnaces, refined in argon–oxygen decarburization vessels, poured into large electrode molds, electroslag remelted, then hot forged and hot rolled to plates and bars. Wrought sheets are typically produced by cold rolling. Wires for structural welding are drawn down from hot rolled coil feedstock.

Many intermediate solution anneals are necessary during the cold rolling and cold drawing operations, because cobalt alloys exhibit high work hardening rates. Solution annealing is normally carried out at temperatures in excess of 1100°C and is followed by rapid cooling to prevent the precipitation of grain boundary carbides.

With regard to castings of the (high-carbon) cobalt wear alloys, these are typically made using the investment casting process or the resin shell casting process. Normally, small quantities of remelt bar or remelt ingot are taken from larger arc-melted heats of material for this purpose. Investment cast molds, made by the lost-wax process, are usually clamped to high-frequency furnaces containing small quantities of molten remelt material, and rolled over to produce the castings. Pouring temperatures are in the vicinity of 1650°C.

Weld overlay consumables of the same materials are produced by several methods, depending on the type of consumable desired. Straight lengths of wire, for gas tungsten arc (TIG) and shielded metal arc welding (with a suitable flux coating), are typically made by continuous casting. Coiled wires for gas metal arc (MIG) welding are made by wrapping the alloying ingredients in sheaths of cobalt (actually a cobalt–iron alloy, made by powder metallurgical techniques). Powders for plasma transferred arc welding are made by gas atomizing.

Bibliography

Massalski T B, Okamoto H, Subramanian P R, Kacprzak L 1990 *Binary Alloy Phase Diagrams*, 2nd edn. ASM International, Materials Park, OH, Vol. 2, pp. 1214–15

D. Klarstrom and P. Crook

Colloid Casting

Colloidal casting is used to produce complex ceramic shapes from a broad range of raw materials including silicon carbide, silicon nitride, clay, and alumina. The three primary methods of colloidal casting include slip casting, pressure casting, and centrifugal casting. Central to each method is the formation of a green body by solid–liquid separation of a slip, which is a fluid suspension of colloidal ceramic particles mixed in liquid (e.g., water). After reviewing each casting method, this article concludes with a discussion of slip formulation guidelines that are common to all colloidal casting methods.

1. Slip Casting

Slip casting, the most common industrial method of colloidal casting, is used to produce tableware, structural tubing, automotive and biomedical components, sanitaryware, artware, etc. (Schilling and Aksay 1991). Slip casting begins with pouring a slip, which is usually aqueous, into a permeable gypsum mold. The pores in the gypsum are much smaller than the ceramic particles in the slip. Capillary suction of the gypsum filters the interparticle liquid from the slip, depositing a layer of densely packed particles against the mold wall. Drain casting involves slip casting for a limited time period to form a cast layer of a desired thickness (Fig. 1). The remaining slip is subsequently drained from the mold. Solid casting is identical to drain casting, except that slip is continually added until a solid cast forms. A less common approach is vacuum slip casting, where a vacuum is applied to a gypsum mold in order to increase the rate of filtration.

Production of highly complex shapes is also possible by slip casting with insoluble mandrels attached to a mold. For example, fugitive wax slip casting is based on the much older technology of investment casting (Ezis and Neil 1979).

1.1 Mechanics of Slip Casting

The pressure at the gypsum–cast boundary is approximately equal to the capillary suction pressure of the gypsum mold, σ, and is approximated by the expression $\sigma = S_g \gamma \cos\,\theta$, where S_g is the specific surface area of gypsum, γ is the surface tension of the interparticle liquid, and θ is the contact angle ($\cos\,\theta = 1$ for water–gypsum). Capillary suction pressure for water in gypsum varies between 0.03 MPa and 1 MPa, depending on the average pore size in the gypsum mold (Nies and Lambe 1956).

Slip casting is generally limited to the production of thin-walled articles (1.5 cm maximum thickness) because of casting rate limitations imposed by the hydraulic resistances of the cast and mold. *In situ* measurements of the cake thickness are usually impossible to perform with the naked eye because of the opacity of the slip. Ultrasonic, *in situ* measurements of cake height have been reported during slip casting in a laboratory setting (Haerle and Haber 1995).

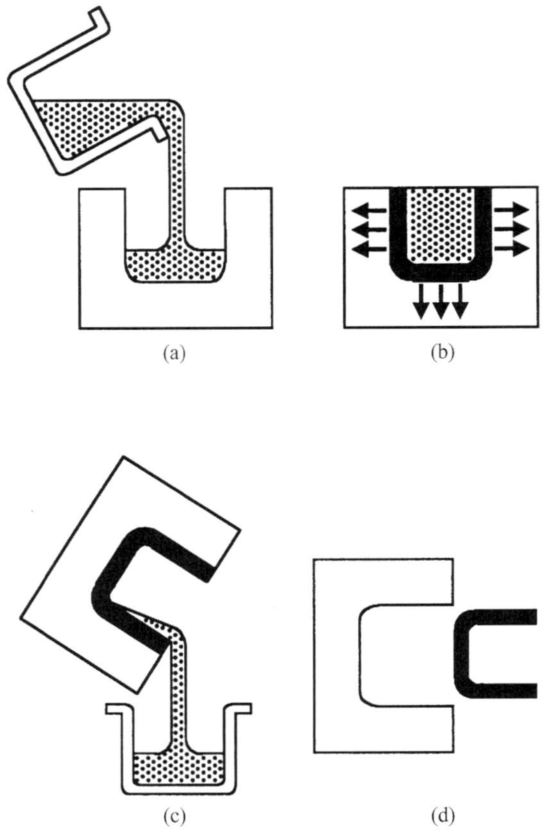

Figure 1
Four consecutive stages of slip casting: (a) the mold is first filled with slip; (b) the mold extracts liquid and the cast layer forms along the mold walls; (c) excess slip drains from the mold; (d) the green body is removed after partial drying.

A more common method entails estimating the cast-layer thickness, ξ_c, based on measurements of the elevation of the slip–air interface, ξ_s:

$$\xi_c = (\xi_s^0 - \xi_s)/n = -\varepsilon_m \xi_m / n \qquad (1)$$

The above model is derived from a mass balance and assumes filtration of an incompressible, constant-density cake onto the planar, horizontal surface of a mold (Fig. 2). ξ_s^0 is the height of the slip–air interface at the start of casting, ξ_m is the depth of the filtrate–air interface in the mold, ε_m is the volume fraction porosity in the mold, and n is a mass balance

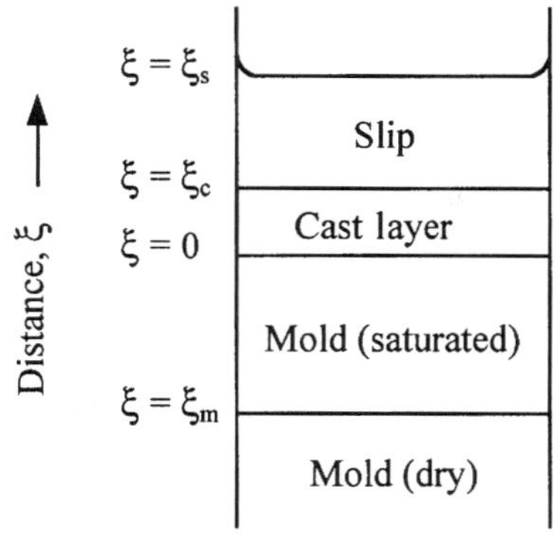

Figure 2
Schematic view of slip casting onto a planar, horizontal surface of a gypsum mold.

factor equal to $(1-c-\varepsilon_c)/c$. In this expression, c is the volume fraction of solids in the slip and ε_c is the volume fraction of voids in the cast layer.

The rate of cast-layer growth can be estimated by a casting kinetic model based on Fig. 2 (Adcock and MacDowall 1957). The model assumes negligible gravity sedimentation of particles in the slip and predicts that the casting rate decreases as a parabolic function of time, t:

$$\xi_c = \{2Pt/[n\eta\varepsilon_m(n\alpha_m + \alpha_c)]\}^{1/2} \quad (2)$$

where η is the viscosity of the filtrate, and α_m and α_c are the specific hydraulic resistances (reciprocal permeabilities) of the mold and cast, respectively. Equation (2) predicts that the casting rate can be increased by (i) decreasing α_m and α_c and (ii) increasing P, c, ε_m, and ε_c. The usual practice of slip casting involves use of dispersed slips containing ultrafine (submicrometer) ceramic particles. In this case, high-density cast layers ($c>0.5$) typically form, having very low permeabilities that dominate the total flow resistance. Under these conditions, α_c can significantly exceed α_m, and Eqn. (2) can be simplified by setting $\alpha_m=0$.

The specific resistance of the cake is strongly affected by particle size and packing density, as shown by the Carman–Kozeny model: $\alpha_c=(5c^2S^2)/(1-c)^3$, where S is the specific surface area of the powder. In the case of spherical particles, S can be approximated as $3c/r$, where $r=$ particle radius.

Useful information on the uniformity of the packing density of a slip-cast layer can be obtained by plotting measurements of ξ_s or ξ_c as a function of $t^{1/2}$. A strictly linear plot indicates a uniformly dense cast layer, because it indicates constancy of the cake microstructure parameters (n and α_c) in Eqn. (2). Typical causes of deviations from linearity and thus microstructure uniformity are associated with one of two factors: (i) the presence of large flow units (e.g., hard agglomerates, large particles) that undergo gravity sedimentation during slip casting or (ii) changes in α_c that might result from compression of the cake or from time-dependent changes in the rheological properties of the slip.

Banno *et al.* (1999) reported a two-dimensional computer simulation method to predict cake growth during slip casting with complicated shape gypsum molds. The model is based on the method of finite differences and simulates cake growth patterns, water penetration patterns, water flow rate distributions, and pressure distributions.

1.2 Gypsum Molds

The gypsum mold is a critical element in slip casting, because the capillary suction of this porous material is responsible for slip dewatering (Dal and Berden 1968). Advantages of gypsum are the ease of its manufacture into a wide variety of shapes, low cost, good dimensional accuracy, and reproducible water absorption properties.

Gypsum molds are made by pouring a plaster–water slurry into a case mold having the same shape as the finally desired greenware. Case molds can be made of gypsum, wood, or plastic. After filling the case mold, the slurry sets into a rigid shape by the chemical hydration of plaster ($CaSO_4\cdot\frac{1}{2}H_2O$) to gypsum ($CaSO_2\cdot 2H_2O$). The setting process involves nucleation and growth of needle-shaped gypsum crystals (2–5 μm in diameter by up to 20 μm in length) that entangle and become interlocked with random orientation.

The pore volume between the gypsum crystals consists of interconnected capillaries (~0.1 μm diameter) that are randomly oriented and filled with excess water during the setting process. Theoretically, only 0.186 kg of water are required to hydrate 1 kg of plaster; however, in practice more water (0.64–0.8 kg water per kg plaster) is added to impart fluidity to the plaster slurry. Increases in the consistency (the water-to-plaster weight ratio) increase the pore volume of the hydrated gypsum, which in turn decreases the strength of a mold and increases its water absorption capability. The slurry volume expands by as much as 0.3% during setting, and this expansion decreases as the slurry consistency is raised. The freshly set mold is separated from the case mold, and the excess pore water is then evaporated.

Careful control of the mixing process is needed to ensure repeatable and reliable mold properties (Lambe 1955, 1963). Increases in the mixing water temperature reduce the setting time and increase the mold compressive strength. The solubility of gypsum in water has a maximum between 21°C and 37°C, and it is recommended that mixing take place in this temperature range. Increasing the mixing energy increases the mold compressive strength and decreases the water absorption capacity of a mold. Whiteside (1966) reported mold strengths that changed from 9.6 MPa to 13.8 MPa for samples that were mixed for 2.3 min and 9 min, respectively.

Metal ion impurities in the mixing water and slips should be minimized in order to avoid chemical reactions with gypsum that produce soluble salts (sodium sulfate, magnesium sulfate). These salts migrate to the surface of a mold during drying (efflorescence), producing areas on the mold surface that cause adhesion of greenware to the mold.

Careful pouring of the plaster slurry is necessary to avoid air bubble entrapment. Agitation of the filled case mold is used to avoid bubbles near the mold surface. Cracks or subsurface bubbles in a mold can result in mold surfaces having nonuniform suction. In turn, this can produce defects in greenware.

Case molds are precoated with a soap film that promotes mold release. Neutral or potassium soaps are more desirable than highly alkaline soaps, which are sources of sodium. The best soaps are those that produce an insoluble film (e.g., calcium stearate).

Trial-and-error methods are usually necessary to bring a new mold to a proper water concentration in order to control the rate of water absorption during slip casting. If a slip is dewatered too rapidly, regions of the cast layer adjacent to the mold can shrink excessively. In turn, this reduces the casting rate, it may cause surface cracks in the green body, and it can also cause premature release of the green body from the mold. If a slip is dewatered too slowly, the cast layer does not build up at a satisfactory rate and is likely to be too fluid and stick to the mold. An intermediate condition is usually achieved by a "conditioning" process that involves rewetting a dried mold until it reaches a moisture concentration of 5–15% by weight.

The amount of free water in a gypsum mold significantly affects the rate of water absorption. When a dry mold is exposed to water, absorption rates are initially rapid ($\sim 0.45\,g\,cm^{-2}\,min^{-1}$) and then parabolically decrease to lower values. The absorption rate of new molds increases during the first three cycles of immersion and drying because of the dissolution of gypsum from the pore walls.

Once casting is completed, a given part begins to dry and shrink away from the mold. This shrinkage is necessary to release the part from the mold. Mold release can be aided by coating the walls of a mold with a release agent such as silicone oil, olive oil, or a 5% solution of sodium alginate in water.

Immediately after slip casting, excess water is evaporated from the molds using a drying oven or a drying room. Optimum drying conditions entail the use of uniform temperatures and air circulation. Gypsum molds should not be heated above 49°C, the temperature at which gypsum dehydrates and permanently loses its strength.

2. *Pressure Casting*

Pressure casting is an automated production operation that resembles slip casting, except that the slip is pressurized to increase the filtration rate (Blanchard 1988). Two main types of pressure-casting equipment include medium- (0.3–0.4 MPa) and high-pressure (up to 4 MPa) systems. The materials of construction and tolerances of mold fit become more expensive as the pressures increase. Medium-pressure systems are used on large, high-production articles such as sanitaryware. In this case, the molds consist of two mating halves that are made of either gypsum or porous plastic. A series of up to 15 molds is assembled and then stacked together before slip is simultaneously pumped into each mold. After the casting process is completed, the mold halves are separated and the green body is removed. The saturated molds are subsequently purged with compressed air to remove the liquid occupying the pores within each mold. The entire production cycle takes approximately 75–90 min with a 15-mold bench.

Higher pressure casting can become advantageous for lower production levels and smaller shapes (e.g., tableware). Production levels are usually lower than with medium-pressure systems, because the equipment typically operates with a single mold. High-strength plastic molds are needed for high-pressure casting, because gypsum molds tend to fracture.

Pressure casting has several economic advantages over slip casting. A major problem with slip casting is the slow filtration rate, which requires a large number of gypsum molds for commercial production. Another problem that is eliminated is the drying of plaster molds, a slow process involving large amounts of floor space. Pressure casting has a shorter production cycle, which requires less floor space and less labor. Fluctuations in the quality of greenware are reduced because of the automated nature of pressure casting.

Unlike gypsum, plastic molds are not as prone to surface wear. This improves the degree of fit with mold parts, decreases the surface roughness of greenware, and increases the consistency of edges and sharp borders. Plastic molds are also less prone to chemical attack from metal ions contained in slips. Mold conditioning is also avoided, because the slip pressure is not affected by variations in mold saturation. Another advantage of pressure casting over slip casting is that the greater pressures result in higher green densities with less moisture. Hence, drying and sintering shrinkages are reduced, and dimensional tolerances for near-net-shape processing are improved. One disadvantage is that plastic molds are more expensive than gypsum molds.

3. *Centrifugal Casting*

Centrifugal casting is the least common industrial method of colloidal casting and involves solid–liquid separation by spinning a mold in a laboratory-scale centrifuge (Lange 1986, Bachmann *et al.* 1989). These molds are not necessarily porous and can be fabricated from various metals and plastics. Unlike slip casting or pressure casting, the interparticle liquid in centrifugal casting is removed through the top of a cake, which tends to be lower in compaction stress, lower in packing density, and thus higher in permeability than regions of a cake that are more distant from the center of rotation. With centrifugal casting, it is possible to achieve more rapid casting of much thicker shapes than those typically made by other colloidal casting methods.

During centrifugal casting, significant gradients in compaction stress typically exist within a cake at different radial distances from the center of the centrifuge. These spatially varying compaction stresses can produce packing density gradients in a cake depending on the cake compressibility (Schilling *et al.* 1996). This compressibility is largely controlled by slip processing conditions that regulate interparticle friction and adhesion (e.g., variations in the chemistry of surface-adsorbed polymers or ions). Flocculated slips tend to produce compressible cakes that are more prone to the formation of packing density gradients and consequent shrinkage cracks during drying and debinding. In contrast, dispersed slips tend to produce incompressible cakes that are more likely to produce uniform packing densities and crack-free greenware (Lange 1989).

Composites can be made by centrifugal casting slips that contain a mixture of different types of particles (Chang *et al.* 1991). Differential sedimentation velocities are important to avoid during centrifugal casting of composites, as differential sedimentation can produce segregated microstructures that are prone to cracking during drying, debinding, and sintering. This problem is especially severe when using dispersed slips where particles do not agglomerate and can independently move past one another during centrifugation. Flocculated slips are preferable, because they can form a continuous network of particles that traps heavy particles from differential sedimentation (Chang *et al.* 1991). A trade-off with the use of flocculated instead of dispersed slips is that flocculated slips tend to consolidate into compressible cakes that are more prone to forming packing density gradients, as mentioned above.

Marshall *et al.* (1991) demonstrated the formation of layered microcomposites by the sequential centrifugal casting of different slip formulations onto the same cake. Layer thicknesses can be made as small as a few micrometers by controlling the speed of centrifugation and the solids concentration in the slip.

4. *Slip Control*

A major requirement of colloid casting is the control of suspension chemistry to produce low-viscosity slips that are highly concentrated in powder. Low viscosities (<1 Pa s) are needed to allow pouring a slip into a mold. High solids concentrations (up to 60 vol.%) are important because they increase the rate of filtration during slip casting and pressure casting (Eqn. (2)). Typical methods for achieving low-viscosity slips with high solids concentrations entail the use of polydisperse particle size systems and repulsive interparticle forces that arise from surface-adsorbed ions or polymers (Lange 1989).

Constancy of slip properties is essential for achieving reliable cast-layer properties. Critical parameters include rheological properties of the slip, sedimentation rate, casting rate, drying shrinkage, and mold release behavior. The slip must be easily reproduced, and its casting behavior should be insensitive to slight variations in slip conditions such as the solids concentration, chemical composition, mixing conditions, and storage time. Trial-and-error methods are usually needed to determine optimum conditions. The slip must be free of gas bubbles and solid contaminants that may become fracture-initiating defects in the green or sintered part. Slips are usually aged to keep properties constant, then final viscosity measurements are performed before vacuum de-airing and casting.

Organic polymers are common slip ingredients that serve many functions including dispersants, viscosity modifiers, plasticizers, and binders (in reference to their ability to improve green-body strength). Aqueous suspending media are typically used to dissolve these polymers. Binders are usually 10000 to 30000 molecular weight polymers such as acrylics (e.g., poly(acrylic acid), poly(methacrylic acid)), vinyls (e.g., poly(vinyl alcohol)), alginates, and various cellulosics. Plasticizers are typically small to medium-sized molecules that decrease cross-linking among binder molecules and make castings more pliable (e.g., polyethylene glycol, butyl, benzyl, phthalate). Shanefield (1995) compiled an extensive review of organic additives for ceramic slips.

A concern with the use of organic polymers in slips is that they are transported into the capillaries of slip/pressure casting molds and may be difficult to remove. This can change the mold permeability, suction behavior, and shorten the life of the mold. The use of organic chemicals may also pose health hazards. There are also concerns associated with the burnout of organics from ceramic greenware, e.g., the production of toxic gases, the formation of cracks and gas bubbles in a ceramic part, and the formation of carbonaceous residues in sintered microstructures. It is also important to control the particle size distribution in order to maximize the green density and thereby minimize dimensional shrinkages during drying, debinding, and sintering. Several packing models are available to aid in formulating mixtures of different sized particles to achieve maximum packing efficiency (Reed 1995).

Within a given cast layer, the packing density should be uniform to avoid nonuniform shrinkages that cause warpage and cracks during drying, debinding, and sintering (Pujari and Tracey 1993). To understand the development of packing density gradients it is important to consider that spatial variations of compaction stress exist during cake formation by all colloidal casting methods (Schilling *et al.* 1996). It has already been mentioned that significant gradients in compaction stress exist within a centrifuge cake at different radial distances from the center of the centrifuge. In slip casting and pressure casting, high levels of compaction stress exist at the mold–cake interface, whereas much lower compaction stresses exist at the cake–slip interface. Whether or not these spatially varying compaction stresses produce packing density gradients depends on the compressibility of the cast layer, which in turn depends on chemical properties of the slip that control interparticle friction and adhesion.

Differential sedimentation velocities of large vs. small particles is a common cause of nonuniform microstructures during colloidal casting. Differential sedimentation results in localized, segregated zones of these large particles, and in turn this produces cracks and warpage during drying, debinding, and sintering. This problem can be alleviated by preparing a slip as follows: (i) narrow the particle size distribution of the starting powder so it does not contain large particles that are prone to rapid sedimentation, (ii) rigorously mix or ultrasonicate a dispersed slip (repulsive interparticle forces) to break apart weak agglomerates, and (iii) subsequently remove large particles and agglomerates by sedimentation or centrifuging the slip prior to casting. Another approach to minimize segregation of heavy particles is to reduce the time available for segregation during casting. An easy way to accomplish this is to raise the casting rate by increasing the solids concentration in a slip (Eqn. (2)).

Different cast microstructures can be produced by adjusting the interparticle forces in the slip (Aksay and Schilling 1984). Dispersed slips generally result in slow casting of densely packed microstructures having low permeability. In contrast, attractive surface forces result in flocculated slips that can be slip cast or pressure cast more rapidly because they form cakes that are more porous and permeable.

Slips containing mixtures of different particle chemistries are used in the colloidal casting of composite materials. Special care must be exercised to avoid spatial variations of mixedness in the cast layer, mixedness variations that can produce differential shrinkages that lead to cracks and warpage during drying, debinding, and sintering. Chang *et al.* (1991) illustrated how segregation of alumina and zirconia particles could be avoided through the use of flocculated instead of dispersed slips during centrifugal casting. Hoffman *et al.* (1989) showed another example of segregation during slip casting of a mixture of equiaxed Si_3N_4 particles and SiC chopped fibers. During slip casting, the fibers preferentially align themselves parallel to the surface of the mold–cake interface. This preferential alignment is observed in regions adjacent to the mold–cake interface, whereas the fibers are randomly oriented in regions of the cast layer that are more distant from the mold–cake interface. Drying cracks form at the boundary of these two regions as a result of differential shrinkage. Additional research is needed to explain how changes in colloidal casting conditions influence spatial variations of mixedness in composite materials.

Bibliography

Adcock D S, MacDowall I C 1957 The mechanism of filter pressing and slip casting. *J. Am. Ceram. Soc.* **40**, 355–62

Aksay I, Schilling C H 1984 Colloidal filtration route to uniform microstructures. In: Hench L L, Ulrich D R (eds.) *Ultrastructure Processing of Ceramics, Glasses, and Composites.*. Wiley New York, pp. 439–47

Bachmann P K, Geitner P, Krafczyk E, Lydtin H, Romanowski G 1989 Shape-forming of synthetic silica tubes by layerwise centrifugal particle deposition. *Am. Ceram. Soc. Bull.* **68**, 1826–31

Banno T, Saburo S, Oda K 1999 Application of a two-dimensional computer simulation to cake growth in slip casting with complicated-shape gypsum molds. *J. Am. Ceram. Soc.* **82**, 3401–6

Blanchard E B 1988 Pressure casting improves reliability. *Am. Ceram. Soc. Bull.* **67**, 1680–3

Chang J C, Velamakanni B V, Lange F F, Pearson D S 1991 Centrifugal consolidation of Al_2O_3 and Al_2O_3/ZrO_2 composite slurries vs. interparticle potentials: particle packing and mass segregation. *J. Am. Ceram. Soc* **74**, 2201–4

Dal P H, Berden W H 1968 The capillary action of plaster molds In: Stewart G (ed.) *Science of Ceramics.* British Ceramic Society, Vol. 4, pp. 113–31

Ezis A, Neil J T 1979 Fabrication and properties of fugitive mold slip-cast Si_3N_4. *Am. Ceram. Soc. Bull.* **58**, 883

Haerle A G, Haber R A 1995 Ultrasonic real-time monitoring of cake structure during slip casting. *J. Am Ceram. Soc.* **78**, 819–22

Hoffman M J, Nagel A, Greil P, Petzow G 1989 Slip casting of SiC-whisker-reinforced Si_3N_4. *J. Am. Ceram. Soc.* **72**, 765–9

Lambe C M 1955 Plaster molds for mechanized forming of clay ware. *Am. Ceram. Soc. Bull.* **34**, 251–5

Lambe C M 1963 Preparation and use of plaster molds. In: Kingery W D (ed.) *Ceramic Fabrication Processes.* MIT Press, Cambridge, MA, pp. 31–40

Lange F F 1986 Forming a ceramic by flocculation and centrifugal casting. *US Pat.* 4 624–808

Lange F F 1989 Powder processing science and technology for increased reliability. *J. Am. Ceram. Soc.* **72**, 3–15

Marshall D B, Ratto J J, Lange F F 1991 Enhanced fracture toughness in layered microcomposites of Ce–ZrO_2 and Al_2O_3. *J. Am. Ceram. Soc.* **74**, 2979–87

Nies B W, Lambe C M 1956 Movement of water in plaster molds. *Am. Ceram. Soc. Bull.* **35**, 319

Pujari V K, Tracey D M 1993 Processing methods for high reliability silicon nitride heat engine components. *ASME*

Trans. Proc. Int. Gas Turbine and Aeroengine Congress. ASME 93-GT-319, Cincinnati, OH
Reed J S 1995 *Principles of Ceramic Processing*, 2nd edn. Wiley, New York
Schilling C H, Aksay I A 1991 Slip casting: a review. *Engineered Materials Handbook, Ceramics & Glass.* American Society for Metals, Metals Park, OH, Vol. 4, pp. 153–60
Schilling C H, Huss S P, Gray J N 1996 Regulating density gradients in slip consolidation: a new methodology. In: Henkes V E, Onoda G Y, Carty W (eds.) *Science of Whitewares.*. American Ceramic Society Westerville, OH, pp. 117–31
Shanefield D J 1995 *Organic Additives and Ceramic Processing.* Kluwer, Boston
Whiteside E L 1966 Quality control in the plaster mold shop. *Am. Ceram. Soc. Bull.* **45**, 1022–6

C. H. Schilling

Composites, Joining of

Composite materials, be they metal matrix composites (MMCs), polymer matrix composites (PMCs), or ceramic matrix composites (CMCs), are joined by the same procedures that are well-known for joining isotropic materials such as metals or polymers. This is to be expected since the process of transferring loads between joined components sets up stress patterns which are more dependent on the component/joint configuration than on the material used to fabricate the article.

Of course, composites often display directionality of mechanical properties, especially those fabricated from long or continuous fibers and used in laminated form e.g., graphite fiber-reinforced epoxies used in aerospace applications. Such directionality, which usually results in low strength through the thickness of plate-like constructions, causes particular problems and the emphasis when designing load-carrying joints is on minimizing the stresses in these weak directions.

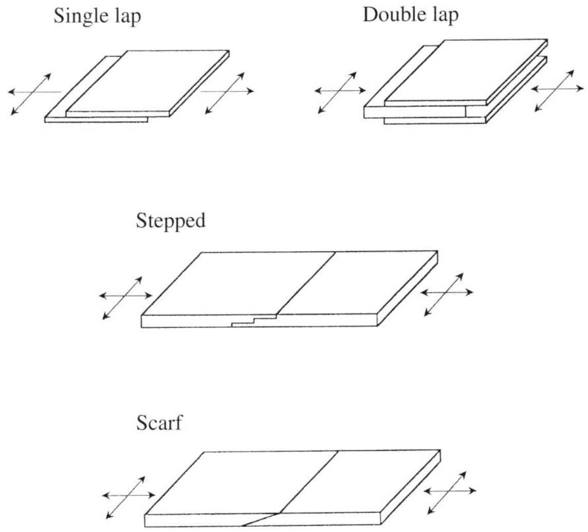

Figure 1
Lap joint configurations for in-plane loading.

1. *Joint Configurations*

1.1 *Standard Configurations*

The type of joint most often encountered is the lap joint, which has the four basic forms shown in Fig. 1; single lap, double lap, stepped lap, and scarf. Such joints are mainly used between fiber-reinforced plastic (FRP) laminates, but in the stepped and scarf forms could be between an FRP laminate and a metal plate.

Flange joints are also commonly encountered, although there is a shortage of generally available information. Such joints would be used to join sections of pipes (tubes), cylinders (e.g., turbojet casings), or the (usually flat) panels that form the sides of containers, for holding water, etc. Tubular components can also be joined by cylindrical forms of the flat lap joints.

1.2 *Special Configurations*

Nonstandard configurations are reserved for particular applications, e.g., wedge joints are used for securing turbine blades or propeller blades to the turbine disc or propeller hub, respectively. As the rotational speed of the device increases so does the corresponding radial force on the blade and the constraint afforded by the wedge produces an ever tightening joint. The same principle is encountered in the wedge grips of tensile testing machines.

Another special configuration, which has been used for securing helicopter rotor blades to the rotor hub, is the "race track" or "wrapped" joint. Here, continuous fibers are positioned around the hole (Fig. 2) in which the securing pin is located, the forces on the joint being such as to load the fibers in tension, an ideal uni-axial stress state.

2. *Load Transfer*

The purpose of a structural joint is to transfer load from one side of a joint to the other. In so doing a complex state of stress, which varies throughout the joint, will be set up. The main points can be illustrated by reference to the single cover joint (end-to-end single lap joints) shown in Fig. 3. The load in plate A has to be transferred to plate B via cover plate C. As a consequence the joining medium (adhesive in Fig. 3(b)) will deform in shear, with maximum stresses occurring at the ends of the overlaps. Because the line of action of the force in A and B is not aligned with that in C, overall bending, as shown in Fig. 3(c) will occur. The ends of the overlaps will move apart, setting up through-thickness tensile (so-called, "peel") stress in the joining medium and the plates. It is these stresses mentioned above that are likely to be the cause of failure in FRP laminates.

With flange joints the usual consideration is to maintain a leak-proof configuration by ensuring that the mating faces of the flanges do not separate. Such issues are well understood from many decades of joining steel pipes. With composites, the usual problem is to prevent failure either at the corner of the flange due to the gross bending, or due to the connecting fasteners pulling through the material (Godwin *et al.* 1986).

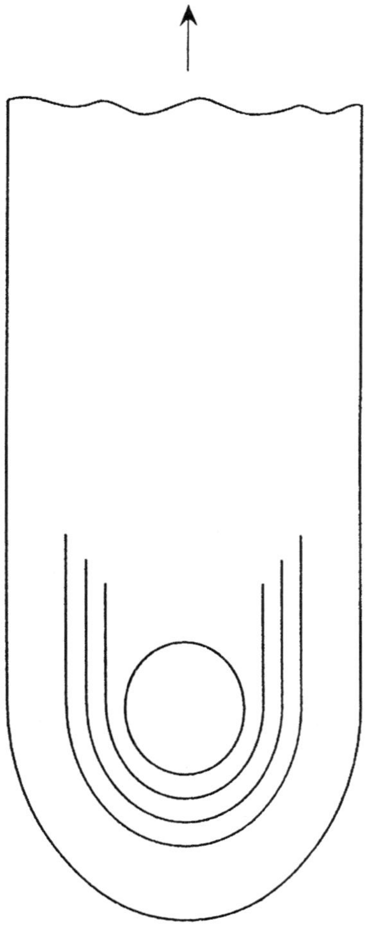

Figure 2
"Wrapped" joint.

3. *Joining Polymer Matrix Composites*

Polymer matrix composites (PMCs), or FRPs, constitute the bulk of the commercial market. The majority of FRPs will have a thermosetting matrix and their popularity is due to their ease of processing. These materials are normally used in the form of laminates, with continuous or long fibers, and hence joints are usually of lap configuration. FRPs with thermoplastic matrices are also gaining popularity, although these are more commonly produced with short fibers by processes such as injection molding and hence, rather than being plate-like, will have a bulk form. Joining procedures are then similar to those used for bulk isotropic materials, or for MMCs (see Sect. 4).

Recently introduced materials for aerospace use includes laminates of aluminum alloy sheets bonded to sheets of aramid fiber/epoxy (ARALL) or to sheets of glass fiber/epoxy (GLARE) (Wu *et al.* 1992). Such hybrid laminates would be joined in the same way as conventional FRP laminates.

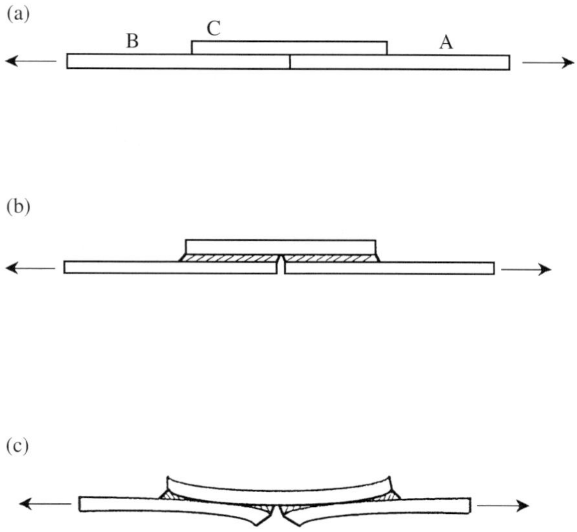

Figure 3
Load transfer across a single cover joint.

3.1 Adhesive Bonding

Adhesive bonding (or gluing) is used to join (pre)-cured laminates. A layer of adhesive is inserted between the laminates in the joint region and allowed to cure. So-called "cobonding" is a similar procedure, but here the joint is between uncured laminates, and the adhesive and matrix cure in the same operation. Slightly different, but producing identical configurations, is "cocuring" when there is no additional adhesive layer between parts and the matrix resin acts as the glue during the cure process.

The bonding procedure is perfectly straightforward and well-understood (Matthews 1994). However, care is needed to prepare the mating surfaces by removing all traces of mold release agent, or other contaminant, from the composite. If the composite is to be bonded to a metal then the latter will also need pretreatment to remove grease, etc., and any surface oxide layer (for aluminum and titanium). Adhesives are normally epoxy resin, but could be acrylic, phenolic, or polyurethane. They come in liquid, paste or film form and cure at room temperature, 120°C or 170°C.

Of the standard configurations the single lap is the weakest, followed by the double lap, then the stepped lap, with the scarf being the strongest joint (Matthews 1994). Essentially this is due to the eccentricity of load paths being greatest for the single lap and zero for the scarf. Indeed, by making the angle sufficiently small, it is possible to make a scarf joint stronger than the material being joined (efficiency of 100%). The only drawback is the need to maintain a "knife edge" at the corner of the scarf and this is difficult to produce, and avoid being broken off, in wide joints (i.e., items bigger than laboratory test specimens). The stepped joint is difficult and costly to fabricate and so would only be used in the most demanding applications.

The performance of single and double lap joints is enhanced by increasing the length of the overlap region. Further improvements can be brought about by tapering the thickness (being a minimum at the overlap end) of the laminate. Performance is also enhanced by using a ductile adhesive and having the largest possible in-plane stiffness of the laminate, consistent with other structural constraints Hart-Smith 1972).

Typical applications of bonded FRP joints include: secondary and tertiary (control surfaces, fairings) as well as primary (tail plane-to-fuselage) aircraft structures (Anon 1990, Hoskin and Baker 1986), the chassis of Formula 1 race cars, satellite structures, ship hulls, the doors of truck cabs, suspension bridge decks, and indeed any situation where a smooth profile is desirable and/or drilling holes for fasteners is not feasible (Phillips 1989).

3.2 Mechanical Fastening

Mechanically fastened joints are used in aircraft (e.g., spar-to-skin joints in wings), ships (frame-to-hull joints), and race cars (engine-to-chassis joints). In the aircraft case they are a necessary feature of the overall assembly process (Anon 1990). Mechanical fastening is a suitable procedure when bonding is either difficult (e.g., with thermoplastics) or its durability is questionable. Mechanical fasteners are ideal if disassembly is required, and could be an acceptable approach for (temporary) repair. Of course, fasteners are heavy so a weight penalty will be incurred.

Typical fasteners are screws (only used for the most lightly loaded joints), rivets (limited to laminates no thicker than 6 mm), and bolts (Matthews 1994). The latter produce the strongest joints. Of course, it is necessary to drill holes in the laminate to receive the fasteners, and these holes will give rise to stress concentrations, which inevitably limit the strength of the joint. Drilling the holes requires diamond or carbide tipped bits and controlled feed rates to prevent ply separation on the back face of the laminate.

Mechanical fasteners are mainly used for single lap joints (rivets and bolts), double lap joints (bolts), and, of course, for flanges (bolts). The fiber/matrix combination is very important; carbon (graphite) fiber/epoxy (CFRP) gives the strongest joints, followed by glass fiber/epoxy (GFRP), with "Kevlar" fiber/epoxy (KFRP) being the weakest. A joint with optimum dimensions containing one fully tightened bolt could have an efficiency of around 40% (Hart-Smith 1987). The obvious way to overcome this low performance is to increase the thickness of the laminate in the region of the joint.

Other parameters that can influence performance are the dimensions of the joint, i.e., the fastener diameter, thickness of the laminate, distance of the hole center from the end of the joint, and width of the joint (for one fastener) or the pitch between fasteners for an array of fasteners (Matthews 1994).

3.3 Welding

Because of their low surface energy, thermoplastic matrix composites (TPMCs) are difficult to bond, although with appropriate pretreatment, such as corona discharge, bond strengths are equivalent to those obtained with thermosets (Kinloch *et al.* 1992). The welding process involves melting of the matrix polymer alone and/or an intermediate layer of unreinforced polymer. The general characteristics of welded joints will be similar to those of bonded joints.

The processes that have been considered include: hot plate, vibration, ultrasonic, resistive implant and induction welding (Wise 1992), and, more recently, microwave welding. Induction welding appears to have the most potential for large-scale use and has been employed successfully to fabricate wing components on Airbus airliners (Clements 2000). With all these processes it is possible to weld small specimens, the difficulty is in applying the procedures to real components when large areas have to be heated and consolidated.

4. Joining Metal Matrix Composites

Although not so widely used as PMCs, MMCs have been employed in some key areas, and are becoming more popular. Applications include: automotive drive shafts, valves, pistons, cylinder liners and brake discs; mountain bike frames and golf clubs; struts in space craft; electronic instrument racks; and yacht fittings (Ellis 1994, Koczak *et al.* 1993).

MMCs are normally reinforced by particles or short fibers and hence do not display the anisotropy seen with long fiber reinforcement. They could, clearly, be joined by mechanical fastening or adhesive bonding (provided the latter is compatible with the operation temperature).

However, the main attraction is to be able to use conventional metallurgical techniques such as fusion, forging, and brazing. It is important to ensure that the joining process does not react adversely with the MMC so that properties are reduced in the joint region. The matrix metal is a key determinant of performance, in the latter respect; aluminum (Al) is the most commonly encountered, but titanium (Ti), magnesium (Mg), and other metals have been used (Koczak *et al.* 1993).

4.1 *Fusion Welding*

During welding the matrix will be above its melting point, but the reinforcement (typically silicon carbide, SiC, or alumina, Al_2O_3) will remain solid. The defects normally found when welding unreinforced alloys, such as solidification cracking, may still be encountered with MMCs, but with the latter additional problems occur, i.e., the higher viscosity of MMC melts; segregation on solidification; reinforcement/matrix interactions; evolution of occluded gases (Ellis 1994).

There are several processes available for fusion welding of MMCs: tungsten inert gas (TIG), metal inert gas (MIG), electron beam (EB), laser beam (LB), resistance, capacitor discharge, plasma (Ellis 1994). TIG and MIG processes should be used when weld fillers are needed. EB and LB methods are attractive because they have short thermal cycles and low heat input.

4.2 *Forging and Solid State Welding*

Since melting is largely absent, solid state processes avoid the problems encountered with fusion welding of MMCs.

Diffusion bonding is one main category. In this approach the components are held under load at elevated (below melting) temperature in a protective atmosphere or vacuum. Solid phase diffusion bonding (SPDB) is one variant of the process, although transient liquid phase diffusion bonding (TLPDB) is a more successful approach.

Friction welding is another solid state joining process. Friction is created by moving one part against the other, or by a separate tool, using rotary (stir welding), linear, or orbital motion. After sufficient time has elapsed the motion is stopped and the parts brought together under load for the forging stage of the weld. Drive shafts are joined by friction welding. With some MMCs the particulate or fibrous reinforcement close to the bond line may be fragmented.

4.3 *Brazing*

Since brazing does not involve melting of the matrix, problems of matrix/reinforcement reactions and reinforcement destruction are reduced. Dissimilar materials can be joined by brazing and heat induced distortion is minimized.

5. *Joining Ceramic Matrix Composites*

CMCs are used in high temperature applications such as turbine blades, piston engine valves, and turbo charger components. It may be necessary to join CMCs to themselves or to a metal (Fernie *et al.* 1991). A particular problem is the brittleness of CMCs, which means that tensile stresses in the joint must be kept to an absolute minimum. The use of mechanical fasteners is normally not feasible, because of the associated stress concentrations, and the difficulty of making holes in such hard materials.

One approach is to use shrink fitting, in which differences in thermal contraction (on cooling from elevated temperature) are exploited to form a tight joint. This method is used in the automotive industry to produce spark plugs (which have an alumina insulator).

Friction welding has also been used to join CMCs to metals, as have SPDB and TLPDB. Brazing can also be used, although this may be difficult because ceramics are not easily wetted by the braze metals. Finally it is possible to join CMCs using ceramic adhesives. However, such materials have very low toughness so joint strengths are generally low.

6. *Concluding Remarks*

All forms of composite can be satisfactorily joined. Adhesive bonding and mechanical fastening are the main methods for PMCs, and these are now mature technologies with well-established design rules.

MMCs can be joined by a variety of methods, but care is needed to avoid unwanted reactions, particularly with fusion processes, and destruction of the reinforcement in the joint region. CMCs are the most difficult type of composite to join and development is needed. Shrink fitting can give an acceptable joint in the right circumstances.

Bibliography

Anon 1990 Case studies. In: Middleton D H (ed.) *Composite Materials in Aircraft Structures*. Longman, Harlow, UK, Chap. 14

Clements L L 2000 Thermoplastic-matrix components lightens commercial transport. *High Performance Compos.* **8**, 27–31

Ellis M B D 1994 *Joining Metal Matrix Composites—A Review*. Research Report 489/1994. The Welding Institute, Abington, Cambridge

Fernie J A, Threadgill P L, Watson M N 1991 Progress in joining of advanced materials. *Weld. Met. Fabric.* **59**, 179–84

Godwin E W, Matthews F L, Kilty P F 1986 The design of bolted flange joints in GRP. *Plastics Rubber Process. Applic.* **6**, 161–7

Hart-Smith L J 1972 Design and analysis of adhesive-bonded joints. In: *USAF Conf. Proc. AFFDL-TR*-72-130. US Air Force Flight Dynamics Laboratory, Dayton, OH, Paper 6059A

Hart-Smith L J 1987 Design and empirical analysis of bolted or riveted joints. In: Matthews F L (ed.) *Joining Fibre Reinforced Plastics*. Elsevier Applied Science, Barking, UK, Chap. 6

Hoskin B C, Baker A A 1986 Aircraft applications. In: Hoskin B C, Baker A A (eds.) *Composite Materials for Aircraft Structures*. American Institute for Aeronautics and Astronautics, New York, Chap. 13

Kinloch A J, Kodokian G K A, Watts J F 1992 The adhesion of thermoplastic fibre composites. *Philos. Trans. R. Soc. London, Ser. A* **338**, 83–112

Koczak M J, Khatri S C, Allison J E, Bader M G 1993 Metal-matrix composites for ground vehicle, aerospace, and industrial application. In: Suresh S, Mortensen A, Needleman A (eds.) *Fundamentals of Metal Matrix Composites*. Butterworth-Heinemann, MA, Chap. 16

Matthews F L 1994 Bonded and mechanically fastened joints. In: Hollaway L (ed.) *Handbook of Polymer Composites for Engineers*. Woodhead Publishing, Cambridge, Chap. 8

Phillips L N 1989 Applications. In: Phillips L N (ed.) *Design With Advanced Composite Materials*. The Design Council, London, Chap. 7

Wise R 1992 Approaching total joining technology for carbon fibre reinforced plastics. *Weld. Inst. Bull.* **1**(January/February), 4–6

Wu H F, Bucci R, Slagter W, Vogelsang B 1992 Bearing strength characteristics of fiber-reinforced aluminum laminates. In: *Proc Composites Testing and Standardisation*. European Association for Composite Materials Bordeaux, France, pp. 453–70

F. L. Matthews

Conducting Polymers, Solution- and Gel-processing of

Semiconducting polymers are commonly macromolecules comprising extended sequences of conjugated bonds, that are useful by themselves for, among other things, numerous optical and semiconductor applications. They may also be rendered into materials that exhibit relatively high levels of electrical conductivity by partial oxidation, reduction, or protonation. These relatively new materials are of potential use for antistatic coatings and films, (flexible) electrodes, electrochromic windows, photovoltaic devices, light-emitting diodes, etc. (Skotheim *et al.* 1998).

It is the very same conjugation which makes these macromolecules unusual, versatile, and of attractive use, that is the origin of the difficulties that were initially encountered in attempts to process the early, more simple forms of these polymers (e.g., polyacetylene, poly(*p*-phenylene)) into useful articles. The latter is due to the fact that conjugated polymers often are relatively rigid macromolecules, which are typified by low melting entropies, commensurate with high melting temperatures. This last property is particularly well-illustrated by Fig. 1, in which the melting temperatures of crystalline, oligomeric *p*-phenylenes, and 2,5-thienylenes are displayed as function of the number of monomer repeat units (*n*). As is evident from these data, both families of oligomers already at very low degrees of polymerization exhibit high melting temperatures that exceed their thermal decomposition temperature. Hence, processing polymeric versions of these materials, with associated much higher melting temperatures, by the standard melt-processing schemes in use for most common industrial polymers, has not been successful. In addition, the nonpolar chemical nature of the above, exemplary polymers permits only modest enthalpic interactions with potential solvents (much like polyethylene), which prevents simple solution processing as industrially performed with, for example, the rigid poly(*p*-phenyleneterephthalate) better known as the aramid polymer that constitutes the high-strength fiber Kevlar®. The latter polymer has an estimated melting temperature of 650 °C, but is nonetheless soluble at room temperature in concentrated sulfuric acid due to very strong polymer–solvent interactions.

Fortunately, as summarized here, since the 1980s a number of ingenious schemes have been developed that now permit processing of a broad spectrum of conjugated, (semi)conducting polymers into a wide variety of useful forms. This article is concerned with the physicochemical aspects of these processing technologies. It does not address the particular details of the mechanics and equipment needed or used, which are of limited complexity and can be readily

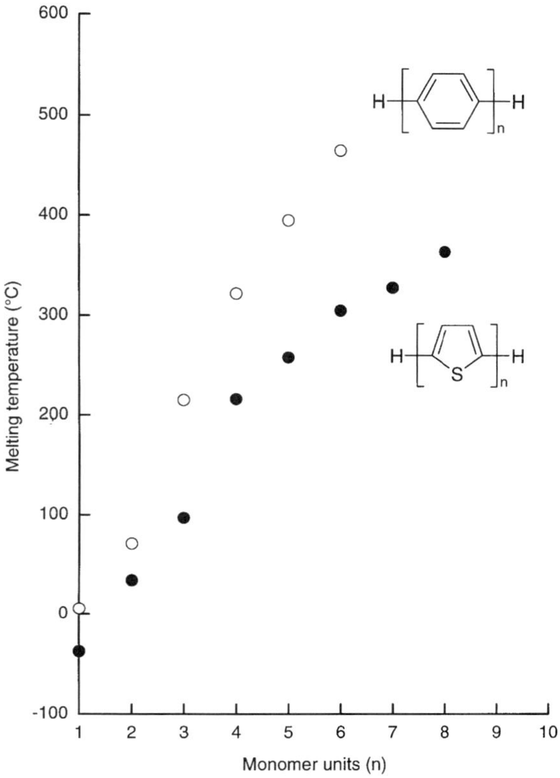

Figure 1
Melting temperatures of conjugated oligomers as a function of the number of monomer repeat units, ○ *p*-phenylenes, and • 2,5-thienylenes (Rothe 1999).

selected and found with the aid of standard texts about polymer processing.

1. *Processing Routes*

A cartoonist's overview of processing routes currently employed for a variety of conjugated polymers is shown in Fig. 2 (after Moulton and Smith 1993).

Route I illustrates the process of direct chemical conversion of monomers into products comprising intractable conjugated macromolecules. This technology is suitable mostly for coatings and thin films, or colloidal particles which subsequently require further processing into final articles. Interestingly, it is the accidental discovery of polyacetylene films directly formed from the monomer during polymerization that eventually led to the birth of the field of conducting polymers (Shirakawa and Ikeda 1971, Frommer and Chance 1986). A number of commercially available products exist that follow this route. For example, monomers for *in situ* synthesis of thin films and coatings (3,4-ethylene dioxythiophene, Baytron M, Bayer), coated, conducting fabrics

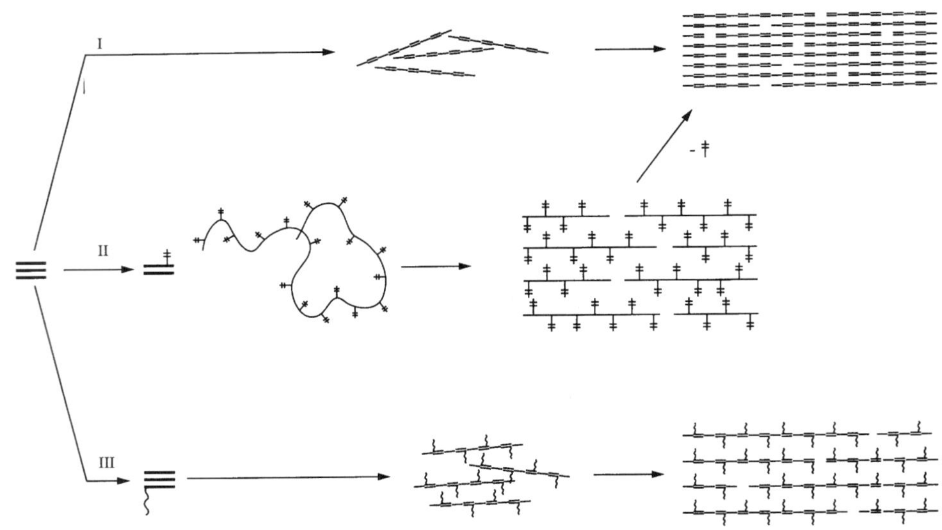

Figure 2
Schematic representation of processing routes for conjugated polymers. (I) Direct synthesis of intractable conjugated macromolecules. (II) Synthesis of tractable, nonconjugated "precursor" macromolecules comprising appropriately selected leaving groups, which are processed with standard techniques, and subsequently converted into the conjugated polymer by abstraction of the substituent. (III) Synthesis of conjugated macromolecules derivatized with solubilizing side groups, followed by standard solution-, gel-, or melt-processing (after Moulton and Smith 1993).

(Contex, Milliken), and colloidal suspensions of polyaniline (Ormecon; although this polymer, when properly synthesized, can be solution processed, see below). The process has the major drawback that it virtually permits no simple control over the organization of the solids formed, at any length scale.

Route II in Fig. 2 attempts to depict the so-called precursor route, which consists of (i) synthesizing tractable, nonconjugated, "precursor" macromolecules comprising appropriately selected leaving groups; (ii) processing these polymers according to standard (mostly solution-based) techniques; and (iii) converting the precursor polymer into the final (intractable) conjugated species by abstraction of the leaving group. Two classical examples of the chemical rearrangements involved in the conversion of the precursor to the conjugated species are displayed in Fig. 3 for polyacetylene (after Edwards and Feast 1980) and poly(*p*-phenylene vinylene) (after Wessling and Zimmerman 1986, see also Bradley 1987). Although initially often employed for the formation of thin films for use in, for example, light-emitting polymer diodes, the precursor method suffers from the major difficulty that complete removal of the leaving groups is not trivial; of course, any remaining substituents interrupt the conjugation in the polymer backbone, which may adversely affect some of its properties (see *Conjugated Oligomers and Polymers, Synthesis of*).

Route III in Fig. 2 indicates the synthesis of tractable, conjugated macromolecules that can be processed according to standard solution-, gel-, or even melt-processing routes. This extremely useful and commercially relevant property is induced by covalently attached side groups that dramatically increase the melting entropy of the macromolecules, or provide sites for strong interactions with solvents, or both. Efficient increase in melting entropy and resulting reduction in melting temperature is readily achieved by the derivatization of conjugated polymers with linear or branched flexible side chains. This is illustrated by the data in Fig. 4, which shows the melting temperatures of poly(3-alkylthiophene)s as a function of alkyl side chain length (after Elsenbaumer *et al.* 1986, Moulton and Smith 1993). Indeed, when side chains reach lengths of about three carbon atoms, these polymers can be dissolved at elevated temperatures in xylene, for example; increasing the length to eight carbon atoms and results in poly(3-alkylthiophene)s that are soluble at room temperature and can also be processed from the melt without decomposition. Conjugated polymers comprising appropriate polar side groups can readily be dissolved in common polar solvents, or even water. The above concepts are widely and successfully used for a broad range of polymers, including poly(phenylene vinylene)s, polyfluorenes, and many others (Skotheim *et al.* 1998, Kraft *et al.* 1998). The

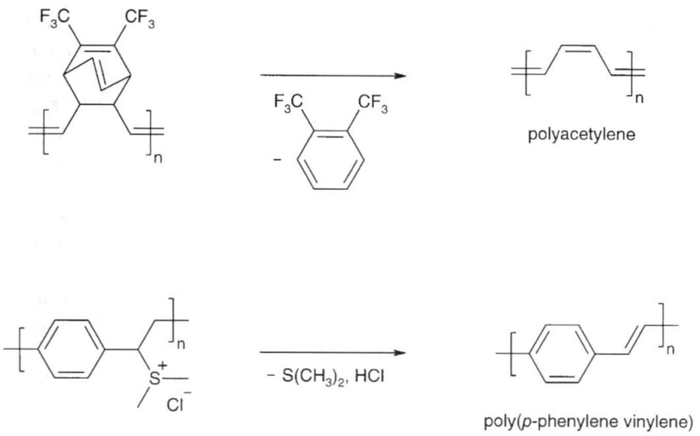

Figure 3
Reaction schemes of conversion of "precursor" polymers into (intractable) conjugated polymers for polyacetylene (after Edwards and Feast 1980) and poly(*p*-phenylene vinylene) (after Wessling and Zimmerman 1986, see also Bradley 1987).

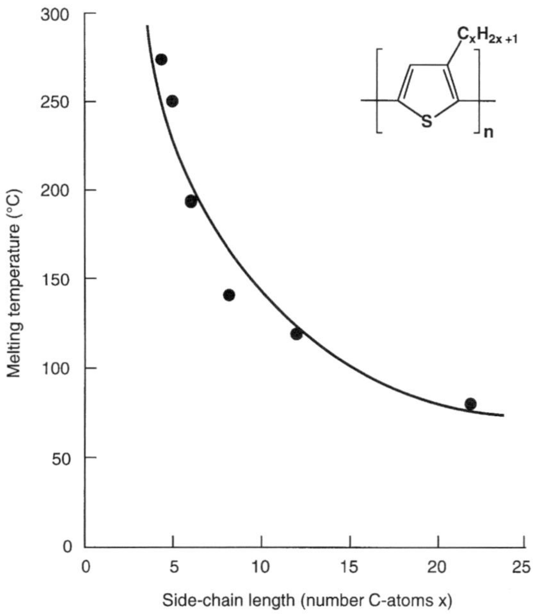

Figure 4
Melting temperatures of poly(3-alkylthiophenes) as a function of side chain length (after Elsenbaumer *et al.* 1986, Moulton and Smith 1993).

principal negative aspects of Route III appear to be that, by design, the density of conjugated moieties in the polymer systems is reduced and that electronic intermolecular interactions in at least one dimension (perpendicular to the main chain) are hampered.

One additional concept that results in increased processibility, which is not shown in Fig. 2, is the reduction of the length of the conjugated sequence in the polymer chain and connecting these functional moieties through, for example, flexible spacers. Although this route is attractive and of use for a number of optical properties, such as photo- and electroluminescence, it is not desirable when high levels of electrical conductivity are sought, because the latter property most often rapidly decreases with decreasing conjugation length.

Finally, attention should be drawn to polyaniline (PANI) which may occur in various states of oxidation (Geniès *et al.* 1990), and is a polymer—when properly synthesized—that is soluble without modification and in high molecular weight form in, for example, *N*-methylpyrrolidinone, as well as a variety of acids, including, *m*-cresol and acetic acid, according to a mechanism not unlike that of dissolution of aramids in sulfuric acid.

It should be noted that the above processing route III generally applies to forms of conjugated polymers that are not oxidized or reduced (i.e., that are *not* in their electrically conducting form); those that are typically exhibit increased chain stiffness due to electron delocalization, and are substantially less tractable. Exemplary, the highly soluble poly(3-alkylthiophene)s no longer are blessed with that property when oxidized, for example with $FeCl_3$, to their conducting counterparts. Thus, although the above route III is useful and adequate for processing conjugated polymers for optical and many semiconductor applications, it requires the additional, often cumbersome, step of postshaping oxidation, reduction, or protonation (often referred to as *doping*) to yield electrically conducting products. This, of course, severely limits the types of product that can be manufactured (generally only shapes of limited thickness).

The latter major technological hurdle has been overcome only in recent years by a scheme that first was demonstrated for polyaniline (Cao *et al.* 1992). In this process, the nonconducting polyaniline emeraldine base form is converted into a soluble, electrically conductive salt through protonation with an acid, the counterion of which induces processibility similar to route III in Fig. 2 (illustrated in Fig. 5 with dodecylbenzene sulfonic acid). The physicochemical nature of the counterion can be widely varied to impart solubility in a variety of solvents of a broad range of polarity (e.g., xylene, dimethylsulfoxide, chloroform, and even H_2O). This particular scheme is not limited to small molecular protonating acids, but has also been demonstrated with sulfonated polystyrene, e.g., with poly(3,4-ethylene dioxythiophene) (PEDOT). The above processible forms of both conducting PANI and PEDOT are commercially available (as Panipol and Baytron P, respectively), and find increasing use in coatings, films, and parts, and are under development in fiber form.

2. *Solution Processing*

Solution processing of polymers generally refers to the manufacturing step of forming a molecular mixture of the polymer and an appropriate solvent, which most often, but not necessarily, is a liquid. The physicochemical nature or "quality" of the solvent and the concentration of polymer may vary widely. It is selected to: (i) control the viscosity of the solution (of critical import in spin coating), (ii) to adjust the dissolution temperature (naturally to below the thermal decomposition temperature of

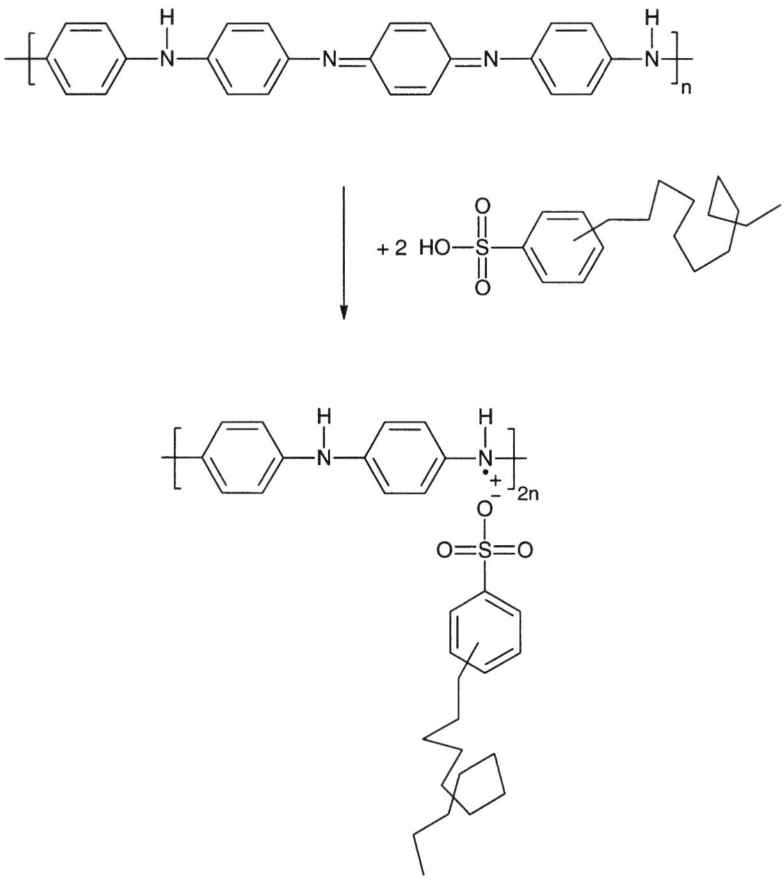

Figure 5
Conversion of (nonconducting) polyaniline emeraldine base into the soluble, electrically conductive salt form through protonation with dodecylbenzenesulfonic acid. The counterion of the latter induces processibility similar to route (III) in Fig. 2 (after Cao *et al.* 1992).

Figure 6
Optical micrograph of lyotropic solution of 21%(w/w) polyaniline protonated with camphor sulfonic acid in *m*-cresol (crossed polarizers, magnification 50 ×) (after Cao and Smith 1993).

the polymer), (iii) to induce liquid crystallinity to enhance macromolecular orientability (in the case of relatively rigid "solutes"; Fig. 6), and/or (iv) to impart a particular morphology, porosity (for example in membranes) or macromolecular conformation in the final product. As such, processing from solution offers a broad spectrum of options to create a wide range of products. In addition, solution processing is the preferred vehicle if blends of polymers are to be created (e.g., low-percolation-threshold blends of conducting polymers with insulating bulk polymers or high-performance, fiber forming polymers; Cao *et al.* 1992, Moulton and Smith 1993), or additives are to be homogeneously dispersed.

After the polymer solution has been produced, it is shaped usually in the form of thin layers, coatings, films, and fibers (i.e., objects of limited thickness). Subsequently, the solvent is removed by either evaporation or extraction with liquids that are nonsolvents for the polymer, upon which the polymer adopts either the amorphous glassy or rubbery state, or forms an ordered (semi)crystalline solid, depending on the regularity of the chain molecules. It is critical to note that the presence or absence of order and orientation, which can be manipulated with the processing conditions, has a very pronounced influence on many of the properties of the final solid polymer. In the case of conjugated polymers, for example, fluorescence may be quenched by macromolecular order, while that same feature dramatically enhances electrical conductivity. With the development of soluble, conjugated macromolecules, as well as soluble, electrically conductive polymers, all the above merits of solution processing can be taken advantage of for these materials.

3. *Gel Processing*

Gel processing is a term that is sometimes used for manufacturing technologies that, at one stage or other, involve the soft-condensed matter state known as gels (Smith 1989). The latter are materials that consist mostly of fluids, but are characterized by certain (mechanical, notably elastic) properties normally associated with solids. As such, gels may be considered as solid matter in which diffusion may take place at a liquid-like pace. Gels can be produced in a great variety of ways and are used in numerous areas of commerce that are outside the scope of this article. Because of their demonstrated use for conjugated polymers, in the present context, gels, and some applications, are discussed that are formed by thermoreversible processes and consist of a macroscopic, mechanically coherent network that is formed and stabilized through interconnected crystalline entities. Such gels can readily be obtained upon cooling of solutions of crystallizable, i.e., sufficiently regular polymers, of polymer concentrations that are above the onset of macromolecular chain overlap, φ^* ($\sim 1/\eta$, where η is the limiting intrinsic viscosity of the polymer). Thermoreversible gels may capture much of the (concentration-dependent) tenuous arrangement of the macromolecules in solution, such as the spacing between chain entanglements, provided that during the formation of the connecting crystals no massive chain disentanglement occurs, i.e., no liquid–liquid phase separation should take place. Interestingly, because of their very nature, gels of conjugated polymers may provide ultra-low percolation threshold electroactive materials ($<1\%$ v/v). Remarkably, this particular feature has not received the attention that it merits.

By contrast, blends of conjugated macromolecules with insulating, gel-forming carrier polymers have been "gel-processed" into highly oriented polymer systems with interesting properties. In these cases, the conjugated polymer typically was co-dissolved at elevated temperatures in decalin with ultra-high molecular weight polyethylene (UHMW PE; average molecular weight $>10^6$ g mol^{-1}), shaped into fiber or film form, gelled by cooling to room temperature, after which the solvent was removed. Not only does this process result in a relatively homogeneous (although not molecular) distribution of the functional polymer in the UHMW PE matrix, it also yields a solid polymer blend of unique ductility. The latter feature derives from the low entanglement density in gel-processed UHMW PE, which permits orienting this material to extraordinary high draw ratios (in the solid state) of 100-fold or more its original length. This deformation has been demonstrated to result in extremely high degrees of macromolecular orientation, which cause dramatic improvement of various properties (Smith 1989), such as mechanical stiffness and tensile strength. Importantly, this exceptional degree of orientation may be imparted also to the conjugated polymer guests, which leads to major enhancement and often useful high degrees of anisotropy of properties such as electrical conductivity, optical dichroism, charge-carrier transport, etc.

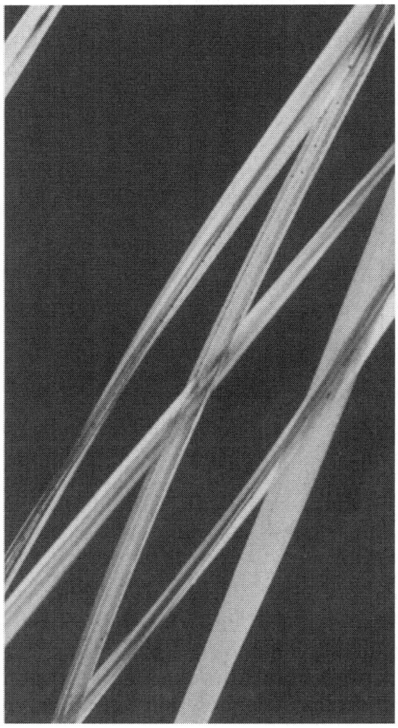

Figure 7
Fiber of 20%(w/w) of polyaniline protonated with dodecylbenzenesulfonic acid in ultra-high molecular weight polyethylene (UHMW PE), gel-spun from decalin (fiber diameter 25 μm) (after Andreatta and Smith 1993).

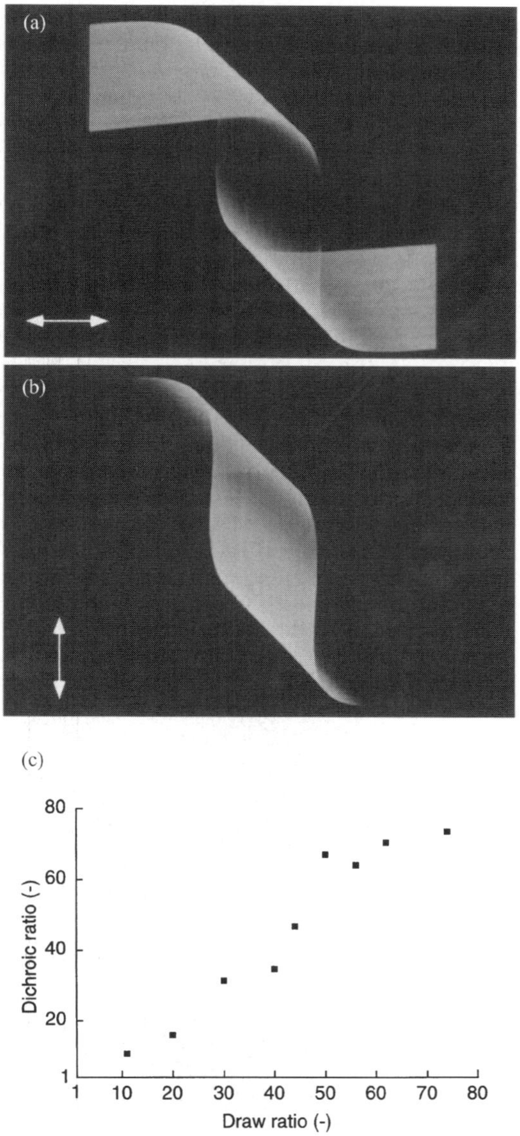

Figure 8
Polarized photoluminescence from a gel-processed, drawn film of 2%(w/w) of a poly(2,5-dialkoxy-*p*-phenylene ethynylene) in UHMW PE. Twisted tape (drawn 80-fold original length) shown through a linear polarizer with its polarization axis oriented, respectively, horizontally (a) and vertically (b) with respect to the draw direction. (c) Dichroic ratio calculated from the peak maxima of emission spectra as a function of the draw ratio (after Weder *et al.* 1997).

Following are two typical examples of materials that are produced with the above-described gel-processing technology. Figure 7 displays a polarized optical micrograph of a high-performance, gel-spun/drawn fiber of 20%(w/w) of polyaniline protonated with dodecylbenzenesulfonic acid in UHMW PE (Andreatta and Smith 1993). This particular fiber was characterized by a bulk electrical conductivity of 0.3 Scm^{-1}, a Young's modulus of 20.6 GPa, and a tensile strength of 1.1 GPa. An example that provides a dramatic illustration of both the intrinsic anisotropic (photophysical) properties of certain conjugated polymers and the capability of gel processing as a technique to maximize this unique feature is given in Fig. 8. Here is presented the polarized photoluminescence from a gel-processed, 80 times drawn film of 2%(w/w) of a poly(2,5-dialkoxy-*p*-phenylene ethynylene) in UHMW PE. The twisted tape is shown through a linear polarizer with its polarization axis oriented, respectively, horizontally (a) and vertically (b) with respect to the drawing direction. Also plotted is the dichroic ratio calculated from the peak maxima of emission spectra as a function of the draw ratio, which approaches the extraordinary high value of about 80 (c) (Weder *et al.* 1997). Such polarized photoluminescent films may find application in, for example, novel liquid-crystal displays.

4. *Concluding Remarks*

The science and engineering of processing conjugated, (semi)conducting polymers has sufficiently developed and matured to the point that one of the initial major drawbacks of these materials, i.e., their purported intractability, has been overcome. Various intelligent processing technologies are now available to shape these interesting materials into a great diversity of forms that range from thin films and coatings to fibers and molded parts, and, importantly, to manipulate their (supra)molecular and segmental order. This latter capability permits control of the properties of these materials, such as level of electrical conductivity, nonlinear optical susceptibility, charge–carrier mobility, and the degree of polarization of absorbed or emitted light, which may span many orders of magnitude. Before truly large-scale industrial application of (semi)conducting polymers will occur, however, it is expected that certain remaining issues need to be satisfactorily and economically resolved, such as their purity, environmental stability, and long-term performance, where the particular set of requirements evidently depends on the end use. It would appear, though, that present niche markets can be well served by the currently available materials.\

Bibliography

Andreatta A, Smith P 1993 Processing of conductive polyaniline-UHMW polyethylene blends from solutions in nonpolar solvents. *Synth. Met.* **55–7**, 1017–22

Bradley D D C 1987 Precursor-route poly(*p*-phenylene vinylene): polymer characterization and control of electronic properties. *J. Phys. D. Appl. Phys.* **20**, 1389–410

Cao Y, Smith P 1993 Liquid-crystalline solutions of electrically conducting polyaniline. *Polymer* **34**, 3139–43

Cao Y, Smith P, Heeger A J 1992 Counter-ion induced processibility of conducting polyaniline and of conducting polyblends of polyaniline in bulk polymers. *Synth. Met.* **48**, 91–7

Edwards J H, Feast W J 1980 A new synthesis of poly(acetylene). *Polymer* **21**, 595–6

Elsenbaumer R L, Jen K Y, Oboodi R 1986 Processible and environmentally stable conducting polymers. *Synth. Met.* **15**, 169–74

Frommer J E, Chance R R 1986 Electrically conductive polymers. In: Mark F, Bikales N M, Overberger C, Menges G (eds.) *Encyclopedia of Polymer Science and Engineering*. Wiley NewYork, Vol. 5, pp. 462–507

Geniès E M, Boyle A, Lapkowski M, Tsintavis C 1990 Polyaniline: a historical survey. *Synthetic Met.* **36**, 139–82

Kraft A, Grimsdale A C, Holmes A B 1998 Electroluminescent conjugated polymers—Seeing polymers in a new light. *Angew. Chem Int. Ed. Engl.* **37**, 403–28

Moulton J, Smith P 1993 Processing technologies of oriented, electrically conductive polymers. In: Lewin M, Preston J (eds.) *Handbook of Fiber Science and Technology III, High-Technology Fibers, Part C*. Dekker, New York, pp. 275–315

Rothe M 1999 Physical data of oligomers. In: Brandrup J, Immergut E H, Grulke E A (eds.) *Polymer Handbook*. Wiley, New York, pp. 97–100

Shirakawa H, Ikeda S 1971 Infrared spectra of poly(acetylene). *Polym. J.* **2**, 231–44

Skotheim T J, Elsenbaumer R L, Reynolds J R 1998 *Handbook of Conducting Polymers*, 2nd edn. Dekker, New York

Smith P 1989 Some industrial applications of soft condensed matter in high-performance polymers processing. In: Riste T, Sherrington D (eds.) *Phase Transitions in Soft Condensed Matter*. NATO ASI Ser. B: Physics 211. Plenum. New York, pp. 353–75

Weder C, Sarwa S, Bastiaansen C, Smith P 1997 Highly polarized luminescence from oriented conjugated polymer/polyethylene blend films. *Adv. Mater* **6**, 1035–9

Wessling R A, Zimmerman R G. 1986. *US Pat.* 3 401 152

P. Smith and C. Weder

Continuous Casting

Continuous casting transforms molten metal into solid on a continuous basis and includes a variety of important commercial processes. These processes are the most efficient way to solidify large volumes of metal into simple shapes for subsequent processing. Most basic metals are mass-produced using a continuous casting process, including over 500 million tonnes of steel, 20 million tonnes of aluminum, and 1 million tonnes of copper, nickel, and other metals in the world each year.

Continuous casting is distinguished from other solidification processes by its steady-state nature, relative to an outside observer in a laboratory frame of reference. The molten metal solidifies against the mold walls while it is simultaneously withdrawn from the bottom of the mold at a rate which maintains the solid–liquid interface at a constant position with time. The process works best when all of its aspects operate in this steady-state manner.

Relative to other casting processes, continuous casting generally has a higher capital cost, but lower operating cost. It is the most cost- and energy-efficient method to mass-produce semifinished metal products with consistent quality in a variety of sizes and shapes. Cross-sections can be rectangular, for subsequent rolling into plate or sheet, square or circular for long products, and even "dog-bone" shapes, for rolling into I or H beams.

Many different types of continuous casting processes exist. Figure 1 is a diagram of a few of the most important ones. Vertical machines are used to cast aluminum and a few other metals for special applications. Curved machines are used for the majority of steel casting and require bending and/or unbending of the solidifying strand. Horizontal casting features a shorter building and is used occasionally for both nonferrous alloys and steel. Finally, thin strip casting is being pioneered for steel and other metals in low-production markets in order to minimize the amount of rolling required.

1. Steel Continuous Casting

Continuous casting is a relatively new process in historical terms. Although the continuous strip casting process was conceived by Bessemer in 1858, the continuous casting of steel did not gain widespread use until the 1960s. Earlier attempts suffered from technical difficulties such as "breakouts," where the solidifying steel shell sticks to the mold, tears, and allows molten steel to pour out over the bottom of the machine. This problem was overcome by Junghans in 1934 by vertically oscillating the mold, utilizing the concept of "negative strip" where the mold travels downward faster than the steel shell during some portion of the oscillation cycle to dislodge any sticking (Wolf 1992). Many other developments and innovations have transformed the

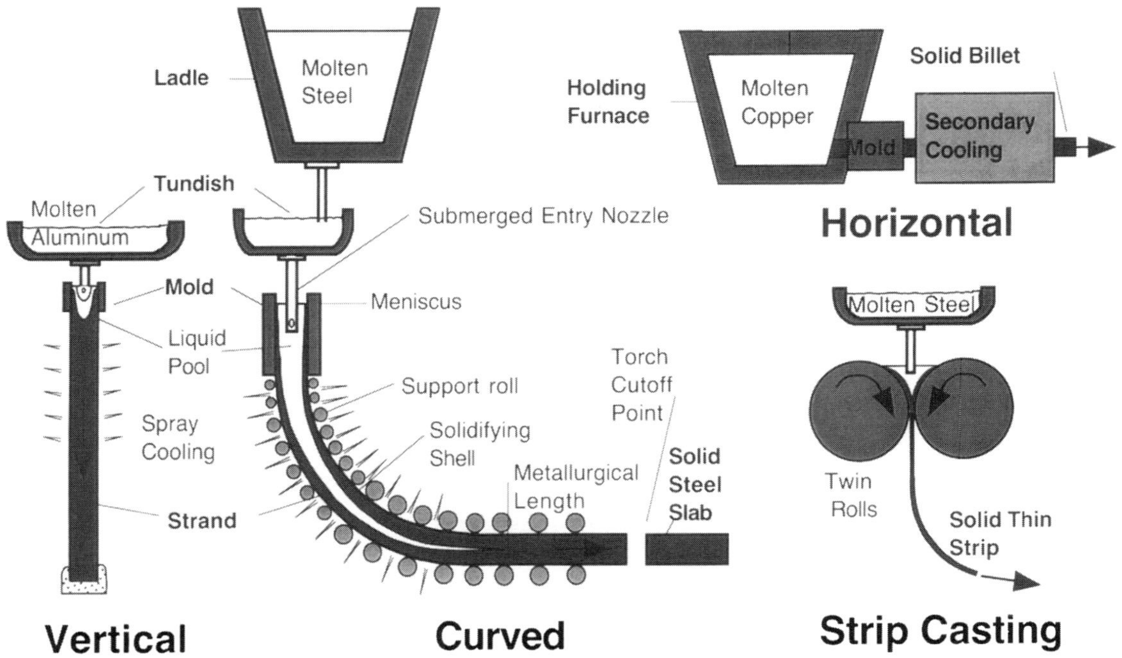

Figure 1
Various continuous casting processes.

continuous casting process into the sophisticated process currently used to produce over 90% of steel in the world today, including plain carbon, alloy, and stainless steel grades (Wolf 1992).

The continuous casting process for steel is shown in Fig. 1 (second frame) and the close-up of the upper mold region is shown in Fig. 2. In this process, molten steel flows from a ladle, through a tundish

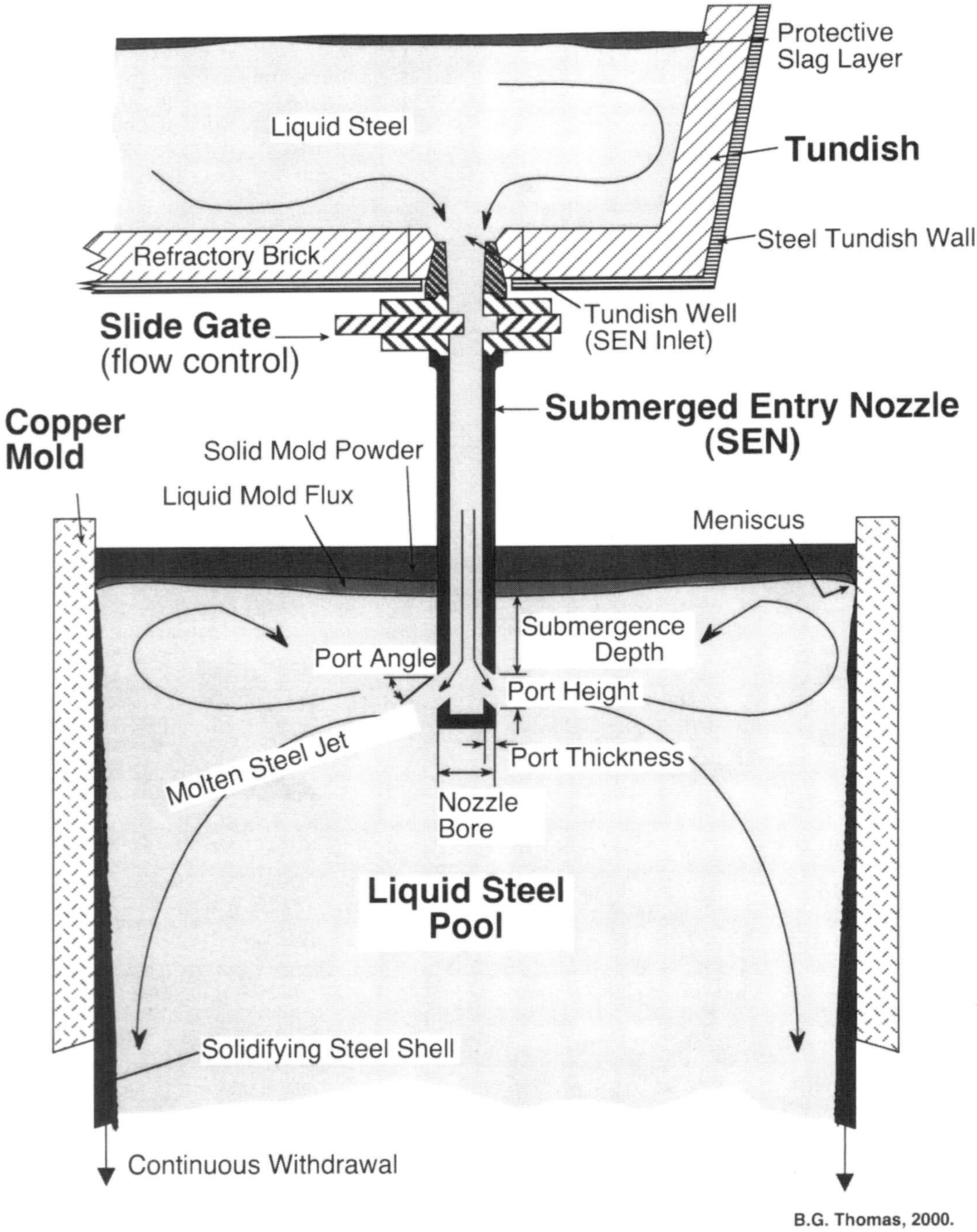

Figure 2
Schematic of mold region of continuous casting process for steel slabs.

into the mold. The tundish holds enough metal to provide a continuous flow to the mold, even during an exchange of ladles, which are supplied periodically from the steelmaking process. The tundish can also serve as a refining vessel to float out detrimental inclusions into the slag layer. If solid inclusion particles are allowed to remain in the product, surface defects such as "slivers" may form during subsequent rolling operations; or local internal stress concentration may occur, leading to a reduction in fatigue life. To produce higher quality product, the liquid steel must be protected from exposure to air by a slag cover over the liquid surface in each vessel and by using ceramic nozzles between vessels. If not, then oxygen in the air will react to form detrimental oxide inclusions in the steel.

Once in the mold, the molten steel freezes against the water-cooled walls of a bottomless copper mold to form a solid shell. The mold is oscillated vertically in order to discourage sticking of the shell to the mold walls. Drive rolls lower in the machine continuously withdraw the shell from the mold at a rate or "casting speed" that matches the flow of incoming metal, so the process ideally runs in steady state. The liquid flow rate is controlled by restricting the opening in the nozzle according to the signal fed back from a level sensor in the mold.

The most critical part of the process is the initial solidification at the meniscus, found at the junction where the top of the shell meets the mold and the liquid surface. This is where the surface of the final product is created, and defects, such as surface cracks, can form if problems such as level fluctuations occur. To avoid this, oil or mold slag is added to the steel meniscus, which flows into the gap between the mold and shell. In addition to lubricating the contact, a mold slag layer protects the steel from air, provides thermal insulation, and absorbs inclusions.

Below the mold exit, the thin solidified shell (6– 20 mm thick) acts as a container to support the remaining liquid, which makes up the interior of the strand. Water or air mist sprays cool the surface of the strand between the support rolls. The spray flow rates are adjusted to control the strand surface temperature with minimal reheating until the molten core is solid. After the center is completely solid (at the "metallurgical length" of the caster, which is 10–40 m), the strand is cut with oxyacetylene torches into slabs or billets of any desired length.

Different continuous casting processes exist to produce cross-sections of different shapes and sizes. Heavy, four-piece plate molds with rigid backing plates are used to cast large, rectangular "slabs" (50–250 mm thick and 0.5–2.2 m wide), which are rolled into plate or sheet. Similar molds are used for casting relatively square "blooms," which range up to 400×600 mm in cross-section. Single-piece tube molds are used to cast small, square "billets" (100–200 mm thick) which are rolled into long products, such as bars, angles, rails, nails, and axles. A new strip casting process is being developed using large rotating rolls as the mold walls to solidify 1–3 mm thick steel sheet.

When casting large cross-sections, such as slabs, a series of rolls must support the soft steel shell between the mold exit and the metallurgical length, in order to minimize bulging due to the internal liquid pressure. Extra rolls are needed to force the strand to "unbend" through the transition from the curved to the straight portion of the path shown in Fig. 1. If the roll support and alignment are not sufficient, internal cracks and segregation may result. These defects will persist in the final product, even after many rolling and other operations, so it is important to control the casting process.

The process is started by plugging the bottom of the mold with a "dummy bar." After enough metal has solidified like a conventional casting onto its head, the dummy bar is then slowly withdrawn down through the continuous casting machine and steady state conditions evolve. The process then operates continuously for a period of one hour to several weeks, when the molten steel supply is stopped and the process must be restarted. The maximum casting speed of 1–8 $\mathrm{m\,min^{-1}}$ is governed by the allowable length of the liquid core and to avoid quality problems, which are generally worse at higher speeds.

After the steel leaves the caster, it is reheated to a uniform temperature and rolled into sheet, bars, rails, and other shapes. Modern steel plants position the rolling operations close to the caster to save on reheating energy. Further information on the continuous casting of steel can be found elsewhere (Wolf 1992, Schrewe 1991, Irving 1993, Cramb 1991, 2001, Continuous casting 1979–1997). The application of computational models to understand and improve this process is discussed elsewhere (see *Continuous Casting: Complex Models*).

2. *Semicontinuous Casting of Aluminum*

Over 90% of commercial aluminum alloys are cast by semicontinuous vertical casting machines, typically as 0.05–0.5 m diameter round sections. This process is similar to steel continuous casting, with the significant difference that it must be stopped periodically when the bottom of the cast "ingot" reaches the floor of the casting pit. Other differences are the slower casting speed (0.03–0.1 $\mathrm{m\,min^{-1}}$), which is required to avoid internal cracks, and the shorter metallurgical length (0.1–1.0 m).

The two most popular types of aluminum continuous casting are the direct-chill (DC) and electromagnetic (EM) processes, which are distinguished by the method for supporting the liquid at the meniscus. The DC process uses water-cooled mold walls, similar to steel casting, while the EM process induces horizontal electromagnetic forces to suspend the metal from ever touching the mold walls. In both processes, the solid skin shrinks away from the mold walls shortly below the meniscus, whereupon the surface is cooled with spray water. Further information on aluminum continuous casting is available elsewhere (Altenpohl 1998, Cast Shop Technology 1997).

3. *Other Continuous Casting Processes*

Copper is often continuous cast using a vertical or horizontal process to produce round billets for subsequent extrusion, forging, or wire drawing (Wilson 1999). A wide variety of other continuous casting processes exist for special applications. Electroslag remelting (ESR) and vacuum arc remelting (VAR) are two forms of vertical continuous casting used for nonferrous metals, superalloys, and specialty alloy sections up to 1.5 m diameter (Mitchell and Auburtin 1997). Heat is supplied from an electrode above the melt and the top surface is protected with a thick slag layer (ESR) or a vacuum (VAR). These processes avoid oxide inclusions and remove impurities such as sulfur to produce highly refined metal with less segregation and fewer defects than found in conventional continuous casting. Their products are more costly, but are needed for critical parts such as found in the aerospace industry.

Bibliography

Altenpohl D G 1998 *Aluminum: Technology, Applications, and Environment*, 6th edn. Aluminum Association, Washington, DC

Cast Shop Technology 1997 In: *Light Metals Conference Proceedings*. Minerals, Metals, and Materials Society, Warrendale, PA, pp. 647–1095

Cramb A (ed.) 1991 Mold operation for quality and productivity. Iron and Steel Society, Warrendale, PA

Cramb A (ed.) 2001 Continuous casting. *The Making, Shaping, and Treating of Steel, 2*. Association of Iron and Steel Engineers, Pittsburgh, PA

Iron and Steel Society Continuous Casting 1979–1997, conference proceedings, Iron and Steel Society, Warrendale, PA, Vols. 1–9

Irving W R 1993 *Continuous Casting of Steel*. Institute of Materials, London, pp. 1–206

Mitchell A, Auburtin P (eds.) 2000 *Proc. 1997 Int. Symp. Liquid Metal Processing and Casting*. Santa Fe, NM

Matsuo M Continuous casting of aluminum. *Journal of Japan Institute of Light Metals* **44**(9), 510–25

Schrewe H F 1991 Continuous casting of steel, fundamental principles and practice. (Stahl und Eisen.) Dusseldorf, Germany, p. 194

Wilson R 1999 *A Practical Approach to Continuous Casting of Copper Alloys and Precious Metals*. Institute of Materials, London, pp. 1–300

Wolf M M 1992 History of continuous casting. In: *Proc. Steelmaking Conference* 75. Iron and Steel Society, Warrendale, PA, pp. 83–137

B. G. Thomas

Continuous Casting: Complex Models

Continuous casting is used to solidify most of the 750 million tons of steel, 20 million tons of aluminum, and many tons of other alloys produced in the world every year. Most previous advances have been based on empirical knowledge gained from experimentation with the process. As computer power increases, mathematical models are becoming increasingly powerful tools to gain additional quantitative insight.

Model applications include basic machine design calculations, identifying and quantifying the mechanisms of various types of defects, troubleshooting the origin of particular defects, and optimizing the various process conditions to increase productivity or minimize defects. This article discusses recent examples of these models with emphasis on the different phenomena that have been modeled in steel continuous casting, comparison with experiments, and practical implications of the findings.

1. *Basic Phenomena and Modeling*

Like most commercial processes, continuous casting involves a staggering complexity of interacting phenomena at the mechanistic level. Some of the important phenomena include:

Fully-turbulent, transient fluid motion in a complex geometry (inlet nozzle and strand liquid pool), affected by argon gas bubbles, thermal, and solutal buoyancies.

Thermodynamic reactions within and between the powder and steel phases.

Flow and heat transport within the liquid and solid flux layers, which float on the top surface of the steel.

Dynamic motion of the free liquid surfaces and interfaces, including the effects of surface tension, oscillation and gravity-induced waves, and flow in several phases.

Transport of superheat through the turbulent molten steel.

Transport of solute (including intermixing during a grade change).

Transport of complex-geometry inclusions through the liquid, including the effects of buoyancy, turbulent interactions, and possible entrapment of the inclusions on nozzle walls, gas bubbles, solidifying steel walls, and the top surface.

Thermal, fluid, and mechanical interactions in the meniscus region between the solidifying meniscus, solid slag rim, infiltrating molten flux, liquid steel, powder layers, and inclusion particles.

Heat transport through the solidifying steel shell, the interface between shell and mold, (which contains powder layers and growing air gaps), and the copper mold.

Mass transport of powder down the gap between shell and mold.

Distortion and wear of the mold walls and support rolls.

Nucleation of solid crystals, both in the melt and against mold walls.

Solidification of the steel shell, including the growth of grains and microstructures, phase transformations, precipitate formation, and microsegregation.

Shrinkage of the solidifying steel shell, due to thermal contraction, phase transformations, and internal stresses.

Stress generation within the solidifying steel shell, due to external forces, (mold friction, bulging between the support rolls, withdrawal, and gravity) thermal strains, creep, and plasticity (which varies with temperature, grade, and cooling rate).

Crack formation, due to internal stresses and composition-dependent ductility problems.

Coupled segregation, on both microscopic and macroscopic scales.

Because of this complexity, no model can include all of the phenomena at once. An essential aspect of successful model development is the selection of the key phenomena of interest to a particular modeling objective and the making of reasonable assumptions. These crucial phenomena should be treated mechanistically, in order to accurately reproduce them. Mechanistic models are based on satisfying the laws of conservation of heat, mass, force, and momentum in an appropriate domain with appropriate boundary conditions. Each phenomenon considered is represented by term(s) in these governing equations. Other phenomena can be ignored or incorporated using empirical constants, obtained through experimentation and model calibration. The equations are discretized using finite difference or finite element methods and are solved numerically with computers, which are becoming increasingly fast and affordable.

2. *Solidification and Heat Transfer*

In the continuous casting process (Fig. 1), molten metal is delivered from the bottom of a transfer vessel (the tundish) into a mold cavity. Here, the water-cooled walls of the mold extract heat to solidify a shell that contains the liquid pool. The shell is withdrawn from the bottom of mold at a "casting speed" that matches the inflow of metal, so that the process ideally operates at steady state. Below the mold, water sprays extract heat from the surface, and the strand core eventually becomes fully solid when it reaches the *metallurgical length*.

Heat flow/solidification models are used for basic design and troubleshooting of this process. These models solve the transient heat conduction equation

$$\rho\left(\frac{\partial h}{\partial t} + \mathrm{v}\cdot\nabla H\right) = \nabla\cdot(k_{\mathrm{eff}}\nabla T) + Q \qquad (1)$$

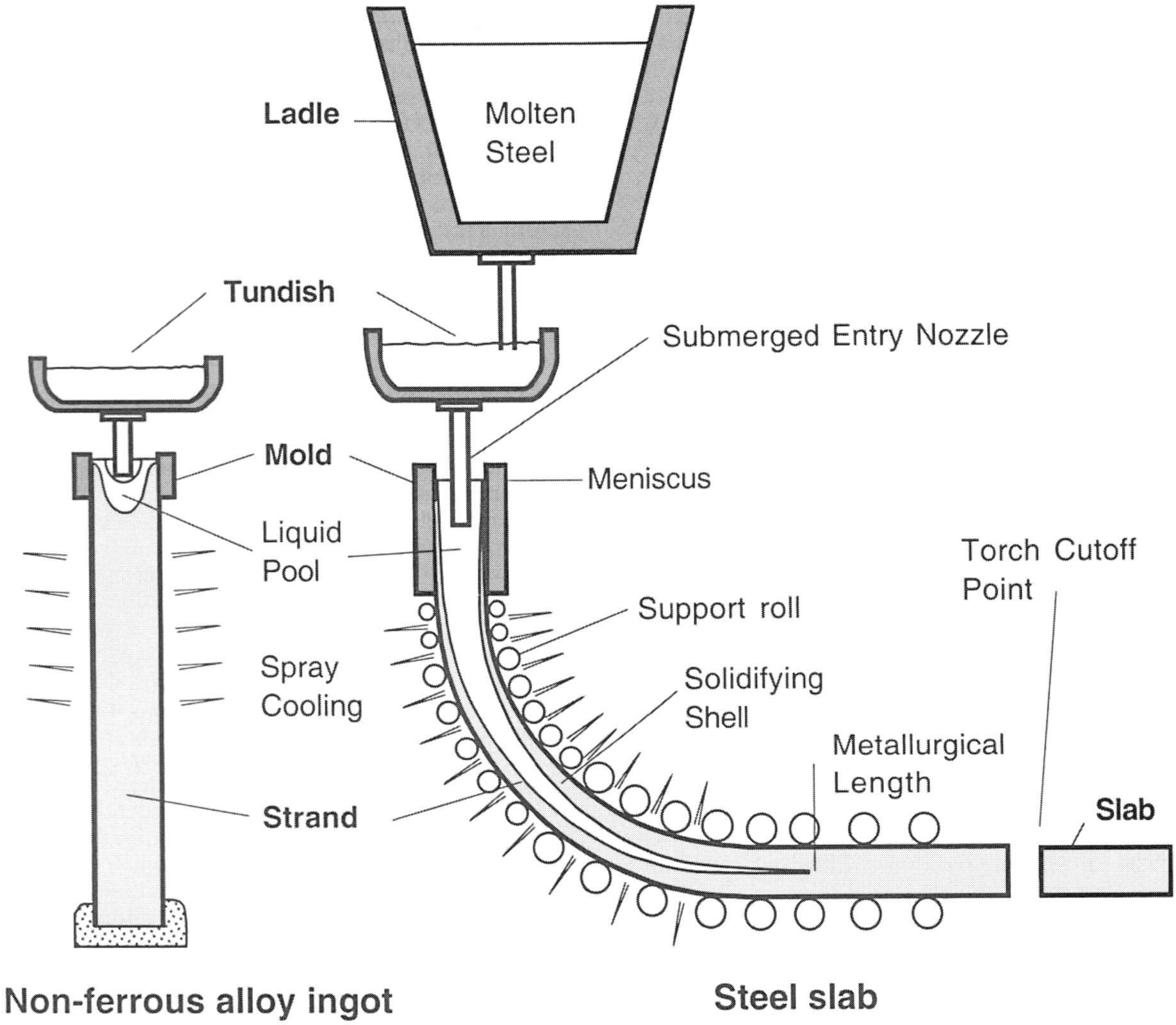

Figure 1
Schematic of continuous casting processes.

Temperature, T, depends on the temperature-dependent material properties of effective thermal conductivity, k_{eff}, and density, ρ, the velocity field, $\mathbf{v}$, various heat sources, Q, and the boundary conditions. Latent heat evolution and heat capacity are incorporated into the constitutive equation that must also be supplied to relate temperature, T, with enthalpy, H.

Axial heat conduction can be ignored in models of steel continuous casting because it is small relative to axial advection, as indicated by the small Peclet number ((casting speed × shell thickness)/thermal diffusivity). Thus, Lagrangian models of a horizontal slice through the strand have been employed with great success for steel. Aluminum continuous casting has a short metallurgical length, owing to its high thermal conductivity and slow casting speed, so this assumption cannot be made. (See *Modeling: Scaling Analysis* for details on how decisions like these are made.)

Heat transfer in the mold region is controlled mainly by heat conduction across the interface between the surface of the solidifying shell and the mold. The greatest difficulty in accurate heat flow modeling is determination of the heat transfer across this gap, q_{gap}, which varies with time and position depending on the thickness, d_{gap}, and the properties of the gas or lubricating flux layers that fill it, such as k_{gap}.

$$q_{\text{gap}} = \left(h_{\text{rad}} + \frac{k_{\text{gap}}}{d_{\text{gap}}}\right)(T_{0\text{shell}} - T_{0\text{mold}}) \tag{2}$$

Where metal shrinkage is not matched by taper of the mold walls, an air gap can form, especially in the corners. This greatly reduces the heat flow. Advanced models simulate the mold, interface, and shell, and use shrinkage models to predict the size of this gap, d_{gap}. In steel slab casting operations with mold flux, such models feature a detailed treatment of the interface, including heat, mass, and momentum balances on the flux in the gap and the effect of shell surface imperfections (oscillation marks) on heat flow and flux consumption (Thomas *et al.* 1998). The coupled effect of flow in the molten metal on delivering superheat to the inside of the shell, and, thereby, retarding solidification, is also modeled. Mold heat flow models can be used to identify deviations from normal operation and thus predict quality problems, such as impending breakouts or surface depressions, in time to take corrective action.

Heat flow models which extend below the mold are needed for basic machine design to ensure that the last support roll and torch cutter are positioned beyond the metallurgical length for the highest casting speed. Below the mold, surface temperature of the strand is maintained by air mist and water spray cooling, while the interior solidifies. Online open-loop dynamic cooling models can be employed to control the spray flow rates in order to ensure uniform surface cooling even during transients, such as the temporary drop in casting speed during a nozzle change. A heat flow model can also be used to troubleshoot defects, e.g., the location of a misaligned support roll that may be generating internal hot-tear cracks can be identified by matching the position of the start of the crack beneath the strand surface with the location of solidification front down the caster calculated with a calibrated model.

3. *Fluid Flow in the Mold*

Flow in the mold is of great interest because it influences many important phenomena, which have far-reaching consequences on strand quality. These effects include the dissipation of superheat (and temperature at the meniscus), the flow and entrainment of the top surface powder layers, top-surface contour and level fluctuations, and the entrapment of subsurface inclusions and gas bubbles. Design compromises are needed to simultaneously satisfy the contradictory requirements for avoiding each of these defect mechanisms.

Some of the important phenomena which govern the continuous casting of steel and determine the quality of the product are illustrated in Fig. 2. Steel flows into the mold through ports in the submerged entry nozzle, which is usually bifurcated. Nozzle geometry greatly affects the flow in the mold and is easy to change, so is an important subject of mathematical modeling.

The jet leaving the nozzle flows across the mold and impinges against the shell solidifying at the narrow face. The flow pattern depends on the casting conditions. It also can fluctuate with time, leading to intermittent defects, so transient behavior is important. Figure 3 shows an instantaneous image of the flow pattern in the mold cavity calculated with a direct numerical solution of the Navier–Stokes equations

$$\rho \mathbf{v} \cdot \nabla \mathbf{v} = -\nabla p \cdot (\mu \nabla \mathbf{v} + \nabla \mathbf{v}^T) + S \qquad (3)$$

The calculations are compared with particle image velocimetry measurements of the flow in a water model, shown on the right side of Fig. 3. These calculations reveal structures in the flow pattern that are important to transient events such as the intermittent capture of inclusion particles.

Although transient calculations can provide detailed insights, the computational cost can be greatly reduced using a turbulence model such as K–ε to approximate the steady time-averaged flow pattern. This model increases the effective viscosity so allows a much coarser mesh. However, it generally requires the solution of two additional conservation equations for the transport of turbulent kinetic energy and its dissipation, which are sometimes inaccurate.

The flow pattern changes radically with increasing argon injection rate (Thomas *et al.* 1994) or with the application of electromagnetic forces (Ishii *et al.* 1996), which can either brake or stir the liquid. These phenomena add significant complexity to the equations that must be solved for the flow model.

4. *Superheat Dissipation*

An important task of the flow pattern is to deliver molten steel to the meniscus region that has enough superheat during the critical first stages of solidification. Superheat is the sensible heat of the liquid metal above the liquidus temperature.

The dissipation of superheat is modeled by solving Eqn. (1) using the velocities found from the flow model (Eqn. (3)). The effective thermal conductivity of the liquid is proportional to the effective viscosity, which depends on the turbulence parameters. The solidification front, which forms the boundary to the liquid domain, can be treated in different ways. Some researchers model flow and solidification as a coupled problem on a fixed grid. Although very flexible, this approach is subject to convergence difficulties and requires a fine grid to resolve the thin porous mushy zone.

An alternative approach for columnar solidification of a thin shell, such as found in the continuous casting of steel, is to treat the boundary as a rough wall fixed at the liquidus temperature using *wall laws* (Huang *et al.* 1992). Figure 4 compares calculations using this approach with measured temperatures in

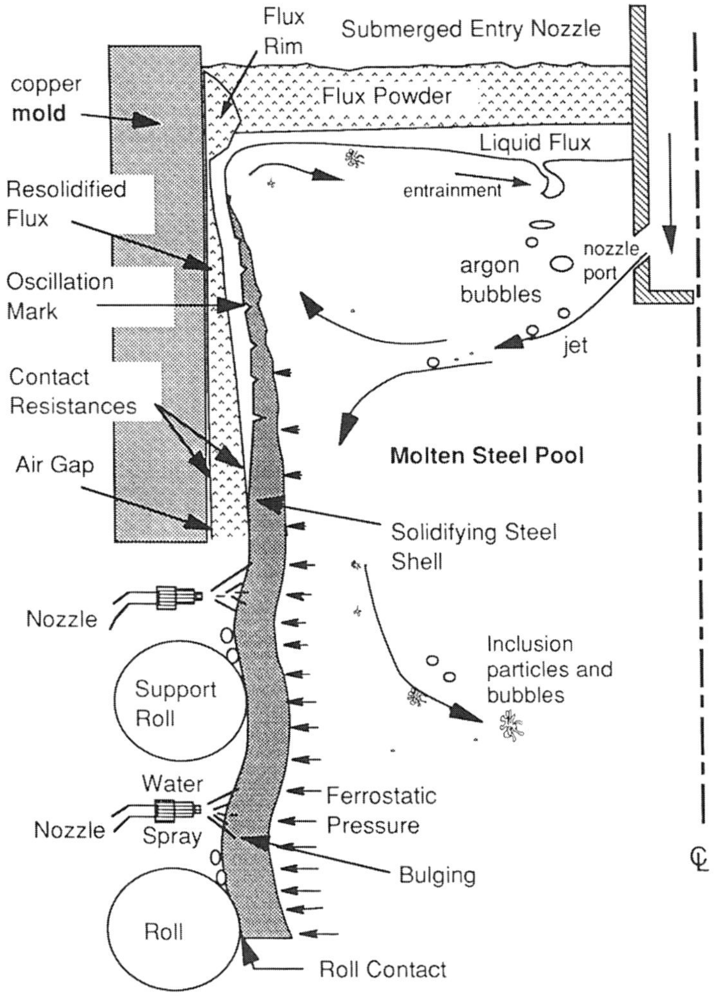

Figure 2
Schematic of phenomena in the mold region of a steel slab caster.

the liquid pool. Incorporating the effects of argon on the flow pattern was very important in achieving the reasonable agreement observed. Figure 4 shows that the temperature drops almost to the liquidus by mold exit, indicating that most of the superheat is dissipated in the mold. Most of this heat is delivered to the narrow face where the jet impinges, which is important to shell solidification.

The coldest regions are found at the meniscus at the top corners near the narrow face and near the submerged entry nozzle (SEN). This is a concern because it could lead to freezing of the meniscus, and encourage solidification of a thick slag rim. This could lead to quality problems such as deep oscillation marks, cracks, and other surface defects. In the extreme, the steel surface can solidify a solid bridge between the SEN and the shell against the mold wall, which often causes a breakout. To avoid these problems, flow must reach the surface quickly. These calculations should be used, for example, in designing nozzle port geometries that do not direct the flow too deep.

5. *Top Surface Powder/Flux Layer Behavior*

The flow of steel in the upper mold may influence the top surface powder layers, which are very important to steel quality. Mold powder is added periodically to the top surface of the steel. It sinters and melts to form a protective liquid flux layer, which helps to trap impurities and inclusions. This liquid is drawn into the gap between the shell and mold, where it acts as a lubricant and helps to make heat transfer more

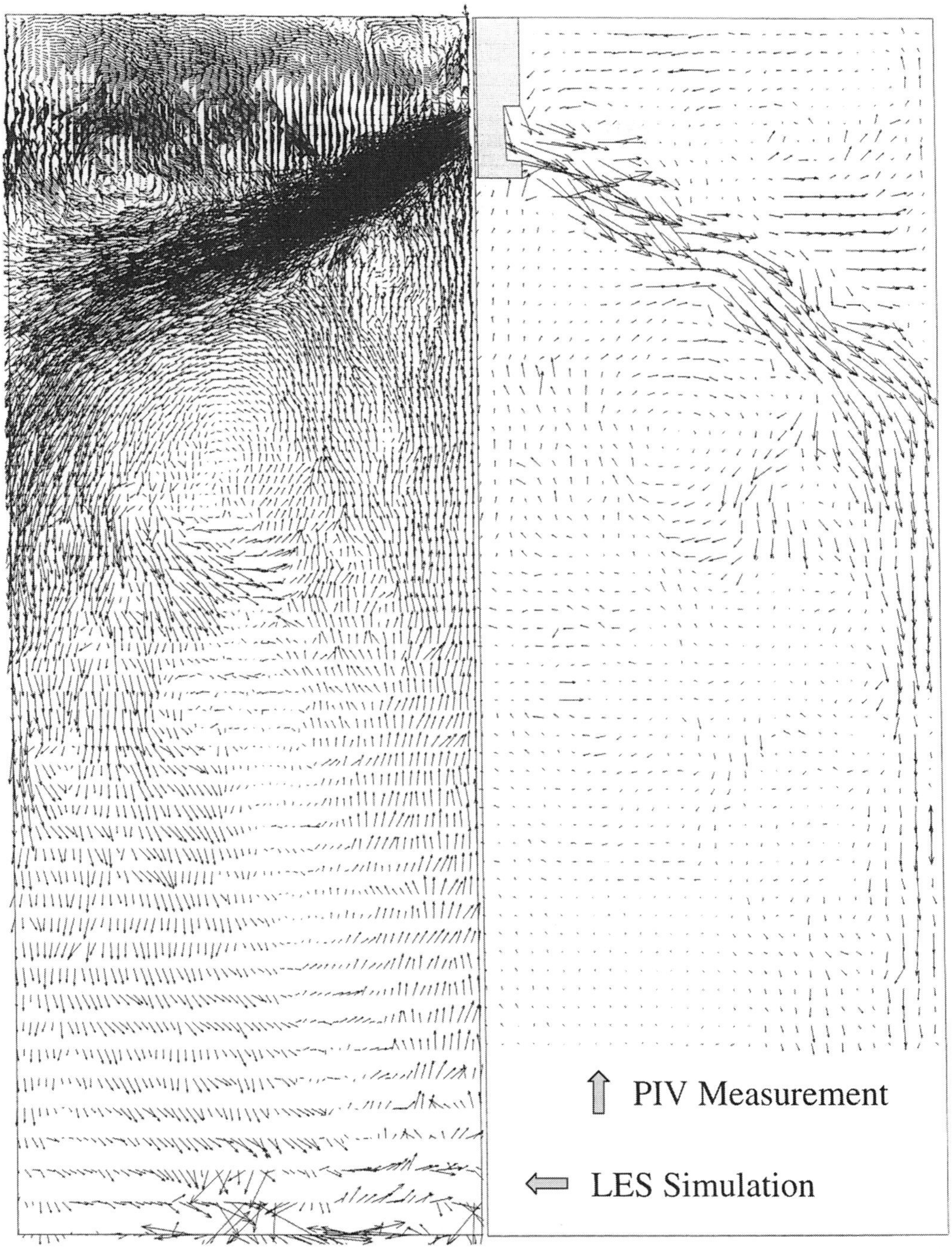

Figure 3
Instantaneous flow pattern in a slab casting mold comparing large-eddy simulation (LES) (left) and particle image velocimetry (PIV) measurement (right) (after Sivaramakrishnan *et al.* 2000).

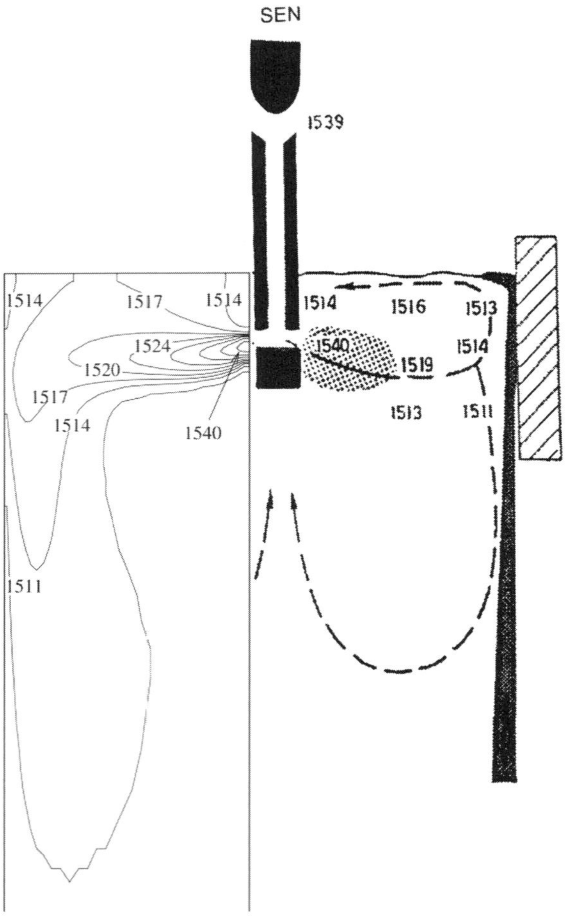

Figure 4
Temperature distribution in mold showing superheat dissipation: (a) prediction, (b) experimental measurement (after Thomas 2000).

uniform. These phenomena are difficult to measure or to accurately simulate with a physical model, so are worthy of mathematical modeling.

Figure 5 shows results from a 3D finite-element model of heat transfer and fluid flow in the powder and flux layers. The bottom of the model domain is the steel/flux interface. Its shape is imposed based on measurements in an operating caster. Alternatively, this interface shape can be calculated by solving additional equations to satisfy the force balance at the interface, which involves the pressure in the two phases, shear forces from the moving fluids, and surface tension, and gravity (Panaras *et al.* 1998). For the conditions in Fig. 5, the momentum of the flow up the narrow face has raised the level of the interface there. The shear stress along the interface is determined through coupled calculations with the 3D steady flow model. The model features different temperature-dependent flux properties for the interior, where the flux viscosity during sintering before melting, compared with the region near the narrow face mold walls, where the flux resolidifies to form a solid rim.

When molten steel flows rapidly along the steel/flux interface, it induces motion in the flux layer. If the interface velocity becomes too high, then the liquid flux can be sheared away from the interface, become entrained in the steel jet, and be sent deep into the liquid pool to become trapped in the solidifying shell as a harmful inclusion. If the interface velocity increases further, then the interface standing wave becomes unstable, and huge level fluctuations contribute to further problems.

The thickness of the beneficial liquid flux layer is also very important. As shown in the model calculations in Fig. 5, the liquid flux layer may become dangerously thin near the narrow face if the steel flow tends to drag the liquid towards the center. This shortage of flux feeding into the gap can lead to air gaps, reduced, nonuniform heat flow, thinning of the shell, and longitudinal surface cracks. Quantifying these phenomena requires modeling of both the steel flow and flux layers.

6. Motion and Entrapment of Inclusions and Gas Bubbles

The jets of molten steel exiting the nozzle may carry argon bubbles and inclusions such as alumina into the mold cavity. These particles may create defects if they become entrapped in the solidifying shell. Particle trajectories can be calculated using the Lagrangian particle tracking method, which solves a transport equation for each particle as it travels through a previously calculated velocity field. The force balance on each particle includes buoyancy and drag force relative to the steel. The effects of turbulent motion can be modeled crudely from a K–ε flow field by adding a random velocity fluctuation at each step, whose magnitude varies with the local turbulent kinetic energy level. To obtain significant statistics, the trajectories of several hundred individual particles should be calculated, using different starting points. The bubbles collect inclusions, and inclusion clusters collide, so their size and shape distributions evolve with time, which affects their drag and flotation velocities and importance. Models are being developed to include these effects (Thomas *et al.* 1997).

Figure 6 shows the trajectories of several particles moving through a steady flow field, calculated using the K–ε model. Most of the argon bubbles circulate in the upper mold area and float out to the top surface. A few might be trapped at the meniscus if there is a solidification hook, and lead to surface defects. A few small bubbles manage to penetrate into the lower

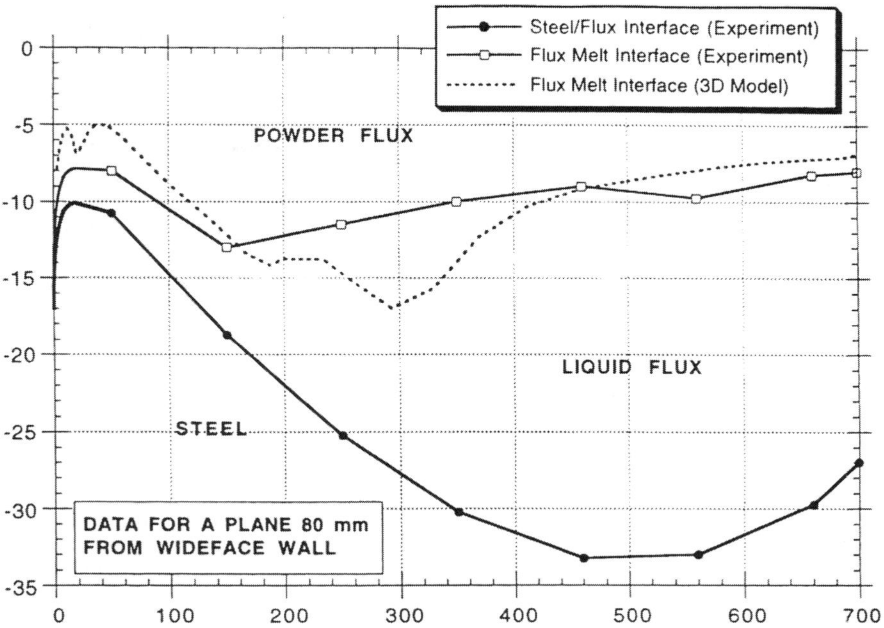

Figure 5
Comparison of measured and predicted melt-interface positions (after McDavid and Thomas 1996).

recirculation zone, where they move similarly to large inclusion clusters. Particles in this lower region tend to move slowly in large spirals, while they float towards the inner radius of the slab. When they eventually touch the solidifying shell in this deep region, entrapment is more likely on the inside radius. Trapped argon bubbles elongate during rolling and in low-strength steel, may expand during subsequent annealing processes to create costly surface blisters and "pencil pipe" defects. Transient models are likely to yield further insights into the complex and important phenomena of inclusion entrapment.

7. *Composition Variation During Grade Changes*

Large composition differences can arise through the thickness and along the length of the final product due to intermixing after a change in steel grade during continuous casting. Steel producers need to optimize casting conditions and grade sequences to minimize the amount of steel downgraded or scrapped due to this intermixing. In addition, the unintentional sale of intermixed product must be avoided. To do this requires knowledge of the location and extent of the intermixed region and how it is affected by grade specifications and casting conditions.

Models to predict intermixing must first simulate composition change exiting the tundish as a function of time. For small tundish operations, mixing in the liquid core of the strand must also be calculated. These calculations can be done using simple lumped mixing-box models and/or by solving the mass diffusion equation in the flowing liquid

$$\frac{\partial C}{\partial t} + \mathbf{v} \cdot \nabla C = \nabla \cdot (D_{\text{eff}} \nabla C) \tag{4}$$

In this equation, the composition, C, ranges between the old grade concentration of zero and the new grade concentration of one. This dimensionless concept is useful because alloying elements intermix essentially equally, owing to the much greater importance of turbulent convection over laminar diffusion. In order to predict the final composition distribution within the final product, a further model must account for the cessation of intermixing after the shell has solidified.

Intermix models have been developed for both slab and bloom casting (Huang and Thomas 1996, Thomas 1997) and are in use at several companies. Such models can be enhanced to serve as on-line tools by outputting for each grade change, the critical distances which define the length of intermixed steel product which falls outside the given composition specifications for the old and new grades. In addition, they can be applied off-line to perform parametric studies to evaluate the relative effects on the amount of intermixed steel for different intermixing operations and for different operating conditions using a standard ladle-exchange operation. Finally, they can

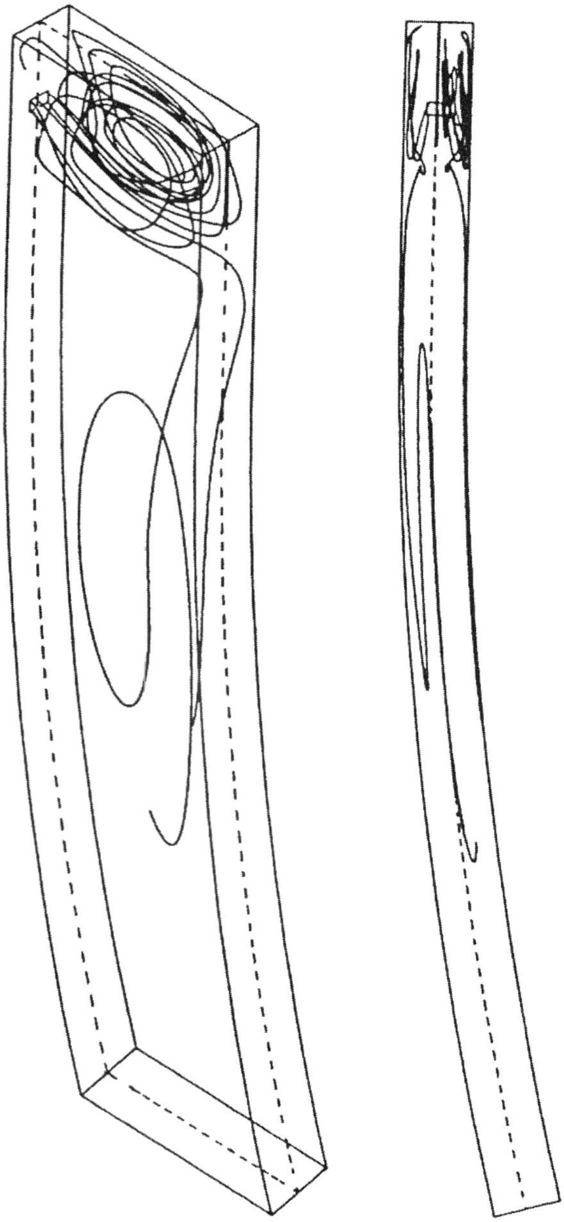

Figure 6
Sample trajectories of five 208 μm solid alumina inclusion particles (after Thomas *et al.* 1997).

be used to optimize scheduling and casting operation to minimize cost.

8. Thermal Mechanical Behavior of the Mold

Thermal distortion of the mold during operation is important to residual stress, residual distortion, fatigue cracks, and mold life. By affecting the internal geometry of the mold cavity, it is also important to the size of the gap that controls heat transfer to the solidifying shell. To study thermal distortion of the mold, 3D elastic-plastic-creep finite element models have been developed for slabs (Thomas *et al.* 1998) and thin slabs (O'Connor and Dantzig 1994 using ABAQUS. In order to match the measured distortion, models should incorporate all of the important geometric features of the mold, which often includes the four copper plates with their water slots, reinforced steel water box assemblies, and tightened bolts. Its four-piece construction makes a slab mold behave very differently from a single-piece bloom or billet mold, which have also been studied using thermal stress models (Samarasekera *et al.* 1982). This type of stress model application will become more important in the future to optimize the design of the new molds being developed for continuous thin-slab and strip casting. In twin-roll strip casting, for example, thermal distortion of the rolls during operation is of the same order as the section thickness of the steel product.

9. Thermal Mechanical Behavior of the Shell

The solidifying shell is prone to a variety of distortion and crack problems, owing to its creep at elevated temperature, combined with metallurgical embrittlement and thermal stress. To investigate these problems, models have been developed to simulate coupled thermal and mechanical behavior of a solidifying steel shell during continuous casting. The equations and solution procedures are documented elsewhere (Dantzig 1989).

An example of the predicted temperature contours and distorted shape of a transverse region near the corner is compared in Fig. 7 with measurements of a breakout shell from an operating steel caster (Moitra and Thomas 1993). The model tracks the behavior of a 2D slice through the strand as it moves downward at the casting speed through the mold and upper spray zones. It consists of separate finite-element models of heat flow and stress generation that are stepwise coupled through the size of the interfacial gap. The heat transfer model was calibrated using thermocouple measurements down the centerline of the wideface for typical conditions. Shrinkage predictions from the stress model are used to find the air gap thickness needed in Eqn. (2) in order to extend the calculations around the mold perimeter. The model includes the effect of mold distortion on the air gaps, and superheat delivery from the flowing jet of steel, calculated in separate models. The stress model includes ferrostatic pressure from the molten steel on the inside of the shell and calculates intermittent contact between the shell and the mold. It also features a temperature-dependent elastic modulus and

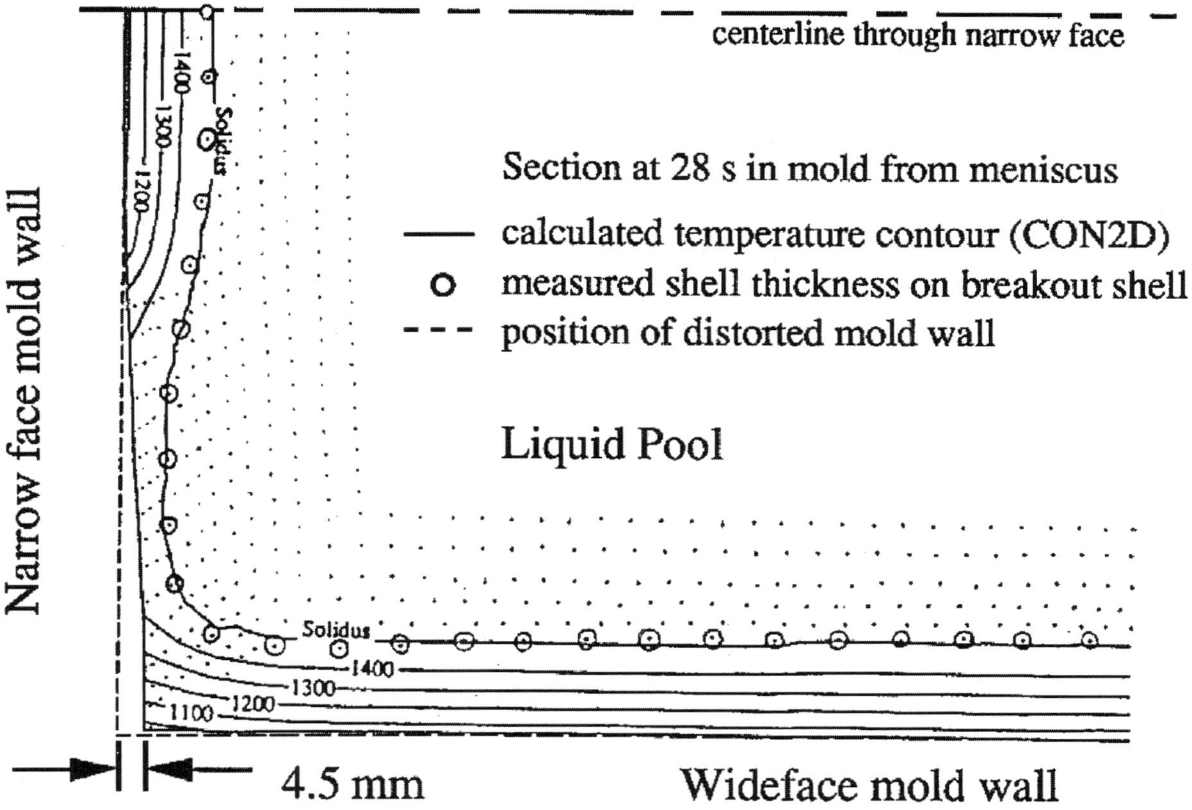

Figure 7
Comparison between predicted and measured shell thickness in a horizontal (x–y) section through the corner of a continuous-cast steel breakout shell (after Moitra and Thomas 1993).

an elastic-viscoplastic constitutive equation that includes the effects of temperature, composition, phase transformations and stress state on the local inelastic creep rate. Efficient numerical algorithms are needed to integrate the equations.

As expected, good agreement is obtained in the region of good contact along the wideface, where calibration was done. Near the corner along the narrow face, steel shrinkage is seen to exceed the mold taper, which was insufficient. Thus, an air gap is predicted. This air gap lowers heat extraction from the shell in the off-corner region of the narrow face. When combined with high superheat delivery from the bifurcated nozzle directed at this location, shell growth is greatly reduced locally. Just below the mold, this thin region along the off-corner narrow-face shell caused the breakout.

Near the center of the narrow face, creep of the shell under ferrostatic pressure from the liquid is seen to maintain contact with the mold, so much less thinning is observed. This illustrates the tremendous effect that superheat has on slowing shell growth, if there is a problem which lowers heat flow.

Figure 8 presents sample distributions of temperature and stress through the thickness of the shell, calculated with this model. To achieve reasonable accuracy, a very fine mesh and small time steps are needed. The temperature profile is almost linear through the shell. The stress profile shows that the shell surface is in compression. This is because, in the absence of friction with the mold, the surface layer solidifies and cools stress free. As each inner layer solidifies, it cools and tries to shrink, while the surface temperature remains relatively constant. The slab is constrained to remain planar, so complementary subsurface tension and surface compression stresses are produced. To maintain force equilibrium, note that the average stress through the shell thickness is zero. It is significant that the maximum tensile stress is found near the solidification front. This generic subsurface tensile stress is responsible for hot tear cracks, when accompanied by metallurgical embrittlement.

This model has been applied to predict ideal mold taper, to prevent breakouts such as the one discussed here (Lawson *et al.* 1994) and to understand the cause

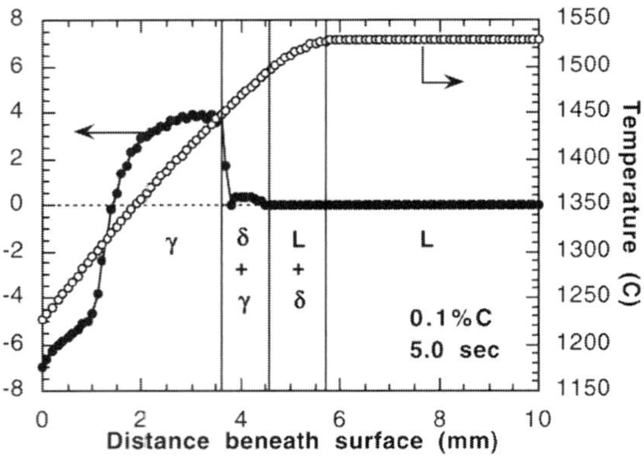

Figure 8
Typical temperature and stress distributions through shell thickness (after Thomas and Parkman 1997).

of other problems such as surface depressions (Thomas *et al.* 1996) and longitudinal cracks in slabs. Crack formation is particularly difficult to model directly. Very small strains can start hot tear cracks at the grain boundaries if liquid metal is unable to feed through the secondary dendrite arms to accommodate the shrinkage. Strain localization may occur on both small scale (when residual elements segregate to the grain boundaries) and on a larger scale (within surface depressions or hot spots). Later sources of tensile stress (including constraint due to sticking, unsteady cooling below the mold, bulging, and withdrawal) worsen strain concentration and promote crack growth. Microstructure, grain size, and segregation are not generally included in a stress model, so modeling these phenomena is difficult. Thus, careful experiments are needed to determine the fracture criteria empirically. A typical experiment applies stress during solidification, which is difficult to control. Thus, detailed modeling of the experiment itself is becoming necessary, in order to extract new fundamental material properties.

Much further work is needed to understand and quantify these phenomena and to apply the results to optimize the continuous casting process. In striving towards these goals, the importance of combining modeling and experiments together cannot be overemphasized.

10. Concluding Remarks

Mathematical models are growing in sophistication and, thereby, are becoming better able to quantify and understand the mechanisms of defect formation in processes such as continuous casting. Despite these advances, much more work is needed to quantify, incorporate, and simplify the key phenomena into better models before the use of mathematical models to design process improvements reaches its full potential and usefulness. In particular, more work is needed on transient phenomena and in better modeling of experiments to extract more fundamental material properties. It is crucial to emphasize that mathematical modeling tools must be used together with the other tools of the process engineer, which include laboratory and plant experimentation, physical models, and knowledge from previous experience.

Emerging continuous casting processes, such as thin-slab and strip casting, present several special challenges, which offer opportunities for the application of mathematical models. With higher casting speeds and thinner mold cavities, these processes are more prone to fluid flow problems. Thinner product (with more surface) puts more demand on solving surface quality problems. The integration of casting and deformation processing together requires better understanding of the interaction between microstructure, properties, and processing. Conventional slab casting has had three decades of development and still is not perfected. For these new processes to thrive, the learning curve towards process optimization to produce quality steel needs to be shorter. As computer hardware and software continue to improve, mathematical models will grow in importance as tools to meet this challenge.

See also: Continuous Casting

Bibliography

Dantzig J A 1989 Thermal stress development in metal casting processes. *Metall. Sci. Technol.* **7**(3), 133–78
Huang X G, Thomas B G 1996 Intermixing model of continuous casting during a grade transition. *Metall. Trans. B* **27**(4), 617–32
Huang X G, Thomas B G, Najjar F 1992 Modeling superheat removal during continuous casting of steel slabs. *Metall. Trans. B* **23**(6), 339–56
Ishii T, Sazhin S S, Makhlouf M 1996 Numerical prediction of magnetohydrodynamic flow in continuous casting process. *Ironmak. Steelmak.* **23**(3), 267–72
Lawson G D, Sander S C, Emling W H, Moitra A, Thomas B G 1994 Prevention of shell thinning breakouts associated with widening width changes,In: *Proc. Steelmaking Conf.*, Warrendale, PA, pp. 329–36
McDavid R M, Thomas B G 1996 Flow and thermal behavior of the top-surface flux/powder layers in continuous casting molds. *Metall. Trans. B* **27**(4), 672–85
Moitra A, Thomas B G 1993 Application of a thermo-mechanical finite element model of steel shell behavior in the continuous slab casting mold. *Proc. Steelmaking*, Dallas, TX. ISS, Warrendale, PA, pp. 657–67
O'Connor T G, Dantzig J A 1994 Modeling the thin slab continuous casting mold. *Metall. Mater. Trans. B* **25**(4), 443–57

Panaras G A, Theodorakakos A, Bergeles G 1998 Numerical investigation of the free surface in a continuous steel casting mold model. *Mater. Trans. B* **29**(5), 1117–26

Samarasekera I V, Anderson D L, Brimacombe J K 1982 The thermal distortion of continuous casting billet molds. *Metall. Trans. B* **13**, 91–104

Sivaramakrishnan S, Bai H, Thomas B G, Vanka P, Dauby P, Assar M 2000 Transient flow structures in continuous cast steel,In: *Proc. Ironmaking Conf.* ISS, Warrendale, PA, pp. 541–57

Thomas B G 1997 Modeling study of intermixing in tundish and strand during a continuous-casting grade transition. *ISS Trans. Iron Steelmak.* **24**(12), 83–96

Thomas B G 2000 Continuous Casting of Steel. In: *Modeling and Simulation for Casting and Solidification: Theory and Applications* Dekker, New York, Chap. 15

Thomas B G, Dennisov A, Bai H 1997 Behavior of argon bubbles during continuous casting of steel. *Proc. Steelmaking Conf.* Warrendale, PA, pp. 375–84

Thomas B G, Ho B, Li G 1998 Heat flow model of the continuous slab casting mold, interface, and shell. *Proc. Alex McLean Symp. Toronto.* Warrendale, PA, pp. 177–93

Thomas B G, Huang X, Sussman R 1994 Simulation of argon gas flow effects in a continuous slab caster. *Metall. Trans. B* **25**(4), 527–47

Thomas B G, Li G, Moitra A, Habing D 1998 Analysis of thermal and mechanical behavior of copper molds during continuous casting of steel slabs. *ISS. Trans. Iron Steelmak.* **25**(10), 125–43

Thomas B G, Moitra A, McDavid R 1996 Simulation of longitudinal off-corner depressions in continuously-cast steel slabs. *ISS Trans.* **23**(4), 57–70

Thomas B G, Parkman J T 1997 Simulation of thermal mechanical behavior during initial solidification. In: Chandra T, Sakai T (eds.) *Proc. Thermec 97 Int. Conf. Thermomechanical Processing of Steel and Other Materials*, Wollongong, Australia. TMS, Warrendale, PA

B. G. Thomas

Copper Alloys: Thermal and Thermomechanical Processing

Copper and its alloys have the ductility to enable them to be processed from cast billet and rod down to thin sheet and wire forms; further, the semifinished products have the formability to be bent, drawn, twisted, or otherwise shaped into finished parts. The final processing steps are designed to put the material into the proper condition to deliver its specified combination of properties: strength, formability, and conductivity. Any one processing step during manufacture alters the properties by changing the microstructure, and therefore the final properties, of the end product.

As with most metals, a considerable effort is devoted to control grain size both in processing and in the final product. A fine grain size promotes strength and eliminates one easy crack path that is provided by coarse grain boundaries. Moreover, a fine grain size in the solidified casting or ingot promotes compositional homogeneity and facilitates homogenizing heat treatments because of the shorter diffusion paths associated with a finer structure. In addition, a finer grain size reduces the potential for cracking along grain boundaries during subsequent hot or cold breakdown processing of the ingot. While the use of second-phase inoculants and rapid cooling via faster heat extraction during solidification can be effective in refining the as-cast grain structure, obtaining fine as-cast grain size is usually determined by the constitution of the alloy.

Copper and its alloys are amenable to being manufactured by powder metallurgy (PM) techniques both to take advantage of net shape production of parts and to permit manufacture of compositions that cannot otherwise be obtained commercially. For example, there are useful alloys that cannot easily be cast, such as the alumina-containing copper dispersion alloys. Or even if able to be cast, there are alloys that are difficult to hot work without cracking of the coarse-grained cast bar condition, such as the copper–nickel–tin spinodal hardening alloys or leaded phosphor bronze alloys for bearing applications. PM processing starts with the premixing of the alloy ingredients as powders and consolidating them by pressing and sintering, i.e., heating to diffusion bond the powder particles. Some products are made into shapes that can then be processed through various cycles of cold working and annealing, and usually continue to exhibit a fine grain throughout their processing history.

Other techniques exist, or are under development, for manufacture of a preform that is subsequently allowed to solidify in a mold or after reheating is then hot forged or pressed. A technique under commercial development is semisolid forming. This involves reheating a cast preform with the correctly adjusted chemistry to be stable in the semiliquid/solid state, with an inherently fine grain, just before and after forging a part to net shape in a closed die.

The majority of copper alloys start as melted and continuously or semicontinuously cast slabs, billets, or rods that are further hot and/or cold processed to finish gauge. Hot working, by definition, means that the microstructure undergoes recrystallization during or just after each hot processing step. The hot worked condition can be cold worked directly with interpass anneals or thermal treatments to provide a recrystallized microstructure needed to restore the ductility required for further cold processing. As the alloy is cold worked, its ductility decreases as hardness increases with continued cold working passes.

By definition, cold working is processing a metal at a low enough temperature such that it will not recrystallize spontaneously during or just after the working step. Copper and its alloys are cold formed routinely at room temperature on all types of commercial processing equipment, rolling, drawing, swaging, forging, etc. At some point the metal cannot be worked further without causing either cracking of the alloy being processed or overloading of the processing equipment. At this point the alloy must be subjected to an appropriate annealing heat treatment to cause softening of the material through the replacement of the cold worked, deformed crystal structure by uniform recrystallization to permit another cycle of cold reduction. Although pure copper can be processed to a great extent, it also must be annealed to remove mechanical and structural directionality and to prepare it for subsequent forming operations.

Cold working (rolling) curves, such as shown in Fig. 1, are a convenient means of showing the relative work hardening of alloys and also to predict or guide the choice of amount of cold work (measured by percentage reduction in thickness) required by a single or multiple series of cold rolling passes to work harden the material to the desired strength. In general, the starting condition is usually fully annealed or recrystallized, and the finer this initial annealed grain size is, the stronger is the product after a given amount of cold reduction in thickness. The increase in yield strength due to grain refinement is usually predicted by the Hall–Petch equation, $\sigma_y = \sigma_0 + Kd^{-1/2}$, where d is the average grain diameter and K is the Hall–Petch constant. In general, a metal with a finer annealed grain size also shows reduced property directionality, better formability, and improved surface appearance in a form part.

The lower stacking fault energy alloys work harden or strengthen more rapidly and must be annealed more frequently to restore ductility. Such alloys must also be processed with more care to avoid or reduce the degree of crystallographic texturing that is induced by the more restricted dislocation

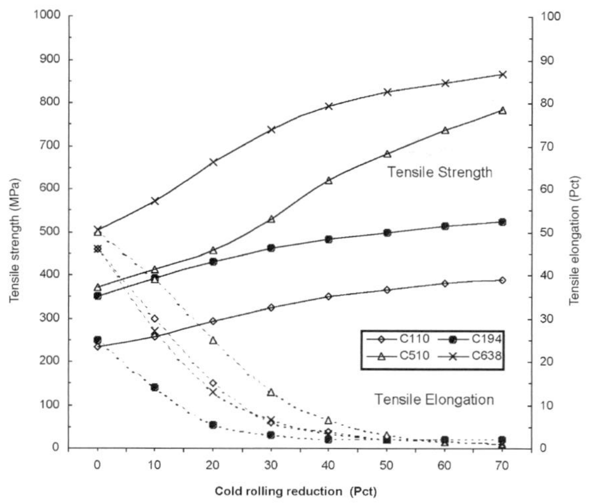

Figure 1
Comparison of the tensile strength and tensile elongation properties as a function of reduction in thickness by cold rolling from a fully annealed initial condition for unalloyed copper (C11000), Cu–2.4 Fe (C19400), Cu–5 Sn (C51000), and Cu–2.8 Al–1.8 Si–0.4 Co (C63800).

motion, leading to the onset of deformation twinning and shear band formation beyond about 40% reduction in thickness. The formation of shear bands—lath-shaped regions of deformed structure that rapidly propagate across several neighboring grains—permits the metal to accommodate the strain demands of continued processing. However, they leave behind cold worked regions that can nucleate cracks during any subsequent metal fabrication operation and can also foster anisotropy or directional mechanical properties, strength or formability, in the plane of the sheet. Processing experience has developed the knowledge to control textural effects through control of processing parameters. In practice, the degree of cold reduction and the temperature/time annealing parameters of the various processing steps are balanced to control the type and to reduce the amount of crystallographic texturing, thus controlling the degree of earing formed at the rim of deep drawn, annealed brass alloy cups, for example.

When a metal is cooled rapidly after casting or is cold processed to effect the desired change in shape or reduction in thickness, it usually contains residual stresses that can lead to part distortion during subsequent machining, stamping, or forming operations. It is usual, therefore, to subject a semifinished or finished copper alloy product to a low-temperature thermal treatment at slightly elevated temperatures, typically 100–300°C. This anneal provides the thermal energy required to allow the dislocation defects to rearrange themselves by diffusion mechanisms into more stable arrays. A relief annealed metal strip will not only exhibit less distortion when formed into parts during subsequent stamping operations, but it will also be more resistant to stress relaxation under load in service. Tension or roller leveling a strip is quite often used to remove sheet distortion or coil set in strip in order to provide a flat product needed by stamping presses. Studies have shown that tension leveling merely redistributes residual stresses rather than eliminating them and will not improve stress relaxation resistance.

For precipitation hardening alloys, the required solution heat treatment above the solvus can usually be done in a strand annealing furnace when the alloy is given an in-process full anneal. Care must be taken to control the grain size during this elevated-temperature treatment and to cool or quench the metal to room temperature rapidly enough to ensure solutionization of the alloying elements that will participate in the subsequent age-hardening treatment. If not quickly cooled or quenched to room temperature, the wrong type, or size, or distribution of precipitates will form. With properly solutionized copper alloys, aging is usually carried out by heat treating at a temperature between room temperature and the solvus in batch annealing furnaces for times of the order of hours. Some alloys that have inherently high aging kinetics can be aged as they are processed more rapidly through a strip or strand annealing furnace.

Another means of providing a more dispersed array of precipitates and enhanced strength is to insert a cold working step between solutionizing and the aging heat treatment. The array of crystalline defects, dislocations, and vacancies produced in this manner not only provides a higher baseline strength, but also offers more numerous nucleation sites for precipitates and paths for higher diffusion rates. The result is that during age hardening, the precipitates form in a finer dispersion and with higher reaction kinetics leading to peak hardness occurring in shorter times and at lower aging temperatures. Moreover, the age hardening of a solution-treated, cold worked structure usually provides a stabilized microstructure with the optimum combination of strength, formability, conductivity, and stress relaxation resistance.

Other processing combinations are used to emphasize one property over another in the final product. For example, a precipitation hardened condition may be cold worked further and given a relief anneal to stabilize the dislocation defect structure. Such a processing sequence will lead to higher strength, but it will be accompanied by reduced formability and stress relaxation resistance than is available with the aged condition. An industrially important variation of thermomechanical processing is the use of an aging treatment at a relatively high temperature to exaggerate overaging to the extent that a homogeneous array of dispersions will be produced in a matrix that has recrystallized into a very fine grain. A subsequent working and relief annealing sequence produces a product with high conductivity and formability at the expense of some strength and some stress relaxation resistance. In general, alternating cycles of annealing and cold working usually lead to the optimum combination of strength and formability, with a reduced amount of crystallographic and mechanical texturing. Warm working a precipitation hardening copper alloy, i.e., applying work to a solutionized alloy at aging temperatures, is usually not commercially effective in achieving combinations of properties that are already obtainable by other, simpler processing sequences.

Two copper-based alloy systems (Cu–Zn at >40 wt.% Zn and Cu–Al at >10 wt.% Al) are capable of undergoing a martensitic phase transformation obtained with beta-phase alloys. But quenched and tempered martensitic copper alloys are generally too weak (Cu–Zn) or too brittle (Cu–Al) for commercial viability. They have not yet been produced with the high strength and toughness combination available with carbon and alloy steels. However, copper alloys known as shape-memory alloys have been developed to make use of the thermoelastic martensitic transformation available in copper–zinc-based alloys, in order to provide cyclical mechanical motion with temperature change, analogous to that seen with the intermetallic Ni–45Ti alloys. These alloys are used in unique thermal actuating applications, such as temperature-actuated safety switches.

See also: Dynamic Recovery and Recrystallization; Primary Recrystallization

Bibliography

Metals Handbook, 10th edn. 1990 Properties and selection: nonferrous alloys and special-purpose materials. ASM International, Materials Park, OH, Vol. 2, pp. 216–427

R. N. Caron

Crystal Growth from the Melt

The most frequently used and most important method of producing bulk single crystals of a given material is by solidification of its own melt—called "crystal growth from the melt" or just "melt growth." Its technical and economic importance is due to the fact that large single crystals (several kilograms up to 300 kg) can be grown very efficiently and with a high yield at a relatively high growth rate (several millimeters per hour up to millimeters per minute). This growth rate is not limited by the transport of crystal species but only by the removal of the heat of crystallization. However, this rule holds only in principle because the temperature gradients in the growing crystal may not exceed certain values (depending on the material), since these gradients may cause crystal imperfections (by thermal stress). This requirement makes it necessary to design very carefully the thermal boundary conditions in the corresponding growth apparatus, which is strongly supported by the aid of computer modeling.

Several techniques of melt growth have been developed and they will be described in the following sections. The main application of melt-grown crystals is in the electronic industry which consumes huge amounts of semiconductor single crystals (nearly 10000 t of silicon and about 600 t of GaAs per year). About 500 t of ruby and 100 t of garnet crystals are produced for laser applications, 100 t of lithium niobate are applied in various electrooptic devices, and more than 1000 t of zirconium dioxide crystals are used as gemstones.

Melt growth cannot be used in cases where the material to be grown has no congruent melting point and where the melting point, T_m, or vapor pressure, p_m, are too high ($T_m > 2000°C$, $p_m > 100$ bar). In these cases crystals have to be grown from melt solutions (fluxes) or from the vapor phase.

The fundamental physical and physicochemical mechanisms which govern melt-growth processes are treated in Sect. 1. For further information on melt-grown crystals see Wilke and Bohm (1988), Hurle (1993–94), and Müller (1998).

1. *Fundamentals of Melt Growth*

1.1 *Thermodynamics and Phase Diagrams*

As melt growth uses the liquid–solid phase transition of an element or a compound, the use of equilibrium thermodynamics and especially phase diagrams is very useful in finding the appropriate growth conditions (e.g., temperature–composition, *T–c*, and temperature–pressure, *T–p*). Details on phase diagrams in correlation with melt growth are given by Rosenberger (1979), Wilke and Bohm (1988), and Rudolph (1998).

Crystallization of a melt, i.e., solidification can only occur if the system deviates from the thermodynamic equilibrium. This means the melt must have a certain temperature, T, below the melting temperature, T_m. This deviation, ΔT, from T_m is called supercooling ($\Delta T = T_m - T$). Then the driving force of crystallization (from the point of view of equilibrium thermodynamics) can be expressed in terms of the difference, ΔG, of the Gibbs energies of the liquid and solid phases:

$$\Delta G = \frac{\Delta H \Delta T}{T_m} \qquad (1)$$

where ΔH usually is the enthalpy released by crystallization. Melt growth processes are characterized by relatively low driving forces (ΔG), as homogeneous nucleation has to be avoided in the supercooled melt region. All melt growth techniques make use of an oriented seed crystal typically of the same material (heterogeneous nucleation). Values of $\Delta H / T_m$ usually fall within the range 10–100 J K^{-1}mol^{-1} and values of ΔT are typically 1–5 K, i.e., ΔG is about 10–100 J mol^{-1}.

1.2 *Growth Kinetics*

On a microscopic scale melt growth occurs by the addition of crystal "units" (atoms or molecules) to the corresponding crystal positions at the growth interface which may be atomically "rough" or "smooth" (e.g., Chernov 1999). Crystal growth from the melt occurs mainly under conditions of an atomically rough interface. This means the top layer of atoms of the phase boundary is continuously fluctuating owing to the relatively low value of the enthalpy of melting, ΔH, compared to the thermal energy, RT_m. For practical purposes of melt growth one may use the criterion of Jackson

$$\alpha_J = \frac{\Delta H \, w}{R T_m u} \qquad (2)$$

to estimate whether conditions of an atomically rough ($\alpha_J < 2$) or smooth ($\alpha_J > 2$) interface can be expected. R is the gas constant, T_m is the melting temperature, w is the number of nearest neighbors of an atom in the growing face, and u is the number of nearest neighbors of an atom in the crystal.

Rough interfaces can grow under conditions of very low supercooling (i.e., driving force). This means the shape of the crystal melt interface is identical to the isotherm of the melting temperature. Only in certain cases, especially in semiconductors, where $\Delta H / RT_m \approx 2$, does the w/u ratio decide whether the given interface tends to be rough or smooth. In such cases parts of the interface can grow with an atomically smooth interface deviating from the curved shape of the isotherm, T_m. Such planar sections of the interface are called facets. Note that in melt growth of dielectrics, especially oxide crystals, where $\alpha_J > 2$, the atomically smooth growth mode becomes predominant.

For an atomically rough interface the kinetically determined crystallization rate, v, can be estimated as:

$$v = \frac{\lambda k \Delta H \Delta T}{h R T_m} \exp(-G_a/RT) = \beta_R \Delta T \tag{3}$$

where G_a is an energetic barrier separating the solid from the fluid which the atoms of the melt have to cross to come into the crystal, h is Planck's constant, k is Boltzmann's constant, λ is the spacing between the layers (of the order of the lattice constant), and β_R is the kinetic coefficient of an atomically rough melt–solid interface. The value of β_R is in the range of some $\text{cm s}^{-1}\text{K}^{-1}$ so that the kinetic growth rate $v > 1\ \text{cm s}^{-1}$. This confirms the earlier statement that melt growth is not limited by growth kinetics but by the transport rate of heat. For further details on growth kinetics see Chernov (1999).

1.3 Transport Processes

(*a*) *Heat transport.* The heat flux across the solid–liquid interface with constant cross-section is described by the following (one-dimensional for simplicity):

$$K_s \frac{dT_s}{dz} = K_l \frac{dT_l}{dz} + \rho_s \Delta H v \tag{4}$$

where K and dT/dz are the heat conductivity and temperature gradient, respectively, in the solid (s) and liquid (l) phase, and ρ_s is the crystal density. Typically, the temperature gradient in the melt is much less than in the crystal which means that the growth rate, v, is more or less determined by the temperature gradient in the crystal. Thus, materials with high thermal conductivity (e.g., metals and semiconductors) can be grown faster than those with low thermal conductivity (e.g., oxides and halides). Another limiting factor of v is the phenomenon of constitutional supercooling (see Sect. 1.3(b) (iii)).

In addition to heat conduction, radiative heat transfer (in transparent crystals and surrounding media) as well as convection (in the melt and gas atmosphere) may have a significant influence on the heat transport. This is especially the case for Czochralski-like techniques (see below). For an extensive discussion of this subject see Müller (1988).

The knowledge of the temperature field is of great importance for optimizing the growth conditions through all growth stages. For example, it controls the concentration of intrinsic point defects. Furthermore, it controls the dislocation density as shown by Völkl (Hurle 1993–94) which is related to the thermal stress caused by curved isotherms in the growing crystal. Computer simulation of heat transport in melt growth configurations is so well developed that it can be used for quantitative predictions of temperature distributions in the growing crystals.

(*b*) *Species transport.* (i) *Segregation and distribution of solute.* In melt growth, species transport is only of relevance for dopants or impurities occurring in dilute concentrations. According to the phase diagram, the equilibrium concentrations c_s (solid) and c_l (liquid) of such species are different, i.e., the equilibrium segregation coefficient $k_0 = c_s/c_l$ is not unity. This causes segregation phenomena during melt growth which can be treated in terms of an effective segregation coefficient $k_{\text{eff}} = C_s/c_l^{(\infty)}$, with $c_l^{(\infty)}$ being the mole fraction of the dopant in the melt far away from the interface. During an actual freezing process the solute is rejected ($k_0 < 1$) or preferentially absorbed ($k_0 > 1$) by the propagating solid–liquid interface, forming an enriched or depleted solute boundary layer in front of it.

The width, δ_s, of this boundary layer is determined by the growth rate, v, and by the diffusive and convective species transport in the melt which is very often difficult to predict. A very popular model, commonly used in melt growth studies, was introduced by Burton, Prim, and Slichter in 1953 (Wilke and Bohm 1988) for the steady state of segregation:

$$k_{\text{eff}} = \frac{k_0}{k_0 + (1 - k_0)\exp(-v\delta_s/D)} \tag{5}$$

The advantage of this model is its easy use for plotting experimental data by fitting δ_s. However, this definition of k_{eff} cannot be used for a quantitative prediction since the value of δ_s is not related physically to diffusion (δ_D) or momentum (δ_v) of boundary layers. This problem is overcome by a definition of $k_{\text{eff}} = (1 + \Lambda)/(1 + \Lambda/k_0)$, with $\Lambda = (v_D \delta_s)/(7.2 v L)$, where v_D is the convective velocity at the edge of the boundary layer, v is the growth velocity, and L is the length of the interface introduced by Müller and Ostrogorsky (Hurle 1993–94). In this case δ_s is a well-defined physical quantity, i.e., the thickness of the solute boundary layer.

From both definitions it follows that k_{eff} equals unity at high growth rates or/and motionless melts (i.e., high values of δ_s). In such cases a high uniformity of the macroscopic distribution within the crystal can be reached. If the melt is stirred, however, the boundary layer is reduced by the melt flow ($\delta_s \to 0$) and k_{eff} equals k_0. In this case the axial distribution of the solute is typically more nonuniform. These extreme cases of axial solute distribution can be described by the equation of Scheil (Wilke and

Bohm 1988) for complete mixing:

$$c_s = c_0 k_{eff}(1 - g)^{k_{eff}-1} \quad (6)$$

where c_0 is the starting concentration and g the solidified fraction; or by an equation given by Tiller for the case of purely diffusive transport (Wilke and Bohm 1988):

$$c_s = c_0\{(1 - k_0)[1 - \exp(-k_0 zv/D)] + k_0\} \quad (7)$$

where z is the axial coordinate and D the diffusion coefficient of the solute in the melt. The Tiller approach, however, can only be used for special cases like melt growth under high magnetic fields or microgravity (Müller 1988).

(ii) *Convection and microsegregation.* Normally, convective transport in the melt determines the distribution of solute, especially of rejected species in front of the growing interface and, hence, the composition of the crystal on the macro- and microscale. Melt convection is classified by the physical mechanisms driving convective flows like buoyancy, forced convection (by rotation, vibration, rotating magnetic field, etc.), and capillary (Marangoni) convection. If the convective flows are unsteady (oscillating), as in most cases of melt growth, the temperature field is also unsteady. Then the solute segregation also becomes nonuniform on the microscale, which can be visualized as so-called striations (Müller 1988). An overview of convection phenomena is given by Müller and Ostrogorsky (Hurle 1993–94).

Convective flows and thus temperature oscillations can be damped very efficiently in conducting melts by applying magnetic fields (Hurle 1993–94).

(iii) *Constitutional supercooling.* Both an enriched ($k_0 < 1$) or depleted ($k_0 > 1$) solute layer, showing a typical diffusion profile at the growing interface, gives rise to constitutional instability if the corresponding liquidus temperatures exceed the actual temperatures in the melt. In this case the melt region is constitutionally supercooled and any microscopic disturbance of the interface is increased. Tiller and co-workers (Wilke and Bohm 1988) were the first to deduce a condition for preventing constitutional supercooling i.e., for morphological stability of the interface:

$$\frac{G}{v} = \frac{mc_l(1 - k_0)}{k_0 D} \quad (8)$$

where G is the temperature gradient in the melt at the interface and m the slope of the liquidus from the T–c phase diagram projection.

Note that Eqn. (8) is an approximation that neglects some stabilizing factors. In particular, a face growing by an atomically smooth mechanism is more stable against disturbances than a rough interface. Further, convection can stabilize the interface morphology by effective removal of the solute layer.

2. *Principles of Melt Growth Techniques*

2.1 *Methods of Crystal Pulling*

(*a*) *Standard Czochralski (Cz) method.* The Czochralski method is the dominant method for the production of bulk single crystals of a wide range of electronic and optical materials (Fig. 1(a)).

The feed material is melted in a cylindrically shaped crucible with a free liquid surface at the top. The heating of the melt is provided by resistance or radio-frequency heaters using an axially positioned moveable pulling rod. It is important to adjust the power of the heater(s) carefully so that a certain portion of the dipped seed is remelted and a melt meniscus is formed. The pull rod, i.e., the seed, is slowly lifted (often under rotation) and the melt crystallizes at the interface of the seed by forming a new crystal portion. The shape of this crystal, especially the diameter, is controlled by adjusting the heating power, pulling rate, and rotation rate of the pulling rod, i.e., the crystal. Typically, the first portion of the crystal is grown with a small diameter to reduce the density of crystallographic defects. This technique is essential for the growth of dislocation-free silicon and is called the "necking procedure."

For economic reasons it is important to grow the cylindrical part with a constant diameter and a length as large as possible. For this purpose an automatic diameter control is necessary, either by control of the meniscus shape (e.g., for silicon) or by weighing the crystal (e.g., for GaAs, InP) or the melt (for oxides). As the gas atmosphere has to be controlled during the growth process, the whole assembly is maintained in a vacuum-tight vessel which is filled with gas (inert gas for semiconductors, oxygen or air for oxides).

The growth of silicon crystals is the most important technical application of the Cz method. Silicon crystals are grown free of dislocations, with diameters of 100–300 mm and masses up to 300 kg.

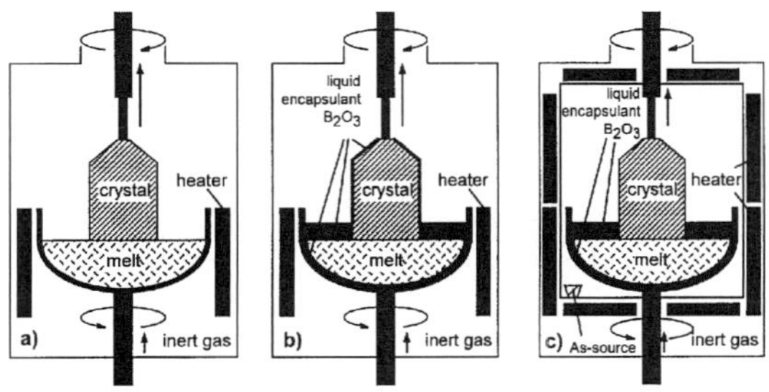

Figure 1
Schematic showing principles of: (a) Cz method, e.g., for silicon; (b) LEC method; (c) VCz method, e.g., for GaAs.

(*b*) *Liquid-encapsulated Cz (LEC) method.* Materials with a high partial pressure of one or more components can be grown by the Cz principle by using a liquid encapsulating the melt surface. Usually boric oxide, but occasionally a halide, prevents the loss of a volatile element by evaporation from the melt if the pressure in the growth chamber exceeds the partial pressures of the components. This is called the LEC method (see Fig. 1(b)) and is the most widely used growth method for group 3–5 compounds and lead chalcogenides. GaAs crystals up to 150 mm in diameter (nearly 30 kg) are grown with a dislocation density of 10^4–$10^5\,cm^{-2}$. The use of B_2O_3 encapsulant also reduces the background impurities effectively owing to its gettering ability. The main disadvantage of the LEC method is the high thermal stress in the growing crystal caused by large temperature nonlinearities and large temperature gradients (about $100\,Kcm^{-1}$) owing to the low thermal conductivity of the liquid encapsulant. This problem can be overcome by the methods described in *Sects. 2.1(c) and 2.4(a)*.

(*c*) *Vapor-pressure-controlled LEC (VLEC or VCz) and hot wall Cz (HWCz) methods.* The VLEC (or VCz) method is a variant of the LEC method which was introduced to reduce the thermal stress in the growing crystal by an *in situ* after-heating of the grown crystal. The after-heating, however, has the consequence that the crystal portion which stands out from the encapsulation layer has such a high temperature that the grown crystal would decompose at the surface by dissociation and be damaged severely. To prevent this dissociation it is necessary to use an inner envelope with hot walls and a vapor pressure source of the volatile component (Fig. 1(c)). This vapor pressure is much smaller than the vapor pressure which would be necessary to avoid a decomposition of the melt but sufficient to prevent a decomposition of the crystal. The VCz method is a complex technique because of the double containment with feed-through of the pulling rod and separately heated vapor pressure source. GaAs, InP, and BGO crystals with reduced defect densities have been successfully grown by the VCz method (e.g., Rudolph *et al.* 1997).

Another variant of the VCz method without using an encapsulant, the hot wall Czochralski (HWCz) method, was also studied for GaAs growth. However, in this case an arsenic pressure of about 2 bar had to be provided in the inner chamber (with hot walls) to avoid a decomposition of the uncovered GaAs melt. Until now the HWCz method has not reached the production stage owing to its technical complexity and material problems.

(*d*) *Magnetic field Cz (MCz) methods.* All variants of the Cz method can in principle be combined with static or time-dependent magnetic devices. MCz methods take advantage of magnetic force fields for damping of temperature fluctuations (unsteady convective flow) or for controlling convective species transport (Hurle 1993–94).

(*e*) *Kyropulous method.* The Kyropulous method has a similar setup to the Cz method (Fig. 1(a)). However, after dipping the seed into the melt the crystal is not pulled from the melt but grows into the melt by a continuous decrease of the heating power. This method is widely used to produce alkali halide crystals for optical components (NaCl, KBr) and alkali iodides for scintillators (NaI, CsI). Crystals with dimensions over 500 mm have been achieved. One major practical problem appears to be seed breakage. For this reason seeds and crucible rotations are usually not used.

(*f*) *Continuous feed Cz (CCz) and double crucible Cz methods.* The uniformity of the grown crystal with respect to the axial doping distribution can be improved if the crystal is grown from a smaller melt portion which is kept at constant volume by supplying the solidified portion from a source. Such a source may be a solid or liquid feed supply (CCz method) or an inner crucible with holes placed in a larger outer crucible ("double crucible" technique). The CCz method was developed for silicon and also to increase the crystal length for one growth run. Its use for production is limited owing to technical and material problems preventing a high yield.

(*g*) *Pulling of shaped crystals from dies.* The diameter and shape of a crystal from Cz pulling techniques is controlled by the shape and position of the melt meniscus at the triple contour crystal–melt–gas (or encapsulant). This control can be improved by using a die. GaAs, GaP, and InP crystals have been grown by the LEC method with a constant circular diameter by using a coracle which floats at the interface on top of the melt. Crystals with profiles of more complex shape, like fibers or tubes, are grown using dies (for further details see *Growth of Shaped Crystals from the Melt*). Shaped crystal growth is used mainly for two applications, silicon ribbons for solar cells (thin-walled, octagonal silicon tubes) and oxide fibers (mainly sapphire and niobates) for laser and nonlinear optics.

2.2 Zone Melting Methods

The importance of zone melting for bulk crystal growth is mainly limited to its use for purification of the feed material but not to the growth process itself. Impurities with a segregation coefficient $k_0 < 1$ can be accumulated very efficiently by multiple passes of melt zones towards the end of the ingot (Pfann 1958).

In special cases of highly reactive melts, the crucible free-zone melting method, the so-called floating zone (FZ) method, is used for the growth of single crystals. An important parameter of the FZ method is the ratio between the surface tension and the density of the melt. It determines the maximum size of the floating melt zone which can be kept stable between the vertically aligned feed rod and growing crystal. The most important application of the FZ method (besides the growth of high-melting-point refractory metals) is the production of silicon crystals with low oxygen concentration ($<10^{16}\,\text{cm}^{-3}$). In this case a relatively small melt zone is heated directly by radio frequency. By the aid of specially shaped heaters the melt zone is stabilized to allow the growth of silicon crystals up to a diameter of 150 mm.

2.3 *Verneuil Process*

One of the oldest (since 1902) and still useful methods to grow oxide crystals from the melt is the Verneuil process. An oxygen–hydrogen flame burner is used to melt the feed which is added as a powder to the gas stream or directly injected into the flame. A thin melt film is formed on top of the growing crystal. The advantage of this process is the low equipment costs; the disadvantage is the extremely large temperature gradients causing a high density of stress-induced crystal defects. Typical applications are the production of sapphire and ruby ($\sim$700 tons per year).

2.4 *Methods of Directional Solidification*

Directional solidification is the basic principle of controlled solidification of a melt, having greatest importance in the field of metallurgical casting. The growth of single crystals by melting a charge in a crucible and freezing the melt from one end (seed) to the other (tail) has been used for many materials for a long time. However, the strong interaction of the growing crystal with the crucible material limits the yield and hence the industrial applicability is limited to only certain cases. The method can be carried out by moving the growth interface in a horizontal or a vertical direction (see Fig. 2). The crystal growth configuration consists typically of a tube furnace which provides a temperature profile with negative gradients parallel to the growth direction (at least in the vicinity of the growth interface but mainly in the major portion of the melt and crystal). The single-crystal seed is positioned at one end of the boat or the lower end of the vertical crucible (see Fig. 2).

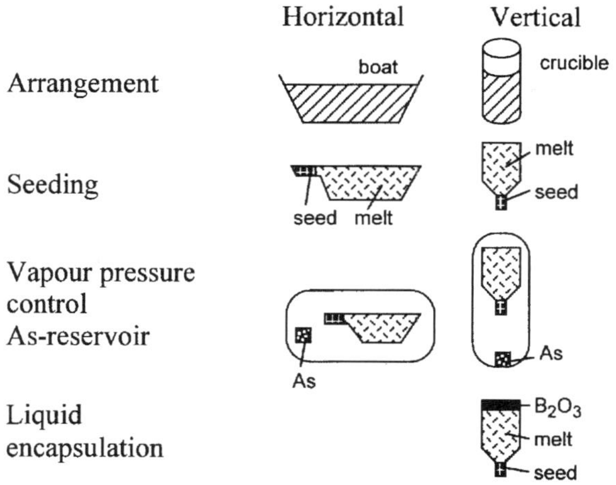

Figure 2
Various principles of directional solidification methods (gradient freeze, Bridgman) and important features, e.g., for GaAs.

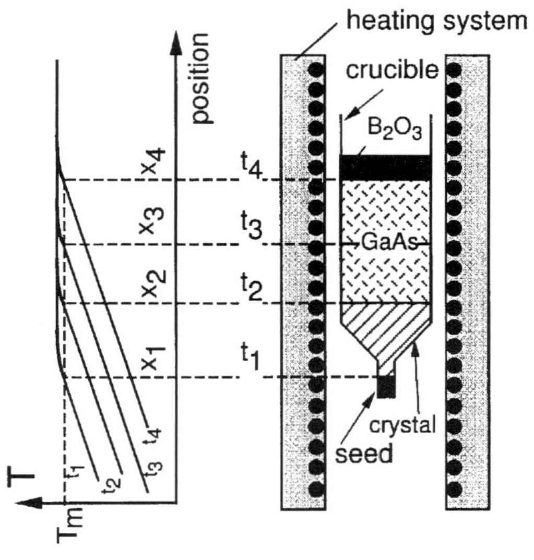

Figure 3
Schematic of the furnace temperature profiles for different time steps (t_1, t_2, t_3, and t_4) with positions x_1, x_2, x_3, and x_4 of the crystal–melt interface (at $T=T_m$) in a vertical Bridgman/gradient freeze configuration (VB/VGF), e.g., for GaAs.

(*a*) *Gradient freeze, Bridgman, and Stockbarger methods.* Crystal growth is carried out by a controlled shifting of the temperature profile relative to the boat or crucible as shown in Fig. 3. This can be achieved in three ways: by mechanical movement of the crucible relative to the fixed furnace (introduced by Bridgman in 1925 and Stockbarger in 1936); by mechanical movement of the furnace relative to the fixed crucible; and without any mechanical movement, the shifting of the temperature profile being achieved only by a controlled changing of the power of the furnace heater (gradient freeze method).

These methods are mainly used for the production of compound semiconductors and optical crystals, e.g., the horizontal Bridgman (HB) (Fig. 2 (left)) and vertical gradient freeze (VGF) (Figs. 2 (right) and 3) methods.

The vertical configurations (VGF, vertical Bridgman (VB)) are preferred for the industrial production of GaAs crystals because of the higher yield of round wafers compared to the HB method. The increasing interest in the use of the VGF technique results from the fact that it uses the simplest principle of melt growth. It has several advantages compared to the Cz methods. It operates under stable hydrodynamic conditions, is suitable for computer modeling and for automatic process control, there is low thermal stress in the growing crystal, round crystals without any diameter control are produced, and the equipment costs are lower.

The VGF method, however, has one severe disadvantage compared to the Cz methods: it does not allow for a direct observation of the growing crystal. Any failure during growth, e.g., by twinning or polycrystal formation cannot be corrected by melting back but is discovered only when the crystal is removed after growth. For further details see Hurle (1993–94).

(*b*) *Heat exchanger method.* The heat exchanger method is a modified directional solidification technique developed mainly for growth of large sapphire crystals with diameters up to 500 mm. However, photovoltaic silicon, CdTe, and BGO crystals are also grown by this method. A crucible with a single-crystal seed centered at the bottom is loaded with a meltstock and placed on top of a helium-cooled heat exchanger. After melting and homogenization, growth is initiated and carried out by increasing the helium flow and decreasing the furnace temperature. After complete solidification the furnace temperature and helium flow are reduced to achieve *in situ* annealing of the boule.

(*c*) *Skull melting.* Some materials like ZrO_2, MgO, CaO, and BaO with high melting temperatures and high chemical reactivity cannot be melted in direct contact with any known crucible materials. In the skull melting technique the charge is heated and melted by radio-frequency heating of the charge in a water-cooled container. As a result of the water cooling, a skull of sintered material is formed and acts as a noncontaminating crucible for the melt. Crystals are typically grown by directional solidification of the molten charge without a seed crystal. The method is used, for example, to grow 50–100 t of cubic zirconia per month for gemstones. Crystal pulling out of a skull-melting crucible is also possible.

See also: Growth of Shaped Crystals from the Melt.

Bibliography

Chernov A A 1999 *Modern Crystallography III*, 2nd edn. Springer, Berlin

Hurle D T J 1993–94 *Handbook of Crystal Growth*. Elsevier/North-Holland, Amsterdam, Vols. 1–3

Müller G 1988 Convection and crystal growth from melt. In: Freyhardt H C (ed.) *Crystals*. Springer, Berlin, Vol. 12

Müller G 1998 Melt growth of semiconductors. In: Fornari R, Paorici C (eds.) *Theoretical and Technological Aspects of Crystal Growth*. Trans Tech, Zurich, Switzerland, pp. 87–108

Pfann W G 1958 *Zone Melting*. Wiley, New York

Rosenberger F 1979 *Fundamentals of Crystal Growth I*. Springer, Berlin

Rudolph P 1998 Elements of thermodynamics for the understanding and design of crystal growth processes. In: Fornari R, Paorici C (eds.) *Theoretical and Technological Aspects of Crystal Growth*. Trans Tech, Zurich, Switzerland, pp. 1–26

Rudolph P, Neubert M, Arulkumaran S, Seifert M 1997 Vapour pressure controlled Czochralski (VCZ) growth—a method to produce electronic materials with low dislocation density. *Cryst. Res. Technol.* **32**, 35–50

Wilke K Th, Bohm J 1988 *Kristallzüchtung*. Deutsch & Thun, Frankfurt am Main, Germany

G. Müller and P. Rudolph

Cutting and Drilling of Metals and Other Materials: A Comparison

Manufacturing technology is one of the key technologies on the world market. In order to keep the market position, it is necessary to realize a high quality product at low market prices in a sufficiently short delivery period. Choosing the most economical manufacturing process is essential for the economic efficiency of a product. In order to show the possibilities of manufacturing including the main process characteristics, some of the most important cutting and nonconventional processes will discussed in this article.

1. Cutting Processes

Cutting processes form the core of industrial manufacturing processes. Apart from economical aspects, this is due to the obtainable surface qualities and the geometrical, dimensional, and positional accuracy combined with almost unlimited machining possibilities. Therefore, considerable competitive advantages are expected for cutting processes.

Cutting processes transform workpieces into finished products by mechanically removing material. Cutting requires an active contact and the interaction of the two partners, tool and workpiece, during machining. The final shape of the workpiece is determined by the geometry of the tool, of which the cutting edge geometry can either be defined or undefined, and by the relative motion between tool and workpiece.

The following parameters are significant for cutting processes; cutting speed, feed, cutting depth, workpiece material, cutting tool material, tool geometry, and cooling conditions.

1.1 Cutting with a Geometrically Defined Cutting Edge

Cutting with a geometrically defined cutting edge is the removal of material using a tool with a defined number of cutting edges, a defined geometry of the wedge, and a defined position of the cutting edges. Such cutting processes are classified according to geometry of the target surface or according to the kinematics of tool and workpiece. Some of the processes, most of those applied in industry, are turning, milling, and drilling.

Cutting processes allow machining of a wide range of materials, most of which are metals and polymers. In certain cases machining of brittle hard materials is also possible. Furthermore, the ongoing developments in the field of cutting materials and machine tools lead to an increase in productivity and new possibilities to machine innovative materials. Even nickel-based alloys that are difficult to machine can be machined. This is done at cutting speeds of $\sim 30\,\mathrm{m\,min^{-1}}$ using hard metal as cutting tool material. An increase in cutting speed of up to 150 m $\mathrm{min^{-1}}$ can be achieved using crystalline boron nitride (CBN) as cutting tool material. Plastics can also be machined using cutting. More and more fiber-reinforced plastics are now being applied because of their high specific strength and stability, as well as other positive characteristics. In spite of problems occurring due to the bleeding of fibers or due to the low thermal conductivity of fiber-reinforced plastics, these materials can be machined best with polycrystalline diamond. In this case, the advantages of diamond, such as the high abrasive wear and high thermal conductivity, can be implemented.

(a) Cutting tool materials and their application. Cutting tool materials have a decisive influence on the cutting process. They have an impact on tool life and, thus, on manufacturing times and costs for tools, machines, and labor. Due to a direct contact with the workpiece, cutting materials are exposed to high mechanical and thermal loads during the machining process. The demands placed on cutting materials can be summarized as follows: hardness and compressive strength, bending strength and toughness, edge strength, inner bonding strength, heat resistance, oxidation resistance, low tendency towards diffusion and adhesiveness, abrasion resistance, and reproducible wear behavior.

The development of new cutting tool materials in the cutting manufacturing process has lead to a rapid increase in cutting speeds, and hence to an improved efficiency (Fig. 1).

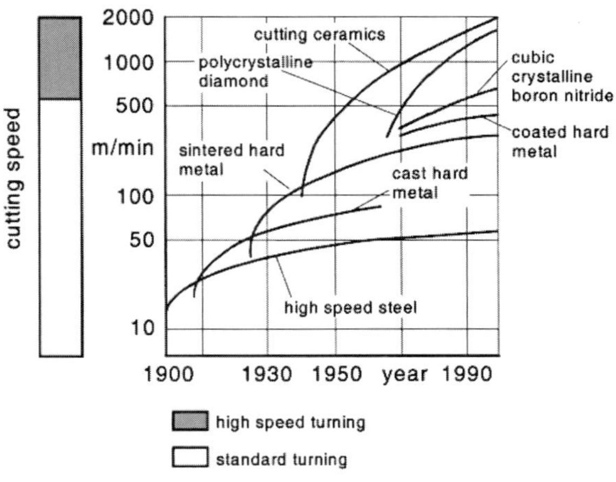

Figure 1
Development of cutting speeds during turning of different cutting materials.

In the following, the most important cutting tool materials, their specific characteristics and their main fields of application are stated.

Tool steels. Tool steels are classified into non-alloyed and alloyed. By adding carbide forming alloying elements, the wear resistance, retention of hardness, and heat resistance can be increased. In industrial manufacturing with modern machine tools, tool steels no longer play an important role because of their comparatively low red hardness of ~200–300°C. They are mainly used for machining steel with manual tools with low cutting speeds.

High-speed steels. High-speed steels are high-alloy steels composed of tempered martensite as base material and embedded double carbide based on molybdenum and tungsten. Further alloying elements, dissolved in the matrix, are: carbon to increase the hardness and the wear resistance; chromium for an increase in the through-hardening capacity; molybdenum, cobalt, and tungsten to increase the heat resistance, tempering resistance, and toughness; and vanadium in order to improve the wear behavior. Currently, many high-speed steels coated with hard materials such as titanium nitride and titanium carbide are used to improve the wear characteristics and to prolong the tool life. High-speed steels are applied for the machining of steel and casting alloys with high material removal rates.

Hard metals. Hard metals are sintered material systems including metal carbides such as tungsten carbide, titanium carbide, tantalum carbide, or niobium carbide as hard particles and a binder metal determining the toughness such as cobalt, nickel, or molybdenum. Depending on the composition of the alloying elements, hard metals can be applied mainly for the machining of die long-chipping steels, ferrous casting alloys, chilled casting, hardened steel, nonferrous metals, and plastics. As in the case of high-speed steels, coated hard metals are also applied to increase the wear resistance and chemical resistance as well as to decrease the cutting temperature.

Ceramics. Cutting ceramics are cutting tool materials made of aluminum oxide or silicon nitride. With the use of cutting ceramics, cutting speed and tool life can be still further increased. The main field of application for cutting ceramics is the machining of cast iron, case hardening steels, and heat treatable steels. Apart from a good wear pattern, they show a high heat resistance, a low tendency to diffuse, and a high oxidation resistance. Cutting ceramics allow machining of steels with a hardness of up to 64 HRC. Disadvantages of cutting ceramics are the small bending strength and the high sensitivity to shock loads and thermal shocks in the case of aluminum oxide, and the chemical affinity for iron in the case of silicon nitride.

Diamond. Diamond is the hardest cutting tool material which it can be applied in either the monocrystalline or the polycrystalline state. Monocrystalline diamond is used for fine machining and polycrystalline diamond enables machining with higher cutting depths. The typical applications of diamond tools are the machining of nonferrous metals, plastics, and other nonmetal materials (e.g., hard rubber, graphite, ceramics, glass, stone, and asbestos).

In the automobile industry polycrystalline diamonds are increasingly applied for machining of Al–Si alloys. Due to the affinity of carbon for ferrous materials, diamond cannot be used to machine these metals.

Boron nitride. Next to diamond, cubical CBN is the hardest cutting tool material. It has a relatively high tensile strength and bending strength, a high thermal stability, and a high thermal resistance to iron. The superiority of CBN over other cutting tool materials such as hard metals and high-speed steels is obvious during the machining of steel at a hardness of 54–68 HRC or difficult-to-machine materials such as high-temperature cobalt and nickel-based alloys. By machining with CBN, surface qualities equivalent to those of fine machining processes can be realized.

(*b*) *Turning.* Turning is cutting with a geometrically defined cutting edge with a close, mostly circular cutting motion and a random feed, which is mostly diagonal to the direction of the cut. During turning, the cutting motion is carried out by the workpiece, while auxiliary motions such as the feed and the engagement are carried out by the tool. Generally, cylindrical turning, face turning, and surface turning are distinguished. Turning is used for the production of rotationally symmetrical components of all kind, including drillings and tapped holes.

In most cases, turning is a two-step process. During the first step, termed *roughing*, as much material as possible is machined, in order to achieve a high accuracy of IT 7 to IT 8 (IT: tolerance class according to ISO 286); in the second step, termed the *finish turning* under optimum conditions, an accuracy of IT 6 can be achieved. The surface roughnesses obtained during turning are in a range of 1.6–6.3 μm.

(*c*) *Drilling.* Drilling is cutting with a geometrically defined cutting edge with a circular cutting motion, with the axis of revolution of the tool and the axis of the produced inner surface being one and the same. The feed motion follows the direction of the axes. Fields of application for drilling include drillings in solid material, counterboring of existing drillings, sinking for the production of rotationally symmetrical geometries on workpiece surfaces, screwing, and rubbing to obtain high surface quality and accuracy.

Drilling with a twist drill can be considered as roughing. The obtained accuracy is in a range between IT 11 and IT 12. The surface roughness is ~80 μm. An improved surface quality and better accuracy can be achieved with finishing processes like rubbing, counterboring, and precision drilling. During finishing, it is possible to obtain an accuracy of up to IT 7 and a peak-to-valley height of up to 4 μm.

(*d*) *Milling*. Milling is cutting with a geometrically defined cutting edge with a multi-tooth tool. The circular cutting motion is carried out by the tool. The feed motion is also carried out by the tool and is either vertical or in a raking angle to the axis of revolution of the tool. Milling is applied for the machining of prismatic components. In milling, it is possible to select almost any surface geometry of the workpiece to be produced, even a machining of sculptured surfaces is possible.

Depending on the nature of the applied milling process, the accuracy obtained during milling is between IT 6 and IT 8 at a surface roughness of 10–30 μm. The best work results can be obtained using front milling. The accuracy achieved with this process is IT 6 and the surface roughness 10 μm.

1.2 Cutting with a Geometrically Undefined Cutting Edge

Cutting with a geometrically undefined cutting edge is the removal of material using a tool, of which the number of cutting edges, the geometry of the wedge, and the position of the cutting edges are not defined.

Machining processes with a geometrically undefined cutting edge are processes in which high surface quality and high dimensional accuracy can be obtained. So far, these processes have mainly been applied in finishing. However, they may also represent an economic alternative for certain machining tasks such as the manufacturing of narrow, deep grooves of hardened components, since the obtainable material removal rates for these applications are higher than those of other cutting processes. The group of machining processes with a geometrically undefined cutting edge comprises basically honing, lapping, and grinding. Honing is a technology for finishing, especially short stroke honing, or for the manufacturing of highly precise geometries, particularly long-stroke honing. Applications for honing are the machining of connecting rod eyes, cylinder liners, rolling and ball bearings, crank shafts, piston pins, and the finishing of cog wheels. During lapping, all materials can be machined and highly precise geometries can be obtained with isotropic surface structures. Lapping is for instance applied to manufacture ceramics and conventional roller bearings, ceramic sealing faces on control elements, and gauge blocks as well as read/write heads of hard disk drives.

Grinding is the process within this group which is mostly applied in industry. Grinding is cutting with a geometrically undefined cutting edge using a multi-cutting edge tool. In this process, the geometrically undefined cutting edges are formed by a number of bound abrasive grains from natural or artificial abrasives, such as Al_2O_3, SiC, CBN, or diamond. The material is removed with cutting speeds up to 200 m s^{-1} in order to improve or change the shape, dimensional accuracy, position, and surface of the workpiece.

Grinding is mainly applied for machining of hardened materials with high demands on the surface quality (Table 1). Even brittle, hard materials, such as ceramics, can be machined. The properties of ceramics such as high hardness, wear resistance, and lack of ductility restrict the number of potential processes considerably, when short tool lives, poor surface qualities, and subsurface damage widely exclude machining processes such as turning, milling, and drilling.

The cutting speed, feed speed and feed are important kinematic parameters of the grinding process. An increase in cutting speed leads to lower grinding forces and better surface qualities, but also to a higher temperature in the contact area and, hence, to an increase of subsurface damage. In order to avoid these disadvantages, more effort has to be made concerning cooling. An increase in the speed with constant parameters shortens the cutting time and results in higher grinding forces, a higher wear, and an increase in temperature. The workpiece subsequently shows lower accuracy and lower surface quality as well as higher subsurface damage.

2. *Nonconventional Machining Processes*

2.1 Electric Discharge Machining

Electric discharge machining (EDM) is a manufacturing process that enables the machining of all electrically conductive materials independent of their hardness or strength. The main application for EDM is the manufacturing of complex 3D geometries, especially the manufacturing of dies and molds. The process is based on the erosive effect of electrical discharges that are ignited between the tool and

Table 1
Obtainable accuracy during grinding.

	Dimensional accuracy	Surface quality R_t (μm)
Surface grinding	IT 8–IT 9	3–8
Profile grinding	IT 4–IT 5	2–4
External grinding	IT 6–IT 8	5–10
Internal grinding	IT 8–IT 10	10–20
Centerless grinding	IT 4–IT 6	2–4

workpiece electrode. Every single spark melts and evaporates a microscopic part of the material which is removed by mechanical forces during the collapse of the discharge channel. An eroded surface is, hence, a superposition of discharge craters, whose topography has a direct influence on the global surface formation (Fig. 2).

Depending on the amount of the transferred energy, the thermal exposure of the machined surface also leads to a thermally influenced subsurface zone. During extremely high discharge energies, the influence on the subsurface can reach up to several millimeters into the bulk material. The use of high frequency generators minimizes the influence on the subsurface and enables a surface roughness down to an arithmetical average peak-to-valley height of $R_a = 0.1\,\mu m$.

A disadvantage of EDM is the technology related multistage process. In the first step, an electrode with the desired geometry has to be machined. In the next step, the desired geometry is created by the shape of the tool into the material at a high erosive power. This is followed by several finishing steps at a successively diminished discharge energy.

Furthermore, difficulties arise concerning productivity and quality during EDM machining of nonmetals such as electrically conductive advanced ceramics and polycrystalline diamond. Due to the microscopic inhomogeneity of these materials, there is an attack on grain boundaries or crystal boundaries during the machining with low discharge energies, which leads to a deterioration of the surface quality.

2.2 Laser Beam Cutting

During laser beam cutting, the material is locally heated by transferring the energy of the laser beam to the interacting zone. Using the cutting gas flow from the gas nozzle the material is disposed of as vapor, melting charge, or slag either directly or after a reaction with the cutting gas. In comparison with other thermal cutting processes, the edge in laser beam cutting is marked by almost vertical generated surfaces, a low surface roughness, and a small heat affected zone in the ranges along the edge.

In this field of technology, the process types of laser beam torch cutting, laser beam melt cutting, and laser beam sublimation cutting are distinguished. A characteristic of laser beam sublimation cutting is the removal of material in the gaseous state, of laser beam melt cutting in the liquid state, and of torch cutting as slag after an exothermal reaction with oxygen. The additional energy allows the increase of the cutting speed up to eight-fold higher compared to melt cutting. Laser beam melt cutting is applied to produce an oxide-free cutting edge as is necessary for the machining of aluminum sheets and high-alloy steels.

The absorption rate is decisive for the laser beam cutting process. Depending on the laser source used, certain materials (e.g., copper) have an extremely low absorption rate concerning certain laser beams. The main influences on laser beam cutting are shown in Fig. 3.

Lasers are industrially used for machining paper, wood, rubber, and plastic, and, furthermore, iron and nonferrous metals and non-metals of any hardness such as glass and ceramics. Economic machining is possible up to the following maximum values, steels 20 mm, high-alloy steels 10 mm, nonferrous metals 6 mm, ceramics 6 mm, plastics 50 mm, and composites 30 mm. The cutting speeds that are significant for industrial manufacture cannot be considered independent of the applied laser power and

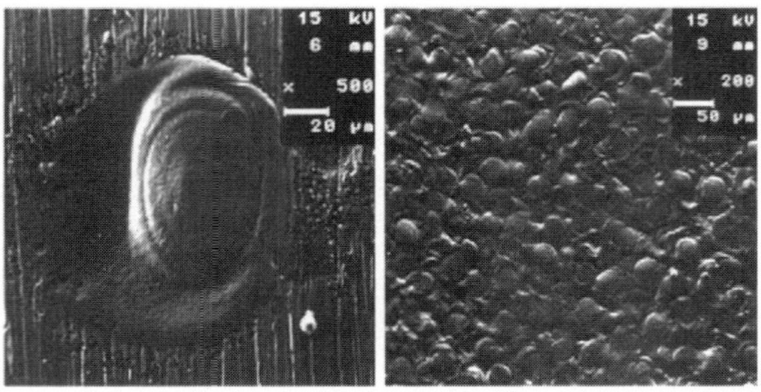

Figure 2
Single discharge crater and eroded surface of a hot-work die steel.

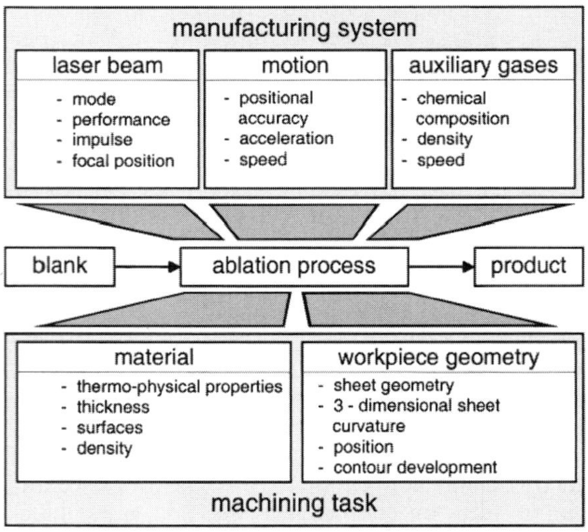

Figure 3
Influence on laser beam cutting.

Table 2
Cutting powers for a CO_2-laser for different materials.

Material	Workpiece thickness (mm)	Laser power (W)	Cutting power ($m\ min^{-1}$)
Structural steel	2.0	1000	9.00
Structural steel	10.0	1000	0.60
High-alloy steel	2.0	1000	3.50
High-alloy steel	8.0	1000	0.50
Aluminum	2.0	1000	1.20
Aluminum	6.3	2000	0.75
Copper	2.0	2000	0.50
Copper	3.0	2000	0.25
Polyethylene	2.0	500	10.00
Polyethylene	15.0	500	0.50
Polyvinyl chloride	3.0	400	6.00
Polyvinyl chloride	10.0	500	2.00

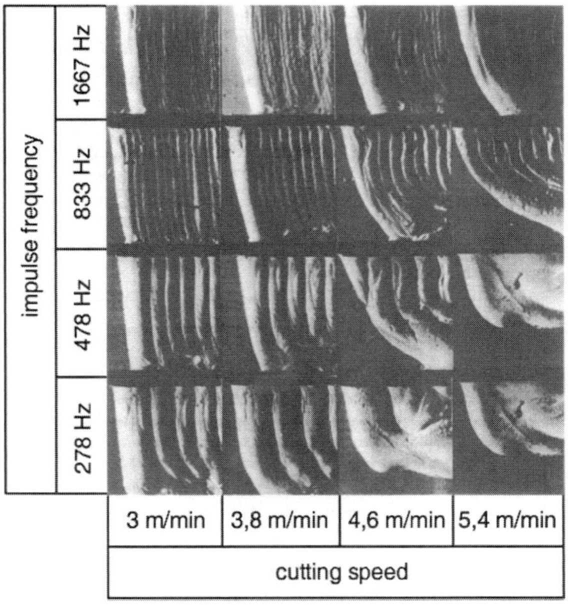

Figure 4
Formation of fronts during the separation of mild steel at a sheet thickness of 1 mm and a laser power of 750 W.

the thickness of the material. Table 2 provides a few reference values for the use of a CO_2-laser.

The typical course of a cutting edge of mild steel with a laser power of 750 W is displayed in Fig. 4. The best surface qualities can be obtained at higher pulse frequencies or especially during continuous wave operation.

In comparison to other cutting technologies, laser beam cutting has the following advantages: no tool wear because of contact free machining and no refurbishing; manufacturing of workpieces without further need of finishing; no mechanical load on the workpieces during machining, machining of materials independent of their hardness and solidity; high accuracy; high machining speed, good reproducibility and low thermal load on the material; and high flexibility and efficiency due to an optimum economic efficiency arising from the possibility of optimum automation.

2.3 Water Jet Cutting

Water jet cutting is characterized by fluid jet systems which are accelerated and forced to collide on the workpiece surface. The machining performance is enhanced by adding abrasive particles to the water jet. The application range for water jet cutting comprises the cutting of a variety of materials, starting with soft materials such as leather, plastics, or paper to metals, glass, and stone. A pressure of up to 1000 MPa is used to accelerate water through a nozzle with a diameter of 0.1–0.5 mm. In practice, however, mostly pressures of approximately 400 MPa are used. By using water jet cutting, steel parts can be machined with a cutting speed of 90 mm min^{-1}, aluminum parts with 150 mm min^{-1} and fiber-reinforced plastic parts with 300 mm min^{-1}.

The material removal effect of water jet cutting is due to the effect of the kinematic energy in the beam on the workpiece surface. The liquid flow is dammed up, hence forces are liberated in the area of contact with the workpiece to exert mechanical stresses on the material. Depending on the fragmentation characteristics of the material, the removal either takes place by divergence, transmission and union of micro fissures, or by pinching off plastically deformed material particles.

3. Concluding Remarks

A comparison of different manufacturing processes cannot be limited to quantifiable values, such as the obtainable accuracy or the surface quality of the finished part. Other boundary conditions and considerations influence the decision for the use of a process. Most attention is paid to technological aspects. Initially, the question arises as to whether it is possible to machine a certain material using a certain technology. Eroding of dielectric materials, for example, can only be used for machining of electrically conductive materials.

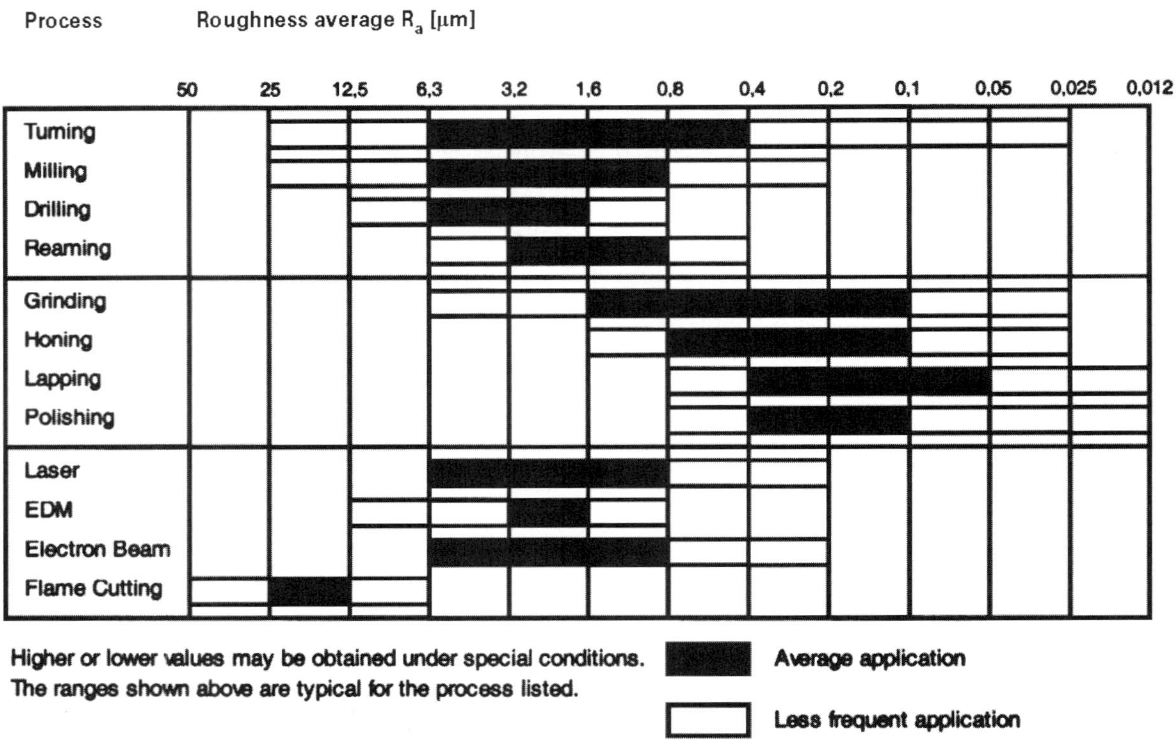

Figure 5
Surface quality with different manufacturing processes.

There are further technological characteristics, such as workpiece damage caused by the machining, residual stresses, structure, or compulsory surface structures. Hence, the comparison of different manufacturing processes is only possible to a limited extent, as not every manufacturing process is appropriate for certain machining operations. An important criterion for the selection of a process is the obtainable surface quality. An allocation of the obtainable surface qualities and the selected manufacturing processes is displayed in Fig. 5. In order to compare different manufacturing processes, further factors have to be considered, such as economic efficiency, material-linked factors, environmental protection, and industrial safety.

Bibliography

Chryssolouris G 1991 *Laser Machining, Theory and Practice*. Springer, New York

Guitrau E P 1997 *The EDM Handbook*. Hanser Gardner Publications, Cincinnati, OH

König W, Klocke F *Fertigungsverfahren Band 2: Schleifen, Honen, Läppen*. VDI, Düsseldorf, Germany

Shaw M C *Principles of Abrasive Processing*. Clarendon, Oxford

Spur G *Handbuch der Fertigungstechnik Band 3/1. Spanen*. Hanser, Munich, Germany

Uhlmann E *Tiefschleifen hochfester keramischer Werkstoffe*. Hanser, Munich, Germany

E. Uhlmann and T. Frost

Cutting Fluids for Machining

Cutting fluids are important process aids in the manufacture of nearly all metal components and end products. The fluids are applied to the interface between a cutting tool and the metal workpiece to provide friction reduction, heat removal, and flushing of the removed metal debris. The fluid may also provide corrosion protection for machinery and the workpiece. These functions of cutting fluids can have a direct effect on tool life and the machined component quality. The effective use of cutting fluids is an important factor in the productivity of highly automated machining operations that can typically produce millions of components per year.

The development and widespread use of cutting fluids for machining began with the Industrial Revolution and mass production machining practices. A historical overview of cutting fluid development is given in Brown (1999 pp. 627–31). Since the beginning of the twentieth century the growth of the petrochemical industry has provided an array of raw materials with which cutting fluids can be formulated. The availability of these materials has given cutting fluid manufacturers the ability to produce a myriad of products, each formulated with the goal of providing machining performance and cost advantages. A highly competitive cutting fluid industry has evolved with a diversity of products and manufacturers, driven by the large economic incentives for improving production machining performance and costs. Beginning in the 1960s, scientists started to recognize concerns with worker exposure and ecological effects of cutting fluids.

The diversity of products and manufacturers has resulted in challenges for those charged with the selection and optimization of cutting fluids. Fluid formulations are typically considered to be proprietary by manufacturers, making it difficult to develop an understanding of the relationship between fluid compositions and machining performance. Environmental and worker health and safety concerns have also driven the need to develop technical information on the selection and use of cutting fluids. Regulations in North America and Western Europe require the disclosure of components in cutting fluids if they are known to be health, safety, or environmental hazards. Some large end users have taken a further step to require a full disclosure of the formulation ingredients by suppliers. Before the fluids can be used, they are subjected to proprietary internal toxicological and environmental compliance reviews.

Three major factors will probably shape cutting fluid development requirements for the foreseeable future: machining performance, environmental conservation, and worker health and safety effects.

1. Selection of Cutting Fluids

Cutting fluids are a complex blend of components that must meet a number of criteria and function under a wide range of conditions. Historical information often plays a major role in the decision of which type of fluid to use for a machining process. The foremost consideration must be compliance with health, safety, and environmental testing and regulations (National Center for Manufacturing Sciences [NCMS] 1997, Section 1). Provided these compliances are met, appropriate laboratory and machining tests are used to evaluate a fluid for compatibility with equipment and workpiece materials, maintainability characteristics, and machining performance. The available bench tests in these areas can often be inadequate predictors of fluid performance in the plant environment. Plant trials are still the ultimate test of a fluid, but are time consuming, expensive, and difficult to conduct in a controlled fashion. The potential for shutting down a critical machining line is also an unpleasant and costly possibility. A reliable protocol for screening fluids prior to full-scale plant trials is desirable.

1.1 New Product Testing and Specification

There are few industry standard specifications, based on tests conducted on the unused fluid, available for reliably defining fluid performance in the plant environment. Fluids that have proven performance in the same or similar production operations are often be used to set minimum relative performance benchmarks. The (former) American Automobile Manufacturers Association (AAMA) developed a standard that defines several key performance parameters and identifies several other parameters for which test methods are needed for defining fluid performance (AAMA 1998). Where meaningful standard tests exist, parameters include rust and corrosion prevention, foaming tendency, thermal stability, extreme pressure properties, seal compatibility, physical properties (flash point, volatility), emulsion stability, friction-reducing properties, filterability, and various chemical properties. Other key parameters not adequately defined by industry standard tests include machining performance, misting tendency, bioresistance, and gumming tendency.

The NCMS has published a guide for fluid evaluation (NCMS 1997). The guide covers a broad scope of considerations in the selection and use of cutting fluids, including health and safety, environmental, system compatibility, machining performance, and in-process maintenance considerations. This document also includes a compendium of standard and nonstandard criteria test methods.

Some end users are trying to develop tests to predict reliably machining performance and misting tendency. Dasch *et al.* (1998) have described the construction and function of a small, dedicated machining center for the purpose of evaluating cutting fluids. This apparatus is equipped with extensive monitoring and data acquisition equipment. Air monitoring equipment has been used to measure fluid volatile organic carbon (VOC) and mist emissions. Horsepower, forces, torque, vibration, acoustic emissions, tool wear, and workpiece surface finish are measured. The fluid bath condition (pH, concentration, alkalinity, and biological activity) is regularly monitored. This machining center has been used with reasonable success to discriminate among several candidate fluids. It is unlikely that a given standard machining test with a fixed passing limit will be acceptable for predicting the plant performance of a given fluid. Rather, tests conducted on materials identical to those used in the plant, at spindle speeds and metal removal rates similar to those used in production could be very helpful in rank ordering candidate fluids, particularly when the plant fluid is used as a reference. If a candidate fluid shows lower forces and better tool life than the plant fluid, plant trials on this fluid may well be justified. Fluids that have demonstrated required performance in all of these initial tests should be evaluated in an in-production machining trial as a final step in the test regimen.

1.2 Cutting Fluid Compositions

There are three basic types of cutting fluids: straight oils, emulsions, and synthetics (chemical solutions). Each of these types can be further categorized based on the requirements and severity of the machining application for which they are formulated. The categories are typically based on the presence of additives for friction reduction and extreme pressure lubrication. For example, the reader is referred to a specification of cutting fluids based on formulated properties (AAMA 1998).

(a) Additives. A number of additives have been developed to enhance the characteristics and performance of cutting fluids. The additives include corrosion inhibitors, boundary lubricants, extreme pressure lubricants, biocides, antifoams, emulsifiers, dispersants, and viscosity index modifiers. Some of these additives are discussed below, relevant to the different types of cutting fluids. However, a detailed discussion of the myriad of cutting fluid additives is not possible within the scope of this article. Additives used to enhance the lubrication properties of cutting fluids deserve special mention, as they are important in understanding the formulation and classification of cutting fluids based on machining requirements.

Boundary lubrication is an important mechanism for reducing friction and the resultant heat and forces at the tool–workpiece interface. In this type of lubrication, a molecular film forms and adheres to the surfaces by physical or chemical adsorption. Boundary additives are organic compounds with strong dipole moments that interact with the fresh metal surfaces created during machining. Additives include fatty acids, alcohols, amines, and esters. These are usually derived from animal or vegetable sources and have long been used as lubricity additives in cutting fluids. Concerns associated with the use of some of these additives and the maintenance of water-based fluids are discussed in Sect. 3. Synthetic polyalkylene glycols (and their ester derivatives) have been investigated as alternative lubricity additives in cutting fluids (Brown 1999).

Extreme pressure (EP) additives are different from boundary lubricants in that they chemically react with tool and workpiece fresh metallic surfaces at elevated temperatures during the cutting event. The reaction products form low-shear-strength films that isolate the tool from the workpiece and chip, thereby reducing friction and wear. Three different types of EP additives are in common use based on sulfur, chlorine, and phosphorus chemical compounds. The development and use of EP additives for metal cutting operations has been common with similar additives for other types of mechanical (gear lubricants) and metalworking (draw lubricants) operations. Technical evidence and experience suggests that these compounds function by the same basic chemical mechanism when machining iron-based alloys: the compounds react in the cutting zone to form iron sulfides, chlorides, and phosphates lubricating film deposits. These film deposits have different temperature domains where they are effective. In addition, metals other than iron-based ones may not react and form similar lubricating films. The choice of additive or combination of these additives is dependent on the severity of the machining process and the workpiece material in question. Lubrication in metal cutting operations involves different mechanisms, depending on the chemistry of the tool–workpiece interface and the mechanics of the machining operation. In addition, the lubrication mechanism can vary significantly during a machining cycle depending on the cutting speeds and forces involved.

Sulfur EP additives are typically classified as reactive and nonreactive based on their ability to react with copper and form corrosion and staining under specified temperature test conditions (AAMA 1998). Reactive additives readily react with copper at 38°C, while the more strongly combined forms of sulfur are essentially nonreactive under these conditions. Reactive additives include elemental sulfur, sulfurized fats, and olefins containing labile polysulfide moieties. Nonreactive additives include monosulfides and disulfides. In general, the use of active sulfur compounds is limited in metal cutting operations to prevent corrosion and staining of the machine tool or workpiece away from the cutting zone. In addition, active sulfur compounds tend to produce strong odors and are objectionable for the workplace environment. The heat created at the chip–tool interface is usually sufficient to create the EP effect with nonreactive sulfur additives, making these additives preferable in metal cutting applications. Sulfur additives are commonly used in severe machining operations using straight oil cutting fluids where the chip–tool interface reaches high temperatures. The sulfur EP effect is believed to be effective from about 200°C to about 500°C.

Chlorine additives are typically chlorinated paraffins, olefins, and esters. In general, they are more reactive than "nonreactive" sulfur additives. They typically form EP lubricating films from about 100°C up to about 400°C. Sulfochlorinated additive hydrocarbon and ester compounds have been developed to combine the EP effects of sulfur and chlorine in one additive package. These additives can function over a range of temperatures that may occur in a typical machining cycle. Concerns do exist regarding the use of chlorinated additives. Additives that contain free chloride ionic residue or react to produce free chloride away from the tool–chip interface can be highly corrosive. Evaluating fluids containing these types of additives for corrosion effects should be a consideration (see Sects. 1 and 2).

Phosphorus compounds are used as EP additives, usually in the form of phosphate organic esters. They are typically effective in the range 60–300°C. Phosphate esters do find application in heavy-duty straight oil cutting fluids. They are less often used in water-based fluids as they can be hydrolytically unstable and have undesirable effects on the fluid properties. Phosphate esters can also be a concern from a health and environmental standpoint (Marino and Placek 1999).

(*b*) *Straight oils.* Straight oil cutting fluids are usually derived from several viscosity grades of severely solvent refined or severely hydrotreated petroleum oils. Occasionally, oils derived from animal, marine, or vegetable sources or synthetic lubricants are also formulated into straight oils. These products are not designed to be diluted with water and are used when the machining application requires a high degree of lubrication, and cooling of the workpiece and tooling are less important. Broaching and tapping operations done at low speed with large cutting forces are examples of operations where straight oils would be used. Low-viscosity oils are typically used in light-duty applications such as polishing, honing, finish reaming, and grinding applications where dimensional sizing and surface finishes are critical.

Products designed for light-duty applications may contain little or no additives beyond dyes or scenting agents. Those designed for more severe applications contain substantial amounts of boundary and EP additives. Additives to neat oils normally involve sulfurized or chlorinated compounds dissolved in an oil carrier. Fats or lard oils and phosphate esters are also added for enhanced performance.

(*c*) *Emulsions.* Emulsions are stabilized mixtures of oil and water designed to combine the strong heat extraction properties of water with the lubricating qualities of oils. Emulsions include soluble oils and semisynthetic types that are formulated as oily concentrates. These concentrates are diluted with water; typically dilutions from 10 to 40 parts of water to one part concentrate are used in metal cutting and grinding operations. Soluble oils contain a high content of mineral oil and form a milky white fluid when diluted with water for use. Semisynthetics contain lesser amounts of mineral oil with a higher proportion of surfactants and other synthetic additives. They are typically translucent when diluted with water for use. Surfactants (emulsifiers) are added to provide emulsion stability. Sodium sulfonates, alkyl ethoxylated phenols, fatty amides, and sodium salts of fatty acids (soaps) are commonly used. Such compounds are molecular species having a linear structure in which one end is hydrophilic and the other lipophilic. When present in an oil–water mixture, the surfactants are aligned on the surface of the oil droplets and effectively form a transition at the oil-water interface. Another important aspect of this configuration is that the polar molecules produce an electrostatic repulsive charge between oil droplets. This repulsive charge interferes with the normal coalescence process by which oil droplets increase in size and lose the ability to remain suspended.

Typically, the emulsifiers used in emulsions also function to some degree as corrosion inhibitors. Some products contain additional film corrosion inhibitors such as alkanol amines, fatty acid amides, amine salts of boric acid, and triazoles (nonferrous corrosion protection).

Boundary and EP additives for increased performance in severe operations are dissolved in the oil phase prior to emulsification in much the same manner and proportions as previously described for straight oils. These additives provide essentially the same functions but with the dramatically improved heat extraction properties afforded by the water phase.

Glycol coupling solvents, chelating agents (e.g., a salt of ethylenediamine tetraacetic acid (EDTA)), defoamers, and biocides are also commonly included.

(*d*) *Synthetics.* Synthetics are also known as chemical solutions. The common definition requires that they contain no petroleum oil and are formulated with chemicals that are soluble in water and form clear solutions. Synthetic fluids are formulated as water-based concentrates. These concentrates are diluted with water; typically dilutions from 10 to 40 parts of water to one part concentrate are used in metal cutting and grinding operations. Typical synthetic concentrate formulations may include polyalkylene glycols, amides, esters, phosphate esters, and sulfurized fatty acids as lubricants. Amine salts of carboxylic and boric acids, and alkanolamines are typically included as corrosion inhibitors. Glycol coupling solvents, chelating agents (e.g., EDTA, NTA), defoamers, and biocides are also commonly included.

1.3 Health, Safety, and Environmental Considerations

There are several environmental and health and safety concerns associated with the use of cutting fluids (Organization Resource Counselors [ORC] 1999). Exposure to cutting fluids has been implicated in worker health concerns, including dermatitis, respiratory problems, and cancer. However, the changing nature of the fluids and the relative levels of worker exposure confound many of these health effects studies over time (the reader is referred to several articles on health effects in AAMA 1998). Since the 1960s growing knowledge and emphasis in the health and environmental sciences has driven these changes. For example, the use of severely refined petroleum oils in all cutting fluid applications (including straight oils as well as emulsions) has become a standard requirement to eliminate carcinogenic polycyclic aromatic compounds (Howell *et al.* 1996, Skisak 1996, AAMA 1998 pp. 8, 11). The following discusses some of the other effects health and safety and environmental drivers have had on cutting fluid practices to the end of the 1990s.

(*a*) *Health and safety.* Dermatitis is probably the most common worker ailment. A common mode of action by cutting fluids is simply the removal of oils from the skin or "de-fatting." The incidence of worker dermatitis can be minimized by the use of barrier skin creams, impervious gloves, and means to avoid direct skin and clothing contact with cutting fluids.

The generation of smoke and fluid aerosol (mist) from the use of cutting fluids in a machining facility is a primary mechanism for worker exposure and has resulted in concerns with respiratory health effects in workers (ORC 1999). Effective mist and smoke containment and ventilation engineering controls can be designed as part of the machining equipment. In addition, mist-suppressant additives have been developed (ORC 1999). As a result, the relative mist concentration in plants is as much as an order of magnitude lower than it was in the early 1980s. The study of health effects associated with the inhalation of cutting fluid aerosols harboring microbial material has become an active area of research (Kreiss 1998, Milton 1998). Proposed regulations are directing efforts to further reduce worker exposure to fluid mist, and formulation of rules by the US Occupational Safety and Heath Administration will likely set the permissible exposure limit (PEL) for mist in the range 0.5–1.0 $mg m^{-3}$ early in the twenty-first century.

The use of straight oil fluids poses special concerns. Special fire hazard control equipment is usually required at machining operations using straight oils. The large amount of carry-off on workpieces and chips and the covering of machining equipment also result in increased fire hazards, slip-fall injuries, worker skin exposure, and cleaning requirements. There will continue to be significant motivation to find alternative cutting fluids and machining process methods to avoid the use of straight oil cutting fluids.

Specific types of chlorinated paraffins have been eliminated from use in cutting fluids based on evidence that they are potential carcinogens and toxic to the environment (Howell *et al.* 1996, p. 16).

In the late 1960s *N*-nitrosoamines, known animal carcinogens, were first discovered in some water-based fluids. It was determined that the nitrosoamines were formed from primary and secondary amine additives (e.g., diethanolamine) and sodium nitrite, a commonly used corrosion inhibitor present in these fluids. Subsequently, the use of nitrite inhibitors in cutting fluids was curtailed (Howell *et al.* 1996).

Water-based fluids are often treated with biocides during use to provide protection from microbial degradation. Depending on the nature of the fluid formulation and in-use maintenance practices, biodeterioration can cause a number of workplace concerns including emulsion destabilization, slime production, malodors and worker respiratory complaints. Biocides used to control biodeterioration are toxic by design and there can be health concerns associated with their handling and misuse (Passman 1996).

(*b*) *Waste treatment and disposal.* Straight oils can provide a long service life provided proper maintenance and filtration practices are used (see Sect. 3). During use, containment and handling procedures are required that segregate the oil and oil-contaminated materials (metal chips/swarf, filtration media) to prevent soil and groundwater contamination and to facilitate effective oil reclamation. Oxidation and contamination with tramp oils, water, and particles can eventually degrade the machining performance properties of the oil. Waste cutting oils are largely reclaimed using collection and re-refining procedures similar to those used for waste motor and lubrication oils. Waste oil residuals that cannot be effectively reclaimed are typically incinerated as a fuel oil. Precautions are required to prevent the contamination of waste oils during storage and transport that may make the oil a hazardous waste. In particular, contamination with polychlorinated biphenyls (PCBs) and other hazardous chlorinated organics can be a concern. The disposal of cutting oils containing chlorinated EP additives requires special consideration. In some European countries, all waste oils containing significant concentrations of chlorinated organics, including EP additives can be considered hazardous. In the USA and other countries, it may be necessary to document by sourcing and chemical analysis that only nonhazardous forms of chlorinated paraffins and other chlorinated additives are present to avoid hazardous waste classification. In addition, oils containing such expensive additives are now more often reclaimed for reuse at considerable cost savings over virgin fluids.

The disposal of water-based fluids has become an important consideration in their selection and use. Environmental conservation and wastewater treatment standards can be expected to drive the need for fluids with advantages in long service life (see Sect. 3) and ultimate waste treatment and disposal. Five methods are typically used for the waste minimization and disposal of spent water-based fluids: contract waste hauling, evaporation, chemical treatment, membrane separation, and biological treatment (ORC 1999). The selection of disposal method(s) depends on several factors, including disposal costs, the size and location of the machining plant, the size and capabilities of publicly owned wastewater treatment works, and local government waste discharge regulations (NCMS 1997). The type of disposal method and regulations can dictate the nature and selection of fluids. For example, synthetic fluids, certain types of emulsions, and fluids containing chelating agents may not be amenable to chemical treatment methods for meeting wastewater discharge regulations.

2. *Criteria for In-production Evaluation*

As mentioned above, in-production evaluation must be considered the critical and final step in the cutting fluid selection and performance-evaluation process. There are substantial financial and productivity improvement opportunities to be achieved through the effective selection and implementation of a cutting fluid system. Returns on investment can often exceed factors of three or four in virtually any facility, regardless of the size or the level of automation involved. Direct and indirect cost data are necessary to establish a baseline as well as provide a structure for cost comparisons with cutting fluid system modifications. Generally, the direct costs are readily defined, including fluid purchase, sump-side chemical maintenance additions (e.g., biocides, pH adjusters, emulsifiers, lubricity agents), and disposal of the spent fluid. However, indirect costs associated with fluid performance may be the most significant, but difficult to quantify in a plant environment. Indirect costs can be associated with part production, machining equipment, and the fluid handling and distribution system.

2.1 Production Evaluation

The critical function of a fluid is as a process aid in the manufacture of parts within quality tolerances at production rates of metal removal. The dimensional and surface finish qualities are directly affected by tool wear and chip morphology, and these in turn can be directly affected by cutting fluid performance. It can be a difficult task to distinguish and quantify cutting fluid effects in the presence of other important variables affecting tool performance, including tool geometry and alignment, machining speed and feedrate, and workpiece material variation. However, attributing indirect expenses associated with tool purchasing and reconditioning costs and production downtime for tool change and fluid replacement can be a very significant item for fluid evaluation.

Cutting fluids are often required to provide corrosion and staining protection to the workpiece and machining equipment. Indirect costs associated with part scrapping and rework, and machine maintenance requirements can be an important aspect of fluid evaluation. Corrosion concerns may result from changes in the fluid during usage that are not reflected in laboratory prescreening procedures. For example, certain fatty acid corrosion inhibitors may be lost because of hard water precipitation or certain corrosive by-products from chlorinated paraffins may form under in-use conditions. Corrosion problems may also result from improper or inadequate fluid maintenance practices (see Sect. 3).

Fluid performance can also be observed to affect production operations significantly in other ways that may be difficult to quantify: The gradual buildup of crystalline or gummy deposits on slide ways, actuators, and tool holders can cause equipment downtime and added costs for equipment cleaning and maintenance. The formation of gelatinous residues can clog chip transport and fluid filtration systems resulting in added costs for cleaning, ineffective filtration, and excessive disposable filter media replacement costs. Some cutting fluids develop excessive foam in production resulting from the selective loss of antifoam agents in the formulation. This can cause containment spills and severe concerns with floor safety and housekeeping requirements. In some cases, the foam can form entrained air in the fluid delivery system, resulting in excessive pump wear from cavitation and thermal stress fatigue to sensitive cutting tool elements.

As mentioned above, worker health and safety concerns are extremely important in the evaluation of fluids and fluid maintenance practices. Worker concerns with smoke, mist, malodors, dermatitis, biodegradation, and general workplace cleanliness and safety can be important considerations in determining the suitability of cutting fluid practices in a plant.

3. *Cutting Fluid Maintenance Practices*

An effective cutting fluid maintenance program must include a protocol of monitoring and corrective responses to prolong fluid functional life. The maintainability and the ability to achieve long-term performance should be an important consideration in the fluid selection and evaluation process. An effective program can provide significant benefits to direct and indirect costs, including the environmental benefits of reduced disposal requirements (NCMS 1997).

Chemical and physical test methods are used in-process to identify and monitor properties that are important attributes of fluid performance. These test data can be used to determine process control limits for the various fluid properties and provide a warning that corrective actions may be necessary. The data can also aid in troubleshooting the causes of production problems and whether they are caused by fluid, mechanical, or other factors. Generally, monitoring procedures are needed to measure the concentration of the fluid and its functional components, the concentration and effect of the various materials that may contaminant the fluid, and in-process checks of fluid performance (e.g., tool life, corrosion protection).

Control responses require several methods to manage and avoid the effects that tend to degrade the properties and performance of a fluid over time. Examples of concentration controls include the addition of concentrate and water to water-based fluids and lubricity additives to straight oils to make up for losses during use. Water-based fluids often require other control additions including the use of deionized water for makeup, pH adjusters, emulsion stabilizers, and biocides.

Procedural repairs and changes can be an important part of an effective control program. For example, machinery design and maintenance practices may often be required that prevent the contamination of the cutting fluid with deleterious hydraulic and lubricating (tramp) oils and process and floor cleaners.

Mechanical clarification to remove chips and fines from the fluid is critical to prolong fluid lifetime as well as to reduce tool wear and part damage from recirculating particles. The fresh metal surfaces of the chips and fines are cationic in nature and chemically reactive. If they are not removed, they can oxidize and, particularly in the case of emulsion fluids, selectively strip functional additives from the fluid. In addition, very fine abrasive particles can interact at the tool–workpiece interface. Proper filtration is essential to ensure that harmful particles a few micrometers in size are removed with adequate efficiency to maintain a sufficiently low equilibrium concentration for the machining process (Joseph 1985). The buildup and migration of larger chips and swarf can clog the fluid delivery system. This can be a particular problem when hollow cutting tools are involved, having narrow fluid delivery passages.

4. *Future Directions in Cutting Fluids*

There are strong economic and environmental incentives for reducing the use of cutting fluids. Precision near-net part casting and forming operations will play a role in significantly reducing metal cutting requirements and fluid usage. The development of new "dry" machining materials and processes may also reduce fluid use. However, cutting fluid applications are likely to remain an important consideration in many metal removal operations for the foreseeable future.

There will be a continuing need to develop standardized test methods that can be used to evaluate fluid performance characteristics. These methods include screening tests that can be applied to new formulations to evaluate parameters such as chemical and biological stability, residue formation, and machining performance (tool wear). Methods are also needed to evaluate fluid performance for critical machining operations in a full production environment. Data from fully instrumented machining centers capable of measuring cutting tool performance under full part production conditions combined with design of experiment and statistical methods could be invaluable in quantifying the effects of cutting fluid *factors* on key *responses* of the production machining system (Peterson 1997, Chap. 12).

It is likely that environmental considerations and total life cycle management will be increasingly important in the formulation, use, and disposal of cutting fluids. It is merely a question of time as to when regulations and economics will dictate a zero-discharge policy for virtually all manufacturing operations. Cutting fluid formulations and their associated mechanical and reprocessing systems must be designed with this objective in mind.

See also: Cutting-tool Materials; Grinding

Bibliography

American Automobile Manufacturers Association (AAMA) 1998 *520 Lubricants, Industrial Oils and Related Products (Class L) Part 9: Type M (Metal Removal)—Specification.* AAMA, Detroit, MI (for updates of this specification see Ground Vehicle Standard MS1008, SAE International, Warrendale, PA (http://www.sae.org))

Brown R L 1999 Metalworking fluids. In: Rudnick L R, Shubkin R L (eds.) *Synthetic Lubricants and High-Performance Functional Fluids.* Marcel Dekker, New York, Chap. 27

Dasch J M, Ang C C, D'Arcy J B, Powell C A, Rogers J D, Smolenski D J, Tung S C, Gagnon E G, Lee J M, Knowles D, Beattie P 1998 General Motors clean machining program. In: *Proc. Symp. on the Industrial Metalworking Environment: Assessment and Control of Metal Removal Fluids.* American Automobile Manufacturers Association, Detroit, MI, pp. 336–43

Howell J K, Lucke W E, Steigerwald J C 1996 Metalworking fluids: composition and use. In: *Proc. Symp. on the Industrial Metalworking Environment: Assessment and Control.* American Automobile Manufacturers Association, Dearborn, MI, pp. 13–20

Joseph J J 1985 *Coolant Filtration.* Joseph Marketing, East Syracuse, NY

Kreiss K 1998 Hypersentivity pneumonitis from inhaled antigens. In: *Proc. Symp. on the Industrial Metalworking Environment: Assessment and Control of Metal Removal Fluids.* American Automobile Manufacturers Association, Detroit, MI, pp. 55–8

Marino M P, Placek D G 1999 Phosphate esters. In: Rudnick L R, Shubkin R L (eds.) *Synthetic Lubricants and High-Performance Functional Fluids.* Marcel Dekker, New York, Chap. 4

Milton D K 1998 Biological responses to endotoxin in the airways and alveoli In: *Proc. Symp. on the Industrial Metalworking Environment: Assessment and Control of Metal Removal Fluids.* American Automobile Manufacturers Association, Detroit, MI, pp. 59–65

National Center for Manufacturing Sciences (NCMS) 1997 *Metalworking Fluids Optimization Guide.* NCMS, Ann Arbor, MI

Organization Resources Counselors (ORC) 1999 *Metal Removal Fluids: A Guide to Their Management and Control.* ORC, Washington, DC, (http://www.orc-dc.com)

Passman F J 1996 Biocide toxicity: a comparison of the toxicological properties of common metalworking fluid biocides In: *Proc. Symp. on the Industrial Metalworking Environment: Assessment and Control.* American Automobile Manufacturers Association, Dearborn, MI, pp. 82–7

Peterson D S 1997 *Practical Guide to Industrial Metal Cleaning.* Hanser Gardner, Cincinnati, OH

Skisak C M 1996 Metalworking fluids, base oil safety. In: *Proc. Symp. on the Industrial Metalworking Environment: Assessment and Control.* American Automobile Manufacturers Association, Dearborn, MI, pp. 44–9

R. S. Marano and D. J. Smolenski

Cutting-tool Materials

Machining of metals with a cutting tool harder than the work material is a common manufacturing operation in the production of a variety of parts (Komanduri 1997). The unwanted material is removed from the workmaterial in the form of chips, using geometrically defined, single- or multiple-point cutting tool on a machine tool. Various machining processes include turning, drilling, milling, boring, threading, tapping, and broaching. The process can be either continuous or intermittent. Each of them would operate under different set of machining conditions, namely, speed, feed, and depth of cut. Consequently, the requirements of the tool material would differ from one operation to another. Materials used for grinding, polishing, lapping, etc., that use abrasives whose cutting edges are not geometrically defined (i.e., random geometry) and whose geometry changes continuously as the process progresses (also randomly) are outside the scope of this article. Similarly, tool material requirements for processes that use either a noncontacting "soft" tool as in electrical discharge machining (EDM), chemical machining (CM), or no tool at all as in laser machining are considered outside the scope of this article.

1. General Considerations

Machining of metals involves extensive plastic deformation (shear strain of $\sim$2–8) of the work material ahead of the tool in a narrow region; high tool temperatures ($\sim$1000°C); freshly generated, chemically active surfaces (underside of the chip as well as the machined surface) that can interact extensively with the tool material resulting in wear; and high mechanical and thermal stresses (often cyclic) on the tool. The performance and life of a cutting tool depend upon the cutting conditions as well as the combination of tool and work material used. The exact nature and extent of these effects depend on the forces, mechanical and thermal stresses, temperature distribution, and chemical interactions between the tool-workmaterial–environment imposed by the cutting process.

A successful cutting tool material must resist these severe conditions and yet provide a sufficiently long tool life to justify its use economically. The specific energy (i.e., the energy consumed per unit volume of the material removed) involved in machining is significant (0.2 to 10×10^6 in lb/in^3). Also, a considerable amount of unwanted material is removed from the workmaterial in the form of chips which is considered as scrap. While the percentage volume of the chips generated (scrap) is not generally a large fraction of the part ($\ll$25%), in some applications, as in the machining of various components used in the frame of an aircraft (generally, aluminum alloys), most of the material is removed in the form of chips (to reduce the weight of the parts significantly) leaving the weight of the part only to be a small fraction (10%) of the total weight of the raw stock before machining. In spite of these limitations, the machining process is a very versatile operation due to the fact that components of various shapes and sizes can be easily made economically without the need for expensive dies or requiring high power, expensive presses as in metal forming.

The cutting tool is a very important component of the machining system. Consequently, the tool material significantly affects the productivity of the machining operation. Other important elements include cutting conditions, tool geometry including the chip-grooves on the tool inserts for ease of breakage and disposal of chips, and the characteristics of the work material (chemical composition, crystalline and microstructure, and metallurgical condition), nature of the parts produced (geometry, accuracy, finish, and surface-integrity requirements), machine tool (adequate rigidity with high horsepower, and wide speed and feed ranges), and support system (operator's ability, sensors, controls, method of lubrication, and chip control). Thus, in the development of any new tool material these factors should be taken into account. This may require the use of available enabling technologies for the synthesis and production of new tool materials. For example, mechanical chip breakers were used till early 1970s for chip breakage and disposal when using cemented carbide tools. But currently, mold-in chip breakers are very common on these inserts. Similarly, chip groove geometry was difficult to incorporate in ceramic tools limiting their application. Grinding of the chip groove geometry in the tools was too expensive. But new ceramic processing technology has enabled production of tool inserts with built-in chip groove geometry, thereby extending their applications.

A tool material must meet several stringent requirements imposed by the cutting process including high hot hardness, adequate mechanical strength, high thermal/mechanical shock resistance, chemical inertness towards the workmaterial and/or the environment under the conditions of cutting, adequate stiffness, adequate wear resistance, suitable thermal properties, ease of fabrication and reshaping, availability, and low cost. The relative importance of each depends, however, upon the type of machining operation—continuous versus intermittent cutting; roughing versus finishing; condition of the machine tool; volume of production (small batch versus mass production); chip handling and disposal; use of a cutting fluid—whether permitted or not; geometry, finish, accuracy, and surface integrity requirements; tool-lifetime economics; other constraints such as delivery schedules and bottle-necks.

A wide range of cutting tool materials with a variety of properties, performance capabilities, and cost is available. These include high carbon steels and low/medium alloy steels, high speed steels (HSS), cemented carbides (WC-based and solid solutions of multicarbides of W, Ti, Ta in a Co binder, or TiC in a Ni–Mo binder), coated tools (both on cemented carbides and on HSS), ceramics (alumina-based and silicon nitride-based), sintered polycrystalline diamond and sintered polycrystalline cubic boron nitride (cBN) by high temperature–high pressure (HT-HP) synthesis, thin diamond coatings on cemented carbides and ceramics by the low pressure–high temperature (LP-HT) synthesis, and single crystal natural diamond. Diamond and cBN because of their extremely high hardness are referred to as *superhard materials*.

In general, in going from carbon steels to diamond, an increase is observed in the hardness and wear resistance; cutting speed capability; edge deformation resistance; bulk plastic deformation resistance; obtainable finish and accuracy; and cost; while the thermal shock resistance and ease of fabrication decrease. Since in each class of tool material a wide range of composition and properties is obtainable, overlapping of characteristics exist in many cases and exceptions to the rule are very common. What follows is a brief description of various tool materials together with their properties, applications, and limitations.

2. *Carbon Steels and Low-to-medium Alloy Steels*

Low-to-medium alloy steels contain elements, such as Mo and Cr for hardenability (solid solution strengthening), and W and Mo for wear resistance (carbide formation). They, however, lose their hardness rapidly when heated above 150–340°C because of limited high temperature solid solution strengthening elements in the alloy. Their abrasion resistance is also limited because of the low volume fraction of hard, refractory carbide phase present in these alloys.

3. *High Speed Steels (HSS)*

High speed steels (HSS) are a class of high alloy steels with significant amounts of W, Mo, Co, V, and Cr in addition to Fe and C. Extensive solid solution alloying of these elements with the ferrite matrix strengthens the matrix and increases the hot hardness. High hardness and wear resistances are obtained by the presence of uniformly distributed refractory hard carbides of W, Mo, V, and Cr. Vanadium also acts as a grain growth inhibitor.

Tool steels are broadly classified as the T-type and the M-type depending on whether tungsten (T) or molybdenum (M) is used as the principal alloying element (see Table 1). Shortage of W, a strategic material, led to the successful search for an alternative element, Mo, in place of W. The two types of tool steels (T- and M-) are used interchangeably for they possess essentially the same properties and have comparable cutting performance. In general, M-type tool steels are more popular (~85% of all tool steels) as they are less expensive (~30%) than the corresponding T-type steels. The latter is due to greater availability and lower cost of Mo and the fact that only about one half as much in weight of Mo as W is needed to yield the same properties and performance.

HSS are available in cast, wrought, and sintered form (using power metallurgy technique). Carbide segregation, formation of carbides with significant variability in size and distribution in the matrix, highly oriented grains are some attributes of the cast/wrought alloys (Fig. 1(a)). Such a tool steel material is more difficult to shape by grinding, has inconsistent properties, and lack of performance reliability. To alleviate these problems a technique has been developed that involves atomization of the prealloyed molten tool steel into fine powder followed by consolidation under hot isostatic pressure (HIP). The resulting microstructure has more uniform distribution of fine carbide particles in the matrix (Fig. 1(b)). The hot hardness and performance at elevated temperatures were maintained the same without the Co addition in these alloys by an appropriate increase in the Mo–W and/or V content by solid-solution strengthening. Tool steels made in this manner grind more easily, exhibit more uniform properties, and perform more reliably in cutting. Tool steels up to a hardness of R_a of 70 can be obtained with this method using high Co (up to 20%) and high V (also up to 20%) that cannot be made by conventional casting or hot forming. Because of the fine carbide size, tools made of this material exhibit significant edge strength and maintain edge sharpness during such cutting operations as in end milling. Consequently, HSS material made by this process is extensively used from multipoint tools, such as drills, milling cutters, taps to many expensive form tools, such as broaches, shaper cutters for gears, various dies for metal forming applications. Table 1 gives a summary of the chemical composition of selected HSS tool materials.

Two new classes of HSS were developed recently with a submicron (0.1–0.5 μm) grain size of TiC or TiN (40–55%) in a steel matrix (45–60%). The percentage of the hard phase, TiC or TiN, in these alloys is higher than in conventional HSS but lower than the lowest limit of the cemented carbide. It is thus harder than any conventional HSS but tougher than most cemented carbides. Small amounts of Cr (3–17.5%), Mo (0.5–4%), Ni (0.5–12%), Co (5–5.7%), Ti (0.5–0.7%), and C (0.4–0.85%) are added to provide solid solution strengthening as well as hot hardness of the matrix material. The material is produced by initially compacting TiC or

Table 1
Nominal chemical composition of reprexentative HSS tool grades.

HSS grade	Nominal chemical composition (%)							Remarks
	C	W	Mo	Cr	V	Co	W_{eq}^{a}	
General-purpose grades (no cobalt)								
T1	0.7	1.8		4	1		18	Used in ~90% of the general-purpose HSS-grade applications
M1	0.8	1.5	8	4	1		17.5	
M2	0.85	6.0	5	4	2		16	
M10	0.9		8	4	2		16	
Higher V-content grades (no cobolt)								
M3	1.2	6	5	4	3		16	Used for abrasive-resistance applications. C content is increased by 0.1% for each 1% increase in V. More difficult to grind
M4	1.4	5.5	4.5	4	4		14.5	
M7	1.0	1.75	8.75	4	2		19.25	
Co HSS grades								
T5	0.8	18		4	2	8	18	Used for higher hot hardness or higher temperature-deformation resistance. More difficult to machine, heat treat and grind
M34	0.9	2	8	4	2	8	18	
M35	0.8	6	5	4	2	5	16	
M36	0.8	6	5	4	2	8	16	
Higher V and higher Co grades								
T15	1.5	12		4	5	5	12	Used for combined abrasive resistance and higher temperature-deformation resistance. Extremely difficult to grind. More expensive
M43	1.25	1.75	8.75	3.75	2	8.25	19.25	

[a] $W_{eq} = 2(\%Mo) + \%W$

TiN powder in a steel die. The resulting porous compact is sintered at high temperature and subsequently infiltrated with molten steel under vacuum. The microstructure of this material in the annealed condition shows a well rounded carbide grains in a spherodite steel matrix and in the quenched condition in a fine martensite matrix. The new HSS tool materials combine the hardness (consequently, the wear resistance) of cemented carbides with the heat treatability of HSS. Thus, they can be machined in the annealed condition and subsequently heat treated to the desired hardness and microstructure. The hardness of TiC in a steel matrix material is ~69 R_a in the annealed condition and after heat treatment ~86.5 R_a. The elastic modulus is in the range of 303 GPa (44×10^6 psi) which is ~50% over HSS. The transverse rupture strength (TRS) is ~50–75 GPa (300–400 ksi). An additional feature of these materials is that they can be easily brazed or welded to the steel substrate without any danger of cracking due to comparable thermal expansion coefficients.

HSS tools are used mostly for low speed, heavy duty applications where built-up edge, adhesion of the chips to the tool, and high friction are the primary concerns rather than high tool temperatures. Consequently, TiN coated (~5 μm) HSS tools, to be discussed shortly are gaining popularity especially for the more expensive tools. The coatings are deposited by a low temperature (450°C) physical vapor deposition (PVD) technique. Improvement in tool life from three to 10 times is feasible by such a coating in the speed range capability of HSS tools. TiN coating besides providing low frictional conditions also has a golden color that is aesthetically very attractive. Such a coating at the same time can also facilitate in the estimation of the extent of tool wear due to color difference between the coating and the substrate.

Since HSS tools are tough, they can be used in continuous as well as in interrupted cutting. The submicron HSS tool materials can be used for machining aluminum and other soft materials at very high speeds using a high positive rake angle (~40°) tool thus reducing the forces significantly. Most form tools, such as milling cutters, drills, broaches, gear shapers, and hobs are made of HSS. Current trends in the HSS tool development are the elimination of undesirable elements in the steels and production of a more consistent alloys for increasing reliability in service of the parts made of this material.

4. *Cemented Carbides*

Cemented carbides form an important class of tool material containing a large volume fraction (⩾90%) of refractory carbides in a metal binder. By varying the size and composition of the carbides as well as the

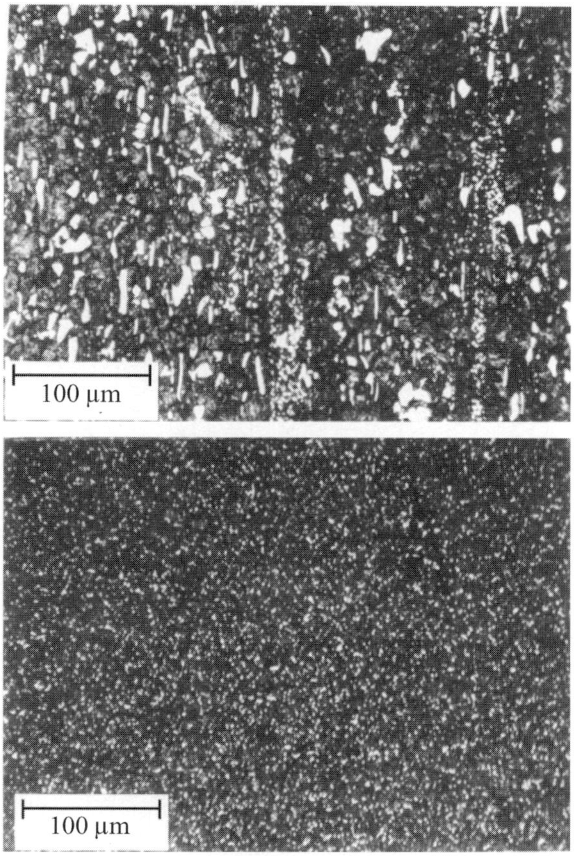

Figure 1
Optical photomicrograph of a quenched and tempered AISI T-15 tool steel showing the microstructure (Picral etch) (a) produced by the conventional cast/wrought technique showing carbide segregation, formation of carbides with significant variability in size and distribution in the matrix, highly oriented grains, (b) produced by the consolidation of the atomized tool steel alloy power by hot isostatic pressing showing a microstructure with more uniform distribution of fine carbide particles in the matrix.

amount of the binder phase, cemented carbide tool materials of different hardness and toughness values can be obtained. Cemented WC in a Co binder is used for machining nonsteels (i.e., nonferrous alloys as well as cast iron) while solid solutions of multicarbides of W, Ti, and Ta or Nb in a Co binder are used for machining ferrous alloys.

Cemented carbides offer several beneficial features as compared to HSS tools, including higher speed capability (3–6 times), higher hot hardness, superior wear resistance, greater refractoriness, better chemical stability, greater stiffness (Young's modulus, E, of cemented carbide is about three times that of steel), greater versatility, and reduced tendency for adhesion. The greater stiffness arises from carbide phase being the continuous phase in cemented carbide while the metallic phase predominates in HSS. The greater versatility results from the ability to change the amount of the binder content as well as the chemistry and size of the carbides. Further, a thin ($\sim 5\,\mu m$) refractory, hard coatings (single or multiple) can be deposited for improved wear resistance without affecting the substrate adversely. Cemented carbides have some drawbacks including lower toughness, lower resistance to mechanical and thermal shock, and higher cost to process and fabricate. Table 2 gives the nominal chemical composition and properties of some representative (nonsteel and steel) grades of cemented tungsten carbide tool materials. The nonsteel grades are straight WC that can be recognized by the angular carbide morphology (Fig. 2(a)) while the steel grades are multicarbides (Fig. 2(b)) which have a rounded morphology. As one progresses from a roughing grade to precision machining grade, the toughness or transverse rupture strength decreases, edge and general wear resistance increases, and the Co content and/carbide grain size decreases.

Cemented WC tool materials with submicron grain size (~ 0.1–$0.5\,\mu m$) were developed to increase the toughness and edge strength. Fine dispersion of small amounts ($\sim 0.5\%$) of submicron chromium carbide is added to restrict the grain growth. The transverse rupture strength (TRS) of the submicron grain sized cemented tungsten carbide can be up to 2.76 GPa (or 400 ksi) which is close to that of HSS but the hardness of this carbide can be significantly higher, namely, 91.5 R_a compared to 70 for HSS.

Another type of cemented carbide tool material (cermet) for high speed (>300–$400\,m\ min^{-1}$) finish machining (low feed) of steels is based on TiC in a Ni–Mo binder. ASTM refers cermet (ceramic–metal) as an acronym to designate a heterogeneous combination of metal(s) or alloy(s) with one or more ceramic phases in which the later constitutes approximately 15–85% by volume and in which there is relatively little solubility between the metallic and ceramic phases at the preparation temperature. Although strictly cemented tungsten carbide also come under this category, the common practice is to consider only Ti-based materials as cermets. Further advancements of this cermet are based on the following four concepts: (i) Improvements through the additions of other carbides, such as MoC, TaC, and WC, (ii) significant additions of TiN to TiC either as separate phases or as titanium carbonitride (TiCN) resulting in TiC–TiN cermets, (iii) modification of the composition of the cermet by adding Al to the Ni–Mo binder alloy which would precipitate fine intermetallic compound Ni_3Al particles in the binder phase for improving the hot strength (similar to the strengthening effect found in nickel base superalloys),

Table 2
Nominal chemical composition and properties of some representative (nonsteel and steel) grades of cemented-carbide tools for roughing, general-purpose and finishing operations.

	Nominal chemical composition (%)				Grain size	Density ($g\,cm^{-3}$)	Hardness R_a	Transverse rupture strength (GPa)	Elastic modulus E (GPa)	Impact strength (N m)	Compressive strength (GPa)	Tensile strength (GPa)	Relative abrasion resistance (volume loss) (cm^{-3})	Thermal conductivity ($cal\,cm^{-1}\,s^{-1}\,°C^{-1}$)	Thermal expansion ($10^{-6}\,K^{-1}$)
	WC	TiC	TaC	Co											
Nonsteel grades (C–1 to C–4)															
Roughing	94			6	Coarse	15.0	91	2.2	640	1.4	5.2	1.5	15	0.29	4.3
General-purpose	94			6	Medium	15.0	92	2.0	650	1.4	5.5	1.5	35	0.24	4.5
Finishing	97			3	Fine		92.8	1.8	610	1.0	5.9	1.8	60		4.3
Steel grades (C–5 to C–8)															
Roughing	72	8	11.5	8.5	Coarse	12.6	91.1	1.7	560	0.9	5.2		8	0.12	5.8
General-purpose	71	12.5	12	4.5	Medium	12.0	92.4	0.7	570	0.8	5.8		7	0.084	5.2
Finishing	64	25.5	4.5	6	Medium	9.9	93.0	1.4	460	0.5	4.9	0.5	5		5.9

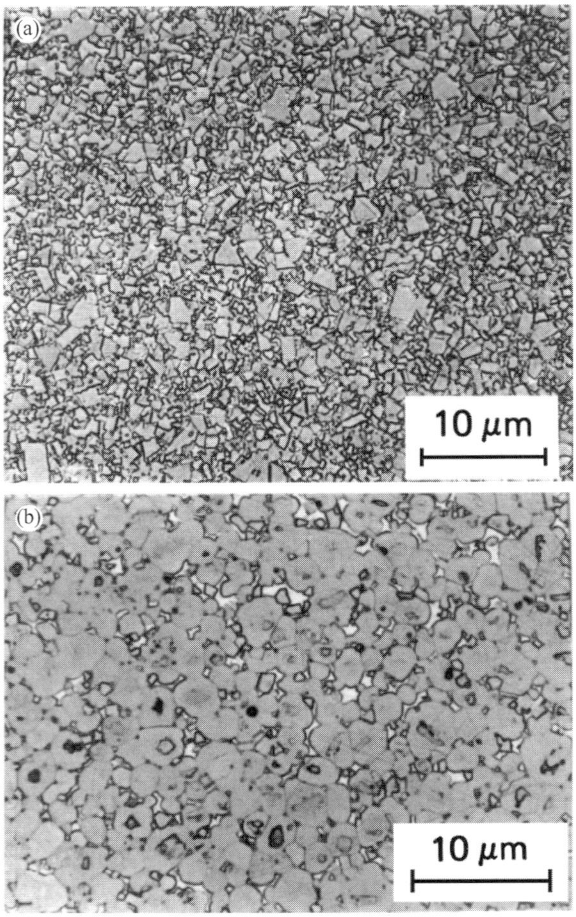

Figure 2
(a) Micrograph of a nonsteel grade, straight WC showing an angular carbide morphology and (b) micrograph of a steel grade multicarbide (W, Ti, Ta(Nb)) showing a rounded carbide morphology.

and (iv) a coating of TiN preferably by PVD on the cermet tools. With these additions the hot hardness, transverse rupture strength (TRS), oxidation resistance, and thermal conductivity can be significantly increased. The higher TRS provides better edge strength and chipping resistance while higher thermal conductivity provides thermal shock resistance both of which have limited the application of this material for a long time. Cemented TiC in a Ni–Mo binder is used for selective higher speed ($\sim$350–500 m min^{-1}) finishing of steels and malleable cast iron. Figure 3 shows a fractograph of a TiC in a Ni–Mo binder cermet showing essentially the brittle nature of this material.

The TiC–TiN cermets or TiCN cermets, especially with coatings, cover the machining range between cemented carbides at one end and ceramics on the

Figure 3
Fractograph of a TiC in a Ni–Mo binder cermet showing essentially the brittle nature of this material.

other. This is mainly because of their high hot hardness (and consequently high deformation resistance), high wear resistance (both flank and crater wear), high resistance to metal build-up, and higher thermal conductivity and thermal shock resistance. Consequently, improved surface finish on the part, consistent part tolerances, high-speed capability, and longer tool life results. Different grades of TiC–TiN were developed to cover roughing ($\sim$12% binder), general purpose ($\sim$10% binder), and finishing ($\sim$9% binder). In addition, coated grades with high binder content ($\sim$18% binder) were developed for interrupted cutting, such as milling.

In selecting a carbide grade for a given application, the following general guidelines should be followed: the grade with the lowest Co content and the finest grain size consistent with adequate strength to eliminate chipping should be chosen; straight WC grades can be employed if cratering, seizure, or galling is not experienced and for work materials other than steels; to reduce cratering and abrasive wear when machining steels, TiC grades are preferred; and for heavy cuts in steel where high temperatures and high pressures generated at the cutting zone deform the cutting edge plastically, a multicarbide grade [W–Ti–Ta(Nb)] with a low binder content should be used.

Cemented carbides are especially effective at higher speeds, generally in the range of 45–180 m min^{-1}. This speed can be much higher (<300 m min^{-1}) with materials that are easier to machine, e.g., Al alloys, and much lower ($\sim$50 m min^{-1}) with materials that are more difficult to machine, e.g., Ti alloys. In interrupted cutting applications, edge chipping is prevented by appropriate choice of cutter geometry and cutter position with respect to the workpiece in such a way as to transfer the point of application of the load away from the tool tip. Finer grain size and higher Co content improves toughness in straight WC–Co grades and are considered desirable in materials used for interrupted cutting. Because of the high hardness, cemented carbides tools can be finished only by diamond grinding.

Cemented carbide tools are produced in insert form (sometimes with a molded-in chip groove geometry for ease of chip disposal and to reduce forces) by cold pressing to shape and nominal size, followed by liquid-phase vacuum sintering. As the chip grooves are built-in, no additional costs are involved in producing them separately. The inserts are finished by diamond grinding. Cemented carbide inserts are generally of the throw-away type although they can be reground and reused, if need be. They are used for practically all machining operations including turning, milling, drilling, broaching, and gear shaping. Like the HSS tool technology, the cemented carbide tool technology is also a matured technology.

5. *Coated Tools*

A thin ($\sim$5 μm), hard, refractory, binderless coating of TiC was introduced in the late 1960s as an effective barrier between the chip and the tool and for reducing the crater wear of cemented carbide tools significantly. This led to the development of other single and multiple coatings of various carbides, oxides, and nitrides on different cemented carbide substrates as well as TiN on HSS with great success. Chemical vapor deposition (CVD) was used for the cemented carbide tools with the substrate heated to $\sim$950–1050 °C and physical vapor deposition (PVD) at lower temperatures ($\sim$400–450 °C) for HSS tools.

The basis for the development of thin coatings on cutting tools is the recognition that the requirements at or near the surface of the tool are different from the bulk. The surface has to be hard to be wear resistant, refractory, and chemically stable for an effective diffusion barrier between the constituents of the tool and the work material. In contrast, the bulk of the tool should be tough enough to withstand high-temperature plastic deformation (at the tool tip and in bulk) under the conditions of cutting. These needs have led to the design of engineered tool materials. To be effective, the coatings should be bonded metallurgically to the substrate with a graded interface, thick enough to provide long tool life but thin enough not to act in a brittle manner, free from any binder, of fine grain size with no porosity, free from the tendency of the chip to adhere or seize to the tool face, easy to deposit in bulk quantities and inexpensive.

Several hard, refractory coatings including single coatings of TiC, TiN, Al_2O_3, HfN, or HfC, and multiple coatings of Al_2O_3 or TiN on top of TiC have been developed using the CVD technique. To deposit a coating of TiC, the tools are loaded in a chamber and heated to a temperature of ~1000°C. $TiCl_4$ and CH_4 gases are introduced in the chamber and reacted in a hydrogen atmosphere. The hard reaction product, namely TiC, deposits on the tool and forms a metallurgical bond with the substrate at high temperature. In the case of TiC coating, depletion of carbon from the substrate at high temperature results in the formation of a brittle η-phase, $Co_xW_yC_z$ and associated microporosity at the coating-substrate interface. This brittle η-phase can reduce the transverse rupture strength (TRS) of coated tools by as much as 30% of the substrate's TRS resulting in premature tool failure especially in interrupted cutting or in roughing. This can be a serious limitation in interrupted cutting and hence methods to overcome this or augment the strength by other methods are essential for the success of the coated tools. Enrichment of cobalt at and near the rake face of the tool, use of lower temperature PVD technique which effectively eliminates the formation of the brittle η-phase, use of top most coating of TiN by PVD to introduce residual compressive stresses are some of the attempts in this direction to overcome this deficiency.

For a coating of Al_2O_3 a gaseous mixture of hydrogen, water vapor, and an aluminum halide, such as aluminum trichloride is used. The water vapor is most conveniently formed by reacting hydrogen with CO_2 in the deposition chamber to form CO and water vapor. H_2 is found to be necessary to ensure oxidation of aluminum at the carbide interface and to form a dense, adherent coating. To enhance adhesion an interlayer of a transition metal which is both a carbide former and an oxide former, such as Ti, Ta, Hf, Zr, or Nb, is selected. For example, the tool is first coated with a thin layer of TiC by CVD process described earlier. It is then oxidized to form TiO at the surface to which Al_2O_3 coating is deposited. Interlayers are thus provided for compatibility between oxides and carbides by a gradual transition or by forming oxi-carbides. This way thermal expansion mismatches and chemical incompatibility are minimized.

TiN coating, similarly, is deposited by reacting $TiCl_4$ and N_2 or NH_3 in a hydrogen atmosphere and TiCN coating is deposited by reacting $TiCl_4$, CH_4, and N_2 in a hydrogen atmosphere. A coatings of 5 μm thickness, found to offer the best performance, requires about 8 hours of processing time. The purpose of multicoatings is to build the thickness for prolonging tool life without the detrimental effects of a thicker, harder single coating. In addition, multiple coatings allow a stronger interface and provide a variety of protections when machining under different cutting conditions. Figure 4 is a micrograph of a representative multicoating of alternate layers of TiN–Al_2O_3 followed by a thick interlayer of TiCN on cemented WC substrate.

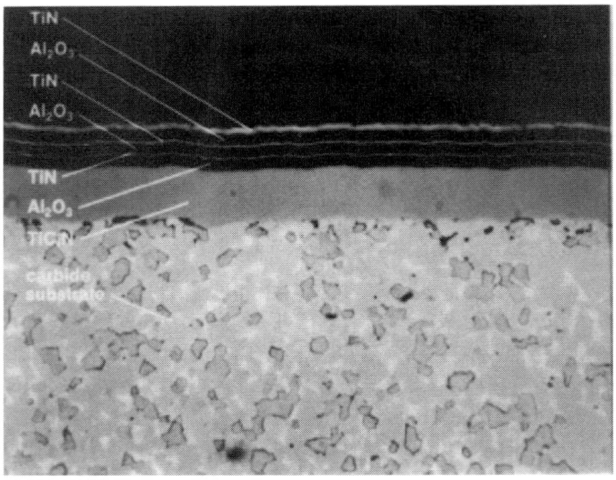

Figure 4
Micrograph of a representative multicoating of alternate layers of TiN–Al_2O_3 followed by a thick interlayer of TiCN on a cemented WC substrate.

Physical vapor deposition (PVD) technology includes vapor deposition, various types of sputtering (DC, radio frequency (RF), reactive, magnetron, and plasma enhanced magnetron), and ion plating. The mean free path in PVD is large and the vapor species arrive at the substrate without extensive gas collision. Consequently, it is a low temperature, line of sight process. Because of lower substrate temperatures adhesion of the coating to the substrate can be a problem. However, unlike in CVD, the formation of η-phase is practically eliminated due to lower substrate temperatures. Also, PVD process can introduce compressive residual stresses beneficial to adhesion of the coating to the substrate and the overall performance of the coated tools. The trend now is to use PVD technology at least to deposit the first layer on the tool substrate. Magnetron sputtering and ion plating are some of the common PVD techniques currently used in the cutting tool coating industry.

Various multicoatings used on cutting tools include TiN, TiN–TiCN–TiN, TiN–TiCN, TiN–Al_2O_3–TiC–TiCN, TiN–TiCN–TiC, TiN–TiCN–TiC, TiN–Al_2O_3–TiCN, Al_2O_3–TiC, Al_2O_3–TiC–TiCN, TiN–Al_2O_3–TiC, TiN–Al_2O_3–TiN–Al_2O_3–TiN–Al_2O_3–TiCN engineered on selected cemented carbide (different Co content and straight WC or multicarbide) as well as cobalt-enriched substrates. In addition, one grade of TiN–Al_2O_3 coating on a Si_3N_4 substrate is also available.

Currently CVD coating technology has reached a stage wherein turnkey systems are available for coating of a variety of materials. Of the various types of cemented carbide tools used today in industry, some 65% are coated especially in the USA, Western Europe, and Japan. This percentage is expected to increase considerably (as high as ~80%) as we go into the 21st Century with advances in coating material concepts, the technology of coating, and the need to improve productivity without effecting the environment.

6. Multiple Nanocoatings

Since the properties of extremely thin coatings can significantly be enhanced by reducing the microstructural or spatial scale of a material system to nanometer dimensions, numerous multiple nanocoatings of alternate hard/tough or hard/lubricating coatings are being investigated to improve the speed capability of cutting tools especially for dry machining or "green" manufacturing. In this approach, a crack initiated in any hard nanolayer is stopped when it reaches the tough nanolayer thus facilitating higher toughness. The number of layers can be several hundreds to several thousands in contrast to a few layers of micrometer thickness currently being used on cutting tools by CVD technique.

The number of material systems that can be used for nanocoatings is virtually unlimited. Any refractory hard material, such as carbides, borides, nitrides, or oxides can be used as the hard material. Compatible metal can be used as the tough material, e.g., Mo, W, graphite, MoS_2, or other layered structured materials can be used as a solid lubricant. Some examples of material systems for nanocoatings are given in the following.

6.1 *In situ formation of oxidation-protective and low-friction layers*

Hard carbide/hard carbide systems: B_4C/SiC, HfC/B_4C, HfC/SiC

Hard carbide/metal co-sputtered or layered systems: B_4C (HfC, SiC)/W (Al, Cr, Ti, Si, Zr)

6.2 *Low friction, low stress coatings*

Layered lattice (solid lubricant)/metal systems: MoS_2/Mo, MoS_2/Ag–Mo, TaS_2/Ta, WS_2/W

6.3 *Low friction, hard coatings*

Hard carbide, oxide, nitride, boride/layered lattice (solid lubricant systems)

Numerous multiple nanocoatings are deposited by PVD technique, chiefly, magnetron sputtering, or pulse laser deposition (PLD) using multiple targets and rotating the tool with respect to the targets such that the tool sees different sputtering targets alternately. The thickness of the coating is determined by the rotational speed of the tool and the sputtering rate of the coating materials.

7. Ceramics

Advanced ceramic tool materials have been introduced for high-speed finishing at light feeds and for high removal-rate machining at low speeds and large depths of cut for machining of some difficult-to-machine steels and highly alloyed cast irons. Two classes of ceramic tools are used, one is alumina-based and the other silicon nitride-based (also called nitrogen ceramics). Ceramic tools are highly refractory (higher melting point), more wear resistant (harder), chemically more stable than HSS and cemented carbides. They are, however, far less expensive than diamond and cubic boron nitride (cBN). Because of their high temperature stability, ceramics perform exceptionally well at high cutting speeds. The main limitations of ceramic tools are relatively low transverse rupture strength (TRS), low fracture toughness, poor thermal and mechanical shock resistance, and a propensity to chipping failure. Notching at the depth of cut line (known as DCL notching), microchipping of the tool edge, chipping due to thermal and/or mechanical cyclic stresses during interrupted cutting, and gross fracture of the tool are the predominant modes of wear. Interrupted cutting is especially severe on ceramic tools because of repeated entry and exit of the cut. Tool edge preparation is particularly important for this reason. Performance of ceramic tools can be improved significantly by using a rigid, high precision machine tool.

Alumina-based ceramics are either pure or alloyed with appropriate amounts of the oxides of Ti, Cr, Zr, TiC, and SiC to increase fracture strength as well as thermal and mechanical properties. Small amounts of MgO or Y_2O_3 are added as sintering aids. The tools are produced by cold pressing followed by vacuum sintering, hot pressing, or hot isostatic pressing (HIP'ing). Higher strength and greater densification (~99.5%) are obtained by the later two processes. High purity and fine grain size (1–3 μm) are paramount for increased strength. Submicron (0.1–0.5 μm) alumina tools were developed to further improve strength, especially the edge strength (Figure 5).

Alumina–TiO ceramic introduced first in the 1950s was characterized by a grain size of ~3 μm and reasonably uniform microstructure; the TiO constitutes ~10%. A density of ~90–95% theoretical, a hardness of ~93–94 HRA and a TRS > 550 MPa (~80000 psi) were achieved with cold pressing followed by sintering. The next advancement was the development of pure alumina and alumina–TiC dispersion-strengthened ceramics containing ~30 wt.% TiC and small amounts of yttria as a sintering agent, resulting in a density close to 99.50% theoretical. High purity, fine grain size, and elimination of porosity are the principal reasons for the high TRS (700–900 MPa or ~100–130 ksi). The fracture

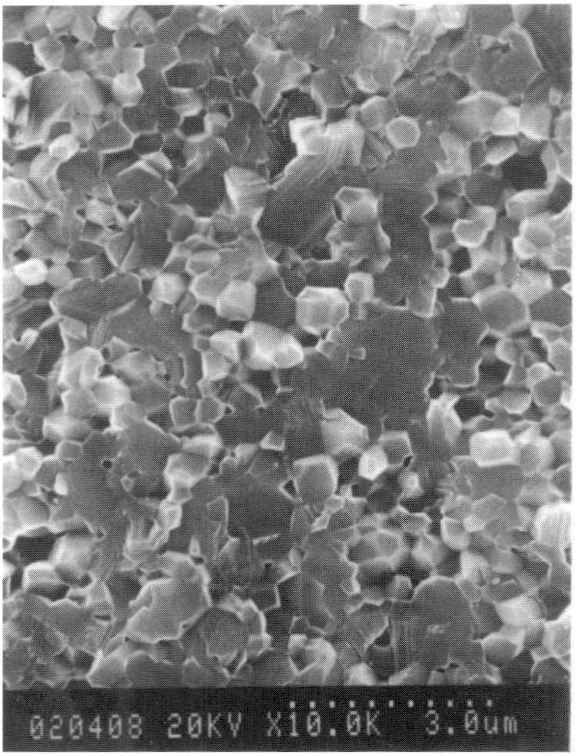

Figure 5
SEM micrograph of a submicron sized (0.1–0.5 μm) pure (no binders or grain growth inhibitors) alumina for tools with improved strength, especially the edge strength.

toughness ($\sim$3.3 MPa $m^{1/2}$) and TRS are improved through obstruction of the cracks, crack deflection or crack branching caused by the dispersed hard TiC particles. The lower thermal expansion coefficient and higher thermal conductivity of the composite compared to straight alumina increases the thermal shock resistance and thermal shock cycling experienced in interrupted cutting. Figure 6 is a micrograph of a hot-pressed alumina–TiC ceramic, showing a white TiC phase and a dark alumina phase with some fine porosity.

The second class of ceramic tool material is based on Si_3N_4 either nearly pure Si_3N_4 (except for some minor additions of sintering aids) or with various additions of alumina, Y_2O_3, and TiC. Si_3N_4, a covalently bonded material, possesses higher elevated temperature strength, thermal stability, low thermal expansion coefficient, higher thermal conductivity, and higher fracture toughness than alumina. However, it is extremely difficult to fabricate pure Si_3N_4 into a dense form (with no porosity) by conventional sintering or hot pressing techniques without the sintering aids, such as MgO, Y_2O_3, TiO, and Cr_2O_3. At the densification temperature, the sintering aids

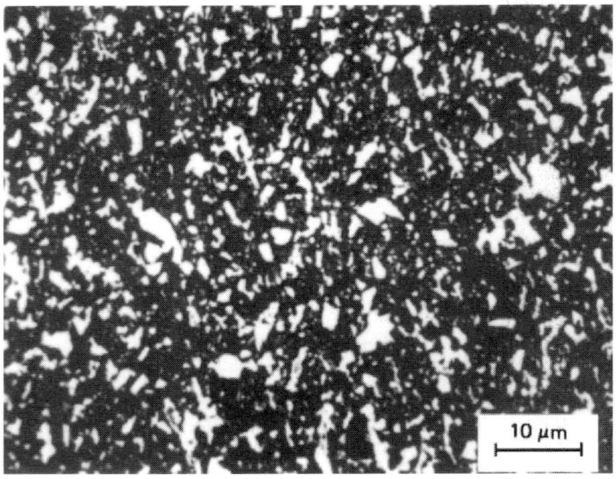

Figure 6
Optical micrograph of a hot-pressed alumina–TiC ceramic showing a white TiC phase and a dark alumina phase with some fine porosity.

Figure 7
SEM micrograph of a silicon nitride tool showing a microstructure of interlocked, elongated grains of β-Si_3N_4.

combine with SiO_2 to form a glassy liquid phase; α-Si_3N_4 particles dissolve in this liquid phase and precipitate out as β-Si_3N_4. The β-Si_3N_4 nuclei grow as elongated grains and form an interlocked grain structure (Fig. 7). The higher fracture toughness ($\sim$4.1 MPa $m^{1/2}$) and higher strength of Si_3N_4 are attributed to this elongated grain structure. Si_3N_4 is characterized by a two-phase structure consisting of β-Si_3N_4 crystallites and an intergranular bonding phase. The hardness and toughness of Si_3N_4 can be further increased by the additions of yttria ($\sim$6%), alumina ($\sim$2%), and a fine dispersed phase of TiC,

TiN, or HfC or combination of TiN and TiC. The toughness is increased via crack interactions with the dispersoid as well as crack deflection of the composite material. Si_3N_4 tools are recommended for high speed (~450–1500 m min^{-1}) machining of gray cast iron in the auto industry.

Another Si_3N_4 based ceramic tool is a complex compound of Si–Al–O–N (SiAlON). It was found that oxygen (O^{2-}) could be substituted for nitrogen (N^{3-}) in the β–Si_3N_4 crystal provided aluminum (Al^{3+}) is simultaneously substituted for silicon (Si^{4+}) to maintain charge neutrality. SiAlON tools are currently produced by sintering. The powder charge consists of a mixture of Si_3N_4, AlN, Al_2O_3, and Y_2O_3; the latter is added as a sintering aid for densification. The powder mix is first ball-milled, dried, pressed to shape by cold pressing, and subsequently sintered at a temperature of ~1800°C under isothermal conditions for ~1 hour before it is allowed to cool slowly. Y_20_3 reacts with β-Si_3N_4 to form a silicate that is a liquid at the sintering temperature. The resulting Si_3N_4 material thus has a glassy intergranular phase (Fig. 8). This material offers significant improvements in tool life, consistency in tool performance (more reliability), and higher removal rates at reasonable cutting speed (90–125 m min^{-1}) when machining nickel base superalloys.

A new whisker reinforced ceramic composite cutting tool material (SiC whisker-reinforced alumina) possessing improved fracture toughness (K_{IC}~8.5–9.0 MPa $m^{1/2}$) was introduced for machining of nickel–iron base superalloys used in aircraft engines. Whiskers of SiC possess very high tensile strength (7 GPa) and have higher thermal conductivity and higher coefficient of thermal expansion than alumina. Consequently, the composite exhibits higher strength, fracture toughness, and thermal shock resistance. SiC whiskers (0.5–1.0 μm in diameter and 10–80 μm long) ~20% volume fraction are mixed homogeneously with micron sized fine powder of alumina. The mixture is then hot pressed to over 99% of the theoretical density at a pressure in the range of 28–70 MPa and temperature in the range of 1600–1950°C for pressing times varying from ~0.75 to 2.5 hours. Figure 9 is a SEM fractograph of SiC whisker reinforced alumina showing long micrometer sized SiC whiskers in an alumina matrix. The improved toughness is explained thus. As the cracks propagate through the ceramic matrix, bonds between the matrix and the whiskers are broken. However, the SiC whiskers, because of their inherently high strength (7 GPa) remain essentially intact. Consequently, whisker pullout occurs due to the separation of the matrix from the whiskers. Interfacial shear stress resisting the whisker pullout absorbs a substantial amount of this fracture energy and inhibits crack propagation.

Ceramic tools are inherently more brittle than cemented carbides, and tool geometry of −10° rake is

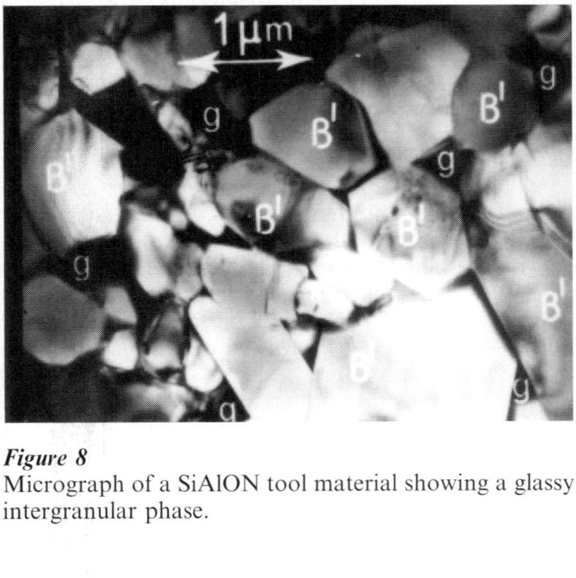

Figure 8
Micrograph of a SiAlON tool material showing a glassy intergranular phase.

Figure 9
SEM fractograph of SiC whisker reinforced alumina showing long micrometer sized SiC whiskers in an alumina matrix.

recommended instead of -5° rake for cemented carbides. In interrupted cutting, attempts should be made to shift the point of application of the load away from the cutting edge to minimize chipping. Suitable edge preparation involving honing a small radius or a small negative land on the rake face is also recommended. The work materials recommended include hardened steel, chill-cast iron, and superalloys (Ni-base and Co-base). Improved design (rigidity and precision) and performance (speed and power) of the machine tools will further improve the efficiency of machining with ceramics.

Alumina based ceramic tools are not suitable for machining aluminum, titanium, and similar materials because of the strong tendency to react chemically. Similarly, Si_3N_4-based ceramic tools are not recommended for machining steels. Ceramics are also not generally suited for low speed and intermittent cutting operations because of failure by chipping unless used on extremely rigid high precision machine tools. Poor thermal shock resistance forbids intermittent application of cutting fluids. Either heavy flooding or no coolant at all is recommended for machining with ceramic tools for this reason. Nevertheless, the ceramic tools have great potential for increased productivity, consistent with economic and technological factors.

8. *Single and Polycrystalline Diamond*

Diamond, the hardest of all known materials (Knoop hardness 8000 kgf mm^{-2}) has been in use for cutting since antiquity. It is the cubic metastable form of carbon and is available both in the natural (single-crystal) and synthetic (polycrystalline sintered body) forms for cutting tool applications. Beneficial features of diamond besides its high hardness include its ability to form a sharp edge by cleavage (single-crystal natural diamond); good thermal conductivity; low friction; nonadherence to most work materials; ability to maintain a sharp edge for a long period of time (long tool life), especially when machining soft materials like copper and aluminum; and high wear resistance. The chief drawbacks of diamond, besides high cost, are its extreme brittleness resulting from easy cleavage, its limited chemical stability at high temperatures and low pressures where it is unstable and reverses back to its stable hexagonal graphite form, its strong tendency to react with elements of groups IVB, V, VI, and VIII (e.g., Ti, Zi, W, Co, Fe, and Ni) of the periodic table, and the difficulty and expense involved in tool fabrication and resharpening after use.

Good quality, single crystal natural diamond tools are used in ultraprecision machining that no other tool material can compete and maintain a sharp edge with minimal wear for prolonged periods of cutting. These include machining of Al and Cu front-surface mirrors for laser applications and microtome knives for biological applications. The lower-quality industrial diamonds are used for the machining of abrasive Al–Si alloys and cast iron for the automotive industry, copper commutators for the electrical industry, polymer and glass–epoxy composites in the aircraft industry, and concrete for the construction industry, and cold pressed and sintered carbides and oxides in the metal cutting and metal forming industries. Easy cleavage, unknown amounts of impurities and imperfections in natural diamond tools result in variability and unreliability in performance. In addition, limited supply, increasing demand for the strategic material and high cost have necessitated an intense search for an alternative dependable source of diamond. This search led to ultrahigh-pressure (~ 50 GPa)–high temperature ($\sim 1500\,^\circ$C) (HP-HT) synthesis of diamond from graphite in the presence of a catalyst–solvent material (e.g., Ni, and Fe) in the mid1950s and the subsequent development of polycrystalline sintered diamond tools in the late 1960s.

The polycrystalline diamond tools consist of a thin layer (~ 1.0–1.5 mm) of fine grain size diamond particles sintered and metallurgically bonded to a cemented carbide base, which provides the necessary elastic support base for the hard and brittle diamond layer above it. An assortment of tool geometries can be made (see Figure 10(a)). These tools are formed under much the same HP-HT conditions as those used for the synthesis of diamond. In the process of making sintered diamond tools, fine diamond powder (1–30 μm) is packed on a cemented carbide base and sintered in a press under the appropriate conditions of pressure and temperature (in the diamond-stable region). Complete consolidation and extensive diamond-to-diamond bonding (see Figure 10(b)) during sintering takes place as a result of high stress concentration at the sharp corners of the diamond crystals. In addition to the diamond-to-diamond bonding, good metallurgical bonding is established between the diamond layer and the cemented-carbide support base in this process. However, some of the binder phase and other impurities from the underlying cemented carbide can diffuse into the diamond layer above it. This can affect the high temperature performance of the polycrystalline diamond. In order to eliminate this effect, the metallic impurities in the polycrystalline diamond (also cBN) are leached and the resulting diamond (or cBN) layer is subjected to HP-HT conditions to obtain a much denser, impurity-free material with superior performance. Sintered diamond (or cBN) tools are then cut to shape by laser cutting or electrical discharge machining (EDM) followed by grinding, polishing, and lapping to size, finish, and accuracy.

Sintered diamond tools are made in an assortment of shapes (squares, rounds, triangles, and sectors of a circle of different included angles) and sizes. The main advantages of sintered polycrystalline diamond

Figure 10
(a) Photograph of an assortment of tool shapes of polycrystalline diamond (or cBN) consisting of a thin layer (~1–1.5 mm) of fine grain size particles sintered and metallurgically bonded on to a cemented carbide base and (b) micrograph of a polycrystalline diamond showing complete consolidation and extensive diamond-to-diamond bonding during sintering.

tools over natural single-crystal diamond tools are greater toughness, higher wear resistance and random orientation of the diamond crystals. In addition, the availability of sintered diamond tools is not dictated by nature or some artificial control; thus, such tools can be manufactured to meet strategic needs.

Sintered polycrystalline diamond tools are much more expensive than conventional cemented carbide or ceramic tools because both the processing technique and the finishing methods for diamond tools are expensive. In spite of higher cost, these tools are used for special applications because of higher productivity, longer tool life, and the overall economics of part manufacture. Being inherently brittle, diamond tools are used with a negative rake (~5°) geometry and 90° lip angle. When wear is negligible, as in the machining of Cu and Al front surface mirrors, a positive rake geometry can be used.

9. Low Pressure Diamond Coatings

A thin coating (~2–5 μm) of diamond on cemented carbide and ceramic cutting tool can prolong tool life considerably for such applications where currently expensive polycrystalline diamond tools or uncoated cemented carbide tools are used resulting in poor tool life as in the machining of abrasive Al–Si alloys, oil and gas drilling. Low pressure, metastable diamond synthesis is an attractive alternative for thin coatings (1–5 μm) (in contrast to a thick layer 1.0–1.5 mm by HP-HT technique). Many processes were developed recently for thin coatings of diamond. They are (i) activation by high temperature-hot filament, laser heating, arc discharge and arc plasma jets, and combustion; (ii) activation by electric or electromagnetic discharge-microwave and RF discharge, d.c. and a.c. glow discharge; (iii) plasma jet methods, RF, microwave, and d.c.; and (iv) combined activation-hot filament + microwave; hot filament + d.c. discharge, and hot filament + bias.

In the CVD diamond synthesis, hydrocarbon gas, typically methane (CH_4~1–5%) and hydrogen (H_2~95–99%) are used. In addition, oxygen is used at times to produce improved diamond coatings. Methane provides the source of carbon. High temperatures are needed to convert the molecular hydrogen to atomic hydrogen (>2000°C) that preferentially etches graphite without interfering with the diamond growth. The growth rates can vary from ~1 μm h^{-1} to 1000 μm h^{-1} depending on the process used. It may be noted that the highest deposition rate achieved by LP-HT is still an order of magnitude lower that with the HP-HT synthesis. Low cobalt (3–6% Co) bonded cemented carbides and silicon nitride substrates are the candidate materials for thin diamond coatings. Adhesion of diamond coating to the tool substrate is a challenging problem in view of diamond's unique properties and lack of compatibility with most tool materials. Hence, substrate preparation prior to diamond coating is essential for good bonding of diamond to the substrate at the interface. This includes etching of the Co from the surface of the substrate or mechanical treatment of the surface prior to coating. Figure 11 is scanning electron micrograph of a diamond coating on a silicon nitride cutting tool.

10. Cubic Boron Nitride

Cubic boron nitride (cBN), next only to diamond in hardness (Knoop hardness ~4700 kgf mm^{-2}), is an unusual material in that it does not exist in nature. Synthesis of cBN involves the structural transformation of boron nitride from hexagonal to cubic form, similar to diamond, at high temperature–high pressure (HP-HT) in the presence of a suitable catalyst–solvent (typically alkaline earth metals and their compounds). Though not as hard as diamond, cBN

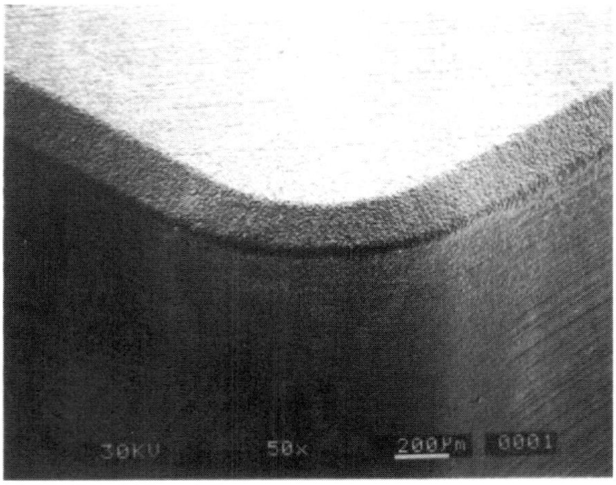

Figure 11
SEM micrograph of a diamond coating on a silicon nitride cutting tool by the LP-HT technique.

is less reactive to ferrous alloys. Consequently, many hard and difficult-to-machine materials such as hardened (R_c up to 65) alloy steels, nickel base superalloys, and hard, chill-cast irons can be machined at higher speeds and higher removal rates (3–5 times those of carbides) with good accuracy, finish, and surface integrity (both metallurgical integrity and residual stresses).

Sintered polycrystalline cBN are fabricated using similar methods to those for diamond tools and are available in similar shapes and sizes. Their costs are comparable because of similar processing and fabricating techniques but are significantly higher than the costs of cemented carbide or ceramics tool.

cBN tools have established a clear superiority in performance and productivity gains for certain machining applications, such as machining of hardened steels ($R_c > 40$), nickel base superalloys over cemented carbide or ceramic tools. However, the development of some of the newer ceramic tools, such as hot pressed alumina–TiC, SiAlON, and SiC whisker reinforced alumina are challenging this with lower costs and reasonable performance. cBN, however, maintains its superior performance over ceramics in the machining of extremely hard steels and hard chill cast irons where ceramics show poor tool life. Hard turning of bearings steels to final dimensions without the need for machining in the soft state, heat treatment to the final hardness, and final grinding is one important application of cBN. The potential of cBN can be increased significantly by using a high precision, rigid machine tool with adequate speed and power capabilities. Many unique applications of cBN have been identified in this field. cBN technology is rapidly advancing in the processing and fabrication areas to be cost effective with competing tool materials. Some developments in this direction include the fabrication of cBN tools (2–3 mm thick) without the cemented carbide substrate to take advantage of both sides for machining, production of significant volume fraction of nonmetallic second phase, such as TiC or TiN to increase toughness. The latter is targeted for intermediate hard steels (R_c 40–50) where toughness of the tool is increased at the expense of some hardness reduction.

The two predominant wear modes of cBN tools at normal speeds are notching at the depth-of-cut line and microchipping. These tools are not recommended for interrupted cutting as they are brittle, easily chipped, and expensive. Also, cBN is not stable at higher temperatures (i.e., at higher speeds) for it tends to transform to its stable hexagonal form (similar to graphitization of diamond) and react with air to form B_2O_3. Consequently, cBN tools are not currently recommended either for very low-speed cutting applications (not economical) or very high-speed (over 350 m min^{-1}) applications. Aside from these limitations, these tools are capable of very high removal rates when used with precision machine tools with adequate power and stiffness. In fact, contrary to intuition, they perform better under heavier cuts than with lighter cuts. Being inherently brittle, cBN tools perform better with a negative rake ($\sim -10°$) tool and a proper edge preparation.

Bibliography

Exner H E 1979 Physical and chemical nature of cemented carbides. *Int. Met. Rev.* **243**, 149

Komanduri R 1997 Tool Materials In: Howe-Grant M (ed.) *Kirk-Othmer Encyclopedia of Chemical Technology*. 4th edn., Vol. 24, pp. 390–455

Brooks K J A 1992 *World Directory and Handbook of Hard Metals*. 5th edn., An Engineer's Digest Ltd, UK

Doi H 1984 Advanced TiC and TiC–TiN base cermets. In: Almond E A, Brookes C A, Warren R (eds.) *Science of Hard Materials*. Hilger, Bristol, UK, pp. 489–523

Lux B, Haubner R 1989 Low pressure synthesis of superhard coatings. *Int. J. Refractory Metals Hard Mater.* **9**, 158–74

Moustakas T D 1989 *US Pat.* 4 804 583

Santhanam A T, Quinto D T 1994 *Surface Engineering of Carbide, Cermet, and Ceramic Cutting Tools*. ASM Handbook, Vol. 5, pp. 900–8

Santhanam A T, Tierney P, Hunt J L 1990 *Cemented Carbides*. ASM Handbook, 10th edn., Vol. 2, pp. 950–77

Schwarzkopf P, Keiffer R 1960 *Cemented Carbides*. Macmillan, New York

Schwarzkopf P, Keiffer R 1953 *Refractory Hard Metals*. Macmillan, New York

Stevenson R W 1984 *Cemented Carbides*. ASM Handbook, 9th edn., pp. 773–83

Lee M, Richman M H, Stanislav J 1971 The effect of a coated thin surface layer on the wear of cemented carbide tools. *Proc. 1st Int. Cemented Carbide Conference*. Chicago, USA, Feb. 1–3, 1971, Vols. I and II, Cemented Carbide Producers Association and the Society of Manufacturing Engineers (SME), FC71-929

R. Komanduri

D

Deformation Processing and Primary Shaping of Ceramics and Polymers: Comparison with Metals

Metals, ceramics, and polymers are used as engineering materials for a large number of industrial applications. The development and availability of metals, ceramics, and plastics with optimized properties has led to comparable material values in many cases (see Table 1).

This chapter deals with the material properties of metals in comparison with those of polymers and ceramics, and describes characteristic differences. In this connection, the focus is on the behavior under mechanical and thermal stresses, and the influence of these properties on deforming processing technologies. The most important physical processes which determine the behavior of the material are explained. Powder metallurgical fabrication processes for ceramics, and casting processes for plastic shaping will illustrate primary shaping. In this connection, the focus is on the fabrication and applications of fiber-reinforced plastics.

1. Physical and Thermal Properties

The main physical and thermal properties are density, melting behavior, thermal conductivity, and thermal expansion. The density of materials depends on several parameters, such as closeness of packing, atomic weight and size, and the porosity of the microstructure. Different materials usually show different melting characteristics such as congruent or incongruent melting, sublimation, or decomposition. The melting or sublimation behavior is largely determined by the strength of the atomic bond. Materials with strong primary and atomic structural bonds tend to have high melting temperatures. Therefore, materials with weaker primary bonds as the major kind of linkage, such as van der Waal bonds, show lower melting temperatures.

Weakly bonded alkali metals and monovalent ionic ceramics have low melting temperatures, stronger bonded transition metals have very high melting temperatures, multivalent ionic ceramics with increasing covalent bond behavior have increased melting temperatures, strongly bonded covalent ceramics have very high melting or dissociation temperatures, and strongly bonded metals have very high melting temperatures. Organic materials have low melting or decomposition temperatures because of the weak van der Waal bonds. Another determining parameter is the type of structure. Linear structures, such as thermoplastics melt, whereas network structures, such as thermosetting resins, decompose or degrade. The melting temperature of thermoplastic alloys can be increased by cross-linking or chain branching.

The upper operating temperature of a polymer is determined by the glass or melting transition temperature. However, the transition temperatures can be lowered by the use of plasticizers. As polymers

Table 1
Comparison of typical mechanical properties of several ceramics, polymers, and metals.

Material		Composition	Density ($g\,cm^{-3}$)	Melting temperature (°C)	Elastic modulus (GPa)	Vickers hardness ($kg\,mm^{-2}$)
Ceramics	α-Aluminum oxide	α-Al_2O_3	3.987	2050	380	2370
	Silicon carbide	SiC	3.220	2400	414	3300
	Silicon nitride	Si_3N_4	3.190	1800	304	2600
	Zirconium oxide	ZrO_2	5.827	2550	138	1410
Metals	Aluminum	Al	2.700	660	69	~60
	Magnesium	Mg	1.740	450	45	~40
	α-Iron	Fe	7.875	1534	197	75
	Steel (1040)	Fe-based alloy	7.850	1400	235	~140
Polymers	Polystyrene	Styrene polymer	1.050	70	3.3	~10
	Teflon	Polytetrafluorethylene	2.200	290	7.5	~10
	Plexiglass	Polymethylmethacrylate	1.200	120	3.0	~10

already start to flow or become sticky below the transition temperature, the upper operating temperature is usually 22–28°C lower than the transition temperature. At increasing temperatures, polymers tend to behave like viscous liquids, which has a negative effect on the mechanical properties of the polymer. Continuous exposure to high temperatures causes oxidic degradation, which leads to higher brittleness and change of color. The oxidic reactions have a more than negative effect on materials with a double bond in the main chain, such as isoprene and butadiene. Therefore, antioxidants are usually added to most commercial products made of polymers. Although the flammability of polymers can be reduced by the use of additives, polymers still release toxic fumes when they burn.

The amount of heat transfer is determined by the amount of applied heat energy, the nature of the heat carriers, and the amount of dissipation. In metals, heat is transported by electrons. Because of the nature of metallic bonds, these electrons are released from their atoms, and are therefore relatively free to move in all directions. The large number of heat carriers and the mean free path are the reasons for the high thermal conductivity of pure metals. Alloying reduces electron motion, which results in a decreased thermal conductivity.

Most organic materials have a low thermal conductivity (see *Thermal Conductivity*) because of the covalent bonding, the large molecular size, and the lack of crystal structures. Foamed polymers are excellent thermal insulators. The thermal conductivity of ceramics varies greatly. The primary carriers of thermal energy are photons and radiation. The highest thermal conductivity is achieved in regular arrangements; that is, structures consisting of a single element, structures made up of elements of similar atomic weight, and structures without impurities.

The thermal expansion coefficient of the material, i.e., the change in dimensions, results from an increasing amplitude of vibration between the atoms in the material at increasing temperatures. The level of thermal expansion is determined by the strength of the bonds and by the arrangement of the atoms. Metals with weak bonds and a closely packed structure have very high thermal expansion coefficients. Plastics can have even higher thermal expansion coefficients, e.g., polyethylene, which has a thermal expansion coefficient that is about four times that of aluminum and about 10 times that of steel.

2. *Mechanical Properties*

The most important mechanical properties of plastics, ceramics, and metals are their elasticity and their behavior under applied stress (see Fig. 1). The elastic modulus E is determined by the strength of the atomic bonds. The stronger the atomic bonds, the higher the stress required to increase the space between the atoms, and thus the greater the modulus of elasticity. Ceramics with weak ionic bonds have low E values, e.g., NaCl. Ceramics with strong covalent bonds have high E values, e.g., TiC.

The behavior is determined by the creation/movement of dislocations, and by stress concentration on microstructural defects. If the applied stress reaches the yield stress for dislocation motion first, plastic deformation occurs. If the local stress reaches the critical fracture stress on a microstructural defect first, brittle fracture occurs. Typical ceramic fractures occur in a brittle mode with only initial elastic deformation before fracture, while typical metal

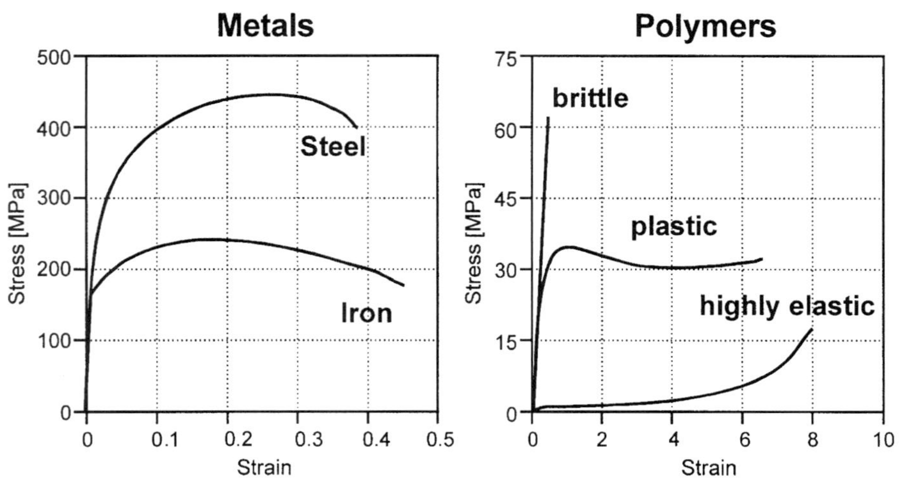

Figure 1
Stress–strain curves of different polymer types in comparison with metals.

fractures occur in a ductile mode with initial elastic deformation followed by a yield point and plastic deformation. The metal is elongated and decreases in cross-section during plastic deformation. Fracture occurs when the cross-section decreases to a degree that the applied stress can no longer be sustained.

The distinctive mechanical properties of polymers are the main reason for their widespread use in everyday life. Polymers can be both rigid and flexible; models to describe their mechanical properties contain both terms. It is well known that polymers over short time periods or at low temperatures show properties associated with glass, while polymers, over long time periods or at high temperatures show properties associated with viscous liquids. However, polymers also have many properties which are normally not associated with viscous liquids, such as die swell and second normal stresses. Furthermore, solid polymers can undergo gradual changes in shape, and they respond to stresses generated during the manufacturing process in the course of many years.

Mechanical properties are determined by a wide range of parameters, such as tensile strength, impact resistance, tear resistance, and hardness. In polymers, as in most materials, sample failure is generally due to inhomogeneity within the polymer sample, such as microscopic cracks or dust particles. If the temperature is low enough, or if it changes rapidly over a short time, all polymers undergo a ductile-to-brittle transition. Another phenomenon important to polymers is the ability of some of them to crystallize during applied stress. This stress-induced crystallization becomes visible in the form of a steep increase in stress at high (>50%) strain levels in the tensile curve. Creep and warping are a result of the same phenomenon, but creep refers to shape changes in response to an external stress field, while warping occurs because of an internal stress field that remains after processing. Although polymers have many characteristics associated with solids, they change over long time scales.

The hardness of materials depends on the tensile strength and on the toughness. Therefore, most ceramic materials, e.g., Al_2O_3 and SiC, have a very high hardness, which makes them an appropriate material for cutting and machining of metals. For many metals, hardness and tensile strength are approximately proportional to each other.

3. *Primary Shaping and Processing of Ceramics*

The basic material for most ceramic fabrication processes is powder. Processes like pressing, slip casting, extrusion, injection molding, tape forming, and green machining are used to give the final, or near-final, product shape (see Fig. 2). The shapes resulting from these forming processes essentially consist of powder compacts which are densified by

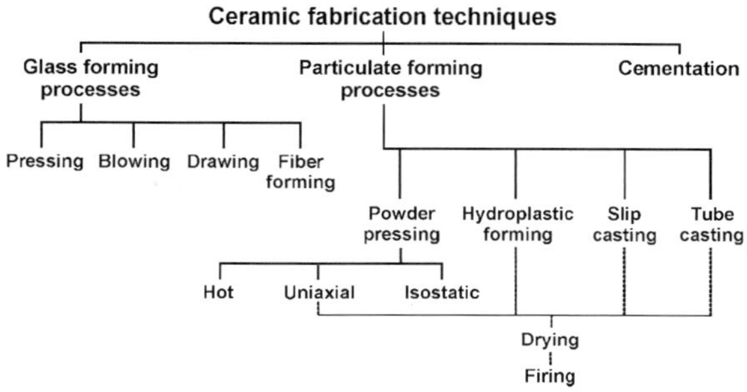

Figure 2
Classification scheme of ceramic-forming techniques.

high-temperature processing to obtain an adequate strength, and several other properties. Some processes combine forming and densification in a single step, like, for example, hot-pressing, chemical vapor deposition, the use of liquid particle spray, and cementitious bonding. Ceramic components requiring small tolerances must be machined after densification.

Since the 1950s, scientists and engineers have acquired a much better understanding of ceramic materials and ceramic processing. They have found out that unique properties of ceramics can be achieved by refining naturally occurring materials, or synthesizing new compositions. These refined or new ceramics are often referred to as modern ceramics, whose special feature is a highly controlled composition and structure. As a result, modern ceramics are appropriate for highly demanding applications. Types of modern ceramics are:

oxide ceramics (such as Al_2O_3, ZrO_2, ThO_2, BeO, MgO, and $MgAl_2O_4$)

magnetic ceramics (such as $PbFe_{12}O_{19}$, $ZnFe_2O_4$, and $Y_6Fe_{10}O_{24}$)

ferro-electric ceramics (such as $BaTiO_3$)

nuclear fuels (such as UO_2 and UN)

nitrides, carbides, and borides (such as Si_3N_4, SiC, B_4C, and TiB_2).

Al_2O_3, MgO, SiC, and Si_3N_4 are made of powder. The characteristics of the ceramic powder are determining parameters for the final properties. Purity, for instance, greatly influences high-temperature properties, such as strength, stress-strain behavior and oxidation resistance. As closer packing can be obtained by the use of different particle sizes, a wide range of particle sizes is required. The powder will then be formed to the final shape by the use of several techniques. Pressing is done by placing the powder into a die and by applying pressure in order to compact the particles.

Common methods are isostatic and nonaxial pressing. The casting of ceramics is done at room temperature. The ceramic particles are suspended in a liquid, and cast into a porous mold which removes the liquid and leaves a particulate compact in the mold. According to the viscosity of the ceramic liquid suspension, the mold, and because of the procedures used, there is a wide range of varieties to this process. The most common one is slip casting. Tape casting has been developed to allow low-cost mass fabrication of thin ceramic sheets. The slight difference between tape and slip casting is that the slip is spread onto a flat surface rather than being poured into a shaped mold.

Extrusion is an extensively used plastic-forming process that has been adopted by the manufacturers of ceramic components (furnace tubes, insulators, bricks, pipes, tiles, tubular capacitors, catalyst supports, magnets, heat-exchange tubes, and others). This process consists of forcing a highly viscous and doughy mixture of powdered ceramics, plastics, and additives through a die.

Although injection molding is very different from extrusion, the cross-section of the mold is approximately that of an extruder. In this process, a mixture of ceramic powder, thermoplastic polymer, plasticizer, wetting agent, and anti-foaming agent are preheated in a barrel to a temperature at which viscosity of the polymer is low enough to allow flowing under applied pressure. Furthermore, the heated material is compacted by a ram or plunger. Then the viscous material is forced through an orifice into a narrow passageway that leads to the shaped tool cavity. At this point, the mixture is much more fluid than an extrusion mixture, and is unable to form a self-supporting shape. The strand piles up until the cavity is full and the material has "knit" or fused together under the pressure and temperature to produce a homogeneous part. The shaped tool is cooler than the injection-molding mix so that the mix becomes rigid in the tool cavity. The part can be removed from the tool as soon as it is rigid enough to handle without deformation. Cycle times can be rapid, providing the potential for injection molding as a high-volume, low-cost process for fabrication of ceramics into complex shapes.

4. *Shaping Processes of Polymers*

The most common method of giving the desired geometry and dimensions to a polymeric material is to shape it in its softened or molten state. Depending on the complexity of the shape and the desired dimensions (thickness, width, radii), the choice of processing will vary: it may be a process which produces a simple flat product (film, sheet, slab), a process which produces a profile (rod, tube, or solid profile with a certain cross-section and surface texture), or a process which produces a complex or hollow product.

Flat films, sheets, and slabs are the simplest shapes of plastic products. The methods used to manufacture flat goods from molten polymers are either calendering or extrusion. Flat goods with a limited length may be produced by compression molding.

Parts with different shapes and sizes are produced by shaping a molten or softened polymer under externally applied stress. There is a great variety of processes for the production of molded parts, according to the required shape and dimensions, the number of pieces, costs of purchase for the equipment, and operating costs. Common processes are: compression molding, transfer molding, liquid resin molding, resin transfer molding, injection molding, reaction injection molding, blow molding, thermoforming, and rotational molding.

Most plastic products do not require further forming since they are already produced in the first step with the required shape, dimensions, and mechanical properties. However, certain products need further processing, such as deflashing, machining and drilling, cutting, calibration, slitting, stretching, painting, surface treatment, metallization, and printing.

Plastics are very versatile and can be given various chemical and physical parameters for specific applications. Multiple compositions are classified based on the monomers and the reaction sequence used to form the polymer. Both thermoset plastics and thermoplastic polymers can be used for the development of reinforced plastic objects. For the most part, fiberglass (see *Fiberglass*) is used as a reinforcing agent, but carbon and aramid fibers are also used. Common fillers and extenders include talc, mica, glass flakes or glass, spheres, sawdust, and calcium carbonate.

Reinforced plastics are fabricated either by lamination, resin injection, spray up, preform molding, resin transfer molding, or sheet molding compound processes. In modern engineering, reinforced plastics have started to replace other materials in applications which require a high tensile strength, but a lower weight, for example in the automotive industry. Although a sheet molding compound (SMC) is four times lighter than steel, it can show up to three-fifths of the tensile strength of steel (comp. Table 2). These and many other advantages of reinforced plastics make them one of the most important materials for modern engineering.

In lamination, an open-faced mold is used to produce a finished part with only one surface finished to high quality. These parts have no shape or size limitation except those imposed by handling considerations. Typically, a polyester resin is used for most commercial applications, because of its low cost. However, epoxies are used in many aircraft applications where performance is of primary concern. In general, the physical properties of laminates are improved by an increased fiber length, and optimum properties are obtained by specific resin-to-fiber

Table 2
Comparison of mechanical properties of SMC and steel.

Property	Low-profile SMC	Structural SMC	SAE 1008 steel
Tensile strength (MPa)	65–90	124–204	330
Tensile modulus (GPa)	8.3–12.4	12.4–20	207
Specific gravity	1.8–2.0	1.85–2.15	7.86

ratios. A higher pressure during the curing process causes improved physical properties; therefore, pressing is used even for very large parts. Mats, woven cloth, rovings, woven rovings, chopped strand mats, etc., can all be employed in the lay-up process. The first step is to apply a releasing agent to the mold surface, followed by a gel coat if a smooth, colored surface is required. The gel coat is a viscous coating, containing pigments, ultraviolet stabilizers, as well as other additives, and is applied by spraying to a thickness of 0.5–1 mm. Then, a thin mat is applied and coated with resin, using brushes or rollers. Successive layers are applied until the desired thickness is achieved, and the resin has been applied. The entire surface is rolled to remove trapped air and to wet the fibers completely. Although hand lay-up requires skill and is labor-intensive, it is still used extensively in the fabrication of low-volume products.

The hand or spray lay-up is an open-mold system. The mold may be a simple wood or plaster unit, or a cast or sheet metal construction requiring permanence. The size may vary from a 300 mm box up to 20 m for a yacht body. The mold surface must be waxed intensively and then be sprayed with a polyvinyl alcohol solution to prevent the laminate from sticking to the surface. To make a smooth surface, a gel coat is applied. The thickness of the gel coat, which is either brushed or sprayed, is approximately 0.5 mm. The gel is an activated polyester resin that has a high filler level and is stabilized for weathering resistance.

Resin injection molding (RIM) as an intermediate volume process uses tooling equipment which is lower in cost than that used in conventional compression or injection molding techniques. Molding cycles take a much longer time than those for compression and injection methods, but production rates exceed those of the open-mold techniques by far.

The closed mold, made from plastics, spray metal, or cast aluminum, produces parts with smooth surfaces on both sides. A gel coat may be used to obtain the desired color, to improve weathering resistance, to increase corrosion resistance, and to enable post-finishing operations. The mold has cooling coils for exothermal control, a perimeter gasket to prevent resin and pressure leakage, and vents to bleed off air and determine part fill. The resin compound, including catalyst and filler, is pumped into the mold through an injection port. The curing cycle is from 10 to 20 minutes because the resin is injected rapidly, and thus allows faster gelling and curing in laminating systems. This closed-mold technique also enables foam encapsulation and the use of ribs and inserts.

In the spray-up process, a mixture of chopped fibers and matrix resin can be applied to a gel-coated mold using an airflow and a spray gun specifically designed for this purpose. The catalyst system and resin are normally mixed in the cavity of the spray gun. Pressurized vessels which eliminate air pressure in the gun can also be used. Hand-held guns allow a free distribution pattern for chopped glass and resin to optimize the properties of the final product. Robotic spray guns can be employed to automate the filling of large molds or mold clusters for smaller parts. Hand-spray lay-up techniques are common for fiberglass shower stalls, tubs, pools, camper tops for pickup trucks, temporary shelters, boats, canoes, surfboards, and signs. These techniques even enable the fabrication of very large pieces.

In preform molding, glass preforms are used in compression molds for parts finished on both surfaces. Chopped fiber with a binder is sprayed from a gun or tube onto a screen through which air flows. The fibers are on the surface as the binder cures, creating a preform that is trimmed to size. Normally the excess fibers of the preform are removed after compression molding by manually trimming the final part. If a water jet is used prior to molding, the preform may have to be dried.

Steel rule dies and laser trimming are alternative methods. The preform is inserted into a mold, resins are applied to the surface, and the part is molded. This technique is used where deep draws and difficult contours make it hard to achieve the desired properties with SMCs, or bulk molding compounds. Farm tractor bodies, industrial housings, truck cabs and hoods, concrete forms, bus seating, boats, etc., are molded by this technique.

The slurry process for making preforms uses chopped glass roving in water. Chopped fiber and a fibrous binder are mixed in a water bath. A screen in the shape of the preform is dropped to the bottom of the bath. When the screen is pulled up, water running through the screen deposits fiber on it. The size and placement of holes in the screen control the fiber distribution and location.

Resin transfer molding (RTM) combines the versatility of hand lay-up with the speed and low volatile emissions of closed-mold processes. In this process, the resin is injected directly into the closed mold, with no direct exposure of the resin to the air. The reinforcement is prepositioned in the mold, although in some cases the reinforcement fibers can be added to the resin before injection. High-quality surfaces can be achieved at relatively low mold temperatures (60–90 °C) and low pressures (14–28 kPa), which have substantial impact on the reduction of the tooling cost. Steel tooling is not required, since the temperature and pressure involved are relatively low. This system offers advantages for short runs and prototyping.

In the SMC manufacturing process, liquid thermosetting resin, glass fiber, and filler are converted to a sheet molding compound, which is normally about 3–5 mm thick and 1 m wide. The length of the sheet is determined by molder preference. The sheet is handled on a roll or in bins. Before SMC can be used for molding, it must age or mature. During this maturation time, the residual polymer acid functions react with the calcium oxide or other metal oxide thickeners, which results in a thickening of the low-viscosity resin. The thickened SMC prevents resin paste from being squeezed out of the glass-fiber matrix, making it easier to handle and mold. Approximately three to five days are required to reach the desired viscosity.

During manufacturing of SMC, all ingredients, except the glass fibers, are blended together to form a thick paste. The paste is applied to top and bottom carrier films with doctor boxes that control the thickness of each carrier film. A rotary chopper is used to cut glass fiber roving to length. Normally, fibers are chopped to 25 mm, but the length can vary from 5 to 50 mm, depending on the application. The glass fibers are deposited randomly on the paste-coated carrier film and are generally oriented more or less parallel to the direction of film travel. The amount of deposited glass fibers is controlled by the speed of the rotary chopper and the speed of the carrier film.

When the SMC is ready for molding, the material is spread onto a cutting table and cut into pieces of predetermined size and shape. Cutting is usually done manually by the press operator with a template and mat knife. Some large volume applications have an automated slitter table. Stacked and assembled pieces are arranged into a charge pattern that has been determined to provide the optimum shape and volume to fill the mold cavity. It is important to trim and assemble the charge pattern as accurately and consistently as possible to avoid process variation. A charge pattern for a Class A part usually covers 20–40% of the mold surface. For structural parts, a charge pattern may cover up to 90% of the mold surface. The mold is heated to 120–160 °C with pressures for low profile SMC of 5500–8300 kPa. The curing time for SMC varies from 30 to 150 s. The curing time is mainly determined by the formulation of the SMC and the part thickness. Initially, the resin system has a low viscosity in order to ensure complete mixing of the additives and wettability of the glass reinforcement. The viscosity increases while the sheet compound is being formed.

Bibliography

Callister W D Jr 1997 *Materials Science and Engineering: An Introduction*, 4th edn. Wiley, New York

Dieter G E 1986 *Mechanical Metallurgy*, 3rd edn. McGraw-Hill, New York

Moore G R, Kline D E 1984 *Properties and Processing of Polymers for Engineers*. Prentice-Hall, New York

Nielsen L E 1994 *Mechanical Properties of Polymers and Composites*, 2nd edn. Dekker, New York

Reed J S 1988 *Principles of Ceramic Processing*, 2nd edn. Wiley, New York

Richerson D W 1992 *Modern Ceramic Engineering*, 2nd edn. Dekker, New York

H. Haferkamp

Die Casting (Permanent Mold)

The die casting process is a name given to metal casting processes that utilize metal molds or dies. There are really several distinct processes included under the general name. The die casting process actually has three main sub-processes. These are: (i) permanent mold casting, also called gravity die casting, (ii) low-pressure die casting, and (iii) high-pressure die casting. In North America, the term die casting is used to mean high-pressure die casting, but the general term embraces all three subcategories.

The three processes differ mainly in the amount of pressure that is used to force the molten metal into the die. In permanent mold casting, the molten metal is poured into the mold and flows only at the force of gravity. The low-pressure process normally utilizes air pressures up to 1.5 MPa to force the molten metal into the die, although for special products the pressure can be as high as 7 MPa. The high-pressure die casting process subjects the molten metal to hydraulic pressures as high as 140 MPa. As this pressure on the metal is increased there is a corresponding reduction in the time required for the molten metal to fill the die.

The die casting process involves pouring or injecting molten metal into hardened steel dies, where the metal solidifies to the desired casting shape. In the casting processes that utilize sand or plaster for molds, the mold is destroyed by the molten metal heat or when the casting is extracted from the mold. In all of the die casting processes, the dies can withstand the casting heat and are constructed with movable sections that allow easy removal of the solidified casting. Therefore, these molds are reusable and can be used for producing many thousands, or even millions, of castings. The advantages of such a process include high production rates, exceptional dimensional repeatability, low part costs, and less machining due to reduced casting finish stock. The disadvantages are design limitations due to steel dies, higher initial die cost, and longer lead time for die construction and for changes to the die caused by a casting design change.

1. Permanent Mold Casting

In permanent mold casting, the molten metal is poured into the steel die and flows only at the force of gravity. For the most part, permanent mold castings are produced by pouring the molten metal into the top of a die that has been made in the desired shape of the casting. There are many deviations from this simple approach in actual casting practice. Sometimes, the metal is poured into the mold at the top, but a runner is cut into the die that conveys the molten metal to the side or bottom of the casting cavity itself, so that the metal flow during filling is from the bottom or side of the mold. This is done to optimize casting conditions for the part being produced. Other times the metal may be poured into a basin beside the die and then the entire unit is tilted to achieve controlled filling of the mold. The flow of metal into the die cavity and the flow of heat from the metal to the die during solidification are two of the main criteria for successful casting production by the permanent mold process.

Dies for the permanent mold process are produced from hardened steel. The heat transfer rate is higher than from sand or ceramic molds used in other processes and this provides advantages in casting cycle time. Often, a ceramic coating is applied to the steel to protect it from the molten metal and to control the heat transfer rate from the molten metal to the die at only a slight increase in cycle time. The solidification of the molten metal must be controlled to prevent solidification shrinkage cavities in the final product. Also, the solidification rate is managed to optimize the microstructure of the resulting casting to achieve the design objectives. For high production castings, the dies are often water-cooled, further decreasing the cycle time and reducing the cost for each part.

Sometimes the casting design requires undercuts or hollow cavities in the casting that cannot be produced with conventional steel dies. In these cases, the die designer may choose a loose piece or an expendable core for the permanent mold process. Both of these options increase die and casting costs, but often the function of the product requires these features in the casting. When expendable cores are used, the process is called semipermanent mold casting. This subcategory of permanent mold casting refers to processes that utilize one or more expendable cores that are placed in the steel die before the production of each casting. The expendable core is then removed by vibration or heat treatment after the casting is removed from the die. Expendable cores for semipermanent mold casting are normally produced from sand with a binder to give them strength, similar to the cores utilized in sand casting processes.

The advantages of the permanent mold and semipermanent mold casting processes are: reasonable piece costs resulting from the high production rates achieved with metal molds (especially water-cooled molds) compared to sand and investment casting, and lower investment required for equipment when compared to low-pressure and high-pressure die casting. In addition, the use of expendable cores in semipermanent mold casting permits great design flexibility for castings.

One of the disadvantages of the permanent mold casting process is that metal dies are more expensive than patterns for sand casting or investment casting so the process is not economical for short runs. At low volume, it is difficult to overcome the high initial tooling cost and compete based on casting cost. In addition, since castings are filled with liquid metal under only the pressure of gravity, castings sections tend to be thicker in permanent mold casting than in the low-pressure and high-pressure die casting processes. Also, where material properties are critical, emerging concepts in high-pressure die casting, such as squeeze casting and semisolid metal processing, are creating new competition for permanent mold casting.

Typical parts produced in the permanent mold process include automotive parts such as aluminum pistons, steering knuckles, brackets, wheels, and pump impellers. Parts are also produced in zinc, brass, copper, lead, and even gray iron. Since the process has great design flexibility and is compatible with so many metals, the types of products that can be produced are almost unlimited.

2. *Low-pressure Die Casting*

Low-pressure die casting was developed in an attempt to eliminate hand ladling of the hot metal. This process also utilizes metal molds to produce castings, but the molten metal is pressurized to achieve faster or better-controlled filling of the mold. The process utilizes pressures up to 7 MPa for special products such as automotive wheels, but typically the pressure used is below 0.5 MPa.

In low-pressure casting, the molten metal is contained in an insulated crucible or furnace which is pressure sealed. A tube passes vertically down through the furnace, with its lower end immersed in molten metal and its top flange sealed against the furnace lid. The die is mounted on the machine, over the furnace, and sealed to the open end of the tube. When air pressure is applied to the furnace, it displaces the molten metal, causing it to travel up the tube and into the die cavity as seen in Fig. 1.

The cooling process takes place in a sequence starting at the extremities of the die, working back to the feed head, which is the mouth of the tube. Upon solidification of the casting, the unwanted metal is returned to the furnace by relaxing the air pressure; the die is opened, the casting is removed, and the casting cycle is repeated. Sometimes a vacuum is applied to the mold cavity to cause the metal to flow. Atmospheric pressure on the metal then forces the molten metal up into the bottom of the die and fills the part.

Using pressure to fill the die cavity permits faster filling of the die than by gravity casting. Faster fill times are important because the molten metal loses temperature to the die as it flows past the die steel and the hotter the metal the thinner the section thickness that can be filled. For some products, the fill rate must be varied depending on the casting section thickness and very sophisticated computer-controlled, low-pressure machines are utilized. This additional capability is not free, however, and is not

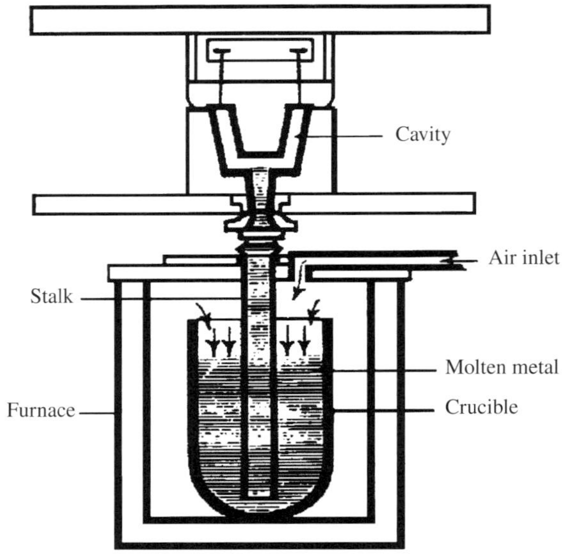

Figure 1
Low-pressure die casting.

cost effective for every product. In that sense, low-pressure die casting is an enhanced process that is generally used for parts with premium requirements. This is the case for many automotive engine components such as heads, lower crankcases, and manifolds that require expendable cores because of the intricate internal passageways that are required. These parts must also be leak free. Many of these parts are produced by the low-pressure die casting process because of their size, complexity, and volume. Gravity and low-pressure, semipermanent mold casting are the primary methods used to produce these complex aluminum automotive components throughout the world. The decision to use gravity or low-pressure casting depends on product requirements and casting volume.

Dies for low-pressure die casting are also produced from hardened steel, but they require more complex design and engineering to assure that pressurized molten metal is contained within the casting system. Detailed metal feed systems and connections to the molten metal bath are required as well as seals in the die where a vacuum is used. These features increase the cost and lead-time of dies when compared to gravity permanent mold dies. The dies are still less costly than those used in high-pressure die casting, however.

The casting machines for the low-pressure process can also become fairly sophisticated as well, requiring higher capital investment for sophisticated pressure or vacuum control systems. These higher costs are offset by the higher value of complex castings that can be produced, as well as by higher production rates and improved casting fill capability, and they still cost much less than machines for high-pressure die casting.

The advantages of low-pressure die casting are that casting quality is very good, especially in heavy wall sections, and casting yield is 90–95% since no runners, risers, or gates are used. This results in much less trimming to remelt and corresponding savings in fuel cost. There is a much greater alloy choice than with high-pressure die casting, permitting the use of heat treatable alloys. The use of expendable cores in low-pressure die casting allows this process to cast parts that are impossible to produce in the high-pressure process.

The disadvantages of low-pressure die casting are that it has a lower production rate than high-pressure casting and it is not normally suitable for casting smaller parts. The minimum wall thickness, while smaller than permanent mold casting, is still larger than for high-pressure die casting. Tooling cost is also higher than for the permanent mold process.

3. *High-pressure Die Casting*

High-pressure die casting is a process in which molten metal is forced under pressure into a securely locked metal die cavity, where it is held by a powerful press until the metal solidifies. After solidification of the metal, the die is unlocked, opened, and the casting ejected. After removal of the casting, the die is closed and locked again for the next cycle. The injection of metal into the die cavity is completed in a fraction of a second. Often, while the molten metal is still held in the die, extremely high pressure is applied (called intensification pressure). This high pressure compresses any gas entrapped in the metal and feeds additional metal into the cavity to compensate for the shrinkage of the metal as it solidifies.

Two types of systems are used for injecting the molten metal into the die. The hot chamber system is used with metals such as zinc, magnesium, and lead. The injection system of a hot chamber machine is immersed in the molten metal bath of the melting furnace. As the shot plunger moves, it forces metal through the nozzle and into the die.

The cold chamber system is used for metals that melt at high temperatures, such as aluminum, brass, and magnesium. Magnesium parts can be produced using both systems, though usually small parts are produced in hot chamber machines and large parts in cold chamber machines since hot chamber machines are limited in size. There are also two injection systems used in the cold chamber process, horizontal and vertical injection.

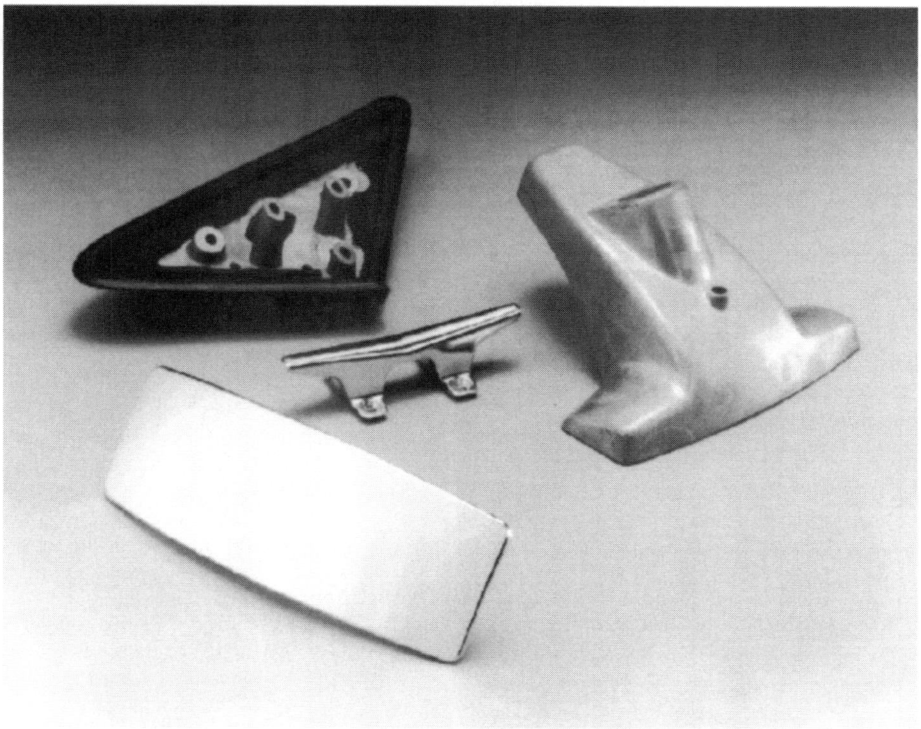

Figure 2
Typical zinc die castings.

In the cold chamber process, the molten metal is poured, by hand or by automatic means, into a port of the cold chamber sleeve. A hydraulically operated plunger advances through this steel sleeve, sealing off the port, and forcing the metal into the die at high speed and pressure. After solidification of the casting, the plunger is retracted, the die opened, the casting ejected, and the system is then ready for the next shot cycle. Higher pressure is used in this system than the hot chamber process. The production rate of a hot chamber machine is higher than that of a cold chamber machine because of the shorter time required during the pour operation. Typical zinc castings produced by the high-pressure die casting process are shown in Fig. 2.

In addition to the conventional high-pressure die casting processes, several enhancements to the process have been developed in recent years. These enhancements include the use of vacuum systems to reduce entrapped gas, slower fill processes to eliminate turbulence during fill and permit the use of heat treatment to enhance mechanical properties of the castings, and the application of semisolid metal processing to produce pressure tight parts not normally able to be produced in high-pressure die casting. Each of these processes utilizes the fundamentals of high-pressure die casting, but with additional capabilities to produce high integrity parts. They also have developed unique names for the purpose of distinguishing themselves from the conventional processes of hot chamber and cold chamber die casting.

Vacuum die casting utilizes a vacuum system on the die cavity to remove gas from the cavity prior to injection of the molten metal. The result is a high integrity part with very low levels of porosity and high mechanical properties. The process is used to produce critical components in light alloys, such as structural and safety components for motor vehicles. The process has higher costs than conventional die casting, but the parts produced justify the additional cost and they cannot be produced successfully in conventional die casting.

Slow fill die casting is often called squeeze casting. This process is widely practiced throughout the world today for producing parts that must be heat treated to achieve the required mechanical properties. Many automotive components that have been converted

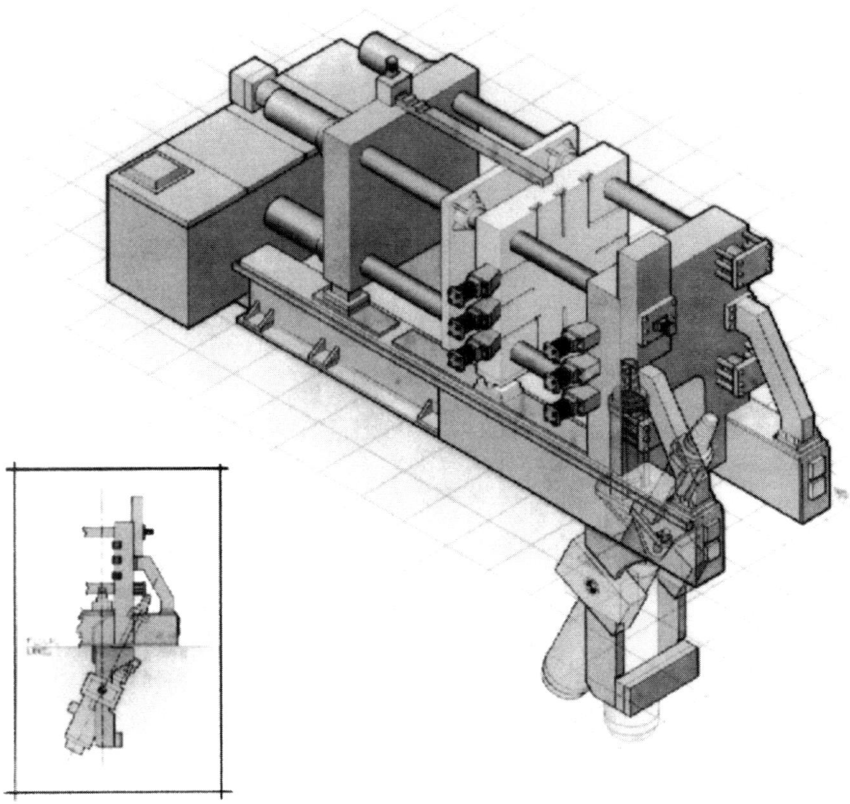

Figure 3
Vertical shot horizontal squeeze casting machine.

from iron castings or weldments of castings and/or stampings are now produced in light alloys in the squeeze casting process. This process utilizes the advantages of the low-pressure die casting process, controlled filling, and directional solidification of the molten metal, as well as the advantages of high-pressure die casting, high-pressure solidification, and fast cycle times. Parts produced in the squeeze casting process include steering and suspension components, e.g., alloy wheels, steering knuckles, and control arms, and air conditioning parts, e.g., compressor scrolls. Parts are produced with both vertical and horizontal shot systems. A schematic diagram of a vertical shot horizontal squeeze casting machine is shown in Fig. 3.

Semisolid metal processing (SSM) is beginning to share some of the same markets as squeeze castings, as well as some of the smaller automotive components such as link arms, fuel rail parts, and drivetrain parts. In this process, the metal injected into the die is only about 50% liquid. The process offers distinct advantages in that the feed metal fills the cavity in a manner that is less turbulent than in conventional die casting. Furthermore, the casting is partially solidified at the onset of the process, thus the solidification "journey" is a much shorter one. The technical capabilities of the process include high integrity of the castings, high mechanical properties, the ability to cast thin walls, good dimensional accuracy, surface conditions similar to high-pressure die casting, and suitability for heat treatment and welding. Semisolid casting and SSM materials are covered in detail in *Semisolid Processing*, and *Casting of Semisolid Metals: Engineering Applications*

The advantages of high-pressure die casting include a higher production rate than with gravity or low-pressure casting. Also, the ability to produce castings with close dimensional control greatly reduces machining operations. Die castings have good surface finish, which is a prime requirement for plating, and much thinner wall thickness is possible reducing overall casting weight. Dies have a long life, reducing unit part costs, and more complex parts can be produced, thereby reducing the number of components required in an assembly.

The disadvantage of high-pressure die casting is that it is best suited to high volume parts. High tooling costs make short production runs uneconomical. Also, the internal porosity prevalent in conventional high-pressure die castings makes producing pressure tight parts difficult, often requiring the use of vacuum die casting, squeeze casting or SSM casting. There are a limited number of alloys suitable for die casting and this restricts the heat treatment or welding of the finished castings. Iron or steel alloys are normally not die castable. There are restrictions in die casting on the casting size and wall thickness which eliminate the possibility of die casting some parts, but this has been reduced with the development of the new high integrity processes of vacuum die casting, squeeze casting, and SSM casting. Die casting machine and maintenance costs are higher than for other casting processes.

4. Concluding Remarks

The term die casting is used to describe processes that utilize metal dies, or molds, to produce parts from various metals. These processes include gravity permanent mold casting, in which the liquid metal is poured into the die; low-pressure die casting, in which the metal is forced into the mold with air pressure; and high-pressure die casting, in which a hydraulic ram is used to inject the molten metal into the die at extremely high pressures. In North America, the term die casting is used to describe high-pressure die casting, while in most of the world the term die casting may be used to describe all of these processes that utilize metal molds.

The die casting processes provide a great deal of flexibility for the production of metal components. Permanent mold and low-pressure die casting can produce parts from almost any metal in almost any shape. In the case of extremely high volume parts, high-pressure die casting provides high production rates and good dimensional repeatability. In recent years, enhanced high-pressure die casting processes such as vacuum die casting, squeeze casting and semisolid metal processing have created new opportunities for utilizing the advantages of high-pressure die casting for structural or leak free components.

Bibliography

Bradney D D 1994 *The NFFS Guide to Aluminum Casting Design: Sand and Permanent Mold.* Non-Ferrous Founders' Society, Des Plaines, IL

Corbit S, DasGupta R 1999 Squeeze cast automotive applications and squeeze cast aluminum alloy properties. In: *SAE Technical Paper Series, International Congress and Exposition.* Society of Automotive Engineers, Inc., Warrendale, PA

Herman E A 1982 *Die Casting Handbook.* North American Die Casting Association, Rosemont, IL

Jorstad J 1999 Selecting parts and material for conversion to SSM. In: Nussbaum A I (ed.) *Semi-Solid Metal World Users' Conference.* Ormet Primary Aluminum Corporation, Hannibal, OH

Midson S P, Young K P 1999 Technical support for high integrity processes. *Diecasting World* (June), 16–18

Prince Machine Corporation 1980 *Low-pressure Manual.* Prince Machine, MI, USA

W. A. Butler

Directional Recrystallization

When a cold-worked metal is annealed, *recrystallization* results from the passage through the material of high-angle grain boundaries. This process involves the nucleation of new, strain-free grains in the deformed material, which then grow until all the latter is consumed. Such a sample will usually then consist of an array of equiaxed grains, but there are cases where unconventional alloying or processing leads to the development of a microstructure of columnar grains aligned along a particular sample axis, which is known as a *directional recrystallization* (DR).

The majority of reported examples of DR are found in oxide-dispersion-strengthened (ODS) alloys, which have been developed for their exceptional creep resistance. Improved high-temperature ductility and creep rupture behavior is found in alloys with a high grain aspect ratio (GAR, the ratio of maximum to minimum grain dimension) when compared with equiaxed material. DR treatment has led to rupture times being increased by more than two orders of magnitude in such materials (Singer and Artz 1986).

1. The Driving Pressure for DR in ODS Alloys

The driving pressure (P) for boundary migration can originate from stored dislocations and from the reduction in total grain-boundary area. ODS alloys are usually produced by *mechanical alloying* (see *Mechanical Alloying*), in which mixtures of metal powders and fine refractory oxide powders are milled together, compacted, and extruded at high temperature. At this point they have a relatively low ($\sim 10^{13}\,m^{-2}$) density of dislocations (ρ) and an ultrafine ($\sim 0.2\,\mu m$) grain size (D), which is stabilized by the dispersed oxide phase. The driving pressure (P_b) provided by the reduction in grain boundary area is given by $\sim 2\gamma/D$, where γ is the grain boundary energy per unit area. Taking $\gamma = 1\,J\,m^{-2}$, this gives a typical value for P_b of 10 MPa.

The driving pressure (P_d) due to the reduction in dislocation line length is approximately $\frac{1}{2}Gb^2\rho$, where G is the shear modulus and $\boldsymbol{b}$ the Burgers vector. For nickel-based alloys, typical values of P_d would be 3×10^{-2} MPa, which indicates that the driving pressure for DR mainly originates from the reduction in grain boundary area, thus identifying it as a process of *secondary recrystallization (SRX)*, alternatively called abnormal grain growth.

2. Processing for DR

In order to produce an elongated grain structure, three possible annealing treatments may be used, namely isothermal annealing, static annealing in a temperature gradient, or annealing in a moving temperature gradient (*zone-annealing*). While some success has been achieved using isothermal annealing to produce elongated grain structure in ODS materials, the preferred technique for both ODS and non-ODS alloys is zone-annealing. A local heating element is moved along the specimen, causing a hot zone to traverse its length. An example of a microstructure produced by this process is shown in Fig. 1, which is a longitudinal section through a partially zone-annealed forging. The inclined recrystallization interface shows that the transformation temperature varies across the specimen, which reflects the original strain gradient in the forging.

An important associated feature of material which has undergone DR is the presence of a strong *fiber texture* most nickel-based ODS alloys exhibit a <110> texture.

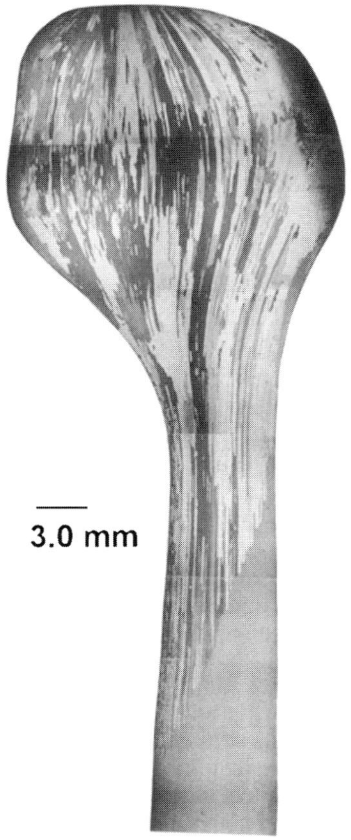

Figure 1
Longitudinal section through a partially zone-annealed forging of MA 6000 showing an inclined recrystallization interface reflecting the strain gradient in the forging (after Marsh and Martin 1991).

2.1 Nucleation of DR in ODS Alloys

Mujahid and Martin (1994) showed by quenching experiments that the nucleation of SRX takes place at a temperature significantly higher than that at which the steady-state interface (Fig. 1) migrates along the specimen. They suggest that thermally activated unpinning of clusters of grains with <110> texture takes place, and these form nuclei with the requisite size advantage to grow into the matrix.

2.2 Growth of DR Grains

It has long been known that massive grain coarsening, even growth of single crystals, can be achieved by passing material through a steep temperature gradient. The temperature gradient ensures that only a small region of the material at any one time will be above the SRX temperature, and the fact that the nucleation temperature is higher than the growth temperature is beneficial for successful DR, in that it provides a "buffer region" below the interface where nucleation is suppressed.

Hotzler and Glasgow (1982) have shown that below the SRX temperature stored energy is gradually annealed out by normal grain growth, resulting in a reduction in driving force for SRX. This imposes a minimum temperature gradient below which zone annealing will not be effective in producing elongated grains, and which in practice sets a lower limit to the rate of zone travel.

If the zone speed is too high the transformation will be incomplete, as the moving interface will not be able to keep up with the hot zone, and layers of fine material may remain in the material. As shown in Fig. 2 (Gessinger 1984), for the case of two mechanically alloyed superalloys, the GAR increases with decrease in zoning speed. In order to maximize the GAR, most mechanically alloyed products are zone-annealed with speeds less than $100\,\mathrm{mm\,h^{-1}}$.

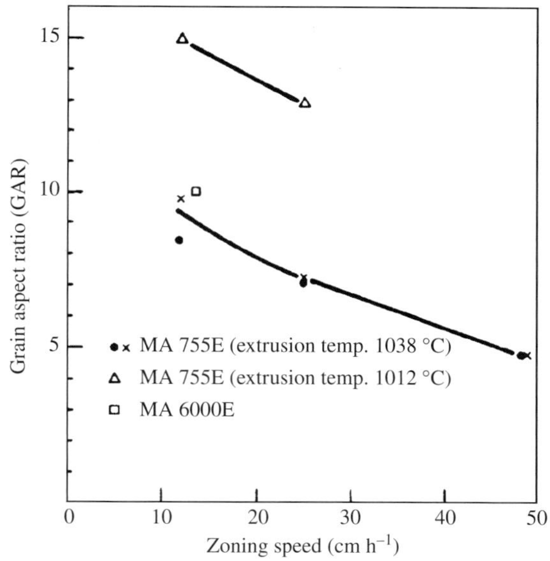

Figure 2
Showing the effect of zoning speed on the grain aspect ratio of a number of mechanically alloyed materials (reproduced by permission of Butterworth-Heinemann from *Powder Metallurgy of Superalloys*, 1984).

3. Computer Simulation of DR

Godfrey and Martin (1995) have developed a Monte Carlo method of simulating the development of a DR microstructure with strong crystallographic texture during SRX in a temperature gradient. Although their model did not include the effect of a dispersed second phase (as is encountered in ODS alloys), it appears qualitatively to reproduce some features of experimentally observed zone annealing, notably the breakdown of the interface into equiaxed grains at speeds which are too slow or too high. This agrees with both experimental observation and theoretical considerations, although there are many unquantified variables in the model.

4. Conclusion

The mechanism of directional recrystallization of hot worked ODS alloys is normally one of secondary recrystallization, or abnormal grain growth, due primarily to the initial ultrafine grain size. DR is accompanied by the development of a strong fiber texture, although the details of its nucleation are still unclear.

In view of the dramatic enhancement of creep life brought about by DR, it is likely to remain an important stage in the processing of ODS materials.

Bibliography

Gessinger G H 1984 *Powder Metallurgy of Superalloys*. Butterworths, London

Godfrey A W, Martin J W 1995 Some Monte Carlo studies of grain growth in a temperature gradient. *Philos. Mag. A* **72**, 737–49

Hotzler R K, Glasgow T K 1982 The influence of γ' on the recrystallization of an oxide dispersion strengthened superalloy—MA 6000E. *Metall. Trans. A* **13A**, 1665–74

Marsh J M, Martin J W 1991 Micromechanisms of texture development during zone annealing of MA 6000 extrusions. *Mater. Sci. Technol.* **7**, 183–8

Mujahid M, Martin J W 1994 The development of microstructures of high grain aspect ratio during the zone annealing of ODS superalloys. *Mater. Sci. Technol.* **10**, 703–10

Singer R F, Artz E 1986 Structure, processing, and properties of ODS superalloys. In: Betz W, Brunetaud R, Coutsouradis D, Fischmeister H, Gibbons T B, Kverenes I, Lindblom Y, Marriott J B, Meadowcroft D B (eds.) *High Temperature Alloys for Gas Turbines and Other Applications.* Reidel, Dordrecht, The Netherlands, pp. 97–126

J. W. Martin

Dynamic Recovery and Recrystallization

During deformation at high temperature, dynamic recovery (DRV) and dynamic recrystallization (DRX) reduce the strain hardening that usually arises from interactions of dislocations whose motion provides the plastic strain except for minor contributions from grain boundary (GB) sliding or diffusion creep. DRV involving dislocation annihilations and rearrangement into low-energy subgrain boundaries (SGB) occurs at all strains in most crystalline materials over a broad range above $0.4\,T_M$ (melting point, K). DRX, usually involving dislocation elimination by GB migration in the formation of new grains, occurs after a critical strain over a range extending above $0.6\,T_M$. Both mechanisms are named in relation to the equivalent static restoration mechanisms that occur normally during annealing after deformation. Recovery tends not to alter the grain structure which is being elongated by the plastic strain whereas recrystallization (see *Primary Recrystallization*) develops new grains classically by discontinuous nucleation and growth although there are additional forms to be described: continuous, geometric, and rotation, which are allied with DRV.

1. Characteristic Flow Curves

As temperature rises, true stress–strain curves exhibit decreasing stress levels with a tendency towards a plateau, rising ductility, and increasing dependence on strain rate $\dot{\varepsilon}$. In hot working, ideally at constant $\dot{\varepsilon}$ and T, DRV gives rise to monotonic hardening to steady-state plastic flow in which dislocation density remains constant; examples of this are aluminum (f.c.c. crystals of high stacking fault energy, SFE) and α-iron (as well as other b.c.c. metals) (Fig. 1). Similarly during creep at constant stress σ, DRV leads to a minimum constant $\dot{\varepsilon}$ after a transient in which a hardening substructure develops. In most other materials including nickel, copper, γ-iron (f.c.c. alloys with low SFE) at high strain rates or stresses, the dislocation density becomes sufficiently great to initiate the nucleation and growth of new grains, which causes the stress–strain curve to develop a peak and soften to a steady-state regime; in creep this appears as sudden accelerations in rate followed by a new hardening transient. During the DRX steady state, new grains repeatedly nucleate and grow distributed throughout the volume undergoing dislocation deformation and dynamic recovery, the flow stress being dependent on the average dislocation density of the grains. The steady-state, or peak, flow stress is related to T and $\dot{\varepsilon}$ in the Zener-Hollomon parameter ($Z = \dot{\varepsilon}\exp(Q/RT)$, where Q is an activation energy and R is the gas constant related to the distribution of atomic thermal vibrations) by several functions: at low σ, (power law) $A'\sigma^{n'}$; at high σ, $A''\exp\beta\sigma$; and across a broad σ range, $A\,(\sinh\alpha\sigma)^n$.

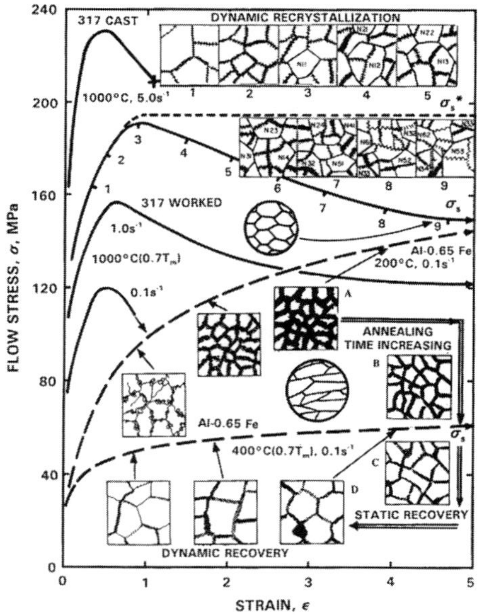

Figure 1
Stress–strain curves for Al-0.65Fe and 317 stainless steel at about $0.7\,T_M$. The aluminum alloy strain hardens monotonically to a steady state as a result of DRV which is high at 400°C but low at 200°C, $0.5\,T_M$ as illustrated by substructure diagrams; SRV, illustrated in sequence A to D indicated by double arrows at bottom right, would require many hours. The 317 curves display effects of $\dot{\varepsilon}$ and of work softening due to DRX which is illustrated by schematic diagrams; the as-cast is strengthened by ferrite particles that stimulate nucleation (courtesy *J. Mater. Proc. Tech.*).

2. Substructure Evolution in Dynamic Recovery

The most characteristic feature of DRV is the development of a stable substructure associated with mechanical steady state which is related to the stress or the combination of T and $\dot{\varepsilon}$ in Z (Fig. 2). As, σ, $\dot{\varepsilon}$, or Z declines or T rises, the subgrains have larger diameter (d), lower internal dislocation density (ρ_i) and greater dislocation spacing (s) and orderliness in the walls (same quantitative relation in hot working and creep). For aluminum, d ranges from about 4 μm at 400°C, $1\,s^{-1}$ to about 30 μm at 400°C, $10^{-6}\,s^{-1}$. Associated with the walls of low-energy dislocation arrays, there is a tilt or rotation (several degrees) which can be observed after etching in α-iron alloys and anodizing of aluminum alloys (polarized light). The misorientations can be measured by diffraction (Kikuchi lines) in scanning electron

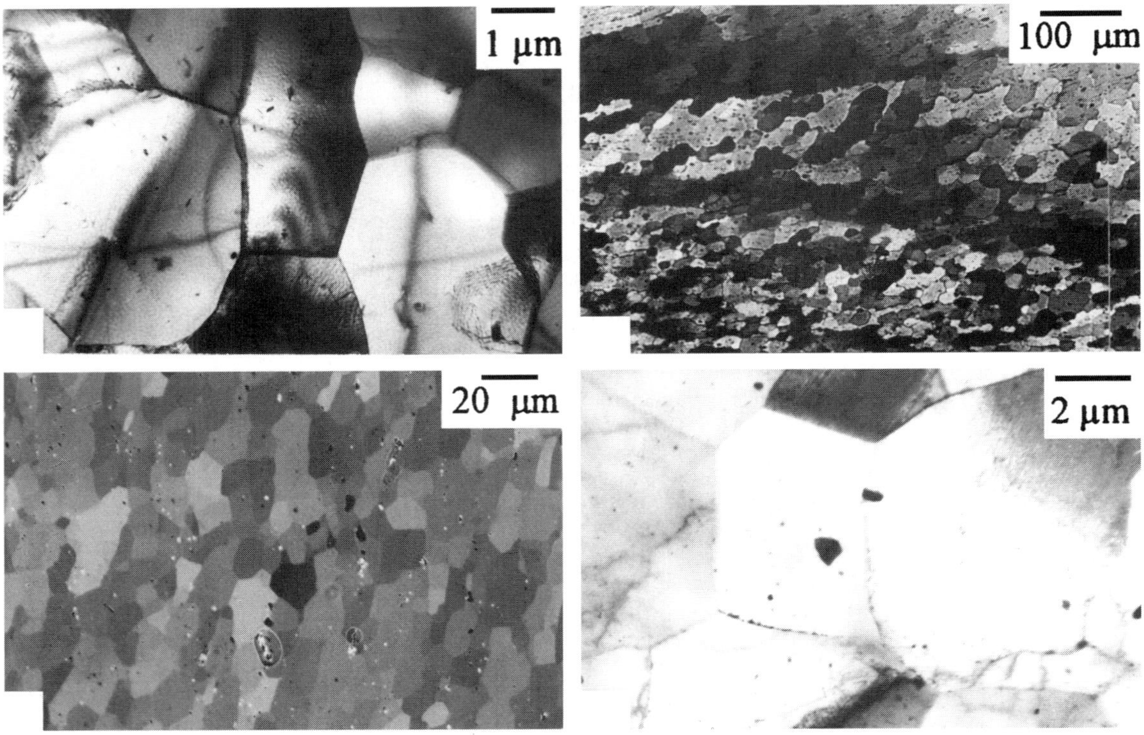

Figure 2
In high SFE aluminum deformed (a) in compression into steady state ($\varepsilon = 0.7$) at 400°C (0.72 T_M), 1.4 s^{-1}, subgrains with neat SGB are visible in TEM. In torsion ($\varepsilon = 40$) at 400°C, 0.2 s^{-1}, (b) polarized optical microscopy exhibits subgrains without visible GB in gage (bottom) but within grains ($D_0 = 2000$ μm) in transition to shoulder (top); (c) in SEM EBSI ($\varepsilon = 20$, $D_0 = 100$ μm) subgrains tend to mask the GB and (d) in TEM ($\varepsilon = 40$, 100 μm) well-polygonized subgrains are visible.

microscopy (electron back scattering) (SEM-EBS), or in transmission (TEM) where individual dislocations can be imaged to measure ρ_i or s.

In the steady state all the substructural features remain dynamically stable. While the grains elongate, the subgrains remain equiaxed as a result of subboundary disintegration, migration, annihilation, and reformation. When $\dot{\varepsilon}$, T, or σ are changed there is a mechanical transient somewhat similar to the initial straining in which the substructure evolves to the new steady-state microstructure. Theories relating T, $\dot{\varepsilon}$, and σ have commonly been based on the internal dislocation network ρ_i; this works because all the parameters d, ρ_i, and s scale together in the steady state (in cold working strength is related to d and s). In the "composite" model the mechanical parameters are related to all the substructural features by providing equal strain rates in the hard subboundary regions with the help of a high internal forward stress, and in the soft subgrains with a low internal back stress. The model is capable of handling pure metals and solute alloys (e.g., Al–5Mg) in which there is solute atom drag on the dislocations. The enhancement of DRV as T rises effectively reduces σ and greatly increases ductility (Fig. 3).

In polycrystal deformation, there is the need for grains to deform on five independent slip systems, being specified by orientation with the possibility of alternative symmetrical systems. Due to solidification mosaic regions, or to different stress concentrations from neighbors, a grain (~200 μm) may break up into deformation bands (~50 μm) which are usually smaller as T declines. Slipping on different systems the bands rotate differently so that transition boundaries between them increase in misorientation ψ in relation to the density of geometrically necessary (ψ creating) dislocations (usually in a single wall unlike the multiple ones in cold work). The transition boundaries become permanent and increase in misorientation as strain rises as well as elongating like the grain boundaries. As T drops through warm to cold working, the deformation bands themselves break up into blocks as microbands (created on new slip systems) cut through the old cell walls; these block walls may also become permanent, increasing in misorientation with rising strain. In hot working,

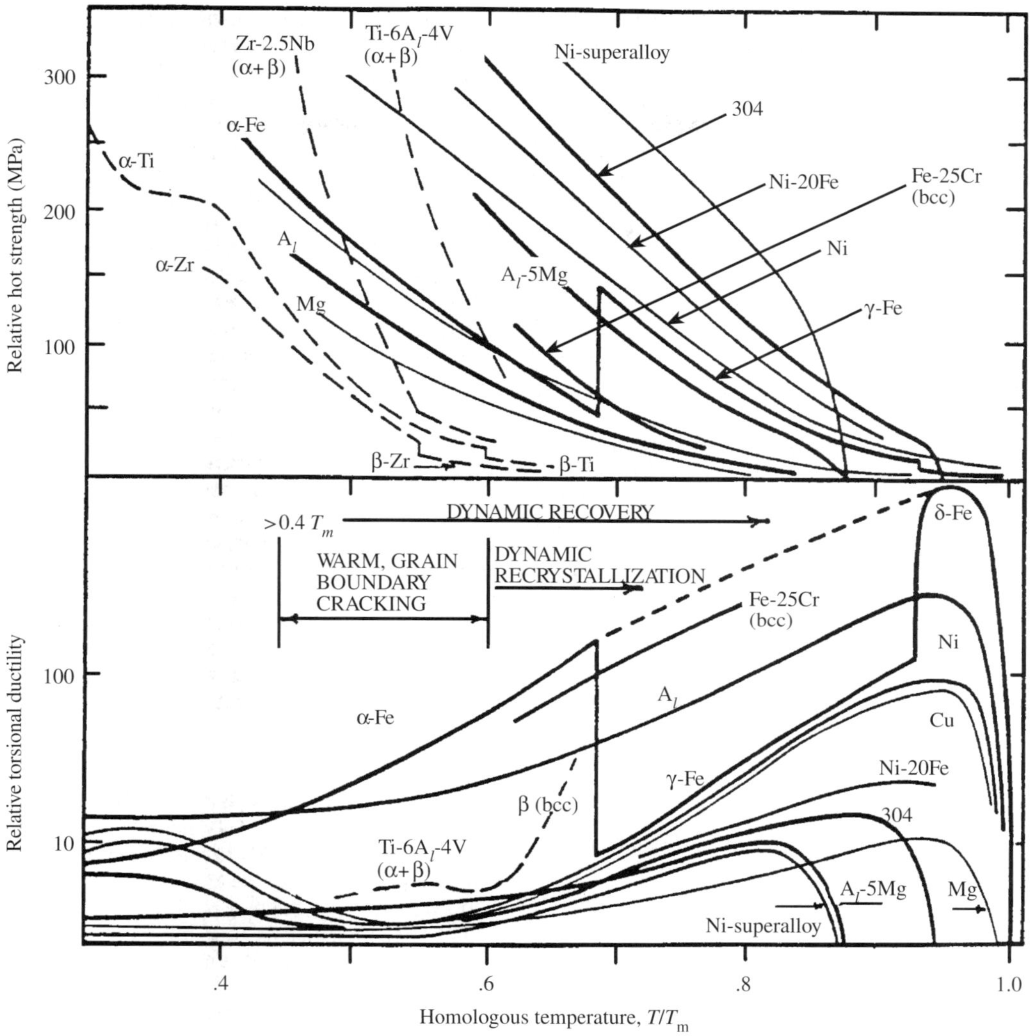

Figure 3
In αFe, Fe–25Cr, β-Ti, β-Zr and Al alloys, the high ductility results from DRV. For low SFE f.c.c. alloys (γ-Fe, Ni, Cu), DRX furnishes high ductility for $T/T_M > 0.6$, i.e. above the trough near 0.45 from intergranular cracking. Higher strengths are exhibited by low SFE f.c.c. alloys especially when solute and precipitates are present. The Zr and Ti alloys are displaced to lower T/T_M and exhibit strong T sensitivity in the $\alpha + \beta$ structure (courtesy of TMS).

the bands ideally do not break up nor are there microbands since the substructure completely rearranges in each strain interval equalling the initial transient and the intercrystal constraints are reduced by GB sliding, migration and diffusion. The rotation of the bands or blocks according to the Taylor theory leads to differently oriented regions contributing the microtexture which adds up to the macrotexture. The regions in hot working are larger than in cold working yet the rolling textures (see *Deformation Textures*) are essentially the same and so are the recrystallization textures (see *Annealing Textures*) from subsequent annealing or slow cooling.

2.1 Static Restoration in Annealing or After Hot Working

On annealing metal cold worked to $\varepsilon = 2$ for static recovery (SRV), the dislocations inside the cells and redundant ones in the walls combine and annihilate, resulting in conversion to subboundaries. At longer times, the cells grow larger partly by wall migration and partly by coalescence as weak walls disintegrate losing dislocations to stronger walls. SRV proceeds under the internal stresses with declining rate as dislocation density diminishes; thus, there is no steady state as in DRV. Upon annealing metal cold worked to a fixed strain, SRV takes a very long time to reach the level of recovery which DRV driven by the applied stress attains by that strain at the annealing temperature (Fig. 1).

Moreover, SRV is frequently curtailed by static recrystallization (SRX) for which it has prepared the nuclei by developing regions with low ρ_i and high ψ walls which relax into mobile grain boundaries unaffected by stress. While SRV tends to heterogenize the substructure, DRV tends to homogenize it because of the continuing deformation. On holding after hot deformation or between stages, the DRV substructure undergoes SRV and develops SRX nuclei in a density proportional to the substructure and in a time easily surpassed in steady state without inducing DRX. The time for, and the grain size developed by, SRX decreases with rising ε during the work hardening transient but becomes constant during steady state.

2.2 *Ultrahigh Strains*

As high-temperature strain proceeds, the grain boundaries lengthen in the elongation direction and become serrated with an amplitude and half wavelength almost equalling the subgrain diameter as a result of local migration to absorb sub-boundaries (Figs. 2(b), 2(c), 2(d)). This evolution is initially easily followed by optical microscopy but as the grains reach a thickness of about 2.5 d opposing serrations frequently meet, pinching off the grains into short segments. The general effect occurs at lower strains for smaller initial grains and for larger subgrains (higher T or lower $\dot{\varepsilon}$ or Z). The process begins at lower strains in the sharp ends of grains at triple junctions (like zipping up the arms of a Y) which leads to some decrease of grains across the section; there is also a net migration into grains of higher dislocation density. The grains are divided into progressively shorter segments which are hard to distinguish from the subgrains, leading to a phenomenon named *geometric DRX* because it gives the appearance of new grains containing substructure although it arises from DRV, plastic shape change of grains, and only local random migration of the original grain boundaries. In aluminum at 400°C, $0.2\,s^{-1}$ ($d = 6\,\mu m$), the effect is found in 100 μm grains at true strains near 5 and in 1000 μm grains only at true strains exceeding 50. The GB, transition bands and subboundaries can be distinguished by diffraction techniques in TEM or SEM-EBS. In orientation image mapping of walls greater than 1 μm spacing and 1° angle, there are often additional short segments with 5–15° angles, which indicates that the ideal substructure theory is not perfectly followed as T declines into the warm range (0.4–0.6 T_M).

2.3 *Effects of Alloying*

In aluminum–magnesium alloys, the substructure is similar to aluminum at the same stress but at the same Z (Q being about equal for diffusion of aluminum or of magnesium in aluminum) the TEM subgrains are about one-fifth the size they are in aluminum consistent with the stress of about five times higher (Fig. 3). The transient is inverted, i.e., there is a softening as the stress declines from that necessary to initially accelerate the dislocations away from magnesium atmospheres; the dislocations initially form in layers which later develop cross-links and finally equiaxed cells. The strain to the substructure steady state is almost double that for the mechanical one. The polarized optical microstructures exhibit crystallites about five times the TEM subgrain size and serrations subtending about five subgrains. As a result the serrations are much more meandering, or billowy, leading to pinching off at their base (partly due to GB sliding), forming what looks like new grains, termed *rotation DRX*. This is also found in magnesium alloys and in some organic and mineral crystals where subgrains form from multiple slip systems only near the grain boundaries due to stress concentrations. The pinched-off crystallites form at intermediate strains of 1–2 (about 400°C, $1\,s^{-1}$ in Al–5Mg) and do not grow like nuclei since the substructure density inside is similar to that in the surroundings. At larger strains, they contribute to earlier geometric DRX and become indistinguishable from it. Other solute elements in aluminum, such as copper, zinc, and silicon are much less effective strengtheners, partly because they are caught up in large overaged precipitates in the hot working regime. In ferritic steels, chromium, molybdenum, nickel, aluminum, and silicon moderately reduce the level of DRV. The presence of fine particles $< 0.2\,\mu m$ (Mn or Fe dispersoids in Al) stabilizes the substructure at a size defined by particle distribution and raises the strain hardening. Large particles $> 2\,\mu m$ (Fe, Si intermetallic constituent phases) cause turbulent flow in the surrounding region leading to dense substructures.

3. *Dynamic Recrystallization*

In classical discontinuous dynamic recrystallization, the grain size is smaller (and the critical strain ε_c is higher) as Z or $\dot{\varepsilon}$ increase and T decreases and is independent of strain during steady state. When the characteristic grain size is smaller than the initial ones, nucleation starts at $\varepsilon_c = 0.6\,\varepsilon_p$, so that by the peak strain ε_p DRX is about 30% complete, becoming about 95% at steady state ε_s. The grains form as necklaces at the initial grain boundaries and then repeatedly as layers at the interface until the old grains are consumed. The new grains fill up with substructure, which stops growth, defines the grain size and finally leads to overlapping cycles of DRX (Fig. 4). The flow stress depends on the dislocation density as defined by elimination from migrating GB and by DRV, which also defines the nucleation strain and the grain size.

In polycrystals, DRV subgrains form first near the GB producing serrations; some serrated regions reach sufficient misorientation that they become nuclei leading to necklace formation. When subgrains do not easily form at grain cores, serrated regions rotate and pinch off leading to *rotation DRX* only in the GB mantle. As the initial grain size increases under constant T and $\dot{\varepsilon}$, the critical strain increases but the characteristic grain size remains unaltered. In single crystals, the critical condition is defined by the dislocation density and hence by the stress; in the absence of GB nucleation sites, considerably greater stress and strain levels are required than for polycrystals. DRX is likely to nucleate at a transition boundary and proceeds by repeated twinning to faster growing orientations. When the initial grain size is smaller than the characteristic size for the conditions, the flow curve is cyclic with declining amplitude; this is related to several cycles of DRX each with increased grain size.

Dynamic recrystallization has many characteristics of the static variety SRX; it follows Avrami kinetics for nucleation and growth and its activation energy is similar to that for grain boundary migration. However, ε_c is much higher than that for SRX because reinsertion of substructure within potential nuclei requires a higher matrix dislocation density as thermodynamic driving force. Unlike static grains, which stop growing upon impingement with similar grains, the rising dislocation density in the expanding new grains reduces the driving force and stops growth. The critical strain for DRX rises with

Figure 4
In low-SFE Cu deformed in torsion at 800°C (0.8 T_M) 11 s^{-1} into steady state ($\varepsilon = 30$) (a) the distorted DRX grains are illustrated in SEM-EBSI and (b) the reformed substructure within them are exhibited in TEM with two nuclei near the center. In Cu rolled at 20 s^{-1} in one pass to $\varepsilon = 2.3$ ($<\varepsilon_c$ DRX), (c) subgrains and some SRX grains appear after finishing at 600°C (0.65 T_M) and (d) dense cellular substructure at 490°C (0.55 T_M).

increasing $\dot{\varepsilon}$ in the hot working domain ($>10^{-2}\,s^{-1}$) whereas that for SRX decreases. Moreover, the critical strains for both DRX and SRX increase with declining $\dot{\varepsilon}$ in the creep domain ($<10^{-4}\,s^{-1}$) because of the slow accumulation of dislocation density. Although the rate of DRX speeds up as $\dot{\varepsilon}$ rises ($\varepsilon_s/\dot{\varepsilon}$ decreases), the strain to steady state rises. When deformation is stopped, some dynamic nuclei continue to grow and reach a larger size (no new impeding internal substructure) in metadynamic recrystallization. Subsequently, SRX can arise because of the substructure in retained DRX grains and gives rise to still larger grains (Fig. 4). In multistage rolling, the individual pass strains are not high enough to induce DRX, but are usually sufficient to induce SRX during the intervals leading to beneficial grain refinement. At low T with short intervals preventing SRX, cumulative strain may be high enough to induce DRX.

3.1 *Effects of Alloying*

The presence of solute or of second-phase particles which impede grain boundary migration lead to higher critical and peak strains so that DRX may not reach completion before failure ensues. In austenite, very high chromium, nickel, and molybdenum solute contents retard nucleation and growth, thus reducing ductility; likewise fine microalloy carbonitrides impede DRX as they do SRX, usually resulting in pancaked grains after finish rolling. Contrastingly, in stainless steels, segregated ferrite particles are sufficiently large ($>2\,\mu m$) and rigid to cause the surrounding matrix to develop a high-density substructure through additional compensatory and turbulent plastic flow, thus resulting in discontinuous, *particle-stimulated nucleation* (*PSN*) *DRX*. Under conditions where the substructure is strongly pinned by particles (Al–Cu–Mg, Al–Li–Zr, and Al–10Mg–Zr alloys), and notably where the substructure density has been built up by high $\dot{\varepsilon}$, low-T thermomechanical processing, straining at a relatively low $\dot{\varepsilon}$, and high T can lead to most of the dislocation walls making a conversion to high-angle (but pinned) grain boundaries capable of sliding, so that superplastic deformation ensues. It appears that the shear stresses enhance the conversion mechanism since the above *continuous DRX* is much more rapid than continuous SRX in simple annealing. Although PSN SRX commonly occurs in many aluminum alloys with constituent particles (Mn, Cr, Fe, Si, Al intermetallics), PSN DRX only occurs in Al–Mg–Mn alloys where the magnesium slows the DRV around the manganese constituent particles thus maintaining the nucleating substructure.

4. *Hot Workability*

Dynamic recovery and recrystallization play a significant role in lowering the flow stress and thus reducing the forces in mechanical shaping processes, notably when the work pieces are large as in primary rolling, forging, and extrusion (Fig. 3). They also reduce the stresses on tooling, thus extending its life; however, this is countered by thermal damage. The reduction in stress improves ductility in so far as stress concentrations at any notch or defect are reduced. The effects for different metals are similar at the same fraction of the base element melting temperature T_M, which indicates the extent of thermal vibration or activation; however, the high DRV alloys are markedly softer than low DRV alloys which is made evident upon a change of phase, such as b.c.c. iron below 910°C, being softer than f.c.c. iron above that (Fig. 3). Alloying lowers the upper T of the potential working range at which the material can no longer retain its cohesion under negligible tensile stress. High-temperature processing ensures good ductility in as-cast feed stock in which large irregular grains and segregation of second phases at the boundaries enhance cracking.

The ductility is notably increased by the restoration mechanisms in the domain where grain boundary sliding becomes operative ($>0.45\,T_M$) (Fig. 3). Failure is initiated by intergranular cracks appearing at triple junctions due to the differential sliding on boundaries oriented at different angles to the tensile axis. Intense DRV in aluminum and α-iron reduces the flow stress inside the grains so that accommodating flow occurs at triple junctions relieving the induced stress concentrations. For metals of high recoverability ultraductility in torsion with no inherent instability gives rise to as much as 150 true strain, equivalent to more than 6000% elongation which is much more than is ever observed in superplasticity. This is not effective in tension because of the necking instability, which occurs earlier as the strain hardening rate is reduced relative to the stress. In contrast, in the aluminum–magnesium alloys where the solute drag gives a strain rate sensitivity of about 0.33, the tensile ductility is raised to about 300%; however, the reduced level of DRV decreases the torsional ductility by a factor of 10 at 5% magnesium.

A material that has limited DRV usually exhibits a noticeable decline in ductility in the GB sliding range due to severe boundary fissuring. The ductility rises markedly as DRX becomes more rapid since the associated migration of boundaries eliminates triple point stress concentrations and separates boundaries from any pores (Fig. 3). Static recrystallization between stages has a similar beneficial effect. Alloying as solute or particles, which slows down GB migration, diminishes the ductility; the segregation of nonequilibrium phases to boundaries of large grains gives rise to very severe conditions. Alloying or impurities reduce ductility in DRV metals to a much smaller degree since the action of pinning dislocations is dispersed throughout the entire grain and not simply concentrated in pinning the boundaries to impede DRX. However, the presence of large particles of phases that crack or melt, being almost always at the GB, encourages cracking there beyond the remedy of DRX or DRV so they must be reduced by melt and solidification control. Intentional heavy alloying, particularly to raise high-temperature creep resistance as in nickel-base superalloys, reduces the ductility severely through DRV and DRX attenuation, thus raising the cool end of the hot working range while simultaneously lowering the upper end through melting.

Bibliography

Blum W, McQueen H J 1996 Dynamics of recovery and recrystallization. Aluminum alloys, physical and mechanical properties. *Mater. Sci. Forum* **217**, 31–42

Humphreys F J, Hatherly M 2004 *Recrystallization and Related Annealing Phenomena.* 2nd edn., Elsevier, London

McQueen H J 1987 Initiating nucleation of dynamic recrystallization primarily in polycrystals. *Mater. Sci. Eng. A* **101**, 149–60

McQueen H J 1991 Micromechanics of dynamic softening in Al alloys during hot working. In: Langdon T G, Merchant H D (eds.) *Hot Deformaton of Aluminum Alloys.* TMS-AIME, Warrendale, PA, pp. 31–54

McQueen H J, Evangelista E, Ryan N D 1990 Dynamic recrystallization and recovery: mechanical and kinetic behavior, nucleation and growth mechanisms. In: Chandra T (ed.) *Recrystallization ('90) in Metals and Materials.* AIME, Warrendale, PA, pp. 89–100

McQueen H J, Ryan N D, Zaripova R, Farkhutdinov K 1996 Comparison of hot working behavior of austenitic and ferritic stainless steels. *Proc. Iron and Steel Society 37th Mechanical Working and Steel Processing Conf.* AIME, Warrendale, PA, pp. 883–94

Sakai T, Jonas J J 1984 Dynamic recrystallization: mechanical and microstructural considerations. *Acta Metall.* **32**, 189–209

Sellars C M 1978 Recrystallization of metals during hot deformation. *Philos. Trans. R. Soc. London Ser. A* **288**, 147–58

H. J. McQueen

E

Electroceramics, Colloidal Stability and Processing of

The term *electroceramics* refers to a class of ceramic materials that exhibit useful electrical properties. The properties of electroceramics, like many other materials, depend upon composition and microstructure. What is also of considerable interest is that the properties of many electroceramics can be "tuned" or altered by providing an external stimulus such as temperature, stress, chemistry of the surrounding medium, etc. Therefore, many electroceramics are considered to be active or field-controllable materials. Electroceramics are useful for passive devices, as well as for sensor and actuator applications. Technologically important examples of electroceramics include materials such as alumina (Al_2O_3), beryllia (BeO), barium titanate ($BaTiO_3$), barium polytitanates (e.g., $BaTi_4O_9$. $Ba_2Ti_9O_{20}$), barium zinc tantalate (BZT), lead zirconium titanate (PZT), silica (SiO_2), and ceramic superconductors based on yttrium barium copper oxides ($YBa_2Cu_3O_{7-x}$) and other compositions. The properties of electroceramics that may be important for different applications include:

- Resistivity (ρ) or conductivity (σ).
- Dielectric constant (k).
- Dielectric losses (tan δ) or Q ($\sim 1/\tan\delta$).
- Temperature and frequency dependence of the dielectric constant (k) and tan δ.
- Piezoelectric, pyroelectric, electro-optic coefficients, as applicable.
- Changes in the conductivity/resistivity as a function of the surrounding atmosphere or chemical species adsorbed on the surface.
- Dielectric breakdown strength.

Electroceramics are used for the preparation of conductors, inductors, resistors, capacitors, microwave dielectric resonators, piezoelectric transducers, and other devices. For devices made from electroceramic materials, it is important to consistently and economically achieve some of the above-mentioned properties that are "key" to a particular application, so that certain specification levels are met. For example, for multilayer capacitors based on electroceramics such as barium titanate, it will be important to achieve a high dielectric constant, a low loss tangent, and temperature stability. For capacitors, a high dielectric constant is necessary for higher volumetric efficiencies. In addition to the desirable levels of electrical properties, devices manufactured from electroceramic materials should also exhibit adequate levels of mechanical and other properties.

In this article, we first summarize the so-called conventional process for the processing of electronic ceramics. Following this, we will present a brief description of some of the chemical routes for the synthesis of electroceramic powders. The reader may find a more detailed discussion of some of these processes in other articles in this encyclopedia. In this article we will not discuss the processing of thin films of electroceramic materials, such as thin films prepared using sol–gel processing or chemical vapor deposition techniques. The focus of the article is on the synthesis and processing of oxide powders. A detailed discussion of the chemical synthesis of nonoxide (e.g., nitride) ceramics can be found in Oxide and Nonoxide Ceramics, Synthesis of.

1. *Traditional Synthesis of Electronic Ceramics*

The so-called traditional or "oxide-mix" process used for the manufacturing of electroceramic powders often results in relatively coarse ceramic powders, since the raw materials used are coarser in nature (average particle size $>10\,\mu m$). The ceramic powders derived using conventional routes also tend to be chemically inhomogeneous. In some situations this can lead to relatively inferior electrical properties. The traditional process is one of the most economical ways of manufacturing devices/components based on electronic ceramics. Typical examples of materials that are used as oxides include TiO_2, ZrO_2, Al_2O_3, PbO, and SiO_2. For the alkali and alkaline-earth groups, the elements' carbonates (e.g., $BaCO_3$, $SrCO_3$, etc.) are used as starting materials. The properties of the raw materials used, such as the average particle size, particle size distribution, specific surface area, chemical composition, and types and levels of impurities are very important and need to be tested regularly. It is also important to make sure that the properties of incoming raw materials are consistent, i.e., the properties do not change significantly from lot to lot. The raw materials are blended in appropriate proportions. Typically, wet ball milling is used to conduct the mixing process. In the ball milling process, alumina or zirconia grinding media are used. In this process, there is always a need to ensure that contamination of the final powder being mixed is kept to a minimum. Following the ball milling step, the mixed powders are then "calcined." During calcination, such precursors as carbonates decompose into their respective oxides. The particles of different precursors also begin to react and form phases of interest for the final application. For example, in $BaTiO_3$ powder manufacturing, $BaCO_3$ decomposes and reacts with TiO_2 to form intermediate phases and ultimately $BaTiO_3$. Calcined powders are often milled again to break up any agglomerates that may have formed. Following this, the powders may also be appropriately spray dried and then shaped using dry pressing, isostatic pressing, tape casting, extrusion, or slip casting. The result of the shaping operation is the so-called "green body" or "green ceramics." Often the ceramic materials in the green state undergo machining (cutting, dicing, drilling, etc.) and other operations (e.g., electroding). Following this, the ceramics are sintered under conditions in which temperature, time, heating and cooling rates, and atmosphere are controlled. The resultant product is then processed for connecting leads for mounting and any encapsulation or packaging. The most important electrical properties specified by the customer are tested continuously.

2. *Chemical Processing of Electroceramics*

The rationale for developing chemical routes for processing electroceramic materials is that such routes enable the preparation of finer, chemically homogeneous, and sinterable powders at lower temperatures. In addition to this, chemical techniques also prove to be quite useful in the precise control of dopants, as well as cation stoichiometry. In this section, we will elaborate on these advantages associated with chemical techniques. Many technologically useful electroceramic compositions are multicomponent. For example, cordierite ceramics contain MgO, Al_2O_3, and SiO_2. The so-called PZT ceramics contain at least three components, namely, PbO, ZrO_2, and TiO_2. Often other components (such as niobium oxide or lanthanum oxide) are added to tailor the electrical and mechanical properties for specific applications. For such multicomponent compositions, it is important to make sure that the spatial distribution of the components is such that it leads to the best possible electrical properties. Similarly, the electrical properties of these materials are often quite sensitive to the chemical composition and microstructure. Small levels of components, known as dopants, also often need to be added to achieve a particular property. For example, in the processing of dielectric materials based on barium polytitanates or magnesium titanate, used for the fabrication of microwave components, a small amount of manganese is often added to compensate for the reduction of Ti^{4+} to Ti^{3+}. If titanium undergoes reduction, typically the dielectric losses increase. Preventing titanium reduction ensures that the quality factor ($1/\tan\delta$) of the dielectric components remains at an acceptable level. Since the concentration of manganese needed is small, it is often added chemically as manganese nitrate solution rather than as an oxide powder. This ensures that the manganese is uniformly distributed throughout the powder mix.

In some applications of electroceramics, the need for resorting to the use of chemical synthesis techniques could lead to lower sintering temperatures. For example, in multilayer capacitor applications there is a need to lower the sintering temperatures associated with barium titanate formulations so that base metals, such as copper or nickel, can be used as electrode materials instead of using expensive, precious metal-based electrodes such as silver and palladium. For multilayer capacitors, almost 80% of the cost arises from the use of precious metals as electrodes. Therefore, there is a significant economic incentive to develop powder formulations that can be sintered at lower temperatures, so that base metal electrodes can be used.

Another application of chemically derived electroceramics is for thin films. Many electronic ceramics are finding applications as thin films integrated onto semiconducting substrates. For example, PZT thin films have been extensively researched for nonvolatile memory, microelectromechanical systems (MEMS), and other applications. Similarly, recently there has been considerable interest in the use of barium strontium titanate (BST) thin films for frequency-agile components. The dielectric constant of such materials can be changed using an applied electric field. Chemical routes, such as sol–gel processing and metallorganic chemical vapor deposition (MOCVD), are being investigated for the deposition of thin films of such multiple-component systems.

One of the major disadvantages associated with the use of chemical techniques for electroceramics is the cost of the ceramic powders generated or the precursors used. Chemically derived powders tend to be considerably more expensive than those derived using conventional processes, since the process uses high-purity chemical precursors. The yield of oxide product, on the basis of per unit mass of the precursors used, is also relatively small. Therefore, the use of chemically derived electroceramics, especially in the form of powders, is justified only when the gain or enhancement in the performance of the final product is significant. It is to be noted that the processing of electroceramic powders also often involves processes such as tape casting or slip casting. In these processes, many polymeric materials play a key role as a dispersant, binder, and plasticizer. For a review of the colloidal chemistry of ceramic processing see Lewis (2000).

3. *Important Chemical Synthesis Routes for the Synthesis of Electroceramic Powders*

3.1 *Oxalate Process*

One of the most successful and commercial chemical processes for the production of electroceramic powders is the so-called oxalate process. This process is particularly useful for obtaining ceramic powders with relatively precise cation stoichiometry. We exemplify the use of this process through its application for barium titanate ($BaTiO_3$). For the manufacture of $BaTiO_3$, titanium tetrachloride ($TiCl_4$) and barium chloride ($BaCl_2$) are dissolved in water and reacted with oxalic acid ($H_2C_2O_4$) to form crystals of barium titanyl oxalate $BaTiO(C_2O_4)_2.4H_2O$, with a very well-defined Ba/Ti stoichiometry. When this precursor powder is heated to 700–900°C for a few hours, fine-particle barium titanate powders are obtained. The decomposition process is well characterized and has been shown to involve formation of carbonate-like intermediates. The weight loss in the conversion of barium titanyl oxalate precursor powders to barium titanate is approximately 60%. The purity of oxalate-derived powders is considerably higher than that of barium titanate powders prepared by the traditional process based on using $BaCO_3$ and TiO_2 (see Table 1).

In Table 1, notice the low levels of impurities in the oxalate-derived powders. Sodium is often considered to be one of the most undesirable impurities, since it can contribute to ionic conductivity in capacitor and other dielectric formulations. Also, the cation stoichiometry of oxalate-derived powders (Ba/Ti, in this case) is controlled very well, and this is one of the advantages associated with the use of most chemically derived electroceramic powders. The homogeneity of oxalate-derived $BaTiO_3$ ceramics is also reflected in their dielectric properties. Typically, oxalate-derived barium titanate formulations exhibit higher values of the dielectric constant k. For example, the Ferro Corporation has developed oxalate-derived $BaTiO_3$ that exhibits a dielectric constant of 4000 to 5000 for the so-called X7R dielectric formulations. As a comparison, with comparable solid-state-derived $BaTiO_3$, a dielectric constant of around 2500 can be achieved. Similarly, the oxalate-derived $BaTiO_3$ is more reactive and makes complete $(BaCa)(Ti_xZr_{1-x})O_3$ solid solutions much easier to achieve, and at lower temperatures. A scanning electron micrograph (Fig. 1) of barium titanate powders manufactured by the Ferro Corporation illustrates the morphology and particle size of oxalate-derived powders.

Table 1
Comparison of impurities in oxalate-derived and traditionally processed barium titanate powders.

Type of impurity (wt.%)	SiO_2	Al_2O_3	SrO	Na_2O
Oxalate-derived barium titanate	0.01	0.01	0.02	<0.02
Traditional synthesis-derived barium titanate	0.12	0.12	1.00	0.13

3.2 *Pechini Method*

This method was originally described by Pechini in a patent, and since then has been widely investigated by many researchers, including Eror and Anderson (1986) for the synthesis of multiple-component oxide materials with well-defined cation stoichiometry. In this process, an α-hydroxycarboxylic acid such as citric acid and a polyhydroxy alcohol such as ethylene glycol are used. At a temperature of ~100–150°C, an esterification reaction occurs between the acid and the alcohol. This reaction results in the formation of a polymeric, resin-like product and elimination of water as a by-product. Many metal ions form stable complexes with citric acid. Therefore when precursors containing citric acid complexes are used, the polymeric resin can accommodate several transition metal cations such as titanium or niobium, as well as alkali and alkaline-earth cations. The resin containing these cations in the intended stoichiometry is then subjected to heat treatment at temperatures between 300 and 700°C. This leads to the formation of ultrafine, chemically homogeneous ceramic oxide powders. Similarly to the oxalate process, one of the advantages of using the Pechini process is the control over stoichiometry. The process is also very versatile, since cations of many metals can be incorporated. Since the precursors are mixed at a nearly atomic scale, the temperatures needed to form

Figure 1
A representative micrograph of oxalate-derived barium titanate electroceramic powders made by Ferro Corporation (courtesy of Dr. M. Chu, Ferro Corporation).

the crystalline ceramic phases of interest tend to be lower than those required in conventional processing. From a large-scale production viewpoint, one of the potential problems associated with this process is that the burning-off of the polymer resin can lead to emissions of carcinogenic polycyclic aromatic hydrocarbons (PAHs). For a review of this process, the reader is directed to publications by Lessing (1989) and Kakihana and Domen (2000).

3.3 Hydrothermal Synthesis

In this process, precursors for metal oxides are typically dissolved in an aqueous medium. Colloidal particulate precursors such as titania, silica, or alumina sols can also be used. A precursor mix of the intended composition is then subjected to higher temperatures and pressures in a sealed reactor. During the hydrothermal synthesis, the colloidal and dissolved species undergo chemical reactions leading to the formation of ultrafine particles. Katsuki *et al.* (1999) and Moon *et al.* (1999) have investigated the application of the hydrothermal synthesis technique for many electroceramics, including barium titanate. Many ceramic oxide materials have been prepared using hydrothermal techniques. The temperature needed for hydrothermal synthesis is usually between 100 and 300°C. In 1989, the Cabot Corporation introduced hydrothermally derived barium titanate powders for capacitor applications. Some of the advantages of hydrothermally derived barium titanate powders are their ultrafine size, the possibility of developing particles with a dopant coating, and also the ability of these powders to be processed further in the form of aqueous suspensions. In recent years, this process has also been applied for the preparation of thin films of such ceramics as barium titanate and lead zirconium titanate (PZT). In another innovation for the process, Katsuki *et al.* (1999) have used a combination of microwave energy and hydrothermal conditions for the synthesis of many ceramic materials.

3.4 Other Techniques

Many other chemical processes have been developed for the synthesis and processing of electroceramic materials. The sol–gel process makes use of metal alkoxides, colloidal particles and other precursors. It is a versatile process for the synthesis of powders, dense abrasive minerals, porous materials, and thin films. For further information, the reader is directed to the article Ceramics and Glasses, Sol–Gel Synthesis of. For electroceramics, the ability of the sol–gel process to form dense, fine-grained, and often oriented thin films is of significant interest. Sol–gel processed barium strontium titanate (BST) and lead zirconium titanate (PZT) thin films for non-volatile memory and dynamic random access memory applications have been investigated by many researchers. Similarly, metallorganic chemical vapor deposition (MOCVD) is another technique that has been shown by Dey and Allury (1996) to be quite useful for low-temperature deposition of electroceramic thin films. There are also other interesting processes that have been proven useful for specific electroceramic compositions. For example, electrophoresis – the movement of charged particles in a dispersion under the application of an electric field – has been explored for the deposition of thin films. A particularly noteworthy technique in this regard is electrochemical. Electrochemical techniques are widely used for the deposition of metals and alloys. In some very interesting work, Switzer *et al.* (1994) have demonstrated that electrochemical deposition can be used for the deposition of thallic oxide-based ceramic superlattices. Some of these techniques, such as this electrochemical deposition of ceramic superlattices described by Switzer *et al.* (1994), and the use of self-assembling monolayers (SAMs) for templating and microcontact printing of inorganic structures, appear to have the potential to advance the science, technology, and applications of the chemical processing of electroceramic materials and need to be explored further.

Bibliography

Dey S K, Allury P V 1996 *MRS Bull.* **21**, 44

Eror N G, Anderson H U 1986 In: Brinker C J, Clark D E, Ulrich D R (eds.) *Better Ceramics Through Chemistry II.* Materials Research Society, Pittsburgh, PA, Vol. 73, pp. 571

Jameel M, Reid R C, Wagner L P 1989 *US Pat.* **4**, 863–83

Kahikana M, Domen K 2000 *MRS Bull.* **25**, 27

Katsuki H, Furuta S, Komarneni S 1999 *J. Am. Ceram. Soc.* **82**, 2257

Lessing P A 1989 *Ceram. Bull.* **68**, 1002

Lewis J A 2000 *J. Am. Ceram. Soc.* **83**, 2341

Moon J, Kerchner J A, Krarup H, Adair J H 1999 *J. Mater. Res.* **14**, 425

Pechini M P 1967 *US Pat.* **3**, 330–697

Switzer J A, Hung C J, Brefogle B E, Shumsky M G, Van Leeuwen R, Goldenz T D 1994 *Science* **264**, 1573

P. P. Phulé

Electroceramics: Rapid Prototyping

Solid freeform fabrication (SFF) refers to various rapid prototyping techniques where 3D components are directly built, layer-by-layer, from computer data description or computer-aided design (CAD) files. The fabrication process starts with creating a CAD file of the 3D part, followed by mathematical slicing of the CAD part into several thin layers. Such data description (tool path) is then used to build the part in a layer-wise fashion without using molds, dies, or hard tooling. Due to the reduced lead-time and lower cost in developing new products, the SFF techniques are valued by the commercial sectors. Since 1980, several rapid prototyping techniques have been developed. Many of these techniques such as stereolithography (SLA) (Hull 1990), selective laser sintering (SLS) (Beaman and Deckard 1990), fused deposition modeling (FDM) (Comb and Priedman 1993), and 3D printing (Sachs *et al.* 1992) are primarily designed to manufacture plastic parts. During the late 1990s, some of the SFF techniques have been modified for direct production of functional metal and ceramic parts. Among these techniques, fused deposition of ceramics (FDC) has been specifically developed for the manufacturing of ceramic components. Following is a brief description of a few commonly used prototyping techniques (Section *1*), followed by some of electroceramic components (Section *2*) made by these techniques.

1. Rapid Prototyping Techniques

1.1 Stereolithography

Stereolithography (SLA) is considered to be the first SFF technique developed by 3D Systems, Inc. in 1988 (Fig. 1) (Hull 1990). The SLA process is based on photo-polymerization of liquid monomers using ultraviolet (UV) radiation. A UV laser is scanned over a layer of the liquid monomer to cure the monomer in selected areas as dictated by the part geometry (or the computer-generated tool paths). After completion of one layer, another layer of resin is coated on top of the cured layer, and the process is repeated until the part is completed.

The SLA process has been modified to produce ceramic parts (Griffith and Halloran 1996a). A ceramic particle-filled photocurable resin is used in place of the virgin resin to make ceramic components. When a laser beam interacts with a liquid suspension, it polymerizes the resin over a finite area (i.e., particular depth and width). A variety of ceramic materials such as silica, alumina, silicon nitride, and hydroxyapatite has successfully been used in the SLA of ceramics (Griffith and Halloran 1996b). Key advantages of SLA of ceramics are the possibility of

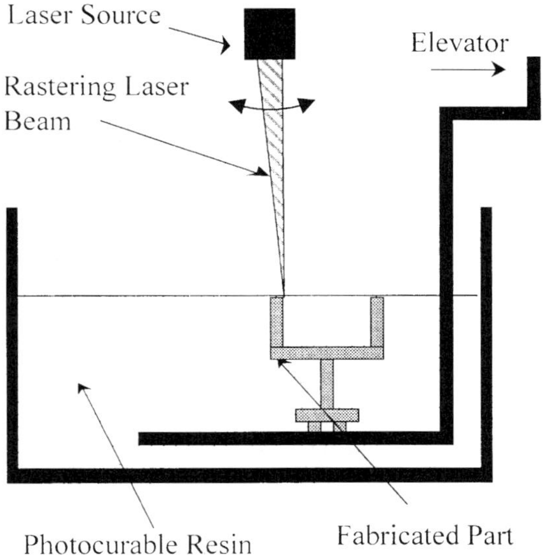

Figure 1
Schematic showing the operation of a typical stereolithography (SLA) apparatus.

fabricating ceramic parts with fine features (about 100 μm) and excellent surface finish.

1.2 Laminated Object Manufacturing

In laminated object manufacturing (LOM), parts are built by lamination of sheet feedstock, which is laser cut along contours defined by the tool path (Klosterman *et al.* 1996). In LOM of ceramics, ceramic tapes are used as the feedstock. The tapes are sprayed with a solvent and then stacked. The solvent acts as an adhesive, thus bonding the tapes together. A laser is then used to cut the tape according to the tool path file. A heated roller is used to further bond the tapes through thermocompression. The area of the tape outside the build contour is cut in a crosshatch fashion. At the end of the build, the part is manually removed by cutting the crosshatched region from the main part.

Computer aided manufacturing of laminated engineering materials (CAM-LEM) is a similar technique to LOM, but uses a powder-filled sheet feedstock, which is laser cut using a low-powered CO_2 laser along contours defined by the tool path file. The cut sheets are then robotically stacked to form the 3D object, which undergoes subsequent postprocessing to obtain the fully dense functional component. Some of the materials used in CAM-LEM include alumina, PZT (lead zirconate titanate), silicon nitride, and stainless steel (Liu *et al.* 1999).

1.3 Direct Write

Direct write, developed at Sandia National Laboratories (Dimos *et al.* 1997), has been used to fabricate multifunctional multimaterial integrated ceramic components. These components involve a wide variety of devices such as capacitors, inductors and conductors embedded in a ceramic substrate. In this method, ceramic- or metal-loaded slurries are printed onto a substrate, according to the tool path. The painted layer is then dried prior to application of the next layer. Slurries of silver (conductor), RuO_2 (resistor), lead zirconate titanate (dielectric and thick-film actuator) have been successfully printed using this technique (Dimos *et al.* 1997). Cofiring the green part produced by the direct write method yields the final integrated component. The direct write method uses a commercial deposition technology called Micropen (Ohmcraft, Inc., NY, USA) which is an automated printing device for ceramic slurries.

1.4 Fused Deposition of Ceramics

Fused deposition of ceramics (FDC), developed at Rutgers University (Agarwala *et al.* 1996, Danforth *et al.* 1998), is based on fused deposition modeling (FDM) technology. In the FDC process, ceramic-loaded polymer filaments are used to fabricate green ceramic components (Fig. 2). Feedstock in the form of a spooled filament is fed into a heated liquefier via a set of computer-driven rollers. The liquefier motion is computer controlled in the *x–y* plane, where the material is extruded through a fine nozzle on a fixtureless platform. An FDC part is built in a layerwise fashion similar to the other rapid prototyping techniques. When the deposition is completed, the part is heated to remove the organic and to obtain high density (sintering) at elevated temperatures.

A variety of structural, bioceramic, and electroceramic components have been fabricated using the FDC process. These include silicon nitride, alumina, hydroxyapatite, lead-zirconate-titanate (PZT), lead-magnesium niobate (PMN), bismuth titanate, as well as stainless steel. Normally, in the FDC process, single-material components are made. The fused deposition of multimaterials (FDMM) process has been developed to fabricate functional components with up to four different materials (Jafari *et al.* 2000).

1.5 Sanders Prototyping

The Sanders prototype (SP) technique is an indirect SFF method where initially a sacrificial wax mold with a negative of the desired ceramic structure is made. The mold is then infiltrated with ceramic slurry to produce the desired part.

The SP process uses a liquid-to-solid inkjet plotter that deposits the polymer on a movable Z-platform (Fig. 3). Two molten polymers, i.e., a thermoplastic building wax (used to make the mold) and a supporting wax (used to support overhangs and cavities in the mold), are fed to heated movable inkjet print heads capable of motion in the *x–y* plane. The piezoelectric-driven inkjet heads deposit the build

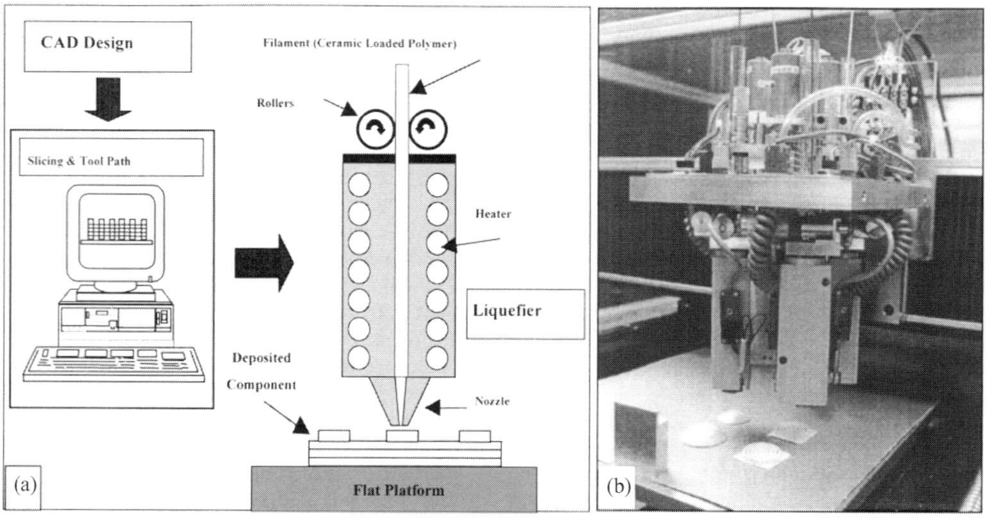

Figure 2
(a) Schematic of fused deposition of ceramics (FDC) process, including CAD file design, slicing and tool-path generation, and deposition on a platform in layer-by-layer fashion. (b) Photograph of fused deposition of multimaterials (FDMM) system with the capability of making components with up to four different materials.

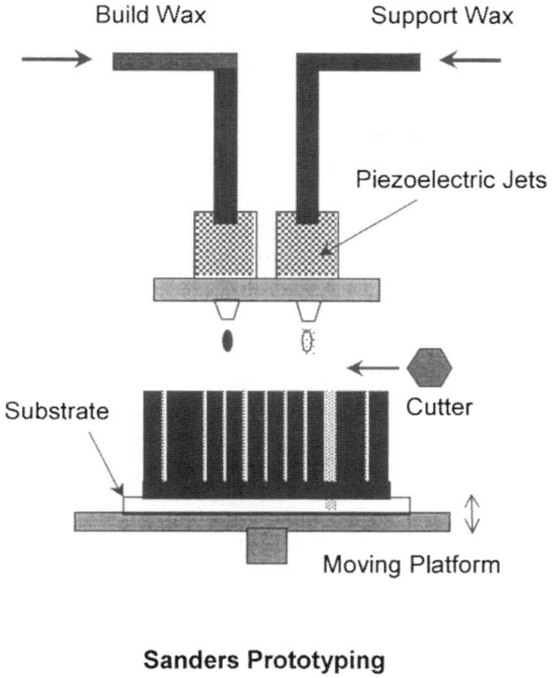

Figure 3
Schematic of the Sanders prototype using inkjet-type deposition process.

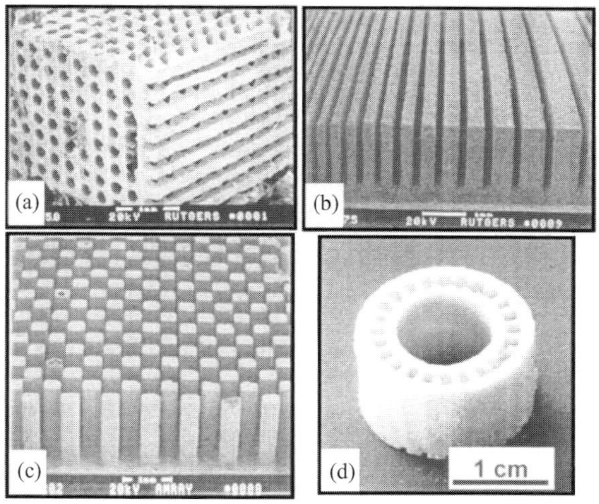

Figure 4
Variety of piezoelectric ceramic structures with different connectivities fabricated by Sanders prototyping (indirect method). (a) Honeycomb with 3-3 connectivity, (b) wax mold with Gaussian gradient, (c) 1-3 stagger rods, and (d) radial structure with 3-1 design.

and support materials as very fine droplets ($\leqslant 75\,\mu m$) at a high rate of 6000 drops/min. The molten droplets ejected from the inkjets solidify upon impact with the cooler mold surface. Any spacing between build polymer walls in a layer is filled with a support wax to provide support to the mold walls and to avoid the filling of the voids with build material during the milling operation. After the deposition of each layer, the cutter mills the mold surface to make it smooth and horizontal for the deposition of a next build layer. After the mold is built, the support polymer is dissolved using an organic solvent that does not attack the build polymer. The mold is then infiltrated using ceramic slurries to fabricate green ceramic components.

2. *Electroceramic Components Developed by Rapid Prototyping*

A vast number of electroceramic components have been fabricated by SFF techniques (Janas and Safari 1995, Safari *et al.* 2000, Safari *et al.* 1998a, Safari and Danforth 1998, Safari 1994, Bandyopadhyay *et al.* 1997, Bandyopadhyay *et al.* 1998). SFF processes allow rapid manufacturing of new designs and the possibility of frequent design alterations and tailoring of electromechanical properties by changing the shape and spatial distribution of the ceramic material.

This section focuses on the electroceramic parts such as piezoelectric ceramics and ceramic/polymer composites, various actuators and sensors, photonic band gap structures, and electroceramics with grain-oriented microstructures (Figs. 4 and 6–10). These electroceramic components have been developed using two different routes: (i) "direct route" by fused deposition of ceramics (FDC), and (ii) "indirect route" or lost mold process, where sacrificial polymer molds were manufactured initially, and then infiltrated with a PZT (or any desired) slurry to make green ceramic parts.

In the indirect route, sacrificial polymer molds were fabricated via the FDM or SP processes. Commercially available investment casting wax (ICW-06) and the SP build wax were utilized in the FDM and the SP processes. For PZT components, a water-based ceramic slurry with 52 vol.% PZT loading was developed. After slurry infiltration and drying, the thermoplastic polymer mold was evaporated using a specific thermal cycle during the early portion of the binder burnout step.

All the green parts made by the indirect and direct techniques were subjected to a careful heat treatment cycle of binder removal to 550°C and sintering at high temperatures (1285°C for PZT), followed by embedding in an epoxy resin (for the composites only), poling, and electrical measurements.

2.1 Electromechanical Components Made by the Indirect Route—Piezoelectric Ceramic/Polymer Composites

Piezoelectric ceramic/polymer composites provide many advantages for ultrasonic transducers, including a low acoustic impedance, high electromechanical coupling coefficients, formability into complex shapes, suppression of lateral modes, and tailoring of material properties. In order to reduce the grating lobes of a transducer, the piezocomposite architecture can be altered. Some examples of ceramic/polymer composites with unique structures are briefly explained below.

(*a*) *Honeycomb composites.* Piezoelectric ceramic/polymer composites with 3-3 connectivity have been manufactured using Sanders Prototyping (Safari *et al.* 1998a, Safari *et al.* 1998b). The SP process was used to form a number of "3D Honeycomb" composites. Figure 4(a) shows the scanning electron micrograph of a PZT 3D honeycomb structure. The structure shows the uniformity and the regularity of the holes that are connected in all three directions. The volume fraction of the PZT structure is approximately 25% with a hole diameter of 200 μm and center-to-center spacing of about 350 μm. The molds of these structures were made by making a 3-3 polymer ladder structure with a 90° rotation between roads of consecutive layers.

Molds were also made with various orientations other than 90° between consecutive layers to study the effects of orientation on the electromechanical properties of the 3D honeycomb structures. It has been found that the composites processed by rapid prototyping show improved electromechanical properties, probably due to controlled regularity and the phase periodicity of the ceramic phase in those structures.

(*b*) *Volume fraction gradient piezoelectric composites.* Volume fraction gradient (VFG) piezoelectric composites with a 2-2 connectivity were fabricated by SP (Panda *et al.* 1998). Many mathematical functions including Gaussian, linear and exponential gradient were used to describe the gradient. Figure 4(b) shows a sacrificial mold with Gaussian gradient obtained using the SP technique. After infiltrating PZT slurry into the molds and heat treatment, the sintered structures were embedded in spurr epoxy. The increase in the spacing between ceramic walls from the center to the edges follows distinctly different functions (Table 1) (Panda *et al.* 1998).

The vibration profiles of the 2-2 composites with and without VFGs, as a function of the distance from the center, are shown in Fig. 5. These measurements were taken by measuring the displacement of the composites using a photonic sensor. The pressure outputs of the VFG designs were compared with a regular 2-2 composite, which was a diced PZT-5H ceramic structure having a wall width of 220 μm and a kerf of 440 μm. The regular design had ~33 vol.% total ceramic content with a d_{33} and k_t of 340 pC/N and 63%, respectively. As expected, the amplitude of the vibration for the regular 2-2 composite did not vary much with the distance from the center, because this structure was the same everywhere. On the other hand, all the VFG distributions showed a decrease in the vibration amplitude, as distance from the center increased. The Gaussian VFG was designed to have slow ceramic vol.% fall with distance from the center. The vibration amplitude followed a similar trend, initially remaining constant and then dropping by only −4 dB at a distance of 5 mm. Thus, this distribution was not very different from the regular 2-2 composite out to 5 mm from the center. The linear gradient showed a very smooth decline in the vibration output, linearly falling to about −8 dB at a distance of 5 mm. Th exponential distribution had a vibration profile that was slightly lower in amplitude than the linear distribution (Panda *et al.* 1998).

(*c*) *Staggered composites.* Figure 4(c) shows a micrograph of a sintered staggered rod structure. This structure was made using the SP method. In this design, the square rods do not have their faces parallel to their nearest neighbors. In one individual row, the structure consists of 350 μm-wide square rods separated from each other by a gap of ~450 μm (a center-to-center distance of 800 μm). The rows next to it were staggered, by a distance

Table 1
Electromechanical properties of Gaussian and linear volume fraction gradient (VFG) composites.

Distribution type	Ceramic wall width (μm)	Spurr epoxy width (μm) Center	Spurr epoxy width (μm) Edges	Total vol.% ceramic	K	k_t (%)	k_p (%)
Gaussian VFG	115	85	500	34	900	68	26
Linear VFG	115	95	550	28	825	66	28

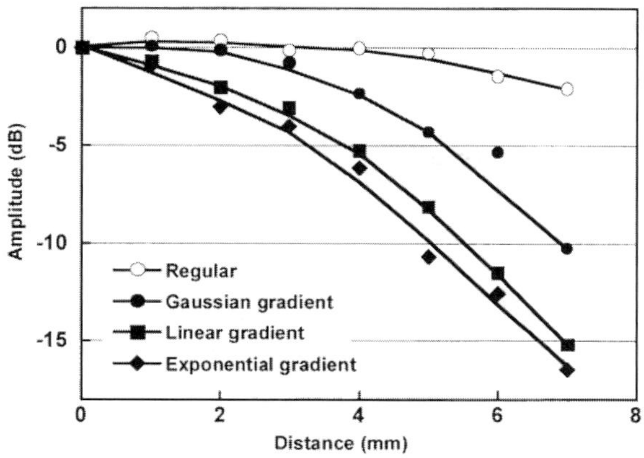

Figure 5
Vibration amplitude profiles comparing 2-2 regular and VFG composites.

equal to half the center-to-center distance (400 μm) between the rods in a row. The center-to-center nearest-neighbor distance between rods of two adjacent rows was ~575 μm. This structure had a ceramic content of ~23 vol.%, and a dielectric constant of 425. The d_{33}, k_t, and k_p of the composite were 290 ± 15 pC/N, 68%, and 29%, respectively. Thus, the k_t was high, even for this coarse structure.

(*d*) *Radial composites.* In current naval undersea towed-array systems, the hydrophone element consists of a monolithic ceramic tube poled in the radial direction. Like their planar counterparts, monolithic ceramic tubes suffer from similar deficiencies, such as high weight, low sensitivity, and low figure of merit ($d_h g_h$). Until recently, the fabrication of functional piezoelectric composites with radially oriented connectivity has been difficult or impossible using conventional processing. With the advent of rapid prototyping, these types of composites can be fabricated with relative ease.

Three different connectivities i.e., 2-2-, 3-1-, and 3-2-type composites, have been studied (McNulty 1999). For the purpose of simplicity, cylindrical coordinate systems are being used to designate the connectivity of each composite design. For 2-2-type composites, both the ceramic and polymer phases are continuous in the radial (r) and z directions. In the case of 3-1-type composites, the ceramic phase is continuous in all three directions (r, θ, z), while the polymer is continuous in only the z direction. Finally, for the 3-2-type composites, the ceramic phase is continuous in all three directions, while the polymer phase is continuous in the z and θ directions. Radially oriented composites have been made using both SP and SFF processes. Figure 4(d) shows the photograph of a tubular 3-1 PZT ceramic preform made using the SP technique. In this structure, the inner and outer ceramic rings are approximately 1 mm in thickness, and the ceramic sheets connecting the inner and outer rings are approximately 1 mm thick by 3 mm wide.

2.2 *Electromechanical Components Made by FDC (Direct Route)*

(*a*) *Curved transducers.* Novel piezoelectric ultrasonic transducers for medical imaging applications were built using the FDC technique. Curved ceramic skeletons for 2-2 composite transducers were designed and built (Fig. 6(a,b)) with different radii of curvature (Lous *et al.* 2000). In the built green parts, the ceramic plates were 550 μm thick and the spacing between the plates was 915 μm. In other designs, samples were made with plates as small as 330 μm, with a spacing of 760 μm. Figure 6(c) shows a micrograph of the surface roughness associated with the layer-wise building of the FDC technique. SEM of the surface and fracture surfaces of the sintered samples showed that the FDC roads bonded well together with no delamination, and a dense microstructure.

(*b*) *Ladder composites.* Ladder structures with fine dimensions were fabricated using the FDC process. Figure 7(a) shows a view of the 3-3 ladder structure of sintered PZT ceramic with volume fraction of approximately 70%. The volume fraction of ladder structures can easily be altered by changing the width and the spacing between the ceramic.

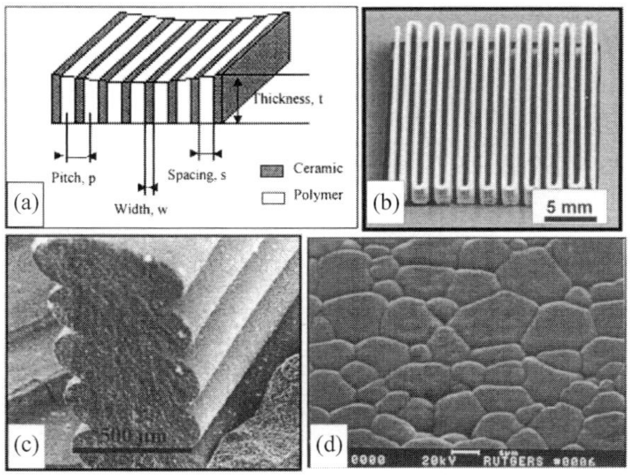

Figure 6
(a) Schematic of a 2-2 curved transducer with alternate ceramic and polymer plates, (b) a curved transducer made by FDC, (c) a SEM micrograph showing the good bonding between individual rods, and (d) a typical dense microstructure of the sintered FDC part.

Other ladder structures with different orientations were also made by FDC (e.g., Figure 7(b)). The effect of orientation on the d_{33} coefficient (structure) is plotted in Figure 7(c).

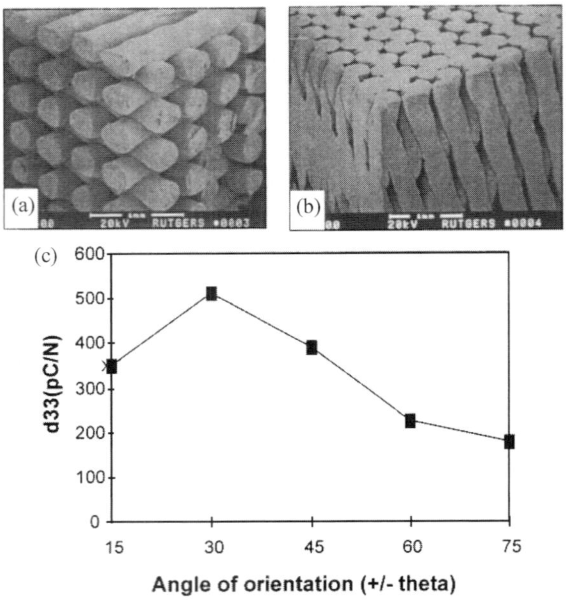

Figure 7
(a, b) SEM micrographs of ladder structures (90° and 15° angles) and (c) a plot showing the variation of d_{33} versus the angle of orientation.

(*c*) *Spiral actuators.* New spiral actuators (Fig. 8) were fabricated in order to obtain high displacement via utilizing the d_{31} coefficient (Mohammadi *et al.* 1999). When an electric field was applied to a spiral actuator in the poling direction (across the width), the spiral tip displaced in both tangential and radial directions. The tangential displacement was in the millimeter range, whereas the radial movement was relatively small—of the order of a few microns. Figure 8 shows a typical tangential displacement of a spiral actuator as a function of d.c. driving voltage in comparison with the transverse displacement of a long strip of the same length, width, and height as the spiral. The spiral consisted of 5 turns with an external diameter of 32 mm. The PZT spiral actuator exhibited a displacement of 1900 μm, which is more than 12 times larger than that of the long strip (displacement of 153 μm) under the same voltage. This suggests that the d_{31} effect (transverse strain) is of minor importance for the observed displacement in spiral actuators. The actuator exhibited a much larger displacement when driven in a.c. mode at resonance frequency, reaching to about 7500 μm at 100 V a.c. Such amplification is very important for applications in piezoelectric motors, where the actuators operate in resonance mode.

(*d*) *Telescoping actuators.* Another monolithic ceramic actuator with telescoping structure was fabricated by the FDC method. The device consists of interconnected PZT ceramic tubes (Fig. 9) bearing a common center with internal displacement amplification under electric field. Upon the application of an electric field, the tubes elongate and shrink alternatively in the tube height (z) direction,

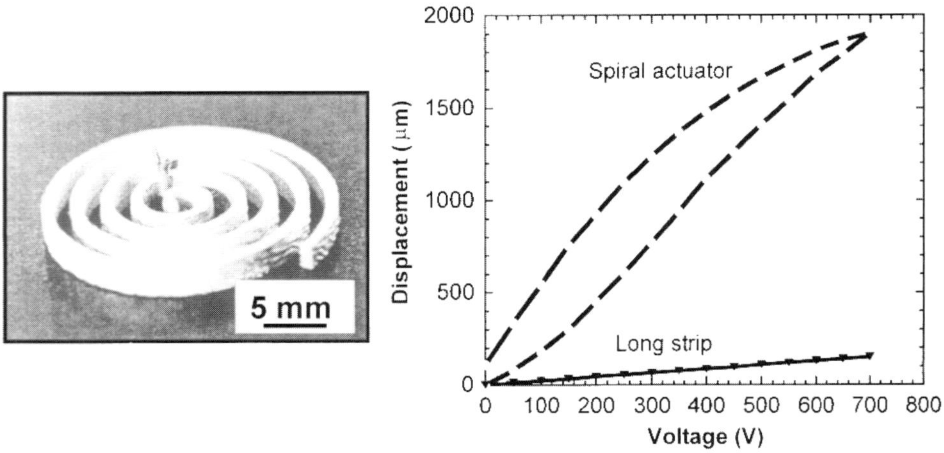

Figure 8
A spiral actuator and its tangential displacement as a function of driving voltage in comparison with the d_{31} response of a straight PZT strip of the same dimensions. The effective length, width, and height of the spiral are 280 mm, 0.6 mm, and 2.75 mm, respectively.

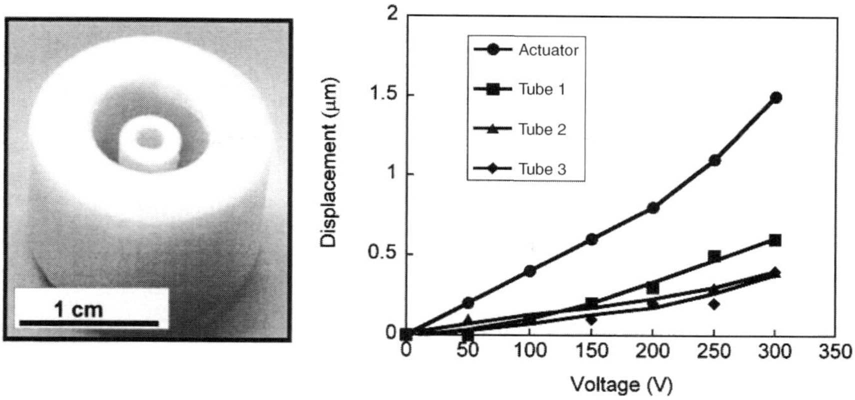

Figure 9
A photograph of a telescoping actuator and its displacement, which is the sum of each tube's displacement.

yielding a net additive displacement in the z direction. A telescoping actuator with three interconnected PZT ceramic tubes and a common center was examined. Under applied electric field, the inner and outer tubes elongate while the middle tube shrinks. The telescoping actuation of the device is the summation of displacement of all the individual tubes, as plotted in Fig. 9. Hence, the displacement can be further enhanced by increasing the number of the tubes. In addition, the displacement and force of telescoping actuators can be enhanced by tailoring of the dimensions (thickness, height, and diameter) of the tubes.

(*e*) *Tube actuators.* PZT tube and tube arrays (Fig. 10(a)) were fabricated for actuator, broadband resonant transducer, and sensor applications. Single tubes and tube arrays with different dimensions were fabricated to optimize the field-induced strain in the axial (tube height) direction. Moreover, tube actuators with zigzag wall design (bellows, Fig. 10(b)) with 30°, 45°, and 60° angles have been designed and fabricated. In comparison with the straight tubes, bellows show up to 50% larger radial displacements, whereas the displacement along the height was about 20–30% lower in bellows than that of the straight tubes for the particular dimensions examined.

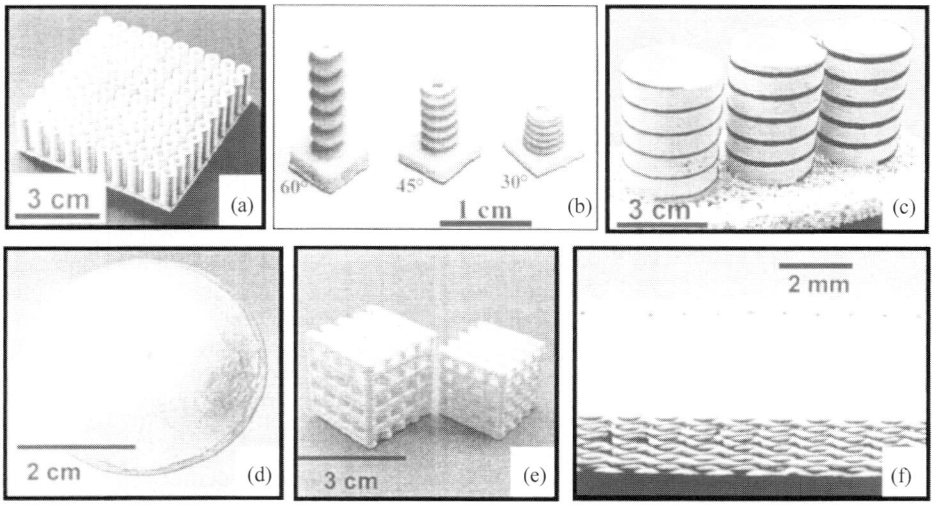

Figure 10
Photographs of various PZT transducers made by the fused deposition of ceramics (FDC) process. (a) Tube array actuator, (b) bellows actuators, (c) tube actuators with helical electrodes, (d) dome actuator, (e) photonic band gap structures of alumina rods, and (g) a structure of hard (dark) and soft (light) PZT made by the FDMM process.

Tube actuators with helical electrodes (Fig. 10(c)) were also fabricated to determine the possibility of enhancing field-induced displacement along the tube height. The helical electrodes allow one to pole the tubes in the height direction, therefore, the higher d_{33} coefficient can be used as opposed to the d_{31} coefficient in the radially poled tube actuators. Tube actuators with helical electrodes can potentially have over two times larger axial displacements relative to the conventional tubes.

Tubes with helical electrodes were made of a piezoelectric material (e.g., PZT, PMN-PT, etc.) and silver–palladium electrode. Two different CAD designs were considered. In the first design, the helical electrodes were located within the tube wall. Therefore, both ceramic and metal electrode were deposited in each layer using the Fused Deposition of Multi-Material (FDMM) system. For the second design, a two-step deposition was used; initially, simple tubes were made by FDC, and then helical electrodes were deposited on the surface in the next deposition step.

(f) Dome actuators. Novel designs, such as Moonies and RAINBOW actuators, have recently been proposed and fabricated for strain amplification. The improved actuation results from curved geometry. Using the FDC process, a new actuator with curved surface (i.e., Dome actuator, Fig. 10(d)) has been developed. The radius of curvature and other dimensions were varied in order to study the axial displacement, resonance behavior, and electromechanical hysteresis. Displacements of up to 7 μm were achieved from the domes with 23 mm and 1.4 mm in diameter and height, respectively.

2.3 Net Shape Components With Oriented Microstructures

Recently, a novel technique referred to as templated grain growth (TGG) has been implemented to create oriented grain structures in a variety of ceramics (Holmes *et al.* 1979, Horn *et al.* 1999). Grain-oriented microstructures are beneficial due to the improved mechanical and/or electrical properties that can be obtained as a result of crystal anisotropy. In order to achieve primary orientation in the TGG process, a small percentage of seed particles must be initially incorporated and oriented within the fine powder matrix. Thus, a fabrication method with shear mechanism is required. Tape casting has normally been employed, where the slurry, including 5–15 vol.% seeds, is sheared to align the seeds in the casting direction. In the FDC process, the material to deposit is sheared both during filament fabrication and deposition of material through fine nozzles. Therefore, the FDC process is suitable for aligning

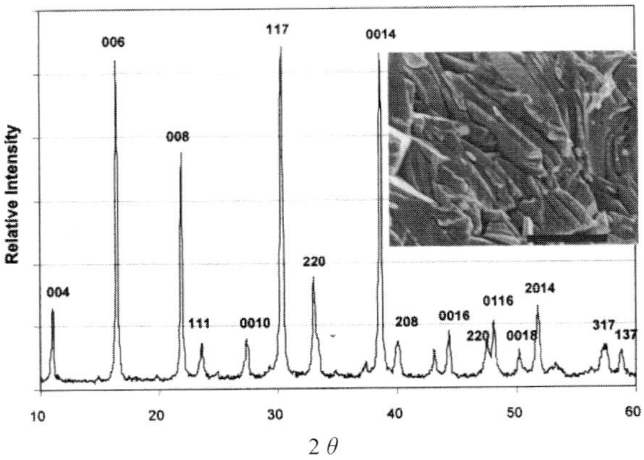

Figure 11
X-ray diffraction from the surface of an FDC-BiT sample, illustrating the high intensity of (00L) planes resulting from grain orientation. Fracture surface of the sample also shows partial grain orientation (5 μmbar).

seed particles. We have implemented the combination of the FDC process and TGG method to fabricate net shape electromechanical components with oriented microstructures.

The possibility of FDC-TGG was explored in a bismuth titanate (BiT) system. BiT filaments with 5 vol.% platelet seed were developed and used to fabricate net shape parts. Examination of the green samples showed that the BiT seeds were aligned during the deposition process. The x-ray diffraction (Fig. 11) and SEM of the TGG-sintered samples showed that partial orientation was obtained in the net shape BiT components (Allahverdi *et al.* 2000).

2.4 Photonic Bandgap Structures

Photonic bandgap (PBG) structures are periodic dielectric structures with alternate high and low permittivity materials in order to obtain an electromagnetic stop-band in a desired direction (Yablonovitch 1993). A variety of potential applications such as thresholdless lasers, high-quality single-mode LEDs, and microwave antennas are already being explored.

We have implemented rapid prototyping to make a variety of PBG parts with different dimensions according to the results obtained form the simulation work. Figure 10(d) shows an alumina PBG structure with square-section rods. Simulation works suggest that such a PBG structure will provide a band gap in the frequency range of 8–12 GHz.

See also: Electroceramics, Colloidal Stability and Processing of

Bibliography

Agarwala M K, Bandyopadhyay A, van Weeren R, Whalen P, Safari A, Danforth S C 1996 Fused deposition of ceramics: rapid fabrication of structural ceramic components. *Am. Ceram. Soc. Bull.* **75**(11), 60–5

Allahverdi M, Jadidian B, Ito Y, Safari A 2000 Fabrication of bismuth titanate components with oriented microstructures via FDC and TGG. *Proc. 12th IEEE Int. Symp. Applications of Ferroelectrics*. Hawaii. IEEE, New York

Bandyopadhyay A, Panda R K, Agarwala V F, Danforth S C, Safari A 1997 Processing of piezocomposites by fused deposition technique. *J. Am. Ceram. Soc.* **80**(6), 1366–72

Bandyopadhyay A, Panda R K, McNulty T F, Mohammadi F, Danforth S C, Safari A 1998 Piezoelectric ceramics and composites via rapid prototyping techniques. *Rapid Prototyping J.* **4**(1), 37–49

Beaman J, Deckard C R 1990 *Selected Laser Sintering with Assisted Power Handling*. US Patent No. 4 938 816

Comb J W, Priedeman W R 1993 Control parameters and materials selection criteria for rapid prototyping systems. In: Marcus H L, Beaman J J, Barlow J W, Bourell D L, Crawford R H (eds.) *Proc. Solid Freeform Fabrication Symp.* University of Texas, Austin, Texas, Vol. 1992, pp. 86–91

Danforth S C, Agarwala M, Bandyopadhyay A, Langrana N, Jamalabad V R, Safari A, Weeren R 1998 *Solid Freeform Fabrication Methods*. US Patent No. 5 738 817

Dimos D, Yang P, Garino T J, Raymond M V, Rodriguez M A 1997 Direct-write fabrication of integrated, multilayer ceramic components. In: Bourell D L, Beaman J J, Marcus H L, Crawford R H, Barlow J W (eds.) *Proc. SFF Symposium*, University of Texas, Austin, Texas, Vol. 8, p. 33

Griffith M L, Halloran J W 1996a Free form fabrication of ceramics via stereolithography. *J. Am. Ceram. Soc.* **79**(10), 2601

Griffith M L, Halloran J W 1996b Curing behavior of ceramic resin for stereolithography. In: Bourell D L, Beaman J J, Marcus H L, Crawford R H, Barlow J W (eds.) *Proc. SFF Symp.*, University of Texas, Austin, Texas, Vol. 7, p. 403

Holmes M, Newnham R E, Cross L E 1979 Grain-oriented ferroelectric ceramics. *Am. Ceram. Soc. Bull.* **58**(9), 872

Horn J A, Zhang S C, Selvaraj U, Messing G L, Troiler-McKinstry S, Yokoyama M 1999 Fabrication of textured $Bi_4Ti_3O_{12}$ by templated grain growth. *J. Am. Ceram. Soc.* **82**(4), 921–6

Hull C W 1990 *Method for Production of 3-D Objects by Stereolithography*. US Patent No. 4 929 402

Jafari M A, Han W, Mohammadi F, Safari A, Danforth S C, Langrana N 2000 A novel system for fused deposition of advanced multiple ceramics. *Rapid Prototyping J.* **6**(3), 161–75

Janas V F, Safari A 1995 Overview of fine-scale piezoelectric ceramic/polymer composite processing. *J. Am. Ceram. Soc.* **78**(11), 2945

Klosterman D, Chartoff R, Priore B, Osborne N, Graves G, Lightman A, Han G, Pak S, Weaver J 1996 Structural composites via laminated object manufacturing. In: Bourell D L, Beaman J J, Marcus H L, Crawford R H, Barlow J W (eds.) *Proc. SFF Symp.*, University of Texas, Austin, Texas, Vol. 7, p. 356

Liu Z, Suppakarn N, Cawley J D 1999 Coated feedstock for fabrication of ceramic parts by CAM-LEM. In: Bourell D L, Beamen J J, Marcus H L, Crawford R H, Barlow J W (eds.) *Proc. SFF Symp.*, University of Texas, Austin, Texas, Vol. 10, p. 393

Lous G, Cornejo I, McNulty T F, Safari A, Danforth S C 2000 Fabrication of piezoelectric ceramic/polymer transducer by fused deposition of ceramics. *J. Am. Ceram. Soc.* **81**(1), 124–8

McNulty T F 1999 Prototyping of radially oriented piezoelectric ceramic-polymer tube composites using fused deposition and lost-mold processing techniques. Ph.D. Thesis, Rutgers University

Mohammadi F, Kholkine A L, Jadidian B, Safari A 1999 High-displacement spiral piezoelectric actuators. *Appl. Phys. Lett.* **75**(16), 2488–90

Panda R K, Kholkine A L, Safari A 1998 Design and fabrication of volume fraction gradient piezoelectric composites for medical imaging applications. *Proc. 11th IEEE Int. Symp. Applications of Ferroelectrics*. ISAF-XI, Montreux, Switzerland. IEEE, New York, pp. 265–8

Sachs E M, Cima M J, Williams P, Brancazio D, Cornie J 1992 Three-dimensional printing: rapid tooling and prototypes directly from a CAD model. *J. Eng. Ind.* **114**, 481–8

Safari A 1994 Development of piezoelectric composites for transducers. *J. Phys. III France* **4**, 1129

Safari A, Danforth S C 1998 Development of novel piezoelectric ceramics and composites for sensors and actuators by solid freeform fabrication. *Proc. 11th IEEE Int. Symp. Applications of Ferroelectrics*. IEEE, New York, pp. 229–34

Safari A, Danforth S C, Bandyopadhyay A, Janas V F, Panda R K 1998a *Oriented Piezoelectric Ceramics and Ceramic/Polymer Composites*. US Patent No. 5 796 207

Safari A, Danforth S C, Jafari M A, Allahverdi M, Jadidian B, Mohammadi F, Vankataraman N, Rangarajan S 2000 Fabrication of advanced functional ceramics by fused deposition technique. *Proc. 9th Eur. Conf. Rapid Prototyping and Manufacturing*, Athens, Greece, pp. 247–63

Safari A, Danforth S C, Panda R K, Lous G, Cornejo I, Kholkine A 1998b Novel piezocomposites for ultrasonic transducer applications. *Proc. IEEE Ultrasonic Symposium*. IEEE, New York, Vol. 1, pp. 615–8

Yablonovitch E 1993 Photonic band-gap structures. *J. Opt. Soc. Am.* **10**(2), 283–95

A. Safari and M. Allahverdi

F

Ferrous Metals, Heat Treatment of

Ferrous metals are a very large group of alloys that consist mostly of iron, may contain carbon as a major alloying element, and may contain one or more other alloying elements. This article concentrates on the heat treatment of wrought carbon steels (Krauss 1990), ranging in carbon content from 50 ppm in ultra-low carbon sheet steels to 1.00 mass.% in bearing and low alloy tool steels. Many of the heat treatments described are also applied, with appropriate modifications to account for differences in alloying, to other types of ferrous alloys, such as cast irons, stainless steels, and high-alloy tool steels (Krauss 1990). Some heat treatments are applied to enhance in-process properties during manufacture, while other heat treatments are performed at the end of processing to produce the properties required for a given application. Details and recommended processing schedules for the heat treatment of specific steels are provided in handbooks and other literature (ASM International 1995).

The object of any heat treatment is to provide a microstructure that produces a desired combination of properties. Microstructures consist of arrangements of phases, each with a unique crystal structure, chemistry, and properties. In steels, quite different phases are stable at various temperatures. In steels, at high temperatures, the dominant phase has a f.c.c. crystal structure and is termed *austenite*. Austenite has a fairly high solubility for carbon, with a maximum of about 2 mass.% at 1150 °C. At low temperatures, including room temperature, the dominant phase is b.c.c. and is termed *ferrite*. Ferrite has a low solubility for carbon, and as a result a carbide phase, with an orthorhombic crystal structure, termed *cementite*, generally coexists with ferrite. In iron-carbon alloys, the equilibrium temperature where austenite begins to form is 727 °C. The temperatures which mark the transition from one phase or several phases to another phase or combination of phases are called *critical temperatures*. Heat treatments applied below the temperature where ferrite and cementite are replaced by austenite are, therefore, called *subcritical heat treatments*.

1. *In-process Heat Treatments of Carbon Steels*

In-process heat treatments are applied to steels to increase ductility for forming, to reduce hardness for machining and/or forming, or to provide a uniform microstructure for subsequent finish heat treatments. Subcritical heat treatments include recrystallization annealing and stress relief heat treatments. For example, low-carbon sheet steels are cold rolled and, therefore, their ferritic microstructure is highly deformed and of low ductility. As a result, a subcritical annealing treatment that causes the strained ferrite crystals or grains to recrystallize is applied, and ductility is restored for subsequent forming operations. The application of this type of processing to sheet steels defines the class of steels referred to as cold rolled and annealed steels. Many manufacturing operations introduce residual stresses, and subcritical annealing treatments are frequently applied to reduce such stresses which may cause distortion and cracking during subsequent processing.

Other in-process heat treatments involve heating above temperatures at which ferrite is stable, therefore resulting in the formation of austenite in low or medium carbon steels, or austenite and cementite in high carbon steels. On cooling the austenite transforms to various combinations of ferrite and cementite, one of which consists of parallel arrays of ferrite and cementite termed *pearlite*. Normalizing is a heat treatment applied to forgings that may have coarse and nonuniform microstructures. Normalized steels are heated just above the upper critical temperatures and air cooled to produce a fine pearlitic microstructure. Full annealed steels are heated to lower temperatures in the austenite phase field and furnace cooled to produce a coarse pearlite of high ductility and low hardness. Spheroidizing heat treatments, which produce a dispersion of spherical cementite particles in ferrite, are applied both above and below the austenite-ferrite critical temperature. The spheroidized microstructure has the lowest hardness and highest ductility of any microstructure that may be formed in a given steel.

2. *Through Hardening Heat Treatments Applied to Steels*

Many applications require high strength, good fatigue resistance, and good wear resistance. When carbon steels are heated to temperatures where austenite is stable, a heat treatment step referred to as *austenitizing*, and the austenite is subsequently cooled rapidly or quenched, carbon atoms cannot diffuse to produce microstructures of ferrite and cementite, and a hard phase termed martensite forms. The crystal structure of martensite in steels is body-centered-tetragonal, the tetragonality introduced because the carbon atoms are trapped between the iron atoms of a body-centered structure. Martensite is the hardest phase that can be produced by heat treatment in a given steel, and, therefore, the process of producing martensite is referred to as *hardening*. The ability of a steel to form martensite, or its hardenability, depends on chemical composition, cooling rates, and section size, and many steels and heat treatments are selected to produce martensite or a significant amount of martensite through the cross-section of a part. Both hardness and hardenability increase with increasing carbon content, and, therefore, medium-carbon (0.3–0.6 mass.% carbon) and high-carbon (0.7–1.5 mass.% carbon) steels are most frequently hardened.

As-quenched martensitic microstructures, however, have low toughness or fracture resistance, and, therefore, are subjected to a subcritical heat treatment referred to as tempering. The tempering permits carbon atoms to diffuse and relieve the supersaturation of carbon atoms in the as-quenched martensite. Various mixtures of ferrite and carbides, depending on tempering time and temperature are produced. The higher the tempering temperature, the coarser the ferrite-carbide mixtures and the lower the hardness. Generally the lower the hardness the higher the ductility and fracture resistance. However, tempered martensitic microstructures in steels with high concentrations of impurity elements such as phosphorus are sensitive to various types of embrittlement. Tempered martensite embrittlement develops between 250 °C and 350 °C, temper embrittlement develops around 500 °C, and these tempering temperature ranges should be avoided in susceptible steels.

3. *Surface Modification Heat Treatments Without Chemical Change*

In many applications only the surface of a steel part is subjected to high stresses. As a result, only the surface needs to be hardened. Such surface hardening is accomplished by various modes of heating only the surface of the steel, including flame, induction, laser, and electron beam heating. The more recently applied techniques of laser and electron beam heating heat to much shallower depths than do flame and induction hardening. Each type of heating produces a surface layer of austenite which on quenching transforms to martensite which in turn is tempered to a given hardness. The interior or core of the part remains unheated and retains the microstructure produced by previous processing. The hardened surface or case is generally in a state of residual compressive stress produced when the expansion that accompanies the transformation of austenite to martensite is restrained by the unheated core. The compressive stresses effectively increase resistance to bending fatigue. Another advantage of the surface heating techniques is that only selected areas subjected to the highest stresses can be hardened, and other areas that are not highly stressed need not be heat treated. Since there is no composition change, generally medium carbon steels which are capable of forming martensite of high hardness are subjected to surface heating.

4. *Thermochemical Surface Modification Heat Treatments of Steels*

Another approach to surface hardening consists of changing surface chemistry. Carburizing is a widely used example of such a thermochemical process. Low-carbon steels, containing around 0.2> mass.% carbon are austenitized and exposed to a carbon-containing atmosphere. The carbon diffuses into the austenite, a surface carbon content about 0.8 mass.% carbon is maintained, and a case with a carbon gradient between the surface and that of the low carbon core is produced. The case depth is a function of time and temperature. For example, a case depth of about 1 mm is produced by carburizing for about 4 hours at 930 °C. The austenite is then quenched to martensite, with a very high surface hardness typical of a high-carbon steel, and tempered at low temperatures in order to preserve as much of the as-quenched hardness as possible. In addition to high case hardness, carburizing also produces beneficial residual surface compressive stresses.

Nitriding is another example of a thermochemical surface heat treatment applied to steels. In contrast to carburizing, nitriding is performed at subcritical temperatures, around 500 °C, often on hardened steels tempered to produce fine mixtures of ferrite and carbides. Ammonia atmospheres provide the nitrogen which diffuses into the ferritic microstructure and creates very high hardness. Nitriding requires much longer times because of the lower processing temperatures relative to carburizing, but, because the transformation of austenite to martensite is not involved, distortion and dimensional changes are minimal.

Combinations of carburizing and nitriding are also frequently applied for specific applications. Carbonitriding involves the diffusion of carbon and nitrogen into austenite, with surface hardening produced by martensite formation, while nitrocarburizing techniques involve the diffusion of carbon and nitrogen into ferrite at subcritical temperatures. Traditionally, the various thermochemical surface hardening processes have been applied with gas atmospheres, but salt baths are also used. More recently vacuum carburizing, in which a vacuum is produced prior to the introduction of a carburizing gas, and plasma carburizing and plasma or ion nitriding, in which the work pieces are made cathodic, are increasingly being used. Benefits of the latter processes include shorter processing times, cleaner surfaces without oxide formation, and cleaner environmental operation.

5. *Coating Surface Heat Treatments Applied to Steels*

Another type of surface modification heat treatment applied to steel is the application of thin coatings or layers of very hard ceramic compounds such as titanium nitride. In contrast to the thermochemical surface modification techniques described above, chemical elements do not diffuse into austenite or ferrite, but combine to produce very thin compound layers, often only $<10\,\mu m$ in thickness, on work piece surfaces. Nevertheless, because of the very high hardness of the compound layer, wear resistance is substantially improved. The coatings are produced by physical vapor deposition (PVD) processes in which atoms or molecules, generated by evaporation, sputtering, or ion plating, are deposited on a substrate in a high vacuum chamber. Chemical vapor deposition (CVD) processes, in which reaction products of a chemical reaction in the vapor phase are deposited on a substrate, are also used to produce hard coatings on steels.

Bibliography

ASM International 1995 *Heat Treater's Guide: Practices and Procedures for Irons and Steels*. ASM International, Materials Park, OH

Krauss G 1990 *Steels: Heat Treatment and Processing Principles*. ASM International, Materials Park, OH

G. Krauss

Forming Procedures in Polymer Component Manufacturing

Forming procedures for thermoplastics have become an important sector of the plastics processing industry. They are widely used for producing packaging material, consumer items, or even technical parts. Products manufactured by these techniques range from mass-produced articles like cups ($10^6\,h^{-1}$) to large-size articles like swimming pools ($1.5\,h^{-1}$). Moreover, the forming of polymers is an important procedure for automotive parts such as dashboards.

Starting with simple machines and molds at the beginning of the 1950s, the forming procedures have progressed to completely automated industrial processes with complex machines and tools of high precision. Although forming in general might be achieved using different processing techniques, the forming of thermoplastics usually involves vacuum or compressed air in addition to mechanical stretching aids. In contrast to molding procedures, which are carried out at the polymer's melting temperature, the forming is done when the material is thermoelastic (compare *Sect.* 1.2). As a result the semifinished product is subjected to a rapid biaxial stretching, which leads to an improvement of the mechanical properties.

Initially the forming procedures, which are commonly subdivided into thermoforming and stretch blow molding—a variety of the blow molding process—were developed on the basis of pure empirical approaches. Nowadays fundamental principles of science are used to understand the processes and material properties. Therefore, this article outlines the physical basis of thermoforming and stretch blow molding before the process engineering is described.

1. Physical Basis

Both procedures are based on the forming of semifinished polymer products under the influence of heat. Therefore one may structure them into the process steps: heating, forming, and cooling.

1.1 Heating

To enable the forming of thermoplastics the mechanically rigid-to-brittle semifinished product has to be converted into a thermoelastic state. For most thermoplastic materials this range is between 80 and 160 °C. In general, the surface temperature should not exceed a maximum in order to avoid any surface damage due to degradation, and the material temperature has to be above a minimum for forming to occur. In addition, the temperature profile over the polymer surface and along the cross-section has to be precise and reproducible to allow homogeneous forming. Consequently, the requirements for a heating system are stringent. Three heating methods are available:

(i) contact heating (metal plates or rolls);
(ii) convection heating (circulating air unit);
(iii) radiation heating (infrared radiation).

(*a*) *Contact heating.* Using a contact heating system the surface of the semifinished product reaches the temperature of the heating plates almost immediately. However, the temperature rise of the core is caused by heat conduction and takes some time because of the low thermal conductivity of polymers. This is unimportant for thin, semifinished products like films used for blister packs, but inevitably the contact heating is of limited suitability for variable-shaped or three-dimensional semifinished products.

In addition, the heating plates have to be coated with anti-adhesive to avoid the polymer sticking to the plates; unfortunately this coating impedes the heat transfer. Therefore, and in order to shorten the cycle time, the temperature of the heating plates is adjusted to above the forming temperature at about 200 °C. This leads to a high temperature gradient across the thickness of the semifinished product, which is partly compensated for in the subsequent forming process.

(*b*) *Convection heating.* The main advantage of convection heating is that such systems are nearly independent of the size and geometry of the semifinished product. Because of the low thermal conductivity and low heat transfer coefficients of polymers the surface temperature rises slowly during prolonged gentle heating and the temperature gradient across the semifinished product thickness is small. Therefore extremely thick plastic plates are often heated in this way.

(*c*) *Radiation heating.* Owing to its universality, infrared radiation heating is the most frequently used heating system. Infrared radiation is not only absorbed at the polymer surface but it is also effective in the inner part of the material and this partly overcomes the problem of the low thermal conductivity of polymers. As a result, high radiation doses can be used to realize short heating times without any damage of the surface due to thermal degradation.

For plastics processing ceramic and quartz radiators are mainly used, though halogen radiators are becoming more widespread. Most radiators are seldom used. The mass of the different radiator systems have a negligible effect on the energy input. In contrast to ceramic radiators, which react the most slowly, halogen radiators respond very quickly, and this facilitates control of the radiator temperature from cycle to cycle.

In several areas of heat engineering the calculation of the exchange of radiation between two solids is based on the assumption of black or gray bodies. This approach is limited for polymers, as they show a distinct dependence of radiation-specific parameters (reflection, absorption, transmission) on radiation wavelength. This makes a calculation of the temperature rise due to radiation quite difficult. These calculations are necessary if the advantages of infrared radiation heating are to be exploited. Basic relations to do so can be found in Throne (1987).

Polymers show a characteristic absorption of infrared radiation mainly in the wavelength range of 2.5–15 μm. Their ability to absorb is based on the excitation of molecules and molecular bonds that are in equilibrium to create resonant frequencies. The absorption of infrared radiation increases the rotational and vibrational energy of the molecules and this leads to a rise in temperature. Because of the similarities in chemical structure most polymers have similar absorption profiles. These are very high, in the range 3.3–3.5 μm, due to the valence oscillations of the carbon–hydrogen bonds.

At the interface between materials of different optical density, for example, air and polymer, part of the incident radiation is refracted into the material and the rest is reflected. The surface reflection ability of polymers, even for apparently glossy surfaces, is low in the infrared range and amounts to 4–5% in a spectral range between 0.6 and 200 μm. The rest of the radiation which is neither absorbed nor reflected is transmitted. Although the coloring of polymers influences strongly the transmission of visible light, its influence is weak in the infrared range.

1.2 Temperature-dependent Behavior of Thermoplastics

The technological behavior of macromolecular thermoplastics is characterized by a wide transition area between the solid and liquid states, i.e., the thermoelastic range. The Young's modulus (E) and the tensile strength (σ) yield information about the deformation resistance (Fig. 1). A measure for the deformation ability is the elongation at rupture (ε_B). For amorphous thermoplastics the maximum of the deformation ability (maximum ε_B) can be found in the thermoelastic range, where the deformation resistance is low (Fig. 1(a)). Therefore, amorphous materials like polystyrene (PS) or polyvinylchloride (PVC) can be formed over a wide temperature range and are very well suited to forming procedures, as they are easy to process.

In general, semicrystalline thermoplastics are not formed in the area of maximum strain, which is above the crystallite melting temperature, as the tensile strength is too low in this range (Fig. 1(b)). For this reason these polymers are formed at just below the crystallite melting range, thereby achieving large improvements in the mechanical properties within the formed component since high molecular orientations are introduced into the material. The Young's modulus is still quite high in this temperature range. Therefore the force needed for forming is also high and the solid-phase pressure forming technique has to be used (Sect. 2.2). Two of the most important examples of semicrystalline thermoplastics processed are polypropylene (PP) and polyethylene terephthalate (PET).

In contrast to uniaxial stretching, which is usually practiced to improve the mechanical properties of synthetic fibers (e.g., Kevlar®), the biaxial stretching while forming an amorphous or semicrystalline thermoplastic leads to more or less balanced properties. This is very important for polymer films as stiffness and strength only increase in direction of the stretching. Perpendicular to this direction the mechanical properties degrade, thereby leading to a splice tendency if the material is loaded transversely to the oriented molecules. Therefore, the anisotropy known from synthetic fibers, which affects mechanical properties as well as characteristics like thermal expansion and conductivity, is not very pronounced for a biaxially stretched thermoplastic.

1.3 Forming

All forming procedures have in common that the heated semifinished product is subjected to a biaxial stretching, which is considered to be continuous and isothermal as long as the material has not touched the tool. As the surface area grows the thickness decreases. Immediately after the contact with the tool's surface the material starts to cool down and simultaneously the deformation resistance rises rapidly. Where the material is in contact with deformation aids like stretching rods, which have a much higher temperature than the tool, it is not usually possible to define how much further stretching of the material in the contact zones is possible. However, this phase of the process can be analyzed since the geometry of the component, the force transfer, and the cooling behavior are usually easily formulated.

As thermoplastics behave like rubber–elastic in the thermoelastic range, one can assume that the theory of deformation of elastomers is valid. Accordingly, the deformations are considered to be large and reversible and the stress deformation behavior is nonlinear. Therefore the Mooney/Rivlin approach may be applied (Rivlin 1984).

1.4 Cooling

The cooling phase starts with the first contact of the material with the surface of the tool. The heat transition at the contact interface between the tool and

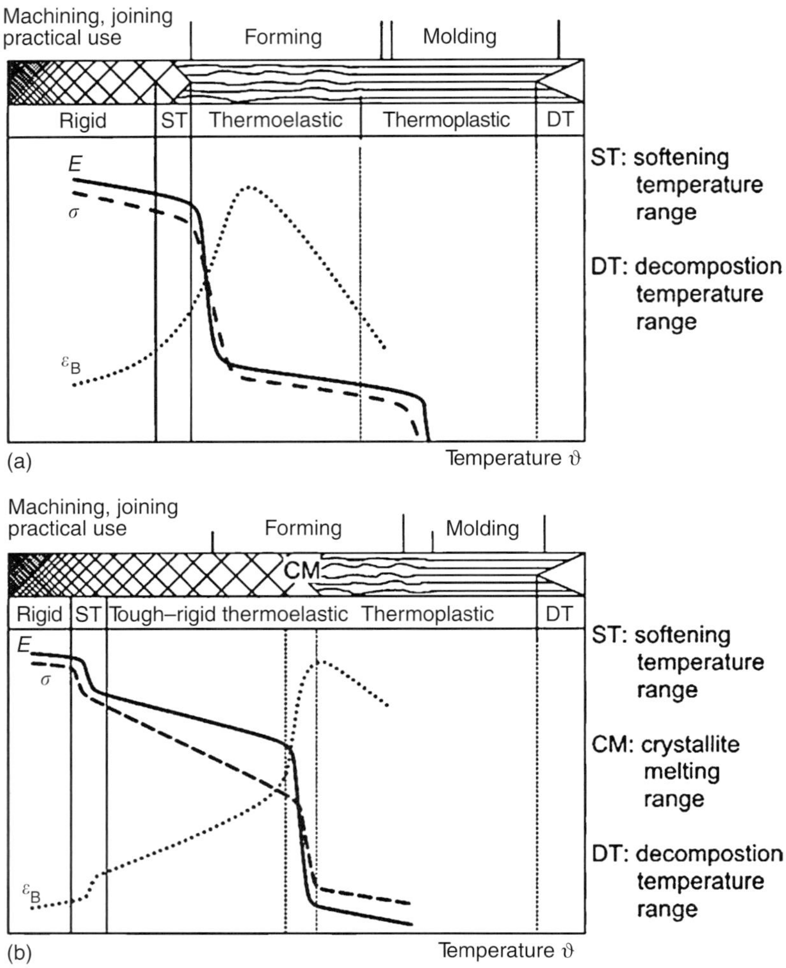

Figure 1
Temperature-dependent behavior of thermoplastics: (a) amorphous thermoplastics; (b) semicrystalline thermoplastics.

the material can only be given as an order of magnitude, which is assumed to be approximately $\alpha_{cont} = 1.000\,Wm^{-2}K$. At the opposite surface the heat transition results from convection and radiation. In general this convection is intensified by fan or spray cooling (water vapor), for which the following total heat transfer coefficients (including convection and radiation) might be used:

free convection: $\alpha_{tot} = 5\,Wm^{-2}K$;
fan cooling: $\alpha_{tot} = 23\,Wm^{-2}K$;
spray cooling: $\alpha_{tot} = 80\,Wm^{-2}K$.

Due to the boundary conditions (for instance, asymmetrical cooling) analytical descriptions of this process phase are nearly impossible. Only rough approximations may be applied after the boundary conditions have been simplified.

An empirical analysis of numerous cooling procedures with different materials and all three kinds of cooling lead to curves of the mean temperature $\bar{T}$ displayed in Fig. 2 in a dimensionless way. Using this diagram an estimation concerning the cooling time is possible.

2. *Process Engineering of Thermoforming*

Thermoforming is one of the oldest techniques used in plastics processing. It is known that from the end of the Middle Ages American aborigines produced thermoformed cups out of keratin from animal horn (Throne 1987). The flat materials were plasticized using boiling water, stretched over a mold, and finally cooled using cold water. In the eighteenth century the

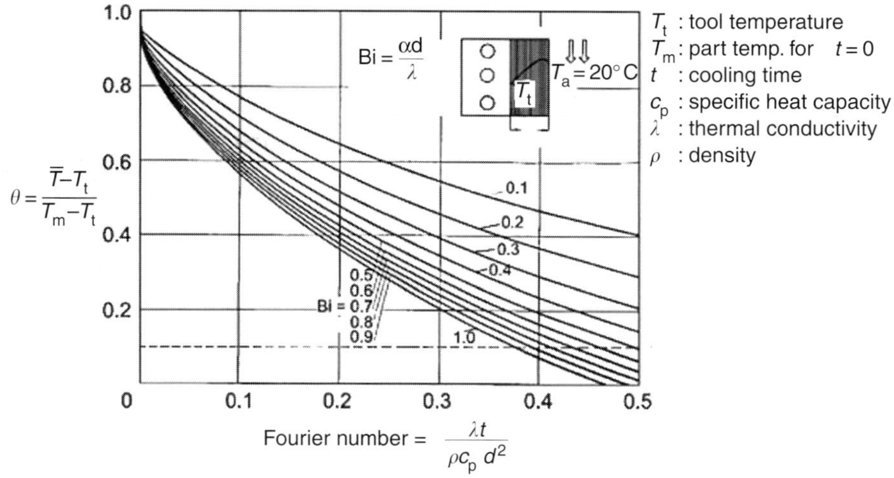

Figure 2
Characteristic curves for cooling in the mold.

first products were thermoformed by a comparable technique from naturally derived cellulose. As industrialization progressed these materials were substituted by celluloid and cellulose acetate. In the 1930s the old method was succeeded to produce the first sheets of PVC, PS, and polymethyl methacrylate (PMMA) which were suitable for thermoforming. Nowadays mainly acrylonitrile butadiene styrene (ABS), PVC, PS, PET, PP, and polyethylene (PE) are thermoformed.

2.1 Machine Technology

The first thermoforming units were versatile machines that served to form all kinds of components, from the margarine tub to the inner receptacles of refrigerators. Later, specialized machine types were developed that satisfied the increased demands concerning output rates and usability in a better way than the versatile machines (Throne 1987). There are two basic systems of machine design: single-station machines and multiple-station machines.

The single-station machine shown in Fig. 3(a) combines heating and forming in the same station and the complete cycle time is the sum of the times of each process step. The multiple-station machine allows heating and forming in separate chambers (Fig. 3(b)). Therefore the slowest process step determines the total cycle time.

Furthermore, sheet machines, machines with material supply from a roller, skin- and blister-packaging machines, food packaging lines and inline units are distinguished.

Usually sheet machines are fed from a stack of polymer sheets having a thickness from 0.1 to 12 mm; occasionally even thicker semifinished products are processed. Although it is easier to supply the material from a roller than from a sheet stack, the thickness of the semifinished product is limited to 2 mm (Fig. 3(b)).

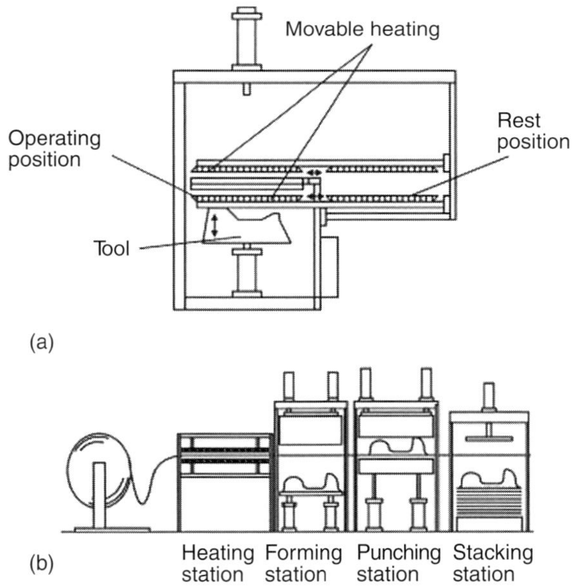

Figure 3
Setups of thermoforming machines: (a) single-station machine; (b) multiple-station machine.

Skin-packaging machines cover objects skin-tight with transparent film. Not only small and lightweight articles are packed in this way but also heavy machine parts up to a weight of 50 Kg. Blisters are transparent, thin-walled cavities which are filled with the product (usually pharmaceutical preparations) and afterwards bonded to a base. The production of blisters may be done using all kinds of thermoforming machines. For small quantities the filling of the blisters is done manually; for higher quantities automatic feeding units are used, which are able to insert up to 8000 tablets per minute.

A very special type of problem is the packaging of liquid or pasty foods by fully automated packaging lines. The requirements of the food industry demand the cleaning of machine parts with aggressive, liquid agents and this necessitates the use of highly alloyed steel or costly coatings. In general all steps, starting with the forming of the container, its filling and sealing with polymer-coated aluminum foil, up to the separation of the packaging are all carried out in one machine body (form–fill–seal lines).

The so-called inline-thermoforming units are supplied directly from an extrusion machine with a warm film or sheet (below melting temperature). In addition to single-layered semifinished products, multilayered ones may also be produced, if an appropriate die and several extruders are used. The main advantage of these inline units is the processing out of one primary production process and the ability to influence the quality of the semifinished product shortly before forming. Its disadvantage is that a defect in one component of the system could lead to a complete halt in production.

2.2 *Forming Techniques*

In principle one differentiates negative forming into a cavity from positive forming over an elevated contour. Apart from these basic forming techniques several variations and combinations have been developed, which are all aimed at achieving an even wall thickness distribution. The decision for either negative or positive forming depends on which side of the molded part has to show exact dimensioning. Only the surface that is in contact with the tool is true-to-size; the opposite side shows deviations which result from thickness differences (Fig. 4). However, from an optical point of view this side has the more glossy surface. Therefore, bathtubs are formed into a cavity.

(*a*) *Negative forming.* During negative forming with vacuum assistance the semifinished product is sucked from the clamping plane into the cavity with a pressure difference of 0.08–0.093 MPa. This simple technique is only applicable for components with a small draw ratio (height/diameter), since the deeper the cavity the thinner the material becomes at the edges and corners.

A further development of this technique is the solid-phase pressure-forming technique shown in Fig. 4(a). As the forming pressure reaches 0.4–0.7 MPa, occasionally even 2.5 MPa, the forming can be done at lower temperatures, thus saving heating time and energy. Usually the wall thickness distribution is not improved while using compressed air, but edges and corners are formed more precisely and the tight contact of the molded part and the tool improves the cooling efficiency and hence the cycle time. Materials having a higher deformation resistance are formed more easily.

In order to improve the wall thickness distribution rams are used to achieve a mechanical prestretching. The ram, which is made of temperature-resistant plastic or heated aluminum, ensures that the bottom section of the thermoformed component has sufficient material available so that the side walls and the bottom are formed more uniformly (Fig. 4(b)).

The thickness distribution is affected by the material temperature, ram geometry, ram speed and temperature, depth of immersion and the time sequence of ram motion, mold motion, and introduction of differential pressure. Under optimum conditions good material distributions are achievable up to a draw ratio of 2:1.

Further improvements in part quality are possible if the semifinished product is preblown before prestretching by the ram (Fig. 4(c)). The last forming step might be done by either compressed air or vacuum as usual.

(*b*) *Positive forming.* While negative forming produces components with a thin bottom and thin edges, positive forming results in components with a thick bottom and thin walls. The mold moves upwards, thereby prestretching the clamped material, before differential pressure is applied (Fig. 5(a)). In order to avoid thicker areas the semifinished product is usually preblown. Using this technique the material is stretched in the middle twice as much as along the edges. Afterwards the motion of the ram stretches the side walls, and forming is completed using a vacuum. Using this method a very even thickness distribution is achievable for parts having almost vertical side walls (Fig. 5(b)). Dome-like geometries having an optically perfect surface may be formed by blowing without a cavity or ram.

Figure 5(c) shows the process of compression molding that has been adopted from metal forming. Instead of a female mold this technique uses a cellular foam plate and differential pressure is not required.

3. *Processing Technology of Stretch Blow Molding*

Plastics are gaining growing importance for the production of hollow articles and bottles for foods, beauty aids, and household chemicals. Recyclable plastic bottles for the food industry are almost exclusively produced by stretch blow molding. The specific processing technique requires only a little material to achieve high strength, stiffness, and

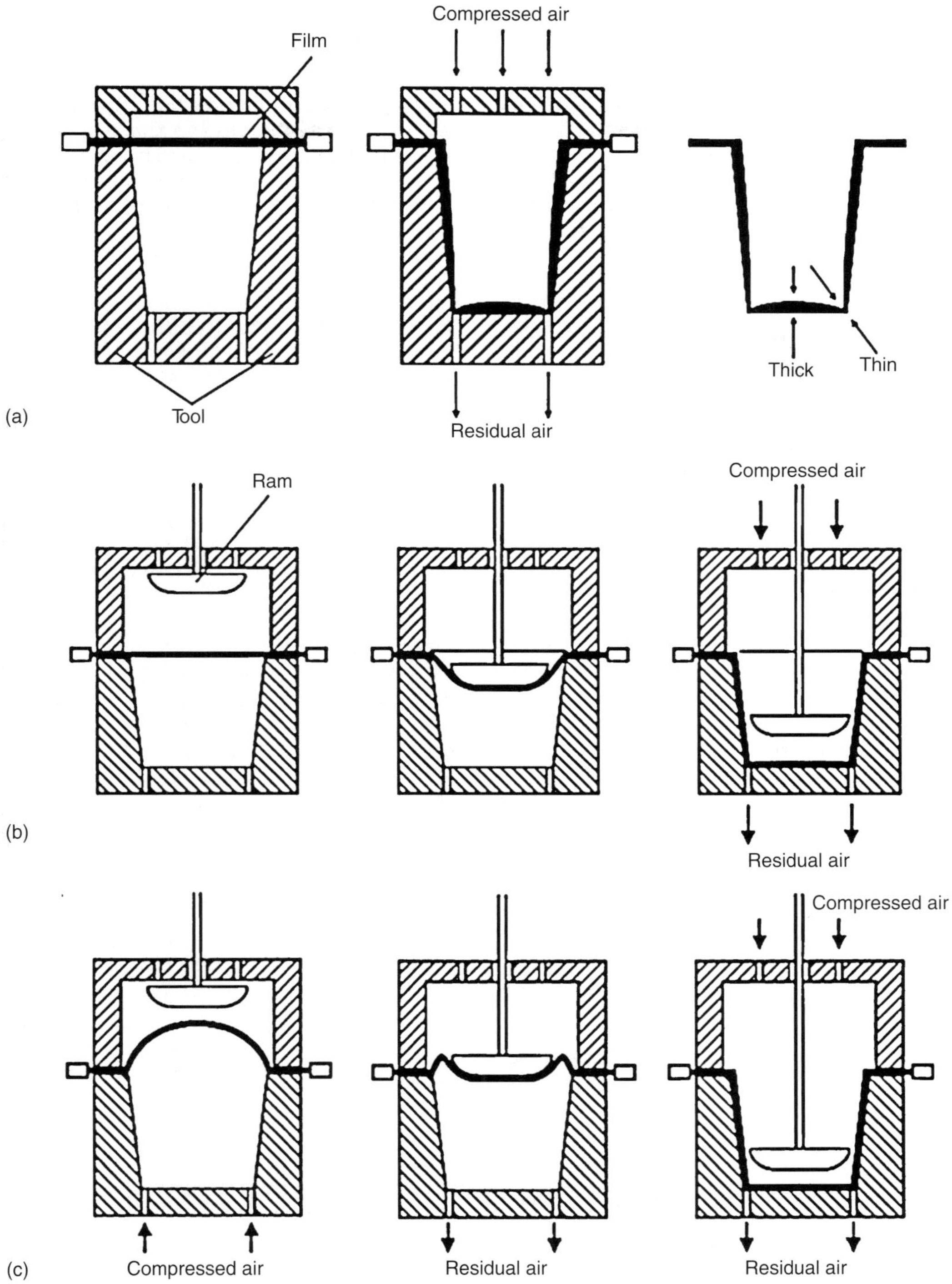

Figure 4
Negative forming: (a) pressure forming; (b) mechanical prestretching; (c) preblowing.

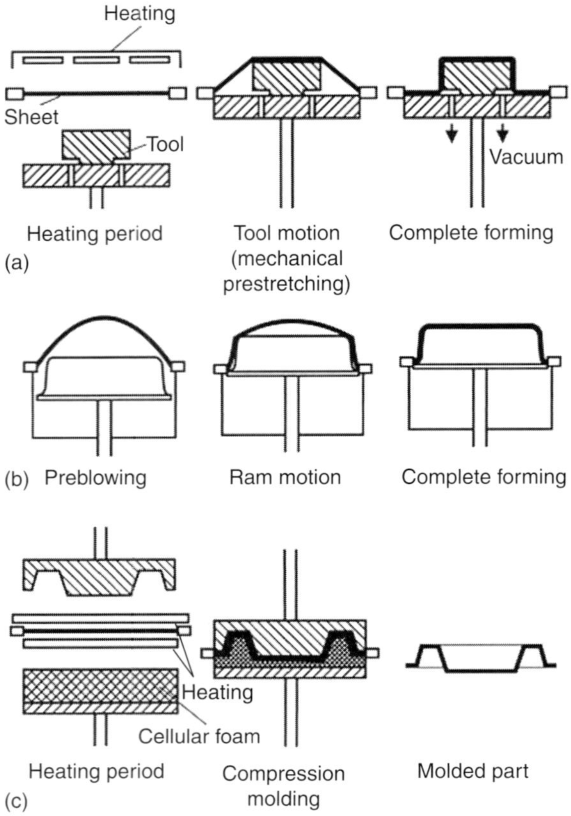

Figure 5
Positive forming: (a) simple positive forming; (b) preblowing; (c) compression molding.

impact strength. Additionally the molded part has a glossy and transparent appearance and a low gas permeability. Amorphous thermoplastics are as easily stretch blow molded as semicrystalline ones. Predominantly PET, PP, PVC, and PC are processed but PE and polyacrylonitrile (PAN) are also processed in minor quantities (Rosato and Rosato 1989).

In stretch blow molding a so-called preform is heated up to its thermoelastic temperature range before a mandrel stretches the hollow part in the axial direction and compressed air forms the component in the radial direction. As with thermoforming, the wall thickness distribution is the most important quality characteristic, and it is dependent on the geometry of the preform, the required bottle geometry, and process parameters such as the temperature distribution in the preform just before forming.

With regard to the processing method there are two different concepts: the one-step process or "method of first heat"; and the two-step process or "method of second heat."

The methods differ in the temperature history of the processed thermoplastics, as can be seen in Fig. 6. Regardless of the processing method the preform is almost exclusively injection-molded. The opening mechanism of the component, and if necessary also the thread of a screw top, is produced very accurately.

3.1 One-step Stretch Blow Molding

The one-step process shown in Fig. 6 differs from the method of second heat only in that there is neither a temporary nor a local separation of preform

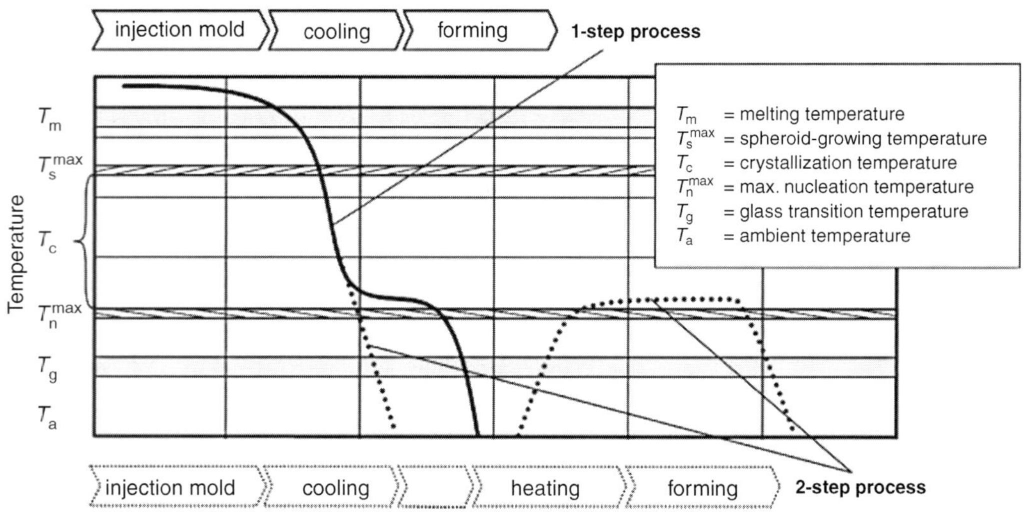

Figure 6
Temperature control during stretch blow molding.

manufacturing and further processing. Immediately after injection molding the preform is cooled down to the thermoelastic temperature range and demolded. It is then transferred either to a thermal conditioning zone or directly to the forming tool.

Thermal conditioning is necessary for large and complex components, as they require a precise control of the preform temperature profile. The conventional conditioning technique is contactless tempering by means of strip heaters (Fig. 7(a)) where the heat exchange is achieved by heat radiation. Contact heating is achieved by pressurizing the preform and thereby pressing it against the surface of a conditioning tool, which is tempered by several separate oil circuits (Fig. 7(b)). Both methods may also incorporate a tempered mandrel inside the preform, if, apart from an axial and a radial temperature profile, a circumferential profile is also needed. Infrared radiation can also be used for contactless tempering (see Fig. 7(c)).

After thermal conditioning the preform is transferred to the blow mold, where it is biaxially stretched. First the preform is stretched axially by the mandrel. Simultaneously a low prepressure acts on the inner surface of the preform to prevent necking. The forming pressure, which is several times higher than the prepressure, is then applied to complete axial stretching and leads to the radial forming of the part. After cooling down by the cavity wall the part is demolded.

3.2 *Two-step Stretch Blow Molding*

The two-step process is characterized by a temporary, and usually local, separation of the injection molding process and the tempering and stretching process. The preforms are cooled down after injection molding to ambient temperature, and may be stored and subsequently heated again before stretching. The processing procedure is shown in Fig. 8.

The cold preform runs through one large or several small heating zones. To ensure an even heating over the circumference, the preform rotates during the heating period. The heating is supplied by quartz radiators having a color temperature of approximately 2400 K. As a considerable part of the radiation is absorbed at the surface of the preform, it is simultaneously cooled by fans to avoid degradation or crystallization of the surface layers. The subsequent compensation period guarantees a homogeneous temperature distribution in the preform by thermal conduction. The stretch procedure does not differ from that in the one-step process.

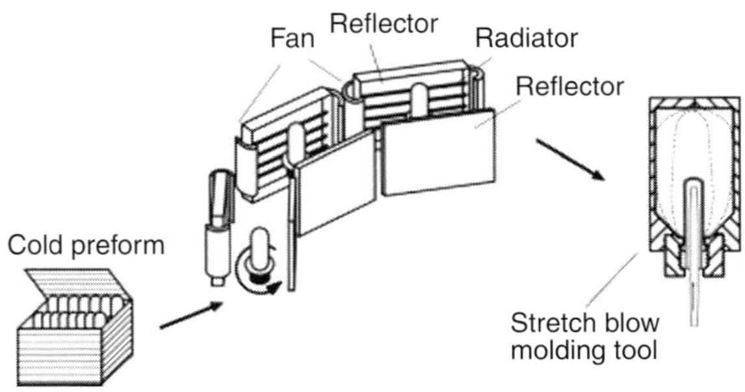

Figure 8
The two-step process.

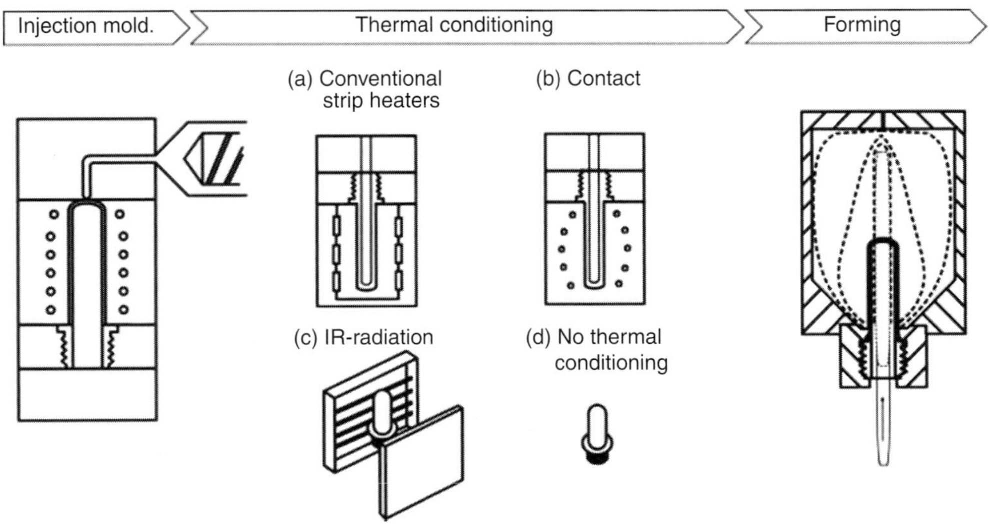

Figure 7
The one-step process.

Bibliography

Bird R B, Stewart W E, Lightfoot E N 1960 *Transport Phenomena*. John Wiley & Sons, New York

Bruins P F 1973 *Basic Principles of Thermoforming*. Gordon and Breach Inc., New York

Hartwig K, Michaeli W 1995 Modeling and simulation of the injection stretch-blow molding process. In: *Proc. Ann. Technical Conference (ANTEC)*, Society of Plastics Engineers, Boston, MA, pp. 925–9

Michaeli W 1995 *Plastics Processing—An Introduction*. Carl Hanser Verlag, Munich

Mooney M 1940 A theory of large elastic deformations. *J. Appl. Phys.* **11**, 582–92

Neitzert W A 1967 *Vacuum-forming of Thermoplastic Films* (in German). Zechner und Hüthig Verlag, Speyer

Rivlin R S 1984 Forty years of non-linear continuum mechanics. In: *Proc. 9th Int. Congress on Rheology*, Mexico, pp. 1–29

Rosato D V, Rosato D V 1989 *Blow Molding Handbook*. Carl Hanser Verlag, Munich

Schwarzmann P 1997 *Thermoforming in Practice* (in German). Carl Hanser Verlag, Munich

Throne J L 1987 *Thermoforming*. Carl Hanser Verlag, Munich

E. W. Michaeli and J. Kopp

Freeform Fabrication

Solid freeform fabrication is a family of rapid prototyping methods that build a material layer-by-layer to form a final shape. This approach allows more freedom of design than conventional molding, for instance a loose ball inside a (nearly) closed box or a set of movable meshed gears can be formed (Fig. 1).

Much biological growth is also a layer-by-layer process. This suggests that freeform methods could also be applied to processes where solids are formed by chemical reaction. Conventional molding methods are incompatible with the large shrinkages of most chemical solidification reactions, so that only fibers and films can be made. If the structure is built in layers, the shrinkage can be accommodated as the layers form. Also, as in biology, it becomes natural to design parts from several different materials and to connect these via compositional gradients rather than sharp interfaces.

Current methods of solid freeform fabrication are targeted at making prototypes, mostly from simple polymeric materials. Here, form and fit is more important than mechanical performance. There has been considerable effort recently to extend the range of materials to make fully functional prototypes in plastic, metal, or ceramic. Beyond this there are research efforts exploring approaches to making new combinations of materials with superior properties.

Figure 1
Freeforming allows assemblies with moving parts to be made as a unit. "Braingear" shown with water-soluble support structure and after removal of support (courtesy of Advanced Ceramics Research Corp., Tucson).

1. Methods

In essence, any method of converting a small volume of liquid or powder to a solid can be used to build 3D parts. In addition, any method that allows sheets to be cut, stacked, and fused can build parts. This article surveys the major approaches but many more are possible. Freeforming can be compared to printing in that there is a range of methods from pencil and paper, through inkjet printing to offset lithography. The advantages and disadvantages of each cause them to be preferred for particular purposes (Calvert and Crockett 1997).

1.1 Stereolithography

Stereolithography was developed by 3D Systems of Valencia, CA, USA after parallel efforts in several laboratories (Kodama 1981, Jacobs 1992). As sketched in Fig. 2, a He–Cd or argon-ion laser traces a shape at the surface of a bath of photopolymerizable monomer. A supporting platform then drops by one layer thickness, a further layer of monomer is swept across the newly solid surface, and the process repeated. After a few hours, the platform is raised and the completed part removed.

The desire to make investment casting molds from stereolithography parts has led to the Quickcast process where the interior of the part is not solid but formed as a lattice of struts. Excess liquid can be drained out of a hole in the base after forming. In principle, this more compliant part can be used to make a ceramic mold that will not crack due to differential expansion as the resin is burnt out.

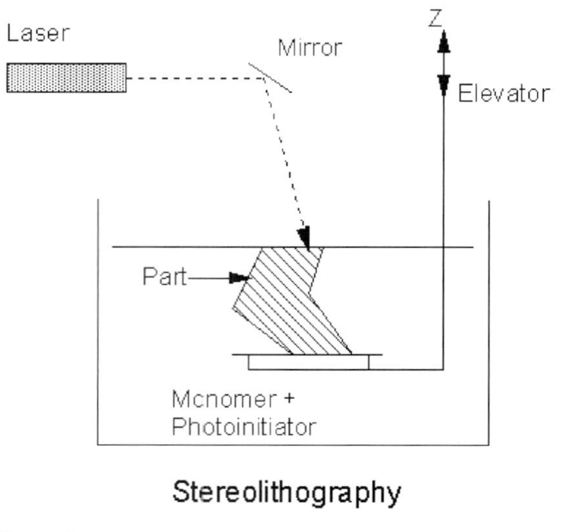

Figure 2
Stereolithography. A moving mirror tracks a laser across the surface of a bath of monomer. As each layer is written, the elevator drops and a new film of monomer is swept across the existing solid surface.

The original monomers were combinations of multifunctional acrylates that formed cross-linked resins. Newer photopolymerizable epoxies offer better properties (Pang 1993). The lower shrinkage of the epoxies during postgelation polymerization leads to lower distortion.

1.2 Selective Laser Sintering

DTM Corporation of TX, USA has developed a selective laser sintering method. The principle is similar to stereolithography but the liquid monomer is replaced by polymer powder which is fused by a CO_2 laser (Marcus *et al.* 1990, Nelson *et al.* 1993). A layer of heated powder is spread on a piston and selected areas are fused, the piston is then lowered slightly, and a new layer of powder spread and fused. The fused, but still porous, part is finally removed from the powder bed by raising the piston. Layer thickness is typically 25–30 μm with a rate of building of $5\,cmh^{-1}$ and an accuracy of about 400 μm. Precision is closely linked to the size of the powder particles (Nelson *et al.* 1995).

Bronze–nickel parts have been formed to fractional densities of 65–80% (Agarwala *et al.* 1993). A number of high temperature materials have also been formed by fusing polymer-coated powders of silicon carbide (Vail *et al.* 1993), other ceramics, and metals (Badrinarayan and Barlow 1995). The porous parts can be converted to high density by hot isostatic pressing (HIPping). There are also schemes for forming composites by partially sintering and impregnating the porous body with liquid metal (Deckard and Claar 1993).

1.3 Fused Deposition Modeling

Stratasys Corporation of Minneapolis, MN, USA, developed fused deposition modeling where a 1 mm strand of wax is fused in a write head attached to an x–y drive, like an inverted pen plotter (Comb *et al.* 1994). The resulting wax parts can be used for investment casting but are rather fragile. (In the investment casting process, a machined metal model is used to make a reusable rubber mold. Wax is cast into this mold and the wax model is coated with refractory cement. The wax is melted out and metal cast into the cement mold, which must then be broken to retrieve the cast metal part.) Steps on the surface, corresponding to the layers, are often noticeable. Polymer materials are also available including a low-melting form of nylon and acrylonitrile-butadiene-styrene plastic (ABS). Layer thicknesses range from 25 μm to 1 mm with a precision on a 30 cm part of 125 μm. The process tends to leave voids amounting to 10–25 vol.% (Weeren *et al.* 1995).

Unlike stereolithography or selective laser sintering, a support material is needed for fused deposition modeling of cantilevered shapes. Water-soluble polymers are available for this or conventional material can often be used with fine break-away points.

1.4 3D Printing

3D printing was developed at MIT and has been licensed to Soligen Inc, who use it to make molds for metal casting. An inkjet system writes a binder, such as a polymer latex or silica colloid, into a bed of ceramic powder (Sachs *et al.* 1992). A layer of powder 100–150 μm thick is spread and then binder sprayed into the zone to be solidified as a stream of droplets. The method is excellent for coarser ceramic powders but is more difficult with fine, submicron powders where capillary forces become significant and spreading the powder is difficult. The method has also been applied to building parts by printing binder into a bed of starch (Z Corporation). There are also commercial systems for inkjet printing wax parts.

1.5 Shape Deposition Manufacturing

In shape deposition manufacturing, thick wax or polymer layers are deposited and then machined using a modified computer-controlled milling machine. The machining step allows the edges of individual layers to be profiled, which eliminates the layer-to-layer stepping found in most freeforming methods. Where necessary, a water-soluble material is used to support overhangs. This method may produce parts directly or be used to make a mold. After removal of the support, the wax mold can be filled with polyurethane, epoxy, or a gel-casting ceramic suspension to make the final part (Cooper *et al.* 1999). This method is complex but offers better surface finish and resolution than most freeforming methods.

1.6 Other Methods of Freeforming

Other approaches include a system of cutting and stacking ceramic green tapes (Griffin *et al.* 1994, Cawley *et al.* 1995) or butchers paper and adhesive (Helisys Inc, Torrance CA, USA; Feygin *et al.* 1992). Metal parts have been formed by laser fusion of stainless steel powder carried in a gas stream (Sandia National Labs, LENS process (Keicher *et al.* 1995, Brooks *et al.* 1999) and by deposition of liquid metal (Hartmann *et al.* 1994). There are also efforts to make shaped parts by related gas-phase deposition methods, some of which are mentioned below.

Several groups are developing inkjet printing of functional materials, particularly ceramics. The method offers high resolution and thin layers but the build rates are slow. The print system is generally limited to low viscosity fluids, so the powder content of the slurry is low (Blazdell and Evans 1999).

These systems all have their advantages and drawbacks for parts production; the interest here is in their implications for new materials.

2. Properties of Freeformed Parts

Precision is a major issue in freeforming where parts need to be comparable with those produced by machining or molding. The resolution of most of the processes, in terms of the dot size within a layer, is about 100 μm. In stereolithography this relates to beam spreading and to free-radical diffusion during cure. In fused deposition modeling this relates to the size of the nozzle. Layer thickness, the resolution in the z direction, can often be less than this, down to 25 μm. However, thinner layers mean longer build times. Surface finish can almost certainly be improved by changing resolution at the surface, by multi-axis machines that rotate the part as it is built, or by adding a small cutting tool to the equipment (Shi and Gibson 1998). Precision is important in machines that must be assembled from parts. It is not so important in biology where an animal is built as an integrated unit. A materials solution to the resolution question would come from a method to deposit either very fine drops very rapidly or varying amounts of material.

Mechanical properties are still a limiting factor in most commercial freeforming systems. Waxes build well but are weak, as are most bonded or partly fused powders. Polymers for fused deposition modeling tend to have low molecular weights to allow flow through fine nozzles. The rapid freezing of the deposited line also limits interlayer adhesion. In most freeformed materials, strength across the layers tends to be lower than that within the layers. An ideal system would probably be a chain-extendable wax.

3. Applications

There are many cases where freeform methods have special advantages for making parts from conventional materials. This will be the case where each part has a unique design or where the shape cannot be made by a conventional method.

The original use of freeformed parts was to allow designers to demonstrate proposed shapes. There was a need to test the function of these parts, in a wind tunnel for instance, and this required improved properties. However, freeformed plastic parts still do not have the strength of conventional molded parts. An alternative is to freeform a mold and to use that to make a small number of test parts, thus, freeforming competes with machining.

Ceramics lend themselves to freeforming because conventional processing is slow and expensive and parts are often needed in quite small numbers. Most of the conventional freeforming processes can be modified to form ceramics. Silicon nitride parts have been made by fused deposition modeling, 3D printing, and by gelling a slurry of monomer and ceramic powder cast into wax molds made by shape deposition manufacturing (Gasdaska *et al.* 1999). For a high performance ceramic made by any process the final part must be machined, at great expense, to remove surface flaws that reduce the strength and to ensure a precise fit. The properties of freeformed ceramics after sintering and machining compare well with those of conventional material (Denham *et al.* 1998, Cawley 1999, Halloran 1999, Lenk 2000, Morissette *et al.* 2000).

Powder metallurgy is used as a cheaper alternative to machining for metals that cannot be easily molded. However, the need to press the powder limits the parts that can be made to shapes with axial symmetry. Freeforming methods can be used to make similar parts, via building and sintering a powder preform (Sercombe and Schaffer 1997) or by direct powder fusion in the LENS process described above.

Freeforming of medical implants is being explored because each patient is different and tomography data could be used to define the required dimensions. So far, the material properties are not good enough to make a convincing case for freeforming. An ideal machine would also be able to process a range of materials for different implant sites. Freeforming has been applied to the ultra-high molecular weight polyethylene used in hip implants (Rimell and Marquis 2000), to calcium phosphate ceramics analogous to bone mineral (Vail *et al.* 1999), and as molds for cast metal implants with porous surface textures (Curodeau *et al.* 2000).

One example of a shape that cannot be made in other ways is a mold with cooling channels below the surface and conforming to the shape of the cavity (Sachs *et al.* 2000).

The desktop modeler, to sit alongside the laserjet printer and produce strong 3D parts, is still not feasible but should become so, once a suitable material is developed. One approach would be to combine an oligomeric wax with a photochemical postcuring system to yield a tough resin. A number of groups have worked on 3D versions of the laserjet process.

4. New Materials

4.1 Porous Materials (with Special Reference to Biological Materials)

Biological materials are characterized by the exploitation of layered and graded structures to produce intriguing combinations of toughness and stiffness in hard and soft tissues. Since the stiffness of biological materials often derives from organic or mineral fibers and plates, their orientation can align the best properties with the high stress direction. There are also many examples of layered fibrous composites that approach orthotropic stiffness (isotropic stiffness in the plane).

In synthetic materials, pores are generally regarded as flaws which need to be eliminated. Porosity in biological materials is used to reduce the weight of bones and spines (Karam and Gibson 1994) (Gibson and Ashby 1988). As in the skull, pores also introduce toughness by allowing a crushing response to impact. Freeform methods allow specific pore distributions to be introduced into materials (Hattiangadi and Bandyopadhyay 2000, Bose *et al.* 1999).

Porous structures based on a repeated cell made of connected struts can be designed to have a negative Poisson's ratio (Sigmund *et al.* 1998). Porous hydroxyapatite bone implants have been produced to allow ingrowth and tissue bonding (Panda *et al.* 1998, Richter *et al.* 1999).

Pores have been introduced to optimize the response of piezoelectric ceramics. Open or polymer-filled channels can reduce the effective modulus and so increase the stress sensitivity of the material (Safari and Danforth 1998, 1999).

4.2 Controlled Orientation

Many biological polymers form by extracellular self-assembly with controlled orientation. The basic materials thus tend to have high stiffness and strength in only one direction. Layered structures allow orientation distributions that can be matched to local stress axes. Extrusion freeforming methods for short fiber reinforced polymers align the fibers parallel to the motion of the write-head (Calvert and Crockett 1997, Peng *et al.* 1999). This also allows local orientation to be controlled and optimized within a part. Thus, stiffness or other properties can be optimized for local stress. Stereolithography of liquid crystal monomers in a magnetic field can be used to make structures with controlled orientation (Chartoff *et al.* 2000). Liquid crystal polymer-thermoplastic composites have also been used in fused deposition modeling (Gray *et al.* 1998). There have been attempts to incorporate long fibers into stereolithography (Ogale *et al.* 1991, Charan *et al.* 1994).

4.3 Multimaterial Combinations

Biological growth also allows for different materials to be incorporated into a single unit, with graded interfaces in place of adhesive joints or rivets. Many freeforming methods also allow several materials to be combined. One simple approach is to incorporate existing inserts into the part as it is built (Khandelwal and Kesavadas 1999).

Extrusion processes can form graded materials by the progressive mixing of two liquid feeds, such as for tungsten and alumina (Calvert *et al.* 1994) and for alumina and aluminum (Sercombe *et al.* 1999). Toughening regions, layers, or lines can also be introduced, for instance, by incorporating metal regions into a ceramic structure. In this case, the metal and ceramic must sinter at the same temperature and with similar volume changes, to avoid high stress concentrations.

Gradients can also be formed by making a material with progressive porosity and then back filling the pores with a second material. This has been carried out for soft and hard polymers (Crockett and Gervasi 1999) and for alumina and aluminum (Cesarano *et al.* 1998, Cesarano 1999). Such back filling of porosity is also widely used to make dense parts via selective laser sintering (Griffith *et al.* 1997, Sand *et al.* 1999).

One problem with multimaterial structures is that there is relatively little experience in their design outside of the rather specialized case of glass–metal seals. Thus, it is not yet clear what we can do with these possibilities; there is little previous experience because, outside of special applications, such as armor, there have been few reasons to design using combinations of materials. Likewise, practical uses of oriented materials mainly occur with composite plates and struts, not in complex shapes. Carefully designed layered structures should be capable of stopping or deflecting cracks in brittle materials and of providing new combinations of stiffness and strength. However, shrinkage, which occurs during manufacture and thermal stresses, arising from differential expansion, is crucial. A number of groups have addressed the complexity of multimaterial design and of developing computer representations for such structures where material choice would effectively be a "color" in a 3D design (Weiss *et al.* 1997, Jackson *et al.* 1999).

4.4 *Chemical Synthesis of Materials*

Another characteristic of biological growth is that the liquid–solid transformation is a chemical change from soluble to insoluble material, usually with the large volume change from solution to dense solid. It seems reasonable that this process will accommodate the volume change without the cracking of the solid, if it occurs as growth of a series of thin layers, much like a solid can be formed by deposition from the vapor phase. Efforts to produce glasses or inorganic–organic hybrids by a layer-by-layer reaction have not been very successful and have cracked badly. However, parts could be made by building the whole as a porous semi-solid mesh and then transforming the whole part (Calvert *et al.* 1998). Stiff and strong bone-like composites have been produced by a process of forming gels followed by mineralization with calcium carbonate or phosphate (Calvert *et al.* 1998).

Materials with small volume changes, such as epoxy and acrylate polymers can readily be formed by stereolithography. Since cross-linked polymers tend to be brittle, thermal polymerizations of linear acrylate were attempted in extrusion freeform fabrication but encountered problems of monomer evaporation since a higher degree of reaction is needed to form a solid than for gelling polymers and so more heat of polymerization is released. The conversion of caprolactam to Nylon 6 does have a low heat of polymerization. This polymer was made successfully by extrusion freeforming of the monomer (Lombardi and Calvert 1999).

There have been various studies applying chemical vapor deposition and laser synthesis to freeforming of ceramic parts (Maxwell *et al.* 1995, Loschau and Lenk 1999, Sun and Shaw 1999, Slocombe and Li 2000). This method particularly lends itself to the growth of thin films, rods, and interconnected frameworks of fine struts.

4.5 *Microstereolithography*

Conventional photoresist methods can readily be adapted to use a laser for curing instead of exposure through a mask. The combination of highly focused beams and thin films allows higher resolution that can be achieved with conventional stereolithography (Bertsch *et al.* 1999, Zhang *et al.* 1999). Dyes with strong two-photon absorption can give curing with much higher resolution that can then be used to build very fine structures as shown in Fig. 3 (Cumpston *et al.* 1999).

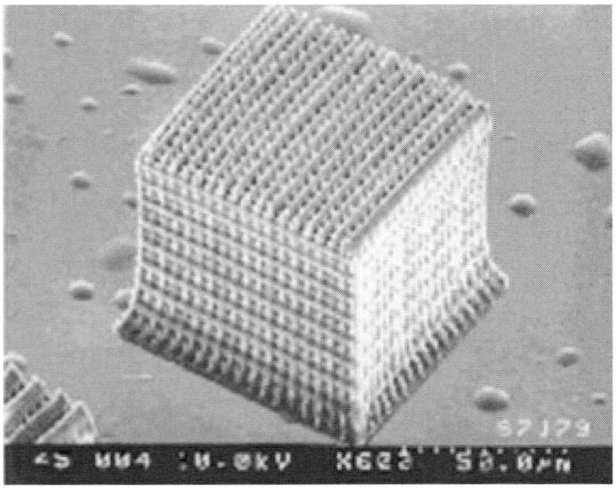

Figure 3
Micron-scale "logpile" formed by stereolithography using two-photon initiators for higher resolution (courtesy of J Perry, University of Arizona, Tucson).

There is currently much interest in direct writing methods to form electrical circuits. Methods such as inkjet printing and micropen deposition give a resolution of 50–100 μm (Dimos *et al.* 1997). Microcontact printing can give much higher resolution but it is more of a multistep process (Xia and Whitesides 1998).

There are several needs driving the development of these processes. Integrated circuit manufacture involves very many steps so that the production of a single chip can take months. There is a need for rapid production of prototype integrated circuits. "Thick film" electronics allows resistors, capacitors, and inductors to be silk-screen printed onto ceramic substrates and then fired. Printed circuit boards are prepared by a photolithography and etch process. A flexible rapid prototyping system could replace these methods for short production runs. There is also a desire to build electronics on flexible plastic substrates, which will require electrical circuit elements to be formed at low temperatures. The expansion of optoelectronics is also driving interest in incorporating optical elements, including light emitting diodes, photodetectors, waveguides, and switches.

Where mechanical parts are strongly 3D and usually made of one material the circuit structures will normally be much thinner than their width but will contain multiple materials. In the long term, clearly the mechanical and electrical functions should be combined into a single monolithic unit. Embedded electronics would remove much of the vulnerability to vibration, corrosion, and mechanical shock.

5. *Future of Freeforming*

The various freeforming methods are widely used in industry for prototyping and will be used for making molds for short-run manufacturing. As the technology becomes more established and the materials stronger, the methods will be used more in smaller businesses such as for sculpture, 3D display, and medical implants. Eventually desktop manufacturing could replace many machined metal parts. Growing familiarity with the approach can be expected to lead to a much wider application to new materials and structures.

Bibliography

Agarwala M, Bourell D L, Wu B, Beaman J J 1993 Evaluation of the mechanical behavior of bronze–Ni composites produced by selective laser sintering. In: Marcus H L, Beaman J J, Bourell D L, Barlow J W, Crawford R H (eds.) *Proc. Solid Freeform Fabrication Symp.* University of Texas Press, Austin, TX, pp. 193–203

Badrinarayan B, Barlow J W 1995 Effect of processing parameters in SLS of metal-polymer powders. In: Marcus H L, Beaman J J, Bourell D L, Barlow J W, Crawford R H (eds.) *Proc. Solid Freeform Fabrication Symp.* University of Texas Press, Austin, TX, pp. 55–63

Bertsch A, Lorenz H, Renaud P 1999 3D microfabrication by combining microstereolithography and thick resist UV lithography. *Sensor Actuat. A* **73**, 14–23

Blazdell P F, Evans J R G 1999 Preparation of ceramic inks for solid freeforming using a continuous jet printer. *J. Mater. Synth. Process.* **7**, 349–56

Bose S, Suguira S, Bandyopadhyay A 1999 Processing of controlled porosity ceramic structures via fused deposition. *Scr. Mater.* **41**, 1009–14

Brooks J, Robino C, Headley T, Goods S, Griffith M 1999 Microstructure and property optimization of LENS deposited H13 tool steel. In: Marcus H L, Beaman J J, Bourell D L, Barlow J W, Crawford R H (eds.) *Proc. Solid Freeform Fabrication Symp.* University of Texas Press, Austin, TX, pp. 375–82

Calvert P, Crockett R 1997 Chemical solid free-form fabrication: making shapes without molds. *Chem. Mater.* **9**, 650–63

Calvert P, Crockett R, Lombardi J, O'Kelly J, Stuffle K 1994 Extrusion methods for solid freeform fabrication. In: Marcus H L, Beaman J J, Bourell D L, Barlow J W, Crawford R H (eds.) *Proc. Solid Freeform Fabrication Symp.* University of Texas Press, Austin, TX, pp. 50–5

Calvert P, O'Kelly J, Souvignier C 1998 Solid freeform fabrication of organic–inorganic hybrid materials. *Mater. Sci. Eng. C* **6**, 165–74

Calvert P D, Lin T L, Martin H 1997 Extrusion freeform fabrication of chopped-fiber reinforced composites. *High Performance Polymers* **9**, 449–56

Cawley J 1999 Solid freeform fabrication of ceramics. *Curr. Opin. Solid State Mater.* **4**, 483–9

Cawley J D, Wei P, Liu Z E, Newman W S, Mathhewson B B, Heuer A H 1995 Al_3O_3 ceramics made by CAM–LEM. In: Marcus H L, Beaman J J, Bourell D L, Barlow J W, Crawford R H (eds.) *Proc. Solid Freeform Fabrication Symp.* University of Texas Press, Austin, TX, pp. 9–16

Cesarano J 1999 A review of robocasting technology. *Mater. Res. Soc. Symp. Proc.* **542**, 133–9

Cesarano J, King B H, Denham H B 1998 Recent developments in robocasting of ceramics and multimaterial deposition. In: Marcus H L, Beaman J J, Bourell D L, Barlow J W, Crawford R H (eds.) *Proc. Solid Freeform Fabrication Symp.* University of Texas Press, Austin, TX, pp. 697–703

Charan R, Renault T, Ogale A, Bagchi A 1994 Automated fiber reinforced composite prototypes In: *5th Int. Conf. on Rapid Prototyping*. Dayton, OH

Chartoff R P, Ullett J S, Schultz J W 2000 Composite-like structures with controlled thermal expansion from aligned liquid crystal photopolymers. *Polym. Mat. Sci. Eng.* **82**, 361–2

Comb J W, Priedeman W R, Turley P W 1994 FDM technology process improvements. In: Marcus H L, Beaman J J, Bourell D L, Barlow J W, Crawford R H (eds.) *Proc. Solid Freeform Fabrication Symp.* University of Texas Press, Austin, TX, pp. 42–9

Cooper A, Kang S, Kietzman J, Prinz F, Lombardi J, Weiss L 1999 Automated fabrication of complex molded parts using mold shape deposition manufacturing. *Mater. Design* **20**, 83–9

Crockett R S, Gervasi V R 1999 Gradient and discrete region composites produced by solid freeform fabrication techniques. *Mater. Res. Soc. Symp. Proc.* **542**, 25–30

Cumpston B, Ananthavel S, Barlow S, Dyer D, Ehrlich J, Erskine L, Heikal A, Kuebler S, Lee I, McCord – Maughon

D, Qin J, Rockel H, Rumi M, Wu X, Marder S, Perry J 1999 Two-photon polymerization initiators for three-dimensional optical data storage and microfabrication. *Nature* **398**, 51–4

Curodeau A, Sachs E, Caldarise S 2000 Design and fabrication of cast orthopedic implants with freeform surface textures from 3-D printed ceramic shell. *J. Biomed. Mater. Res.* **53**, 525–35

Deckard L, Claar T D 1993 Fabrication of ceramic and metal matrix composites from selective laser sintered ceramic preforms. In: Marcus H L, Beaman J J, Bourell D L, Barlow J W, Crawford R H (eds.) *Proc. Solid Freeform Fabrication Symp.* University of Texas Press, Austin, TX, pp. 215–22

Denham H B, Cesarano J, King B H, Calvert P 1998 Mechanical behavior of robocast alumina. In: (eds.) *Proc. Solid Freeform Fabrication Symp.* University of Texas Press, Austin, TX, pp. 589–596

Dimos D, Yang P, Garino T J, Raymond M V, Rodriguez M A 1997 Direct-write fabrication of integrated, multi-layer ceramic components. In: Bourell D L, Beaman J J, Crawford R H, Marcus H L, Barlow J W (eds.) *Proc. Solid Freeform Fabrication Symp.* University of Texas Press, Austin, TX, pp. 33–40

Feygin M, Hsieh B, Melkanoff M A 1992 Laminated object manufacturing (LOM): a new tool in the CIM world. *IFIP Trans. B* **3**, 457–64

Gasdaska C J, Clancy R, Jamalabad V, Dalfonzo D 1999 Solid freeform fabrication of silicon nitride ceramics. *Mater. Res. Soc. Symp. Proc.* **542**, 79–84

Gibson L J, Ashby M F 1988 *Cellular Solids: Structure and Properties*. Pergamon, Oxford

Gray R W, Baird D G, Bohn J H 1998 Thermoplastic composites reinforced with long fiber thermotropic liquid crystalline polymers for fused deposition modeling. *Polym. Compos.* **19**, 383–94

Griffin C, Daufenbach J, McMillin S 1994 SFF of functional ceramic components using a laminated object manufacturing technique. In: Marcus H L, Beaman J J, Bourell D L, Barlow J W, Crawford R H (eds.) *Proc. Solid Freeform Fabrication Symp.* University of Texas Press, Austin, TX, pp. 17–24

Griffith M L, Harwell L D, Romero J T, Schlienger E, Atwood C L, Smugeresky J E 1997 Multi-material processing by LENS. In: Marcus H L, Beaman J J, Bourell D L, Barlow J W, Crawford R H (eds.) *Proc. Solid Freeform Fabrication Symp.* University of Texas Press, Austin, TX, pp. 387–93

Halloran J W 1999 Freeform fabrication of ceramics. *Br. Ceram. Trans.* **98**, 299–303

Hartmann K, Krishnan R, Merz R, Neplotnik G, Prinz F B, Schultz L, Terk M, Weiss L E 1994 Robot-assisted shape deposition manufacturing. In: *IEEE Int. Conf. Robotics and Automation*. IEEE, New York, pp. 2890–5

Hattiangadi A, Bandyopadhyay A 2000 Modeling of multiple pore ceramic materials fabricated via fused deposition process. *Scr. Mater.* **42**, 581–8

Jackson T, Liu H, Patrikalakis N, Sachs E, Cima M 1999 Modeling and designing functionally graded material components for fabrication with local composition control. *Mater. Design* **20**, 63–75

Jacobs P F 1992 *Rapid Prototyping and Manufacturing*. Society of Master Engineers, Dearborn, MI

Karam G N, Gibson L J 1994 Biomimicking of animal quills and plant stems. *Mater. Sci. Eng. C* **2**, 113–32

Keicher D M, Jellison J L, Schanwald L P, Romero J A, Abbott D H 1995 Towards a reliable laser powder deposition system through process characterization. In: *27th Int. SAMPE Tech. Conf.* SAMPE, Albuquerque, NM, p. 1029

Khandelwal D, Kesavadas T 1999 A computational approach to multi-material solid freeform fabrication using simulated annealing. *J. Mater. Proc. Manuf.* **7**, 405–15

Kodama H 1981 Automated method for fabricating a three dimensional plastic model with a photohardening polymer. *Rev. Sci. Instrum.* **52**, 1770–3

Lenk R 2000 Rapid prototyping of ceramic components. *Adv. Eng. Mater.* **2**, 40–1, 44–7

Lombardi J L, Calvert P 1999 Extrusion freeforming of Nylon 6 materials. *Polymer* **40**, 1775–9

Loschau W, Lenk R 1999 Rapid prototyping of complex-shaped parts of Si/SiC-ceramics by laser sintering. In: *EUROMAT 99. Biann. Meet. Fed. Eur. Mater. Soc. (FEMS)*. Wiley-VCH, New York, pp. 145–9

Marcus H L, Beaman J J, Barlow J W, Bourell D L 1990 Solid freeform fabrication: powder processing. *Ceram. Bull.* **69**, 1030–1

Maxwell J, Pegna J, Hill E 1995 Gas-phase laser-induced pyrolysis of tapered microstructures. In: Marcus H L, Beaman J J, Bourell D L, Barlow J W, Crawford R H (eds.) *Proc. Solid Freeform Fabrication Symp.* University of Texas Press, Austin, TX, pp. 143–50

Morissette S L, Lewis J A, Cesarano J, Dimos D B, Baer T 2000 Solid freeform fabrication of aqueous alumina-poly(vinyl alcohol) gelcasting suspensions. *J. Am. Ceram. Soc.* **83**, 2409–16

Nelson C, McAlea K, Gray D 1995 Improvements in SLS part accuracy. In: Marcus H L, Beaman J J, Bourell D L, Barlow J W, Crawford R H (eds.) *Proc. Solid Freeform Fabrication Symp.* University of Texas Press, Austin, TX, pp. 159–68

Nelson J C, Xue S, Barlow J W, Beaman J J, Marcus H L, Bourell D L 1993 Model of the selective laser sintering of bisphenol-A polycarbonate. *Ind. Eng. Chem. Res.* **32**, 2305–17

Ogale A A, Renault T, Dooley R L, Bagchi A, Jara-Almonte C C 1991 3-D photolithography for composite development. *SAMPE Q* **23**, 28–38

Panda R K, Teung P, Danforth S C, Safari A 1998 Fabrication of calcium phosphate based ceramics with controlled porosity for bone scaffolds. *Bioceram. Proc. Int. Symp. Ceram. Med.* **11**, 727–30

Pang T H 1993 Stereolithography epoxy resin development: accuracy and dimensional stability. In: Marcus H L, Beaman J J, Bourell D L, Barlow J W, Crawford R H (eds.) *Proc. Solid Freeform Fabrication Symp.* University of Texas Press, Austin, TX, pp. 11–26

Peng J, Lin T L, Calvert P 1999 Orientation effects in freeformed short-fiber composites. *Composites A* **30**, 133–8

Richter P, Thomas M, van Deventer T 1999 Macroporous hydroxyapatite bioceramics by solid freeform fabrication: towards custom implants. *S. Afr. J. Sci.* **95**, 325–6

Rimell J T, Marquis P M 2000 Selective laser sintering of ultra high molecular weight polyethylene for clinical applications. *J. Biomed. Mater. Res.* **53**, 414–20

Sachs E, Cima M, Williams P, Brancazio D, Cornie J 1992 Three dimensional printing: rapid tooling and prototypes directly from a CAD model. *J. Eng. Ind.* **114**, 481–8

Sachs E, Wylonis E, Allen S, Cima M, Guo H 2000 Production of injection molding tooling with conformal cooling channels using the three dimensional printing process. *Polym. Eng. Sci.* **40**, 1232–47

Safari A, Danforth S C 1998 Development of novel piezoelectric ceramics and composites for transducer applications. *Proc. SPIE-Int. Soc. Opt. Eng.* **3341**, 184–95

Safari A, Danforth S C 1999 Solid freeform fabrication of novel piezoelectric ceramics and composites for transducer applications. *Ferroelectrics* **231**, 731–45
Sand C, Adler J, Lenk R 1999 A new concept for manufacturing sintered materials with a three dimensional composition gradient using a silicon carbide–titanium carbide composite. *Mater. Sci. Forum.* **308–11**, 65–70
Sercombe T B, Schaffer G B 1997 Sintering of a non-conventional aluminium powder alloy. In: *Adv. Powder Metall. Particulate Materials: Proc. 1997 Conf. on Powder Metallurgy and Particulate Materials*
Sercombe T B, Schaffer G B, Calvert P 1999 Freeform fabrication of functional aluminium prototypes using powder metallurgy. *J. Mater. Sci.* **34**, 4245–51
Shi D, Gibson I 1998 Surface finishing of selective laser sintering parts with robot. In: Marcus H L, Beaman J J, Bourell D L, Barlow J W, Crawford R H (eds.) *Proc. Solid Freeform Fabrication Symp.* University of Texas Press, Austin, TX, pp. 27–35
Sigmund O, Torquato S, Aksay I A 1998 On the design of 1-3 piezocomposites using topology optimization. *J. Mater. Res.* **13**, 1038–48
Slocombe A, Li L 2000 Selective laser synthesis of TiC–Al_2O_3 composite through controlled solid state combustion. In: *Proc. Laser Materials Processing Conf. 1999*. Laser Institute of America, p. 87
Sun L, Shaw L 1999 Solid freeform fabrication of in situ SiC/C thermocouples in macrocomponents. *Metall. Mater. Trans. A* **30**, 2549–51
Vail N K, Barlow J W, Marcus H L 1993 Silicon carbide preforms for metal infiltration by SLS of polymer encapsulated powders. In: Marcus H L, Beaman J J, Bourell D L, Barlow J W, Crawford R H (eds.) *Proc. Solid Freeform Fabrication Symp.* University of Texas Press, Austin, TX, pp. 204–14
Vail N K, Swain L D, Fox W C, Aufdlemorte T B, Lee G, Barlow J W 1999 Materials for biomedical applications. *Mater. Design* **20**, 123–32
Weeren R V, Agarwala M K, Jamalabad V, Bandyopadhyay A, Vaidyanathan R, Langrana N, Safari A, Whalen P, Danforth S C, Ballard C 1995 Quality of parts processed by fused deposition. In: Marcus H L, Beaman J J, Bourell D L, Barlow J W, Crawford R H (eds.) *Proc. Solid Freeform Fabrication Symp.* University of Texas Press, Austin, TX, pp. 314–21
Weiss L, Merz R, Prinz F, Neplotnik G, Padmanabhan P, Schultz L, Ramaswami K 1997 Shape deposition manufacturing of heterogeneous structures. *J. Manuf. Syst.* **16**, 239–48
Xia Y, Whitesides G 1998 Soft lithography. *Annu. Rev. Mater. Sci.* **28**, 153–84
Zhang X, Jiang X N, Sun C 1999 Micro-stereolithography of polymeric and ceramic microstructures. *Sensor Actuat. A* **77**, 149–56

P. Calvert

G

Gel Casting

Gelcasting is an advanced ceramic powder compact forming process capable of fabricating complex near-net shapes. The principal gelcasting process was originally developed by Oak Ridge National Laboratory (ORNL) scientists during the 1980s and is based on free-radical polymerization of acrylamide. It has the capabilities for short forming time, uniform powder packing densities, high green strength, and short binder pyrolysis times.

The traditional ceramic powder compact forming processes of slip casting, isostatic or uniaxial pressing of spray-dried powder, and injection molding are widely used manufacturing processes for a variety of industrial (e.g., whitewares, sanitary ware) and technical (e.g., electronics, structural) applications. Although relatively mature and widely used, these forming processes have significant limitations for specific applications, especially for use with nonclay (powder systems not exhibiting plastic properties) materials and fine ($<1\,\mu m$) particle size powders. Slip casting limitations include the need for molds with porous dewatering surfaces, which is a difficulty for complex shaped molds for parts such as turbine wheels, significant powder packing density gradients for fine powders, long forming times when using submicron powders due to the very small pore channels, and low green strength. Injection molding of fine, relatively monodisperse powders requires 40–60 vol.% wax/plastic binder as the carrier medium. The high binder concentration and small pore size results in the need to pyrolyze the binder very slowly (up to 2–3 weeks for some components) and significantly limits part thickness. In isostatic or uniaxial pressing of spray-dried powders, the small amount of binder results in relatively weak powder compacts, requiring careful handling. Pressing of fine powders can also result in relatively nonuniform powder compaction, generating packing density gradients and leaving vestiges of the spray-dried granule agglomerates in the powder compact. With the progress in the development of high-performance ceramic materials and their requirement for fine monodisperse powders and defect-free powder compacts, there is a significant need for improved ceramic powder forming processes.

1. *Gelcasting Technology*

The first gelcasting technology was developed during the 1980s by ORNL scientists (Omatete *et al.* 1991) to address the need for an improved process for making complex-shaped ceramic powder compacts. The basic concept of gelcasting is that a fluid slurry of ceramic powder and a solvent/monomer solution is poured into a mold and rapidly polymerized to form the powder in the shape of the mold. The part is removed from the mold, the solvent removed by drying, and the remaining polymer removed by pyrolysis. The part is then processed as any other typical ceramic powder compact through densification. The initial ORNL gelcasting system was based on free radical-initiated polymerization of acrylamide monofunctional monomer. The gel system consisted of acrylamide, methylene bisacrylamide (MBAM)—a difunctional monomer cross-linker, and ammonium persulfate (APS)—the free radical initiator. Tetramethylethylenediamine (TEMED) was typically added as a catalyst to break down the APS and speed up the cross-linking reaction. The total amount of monomer and cross-linker used is 2–4 wt.% of the slurry ceramic powder content. The cross-linker/monomer ratio used ranges typically from 1:2 to 1:8, depending on the specific powder system used, its unique chemical interactions with the gel constituents, and the desired balance of wet gel powder compact strength and dried (green) part strength. Increasing monomer content and a higher ratio of cross-linker/monomer increases gel strength. Conversely, increasing cross-linker/monomer ratio decreases dried powder compact strength. Very small amounts of initiator and catalyst are required, typically ~ 0.01 wt.% initiator and ~ 0.001 wt.% catalyst.

The resulting cross-linked polymer–solvent gel contains approximately 85–90 wt.% solvent and 10–15 wt.% polymer. The polymerization process is started by heating the slurry to 35–85 °C, although the polymerization process can also be initiated chemically at room temperature depending on the concentrations of initiator and catalyst. The gelation process can be very rapid, in the order of minutes for relatively small parts. The gelled part is then removed from the mold and dried to remove the solvent system. The 2–4vwt.% polymer that remains in the powder compact is then removed by oxidative pyrolysis at temperatures up to 600 °C in air. After pyrolysis the powder compact is ready for densification by the appropriate means for that material system.

The acrylamide-based gelcasting system was very effective and robust as a powder compact forming process. It resulted in rapid forming (8 cm diameter Si_3N_4 turbine wheels gelled in less than 15 minutes), high green strength—up to 3–4 MPa (Nunn *et al.* 1994), rapid drying, and rapid pyrolysis due to the open pore channel network and the need to remove only 2–4 wt.% polymer. The acrylamide system did have two major factors limiting its ultimate potential as a viable forming process: acrylamide is a neurotoxin which requires very careful handling not usually desirable in manufacturing; and it forms carbon–nitrogen bonds during polymerization which are very difficult to break, requiring oxidative pyrolysis to 600°C or above. Two low-toxicity gelcasting systems have subsequently been developed by ORNL, which have been commercialized by ceramic manufacturers. Methacrylamide (MAM) is an effective replacement for acrylamide, and again typically utilizes MBAM as the cross-linker, APS as the initiator, and TEMED as a catalyst (Janney *et al.* 1998). The polymerization is again triggered by heating the slurry to 35–50°C. Another variant of the acrylamide gelcasting technology has been developed: hydroxymethacrylamide (HMAM) (Omatete and Nick 1999). HMAM self-cross-links and does not require a cross-linker. Experimentation has shown that HMAM can be initiated by APS, but exhibits relatively weak gel strengths. A more effective initiator, azobis[2-(2-imidazolin-2-yl)propane] · 2HCl (AZIP), was discovered that generates very strong gels with HMAM. However, AZIP requires a higher temperature to begin dissociating, so the HMAM–AZIP gel system requires gelation temperatures of 60–85°C. The HMAM system has shown an advantage over the MAM–MBAM system in that higher slurry solids contents are achievable for the same desired slurry viscosity, presumably because HMAM also further enhances powder dispersion. Both the MAM–MBAM and HMAM gelcasting systems are in commercial use.

2. *Gelcasting Process*

The gelcasting process uses similar equipment and process steps to those used in standard ceramic powder compact forming processes such as slip casting. The generic flowchart for the gelcasting process is shown in Fig. 1 and the individual steps are discussed below.

2.1 Slurry Preparation

Water is the solvent of choice for most gelcasting applications because of the ability to strongly enhance dispersion of ceramic powders using the polar nature of the water molecule to generate electrostatic repulsion, principally by pH adjustment. Organic solvents have also been effectively demonstrated. Steric repulsion using long-chain polymers is the primary powder dispersion mechanism for organic solvents and it is also used with water-based systems. The acrylamide, MAM, and HMAM gelcasting monomers also provide some component of dispersion (they are similar in steric repulsion capability to the polyacrylate dispersants used in ceramic powder processing) and thus serve an added benefit. Most slurries for gelcasting have been prepared using the traditional ball milling process, which provides effective mixing of powders and allows aging time for hydroxylation of particle surfaces to achieve stable slurry dispersions. Ball milling also allows for controlled particle size reduction, if required. High-energy milling techniques such as turbomilling and attrition milling have also been used effectively to prepare slurries, but care must be taken to allow suitable time for slurry stabilization to take place. The monomer and cross-linker gelcasting system components are added at the slurry preparation step to assure they are intimately mixed with the powder in the slurry and also to provide added dispersion, as described above. Because the slurry is subsequently poured or pumped into a mold, low viscosity (<500 mPas) is desirable. Since the solids content of the slurry becomes essentially the resulting gelcast part powder packing density, maximizing solids content while keeping the slurry viscosity low enough to be readily pourable is a typical optimization requirement. High solids contents are achievable with coarse powders (larger than $\sim 5\,\mu m$) and powders with graded particle size distributions to optimize packing and rheology. Some fine powders can be dispersed effectively to achieve high solids contents, e.g., 62 vol.% for 1 μm Al_2O_3 (Young *et al.* 1991), but maximum achievable solids contents below 50 vol.% are typical of many fine powders due to high surface area and surface chemistry/dispersivity limitations. Omatete and Nick (1999) demonstrated that for a 0.8 μm-sized silicon nitride powder, a maximum solids content of only 45 vol.% was achievable before exceeding maximum slurry flowability (<500 mPas) for effective mold filling. After mixing/milling, the slurry may be deaired to remove entrapped air generated by the slurry preparation process. The initiator and catalyst (if used) are added after the slurry has been prepared. Since APS, AZIP, and TEMED are all strongly acidic, they are added to basic pH slurries while the slurries are rapidly mixing to prevent coagulation and spot-gelation.

2.2 Mold Materials

Gelcasting molds require impervious surfaces, good part release characteristics, and tight sealing. Many mold materials have performed very effectively with the gelcasting technology. Suitable metal mold materials include aluminum and stainless steel. Anodized aluminum offers a more wear- and corrosion-resistant surface, especially with basic pH

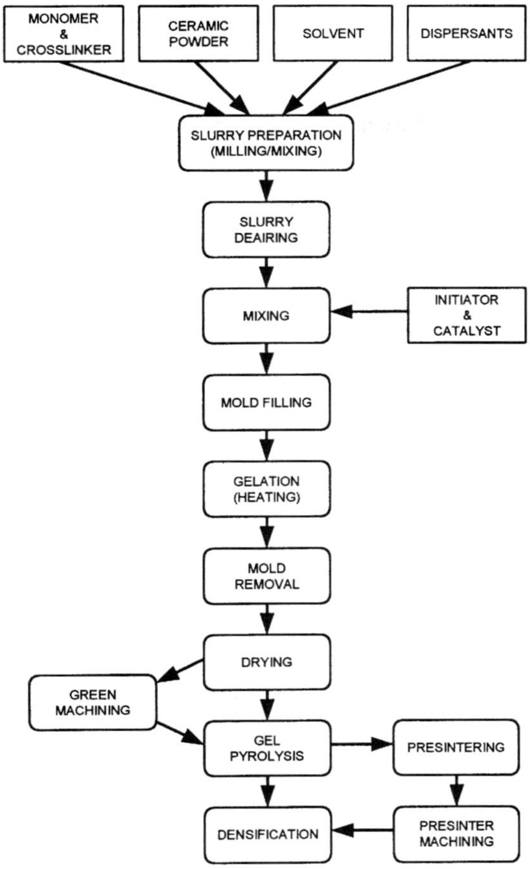

Raw Materials
- Monomer/Crosslinker Additions (2-4 wt%)
- Water is Solvent, Organic Solvents Also Demonstrated

Slurry Preparation
- Ball Milling, Attrition Milling Typically Used

Slurry Deairing
- Short (1-10 min.) Evacuation of Slurry to Remove Entrained Air

Initiator/Catalyst Addition
- Typical Initiator Addition ~ 0.01 wt%
- Catalyst Used for Specific Gel Systems

Mold Filling
- Fluid Slurry is Poured into Mold
- Slurry Working Time Controlled by Initiator/ Catalyst Concentration and Slurry Temperature

Gelation
- Gelation Initiated by Heating to 35°C-85°C
- Typical Gelation Times – 5 min. to 120 min., Depending on Part Thickness/Size

Drying
- Standard Drying Practices Used for Powder Compacts

Green Machining
- Specific Gel Systems Allow Effective Green Machining by Milling/Turning Techniques

Gel Pyrolysis
- Gel Removed by Pyrolysis in Air at 400°C-600°C
- Short Cycle Times (Only 2-4 wt% Binder)

Presintering/Densification
- Standard Densification Practices Used

Figure 1
Flowchart and details of typical ceramic powder gelcasting process.

slurries, and has improved part release compared to unanodized aluminum. Aluminum and steel can be readily machined into complex molds. Glass has been shown to be an effective mold material, due to its combination of very smooth surface and resistance to corrosion, but is difficult to generate complex molds from and too fragile for typical handling. Plastics such as PVC have made effective gelcasting molds, and they take advantage of what is many times a disadvantage for acrylamide-based gelcasting systems—that oxygen inhibits the free-radical polymerization process. PVC has a slightly oxygenated surface and thus generates a small (<0.1 mm) ungelled layer (typically referred to as a reaction layer) on the surface of the part which acts as a lubricant allowing part removal. Many plastics and waxes evaluated for mold materials have much more oxygenated surfaces and generate significant surface reaction layers (>1 mm) which is undesirable because these thicker reaction layers spall off and generate cracks and stresses during drying and pyrolysis. Sensitivity to oxygen can be adjusted with the monomer/cross-linker ratio. For example, in the MAM/MBAM system, higher cross-linker concentrations increase reaction layer thicknesses. Because of the sensitivity to oxygen during gelation, molds should be designed to contact all part surfaces that are desired to be near-net-shape. Relatively fluid slurries are used in gelcasting, so molds must be very close fitting to prevent leaking. Since molds are heated to initiate the gelation process, it is usually desirable also to use mold materials with high thermal conductivity to minimize process heating and cooling times. The effective use of dissolvable or meltable cores (analogous to the lost-wax process for metal casting processes) has also been demonstrated for gelcast parts. Waxes dissolvable in solvent and low-melting metal alloys have been used to form gelcast powder compacts with complex internal features (Pollinger 1997) such as gas turbine combustor cans. The cores must be removed prior to drying of the gelcast part due to the typical 1–3% shrinkage that occurs during drying.

To enhance part removal from molds, release agents are frequently used. A number of mold releases have been evaluated, with release agents used for injection molding of plastics being typically effective. Mold release agents vary in effectiveness depending on particular ceramic powders, gelcasting systems, and mold materials, thus some trial and error is usually required when evaluating release agents for each application. Materials that have demonstrated effective release for specific systems include cooking spray, petroleum jelly, and spray lubricants. The ability of a mold release to provide effective part release must also be balanced with the fact that many release agents contain potential contaminants and that they function by preventing wetting by the gelcasting slurry, thus increasing the potential for incompletely filled molds and air pockets.

2.3 Gelation

Gelation itself is triggered by heating the molds. Since the temperatures to which the molds must be heated are moderate (35–85°C), there is a variety of heating techniques that have proven effective: static air ovens, forced air ovens, water baths, water jacketed molds, and infrared heating of molds. The mold material and heating method should be selected based upon the desired gelling time and size of component, which controls the amount of heat flux needed. An important characteristic of the gelled part is that the gelcasting process results in powder compacts with very uniform powder packing densities. The gelation process "freezes in place" the homogeneously distributed powder in the slurry.

2.4 Solvent Removal

After gelation the solvent is removed by standard drying practices. For large particle size powder compacts with corresponding large pore sizes, static or circulating air drying suffices. For finer particle size ($<1\,\mu m$) powder compacts, drying must be performed more carefully due to submicron pore channels and the potential to crack parts as a result of internal solvent thermal expansion and/or increasing vapor pressure before open pore channels to the surface are achieved. In this case, controlled humidity dryers are typically used. Because the gel provides enhanced strength to the powder compacts during drying, they can handle more rapid drying rates then corresponding slipcast parts. This is balanced with the fact that the gel shrinks as the water is removed from its structure. Gelcast parts with submicron powder and ~45–50 vol.% powder packing can shrink 1–3% during drying, which generates large stresses if not carefully controlled (Ghosal *et al.* 1999). Example drying times for silicon nitride parts using 0.5 μm powder and with 50 vol.% packing are ~20 hours for an 8 cm diameter turbine wheel, and ~80 hours for a 15 cm diameter wheel. The dried parts have also been shown to be readily green machined (Nunn and Kirby 1996) due to the high dried gel strength.

2.5 Polymer Removal

The polymer remaining after drying of the gel is typically removed by oxidative pyrolysis. Where the initial acrylamide monomer gel system required 600°C or greater to achieve complete polymer removal, the MAM–MBAM and HMAM systems pyrolyze at much lower temperatures due to their weaker bonds and the polymers can be removed at temperatures as low as 400°C in air. Since the part has open pore channels because of previous removal of the solvent which constitutes 85–90% of the gel, the pyrolysis can be performed very rapidly, usually in less than a day. For larger and thick cross-section components, pyrolysis times are lengthened to assure complete removal of polymer from the interior. For those materials requiring inert pyrolysis to prevent oxidation, nitrogen and vacuum pyrolysis at temperatures up to 600°C show that ~0.5 to 1 wt.% carbon residue is left. After polymer pyrolysis, gelcast components are then treated as any other powder compact for subsequent densification.

3. Gelcasting Capabilities

The acrylamide-based gelcasting technology results in a number of demonstrated capabilities as a powder compact forming process. The gelation process is rapid, for example an 8 cm diameter part can be gelled in 10 minutes. The process results in very uniform powder packing densities, minimizing stresses and potential cracking during subsequent drying and pyrolysis and maintaining near-net-shape during densification. A variety of mold materials, many of them relatively inexpensive and easily machined into tight tolerances, can be used. Green strengths on the order of 3–4 MPa allow for robust handling, the ability to green machine, and faster drying. Pyrolysis is rapid because 85–90% of the gel system is removed during drying, leaving an open pore network and only 2–4 wt.% polymer. The combination of ease of mold filling by fluid slurry, tight tolerance mold capability, uniform packing densities, and high as-gelled and green strength results in the ability of the gelcasting process to form readily complex and near-net shapes. The mold filling process, the chain reaction capability of the free radical polymerization to initiate gelation uniformly, the high green strength, uniform packing density, and removal of most of the gel system by low stress evaporation also results in the ability to form and process both large and thick cross-section powder compacts. The gelcasting process is also automatable, as has been demonstrated for Si_3N_4 turbine wheels (Nick *et al.* 1999).

4. *Gelcasting Limitations*

Although the acrylamide-based gelcasting technology and its improvements have a number of desirable powder compact forming capabilities, there are number of potential limitations that must be considered. Because low-viscosity slurries are used, sealing of molds is a potential issue. The mold tooling must be tightly sealed, which is potentially difficult with complex shape molds. Many mold materials have some oxygenated surface content and can cause inhibition of gelation and formation of reaction layers. Higher cross-linker content increases sensitivity to oxygen-induced inhibition of gelation, which can limit the ability to generate high gel strengths. Many mold releases will provide effective release but generate high wetting angles with the slurry, making complete mold filling without air entrapment a potential problem. Although uniform powder packing densities are achieved, high surface area/fine particle size powders present a significant problem in that they generate high slurry viscosities at relatively low solids contents. The resulting low powder packing densities present potential problems of extensive shrinkage during drying, pyrolysis and densification which can cause significant stresses and cracking, part distortion, and reduced dimensional reproducibility. Additionally, low powder packing densities reduce the driving force for densification and can inhibit achievement of full density and desired microstructures. Finally, inert pyrolysis cannot completely remove the polymer, leaving residual carbon char.

5. *Other Gelcasting Systems*

Another significant gelcasting technology being developed is the polymerization of existing polymers in solution. Morrisette and Lewis (1999) have developed and demonstrated the effective gelcasting of alumina powder aqueous slurries by cross-linking poly(vinyl alcohol) (PVA) with organotitanates to form strong gels at room temperature. PVA has the added powder processing benefits of being extensively used in current powder processing applications as a binder for spray drying and in being an effective powder dispersant in water suspensions. The gelation process occurs rapidly at room temperature, so slurries are cooled after cross-linker addition to temperatures below room temperature (5–20°C) to slow the gelation process, and provide workable low-viscosity slurries for mold filling.

Direct coagulation casting (DCC) is a new forming technology that also converts an aqueous slurry into a ceramic green body in a nonporous mold—similar to gelcasting, but in this case by coagulation of the powder suspension. Two approaches have been developed to achieve *in situ* suspension coagulation, using enzyme-catalyzed reactions: (i) to change suspension pH by forming acids or bases to drive pH towards the powder suspension isoelectric point (IEP); or (ii) to form salts to reduce double layer thickness. Baader *et al.* (1995) demonstrated the coagulation casting of alumina through the decomposition of urea to generate ammonia and increase suspension pH. The alumina powder, which has an IEP of ~8.5, was first dispersed in water with a small amount of urea at pH 4 to generate a low-viscosity suspension for casting. Urease enzyme was then mixed into the suspension and the casting mold filled. The urease catalyzed the decomposition of the urea to generate ammonia and shift the pH higher, to pH 9. The alumina suspension is then effectively coagulated into a green body as the pH approaches the alumina IEP. The rate of urea decomposition can be controlled by temperature (decreasing the suspension temperature from 25°C to 5°C) and increased working time (no appreciable increase in viscosity) from 10 to 60 minutes, and decreasing the urease concentration by an order of magnitude increases both working time and coagulation time by an order of magnitude. Granule *et al.* (1995) also demonstrated the use of *in situ* salt generation to coagulate silicon carbide aqueous suspensions. In this case silicon carbide has an IEP of two and well-dispersed suspensions are generated at pH 9–11. It is very difficult to decrease pH to such a low value to coagulate the suspension. Here the urease enzyme-catalyzed decomposition of urea was used to form ammonium carbonate salt to decrease the silicon carbide particle double layer thicknesses to achieve coagulation. Approximately $2\,mol\,l^{-1}$ salt concentration was needed to achieve complete destabilization of the silicon carbide suspension, while the suspension pH stayed at about pH 9–10. For both of these coagulation processes, high solids content suspensions result in higher green strength parts and binder burnout is not needed because less than 1 vol.% organic material is added to generate the coagulation process.

6. *Summary*

Gelcasting is an advanced ceramic powder compact forming process that has demonstrated the capability for fabricating complex near-net shapes. The principal gelcasting process of polymerization of acrylamide monomer using a cross-linker and free radical initiator has been improved through the development of a low toxicity monomer (MBAM) and a self-cross-linking monomer (HMAM). The capabilities of these gelation processes include short forming time, uniform powder packing densities, high as-gelled and green strength, short binder pyrolysis times, and the ability to form both thick cross-section powder compacts and complex shaped compacts with both thin and thick sections. Limitations include difficulty in sealing complex molds against leaking of the low-viscosity slurries, the potential for low powder packing densities with fine powders, the sensitivity of the polymerization process to oxygen exposure, and incomplete polymer pyrolysis in nonoxidative environments. Gelcasting technology has been demonstrated for a variety of ceramic materials and commercial applications have been initiated.

Bibliography

Baader F H, Will J, Tieche D, Graule T, Gauckler L J 1995 High strength alumina produced by direct coagulation casting. *Ceram. Trans.: Ceram. Proc. Sci. Technol.* **51**, 463–7

Ghosal S, Emami-Naeini A, Harn Y, Draskovich B S, Pollinger J P 1999 A physical model for the drying of gelcast ceramics. *J. Am. Ceram. Soc.* **82**, 513–20

Granule T J, Si W, Baader F H, Gauckler L J 1995 Direct coagulation casting (DCC): fundamentals of a new forming process for ceramics. *Ceram. Trans.: Ceram. Proc. Sci. Technol.* **51**, 457–61

Janney M A, Omatete O O, Walls C A, Nunn S D, Ogle R B, Westmoreland C G 1998 Development of low-toxicity gelcasting systems. *J. Am. Ceram. Soc.* **81**, 581–91

Morrisette S L, Lewis J A 1999 Chemorheology of aqueous-based alumina–poly(vinyl alcohol) gelcasting suspensions. *J. Am. Ceram. Soc.* **82**, 521–8

Nick J J, Newson D D, Masseth M, Monette M 1999 Gelcasting automation for high volume production of silicon nitride turbine wheels. *Ceram. Eng. Sci. Proc.* **20**, 225–31

Nunn S D, Kirby G H 1996 Green machining of gelcast ceramic materials. *Ceram. Eng. Sci. Proc.* **17**, 209–13

Nunn S D, Omatete O O, Walls C A, Barker D L 1994 Tensile strength of dried gelcast green bodies. *Ceram. Eng. Sci. Proc.* **15**, 493–8

Omatete O O, Janney M A, Strehlow R A 1991 Gelcasting—a new ceramic forming process. *Am. Ceram. Soc. Bull.* **70**, 1641–9

Omatete O O, Nick J J 1999 Improved gelcasting systems. *Ceram. Eng. Sci. Proc.* **20**, 241–8

Pollinger J P 1997 Status of silicon nitride component fabrication processes, material properties and applications. Am. Soc. Mech. Eng., Turbo Expo '97, Paper GT-321

Young A C, Omatete O O, Janney M A, Menchhofer P A 1991 Gelcasting of alumina. *J. Am. Ceram. Soc.* **74**, 612–8

J. P. Pollinger

Gold Casting Alloy, New, for Dental Applications

Introduction

Man's attempt to replace lost teeth dates to the seventh century BC, when the Etruscans tried animal teeth bound with gold alloy bands. Historically, a number of materials have been used to carve replacement teeth, including ivory and wood. Despite these efforts the inability to replicate the environment of the mouth and the poor fit of the restorations resulted in loss of masticatory function.

The fabrication of dental prosthetic devices was revolutionized in the early 1900s with the adaptation of the lost-wax method of casting to the fabrication of dental inlays, crowns, and bridges. This invested pattern methodology developed by H. William Taggert solved the ill-fitting problem. The excellent fit provided by the method allowed the restorations to be cemented in place and function like natural teeth. (The cementation process defines the classification of these prosthetic devices as "fixed" prosthetics as opposed to "removable" prosthetics such as dentures or partial dentures.)

1. Evolution of the Prosthetic Fabrication Process

In the fabrication of a single crown, the lost-wax method involves several steps:

Step 1. The dentist prepares the natural tooth by grinding away the diseased portion leaving a "stump." The unique outer surface of the remaining tooth structure will mate to the inner surface of the prostheses.

Step 2. An impression of the "stump" and adjacent gingival regions is taken using a nonaqueous elastomer material and sent to a dental laboratory. The impression material now has a negative image of the area of interest in the patients' mouth.

Step 3. The dental lab pours a stone model into the impression. The stone hardens creating a positive replica of the area of interest in the patient's mouth.

Step 4. Wax is built up on the model by hand for the portion of the tooth that will be replaced by alloy. A sprue is attached to the wax pattern.

Step 5. The wax pattern and sprue are invested in a refractory material that hardens and is placed in a furnace. The wax is melted and vaporized leaving a void.

Step 6. The dental alloy is cast into the void thus duplicating the wax pattern.

The inside contours of the cast alloy from Step 6 matches the outer surface of the prepared "stump" in Step 1. This allows the restoration to "fit" and restores masticatory function. The prostheses feels and functions like the natural tooth. A more detailed description of the lost-wax process can be found in Anusavice (2003).

The alloys first used in the early 1900s were high-gold alloys (75%+) containing silver and copper. These materials were easy to cast by torch, worked well with available gypsum-based investments, and exhibited excellent biocompatibility and corrosion resistance in the mouth. At first the entire restoration was made of the gold alloy. Later acrylic facings were mechanically attached to the outer surfaces of the gold teeth in order to make the restorations appear more life-like when the patient smiled. The esthetics of fixed prosthetics was advanced further in the 1960s with the introduction of a porcelain-fused-to-metal (PFM) system developed by Weinstein *et al.* (1962). The left-hand side of Fig. 1 shows a cross-sectioned PFM crown, the alloy portion is a thin-wall casting fabricated using the same lost-wax method as before but the bulk of the restoration is made by firing successive layers of porcelain onto the alloy. The alloy provides the fit and imparts resiliency and strength to the restoration while the porcelain

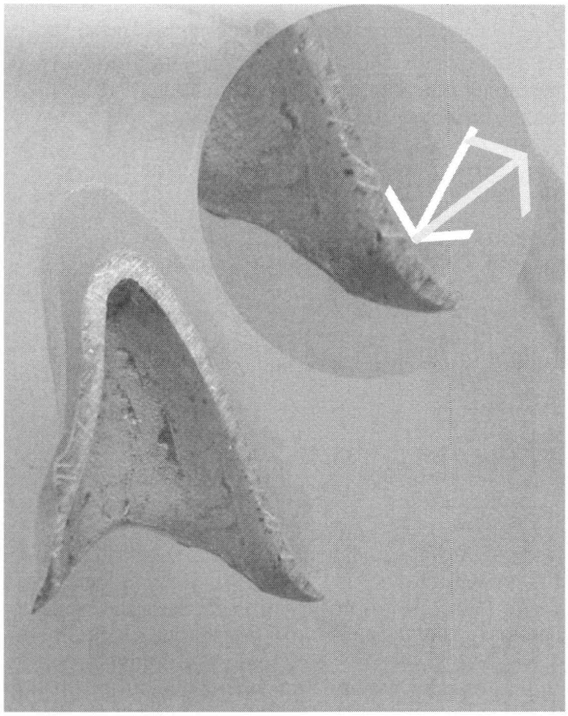

Figure 1
An example of a porcelain-fused-to-metal crown made from the developed alloy. The bottom of the crown sits below the gum line. The enlarged inset shows how the light reflected off the alloy surface will be yellow thus complementing the natural gum tissue (See colour plate section).

imparts life-like esthetics. This type of restoration quickly became the restoration of choice and remains so today. In order to accommodate the firing of porcelain the traditional gold–silver–copper alloys could not be used for the PFM processing technique. A recent review article by Knosp *et al.* (2003) discusses the use of gold in a wide range of dental applications. This article will focus on alloys for the PFM processing technique. The nature of these alloys is discussed next.

2. *The Requirements for a Porcelain-fused-to-metal Alloy*

2.1 Chemical Properties

In order to be biocompatible, the alloy must consist of elements that are nontoxic. The alloy must resist corrosion under both acidic and caustic pH conditions. The alloy may be nonreacting (noble) like gold- and palladium-based alloys or passivating like nickel–chromium and cobalt–chromium alloys where a protective oxide film is formed. Some typical alloy compositions and their properties are shown in Table 1.

Issues surrounding the leaching of base metals into the soft gum tissue, and galvanic reactions between dissimilar alloys in the mouth have been reviewed by Lang *et al.* (1986).

2.2 Mechanical Properties

Biting forces in the mouth vary from 90 N in the anterior (front) of the mouth to almost 900 N in the posterior regions as illustrated in Fig. 2. The amount of stress applied to a restoration is dependent upon the restoration's size and placement in the mouth. Single units will be subject to less stress than bridgework (multiple units). Alloys for the PFM technique generally comply with the International Standard ISO 9693 that requires the minimum proof stress to be 250 MPa at 0.2% offset. Proof stress is used as a criterion because under stress, the alloy must remain in the elastic region in order to function properly. If any plastic deformation occurs, then the restoration will no longer fit. The proof stress of the alloy must increase as the amount of tooth structure being replaced increases and as the number of teeth being replaced increases. The ISO standard minimum value assumes that the alloy will be able to function anywhere in the mouth regardless of the size of the restoration. The standard also requires a minimum elongation of 3%.

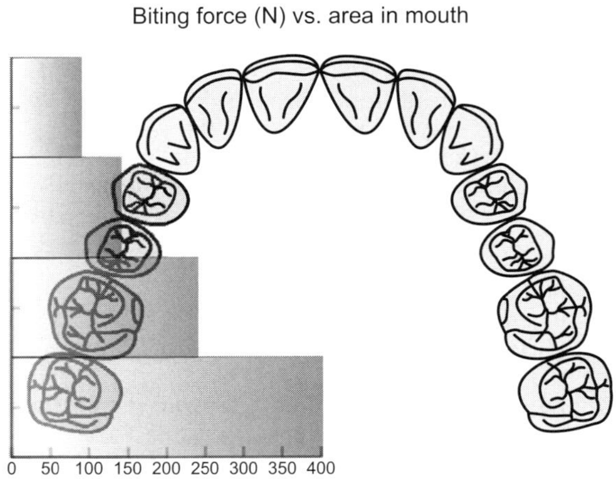

Figure 2
The biting forces increase considerably from the front (anterior) section of the mouth to the rear (posterior) section of the mouth.

Table 1
Typical dental alloy compositions and properties for the porcelain-fused-to-metal processing technique.

Major elements	Minor elements	Color	Solidus (°C)	Liquidus (°C)	0.2% Proof strength (MPa)	Elastic modulus (GPa)
99.7 Au	In, Zn	Gold	1030	1070	14	77
84.4 Au, 9.5 Pt	Mn, Zn, In, Ir	Very yellow	1050	1160	510	90
86.5 Au, 10.4 Pt	Rh, Ta, Mn, Zn, In, Ir	Yellow	1045	1115	524	94
87.4 Au, 4.5 Pt, 5.9 Pd	Ag, Fe, Sn, In	Pale yellow	1150	1180	448	98
76.6 Au, 9.9 Pt, 9.3 Pd	In, Ag, Fe, Sn	White	1150	1230	630	101
65 Au, 26 Pd	In, Ga, Ru	White	1140	1255	552	120
80 Pd	Ag, In, Ga, Ru	White	1100	1290	631	130
55 Pd, 34 Ag	In, Sn, Zn, Ru	White	1170	1265	724	125
54 Ni, 22 Cr	Mo, Fe, Nb, Ta	White	1220	1230	359	160
60 Co, 31 Cr	Mo, Si	White	1240	1350	448	160

The first entry is the new alloy discussed.

The PFM fabrication method requires the alloy portion to be a thin wall casting of ~0.2–0.4 mm in thickness. The higher the alloy elastic modulus the thinner the casting can be made. This is important because maximizing the porcelain thickness within the finite room available in the patient's mouth improves the esthetics of the restoration. The elastic modulus of the alloy is also important during mastication. How well the alloy absorbs stress is critical during uneven loading. This occurs in the posterior regions, especially on molars and on longer span bridgework.

2.3 *Thermal Characteristics*

The technology of casting used by dental laboratories throughout the world consists of a gas and oxygen torch. This limits the liquidus temperature of the alloy to a maximum of ~1300 °C.

A joining operation may be performed on the metal framework before the application of porcelain. The solidus temperature of the alloy needs to be above the melting range of the filler material used. In dental terminology the joining operation is called soldering.

The dental porcelain is applied as a paste onto the alloy and then the assembly is heated to ~960 °C at a rate of 100 °C min^{-1}; held for 1 min and then cooled. The atmosphere in a dental porcelain furnace under full "vacuum" contains ~10% oxygen (dental porcelain furnaces employ a mechanical roughing pump). Several layers of porcelain are applied so that the cycle is repeated at least 3 times. Each subsequent firing is done at the same or lower temperature. The thermal cycling requires the alloy to have sufficient high-temperature creep resistance. If any distortion occurs during the firing then the restoration will not fit. Distortion due to thermal treatment is called "sag."

The thermal expansion of the alloy must be slightly above the expansion of the porcelain that is to be baked on the alloy so that the porcelain will be under a compressive stress at room temperature. Since there is a variety of porcelains and alloys on the market, the user must be familiar with the compatible pairs. Each porcelain brand has a range of compatible alloy expansion values as shown by the samples in Fig. 3. If the alloy has a thermal expansion value outside of the compatible range then the porcelain will fracture. For this reason the dental lab will choose alloys dependent upon the type of porcelain used.

Porcelain adherence is achieved by the formation of a chemical bond during the firing of the first porcelain layer. The alloy must form a thin, adherent metal oxide that can be dissolved by the porcelain. For the noble metals, additions of tin, indium, and gallium are used to generate a thin oxide layer on the alloy. Control of the alloy oxidation is a problem for the nonprecious alloys. Generally this is accomplished by sufficient additions of chromium to form a continuous chrome oxide layer.

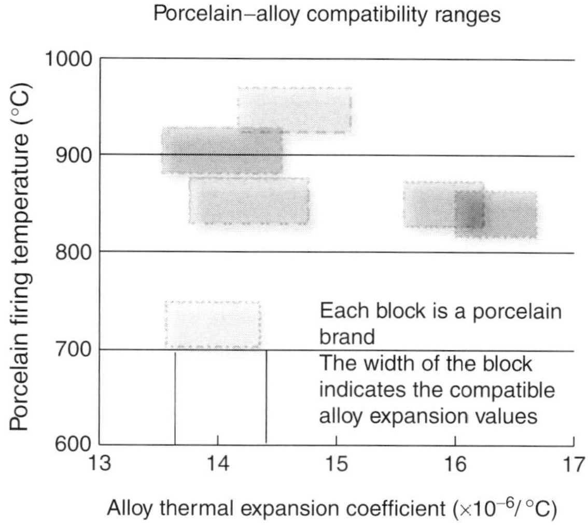

Figure 3
Porcelain–alloy compatibility ranges. Each block represents a porcelain brand and is positioned vertically dependent upon its fusion temperature. The width of the block indicates the alloy thermal expansion values that are compatible with the porcelain (See colour plate section).

2.4 *Esthetics and Alloy Color*

The ultimate goal in the restoration of teeth is to have the prosthesis appear life-like as well as function like natural teeth. The PFM processing technique achieves this goal to varying degrees. Although the alloy is covered with successive layers of porcelain, light travels through the porcelain and is reflected back out. If the surface of the alloy is silver (white) in color then the entire light spectrum is reflected. This includes the blue region, which when passing through the pink gum tissue produces a dark unsightly outline. As one can see from Table 1 most of the alloys are silver in color except for the high-gold alloys containing low amounts of platinum and palladium. Sometimes functional requirements cannot be sacrificed in order to attain a desired esthetic result, that is, for a large restoration, the alloy must be especially strong. Additions of platinum and palladium strengthen a high-gold alloy but also degrade the yellow color of pure gold.

The right-hand side of Fig. 1 is a magnified cross section of a crown illustrating the situation when the surface of the alloy is yellow in color. Incident light reflecting off the alloy surface is reflected back as yellow light. The light from the blue portion of the spectrum is absorbed. The reflected light produces very warm porcelain shades complementary to the gum tissue. This is especially true at the margin line (bottom of crown) where it can be seen that the porcelain thickness is reduced. As the color of the alloy approaches that of pure gold, the dark gum line decreases and, at the color of pure gold, it disappears. While the inertness of gold provides excellent biocompatibility and ease of fabrication, it is the color of gold that bestows unparalleled esthetics when used for the PFM technique. The richer the alloy color the better is the result.

The color of the alloy oxide may also be a problem since it can interfere with the porcelain shades and in severe cases bleed through the porcelain.

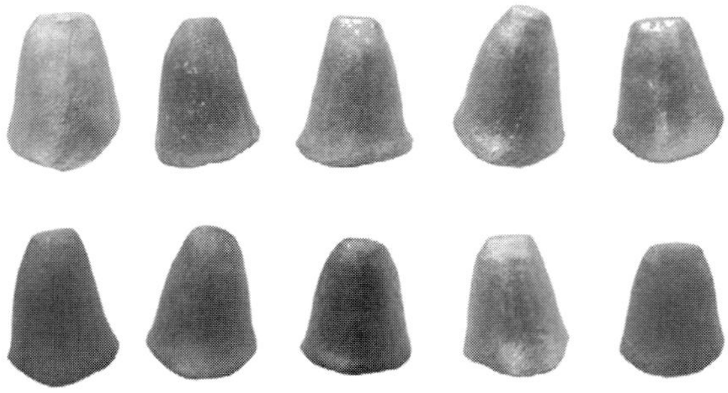

Figure 4
Alloy oxide colors resulting from varying the nature and content of base elements in a 99% gold alloy. The final alloy developed is shown on the top left (See colour plate section).

3. *Alloy Development*

The quest for better esthetics in dental restorations has resulted in many different materials both metallic and nonmetallic, methods of fabrication, and processing techniques being introduced onto the marketplace.

The most reliable and cost effective of these, however, is still the PFM processing technique. How then can the esthetics of gold be best utilized?

3.1 Strength Considerations

Developing an almost pure gold alloy that would produce the desired life-like esthetics compromises the strength of the alloy. Of course, the alloy still had to produce a functional prosthetic device.

For most people their "smile-line" shows only the upper front teeth and the juncture of the tooth with the gum is seen in the front six teeth.

This anterior portion of the mouth is the area of the jaw that experiences the least force. As discussed above, the front teeth generally experience a force of only 90 N. Pure gold can easily accommodate this stress level. The size of the restoration must be limited, however, in order to avoid any significant torque on the alloy. Bridgework consisting of multiple units naturally develops a torque due to the uneven loading that occurs during mastication. By limiting the restoration to primarily anterior single units or short three-unit anterior bridges the applied stress remains uniform, enabling the almost pure gold to function as a prosthetic material.

3.2 Porcelain Adherence and Oxide Color

Figure 4 shows the dramatic color difference resulting from varying the nature and content of base elements in a 99% gold alloy. By adding small quantities of indium and zinc it was found that the alloy developed an almost transparent continuous surface oxide (Fig. 4—top left).

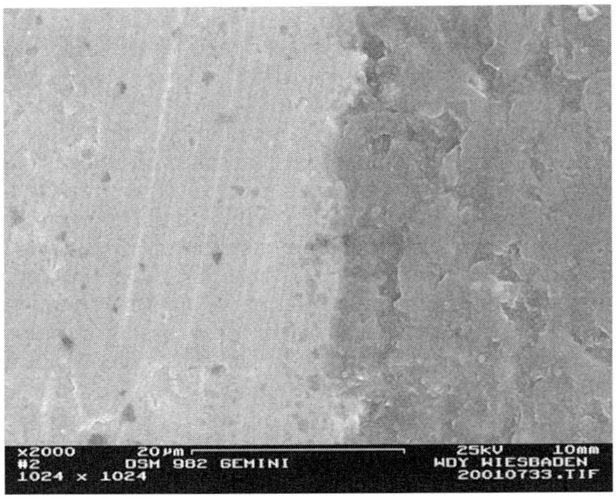

Figure 5
An SEM photomicrograph of the gold alloy–porcelain interface showing very good wetting of the dental porcelain (right side of figure) onto the alloy surface.

The total base metal addition is less than 0.4 wt.%. This quantity is sufficient to allow the glass in the dental porcelain to wet the alloy and form a strong chemical bond. Figure 5 shows an SEM photomicrograph of the gold alloy–porcelain interface. The porcelain (right side of the figure) wets the alloy surface and bonds with the alloy.

3.3 Porcelain Compatibility

Concerning porcelain compatibility, all porcelains on the market appeared to be compatible with the gold alloy, despite the high thermal expansion of pure gold. It is believed that the high heat transfer rate of pure gold keeps the alloy at a much lower temperature than the porcelain during the cooling portion of the porcelain firing cycle. The softness of the gold may also accommodate any stress that is developed by the porcelain as it cools.

3.4 Biocompatibility

One major benefit of having a high gold content is the exceptional biocompatibility of pure gold. In addition to the nontoxic nature of gold, the noble nature of gold may impart additional benefits to the alloy, such as low plaque accumulation. The accumulation of plaque is harmful to the gum tissue. By using an alloy that accumulates less plaque, the gum tissue stays healthy longer.

Internal studies (Munoz and Bernal unpublished) have demonstrated that the commercial alloy resulting from this development effort (GOLDTECH BIO2000®) (Cascone 1999) performed better than another commercial material known for its ability to resist plaque.

4. Summary

The development of an almost pure gold alloy for the PFM processing technique required a restriction in the application of where the alloy was to be used in the mouth. By restricting its use to the anterior portion (bi-cuspid to bi-cuspid) and the size of the restoration, single crowns, or three unit bridges, the alloy has sufficient strength to withstand normal mastication forces. The selection and mixture of base elements indium and zinc provided porcelain adherence and an almost transparent oxide color. Thus the material has the most desirable gold color ensuring an esthetic restoration.

For the dental laboratory technician the new alloy provides an expedient way to produce esthetic restorations. The alloy uses the normal lab processing methods so no additional training or capital expense is required. Casting three-unit bridges is done in a single step, unlike other noncasting high-gold systems.

The high heat transfer rate of gold endows GOLDTECH BIO2000® with universal porcelain compatibility, a very unique property.

For the dentist, GOLDTECH BIO2000® provides the esthetics of all-ceramic restorations with the reliability of the PFM technique. This also means that less tooth structure is cut and it is easier to cement in place. The patient gets a restoration with exceptional esthetics and, due to the low surface tension of the alloy, less plaque accumulation than other gold systems.

Bibliography

Anusavice K (ed.) 2003 *Phillips' Science of Dental Materials*, 11th edn. WB Saunders Co., St. Louis, MO

Cascone P 1999 *US Patent* 5922276

International Standard ISO 9693 Dental ceramic fused to metal restorative materials. International Organization for Standardization, Switzerland

Knosp H, Holliday R, Corti C 2003 *Gold in Dentistry: Alloys, Uses and Performance Gold Bulletin* **36**(3), 93–102

Lang B R, Morris H F, Razzoog M E 1986 *International Workshop Biocompatibility. Toxicity and Hypersensitivity to Alloy Systems Used in Dentistry*. The University of Michigan School of Dentistry, Ann Arbor, MI

Munoz C A, Bernal G In-Situ Clinical Evaluation of the Plaque Retention of Four Prosthetic Materials, Case Study by Loma Linda University School of Dentistry. Unpublished study conducted at Loma Linda University. Copies available from Argen

Weinstein M, Katz S, Weinstein A B 1962 *US Patent* 3052982

P. Cascone

Green Machining

The topic of green machining of ceramics is not widely documented in the technical literature. Green machining parameters tend to be sensitive to a combination of material properties, machine tool characteristics, and the preferences and experience of the machinist. The process also tends to fall under the category of trade secrets and proprietary information. However, there are some general aspects of green machining that do apply to the fabrication of many ceramic articles, and these are discussed here. By definition, this fabrication step refers to the machining of ceramic materials in their unfired or "green" state. The process is frequently required due to limitations in the ability to incorporate complex or multidimensional features into ceramic components during the shape-forming process. Economics as well as technology can drive these limitations. Some instances where green machining is pursued for economic reasons include short production runs, where the expense of tooling and equipment to form net-shape parts exceeds the cost of green machining, or where the design and delivery time for net-shape tooling is greater than the customer is willing to tolerate. Technical reasons include features that are difficult to incorporate during die pressing, molding, or casting, such as threads, holes, or slots (e.g., especially features perpendicular to the die-pressing direction). Of course, one of the greatest economic drivers for green machining is the increased difficulty and expense associated with hard grinding of densified ceramics. The time required for hard grinding, and the rate of tool wear, is typically at least an order of magnitude greater than for green machining.

1. Importance of Green Ceramic Characteristics

The importance of various characteristics of the green ceramic body cannot be overemphasized. The binder system in particular has a great influence on the difficulty of machining, quality of the machined surfaces, and wear rate of tool bits. The variability in green density from part to part is also a critical parameter since it corresponds directly to the variability in shrinkage that will occur during sintering. The level of confidence in controlling the shrinkage variability dictates how much material can be removed, and how much grindstock must be left on the part for hard grinding after sintering.

Numerous binder systems are employed to consolidate ceramic powders and retain the green shape prior to sintering. They are generally organic compounds such as poly(ethylene glycol) (PEG), poly(vinyl alcohol (or acetate)) (PVA), acrylics, or preceramic polymers. The binder is critical to providing sufficient green strength for machining. Selection of the binder must also include consideration of its stability to temperature and humidity, and its spring-back or creep behavior during and after machining. Aqueous binder systems can absorb enough atmospheric water to change the dimensions of the green body, whereas some preceramic polymer binders are used with nonaqueous solvents and are relatively stable with humidity. In some cases the green bodies are stored in an environmentally controlled room with a stable temperature and humidity level. Yet another problem that can be encountered with certain binders is sticking of the machining debris to the tool bit and other surfaces. Ideally the machining debris should consist of small chips that can be easily vacuumed away. Very fine dust particles are difficult to clean up, while particles that are very large are usually associated with damaged surfaces.

Another option for adding the mechanical integrity needed for green machining is partial sintering, or bisque firing, of the green body. In this case, enough presintering of the ceramic must take place to provide green strength, since the binder is typically decomposed during bisque firing.

2. Green Machining Methods and Parameters

Green machining of ceramics has been demonstrated to be a versatile process, in that many types of machining procedures are feasible. Grinding, slicing, drilling, broaching, turning, and milling have all been successfully used to introduce many features into green bodies. The introduction of computer numerically controlled (CNC) machine tools has especially expanded these capabilities. Laser machining is also used for very detailed work such as vias in substrates for electronics, or scribing parts that will later be separated; laser machining is a less-developed and more expensive machining method.

Parameters such as feed rate, chucking/gripping loads, lathe speeds, removal rates, etc., are substantially different for green ceramics compared to those for metals or fired ceramics. The softer nature of green ceramics dictates a less aggressive approach, and requires a stiffer set-up and machine to maintain accuracy due to the relatively low elastic modulus of the green ceramic. Examples of table speeds and feed rates for a typical green alumina material suggested by Larson (1991) are shown in Table 1. In general, as the amount of material removed with each pass is increased, a lower surface speed should be used to allow more time for the removal of debris from the surface, and also to prevent overheating of the part and the tool bit or grinding wheel. The cutting angle between the green body and the tool bit will affect the way material is removed from the surface; usually a small cutting angle results in less damage or spalling of the surface. A pneumatic rather than a mechanical chuck can be used for fixturing the parts during machining to afford greater control over the load

Table 1
Typical feed rates and surface speeds for green machining.

Depth of cut		Feed rate		Surface speed	
(mm)	(in)	(mm rev^{-1})	(in rev^{-1})	(m min^{-1})	(sfm)
0.13	0.005	0.05	0.002	76	250
0.38	0.015	0.13	0.005	46	150
0.76	0.030	0.25	0.010	30	100
1.50	0.060	0.38	0.015	23	75

Source: Larson (1991).

applied to the part, especially when low loads are necessary to prevent damage to the parts. Extreme speeds during lathe operations may cause parts to fly apart if the strength of the binder is exceeded. In the case of hollow or cylindrical shapes, a mandrel can be inserted to provide extra support.

Owing to the abrasive character of the ceramic powders in green bodies, the wear of tool bits and grinding wheels must be monitored frequently. The most popular tool bit materials are tungsten carbide and diamond. The former is typically cobalt-bonded, while polycrystalline diamond or diamond grit may be bonded with metal, resin, or other adhesives. Although diamond tooling is about 10 times more expensive than carbide tooling, its lifetime and the reduced frequency of tool bit changes usually justifies the expense. Conventional steel tool bits do not tolerate abrasive ceramic powders well, and changes in tool bit size due to wear result in dimensional inaccuracies.

Inspection of green machined surfaces generally includes looking for chipping, pull-outs, grooving, and other undesirable artifacts. The extent of these types of damage can be reduced by modifying binder properties, machining parameters, or by adding more generous tolerances or a larger radius to sharp corners and complex features.

Finally, the issue of dust generation during green machining must be addressed. Dust collection systems, consisting of an industrial vacuum and dust filtration unit, are required. The machining debris not only impedes the accuracy and efficiency of machining, but will result in an increase in tool wear and damage to machine tools if not removed. It can also be a significant industrial hygiene issue if not addressed appropriately.

3. *Trends*

Developments in ceramic processing have the potential to reduce the need for green machining. Rapid prototyping technology (see *Rapid Prototyping*), for example, can produce components with internal cavities and complex shapes that are impractical to make with conventional powder consolidation methods, regardless of the availability of green machining. Injection molding (see *Injection Molding*) of ceramic parts can also yield near net-shape components that do not need to have features introduced by green machining prior to sintering. It is clear that the number of green body properties and machining parameters that can affect green machining behavior are substantial. Although this fabrication process may at first be considered trivial, it is clearly a process that requires significant levels of engineering expertise.

Bibliography

Islam M U, Campbell G 1993 Laser machining of ceramics: a review. *Mater. Manuf. Process.* **8**(6), special issue on advanced materials

Larson D 1991 Green machining. In: *Engineered Materials Handbook: Vol. 4. Ceramics and Glasses.* ASM International, Materials Park, OH, pp. 181–5

Lenk R 1995 Rapid prototyping von Keramischen Funktionsmustern. In: *1995 Annual Report of the Fraunhofer Institute.* Keramische Technologien und Sinterwerkstoffe, Dresden, Germany

National Bureau of Standards 1978 *The Science of Ceramic Machining and Surface Finishing*, Special Publication no. 562. National Bureau of Standards, Gaithersburg, MD

Richerson D W 1982 *Modern Ceramic Engineering.* Dekker, New York

Sheppard L M 1999 Green machining. *Ceram. Ind.* **149**(6), 65–9

T. E. Easler

Grinding

Grinding, defined here as particle size reduction by mechanical means, is an important step in the processing of ceramic powders. Typically, the objective is to produce powder with a specified particle size distribution appropriate for subsequent processing and fabrication operations. In some cases, blending of different components is an important secondary goal. Primary considerations include minimizing contamination and the efficient use of equipment and raw materials. Energy utilization can be an important factor in large-scale operations, especially for ultrafine grinding.

Size reduction in grinding processes results from fracture of particles under the action of applied stress. In the grinding of brittle solids, such as most ceramic powders, the applied stresses are usually compressive although some component of shear may also often be present. However, the actual fracture probably occurs at locations within the particles where shear or even tensile stresses are high. Fracture is initiated when the stress at some location in a particle exceeds its strength. The release of stored strain energy following initial failure leads to secondary cracking and catastrophic disintegration of the particle. Consequently, the result of brittle fracture is normally the production of a complete suite of fragments whose sizes range from quite close to that of the original (parent) particle to very fine dust. The distribution of sizes resulting from each single fracture in effect limits the product size distribution that can be obtained from the process. It is not possible to produce a narrower distribution by grinding alone.

Fracture of a particle requires that the applied stress exceeds its strength. If the stress is greater than the strength of the whole particle, massive fracture occurs as described above. It is also possible, however, for local stresses to exceed the strength of the material, at edges or corners of irregularly shaped particles, for example. In such cases, fracture itself might also be localized, leading to size reduction by attrition rather than by massive fracture. The products of this kind of breakage are the original (though slightly reduced) particle and a quantity of much finer material.

In grinding machines, stress is applied to the particles through the mechanical action of the machine. The magnitude of the stress acting on any individual particle varies, as does the strength of the particle itself. Consequently, some particles experience massive fracture, some undergo attrition, and some escape entirely. The relative proportions of these vary with the average magnitude of the applied stress. The rate of breakage depends, therefore, not only on the frequency of stress application but also on the magnitude of the stress, since higher stresses increase the probability of breakage in each stressing event. The products of breakage also depend on the stress magnitude. Higher stresses lead to more of the particles undergoing massive fracture as opposed to attrition, and higher levels of stored strain energy cause more intense cracking and fragmentation of the particles.

1. Breakage Characteristics and Mill Performance

The performance of a grinding process can be characterized by the specific rates of breakage and the primary breakage distributions. For particles in a given size class, the specific rate of breakage is defined as the rate, relative to the amount present, at which they are broken out of that size class. Since the strength of particles generally increases with decreasing size, while the volume stressed in each event decreases, breakage rates typically decrease with decreasing size. For a fixed milling condition, the forces applied to particles of different size are about the same. However, because of their smaller cross-section, fine particles are subject to greater stress and are, therefore, more likely to be broken when stressed. Larger particles, on the other hand, receive lesser amounts of stress and are less likely to be broken. The overall result is that with increasing particle size breakage rates first increase but then reach a maximum as they become less likely to suffer massive fracture at the available stress levels. Further increase in size then leads to a decrease in the breakage rate.

The primary breakage distribution is defined, in incremental form, as the mass fraction of the fragments resulting from a single breakage event that fall into a particular size class. In other words, it is the size distribution obtained when an "average" particle is broken. The distributions are often presented in cumulative form, i.e., as the fraction of fragments *finer* than a given size. Experimental studies have shown that breakage distributions vary with the energy available to the breakage event. High-energy breakage, such as that resulting from high-speed impact, produces very fine fragments, substantially smaller than the original particle. Lower energy levels, just sufficient to cause massive fracture, generally yield coarser distributions with fragments ranging in size up to perhaps one third of the mass of the original. Examples of these two types are given in Fig. 1.

Breakage distributions are often found to be normalized in the sense that, on a relative basis, they are independent of the size being broken. Breakage of large particles, where the applied stress may be insufficient to cause massive fracture, tends to be accompanied by an increase in the relative importance of the attrition mode and the distributions often exhibit a bimodal character. These breakage

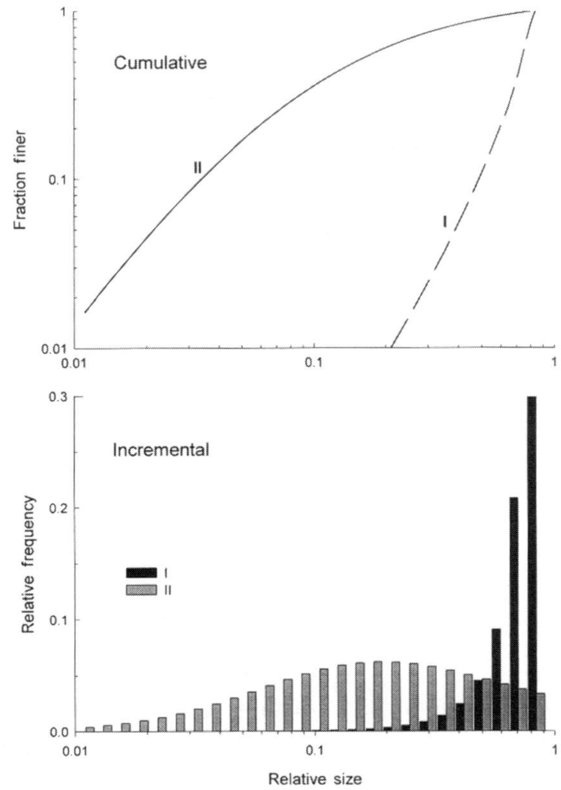

Figure 1
Examples of primary breakage distributions: curve I corresponds to low-energy breakage; II represents high-energy breakage.

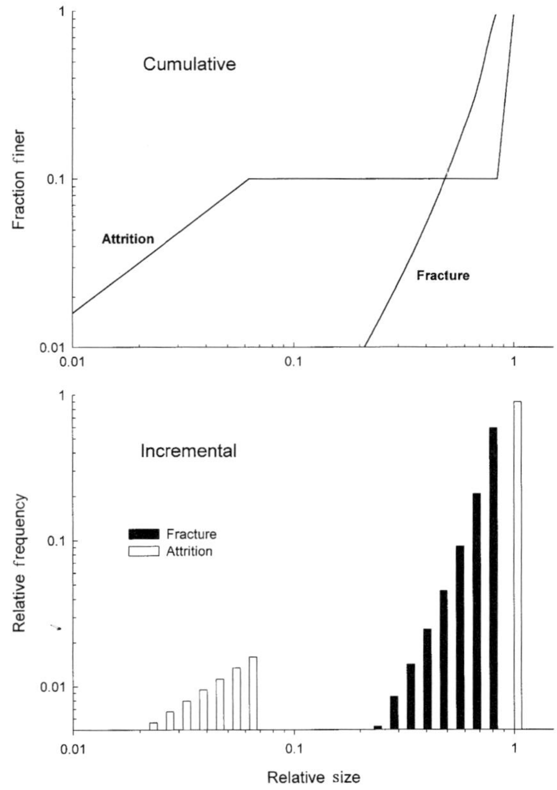

Figure 2
Effective primary breakage distributions for fracture (low energy) and attrition. Note that attrition fragments include the bulk of the original particle. The latter disappears entirely in the case of fracture.

distributions are not usually normalized with respect to those associated with the breakage of smaller particles, i.e., that undergo massive fracture. A comparison of effective breakage distributions corresponding to fracture and attrition is given in Fig. 2.

Deaggregation processes, where the feed particles consist of aggregates of much finer material, also give rise to bimodal breakage distributions. Individual breakage events produce some relatively large fragments of the original aggregate together with a substantial amount of much finer material consisting of very small aggregates (binary, ternary, etc.) and primary particles. Since deaggregation rates are usually much higher than breakage rates for (solid) primary particles, very little *particle* breakage occurs.

In principle, knowledge of the relationships between process variables, including mill size, and the breakage parameters provides the basis for process design and scale-up. In practice, of course, the parameters depend on both machine and material and some degree of empirical testing is always required. Nevertheless, a general understanding of how breakage parameters vary with conditions can be of substantial value as a guide to the use of such tests in equipment selection and process optimization.

2. Grinding Machines

Crushing and grinding equipment is available that can reduce particles in size from coarse (tens of centimeters) to very fine (submicron). The discussion here will be limited to grinding in the finer sizes, e.g., millimeters to micrometers and less. It should also be noted that the reduction ratio (feed size:product size) is limited for most machines and seldom exceeds about 10:1. In order to achieve higher reduction ratios, it is usually necessary to grind in stages.

Equipment suitable for fine grinding of brittle solids such as ceramic powders generally falls into one of several classes. Important examples include media mills, impact mills, and fluid-energy mills.

2.1 Media Mills

Media mills are the most widely used machines for fine grinding of powders. These machines contain individual loose grinding elements—the grinding media—which can consist of balls, rods, pebbles, beads, etc. The action of the mill leads to motion of the media elements, which collide with each other and with the walls of the grinding chamber. Grinding occurs when particles are caught, "nipped," in these collisions. Depending on the mechanisms used to induce media motion, this class of machines can be further subdivided into tumbling mills, vibratory mills, and stirred mills.

For each of these, the breakage characteristics are determined by the frequency of media collisions and by the energy associated with each collision. Collision frequencies depend on the motion of the mill and the number of media elements contained in it. The collision energy is also related to mill motion as well as to the mass of the individual elements. It follows that smaller media provide increased collision frequency but reduced breakage energy associated with each collision. As a consequence, breakage rates in media mills typically exhibit the kind of pattern illustrated in Fig. 3. For a fixed size of media, the rates initially increase with particle size because a greater volume is stressed in each collision. For very small particles impacted by coarse media the impact energy is sufficient to ensure breakage, while in the case of coarse particles the impact energy may be insufficient and only a fraction of such impacts will lead to breakage. Efficient operation of media mills depends on the proper matching of the media size to the size of the particles being ground. Size ratios of about 20:1 (media:particles) generally appear to be appropriate. Media density determines the impact energy and thereby the location of the maximum in the breakage rate–size relationship. Using a range of media sizes can extend the range of particle sizes that can be ground effectively. However, in order to obtain a large reduction ratio, it is usually best to grind in stages using progressively finer media in different mills or in different compartments of a continuous mill.

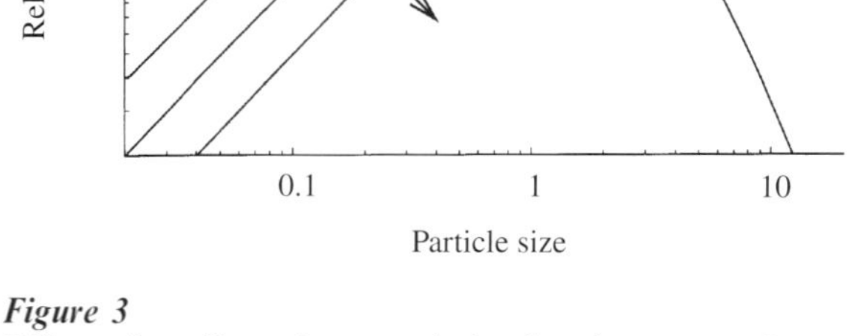

Figure 3
Effect of medium size on relative breakage rates for different particle sizes.

Since media mills generally operate by applying compressive stresses to particles, the relationships between breakage parameters and particle size tend to follow very similar patterns for different machines and operating conditions. Actual magnitudes, of course, vary substantially. For batch grinding systems, the set of breakage rates and breakage distributions for the different particle sizes present (or that can be produced) can be used to predict product size distributions as a function of grinding time. In the case of continuous grinding, additional information on the transport of particles through the mill is needed. Procedures for making such predictions are beyond the scope of this article and are well documented in the modern grinding literature (see the Bibliography). However, a useful, though very approximate, relationship that can be derived from a simplified treatment of the breakage process is that the required grinding time varies roughly with the reciprocal of the desired product size. Specifically:

$$t \approx \frac{x_f}{x_p S_f} \tag{1}$$

where, x_f is a characteristic size (e.g., median size, x_{50}) of the feed material, x_p is the corresponding size of the product, and S_f is the breakage rate for the feed particles.

The grinding energy in a properly designed media mill with appropriately sized media is usually sufficient to ensure massive fracture of the particles. However, the grinding action does not normally include high-speed impacts, so the breakage distributions typically take the form shown by curve I in Fig. 1.

(*a*) *Tumbling mills.* Tumbling mills are the simplest and most commonly used type. Typically they consist of a horizontal cylinder, partially filled with the media and rotated about its horizontal axis, as illustrated in Fig. 4. Essentially, these mills operate by raising the media through rotation of the mill and allowing them to fall back under gravity. Thus the grinding energy is supplied by, and therefore limited by, gravity. The motion of the media is also controlled by gravity. As the rotational speed is increased, the motion changes from a cascading type where individuals roll and bounce over each other, to the cataracting condition which involves projection of media out of the bed after which they remain in free flight for a period before falling back into the bed. Eventually, a speed is reached at which the outer layers of media begin to centrifuge. This is known as the critical speed of the mill and it can

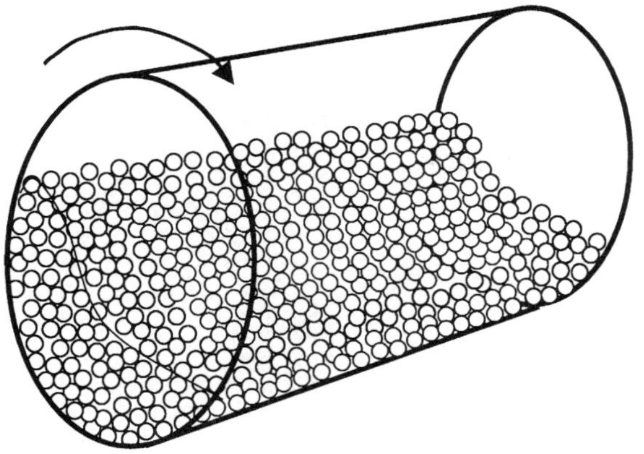

Figure 4
Tumbling ball mill (schematic).

be calculated, by equating gravity and centrifugal forces, from

$$N_c = \frac{60}{\pi}\sqrt{\frac{g}{2(D-d)}} \quad (2)$$

where N_c is the critical speed (rpm), g is the gravitational constant, D is the mill diameter, and d is the media diameter. Tumbling mills are normally operated at about 70–80% of the critical speed, which corresponds to maximum power input and is probably related to the transition between the cascading and cataracting regimes.

Media filling levels of about 50% of the mill volume are common in batch grinding; somewhat lower levels are required in continuous mills to allow for feed and discharge of powder. Powder loading is generally considered to be optimum when the voids in the media bed are just filled, i.e., the bulk volume of powder should be about 40% of the bulk volume of the media. Lower loadings allow too much direct media–media contact and wear while higher levels favor "cushioning" of media–media impacts and energy loss in rearrangement of particles. Fine grinding is very often carried out wet using slurry concentrations of up to 40% solids by volume. Wet grinding is generally considered to be more "efficient" than dry grinding, probably because of improved dispersion of the particles and reduced tendency for aggregation of fines. Grinding aids, typically surface-active agents, are available commercially for both wet and dry grinding. They probably function by improving particle dispersion.

Because of their wide application, design and scale-up criteria for tumbling mills are well established and standard test procedures, such as the well-known Bond Test, are available for evaluating specific materials. Tumbling media mills are quite effective for grinding to sizes in the 10–50 μm range, but often require inordinately long grinding times for size reduction into the micron range.

(*b*) *Centrifugal mills.* Centrifugal mills are, for the most part, a special case of tumbling-media mills, the difference being the addition of a centrifugal component to increase the effective gravitational force. The typical arrangement involves a planetary type of motion in which the mill shell rotates about its own axis, which is itself rotated about a central axis. In effect, the mill is equivalent to a simple tumbling mill operating in a centrifugal, rather than a gravitational field. In contrast to conventional tumbling mills, rotational speeds, which determine media collision frequency, are not limited by the onset of centrifugation when the centrifugal force due to rotation of the shell equals or exceeds that of gravity (i.e., at the critical speed). As the mill speed increases, the effective gravitational force also increases. Provided the ratio of mill speed to orbital speed falls within a specified range, determined by the geometry of the planetary system, media motion in the mill is essentially independent of the actual speed. Practical operating speeds are limited only by mechanical design constraints. The principal advantage of centrifugal mills is that substantially higher grinding forces can be obtained than with the conventional types, even with small mills and fine media. Consequently, mill capacity is substantially increased and grinding can be extended to finer sizes. The main disadvantages are in mechanical design and questions of reliability. For these reasons, centrifugal mills have found only rather limited applications in commercial-scale grinding.

(*c*) *Stirred-media mills.* Stirred-media mills generally consist of a cylindrical, stationary vessel filled almost completely by grinding media and agitated by means of an internal impeller. In contrast to the tumbling mills, in which shearing of the media bed and, consequently, the grinding action is largely confined to the surface layers of the bed, the entire charge in a stirred mill is in motion at all times. As a result, the inherent capacity of these mills is significantly higher than that of their conventional counterparts. Mills agitated at low speeds are usually oriented vertically since the grinding force relies, to some extent, on gravity. In high-speed mills, centrifugal forces dominate over gravity and orientation is more a matter of convenience. For the most part, the grinding energy is derived from the kinetic energy of the media. Increasing the agitation rate increases both the media collision frequency and the impact energy, which leads to higher breakage rates and also extends the range of particle sizes that can be broken. Because of the high energies involved, stirred mills using fine media are especially suitable for grinding into submicron size ranges. As with the other kinds of media mills, careful matching of the media size to the size of the particles being ground can be critical to efficient operation. Powder loading so as just to fill the void space between the media elements is appropriate for these mills also. Wet grinding is normal, both to ensure particle dispersion and to assist in the removal of the excess heat developed in the process. For grinding to very fine sizes, it is common practice to circulate the slurry through a heat exchanger. Ceramic liners, agitators, and media can be used to minimize contamination due to wear.

(*d*) *Vibratory mills.* Vibratory mills use oscillatory motion of the mill shell to agitate the media. As for the stirred mills, the active grinding zone encompasses the entire mill volume. The grinding energy is supplied by the inertia of the media and is not limited by gravity. In principle, high energy can be supplied to quite fine media, making these devices attractive for ultrafine grinding applications. By very careful matching of media size, powder size, and energy input (based on vibrational amplitude and frequency) it should be possible to achieve quite high grinding efficiencies. Unfortunately, mechanical design for reliability and low maintenance is not simple. Problems in these areas have tended to limit their large-scale application.

2.2 Impact Mills

Impact mills induce particle breakage by collision with moving parts of the machine itself. A widely used example is the hammer mill which typically consists of a set of hammers rotating, usually at high speed, in a cylindrical case as shown schematically in Fig. 5. The shell of the case generally takes the form of a grate through which product particles can exit the mill. Breakage can result from direct impact between particles and hammers, by shearing of large particles between the hammers and the grate, or by particle–particle collisions in the highly turbulent environment of the mill. The classifying action of the grate is limited to quite coarse particles and product sizes are typically considerably finer than the grate opening size.

Impact velocities in hammer mills are normally very high which leads to fine breakage distributions of the form shown as curve II in Fig. 1. Because of the high impact velocities employed, large particles are broken very effectively. At finer sizes, however, breakage is probably limited by aerodynamic factors; entrainment of the particles tends to reduce the probability of impact as well as the severity of those impacts that do occur. As a consequence, high-speed, mechanical-impact mills are ideal for reducing a relatively coarse feed to a considerably finer size in a single stage. They cannot, however, be used to reduce a complete batch of material into the micron size range.

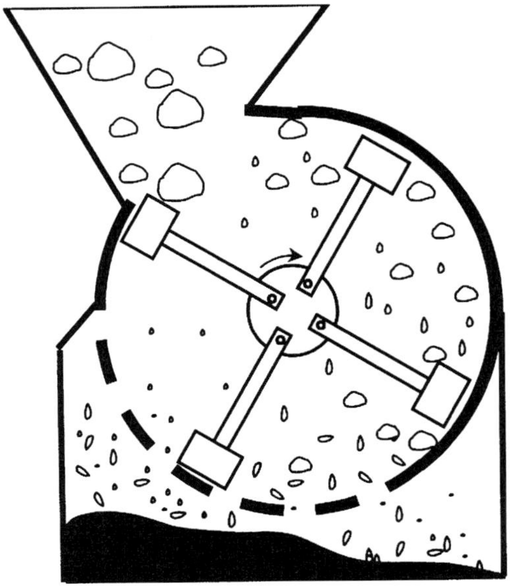

Figure 5
Hammer mill (schematic).

2.3 Fluid-energy Mills

Fluid-energy mills, also known as jet mills, rely on collisions in a stream of particles entrained in a high-velocity fluid, typically air or steam, to effect breakage. In the "pancake" type, the particle-laden stream is injected through peripheral jets into a flat, cylindrical chamber at extremely high velocity. The highly turbulent environment so generated leads to high-velocity impacts between particles and with the walls of the chamber. The systems are generally designed so as to remove fine particles with the fluid while coarser material is retained for further breakage. An alternative arrangement is the opposed jet type in which particles enter the chamber through jets arranged in direct opposition to each other, the idea being to promote breakage by particle–particle collisions rather than by impact against chamber walls. Designs of this kind are intended to reduce contamination due to wear of machine surfaces.

Fluid-energy mills are commonly used to produce particles in the 1–10 μm size range. Product size distributions are often relatively narrow which may be due in part to the loss of extremely fine (submicron) material in the product collection system.

3. Grinding Circuits

The efficiency of grinding systems and control over product size distribution can generally be improved through the use of a grinding/classification circuit in which the mill itself is connected to a classifier which removes product-size material from the circuit and returns oversize particles to the mill for further grinding. If undersize (i.e., product) particles can be removed from the system as quickly as possible, energy expenditure on over-grinding can be minimized and a more closely sized product can be obtained.

An example, based on computer simulation, illustrating the potential benefits of closed circuit grinding to obtain a product having a mass median particle size of 0.3 μm is given in Fig. 6. The curves labeled "open-circuit" represent the distribution obtained by grinding without classification while those labeled "closed-circuit" indicate the product size distribution that could be produced with an identical grinding mill operating in closed circuit with a classifier set to separate at a size of 0.5 μm. The circulation ratio, i.e., the ratio of recycle rate to fresh feed rate, would be 4.7 in this case. In other words, the actual feed to the mill would consist of one part fresh feed to 4.7 parts oversize material returned by the classifier. The substantial narrowing of the size distribution by including the classifier in the circuit is readily apparent. In addition, the production rate of the closed circuit is estimated to be 45% higher than that of the equivalent open circuit.

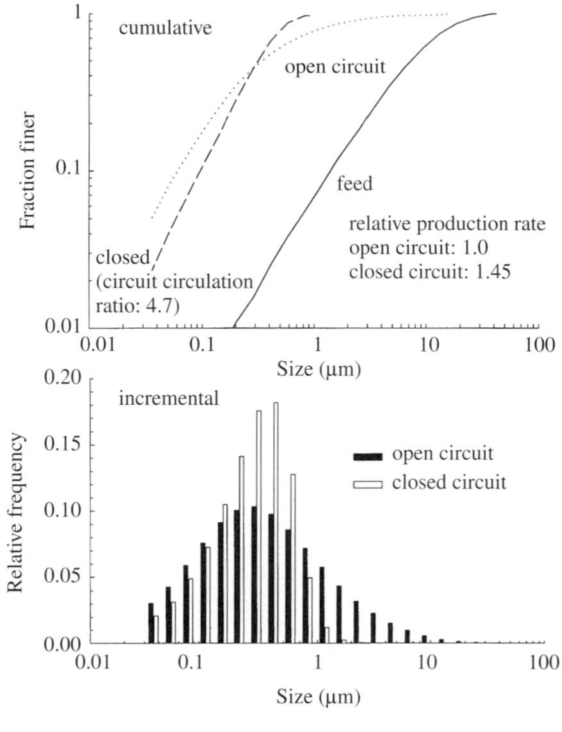

Figure 6
Simulation of fine powder production (mass median size: 0.3 μm) by open and closed circuit grinding.

Essentially all types of grinding mills can be incorporated into a closed circuit. Media mills can be operated continuously and connected to an external classifier. Impact mills and fluid-energy mills generally include some kind of internal classification action arising from selective product removal and can be linked to an external classifier as well. Closed-circuit grinding can also be applied to batch milling by periodically removing the mill contents, classifying them and returning oversize particles to the mill along with make-up fresh feed. Obviously, this approach is labor-intensive, but may be a practical alternative for small-scale production of high-value products. The procedure is also useful for closed-circuit evaluation and design and forms the basis for the Bond procedure used in industrial ball-mill circuit design.

4. *Equipment Selection*

The choice of a particular kind of mill for a specific grinding application depends on the material to be ground, the particle size of the feed and the desired product, the required capacity (throughput), and very often the amount of product contamination that can be tolerated. The various media and impact and fluid-energy mills described above can all be used to reduce feed particles in the 0.1–1 mm size range to products finer than 10 μm. Media mills generally provide the greatest flexibility with respect to product size range since grinding time can be controlled independently. Tumbling mills are effective for low-cost grinding in the coarser size ranges while centrifugal, vibratory, and stirred-media mills are more appropriate for the finer sizes. Impact mills and fluid-energy mills usually cause less contamination of the product, but are less easy to adapt to different product specifications. Scale-up criteria are widely available for tumbling ball mills but considerably less so for the other types.

Classification systems suitable for incorporation into grinding circuits include a variety of air classifiers capable of separation at sizes down to a few micrometers (albeit at increasingly reduced capacity and sharpness). Simple hydrocyclones can be used for wet classification down to about 10 μm and are highly effective for removing coarse particles from the product stream. However, entrainment of the finest particles in the liquid leads to misplacement of up to 30% of the fines into the coarse (recycle) stream.

5. *Limitations of Fine Grinding*

Size reduction by grinding is inherently constrained by the primary breakage distribution of the material being ground, which represents, in effect, the narrowest product size distribution that can be obtained from the system. Narrower distributions, e.g., "single size" fractions, can only be produced by separation of the product into discrete fractions.

Limits on overall product fineness can arise for several reasons. The asymptotic behavior of the process as exemplified in Eqn. (1) can lead to an apparent limit in that additional grinding seems not to produce significant further reduction in size. Agglomeration of fine particles, especially in dry grinding, can lead to real size limits since the grinding energy is expended in deformation of agglomerates rather than in particle breakage. There are also theoretical and experimental grounds for believing that there is a true limit to the size of particles that can be broken or produced by mechanical action. This "absolute" grinding limit seems to lie in the 10–30 nm range for solid particles of a brittle material.

Bibliography

Austin L G 1971 A review introduction to the description of grinding as a rate process. *Powder Technol.* **5**, 1–17

Austin L G, Klimpel R R, Luckie P T 1984 *Process Engineering of Size Reduction: Ball Milling*. AIME, New York

Cho H, Waters M A, Hogg R 1996 Investigation of the grind limit in stirred-media milling. *Int. J. Miner. Process.* **44-5**, 607–15

Crabtree D D, Kinasevich R S, Mular A L, Meloy T P, Fuerstenau D W 1964 Mechanisms of size reduction in comminution processes. *Trans. SME/AIME* **229**, 201–10

Herbst J A, Fuerstenau D W 1980 Scale-up procedure for continuous grinding mill design using population balance models. *Int. J. Miner. Process* **7**, 1–31

Hogg R 1981 Grinding and mixing of nonmetallic powders. *Am. Ceram. Soc. Bull.* **60**, 206–11

Hogg R 1999 Breakage mechanisms and mill performance in ultrafine grinding. *Powder Technol.* **105**, 135–40

Kwade A, Blecher L, Schwedes J 1996 Motion and stress intensity of grinding beads in a stirred media mill, Part 2. Stress intensity and its effect on comminution. *Powder Technol.* **86**, 69–76

R. Hogg and H. Cho

Growth of Shaped Crystals from the Melt

Crystals for modern engineering applications usually need to have the form of plates, rods, or tubes. Sometimes the design detail can be more complicated. When starting from cylindrical boules there is a loss of expensive material (often up to 90%) as well as the formation of structural defects during the machining needed to obtain the desired shape. Therefore, crystals of specified size and shape with controlled defect and impurity structure that allows one to use them as final products without additional machining are a great advantage.

Both theoretical and practical aspects of shaped crystal growth have been developed (Fig. 1). The main information concerned with all these aspects can be found elsewhere (Tatarchenko 1993, 1994, Cullen *et al.* 1980, Kalejs *et al.* 1986, Kalejs and Surek 1989).

1. Capillary Shaping Techniques

The problem of shaped crystal growth seems to be simply solved by profiled container crystallization just as in the case of casting. Indeed it is possible to realize this idea in the Bridgman technique (see *Crystal Growth from the Melt*) for growth, for instance, of crystals of silicon or CaF_2 with different shapes of crucible. However, such a solution is not always acceptable. Crucible material should satisfy a whole set of requirements: it should neither react with the melt nor be wetted by it. Even if these requirements are satisfied, perfect crystal growth is not ensured.

Therefore, the techniques of crystal lateral surface shaping without contact with container walls have to be considered as the best methods for shaped crystal growth. The main techniques of this type of growth are the Czochralski technique (CzT, see *Crystal Growth from the Melt*), the Verneuil technique (VT), and the floating zone technique (FZT, see *Crystal Growth from the Melt*). For all these techniques the final shapes and the dimensions of the crystals are controlled by the shapes and the dimensions of the meniscus of melt at the crystal–melt interface. The meniscus shape is in turn controlled by the melt surface tension–capillary forces. Therefore, these techniques are called capillary shaping techniques (CSTs).

2. Techniques of Pulling from Shaper

CSTs allow special shapes of the meniscus to be obtained. Sometimes, electromagnetic, ultrasonic, or other external factors are used to control the meniscus. Normally CSTs allow one to obtain only very simply shaped crystals (cylinders, plates, or tubes) and they have to be modified for the growth of more complicated shaped crystals.

Some possible schemes for the modification of the CzT are presented in Fig. 2. They differ from the standard CzT in two aspects: first, the presence of a shaper at the melt and second, the crystal can be pulled upwards as well as downwards. The following gives a brief history of shapers. The use of dies (plates, holes) was first attempted by Gomperz (1922), but it was only because of the work carried out by Stepanov (1959) and others (Tatarchenko *et al.* 1969, Tatarchenko 1973, 1993, 1994) that the technique was developed for various materials. In the literature all techniques of crystal growth from a shaper are called Stepanov techniques (STs). Subsequently, edge defined film fed growth (EFG) (LaBelle 1980) and many others were proposed. It is evident authors prefer to use their own names without serious analyses of the peculiarities of each technique in comparison to others. It has been suggested (Tatarchenko 1993, 1994) that all these techniques should bear the same name: techniques of pulling from shaper (TPS).

In fact, the main principle of all these techniques is that of limiting the free surface area and thus the intensity of the liquid free surface perturbations. As has been shown (Tatarchenko *et al.* 1969, Tatarchenko 1993, 1994), only two possibilities can be used for the shaping: the edges or walls of a shaper. This means that, in the theoretical analysis, two different boundary conditions have to be used: (a) fixing a line on the surface of the meniscus by the edges of shaper or (b) fixing the angle of wetting between the melt and the wall of the shaper. Special design features can characterize each scheme. (a) Pulling up or down; (b) using for the shaping only the walls (Figs. 2(a), (f)), only the internal edges (Figs. 2(b), (d)), only the external edges (Figs. 2(e), (j)), or a combination (Figs. 2(c), (g), (h), (i)) with positive (Figs. 2(b), (g)) or negative (Figs. 2(d), (e), (h), (i), (j)) pressure of the melt; (c) using a capillary feeding of the melt (Figs. 2(d), (e), (i), (j)), etc. It is necessary to emphasize that the negative pressure of the melt, especially with capillary feeding, is possible only if the shaper material is wetted by the melt.

The dimensions and the shape of the specimens being pulled by TPS depend on the following factors: first, shaper design; second, the pressure of feeding the melt to the shaper; third, the crystallization surface position with respect to shaper. The seed shape, as a rule, is not important because it influences only the length of the transition zone between the beginning of process and stable growth.

3. Variable Shaping Technique (VST)

The VST allows one to change completely the cross-section of the crystal (periodically if necessary)

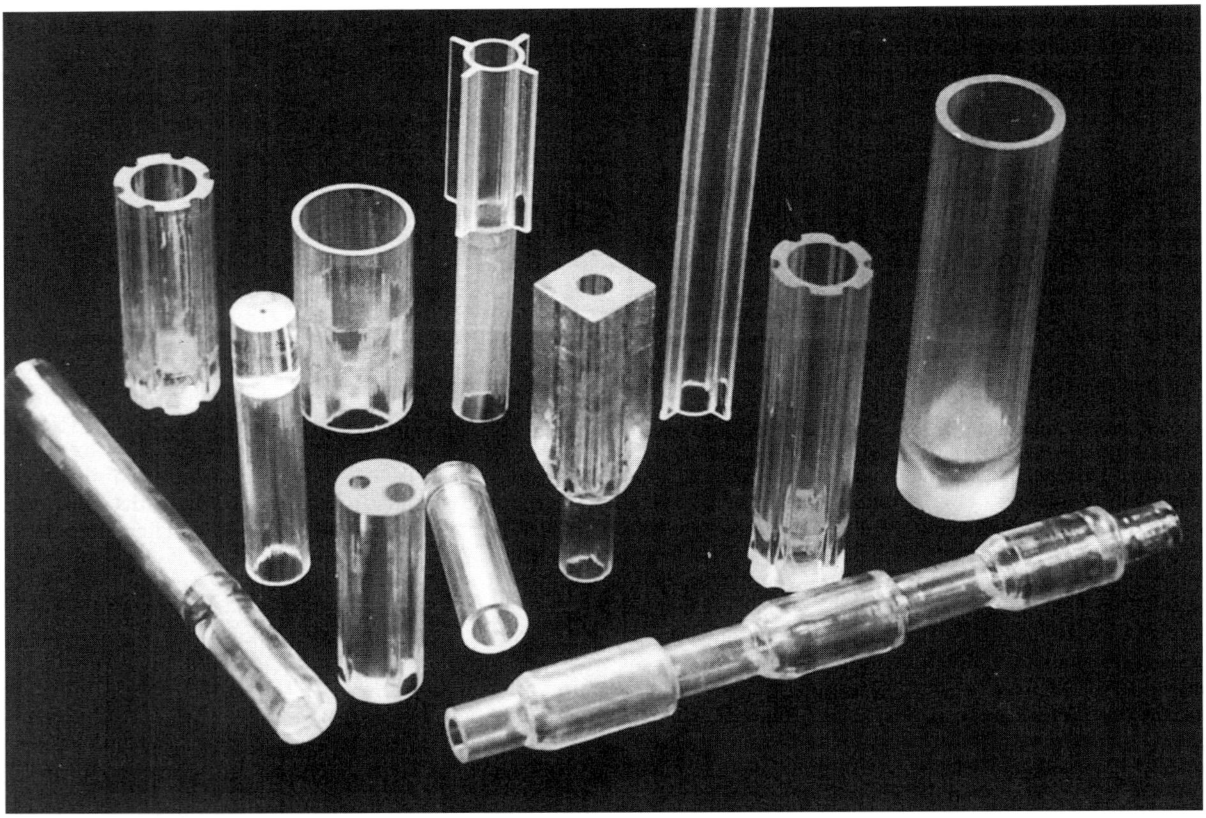

Figure 1
Monocrystalline sapphire profiles grown by TPS and VST by the authors with collaborators in the Solid State Physics Institute of Academy of Sciences of Russia.

during pulling (Fig. 1) and was developed by Borodin *et al.* (1986). The VST is based on controlling the melt mass flow towards the crystallization interface when passing to a new specific cross-sectional shape. In this case, different shaping elements on the same shaper, its free edges or walls, operate in turn.

4. *Dynamic Stability of Crystallization—The Basis of Shaped Crystal Growth*

As a crystal is not restricted by the crucible walls, its cross-section depends on growth parameters. Any deviations of growth rate, temperature conditions, etc., result in crystal cross-section changes. Therefore, the problem of crystal cross-section stabilization arises.

The modern system of diameter control by crystal weighing or detection of interface position, allow one to obtain cylindrical crystals by CSTs. However, another approach is possible: to analyze theoretically the dynamic stability of real crystal growth systems and to select, on the basis of this analysis, the stable regions. In the dynamically stable system, any perturbations of parameters tend to attenuate. As a result it has become possible to obtain crystals of specified shape and of controlled cross-section without special regulation.

5. *Lyapunov Set of Equations*

A comparative analysis of dynamic stability of crystallization processes was first carried out for the CzT and TPS (Tatarchenko 1973). The Lyapunov approach of stability analysis (Korn and Korn 1961, p. 282) that has now been developed in detail for many growth techniques (Tatarchenko 1993, pp. 19–43; 1994, pp. 1016–24) is recommended in this respect.

In the Lyapunov approach, all crystallization techniques are characterized by a finite number, n, of main variables (degrees of freedom), X_i, that can vary arbitrarily during crystallization. Thus, for the CzT they will include crystal dimension and crystallization front position, i.e., this technique has two degrees of freedom ($n=2$). For the FZT $n=4$: crystal dimension, melted zone volume, position of crystallization surface, and position of melting surface.

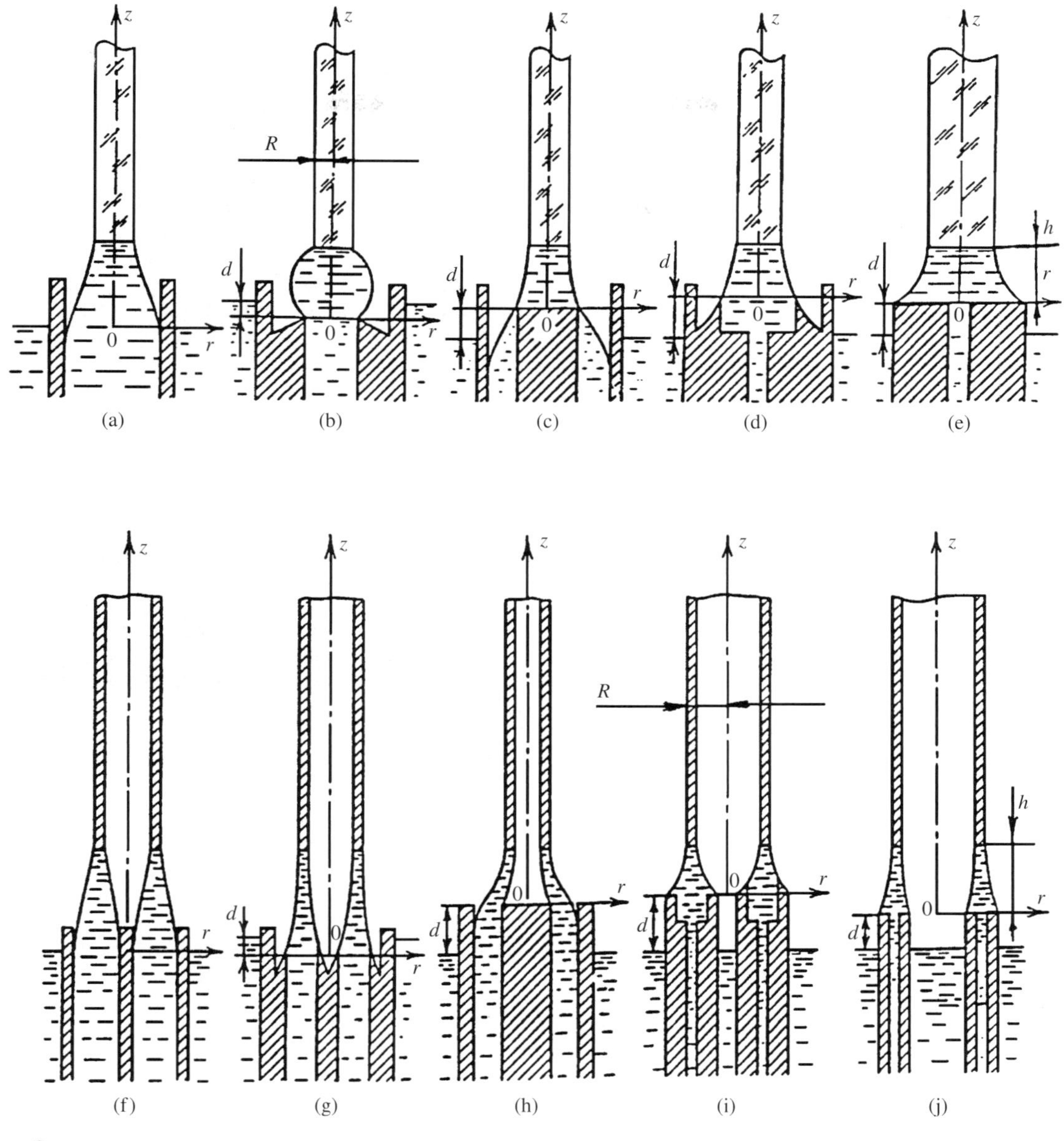

Figure 2
Schematics of pulling from shaper of (a–e) a crystalline rod and (f–j) a tube at various melt pressures, *d*. *R*, crystal radius; *h*, crystallization front height.

To realize a mathematical analysis of stability, a set of equations given by the derivative of each degree of freedom, X_i, with respect to time t as a function of the other n degrees of freedom, X_1, ... X_n, and parameters, C (temperature of melt, velocity of pulling, etc.), of the process has to be obtained:

$$\frac{dX_i}{dt} = f_i\left(X_1, X_2, \ldots X_n, \frac{dX_1}{dt}, \frac{dX_2}{dt}, \ldots \frac{dX_n}{dt}, t, C\right); \quad i = 1, 2, \ldots n \tag{1}$$

To find the explicit function f_i, a set of simultaneous equations must be solved. The set of equations includes: (a) the melt or solution flow equation (the Navier–Stokes equation) with the boundary conditions on the meniscus free surface (the Laplace capillary equation); (b) the continuity equation (the law of mass conservation); (c) the heat transfer equations for the liquid and the solid phases, with the equations of heat balance at the crystallization front and at the melting front as boundary conditions (the law of energy conservation); (d) the diffusion equation (impurity mass conservation); and (e) the growth angle constancy condition. The set of these equations is general for all the crystallization techniques under consideration (but it does not mean that each time all of them are used), while the specific features of each technique and crystallization scheme are accounted for by the set of boundary conditions and real values of the parameters included in the equations.

The condition $dX_i/dt = 0$ corresponds to the growth of crystals of constant cross-section with stationary crystallization front position and other parameters of crystallization (the system at the conditions of equilibrium):

$$f_i(X_1^0, X_2^0, \ldots X_n^0, t, C) = 0 \tag{2}$$

Only stable solutions of Eqn. (1) are of interest here. According to Lyapunov (Korn and Korn 1961, p. 284), the solutions of Eqn. (1) are stable if they are stable for the linearized set of equations

$$\frac{\partial X_i}{\partial t} = \sum_{k=1}^{n} \frac{\partial f_i}{\partial X_k} \delta X_k \tag{3}$$

where $\delta X_k = X_k - X_k^0$, and all partial derivatives are taken with $X_k = X_k^0$. The stability of Eqn. (3) is in turn observed when all the roots, S, in the characteristic equation have negative real components:

$$\det\left(\frac{\partial f_i}{\partial X_k} - S\delta_{ik}\right) = 0 \tag{4}$$

where δ_{ik} is the Kronecker symbol. The equilibrium will be unstable if Eqn. (4) has at least one root with a positive real component, and unless an imaginary number can be found among the roots, additional study including an approximation for the nonlinear terms in Eqn. (3) is required.

The results of stability analysis of CSTs show that the capillary stability (that depends on a scheme of formation of melt meniscus) and the heat stability (that depends on a temperature distribution at the zone of growth) play the main role in terms of dynamic stability.

6. *Static Stability of Meniscus*

Together with the dynamic stability of the crystal growth process, a static stability of the meniscus has to be provided. Static stability means that the melt meniscus exists for all values of crystallization parameters. An analysis of static stability of the meniscus can be realized on the basis of the method described by Funk (1970, p. 132).

7. *Growth of Shaped Crystals by the CzT*

An analysis of dynamic stability of the CzT shows that capillary stability is absent, therefore shaped crystal growth can be realized if heat stability is provided. Practically, this means that cylindrical crystals can be grown when the melt is superheated and the radius of the crystal is bigger than the melt capillary constant $a = (2\gamma/\rho g)^{1/2}$, where γ is the melt surface tension coefficient, ρ the melt density, and g the gravitational acceleration. Examples of shaped crystal growth by the CzT are provided by cylindrical crystals of silver (Bachmann *et al.* 1970) and of silicon tubes of very large diameter (Alioshin *et al.* 1990).

8. *Growth of Shaped Crystals by TPS*

An analysis of dynamic stability of TPS (Tatarchenko 1993, 1994) shows that both capillary and heat stability can be achieved if the shaper and zone of growth are well designed. However, this remains a real art. Table 1 gives a list of shaped crystals with the corresponding profiles, the shaping schemes, and relevant references (Antonov *et al.* 1977). More complete bibliographies are contained in Cullen *et al.* (1980), Kalejs *et al.* (1986), and Kalejs and Surek (1989).

9. *Growth of Shaped Crystals by the VT*

An analysis of dynamic stability of the VT shows that capillary stability exists only for crystals with a diameter less than two capillary constants a (e.g., for sapphire, $a = 6$ mm). Cylindrical crystals with diameters larger than 12 mm can be grown only if heat stability is provided. Practically, this means that it is necessary to obtain a special thermal field in the zone of growth. Crystalline sapphire and ruby tubes have been successfully grown under the dynamically stable conditions (Tatarchenko 1993, 1994).

10. *Some Advantages of Shaped Crystal Growth*

Besides the above-mentioned economizing in materials and the possibility of avoiding machining, shaped crystal growth has three main advantages with respect to other techniques: first, the crystals can have a more perfect structure; second, the distribution of impurities along the axis of the crystal can be more uniform; and third, the distribution of impurities on the cross-section of crystal can be controlled.

Any deviations of growth rate, temperature conditions, etc., result in crystal cross-section changes. For example, it is found experimentally that an increased number of inclusions and nonuniform impurity distribution are observed at pinch locations.

Table 1
Controlled profile crystal growth for different materials.

Substance	Crystal	Part of Fig. 2	Profile	Shaper material	Shaper–melt interaction	Author	Year
Semiconductors	Ge	a, b	Ribbon	Graphite	Not wetted	Tsivinski *et al.*	1963
		h	Tube	Graphite	Not wetted	Antonov *et al.*	1966
		b	Various profiles	Graphite, etc.	Not wetted	Stepanov *et al.*	1968
		e, h	Tube, ribbon	W	Wetted	Koptev *et al.*	1968
	Si	b	Ribbon	BN, quartz	Not wetted	Boatman *et al.*	1967
		e	Ribbon, tube	Graphite	Wetted	Ciszek	1972
		c	Ribbon	Graphite	Wetted	Tatarchenko *et al.*	1977
	GaAs	b	Ribbon	Graphite, BN, SiTi	Not wetted B_2O_3	Goundry	1966
	InSb	b	Ribbon	Graphite	Not wetted	Nosov *et al.*	1969
Dielectrics	CsI, LiF	c, h	Various profiles	Quartz	Wetted	Golzman	1958
	LiF	e	Ribbon	Ni	Wetted	Cocks *et al.*	1972
	LiF–NaCl	e	Ribbon	Ni	Wetted	Pollock *et al.*	1974
	AgCl	e	Ribbon, tube	Mo, glass	Wetted	Koptev	1976
	α-Al_2O_3	e	Various profiles	Mo	Wetted	Mlavsky *et al.*	1970
	β-Al_2O_3	e	Tube	Ir	Wetted	Cocks *et al.*	1974
	$LiNbO_3$	e	Plate	Pt, Ir	Wetted	Fukuta *et al.*	1975
	$Gd_3Ca_5O_{12}$	e	Plate	Ir	Wetted	Swartz *et al.*	1974
	$MgAl_2O_4$	e	Plate	Re	Wetted	Hurley *et al.*	1975
	Mn_2SiO_4	e	Rod	Pt	Wetted	Finch	1975
Metals	Pb, Zn, Sn, Cd, Al, Bi	b	Wire	Mica	Not wetted	Gompertz	1922
	Bi	b	Rod	Quartz	Not wetted	Kapitza	1928
	Metals, alloys	b	Various	Cast iron, etc.	Not wetted	Stepanov	1941
	Sb, Hg, Ga	b	Rod	Glass	Not wetted	Tsivinski *et al.*	1968
	Fe	c, h	Rod, tube	BNC	Not wetted	Tatarchenko *et al.*	1971
Composites	Al–steel, etc.	b	Various	Cast iron, etc.	Not wetted	Hohlov *et al.*	1976

Pinches themselves do not seem to cause defect formation, but crystal dimension change is indicative of growth conditions (mainly crystallization rate) deviating from equilibrium. From this point of view, a constant cross-section crystal growth may guarantee, to a certain extent, the absence of defects caused by sudden changes of crystallization rate.

If the impurity distribution coefficient is not equal to unity, during the growth the concentration of impurities changes along the axis of crystals. A solution to this problem is to replenish the melt at the zone of growth. This is obtained, for instance, by application of a floating (double) crucible in the CzT. However, the capillary feeding in TPS gives a more effective result. At the same time, the application of special systems of capillaries for the melt feeding allows one to control the distribution of impurities at the cross-section of crystals (Tatarchenko 1993, p. 242).

Bibliography

Alioshin A A, Bletscan N I, Bogatyriov S I, Fedorenko V N 1990 Silicon furnace components for microelectronic applications fabricated from shaped silicon tubes. *J. Cryst. Growth* **104**, 130–5

Antonov P I, Nikanorov S P, Tatarchenko V A 1977 The growth of controlled profile crystals by Stepanov's method. *J. Cryst. Growth* **42**, 447–52

Bachmann K J, Kirsch H J, Vetter K J 1970 Programmed Czochralski growth of metals. *J. Cryst. Growth* **7**, 290–5

Borodin V A, Sidorov V V, Steriopolo T A, Tatarchenko V A 1986 Variable shaping growth of refractory oxide shaped crystals. *J. Cryst. Growth* **82**, 89–94

Cullen G W, Surek T, Antonov P I (eds.) 1980 Shaped crystal growth. J. Cryst. Growth 50

Funk P 1970 *Variationsrechnung und Anwendung in Physik und Technik*. Springer, Berlin

Gomperz E V 1922 Untersuchungen an einkristalldrahten. *Z. Phys.* **8**, 184–90

Kalejs J P, Surek T (eds.) 1989 Shaped crystal growth. J. Cryst. Growth 104

Kalejs J P, Surek T, Tatarchenko V A (eds.) 1986 Shaped crystal growth. J. Cryst. Growth 82

Korn G A, Korn T M 1961 *Mathematical Handbook for Scientists and Engineers*. McGraw-Hill, New York

LaBelle H E 1980 EFG, the invention and application to sapphire growth. *J. Cryst. Growth* **50**, 8–17

Stepanov A V 1959 A new technique of production of sheets, of tubes, of rods with different cross section from a melt. *Zhurnal Tekhnicheskoi Fisiki*. **39**, 1302–7 [Transl. in *Sov. Phys.—Tech. Phys.*]

Tatarchenko V A 1973 Influence of capillary phenomena on the stability of the crystallization process during the pulling of shaped specimens from the melt. *Fiz. and Khim. Obrabotki Mater.* **6**, 47–50 [*Sov. Phys. Chem. Mater. Treatment*]

Tatarchenko V A 1993 *Shaped Crystal Growth*. Kluwer, Dordrecht, The Netherlands

Tatarchenko V A 1994 In: Hurle D T J (ed.) *Handbook of Crystal Growth*. North-Holland, Amsterdam, Vol. 2b

Tatarchenko V A, Saet A I, Stepanov A V 1969 Boundary conditions of capillary shaping at crystallization from melts. *Izvestiya Akademii Nauk SSSR, Seriya Fizicheskaya* **33**, 1954–9 [Transl. in *Bull. Acad. Sci. USSR, Physical Series*]

V. A. Tatartchenko

H

Heat Treatment

Heat treating is a process in which metal is heated and cooled under tight controls to improve its properties, performance, and durability. Heat treating can be used to control the microstructure of the metal to make it easier to process, more formable, or more machinable; or to impart improved properties and performance to make it more durable, more corrosion or wear resistant, stronger, or more fatigue resistant.

All metals and alloys can be heat treated. Aluminum alloys are annealed or solutionized, quenched, and aged hardened. Copper, brass, and bronze alloys are also heat treated by annealing as well as age hardening. Steels are heat treated by annealing, quenching, and tempering, or age hardening. In this article these important heat-treating processes of annealing, solutionizing, age hardening, and controlled cooling will be presented in terms of alloy systems and process design and control requirements. The discussion is organized in terms of phase diagram features and phase transformation morphology and kinetics.

1. Single-phase Alloys

Single-phase alloys like OFHC copper, α-brass, and austenitic stainless steels are strengthened by cold work. Frequently these alloys are cold worked and require annealing to continue cold working or to control their mechanical properties. The annealing of alloys that do not undergo a phase transformation on heating or cooling is discussed below.

During annealing of single-phase alloys the cold worked metal undergoes three distinct structural changes (Reed-Hill and Abbaschian 1994): recovery, recrystallization, and grain growth. At relatively low temperatures these alloys recover many of their physical properties that changed as a result of cold working. The electrical conductivity and heat capacity are restored to their well-annealed values. This recovery is due to the increased mobility of the dislocations that were generated during cold work as the temperature increased. With this increased mobility these dislocations can move and be destroyed. The decrease in dislocation density allows the physical properties to be recovered. No change in grain morphology or size occurs during the recovery process.

As the temperature of the cold worked alloy increases recrystallization can occur. In the cold worked alloy the grain morphology has become distorted, typically elongated, during the bulk deformation of the metal. These grains may have become lenticular or rod-like in shape. In these cases the grain boundary surface area and therefore grain boundary surface energy may be much greater than an alloy with equiaxed grains. During recrystallization new equiaxed grains are formed at the triple points of the intersection of the deformed cold worked grains. This recrystallization process occurs by nucleation process and growth processes and takes time to complete.

Figure 1 shows the fraction of a cold worked copper alloy that has recrystallized as a function of time for several temperatures. The shape of these curves is usually termed sigmodal and can be represented mathematically by the Mehl-Avarami equation:

$$F = \exp(-kt^n) \qquad (1)$$

where k and n are temperature-dependent constants and F is the fraction of the structure that has recrystallized.

While not apparent from Eqn. (1) and Fig. 1 each alloy system will have a practical recrystallization temperature. This temperature can be calculated

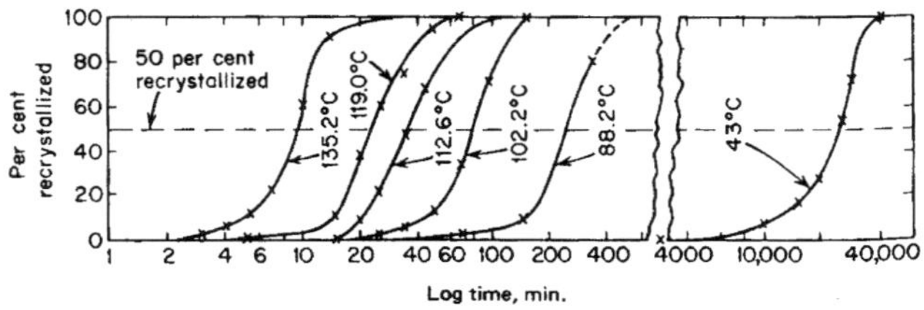

Figure 1
Isothermal recrystallization curves for pure copper cold-rolled 98 percent.

from the Arrhenius-type equation that relates the time to achieve a fraction recrystallized (i.e., 90% recrystallized) as a function of temperature:

$$1/t = (1/t_0)\exp(-Q/RT) \qquad (2)$$

where t is the time required to achieve 90% recrystallization, t_0 is a constant, Q is the activation energy for the process, R is the gas constant, and T is the absolute temperature. Recrystallization is frequently discussed in terms of the recrystallization temperature for the alloy. Typically the recrystallization temperature is approximately 40% (it may vary between 30% and 70% of the absolute melting temperature of the alloy) of the absolute melting temperature of the alloy. The recrystallization temperature is the temperature at which an alloy with a certain fraction of cold work will recrystallize in a reasonable time.

After the new equiaxed recrystallized grains have formed they will grow. After recrystallization is complete the grain diameter of the alloy will increase with time:

$$D(t)^2 - D_0^2 = Kt \qquad (3)$$

where D_0 is the initial grain size, $D(t)$ is the grain size at some time t, and K is the grain growth proportionality coefficient, being an exponential function of temperature:

$$K = K_0 \exp(-Q/RT) \qquad (4)$$

where K_0 is a constant that depends on the alloy system composition.

As clearly seen in Eqns. (1)–(4) the annealing behavior of cold worked alloys (i.e., percent recrystallization and grain size) is a function of the annealing temperature and time. Typically for annealed alloys the percent recrystallization and grain size are part of the specifications. Therefore, in practical terms the important heat treatment process parameters that must be controlled to obtain the specified percent recrystallization and grain size are temperature and time. In general, the control strategy for annealing heat treatment focuses on measurement and control of temperature and time.

2. *Two-phase Alloys*

2.1 Age Hardening: Aluminum–Copper Alloys (Solutionizing, Quenching, Aging)

Age or precipitation hardening heat treatment can be used to very effectively and predictably strengthen two-phase alloys that can be heated into a single-phase region on the phase diagram presented in Fig. 2, but can transform into a two-phase microstructure on cooling owing to the solubility decrease of the solute.

The alloy containing 4 wt.% copper can be used to describe the age hardening heat-treatment process. First the alloy that may have either a cast or wrought structure is heated to a temperature in the single-phase α region but below the eutectic temperature

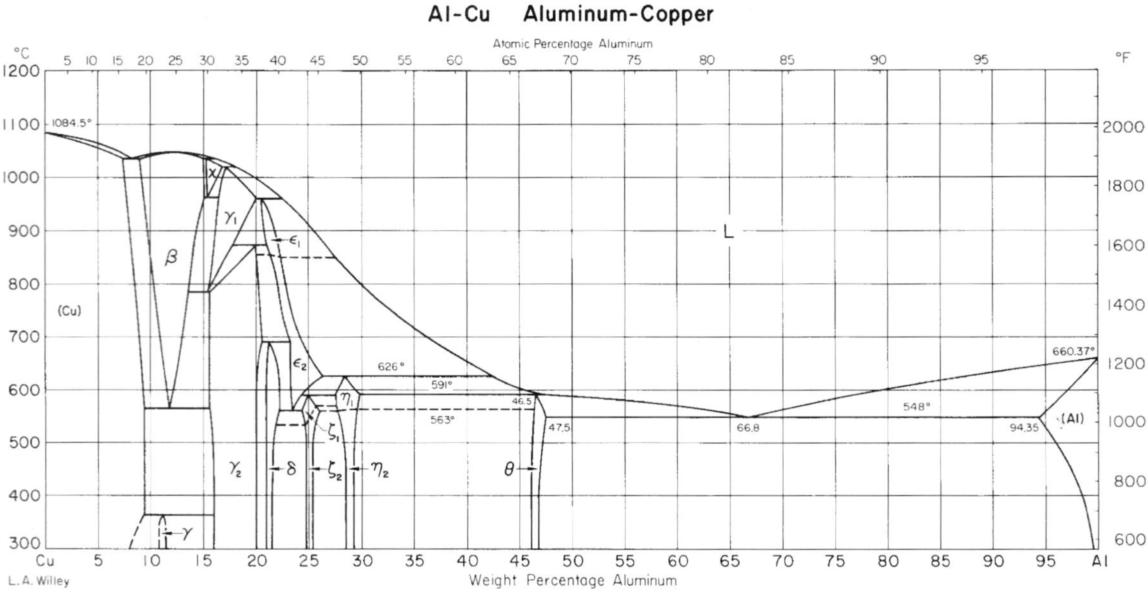

Figure 2
Aluminum–copper equilibrium phase diagram.

for the alloy system. This temperature control is necessary to solutionize the single-phase structure without the danger of incipient melting owing to composition segregation in the alloy. The time required for the solutionizing heat treatment is a function of the starting microstructure of the alloy in terms of the composition segregation in the alloy and the size and distribution of any second phases. The larger the second phase precipitates and the larger the amplitude and wavelength of composition variations the longer the time required for the solutionizing heat treatment. This segregation is more severe for cast alloys, where interdendritic eutectic structures can develop, than for wrought alloys. In addition to the solutionizing process described above the alloy will also be annealed and undergo recovery, recrystallization, and grain growth.

After the solutionizing heat treatment has been completed, the alloy must be quenched from the solutionizing temperature to retain the single-phase structure. This quenching is usually accomplished by immersion in water. While water quenching cools the alloy rapidly enough to avoid any decomposition of the matrix phase this rapid quench can cause residual stresses and distortion of the part being heat treated. Quenching in gases like argon, nitrogen, or helium may allow a more controlled cooling rate that would reduce residual stresses and distortion while avoiding the decomposition of the solutionized phase. After quenching the alloy will exhibit the properties of a well-annealed structure including high ductility and low strength and hardness. These alloys must be aged to increase their strength. Some alloys can be naturally aged at room temperature, while many alloys require aging at a higher temperature in the two-phase region of the phase diagram. The typical aging behavior of an age hardenable aluminum alloy is shown in Fig. 3, where the yield strength of the alloy is plotted as a function of aging time for several temperatures.

As seen in Fig. 3, the strength of the alloy increases with time to a maximum value followed by a slow decrease in strength. At higher temperatures the time to reach the maximum value is shorter than at lower temperatures while the maximum value of the strength is higher at low temperatures. Initially during the aging process the second phase nucleates in the matrix of the solutionized phase. This second phase, Al_2Cu in the case of the aluminum–copper system, precipitates coherent with the matrix. This coherent precipitation of a very small phase is very effective in developing a strain field around the precipitate. This strain field acts to inhibit dislocation motion and therefore increases the strength of the alloy. However, with time these precipitates grow by diffusion of the solute elements and become incoherent with the matrix. In addition, as the precipitates grow their number decreases as the large second-phase precipitates consume the smaller precipitates. This process reduces the strain fields around the second-phase precipitates and therefore reduces the strength of the alloy.

The aging kinetics as a function of temperature can be modeled using the same Arrhenius-type equation as Eqn. (2). These processes can both be modeled using this equation when the processes are diffusion controlled.

2.2 *Eutectoid Transformations*

The next phase diagram feature is best represented by the Fe–Fe_3C metastable phase diagram, as shown in Fig. 4. The feature of discussion in this section is the invariant eutectoid transformation of austenite (γ) to ferrite (α) and cementite (Fe_3C) at 723°C: $\gamma \rightarrow \alpha + Fe_3C$.

During the heat treatment of these steel alloys the alloy is heated into the austenite single-phase region and solution annealed. As discussed in the previous section this heat treatment is conducted for a time sufficient to transform the entire microstructure into a homogenous single-phase solid solution. In this case typically the starting microstructure contains no austenite and the ferrite and carbides must transform and dissolve to form the single-phase austenite solid solution. As is discussed in detail in other articles of this encyclopedia the phase transformations that occur on the cooling of the austenite control the microstructure and therefore the properties of these steels.

The decomposition of austenite to a variety of ferrite and carbide microstructural features is a function of the cooling rate. The time–temperature–transformation (TTT) diagram for the eutectoid steel is shown in Fig. 5, where the relationships between the cooling curve (i.e., temperature vs. time) and the phase transformations and microstructural features that develop are presented. For the transformations of austenite to ferrite and carbide slow cooling creates a microstructure of alternating layers or plates of ferrite and cementite. This microstructural feature is called pearlite and represents the typical morphology that develops when a single phase of either solid or liquid transforms to two different phases. Steels with this microstructure are ductile and tough with limited strength.

Alloys that are cooled to a temperature below the "nose" of the TTT diagram and held to isothermally transform, will transform to a finer structure called bainite. Bainite is generally stronger and less ductile than pearlite. In general for these diffusion-controlled phase transformations from austenite to ferrite and cementite the finer the features in the microstructure the stronger and less ductile the steel.

3. *Martensitic Transformations*

On rapid cooling many alloys including steels and titanium alloys will transform to an unstable phase

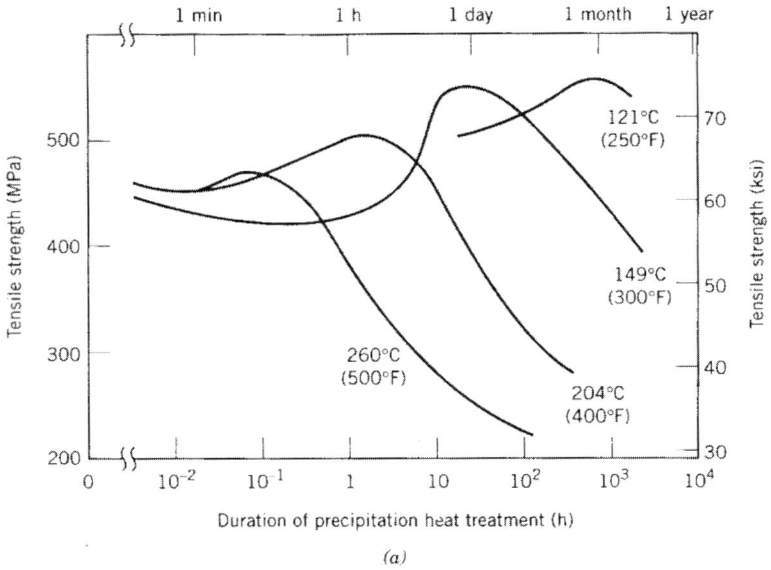

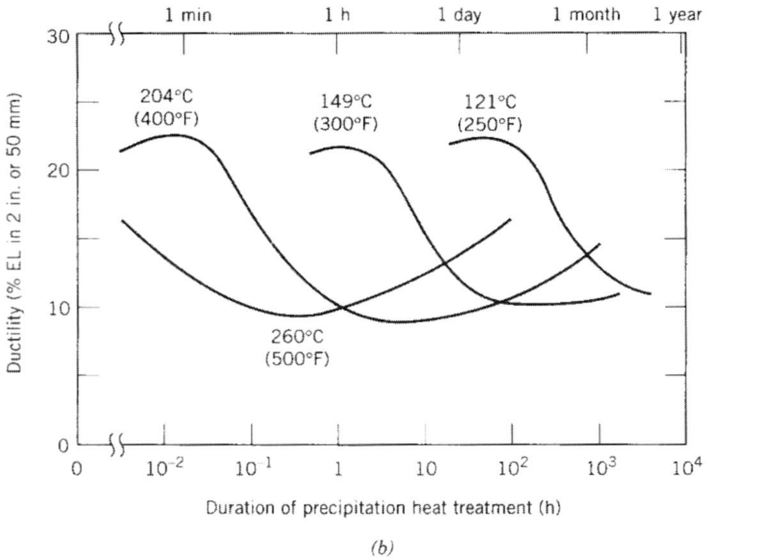

Figure 3
Precipitation hardening characteristics of aluminum alloy 2014: (a) tensile strength, (b) ductility.

by a martensitic-type phase transformation. These martensitic phase transformations do not occur by a diffusion-controlled mechanism but by a shear transformation with the atoms changing their relative position in the lattice by only slightly shifting position. Martensitic steel has a body-centered tetragonal crystal structure with a c/a ratio greater than unity. This transformation is dependent on temperature and occurs at cooling rates rapid enough to avoid the diffusion-controlled transformations. The onset and completion of martensite formation are presented as the isothermal lines on the TTT diagram of Fig. 5 for the eutectoid steel.

Martensitic structures possess very high strength, low ductility, and very low toughness. To increase the toughness these as-quenched martensitic alloys must be given a tempering heat treatment. Tempering is accomplished by heating the as-quenched martensite to a temperature below the eutectoid temperature for the alloy but high enough to allow some

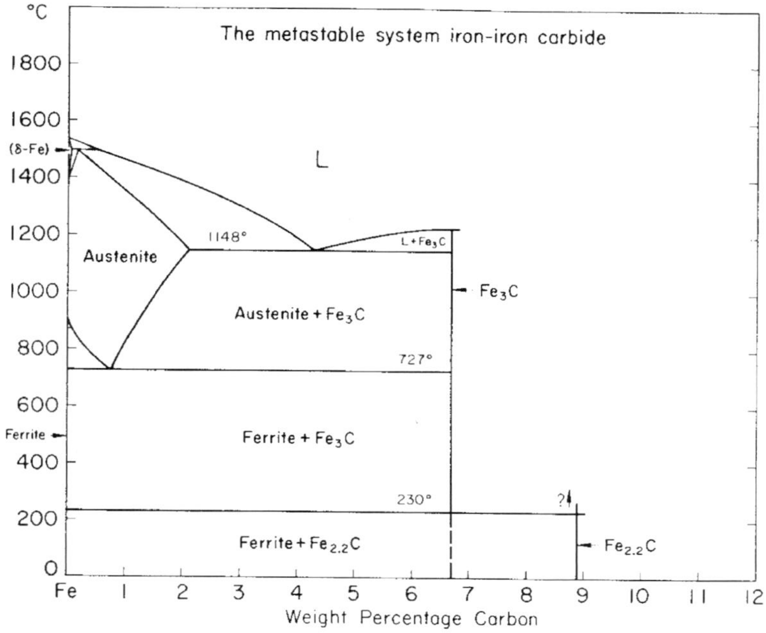

Figure 4
Iron–iron carbide metastable equilibrium phase diagram.

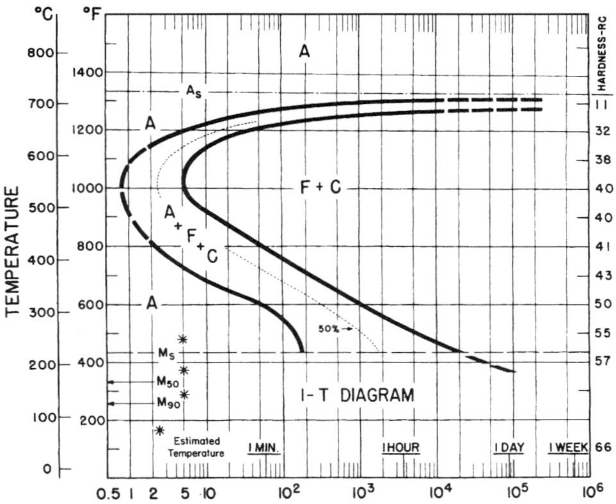

Figure 5
Time–temperature transformation diagram for eutectoid steel.

controlled phase transformation in the alloy. For a steel alloy with as-quenched martensite with a c/a ratio of 1.15, the tempering occurs by the decomposition of the martensite to a carbide containing iron and other carbide-forming elements like chromium and martensite with a lower c/a ratio. In many cases this precipitation of carbides may actually increase the strength of the steel slightly while greatly increasing the toughness. This transformation mechanism is similar to the reactions that occur in the aging heat treatment of precipitation hardening alloys.

4. Summary

The heat treatment of metallic alloys can be effectively used to obtain the mechanical properties that are desired for the alloy. Many of the phase transformations that occur are controlled by the temperature and time of the heat treatment. The precise control of temperature and time is necessary to produce high-quality parts and components with the specified properties.

The cooling rate from a high-temperature heat treatment is also a very important process parameter. As discussed for steels, slow cooling will yield ductile and tough pearlitic microstructures. Quenching will produce high-strength and low-toughness martensitic microstructures that may be tempered to produce steels with a combination of high-strength and high-toughness tempered martensitic microstructure.

It is the understanding and control of temperature and time of heat treatments and the cooling rates from heat treat temperatures that will ensure the production of high-quality alloy components.

Bibliography

ASM 1973 *Metals Handbook: Vol. 8. Metallography, Structure and Phase Diagrams.* American Society for Metals, Metals Park, OH

Callister Jr. W D 1997 *Materials Science and Engineering: An Introduction*, 4th edn., Wiley, New York

Reed-Hill R E, Abbaschian R 1994 *Physical Metallurgy Principles*, 3rd edn., PWS, Boston

United States Steel 1963 *I–T Diagrams Isothermal Transformation of Austenite in a Wide Variety of Steels.* United States Steel, Pittsburgh, PA

R. D. Sisson Jr.

Hot Rolling

A revolution is in progress in the materials industry as a result of strong global competition, spurred on by increasing concern for the environment. The two thrusts driving the materials revolution are the need to tailor materials to meet specific properties at low cost, and the need to minimize the impact of materials production on the environment. Although metals and their alloys have not always been included in the lexicon of Advanced Materials, the technologies that have been brought to bear on the production of a wide range of metals are indeed highly sophisticated. Unprecedented advances have been made in the processing of metals to yield superior properties at lower costs, which in combination with recyclability, assures their dominance on the global scene in the next century. In 1998, the worldwide production of crude steel, which constitutes ~95% by volume of the total output of metals, was 760 million tonnes and nearly half of the output was in the form of sheet and strip. Steel sheet and strip has a wide range of application in numerous sectors, such as automotive, bridges, pipelines, packaging, tanks, utility poles, and in industrial and residential construction. In tonnage, aluminum is second in volume to steel, and primary aluminum production is estimated to be ~17 million tonnes. Over 85% of fabricated aluminum products are rolled sheet and foil. A major portion of aluminum sheet is utilized to produce can stock for beverage containers, and it is also used in the construction, transportation, printing industry, and industrial distribution markets. Other applications include the production of foil for household and commercial packaging. Impressive gains have been made in the area of recycling of these two metals. The North American steel industry's overall recycling rate is ~68%, with the rate approaching 96.5% for automobiles. On a worldwide basis, 80% of the beverage cans are made of aluminum with a recycling rate of 55%, while the automotive industry recycles 85–90% of the aluminum used in vehicles.

Both steel and aluminum sheet/strip are produced by hot and cold rolling, continuously cast slabs. In the case of steel, molten metal is produced from pig iron in a steelmaking plant, or from remelted scrap in an electric arc furnace facility. Liquid steel is continuously cast into slabs, which are torch cut into specified lengths. Conventionally cast slabs have a thickness of ~250 mm, while thin slab casting processes, produce semi-finished slabs, 50–90 mm in thickness. The slabs, which are 6–10 m long, are subsequently reheated in a furnace to 1200 °C prior to hot rolling. Thick slabs are first converted to transfer bar, 25–40 mm thick in a rough rolling mill. The advantage of newer thin slab casting processes lies in the elimination of this step. Finish rolling of the transfer bar, is conducted on a five to seven stand tandem mill, to produce 2–10 mm thick strip or sheet. The material, which exits the finishing roll, stands at temperatures ranging between 800 °C and 1000 °C, is control cooled by water sprays on the runout table and subsequently coiled. The hot band is marketed in the as-rolled conditions for certain applications, or cold-rolled to achieve a further reduction in thickness, better dimensional tolerance, and specific properties. Following cold rolling, an annealing operation is usually conducted to restore the formability of the material. Final processing included prefinishing or precoating. The production of aluminum/sheet/strip begins with the direct chill (DC) casting of large ingots, which are then homogenized. The homogenized ingots processes in a hot breakdown reversing mill to slab gauge and is subsequently hot-rolled, much like steel, in a continuous hot-tandem mill to transfer gauge. The temperature range for deformation in the hot tandem mill is 300–500 °C, and there are usually three to four stands. Following hot rolling the sheet may be annealed and further cold rolled to final gauge. The process may be followed by a final annealing operation depending on the end application.

The automotive industry's demand for materials with properties aimed at reducing vehicle weight has generated intense competition between the aluminum and steel industry. In response to the competition from aluminum in applications that traditionally were the domain of steel, such as the outer body of automobiles, the steel industry initiated the Ultra Light Steel Auto Body (ULSAB) project. This is a $22 million effort involving steel companies from 18 countries aimed at designing an ULSAB from high strength steels by employing special design techniques to optimize the gauge of steel utilized in critical locations. High-strength, low-alloy (HSLA) steels, with yield strengths approaching 600 MPa are widely used in the automotive industry. These low carbon steels contain small concentrations of titanium, vanadium, and niobium (columbium). These elements precipitate out during continuous casting, hot rolling or coiling, significantly increasing the strength of the steel above plain carbon grades. One draw back is the lower elongation that accompanies the increase in strength. Interstitial-free steels, which have extremely low carbon levels, such as 20 ppm, together with small amounts of titanium and nitrogen, which tie up the carbon, are extremely formable, although they have low yield strengths in the range 150–200 MPa. Phosphorous, manganese, and silicon, in various concentrations have been added to these steels to increase their yield strengths, without a serious loss in formability. Bake-hardenable steels have been developed to achieve higher strengths, without the penalty in elongation during forming. These steels are designed to retain a small concentration of carbon in solution, ~20 ppm, during rolling and subsequent forming. During the paint-bake cycle, the carbon in solution causes a strain aging effect which results in as much as 65 MPa increase in yield strength down the line, making it extremely dent resistant and suitable for door outer panels, hoods, and fenders. New steels, with higher strength than the HSLA steels, are under development, and include dual phase steels and transformation-induced plasticity steels (TRIP).

In response to the opportunity to capture a significant fraction of the automotive market, aluminum companies have launched an aggressive campaign. In the USA, the Partnership for a New Generation of Vehicles Program is aimed at developing an aluminum-based vehicle weighing <900 kg with a fuel efficiency >80 mpg. Numerous aluminum alloys are being developed for application in the production of aluminum vehicles, and there are several vehicles on the road. Work has been conducted on 5000 and 6000 series aluminum alloys to achieve the desired strength and formability.

The tolerance range for mechanical properties, such as strength and formability, for automotive sheet are extremely stringent, as are dimensional tolerances of thickness, flatness, and crown. The mechanical properties of cold-rolled material are strongly influenced by the microstructural changes that occur during hot rolling. Traditionally schedules for hot rolling for new grades have been designed by physical simulation of the rolling process in the laboratory to achieve optimum mechanical properties. In translating this knowledge to production, costly on-line experimentation is often required to duplicate the laboratory results, since off-line simulations rarely match the complex thermomechanical history experienced by the sheet or strip during industrial rolling. With the rapid growth of computer power in the 1980s and 1990s and the demand for newer grades of both steel and aluminum alloys with improved properties, mathematical modeling has emerged as an effective tool for simulating microstructural changes that take place during hot rolling. The primary aim of the effort is to predict mechanical properties of the metals with the objective of optimizing the process parameters.

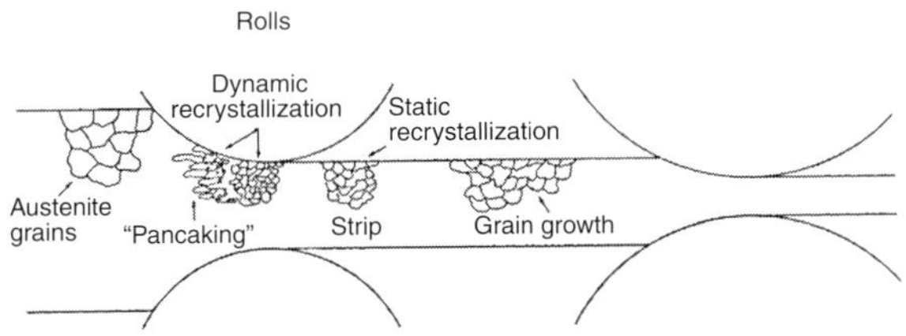

Figure 1
Schematic diagram of microstructural changes occurring during hot rolling.

1. Microstructural Evolution During Hot Rolling

1.1 Microstructural Phenomena

During hot rolling the microstructure and mechanical properties of the material change with its thermomechanical state, determined by composition, percent reduction, strip thickness, strip speed, and heat transfer. For steels, hot rolling is generally conducted at 850–1200 °C, in the austenite phase, while for aluminum it is conducted at 350–500 °C. During each rolling pass, strain is imparted to the material which causes an increase in dislocation density and consequently work hardening. Elongation of the grains along the rolling direction as shown in Fig. 1 also occurs. With increasing internal energy, softening mechanisms, both dynamic and static, begin to operate. Microstructural changes that take place during deformation are termed dynamic, while phenomena that occur between successive passes are termed static. In the austenite phase the primary softening mechanisms are static recovery, and dynamic and static recrystallization, while in aluminum alloys they are dynamic and static recovery and static recrystallization. Recovery is marked by a limited amount of softening, as compared to recrystallization, which involves the nucleation of new, strain-free, grains at grain boundaries. At high temperatures and low strain rates, typical of the roughing mill and the early passes of the finishing mill, dynamic recrystallization can occur during rolling of steels, but not aluminum alloys. Between stands, static recovery followed by static recrystallization occurs. In the case of steels some grain growth can occur after recrystallization because of the high temperatures. In HSLA steels, alloyed with niobium, static recrystallization is retarded by the presence of niobium in solution or due to strain-induced precipitation of Nb(CN). This retards static recrystallization in early passes, leading to strain accumulation. When recrystallization occurs between later passes, the recrystallized grain size tends to be much finer because of the high values of retained strain. The fine-grained austenite transforms to fine-grained ferrite during subsequent cooling on the runout table, and consequently higher yield and tensile strength than coarse-grained material. In the case of aluminum, there is no transformation following rolling and the mechanical properties are determined by the statically recrystallized grain size following the last pass.

1.2 Quantitative Microstructural Relationships

Researchers worldwide have directed considerable efforts towards quantification of the microstructural phenomena described in the preceding sections. Static and dynamic recrystallization kinetics can be determined by torsion, compression, or tension tests, at temperatures, strains, and strain rates typically encountered in hot rolling. Of these tests, compression testing is preferred since the material is primarily in a state of compression during rolling. The tests are generally conducted under isothermal conditions for a fixed strain rate and strain. The deformation resistance is also determined from these measurements, and the data is utilized to formulate constitutive equations, which relate flow stress to strain rate, strain, and temperature for a given composition. Transformation kinetics for the austenite-to-ferrite and austenite-to-pearlite reactions have been determined under isothermal and constant cooling-rate conditions using dilatometry. The tests also allow the determination of statically recrystallized grain sizes and in the case of steel the ferrite grain size following transformation. The relationship between final microstructure and mechanical properties, such as yield strength, tensile strength, and elongation and, in the case of aluminum, hot-rolled texture is established through laboratory measurements. Given the range of grades of steel and aluminum commercially processed, and the number of new grades being developed, the experimental work required to quantify all of the microstructural phenomena is substantive.

2. *Mathematical Modeling of Microstructural Evolution*

All of the microstructural phenomena that drive microstructural changes—recovery, recrystallization, precipitation, grain growth, and transformation—are thermally activated and their rates are strongly influenced by temperature. Thus modeling microstructural evolution begins with quantifying the thermal history for each segment of the rolling operation with the aid of computer models. Strain and strain rates also influence many of the microstructural phenomena, which places importance on the deformation history. Beginning with the pioneering work of Sellars and Whiteman (1979), knowledge-based models which link the thermomechanical state of the material to the evolving microstructure, have been developed by research groups worldwide. Figure 2 illustrates the link between the thermomechanical and microstructure models, the industrial process, and laboratory and pilot plant measurements.

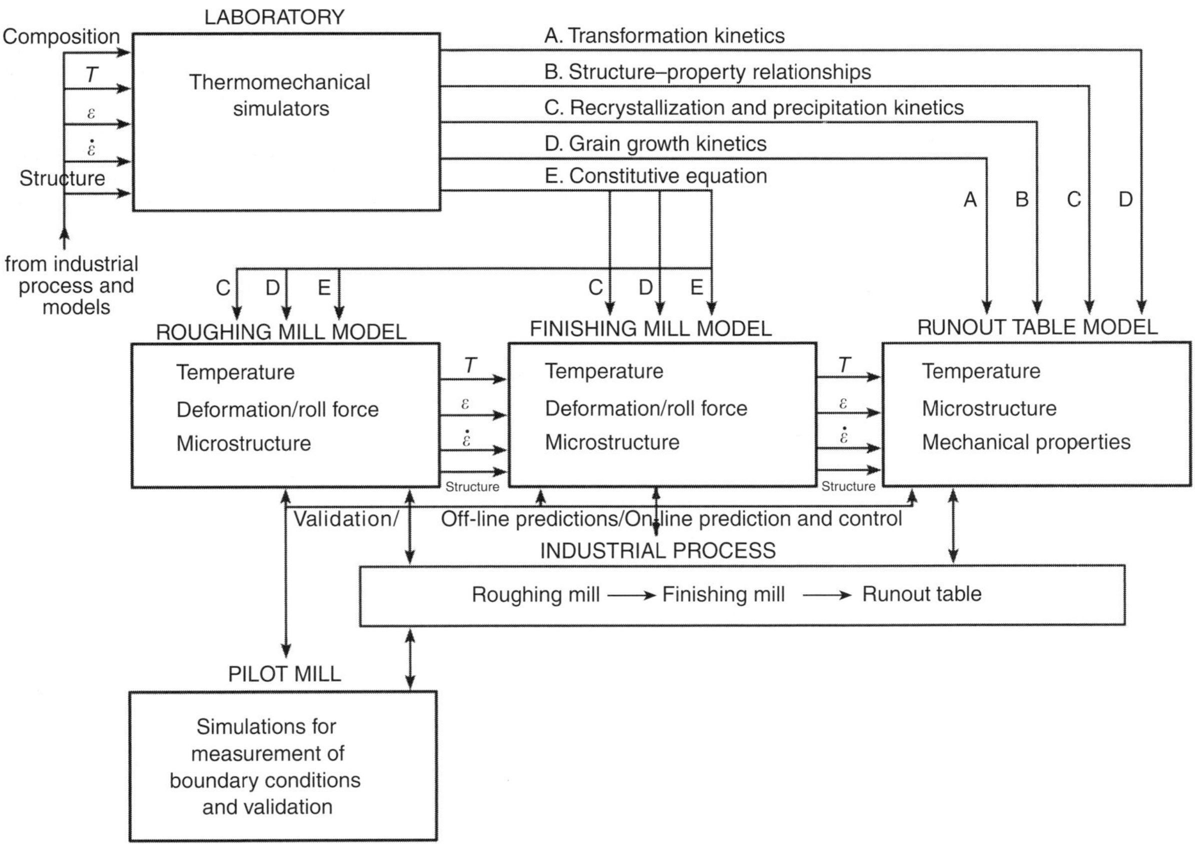

Figure 2
Links amongst thermomechanical and microstructural models, laboratory/pilot plant measurements and the industrial process.

2.1 Heat Transfer and Deformation

Over the past two decades, mathematical models of heat transfer and deformation have been developed to quantify the temperature, strain, and strain rate in the work piece during hot rolling of steel and aluminum. In the case of hot rolling of steel, the slab is reheated in a reheat furnace to hot rolling temperatures. In the ensuing rough- and finish-rolling operations the steel loses heat by radiation to the atmosphere, by conduction to the work rolls, and by convection to water during hydraulic descaling, and interstand spray cooling. The strip is subsequently cooled by laminar water bars or a water curtain system on the runout table to effect the transformation from austenite to ferrite and pearlite. Heat is generated within the slab or strip due to deformation, as a result of friction along the zone of contact with the rolls and during austenite decomposition. Computer models have been developed to take into account the influence of these phenomena on the thermal history of the slab, and are based on a number of simplifying assumptions. Under normal conditions the process operates at steady state, and for a specific rolling schedule, it may be assumed that the sheet or strip temperature at a given location in the mill is time independent. The width of the slab, typically >1300 mm, is significantly greater than the thickness (~250 mm) and hence, conduction along the width direction can be ignored. Conduction along the length of the slab is also negligible in comparison to heat transport by bulk flow of the material owing to high speeds of motion of the material in the mill. Consequently most heat transfer models of the rolling process compute the variation in through-thickness gradients of the slab and strip as a function of process parameters—steel composition, mill layout, strip speeds, rolling schedule, dimensions of strip, work roll geometry, and cooling—and design and operation of the runout table cooling system. In the roll bite the models are often coupled to deformation to properly account for the heat of deformation. Figure 3 shows the prediction of thermal history and austenite grain size of an HSLA-Nb coil during finish rolling from transfer bar of dimension 38 mm to a final strip thickness of 3.6 mm in a seven stand tandem mill with two descale sprays. The effect of the latter and the work rolls on the surface temperature of the strip is evident with chilling effects of ~200°C under the descaler and 400°C in the roll bite. These effects are confined to a narrow (~1 mm) zone near the surface of the strip and are not visible at the centerline, where minor increases (~10°C) associated with deformation heating are evident. At the finish mill exit, the through thickness temperature profile is essentially eliminated. The finish mill exit temperature of 920°C predicted by the model is in good agreement with pyrometer measurements. In general, the model predictions and industrial temperature measurements agree within ±20°C for a wide variety of rolling schedules. For very thin gauge material, final thickness <2 mm, finite-element models that account for through thickness variations in strain and strain rate must be employed to more accurately predict the extent of deformation heating.

There are a number of different finite-element models for predicting the deformation of steel strip during hot rolling. The results of one such model, an Eulerian formulation, is shown in Fig. 4. The two high strain rate zones immediately beneath the surface, close to entry and exit, are responsible for a large temperature rise due to the high local rates of

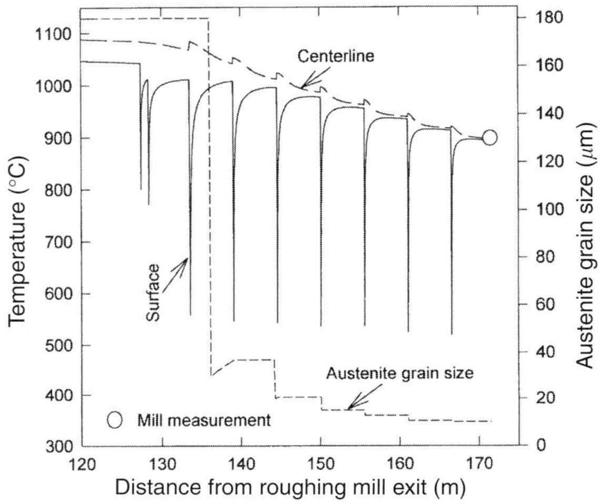

Figure 3
Predictions of thermal history and austenite grain size for an HSLA-Nb steel coil.

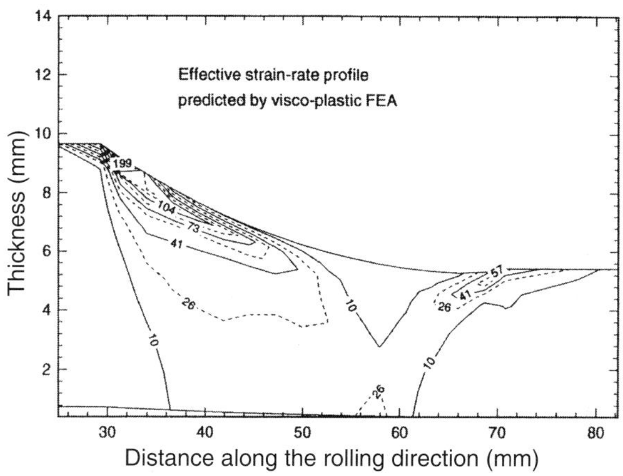

Figure 4
Predicted strain-rate distribution in the roll bite during hot rolling of steel for a reduction of 45 pct.

deformation. This is particularly true for thin strip in the later passes of the finishing mill.

Following rolling in the finishing stands of a hot-strip mill, the strip is cooled on a runout table with laminar water bars, water curtains, and sprays to achieve the desired coiling temperature. The latter is employed as a control variable in the mill to adjust the number of banks that are turned on and the corresponding water flow rate. The runout table is extremely complex from a heat transfer standpoint, and the austenite decomposition is strongly influenced by the thermal history. Studies of heat transfer between water jets and heated steel plates, suggest that immediately underneath the jet, in the impingement zone, the rate of heat transfer is high and transition boiling is the primary mechanism for heat removal. Between the jets, transition boiling, film boiling and convection prevail, depend on the strip surface temperature. Although data for heat transfer under industrial headers is sparse, experimental data from laboratory measurements has been incorporated into mathematical models to simulate runout table cooling.

Figure 5 compares the predicted thermal profiles for two 3.15 mm thick strips of 0.06 pct. carbon with measured coiling temperatures. The temperature predictions compare favorably with measurements and illustrate the difference in cooling intensity required to achieve high and low coiling temperatures. Interestingly, the bottom surface cools more slowly than the top surface on the runout table. Due to gravity, water does not adhere to the bottom surface resulting in lower heat transfer; existing cooling systems do not appear to fully compensate for this effect. In the case of thick strip, these differences in cooling do manifest as differences in microstructure, with the top surface having a finer or more acicular microstructure than the bottom surface.

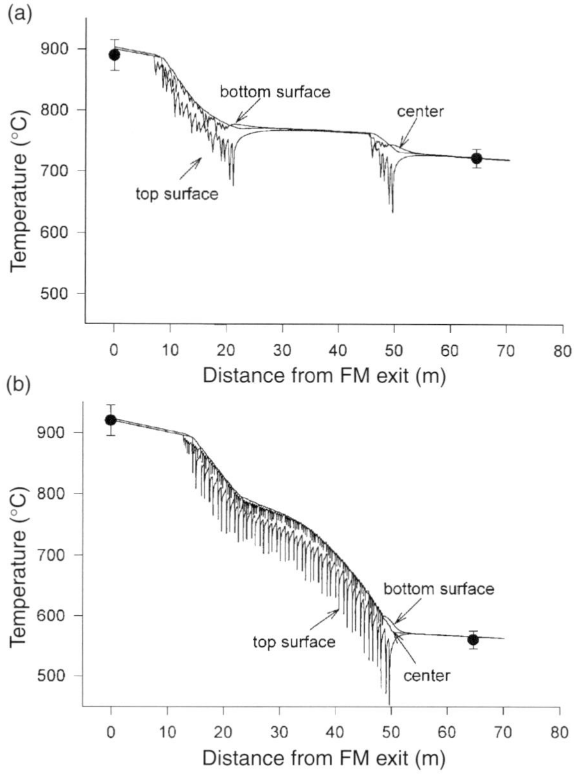

Figure 5
Comparison of predicted and measured temperatures during runout table cooling in the CSP mill, showing the common cooling patterns for (a) high and (b) low coiling temperatures (0.06 C–0.18 Mn).

2.2 Modeling of Microstructure Evolution and Prediction of Final Mechanical Properties

Elaborate integrated models which combine the thermomechanical history with equations describing microstructural phenomena have been constructed and the elusive goal of predicting mechanical properties of hot-rolled strip has become a reality. Although extensive work is required to extend the predictive capability to all grades of steel currently processed, significant advances have been made.

The austenite grain size changes that occur during finish rolling of an HSLA-Nb steel were shown earlier in Fig. 3. The initial austenite grain size has been estimated from transfer bar evaluation to be 180 μm. The austenite grain size is reduced to 20 μm after the second pass due to recrystallization between stands 1 and 2, and deformation in the second pass, with little grain growth between stands 2 and 3. Beginning in stand 3, recrystallization is no longer completed between stands for this grade, because of the retarding effect of niobium in solution on recrystallization kinetics. Figure 6 shows the development of retained strain and fraction recrystallized through this finish rolling process. This is an example of controlled rolling under no-recrystallization conditions, which is usually observed in niobium-containing steels. In this case, there is partial recrystallization, instead of a complete cessation of recrystallization, since the latter tends to be solute drag controlled in hot-strip rolling. Strain-induced precipitation of Nb(CN) is not likely to occur since it requires mill residence times of at least 10 s, which is higher than attained in the finishing train of a hot-strip mill. It is estimated that the finish mill exit microstructure will consist of mixed grains, i.e., finer ~10 μm austenite grains together with larger, ~20 μm, unrecrystallized deformed grains. The prediction of the extent of recrystallization is critical for estimating roll forces and Fig. 6 shows good agreement between predictions and mill measurement.

Similar calculations for grain size evolution during hot rolling of aluminum have also been recently conducted and the predicted fraction recrystallized

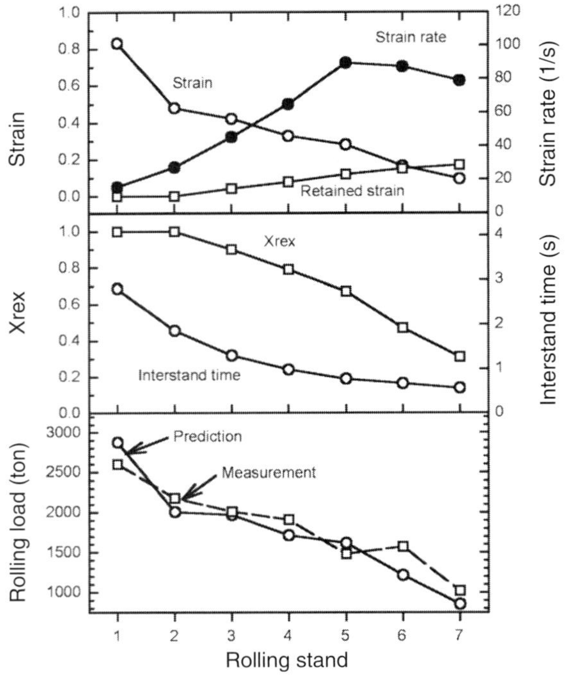

Figure 6
Predictions of strain, strain rate, retained strain, fraction recrystallized and mill loads during 7 pass finish rolling of an HSLA-Nb steel coil.

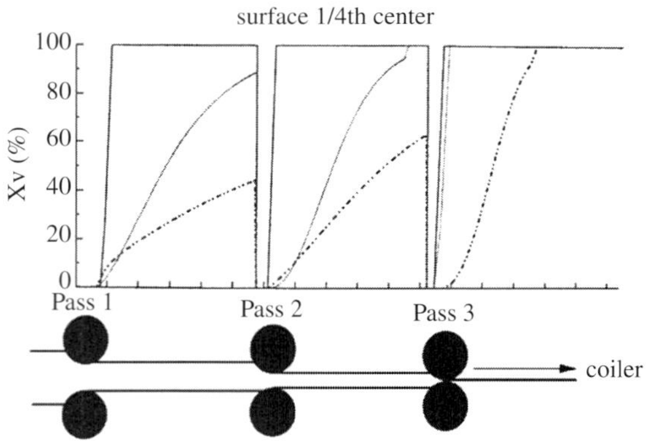

Figure 7
Model-predicted average grain size throughout the hot tandem mill (AA5182).

and grain size evolution for a four stand tandem finish rolling operation for AA5182 alloy (Al–4.5% Mg) are shown in Figs. 7 and 8, respectively. It is evident that the AA5182 alloy only partially recrystallizes between stands, and the surface recrystallizes more quickly than the center. Magnesium is known to reduce the rate of dynamic recovery and

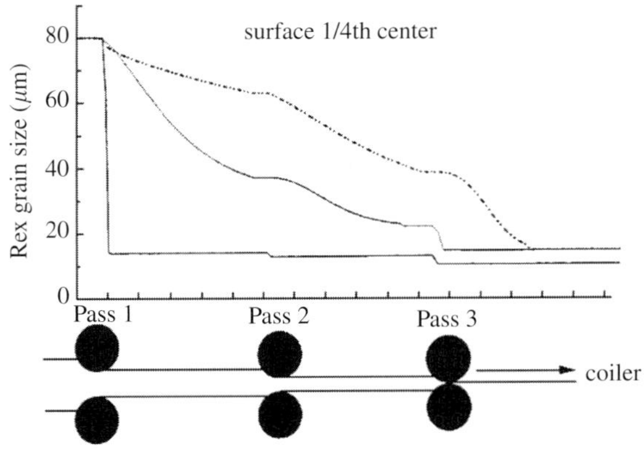

Figure 8
Model-predicted percent recrystallization between stands during hot tandem rolling of an AA5182 alloy.

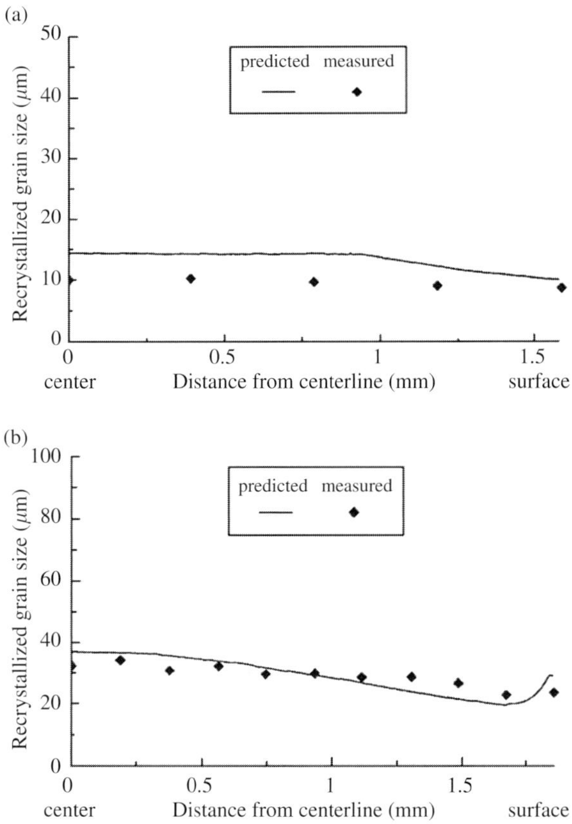

Figure 9
Comparison of model-predicted and experimental measurements of grain size in the strip after hot rolling for (a) AA5182 and (b) AA5052.

rate of work hardening and, hence, the stored energy of the material. The average grain sizes predicted in Fig. 8 are obtained by conducting a weighted average of the grain size of recrystallized and unrecrystallized portions of the strip. The surface of the strip experiences deformation at lower temperatures, higher strain rates and higher strains, and, hence, has a higher stored energy for recrystallization. Model predicted grain sizes compare favorably with industrial measurements in coil as shown in Fig. 9.

Unlike aluminum, steel undergoes a transformation following finish rolling, where the high-temperature austenite phase transforms to a low temperature ferrite/pearlite phase on the runout table during controlled cooling. The final austenite grain size is not the sole determinant of final mechanical properties. The latter is influenced by final ferrite grain size, steel composition, and precipitate size and distribution. As mentioned earlier, the cooling rate driving the austenite to ferrite/pearlite transformation, and the coiling temperature strongly influence metallurgical phenomena and final mechanical properties. Fig. 10 shows microstructure and property prediction for an HSLA-V coil. Good agreement is evident with measurements of ferrite grain size (8 μm), yield strength (390 MPa) and ductility (38% elongation) of industrially processed strips. The through-thickness microstructure profile and the associated properties are also shown.

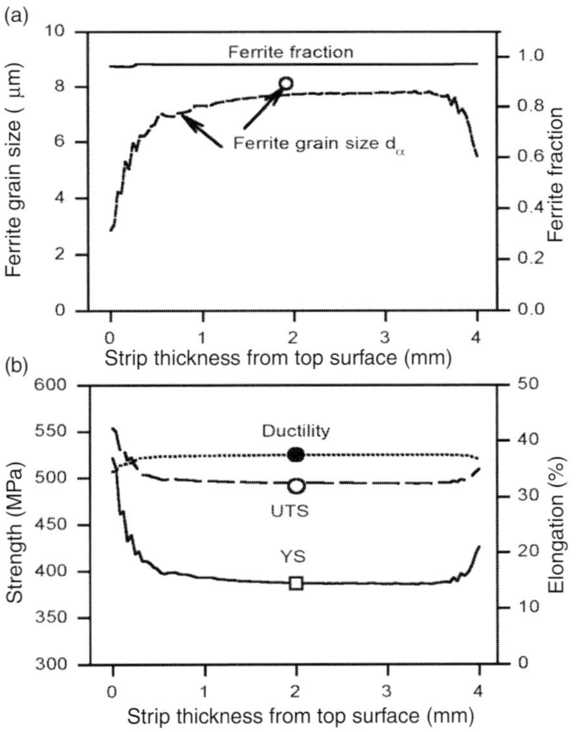

Figure 10
Predictions of (a) ferrite grain size and ferrite fraction and (b) mechanical properties for a 4 mm thick HSLA-V steel coil; symbols indicate measured values.

3. Concluding Remarks

The competitiveness of companies engaged in the production of materials, will be determined by their capability to tailor processes to meet property requirements consistently at low costs. Increasingly this will be achieved through the development of models that draw upon a broad fundamental knowledge base, relying on empirical information only when phenomena are too difficult to quantify, or when the fundamental knowledge simply does not exist. In the case of hot rolling, a marriage of knowledge on transport phenomena, constitutive behavior of the material at elevated temperatures, and structure–property relationships is required for the development of microstructural models. In the short term, models are being developed for off-line and on-line prediction of mechanical properties. In the long term the models will be embedded in expert systems for process control and will serve as a powerful tool for transferring knowledge from research environment to the shop floor.

Bibliography

Chen W C, Samarasekera I V, Hawbolt E B 1993a Mathematical modeling of heat flow and deformation during rough rolling. *Ironmak. Steelmak.* **20**(2), 113–25

Chen W C, Samarasekera I V, Hawbolt E B 1993b Fundamental phenomena governing heat transfer during rolling. *Metall. Trans. A* **24**, 1307–20

Devadas C, Barager D, Ruddle G, Samarasekera I V, Hawbolt E B 1991a The thermal and metallurgical state of steel strip during hot rolling. Part II. Factors influencing rolling loads. *Metall. Trans. A* **22**, 321–33

Devadas C, Samarasekera I V, Hawbolt E B 1991b The thermal and metallurgical state of steel strip during hot rolling. Part I. Characterization of heat transfer. *Metall. Trans. A* **22**, 307–19

Devadas C, Samarasekera I V, Hawbolt E B 1991c The thermal and metallurgical state of steel strip during hot rolling. Part III. Microstructural evolution. *Metall. Trans. A* **22**, 335–49

Jin D Q, Samarasekera I V, Brimacombe J K Modeling inhomo-geneous deformation during hot rolling of plain-carbon steels. *Proc. 38th Mech. Working Steel Conf.* **XXXIV**, Cleveland,OH, p. 329–38

Kumar A, McCulloch C, Hawbolt E B, Samarasekera I V 1991 The thermal and microstructural evolution on the run-out table of a hot-strip mill. *Mater. Sci. Technol.* **7**, 360–8

Kumar A, Samarasekera I V, Hawbolt E B 1992 Analysis of deformation in the roll bite during hot rolling of steel. *J. Mater. Process. Technol.* **30**, 91–114

Militzer M, Giumelli A, Hawbolt E B, Meadowcroft T R 1996 *Metall. Mater. Trans. A* **27**, 3399–409

Militzer M, Pandi R, Hawbolt E B 1996 *Metall. Mater. Trans. A* **27**, 1547–56

Militzer M, Samarasekera I V, Jin D Q, Muojekwu C A 1999 Advances in the prediction of microstructure and mechanical properties of steel strip. *Proc. Chinese Soc. Metals* (in press)
Samarasekera I V, Jin D Q, Brimacombe J K 1997 Critical challenges in the prediction of mech. properties of hot-rolled steel strip. In: Chandra T, Sakai T (eds.) *Proc. THERMEC 97*. Wollongong, Australia
Samarasekera I V, Brimacombe J K 1997 The application of microstructural engineering to the hot rolling of steel. *Proc. 38th Mech. Working Steel. Conf.* **XXXIV**, Cleveland, OH, 313–28
Samarasekera I V, Muojekwu C A, Jin D Q, Brimacombe J K 1998 The state-of-the-art: hot rolling of thin slabs. *Proc. 39th Mech. Working Steel Process.* Conf. **XXXV**, Iron and Steel Society, Indianapolis, IN, 861–73
Sellars C M, Whiteman J A 1979 Recrystallization and grain growth in hot rolling. *Met. Sci* **17**, 187–94
Wells M A, Lloyd D J, Samarasekera I V, Brimacombe J K, Hawbolt E B 1998a Modeling the microstructural changes during hot tandem rolling of AA5XXX aluminum alloys: Part I—Microstructural evolution. *Metall. Mater. Trans. B* **29**(3), 611–20
Wells M A, Lloyd D J, Samarasekera I V, Brimacombe J K, Hawbolt E B 1998b Modeling the microstructural changes during hot tandem rolling of AA5XXX aluminum alloys: Part II—Texture evolution. *Metall. Mater. Trans. B* **29**(3), 621–33
Wells M A, Lloyd D J, Samarasekera I V, Brimacombe J K, Hawbolt E B 1998c Modeling the microstructural changes during hot tandem rolling of aa5xxx aluminum alloys: Part III—Overall model development and validation. *Metall. Mater. Trans. B* **29**(3), 709–19

I. V. Samarasekera

I

Injection Molding

Since the 1980s, considerable attention has been given to the processing and fabrication techniques required to mass produce dimensionally accurate, complex-shape engineering components from ceramic powders. A comparison of forming processes for ceramics shows that injection molding is a very attractive process for near-net shape fabrication of complex-shape structural ceramics. The process can be used to produce monolithic or composite ceramics. Products range from thin (<10 mm) SiC turbine blades, which are accurate to 0.05 mm over the blade profile after a linear shrinkage of 18%, to large cross section (~25 mm) Si_3N_4 turbocharger rotors having relative densities of >99%.

The ceramic injection molding process consists of three characteristic stages in addition to final sintering. These are:

Mixing and compounding. The ceramic powder is dispersed in a suitable binder system to produce a homogeneous suspension with flow properties appropriate for the subsequent injection molding stage.

Injection molding. The ceramic suspension is injected into a mold cavity, which contains the required shape, to produce moldings free of defects.

Removal of binders (also referred to as debinding or thermolysis). The binder system, which enabled flow of the ceramic powder in order to form the shape, is removed from the moldings before sintering the ceramic.

1. *Materials Selection*

Binder systems (sometimes referred to in the literature as the organic vehicle) used in ceramic injection molding formulations contain several components. The major component has a significant influence on the final properties of the formulation. The minor components can provide several benefits. Some are plasticizers, such as phthalates, which reduce the viscosity of the ceramic suspension. Processing aids such as octadecanoic acid and its derivatives act as lubricants, and coupling agents such as silanes and titanates are surfactants to the powder, improving the wetting characteristics between the binders and the ceramic. However, coupling agents sometimes hinder subsequent removal of the binder.

Although emphasis in the literature is on the selection of a suitable binder system, ceramic powder characteristics play a major role in all stages of the process. The volume fraction of ceramic (V_c) dispersed in the binder system has to be optimized carefully. If V_c is too large (>0.7), the viscosity of the formulation will be too high for processing. Also, as V_c increases, more efficient particle packing will result in a smaller inter-particle space for the exit of binder during thermolysis. However, if V_c is too low (<0.5), distortion due to shrinkage and slumping during debinding and subsequent sintering would be unavoidable. A low V_c will also cause handling difficulties after removal of binder, due to low mechanical strength. The powder characteristics also influence the green microstructure, determine the level of iron contamination during processing, catalyze pyrolysis of the binder system, and affect the creation of defects during removal of binder. Optimum ceramic powder characteristics suitable for injection molding, based on a high molecular weight polypropylene-based binder system, include a near-spherical particle shape, a particle size distribution with particles in the size range of 0.2–6 μm (median 1 μm) and a surface area of $3\ m^2\ g^{-1}$. Finer ceramic powders could be injection molded by using low molecular weight binder systems, but these pose particle packing problems.

The use of very low molecular weight binders such as water and alcohol in ceramic injection molding gives rise to mixing and molding problems associated with concentrated suspensions. However, these binders are more likely to diffuse through the small interstices in fine ceramic particle assemblies: but they have low boiling points, which makes them more likely to form voids.

It is also possible to convert the binder system into the ceramic itself. Several preceramic polymers, particularly for SiC, have been investigated, but high shrinkage in the components and low yield of ceramic are major problems.

2. *Preparation of the Suspension*

One of the first tasks in the injection molding of ceramics is the preparation of a suspension by dispersing the ceramic powder in the binder system. This stage causes several ceramic–binder interactions.

Hydrolyzable ceramic powders usually agglomerate during storage. Therefore, it is important to protect these powders from capillary condensation of water at particle junctions, and subsequent surface reaction which produces solid-bridge agglomerates that are difficult to disperse, even by using high-shear mixing. Protection can be achieved by wet milling the powder with silane coupling agents, which interact well with practical binder systems, thus reducing the viscosity of the suspension. Processing aids such as octadecanoic acid derivatives cause acid-base interactions, which result in their being adsorbed onto the ceramic particles, thus coating and protecting the powder from moisture adsorption. This action also has other benefits, such as improved dispersion of the ceramic powder, resulting in better microstructural homogeneity; improved miscibility of the constituents in the binder system; and reduction of the iron contamination due to machinery wear, by lowering the coefficient of friction between the ceramic and metal wall of the mixing, compounding, and injection molding equipment.

Interaction between the ceramic powder and the major and minor binders helps the breakdown of agglomerates which produce strength-limiting defects in the moldings. Fine ceramic particles can agglomerate under a variety of adhesive forces, and the maximum force (F_{max}) required to separate a single spherical particle of radius r_1 in contact with another or similar shape but radius r_2 is given by:

$$F_{max} = 3\pi\eta\dot{\gamma}r_1 r_2\pi \qquad (1)$$

Here η is the viscosity of the suspension and $\dot{\gamma}$ is the shear rate. This expression shows that the force which encourages the separation of particles can be increased by introducing a higher shear stress (τ). Therefore, high-shear mixing is beneficial and thus double-cam mixers, twin-roll mills, and twin-screw extruders are more successful in producing ceramic suspensions with a homogeneous green microstructure than are single- and double-blade mixtures and single-screw extruders. The major causes of defects formed at this stage are undispersed agglomerates of powder and binder, due to inefficient mixing and contamination. The latter type can result from contaminated machinery, the atmosphere or iron contamination.

3. *Flow Properties of Injection Molding Suspensions*

It is essential that the ceramic suspension possess suitable flow properties for injection into the mold cavity without creating defects. Ceramic suspensions used for injection molding are concentrated and with a wide particle size distribution of powder, a V_c of 0.65 can be obtained after high shear mixing. However, as the particle size distribution becomes submicrometer and more monodisperse, a V_c typically only 0.55–0.60 can be incorporated in the suspension. The ceramic powder volume concentration (CPVC), which corresponds to the volume of ceramic powder incorporated in a binder system when the torque during the mixing process reaches a maximum, can be used as an indicator of the appropriate solids loading level. However, it is preferable to establish the curve of relative viscosity η_r (the viscosity of suspension/viscosity of binder system) vs. V_c for the ceramic suspension.

For concentrated ceramic suspensions, equations of η_r vs. V_c contain a parameter V_{max} which depends on the particle characteristics of the ceramic. Recent work has enabled V_{max} for ceramic injection molding formulations to be estimated conveniently by plotting:

$$\frac{1}{\sqrt{\eta_r}-1} \text{ vs. } \frac{1}{V_c} \qquad (2)$$

The gradient/intercept of this linear graph gives the value of V_{max} independently of the shear rate at which η_r is calculated.

Adsorbed layers of water and polymer affect the ceramic loading and are more pronounced for finer ceramic powders. $V_{max} - V_c$ represents the free volume of binder over and above that required to fill the interstices between the packed powder particles. In practice, $V_{max} - V_c$ is 10–15% in a typical ceramic injection molding formulation, in order to prevent seizure of mixing and compounding machinery, and to reduce iron contamination due to wear.

Various techniques have been used to estimate the flow properties of ceramic injection molding formulations. Spiral flow length measurements, which bear direct relevance to the molding situation, have been used, but are time-consuming to perform. The flow properties of ceramic injection molding suspensions using dynamic oscillatory, parallel plate, and capillary rheometry have been investigated. The first method minimizes particle migration effects that could be present in capillary rheometry. However, capillary rheometry resembles the injection molding process on a smaller scale.

Ceramic suspensions suitable for injection molding show Bingham or psuedoplastic flow behavior, and Eqn. (3) describes the quasi-plastic flow behavior of ceramic injection molding suspensions.

$$\tau - \tau_y = k\dot{\gamma}^n \qquad (3)$$

where τ_y is the yield stress of the suspension. k is a constant and for psuedoplasticity $n>1$. If $n<1$ the suspension will be dilatant, and this can result in the separation of the ceramic powder from the binder system. In addition to being psuedoplastic, ceramic injection molding suspensions should possess a viscosity of ~ 1000 Pa s and a low temperature dependence of viscosity at the nozzle temperature and shear rate of the nozzle/mold region (0.3×10^{-3} Pa s°C^{-1} at 108 s^{-1}). Using these criteria, a polypropylene-based binder system that produces void- and crack-free moldings has been identified and is used extensively in industry. This binder system has been modified further by incorporating atactic polypropylene in place of the microcrystalline wax, in order to reduce shrinkage in the molding stage.

A flow property that has been neglected, but which could appreciably influence the formation of defects during injection molding, is the yield stress of the suspension. The suspension can inject into the mold cavity like a thread of ribbon that coils in upon itself if the yield stress is too high (>10 kPa). This induces mechanical stresses in the moldings and causes voids due to entrapment of air. On the other hand, a very low yield stress causes deformation of the molding during removal of binder. Casson's equation given below fits capillary rheometry results extremely well where:

$$\tau^{1/2} = (\tau_y)^{1/2} + k_1(\dot{\gamma})^{1/2} \quad (4)$$

This allows the estimation of τ_y. Although $\tau_y \ll \tau$, τ_y increases with V_c and at a certain V_c, $\tau_y \rightarrow \infty$ (Fig. 1). This behavior is similar to the η_r vs. V_c relationship described earlier. The τ_y vs. V_c graph enables the adjustment of the solids loading to obtain an acceptable yield stress and viscosity.

4. *Shape Forming by Injection Molding*

Reciprocating screw and plunger-injection molding machines, widely used in the polymer processing industry, are also used with some modification for ceramic suspensions. Reciprocating screw machines provide better metering and injection pressure control, lower cycle time, flexibility in part size, and superior homogeneity and reproducibility of the molded component. However, plunger-type machines are, in general, cheaper and have a longer machine life due to less wear.

The type of injection molding machine used may also depend on the binder system. For wax binder systems, which produce suspensions with lower viscosities of about 100–200 Pa s below 100°C, a type of plunger machine known as a wax injector could be used. Here the suspension is transferred to the injection chamber just prior to flow in the mold, but otherwise it is held in another agitated heated chamber. For thermosetting binder systems, transfer molding is used, and a plunger or a screw transfers the suspension into a heated mold where the thermosetting resin cross-links.

Low-pressure (0.2–0.6 MPa) injection molding, which is limited to low viscosity (~ 100 Pa s) ceramic suspensions, has been successfully used to injection mold non-oxide ceramics. The binder system contains low molecular weight binders such as wax or cyclohexane. In the latter instance, the mold temperature was -78°C and the molding required freeze drying. Also, the flow rate of the suspension is of crucial importance. Excessive flow rates cause turbulence which gives rise to void formation. However, it is claimed that components with larger section sizes ($\gg 25$ mm), can be made using low-pressure injection molding. Lower pressures also allow the use of less expensive molds and result in less wear of machinery and molds.

The full benefit of high-shear mixing and high-pressure molding can be exploited when high molecular weight binder systems are used. However, when typical high molecular weight polymers like polypropylene are filled with ceramic powder, an appreciable decrease in heat capacity and increase in thermal conductivity occur. These cause rapid cooling of the ceramic suspension in the runners and the mold. Therefore, optimization of pressure, temperature and injection speed is essential to injection mold ceramics. In particular, mold cavity temperature and pressure should be carefully controlled. Molding trials performed using a reciprocating screw machine revealed that the mold cavity filling rate is determined by injection speed, rather than by injection pressure, while the weight of the molding is more affected by the hold pressure than by the injection speed. A shallow metering screw was also found to reduce viscosity fluctuation and resulting inhomogenities. Mold design is also important for the production of defect-free moldings. The main parameters in mold tool design are positioning and geometry of the gate and runners, part geometry, ease of automation, wear, and cost.

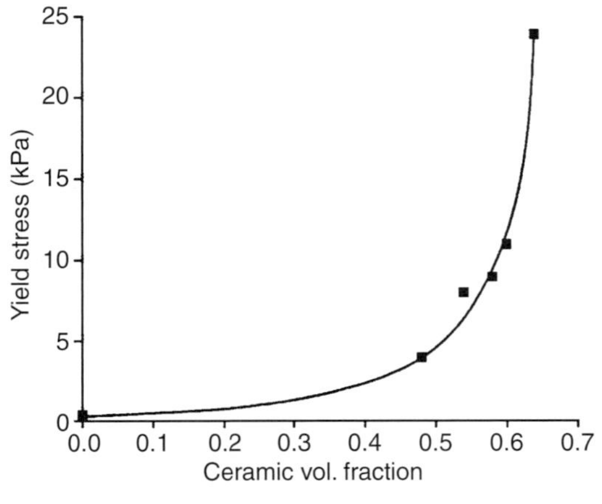

Figure 1
Yield stress variation with ceramic volume fraction for alumina–polypropylene suspensions.

Solidification in the mold cavity should be controlled carefully. The control of sprue solidification time and, therefore, its determination by computer simulation is a key factor in controlling molding defects, especially in thick sections. Immediately after injection a solid layer forms at the mold surface. Thickness of this layer increases as solidification continues. The machine hold pressure continues to fill the mold cavity, compensating for shrinkage, and this occurs until the center of the sprue solidifies. However, in thicker sections the sprue solidifies before the entire molding is solid, and a pocket of molten suspension at the thickest section is isolated. During solidification of this molten pool of suspension, shrinkage occurs without the compensating influence of hold pressure, and the resulting stresses cause cracking.

Therefore, avoidance of shrinkage-related defects in ceramic moldings by controlling solidification is of vital importance. Several solutions are available. The sprue and runners can be kept molten, while the larger sections of the molding solidify, by using electrically heated inserts. A heated sprue and the use of a combined injection/compression molding device are other alternatives. However, in all these methods the incorporation of suitable inserts in the sprue/mold area makes machine/tooling design and process control extremely difficult. One special technique requires the use of a valve that sits between the machine nozzle and the mold (Fig. 2). This valve is used to oscillate the pressure applied on the suspension between wide limits. The resulting heat dissipation caused by this mechanical action prolongs the solidification of the sprue. The valve is computer-controlled, and the frequency and duration of oscillation and intensity of pressure applied can be controlled carefully to eliminate the defects observed in the thick section of the molding (Fig. 3). An on-line compounding injection molding machine, incorporating the co-rotating, intermeshing twin-screw extruder and a reciprocating screw injection molding machine fitted with the oscillating pressure valve, has been developed.

Voids and cracks are the major defects at the injection molding stage and their avoidance depends mainly on mold cavity pressure and temperature. As cavity pressure increases, voids are avoided, but cracks could still appear. However, excessive pressure should not be used as this can cause distortion of the moldings during initial heating of the binder removal stage.

A major problem during compounding/injection molding is the iron contamination by the ceramic suspension, due to wear of machinery. Check ring type non-return valves, nozzle/nozzle seat could suffer appreciable wear, which affects pressure transfer during injection molding. The oscillating pressure valve could also suffer extensive wear damage. Such wear problems have led to detailed investigations of ceramic injection molding machine parts. In particular, improvements in machinery for low-pressure

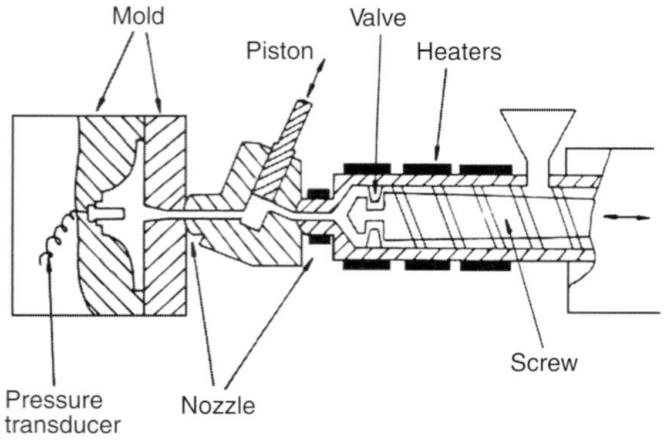

Figure 2
Schematic diagram of the oscillating pressure equipment in a ceramic injection molding machine.

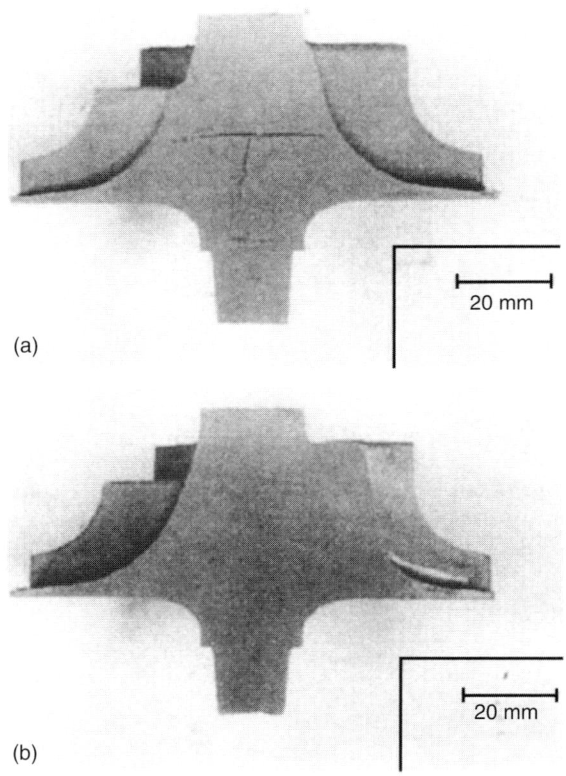

Figure 3
Sections of silicon nitride (50 vol.%) made using a polypropylene-based binder system with (a) static hold pressure and (b) oscillating pressure, both at a mold temperature of 70°C.

injection molding have been made already. It is very clear that the mild steel machine parts of polymer processing machinery are unsuitable for the processing of ceramic suspensions and suitable modifications, with overall machine design being taken into account, still require attention.

5. *Binder Removal*

This stage of the process is widely considered to be the most likely to introduce defects into the moldings, hindering this processing route from being used to make parts with a large cross section (>25 mm). However, certain defects such as residual stress-induced cracking in thick sections, which really originate at the molding stage, become evident only after removal of binder or even after final sintering. Debinding usually involves long cycle times (e.g., 604800 s) and as the section size of the molding increases, avoiding strength-limiting defects during this stage becomes significantly more difficult. Defects such as slumping, bloating and void formation occur during the early stages of binder removal, while cracking occurs later on. In non-carbide ceramics, carbon residue after removal of binder could be a problem. Other residue types such as titanium, aluminum, silicon, and calcium could also be present, mainly as the residue of polymeric binders.

Techniques for the removal of binder can be classified into two major types: extraction by slow pyrolysis to about 500°C (thermolysis), and solvent extraction. Thermolysis is the preferred method, and on heating several mechanisms facilitate the removal of binder. Firstly, evaporation or sublimation can occur and low molecular weight constituents are removed from the moldings without further breakdown of their chemical structure. Secondly, thermal or oxidative degradation causes chain scission in high molecular weight binders, and the resulting decomposition products are removed. Thirdly, low molecular weight binders such as waxes can be extracted by capillary action. To facilitate capillary action, the moldings can be packed in a suitable ceramic powder bed during pyrolysis. Solvent extraction is usually carried out on a two-component binder system, in which one constituent is soluble in the solvent used for extraction. Application has been largely limited to suspensions containing coarse metal powders. It is claimed that solvent extraction provides a faster rate of binder removal. However, compared with thermolysis, the operation of this method under modern safety regulations requires expensive plant and accessories.

The binder removal process depends on several parameters. The atmosphere under which removal of binder is carried out, properties of specific constituents and the overall composition of the binder system, ceramic powder characteristics, and component geometry are the major factors that affect the process. Much research has been devoted to empirically determining a temperature ramp for the successful removal of a specific binder system containing one type of filler. Rate-controlled thermal extraction, where the temperature increase is governed by weight loss of binder, is reported to have been successful with injection molded Si_3N_4 turbocharger rotors. However, most thermoplastic binder systems commonly used in ceramic injection molding suspensions lose the major binder during a very narrow temperature range when pyrolyzed in air or an inert atmosphere, and this makes control of temperature using weight loss extremely difficult. Thermogravimetry and microstructural information have been used to define a thermal profile for specific injection-molded articles. A method of choosing binder system constituents to achieve a linear weight loss during pyrolysis has been developed, but it is unclear whether this would give rise to defect-free components.

Several types of atmosphere can be used for successful debinding. Pure oxygen, air, pressurized nitrogen, and vacuum have all been used successfully. The case-specific nature of binder removal has led to a method which requires the development of temperature-heating rate diagrams for molded components. Empirical studies have led to two vital, experimentally verified facts about the removal of binder. Firstly, the heating rate up to the softening point of the formulation is rapid in comparison with that used during evaporation or degradation of binder constituents. Secondly, the time required for removal of binder is dependent on the square of the section size.

More recently, modeling the removal of binder from filled suspensions has got under way, in order to uncover the fundamentals of the process. Some models deal with diffusion of liquid binder or vapor in a porous body, but this does not address the criteria for defect formation. Others have developed models that deal with stresses caused by the expansion effects of the binders due to heating during removal but assume a rigid skeleton of ceramic powder. This assumption is inconsistent for ceramic suspensions, where particle movement and shrinkage during removal of binder are clearly established experimentally. Mass transport in ceramic suspensions has also been modeled comprehensively, but the use of such a model to predict the formation of defects during the various stages of binder removal is not fully discussed.

Steady-state and transient diffusion models were developed in the 1990s for cylindrical, spherical and flat-plate geometries. Thermal degradation produces decomposition products which initially dissolve in the parent polymer and then diffuse in the continuous phase (polymer melt) to free surfaces where they evaporate. If the vapor pressure over solution reaches ambient pressure, bubbles form, the molding bloats and is rendered scrap. Transient diffusion models based on this criterion have been validated experimentally using a polyalphamethylstyrene–50 vol.% alumina system for cylinder and plate geometries, and have been extended recently to incorporate optimized heating schedules.

Either continuous porosity or a shrinking unreacted core can develop during debinding, representing two extremes of porosity configuration between which all real cases lie. The modeling has been modified to predict upper and lower bounds for the critical rate of heating, corresponding to these two configurations and, subsequently, intermediate scenarios have also been modeled. These models predict the maximum permissible heating rate (Z_c, also referred to as the critical heating rate) before boiling occurs in the bulk at the center of the thickest section. Therefore, the variation of Z_c for different geometries can be estimated and, as expected, shows a steep increase with section thickness (Fig. 4).

It is possible to perturb the models in a systematic way to find out how changes to the polymer and degradation product properties affect the maximum permissible heating rate. Elevated pressure acts to suppress boiling and allows an increase in critical heating rate, as predicted by the models. Commercial pressure ovens are available for this application. Six parameters influence Z_c significantly, and these have been identified by perturbing the models. These six parameters can be classified into three groups: (i) those which influence the rate at which degradation products are produced; (ii) those which determine the diffusion coefficient through free volume theory; and (iii) those which determine the boiling point of the solution. It is now possible to optimize these parameters and to synthesize binders suitable for ceramic injection molding.

The fracture strength of ceramic injection moldings before debinding is about 15 MPa. Due to significant difficulties faced during removal of binder from thick sections, it is not established whether the mechanical properties and Weibull Modulus of sintered components produced using the injection molding processing route compare favorably with those obtained from alternative processing routes. However, modifications of the ceramic injection molding process such as the quick set injection molding process can produce very high strengths—in excess of 1000 MPa in Si_3N_4.

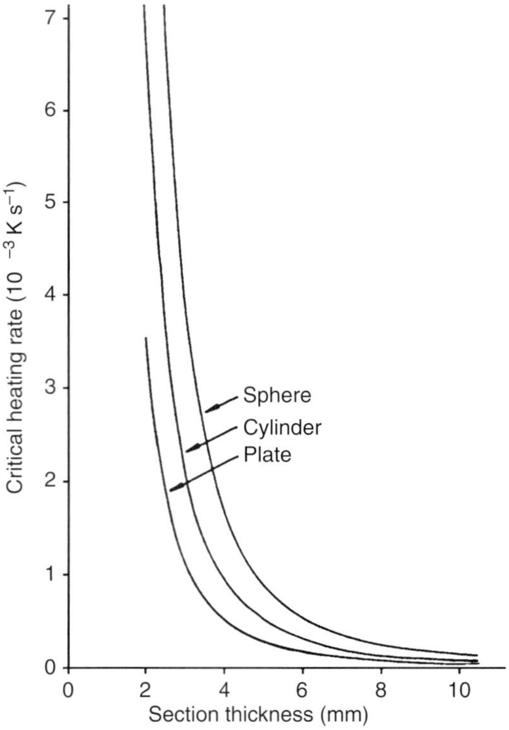

Figure 4
Z_c values predicted by the transient diffusion model in a body having distributed porosity.

Bibliography

Adams R W, Householder W B, Sundback C A 1991 Applicability of quickset injection moulding to intelligent processing of ceramics. *Ceram. Eng. Sci. Proc.* **12**, 2062–71

Carlsson R 1989 Shaping of engineering ceramics. *Mater. Des.* **10**, 10–4

Edirisinghe M J 1997 Thermolysis of plastically formed ceramics: modeling and computer simulation. *Mater. Manuf. Process.* **12**, 609–28

Edirisinghe M J, Evans J R G 1986a Review: fabrication of engineering ceramics by injection moulding 1. Materials selection. *Int. J. High Technol. Ceram.* **2**, 1–31

Edirisinghe M J, Evans J R G 1986b Review: fabrication of engineering ceramics by injection moulding 2. Techniques. *Int. J. High Technol. Ceram.* **2**, 249–78

Edirisinghe M J, Evans J R G 1987a An oscillating pressure unit for ceramic injection moulding part A. Materials and design. *Mater. Des.* **8**, 248–88

Edirisinghe M J, Evans J R G 1987b Properties of ceramic injection moulding formulations 1. Melt rheology. *J. Mater. Sci.* **22**, 269–77

Edirisinghe M J, Evans J R G 1987c Properties of ceramic injection moulding formulations 2. Integrity of mouldings. *J. Mater. Sci.* **22**, 2267–73

Edirisinghe M J, Shaw H M, Tomkins K L 1992 Flow behaviour of ceramic injection moulding suspensions. *Ceram. Int.* **18**, 193–200

Edirisinghe M J, Zhang J G, Evans J R G 1988 An oscillating pressure unit for ceramic injection moulding part B. Performance. *Mater. Des.* **9**, 85–93

Evans J R G, Edirisinghe M J 1991 Interfacial factors affecting the incidence of defects in ceramic mouldings. *J. Mater. Sci.* **26**, 2081–8

German R M 1990 *Powder Injection Moulding*. Metal Powder Industries Federation, NJ

Krug S, Evans J R G, ter Maat J H H 1999 Aetiology of defects in large ceramic mouldings. *J. Am. Ceram. Soc.* **82**, 2094–100

Mutsuddy B C 1987 Mechanical properties of injection moulded ceramics. *Powder Metall. Int.* **19**, 43–6

Mutsuddy B C, Ford R G 1995 *Ceramic Injection Moulding*. Chapman and Hall, London

Peltsman I, Peltsman M 1982 Improvements in machinery for hot moulding of ceramics under low pressure. *Ceram. Eng. Sci. Proc.* **3**, 865–8

Pinwill I E, Edirisinghe M J, Bevis M J 1992 Development of temperature-heating rate diagrams for the pyrolytic removal of binder used for powder injection moulding. *J. Mater. Sci.* **27**, 4381–8

Shaw H M, Edirisinghe M J 1995 Shrinkage and particle packing during removal of organic vehicle from ceramic injection mouldings. *J. Euro. Ceram. Soc.* **15**, 109–16

Song J H, Evans J R G, Edirisinghe M J, Twizell E H 2000 Optimization of heating schedules in pyrolytic binder removal from ceramic mouldings. *J. Mater. Res.* **15**, 449–57

Woodthorpe J, Edirisinghe M J, Evans J R G 1989 Properties of ceramic injection moulding formulations 3. Polymer removal. *J. Mater. Sci.* **24**, 1038–48

Zhang J G, Edirisinghe M J, Evans J R G 1989 A catalogue of ceramic injection moulding defects and their causes. *Ind. Ceram.* **9**, 72–82

M. J. Edirisinghe

Investment Casting (Disposable Mold)

"The investment casting process" or what is frequently referred to as the "lost wax process" can be traced back to at least 4000 BC. The process uses a wax pattern to define the component geometry and has numerous advantages over more conventional casting processes, particularly in the degree of detail that can be achieved. Evidence for early use of the process can be found in China, India, Egypt, Africa and particularly in Greece and other parts of Europe. The early Colombians and Aztecs used the process extensively in South America. The lost wax process is used extensively today for producing sculptures in various materials particularly bronze and is favored because of the ease with which very fine detail and surface finish of the desired quality can be achieved. In modern times the lost wax process was adapted for dental work where accurate castings for fillings and crowns could be produced in gold and other metals. A centrifugal casting process is frequently used for these small fine detail parts and centrifugal casting is also used by the jewelry industry, again because of the ease of which fine detail can be produced.

Rapid growth of the investment casting process occurred during World War II. It was realized that cobalt base alloys, which had been developed by Austenal Laboratories which was then operating in New York City, could be used to produce turbine components for turbochargers, which was of interest to aircraft engine manufacturers. The Vitallium alloy, developed by Austenal, was found to be ideal for GE's supercharger buckets and was rapidly incorporated into production. GE ultimately made some slight modifications to the composition of Vitallium to improve its properties and by the time World War II was over some 35 million blades had been produced.

The investment casting process is shown schematically in Fig. 1. The process consists of first

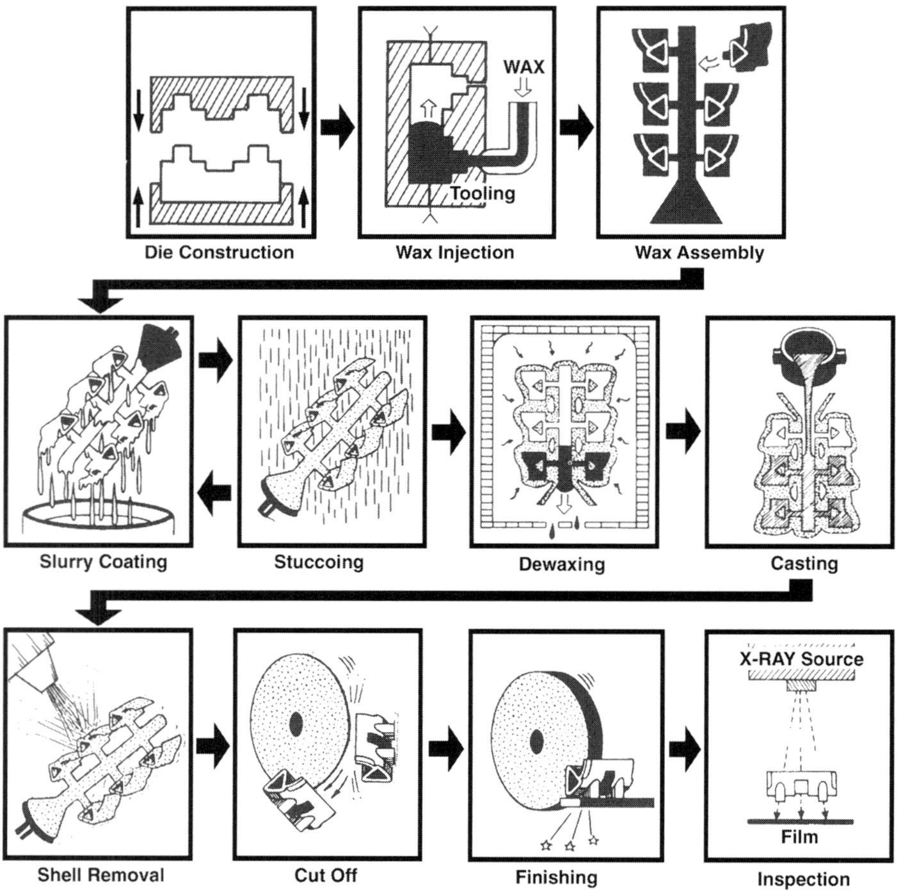

Figure 1
Schematic diagram showing the processes involved in investment casting.

producing a wax pattern by injecting wax into a die. This produces a model of the component to be cast. As the first step in the process, it is imperative that the dimensions of the wax pattern and other important features required in the final component be reproduced. Much effort has been expended on developing improved wax pattern materials that are both easy to inject, have good surface finish, and also mimic the final dimensions required as accurately as possible. The individual wax patterns are then fastened by a feed gate to a pour cup. The dimensions and location of the feed gate are very important to achieve final desired product quality. In today's technology, sophisticated, computerized modeling systems are available to guide the gating design to produce an optimum quality casting. Much effort is also expended in maximizing the number of parts per mold since this has significant effect on product cost. The wax assembly is then dipped into a ceramic slurry, drained and then coated with a fine ceramic sand. After drying, this process is repeated using progressively coarser grades of ceramic material until a self-supporting shell has been formed. Thickness of the shell depends on the details of the process but is generally of the order of 3 mm to 15 mm thick. The coated cluster is then placed in either an air furnace or a steam autoclave where the wax pattern melts out through the gates, leaving a cavity of the desired dimensions to produce the casting.

After the ceramic shell has been produced and inspected, molten metal is poured into the cavity using one of several processes which may involve either centrifugal casting, vacuum casting, or inverse gravity casting. In the latter process, the mold is inverted over the molten metal and a vacuum applied through the porous shell which draws the molten metal up into the shell. The final step is then to remove the shell, again by one of several processes that might involve caustic dissolution of the shell, water blasting, or shot blasting mechanically. Finally the parts are cut off from the gates and various metal finishing steps are used to produce a finished component, which can then be inspected by x-ray inspection or other methods as appropriate.

It is possible also to produce hollow components by employing a ceramic core, which is incorporated into the wax pattern during the initial wax injection step. The leaching of the ceramic core from the interior of the component then results in a hollow metallic shape. This process is shown in Fig. 2 for the production of hollow airfoils.

Other variants of the investment casting process can be applied to produce particular kinds of components. For example, a directional solidification process is frequently employed to produce parts for the aircraft engine industry where very high temperatures are anticipated and there is a desire to eliminate grain boundaries transverse to the principal stress axis. In the most sophisticated processes, components can be produced which contain one single grain and grain boundaries are entirely eliminated. Single crystal components for land-based turbines are now being produced which weigh as much as 20 kg.

Investment cast components are now used today in a wide variety of applications including automotive,

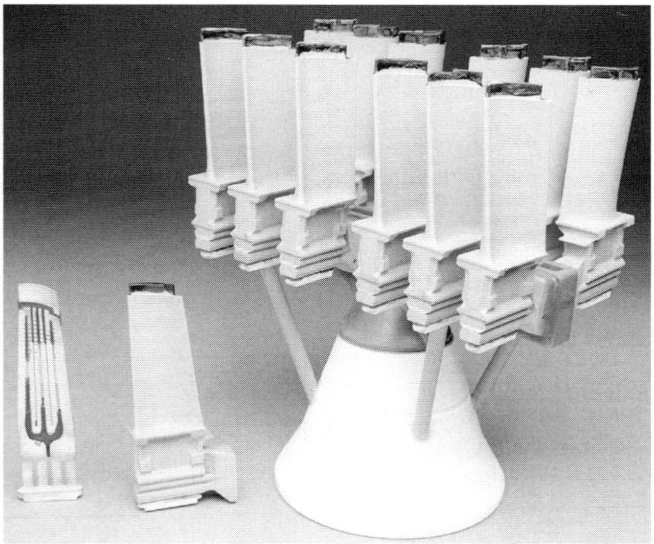

Figure 2
Hollow airfoils produced by investment casting.

aircraft structures, gas turbine components, and also for more traditional sculptures used for decorative purposes. Some of the larger sculptures are produced in several pieces and subsequently welded together to form a monolithic structure. The largest investment castings for such purposes may weigh as much as several thousand pounds. Some of the advantages of the investment casting process include low initial tooling cost and the ability to reproduce very fine details. The ability to produce rapid prototypes from such methods as stereolithography, where no mold is required to produce patterns, is also important. This enables prototype hardware to be produced from electronic data, and is extremely useful in quickly producing prototype hardware for test purposes. In this way, design changes can easily be made prior to producing hard tooling and significant reduction in lead times can be achieved.

There are numerous sources for investment castings. Currently there are approximately 350 investment casting companies in the USA with many more worldwide. Many have particular areas of specialization, such as castings for a particular industry or from a particular alloy. The process also has the advantage that it can produce, economically, either very small numbers, often in the art industry as low as one or two, to very high numbers of components as in the automotive industry where millions of a particular component can be produced per year, at very competitive costs.

N. E. Paton

J

Joining of Metals

Joining techniques are divided into different groups. They are based on different physical effects and three different types of joints are used: material joint, pressure joint, or geometrical joint. Local deformation is a combination of joining by pressure and by geometry and it is used in different joining techniques, such as rivets, clinching, toxing, and all welding techniques without fused bonding zones.

1. Explosive Welding

Explosive welding is mainly used for cladding processes. Nearly all kinds of metals and alloys allowing more than 5% of strain may be joined by this technique (Andernach und Bleck). Titanium explosive cladded tank bottoms have successfully withstood testing by use in chemical apparatus.

In explosive welding, a compression force created by detonation of explosives is used to join overlapping metal sheets. The joining parts are arranged towards each other at an angle of 1–15°, depending on the material and method, and are prepared with a layer of explosive on the top. After ignition the joining areas are moved against each other at high speed. Joining happens continuously by local plastic deformation of the contact area (Ruge 1993) (see Fig. 1).

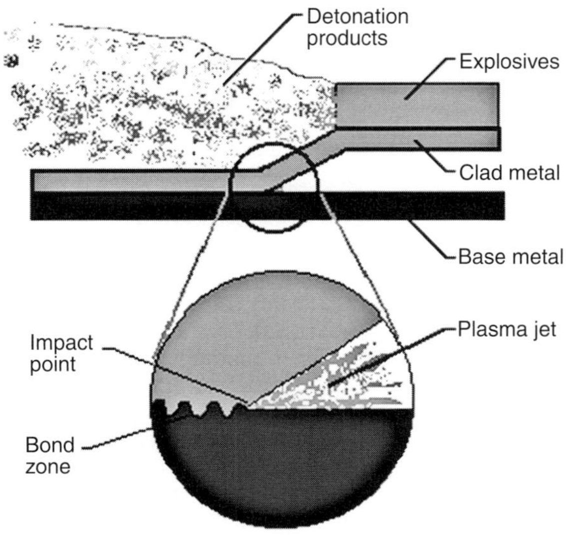

Figure 1
Function principle of explosion welding (after High Energy Metals Inc.)

The thickness of the cladding may vary between 0.1 and 30 mm and the detonation velocity is between 1200 and 7000 $\mathrm{ms^{-1}}$. The top plate hits the bottom plate with a speed of 100–1000 $\mathrm{ms^{-1}}$, and pressures are in the range of 10 to 100 kbar. Under certain conditions a superheated layer of material forms in the contact area. The joint is then caused by a deformation-induced melting bath. The melt layer is quite thin; for example, when cladding an aluminum alloy it is about 0.5–4μm (Kreye 1985, 1997). The dilution of the molten zones results in plane or corrugated interfaces.

In contrast to surface-layer welding, explosive welding causes no change in microstructure, and corrosion resistance of the layers is not affected (Pocalco 1987).

2. Friction Welding

Friction welding is a pressure welding process which generates the necessary temperature at the weld interface by mechanical friction.

2.1 Friction Welding with Rotating Workpiece

The friction is usually caused by one freely rotating and one fixed workpiece as axial pressure is applied (Spindler 1978). This causes destruction of local welds and, at the end of weld cycle, hot torsion of the plasticized joint area. Locally, complete fusion may occur but this is not necessary for the formation of good welds. When interfaces are heated to the proper temperature, the drive motor is shut off and the workpieces are pressed together at a predetermined load.

An even, lip-shaped bead ,which is typical of the process, is built up. It can be removed by special additional devices in the welding machine in the hot state by turning or shearing. Application of friction welding allows the joining of a number of material combinations which cannot be joined by fusion processes. Examples are joints of steel–aluminum, magnesium–aluminum, magnesium–steel, and various ceramics. A sample of a steel–aluminum joint is shown in Fig. 2.

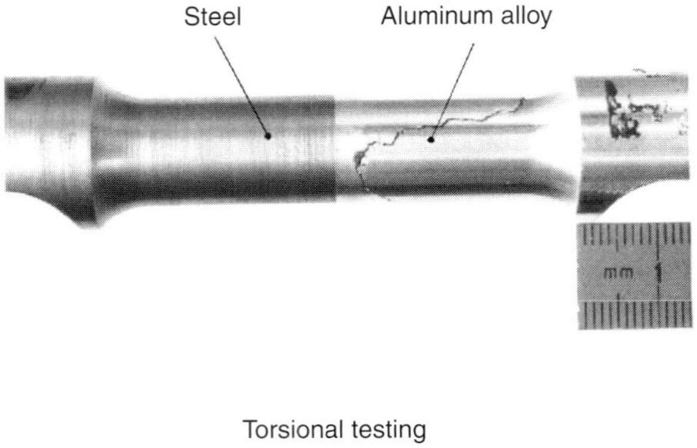

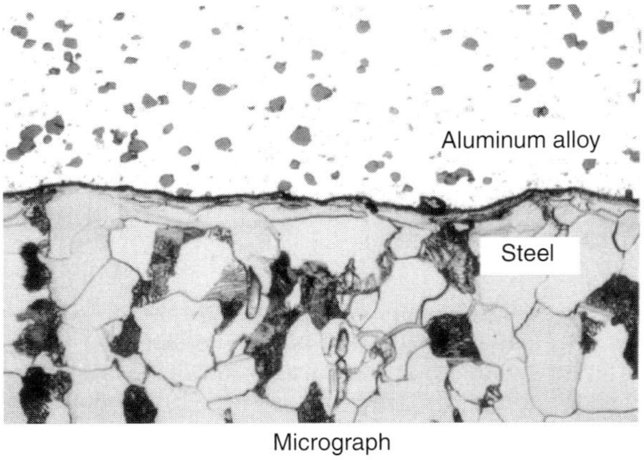

Figure 2
Friction-welded aluminum–steel joint (after Suthoff 1999).

2.2 Friction Stir Welding

For friction stir welding, a rotating mandrel-like tool is pressed between the joining line of two workpieces, which are then butt-joint welded along the joining line (Fig. 3). The parts to be joined have to be fixed to each other and clamped to a ground plate to avoid slipping during processing. The wear-resistant tool frictionally heats the workpiece material (Russell 1997). This heat plasticizes the base material without melting it. The joining temperature for aluminum—the favorite material for this process—is at about 500–550°C. Plasticized material is stirred in front of the tool pin device to the top in a direction contrary to the welding direction. As material flow is limited by the shoulder of the tool and by the fixing plate to the top and bottom sides, a joining process is caused along the joining line (Rhodes *et al.* 1997, Schofer 1999). The workpiece thickness varies between 1.6 mm and 30 mm. Butt and overlap joints are common weld shapes. Other joint preparations are feasible if the necessary high force input can be generated.

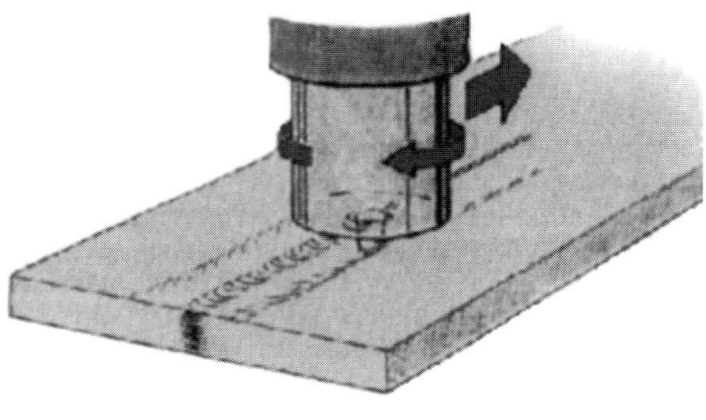

Figure 3
Principle of friction stir welding (after Schofer 1999).

3. Mash Seam Welding

In conventional mash resistance welding, wheel electrodes run over a small overlap of two clamped metal sheets. By the application of an electrical current the sheet edges are plasticized and with pressure from the electrodes the workpiece almost matches the wall thickness of a single sheet. The process results in smooth and less reinforced seam surfaces. To prevent sheets from slipping during manufacturing a sturdy clamping device or tacking of the blanks is necessary.

A process modification is when the forging is not controlled by wheel electrodes but by separately arranged pressure rolls leveling the seam during the hot state (Ruge 1993).

4. Cold-pressure Welding Techniques

Cold-pressure welding is suitable for joining similar or dissimilar metals when the carefully cleaned surfaces are deformed under pressure such that foreign surface layers are cracked. The metal areas which are exposed by this and which have not yet had atmospheric contact are brought close enough to effect their atomic binding forces.

Degreasing and removing of existing oxide layers is necessary before welding. As the process is carried out at room temperature, no nondesirable brittle intermediate layers are formed as they are generated in fusion welding processes of dissimilar metals, e.g., copper and aluminum (Ruge 1993).

Cold-pressure welding is used for a variety of applications. In lap welding, working pieces are put one

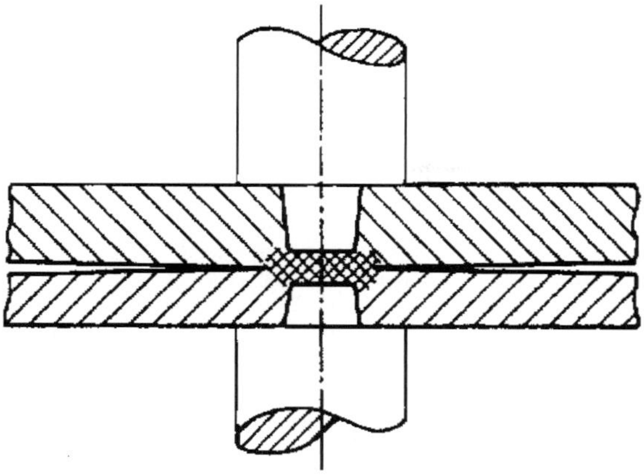

Figure 4
Lap weld (after Ruge 1993).

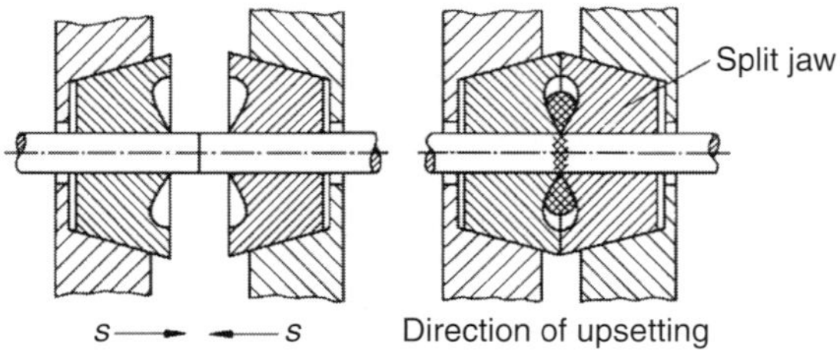

Figure 5
Butt welding (after Ruge 1993).

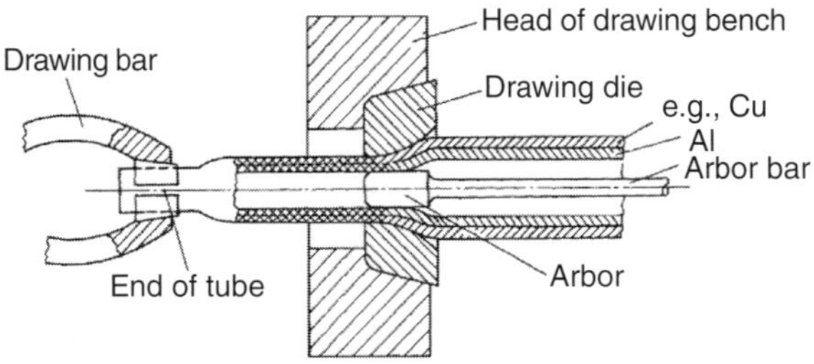

Figure 6
Cold-pressure welding with drawing processes (after Ruge 1993).

on top of another and deformed between two plungers. This results in a point- or line-shaped joint. A schematic of the principle of lap welding is shown in Fig. 4.

In butt welding, the joining parts are, after appropriate cleaning, fixed into a holder and placed with their end faces towards each other and then plasticized. A flash is formed by upsetting (Fig. 5).

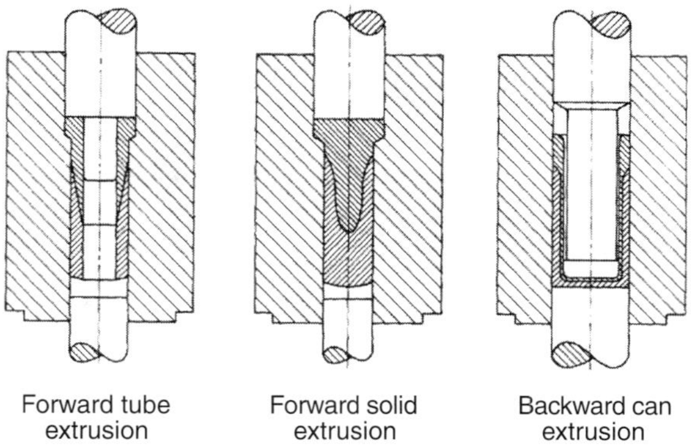

Figure 7
Alternatives to cold-pressure welding with extrusion processes (after Ruge 1993).

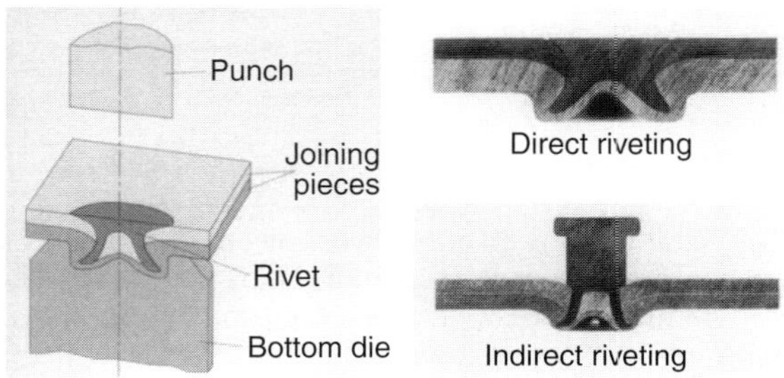

Figure 8
Function principle of riveting using a half-hollow rivet; examples of joining element structure with direct and indirect riveting (after Budde and Pilgrim 1995).

The welding joint has a higher strength than the base material due to strain hardening.

During cold-pressure welding with drawing processes, composites with tubular as well as round massive shapes can be manufactured. This process is also suitable to generate material composites which cannot be made by fusion technology. One example is the aluminum–copper pipe connection. The process method is shown in Fig. 6.

Cold-pressure welding with extrusion processes can be divided into three basic kinds of process: forward tube extrusion; forward solid extrusion; and backward can extrusion.

The process scheme of these alternatives is shown in Fig. 7. The advantage of forward tube extrusion lies in the absence of necking and compression flash at the welding area. In addition, the welding process and working piece molding can be combined into one operation (Ruge 1993).

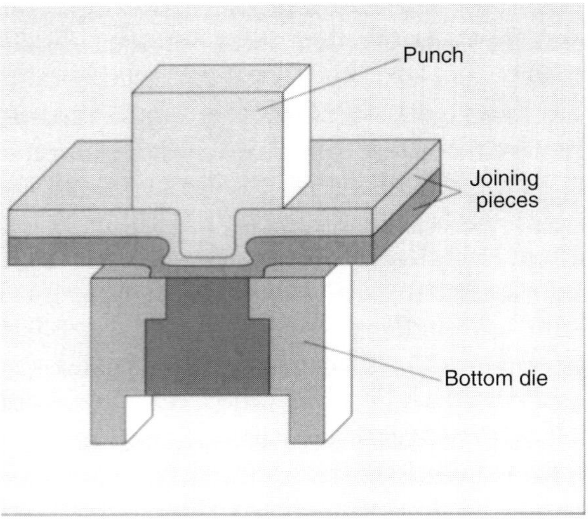

Figure 9
Function principle of clinching process (after Budde and Pilgrim 1995).

5. *Mechanical Joining*

5.1 Riveting

Using the riveting process, workpieces are joined by means of an accessory joining part (Fig. 8). The prepunching process, necessary in conventional riveting, is nowadays replaced by a rivet-cutting operation.

Riveting can be carried out by indirect riveting or direct riveting. Indirect riveting uses accessory joining parts (the rivet) for form-fit and nonpositive combining of the joining elements. Direct riveting also uses additional joining parts for the process but, in contrast to indirect riveting, this is developed as function element (Budde and Pilgrim 1995).

5.2 Clinching

Clinching is a mechanical bonding technology for materials of sheet metal and profile parts using overlapped joining with piercing and nonpiercing tools. By cold heading, which causes flattening or extrusion, a form-fit and nonpositive joint is generated. In contrast to riveting, no additional accessory joining part is needed. Figure 9 shows the function method of clinching.

Bibliography

Andernach und Bleck GmbH Website http://www.andernachu.bleck.com/willkommen/

Budde L, Pilgrim R 1995 *Stanznieten und Durchsetzfügen*. Die Bibliothek der Technik. Verlag moderne Industrie, Vol. 115

High Energy Metals Inc. Website http://www.highenergymetals.com/page2.html

Kreye H 1997 Melting phenomena in solid state welding processes. *Welding J. Res.* **5**, Suppl. 56, 154s–8s

Kreye H, Hammerschmidt M, Steigleder B, Woidneck C-P 1985 Über den Bindemechanismus beim Explosivschweißen. *Schweißen Schneiden* **37**, 297–302

Pocalco A 1987 Fabrication of explosion welded titanium-clad composites. *Welding J* **66**, 24–30

Rhodes C G, Mahoney M W, Bingel W H, Sparling R A, Bompton C C 1997 Effects of friction stir welding on microstructure of 7075 aluminum. *Scripta Materialia* **36**, 67–9

Ruge J 1993 *Handbuch der Schweißtechnik—Verfahren und Fertigung Dritte Auflage*. Springer Verlag, Berlin, Vol. 2

Russel M J, Shercliff H R, Threadgill P 1997 *Development and Modelling of Friction Stir Welding*. Cambridge University Engineering Department Technical Report. .

Schofer E 1999 *Fügen zukunftsweisender Werkstoffe—Fortschritte, Erfahrungen, Tendenzen 6*. Aachener Schweißtechnik Kolloquiumpp, Shaker Verlag, Aachen, 71–8

Spindler D E 1978 *Inertia Friction Welding*. In: *Proc.* American Welding Society, Miami, FL

Suthoff B 1999 *Optimiertes Reibschweien zukunftsweisender Verbunde 6 (Optimized Friction Welding of Future Composites). Aachener Schweitechnik Kolloquium,* Shaker Verlag, Aachen

U. Dilthey

L

Laser Cutting

By the end of 1999 more than 20 000 high-power (>1000 W) industrial carbon dioxide (CO_2) lasers were integrated with two- and three-axis motion systems and used as sheet-metal cutters. Another 3500 units with power <1000 W were being used in nonmetal cutting applications. High-power (>500 W) solid-state (Nd:YAG) lasers are gaining importance in three-dimensional sheet-metal cutting applications. As a result laser cutting is the single largest market for high-power lasers in manufacturing operations.

The noncontact and rapid cutting rate characteristics of the process lead to cost-effective processing, which typically results in a profitable operation. Since profits and productivity are the keys to successful manufacturing, it is easy to understand why the laser has become a commonly used tool in cutting operations.

Because of technological differences between the cutting of metals and nonmetals, it is necessary to treat the details of each separately, although the systems employed for both material groups may be quite similar.

1. Metal Cutting

The process by which lasers cut metals is simple in concept. The photon energy in a focused laser beam is absorbed in a thin surface layer of the metal. This absorbed energy is converted into heat, raising the metal to its melting point and, in some situations, to the point of vaporization. A hole with molten walls is produced through the metal in microseconds. Movement of this hole along a cutting path, while laser energy continues to be delivered, produces a narrow cut in the metal. A continuous jet of gas, coaxial with the laser beam, blows the cut byproducts through this hole. The width of this cut (kerf), the quality of the cut edges (squareness and striations), and the appearance of the cut underside (dross) are determined by choice of laser, beam quality, delivered power, and type of motion employed (beam, workpiece, or combination). Figure 1 schematically identifies the factors involved in producing a high-quality cut.

1.1 Cut Thickness

Beam power vs material thickness or rate is a straight-line function for most ferrous metals cut by lasers up to about 15 mm in thickness. A simple relationship is that cut thickness scales with power and is inversely proportional to speed.

Thickness of cut is attained with some degradation in cut edge quality, especially in ferrous metals thicker than 6 mm. The maximum thickness possible, therefore, is a function of power, rate, and a compromise on cut quality. Currently, a thickness of 25 mm is considered as a practical maximum for steels. In certain alloyed metals the Nd:YAG laser produces quality cuts in thicknesses up to 25 mm. However, these cuts are produced at extremely slow rates compared to kilowatt CO_2 laser cuts.

1.2 Surface Reflectivity

At room temperature, untreated steel surfaces reflect 93% of CO_2 laser light (10.6 μm) and 75% of Nd:YAG laser light (1.06 μm). To overcome the far infrared CO_2 light reflectivity factor, current metal cutters rely on the high energy density generated by 1000–3500 W CO_2 lasers operating with high beam quality. The key is the small amount of absorbed power (7–8%), which couples into

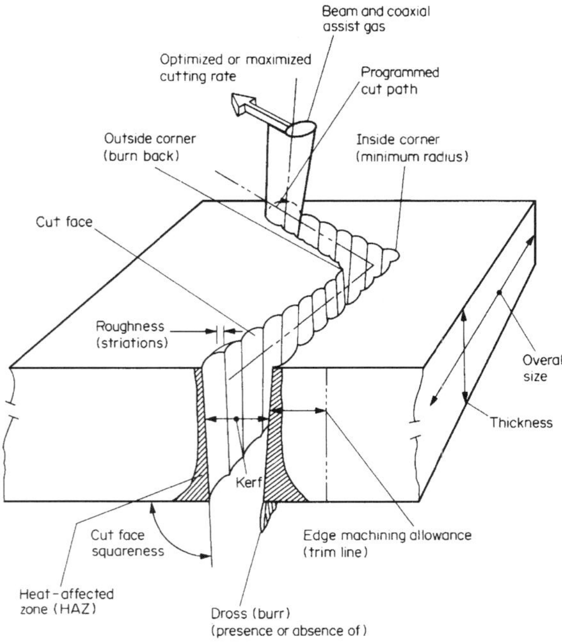

Figure 1
Laser cutting quality factors.

the metal, forming a molten puddle. This molten metal has enhanced beam absorption characteristics that couple more than 50% of the beam power, allowing the required hole to be formed in microseconds. The shorter wavelength laser, Nd:YAG, inherently couples more energy into the metal leading to the cutting mechanism, albeit at a slower process rate.

1.3 Beam-assisted Techniques

Exothermic gas-jet cutting is a process in which a flow of oxygen is directed to the metal surface in a stream concentric with the beam. The heat of the laser-produced molten puddle causes the oxygen to ignite and burn through the thickness of the metal. This exothermic reaction is confined to the focused laser spot diameter, differentiating this process from other gas-cutting techniques. The gas pressure ejects the molten metal from the cut kerf. In typical mild steel applications, the gas flow is at subsonic levels.

Oxygen-assist laser cutting is the standard process for mild steel applications. Control of gas pressure, shape of the gas stream, and positioning of the gas nozzle orifice above the metal surface are critical factors.

In highly alloyed steels, such as stainless steel, expelled metal may cling to the cut backside. This material (dross) must be removed mechanically. Newly developed endothermic techniques using pulsed CO_2 laser beams and high-pressure (15 bar) nitrogen gas jets can reduce or eliminate the dross. Cutting 2–3 mm thick stainless steel with high-pressure nitrogen produces excellent cut quality.

1.4 Kerf Width

The narrowness of the kerf (material cut away) is a major advantage of the laser process. A narrow kerf allows cut patterns to be nested as close as one beam diameter apart, an effective way to reduce scrap metal. In 2 mm thick mild steel, kerf widths can typically be 0.1 mm. Sharply contoured and profiled cuts can be made, especially in sharply profiled shapes.

Kerf width is a function of beam quality, gas pressure, processing rate, and motion parameters. These factors are interrelated, generating a high number of process variables.

1.5 Cut-edge Conditions

In thin metals, 1 mm or less, square side cuts can be obtained when cut parameters are optimized. As metal thickness increases, squareness will degrade. A general rule of thumb is to expect a taper of 1/1000 of the sheet thickness up to 6 mm thickness.

Striations resulting from the laser melting process are generally present. The coarseness of these striations increases with increasing metal thickness. Continuous wave (CW) CO_2 laser cutting can produce surface roughness values 8–15 μm in 1.6 mm thick cold-rolled steel and 30–35 μm in mild steel. For stainless steels, surface roughness of 30–50 μm in thin-gauge sheets is routine when using oxygen assist. Nitrogen gas can reduce this roughness to 10 μm in 2 mm thick applications.

Careful control of beam, gas, and process rate parameters can produce small increases in the heat-affected zone (HAZ) of mild steel. An HAZ of 0.5 mm with continuous wave CO_2 lasers and 0.15 mm with pulsed CO_2 lasers is typical in 3 mm thick mild steel.

Auxiliary optical systems produce a circularly polarized laser beam which is favorable for cutting straight-sided profiled shapes.

1.6 Laser Metal Cutting Advantages

A long list of benefits makes laser sheet-metal cutting a relatively easy process to justify. Minimal kerf is a significant advantage for the laser over most other cutting processes, except abrasive fluid jet. This feature allows for maximizing sheet metal usage by reducing scrap. The laser process is noncontact, eliminating part distortion. The focusing optics are always in place, so there are no tool set-up and storage costs.

The ability to digitize part drawings and cut under computer numerical control (CNC) or with a personal computer (PC) reduces order cycle time to a matter of hours, not days. This productivity improvement is considered a prime reason for laser system purchase.

Laser-cut edges are superior to edges produced by most other cutting processes in every respect. Concentrated laser beam energy produces little or no thermal effects. Piece parts cut by the laser do not require subsequent machining operations.

1.7 Cutting Rates

Cutting rates vary with the metal being cut, the thickness of these metal, and the profiles of the part to be cut. Straight line cutting rates of 150 mm s^{-1} are typical of current 1.5 kW CO_2 laser systems. Profiling can reduce this rate to 21 mm s^{-1}. However, these rates are significantly higher than those for other thermal cutting processes.

2. Nonmetal Cutting

Nonmetal cutting has three requirements: (a) a focused beam of energy at a wavelength that will be absorbed by the material to be cut so that melting, chemical degradation, or vaporization can occur; (b) a concentric jet of gas (usually compressed air) to remove cut by-products; and (c) a means of generating cuts in straight lines or curved patterns. One result of the process, residual thermal effects in heat-sensitive materials, is a greater problem than in metal cutting. Therefore, the laser may not always be applicable to cutting all nonmetals.

2.1 *Material Considerations*

Because the nonmetal cutting process requires moderate amounts of power, the only limitation on cut thickness is the quality of the cut. In practice, the majority of cutting applications occur in material less than 12 mm thick. For quality cuts, this thickness is reduced to 6 mm or less. All nonmetals that can be thermally cut will absorb CO_2 light almost completely. Any beam reflectivity losses are negligible. Some heat-sensitive materials can be successfully cut if the edge conditions are acceptable. Nd:YAG lasers can be used to cut materials that will absorb the 1.06 μm wavelength.

2.2 *Gas-assist Techniques*

Nonmetal cutting applications require a gas jet to remove process by-products from the cut area. Oxygen-free gases are used when the products combust. Many nonmetals produce potentially hazardous fumes when thermally cut, so care should be taken to exhaust by-products from the cutting area.

2.3 *Kerf Width*

Beam quality and focus, gas assist, material thickness, and process rate determine kerf width. Narrow kerf is a feature of nonmetal cutting and it is especially important in the cutting of compactly nested parts, such as those produced in fabric cutting.

2.4 *Cut-edge Conditions*

Sharp-edged cuts are typically produced in nonmetal cutting applications. Some thermally sensitive materials exhibit slightly rolled cut-edge surfaces resulting from the thermal effects of the process. Combustible materials, such as wood or paper, may exhibit a discoloration graduating to an ash residue as thickness increases. Many woven fabrics cut by laser beam exhibit a desirable fused edge, which resists unraveling.

Nonmetals react in a variety of ways to laser-generated heat, so it is difficult to generalize on edge roughness. Some materials, such as acrylic and Plexiglas, exhibit a completely clear edge after cutting. Others, such as epoxy, produce a gummy surface. A general rule is that thermally sensitive materials will usually exhibit less than perfect edge effects.

Because the nonmetal cutting process utilizes all of the laser-generated heat, and because the materials being cut are generally poor thermal conductors, residual heating effects are not a serious problem. However, for those materials that are thermally sensitive, some minor subsurface degradation can be expected.

Circularly polarized output beams contribute to kerf width control, regardless of the direction of the cut. This feature allows perfectly square-edged cuts to be made.

2.5 *Nonmetal Cutting Advantages*

Many of the cutting advantages cited in Sect. 1.6 also apply to nonmetal cutting, the most important of these being the noncontact aspect of light beams. Cutting fragile, brittle, or very thin materials, without mechanical contact or force applied to the material surface, is an advantage that shows up in manufacturing cost reduction.

Another major factor in nonmetal cutting is the extremely narrow kerf obtained, which provides an advantage in material utilization. A narrow kerf allows the beam path on one piece to be separated from the next by one beam diameter. This contributes to reduced scrap, which can be a positive economic factor.

Of all the nonmetal cutting techniques, except fluid jet, the laser is the most versatile. Essentially all nonmetals can be cut by the beam from a CO_2 laser. The same beam used to cut wood can cut plastics, asbestos, rubber, glass, paper, cloth, ceramics, fiberglass, and composites. The same laser used to cut these nonmetals can, with sufficient power, cut metals. No other cutting process offers this versatility.

2.6 *Cutting Rates*

Cutting rates are a function of material, thickness, and beam power, and are thus too varied to be detailed in this summary. A general rule of thumb is that cutting rate and cut thickness are linear functions of output power. Typical rates for 1 mm thick nonmetals are in the range 10–25 ms^{-1}.

3. *Summary*

Laser cutting has been proven as a technically superior, cost-effective alternative in a broad spectrum of manufacturing requirements. Precision high-quality cuts, at very rapid rates, are hallmarks of this widely accepted process.

Bibliography

Belforte D A 1998 Laser metal cutting. *Ind. Laser Rev.* **13**(8), 25–7

Belforte D A 1998 Non-metal cutting. *Ind. Laser Rev.* **13**(9), 11–3

Migliore L 1996 In: Migliore L (ed.) *Laser Materials Processing*. Dekker, New York, pp. 139–64

Powell J 1998 *CO_2 Laser Cutting*. 2nd edn. Springer, Berlin

Steen W M 1998 *Laser Materials Processing*. Springer, Berlin, pp. 69–107

D. A. Belforte

Laser Forming of Metals

The process of laser forming has been investigated widely since 1987, especially regarding its extraordinary potential for rapid prototyping in sheet metal working. The reason for the increasing interest in laser forming is a specific advantage of the process, namely the possibility of producing sheet metal parts with nearly unlimited geometrical complexity without using special tools or dies. This advantage of flexible forming was already known from flame bending, which has been used for diverse straightening tasks in industrial applications since the early 1920s. Inspired by the development of reliable and powerful laser systems for production engineering tasks, the idea of using lasers as controllable heat sources for thermal bending operations arose. Starting from basic investigations that enabled an overview of the geometrical possibilities and future applications of laser forming, research soon focused on analyzing the underlying forming mechanisms. These investigations led to numerous ideas for applications of laser forming, such as rapid prototyping, laser straightening, and laser adjustment.

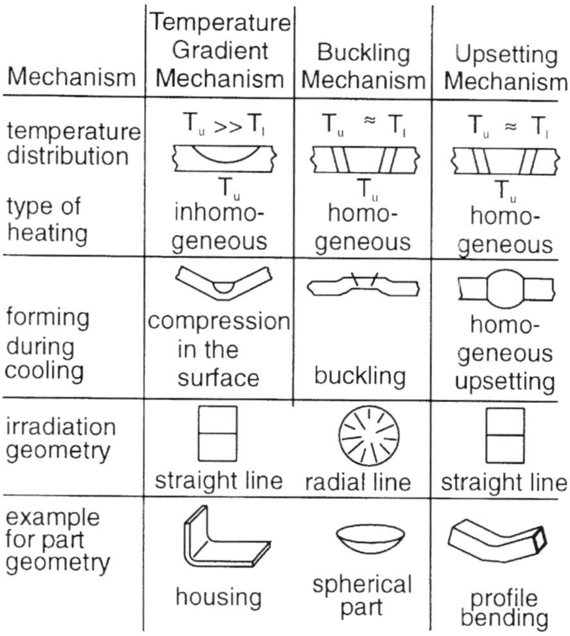

Figure 1
Mechanisms of laser forming.

1. Principle of Laser Forming

A temperature gradient is generated by locally heating a sheet metal part with a laser. The increased temperature leads to a thermal expansion of the heated area. The surrounding structure counteracts the thermal expansion. This obstructed thermal expansion results in compressive stresses. In addition to the reduced yield stress owing to increased temperature, the compressive stresses cause a local compressive plastic strain. The plastic strain gradient results in the forming of the part. Depending on the temperature field compared with the sheet geometry, a separation into several mechanisms can be made (Fig. 1).

In the temperature gradient mechanism a significant temperature difference between the upper and the lower side of a sheet metal part is generated by a relatively short laser pulse or a fast laser scan along the bending line. A bending towards the laser with a typical value in the range 0.001–3° can be manufactured with one single laser spot or scanning line.

In the buckling mechanism, as well as in the upsetting mechanism, a relatively homogeneous temperature field perpendicular to the sheet is generated by a high thermal conductivity or relatively long laser pulses as well as a low scanning speed in continuous-wave (CW) mode. In the buckling mechanism the buckling stiffness of the sheet is relatively low whereas the lateral expansion of the temperature field is large. This causes buckling of the part. The bending angle is usually in the range 0.1–15°. The direction of the bending can be towards or away from the laser depending on the precurvature of the part.

The upsetting mechanism is dominant when, next to the homogeneous temperature field in the sheet thickness, the lateral expansion of the heated area is smaller than the sheet thickness. This causes a shortening of the structure. In aligning processes a shortening in the region of 10 μm can be processed.

2. Process Parameters

All kinds of industrial lasers can be used for laser forming. For macroscopic applications CO_2 lasers are dominant owing to their power. Nd:YAG and excimer lasers are suitable for adjustment tasks. For special applications frequency doubled Nd:YAG lasers, fiber lasers, and diode lasers become more dominant. There is no limitation of laser forming by the sheet thickness. Thin metal foils in the region of 10 μm can be formed with short-pulse lasers. Thicker sheets of a few millimeters can be formed with Nd:YAG or diode lasers. For thick sheets of several centimeters high-power lasers such as CO_2 lasers are used.

The main influences on the results can be divided in three groups: laser parameters, material properties, and geometrical properties. First, the absorbed energy has to be higher than the threshold energy. The threshold energy is the energy where plastic deformation starts, and it is dependent on the material

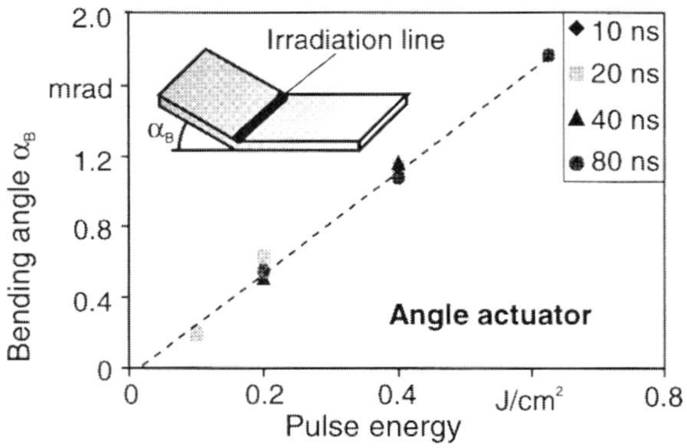

Figure 2
Influence of laser pulse energy on bending angle.

properties and the stiffness of the geometry. The stiffness of the part has to be high enough to stop counter-deformation, but low enough to allow forming in the opposite direction, the direction of the laser forming. With higher laser power a linear correlation between laser power and the forming effect occurs (Fig. 2).

In the temperature gradient mechanism the bending effect is linear correlated to the reciprocal value of the sheet thickness squared. The effect of the sheet thickness for the other two mechanisms is not that strong and can be expressed by a linear correlation between the forming effect and the reciprocal value of the sheet thickness.

The beam geometry has two effects. First, increasing the beam diameter increases the forming effect owing to the increased forming zone. Second, the energy density is lowered, which decreases the maximum temperature and therefore the maximum plastic strain. The balance of both effects determines the forming effect.

In principle, the surface does not affect the process. However, the absorbed energy, which is relevant for the balance between the mechanisms and the effect of the forming, is strongly dependent on the material, laser wavelength, and surface roughness and coating. The absorptivity of the sheet can be increased by using special coatings, e.g., graphite or TiN. The surface is not modified by the laser forming process because of the low temperatures involved, which are usually lower than the melting temperature. However, for large forming results the surface can be influenced by the coating, oxidation of the sheet, and quality after the laser forming.

The most important material property is the thermal expansion coefficient. A higher thermal expansion coefficient results in a higher thermal expansion at the same temperature and therefore in a larger forming result. The thermal conductivity influences the temperature field; a lower thermal conductivity reduces the heated volume, which reduces the forming effect. However, a reduced heated volume causes higher maximum temperatures and therefore larger effects.

3. Applications

3.1 Rapid Prototyping

One possible application of laser forming is the rapid prototyping of metallic components from sheet metal or profiles. The costs of conventional tools and dies for producing only a small number of functional prototypes are in most cases far too high, and the process too time-consuming, for an economical production. These drawbacks can be avoided by using laser forming as a flexible forming method for rapid prototyping. As laser forming is an incremental process, many successive irradiations have to be processed in order to generate a high degree of forming, e.g., a large bending angle. Therefore, a suitable irradiation strategy and an online measurement of component geometry are always necessary for rapid prototyping of complex-shaped parts by laser forming. A suitable closed-loop control system for incremental laser bending of profiles is presented in Kraus (1997). The advantages of using laser forming for prototyping of profiles are the high flexibility of the process, e.g., if slight modifications of the geometry are necessary, and the comparatively high processing speed of several hours in comparison to several days or weeks for conventional processing.

3.2 Laser Straightening

In addition to investigations concerning the rapid prototyping capabilities of laser forming, the potential of the process to become a substitute for flame bending has also been investigated. A study on laser straightening of welded car body parts shows that the distortions introduced by welding of the parts can be reduced successfully by local laser forming. The irradiation strategies developed for rapid prototyping of profiles can also be used for these laser-straightening tasks.

3.3 Laser Adjustment of Preassembled Systems

In the production of electrical, optical, and micromechanical parts, an accuracy in the submicrometer range is often necessary. In the process of laser adjustment of preassembled systems, the functional element is fixed on an actuator, and the actuator is fixed on the surrounding area. The first step in this process sequence, the joining and assembly of the functional element and the actuator, can be done in a fast and inaccurate manner. In the second step, the laser adjustment step, a controlled displacement of

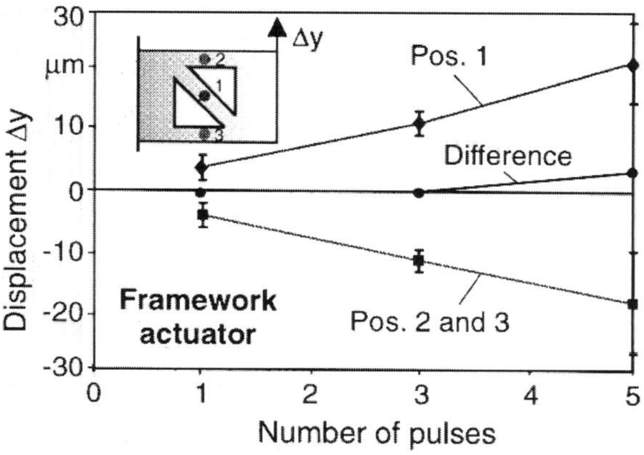

Figure 3
Displacement of the framework actuator.

the functional element is done by irradiating particular points of the actuator with a Nd:YAG or excimer laser. An accuracy of 0.1 mrad in two directions can be reached in less than 20 s.

With special actuators a movement in two or more directions is possible (Fig. 3). Irradiating position 1 of the framework actuator results in a movement in the positive *y*-direction, which can be reduced by irradiating positions 2 and 3. Using this principle the tolerance field of a system can be reduced to the required value.

4. Conclusion

The future trends in laser forming can be seen in production areas where small forming effects are necessary. Examples of these are the adjustment and straightening applications for the formation of small, complex shapes for micro parts such as sliders for hard disk drives.

Bibliography

Frackiewicz H, Mucha Z, Trampczynski W, Baranowski A, Cybulski A 1987 A method of bending metal objects. *Eur. Pat.* 0 317 830 A2

Geiger M, Vollertsen F, Amon S 1991 Flexible Blechumformung mit Laserstrahlung. *Blech Rohre Profile* **38**(11), 856–61 (in German)

Kraus J 1997 *Laserstrahlumformen von Profilen* Maisenbach, Bamberg, Germany (in German)

Magee J, Watkins K G, Steen W M 1998 Advances in laser forming. *J. Laser Appl.* **10**(6), 235–46

Thomson G A, Pridham M S 1997 Controlled laser forming for rapid prototyping. *Rapid Prototyp. J.* **3**(4), 137–43

Vollertsen F 1996 *Laserstrahlumformen—Lasergestützte Formgebung: Verfahren, Mechanismen,* Modellierung, Meisenbach, Bamberg, Germany, (in German)

M. Geiger, B. Müller, and T. Hennige

Liquid Phase Sintering: Ceramics

Liquid-phase sintering (LPS) has been used as an important densification process to manufacture various ceramic components from shaped powder compacts. Two major advantages of LPS are enhanced sintering kinetics and additional tailorability of properties. A number of commercial ceramic products takes advantage of these benefits, including Al_2O_3, ZrO_2, TiO_2, $Pb(Zr,Ti)O_3$, ZnO, mullite, SiC, Si_3N_4, and SiAlON, for structural, electronic, chemical, and thermal applications. LPS is being further exploited as an enabling process to design multifunctional ceramic components.

In a typical LPS process of ceramics, at least two solid powders are first homogeneously mixed by dry or wet mixing. The mixed powder is formed into a green compact of 50–60% relative density by various forming methods, e.g., uniaxial die pressing, cold isostatic pressing, slip casting, injection molding, etc. The structure of ceramic green compact is best described as random loose or dense packing. A liquid phase, typically 1–20 vol.%, is formed upon heating the mixed powder compact. While the mixture of solid particles and liquid sinters together, porosity of a powder compact gradually diminishes to form dense ceramic parts.

Three general requirements for LPS (Kingery 1950) are: (i) presence of liquid at sintering temperature; (ii) good wetting of solid with a liquid; and (iii) appreciable solubility of solid in liquid. Unlike LPS of metals, the liquid in ceramic systems is different in its viscous grain boundary phase, limited mutual solubility, and slow reactions between constituents. Hence, the LPS of ceramics is significantly different from that of metals (German 1985).

The presence of a small amount of liquid phase during sintering greatly differentiates the behavior of LPS from solid-state sintering (SSS). LPS can be a dominant process even with <1 vol.% liquid in the system during the sintering. Although LPS of technical ceramics is somewhat similar to the firing of pottery or porcelain it differs in the amount of liquid during sintering. Densification of a system with more liquid can be best described as viscous composite sintering or viscous glass sintering.

1. Densification Stages and Driving Force for LPS

Densification during LPS is divided into three distinctive stages as defined by three different rate-controlling processes. With heating, the densification mechanism progressively changes from grain rearrangement to solution-precipitation, and then to final pore (or vapor phase) removal stages. There is significant overlap between these stages in real powder compacts as shown schematically in Fig. 1.

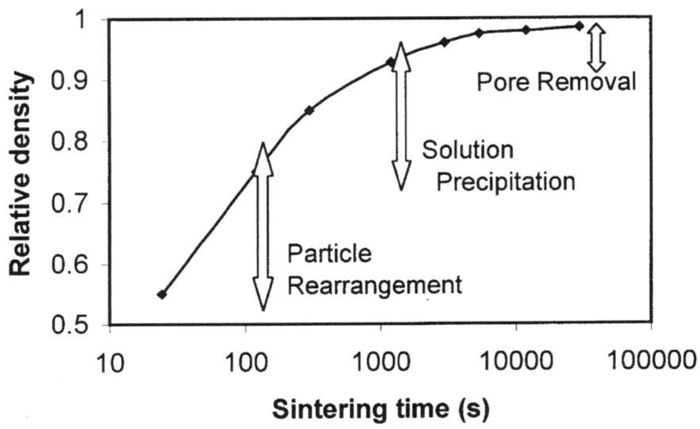

Figure 1
A densification curve as a function of sintering time showing the three stages of liquid-phase sintering and the possible overlap between stages.

Densification during LPS is driven by the thermodynamic driving force to minimize the interfacial free energy of a system. In general, the change in free energy when going from one configuration to another in a solid–liquid–vapor system is

$$\Delta G = \Delta A_{sv}\gamma_{sv} + \Delta A_{ss}\gamma_{ss} + \Delta A_{sl}\gamma_{sl} + \Delta A_{lv}\gamma_{lv} \quad (1)$$

where ΔA_{sv}, ΔA_{ss}, ΔA_{sl}, and ΔA_{lv} are changes in the various interfacial areas and γ_{sv}, γ_{ss}, γ_{sl}, and γ_{lv} are their corresponding interface energies (subscripts s, l, and v represent solid, liquid and vapor), respectively. If good wetting of the solid by a liquid is assumed, ΔA_{sv} and ΔA_{ss} are unimportant. Also, when there is no grain growth, ΔA_{sl} is negligible. Therefore, $\Delta A_{lv}\gamma_{lv}$ is the primary and most important component of the driving force for LPS.

2. Particle Rearrangement

In the initial stage of LPS, a number of serial and parallel processes occur including melting, wetting, spreading, and liquid redistribution. Both the solid and liquid are subject to rearrangement due to unbalanced capillary forces around solid particles giving rise to shear and rotational movements of particles. Liquid films between particles act as a lubricant. The rearrangement of particles proceeds in the direction of reducing porosity from the loose random packing to the dense close packing (Shaw 1986). As density increases, particles experience an increasing resistance to further rearrangement by physical constraints from neighboring particles and form a close packed structure.

Rearrangement can be modeled as a viscous flow process with a geometric model of a liquid film between solid particles. Assuming a Newtonian liquid, the deformation rate of particles with a liquid film is proportional to the shear stress to reduce the liquid–vapor surface area (Kwon 1996). Accordingly, the densification rate is given by

$$\frac{\mathrm{d}(\Delta\rho/\rho_0)}{\mathrm{d}t} = A(g)\frac{\gamma_{\mathrm{lv}}}{\eta}r_{\mathrm{s}}^{-1} \tag{2}$$

where ρ is the relative density, t is time, $A(g)$ is a geometric factor, η is the liquid viscosity, and r_s is the radius of the solid particle. The term $A(g)$ decreases with increasing volume fraction of solid, liquid, and relative density. At approximately 25–35 vol.% liquid, full densification can be achieved by rearrangement alone.

3. *Solution-precipitation*

When the rearrangement of particles becomes too slow or stops, additional densification can be achieved by the dissolution of solid at grain contacts, resulting in the center-to-center approach of particles. Figure 2 schematically shows a two-sphere model with a liquid meniscus at the contact. The solid solubility in the liquid is proportional to the normal traction at the grain contact arising from the capillary forces (Laplacian forces) pulling the solid particles together to minimize $\Delta A_{\mathrm{lv}}\gamma_{\mathrm{lv}}$ in Eqn. (1) (Heady and Cahn 1970). The effective stress at a particle contact can be as high as 10 MPa for particles with a diameter of $\sim 1\ \mu$m.

The dissolved solute from the grain contact transfers to the uncompressed side of the grains by diffusion through the liquid phase followed by precipitation. The dissolution rate of the solid decreases as the contact area increases because there is a gradual reduction of effective stress at the contact. Accordingly, the densification rate decreases as the density of a powder compact increases.

With appropriate geometrical models for grain, liquid, and pore structures, parametric relationships

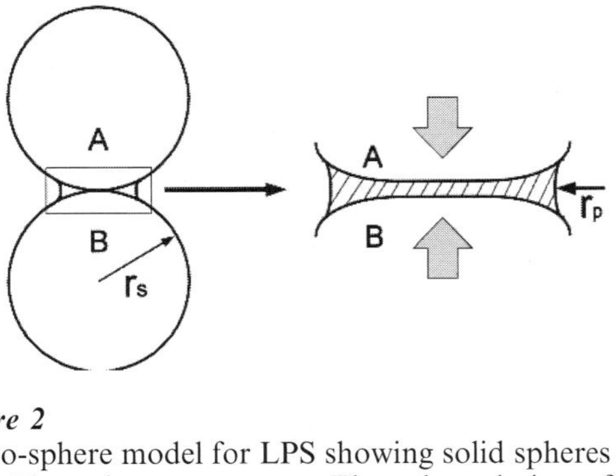

Figure 2
A two-sphere model for LPS showing solid spheres with a liquid meniscus at contact. The enlarged view of the grain contact shows a flattened contact separated by a liquid film. Note that the spheres are also completely wetted by the liquid in a satisfactory LPS system.

for the densification rate can be derived. Assuming pores are located at the edges or corners of a tetrakaidecahedron (Wray 1976), the driving force can be determined by the geometry of the solid–liquid–vapor assemblage (Kwon and Messing 1991).

In general, there are two rate limiting processes for solution-precipitation. When material transport is limited by diffusion through the liquid phase, the densification rate is

$$\frac{\mathrm{d}(\Delta\rho/\rho_0)}{\mathrm{d}t} = B(g)\frac{\delta D_{\mathrm{b}}c_{\mathrm{l}}\gamma_{\mathrm{lv}}\Omega}{kT}r_{\mathrm{s}}^{-4} \tag{3}$$

where $B(g)$ is a geometric factor, δ is the thickness of the liquid boundary, D_b is the grain boundary diffusion constant of solute, c_l is the solubility of the solute, Ω is the molecular volume of the solute, and kT has its usual meaning. If the material transport is interface-reaction controlled the densification rate is

$$\frac{\mathrm{d}(\Delta\rho/\rho_0)}{\mathrm{d}t} = C(g)\frac{Kc_{\mathrm{l}}\gamma_{\mathrm{lv}}\Omega}{k_{\mathrm{B}}T}r_{\mathrm{s}}^{-2} \tag{4}$$

where $C(g)$ is a geometric factor and K is the interface-reaction rate constant.

It is interesting to note that both Eqns. (3) and (4) show that densification strongly depends on r_s. The solution-precipitation kinetics of an alumina–glass system, densified by isothermal LPS, was analyzed by Kwon and Messing (1989) and the rate controlling mechanism was identified by evaluating the particle size dependency of the densification rate, based on the above models.

Transmission electron microscope (TEM) studies and modeling of the structure of the thin liquid film in grain boundaries has increased the understanding of the nature of grain boundaries in LPS systems (Clarke 1987, Kleebe *et al.* 1993). The morphologies of continuous liquid intergrain boundaries, typically 1–2 nm thick, have been documented in various liquid-phase sintered ceramics by high resolution TEM.

4. *Removal of Closed Pores*

As contacts between solid particles flatten, the densification rate by solution-precipitation decreases. At a later stage of solution-precipitation the interconnected pore channels pinch off to form isolated (closed) pores to reduce $\Delta A_{\mathrm{lv}}\gamma_{\mathrm{lv}}$ of the system. The final stage of LPS starts after pore closure at a relative density range from 0.9 to 0.95. The closed pores usually contain gaseous species from the sintering atmosphere and vapor from the liquid. After pore closure, the driving stress for densification is

$$S_{\mathrm{D}} = \frac{2\gamma_{\mathrm{lv}}}{r_{\mathrm{p}}} - \sigma_{\mathrm{p}} \tag{5}$$

where σ_p is the gas pressure inside a pore. If r_p and σ_p remain small (i.e., $S_D>0$) densification will proceed (Kang *et al.* 1989).

If gaseous species are entrapped in closed pores, the diffusion of gases out of the pores to the surface of the powder compact can control the rate of densification. Unlike solids, the solubility of gaseous species in liquids is very high. In addition, the liquid vaporizes, transfers, and condenses to other pore ;sites within a pore. Hence, the sintering atmosphere can significantly influence final-stage densification. In order to achieve full density the temperature–time cycle for sintering must be optimized for the component size to facilitate gas transport from the pores. The change of sintering atmosphere in the final stage is effective in achieving a higher final density due to an increased chemical driving force for the transfer of entrapped gases from pores. Complex chemical reactions between the sintering atmosphere and the powder compact can also occur at high temperatures to form gaseous reaction products within closed pores. Such gas-forming reactions are partially responsible for the incomplete densification by LPS in some cases.

Pressure-assisted sintering techniques, e.g., hot pressing and hot isostatic pressing, may be used to lower the sintering temperature, achieve a higher final density, and produce a more homogeneous microstructure. The applied stress (or pressure) enhances the driving forces for densification in all three LPS stages.

Several processes can occur simultaneously during the final stage of LPS including growth and coalescence of grains and pores, dissolution of solid into liquid, phase transformations, vaporization of solid and liquid, and the formation of reaction products between liquid and solid. These are not necessarily essential densification processes, but important to achieve a desired microstructure, as described below.

5. *Grain Growth*

If $\Delta A_{lv}\gamma_{lv}$ in Eqn. (1) approaches its minimum value in the final stage of LPS the free energy of system can be further reduced by minimizing $\Delta A_{sl}\gamma_{sl}$, resulting in grain growth. While the dissolution of the solid in a liquid contributes to densification, a differential dissolution between different shapes and sizes of particles results in grain growth by Ostwald ripening. The dissolved solute from small or sharp corners of particles tends to reprecipitate on coarser particles. Hence, coarse particles grow as fine particles are consumed. In general, the grain growth kinetics of spherical particles can be described by

$$r_{st}^n - r_{s0}^n = k(G)t \qquad (6)$$

where, r_{st} and r_{s0} are the radii of grains at time t and 0, respectively, and $k(G)$ is the rate constant for grain growth. The grain growth exponent, n, varies depending on the rate-controlling mechanism, where $n=3$ for diffusion control and $n=2$ for interface-reaction control.

6. *Reactive and Transient Liquid-phase Sintering*

There are many LPS systems in which a chemical reaction takes place during sintering. The chemical driving force for various reactions can be much higher than the driving force from interfacial energies in some multiphase LPS systems. The active control of the temperature–time cycle can be used either to promote or limit certain reactions based on phase equilibria. The material can also react with gases in the atmosphere to form gas, liquid, or solid reaction products.

Transient liquid phase has been used to enhance the densification kinetics in some special systems. The transient liquid can be removed by evaporation or remain in the system by forming a solid solution or crystalline reaction products. Accordingly, transient LPS can be an alternative route to fabricate fully dense polycrystalline ceramics, if solid-state sintering is difficult.

7. *Design of Microstructure for Applications*

There exist few real ceramic systems that closely follow the sphere models depicted in Fig. 2. In real powder compacts, which have different particle shapes, sizes, and phase distributions, a number of complex processes may occur that significantly deviate from such idealized processes. For instance, Figs. 3 and 4 show typical microstructures of liquid-phase sintered silicon carbide and silicon nitride, respectively. Note that grains in both figures are jagged and elongated with limited round shapes.

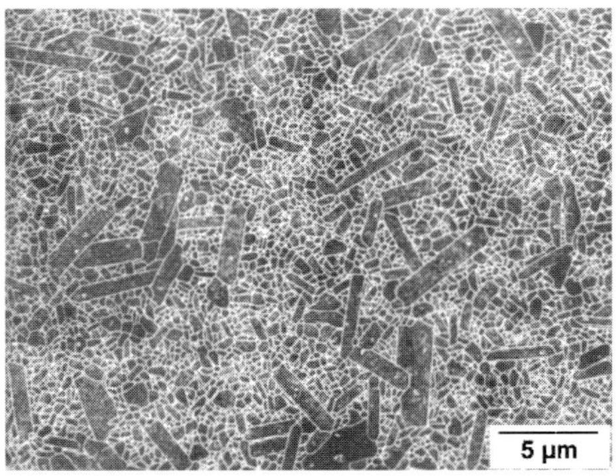

Figure 3
Microstructure of SiC by the controlled grain growth for an improved fracture toughness (courtesy of Y.-W. Kim, Seoul City University).

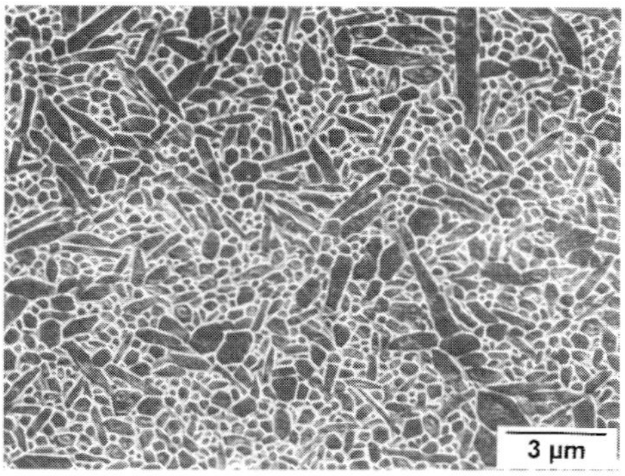

Figure 4
Typical microstructure of liquid-phase sintered silicon nitride revealed by plasma etching; optimized for a high rolling contact fatigue life (courtesy of V. Pujari, Saint-Gobain Ceramics and Plastics, Inc.).

In order to tailor the microstructure of components for targeted applications the known mechanisms of LPS can be used to optimize the microstructural features. The material system may be defined by information from phase diagrams, thermochemical reactions, solubility, viscosity, diffusivity, and vapor pressure. Processing parameters such as temperature, atmosphere, time, and heating and cooling rates can also be determined from data- and experience-bases. Unknown input data can be estimated using physical and thermochemical packages based on available models.

For instance, a reduction of sintering temperature of alumina substrates can be obtained using the principles of LPS and phase diagram data. The microstructure for the Si_3N_4 rolling-element bearing components can be optimized to achieve an improved fatigue life and crack resistance. The introduction of β-SiC seeds in a fine α-SiC grain matrix can be an effective means to control the β- to α-SiC transformation and grain growth resulting in a tough microstructure (Kim *et al.* 1997). The properties of ZnO varistor can be tailored by dopants to achieve the desired nonlinear properties (Gupta 1990). A special heat treatment can be used to form favorable crystalline phases in liquid grain boundaries at the end of sintering. A separate heat treatment run can also be used to tailor further the microstructure for optimized mechanical or electrical properties.

8. *Future Perspectives*

The development of ceramic alloys to take advantage of various functional properties of ceramic materials is in its infancy compared with the well-established metallic alloys. A number of advances in LPS have been beneficial in reducing the alchemy of ceramics in the future. For example, the chemical variation within a single grain was observed as the core–rim structure in SiC and Si_3N_4 ceramics. Functionally gradient ceramics can be prepared by a directional reaction of liquid. Computer simulations have been employed to predict and design microstructures. LPS has also been used as an alternative tool to process nanoceramic alloys. A small amount of liquid forming additive is intentionally added to facilitate superplastic formation.

The optimization of material system and processing parameters are not a simple task in real life due to various unknown factors, especially with new or lesser known material systems. Improved knowledge of physical and chemical factors influencing LPS processes would help to enhance the process design and control in developing new multifunctional ceramics to build smart devices, components, tools, machines, and systems.

Bibliography

Clarke D R 1987 On the equilibrium thickness of intergranular glass phases in ceramic materials. *J. Am. Ceram. Soc.* **70**, 15–22

German R M 1985 *Liquid Phase Sintering*. Plenum, New York

Gupta T K 1990 Application of zinc oxide varistors. *J. Am. Ceram. Soc.* **73**, 1817–40

Heady R B, Cahn J W 1970 An analysis of the capillary forces in liquid-phase sintering of spherical particles. *Metall. Trans.* **1**, 185–9

Kang S-J, Greil P, Mitomo M, Moon J-H 1989 Elimination of large pores during gas-pressure sintering of β'-sialon. *J. Am. Ceram. Soc.* **72**, 1166–9

Kim Y-W, Mitomo M, Hirotsuru H 1997 Microstructural development of silicon carbide containing large seed grains. *J. Am. Ceram. Soc.* **80**, 99–105

Kingery W D 1950 Densification during sintering in the presence of a liquid phase. I. Theory. *J. Appl. Phys.* **30**, 301–6

Kleebe H-J, Cinibulk M K, Cannon R M, Ruhle M 1993 Statistical analysis of the intergranular film thickness in silicon nitride ceramics. *J. Am. Ceram. Soc.* **76**, 1969–77

Kwon O-H 1996 Liquid-phase sintering. In: Brook R J (ed.) *Processing of Ceramics. Part II. Materials Science and Technology* VCH, New York, Vol. 17B, Chap. 14

Kwon O-H, Messing G L 1989 Kinetic analysis of solution-precipitation during liquid-phase sintering of alumina. *J. Am. Ceram. Soc.* **73**, 275–81

Kwon O-H, Messing G L 1991 A theoretical analysis of solution-precipitation controlled densification during liquid phase sintering. *Acta Metall. Mater.* **39**, 2059–68

Shaw T M 1986 Liquid redistribution during liquid-hase sintering. *J. Am. Ceram. Soc.* **69**, 27–34

Wray P J 1976 The geometry of two-phase aggregates in which the shape of the second phase is determined by its dihedral angle. *Acta Metall.* **24**, 125–35

O.-H. Kwon

Liquid Phase Sintering: Metals

The formation of a liquid phase is typical in sintering many engineering materials, both by intention and due to impurity effects. Liquid phase sintering has been practiced since 4000 BC, largely in the fabrication of pottery and glass-bonded ceramics, such as porcelain. Today, many products are processed by liquid phase sintering including dental abrasives, cutting tools, dental amalgams, automotive transmission gears, and refractory ceramics (German 1985).

In most liquid phase sintering systems there is only a weak chemical reaction between the constituents. Thus, surface tension from the liquid phase is a significant factor in determining the sintering rate. For such cases, the three main concerns are solid solubility in the liquid, wetting of the liquid on the solid grains, and solid phase diffusion in the liquid (German 1996). Usually the liquid wets the solid particles and has solubility for the solid. The resulting capillary force compresses the structure and allows densification when solid diffusion occurs in the liquid. Diffusion rates in liquids are relatively high, resulting in both densification and microstructural coarsening.

Liquid phase sintering densification occurs in stages as sketched in Fig. 1 (Kingery 1959). Initially, mixed powders are heated to a temperature where liquid forms. Solid-state sintering during heating can yield considerable densification prior to formation of the first liquid. Subsequent densification depends on the amount of liquid phase. Three stages of densification are encountered after the liquid forms—rearrangement, solution-reprecipitation, and final stage sintering. Figure 2 summarizes these stages by plotting the density vs. the volume of liquid phase. If there is a high liquid level, then full density can be achieved via rearrangement upon liquid formation. Alternatively, at low liquid contents the solid skeleton slows densification. A limiting case is no liquid phase and total densification by solid-state sintering.

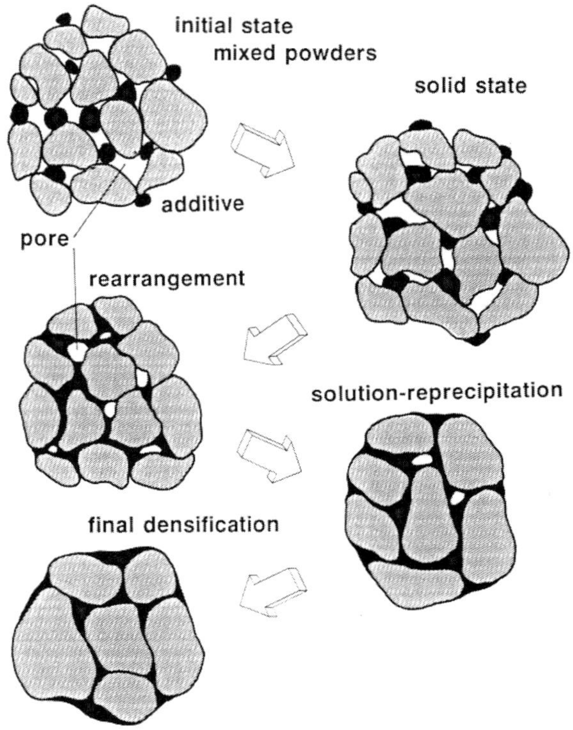

Figure 1
A schematic diagram of classic liquid phase sintering involving mixed powders which form a nonreactive liquid on heating, allowing particle rearrangement and densification by solution-reprecipitation and solid skeleton sintering.

In contrast, some systems swell during heating since the liquid dissolves into the solid (Savitskii 1993). This is evident in the cobalt–aluminum system where molten aluminum diffuses into the solid copper and increases porosity. Consequently, high sintering temperatures are required to offset this swelling reaction. Alternatively, aluminum-based alloys with copper additions exhibit densification due to a high solubility of solid aluminum in copper-rich liquid.

1. *Phase Diagram Features*

Liquid phase sintering starts from a nonequilibrium condition involving mixed powders of differing compositions. Thermodynamic attributes of the system provide gauges as to the behavior during heating. One key is the solubility of the solid in the liquid. For classic liquid phase sintering, the solid is soluble in the liquid, but the reverse solubility of the liquid in the solid is low. This ensures the liquid is not transient. Figure 3 provides a schematic binary phase diagram corresponding to the composition and temperature where liquid phase sintering is associated with favorable solubility and melting characteristics.

2. *Rearrangement*

When the liquid forms there is a densification burst due to the capillary force exerted by the liquid on the solid particles. The solid grains shrink as they dissolve into the liquid. Rapidly, the solid grains pack to a higher density, releasing liquid to fill pores between grains. For wetting liquids, the solid–liquid surface energy is lower than the solid–vapor surface energy, resulting in a reduced system energy. During rearrangement the compact exhibits a viscous flow

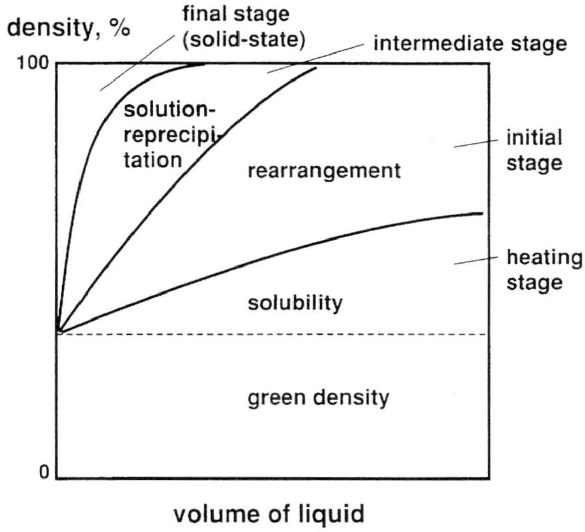

Figure 2
An overview of the densification stages associated with liquid phase sintering, where the sintered density depends on the liquid content. Some densification occurs by solid-state sintering during heating. At high liquid contents rapid densification occurs by particle rearrangement, but with lower liquid contents slower densification invokes a need for solution-reprecipitation and solid skeleton sintering.

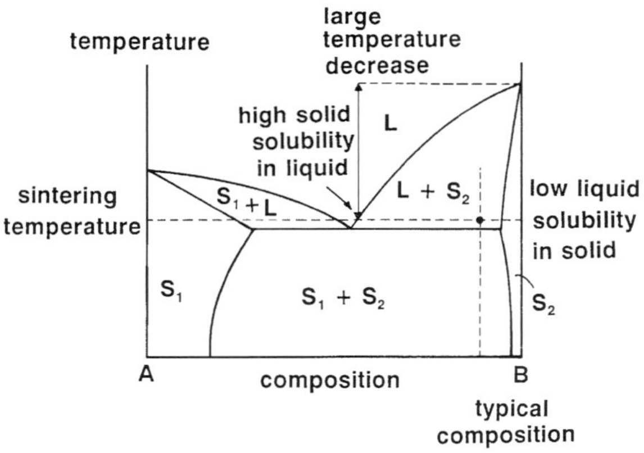

Figure 3
A model binary phase diagram showing the composition and temperature associated with liquid phase sintering in the $L + S_2$ phase field. The favorable attributes for liquid phase sintering include a suppression of the melting temperature, low liquid solubility in the solid, and high solid solubility in the liquid.

response to the capillary action. The elimination of porosity increases the compact viscosity. As a consequence the densification rate continuously decreases. Full density is possible by rearrangement if enough liquid is formed. Since monosized spheres pack to 74% density in theory, at least 26 vol.% liquid is needed to obtain full density by rearrangement. However, rearrangement is inhibited by particle contacts formed through compaction or solid-state sintering during heating.

3. *Solution-reprecipitation*

Concurrent with rearrangement are various other events; although, the kinetics of rearrangement are initially so fast these other events are overshadowed. As densification by rearrangement slows, solubility and diffusivity effects dominate. This second stage of liquid phase sintering is termed solution-reprecipitation. The solubility of a grain in its surrounding liquid varies inversely with the grain size; small grains have a higher energy and solubility than large ones. The difference in solubilities establishes a concentration gradient in the liquid. Material is transported from the small grains to the large grains by diffusion through the liquid. This process is termed Ostwald ripening. The net result is a progressive growth of the larger grains at the expense of the smaller grains, giving fewer grains with a larger average size.

Solution-reprecipitation not only contributes to grain coarsening, but also to densification via grain shape accommodation, allowing a customized fitting together of the growing grains to better fill space. The amount of liquid determines the diffusion distance and the necessary degree of grain shape accommodation. Figure 4 shows the microstructure of a liquid phase sintered W–Ni–Fe alloy exhibiting grain shape accommodation of nearly pure tungsten grains in a solidified matrix alloy. These grains are several times larger than the original particles and are flattened along neighboring faces to allow a high packing density. Although the solid–liquid interface area is minimized by a spherical shape, grain shape accommodation is favored since it eliminates porosity and the high interface energy associated with pores.

4. *Coarsening*

During liquid phase sintering there is grain growth, even as densification slows in the final stage. The maximum density attained depends on the characteristics of the pores and any internal gases trapped in the pores. Furthermore, the skeletal microstructure, which forms after rearrangement, provides rigidity to the compact, and inhibits pore elimination. With continued sintering, the grain size enlarges by solution-reprecipitation. Along with grain growth there are simultaneous changes in the pore size, intergranular neck size, grain–liquid interfacial area, and mean grain separation. Coupled with transport events through the liquid phase, there are simultaneous

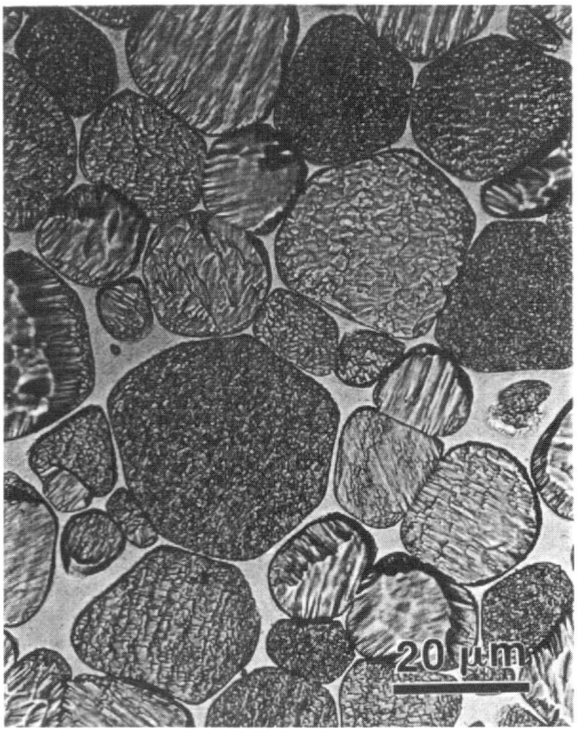

Figure 4
An optical micrograph of a tungsten heavy alloy fabricated by liquid phase sintering to full density, showing the coarsened solid grains, grain shape accommodation, and solidified liquid matrix. The material is a 95% W–3.5% Ni–1.5% Fe alloy sintered at 1470°C for 2 h in hydrogen. The tungsten grain are etched for visual contrast.

solid-state sintering events, but typically liquid diffusion rates are many times faster and dominate all solid-state sintering and coarsening events.

In the final stage, the microstructure approaches a minimum energy solid–liquid configuration. Densification is inversely dependent on the solid–liquid–vapor interface energies. Minimum energy conditions direct solution-reprecipitation events toward grain shape accommodation to annihilate high-energy vapor interfaces. In the final stage the controlling step is sintering densification of the rigid solid skeleton.

5. *Concluding Remarks*

A majority of all sintering is performed in the presence of a liquid phase. In many industrial materials the liquid phase is present by design, but there are many cases where it forms as a byproduct of incidental impurities. Applications range from automotive mechanical components (Fe–2Cu–0.8C) to high thermal conductivity microelectronic ceramics (AlN–$4YF_3$). An important aspect is solubility of the solid in the liquid at the sintering temperature, providing a means to enhance mass flow and densification. The opposite solubility situation gives swelling. In extreme cases, mixed powders can react during liquid phase sintering to form new compounds, e.g., molybdenum and silicon react to form $MoSi_2$. However, most liquid phase sintering systems have a low level of chemical interaction between constituents such that surface energy dominates the microstructural evolution during sintering. To avoid distortion from an excess of liquid, most systems use less than 15 vol.% liquid. Full density is easily obtained with higher liquid contents, but solid–liquid separation can result in nonuniform dimensions and properties. Alternatively, with very low liquid contents, solid grain contact slows final densification.

Bibliography

German R M 1985 *Liquid Phase Sintering*. Plenum, New York

German R M 1996 *Sintering Theory and Practice*. Wiley, New York

Kingery W D 1959 Densification during sintering in the presence of a liquid phase. I. Theory. *J. Appl. Phys.* **30**, 301–6

Savitskii A P 1993 *Liquid Phase Sintering of the Systems with Interacting Components*. Russian Academy of Sciences, Tomsk, Russia

R. M. German

Lost Foam Casting (Disposable Mold)

The lost foam casting (LFC) process employs an expanded polymer (foam) pattern. The true lost foam casting process (also known as expendable pattern casting (EPC)) uses a foam pattern which is coated with a thin layer of ceramic. Unbonded sand is vibrated around the coated pattern and the pattern is displaced by liquid metal during pouring. The lost foam process is used primarily for producing aluminum, gray cast iron, and ductile cast iron parts, although other metals including steel and copper-base alloys have been commercially poured. Some of the best known applications include aluminum cylinder blocks and heads for automotive and marine engines, ductile iron automotive crankshafts, gray iron housings for electric motors, and gray iron fire plugs.

The lost foam process has a number of attractive characteristics. Because the pattern is not removed from the mold prior to pouring, parting lines are eliminated. Cores are not required to produce holes or interior surfaces, and no core fins or core shifts are created. Complex shapes with no draft or even reverse draft can be poured, producing lighter-weight parts with minimal machining requirements. Single castings can replace fabricated assemblies because several foam patterns can be glued together prior to sand compaction and casting. Risers are not ordinarily required to control solidification shrinkage. The process is environmentally friendly, with relatively low emissions compared with most of the conventional casting processes.

1. Metal Flow in Lost Foam Casting

The presence of the pattern and the need to displace the pattern with liquid metal makes the metal filling process vastly different from all other casting processes. The metal front is opposed by the solid foam and the thin layer of liquid and gases produced from the decomposing pattern (Fig. 1). The back-pressure helps to produce a relatively smooth metal front that progresses at a rate determined by how fast the polymer foam collapses and decomposes or by the rate at which the opposing gases are able to escape through the coating. The temperature of the metal front decreases as fill advances. This mold filling behavior helps to reduce casting defects while promoting good directional solidification from the casting extremities to the gating system, eliminating the necessity for risers.

2. The Process

A number of steps are required to produce a lost foam casting (Fig. 2). Careful control of each of these steps is crucial in producing high-quality castings.

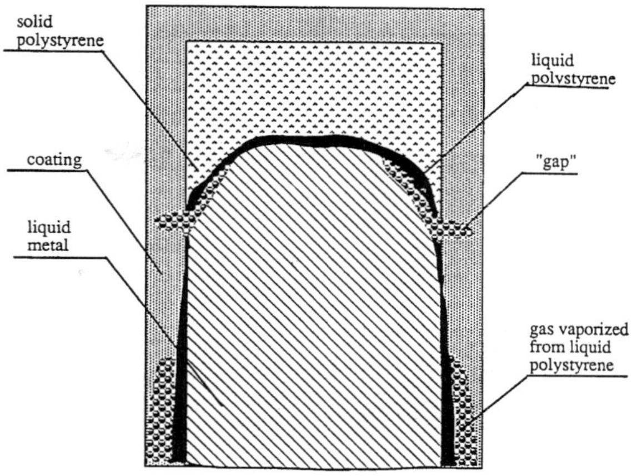

Figure 1
Model for metal fill in lost foam casting. As liquid metal flows from the bottom, the solid polystyrene foam collapses and decomposes, forming a gap containing liquid polystyrene, gaseous decomposition products, and air. The liquid polystyrene is pushed to the coating, where it wets, infiltrates into the coating, and further decomposes. Gaseous products are removed through the porous refractory coating.

2.1 Producing the Foam Beads

Polystyrene, with a molecular weight of approximately $200\,kg\,mol^{-1}$ and containing approximately 5% pentane as the blowing agent, is the most common and least expensive polymer. Solid polymer beads are introduced into a steam chamber and small cells of pentane gas are produced within each bead, producing expanded polystyrene (EPS) beads with a nominal density of $0.02–0.03\,g\,cm^{-3}$ ($1.3–1.8\,lb\,ft^{-3}$). The beads are then stabilized at a controlled temperature and humidity.

Expanded polymethyl methacrylate (EPMMA) has been used for producing iron and steel castings in order to reduce carbon pickup in the metal and lustrous carbon defects on the casting surface. However the high cost and difficulty in blowing patterns with EPMMA has minimized its use. Instead, a copolymer of polystyrene and polymethyl methacrylate is more often used for iron castings.

2.2 Blowing the Pattern

The pattern or pattern components are produced by blowing the pre-expanded beads into an aluminum die. The die includes one or more gun ports through which the beads are introduced into the die cavity, vents through which steam can be introduced and evacuated from the die cavity, and provision for cooling. The dies may be very complex, including slides, so that complex patterns can be made.

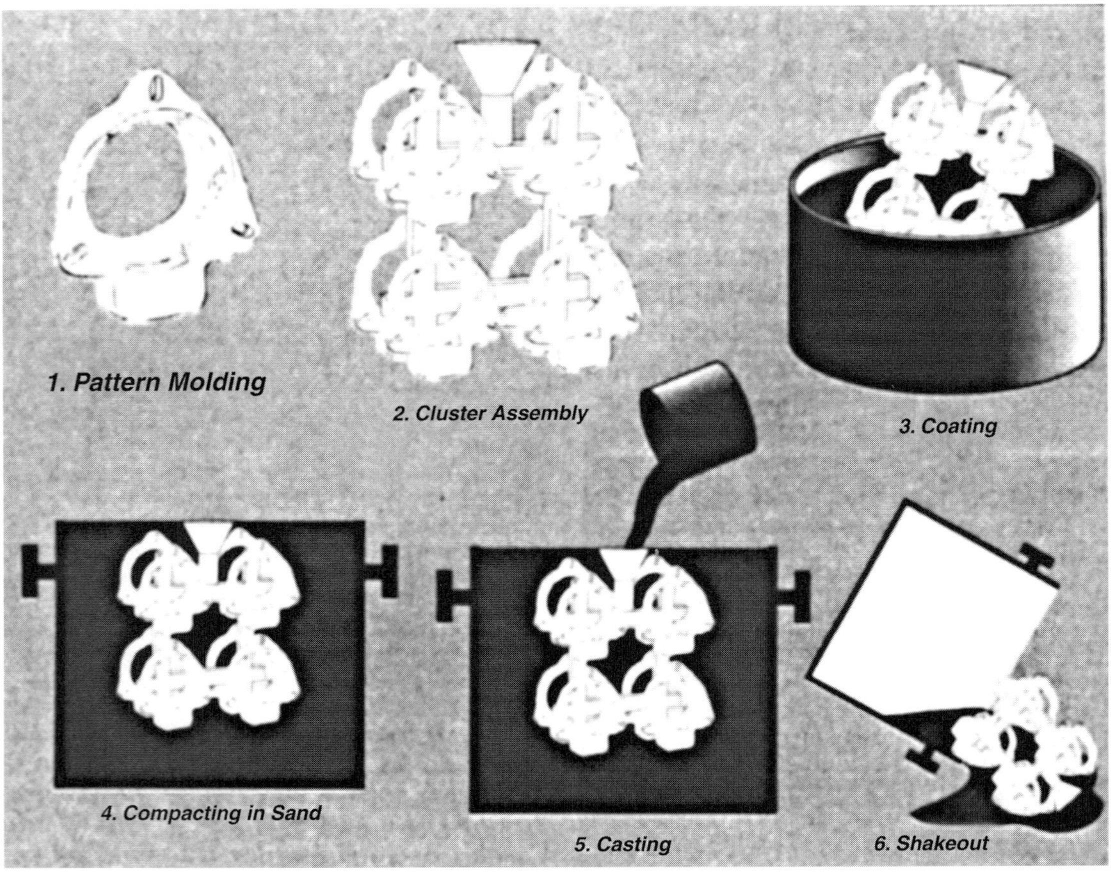

Figure 2
Some of the steps in the production of a lost foam casting. A foam pattern is produced, assembled onto a gating system to produce a cluster, coated with a ceramic, surrounded with unbonded sand, and poured.

In operation, the die is closed and the beads are blown through the gun ports into the die cavity. Steam is introduced through vents on one side of the die, with the steam passing through the die cavity and exiting through vents on the opposite side of the die. In some cases, the direction of the steam flow may be reversed to promote more uniform heating of the beads. The beads expand further to fill up the die cavity, soften, and fuse to one another. After an appropriate steaming time, the die and pattern are water-cooled until the pattern is rigid.

The overall density of the pattern is determined by the density of the pre-expanded beads. However, patterns may have pronounced density gradients through the cross-section of the pattern, with the density much higher at the pattern surface than at the center of the cross-section. The degree to which the beads are fused also varies, with better fusion at the surface than at the centerline. Good fusion at the surface is necessary to produce a smooth pattern and hence casting surface.

2.3 *Pattern Aging and Stabilization*

After the pattern is removed from the die, the dimensions change. The pattern initially increases in size, but this is followed by shrinkage that may continue for weeks or months as pentane diffuses out of the beads and is replaced with air. The patterns may be aged for about 4 hours at 65°C to accelerate these changes and produce a dimensionally stable pattern. Aging also eliminates moisture from the pattern that may lead to casting defects. The dimensional changes that occur during aging must be considered in designing the tooling for producing the patterns.

2.4 *Pattern and Cluster Assembly*

Complex patterns are often produced in several pieces and assembled using a hot melt glue. A typical cluster includes a foam sprue, gating system, and as many patterns as can practically be attached to the cluster. These components are also assembled using hot melt glue. Decomposition of the glue, which is about 40 times denser than the foam, produces a large quantity of liquid and gaseous decomposition products that must be eliminated during pouring. Whenever possible, glue print machines that produce only a thin line of glue at the joint are used.

The gating system must be sufficiently robust that the patterns will remain attached during coating, handling, and compaction. The gating system introduces liquid metal into the pattern at locations that assure complete filling to avoid misruns and proper directional solidification to prevent solidification shrinkage. Unlike conventional "open cavity" casting processes, the gating system does not usually determine the fill time during pouring—the decomposition rate of the foam and/or the properties of the coating act as the choke during fill.

2.5 *Coating the Cluster*

The foam cluster is coated with a thin (approximately 0.3 mm) layer of refractory material. The coating slurry typically contains fine ceramic solids (such as mica) suspended in an aqueous-based binder that hardens on drying. Proprietary additives are used to control the rheological properties of the slurry so that a uniform coating can be applied to the foam.

The coating protects the metal casting surface from the coarse sand used as the molding material (thus producing a good surface finish) and transports air and both liquid and gaseous decomposition products from the metal front into the surrounding sand at a controlled rate during pouring. The coating is applied by dipping the assembled cluster into a large tank of ceramic slurry. The coated cluster is normally dried in a forced air oven before producing the mold.

The viscosity of the coating slurry must be controlled to produce a coating with the desired uniform thickness without dripping. The gas permeability of the coating determines the rate at which air and decomposition gases escape from the liquid metal–foam interface during pouring and consequently determines the rate at which the liquid metal displaces the foam pattern. If the gas permeability is too low, a high back-pressure at the metal–foam interface develops, slows the velocity of the metal front, and leads to misruns. If the gas permeability is too high, the metal front may become turbulent, liquid polymer decomposition products may be trapped within the metal, and pyrolysis defects produced from the liquid polymer may form in the casting. Finally, the thermal conductivity and/or heat capacity of the coating influence the rate of heat removal from the liquid metal during pouring.

2.6 *Compaction*

The lost foam process uses a coarse, unbonded sand aggregate that is vibrated around the pattern as sand is introduced into the flask during molding. Normally round silica sand with a grain fineness number of 30–40 sieve size provides good overall gas permeability to the system. However the large dimensional changes that occur when silica sand heats during pouring may produce stresses that distort the casting. If the stresses are sufficient, the coating may crack during pouring, permitting liquid metal to penetrate through the coating, producing a burn-in defect on the casting surface. Some lost foam foundries use a low-expansion sand such as olivine or synthetic mullite.

During compaction, the sand is introduced into the flask at a controlled rate while the flask is vibrated. The top 5–10 cm of sand in the flask at any time is fluidized and completely surrounds the pattern. Introduction of the sand may be interrupted when the sand level reaches holes or internal cavities in the pattern, permitting sufficient time for these critical areas to also fill with sand. Introducing the sand too quickly does not provide enough time for the sand to properly compact around the pattern.

In some foundries, the sand is thermally reclaimed. During pouring, high molecular weight gases from the decomposing foam condense in the sand nearest the casting surface. The condensed polymer reduces the flowability of the sand and inhibits good compaction. The temperature of the sand must also be controlled. High sand temperatures melt the polymer residue, producing even worse flowability and casting defects.

2.7 *Pouring*

As with other processes, the liquid metal should be poured rapidly enough to keep the gating system and pouring cup full. This assures a constant metallostatic pressure head, which prevents entrainment of gases and dross/slag into the system and helps assure that the sand around the sprue does not collapse and cut off flow of liquid metal to the rest of the gating system.

Most lost foam castings are poured using only gravity. However, steel and many iron castings use a vacuum assist during fill. The vacuum helps to draw the harmful decomposition products through the coating and away from the liquid metal during fill, potentially reducing pyrolysis-related casting defects. Other variations include use of pressure pouring to provide fill.

3. Pyrolysis Defects

The necessity for removing the products of the pyrolysis of the foam pattern may lead to some casting defects that are unique to lost foam casting. Droplets of liquid polymer entrapped within the liquid metal may pyrolyze to a gaseous form and produce internal pores in the solidified casting. In some cases, the bubbles produced from the pyrolyzed liquid polymer may be attached to a surface of the casting, in which case the bubble collapses and produces a fold. Folds, which are thin sheets of carbon and metallic oxide, occur most commonly in aluminum castings. Lustrous carbon on iron and steel castings occurs when liquid polymer trapped between the metal and the coating pyrolyzes to produce a difficult-to-remove carbon film on the casting surface.

Careful control of the processing parameters can help reduce these defects. For example, using a low-permeability coating reduces the rate at which the liquid metal displaces the foam pattern and provides more time for liquid polymer to reach the casting surface and be transported harmlessly into the ceramic coating. Low-density foam patterns reduce the total amount of polymer that must be decomposed, thus reducing pyrolysis defects. Patterns that have large gradients in density or degree of fusion produce an arrow-shaped metal front that pushes pyrolysis products to the coating, where the pyrolysis products can be removed. Designing gating systems such that two streams of liquid metal do not meet, thus trapping pyrolysis products at the junction, is helpful in preventing these defects.

4. Summary

As the twenty-first century begins, the lost foam casting process is approaching maturity. The physical phenomena controlling the process are understood reasonably well. New foams, pattern production methods, and improved coatings have been developed. Techniques by which the characteristics of the foams, coatings, and compaction processes can be monitored and controlled have been devised.

Bibliography

Austin T 1994 *Expendable Pattern Casting, Volume 2: Tooling Manual.* American Foundrymen's Society, Des Plaines, IL

Bates C E, Griffin J, Littleton H 1994a *Expendable Pattern Casting, Volume 1: Process Manual.* American Foundrymen's Society, Des Plaines, IL

Bates C E, Griffin J, Littleton H 1994b *Expendable Pattern Casting, Volume 3: Casting Defects Manual.* American Foundrymen's Society, Des Plaines, IL

Monroe R W 1992 *Expendable Pattern Casting.* American Foundrymen's Society, Des Plaines, IL

D. R. Askeland

M

Machining and Machinability

Machinability describes the totality of all properties of a material that influence the mac'hining process. It must invariably be judged in relations to the process employed, the tool material, and the cutting parameters. The four main variables generally used to assess the machinability of a material are: tool life, cutting force, surface finish, and chip formation. Depending on their properties, there may be wide differences in the machinability of different materials. The choice of the tool and the cutting parameters accordingly opens up different windows of cost-effectiveness for machining any given material. The machinability characteristics of the most important metallic materials (ferrous metals, steel and cast iron, nonferrous metals, aluminum-, nickel-, and titanium-alloys) are discussed and their process limits indicated.

The term machinability is used quite generally to describe the capability of a material to be subjected to cutting processes. The machinability of a material must therefore always be judged in the context of the machining processes involved, the tool material, and the cutting parameters. Machinability is usually judged on the basis of four evaluative criteria (König and Klocke 1997): tool life, cutting force, surface finish of the part, and chip geometry.

1. Essential Parameters

The tool life t of the tool is the most important indicator characterizing the machinability of a material. The tool life t is the service life of a tool up to the point at which it becomes unusable due to wear.

A knowledge of the magnitude and direction of the resultant cutting force F or of its components: the cutting force F_c, the feed force F_f, and the passive force F_p, forms the basis for: designing machine tools, establishing the cutting parameters, predicting the attainable accuracy of the part, and interpreting the phenomena which occur at the contact point. As a rule, the cutting of harder-to-machine materials entails higher forces.

The quality of the surface generated by machining may be a criterion for the design of the cutting process, if cutting is not followed by any subsequent operation. The primary variables influencing surface quality are the cutting parameters and the geometry of the cutting edge.

The shape and size of the chips, together with the way in which they are removed, are especially significant for cutting operations with a limited chip space (e.g., drilling, broaching, and milling) and for automatic lathes, due to the restricted working space and large volume of chips. The degree of upsetting of the chips can also be used as an indicator for chip formation phenomena.

2. Influences on the Machinability of Steel

The machinability of steel materials is influenced by their microstructure and their mechanical properties. The most significant factors affecting microstructure and hence mechanical properties are: carbon content, alloying elements, and heat treatment.

Carbon is the most important alloying element of steel, and is present mainly in the cementite.

Sub-eutectoid steels consist of a mixture of ferrite and pearlite, super-eutectoid steels of secondary cementite and pearlite. The percentages of hard materials are an indicator for the machinability of the steels. The lower the carbon content, the more closely the machinability will resemble that of ferrite (see Fig. 1).

Alloying and trace elements may influence the machinability of steels as a result of changes in the microstructure or through the formation of either lubricating or abrasive inclusions, like graphite, carbides, nitrides, oxides, or intermetallic phases.

Depending on the alloying constituents of the steel, sulfur forms various stable sulfides. Ferric sulfides, FeS, are undesirable, because they have a low melting point and are deposited mainly at the grain boundaries. This causes the feared "red brittleness" of steel. Alloying with manganese prevents this effect.

The elements added to deoxidize steels—aluminum, silicon, or calcium—form hard, nondeformable inclusions, which are then present in the steel, e.g., as aluminum oxide and silicon oxide. They make the material less machinable, especially if the oxides occur in large quantities or in lineate form. Systematic heat treatment can affect the microstructure in terms of the quantity, shape, and arrangement of its constituents, adapting it for the required mechanical properties and machinability.

Three main objectives of heat treatment may be distinguished:

- obtaining a homogeneous microstructure over the entire cross-section;
- obtaining a hardening structure restricted to small segments of the cross-section without altering

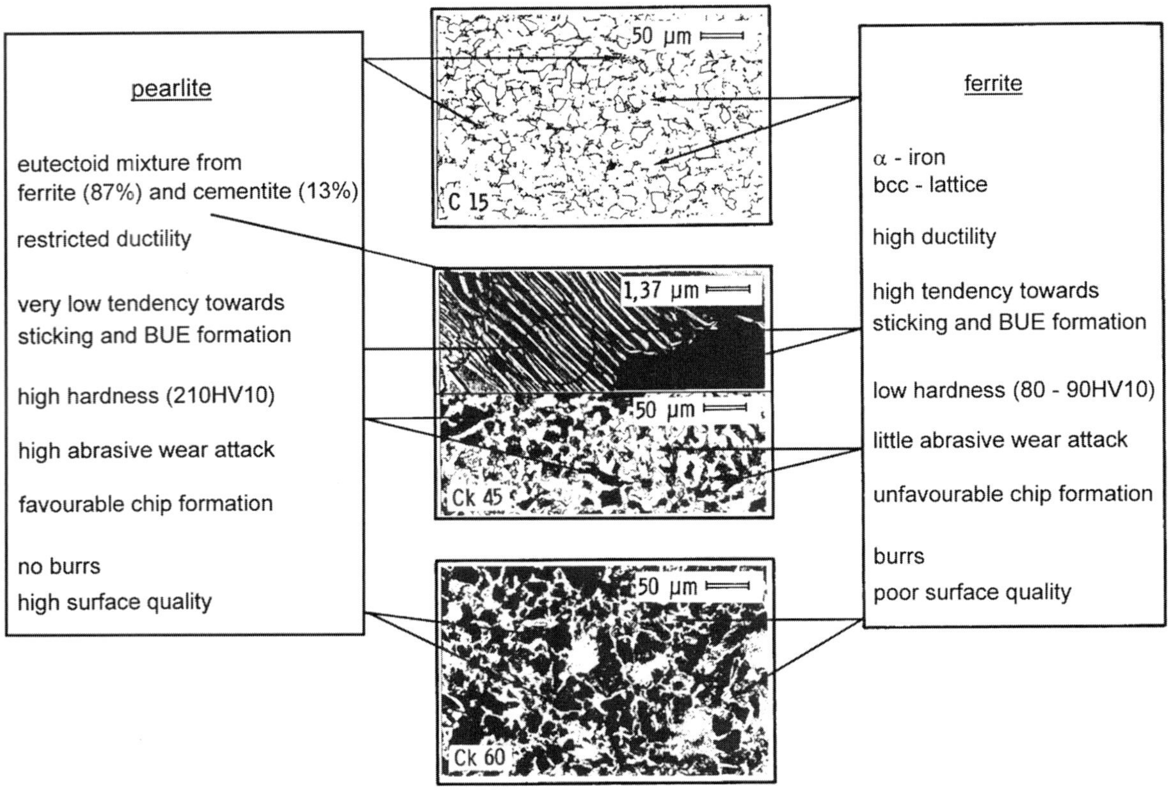

Figure 1
Microstructural constituents of non-alloyed subeutectoid carbon steels.

the chemical composition (especially hardening of the surface zone); and

• obtaining microstructures that differ widely over the cross-section, especially in the surface zone, by modifying chemical composition (carburizing, case-hardening).

Heat treatments of the first class are predominant in the influencing of machinability. In principle, the following heat treatments from this class are available: homogenizing, normalizing, coarse grain annealing, softening, and tempering or annealing.

3. *Machinability of Special Steels*

3.1 Machinability of Free-cutting Steels

Free-cutting steels are materials that, when machined, produce short-breaking chips with little upsetting and smooth surface finishes on the parts, such as to cause little tool wear. These properties are given to free-cutting steels primarily by alloying them with sulfur and phosphorus. Depending on their chemical composition and field of applications, free-cutting steels are employed in various forms (without heat treatment, case-hardened, or tempered). Materials with enhanced S content are also found in the group of nitriding steels (e.g., 30CrAlS5) or the corrosion-resistant steels (e.g., X12CrMoS17 and X12CrNiS18-8).

Phosphorus encourages precipitation of nitrogen and carbon, causing brittleness of the α solid solution (ferrite brittleness). At a shearing plane temperature of 200–400°C, this temperature-dependent brittleness provides more favorable chip formation and shorter-breaking chips. Susceptibility to adhesion in the contact zone is also reduced, and there is a positive effect on surface quality. Free-cutting steels contain a maximum of 0.1% phosphorus.

If, in addition to sulfur, a suitable quantity of manganese is added to the steel (approximately: $\%Mn = 2.5 \times \%S + 0.15$), the greater affinity of the sulfur to manganese rather than to iron results in the almost exclusive formation of manganese sulfide MnS. MnS makes the chips shorter-breaking, improves the surface quality of the part and reduces susceptibility to cutting-edge build-up. However, it also detracts from the strength properties of the steel. Commonly used free-cutting steels are, for example, 9SMn28, 9SMnPb28, 35S20, and 45S20.

3.2 Machinability of Case-hardening Steels

Machining of case-hardening steels takes place almost exclusively before the case-hardening treatment. Since the microstructure of these steels is mainly ferrite, with little pearlite, wear attack on the tool is slight.

High cutting speeds are desirable in order to avoid the range of edge build-up. The best possible surface quality is obtained by using appropriate cooling lubricants, by modifying the tool geometry (positive orthogonal rake) and by reducing the feed rate.

Depending on their alloying elements, the steels are heat-treated to a specific ferrite-pearlite microstructure or a specific strength in order to improve machinability. Coarse-grain annealing is frequently used for alloyed case-hardening steels, to reduce the machining problems caused by adhesion.

Heat treatment has little effect on the cutting parameters that can be employed, provided that cemented carbide tools (usually HW-P10 or coated cemented carbides) are used and the tensile strength of the material is below 650 N mm^{-2}.

Machining is followed by casehardening. After hardening, the hardness values rise as high as 60 HRC. In certain cases, an additional finishing process is necessary due to the distortion induced in the part through the case-hardening operation. Micrograin cemented carbides, mixed ceramics and PCBN tool materials are especially suitable for finishing the highly tempered or hardened steels (>45 HRC). Very high cutting forces are exerted. Because the chip is fully annealed by the significantly higher temperatures, no chip-breaking problems are observed. Very good surface finishes are usually obtained.

Representative case-hardening steels include Ck15, 16MnCr5, 20MoCr4, and 18CrNi8.

3.3 Machinability of Heat-treatable Steels

Heat-treatable steels possess a carbon content between 0.2% and 0.6% C. Their strengths are correspondingly higher than those of case-hardening steels. The principal alloying elements are silicon, manganese, chromium, molybdenum, nickel, and vanadium.

The machinability of heat-treatable steels depends mainly on the nature of their microstructure, and consequently fluctuates across a wide range according to carbon content. In terms of machinability, the microstructure resulting from a specific form of heat treatment outweighs the influence of the alloying constituents. Adding approximately 0.06–0.1% sulfur to a heat-treatable steel produces a significant improvement in machinability for turning, milling, and drilling operations.

In many cases tempering treatment is carried out between the roughing and the finishing processes. Roughing, where high removal rates are the primary factor, is usually performed on materials in the normalized state, whose machinability is characterized by relatively low wear, due to a ferritic-pearlitic microstructure. However, most machining operations on heat-treatable steels take place after the tempering of the material, i.e., at a strength higher than that of the annealed state. High cutting speeds accordingly entail greatly increased tool wear.

Group P cemented carbides are the primary choice for machining heat-treatable steels. These are suitable for virtually all cutting processes. They may also be used for drilling and tapping with extremely long or slender tools.

Heat-treatable steels frequently machined in practice include Ck45, 42CrMo4, 30CrMoV9, or 36CrNiMo4.

3.4 Machinability of Nitriding Steels

The carbon content of nitriding steels is some 0.2–0.45%. They are heat-treatable and are alloyed with chromium and molybdenum (to improve their suitability for through quenching and subsequent drawing) and with aluminum or vanadium (nitriding agents). Nitriding is performed at temperatures in the range 500–600°C, i.e., below the α–γ equilibrium temperature of the material.

These materials are machined before being nitrided, usually in the heat-treated state. The range of applications for nitriding steels is similar to that of case-hardening steels (gears, guides).

3.5 Machinability of Tool Steels

Tool steels (non-alloyed or alloyed) have to withstand different types of stress. In order to provide for these, tool steels are classified as cold-work steels, hot-work steels, and high-speed steels.

Lamellar pearlite and ferrite are present up to a carbon content of 0.8% in nonalloyed tool steels in the forged or rolled state; at higher carbon contents, there is lamellar pearlite and a network of cementite.

In the hardened state, the microstructure in the surface zone consists principally of martensite, gradually giving way towards the interior of the part to transitional microstructures and to fine lamellar pearlite, which can no longer be resolved under the microscope. Additional cementite is included in the basic microstructure of hyper-eutectoid steels.

Nonalloyed tool steels are machined in the soft-annealed state. Non-alloyed sub-eutectoid tool steels can also be machined in the normalized state or as delivered after metal forming. In both cases, relatively poor machinability, with increased susceptibility to adhesion and cutting-edge build-up, is to be expected.

Alloyed tool steels, especially high-alloyed high-speed steels, are considered as difficult to machine, because they tend strongly to adhesion and cutting-edge build-up when machined in the annealed state. Rough, inferior surfaces are generated, and at the point where the tool emerges from the work material there may be light chipping or parts of the chip may remain attached to the surface. Tempering to a greater strength (1200–1400 N mm^{-2}) may offer a partial solution.

Hitherto, it has only been possible to machine these steels in the annealed state, and feasible cutting speeds have increased with the degree of coalescence of the carbides. There is, however, a proportionate rise in the susceptibility to adhesion, so that tool steels with finely dispersed grainy carbides are generally difficult to machine. Cemented carbides of medium toughness containing titanium and tantalum carbide are used to machine tool steels. Certain wear-resistant high-alloyed high-speed steels can also be used to good effect. Acceptable cutting speeds are relatively low, but can be raised by using coated tools.

Tool steels that are available in a pre-tempered state (R_m < 2000 N mm^{-2}) can also be machined using PCBN tool materials, frequently eliminating the grinding operation.

3.6 *Machinability of Stainless, Heat-resistant and Refractory Steels*

Stainless steels characteristically possess good resistance to chemically corrosive agents. They generally have a chromium content of at least 12 wt.%. Stainless steels may be classified as ferritic, martensitic, or austenitic, depending on their microstructural constituents.

Steels with a ferritic microstructure are relatively easy to machine, whereas the austenitic types present great difficulties. Austenitic steels are therefore often machined in the quenched or solution-heat-treated state, characterized by a high susceptibility to chip adhesion, cutting-edge build-up, and cold hardening.

4. *Machinability of Cast-iron Materials*

Cast-iron materials are alloys of iron and carbon with a carbon content exceeding 1.7% (usually 2–4%).

The machining properties of cast iron materials are influenced very strongly by the quantity and nature of the segregated graphite. Graphite segregations both reduce friction between the tool and the part, and interrupt the basic metallic microstructure. This makes the material easier to machine than a graphite-free cast iron or a steel, with short-breaking chips, low cutting forces and longer tool lives.

In cast iron with lamellar graphite (gray cast iron, GGL) the steel-like basic structure is interrupted by graphite lamellae, which invariably produce short-breaking chips. A decrease in the cutting forces is also observed. In general, no burr forms on the edges of the part, but chipping occurs.

The surface zone of a casting is more difficult to machine than the core zone. This is attributable partly to nonmetallic inclusions and partly to the modified nature of the graphite and the microstructure immediately below the casting skin, and to scale. In consequence, there is strong abrasive wear and a wear notch is formed on the cutting edge of the tool. In practice, the cutting parameters are often reduced to allow for this phenomenon.

The classic cemented carbide grades for machining cast iron with lamellar graphite are the K grades. K01 and K05 types and cermets are suitable for finishing and microfinishing. Coated cemented carbides and tool ceramics significantly enhance the removal rate.

The graphite in nodular cast iron occurs in the form of nodular inclusions. The basic microstructures of grades with low strength and good toughness properties consist mainly of ferrite. When machined, these materials may produce helical chips, which are slightly friable due to the lack of chip strength induced by the graphite segregations. Dummy chips may also form at high cutting speeds in operations without cooling lubricant.

The microstructure of vermicular cast iron (GGV) is characterized by worm-shaped graphite inclusions. The reduction of internal stress peaks in the material, of the kind observed at the sharp ends of the lamellar graphite inclusions in GGL, allows good component properties to be combined with good machinability.

When machining hard cast iron (white cast iron), the cutting edge of the tool is subjected to severe stress as a result of the high cementite content in the microstructure. The tool material should possess high resistance to wear and to compressive stress. White cast iron is machined almost exclusively with group K cemented carbides and, especially with high hardness values, tool ceramics (oxidic mixed ceramics), and PCBN tool materials. Chip thicknesses are reduced to minimize the stress on the cutting edge when machining white cast iron. Tool cutting edge angles of 10° to 20° and orthogonal rakes of −5° to +5° are customary for machining white cast iron rollers.

5. *Machinability of Aluminum Alloys*

Aluminum alloys contain one or more of five main alloying elements: silicon, copper, magnesium, zinc, and manganese. Iron, chromium, and titanium are added in small quantities. Specialized alloys may additionally contain nickel, cobalt, tin, lead, or vanadium. Beryllium, boron, and sodium are added as trace agents in order to influence crystallization phenomena.

Non-precipitation-hardenable alloys acquire their strength properties solely from their alloying elements and from their degree of cold hardening (alloys in the AlMg, AlMn groups).

Precipitation-hardenable alloys receive their strength properties from heat treatment. An additional improvement can be obtained through cold hardening, but this method is seldom used (AlMgSi, AlZnMg, AlCuMg, and AlZnMgCu alloys).

Both non-precipitation-hardenable and precipitation-hardenable alloys in the soft state without silicon have a very strong susceptibility to smearing and a tendency to edge build-up even at cutting speeds of up to about $v_c = 300\,\mathrm{m\,min^{-1}}$. Surface quality is poor. If tools with a suitable cutting-edge geometry are used, long, tough ribbon chips are formed, greatly aggravating the difficulties of the machining process. Whenever possible, alloys of this type should not be employed for components that have to be machined.

The machining properties of precipitation-hardened alloys and cast alloys with a silicon content of up to 12% deteriorate with increasing silicon content. Hard, brittle inclusions like Al_2O_3 and the silicon itself improve chip-breaking qualities, but simultaneously increase tool wear. For this reason, cemented carbides should be used as the tool materials. Virtually all K grades are suitable.

Diamond tools are frequently used to generate decorative surfaces, and for mass production. High cutting speeds and small chip cross-sections are used. HSS tools may be advantageous for discontinuous cutting at low cutting speeds with simultaneously high removal rates on alloys with low silicon content.

Tool wear in aluminum machining is generally caused by rounding and simultaneous displacement of the cutting edge. Wear by craters does not occur. Progressive dulling of the cutting edge results in higher cutting forces and increased temperatures at the chip generation point. The temperature may be high enough to produce dummy chips, i.e., plastified material emerges between the flank and the part. This is usually associated with poorer surface quality.

Cast alloys with a silicon content exceeding 12% and piston alloys can be machined effectively in terms of chip shape and attainable surface quality using cemented carbide and polycrystalline diamond tools. The tool cutting edge is exposed to very high abrasive wear attack through the high silicon content. A significant reduction in cutting speed is therefore necessary as silicon content increases.

6. *Machinability of Nickel-base Alloys*

Nickel-base alloys are refractory, corrosion-resistant, and simultaneously very tough. They are therefore the most commonly used modern materials for components subjected to high alternating mechanical loads at working temperatures up to 1100°C.

Owing to their mechanical, thermal, and chemical properties, all nickel-base alloys are classed as difficult-to-machine materials. Generally, the high temperature resistance and low thermal conductivity, as well as the abrasive load from carbides, inclusions, and intermetallic phases lead to impressive thermal and mechanical stress of the cutting edge during machining of such alloys. Nickel-base alloys are therefore usually cut with tools made from high-speed steel and cemented carbide. Due to the high temperatures in the cutting zone the tools work at relatively low cutting speeds, which quite often enhance the formation of build-up edges.

When cutting nickel base-alloys tools with a positive rake angle (turning: $\gamma_0 = 5\text{–}15°$, milling: $\gamma_p = 7\text{–}12°$, $\gamma_f = 15\text{–}25°$) and sufficient clearance angle ($\alpha_0 = 6\text{–}10°$) should be chosen because of the tendency of the material to smear, form build-up edges, and cold-harden. The ductility of the material enhances, depending on the cutting process, the formation of unfavorable long chips. Adopted chip formers on the tools can diminish this problem.

The degree of cold-hardening as well as the depth of the influenced zone of the work piece is mainly depending on the parameters: cutting speed, feed, cutting edge geometry, and wear of the cutting tool. Therefore, small feed and depths of cut should be avoided. This is also very important due to the production-related roundness or chamfer of coated cemented carbides and ceramic grades. Cut-down milling and no reaming should be performed according to the same reasons. A customized cutting fluid can enhance the performance of the cutting tools when machining nickel-base alloys.

7. *Machinability of Titanium Materials*

Titanium and titanium alloys combine low density with high strength ($R_m = 900\text{–}1400\,\mathrm{N\,mm^{-2}}$). They possess good heat resistance up to a temperature of approximately 500°C. They are also resistant to many corrosive media. Titanium materials are classified in four groups (pure titanium, α alloys, $(\alpha + \beta)$ alloys and β alloys).

Titanium alloys are regarded as difficult-to-machine. If they are to be processed cost-effectively, it is necessary to take special note of the physical properties of this materials group. Their strength is high, their rupture elongation of $A_5 = 5\text{–}15\%$ is low (for alloys). The modulus of elasticity is roughly 50% and the thermal conductivity roughly 80% that of steel.

High pressures and temperatures are generated in the tool contact zones. Only a small proportion of the resulting heat is dissipated with the chips. When steel is machined, the chips transport up to 75% of the heat away from the contact zone. With titanium, the equivalent figure is only about 25%. As a result, titanium materials must be machined at comparatively low cutting speeds, because the cutting speed has a much stronger influence on the contact zone temperature than the other machine setting parameters. This applies to both continuous (turning, drilling, broaching, and tapping) and to discontinuous cutting (milling and broaching).

Cemented carbides rich in WC and Co (K grades), high-speed steels with a high cobalt content, and diamond are suitable tool materials for cutting titanium.

8. *Conclusions*

Metallic materials are well known for their different physical and chemical properties. This directly influences their use and the design of the products made from them. The physical and chemical properties influence the machinability of metals in a similar way. The focus of this article is the description of these characteristics and a brief introduction to the cutting technology of these materials.

Bibliography

Kirk D C 1976 Cutting aerospace materials (nickel-, cobalt-, and titanium-based alloys. In: *Tools and Dies for Industry: Proc. Conf. Metals Soc.*, University of Birmingham, UK, pp. 77–8

König W, Klocke F 1997 *Fertigungsverfahren 1–Drehen, Fräsen, Bohren*. Springer, Berlin

N N 1980 *Machining Data Handbook*, 3rd edn. Machinability Center, Cincinnati, OH, Vols. 1 and 2

Pekelharing A I 1974 Built-up edge (BUE). Is the mechanism understood? *Ann. CIRP* **23**, 206–11

Tönshoff H K 1995 *Grundlagen–Spanen*. Springer, Berlin

F. Klocke

Machining and Nonmechanical Shaping: A Comparison

The final shape of a component can be obtained in different ways. Preformed or semifinished workpieces are processed by machining to remove material or by metal forming, which preserves mass and cohesion. This article draws a comparison between cutting and metal forming. With respect to their effective mechanisms, manufacturing processes are classified as mechanical, chemical, and thermal processes. Both cutting and metal forming are mechanical processes. A characteristic that all processes have in common is that they enable direct or indirect influence on size, shape, and surface quality of the workpiece. Some processes even permit an influence on the workpiece properties in the subsurface by means of structural changes. While structural transformations in the subsurface only play a marginal role in cutting, the change of material properties is the objective in metal forming. Material property modifications caused by metal forming are gaining importance. Hence a comparison of the two processes has to include their influence on different material characteristics such as solidity, hardness, structure, and subsurface formation. The forces acting and thermal influences during the process, the attainable accuracy and finish qualities, as well as the resulting by-products are equally of importance. A global consideration also comprises comparisons of economic efficiency, lot size, ecological aspects, and energy balance, as well as analyses of other fields of investigation.

1. *Cutting Processes*

Cutting is a process that modifies the shape of a solid object and removes local cohesion. The significant characteristics of cutting are the variety of producible shapes, a high geometrical accuracy, the ability of the process to be automated, and economic flexibility. During cutting, material is separated mechanically in the form of chips by means of cutting tool edges. Standardization defines cutting with geometrically defined tool edges by position, form, and number of cutting edges, as well as the cutting edge geometry of the tool. Processes of this type include turning, drilling, boring, milling, planing, slotting, broaching, and sawing. In cutting with geometrically undefined tool edges, in contrast, the edges are formed by irregularly shaped and indefinitely arranged hard material grains. Processes of this group include different methods of grinding as well as honing and lapping.

The cutting processes that lead to the separation of chips are divided into different zones and can be described as follows. In the primary shear zone a chip is formed by material shearing. In the second shear zone the frictional forces that are produced on the flank of the tool edge result in a plastic deformation of the material. In the predeformation zone, the material becomes flexible and is deformed as a result of the stresses occurring during chip formation. In the back-up and separation zone high compressive stresses occur and lead to a deformation and subsequent separation of the material. In the case of geometrically defined cutting edges, chip formation and tool edge are determined by geometrically cut angles such as clearance angle, rake angle, and wedge angle. These variables are also discernible on geometrically undefined cutting edges.

Both bound or loose hard material powder particles can be effective in cutting with geometrically undefined cutting edges. Bound tools consist of grains, bonding agent, and pores, while loose tools consist of particles in a paste or a fluid. The cutting angles relevant for chip formation are described statistically by mean values or distributions, since uneven cutting edges are used in most cases. A negative rake angle at the single grain, which is often very high, is typical for these tools. A flexible deformation of the material initially takes place as the surface of the workpiece is affected, later a plastic deformation occurs, and finally chip separation takes place with high friction between single grain and material. A large part of the mechanical energy supplied is converted heat, which predominantly flows into the workpiece, and some of the energy goes into the powder, linkage, and cooling lubricant. The resulting rise in temperature can lead to internal tensions, breaks, or microstructural changes in the subsurface of the material.

2. *Metal Forming Processes*

Metal forming is the direct alteration of form, surface, and material properties of a workpiece while preserving mass and cohesion. The processes of forming use the plasticity of metals for the production of semifinished material and structural parts. Forming is based on the flexible moldability of numerous materials and relies on the ability of the material structure to bear unit deformations along crystalline gliding planes without damaging the material cohesion. Forming saves material as no waste is produced. The objective is to obtain a finished surface in order to avoid expensive finishing. Metal forming processes are classified according to the effective stresses into forming under (i) compressive, (ii) a combination of tensile and compressive, (iii) tensile, (iv) bending, and (v) shearing conditions. Important processes are upsetting, wire drawing, deep drawing, extruding, stretch forming, bending, and forging. The forming process is influenced by the workpiece, the tool, the lubricant, the environment medium, and the machine. It can be described by means of yield stress, deformation size, flow conditions, anisotropy, and the flow curve. Processing heated workpieces is called hot forming, while processing workpieces at ambient temperature is called cold forming. With metallic materials cold forming is usually accompanied by work hardening, which is characterized by the fact that with increasing deformation, yield strength and breaking strength increase and breaking elongation decreases.

The description of forming processes is based on three basic models, the homogeneous shaping of disks, strips, and tubes. The basis is a homogeneous, isotropic material with a constant friction between workpiece and tool, and a constant yield stress. The strip model forming process is described briefly. A strip of a minimum breadth between two parallel delimitation surfaces between the upper and lower forming tool is considered. The upper and lower barrier is regarded as tangential to the contour of the tool. The strip height and the angles described depend on the moving sense of the tool. It is assumed that neighboring cross-sectional areas of the strip remain even and parallel to each other. Only compression stresses are considered, which normally affect the cross-section and delimitation surfaces, the remaining loads are neglected. By multiplying the tensions with the appropriate surfaces the forces affecting these surfaces are obtained. These forces are divided into horizontal and vertical forces and describe the loads at the delimitation surfaces of the strip.

3. *Process Comparison*

Irrespective of the assigned manufacturing, process workpieces are characterized by a variety of features such as shape, size, accuracy, surface, and material. All features are equally important. To describe the shaping processes in forming, material modifications are considered, whereas in cutting the modifications at the tool–workpiece interface are looked at. The significant process differences between cutting and metal forming are shown in Fig. 1.

3.1 Mechanical Properties

Certain stress conditions usually need to be met in order to obtain the necessary material characteristics. The numbers of slip planes of the crystal lattice of pure metals influence their tenacity. Cubic lattices indicate more slip possibilities and thus better tenacity values than hexagonal lattices. However, physical properties determine the extent to which flow movements are obstructed. The static physical properties are influenced by the entire material cross-section, which has an effect on the dynamic characteristics of the surface structure.

Surface properties can be compared best in cutting and forming. Therefore, these characteristics are dealt with in the following. Cutting and cold forming processes only entail surface modifications with a small depth effect. An increase in vibration resistance can be achieved by cold forming. The increase of the transfer density causes cold hardening. The value of the increase depends on the flow direction of the material and on whether the forming process is a homogeneous or a partial one.

In forming, the material is brought into a plastic state at which it is held during the processing. The decisive criterion for the behavior of the workpiece in forming is the yield stress. The flow curve, which depicts cold hardening during plastic deformation, is determined by the yield stress and the forming degree. The yield stress depends on material, forming degree, deformation ratio, temperature, and mean strain. The influence of the material on the flow curve varies with different metals. For alloy steels with a very small molybdenum, nickel, chrome, and carbon content the yield stress constantly increases in cold forming. At forging temperature the carbon content does not influence the yield stress.

In cutting, however, the machined material is mainly influenced in the shear zone. Here thermal loads and shear stress occur. The shear stresses cause material removal, the chip formation. Three basic types of chips can be distinguished: flowing, tearing, and shearing chips. These again can have different

Cutting 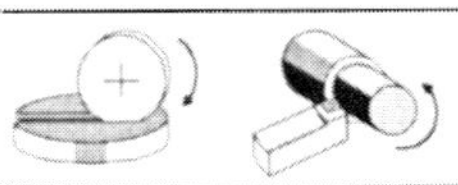		**Metal forming** 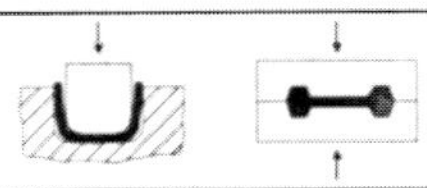	
Pros	*Cons*	*Pros*	*Cons*
⊕ High geometrical accuracy ⊕ High obtainable surface finish ⊕ Variety of producible shapes ⊕ Flexibility regarding lot size and workpiece geometry	⊖ Thermal stresses in the workpiece ⊖ Hazards for workplace and environment due to lubricants, coolants and removal products ⊖ High costs for reprocessing and waste disposal	⊕ Few thermal stresses in the workpiece ⊕ Fast automatic process for high lot sizes ⊕ No removal products ⊕ Low costs for reprocessing and waste disposal	⊖ Necessary cleaning of tool and workpiece ⊖ Necessary finish process ⊖ Hazards for workplace due to noise and vibrations

Figure 1
Comparison of cutting and metal forming.

phases. The flowing chip is a continuously running, connected splinter, which results from plastic deformation in the shear zone. It is usually formed in ductile materials with a homogeneous structure. If, however, the chip formation process takes place on brittle materials, tearing splinters are formed. The material surface is rough and contains fissures. The formation of shearing chips is caused by a combination of the two other separation mechanisms. At first, a noticeable plastic deformation occurs, then the material is prematurely separated. The resultant splinters form individual segments, which may also weld in the splinter flow. This chip category mostly occurs with materials of average ductility. A further separation product mainly occurring with ductile materials is the built-up edge. During processing this built-up edge is constantly formed and disrupted in a cycle. It is formed in the shear zone at the top of the tool by press welding.

The structure in the subsurface is changed by thermal stress. For example, in high-speed grinding workpiece temperatures between 650°C and 1200°C may occur. This leads to oxidation and even combustion of the workpiece, depending on workpiece rate and feed.

3.2 Costs

An economic evaluation of cutting and metal forming reveals that personnel and material costs are equally important, while energy costs in cutting are less significant. The so-called high-performance materials play a key role in manufacturing industry although they only have a small impact on the production costs. These materials include polymers, ceramics, metals, and alloys, as well as composite materials. It is presumed that they will be applied more frequently in the future especially in the aerospace, automotive, and electronic industries.

Metal forming processes in contrast to cutting processes are particularly appropriate for the machining of expensive materials as a very small quantity of waste is produced. No chips must be separated or disposed of. The forming capacity of the material has a great influence on the production costs. The heat required, the adherence to the optimal deformation ratio, and the use of special lubricants and tools have a crucial effect on the costs. A further factor is the disposal of auxiliary materials and effective media. Tool and workpiece have to be cleaned after the forming process. In cutting, coolants and lubricants are usually transported with the chips and have to be reprocessed. No cleaning of tool and workpiece is required.

3.3 Environmental Aspects

Manufacturing has consequences for the workplace and the environment. This results in additional costs and restrictions concerning flexibility. Therefore security systems and job safety measures have to be observed and implemented, including measures against chemical, electrical, and thermal hazards, as well as hazards caused by noise. In manufacturing, the environment is harmed by material emissions in the form of gaseous, liquid, or solid process media or removal products. The legal regulations concerning processing and waste disposal have to be observed. Concerning the impact on the environment, the difference between cutting and metal forming processes is that forming does not produce any chips that have to be disposed of. In cutting, the chips mix with coolants and lubricants. Reprocessing those mixtures is rather costly, because the chips have to be separated from the liquid coolant. Usually the mixtures must be disposed of as special refuse.

In forming, especially in cold forming, two different effects on the workplace and the environment can be observed. On the one hand, noise and vibrations of the drive, the transmission, and the actual transforming process occur. This applies in particular to forging, deep drawing, and extrusion. On the other hand, solid and liquid lubricants, which reduce the friction between tool and workpiece, have to be removed and disposed of after the forming process. Moreover, the temperature of the workpiece also affects the hot forming process. In order to minimize the impact on the operator and other workers, handling systems are used, protective clothing is worn, and the machines are encapsulated.

4. Conventional Machine Tools

Typical machine tools for cutting are universal center lathes and cylindrical surface grinders, and for metal forming are deep drawing and extruding machines. Universal center lathes are usually applied for small-lot production. Turning is the process of cutting with a continuous cutting motion and a feed motion on a level perpendicular to the cutting direction. Here, the axis of rotation of the cutting motion maintains its position relative to the workpiece, independent of the feed motion. In universal center lathes the tool executes the feed motion and the workpiece carries out the cutting motion. Applications are high-speed cutting of materials, like gray cast iron or alloy steel, as well as superfinishing of gray cast iron, carbon steel, and nonferrous metals. Usually tungsten carbides, ceramics, or diamonds are used as cutting materials.

Cylindrical surface grinding can be divided into external peripheral cylindrical and external round peripheral cylindrical plunge-cut grinding. These processes are applied in finishing and used for small-lot sizes as well as mass production. Workpieces of a high accuracy, a surface roughness of 2 μm, and a diameter tolerance of 5 μm can be produced. The grinding tools consist of a multiplicity of bound abrasive grains. There are natural and synthetic cutting materials. Usually brittle hard materials are used such as zirconium oxide, corundum, silicon carbide, boron carbide, boron nitride, and diamond. Normally, the abrasives are held together by bonding agents. These can be inorganic such as ceramics and silicate, organic such as rubber, synthetic resin, and glue, or metallic such as bronze, steel, and tungsten carbide.

In deep drawing, an even sheet metal is formed into a hollow cylinder. The processed sheet metal has an almost constant strength, because it is not clamped laterally. The shape and the surface profile of the tool are imprinted in the workpiece. The process is particularly applied in the production of vehicle parts such as oil pans, interior and exterior forms, vehicle roofs, and transmission parts, and of components such as rinses and baths as well as caps, tins, and cans. Dead or non-dead hot and cold rolled strips are processed, which can be either cast, unalloyed, or alloyed sheet metal.

During the forming process of extrusion molding a block surrounded by a holder is pressed in strands with a full or hollow cross-section. The manufacturing of such strands usually takes place in several steps, until the final form is obtained. A differentiation is made between extrusion with rigid tools and with an active medium. During extrusion with rigid tools, a workpiece block equipped with a stamp is pressed through a matrix. Tool and workpiece are in direct contact. During extrusion with an active medium, the block is pressed through the matrix by the active medium. In the ideal case, a purely compressive stress state develops. These processes are mainly used to machine aluminum and its alloys, lead and tin alloys, copper and its alloys, and various iron, nickel, and titanium materials.

5. *Advanced Processes*

Advanced processes, which in the future will gain in importance in industrial applications, include high-speed cutting, high-speed grinding, and the forming process called internal high-pressure forming. High-speed cutting requires higher static and dynamic rigidities of the machines, high drive powers, improved spindle bearings, cooling lubricant systems, and safety devices, as well as mechanisms for rapid balancing of machining parameters. Other requirements include highly precise feed, roll, or hydrostatic guidance, and loading, unloading, and feed systems, as well as improved controls.

In the field of grinding machines, a clear trend is towards creep feed grinding with boron nitride and diamond grinding wheels at high speed. In the course of complete processing, machines have been developed for outside and internal cylindrical grinding, where both headstocks are arranged parallel on an indexable insert. High-speed grinding is used to increase productivity. When boron nitride grinding wheels are used, cutting speeds of up to $300\,\mathrm{m\,s^{-1}}$ may increase the material removal rates substantially and guarantee a good work result, whereas the increased process temperatures require an adapted cooling lubrication. Moreover, the material removal rate can be increased by controlling the tool during the grinding process. For conventional grinding wheels, grinding with continuous dressing has been developed in which a diamond role is applied. A similar effect can be obtained with diamond or boron nitride grinding wheels by continuous in-process dressing, where dimensional controls are applied to increase the accuracy.

Internal high-pressure forming belongs to the forming processes with an active medium. Hollow semifinished workpieces are used as base material to manufacture components of a complex shape. Forming takes place by means of a liquid loaded by pressure, if necessary combined with a simultaneous effect of outside axial forces. Water emulsions, oils, and elastomers are used as fluid media. The process is used for expanding, expand tossing, bending, interspersing, and calibrating. The range of components manufactured with internal high-pressure forming continues to increase. Body reinforcements are manufactured such as frame constructions in the automotive and railway industries. Also, front wheel carriers and rear axles are manufactured using this process. In pipe construction diverse applications can be found such as the production of exhaust elbow pipes and pipe units. As in the case of the other forming processes, a large variety of metals can be processed.

Bibliography

Beitz W, Küttner K H 1994 *Dubbel: Handbook of Mechanical Engineering*. Springer, Berlin

Czichos H 1996 *Hütte—Die Grundlagen der Ingenieurwissenschaften*. Springer, Berlin

König W 1990 *Fertigungsverfahren*. VDI, Düsseldorf, Germany

Lange K 1985 *Handbook of Metal Forming*. McGraw-Hill, New York

Spur G, Stöferle T 1984 *Handbuch der Fertigungstechnik*. Hanser, Munich, Germany

G. Spur and F. Elbing

Machining and Precision

The purpose of machining is to convert a blank to a finished part productively and efficiently, so as to impart to it a precise geometric form that can be described in terms of dimensions, shape, position, and surface properties. The demands on the precision of machining are increasing continuously. This holds especially for ultraprecise processes and microtechnology (Furukawa and Moronuki 1988, Moriwaki *et al.* 1990, Hatamura *et al.* 1997).

The development of high-strength materials results in an increased power requirement during machining, meaning higher process-related loads on the machine and the tool, and also on the workpiece. Techniques adapted to meet these demands (such as hard turning, laser-supported turning, high-speed cutting (HSC)) are used in response to this.

Using materials with a high degree of internal damping, and compound structures (e.g., welded structures) in the machine design improves machine performance. The use of novel materials is becoming increasingly important in this context, since it is possible in this way to meet better the extreme and sometimes contradictory demands made on the material properties, and to fulfill economic criteria as well. These include:

- new materials adaptable to specific tasks

(i) reaction resin or polymer concrete for rigid and low-vibration column components (thermal inertia, low coefficient of thermal expansion, damping),

(ii) fiber-reinforced plastics for highly stressed components (lightweight construction, damping, low coefficient of thermal expansion),

(iii) structural ceramics for wear- and corrosion-resistant parts (chemical, thermal, and electrical properties, wear resistance);

- coatings primarily to affect the friction properties (Teflon coating of guide means) and wear (coating with hard materials, hardness, wear resistance);
- materials with controlled properties for use as actuators (piezoelectric elements, electro- and magnetorheological fluids).

Figure 1 illustrates the interaction between a workpiece and a machine in the machining process and the effects of this interaction on the precision with which the workpiece is machined. Owing to the deformation behavior of the machine components and of the workpiece itself, the material properties have a significant effect on the precision achieved. Moreover, the machining process to be used is dictated by the material of the workpiece.

The main object of cutting with machine tools is to generate the cutting forces required to remove the chips, while executing the specified relative motion between the cutting edge of the tool and the workpiece, and to do so as precisely as possible under any mechanical and thermal conditions, in order to produce the desired change in shape.

The main criteria for machining are precision and productivity. A high rate of stock removal is necessary for productivity. This can be achieved with large chip cross-sections and high cutting and feed rates. However, the large forces and energy expenditures required for this can make accurate production more difficult.

As control of the geometric and kinematical effects on the machining process and of static and dynamic deformation improves, thermally induced deformations are increasingly becoming the decisive factor for precision in production (Bryan 1990, HPZ 1994).

Even minor changes in component temperatures owing to varying room temperatures or local temperature increases can cause displacements, which are governed by the coefficient of thermal expansion of the structural materials. This can cause workpiece tolerances to be exceeded in the micrometer range.

1. Effects of Manufacturing Process and Workpiece Material on Precision

1.1 Process-related Stress

The material to be machined and the chip-removing conditions largely determine the forces and the temperature increases that occur at the point of machining. In view of the effect on precision, these are referred to as process-related stress. Whereas forces resulting from this process-related stress are a direct cause of deformations, thermally induced deformations in the machine and the workpiece are determined by time—except for those caused by the location and intensity of the heat. For this reason, the precision achieved is not only influenced by the process-related stress, but also by the thermal conditions of the machine environment and by the course of time. This latter point shows that even organizational factors concerning the time of the machining can influence the work results, and they have to be taken into account in precision machining.

Thus, the interface deflections of interest here occur as a reaction to the process-related stress. They are determined by the geometric transfer properties of the system consisting of the workpiece, the tool, the chuck, and the machine, and by the relevant material properties. The material properties of the workpiece and of the machine are defined by material parameters such as strength, Young's modulus, mechanical dissipation factor (η, strain energy divided by loss energy during one cycle), density, Poisson's ratio, thermal conductivity, specific heat capacity, linear thermal expansion coefficient, and hardness. They directly affect the deflections occurring and, thus, the resulting precision of a machining process. Figure 2 is a simple illustration of this relationship.

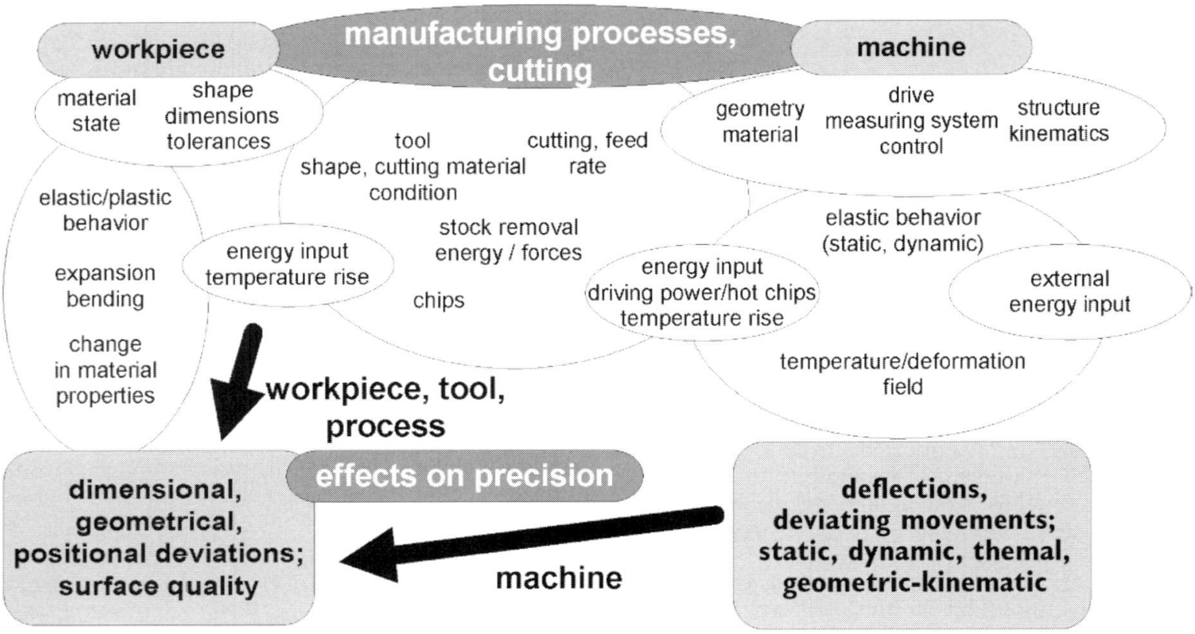

Figure 1
Effects of machining on workpiece quality.

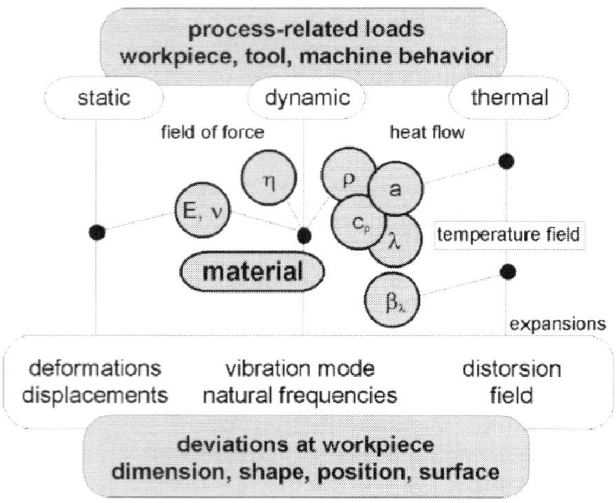

Figure 2
Parameters affecting workpiece precision.

1.2 Thermal Changes in Shape of the Workpiece

In the chip-removal process, mechanical energy is converted to thermal energy in the range of contact of the tool, the workpiece, and the chip. The cutting work is given as the sum of the deformation (shear), chip removal, and friction work (Milberg 1992, Moriwaki *et al.* 1990). Thus, the zones of heat generation are made up of the shearing or chip-removal zone during penetration of the cutting wedge into the workpiece and of the friction zone between the tool and the chip. The thermal energy (Q = heat quantity) is dissipated by thermal conduction in the workpiece, the tool, and the chip (Fig. 3), as well as by radiation and convection.

On the one hand, the heat supplied to the workpiece causes thermal expansion or bending (Fig. 5), which in turn causes changes in dimensions and shape. On the other hand, this heat can also lead to changes in the boundary zones of the workpiece (such as microstructural changes in the material owing to unfavorable settings during grinding).

Both influences can be minimized if optimal chip-removing parameters are selected. For example, the energy input into the workpiece is substantially reduced by HSC, which is rapidly gaining acceptance. With the relatively small chip cross-sections and very high relative speeds, the contact surface between the workpiece and the tool is small and the contact itself is very short-lived.

2. Effect of the Machine Tool on Workpiece Precision

The movements of the workpiece and/or the tool that are required to machine the workpiece (position and rate changes), as well as the conversion of drive and cutting energy, produce undesirable deflections

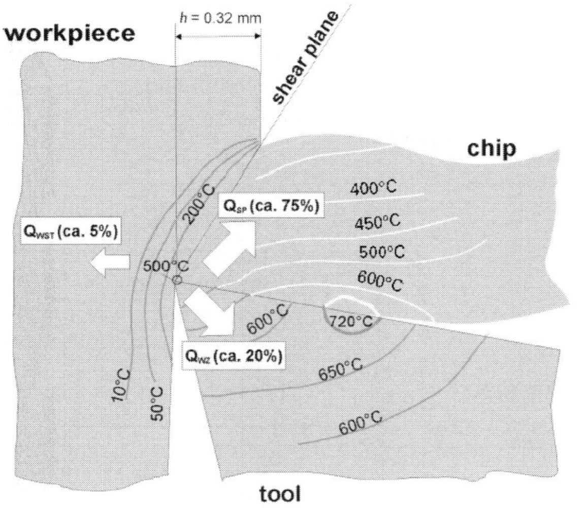

Figure 3
Temperature distribution in the cutting process (according to Milberg 1992). Cutting material: HMP20; workpiece material: steel; cutting rate: $V_c = 60\,\text{m min}^{-1}$; rake angle: $\gamma = 10°$.

influenced by kinematical, static, dynamic, and thermal effects. This places limits on the machining precision that can be achieved.

2.1 *Thermal Deformations and Deflections*

The chain of thermal action of the system consisting of the workpiece, the tool, and the machine as shown in Fig. 4 reflects the causal relationships between the thermal load and deviations in dimensions, shape, and/or position at the workpiece.

(*a*) *Thermal load.* The process-related stress generates heat flows at its effective areas in and at the machine by:

- heat loss of internal heat sources in the machine (drive mechanisms and mobile machine components), and
- temperature and temperature changes of external heat sources (external energy input from machine environment); cf. Fig. 1.

The heat sources consisting of the tool, the workpiece, and the chips, which are introduced by the work process and are located in the work area, are classified either as external or internal heat sources (Shaw 1996). The thermal load can be reduced by high efficiency, cooling measures, or thermal isolation from the environment.

(*b*) *The nonstationary temperature field.* Owing to heat storage and heat transfer in the individual units of the machine tool, the thermal load produces a temperature field that varies in time and place and conforms to Fourier's differential equation:

$$\frac{\partial T}{\partial t} = a\left[\frac{\partial^2 T}{\partial x^2} + \frac{\partial^2 T}{\partial y^2} + \frac{\partial^2 T}{\partial z^2}\right] \quad \text{where } a = \frac{\lambda}{\rho c_p} \tag{1}$$

The thermal diffusivity, a, is the characteristic material parameter here and determines the development of the *temperature field*. The time course of the mean temperature of a unit or of the entire machine is described as the step response of the thermal system by the following equation:

$$\Delta T = \Delta T_B(1 - e^{-t/\tau}) \tag{2}$$

where the temperature difference

$$\Delta T = T - T_U \tag{3}$$

is defined by the ambient temperature (Fig. 5). The steady-state temperature, ΔT_B, is determined by the supplied power, P_0, and the heat dissipation capability, $A\alpha$ (where A is the surface area and α the heat transfer coefficient).

The heat dissipation capability describes the heat exchange with the environment and with the other units by thermal conduction, convection, and radiation:

$$\Delta T_B = \frac{P_0}{A\alpha} \tag{4}$$

The time constant, τ, of the system is obtained from the heat removal capacity and the heat storage capacity, mc_p (where m is the mass and c_p the coefficient of specific thermal capacity). It is a material/design property of the system.

A large time constant is desirable in universal machines. A small time constant may be advantageous in single-purpose machines:

$$\tau = \frac{mc_p}{A\alpha} \tag{5}$$

Measures for lowering and compensating for the nonstationary temperature begin at the heat dissipation capability, the specific thermal conductivity of the material, and the time constant.

(*c*) *The nonstationary distortion field.* A temperature change in a body momentarily changes its volume, V, in direct proportion to its coefficient of volume expansion, β:

$$e_{th} = \frac{\Delta V}{V} = \beta\Delta T = \varepsilon_{x,th} + \varepsilon_{y,th} + \varepsilon_{z,th} = \Delta T(\beta_{l,x} + \beta_{l,y} + \beta_{l,z}) \tag{6}$$

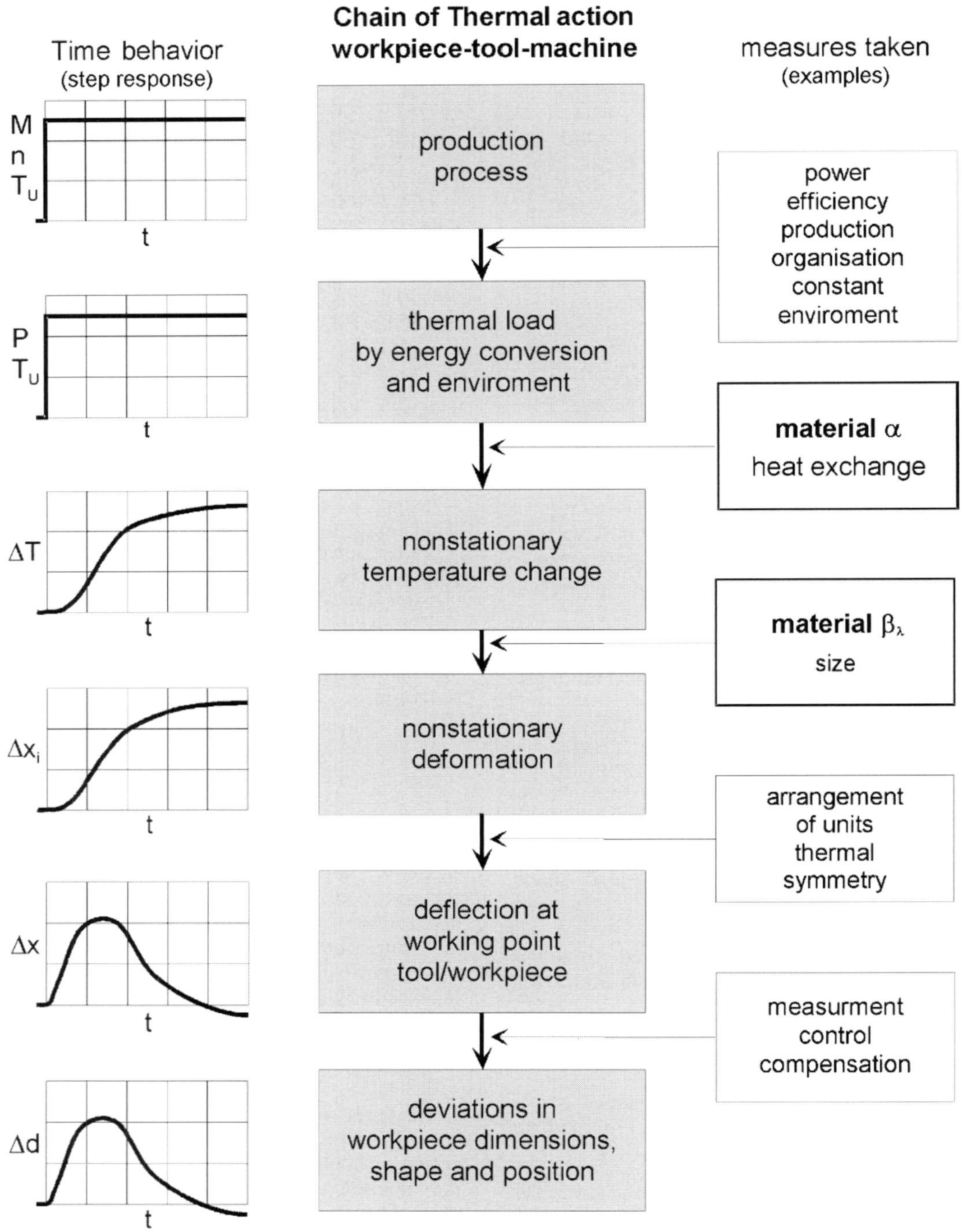

Figure 4
Chain of thermal action of workpiece, tool, and machine tool.

where the thermal volume expansion, e_{th}, is composed of the linear expansions, ε, in the principal directions. Whereas homogeneous bodies expand freely with constant and linear temperature distribution, nonuniform temperatures and irregularities cause thermal tensions and thus elastic deformations. Additional tensions occur if the deformations are externally obstructed or if additional, external

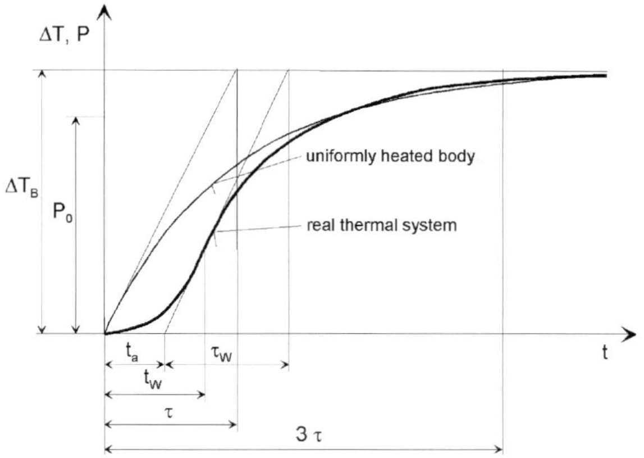

Figure 5
Step response of a thermal system.

forces are applied. The deformation state is described with a generalized version of Hooke's law. The following equation applies:

$$e = e_{\text{elast}} + e_{\text{th}} = \frac{1-2\mu}{E}(\sigma_x + \sigma_y + \sigma_z) + 3\beta_l \Delta T \quad (7)$$

The total expansion in one direction is given as

$$u_x = \int_x \varepsilon_x \mathrm{d}x \quad (8)$$

The greater portion of the thermal deformations can be determined from free thermal expansions or thermal bending. The linear thermal expansion coefficient, β_l, has a decisive effect on the magnitude of the thermoelastic deformations.

(*d*) *Relative deflection at the point of action.* The thermoelastic deformations of the machine units—referring to the tool center point (TCP) of the machine tool—are summed up to all other relative movements between the tool and the workpiece. This summation causes the workpiece deviations that are thermally influenced by the machine. If the machine is designed with this in mind, it is possible for the deformation effects to be approximately neutralized (thermal compensation) and for the displacement vectors to be tangent to the plane of action (e.g., by thermal symmetry). These are common state-of-the-art design principles. The remaining deformations can be further reduced by 60–90% with measurement controls and compensating devices (Weck *et al.* 1995). Software solutions integrated into the CNC control are mainly used for the compensation. Here the control possibilities offered by a position-controlled feed drive are used.

(*e*) *Effects on the workpiece.* The thermally induced deflection of the working point of the machine tool constitutes a deviation from the nominal position or path and is transmitted to the workpiece as a machine error only during the engagement of the tool. A time window that is determined by the process and which defines the workpiece deviation appears in connection with the precision reference, the coordinate system of the undeformed machine as set by the manufacturer, the tool setting, or a reference surface on the workpiece. Owing to the delayed thermal behavior, there is no further direct relationship between the process-related load and the effects on precision.

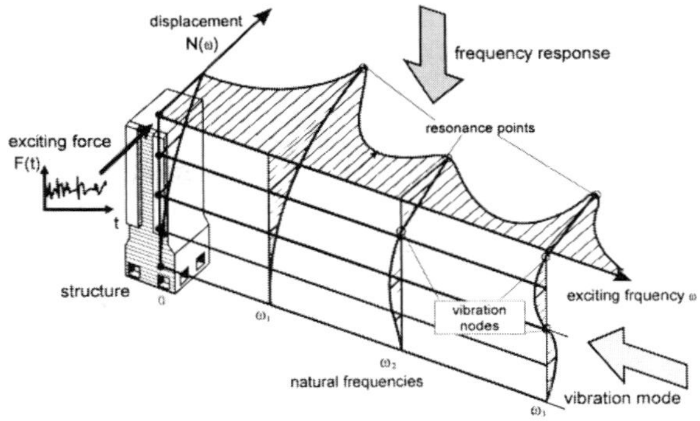

Figure 6
Model analysis of a machine column.

2.2 Static and Dynamic Deformation and Deflection

The static behavior of a machine tool describes its reaction to loads that are constant over time. Analogously, the dynamic behavior includes the reactions to loads that vary with time. In both cases, the relative elastic deflections, which occur between the tool blade and the workpiece, are decisive for the working precision of the machine tool. This shifting of the cutting edge compared with workpiece results from the sum of all deformations of all machine components located within the field of force that have been transformed onto the cutting edge. This includes deformations of the frame components, the joints in between, and the guides for linear and rotary movements. The static and dynamic yielding of the machine as the ratio of the position change to the load change causing it, makes up the characteristics for the deformation behavior at the cutting edge. They depend on the material (E, v, ρ, η), the cross-sectional shape of the components (A, I_x, I_y, I_t, κ), and the structural design of the machine as a whole (serial or parallel arrangement of axes, effective lengths). In addition, the dynamic yielding depends on the oscillation frequency that excites the structure.

The deformation state of an entire machine is described by its vibration modes. Included in this is the static limit case with a frequency of zero.

Using a machine column as an example, Fig. 6 illustrates how the frequency response of the vibration amplitudes is formed from the static elastic centerline, taking the three lowest natural vibration modes into consideration. The magnitude of the amplitudes depends on the location along the column centerline that is being observed. In terms of precision, the aim of the machine design has to be, first of all, to obtain as low a static yield as possible. In addition to this, the vibration amplitudes of the system consisting of the workpiece, the tool, the chuck, and the machine can be selectively reduced with materials having a high internal damping capacity and/or compound structures (such as welded joints).

3. *Summary*

The material properties of the workpieces to be machined, and the properties of the material used in the machine design have a decisive effect on the machining process and on the machine tool performance. The achievable workpiece accuracy and productivity are major factors in production. Developments in tools and engineering on the one hand, and innovations in machine design, drive, measuring, and control technology on the other, are continuously pushing back the limits of precision and productivity, but without eliminating them altogether.

Bibliography

Bryan J 1990 International status of thermal error research. *Ann. CIRP* **39**(2), 645

Furukawa Y, Moronuki N 1988 Effect of material properties on ultra precise cutting processes. *Ann. CIRP* **37**(1), 113

Hatamura Y, Nakao M, Sato T 1997 Construction of an integrated manufacturing system for 3-D microstructure—concept, design and realization. *Ann. CIRP* **46**(1), 313

HPZ 1994 *Hochpräzisionszerspanen mit geometrisch bestimmter Schneide*. Hrsg. Forschungsgemeinschaft Qualitätssicherung Frankfurt am Main, Beuth, Berlin

Milberg J 1992 *Werkzeugmaschinen—Grundlagen: Zerspantechnik, Baugruppen und Steuerungen*. Springer, Berlin

Moriwaki T, Horiuchi A, Obuda K 1990 Effect of cutting heat on machining accuracy in ultra precision diamond turning. *Ann. CIRP* **39**(1), 81

Namba Y, Wada R, Unno K, Tsuboi A, Okumara K 1989 Ultra-precision surface grinder having a glass-ceramic spindle of zero thermal expansion. *Ann. CIRP* **38**(1), 331

Namba T, Yamada Y, Tsuboi A, Unno K, Nakao H, Inasaki I 1992 Surface structure of Mn–Zn ferrite single crystals ground by an ultra-precision surface grinder with various diamond wheels. *Ann. CIRP* **41**(1), 347

Shaw M C 1996 Energy conversion in cutting and grinding. *Ann. CIRP* **45**(1), 101

Weck M, McKeown P, Bonse R, Herbst U 1995 Reduction and compensation of thermal errors in machine tools. *Ann. CIRP* **44**(2), 589

K. Großmann

Machining: Generalities and Economics

Machining is manufacturing that employs cutting or abrasive processes. A tool removes particles of a preshaped part to generate a finished surface or complex shape that cannot be formed by primary shaping or forming processes. The machine moves the tool across the workpiece on a kinematically controlled path. Consequently, the shape of the workpiece is predominantly independent of the profile of the tool. Forces and temperatures during cutting processes are low compared with casting or forming processes. In general, high dimensional accuracy and high surface qualities can be achieved. Productivity in, for instance, parts-per-time units is much lower than in forming or in most primary shaping processes; however, the flexibility of shapes and lot sizes is much higher than with processes that use form tools, such as metal forming and casting processes. Determination of the optimal interface of subsequent processes such as forming and cutting is an important issue, both scientifically and practically. Adaptation, substitution, or integration are the options that can be employed to determine this interface.

1. Cutting Processes: Machining with Geometrically-defined Cutting Edges

In cutting, the number, shape, and position of the cutting edges in relation to the workpiece are both known and can be described (in contrast to abrasive processes, i.e., machining with geometrically undefined cutting edges, e.g., grinding). Figure 1 illustrates important processes within this group (Tönshoff 1995). The processes are distinguished according to the cutting motion direction (cutting speed v_c), the feed motion direction (feed speed v_f), and the resulting effective motion direction (effective speed v_e).

The vectors of feed and cutting speed constitute the working plane (Fig. 1). The angle between these two vectors is the feed direction angle φ, while the angle between the effective direction and the cutting direction is the effective direction angle η. The relationship

$$\tan \eta = \frac{\sin \varphi}{(v_c/v_f) + \cos \varphi} \tag{1}$$

applies to all cutting processes. For turning, drilling, reaming, and sinking $\varphi = 90°$, and therefore

$$\tan \eta = \frac{v_f}{v_c}$$

The mechanical process of cutting particles from the workpiece, i.e., chip formation, can best be described with reference to the orthogonal process (two-dimensional workpiece material deformation). The wedge is determined by the rake angle γ, the clearance angle α, and the cutting edge radius r_β. Penetration of the wedge causes the material to

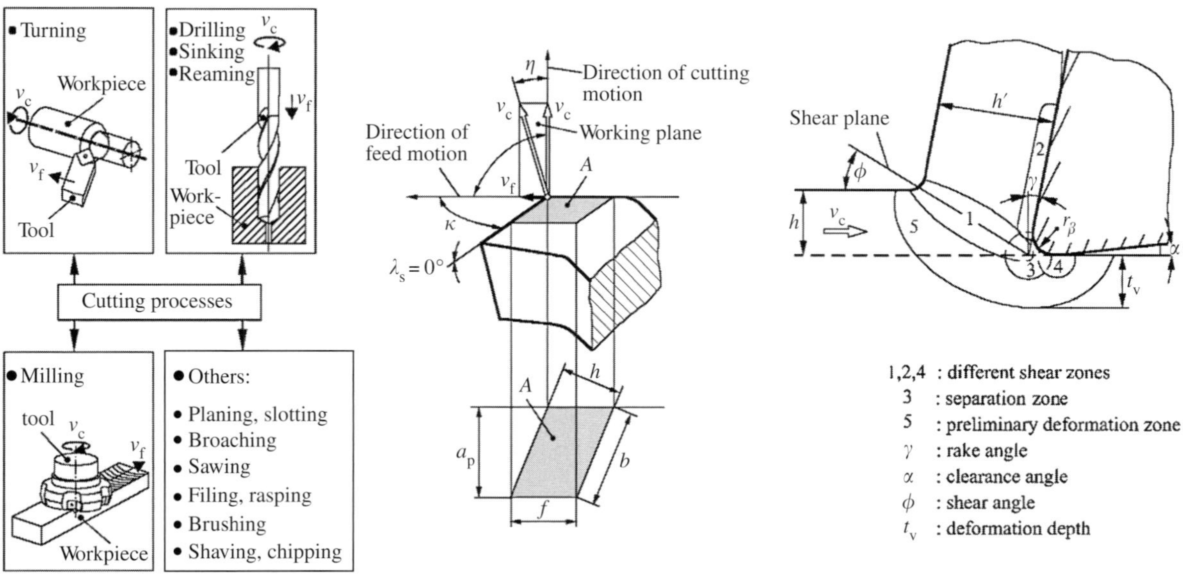

Figure 1
Cutting processes—motion and geometry.

undergo plastic deformation. Figure 1 shows the zones of plastic deformation in which five zones can be distinguished.

The primary shear zone 1 constitutes the actual area of chip formation by shearing. In the secondary shear zones 2 and 4 ahead of the rake face and at the flank, forces of friction act between the tool and workpiece, causing plastic deformation of these material layers near the rake and the flank faces. In the deformation advance zone 5, chip formation causes stresses which lead to plastic and elastic deformation of this zone. In the pressure and separation zone 3, the material is deformed and separated under high compressive stresses.

Because of the shearing mechanism, the chip thickness h in the undeformed condition changes to the chip thickness h', resulting in the chip thickness ratio $\lambda_h = h'/h$. The shear plane consists of the shear angle ϕ with the direction of cutting motion. In the chip formation zone, the cutting energy is

$$E_c = F_c \times l_c \tag{2}$$

where F_c is the cutting force and l_c is the path length due to primary motion.

The cutting energy consists of forming and shearing energy E_ϕ, friction energy at the rake face $E\gamma$, friction energy at the flank E_α, surface energy to form new surfaces E_T, and kinetic energy due to material flow.

The energy converted in the machining of one unit of volume is

$$e_c = E_c / V_W \tag{3}$$

where e_c is the specific energy and V_w the volume of removed material. Like E_c, the individual components of E_c can also be expressed in relation to V_w.

From the specific energy e_c introduced, the specific cutting force k_c can be derived as a material-dependent factor for calculating the cutting force:

$$k_c = F_c / A = F_c / (h \times b) \tag{4}$$

where A is the cross-section of undeformed chip and b and h are the width and thickness, respectively, of the undeformed chip (see Fig. 1), such that

$$e_c = E_c / V_W = P_c / Q_W = (F_c \times v_c) \times (A \times v_c) = k_c \tag{5}$$

where P_c is the cutting power, Q_W is the material removal rate, and F_c is the cutting force.

The energy introduced into the chip formation zone is almost entirely converted into heat (with small energy remnants being transformed into residual stresses in the chip and the workpiece). This produces high temperatures in the cutting zone and also in the tool tip, the latter being subjected to mechanical and thermal stress. This load, along with chemical reactions, causes wear.

The stress on the cutting edges is dependent on various influences. Besides process parameters (such as cutting speed, feed rate, depth of cut, etc.) and environmental influences (such as cooling lubricant), the workpiece material in particular influences tool wear. The different kinds of wear are:

(i) fractures and cracks, which occur in the cutting edge area due to excessive mechanical or thermal stresses;

(ii) mechanical abrasion, which is mainly caused by hard inclusions such as carbides and oxides in the workpiece material;

(iii) plastic deformation, which occurs if the cutting material has inadequate resistance to deformation but sufficient toughness;

(iv) adhesion, which is the shearing off of pressure-welded points between the material and the chip, the point of shearing being located in the cutting tool material;

(v) diffusion, which occurs at high cutting speeds and where there is mutual solubility of the cutting and workpiece materials. The tool material is weakened by chemical reactions, and may become detached or be removed;

(vi) oxidation, which occurs only at high cutting speeds. The cutting tool material oxidizes, due to contact with atmospheric oxygen, and its structure is weakened.

The machining tool is consequently subject to wear. A basic dualism applies to all cutting materials. They should, on the one hand, be hard and wear resistant—but are then less tough and stable under periodically changing loads or under unstable cutting conditions, and in intermittent cutting. On the other hand, the machining tool that can withstand variable mechanical or thermal stresses is then less resistant to wear. To overcome this restricting dualism, various cutting materials are manufactured as composite materials. Tool coatings made of hard, wear-resistant carbides or oxides combine the different functions: physically or chemically vapor-deposited layers provide wear protection, while the underlying, tougher substrate performs the supporting function, even under dynamic load conditions.

Materials used for cutting are plain and alloy steels (still important for manually-guided tools), high-speed steels, cemented carbides, ceramics, and superhard cutting materials (diamond and boron nitride). For a specific tool material/workpiece material combination and a given tool life criterion, tool life depends on the cutting speed according to a power law (Kienzle and Victor 1952):

$$\frac{T}{T_0} = \frac{v_c^k}{C} \tag{6}$$

where T_0 and C are reference values and k is an empirically determined exponent. T_0 is normally set at $T_0 = 1$ min, and C is the cutting speed for which a tool life of $T_0 = 1$ min can be reached.

For economical cutting processes, the optimal cutting speed has to be established according to a time criterion (Fig. 2). The optimal cutting speed is then

$$v_{\mathrm{c,\,opt}} = C(-k-1)_{\mathrm{Wz}}^{1/k} \qquad (7)$$

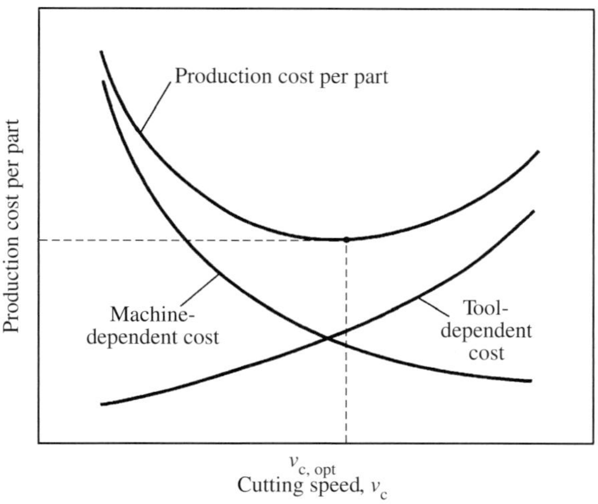

Figure 2
Machining costs as a function of cutting speed.

Optimization of the cutting speed to minimize unit costs takes into account not only the tool changing time, t_{Wz}, but also the tool costs per cutting edge, K_{Wz}, and the hourly machine rate, K_{M}:

$$v_{\mathrm{c,\,opt}} = C(-k-1) \times (t_{\mathrm{Wz}} + (K_{\mathrm{Wz}}/K_{\mathrm{M}}))^{1/k} \qquad (8)$$

1.1 High-speed Cutting and High-performance Cutting

Compared with conventional cutting, high-speed cutting (HSC) uses a decisively increased cutting speed and low feed rates per tooth (Fig. 3) and offers several advantages over conventional cutting (Schulz 1996). Machining with increased cutting speeds leads to reduced cutting forces. The most common benefits are higher material removal rates and a higher surface quality, combined with high shape and form accuracy.

For several metal cutting applications, however, very high cutting speeds are not necessarily advantageous (Tönshoff *et al.* 1999). High-performance cutting (HPC) is a technology which combines the advantages of high-speed machining in finishing, and high material removal rates in roughing, using medium cutting speeds. The rotational frequency of the main spindle is particularly important, especially in roughing. The range in which spindles of high-speed machining machine tools supply maximum power is limited. Due to this, the rotational frequencies that can be chosen during roughing are restricted. The roughing process according to HPC is

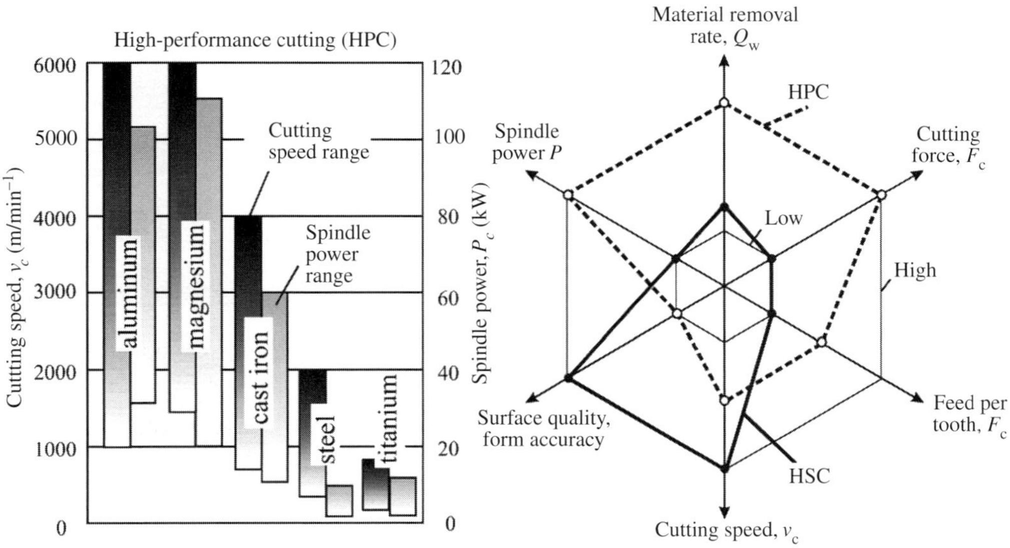

Figure 3
High-performance cutting (HPC) vs. high-speed cutting (HSC).

limited by the power of the main spindle and not by the rotational frequency alone, because the required torque is high and loss of power increases with the rotational frequency. Therefore, an optimal spindle speed for roughing using HPC is not the maximum speed available.

The practicability of high-performance cutting depends to a large extent on the material to be machined. Figure 3 shows the cutting speed and spindle power ranges of HPC together with a comparison with HSC processing.

2. *Abrasive Processes: Machining with Geometrically Undefined Cutting Edges*

Abrasive machining is cutting by the mechanical action of multicutting edges. The cutting edges are formed by grains of hard material whose shape and distribution is irregular (Malkin 1989). Therefore, the cutting edge geometry cannot be described with reference to a single grain and the individual cutting edge is geometrically undefined. The abrasive processes can be distinguished as cutting with bonded grains, such as grinding with rotating tools, belt grinding, or honing, and cutting with nonbonded grains, such as lapping and polishing.

The important cutting edge angles for chip formation—the clearance angle α, the rake angle γ, and the wedge angle β—are only indicated by means of statistical parameters such as mean values or distributions. On average, high negative rake angles and large contact and friction zones are formed between the grain and the workpiece. The cutting edges penetrate only a few micrometers into the material. The undeformed chip thickness distribution depends on the position of the cutting edges in the mixture of grains (microtopography of the cutting edge zone) and the geometry of the workpiece surface to be machined. Not only chip removal but also elastic and plastic deformation without chip removal takes place, which can lead to changes of the surface integrity state. At the beginning of the grinding process elastic effects occur primarily, because of a sharp grinding wheel. With increasing wheel wear, thermoplastic effects arise, and thermally-induced stresses increasingly lead to tensile stresses, while the mechanical load mainly induces compressive stresses.

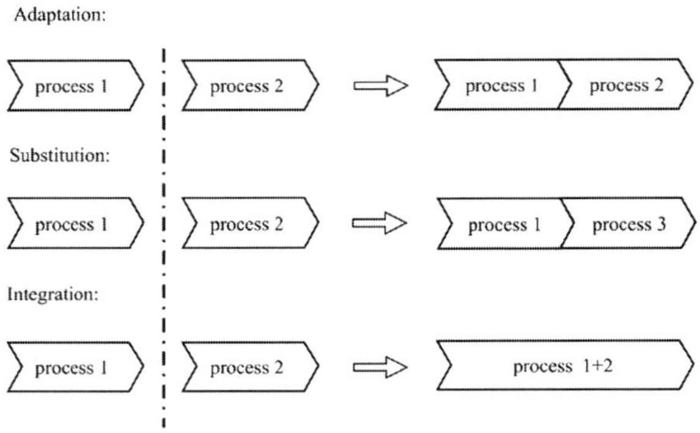

Figure 5
ASI method of rationalization.

	Grinding	Honing	Lapping
Accuracy of shape	+	O	++
Dimensional accuracy	+	++	++
Positional accuracy	+	–	–
Geometrical surface	O	+	++
Favorable subsurface influence	O	+	+
Metal removal rate	++	O	–
Manufacturing costs	–	+	++

– Low   + Large
O Moderate   ++ Very large

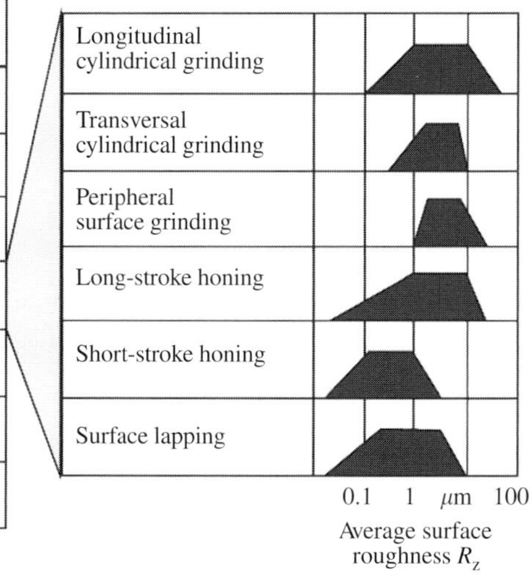

Figure 4
Comparison of abrasive processes.

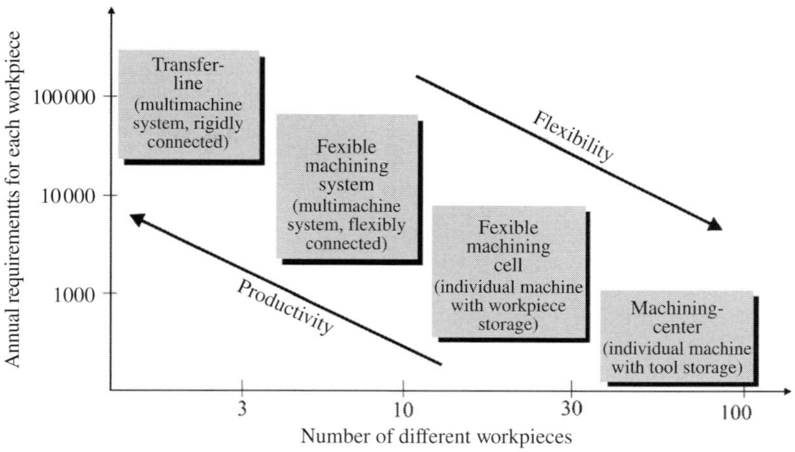

Figure 6
Machining systems and their fields of application.

The increase of normal and tangential forces causes mechanical and thermal loads to increase. Tangential forces determine the related grinding power (P_c). High normal forces between tool and workpiece result due to the predominantly negative rake angles of the cutting edges.

Machining processes with geometrically undefined cutting edges are frequently used as final machining processes for workpieces with high quality requirements. Figure 4 shows a comparison of various precision machining processes with regard to operational results and efficiency.

3. *Machining as a Part of the Process Chain*

Workpieces are manufactured in sequences of operations (manufacturing steps). Rationalization measures to improve economic efficiency and quality, therefore, should not only be applied to individual operations or manufacturing steps, but should also aim at an overall optimum. To achieve this, the possibilities of adaptation, substitution, and/or ntegration (the A–S–I method) may be examined (Fig. 5). Adaptation is the favorable coordination of consecutive processes, e.g., the optimal intersection between metal forming (near net shape) and machining. The development of tools and machine tools may lead to the substitution of one manufacturing process by another, e.g., replacement of grinding by hard turning. Integration of manufacturing steps shortens the sequence of operations, and is often linked to direct cost savings, and certainly to shorter throughput times and reduced material flow costs (indirect costs of complexity). Complete machining of components on multiaxis lathes or machining centers is an example.

4. *Productivity vs. Flexibility*

Quality, time, and costs are the criteria of modern production. The time-to-customer has become a dominant factor in the growing turbulence of markets. For machining processes and manufacturing systems, the time-to-customer takes not only productivity but also flexibility into consideration. Flexibility in the sense of fast change of orders allows a kind of risk management under uncertain boundary conditions given by the market. Flexibility in the sense of applicability of different cutting or abrasive processes allows complete machining of a workpiece. Machining centers and cells are favorable solutions in this respect (Fig. 6). Complete machining shortens process chains, and thus reduces machining and complexity costs and throughput times.

5. *Conclusion*

Cutting and abrasive processes are necessary and ideal complements to forming and casting processes, because of dimensional accuracy and surface finish requirements. Besides quality aspects, economical and ecological reasons may make these flexible and versatile production processes unavoidable. To optimize quality, costs, and production time the best sequences and interfaces in the complete production chain must be found. Therefore, it is necessary not just to look at a single process, but to find a holistic view of part manufacturing.

Bibliography

Kienzle O, Victor H 1952 Die Bestimmung von Kräften und Leistungen an spanenden Werkzeugmaschinen. *VDI-Z* **94**, 299–305

Malkin S 1989 *Grinding Technology: Theory and Applications of Machining with Abrasives*. Ellis Horwood, Chichester, UK
Shaw M C 1996 *Metal Cutting Principles*. Oxford University Press, Oxford, UK
Schulz H 1996 *High-speed Machining*. Carl Hanser Verlag München, Vienna
Tönshoff H K 1995 *Spanen*. Springer-Verlag, Berlin Heidelberg
Tönshoff H K, Winkler J, Andrae P 1999 Tools and process design for high-efficient machining. In: *Proc. 3rd Int. Machining & Grinding*. Cincinnati, OH. Society of Manufacturing Engineers, Dearborn, MI

H. K. Tönshoff

Machining of Metals: Fundamentals

Machining by mechanical chip formation is widely used to remove material from a work-piece in a variety of configurations including turning, milling, drilling, and tapping. It is a process capable of producing precise geometrical shapes with tight tolerances. Machining has a long history. In the eighteenth century it was essential to the Industrial Revolution, as the only process capable of producing cylinder bores to tolerances sufficient for steam engines. Machining is still widely used to remove large amounts of material, of the order of millimeters per pass, from ductile metals. Machining can also be used to form ductile chips in brittle materials. In production of aircraft parts under 75 mm machining tolerances of $\pm 5\,\mu m$ are reasonable. The basic process of chip formation is fundamental to all abrasive processes, e.g., grinding, honing, lapping, and abrasive wear. Machining remains economically important.

Chip removal produces technically and scientifically interesting material responses. In single-point cutting the process concentrates the mechanical power of a machine tool into small volumes of the work-piece by forcing a hard cutting tool with a small edge radius through an outer layer of the work-piece material. Deformations of the work-piece material during chip formation can exceed 300%, at strain rates that have been estimated to be as high as $10^6\,s^{-1}$ (von Turkovich 1970). The tribological conditions at the tool–chip interface can result in macroscopic welding of the material to the cutting tool and dissolution of tool elements into the chip. The geometry of the deformation zones and the resulting microstructures in the work-piece depend on the geometrical and tribological interaction of the cutting tool with the work-piece material. Further understanding of the fundamentals of chip formation is motivated by the desire for higher material removal rates, longer tool life, tighter tolerances, and improved quality of machined surfaces.

1. Fundamental Geometrical Relationships in Chip Formation

The idealized, fundamental geometry of continuous chip formation for orthogonal cutting is shown in Fig. 1. This idealization is plane strain. It is a reasonable approximation for understanding the kinematics of the formation of a continuous chip and to be representative of the formation of the bulk of

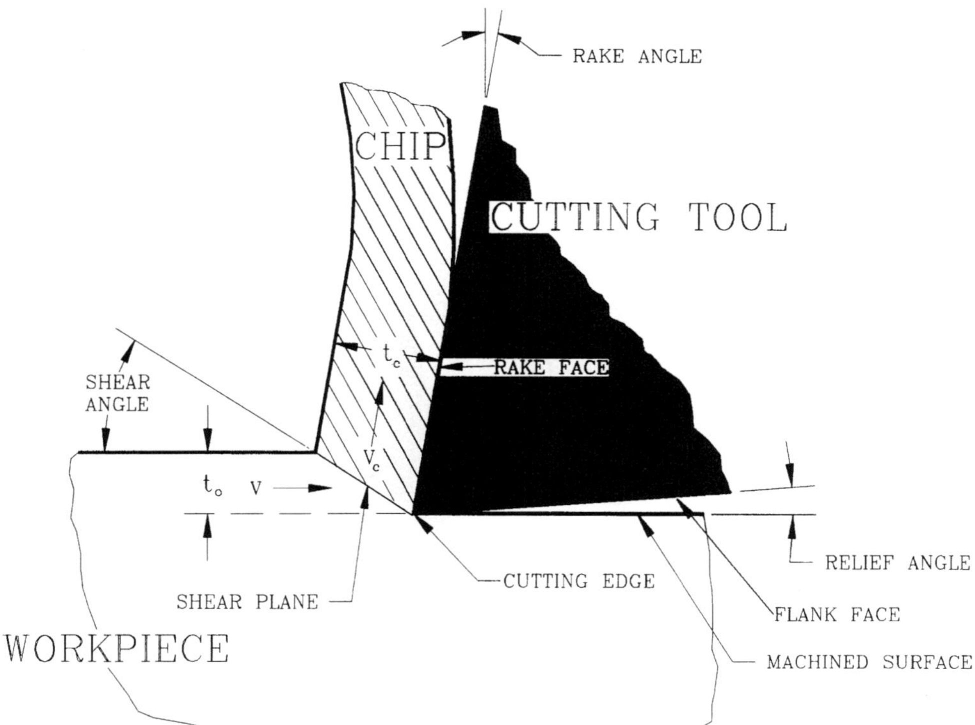

Figure 1
Idealization of orthogonal machining showing the basic geometry of chip formation on the shear plane.

the chip in normal machining. Plane strain conditions exist sufficiently far away from the free surface at the edge of the chip, where plane stress conditions exist, and sufficiently far from the nose of the tool where the flow is complicated by the curvature of the cutting edge.

In this idealization (Fig. 1) the work-piece moves with a speed V against a hard tool with cutting edge radius of zero. The chip forms by shear on the shear plane, extending from the cutting edge to the free surface. In reality there must be a zone of finite thickness in which the chip is formed, and this zone is referred to as the primary shear zone. The chip contacts the tool along the rake face, until the chip curls away from the tool, forming a natural contact length. The region of deformation in the chip adjacent to the rake face of the tool is called the secondary shear zone. In the idealization the tool only contacts the work-piece along a line approximated by the cutting edge. In reality there is a finite area of contact between the tool and the work-piece, around part of the tool nose and along the tool flank. The region of deformation in the work-piece surface adjacent to the tool nose and flank contact is called the tertiary deformation zone. The newly machined surface forms at the cutting edge. It then passes below the relief, or flank, face of the tool where it is separated from the tool by a clearance or relief angle. The shear angle is the angle between the shear plane and the free surface, or the cutting direction. The chip flows up the rake face of the tool with a speed V_c. The rake face of the tool is inclined at an angle α to the normal to the cutting direction. The convention is that the rake angle is positive when the angle between the rake face and the cutting direction is obtuse and negative when it is acute.

The thickness of the material that will form the chip is frequently referred to as the depth of cut in orthogonal machining, t_o, and the chip thickness is t_c. The cutting ratio or chip thickness ratio, r, is defined as

$$r = t_o/t_c \tag{1}$$

From the geometry of Fig. 1 it can be shown that the cutting ratio is related to the shear angle and the rake angle as follows:

$$r = \frac{\sin\phi}{\cos(\phi - \alpha)} \tag{2}$$

and

$$\tan\phi = \frac{r\cos\alpha}{1 - r\sin\alpha} \tag{3}$$

The cutting ratio can be determined from the depth of cut in orthogonal machining and the rake angle, which are generally independent variables and are known, and the chip thickness, which is relatively easy to measure.

The shear strain required to form the chip is derivable from the fundamental geometry of chip formation, as shown in Fig. 2. Consider the unit square in position $A_1B_1C_1D_1$, which moves to position $A_2B_2C_2D_2$, such that C_2 and D_2 are on the shear

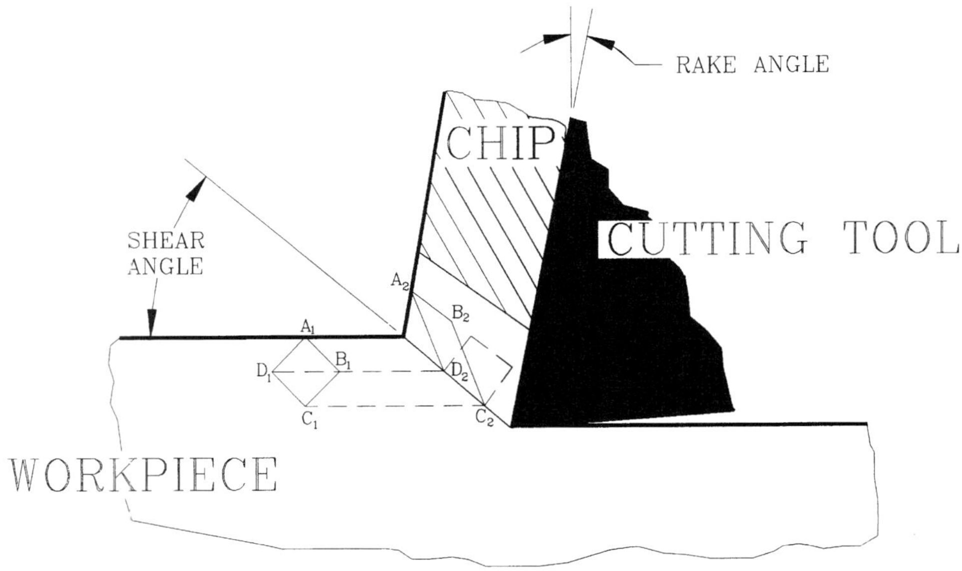

Figure 2
Schematic depiction of the strain of element ABCD as it moves through the shear plane, defined by the shear angle.

plane. It can be shown from geometry that the shear strain is

$$y = \cot\phi + \tan(\phi - \alpha) \tag{4}$$

Using Eqns. (3) and (4), the shear strain can be determined experimentally from chip thickness measurements when the chip formation is approximately continuous. The shear strain can also be determined metallographically, for many continuous and discontinuous chips, from the orientation of the axis of maximum grain elongation in the chip, which is not parallel to the shear plane (Brown 1987).

2. *Forces*

The chip can be considered to be a separate body held in equilibrium by the tool–chip forces and the forces on the shear plane (Merchant 1944). This ignores inertial loading from the chip, which at ordinary machining speeds is significantly lower than the forces of deformation and friction. These forces can be divided into three orthogonal force systems, as shown in Fig. 3. The cutting force, F_c, is parallel to the cutting direction and can be used with the cutting speed to determine the total power consumed in chip removal. The thrust force, F_t, is orthogonal to the cutting force, and therefore does not place a demand on the power, although it is important in the deflection of the tool and the work-piece. The shear force, F_s, represented in Fig. 3 as the force exerted by the chip on the shear plane, can be used to determine the flow stress on the shear plane. The normal force exerted by the chip on the shear plane, F_n, acts to suppress fracture on the shear plane. The tangential and normal forces exerted by the rake face of the tool on the chip, F and N, respectively, are the third system of orthogonal forces, and determine the coefficient of friction, μ. The vector sum of the forces in each of the systems is equal to the total resultant force, R.

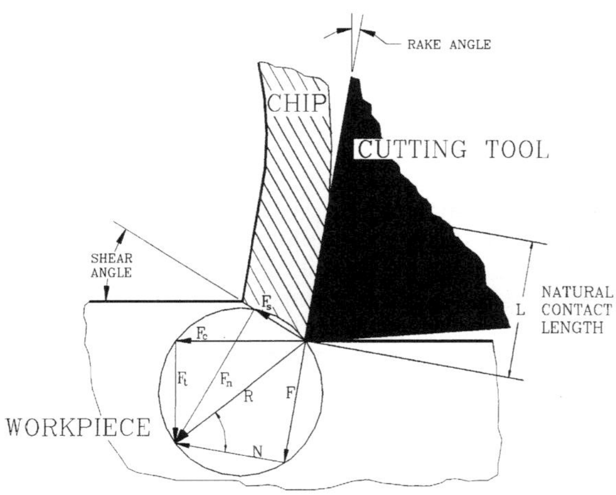

Figure 3
Schematic depiction of three systems of orthogonal forces in chip formation.

The circle diagram in Fig. 3 is a convenient and traditional means of representing systems of orthogonal forces in machining. R is a diameter of the circle. The other force systems consist of two cords that sum to the diameter and are therefore orthogonal to each other. The circle is positioned for clarity in understanding the orientation of the forces. The center of force between the chip and tool is somewhere nearer the center of the natural contact length.

The relationships between the forces can be found by multiplying the force vectors by the appropriate rotation matrix, for example:

$$F = F_c\sin\alpha + F_t\cos\alpha \tag{5}$$

$$N = F_c\cos\alpha - F_t\sin\alpha \tag{6}$$

$$F_n = F_c\sin\phi + F_t\cos\phi \tag{7}$$

$$F_s = F_c\cos\phi - F_t\sin\phi \tag{8}$$

3. *Shear Angle*

The shear angle is of fundamental importance in chip formation. The smaller the shear angle the larger the strain, the machining forces, and the power requirements. There is nothing in the geometry of the tool that dictates what the shear angle should be. The shear angle can be determined *a priori* from the chip thickness (Eqn. (3)), although the chip is not constrained to any particular thickness. The determination of the shear angle is fundamental for understanding chip formation.

The shear angle can be estimated by assuming that it is oriented such that the work to form the chip is a minimum (Merchant 1945). To do this the cutting force is written in terms of the stress on the shear plane, the tool–chip friction, the shear angle, and the parameters of the cutting geometry. Assuming that the tool–chip friction and stress on the shear plane do not vary with the shear angle, the shear angle for minimum cutting force is determined:

$$\phi = \pi/4 - \beta/2 + \alpha/2 \tag{9}$$

where β is the friction angle, such that

$$\mu = F/N = \tan\beta \tag{10}$$

where μ is the average friction coefficient of tool–chip contact. The expression for the shear angle in Eqn.

(9) works reasonably well, despite ignoring strain hardening or the ability of the chip to transmit the deformation forces from the tool to the shear plane.

A slip line field can be used to determine the shear angle (Lee and Shaffer 1951). This approach assumes that the material in the chip that transmits the stress to the shear plane is at the same state of stress as the shear plane:

$$\phi = \pi/4 - \beta + \alpha \tag{11}$$

The Lee and Shaffer solution for the shear angle (Eqn. (11)) results in a lower shear plane and therefore more strain in forming the chip than Merchant's solution (Eqn. (9)), whenever the rake angle is less than the friction angle, which is usual. Therefore, in most cutting situations, for a material that does not strain harden, a larger chip than that predicted by Eqn. (9) would be necessary to transmit the loads for chip formation.

The orientation of the shear plane has also been attributed to the strain-hardening properties of the work-piece in the primary deformation zone, at least to the extent that hardening is not off-set by thermal softening. Strain hardening tends to rotate the shear plane down into the softer, undeformed material in front of it until the lower flow stresses in the softer material are balanced, energetically, by the increased length of the shear plane (Wright 1982, Stevenson 1992). This approach assumes that the chip formation system will work at the lowest work rate for initiating shear instability in severely strain-hardened material, k_1, and that it will operate at this work rate while shearing the softer material, k_0, along a longer shear plane. The longer shear plane will form at an angle ϕ_0 such that

$$\cos(\phi_0 - \alpha)\sin\phi_0 = k_0/k_1 \, [\cos(45 - \alpha/2)\sin(45 + \alpha/2)] \tag{12}$$

Equation (12) is an upper-bound solution for the shear angle, as it ignores tool–chip friction. Interestingly, for many materials it does appear to be a good approximation of the shear angle and/or an upper bound, for others it is less than measured shear angles (Wright 1982). In Wright's work the k_0/k_1 ratio was estimated from published tensile test data, where k_0 is the yield stress and k_1 is the ultimate stress divided by $\sqrt{3}$. Wright noted that these tests achieve neither the strain level nor the strain rates seen in machining. Nevertheless, the strain-hardening properties should be expected to influence any system in which some component achieves such high strains.

4. *Tool–Chip Friction*

The friction at the tool–chip interface is an essential element for understanding the system that forms the chip. Work is required to overcome the friction, and increasing the friction has the additional influence of reducing the shear angle, according to both the Merchant (Eqn. (9)) and Lee and Shaffer (Eqn. (11)) solutions. Reducing the shear angle also increases the work required to form the chip by increasing the area of the primary shear zone (which is proportional to $t_o/\sin\phi$, Fig. 1) and the chip strain (Eqn. (4)). Both these classic solutions for the shear angle presuppose that the tool–chip friction and the rake angle are the only parameters that can influence the shear angle. Therefore, the friction angle in Eqns. (9) and (11) might be better assumed to be a kind of machining constant (Chao and Trigger 1951).

The tool–chip friction in machining occurs under unusual conditions. The chip material is separated from the work-piece and, free from contamination and oxides, is immediately in intimate contact with the tool. Under these conditions the chip material frequently adheres strongly to the tool. Unless there is an independent means of measuring the friction, as it would be under the unusual conditions on the rake face of the tool, Eqns. (9) and (11)) lose their predictive value in an absolute sense. However, the value of the sense of the equations is borne out by the dramatic response of the chip formation system to changes in tool–chip friction. The shear stress at the tool–chip can approximate the shear stresses on the shear plane when the tool rake face geometry is modified so as to restrict the tool–chip contact length below what it would be naturally, i.e., a controlled or restricted contact tool (Brown 1983). The more restricted the contact length, until the point where the chip flows around the restricted contact and remains in contact with the tool, despite the contact restriction, the lower the machining forces. At extremely high machining speeds, 50000 m/min, steel has been observed to melt and the tool chip friction coefficient drops from above 0.25 to 0.01, with melt layer lubrication (Recht 1985).

5. *Discontinuous or Segmented Chips*

Large deformations at high rates can lead to strain localization, and this is seen in machining chips. Rather than the quasi-continuous deformations, as in Figs. 1–3, a segmented or discontinuous chip can form, where narrow regions of large deformation separate larger regions with less strain. There are several kinds of discontinuous chips that may form for different reasons (Komanduri and Brown 1981). High-speed deformation tends to concentrate in regions where it begins when the heat from deformation remains in the region of deformation to raise the temperature and lower the yield stress so that more deformation is favored in the same region. The point at which this kind of catastrophic thermoplastic shear is reached depends on the change in shear stress with respect to temperature and strain, and on the thermal conductivity, density, and specific heat of the material. A ratio of these properties can be used to estimate the relative tendency of materials to form discontinuous chips at lower machining speeds (Recht 1985). Other discontinuities in chip thickness, observed in brass, have been postulated to result from the strain-hardening behavior of the work-piece leading to systematic rotations of the shear plane in both directions (Stevenson 1992).

6. *Energy Partition*

The energy dissipation in machining can be divided into the three deformation regions: primary, in the shear zone; secondary, in the chip adjacent to the tool rake face; and tertiary, in the work-piece adjacent to the tool flank face (Williams 1978). Most of the energy is dissipated in the primary shear zone and most of this goes with the chip in the form of heat. Of the three, the least energy is dissipated in the tertiary shear zone and heats the work-piece. Tools with higher thermal conductivity have been shown to have longer natural contact lengths, and have a thicker secondary shear zone (Balaji *et al.* 1999).

In grinding, where abrasive grains in an agglomeration act as individual cutting tools, long wear flats can develop, increasing the energy dissipation into the work-piece. In especially severe grinding conditions a white etching, martensitic layer can form on the machined surface. The specific energy for material removal in grinding can increase to the specific heat of fusion (Malkin 1989). The energy partition can be calculated from the grinding parameters and thermal properties of the abrasive grain, coolant, and work-piece (Guo *et al.* 1999).

7. *Deformation Mechanics*

Even for the so-called continuous chips the slip process in the primary shear zone may not be homogeneous, but involve cyclic variations in material behavior between predomination by work hardening and predomination by sufficient recovery for constant stress deformation or work softening (von Turkovich 1970). This is consistent with the rippled appearance of the free surface of the chip (Kobayashi and Thomsen 1959). The strain rate in the primary deformation zone can be expressed as a function of the Burgers vector, dislocation density, and dislocation velocity. It has been shown that, making allowance for dislocation parameters such as density, climb rate, and loop length, the width of the primary deformation zone should be about 25 nm and the strain rate is therefore about $8\times10^{6}\,s^{-1}$. Assuming a chip velocity of $1\,m\,s^{-1}$ the total strain should be a reasonable 2, giving support to this dislocation mechanics approach (von Turkovich 1970). Whatever the mechanism, the deformation conditions in machining are severe enough so that there appears to be significant strain gradients and exhaustion of strain hardening during chip formation (Stevenson 1992).

8. *Machined Surfaces*

The topography of the machined surface can be complex. In turning the movement of a round tool nose with a radius r through the work-piece with a feed of s from purely geometric considerations results in a series of cusps with a peak-to-valley roughness of

$$R_t = r - (r^2 - s^2/4)^{1/2} \approx s^2/8r \qquad (13)$$

There is a finite minimum thickness of material that can be removed along the tool nose where the depth of cut thickness diminishes towards zero in many cutting situations, and this can alter the peak-to-valley roughness (Shaw and Crowell 1965). Other deviations from the perfect cusp geometry can be the result of multiple factors including redeposition of material adhered to the tool, cracks and tears resulting from the chip removal and tertiary deformation, vibrations, and microconstituents in the work-piece being dragged along with the tool (Brown *et al.* 1996). Deviations from the cusp geometry may not alter the peak-to-valley roughness, as topographic characterization parameters cannot give a complete description of the geometry. The topography of machined and ground surfaces can be highly complex at fine scales and have been shown to have fractal components (Brown *et al.* 1996).

The process of separating the chip from the work-piece and the subsequent deformation of the freshly created surface in the tertiary deformation zone can have a profound influence on the microstructure and can also influence the composition of the near-surface layers. Hard turning of tool steels can result in a superficial white layer (Chou and Evans 1997). Conventional turning of commercial aluminum alloys can form a nanocrystalline layer, remarkably low in dislocations, for about the first 0.1–0.3 μm, which is followed by another layer, an order of magnitude thicker, which contains localized deformation bands with high dislocation densities (Dupont *et al.* 1988). Machining of pure aluminum can produce a layer of fully recrystallized grains relatively low in dislocations, below which there is a heavily deformed layer. In both cases the large, high-speed deformations and accompanying heat result in a dramatic reorganization of the crystal structure in the superficial layer, which appears to be consistent with the kind of strain saturation hypothesized in chip formation (von Turkovich 1970, Stevenson 1992). Damage–depth profiles have been generated below machined single crystals as a function of CdS crystal orientation using ion beam channeling to show damage depths of a few hundred nanometers depending on machining direction (Lucca *et al.* 1996).

Machining and grinding introduce residual stresses into the near-surface layers (Brinksmeier *et al.* 1982, Lucca *et al.* 1998). There can be high gradients in these residual stresses, making them difficult to measure. In turning, where the tool passes each location once, tensile stresses dominate superficially. With the passage of the tool and the high temperatures resulting from the large, high-speed deformations the superficial material expands, while constrained by cooler underlying material, and yields. Tensile stresses remain after cooling. In milling, where the machined surface is subsequently subjected to re-passage of the tool, plastically deforming the surface, but with significantly less temperature rise, superficial compressive stresses can be introduced. Grinding with sharp, CBN abrasive wheels can also produce superficial compressive stresses (Brinksmeier *et al.* 1982).

Bibliography

Balaji A K, Sreeram G, Jawahir I S, Lenz E 1999 The effects of cutting tool thermal conductivity on tool–chip contact length and cyclic chip formation in machining with grooved tools. *Ann. CIRP* **48**(1), 33–8

Brinksmeier E, Cammett J T, König W, Leskovar P, Peters J, Tönshoff H K 1982 Residual stresses—measurement and causes in machining processes. *Ann. CIRP* **31**(2), 491–510

Brown C A 1983 A practical method for estimating machining forces from tool–chip contact area. *Ann. CIRP* **32**(1), 91–5

Brown C A 1987 Metallographic analysis of strain patterns in machining chips. *Metallography* **20**, 465–83

Brown C A, Johnsen W A, Butland R M 1996 Scale-sensitive fractal analysis of turned surfaces. *Ann. CIRP* **45**(1), 515–8

Chao B T, Trigger K J 1951 Cutting temperatures and metal cutting phenomena. *ASME Trans.* **73**, 777–93

Chou Y K, Evans C J 1997 White layers and thermal modeling of hard turned surfaces. *Manufact. Sci. Technol.* **2**, 75–82

Dupont F, Brown C A, Ilschner B 1988 Microstructure and microhardness in the superficial layers deformed machining 7075 aluminium alloy. *Z. Metall.* **B79**(H2), 74–80

Guo C, Wu Y, Varghese V, Malkin S 1999 Temperature and energy partition for grinding with vitrified CBN wheels. *Ann. CIRP* **48**(1), 247–50

Hill R 1950 *The Mathematical Theory of Plasticity*. Clarendon, Oxford, pp. 206–7, 325–7

Kobayashi S, Thomsen E G 1959 Some observations on the shearing process in metal cutting. *J. Eng. Ind.* **81**, 251–62

Komanduri R, Brown R H 1981 On the mechanics of chip segmentation in machining. *J. Eng. Ind.* **103**, 33–51

Lee E H, Shaffer B W 1951 The theory of plasticity applied to a problem in machining. *J. Appl. Mech.* **18**, 405–13

Lucca D A, Brinksmeier E, Goch G 1998 Progress in assessing surface and subsurface integrity. *Ann. CIRP* **47**(2), 669–93

Lucca D A, Maggiore C J, Rhorer R L 1996 Subsurface damage distribution in ultraprecision machined CdS. *Ann. CIRP* **44**(1), 519–22

Malkin S 1989 *Grinding Technology Theory and Applications of Machining with Abrasives*. Society of Manufacturing Engineers, Dearborn, MI, pp. 105–42

Merchant M E 1944 Basic mechanics of the metal cutting process. *J. Appl. Mech.* **11**, A169–75

Merchant M E 1945 Mechanics of the metal cutting process: II. Plasticity conditions in orthogonal cutting. *J. Appl. Phys.* **16**, 318–24

Recht R F 1985 A dynamic analysis of high-speed machining. *J. Eng. Ind.* **107**, 309–15

Shaw M C, Crowell J A 1965 Finish machining. *Ann. CIRP* **13**, 5–22

Stevenson R 1992 The morphology of machining chips formed during low speed quasi-orthogonal machining of CA 360 brass and a model for their formation. *J. Eng. Ind.* **114**, 404–11

Von Turkovich B F 1970 Shear stress in metal cutting. *ASME J. Eng. Ind.* **92**, 151–7

Williams J E 1978 Some aspects of a three-zone model of machining. *Wear* **48**, 55–71

Wright P K 1982 Predicting the shear plane angle in machining from work material strain-hardening characteristics. *J. Eng. Ind.* **104**, 285–92

C. A. Brown

Machining: Wear of Tools

Machining may be regarded as a tribological system with system elements and input/output variables. In this system, tool wear is the main loss factor decisively affecting the use variable of the system, which is its ability to shape materials. System-related wear variables for cutting tools are presented for the case of machining with a geometrically defined cutting edge. These variables are based on the system-dependent wear mechanisms of abrasion, adhesion, diffusion, tribochemical reaction, and surface scoring. Together with the strength properties of the tool and work materials, the system element "interface material" alters the tribological parameters of the machining process, influencing the finished product. Interface materials are cooling lubricants and their additives and, in the case of machining with a geometrically defined cutting edge, coatings. Such interface materials may increase wear resistance or may modify contact parameters in such a way that the wear attack is reduced.

1. *Machining as a Tribological System*

The structure of any tribological system consists of four basic elements: *body*, *counterbody*, *interface material*, and *ambient medium* (Fig. 1). Interactions between these elements take place in response to load stresses. Within the system, external input variables (the collective load) are transformed into technically useful output variables (use variables). In the process, friction- and wear-induced wear variables occur, manifesting themselves through changes in the surfaces of the body and counterbody.

The interface material frequently consists of a solid or liquid lubricant. The lubricant separates the surfaces of the body and counterbody as they slide against one another. The lubricant can, however, fulfill this function only if the lubricant is attached so firmly to the surface by chemical or physical means that it is not removed completely by the effects of the shearing forces. Interface materials may also be granular, hard, and abrasively active wear particles released from the two frictional surfaces as a result of loading.

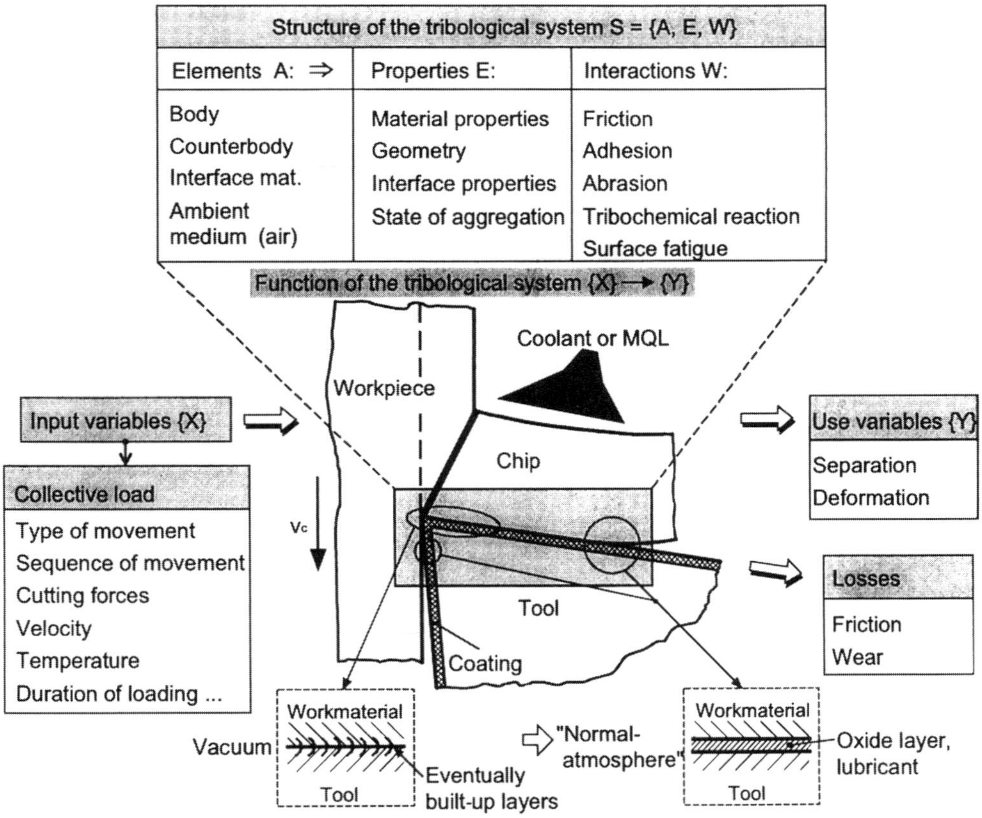

Figure 1
System analysis of cutting with geometrically defined cutting edge.

Machining is an open tribological system with the tool as the body, the work material as the counterbody, and possibly a cooling lubricant as the interface material. In open tribosystems there is a continuous flow of material into and out of the system. New, highly reactive surface areas of the work material are continuously fed to the loaded surfaces of the cutting tool. The functionality of the system is determined by tool wear and by the geometrical and functional interface properties of the part. Both result from conditions in the tool–part contact zone.

2. Wear Mechanisms

The classic analysis of the machining tribosystem considers the effects of the four main wear mechanisms of adhesion, abrasion, tribochemical reaction, and surface rutting according to DIN 50320 on the face and flank of the tool. If the tribological partners are mutually soluble, diffusion wear also occurs at higher cutting speeds. These wear mechanisms are activated by supercritical stresses on sections of the cutting edge resulting from the following: static or dynamic mechanical and thermal overloading; abrasion owing to contact with hard particles; adhesion of work material (shearing of pressure welded material); diffusion phenomena; and scaling.

These phenomena are superimposed on one another over a wide range and both their causes and their effects on wear can only partly be differentiated. They result in visible wear phenomena on the cutting edge, which are more or less pronounced depending on the form and duration of the stress. Figure 2 shows forms of wear occurring mainly on turning tools and affecting the face (crater wear) and on the major and minor flank (flank wear). Depending on the cutting parameters and the work material/tool material pairing, notch wear may occur at the edge of the contact zone of the tool, caused by the abrupt transitions in mechanical and thermal loads at the end of the tool–chip contact zone, to the abrasive nature of the sharp-edged underside of the chip, and to oxidation phenomena.

The frictional phenomena in the contact zones of the tools are comparable with those encountered in dry friction in a vacuum under high mechanical and thermal stresses.

2.1 Mechanical and Thermal Overloading

(*a*) *Chipping.* Large cutting forces result in chipping of the cutting edges or cutting-edge corners if the tool orthogonal wedge angle or the tool included angle is too small or if the tool material is too brittle. Discontinuities in the cut may also cause chipping, especially when machining tough materials with sticking chips. Slight chipping occurs if workpieces contain hard, nonmetallic inclusions, created, for example, during deoxidation of the steel.

(*b*) *Edge cracks.* In discontinuous cutting operations (e.g., milling) the cutting edge is subjected to severe alternating stress, which may lead to fatigue fracture. A rapid sequence of alternating stresses will cause so-called edge cracks, especially when milling with cemented carbide tools.

The rapidly alternating stresses caused by the formation of lamellar chips, e.g., when machining titanium materials, may also lead to the creation of edge cracks.

(*c*) *Ridge cracks.* Ridge cracks are a form of damage to the cutting edge caused by alternating thermal stresses that lead to alternating mechanical stresses as a result of the differing thermal expansion in different temperature zones of the tool material. Stresses of this nature occur mainly in discontinuous cutting.

During the tool contact, the cutting edge heats rapidly to high temperatures. When the tool leaves

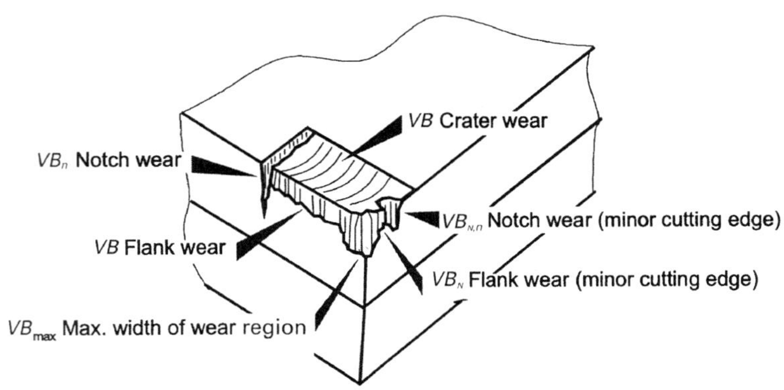

Figure 2
Wear types and parameters.

the material it cools. Among other factors, the difference between the highest and lowest temperature depends on the work material, its thermal conductivity and specific heat, on the cutting parameters, and on the ratio between the distances traversed in the work material and in the air. The use of cooling lubricants in discontinuous cutting has a strong quenching effect, and encourages the formation of ridge cracks, especially on cemented carbides and ceramic tool materials. The course of the ridge cracks follows the isotherms of the temperature field in the tool cutting edge.

(*d*) *Plastic deformation.* Plastic deformation of the cutting edge occurs when the tool material softens at high temperature and begins to flow under the pressure of the cutting forces. Cutting edges made of tool steel or high-speed steel deform in inverse proportion to the difference between the temperature at the cutting edge and the yield temperature of the tool material.

Plastic deformation also occurs in cemented carbides and cermets, but only at higher temperatures (cutting speeds) and higher forces than is the case with tool steels and high-speed steels. With cemented carbides, the higher the percentage of the binding phase (usually cobalt), the greater will be the deformation.

2.2 Abrasion

Particles of tool material detached through the action of external forces in the cutting plane generate mechanical abrasion. The abrasion is caused by hard inclusions in the material such as carbides and oxides.

2.3 Adhesion

The adhesion of work material to the tool material will cause wear if material is welded on to oxide-free surfaces in sufficient proximity, and is then re-separated. The shearing point may lie within the tool material. Increased wear owing to adhesion mechanisms is observed on rough tool surfaces or where there is intermittent contact between the work material and the tool or the flow of material across the tool surfaces is disturbed.

Disturbances in the material flow, e.g., as a result of intensive edge buildup, lead to microchipping, especially at low cutting speeds. Built-up edges are highly hardened layers of the work material that occur as adhesions on the tool. The work material adhering to the cutting edge becomes extremely hard as a result of chip pressure, enabling it to function as a cutting tool in its own right.

Depending on the cutting parameters, particles of the built-up edge periodically slide between the flank and the cut surfaces. If they are very hard, they cause increased flank wear and have a highly adverse effect on the surface quality of the workpiece.

2.4 Diffusion

Diffusion wear must be expected on high-temperature wear-resistant cemented carbide tools used at high cutting speeds if the tool and work materials are mutually soluble. Tool steel and high-speed steel, however, become soft at temperatures where diffusion practically cannot occur (e.g., approximately 600 °C for high-speed steel).

When diffusion wear takes place on cemented carbide tools, iron diffuses into the cobalt binder phase. Cobalt diffuses into the steel. Because of the iron component in the cobalt phase of the cemented carbide, its carbon solubility increases from 0.7% to between 1.5% and 2%. As a result, the tungsten carbides in the cemented carbide dissolve into mixed and double carbides of the forms Fe_3W_3C, $(FeW)_6$, and $(FeW)_{23}C_6$. The carbon released in this process diffuses via the ferriferous cobalt phase into the steel (König and Klocke 1997). The increase in wear is accordingly triggered by a reduction in the wear resistance of the tool material owing to changes in its chemical composition, and primarily by the dissolution of the WC hard material phase.

2.5 Scaling

Scaling (oxidation) is often observed in the areas of the cutting edge close to the contact zone. Its significance depends on the tool material and the temperature of the cutting edge. Cemented carbides of pure tungsten carbide and cobalt oxidize more readily than those with added titanium carbide or other carbides. Scaling can begin at temperatures as low as 700–800 °C. Scaling is of virtually no importance in the case of tool steels and high-speed steels, as their heat resistance is exceeded before severe oxidation can take place.

3. Interface Materials

Interface materials can influence mass and surface wear mechanisms in two ways: by increasing the specific wear resistance or by modifying the contact parameters in such a way that wear attack is reduced. In the machining process, a variety of individual factors influencing friction and wear are superimposed on one another.

3.1 Coatings

Coatings represent a class of solid interface layers in the tribological system of the machining process. Since about 1968 cemented carbides have been coated with hard thin films to enhance their wear resistance and hence their importance. The use of coated carbides and high-speed steels for the machining of various materials is now state-of-the-art technology.

Because of their great hardness and chemical stability, numerous hard materials can improve resistance to abrasion, counteract adhesion mechanisms between the tool and the part, act effectively as a barrier to diffusion, and reduce oxidation of the surface of the tool material. They effectively improve resistance to crater, flank, and notch wear at both low and high cutting speeds.

Multilayer coatings combine the specific functions of individual coating materials. One example is the use of hard–soft coating composites to reduce adhesion and friction in the flute zones of drilling, tapping, and end milling tools.

3.2 Cooling Lubricants

Tribologically, cooling lubricants act in two ways: as an ambient cooling medium and as a lubricating interface material. In order to obtain a lubricating effect under the high pressures per unit area relevant in the machining process, solid layers with high compressive strength and low shearing strength should be present on the metal surfaces. A fundamental precondition is that the lubricant should be able to pervade the contact zone. In the case of increasing edge buildup, these conditions are met through the fluctuation of the built-up edges; the same applies to flutes on drilling tools. Owing to the separation of the chip and the tool material, welded-on material is suppressed or reduced. This effect can be improved by adding sulfur or phosphorus agents to the cooling lubricant, but these become effective only above certain temperatures.

The conditions for the formation of high-pressure lubricating films become gradually less favorable as the cutting speed rises. At the same time, the rise in temperature leads to diffusion processes between the friction partners, a fall in the hardness of the tool material, and, in extreme cases, plastic deformations of the cutting edge, making the cooling function of the cooling lubricant the primary factor.

Bibliography

Czichos H, Habig K-H 1992 *Tribologie-Handbuch: Reibung und Verschleiß. Systemanalyse, Prüftechnik, Werkstoffe und Konstruktionselemente.* Vieweg-Verlag, Braunschweig/Wiesbaden, Germany (in German)

Klocke F, Krieg T 1999 Coated tools for metal cutting—features and applications. *Ann. CIRP* **48/II**, 515–25

König W, Klocke F 1997 *Fertigungsverfahren Bd. 1: Drehen, Fräsen, Bohren.* Springer-Verlag, Berlin

F. Klocke, G. Eisenblätter, and T. Krieg

Magnesium Alloys: Thermal Treatment and Thermomechanical Processing

Magnesium alloys exhibit limited formability at ambient temperatures. This is due to the hexagonal crystal structure with its limited number of operating slip systems. At temperatures below 200°C only basal plane slip occurs and there is consequently a lack of operating slip systems to allow extensive deformation without fracture occuring. Only lithium-containing magnesium alloys are readily formed at room temperature due to their cubic structure. At temperatures exceeding 200°C pyramidal slip occurs and formability is dramatically improved. Compared to aluminum, magnesium alloys need higher specific stresses to deform plastically. Extrusion and rolling speed for a standard magnesium wrought alloy are similar to those of high-strength aluminum alloys. Mechanical properties are comparable to those of medium-strength aluminum alloys. As a consequence of these limitations, the application of wrought magnesium is currently rather restricted. Thermal treatments have been designed to secure proper as-cast grain structure, metal composition, microstructure control by solution treatment, aging, stress relief, and control of the deformation structure by soft annealing or partial annealing.

1. Molten Metal

Control of grain size and metal purity provide examples of thermal treatments carried out in the liquid state.

Although not presently in common use, a grain refining method for aluminum-containing magnesium alloys involves superheating molten alloys to temperatures exceeding 800°C, prior to cooling the melt just before pouring the casting. The mechanisms involved are not well understood; however, it is believed that carbon from the crucible linings plays an important role.

The addition of manganese to magnesium–aluminum alloys at elevated temperatures is used to control the iron content by precipitation of an aluminum–manganese–iron intermetallic compound during controlled cooling of the molten metal.

2. Castings

Thermal treatments are commonly used for sand and permanent mold castings. High-pressure die castings are usually not heat treated. This is mainly due to the fact that the loss of mechanical strength caused by grain coarsening will not be compensated for by artificial hardening. There is also a risk that blisters may form on the surface of the castings due to gas porosity.

The common heat treatments for castings include stress relieving, direct aging T5, solution heat treatment T4, and full age hardening T6. Details on heat treatment schedules are given in *Metals Handbook*, Vol. 4 (1981).

3. Wrought Products

Wrought magnesium alloys mainly develop their strength through work hardening, and are usually applied in the as-fabricated condition, F, or in the T5 condition. Sheet and plate materials can be supplied as-fabricated in the various H1X conditions or in the back-annealed H2X conditions. Soft annealing is commonly performed at temperatures in the range 290–450°C depending upon alloy. Since plastic forming operations are mainly carried out at elevated temperatures, the need for soft annealing is less for magnesium than for metals subjected to cold forming. Practical heat-treatment schedules for wrought magnesium alloys are described in *Metals Handbook*, Vol. 4 (1981).

Bibliography

Metals Handbook, 9th edn. 1981 Vol. 4, *Heat Treatment*. ASM International, Materials Park, OH

H. Westengen

Mechanical Alloying

Mechanical alloying (MA) is a powder metallurgical process which was developed in the mid-1960s by the International Nickel Company (INCO) as a practical solid-state route for introducing and retaining fine refractory oxide particles in high-temperature alloys (Benjamin 1970, 1989). INCO established that it was possible to obtain the desired oxide distribution by using dry, high-energy ball milling of mixtures of raw material powders: mechanical alloying was born. It was quickly found that this solid-state powder processing route enabled intimately mixed, fine-scale composite powders to be produced from a wide range of elemental and master alloy starting powders. The principal aspects of powder processing and handling together with subsequent consolidation and thermomechanical processing which it was important to control in order to ensure reproducible end properties in high-temperature, creep-resistant nickel- (e.g., MA 754) and iron-based (e.g., MA 956) ODS alloys were also determined. Much of this processing knowledge has proved of generic relevance and is as applicable to the control of MA in the many other alloy systems that have subsequently been developed and explored as it was to the early MA ODS alloys.

An enormous amount of research has been performed on the MA technique and the materials systems that it has proved possible to synthesize using it. The broad categories for the latter include intermetallics, ceramics, composites (cermets and metal matrix composites (MMCs)), nanocrystalline materials, amorphous alloys, and self-propagating high-temperature synthesis (SHS) systems. The list is not exhaustive but at least illustrates the range of applicability and versatility of MA. Most of these applications of MA remain at the research stage.

1. *Mechanisms and Processing*

MA in high-energy ball milling equipment is accomplished by processing an initial powder charge usually comprising a mixture of elemental, ceramic (e.g., yttria for ODS alloys), and master alloy powders, all supplied in strictly controlled size ranges. Master alloy powders are used in order to reduce *in situ* oxidation of highly reactive species, such as aluminum or titanium alloy additions during processing. The milling medium normally used in commercial systems is a charge of hardened steel balls, typically 2 cm in diameter. The ball-to-powder weight ratio is chosen carefully for each mill and powder charge combination, but is typically around 10:1 for commercial systems. Given the enormous surface area, both of the initial powders and the fresh powder surfaces generated during MA processing, control of the milling atmosphere and its purity is essential to avoid undue alloy contamination. The principal protective atmospheres employed during commercial milling of MA powders are usually either argon or hydrogen and this protection generally extends both to pre- and post-MA powder handling. Both the purity of these gas atmospheres and the integrity of gas seals on the milling equipment are essential to control contamination, particularly when processing reactive species. For example, levels of oxide contamination in Ni_3Al can double with just a few hours of milling in impure argon. Occasionally, however, deliberate doping of the milling environment has been used to facilitate alloying additions during processing.

The central event during MA is the ball–powder–ball collision within the milling medium during processing. It is repetition of these high-energy collisions which leads eventually to MA of the powder charge. Intimate mixing and eventual MA of the powder charge occurs in a series of identifiable, more or less discrete stages during processing (e.g., Gilman and Benjamin 1983). For ductile–ductile or ductile–brittle combinations of starting powders, MA initially proceeds by the flattening and work hardening of ductile powders and fragmenting of brittle constituents, which is followed by extensive cold welding between powder particles, formation of lamellar structures, and coarsening of the powder particle size distribution. Brittle powder fragments are trapped at cold weld interfaces between the evolving lamellas of the ductile constituents and thus, while continuing to comminute, become dispersed. With continued milling a balance, which is dependent on processing parameters and the composition of the constituents, is established between further cold welding and powder particle fracture, leading to relatively stable powder particle sizes.

This balance between welding and fracture is accompanied both by further decreases in lamella spacings and by folding and mixing-in of lamella fragments to produce microstructures typical of MA (Fig. 1). For ODS alloys, powder constituents are milled to the stage where light microscopy examination reveals that lamella spacings have decreased to below the resolution limit ($\sim 1\,\mu m$). For typical levels of oxide addition (e.g., ~ 0.5 wt.% yttria) this criterion ensures average dispersoid interparticle spacings of $< 0.5\,\mu m$ (Fig. 2). In other systems, milling can progress until true alloying occurs. Surprisingly, MA can also be achieved between essentially brittle powder constituents. The mechanisms by which this occurs are less well understood than in systems incorporating at least one ductile powder component. Nevertheless, granular as opposed to interlamellar mixtures of brittle powder constituents do evolve, typically with smaller, harder fragments progressively incorporated to a very fine scale within the less hard constituents, e.g., alumina–nickel oxide. Moreover, MA of these brittle constituents can progress to true alloying, as has been demonstrated using

Figure 1
Polished and etched metallographic section of ODM 751 FeCrAl alloy powders in the fully MA condition, showing the folded lamellar structures typical of material processed by high-energy ball milling (courtesy of D.M. Jaeger).

lattice parameter measurements on Si–28 at.% Ge progressively milled from constituent powders (Davis and Koch 1987).

Milling of very ductile metals such as aluminum and tin has to be carefully controlled to avoid complete agglomeration of the ductile phase rather than the balance between cold welding and fracture that leads to MA. This is normally achieved by adding precise amounts of organic compounds termed process control agents (PCAs) to the milling environment. Typically waxes or solvents, these compounds that interfere with cold welding progressively break down during milling to become incorporated within the final MA powders (e.g., in aluminum alloys) as fine-scale distributions of carbides or oxides. Similar restrictions to the proclivity for cold welding in ductile powders can be achieved without use of PCAs by milling at low temperatures, e.g., below $-100^\circ C$ for aluminum.

The processing equipment used to effect MA by high-energy ball milling of powders originated in mining and conventional powder metallurgy industries. The range of high-energy ball milling equipment divides, approximately, into two categories: small, high-energy laboratory devices, and larger facilities capable of milling commercial quantities of powder. The former category includes SPEX shaker mills and planetary ball mills. Both devices are capable of rapidly effecting MA, but in quantities of powders up to no more than a few tens of grams. SPEX mills vibrate at up to 1200rpm in three orthogonal directions to achieve ball velocities approaching $\sim 5\,ms^{-1}$. Planetary mills incorporate a rotating base plate upon which are mounted counter-rotating, smaller-radius vials containing the ball/powder charge. The kinetic energy imparted to the ball charge in the planetary mill depends on the base plate and vial radii and angular velocities. Attritor or Szigvari ball mills, depending on their size, can be used either for laboratory or commercial ball milling applications and incorporate a rotating vertical shaft with attached horizontal impellors which stirs a container housing the ball and powder charge. These devices can process batches of up to several kilograms or more of powder through the significant differential movement the impellors generate between the ball and powder charge. Balls can either cascade or tumble when leaving the mill wall during attritor processing, depending on the ball charge and impellor velocity.

The largest commercial devices applied to MA are horizontal ball mills. When these devices exceed several meters in diameter they impart sufficient kinetic energy through ball impacts to effect MA and can process over 1000 kg of powder per batch in larger units. Balls either cascade or tumble during processing in these mills depending on rotational speed (see Fig. 3). The time taken to achieve MA scales approximately inversely to the size of the milling equipment used. Hence, milling which might take minutes to accomplish in a SPEX mill could take hours in an attritor or days in a horizontal ball mill. All of these processing routes, however, have very low energy conversion efficiency, in that only a small fraction of the milling energy expended effects microstructural change contributing to the MA process.

It is worth noting that during MA, powder particles also coat (condition) the ball milling medium, which means that, to avoid cross-contamination of commercial alloys, the repeat use of ball charges is restricted to compositionally similar batches of raw materials.

2. *Modeling*

The approaches taken to modeling the mechanical and dynamical aspects of MA fall into two broad categories: local and global modeling (Courtney 1994, Soni 2000). Local modeling considers the likely fundamental nature of single-collision events, the energy transfer to trapped powder particles, and the way particle welding and fracture events may lead to temporal evolution of key powder parameters such as hardness, particle shape, and size distribution. In order to effect the computations necessary to follow iteratively the evolution of powder particle changes during MA, local modeling requires input of mill-specific parameters such as ball velocity and collision frequencies determined from global modeling approaches. Analytical approaches to local modeling have been based on assumptions of Hertzian collisions between balls leading to transfer of energy to

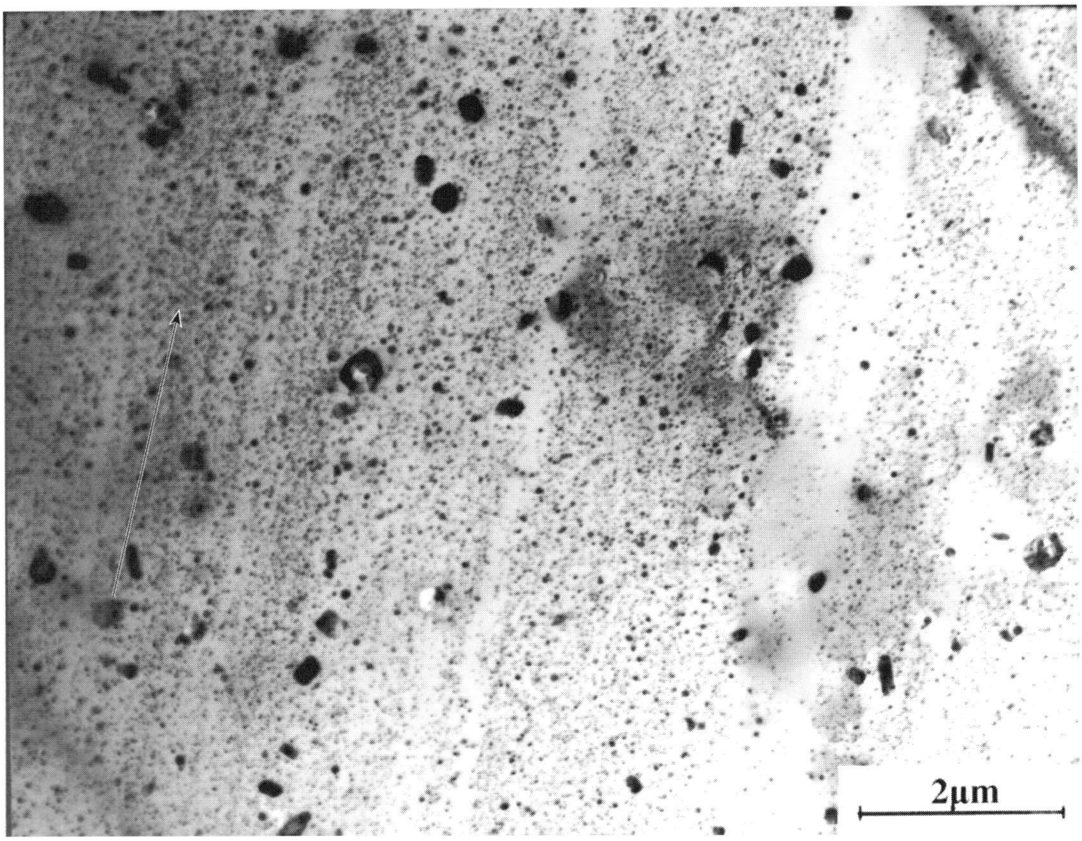

Figure 2
Transmission electron microscope image showing alignment of a fine-scale dispersion of oxide particles in extruded ODS alloy PM2000. The arrow shows the extrusion direction (courtesy of Y.L. Chen).

trapped powder particles that deform in a manner that can be estimated from constitutive equations for plastic deformation. These models predict strain rates in the deforming powders of $\sim 10^3$–$10^4 s^{-1}$ with accumulated strain per collision ranging from a few tenths of 1% (attritor mill) to 1% (SPEX mill).

Calculations of temperature change in MA powders during milling, assuming either adiabatic heating or temperature rise owing to Hertzian pressure, predict modest rises (several hundred kelvin in a SPEX mill) in line with observation. These local models of MA have been extended to consider particle cold welding and particle fracture. The latter events are modeled by assuming a mechanism similar to that of a forging fracture, where this can be taken to originate at the collision impact point and spread radially across the flattened powder under the appropriate conditions of crack initiation and propagation (*J*-integral approach). Using mill parameters estimated from global modeling, local analytical models have been incorporated in computational programs and run to predict the time-based evolution of powder size, hardness, and microstructural refinement occurring during MA.

Comparisons with experimental data obtained from elemental chromium and iron powders processed in a SPEX mill, reveal that the model is capable of predicting the correct shape of curve for the evolution of the average particle size with milling time but underestimates the time to achieve peak average particle size and overestimates its value. Other predictions are for lower hardness and less microstructural refinement than are observed experimentally. Similar predictions for aluminum milled in a SPEX unit again underestimate particle hardness but achieve closer agreement with evolving particle sizes. It is generally accepted that uncertainties in the values of key parameters used both in these and in more phenomenological, local models of MA mean they must be considered qualitative in nature. However, they enable reasonable predictions of trends when either powders or processes are changed systematically.

The requirement for global modeling of MA reflects the inherent complexity of the high-energy

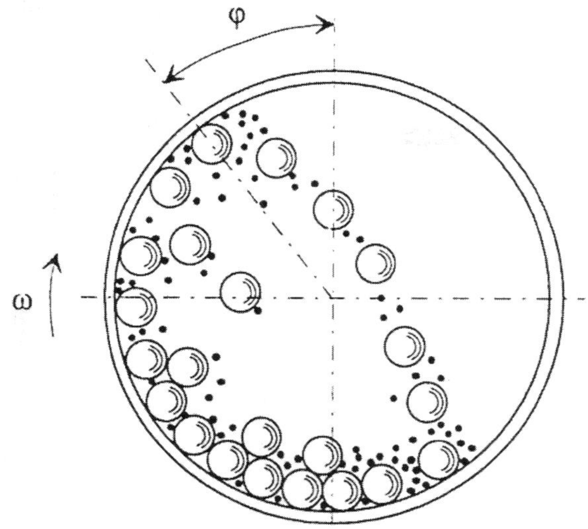

Figure 3
Configuration of a horizontal ball mill, showing the release of the powder and ball charge (at angular position φ) from the inner wall of the mill rotating with angular velocity ω (after Lu *et al.* 1995).

ball milling process, added to which are device-specific characteristics such as positional milling and powder inhomogeneities. For predictive purposes global models of MA seek to define ball impact frequencies, angles, and velocities together with estimates of the effective fraction (that leading to energy transfer likely to be of significance in terms of MA) of milling impact events. Much of this work has been underpinned by experimental validation of patterns of local milling behavior and specific effects of mill design using techniques such as high-speed photography. These models have revealed, for example, that $<1\%$ of ball impacts in a SPEX mill involve the near head-on collisions most effective for ball–powder energy transfer; and that increasing the ball charge increases the impact frequency, while simultaneously decreasing both ball collision velocity (lower mean free path) and the final hardness of milled powders. Similar results have been obtained for analysis of planetary ball mill action, although here it has been established that two different milling regimes (energies) exist, depending on the dimensions and respective angular velocities of support plate and vial. These are ball rolling on the vial inner surface, and ball escape and impact on the vial opposite wall. By independently varying plate and vial angular velocity and thus ball energy, Gaffet (1991) demonstrated that it was possible to transform a powder charge with an average composition of $Ni_{10}Zr_7$ to end states ranging between fully crystalline and fully amorphous, thus underlining the link between process and outcome, or MA and microstructure.

One disadvantage of local/global modeling of the dynamics and mechanics of MA is that these models are not linked to fundamental transformation mechanisms, particularly important for understanding the formation of metastable phases during MA. However, atomistic models of metastable phase formation based, for example, on molecular dynamics are unconnected to the deformation parameters of MA (Khina and Froes 1996). Both phenomenological thermodynamic (e.g., based on heat of mixing) and kinetic (based on multiple layer diffusion couples) models have been used to interpret this metastable phase formation, but further developments of these models are needed to link plastic strain and defect levels directly to the kinetics as well as nucleation and growth mechanisms of metastable phases formed during MA.

3. *Applications*

Despite the fact that MA ODS alloys have been used successfully since high-strength, corrosion-resistant, nickel-based INCO alloy MA 754 bar was used by GEC in the nozzle guide vane assembly of the F404-400 military fighter aircraft engine in 1977, the high cost of these alloys has determined that they remain niche materials. Essentially, the only MA materials that have successfully penetrated commercial markets so far are ODS alloys, principally the nickel- and iron-based variants. Total worldwide production of these materials remains at a level of around several hundred tons per year. Several aluminum-based ODS alloys were developed by INCO from the late 1970s through the 1980s culminating in the commissioning of a production facility in 1988 that was closed in 1992 when INCO withdrew MA ODS aluminum alloys from the market. Potential aerospace end users of INCO high-strength, as-forged, lightweight, stress corrosion-resistant Al–Mg–Li ODS alloy, designated Al-905XL, had to wait seven years for a viable replacement to emerge (designated alloy 5091), produced by Aerospace Metal Composites Ltd.

Nickel-based ODS alloys range from the solid solution-strengthened ODS Ni–20 wt.% Cr alloy MA 754 to alloy MA 758, which is essentially similar but contains 30 wt.% chromium (for corrosion resistance in components used for manufacturing fiberglass from the melt), to the γ' and solid solution (tungsten, molybdenum) strengthened, creep, oxidation, and sulfidation (chromium, aluminum, titanium, tantalum, and tungsten) resistant ODS alloys MA 6000 and MA 760. These higher strength nickel-based ODS alloy variants have found applications as small, highly stressed first- and second-stage turbine blades (MA 6000) and for industrial gas turbine vanes and blades (MA 760). These latter alloys are supplied for service in a very coarse-grained, high grain aspect ratio condition generated by dynamic zone annealing that imparts excellent high-temperature creep resistance. The nickel-based ODS alloys are more difficult to form than their iron-based ODS alloy counterparts and are available in a more limited range of mill forms (e.g., bar, wire-rod, plate, sheet, and precision forgings).

Iron-based ODS alloys (MA 956, PM 2000, and the higher chromium, aluminum variant MA 956HT) which are, essentially, dispersion-strengthened, highly oxidation-resistant variants of FeCrAl alloys, have found a greater variety of specialist applications in a broader range of market sectors than their nickel-based ODS alloy counterparts. As with nickel-based ODS alloys, MA 956 and PM 2000 achieve maximum creep and oxidation resistance when used in service in the high aspect ratio, coarse-grained secondary recrystallized, preoxidized release condition, achieved for iron-based ODS alloys without resort to zone annealing. In contrast, the iron-based MA ODS alloy MA 957, which was developed for use as a low void swelling nuclear fuel cladding tube in fast breeder reactors and is intended for service in liquid sodium at $\sim 700\,^\circ\mathrm{C}$, is supplied with a specified $\sim 30\,\mu\mathrm{m}$ grain size. The excellent fabricability of iron-based ODS alloys ensures they are amenable to a wide range of metal processing techniques.

Other niche applications, for MA aluminum alloys, include race pistons for motorbikes forged from Al–12 wt.% Si, DISPAL-1, produced by reaction-milling techniques, and thin-walled tube for road racing bicycle frames extruded from hot isostatically pressed (HIP) billets of 2124-T4-17.8wt.% SiC_p alloy, manufactured by MA. For both these alloys as well as for the low-density 5091 aerospace alloy, service applications are at ambient, hence the material is supplied in a very fine-grained condition (e.g., $< 1\,\mu\mathrm{m}$ grain size for 5091 alloy).

Elsewhere, essentially small amounts of special composition MA materials have been tested in niche applications, such as the aluminum alloy DISPAL-2 used in electrodes for NaS batteries or for low-distortion aerospace interferometer parts, and the magnesium-based "supercorroding" alloy exhibiting controlled corrosion rates for application as bolting to release deep-sea equipment or as controlled heating/hydrogen sources. In the area of nonstructural applications, a number of specialist coatings have been produced using MA powders. These include Ti–Al powders for RF plasma spraying, Ti–Al–SiC powders for wear-resistant coatings, and sieved, as-MA powders of the standard commercial ODS alloys such as MA 754 and PM 2000 for plasma spraying. A promising emerging MA ODS alloy is the intermetallic composition, yttria-strengthened alloy Fe_3Al, intended for application in components for use in fossil fuel power plants operating above $1100\,^\circ\mathrm{C}$, where the high aluminum content in the alloy is intended to confer enhanced oxidation resistance in aggressive environments.

Future commercial opportunities for MA alloys include the possible use of amorphous Al–Cu–Mg alloys as aerospace structural materials, intermetallics such as Ti_3Al, TiAl, and Cr_3Si for either airframe or high-temperature engine applications, and MA magnesium-containing combinations of Fe/Ni/Co/Ti as potential hydrogen storage materials (Lu and Lai 1998, Soni 2000).

See also: Mechanical Milling

Bibliography

Benjamin J S 1970 Dispersion strengthened superalloys by mechanical alloying. *Metall. Trans.* **1**, 2943–51

Benjamin J S 1989 Mechanical alloying—a perspective. In: Arzt E, Schultz L (eds.) *New Materials by Mechanical Alloying Techniques: Proc. DGM Conf.* Deutsche Gesellschaft für Metallkunde, Germany, pp. 3–18

Courtney T H 1994 Modeling of mechanical milling and mechanical alloying. *Rev. Particulate Mater.* **2**, 63–113

Davis R M, Koch C C 1987 Mechanical alloying of brittle components: silicon and germanium. *Scr. Metall.* **21**(3), 305–10

Gaffet E 1991 Planetary ball milling. An experimental parameter phase diagram. *Mater. Sci. Eng. A* **132**(1–2), 181–93

Gilman P S, Benjamin J S 1983 Mechanical alloying. *Annu. Rev. Mater. Sci.* **13**, 279–300

Khina B B, Froes F H 1996 Modeling mechanical alloying: advances and challenges. *J. Mater.* **48**(7), 36–8

Lu L, Lai M O 1998 *Mechanical Alloying*. Kluwer, Boston

Lu L, Lai M O, Zhang S 1995 Modeling of the mechanical alloying process. *J. Mater. Process. Tech.* **52**, 539–46

Soni P R 2000 *Mechanical Alloying—Fundamentals and Application*. Cambridge International, Great Abington, Cambridge, UK

Suryanarayana C 1999 Mechanical alloying. In: Suryanarayana C (ed.) *Non-equilibrium Processing of Materials*. Pergamon, Oxford, UK, pp. 49–85

Suryanarayana C 2001 Mechanical alloying. *Prog. Mater. Sci.* **46**, 1–184

A. R. Jones

Mechanical Milling

Mechanical milling is a direct descendant of the technique of mechanical alloying, invented in the late 1960s by Benjamin (1970); this is a method involving the ball-milling of constituent powders to create an industrial alloy which is precipitation-hardened and may also be hardened by an oxide dispersion. The process is followed by a consolidation step. The term "mechanical alloying" was invented by a patent attorney for the International Nickel Company. The process, extensively used in industry, especially for nickel alloys intended for high-temperature use, leads to the formation of solid solutions of the participating metallic elements (generally strongly supersaturated) in the form of very fine grains, as well as to a population of finely divided compounds.

The most studied fundamental aspect of mechanical alloying is the metastable extension of solid solubility, which in extreme cases (e.g., iron dissolved in aluminum) can be by a factor of hundreds. A listing of such extensions can be found in a review by Suryanarayana (1999).

Mechanical milling is a research technique as distinct from a production tool. The starting material is most commonly a prealloyed single-phase metallic powder (in most experiments an intermetallic compound), and the interest lies in the creation of metastable configurations: complete or partial destruction of long-range atomic order, or amorphization. It has long been known that substantial plastic deformation will partly disorder a superlattice. Various forms of mechanical working have been used, including particularly extreme cold rolling, but the most practiced technique, as the name implies, is milling in a ball-mill. Here, a population of steel or ceramic balls together with a much smaller mass of powder is tumbled together in a rotating or vibrating container (the two forms of motion are combined in the more sophisticated mills).

Good reviews of mechanical milling, both techniques and findings, are by Koch (1991) and Suryanarayana (1999). There is a comprehensive bibliography of both mechanical alloying and mechanical milling by Suryanarayana (1995).

1. *Disordering*

Hellstern *et al.* (1989) were the first to disorder partially an intermetallic compound (AlRu) by prolonged milling. The Bragg long-range order parameter, S, could only be reduced from 1.0 to ~ 0.7, after which it stabilized at this value (a consequence of competing disordering and reordering reactions). In an important study, Jang and Koch (1990) showed that the less strongly ordered compound Ni_3Al could be disordered completely by milling. They also demonstrated, by means of microhardness measurements on sectioned powder particles after various times of milling, that the yield strength of the compound peaked very sharply for $S \approx 0.5$. Other compounds, such as NiAl or CoZr, cannot be disordered even partially. These differing findings echo the responses of intermetallic phases to irradiation with various particles. The ease of disordering diminishes with a decrease in the ordering energy (the difference in internal energy between a fully ordered and a fully disordered form of the same compound).

Measurements have been made of the change of lattice parameter as an ordered intermetallic compound is progressively disordered by mechanical milling. This was first done for Ni_3Al by Gialanella *et al.* (1992), and it was established that the (cubic) lattice parameter varies linearly with the order parameter, S. This finding was taken further by Cahn (1999), who established that there is a strong correlation between ordering energy and the change of lattice parameter resulting from a complete destruction of long-range order; this correlation is much clearer than that between lattice parameter change and the mismatch of atomic sizes.

2. *Amorphization*

The other major effect of mechanical working, including particularly milling, on the state of an intermetallic compound is the conversion of the crystalline lattice, wholly or partially, into an amorphous (glassy) condition.

The first observation of such amorphization was by a Russian investigator, A. E. Yermakov, in 1981, in the yttrium–cobalt and gadolinium–cobalt systems. This was followed by a systematic investigation by Koch *et al.* (1983) of phases in the nickel–niobium system, which are easy glass-formers. These studies went in parallel with others devoted to solid-state amorphization (see *Amorphization Reactions: Solid State*), in which thin layers of pure metals are allowed to diffuse into each other and, primarily because of the sluggish diffusion of one constituent and consequent difficulty in nucleating the stable intermetallic compound, a metastable glass forms. This was first discovered by Schwarz and Johnson (1983) in lanthanum–gold multilayers and an intense burst of research followed. It has become clear that mechanical working (commonly by rolling of stacks of metal foils) can aid amorphization during a subsequent diffusion anneal (this is called mechanically aided amorphization), or, in some systems, the intense mechanical working alone suffices without the need for an anneal (e.g., aluminum–platinum). This is known as mechanically induced amorphization. The literature concerned with these two categories is reviewed by Cahn and Greer (1996). Amorphization by milling is a subcategory of mechanically induced amorphization.

Amorphization is a stage which follows disordering, and the two are causally linked. This was made particularly clear by some work in the 1980s which studied irradiation-induced amorphization of CuTi and Cu_4Ti_3 (using high-energy electrons) as a function of irradiation temperature; the steady-state degree of residual order was higher for higher irradiation temperatures, and above a critical degree of residual order (and thus above a critical irradiation temperature) amorphization was no longer observed. The details of the relationship between disordering and amorphization are discussed by Cahn and Johnson (1986).

Criteria for amorphization by both irradiation and mechanically aided and induced amorphization have been much debated; there have been disputes between the idea that long-range order needs to be destroyed to a critical degree, and the rival idea that a critical concentration of point defects needs to be attained. These disputes are surveyed by Cahn and Greer (1996). Beke *et al.* (1991) have proposed the most successful theory to date for interpreting such varying responses to irradiation or milling, one which bypasses the above-mentioned disputes.

They estimated the additional elastic (mismatch) energy which is stored in an ordered solid solution when its long-range order is destroyed, whether mechanically or by irradiation; this depends, *inter alia*, on atomic sizes. This energy was then compared with the ordering energy, and it was concluded that amorphization is possible if (i) the (virtual) critical disordering temperature is above the melting temperature, and (ii) the ratio of the elastic mismatch energy to the ordering energy exceeds a critical amount. A comprehensive test against experimental data, for both irradiation and milling, gave widespread though by no means universal agreement.

See also: Mechanical Alloying

Bibliography

Beke D L, Loeff P I, Bakker H 1991 On the elastic mismatch in the order–disorder transformation and solid-state amorphization of intermetallic compounds. *Acta Metall. Mater.* **39**, 1267–73

Benjamin J S 1970 Dispersion-strengthened superalloys by mechanical alloying. *Metall. Trans.* **1**, 2943–51

Cahn R W 1999 Lattice parameter changes on disordering intermetallics. *Intermetallics* **7**, 1089–94

Cahn R W, Greer A L 1996 Metastable states of alloys. In: Cahn R W, Haasen P (eds.) *Physical Metallurgy*. North-Holland, Amsterdam, pp. 1723–830

Cahn R W, Johnson W L 1986 Review: the nucleation of disorder. *J. Mater. Res.* **1**, 724–32

Gialanella S, Newcomb S B, Cahn R W 1992 Order parameter and lattice parameter of $L1_2$ ordered alloys, with special reference to Ni_3Al-based alloys. In: Yavari A R (ed.) *Ordering and Disordering in Alloys*. Elsevier, London, pp. 67–78

Hellstern E, Fecht H J, Fu Z, Johnson W L 1989 Structural and thermodynamic properties of heavily mechanically deformed Ru and AlRu. *J. Appl. Phys.* **65**, 305–10

Jang J S C, Koch C C 1990 Amorphization and disordering of the Ni_3Al intermetallic by mechanical milling. *J. Mater. Res.* **5**, 498–510

Koch C C 1991 Mechanical milling and alloying. In: Cahn R W (ed.) *Processing of Metals and Alloys. Vol. 15: Materials Science and Technology: A Comprehensive Treatment*. VCH, Weinheim, Germany, pp. 193–245

Koch C C, Cavin O B, McKamey C G, Scarbrough J O 1983 Properties of "amorphous" $Ni_{60}Nb_{40}$ by mechanical alloying. *Appl. Phys. Lett.* **43**, 1017–19

Schwarz R B, Johnson W L 1983 Formation of an amorphous alloy by solid-state reaction of the pure polycrystalline metals. *Phys. Rev. Lett.* **51**, 415–18

Suryanarayana C 1995 *Mechanical Alloying and Milling: A Bibliography, 1970–1994*. Cambridge International, Abington, UK

Suryanarayana C 1999 Mechanical alloyingco. In: Suryanarayana C (ed.) *Non-equilibrium Processing of Materials*. Pergamon, Oxford, pp. 49–85

Suryanarayana C 2001 Mechanical alloying. *Prog. Mater. Sci.* **46**, 1–184

R. W. Cahn

Mechanical Surface Treatment

The surface layers of many components used in technical procedures represent the most mechanically loaded workpiece areas. Processes include, for example, quasistatic and cyclic bending or torsion, superposition of homogeneous and inhomogeneous quasistatic and cyclic loadings, such as bending and tension as well as torsion and bending, and the loading of notched components. Furthermore, for friction and wear and for corrosive and combined corrosive–mechanical loading, the direct surfaces and the near-surface regions are the critical parts in respect to failure. Therefore, many mechanical, thermal, and thermochemical surface treatments and surface coating procedures have been developed and are used individually or in combination to improve the strength and corrosion properties of the near-surface regions.

This article covers the mechanical surface work-hardening treatments of metallic materials by shot peening and deep rolling, plus some machining treatments that also produce modified surface conditions. These methods are applied particularly to improve the fatigue strength and fatigue life as well as the corrosion fatigue behavior. The effects of different parameters on the surface state will be regarded and characteristic examples of changes of the surface will be presented. The stability of the mechanically surface-treated condition, especially of the residual stress state during thermal or mechanical loading, is briefly examined. Finally, the consequences of the mechanical surface treatment on some mechanical properties are explained.

1. Surface Treatments

1.1 Shot Peening

Shot peening represents the most important and most frequently used method of noncutting treatment of material surfaces. The process involves the mechanical interaction of hard particles of shot having sufficient kinetic energy with the surface of a workpiece of given geometry, defined chemical composition, and certain mechanical properties. Shot peening is usually carried out by the wheel method (shot is propelled by a motor-driven bladed wheel) or the air blast method (shot is introduced by an air blast machine, either by gravity or by direct pressure). The attainable near-surface changes in the material are dependent on the shot used, the shot parameters and type, and the condition of the workpiece. The principal characteristics of the shot particles are their size, shape, hardness, and specific weight. Typical shot-peening parameters are the shot velocity v (for wheel machines) or the peening pressure p (for air blast machines), the peening intensity, the mass flow $\dot{m}$, the coverage, the exposure time t, and the angle of impingement. The peening intensity i is proportional to the product of $v \times d \times t$ or $p \times d \times t$, respectively, where d is the shot size. The so-called Almen intensity is a technological measure of the shot-peening induced changes of the surface state and is proportional to the peening intensity i. This intensity is expressed as the arc height of an Almen test strip at the saturation coverage.

Shot peening of metallic materials causes inhomogeneous plastic deformation of the near-surface layers. The effect usually extends to about 0.2–0.4 mm below the surface. The most important deformation processes occurring are summarized schematically in Fig. 1. These processes are the Hertzian pressure, with maximum values of the forces below the surface, and the plastic stretching of the near-surface layer, with maximum forces at the surface itself (Wohlfahrt 1982). Compressive residual stresses with characteristic depth distributions are produced together with work-hardening and/or work-softening effects and typical surface topographies are created. The maximum of the compressive residual stresses may be situated below the surface (especially for hard materials) or directly at the surface (especially for soft materials) according to the dominating process. In all materials a peening-induced adiabatic heat production is expected, accompanied by a relatively low rise in temperature. This effect is caused by extremely short impact times of the shot together with plastic deformation to various degrees according to the hardness of the material.

The processes described above are influenced by a complex interaction of a variety of parameters which are divided into two groups: the shot peening conditions and the workpiece conditions. The two parameter groups, however, are largely interactive. The shot peening parameters may have typical influences on the workpiece properties (see, e.g., Sect. 2).

Several models have been developed to describe analytically the surface state, especially the distributions of strains, stresses, and residual stresses during and after the peening process. The models essentially result from considerations not only of the quasistatic penetration but also of the dynamic impact of a single ball in the workpiece. Moreover, the workpiece reactions which are caused by shot peening are considered in terms of repeated deformations including the Bauschinger effect and the cyclic deformation behavior of the material. In most cases, modeling of the residual stresses proves that the distributions are at least qualitatively in agreement with experimental results.

1.2 Deep Rolling

Deep rolling is a widely used mechanical surface treatment, especially for cylindrical components with

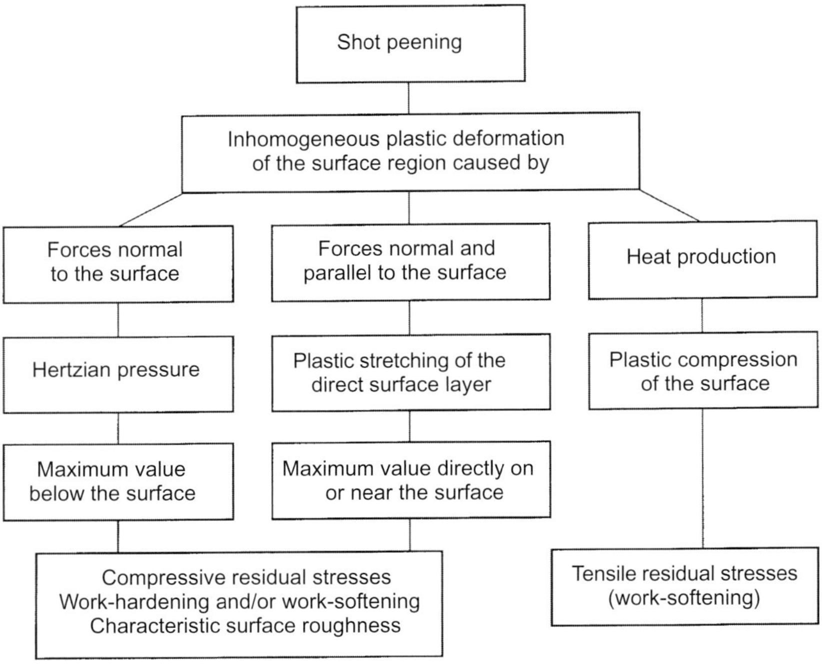

Figure 1
Shot-peening processes and workpiece reactions.

notches or high curvatures. Very often the process can be directly integrated into a production line. The rolling procedure is carried out by specially shaped tools, which have to be well adapted to the geometrical conditions of the components. Single or multiple rolling operations with and without slip are applied. In addition to the contact geometry, the rolling force and the resulting contact pressure are carefully controlled, and constant as well as variable or oscillating rolling forces may be applied.

The basic process is a local inhomogeneous plastic deformation, restricted to near-surface areas, which yields similar consequences to those in the case of shot peening. However, the thickness of the affected surface layers is usually considerably greater and surface roughness is strongly reduced. As a consequence of the applied contact pressure, there is an inhomogeneous triaxial loading stress with a maximum equivalent stress below the surface of the treated component. Since the local yield strength of the material is exceeded, after unloading a triaxial residual stress state is produced with compressive residual stress maxima below the surface of the material.

Strain hardening and improved smoothness of the surface are further important consequences of surface rolling operations. Consequently, all process parameters influencing the plastic deformation history of the rolling operation have to be considered to assess the resulting properties of the near-surface materials.

1.3 Machining

Every machining operation of a metallic material is combined with macroscopic and microscopic inhomogeneous plastic deformation in addition to the local production of thermal energy in the area of the near-surface material. This is the case for machining operations with geometrically determined cutting edges, such as milling or turning, as well as for procedures with geometrically undetermined cutting edges, such as grinding or honing. Consequently, machining operations produce effects comparable to those of shot peening or rolling treatments on the resulting properties of the near-surface material, but with additional more or less pronounced thermal effects. Heating and cooling cycles during machining operations may lead to phase transformations and local softening or hardening effects. Thermally induced plastic deformations are usually compressive and lead to near-surface tensile residual stresses after cooling. Mechanical effects of machining operations lead to strain hardening and to compressive near-surface residual stress states, as already mentioned above. Mechanically- and thermally-induced inhomogeneous plastic deformations together determine the resulting properties of the near-surface materials.

Surfaces after pure cutting operations usually exhibit tensile residual stresses. In practice, however, it is often the consequences of the final squeezing operations that determine the resulting near-surface properties.

Even for a given type of machining operation, such as grinding, the consequences for the resulting states of the near-surface material can be quite different, depending on the type and state of wear of the tool, the cooling and lubricating conditions, and the process parameters applied (Scholtes 1987).

2. *Characterization of the Surface States*

To describe sufficiently the condition of a mechanically treated surface, it is necessary to know the characteristics of the different testing methods and their properties. The best known properties are the microhardness HV, the residual stresses σ^{rs} (usually determined by x-ray diffraction or mechanical methods), the half-width values HW of the x-ray interference lines, and the surface roughness R_t (depth of roughness) and R_a (roughness average). The HW values as a measure of the microstructural work-hardening state and the dislocation density characterize the mean lattice strains and micro-residual stresses, respectively. An example of the depth distributions of a mechanically treated surface for the properties HV, HW, and the residual stresses σ^{rs} in longitudinal and transversal directions obtained by x-ray diffraction and successive electrolytic surface removal as well as a near-surface abnormal microstructure is shown schematically in Fig. 2. In some cases, mechanically- or thermally-induced phase transformations also occur, e.g., the transformation of retained austenite to martensite in the case of hardened steels. Finally, it has to be mentioned that in the case of porous materials, e.g., cast or sintered states, mechanical surface treatments lead to a gradually diminished porosity in near-surface zones (see *Residual Stresses: Measurement by Destructive Testing*; *Residual Stresses: Measurement by Diffraction*; *Stress Measurement by the Thermoelastic Effect*).

2.1 *Roughness*

Since the surface roughness has an influence on the fatigue strength and fatigue life due to micronotch effects, the knowledge of the effect of different shot-peening parameters on the depth of roughness is of great interest. Figure 3 shows the property R_t of the heat-treatable steel AISI 4140 after grinding and shot peening, respectively, as a function of workpiece hardness. In the ground condition R_t is independent of workpiece hardness and has a value of 2 μm. For the hardened condition the depth of roughness is almost the same for both the peened and the ground samples. However, after shot peening with fixed parameters the property R_t increases from the hardened to the normalized condition by a factor of 10. Furthermore, the roughness increases with increasing peening pressure p, shot size d, and hardness of the shot HV_S and decreases with increasing mass flow $\dot{m}$, whereby the changes are lower with increasing workpiece hardness.

2.2 *Residual Stresses*

Figure 4 shows typical depth distributions of residual stresses and x-ray interference line half-width values

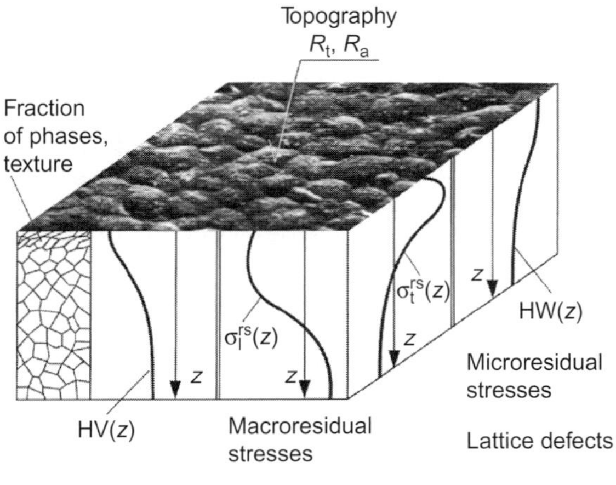

Figure 2
Courses of characteristic properties of surface layers after mechanical surface treatments.

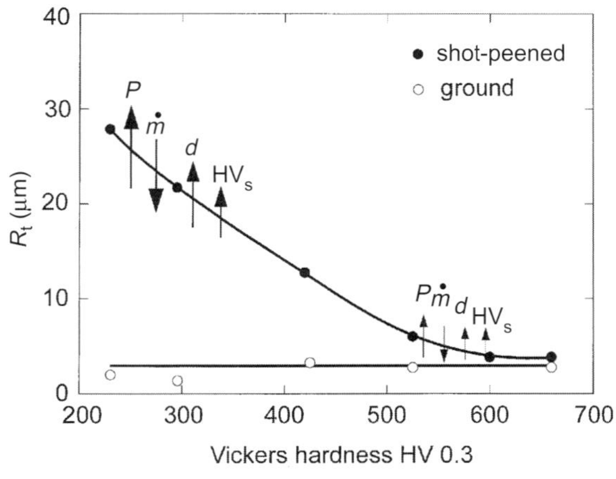

Figure 3
Surface roughness after grinding or shot peening vs. workpiece hardness of AISI 4140 in a quenched and 450°C, 2 h (T450) tempered condition (shot peening: S 170/46 HRC, $p = 1.6$ bar, $\dot{m} = 1.5\,\mathrm{kg\,min^{-1}}$).

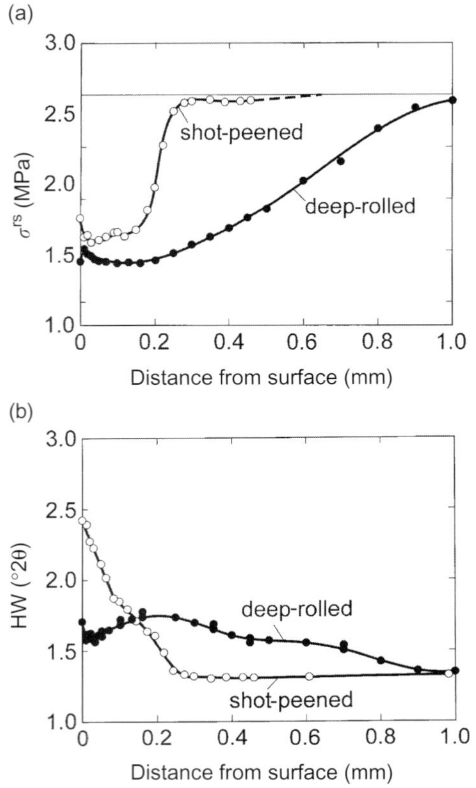

Figure 4
Characteristic depth distributions of (a) residual stresses and (b) x-ray interference line half-width values of shot-peened or deep-rolled quenched and tempered steel AISI 4140 (hardness 320 HV 30).

of a shot-peened or deep-rolled quenched and tempered steel AISI 4140. A characteristic feature is that the thickness of the affected surface layer is higher for deep-rolled compared with shot-peened states. Residual stress depth distributions exhibit maximum compressive residual stresses at a certain distance from surface. Although the amounts of compressive residual stresses are almost the same near the surface, half-width values are completely different, which indicates the presence of differences in local microstructures (see Sect. 2.3).

Examples of residual stress depth distributions of shot-peened samples of the steel AISI 4140 in different heat treated conditions can be seen in Fig. 5, which shows an increasing amount of residual stresses at the direct surface layer for the normalized condition (hardness 230 HV 0.3) up to the hardened and at 450°C-tempered condition T450 (430 HV 0.3). With increasing sample hardness the residual stresses at the surface decrease.

Figure 6 shows a simplified summary of the in fluence of several shot-peening parameters on the

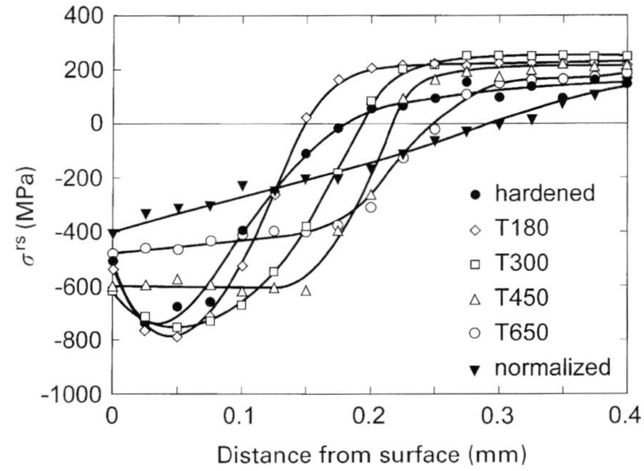

Figure 5
Residual stresses vs. distance from surface of AISI 4140 at different heat treatment conditions after shot peening (S 170/46 HRC, $p = 1.6$ bar, $\dot{m} = 1.5\,kg\,min^{-1}$).

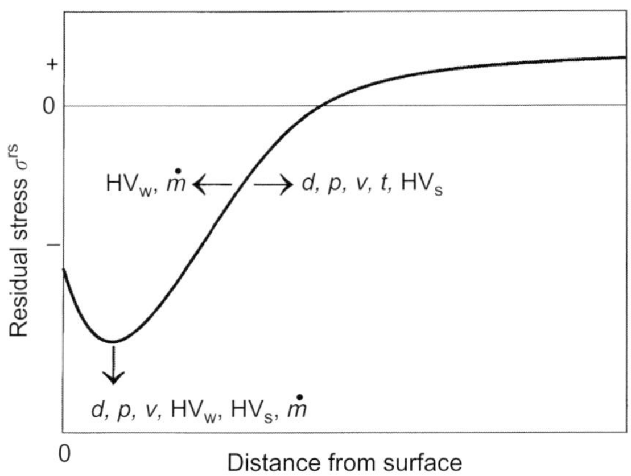

Figure 6
Influence of the peening parameters on the residual stress distribution.

shot peening-induced residual stress distribution (Voehringer 1987b). The arrows show the shift direction of the residual stress distribution when the indicated parameters increase.

2.3 *Microstructure and Work-hardening State*

The half-width values as a measure of the microstructural work-hardening state are shown in Fig. 7 as a function of the distance from the surface for different heat-treated samples of the steel AISI 4140 after shot peening. The curves for normalized and hardened samples and those tempered at 650°C (T650) are similar. However, the HW values for the normalized condition are somewhat smaller than

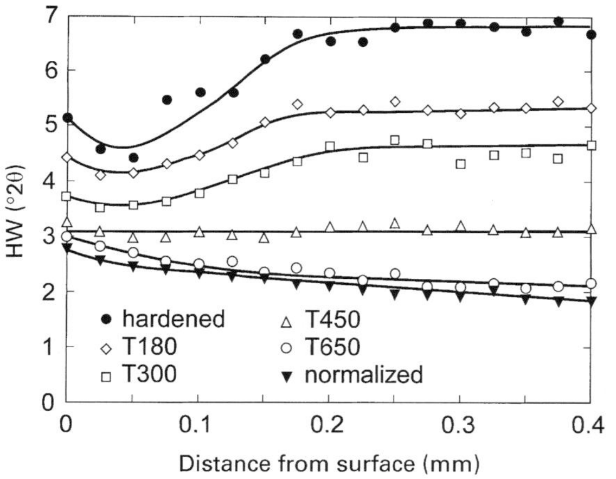

Figure 7
Half-widths vs. distance from surface of AISI 4140 at different heat treatment conditions after shot peening (S 170/46 HRC, $p = 1.6$ bar, $\dot{m} = 1.5\,\mathrm{kg\,min^{-1}}$).

those for the T650 condition. Both curves show increasing values at the surface.

The inhomogeneous peening-induced deformation of the surface region produces in soft and half-hard materials (<500 HV) an inhomogeneous work-hardening which is indicated by increasing half-width values due to increasing dislocation densities and characteristic dislocation arrangements. This work-hardening effect increases with increasing p, v, d, and HV_S (e.g., Voehringer 1987b). However, in relatively hard materials, such as hardened steels or cold-formed materials (>500 HV) with relatively high dislocation densities, the peening process causes decreasing HW values due to work-softening effects. The appropriate work-softening effect is more effective the higher is the hardness. In these states the decrease in the HW values is caused by rearrangements of the dislocations with high density in low-energy structures. In hardened steels solute carbon atoms cause an additional work-softening effect due to stress-induced changes in the positions of the carbon atoms by the Snoek effect. Finally, it can generally be stated that with increasing shot time and multiple coverage, a work-softening effect also occurs which is initiated by fatigue processes such as cyclic deformation, crack formation, and propagation.

The surface topography and the distribution of plastic strains are also important properties of the peened surface. The latter can be estimated, for example, from single ball tests applying a measuring grid. In a normalized SAE 1045, surface plastic strain values in the order of magnitude of 10% were evaluated with this method. Similar strain values are reported by Hakimi *et al.* (1984) who compared the half-width values of shot-peened steels with those of corresponding specimens deformed in compression tests with usual strain rates. Transmission electron microscopy investigations of normalized SAE 5132 (German grade 35 Cr 4) with an exact target preparation show dislocation structures with cells and tangles which can usually be observed at uniaxial deformation, but not at cyclic deformation. The dislocation densities decrease with increasing distance from the surface in a manner similar to the half-width values of the x-ray interference lines.

2.4 *Texture*

Because of the near-surface inhomogeneous plastic deformations, mechanical surface treatments create typical texture states in thin surface layers. Characteristic preferred grain orientations are observed, depending on the surface treatment process and the process parameters applied. Figure 8 shows characteristic texture pole figures of a shot-peened and a milled surface of a normalized steel SAE 1045, which before the surface treatment exhibited a random grain orientation. In order to characterize the type and strength of the texture state the ratio I_{max}/I_{min} of the largest to the smallest number of counts (cts) appearing within the measuring range shown in the pole figures is plotted. In the case of shot peening (Fig. 8(a)), the deformation forces mainly act perpendicular to the surface and tend to create a fiber-like texture with an {110}-fiber axis parallel to the normal of the shot-peened surface. Milling produced a characteristic grain orientation (Fig. 8(b)), whose prominent feature is its asymmetry with respect to both the longitudinal and the transverse direction. This can be explained by the cutting forces exerted by the tool. In the case presented here, due to the axial rake angle of the cutter the resulting cutting forces are inclined at a certain angle to the feed with respect to the longitudinal direction. However, it has to be pointed out that in both cases the textures are relatively weak.

2.5 *Other Properties*

Additional information regarding the peened surface state can be obtained for specific materials by analyzing the microstructure. This method is useful if materials with metastable phases are shot-peened because transformation-induced plasticity can occur, as for example the transformation of retained austenite to martensite in hardened steels or in bainitic cast irons as shown, for instance, in Fig. 9 for bainitic cast nodular iron in two conditions. Furthermore, conclusions regarding the deformation and stress state can be drawn from the existence and the density of deformation twins that are produced in the surface region of peened materials having a sufficiently small stacking fault energy.

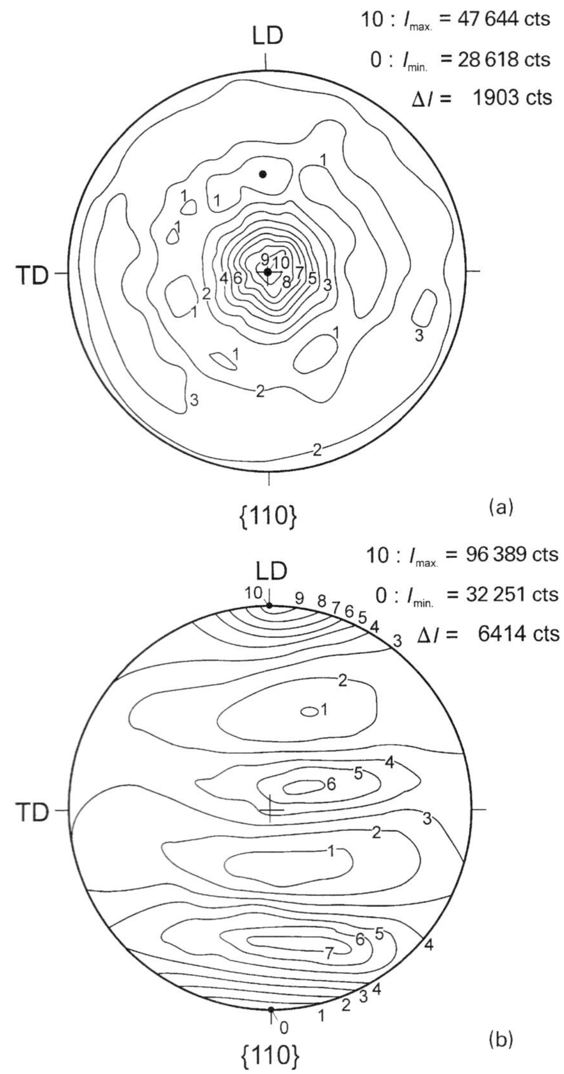

Figure 8
{110}-Texture pole figures of (a) shot-peened and (b) milled specimens of a normalized SAE 1045.

Finally, another method should be mentioned with which the yield strength $R_{s,y}$ of a peened surface layer can be determined. This property can be analyzed using combined mechanical and x-ray measurements with discrete increases of compression loading and unloading up to the critical loading stress $\sigma_{s;crit}$, where the relaxation of residual stresses starts (Voehringer 1987a).

3. *Stability and Relaxation Behavior of the Surface Residual Stresses*

Stability of the mechanically-treated surface state of components is desirable during the application of

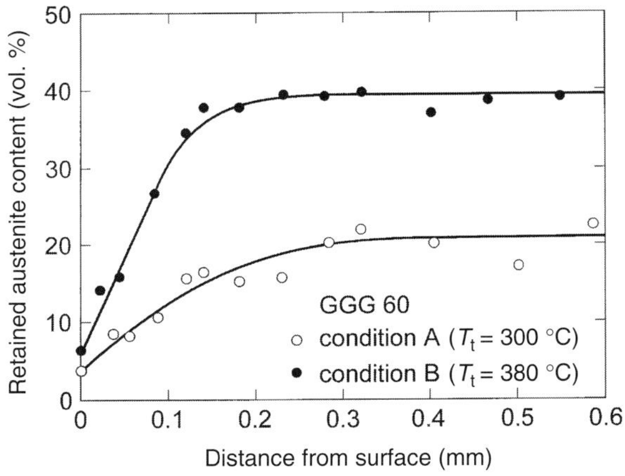

Figure 9
Retained austenite vs. distance from surface of bainitic nodular cast irons in different conditions after shot peening (S330/55 HRC, $p \approx 4$ bar, Almen intensity $I \approx 0.23$ C).

thermal and/or mechanical energy in order to preserve the improvements in the mechanical properties. Therefore, the residual stress relaxation behavior is the most important aspect to be considered. Macro- and microresidual stresses can be reduced or completely relaxed either by heat treatment or under unidirectional or cyclic mechanical loading.

Characteristic examples are represented in Fig. 10 for the relaxation of shot-peening residual stresses of the direct surface of the steel AISI 4140 in a quenched and tempered condition (T450) by different thermal treatments. As can be seen in Fig. 10(a), the thermal relaxation grows with increasing annealing temperature and time. The data points are fitted by curves calculated by an Avrami relationship that enables a quantitative description of the thermal relaxation of residual stresses. In this case the relaxation is caused by a volume diffusion-controlled creep of slip dislocations in the residual stress field of other dislocations. The mechanical relaxation of residual stresses (Fig. 10(b)) starts by quasistatic loading for critical stresses, $\sigma_{s;crit}$, which are lower than the yield strength of the unpeened condition ($R_y = 1300$ MPa). Furthermore, the relaxation is anisotropic in compression and tension ($\sigma^{comp}_{s;crit} < \sigma^{tens}_{s;crit}$). However, the onset of the cyclic relaxation under similar conditions (Fig. 10(c)) occurs for stress amplitudes where $\sigma_{s;crit} > \sigma^{comp}_{s;crit}$ (Holzapfel *et al.* 1998).

Elastic residual strains related to residual stresses can be converted into microplastic strains by suitable deformation processes, such as by dislocation slip or creep. The onset of residual stress relaxation is determined by the critical applied loading stress σ_{crit}, with the meaning of a resistance against relaxation of residual stresses.

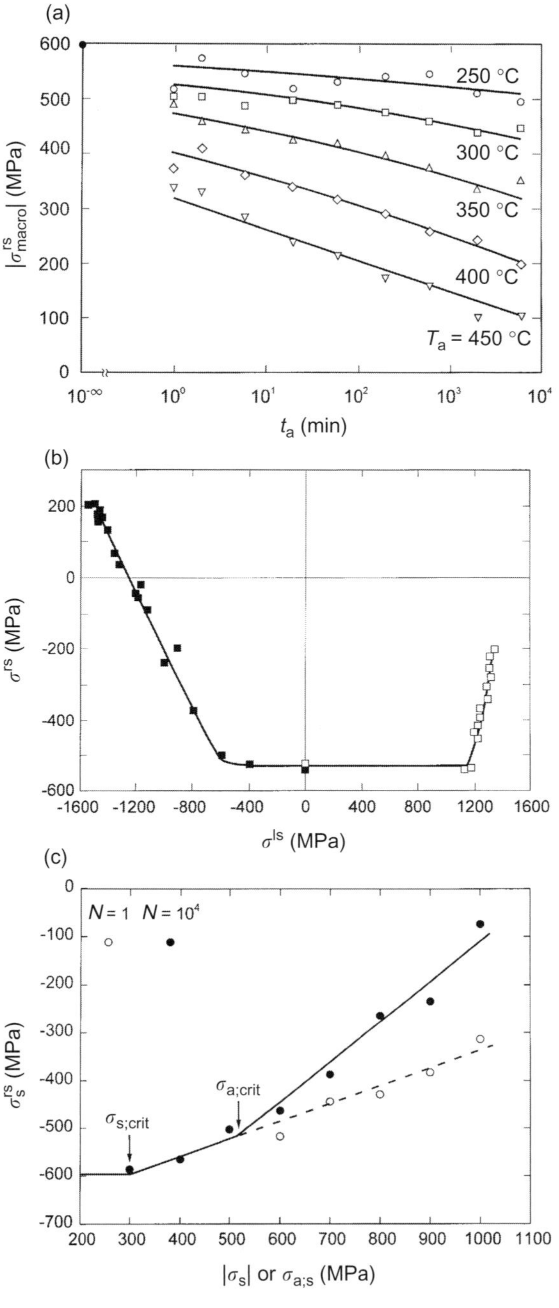

Figure 10
Relaxation of shot-peening residual stresses of AISI 4140 in a quenched and tempered condition (T450) by: (a) thermal treatment; (b) uniaxial tensile and compressive loading; and (c) cyclic bending (S 170/46 HRC, $p = 1.6$ bar, $\dot{m} = 1.5\,\mathrm{kg\,min^{-1}}$).

The applied stress may be zero, have a value that does not vary with time (thermal residual stress relaxation), have a steadily increasing value (unidirectional deformation), or a value varying periodically with time (cyclic deformation).

Since the onset of residual stress relaxation depends on the resistance of the material, σ_i, as well as on the applied stress, σ, all the parameters characterizing the state of the material as well as temperature and time or frequency are of importance. The onset of residual stress relaxation is delayed by the presence of sufficient stable obstacles to dislocation movement which increases σ_i at the loads given, mostly for $T < 0.4T_m$ (T_m = melting temperature in K). The onset and extent of residual stress relaxation is influenced in a complex manner by the combined effects of heat treatment, unidirectional and cyclic stress, and by multiaxial applied and/or residual stress states. The kinetics of residual stress relaxation are essentially determined by the difference of driving forces and resistances. The rate increases with increasing applied load, i.e., as the temperature rises, the time and the magnitude and amplitude of the applied stress, plastic deformation, or number of cycles all increase. The rate often increases as the amount of residual stress increases, and is faster when there are fewer stable obstacles to dislocation glide.

4. *Consequences and Applications*

The near-surface microstructures and residual stress states created by the different mechanical surface treatments have a distinct impact on the strength and lifetime of the components treated (see *Mechanical Surface Treatments for Decorative Purposes*). Increase in fatigue strength by shot peening and deep rolling compared with unfinished conditions is most pronounced for components with high hardness and tensile strength. Thus, the notch-induced loss of fatigue strength can be completely compensated in practical cases to provide adequately high-strength workpiece conditions.

The basic processes influencing fatigue lifetime are cyclic plasticity, crack initiation, and crack propagation. In Fig. 11, for a cyclically softening SAE 1045 and a cyclically hardening magnesium-base alloy AZ31 material, plastic strain amplitudes as a function of the number of cycles are plotted for tension–compression loading with the stress amplitudes indicated. Results for annealed and shot-peened or deep-rolled conditions are shown. It can be seen clearly that mechanical surface treatments reduce considerably the plastic strain amplitudes. The fact that the effect is more pronounced for deep-rolled than for shot-peened states is due to the thicker affected surface layer in the former. Because fatigue damage is clearly correlated with plastic strain amplitude, the benefit of mechanical surface treatments becomes

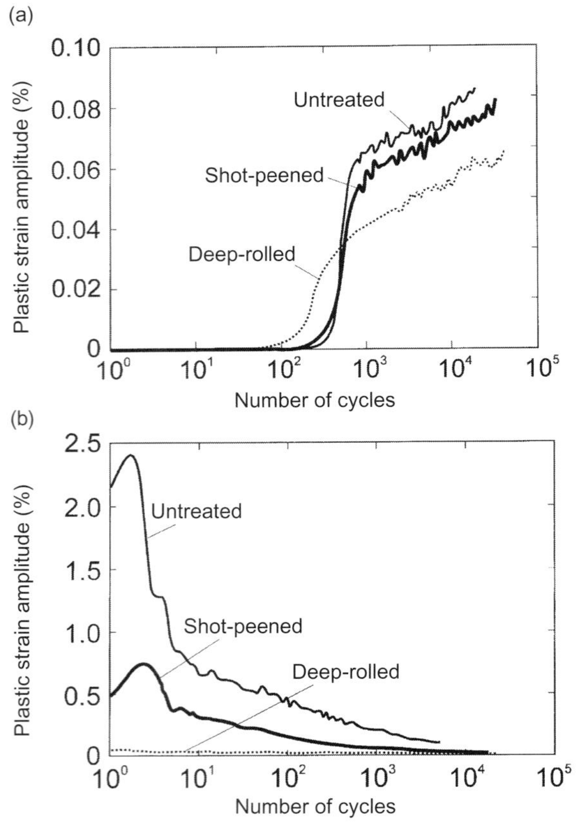

Figure 11
Cyclic deformation curves of different mechanically surface-treated materials for push–pull loading in the LCF regime: (a) plain carbon steel SAE 1045, stress amplitude $\sigma_a = 350$ MPa, Almen intensity $I = 0.120$ mm A, rolling pressure $p = 150$ bar; (b) magnesium alloy AZ 31, $\sigma_a = 125$ MPa, Almen intensity $I = 0.2$ mm A, rolling pressure $p = 100$ bar.

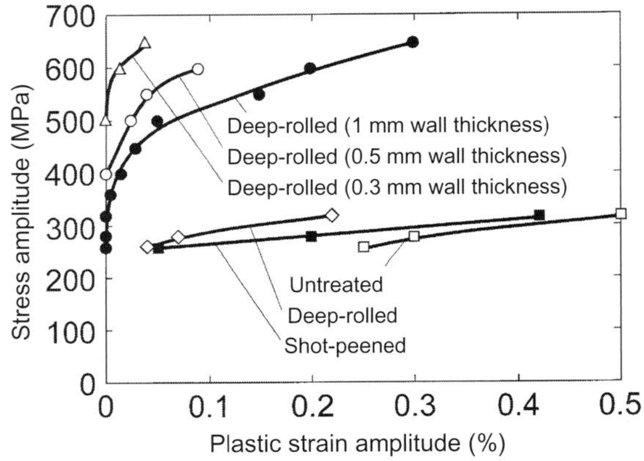

Figure 12
Cyclic stress–strain curves of deep-rolled AISI 304 (rolling pressure 150 bar) using compact and hollow specimens with different percentages of strain-hardened layers.

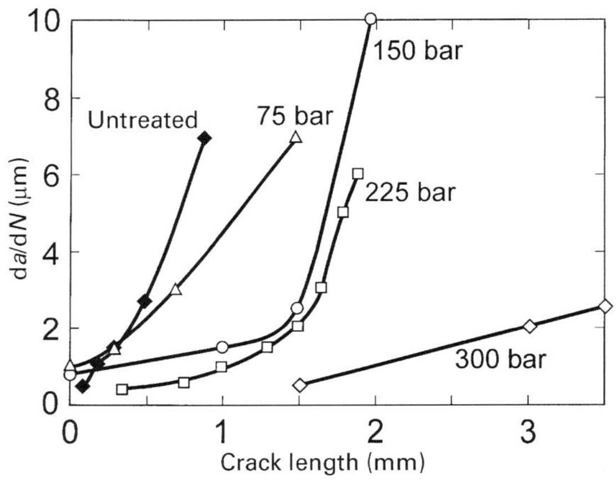

Figure 13
Fatigue crack propagation velocities da/dN in annealed and deep-rolled steel AISI 304 (different rolling pressures); stress amplitude = 320 MPa.

obvious. The influence of mechanical surface treatments on cyclic plasticity can be summarized in cyclic stress–strain curves, correlating plastic strain amplitudes and stress amplitudes for, e.g., half the number of cycles to failure. Figure 12 shows data for an austenitic steel AISI 304 in a shot-peened or deep-rolled condition in comparison with untreated specimens (Altenberger *et al.* 1999). The effects of mechanical surface treatments become more distinct as the ratio between the area of affected surface layer to the cross section of the specimen becomes higher. In the case in Fig. 12, hollow, thin-walled specimens were prepared from compact ones for analysis. It can be seen that cyclic yield strength considerably increases in comparison with compact specimens if only the fatigue behavior of the surface layers of mechanically surface-treated components is investigated.

A major influencing parameter on crack initiation after mechanical surface treatment is the surface topography. In this context, smoothing effects of surface rolling procedures can be very advantageous. In addition to crack formation, crack propagation plays an important role for strength and lifetime. Crack velocity is influenced by both strain hardening and residual stresses in mechanically-treated surface layers. Figure 13 shows crack velocities for stainless steel AISI 304 surface-rolled at different pressures. Crack velocity da/dN, determined by analyzing striations on cracked surfaces, increases with increasing rolling pressure and is considerably diminished compared with untreated specimens (Altenberger *et al.* 1999).

Although the major application of shot peening and deep rolling is related to improvement of the fatigue behavior, other useful applications have been developed—for example, metal forming, improving the resistance to stress corrosion, and surface activation of superalloys by peening for a better adhesive strength of plasma-sprayed thermal barrier coatings.

Bibliography

Altenberger I, Scholtes B, Martin U, Oettel H 1999 Cyclic deformation and near surface microstructures of shot peened or deep rolled austenitic stainless steel AISI 304. *Mater. Sci. Eng. A* **264**, 1–16

Hakimi D, Servant C, Castex L 1984 A study of shot-peened surfaces by X-ray diffractometry and electron microscopy, "in French". In: Fuchs H O (ed.) *Proc. 2nd Int. Conf. on Shot Peening*. The American Shot Peening Society, Paramus, NJ, pp. 249–58

Holzapfel H, Schulze V, Voehringer O 1998 Residual stress relaxation in an AISI 4140 steel due to quasistatic and cyclic loading at higher temperatures. *Mater. Sci. Eng. A* **248**, 9–18

Scholtes B 1987 Residual stresses introduced by machining. In: Niku-Lari A (ed.) *Advances in Surface Treatments*. Pergamon Press, Oxford, Vol. 4, pp. 59–71

Voehringer O 1987 Relaxation of residual stresses by annealing or mechanical treatment. In: Niku-Lari A (ed.) *Advances in Surface Treatments*. Pergamon Press, Oxford, Vol. 4, pp. 367–96

Voehringer O 1987 Changes in the state of the material by shot peening. In: Wohlfahrt H, Voehringer O, Kopp R (eds.) *Proc. 3rd Int. Conf. on Shot Peening*. DGM–Informationsgesellschaft, Oberursel, Germany, pp. 185–204

Wohlfahrt H 1982 Shot peening and residual stresses. In: Kula E, Weiss V (eds.) *Residual Stresses and Stress Relaxation*. Plenum Press, New York, pp. 71–92

B. Scholtes and O. Voehringer

Mechanical Surface Treatments for Decorative Purposes

Generally, mechanical surface treatments such as shot peening or roller-burnishing induce high dislocation densities in near-surface regions. Due to the local plastic deformation, residual stresses are developed and the surface topography is changed. These changes can have contradictory influences on the fatigue strength. The surface roughness determines whether fatigue strength is primarily crack nucleation controlled (smooth surface) or microcrack propagation controlled (rough surface). For smooth surfaces, a work-hardened surface layer is detrimental to crack propagation. Near-surface residual compressive stresses are clearly beneficial, as they can significantly retard crack growth once cracks are present.

The total fatigue life of a component is the sum of both crack nucleation and crack propagation life (Fig. 1). The fraction of the fatigue life that is spent in crack nucleation depends on the stress amplitude: the higher the stress amplitude the earlier cracks nucleate and the higher the fraction of the propagation life to the total life.

1. Fatigue Behavior

To demonstrate the marked effect that surface treatment can have on fatigue performance in titanium alloys, a typical set of S-N curves from rotating beam loading of the Ti–6Al–4V alloy is shown in Fig. 2. The reference EP, electrolytically polished conditions, is compared with mechanical polishing performed with three different particle sizes of 7, 18, and 48 μm. In addition, results after shot peening SP are shown. There is a wide variation in 1×10^7 cycles endurance limits ranging from 475 MPa for coarsest polishing up to 725 MPa for shot peening. For a given microstructure, this variation in endurance limits due to mechanical surface treatments is much greater than that caused by varying the bulk microstructure, i.e., fully lamellar, duplex, or equiaxed for a given surface treatment. Starting with the electropolished reference, the endurance limit of this particular microstructure, roughly 600 MPa, first increases to 700 MPa with mechanical polishing with 7 μm particle size, but then decreases to 550 and 475 MPa after coarser polishing with 18 and 48 μm particles, respectively. The highest endurance limit of 725 MPa was measured after shot peening using cast steel shot.

The change in surface layer properties after mechanical surface treatments is shown in Fig. 3. While transmission electron microscopy (TEM) is suitable to study the change in *dislocation density* after small amounts of plastic deformation, the dislocation density after heavy deformation can best be evaluated by x-ray diffraction (Fig. 3(a)). Here, the half width breadth as a measure of the total dislocation density is shown for the various surface treatments as a function of the distance from the surface. No x-ray interference line broadening could be found for mechanical polishing with the finest particle size of 7 μm, owing to the fact that the x-ray measurement (CuK_α-radiation) integrates over roughly 10 μm while the affected depth is probably only about 1 or 2 μm.

The near-surface plastic deformation can also be clearly seen by optical microscopy (Fig. 3(b)). For a given treatment, the penetration depth measured by x rays (150 μm) is roughly twice that clearly seen by optical microscopy (75 μm).

Residual stresses (macrostresses) were also measured by x-ray diffraction (peak shifting) and

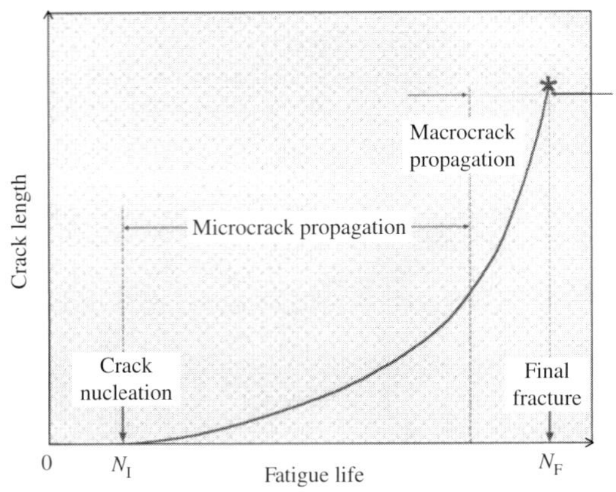

Figure 1
Stages in fatigue life with regard to crack nucleation and propagation (schematic).

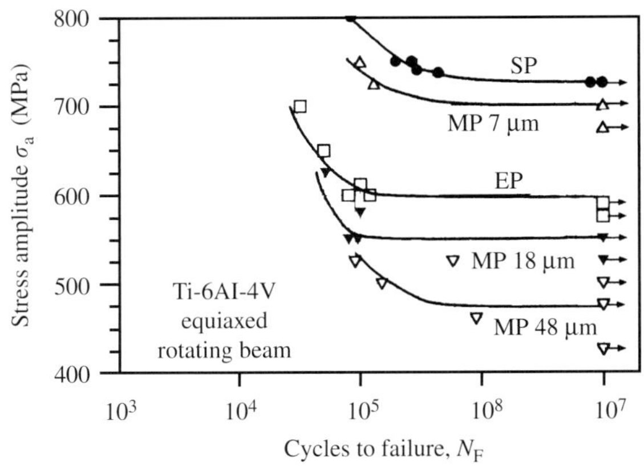

Figure 2
Effect of various surface treatments on fatigue life in Ti–6Al–4V.

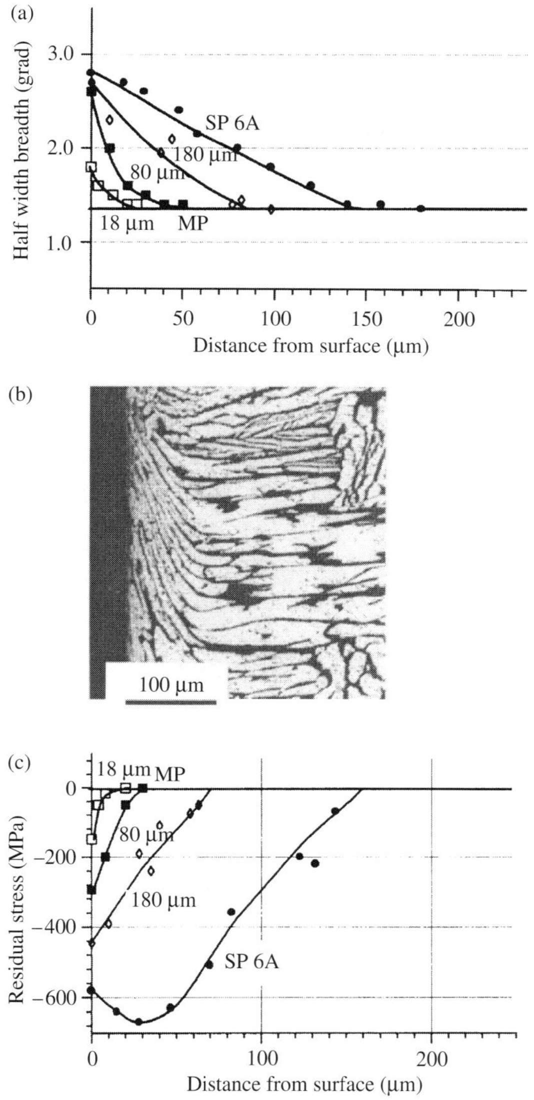

Figure 3
Surface layer properties after mechanical surface treatments. (a) Dislocation density profiles, (b) cross-section, (c) residual stress profiles.

calculated by the $\sin^2\psi$-method. Generally, mechanical surface treatments develop residual compressive stresses in the surface (Fig. 3(c)). For reasons of mechanical equilibrium, these stresses must be balanced by residual tensile stresses in deeper regions. The depth at which the stresses change sign from compression to tension is in good agreement with the penetration of plastic deformation (compare Fig. 3(c) with Fig. 3(b)). After shot peening, maximum compressive stresses are usually found well below the surface.

2. *Separation of Surface Parameters*

Three parameters—surface roughness, dislocation density, and residual stresses affect the fatigue performance. The individual effects of these parameters were separated to determine their influence on fatigue crack nucleation and microcrack growth.

The high surface roughness present after shot peening can be reduced through mechanical polishing followed by electropolishing. After removing a minor amount, the roughness is as low as in the reference (Fig. 4(a)). The profiles of dislocation density and residual stresses are hardly affected by this removal since their penetration into the material is much greater than the scale of the roughness.

Residual stresses can be reduced by plastic deformation or by annealing. In both cases, elastic strains are transformed to plastic strains thus reducing the residual (elastic) stresses. Annealing at various temperatures and holding times was performed on shot peened specimens to study the effect on residual stress decay. A complete stress relief was found after annealing for one hour at 600°C. This treatment is used in the following to separate the effects of residual stresses.

While the residual stresses are completely relieved at this temperature, there is still a high dislocation density in the affected layer after this stress relief. This is seen in the TEM micrographs comparing a fully lamellar microstructure before peening and after peening + stress relieving at 600° for one hour (Fig. 4(c)). A significant decrease of the shot peening induced high dislocation density in Ti–6Al–4V occurs at temperatures above 650°C. At that temperature, the fully lamellar microstructure starts to transform to an equiaxed structure through recrystallization.

As a result of such a parameter separation, there are various conditions that can be compared to evaluate individual effects. The reference EP is characterized by a low surface roughness, no residual stresses, and a low dislocation density. After shot peening all three properties are altered and the surface layer is characterized by a high surface roughness, high residual compressive stresses, and a high dislocation density (SP). If a stress relief is performed (SP + SR), the high surface roughness and the high dislocation densities are still present. If instead of a stress relief, a surface polish is performed after shot peening (SP + EP), both the high residual compressive stresses and the high dislocation densities are maintained while reducing the surface roughness. Finally, both a stress relief and a polish after shot peening can be performed (SP + SR + EP), leaving only the high dislocation density as the main parameter.

For microcrack propagation, two parameters are important, namely dislocation density and residual stresses. The surface roughness is of minor importance since the crack depths of microcracks even in

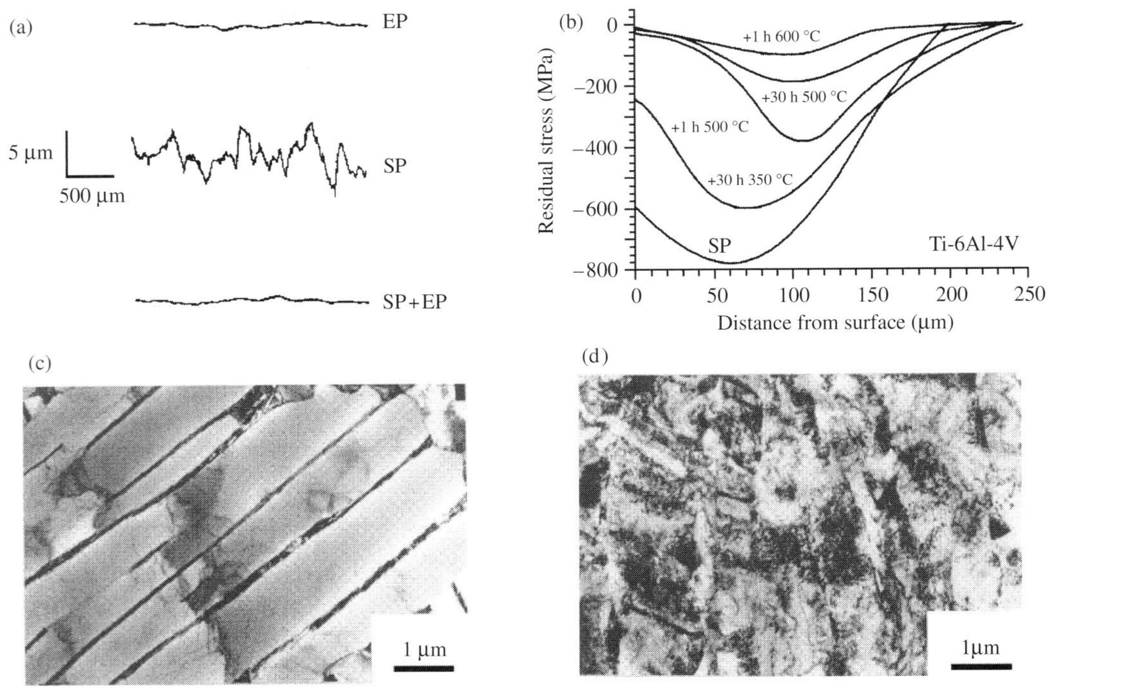

Figure 4
Parameter separation in Ti–6Al–4V (a) Roughness profiles, (b) residual stress profiles, (c) dislocation microstructures (TEM), before shot peened and (d) after shot peening and stress relieving at 1 hour and 600°C (50 µm from as-peened surface).

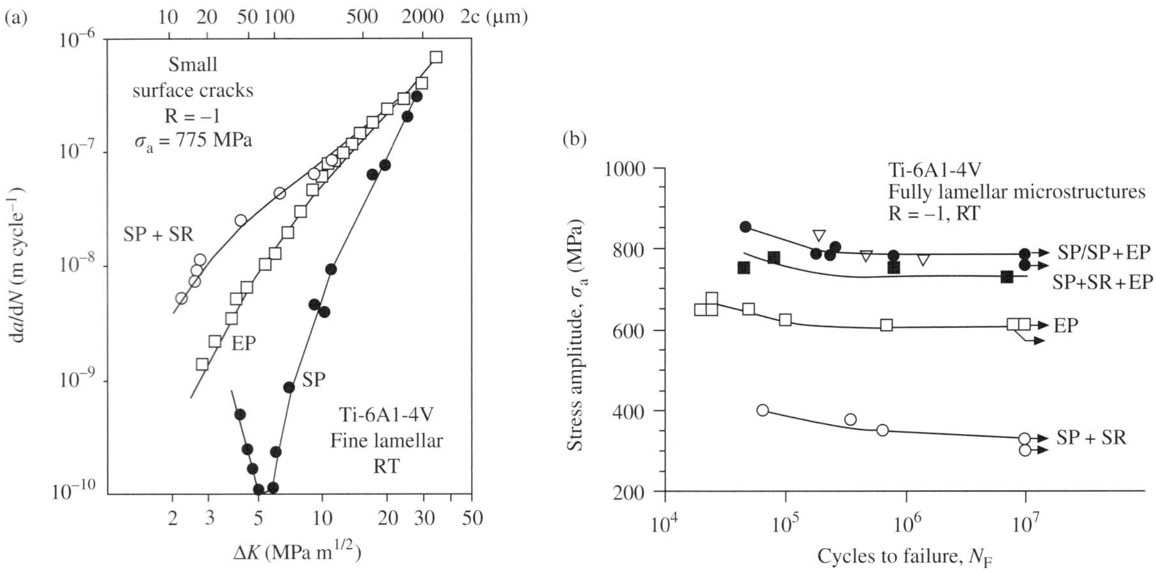

Figure 5
Fatigue performance of the various conditions in Ti–6Al–4V at ambient temperature (a) da/dN-ΔK curves for microcracks and (b) fatigue life.

their early stages are much greater than the maximum roughness.

While the shot peening-induced high surface roughness makes it difficult to determine the crack nucleation life in shot peened specimens, it is obvious that the fatigue life improvement mainly derives from the extension of the crack propagation rather than from nucleation life. Crack growth in the shot peened condition is quite straight and normal to the loading axis. Microcrack growth rates da/dN were calculated by converting surface crack length $2c$ to crack depth a via the ratio $a/2c = 0.45$. For calculating the amplitude of the stress intensity factor, only the tensile portion of the load cycle was considered.

The fatigue performance at ambient temperature of the various conditions in Ti–6Al–4V is shown in Fig. 5. The microcrack growth results are plotted as da/dN-ΔK curves in Fig. 5a. The corresponding crack lengths at the surface are also shown. Compared to the reference, shot peening leads to much lower growth rates—up to two orders of magnitude. There is a minimum in growth rate, probably caused by maximum residual compressive stresses at some distance below the surface. Overall, the beneficial effect of the residual compressive stresses on crack growth clearly overcompensates the detrimental effect of the high dislocation density and low ductility. This is why shot peening improves fatigue life.

With these results in mind, the corresponding S-N curves in Fig. 5(b) can be understood. Shot peening improves the fatigue performance of the reference mainly by inducing residual compressive stresses which hinder crack growth. The degree to which residual compressive stresses can improve the fatigue strength is seen by comparing curves SP + SR with curve SP. For these conditions with a rough surface, the fatigue strength is crack propagation controlled (early crack nucleation at a rough surface) and is, therefore, drastically affected by residual stresses.

3. *Concluding Remarks*

The results also show that even without residual stresses, the fatigue strength can be improved. This is seen by comparing the reference EP with the curve SP + SR + EP. This significant increase of the fatigue strength of roughly 75 MPa is clearly related to the beneficial effect of a work hardened surface layer on crack nucleation. The high dislocation density increases local yield stress and the resistance to crack nucleation. Of course, the surface must be smooth in order that the fatigue life can in fact become nucleation controlled. The effect of surface layer properties on fatigue life is summarized in Table 1.

Table 1
The effect of surface layer properties on fatigue life.

	Crack nucleation	Crack propagation
Surface roughness	accelerates	no effect
Cold work	retards	accelerates
Residual compressive stresses	minor or no effect	retards

A high surface roughness accelerates crack nucleation with no effect on crack propagation. Cold work retards crack nucleation but accelerates crack propagation. Residual compressive stresses have little effect on crack nucleation, but could drastically retard crack propagation.

See also: Mechanical Surface Treatment

Bibliography

Hack J E, Leverant G R 1982 *Influence of Residual Compressive Stress on the Crack Opening Behavior of Part-Through Fatigue Cracks, Residual Stress Effects in Fatigue.* ASTM STP, 776 204

Wagner L, Lütjering G 1984 *Influence of the shot peening parameters on the surface layer properties and the fatigue life of Ti–6Al–4V.* In: Fuchs H O (ed.) Shot Peening. American Shot Peening Society, p. 194

Wagner L, Gregory J K 1993 Thermomechanical Surface Treatment of Titanium Alloy. In: Mittemeijer E J (ed.) *Proc. 2nd ASM Heat Treatment and Surface Engineering Conf. in Europe.* Trans Tech Publications, p. 159

Wagner L, Gregory J K, Gysler A, Lütjering G 1986 *Propagation Behavior of Short Cracks in a Ti-8.6Al Alloy.* In: Ritchie R O, Lankford J (eds.) Small Fatigue Cracks TMS-AIME, p. 117

Was G S, Pelloux R M 1916 *The effect of shot peening on the fatigue behavior of alloy 7075–T6. Met. Trans.* **10A**, 656

T. Dörr and L. Wagner

Metal Forming (Cold): Dislocation Mechanisms and Microstructural Changes

Metals are crystalline except under very special conditions. To about equal parts metals and alloys crystallize in an f.c.c., b.c.c., and h.c.p. crystal structure. There is unambiguous evidence that metals retain their crystal structure during plastic deformation. This conservation principle of crystal plasticity (Fig. 1) has serious consequences for the deformation process and material properties.

1. Deformation Fundamentals: Slip

In principle, there are an infinite number of planes and translation vectors to accomplish an imposed shape change. Due to energetic reasons, crystallographic glide is confined to only a few crystallographic planes (slip planes) and directions (slip directions), usually the most densely packed planes and directions in a crystal. A slip plane and a slip direction constitute a slip system.

The mechanism of crystallographic glide is the motion of crystal dislocations. The glide plane of a dislocation and the direction of its Burgers vector correspond to the slip plane and slip direction, respectively. On an atomistic level, therefore, plastic deformation of a metal requires the generation and motion of crystal dislocations.

Dislocations are crystal defects and, therefore, cause a distortion of the atomic arrangement, which manifests itself in a long-range stress field. Dislocations interact with other dislocations via their stress field. This interaction constitutes a glide resistance, which is macroscopically felt as a yield stress and eventually causes immobilization of moving dislocations. The traveling distance (slip length) of a moving dislocation is usually much smaller than the macroscopic dimension of the formed part, therefore, dislocations are stored in the crystal, and the dislocation density grows with progressing deformation which in turn increases dislocation interaction and thus the flow stress. The relation between flow stress σ and dislocation density ρ can be represented by the Taylor equation

$$\sigma = \alpha M G b \sqrt{\rho} \tag{1}$$

where the dislocation density (m^{-2}) is defined as the total dislocation line length per unit volume—typically 10^{10} (m^{-2}) for annealed and 10^{16} (m^{-2}) for heavily deformed metals, b is the Burgers vector (slip vector), G the shear modulus, M is the Taylor factor which relates the applied stress to the shear stress in the glide system and depends on crystallographic texture (see below). Typically M is of the order of three, and the geometrical constant α is of the order of 0.5.

Any change of the glide resistance of the moving dislocations will affect the flow stress. This is the reason for work hardening (increasing dislocation density), other strengthening mechanisms (grain size, solute atoms, particles, precipitates etc.) and softening by recovery and recrystallization (decreasing dislocation density). It is important to note that the macroscopic deformation behavior can be related to the microscopic properties of the crystal dislocations. The force $\vec{K}$ on a dislocation due to a stress σ applied to a crystal is given by the Peach-Koehler equation

$$\vec{K} = \boldsymbol{\sigma}\, \vec{\mathbf{b}} \times \vec{s} \tag{2a}$$

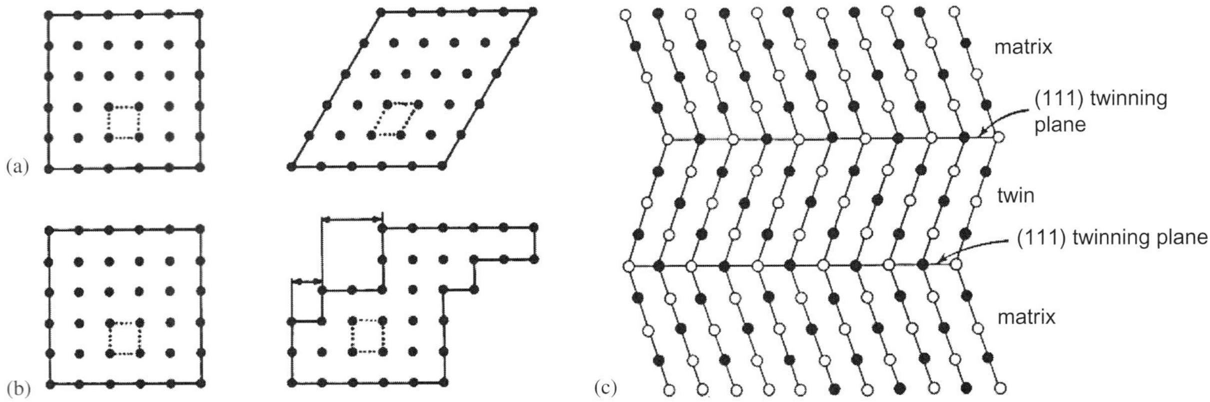

Figure 1
Plastic deformation of crystals (dotted lines = unit cell) (a) with change of crystal structure; (b) by crystallographic glide; (c) by deformation twinning.

or the magnitude of the force K in the slip system due to the resolved shear stress τ

$$K = \tau b \tag{2b}$$

The imposed strain rate is reflected by the dislocation velocity through the Orowan equation

$$\dot{\varepsilon} = \rho_{\mathrm{m}} b v \tag{3}$$

where ρ_{m} is the density of the mobile dislocations moving with an average velocity v. The velocity, of course, depends on the applied stress σ.

2. *Twinning*

While crystallographic glide by dislocation slip is the dominating deformation mechanism in metals, particularly at low temperatures, deformation twinning is also observed, which can accommodate small plastic strains as well. A deformation twin has an atomic arrangement that is the mirror image of the parent crystal. The generation of a twin corresponds to a simple shear deformation by a displacement parallel to the twinning plane (mirror plane). In contrast to dislocation glide the shear strain γ is constant for a given crystal structure, e.g., $\gamma = \sqrt{2}/2$ for cubic crystals. Twinning plane and displacement direction constitute a twin system in analogy to a slip system (Table 1).

Since for a given shear strain the atomic displacement increases with growing distance from the twinning plane (Fig. 1(c)), twins are usually thin, and their formation requires high stresses which are available only at low temperatures, because the material strength increases with decreasing temperature. Deformation twinning is liable to occur if the flow stress is high and the twin boundary energy (approximately half of the stacking fault energy (SFE)) is low. Face-centered cubic metals and alloys with low SFE undergo deformation twinning, like silver at room temperature or copper at liquid nitrogen temperature. As the SFE decreases upon alloying, most alloys are prone to twinning during low temperature deformation, e.g., copper alloys, like α brass (Cu–Zn) where twinning constitutes the major low temperature deformation mechanism.

Most prominent is deformation twinning in materials where dislocation slip cannot fully accommodate the imposed strain, e.g., in h.c.p. metals and alloys with $c/a > 1.63$ owing to their low number of crystallographic slip systems. For an arbitrary deformation a crystal needs to activate five independent slip systems to satisfy the five independent elements of the strain tensor. In h.c.p. metals with $c/a > 1.63$, e.g., zinc, there is only slip on the basal plane and, therefore, there are only two independent slip systems. This is sufficient to deform single crystals to very large strains. However, the individual grains of a polycrystal have to deform compatibly with the adjacent grains to avoid holes or overlap occurring. This can be accomplished only if the individual grains

Table 1
Slip systems and twinning systems of the major metallic crystal structures.

Slip systems

Crystal structure	Slip plane	Slip direction	Number of nonparallel planes	Slip directions per plane	Number of slip systems
f.c.c.	{111}	$\langle 1\bar{1}0\rangle$	4	3	$12 = (4\times 3)$
b.c.c.	{110}	$\langle \bar{1}11\rangle$	6	2	$12 = (6\times 2)$
	{112}	$\langle 11\bar{1}\rangle$	12	1	$12 = (12\times 1)$
	{123}	$\langle 11\bar{1}\rangle$	24	1	$24 = (24\times 1)$
h.c.p.	{0001}	$\langle 11\bar{2}0\rangle$	1	3	$3 = (1\times 3)$
	$\{10\bar{1}0\}$	$\langle 11\bar{2}0\rangle$	3	1	$3 = (3\times 1)$
	$\{10\bar{1}1\}$	$\langle 11\bar{2}0\rangle$	6	1	$6 = (6\times 1)$

Twinning systems

Crystal structure	Twinning plane	Shear direction	Displacement plane	Prototype
f.c.c.	{111}	$\langle 112\rangle$	{110}	Ag, Cu
b.c.c.	{112}	$\langle 111\rangle$	{110}	a-Fe
h.c.p.	$\{10\bar{1}2\}$	$\langle 10\bar{1}1\rangle$	$\{1\bar{2}10\}$	Cd, Zn

can undergo an arbitrary shape change at their boundaries, for which five independent slip systems are required. Therefore, polycrystals of zinc, for example, have to activate mechanical twinning for deformation. Due to crystallographic effects h.c.p. metals and alloys with $1.63 \leqslant c/a \leqslant 1.73$ cannot deform to a large degree by twinning, since twinning ceases to comply with the imposed shape change due to crystal reorientation and the fact that only one sense of twinning shear is crystallographically possible. This is the reason why polycrystalline h.c.p. magnesium ($c/a = 1.63$) and its alloys are not ductile at ambient temperature. Note that hexagonal metals like titanium, beryllium, or zirconium are very ductile even at low temperatures since they have a ratio $c/a < 1.63$, a fact which favors nonbasal slip and thus provides sufficient independent slip systems.

3. Deformation Microstructure

When gliding dislocations become immobilized they are stored in the crystal. The spatial distribution of these stored dislocations can occur at random or in distinct patterns. Most metals and alloys reveal nonstatistical dislocation distributions. Typically dislocations arrange in cellular patterns, i.e., in high dislocation density (cell) walls, which enclose volumes (cells) of considerably lower dislocation densities. For most commercial cold forming processes—like rolling—the cellular dislocation arrangements become superimposed by deformation inhomogeneities, which typically have a band-like shape and can assume macroscopic dimensions to be recognized by the bare eye. In rolled materials, besides microscopic microbands which appear like distinctly framed elongated dislocation cells under the microscope, there are two major types of deformation inhomogeneities, namely deformation bands and shear bands which can be best recognized in micrographs of the lateral surface. Deformation bands are elongated microstructural features that extend parallel to the rolling direction and that gradually accommodate a large misorientation to the matrix. Shear bands are deformation inhomogeneities, which extend under an angle of 30–35° to the rolling direction. Copper-type shear bands comprise clusters of elongated cells, and are confined to grain dimensions. In contrast, brass-type shear bands have (macroscopic) dimensions comparable to the sample thickness. Their internal structure is composed of very small granular features, typically globular grains of nanometer size. The origin and structure of shear bands is far from being understood, but they are generally assumed to be due to local softening (like texture softening) or enforced macroscopic shear in case of obstructed dislocation glide. Brass-type shear bands are typical in alloys with low stacking fault energy, e.g., α-brass. Copper-type shear bands are formed during cold rolling of f.c.c. metals and alloys with higher stacking fault energy.

4. Hardening Behavior

The storage of dislocations during deformation leads to an increased glide resistance for moving dislocations which is experienced as an increased flow stress with increasing strain (Fig. 2). The rise of stress σ with strain ε is referred to as strain hardening or work hardening, and the slope of the hardening curve $d\sigma/d\varepsilon$ is the work hardening coefficient or hardening rate. While the hardening curves of different materials or deformation conditions vary considerably, there is a characteristic shape of the engineering stress–strain curve, namely after yielding a curve with a single maximum, though strength level and hardening rate can be very different.

Of particular interest for forming processes is the hardening behavior for large strain deformation, e.g., sheet rolling (see also *Plastic Behavior of Metals: Large Strains*). A typical true stress–strain behavior for large strain deformation exhibits after yielding some transient with high but decreasing hardening rate until a low but constant rate is attained (Fig. 2), i.e., the stress increases linearly with strain, and eventually a steady state (constant flow stress) is attained. Correspondingly, the large strain-hardening curve exhibits five distinct stages:

Stage I—right after yielding transition to athermal hardening (observed only in single crystals oriented for single slip);

Stage II—athermal hardening with constant high hardening rate $d\sigma/d\varepsilon \approx G/30$ (extended range only in single crystals at low temperatures);

Stage III—dynamic recovery range, hardening rate decreasing linearly with stress;

Stage IV—large strain hardening with constant low hardening rate; and

Stage V—transition to steady state or unstable microstructure, usually not attained during cold forming.

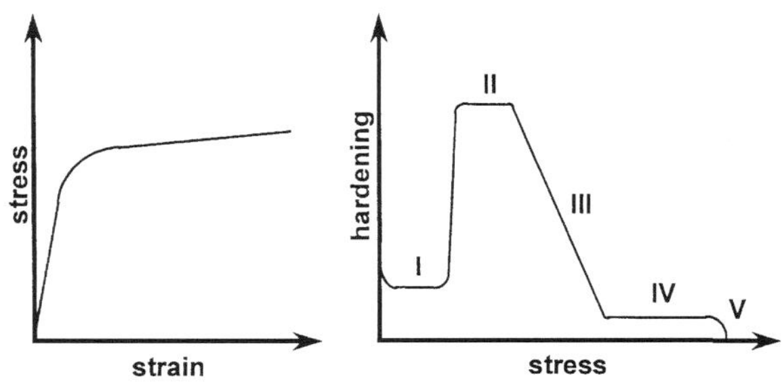

Figure 2
Schematic stress–strain curve and hardening rate vs. stress for large strain plastic deformation of metals and alloys.

The important stages of the flow curves of commercial metals and alloys for industrial cold forming processes are stages III and IV. Their extent and level depends on material composition and processing and is thus history dependent.

The hardening behavior reflects the glide resistance of the moving dislocations and therefore depends on the dislocation structure S which is strongly dependent on processing history, i.e., at strain rate $\dot{\varepsilon}$ and temperature T

$$\sigma = \sigma(\dot{\varepsilon}, T, S) \tag{4}$$

For this particular reason, the macroscopic strain is not a state parameter and, therefore, a prediction of flow stress as empirical function of strain $\sigma(\varepsilon)$ is physically not really correct, although longstanding engineering practice uses it because of simplicity.

In the past the most common engineering representation of the flow curve was the Ludwik or Hollomon equation

$$\sigma = K(\dot{\varepsilon}, T)\varepsilon^n \tag{5a}$$

where n is the hardening exponent, usually in the range of $n \approx 0.3$. Eqn. (5a) gives a poor fit over a long range of strain, since it cannot properly represent stages III and IV. A better representation is obtained by the Voce equation, which follows from a linear decrease of the hardening rate with stress, i.e., stage (iii) behavior

$$\sigma = \sigma_S + (\sigma_0 - \sigma_S)\exp\left(-\frac{\varepsilon}{\varepsilon^*}\right) \tag{5b}$$

(σ_S—steady state stress, σ_0—yield stress, ε^*—characteristic strain) or modified to account also for stage IV

$$\sigma = \sigma_0 + (\sigma_1 + \Theta_1\varepsilon)\left[1 - \exp\left(\frac{-\varepsilon\Theta_0}{\sigma_1}\right)\right] \tag{5c}$$

(σ_1—extrapolated yield stress, Θ_1—slope in stage IV, Θ_0—initial slope).

Eqn. (5c) gives a reasonable fit for a large strain interval of cold-formed metals and alloys, if proper care is taken of the texture dependent Taylor factor. Nevertheless, also the modified Voce equation (Eqn. (5c)) is only an empirical closed form representation of the stress–strain behavior of metals and does not make use of the fact that hardening depends on dislocation structure evolution. Changing deformation conditions—as usual in industrial forming processes—affects structure evolution, which has a serious impact on hardening behavior.

More advanced work hardening models are based on dislocation theory. They follow a common scheme, namely the formulation of a master equation for structure evolution ($d\rho/d\varepsilon$) and a kinetic equation $\dot{\varepsilon}(\sigma)$ to associate structure with glide resistance (Roters *et al.* 2000). These relations are formulated as differential equations and can, therefore, account for transient microstructure evolution.

The total change of dislocation density $d\rho$ in a strain increment $d\varepsilon$ reflects the structure evolution due to production of dislocations $d\rho^+$ and dynamic recovery (e.g., annihilation, dipole formation, lock formation, etc.) $d\rho^-$

$$d\rho = d\rho^+ - d\rho^- = (f_1 - f_2)d\varepsilon \tag{6}$$

The functions f_1 and f_2 have to be determined from dislocation theory. They depend on the particular processes considered, and their actual values vary with dislocation structure. The flow stress is related to the dislocation density and to the strain rate by Eqns. (1) and (3).

The major problem is a proper incorporation of dislocation patterning in the theory. It is a common assumption that the dislocations are arranged in a cell structure, but different approaches have been proposed for the dominant stress and strain rate controlling processes. Due to the non-homogeneous dislocation distribution, the flow stress will be a volume (V) weighted average of the flow stress in cell walls (σ_w) and cell interiors (σ_i)

$$\sigma = \sigma_W V_W + \sigma_i V_i \tag{7}$$

A respective prediction for cold forming of aluminum alloys has shown reasonable agreement with experiment. The quality of prediction and its flexibility to changing deformation conditions has to be paid for by a substantial investment in computer power and computation time.

5. *Texture and Anisotropy*

A simple shear deformation is always accompanied by a rigid body rotation, because the displacement gradient tensor is not symmetrical. Since plastic deformation in metallic materials proceeds by simple shear mechanisms (dislocation glide, mechanical twinning) the grains in a polycrystal change their orientation during metal forming—except for few orientations that are stable under the imposed deformation conditions—and rotate towards the stable orientations. For a given forming process, like rolling, wire drawing, or extrusion, typical orientation distributions, crystallographic textures develop, depending on the deformation mechanism. There are two basic types of cold rolling textures for f.c.c. materials, the copper-type and the brass-type rolling texture (Fig. 3(a)). For simplicity, the texture can be described as superposition of scattered ideal orientations, which are listed in Table 2. The rolling texture of b.c.c. metals is typically made up of two fibers in (Euler) orientation space, namely orientations with ⟨110⟩ parallel to the rolling direction (α-fiber) and orientations with ⟨111⟩ parallel to the rolling plane normal (γ-fiber) (Fig. 3(b)). In wire drawing or extrusion with circular cross-section of f.c.c. metals, fiber textures (i.e., textures with rotational symmetry about a fiber axis) develop where the direction of the wire axis is oriented preferentially parallel to ⟨100⟩ and ⟨111⟩, the relative weight depending on material. Cold drawn wires of b.c.c. metals typically exhibit a ⟨110⟩ fiber texture.

Physical and mechanical properties of crystalline solids depend on crystal orientation. For instance, the tensile yield stress of a copper single crystal pulled in the ⟨111⟩ direction is larger than if pulled in the ⟨123⟩ direction. Correspondingly, the properties of polycrystals depend on their crystallographic texture. An important practical case is the effect of texture on earing during deep drawing of metal sheet. This is due to an influence of the texture on the yield behavior and quantitatively expressed in terms of the R value (Lankford parameter). It is determined by tensile tests on strip specimens machined from the sheet and defined as the ratio of in-plane (w) to through thickness (t) strain

$$R = \frac{d\varepsilon_w}{d\varepsilon_t} \tag{8a}$$

In order to account for in-plane anisotropy, specimens with tensile axis under 0°, 90°, and 45° to the rolling direction are measured to give the average value

Table 2
Major components of rolling and recrystallization textures in f.c.c. metals in terms of Euler angles and Miller indices.

φ_1	Φ	φ_2	{hkl}	⟨uvw⟩	Name
Copper rolling texture					
35	45	0	011	211	Brass-def
62	33.5	63	123	634	S
90	32	45	112	111	Cu
Copper recrystallization texture					
0	0	0	001	100	Cube
63	48	63	122	212	First twin to cube
16.5	26	10.0	148	841	Second twin to cube
76	29	74	148	747	Second twin to cube
Brass rolling texture					
32	45	0	011	211	Brass-def
0	45	0	011	100	Goss
90	66	45	233	311	{223}
0	51	45	111	110	{111}
Brass recrystallization texture					
82	30	34	236	385	Brass rex
90	33	0	035	053	Tilted cube
0	20	0	025	100	Tilted cube
45	38	68	134	132	

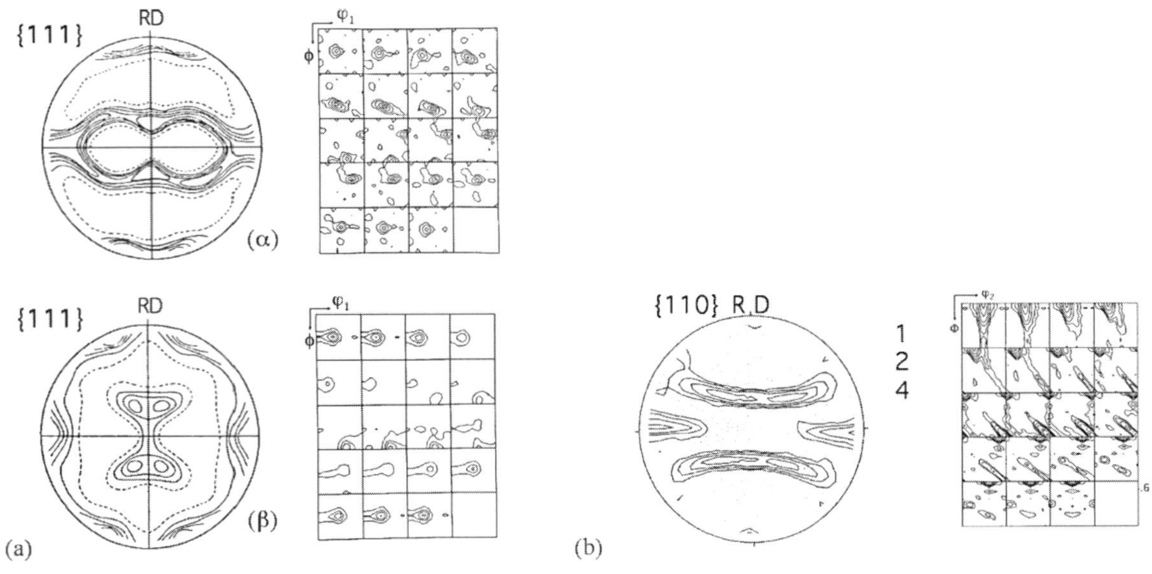

Figure 3
Rolling textures of cubic metals in terms of pole figures and orientation distribution function (ODF) in Euler space. (a) f.c.c. metals (α) Cu-type rolling texture, (β) brass-type rolling texture; (b) b.c.c. metals, e.g., ferritic steel.

$$\bar{R} = \frac{R_0 + 2R_{45} + R_{90}}{4} \tag{8b}$$

and the in-plane anisotropy

$$\Delta R = \frac{R_0 - 2R_{45} + R_{90}}{4} \tag{9}$$

For a random texture $R = 1$ and $\Delta R = 0$ but best deep drawing quality with high forming limit is obtained for $R > 1$ with $\Delta R = 0$, which can be generated by an in-plane fiber texture. The recrystallization texture of deep drawing steels is essentially a $\langle 111 \rangle$ fiber texture and, therefore, most suitable for this forming process. In contrast, aluminum sheet develops a component texture, in the worst case even a single component texture, like the cube texture in recrystallized aluminum sheet. This gives rise to strong earing (Fig. 4). In industrial practice the earing of the cube texture at (0, 90°) is essentially compensated by the earing at 45° due to subsequent cold rolling.

Because of the importance of texture for mechanical behavior a prediction of deformation textures is desirable. Advanced codes (viscoplastic models, relaxed constraints Taylor theory, self-consistent methods, grain interaction codes) are available that are capable of quantitatively reasonable predictions of the deformation texture. None of the approaches, however, takes the formation of deformation inhomogeneities into account.

Forming processes usually do not impose a homogeneous shape change on a material. For instance, during thick section rolling a through thickness texture gradient can develop. In such case texture prediction requires information on the local deformation that can be obtained by coupling advanced simulations to FEM codes, which provide the local displacement gradient tensor. Conversely, the FEM computations can be improved by incorporation of the local texture in the actual plastic flow behavior. (See also *Metal Working: Cold Rolling*.)

6. *Softening: Recovery and Recrystallization*

During plastic deformation, dislocations are introduced to accommodate the imposed shape change. Dislocations, however, are crystal defects that are not stable in thermodynamic equilibrium, thus the crystal strives to annihilate them. At low temperatures the dislocation arrangement is in mechanical equilibrium and is therefore stable. At elevated temperatures, however, thermally activated processes destabilize the dislocation structure and substantially change or even remove the deformation-induced microstructure, which drastically affects microstructure, texture, and mechanical properties (Fig. 5). There are two major restoration processes, recovery and recrystallization. They are termed static processes if they occur subsequent to deformation during annealing, or are referred to as dynamic recovery and dynamic recrystallization if they proceed during deformation. Dynamic recovery also occurs strain induced and is, therefore, integral part of the strain hardening process at finite temperature. Dynamic recrystallization is an important phenomenon during hot working and liable to occur for $T > 0.4T_m$ (T_m melting temperature).

Static recovery encompasses all processes of dislocation annihilation and rearrangement in patterns of lower total energy. Typical recovered microstructures are subgrain structures where the grain interior is subdivided into dislocation-free subgrains, which are separated by low angle boundaries. This is also referred to as polygonization.

Recrystallization proceeds by nucleation of virtually dislocation-free grains, which grow at the expense

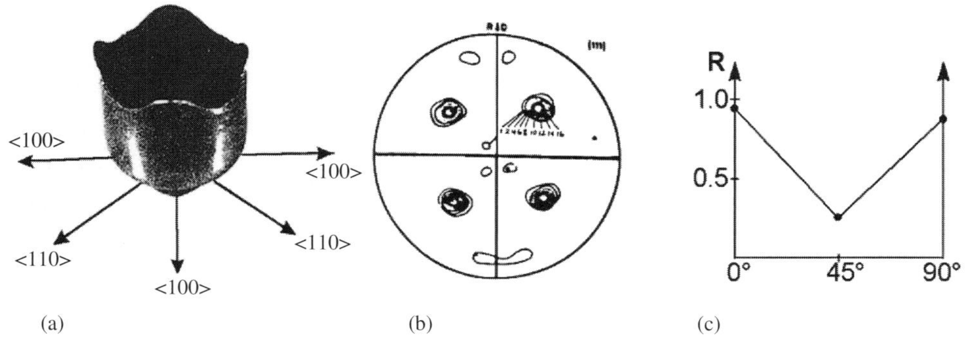

Figure 4
(a) Cup drawn from an Al sheet with a strong cube texture as indicated by {111} pole figure in (b), and respective in-plane variation of Lankford parameter (c).

of the deformed matrix. The moving grain boundaries of the expanding nuclei are capable of absorbing the deformation-induced dislocations so that finally an almost dislocation-free granular structure is generated. Typical dislocation densities of recrystallized polycrystals are $10^{10}\,m^{-2}$ compared to up to $10^{16}\,m^{-2}$ for heavily deformed materials.

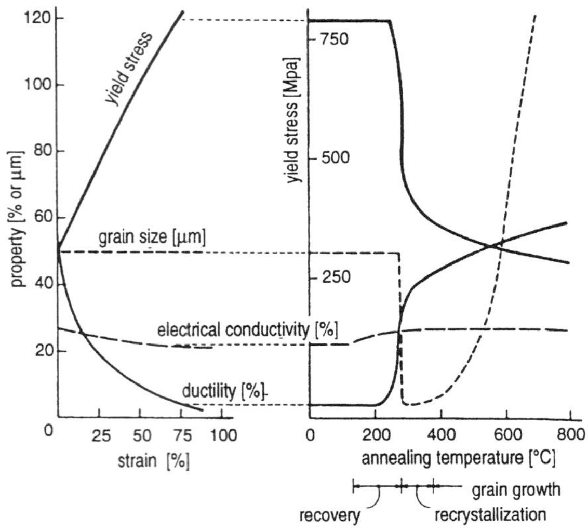

Figure 5
The effect of cold forming and annealing on the properties of a Cu-35%Zn alloy.

While recovery essentially consists of a dislocation rearrangement in the as-deformed structure, a completely new microstructure is formed atom by atom during recrystallization. Correspondingly, both processes have different kinetics and different impact on properties. Invariably either process causes a softening of the material, and fully recovered and fully recrystallized materials have comparable yield stresses. The texture development is very different, however. The deformation texture is essentially retained during recovery or slightly increased in sharpness. During recrystallization a completely new and different texture is generated. The recrystallization texture can be very sharp, like the cube texture in rolled and annealed aluminum, which gives rise to strong earing during deep drawing. Usually, the recrystallization texture comprises several components and can also be very weak, in particular when annealing twins form during the heat treatment. Typically, specific types of deformation textures, as shown in Fig. 3 for rolling, transform to specific recrystallization textures in f.c.c. and b.c.c. materials (Fig. 6). The main recrystallization components are listed in Table 2. However, small changes in chemical composition or processing history can totally alter the recrystallization texture, even if the deformation texture appears unchanged.

Advanced codes for recrystallization structure and texture prediction with high spatial and temporal resolution, e.g., cellular automata or Monte Carlo simulation approaches, are available that can be interfaced to FEM or superimposed to real

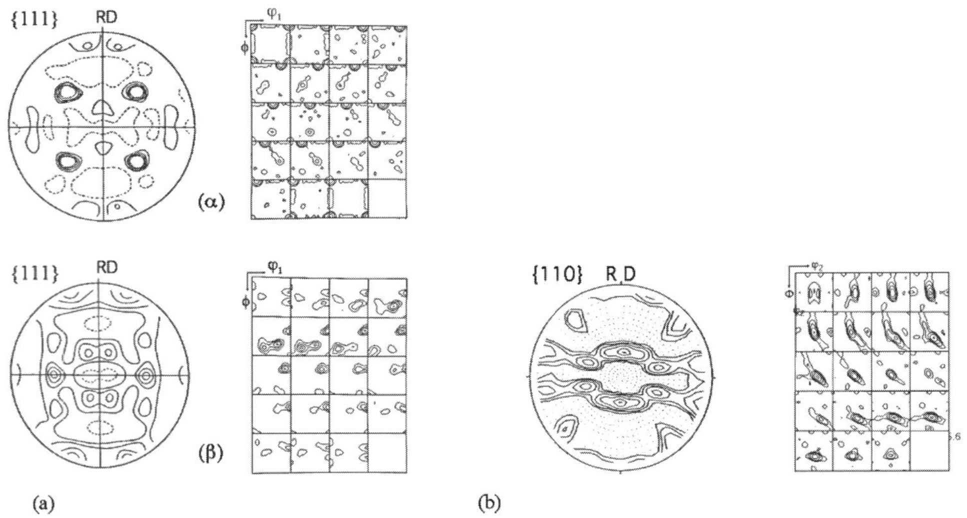

Figure 6
Recrystallization textures of cubic metals in terms of pole figures and orientation distribution function (ODF) in Euler space. (a) f.c.c. metals; (α) Cu-recrystallization texture, cube texture; (β) brass-recrystallization texture. (b) b.c.c. metals, e.g., ferritic steel.

deformation microstructures to account for spatial variation of microstructure (Marx *et al.* 1999).

7. *Conclusions*

The conservation of the crystal structure during plastic deformation of metals causes hardening, crystallographic texture, and propensity to recrystallization during heat treatment, which again is associated with a change of strength, microstructure, and texture. Macroscopically these phenomena affect forming resistance, shape anisotropy, and forming limit. Approaches developed for predicting strength, texture, and microstructure make little use of our understanding of microstructure-property relations and are reliable only in the limits of simple and experimentally tested conditions. Models based on physical mechanisms in conjunction with finite element models require substantial computer power but offer improved predictive capacity for nonstationary and inhomogeneous forming processes.

See also: Laser Forming of Metals

Bibliography

Beckers B, Pomana G, Gottstein G 1999 Interactive texture- and finite-element-simulations for complex deformation processes. In: Khan A S (ed.) *Constitutive and Damage Modeling of Inelastic Deformation and Phase Transformations*. Neat Press, Fulton, MD, pp. 305–8

Gottstein G 1998 *Physikalische Grundlagen der Materialkunde*. Springer, Berlin

Humphreys F J, Hatherly M 1995 *Recrystallization and Related Annealing Phenomena*. Pergamon, New York

Kocks U F, Tomé C N, Wenk H R 1998 *Texture and Anisotropy*. Cambridge University Press, Cambridge

Marx V, Reher F R, Gottstein G 1999 Simulation of primary recrystallization using a modified three-dimensional cellular automaton. *Acta Mater*. **47**, 1219–30

Mecking H 1985 Textures of metals. In: Wenk H R (ed.) *Preferred Orientation in Deformed Metals and Rocks: an Introduction to Modern Texture Analysis*. Academic Press, Orlando, FL

Roters F, Raabe D, Gottstein G 2000 Work hardening in heterogeneous alloys—a microstructural approach based on three internal state variables. *Acta Mater*. **48**, 4181–9

G. Gottstein

Metal Forming (Hot): Microstructural Aspects

The primary reason for heating metal prior to forming is to increase the ease of deformation. However, increasing the temperature also leads to the occurrence of a number of microstructural reactions, particularly in steel. Since the 1950s, there have been attempts to explicitly control these microstructural processes during deformation processing. This approach is often termed thermomechanical processing. Here the aim is to generate superior properties, eliminate heat treatment following deformation, and/or allow a reduction in alloying content.

The following sections review the important microstructure changes that occur during and following the high temperature deformation of steel mainly by means of empirical relations. Many of the concepts also relate to other metals.

1. Processing Parameters

The evolution of the hot deformed microstructure depends upon the processing parameters and the composition of the steel. The most important process parameters are the temperature, time, strain, and strain rate. Generally, the former two have a greater influence than the latter two. Table 1 summarizes the range of process conditions for hot rolling mills. The traditional approach to thermomechanical processing has been to introduce holds, which allow the temperature to drop between different deformation stages, or to change the cooling rate after deformation. The strain for a given deformation does not vary markedly between different types of processes, while the strain rate can vary significantly, but this only has an effect for some processes.

For most of the commercial steels the austenite to ferrite transformation occurs at temperatures between 910°C (for ultralow-carbon steels) and 750°C (for most structural grades). Hence the majority of deformation takes place in the austenite phase field, with a limited number of rolling processes occurring in the ferrite or ferrite + austenite phases.

Examples of two common hot strip rolling mill configurations are shown schematically in Fig. 1. The main steps for conventional slab cast material, in chronological order, are: reheating, roughing deformation, finishing deformation, accelerated cooling, and coiling. Although it is desirous, from a microstructural engineering point of view, to be able to exert control over all the processing parameters, it is evident that there are considerable limitations to the degree to which the conditions can be altered. For instance, in the processing of slab cast material on a hot strip mill there is no way in which heat may be added to the work piece after it has left the reheat furnace. There are also limitations due to the final thickness, the maximum drive speeds, the length of the mill, the placement of the processing units and so on.

These limitations have not prevented thermomechanical processing from making a considerable impact in the production of rolled steel. Furthermore, ongoing mill development has opened, and will continue to open, up new opportunities for microstructure control. A schematic of a thin slab casting

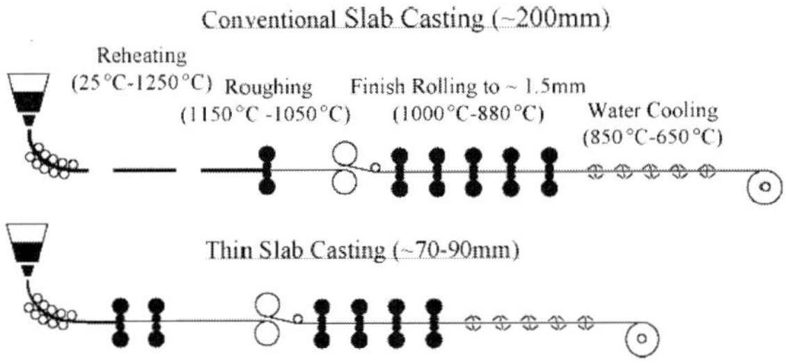

Figure 1
Diagram illustrating the configuration of hot strip mills for conventional slab cast product and thin slab cast material.

Table 1
Typical range of processing conditions for commercial hot rolling mills.

Mill type	Strain per pass	Total strain	Strain rate (s^{-1})	Temperature (°C)	Interpass time (s)
Plate	0.05–0.4	1–4.5	0.3–30	1300–700	5–20 (no holds)
Strip:					
roughing	0.1–0.5	~2.5	0.5–10	1300–950	~10
finishing	0.1–1.3	1.5–3.2	10–100	1050–700	1–5
Structural	0.05–0.8	1–4.5	0.3–30	1250–800	~10
Rod	0.3–0.9	2–5	0.5–300	1150–900	0.07–3
Bar	0.3–0.6	4–7	0.2–2000	1150–1000	0.01–3

and direct hot rolling mill is also shown in Fig. 1. This is the generation of strip mill that follows the conventional mills. In this type of mill, the slab does not cool to room temperature prior to rolling and heat is added after roughing. In addition, in rolling of plate there is a high degree of freedom to affect the temperature-time-deformation history, and therefore the microstructure, because of the nature of the reversing rolling process employed.

2. *Microstructure Evolution*

In most steel products, and indeed most metal products, the important properties are related to mechanical behavior and include hardness, strength, ductility, formability, and toughness. Consequently, the key elements of the microstructure that are of interest are: the grains that make up the primary microstructural constituent (phase), and the grains or particles that make up any other constituents. Typical empirical relationships between the average grain size and the tensile and yield strengths for C–Mn steels are (Hodgson and Gibbs 1992):

$$\text{LYS} = 62.6 + 26\,\text{Mn} + 60\text{Si} + 759\text{P} + 213\text{Cu} + 3286\text{N} + 19.7\Big/\sqrt{\frac{d}{1000}} \quad (1)$$

$$\text{TS} = 165 + 54\text{Mn} + 100\text{Si} + 652\text{P} + 473\text{Ni} + 635\text{C} + 2173\text{N} + 11\Big/\sqrt{\frac{d}{1000}} \quad (2)$$

where LYS is the lower yield strength (MPa), TS is the tensile strength (MPa) and d_α is the ferrite grain size (μm). Other relationships exist relating the other mechanical properties to the composition and key microstructural parameters. In the following sections the evolution of the austenite and the ferrite microstructure is discussed. Particular emphasis will be placed on the interplay between the various microstructural processes.

2.1 *Precipitate Dissolution*

Steel contains many extra elements. Some, like titanium, aluminum, manganese, and niobium, are added intentionally, but others, like sulfur, nitrogen, and oxygen, enter unintentionally during smelting and liquid processing. During cooling after casting most of these foreign elements combine to produce various solid precipitate particles. The most common precipitates in steel are: MnS, AlN, Ti(S,C,N), Nb(C,N), V(C,N), Fe_3C, and BN

In the reheating stage of hot rolling, most of these second phase particles decompose and re-enter into solid solution (a more complicated situation occurs for near net-shaped processing, which will not be considered here). The decomposition of precipitates is important because it determines the level of solute elements present at the start of rolling. This then influences the number and type of precipitation reactions occurring during and after rolling.

In many instances the temperatures during reheating are high and the times are long. Consequently near equilibrium conditions are attained. The most significant processing parameter for dissolution is therefore the reheating temperature. It has been found that the log of the equilibrium constant (solubility product) is a function of the inverse of the temperature:

$$k_s = C_1 - C_2/T \quad (3)$$

where k_s is the solubility product defined as the product of the mass percentage of each element in the precipitate raised to the power of its stoichiometric proportion in the compound (e.g., k_s for Fe_3C is $[Fe]^3 \times [C]^1$), T is the solubility temperature in Kelvin and C_1 and C_2 are constants (see Gladman 1997 for actual values). This equation defines the temperature above which the precipitating species is predicted to dissolve for a given steel composition under equilibrium conditions.

As indicated above many of the precipitate species in steel involve the combination of a substitutional solute element (e.g., Ti, Nb, Al) with an interstitial element (C, N). In some cases this occurs as a 1:1 stoichiometric precipitate (e.g., TiN and TiC). In others, however, carbon and nitrogen are mutually soluble in the precipitate and carbonitrides are produced (e.g., Ti(C,N), Nb(C,N)). The significance of this phenomenon for dissolution is that the carbonitride is generally more stable than the individual carbide or nitride. The carbonitride can therefore exist at temperatures where the carbides and nitrides will not. Knowledge of the precise species that precipitated prior to reheating is, therefore, necessary for the control of dissolution.

2.2 *Grain Growth*

The temperature–time history during casting determines the size of the as-cast grains, which are those at the start of rolling in thin slab processing. However, for most processes the grain size prior to rolling is determined by the reheating operation. During reheating, a fine grain size is initially formed during the ferrite to austenite transformation and then this increases, through grain growth. The migration of grain boundaries occurs through diffusional processes and is therefore thermally activated. It is also important to note that grain growth occurs not just during reheating but also all the while the work piece is at elevated temperature during processing.

The rate of grain growth is determined largely by the initial grain size (and size distribution) and the temperature; smaller grains grow more rapidly than larger ones and faster growth occurs at higher temperatures. In the empirical modeling of the hot rolling of steel, the evolution in grain size (d) due to grain growth following recrystallization is often described by the following expression:

$$d^a = d_o^a + C_3 t \exp\left(\frac{-Q_{growth}}{RT}\right) \quad (4)$$

where d_o is the initial (recrystallized) grain size, t is annealing time, T is temperature, R is the gas constant and a, C_3 and Q_{growth} are constants. Typical values (as for corresponding relations in this article Hodgson and Gibbs 1992) for C–Mn steel in the austenite range are: $a=7$, $C_3=1.5\times10^{27}$ and $Q_{growth}=-400\,kJ\,mol^{-1}$, for $t>1$ s.

Another important factor influencing grain growth is the presence of solute atoms and precipitates. These can, for certain sizes and distributions, impede or even prevent the motion grain boundaries. This phenomenon is critical for many steel grades but particularly during the reheating of aluminum-killed grades (i.e., steels to which aluminum is added to remove oxygen during steelmaking). In these materials, the interaction between grain growth and precipitate dissolution can lead to dramatic changes in grain size.

This is illustrated in Fig. 2 which shows the influence of reheating temperature on grain size for an aluminum-killed steel and a non-aluminum-killed grade. In steels with aluminum and nitrogen the dissolution of AlN precipitates can lead to abnormal grain growth. However, if the temperatures are kept low enough, these steels retain a fine grain structure throughout reheating. Under these conditions, the grain sizes are considerably finer in the Al-killed grade than in the steel without aluminum. This is due to the retarding effect of AlN precipitates. For temperatures higher than 1150°C, AlN fully dissolves and the two grades show a similar grain size. Ti(C,N) is even more effective in retarding grain growth, particularly between rolling passes. In this case the austenite grain size is effectively limited to 20–30 μm, even at 1200°C.

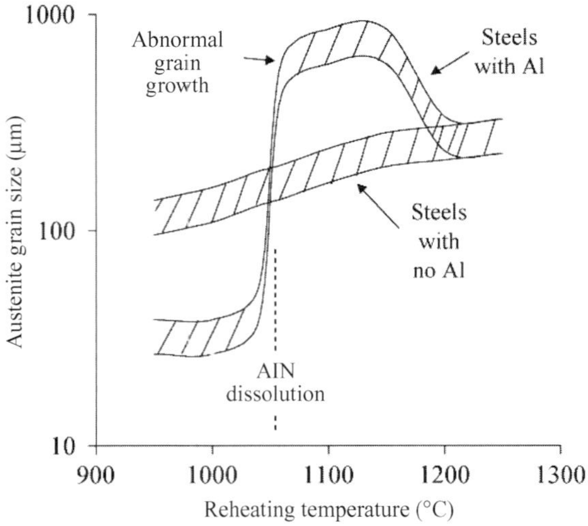

Figure 2
Influence of re-heating temperature on the austenite grain size of C–Mn steels with and without Al addition.

2.3 Formation of the Deformed Structure

From a mechanistic, as opposed to a mechanical, point of view, deformation entails the slip of dislocations along crystallographic glide planes, leading under conventional hot working conditions to a dislocation substructure within the grains. This substructure is further altered both during and after deformation by the thermally activated movement of accumulated dislocations, a process known as recovery. The important parameters of the deformed structure include the dislocation density, the dislocation cell and subgrain size, the misorientations across cell/subgrain walls, and the degree and dimension of other dislocation features such as microbands and shear bands.

The substructure established during deformation can influence the mechanical properties directly if the metal is used in the as-deformed state. However, for steels, this is rarely the case and the major effect is on the subsequent microstructural processes such as precipitation, recrystallization, and transformation. The change in grain shape that occurs during deformation, by the overall macroscopic shape change, is also important.

As would be expected, an important parameter in establishing the deformed sub-structure is the amount of deformation. The nature of the deformation, that is the strain path used to achieve the shape change, is also significant, but this is of secondary importance for the present considerations. The rate and temperature of deformation are, however, of equal if not greater significance than the amount of work. In many instances, the hot deformed substructure reaches a steady state and subsequent deformation ceases to alter the main parameters of the substructure (as defined above). In these cases, the substructure is defined almost solely by the temperature and rate of deformation. The flow curves given in Fig. 3 display a steady-state region at high strains (due to dynamic recrystallization) for which this statement holds.

It is commonly observed that the characteristic dimensions of many deformation-induced structures (e.g., subgrain size) vary with the Zener-Hollomon

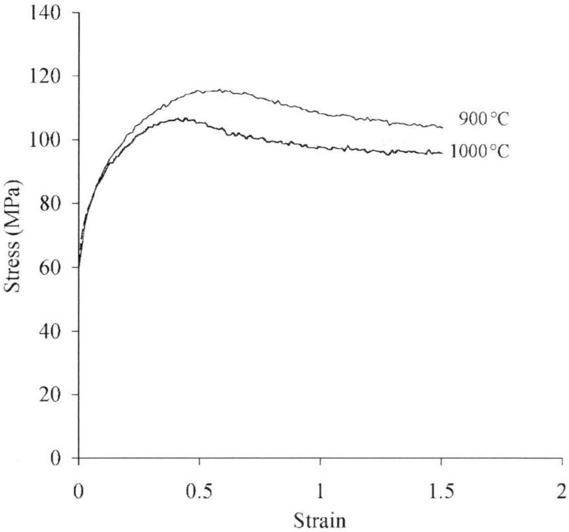

Figure 3
Stress vs. strain curves for a C–Mn steel obtained using torsion testing at a strain rate of $0.3\,s^{-1}$.

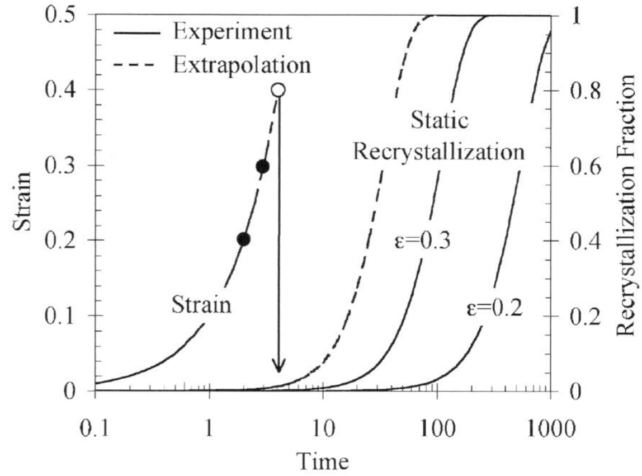

Figure 4
Schematic of the influence of strain on the kinetics of static recrystallization. Dynamic recrystallization will occur for strains greater than 0.4, which corresponds to the point where the deformation time matches the predicted static recrystallization start time (assuming additivity).

parameter or temperature-corrected strain rate, $Z=\dot{\varepsilon}\exp(Q_{def}/RT)$, according to a power law relationship. (For a constant value of Z, the steady-state flow stress during hot working is also constant.)

2.4 *Recrystallization*

The internally stored energy of deformation, which is present in the dislocation structure, can be relieved by the nucleation and growth of new strain-free grains. This process is termed recrystallization and establishes, once complete, a new grain size (and crystallographic texture). According to its definition it takes place following deformation; either between passes of hot working or during the cooling of the work piece to room temperature. As will be discussed below, it can also occur during deformation of austenite.

The rate of recrystallization is a function of the process parameters and increases with increasing strain, temperature, and strain rate. In addition, the grain size before deformation affects the rate of recrystallization with smaller grain sizes leading to higher rates. The recrystallization behavior can be modeled using a simple Avrami type equation to relate the volume recrystallized, X, to the time, t, with the above effects incorporated through a time constant, t_{50} (the time to 50% recrystallization):

$$X = 1 - \exp\left(-0.693\left(\frac{t}{t_{50}}\right)^{b}\right) \quad (5)$$

$$t_{50} = C_4\varepsilon^{-c}d_o^{d}\exp(Q_{rx}/RT) \quad (6)$$

where ε is the strain, d_o is the initial grain size and C_4, b, c, d, and Q_{rx} are constants. Typical values for C–Mn steels with <0.1%V and <0.02%Ti are: $b=2$, $C_4=2.3\times10^{-15}$, $c=2.5$, $d=2$ and $Q_{rx}=230\,kJ\,mol^{-1}$.

One would expect that for conditions leading to very high rates of recrystallization, it would be possible for recrystallization to commence prior to the end of deformation (see Fig. 4). Indeed, this view provides a simple and useful perspective of a process termed (discontinuous) dynamic recrystallization. For a given temperature and strain rate, there exits a critical strain for the onset of dynamic recrystallization. This critical strain is often expressed empirically as a function of the temperature, strain rate, and initial grain size. It can also be predicted quite accurately using the above equations for static recrystallization and assuming that it corresponds to the strain at which 1% "static" recrystallization is reached during the test.

Because recrystallization leads to a lower dislocation density it reduces the flow stress. Work hardening occurs initially but this is halted at a peak flow stress due to the onset of dynamic recrystallization. Dynamic recrystallization is often seen to occur at strains between 0.6 to 0.8 of the strain to the peak stress. The peak stress is frequently followed by a reduction in flow stress until a steady-state value is reached (Fig. 3).

A significant feature of dynamic recrystallization is that it leads to the establishment of a grain size that is uniquely determined by the temperature and strain rate of deformation. That is, it is independent of strain and initial grain size. The dynamically recrystallized grain size for C–Mn steels can be given as:

$$d_{drx} = 1.6 \times 10^4 Z^{-0.23} \quad (7)$$

Interestingly, the grain size after meta-dynamic recrystallization, which is the name given to the static recrystallization process that takes place following deformation in the dynamic recrystallization regime, is approximately 50% coarser than the dynamic recrystallized grain size, irrespective of the deformation conditions (the exponent for Z is the same as for Eqn. (7)).

2.5 Precipitation

As discussed in Sect.2.1, many of the precipitates in steel are dissolved during the reheating step prior to rolling. However, because the temperature drops during and following processing there is a thermodynamic driving force for reprecipitation. As mentioned above, this is important for the direct effects of precipitates on strength and solute level. The kinetics of precipitation are often analyzed using isothermal or continuous cooling transformation "C" curves which are a plot of fraction precipitated against time. These plots show that for austenite the kinetics of precipitation of the various species are generally fairly slow compared with the time of processing, and a number of species do not form precipitates until after the austenite to ferrite transformation (see below). Common examples of this are the precipitation of AlN in aluminum-killed steels, TiC in interstitial free grades and VN in high nitrogen (>100 ppm) grades. It is important to note that the strengthening effect of precipitates in the final product is dramatically increased when they form in the ferrite, due to coherency between the precipitate and the matrix structures.

However, other precipitating species, particularly the carbonitrides of Nb, Ti, and V, are able to precipitate in the austenite under the influence of the deformation. This is termed strain-induced precipitation and results from nucleation at deformation-induced defects such as dislocation structures. One empirical model for the start of strain-induced precipitation of Nb(CN) is (Dutta and Sellars 1987):

$$t = 3 \times 10^{-6}[\mathrm{Nb}]^{-1}\varepsilon^{-1}Z^{-0.5}\exp\left(\frac{2.7 \times 10^5}{RT}\right) \times\exp\left(\frac{2.5 \times 10^{10}}{T^3(\ln k_s)^2}\right) \quad (8)$$

Under certain conditions, precipitation of Nb(CN) can retard and even prevent static recrystallization. For this reason niobium is a commonly used alloying addition in high strength steels. (Titanium and vanadium additions are also used but niobium is most effective, possibly due to its strong influence on grain boundaries while still in solution).

It is of particular interest to quantify the conditions under which precipitation and recrystallization interact. Figure 5 shows the recrystallization curves for a 0.03Nb and a C–Mn steel superimposed on the precipitation start curve, all for the same deformation conditions. The intersection between the precipitation and recrystallization curves gives the temperature below which recrystallization is unlikely to occur for a typical rolling schedule with that pass strain (the recrystallization stop temperature). It can be seen that the recrystallization curve for the niobium steel is displaced towards higher temperatures, compared with the plain carbon grade even prior to the onset of precipitation. This is a result of solute drag. If solute drag had not occurred then the recrystallization stop temperature would be much lower (as given by the intersection with the C–Mn line) and it would not be industrially useful.

It is also important to note that precipitation can interact with recrystallization during annealing after hot-worked material has been cooled to room temperature and cold rolled. In this regard, the precipitation of AlN following hot rolling of steel sheet is important. If the cooling after rolling is carried out at a sufficiently slow rate all the AlN will precipitate prior to cold rolling. Under these conditions it turns out that the AlN precipitates do not interfere with recrystallization any where near the extent to which they do if the aluminum and nitrogen are retained in solution after hot rolling and precipitate out during annealing.

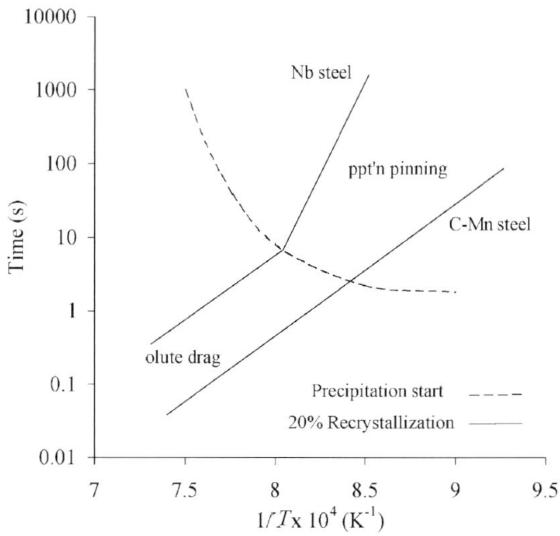

Figure 5
Influence of temperature and time on precipitation of Nb(C,N) and the point of 20% static recrystallization for a Nb grade and a plain C grade.

2.6 Transformation

The final microstructural change in hot deformation processing of steel is the transformation from austenite to the room temperature microstructure. This microstructure may consist of various proportions and morphologies of ferrite, bainite, martensite, pearlite, and cementite. In many cases the microstructure formed during cooling after hot deformation follows the same behavior as those formed during heat treatment. Here, however, there is a strong influence of deformation, which exerts itself primarily on the evolution of ferrite microstructures.

The transformation from austenite to ferrite during cooling occurs by a diffusional process by which ferrite nucleates predominantly on the austenite grain boundaries and on other microstructural heterogeneities such as twins and deformation bands. The ferrite grains then grow until they converge or until the equilibrium fraction is reached. Due to the preferential nucleation of ferrite at particular microstructural sites, a key way to refine the grain size is to increase the density of sites. Consequently, refining the austenite grain size and increasing the amount of retained strain decreases the ferrite grain size. An example of enhanced intragranular nucleation at twins and microbands due deformation is illustrated in Fig. 6.

Increasing the cooling rate also decreases the ferrite grain size by both increasing the number of nuclei due to increased undercooling and also by decreasing the growth rate. At high cooling rates it is also possible to favor intragranular nucleation of ferrite. However, if the cooling rate is too high bainite or martensite will form. One way to promote the intragranular nucleation without forming other phases is to incorporate a period of slow cooling (e.g., by coiling the steel) after an initial period of rapid cooling (e.g., water sprays).

For continuous cooling of a C–Mn steel, the following empirical equation describes the influence of austenite grain size, d_γ, and cooling rate, $\dot{T}$, on the final grain size, $d_{\alpha 0}$:

$$d_{\alpha 0} = C_5 + C_6\dot{T}^{-1/2} + 22(1 - e^{0.015d_\gamma}) \quad (9)$$

where $C_5 = -0.4 + 6.4(\%C + \%Mn/6)$ and $C_6 = 24.2 - 59(\%C + \%Mn/6)$. This expression can be modified to account for retained strain, ε, according to the following equation:

$$d_\alpha = d_{\alpha 0}(1 - 0.45\sqrt{\varepsilon}) \quad (10)$$

This last equation shows that saturation in the level of ferrite refinement is reached by increasing the deformation (in the nonrecrystallization region). It has also been shown that there is a limit to the refinement that can be achieved by increasing the cooling rate. For most structural steels this leads to an apparent plateau in achievable lower ferrite grain sizes of approximately 5 μm.

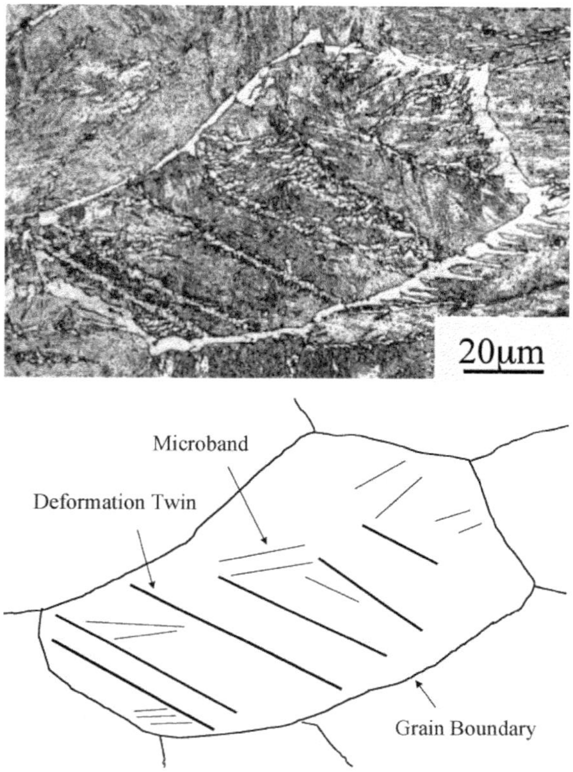

Figure 6
Micrograph of a plain carbon steel quenched within 0.05 s of the end of deformation following torsion testing. It is evident that nucleation of ferrite has occurred inside the austenite grains. Both twins and microbands are sites for nucleation.

However, there are some alternative processing methods to produce much finer grain sizes, approaching 1 μm. One method (Hodgson *et al.* 1998) involves using the opposite logic from that described above. Here the initial austenite microstructure is coarsened prior to rolling. This allows high undercooling at low cooling rates to be achieved. A single deformation step is then employed in which high surface shears are encouraged. This is followed by cooling in air. The final product displays surface layers of ultrafine grains, with a core microstructure similar to that expected in conventional hot rolling.

3. Conclusion

The microstructure evolution during the hot working and cooling of steels is complex and highly dependent on the deformation, temperature, and time history.

There is also a great deal of interaction between the various microstructural events. However, even with this high level of complexity it is still possible to design and implement industrial-scale thermomechanical processes, where the microstructure evolution is manipulated to produce the desired final properties without subsequent heat treatment.

See also: Primary Recrystallization

Bibliography

Barnett M R, Kelly G L, Hodgson P D 2000 Predicting the critical strain for dynamic recrystallization using the kinetics of static recrystallization. *Scr. Mater.* **43**, 365–9

Dutta B, Sellars C M 1987 Effect of composition and process variables on Nb(C,N) precipitation in niobium microalloyed austenite. *Mater. Sci. Technol.* **3**, 197

Gladman T 1997 *The Physical Metallurgy of Microalloyed Steels*. Institute of Materials, London

Hodgson P D, Gibbs R K 1992 A mathematical model to predict the final properties of hot rolled C-Mn and microalloyed steels. *ISIJ Int.* **32**, 1329–38

Hodgson P D, Hickson M R, Gibbs R K 1998 The production and mechanical properties of ultra fine servite. *Mater. Sci. Forum* **284–6**, 63–72

M. R. Barnett and P. D. Hodgson

Metal Forming (Warm): Comparison with Hot and Cold Forming

Warm forming of steel takes place in a temperature range between 600°C and 900°C. A relevant factor for deciding to work in this temperature range is the comparatively low flow stress as well as the great formability potential.

Figure 1 shows the flow stress k_f as a function of temperature for various steels. The decrease of flow stress with increasing temperature is to be explained by the higher mobility of the dislocations as well as by the recovery and recrystallization processes increasingly occurring in parallel.

Figure 2 documents the limiting strain $\varphi_{Gr.}$ (or formability) as well as oxide formation in terms of layer thickness as functions of temperature. The lower limit of warm forming is at approximately 600°C, since it is only from that point up that a significant increase of limiting strain φ_{Gr} can be obtained. The upper limit of warm forming is in the vicinity of 900°C. A beneficial factor is the rise of formability having taken place up to this temperature. A disadvantage is the increasingly occurring oxide formation.

The choice of temperature within these limits is influenced by various factors. Decisive among these are workpiece shape, workpiece material, true strain, the required accuracy, and the workpiece characteristics requested.

Figure 3 shows the influence of forming speed on flow stress for steel grade 100Cr6. At higher forming speeds, flow stress in the range of phase transition (~720–750°C) is higher by a factor of 2–3 than at low forming speeds. This is an indication of the fact that recrystallization, which takes some time, has a positive impact at lower forming speeds.

1. Shape Spectrum and Component Characteristics

The shape spectrum that can be produced by warm forming corresponds largely to that of cold forming. This primarily includes axially and rotationally symmetrical components. In addition, there are also

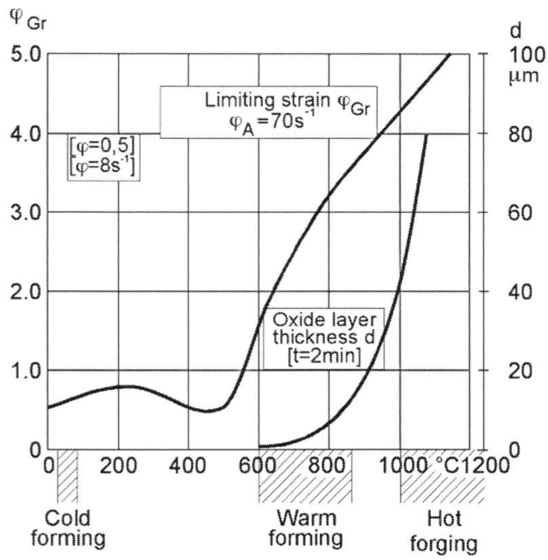

Figure 2
Influence of temperature on limiting strain and oxide formation.

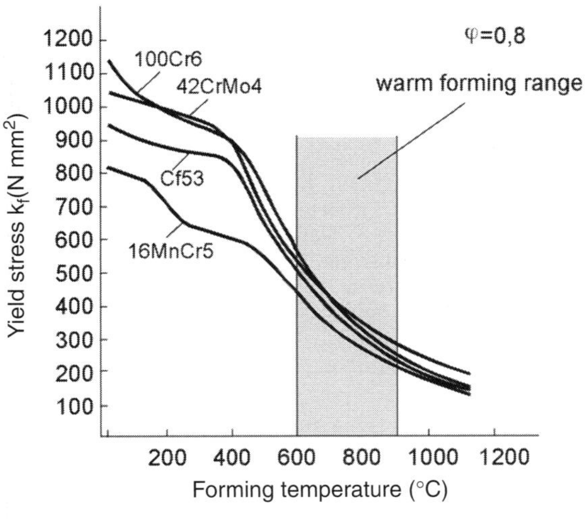

Figure 1
Flow curves of steels 100Cr6, 42CrMo4, Cf53, and 16 MnCr5.

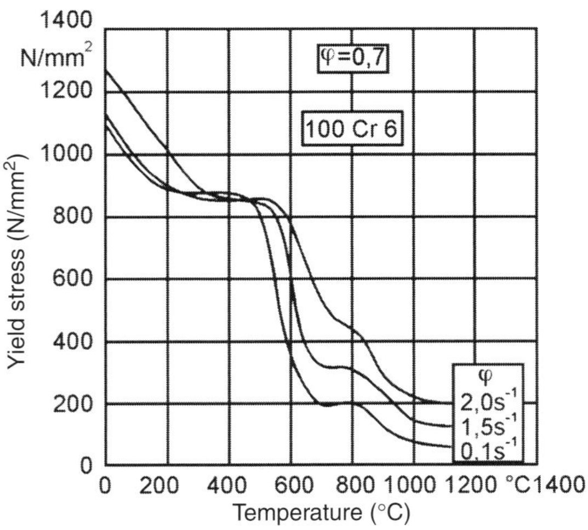

Figure 3
Influence of forming speed on yield stress.

components equipped with gears and gear-like profiles. This is due to the fact that while flow stress decreases with increasing temperature, for higher-alloy steels it still reaches values as they occur in cold forming low-alloy steels.

The component materials used in warm forming are non-alloyed or low-alloy case hardening steels. An advantage is that in addition to these also austenitic stainless steels as well as steels with relatively high carbon percentage can be worked.

The tool technology used in warm forming orients itself largely at tool constructions for cold forming. This means application of female dies equipped with shrink rings as well as split tool constructions at flutes or failure-prone points.

The achievable dimensional and shape accuracies correspond to ISO qualities IT 9–IT 12. They are thus comprised between the accuracies achievable by cold extruding and drop forging. The surface quality of the components is determined by temperature and the lubricant used and is usually comprised $R_t = 10–50\,\mu m$. Further advantages of the process are high material utilization of 85% and the comparatively low energy input of $41\,J\,kg^{-1}$ per finished part. The mechanical characteristics of warm-formed components differ only to a minor degree from those of components heat-treated after cold forming. If no particular demands are made as to the homogeneity of strength distribution within the components, an additional heat treatment can thus be dispensed with.

2. Delimitation as Against Cold and Hot Forming

Hot forming of steel is consecutive to warm forming in the temperature range above 900°C. Here, drop forging is the most common hot forming process. It permits producing components of almost any shape in large quantities and in any grade of steel. However, the intense oxide formation in the temperature range above 900°C deteriorates surface quality ($R_t < 100\,\mu m$) which involves wider dimensional tolerances (IT 12–IT 16) and greater amounts of excess material to be removed by metal cutting operations on functional surfaces. In drop forging, material utilization is between 60% and 80%. The average energy input per kg of finished parts is $46–49\,J\,kg^{-1}$.

The processes of cold forming finds application in the range of room temperature. The most important process is cold extruding. The process limitations are determined by the potential for deformation and by the choice of materials. The spectrum of shapes is largely restricted to rotationally symmetrical components, the choice of materials being limited to non-alloyed or low-alloy structural steels ($C < 0.5\%$). Outstanding advantages of this process are high surface quality ($R_t = 10\,\mu m$) and good dimensional and shape accuracy (IT 7–IT 11). Another advantage is high material utilization from 85% to 90%. The average energy input per kg of finished parts is $41\,J\,kg^{-1}$.

3. Process Chain in Warm Forming

The production sequence in warm forming corresponds largely to that in hot forming. It starts with producing bar sections by shearing. For the coating process, the blanks are preheated in an induction furnace and then are provided with a graphite coat by immersion. After further heating of the blanks to the forming temperature, the forming operation takes place in a press. Depending on further use, the components then are either sand blasted and calibrated or subjected to another heat treatment.

4. Concluding Remarks

The advantage of warm forming in steel is the relatively low flow stress as well as the great formability potential. The shape spectrum corresponds largely to that of cold forming. In addition, there are also components equipped with gears and gear-like profiles. The dimensional and shape accuracies correspond between the accuracies achievable by cold extruding and drop forging.

Bibliography

Herlan Th 1999 Warm forging of straight tooth bevels for the utility vehicle's production. *Adv. Technol. Plast.* **2**, 767–78

Hirschvogel M 1990 Möglichkeiten und Grenzen der Halbwarmumformung. In: *Proc. XIII. Int. Forging Congress*. New Delhi, India

Hirschvogel M 1997 In: *Potentiale der Halbwarmumformung—ein Verfahrensvergleich.* [Potential of warm forming—a comparison of processing methods] Tagungsband 6. Umformtechnisches Kolloquium. Meisenbach-Verlag, Darmstadt, Germany, pp. S195–205

Nagahama T, Enomae S 1992 Cold- and warm-forging press developments and applications. *J. Mater. Proc. Technol.* **35**, 415–27

D. Schmoeckel

Metal Forming at High Energy and High Velocity

In forming operations the maximum strain rate can increase up to $10^3\ s^{-1}$ and the temperature up to the melting point. Table 1 (Alexander 1985) gives an overview of the strain rate of different manufacturing processes. In this table, the highest strain rates are possible in machining. In precision cutting, however, strain rates up to $10^7\ s^{-1}$ are possible depending on the conditions. In all of these processes the strains are moderate. The text will be expanded to high strain and strain rate in plastic deformation in general including machining—this is also a plastic deformation process just like forming—instead of only on forming. This expansion widens considerably the extreme conditions.

A process that combines high strain with high strain rate and temperature occurs in the contact zone of tool and work piece under dry sliding conditions. That holds for forming as well as machining operations. In this context two different dry sliding operations are shown and discussed. They are chosen because of the accessibility of the dry sliding contact zone and the means to estimate the mechanical properties. For this area of high strains and strain rates little or no attention is paid. No tests are developed and for this additional reason two examples are described and discussed. The first is a pin–ring test. The second is the cutting process. Based on the assumption that dry-sliding friction takes place by plastic deformation, analysis of both processes reveals that the strain and strain rate in these zones are much higher than those in the bulk metal after normal mechanical working operations. These zones have very different mechanical and chemical properties compared with materials deformed under normal forming conditions. It is caused by the extreme temperature, strain, and strain rate. Some estimations are made and supported by microscopic observations. These observations show very small grains ($\ll 1\ \mu m$), which are much lower than those found in the bulk after usable forming operations.

Table 1
A typical range of strain rates in manufacturing processes.

Manufacturing process	Strain rule (s^{-1})
Hot forging	100–200
Hot extrusion	10–20
Cold forging and extrusion	10–100
Warm forging	10–100
Creep forming	10^{-4}
Sheet metal forming	10–100
Deposition forming	200–1000
Machining, shearing	>1000
Explosive forming	100–1000
Hot rolling	40–100
Cold rolling	100–1000

It is well known in production engineering that the extent of metal deformation depends on the mechanical operation. The maximum deformation, for instance, in a tensile bending operation is much lower than that in extrusion or wire drawing. In classical material tests, it is known that the maximum deformation in a tensile test is lower than that in a torsion test. Important for this behavior is the hydrostatic stress component ($=\sigma_H$) of the stress field (σ_{ii}), which is defined as:

$$\sigma_H = \tfrac{1}{3}\sigma_{ii} \qquad (1)$$

A high negative stress means that the process takes place under high compressive stresses. This makes it possible for the metal to undergo high deformation without failure. In practice this ideal stress situation occurs very often in mechanical operations in the contact zone between the tool and work piece material during dry-sliding friction.

As mentioned above, two processes will be discussed concerning the friction process between tool and chip in cutting and between pin and ring in dry sliding. For both processes, it will be assumed that the dry-sliding friction takes place by plastic deformation of one of the components.

1. *Models*

1.1 The Deformation of the Chip–Tool Contact Zone in Cutting of Steel SAE 1045

Figure 1 gives a schematic representation of this process. If F_W is the friction force between tool and chip, v_c the chip speed, and dV the volume of material that undergoes the high deformation of the friction process per unit time, the following holds:

$$F_W \times v_c = \int \bar{\sigma} \times d\bar{\varepsilon} \times dV \qquad (2)$$

with

$$\bar{\sigma}^2 = (\sigma_{11}-\sigma_{22})^2 + (\sigma_{22}-\sigma_{33})^2 + (\sigma_{33}-\sigma_{11})^2 \qquad (3)$$

$$d\bar{\varepsilon}^2 = \tfrac{2}{3}(d\bar{\varepsilon}_{11}^2 + d\bar{\varepsilon}_{22}^2 + d\bar{\varepsilon}_{33}^2) \qquad (4)$$

in which $\bar{\sigma}$ is the effective stress and $d\bar{\varepsilon}$ is the effective incremental strain with $d\bar{\varepsilon}_{ii}$ the incremental strain tensor. For $\bar{\sigma}$ = constant, Eqn. (2) can be rewritten:

$$\bar{\varepsilon} = \frac{F_W \times v_c}{dV \times \bar{\sigma}} \qquad (5)$$

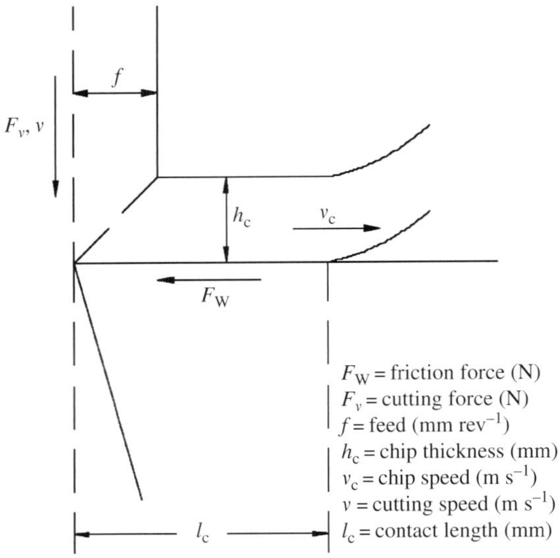

Figure 1
Schematic representation of the cutting process.

If l_c is the contact length between chip and tool, the average effective deformation rate $\dot{\bar{\varepsilon}}$ can be estimated as:

$$\dot{\bar{\varepsilon}} \approx \frac{\bar{\varepsilon} v_c}{l_c} \tag{6}$$

1.2 The Deformation in a Pin–Ring Contact of a Dry-sliding Test

Figure 2 shows a schematic diagram of a pin–ring test set-up. The copper pin (6 × 6 mm) is pressed with a normal force ($=F_N$) against a steel SAE 1045 ring. The ring rotates with a peripheral velocity of v m s^{-1}. The wear of the copper ring (ΔV_p) is the volume per unit time that is transported from the pin to the ring and which adheres to the back of the pin indicated by B. This transport of the pin material takes place by the friction of the pin against the ring ($=F_W$) during the rotation. If in the stationary condition the friction energy is completely consumed by the plastic deformation of the pin, it holds that:

$$F_W \times v = \int \bar{\sigma} \times d\bar{\varepsilon} \times \Delta V_P \tag{7}$$

For $\bar{\sigma}$ = constant, Eqn. (7) gives:

$$\bar{\varepsilon} = \frac{F_W v}{\Delta V_p \bar{\sigma}} \tag{8}$$

The strain rate of the copper material can be approximated by:

$$\dot{\bar{\varepsilon}} \approx \frac{\bar{\varepsilon} v}{l_P} \tag{9}$$

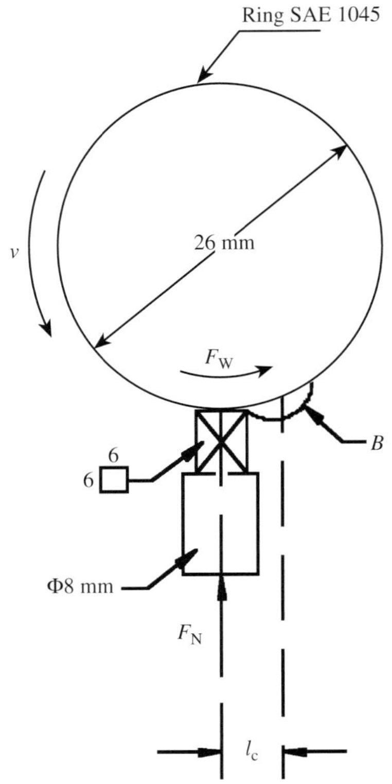

Figure 2
Schematic representation of the dry-sliding process of a copper pin against a steel SAE 1045 ring.

where l_p is the average distance between the middle of the pin and the center of the worn material denoted by B.

1.3 Deformation and Material Displacement

There is no way to determine accurately the high strain in dry-sliding friction. However, it can be estimated as follows. If it is assumed that the deformation takes place by pure shear, a square dxdy (OABC) may deform to OAB′C′ as shown in Fig. 3. The effective strain ($=\bar{\varepsilon}$) can be calculated as:

$$\bar{\varepsilon} = \frac{\tan \gamma}{\sqrt{3}} \tag{10}$$

in which γ is the shear angle. With Eqn. (10) the displacement ($=\Delta x$) of the material in the x-direction is given by:

$$\Delta x = dy \times \tan \gamma = dy \times \bar{\varepsilon} \times \sqrt{3} \tag{11}$$

If the displacement field is known, Eqn. (11) allows us to determine the strain field.

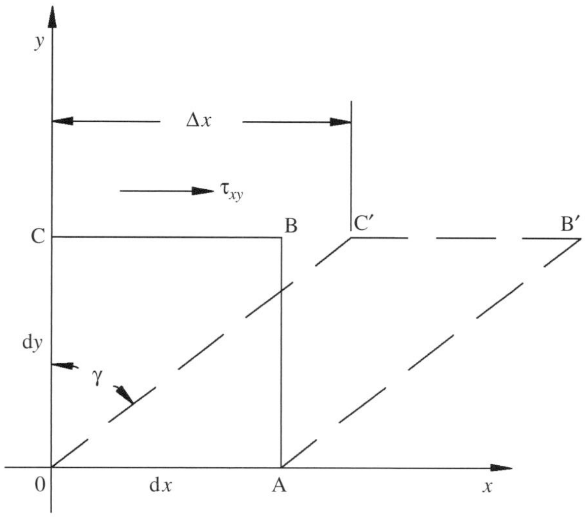

Figure 3
Deformation of a square by a shear stress τ_{xy}.

2. *Results*

2.1 *Machining*

Cutting steel SAE 1045 with a cutting speed of $2\,\mathrm{m\,s^{-1}}$, a feed of $0.2\,\mathrm{mm\,rev^{-1}}$, and a width of cut of 1 mm gives a friction force $F_W = 349\,\mathrm{N}$, a chip speed of $0.8\,\mathrm{m\,s^{-1}}$, and a chip thickness of 0.5 mm. The temperature of the chip is around 400°C. The flow stress at this temperature measured by the Split–Hopkinson bar for a strain rate of $7.5 \times 10^3\,\mathrm{s^{-1}}$ is $800\,\mathrm{N\,mm^{-2}}$ and nearly independent of the strain. The volume dV can be derived from Fig. 4. It shows a micrograph of the instantaneous cutting process made by a quick-stop device. The depth of the deformation field can be estimated at 20 μm with the assumption that the start is identifiable by the bending of the grain boundaries. Using Eqn. (5), one obtains an average effective deformation of 21.8. For these cutting conditions, l_c is 1.4 mm. Entering these values into Eqn. (6) gives $\bar{\dot{\varepsilon}} = 1.2 \times 10^4\,\mathrm{s^{-1}}$.

2.2 *Dry Sliding*

For a normal force of 40 N and a peripheral velocity of $2\,\mathrm{m\,s^{-1}}$, the friction force is 42.8 N and the worn volume is $267 \times 10^{-3}\,\mathrm{mm^3\,s^{-1}}$. If we assume a constant flow stress for copper, meaning it is independent of the strain, of $1000\,\mathrm{N\,mm^{-2}}$ (Hughes *et al.* 1995), we can calculate that $\bar{\varepsilon} = 321$.

For a contact length of the worn material with the ring in the sliding direction of 3 mm, the average distance between pin and worn material B ($= l_p$) is 4.5 mm. With Eqn. (9) a first approximation of the strain rate of $1.4 \times 10^5\,\mathrm{s^{-1}}$ can be calculated.

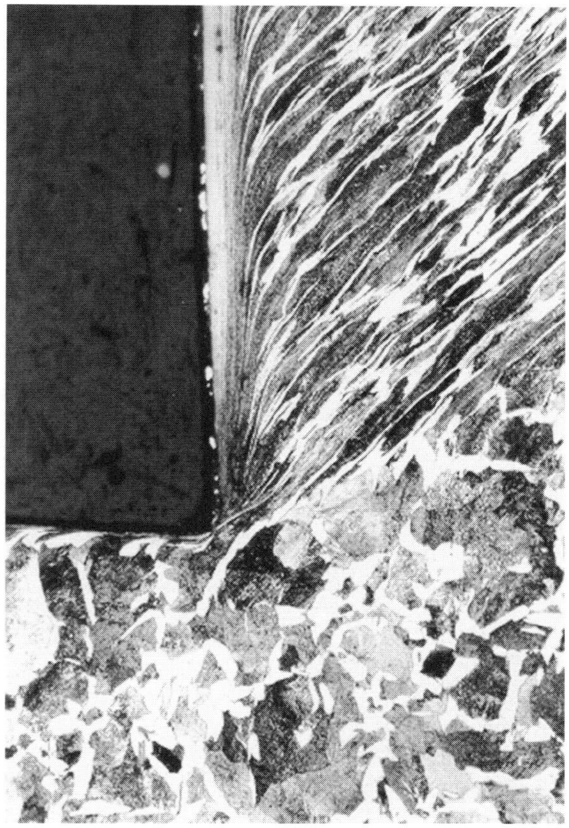

Figure 4
The instantaneous cutting process made by a quick-stop device. The etched structure of SAE 1045 shows the bending of the grain boundaries near the contact surface of tool and chip.

2.3 *Deformation and Material Displacement*

Figure 4 is a cross-section made in the plane of cutting and is a rotated mirror image of Fig. 1. It shows an etched structure of SAE 1045. In the contact zone of the chip and tool a white band is visible. Between this white band and the uniform deformed material of the chip, the grain boundaries of the ferrite and cementite are bent in the direction of the tool tip. The sharp bending of the grain boundaries over a short distance means that the deformation is high. Micrographs made by the scanning microscope reveal that near the contact surface the striations are almost parallel to the surface, which results in very high deformations as concerned by Eqn. (11).

Figure 5 shows the displacement field in the copper pin after dry-sliding friction against steel. It is a cross-section of the pin perpendicular to the contact surface of pin and disk and parallel to the friction force. Near to the contact surface the striations are

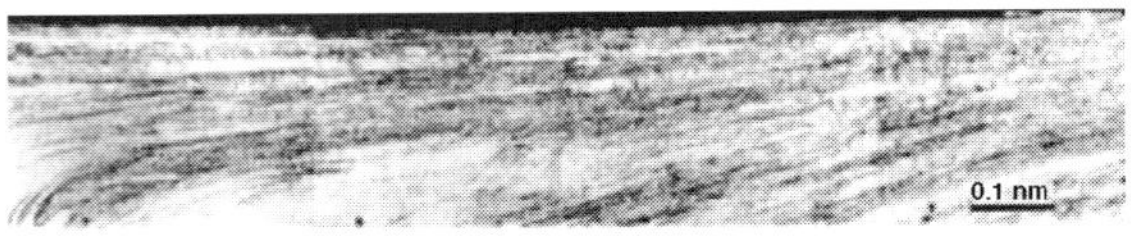

Figure 5
Photograph of the bending of grain boundaries in a worn copper pin taken perpendicular to the contact surface and parallel to the friction force.

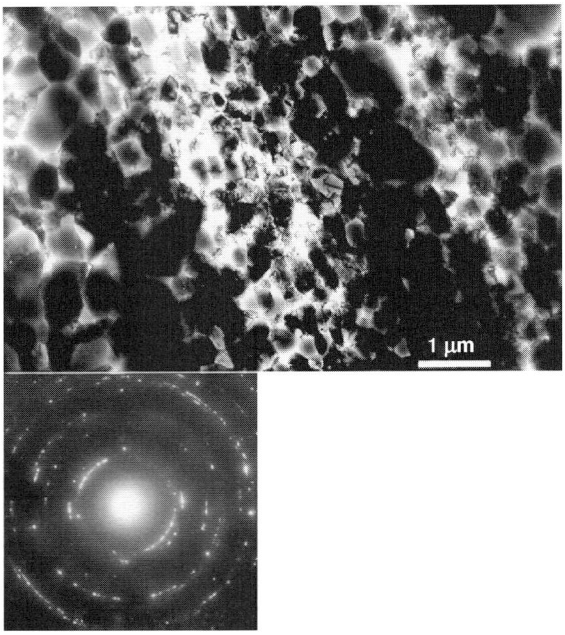

Figure 6
Transmission electron micrograph of the real contact surface of a worn copper pin together with the diffraction pattern (area $3\,\mu m^2$).

nearly parallel to the surface, which means that the material is heavily deformed. A displacement of 1 mm over a thickness of 0.03 mm is common, following Eqn. (11) this means an effective deformation of 19. Figure 6 shows a transmission electron micrograph of a section of the copper pin in the contact zone parallel to the contact zone. It is taken at a place where the strains are relatively low. The different grain sizes are caused by the different wear tracks on the surface. It is clear that the smallest grains are very small, being in the order of 0.2 µm. This is supported by the diffraction pattern, which was taken from an area of $3\,\mu m^2$.

3. *Remarks*

To gain an idea about the level of effective deformation, these processes may be compared with rolling of a strip with an initial thickness of 500 mm and an end thickness of 5 µm. Because the thickness reduction is compensated by the length of the strip and the width is unchanged during the process, we can calculate that the effective deformation is 13.3 after integration of Eqn. (4).

For the calculation of the strain in the contact surface of tool and chip in cutting, we assumed that the strain in the whole contact area was constant (=21.8). In reality it grows from a low value like three to an end value. Consequently the strain in the contact layer has to be 40.6 for a linear distribution.

Figure 6 shows a structure that has been recrystallized. The small grain size, which is abnormal for copper, is obviously caused by the combination of high strain, strain rate, and temperature. This small grain size results in a technical metal with excellent mechanical properties: it has a much higher flow stress and failure strain.

The deformation process in dry sliding can be divided into steps. This results in deformation followed by recrystallization. Afterwards it is difficult to determine how often it has happened. This hampers the determination of the strain of worn material.

The fine structure after recrystallization can explain the high hardness of these layers. Beyond that these layers often have different properties from the bulk material. During etching these layers do not always react. The layer is microscopically visible as a so-called "white layer" or "Beilby layer."

In the literature it is shown that the same phenomenon as described and shown here for two processes and two metals are the same for a number of processes and metals. However more information is necessary to support the idea that dry-sliding friction takes place by plastic deformation of one or both components.

The assumption that dry-sliding friction is a plastic process supports the notion of the transformation of mechanical energy into heat. In the past, too little attention has been paid to this conversion.

4. *Conclusions*

A dry-sliding friction process—high strain, high strain rate, and high temperature—results in a metal with small grain thickness. This normally leads to excellent mechanical properties: a higher flow stress or hardness and a higher failure strain. A sheet thickness of 1 µm with a grain size of 0.2 µm or lower behaves like bulk material in a forming operation. These metals are apt for very small parts in the electronic industry.

Bibliography

Alexander J M 1985 On problems of plastic flow of metals. In: Sawczuk A, Bianchi G (eds.) *Plasticity Today*. Elsevier Applied Science, London

Dautzenberg J H, Zaat J H 1973 Quantitative determination of deformation by sliding wear. *Wear* **23**, 9–19

Dautzenberg J H, van Dijck J A B, Kals J A G 1989 Metal structures by friction in mechanical working processes. *Annals of the CIRP* **38**, 567–70

Hughes D A, Dawson D B, Korellis J S, Weingarten L I 1995 A microstructurally based method for stress estimates. *Wear* **181–3**, 458–68

Rigney D A 1997 Microstructural evolution during sliding. In: *Proc. Materials Solutions '97 on Wear of Engineering Materials*. Indianapolis, IN, USA

J. H. Dautzenberg

Metal Forming at Very Low Strain Rates

A decrease in strain rate during isostructural isothermal forming causes (i) a decrease in flow stress (except when dynamic strain aging or precipitation is present or when stress increase owing to a decrease in adiabatic temperature rise dominates over the stress decrease due to a lowering of strain rate); (ii) an increase in ductility (except where a minimum strain rate is needed for the onset of phenomena like superplasticity, i.e., achieving extreme near-homogeneous elongation in tension of several hundred percent before failure, when maximum formability is present at intermediate strain rates); and (iii) an increase in the friction coefficient if the lubricant film does not break during deformation.

Many microstructural features, e.g., average grain size, grain shape, grain size and shape distributions, preferred orientation (texture), misorientation between grains/grain boundary character distribution, dislocation density, distribution of defects, and compositional inhomogeneities (segregation), strongly influence the mechanical response of a material. The average grain size is the most commonly used microstructural parameter.

1. *Stability of Flow*

At low homologous temperatures ($T/T_m \leqslant 0.3$–0.4, where T is the deformation temperature and T_m is the melting point of the material, both on the absolute scale) strain hardening strongly influences flow. With increasing temperature, strain rate hardening also becomes important. If microstructure is relatively stable (negligible grain growth), strain hardening is insignificant at high homologous temperatures. At intermediate temperatures (warm working) and when grain growth is present at high temperatures, both strain and strain rate hardening are significant.

Instabilities in plastic flow are classified based on strain, strain rate, and/or temperature effects. Uniaxial tensile/compressive flow at constant temperature and grain size is often analyzed using the power law

$$\sigma = K_1 \varepsilon^n \dot{\varepsilon}^m \tag{1}$$

where σ is the true stress, ε the true strain, $\dot{\varepsilon}$ the true strain rate ($=\partial\varepsilon/\partial t$, where $t=$ time; $\dot{\varepsilon}=v/l$, where v is the velocity of cross-head movement and l is the specimen length), n the strain hardening exponent, m the strain rate sensitivity index, and K_1 a temperature-and grain size-dependent material constant. When $m \sim 0$ (low temperatures), the region of stable flow is defined by the condition $\partial\sigma/\partial\varepsilon \geqslant \sigma$ and the limit of uniform flow is reached when $\varepsilon = n$. (This relationship is not obeyed by materials prone to dynamic strain aging.) As for most materials $n=0.1$–0.3, the extent of formability at low homologous temperatures is limited. When strain rate sensitivity is also present, the region of stable flow is defined by the condition $\partial\dot{A}/\partial A \leqslant 0$, where A is the cross-sectional area of a tensile specimen and the overdot denotes a time derivative. Then

$$\frac{n}{\varepsilon} + m \geqslant 1 \tag{2}$$

defines the condition for stable flow.

The grain size dependence of flow stress, σ, is given by the Hall–Petch relationship $\sigma \propto L^{-0.5}$, where L is the grain size. At high temperatures, $\sigma \propto L^a$, where a often lies in the range 0.7–1.2. (The region of transition from an inverse to a direct dependence of flow stress on grain size may be regarded as the boundary between cold and hot working.) While considering stability of flow, structural changes and damage accumulation are taken into account by introducing internal variables in the constitutive equations that are consistent with special kinetic equations. The results depend significantly on the constitutive equation chosen.

2. *Superplasticity*

For a constant load test ($P=$ constant) at high temperature ($n \sim 0$) the time evolution of a local nonuniformity in specimen cross-section δA is given by

$$\delta A = \left(\frac{A_0}{A}\right)^{\frac{1}{m-1}} \delta A_0 \tag{3}$$

where A_0 is the original specimen cross-section and δA_0 is the initial local variation in cross-section introduced, for example, during machining. The above analysis is valid for all situations where $m \neq 0$ and so it is useful for analyzing both conventional and superplastic flow. Equation (3) reveals the very strong dependence on m which explains why large ductility can be obtained even beyond the point of instability in materials for which m is large. Taking a realistic value for δA_0 it is easy to show that large elongations will result if $m \geqslant 0.3$. In this definition the extreme elongation obtainable in environmental superplasticity (applying a small stress in the range of solid-state transformation or along the direction of maximum coefficient of linear expansion of a material with anisotropic expansion properties, deforming a neutron-irradiated material), structural superplasticity (grain size less than 10–15 μm), deformation of hot glass, and heat-softened polymers and metallic glasses will all be viewed as subgroups of the general phenomenon of superplasticity.

A few empirical equations are available that connect the elongation at failure in a tensile test with m.

The equations predict the right trend but the fit is poor. Theoretical analysis of this problem is extremely complicated owing to finite strain behavior and dynamical effects. The *m* value is a complex function (yet to be quantified) of strain rate, temperature, grain size, strain, and prior treatments. As this is the most important practical parameter that determines ductility, its careful experimental evaluation (often in a strain rate jump test) is essential.

The effect of composition (microstructural stability) on formability should be considered along with the effect of temperature (as T_m is composition dependent). Processing history as well as volume, proportion, and distribution of phases affect ductility.

Nonuniform and transient flow of practical interest is mostly ignored in laboratory tests and analyses. The effects of temperature are not fully understood because of microstructural evolution during deformation. The stress state (e.g., tension, compression, or torsion) as well as the presence of a hydrostatic pressure (even as low as 0.3 times the flow stress) significantly shifts the optimal range of superplastic flow. Thus, simple equations like Eqn. (1) and experimental data based only on uniaxial testing are not sufficient to understand superplastic deformation of practical interest. Unlike in conventional plasticity (von Mises solid) where the yield criterion depends on the second stress tensor invariant, for superplastic flow possibly the first stress tensor invariant (hydrostatic part) will also have to be considered. This is yet to be done.

Superplasticity-like phenomena are seen over a wide strain rate range—from flow of geological materials ($\dot{\varepsilon} \sim 10^{-13}s^{-1}$) to diffusional creep ($m = 1$ but superplasticity is yet to be established) to conventional structural and internal stress (environmental) superplasticity ($\dot{\varepsilon} \sim 10^{-6}$–$10^{-2}s^{-1}$). *High strain rate superplasticity* has been established in metallic materials (aluminum and titanium alloys), ceramics, metal matrix composites, and advanced materials like metal ceramic composites and mechanical alloys ($\dot{\varepsilon} \sim 10^{-2}$–$10s^{-1}$). Here the grain size is reduced to below 2 μm (often into the submicrometer range) using adequate techniques like equal channel angular (ECA) pressing, powder metallurgy, physical vapor deposition, mechanical alloying, and consolidation of amorphous and nanocrystalline powders. Although high strain rate superplasticity is yet to be adopted in industry, ECA pressing and powder metallurgy, as methods of producing a very fine grain size, are capable of scale-up to commercial level. Microstructures produced by these two latter techniques are shown in Fig. 1. Another way of exploiting such ultrafine grain sizes is to superplastically form the material at a low temperature, e.g., aluminum alloys at around 200°C and titanium alloys at 250–450°C (*low-temperature superplasticity*). The temperature of forming is also reduced by altering the composition, e.g., hydrogenation of titanium alloys. *Superplastic forming in industry* is done in the strain rate range 10^{-5}–$10^{-2}s^{-1}$. These rates are lower than those of conventional forming processes ($\dot{\varepsilon} \sim 1$–10^3s^{-1}).

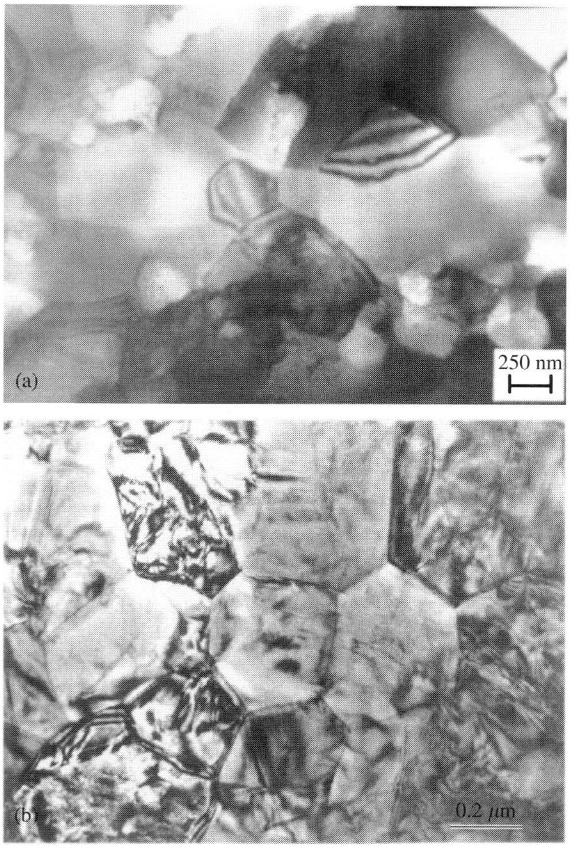

Figure 1
(a) Microstructure of an Al–Mg–Li alloy (Al 1420) produced by equal channel angular extrusion (after Valiev *et al.* 2000). (b) Equilibrated, fine-grained, 5.2wt.% yttria-stabilized tetragonal zirconia produced by powder metallurgy (courtesy of J. Wadsworth and T. G. Nieh).

3. Production of Fine Grain Size

Superplastic titanium alloys (Ti–6Al–4V alloy is by far the most common) are formed at higher strain rates ($\sim 10^{-3}$–$10^{-2}s^{-1}$) than aluminum alloys. Mill-produced (hot rolled in the $\alpha + \beta$ range and finish rolled at 800°C) titanium alloys are superplastic as the grain size is fine (~ 7 μm) and uniform and the two-phase microstructure is bimodal (which confers high grain growth resistance). Titanium alloys do not cavitate during superplastic forming and are easily diffusion bonded.

To produce superplastic microstructures in aluminum alloys two different routes are used. Alloys like SUPRAL 100 (Al–6Cu–0.4Zr) are chill cast from a temperature about 100°C higher than that commonly used for aluminum alloys (to take as much zirconium in solution as possible), homogenized, hot rolled, and cold rolled without intermediate annealing by about 70%. Such alloys acquire a superplastic microstructure during forming where in the early stages a strain rate greater than optimal is applied (strain-assisted continuous recrystallization). Grain growth is minimized by the grain boundary pinning action of Al_3Zr particles. In contrast, alloys like Al 7475 are cast and hot rolled like their non-superplastic variants. Following solution treatment (~500°C) and overaging (~400°C), the alloy is rolled by about 85% at 200°C. Then it is statically recrystallized at about 500°C to obtain an average grain size of about 5 μm. Aluminum–lithium alloys can be processed by either route, but industrially the former (easier) process is adopted. No eutectic aluminum alloys have succeeded commercially, although the Al–Ca–Zn eutectic alloy has shown some promise. Likewise, the Zn–22% Al eutectoid alloy has found some limited applications. Aluminum-, copper-, iron-, and nickel- based alloys cavitate during superplastic deformation.

The initial processing needed to produce the superplastic microstructure increases the cost of aluminum alloys by a factor of 3–5. The two aluminum alloys Al 5083 and Al 2004 together account for more than 90% of commercial superplastic alloys in use.

4. Superplastic Forming and Diffusion Bonding

Five different superplastic forming techniques are in commercial use (Fig. 2). In *simple female forming* (Fig. 2(a)) a heated superplastic sheet is clamped around its edge and stretched into a heated cavity tool using gas pressure. In *drape forming* (Fig. 2(b)) heated and clamped superplastic sheet is stretched into a cavity containing one or more male form blocks. In *male forming* (Fig. 2(c)) gas pressure and tool movement are combined to produce deeper, more uniform thickness parts. (This process is not used for titanium alloys because of engineering difficulties associated with high temperatures.) *Back pressure forming* (Fig. 2(d)) is used with alloys that are prone to cavitation. Here gas pressure is used on both front and back sides of the deforming sheet which produces a hydrostatic confining pressure capable of suppressing cavitation. Gas control creates a positive pressure differential enabling forming to be achieved. The process is applied to female and drape forming methods when less than 0.5% cavitation level is required in structural applications. *Diaphragm forming* (Fig. 2(e)) uses one or two "slave" sheets of a superplastic alloy to guide a smaller unclamped blank to be gradually drawn and draped into tool contact. This process is used when the thickness variation allowed is limited. The choice of process is decided by the size, shape, and performance required of a product as well as economy of operation.

The equipment for superplastic forming is custom-built, computer-controlled male and female installations that cost around US\$0.5–3.0 million, depending on dimensions, clamping forces, and techniques available. Special design features take into account the high temperatures, slow velocities, and requirements of pressure tightness and long hold times. For typical aerospace superplastic aluminum alloy components, the maximum separating force increases from about 400 tons in the absence of a back pressure to about 1100 tons when a back pressure of 0.5 times the flow stress is used. This further rises to 2000 tons if the rate of forming is raised from $4\times10^{-4}\,s^{-1}$ to $1\times10^{-2}\,s^{-1}$. Thus, increasing the temperature and the strain rate of forming and introducing a back pressure increases the cost of equipment and tooling significantly.

Sheet manufacturing techniques can be based on superplastic forming (SPF) or superplastic forming/diffusion bonding (SPF/DB). The advantages are high formability, low flow stress, high design flexibility, improved consistency and interchangeability, extreme shaping accuracy, absence of spring back and residual stresses, capability for processing difficult-to-form materials, optimal use of expensive materials, elimination of one die half because of gas pressure forming and thus reducing tooling, die, and fabrication costs, reducing assembly costs by reducing part counts that in turn leads to weight and fuel savings in service, and improved properties like corrosion resistance and better fatigue and fracture resistance. DB is used to obtain selective reinforcements and self-stiffened structures. Combining SPF and DB in a single step makes the process very attractive for the aerospace industry.

DB, which has proved to be more successful with titanium alloys than with aluminum alloys, can be done before, during, and/or after SPF. Bond properties of titanium alloys are similar to those of the base metal. (However, for achieving good bonds in titanium alloys sometimes pressures similar to those that are applied during SPF will have to be applied at 925 C for 3 h.) In high-strength aluminum alloys a tenacious oxide layer makes good bonding difficult. An aluminum alloy or an amorphous metal reactive layer is used to achieve sound bonds in this case. Diffusion brazing, which is a faster and cheaper technique, is under development. This process can be used if bond properties are allowed to be inferior or dissimilar to the metals that are to be joined. Apart from the usual quality checks, etching to remove the α-phase in titanium alloys and ultrasonics-based nondestructive testing techniques to ensure

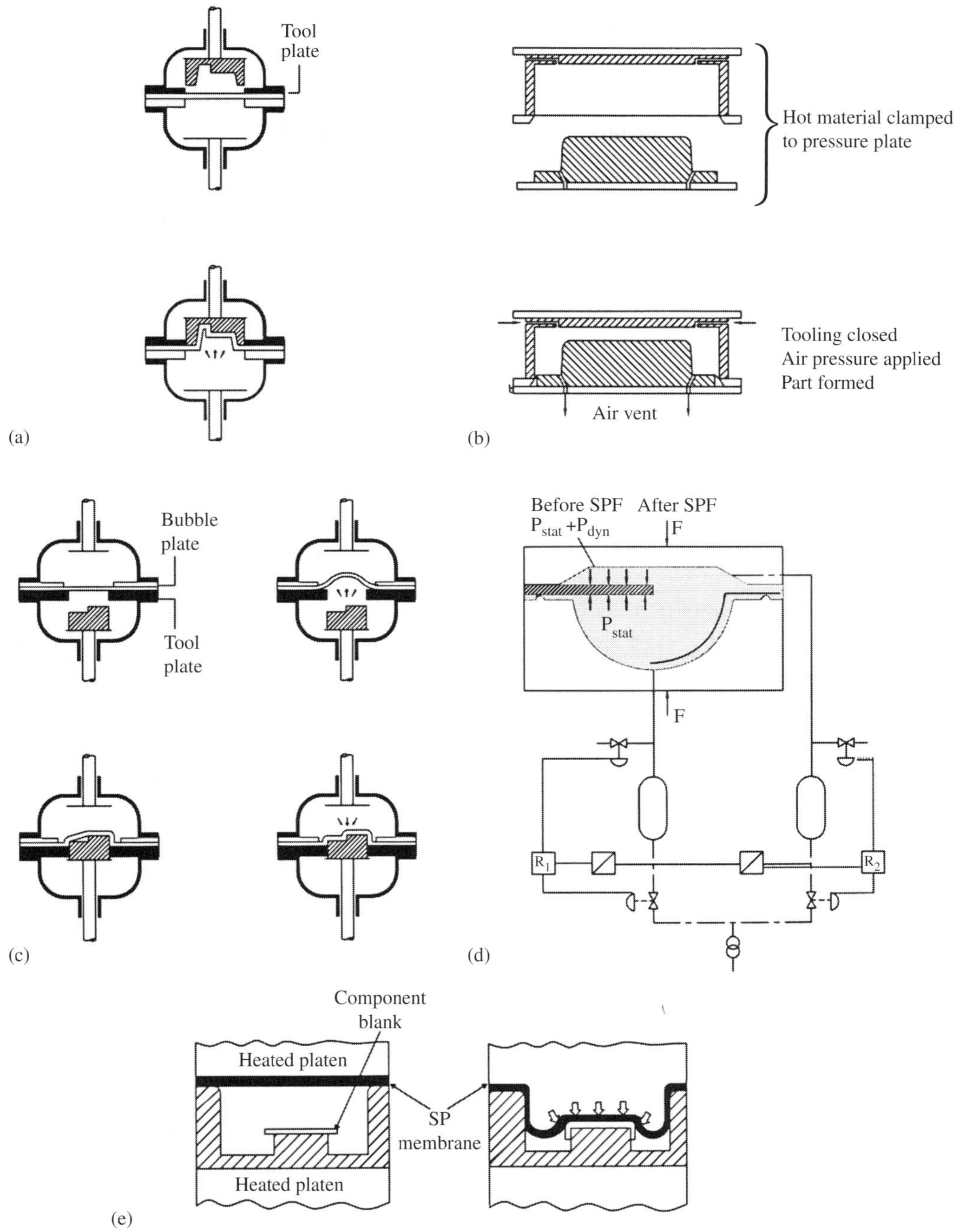

Figure 2
Schematic representations of (a) female forming, (b) drape forming, (c) male forming, (d) back pressure forming, (e) diaphragm/membrane forming.

soundness of bonds are necessary. These requirements add to the costs.

The main disadvantages of SPF are the requirements of high temperature, low strain rate, special material preparation, inability to eliminate thickness variations, higher costs than those of plastic products, and restriction to small and medium batch production. Resistance to new designs and/or

Table 1
Applications of superplastic forming.

Scale	Application: Without requirement for surface finish	With moderate requirement for surface finish	With highest requirement for surface finish
Single component	Satellite components, design studies components	Bodies for heavy vehicles	Racing car parts, sports equipment, e.g., golf club heads, skates
Small-scale production	Tanks, aircraft components	Structures for heavy vehicles, hospital utilities	Formula cars, e.g., tourist cars
Medium-scale production	Balances, interior of automobiles, bus components, computer frames	Aircraft structures, facade of buildings, truck structures, components for textile and paper industry	Luxury cars, e.g., BMW, Mercedes Benz
Large-scale production	Fittings, profile joints, inside of seats	Visible car interiors, e.g., engine cover, components for railway carriages	Car trunk cover, hard tops of cars

new materials, very high cost of developing and qualifying new materials, difficulties in integrating the process into existing technologies, and lack of awareness among designers about the potential of the process are additional problems. When the superplastically formed component is used for high-temperature service, a postforming heat treatment also is necessary.

The main drawbacks of DB are high temperature, long bonding times, need for vacuum or protective atmosphere, special pretreatments needed for aluminum alloys, and special nondestructive testing techniques essential for evaluating the soundness of each bond.

Two-step forming to control localized deformation, preprofiling of initial sheets, membrane hot forming, producing small parts in multiple dies in one run, finite element simulation to eliminate costly try-outs, and use of cheap die materials, e.g., aluminum alloys, cast iron, graphite, for less severe forming conditions are some of the innovative steps being taken to make the technology more attractive.

A few other interesting developments are "creep forming" of advanced composites and thermoplastics using press forming or autoclave forming; titanium matrix composites, intermetallics, e.g., Ti_3Al, TiAl, and intermetallic composites, Y_2O_3-stabilized tetragonal zirconia (YTZP) and 20% Al_2O_3–YTZP all processed through SPF; and solid-state bonding, sinter forging, hot pressing, extrusion, tube forming, punch forming, coining, and hot isostatic pressing of ceramics and hybrid ceramic–metal structures for the aerospace industry processed through SPF/DB. Table 1 lists the areas of applications of superplastic forming.

5. Conclusions

In terms of cost, the main applications of superplastic forming are in the aerospace industry. Architectural products constitute the largest number. Net/near-net-shape technologies of high precision for both bulk (e.g., "Gatorizing" of turbine disks made of superalloys through powder processing) and sheet components, high added value end products, medium- to low-volume batches, situations where the special properties of superplastic materials are required in service and parts in thin gauges, complex shapes, and medium size (0.1–4.0 m^2), all of which exploit SPF or SPF/DB, have decided the application areas. With increased awareness among designers, new and innovative products/applications are likely to emerge.

Bibliography

Baudelet B, Suery M (eds.) 1985 *Superplasticity*. Editions du Centre National de la Recherche Scientifique, Paris

Chokshi A H (ed.) 1997 *Superplasticity in Advanced Materials*, ICSAM-97, Materials Science Forum, 243–245. Trans Tech, Aedermannsdorf, Switzerland

Hamilton C H, Paton N E (eds.) 1988 *Superplasticity and Superplastic Forming*. TMS, Warrendale, PA

Kaibyshev O A 1992 *Superplasticity in Alloys, Intermetallides and Ceramics*. Springer, Berlin

Langdon T G (ed.) 1994 *Superplasticity in Advanced Materials*, ICSAM-94, Materials Science Forum, 170–2. Trans Tech, Aedermannsdorf, Switzerland

Nieh T G, Wadsworth J, Sherby O D 1997 *Superplasticity in Metals and Ceramics*. Cambridge University Press, Cambridge, UK

Padmanabhan K A, Davies G J 1980 *Superplasticity*. Springer, Berlin

Padmanabhan K A, Vasin R A, Enikeev F U 2001 *Superplastic Flow: Phenomenology and Mechanics*. Springer, Berlin
Paton N E, Hamilton C H (eds.) 1982 *Superplastic Forming of Structural Alloys*. TMS–AIME, Warrendale, PA
Sakuma T, Aizawa T, Higashi K (eds.) 1999 *Proc. 1998 Int. Symp. Towards Innovation in Superplasticity II,* Materials Science Forum, 304–6. Trans Tech, Aedermannsdorf, Switzerland
Smirnov O M 1979 *Metal Working in a Superplastic State*. Metallurgy Publishing, Moscow (in Russian)
Valiev R Z, Islamgaliev R K, Alexandrov I V 2000 Bulk nanostructural materials from severe plastic deformation. *Prog. Mater. Sci.* **45**, 171

K. A. Padmanabhan

Metal Forming: Close Die Forging

Closed-die forging is a forming process in which a workpiece that has been previously heated to a high temperature is pressed by means of countermoving form tools (forging dies). The tools that hold the shape of the forged part completely enclose the workpiece during this process and machines provide the necessary high forces and power. The various possibilities in design and the superior material properties are a special characteristic of die-forged parts. With a fiber orientation directed in the way of stresses at dense structures a greater component resistance can be reached than with casting or machining practices (Cartwright 1976). The advantage of this process derives from the fact that in the forming operation the fiber orientation and the expected main course of strain lines can be shaped to be parallel. Figure 1 shows the favorable fiber orientation of a forged part which follows the contours of the workpiece in comparison with cast and machined parts.

Because of these characteristics, die-forged parts are used wherever safety, dependability, and durability are of importance (Gold 1985). This is the case whenever the transfer of forces and movements is crucial, in combination with high static and dynamic strains as well as great wear rates.

Die-forged parts can be found in many areas of technical application. The most important consumers of forged components are the automobile and machine building industries. Products like connecting rods, steering knuckles, steering gear arms, wheel hubs, crankshafts, gear wheels, and turbine buckets are manufactured with closed-die forging technology.

1. *Process Survey and Working Sequence*

The most important process variants are closed-die forging with flash, die forging without flash, and precision forging (Fig. 2). In the die forging process with flash, surplus material is displaced to the outside through a tool gap. In the process without flash no surplus material escapes from the tool.

The accuracy that can be reached with forging with flash is within the tolerances of IT 12–16 (quality number of the ISO tolerance table). By forging without flash the tolerances of the die-forged parts can be increased (IT 8–11) and machining processes can be reduced. Precision forging allows the forging of certain functional and effective surfaces, prefinished in such a way that the accuracy is comparable with finish-machining processes (IT 7–9).

In addition to forming, various other manufacturing processes are included in the work procedure. The whole manufacturing sequence can be divided into separating, heating, forming, possibly trimming, heat treatment and cleaning operations (Fig. 3).

The optimized combination of steel characteristics for specific applications can only be reached with heat treatment. The most important heat treatment processes for die-forged parts of steel materials are normalizing, tempering, tempering from hot-forming temperature, heat treatment for improved machinability, respectively, resistance, and controlled cooling of the forging heat.

The most important nonferrous material for die-forging parts is aluminum. Temper-hardened alloys are mostly used for this application. The thermosetting aluminum alloys obtain their resistance by solution heat treatment followed by quench hardening and aging.

2. *Process Parameters and Forming Behavior*

The result of the die-forging process, i.e., the quality of the component and the efficiency, depends on many different influencing variables. For a technological analysis of the process, a knowledge of the complex interrelations and dependencies of all interacting process parameters is necessary. The important variables of the die-forging process are presented in Fig. 4. They can be divided into input, forming, disturbance, and output variables. The input variables

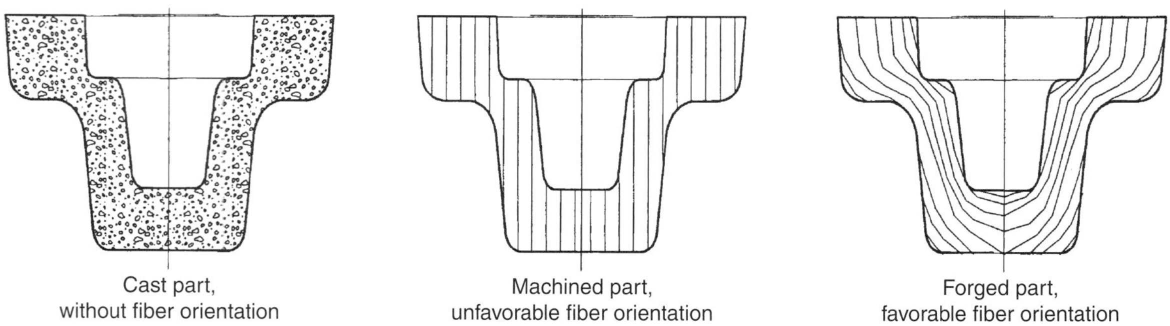

Figure 1
Fiber orientation of parts manufactured by different methods.

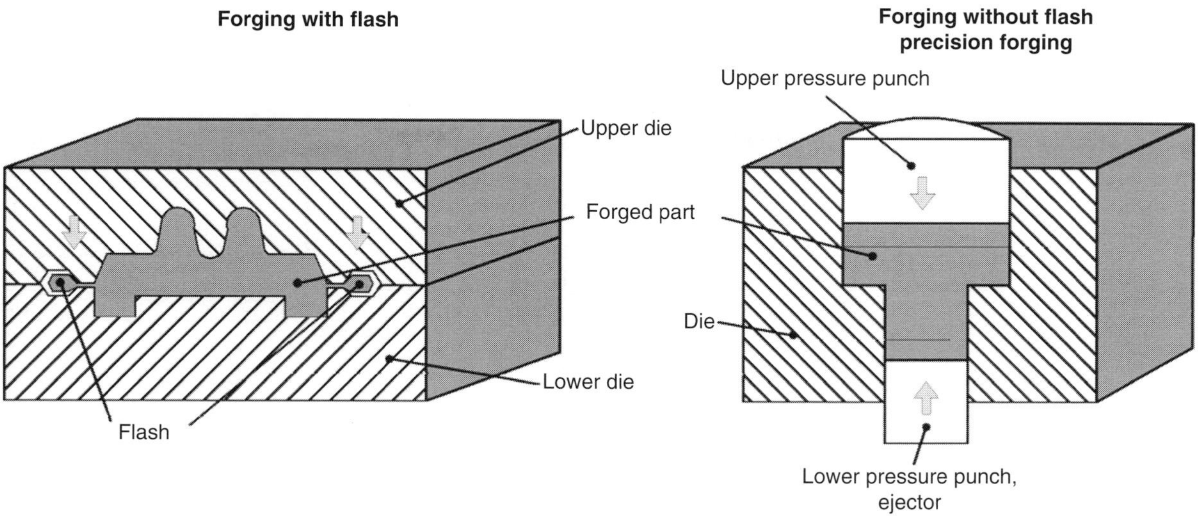

Figure 2
Process variants of closed-die forging (Copyright IFUM).

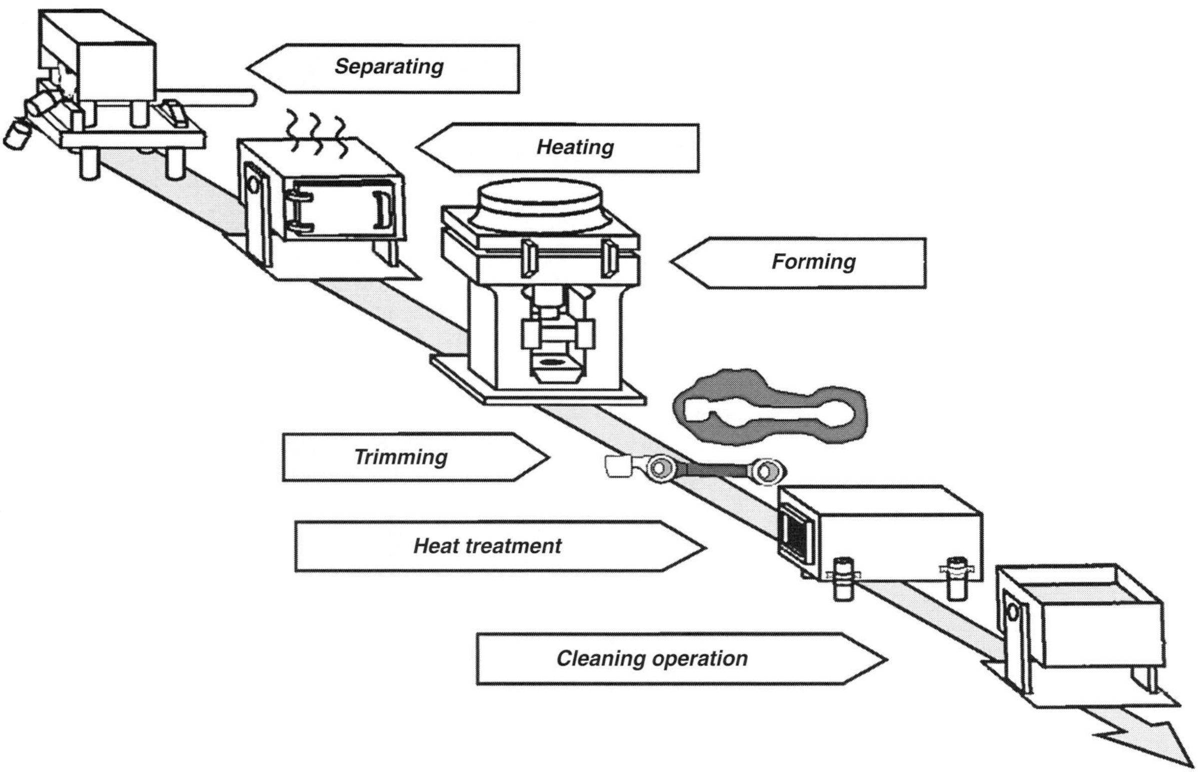

Figure 3
Working sequence in closed-die forging (Copyright IFUM).

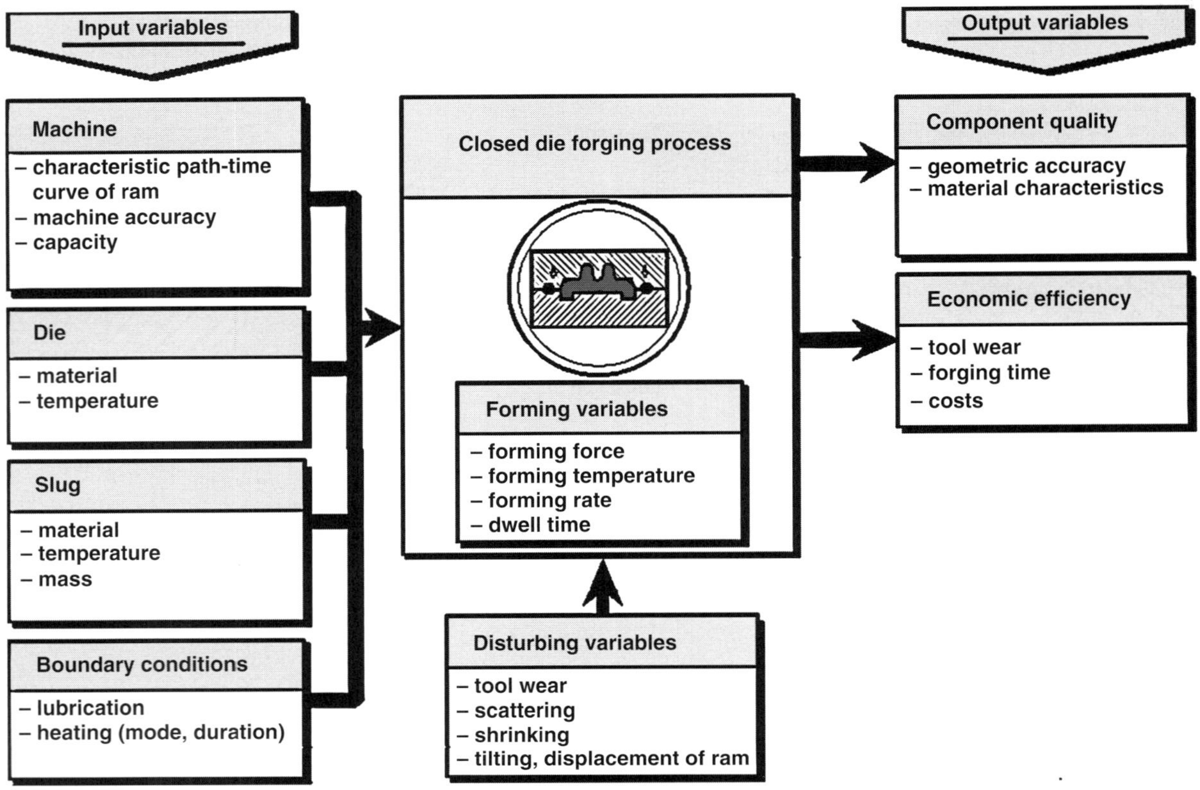

Figure 4
Important variables of the closed-die forging process (Copyright IFUM).

can be assigned to the metal-forming machine tool, tool (die), workpiece (slug), and boundary conditions.

The forming behavior and the resulting need for energy during the forming process vary with the material of the forged part. The behavior depends on temperature, strain, and strain rate, all of which influence the flow stress, plus the workability which shows the ability to tolerate a certain amount of deformation without fractures in the microstructure. Workability and flow stress identify the forgeability of a material.

A major influential variable on the forming behavior of metal materials is the temperature. With rising temperatures the workability improves. The upper limit is determined by solidus temperature, phase transformation, chemical reactions, and coarse grain formation, while the lower temperature limit in forging processes is determined by the point of recrystallization, high rising flow stresses at decreasing temperatures, or phase transformation (Spur and Stoeferle 1984).

3. *Materials for Closed Die Forging*

Forging processes can be applied to all ductile metals. Steel has a special role because there are many different kinds of standard steels available with a wide variety of characteristics. These characteristics can be influenced by deformation and heat treatment on a large scale and are thereby adapted to different applications. The most important characteristics of steel are hardness, tensile strength, elasticity, and toughness as well as fatigue strength, high-temperature strength, wear resistance, corrosion resistance, and machinability. Important for the specific characteristics of different steel materials are the chemical composition and the heat treatment conditions.

Many different steel materials are used for forging parts, from plain carbon steel to low- and high-alloys. The steel groups mostly used for forging parts are unalloyed constructional steels, heat-treatable steels, case-hardening steels, nitriding steels, steels for flame-and induction-hardening, antifriction bearing steels, heat-resistant steels, low-temperature steels, stainless steels, and microalloyed steels (Lange and Poehlandt 1985).

The application of nonferrous metals is more limited than that of steel. Other metals of technical interest are aluminum, magnesium, copper, titanium, nickel, cobalt, and tungsten and their alloys. The conditions under which these nonferrous metals are deformed differs from those for steels, especially with respect to the forming temperature and the deformation resistance (Davis 1998).

Thermosetting wrought alloys are mostly used for forged parts made of aluminum. Naturally hard materials are also used. They show a very good corrosion resistance and a good behavior at low temperatures. Die-forged parts are mainly used in applications where a low density of the component coupled with high fatigue stress is desired, e.g., in the aircraft and automobile industries and for dynamically highly-loaded machining components. Magnesium is used as an alloy, usually in combination with aluminum and zinc. Magnesium alloys show a better resistance to corrosion than aluminum alloys, and they are also often used in the aircraft and automobile industries.

Titanium and titanium alloys show the best combination of strength and density for all construction materials. However, the difficulties with production and processing make titanium materials very expensive so that their application in most cases is limited to airplanes and spacecraft with some exceptions in machine-building and some medical equipment.

Die-forged parts made of special alloys based on nickel, cobalt, or tungsten show great high-temperature strength. Their applications are mainly in turbine disks or buckets. Die-forged parts made of copper and brass are for the most part used in mountings, for electrical engineering, in the automotive industry, and in precision mechanics. They are resistant to corrosion and have good electrical properties.

4. *Tool Materials*

When choosing a proper material for a forging die it is necessary to consider size and design of the die impression, forging temperature, costs for the material, and the load on the tool. The load on the tool can be differentiated into three categories when considering which steels are appropriate:

mechanical loads
thermal loads
wear resistance.

Frequently-used steels for forging dies are: low-alloyed hot-work tool steels like Ni–Cr–Mo–V-alloyed steels which are used for dies facing an impact load; highly-alloyed hot-work tool steels like Cr–Mo–V-alloyed steels which show a good resistance to heat-cracking; and W–Cr–V-alloyed steels which show the greatest heat resistance (retention of hardness) (Lange and Meyer-Nolkemper 1977).

5. *Isothermal Hot Die Forging*

Isothermal hot die forging is one of the variants of die forging. It is rarely used but has advantages in some niche applications. Isothermal hot die forging is die-forming in tools with temperatures nearing the forming temperature. When forming aluminum and magnesium alloys this process is usually used because deformation temperatures are comparatively low (400–500°C). It is also applied to titanium alloys.

6. *Conclusion*

In production, especially in mass production of secure and reliable components with high durability, the closed-die forging technology represents an important branch of the industry. The main methods of closed die forging are forging with flash, forging without flash, and precision forging. Precision forging is currently used infrequently, but it will gain greater popularity in the future because of the near-net-shape character of this forging method. The forming operation and the quality of the forging part is influenced by many parameters. They can be divided into input, forming, disturbance, and output variables.

The most commonly forged materials are steels, but because of the increasing interest in lightweight construction, forging of aluminum and magnesium is regularly a part of research and development activities. In the field of tool technology, reduction of tool wear is at the forefront. Ceramic coatings and inlays for forging tools also show great potential for increasing the life of the tool.

Bibliography

Cartwright B 1976 Ten ways to use grain flow in forgings. *Engineering, London* **216**, 804–5

Davis J R 1998 *Metals Handbook*, 2nd edn. ASM. International, Materials Park, OH

Gold R 1985 The fundamentals of forging. *Precision Metal* **43**, 23, 26–8

Lange K, Meyer-Nolkemper H 1977 *Gesenkschmieden*. Springer-Verlag, Berlin

Lange K, Poehlandt K 1985 *Handbook of Metal Forming*. McGraw-Hill, New York

Spur G, Stoeferle T 1984 *Handbuch der Fertigungstechnik*. Umformen, Carl Hanser Verlag, München, Vol. 2/2

E. Doege, G. Dreyer, and S. Ruesch

Metal Forming: Formability

The meaning of formability or workability is discussed in this article and origins of defects are identified. Some defects, such as wrinkling in deep drawing and necking, can be explained in terms of continuum mechanics. Others, like orange peel, grain coarsening, and mechanical fibering, arise because practical materials are not continua. Many defects concern cracks and fracture, the explanation for which requires consideration of both process mechanics and metallurgy.

1. Formability

Formability of a material concerns whether it is possible, given adequate machine capacity, to perform a given metal forming operation successfully without introducing defects.

There is a great deal of practical knowledge about how to avoid defect formation, such as correct forging temperatures for different metals, correct surface preparation, and prior heat treatment. The controlling parameters invariably are of two sorts, (i) *material-dependent* (e.g., metallurgical "cleanliness," segregation, prior heat treatment, annealing/normalizing cycles, grain size, texture, and anisotropy) and (ii) *process-dependent* (e.g., geometry of deformation, tool and die design, interface friction, rate of deformation, and temperature).

It is generally agreed that the quality of a metal formed product, and the efficiency and economics of the operation, depend on the avoidance of unnecessary inhomogeneity of plastic flow, i.e., minimization of *redundant work* (the work associated with internal plastic strains, imposed by the flow field, which are greater than the homogeneous true strain based upon external shape change alone). Inhomogeneous plastic flow affects residual stress patterns, elastic springback, and the possibility of opening up crack-like defects on exit from forming operations, during subsequent processing, or in service.

In drawing, extrusion, rolling, etc., deformation zone geometry can be characterized by the single parameter Δ defined as the ratio of the mean thickness or diameter h of the workpiece, to the contact length L between tool and workpiece: Δ increases with increasing reduction and increasing die angle (drawing) or increasing ratio of strip thickness to roll radius (rolling). Figure 1 illustrates how Δ is an indicator of uniformity of deformation in 2D forging. Redundant strain is greatest at large Δ. Blazynski (1989) shows how appropriate design of tooling can reduce redundant shearing in plug drawing of seamless tube, rotary piercing, and Assel tube elongation of a thick-walled pipe as the wall thickness is reduced by skew rolling over a freely-rotating mandrel.

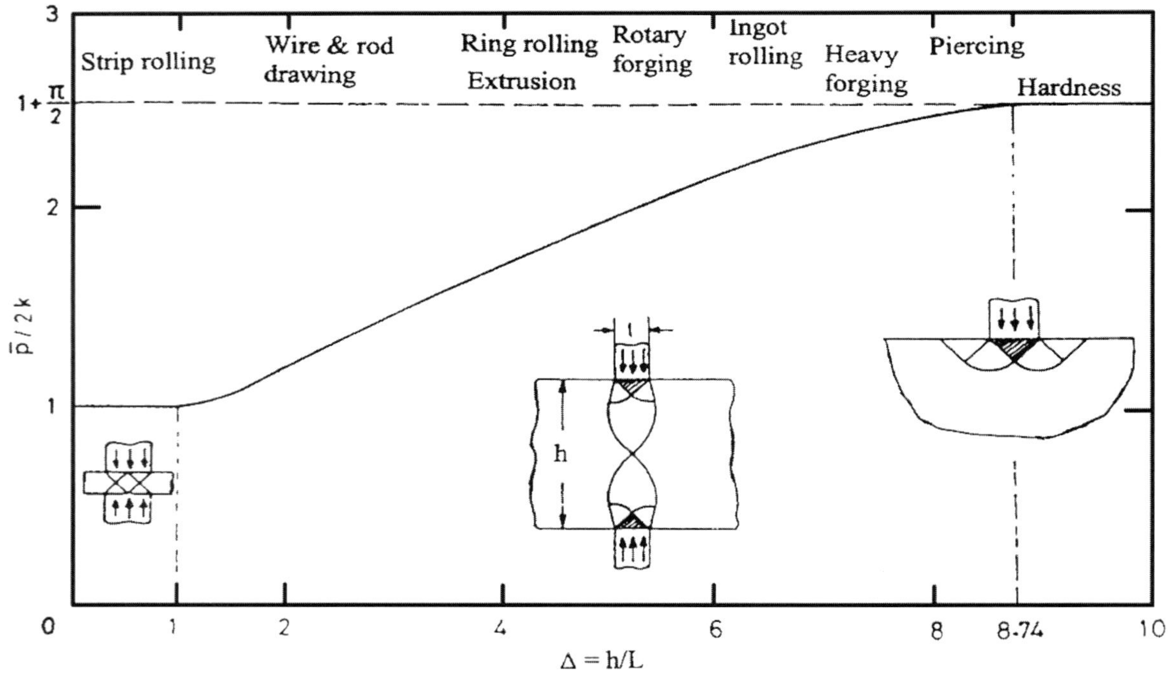

Figure 1
Influence of deformation zone geometry on homogeneity of flow.

Patterns of changing deformation fields like Fig. 1 are found with opposed wedge indentation: when the fields intersect at sufficiently deep penetration, the plastic flow which accommodates vertical motion of the wedges occurs along the axis of the plate, putting the center of the bar into tension, leading to eventual fracture. This is how cutting by pliers takes place and explains why, in other cases, nonuniform flow fields are dangerous for void and crack formation. The V-anvil was introduced as early as 1839 to combat the tendency for crack formation in the center of billets when rounds are forged with flat dies. Similar center-cavities are found in cross-rolling, helical rolling, and cross-wedge rolling and are used in rotary piercing, and the start up of seamless tube production by the Mannesmann process.

2. *Microstructures and Workability*

Practical, commercial metals are not continua as assumed in the sort of analyses used to investigate flow fields and working loads. They consist of grains, each of which deforms in its own anisotropic manner; they contain impurity elements as well as the intended chemistry; and they contain inclusions. While a random aggregate of small crystals behaves in an overall way with characteristic "continuum" properties, the size and behavior of individual grains, microstructural composition (and hence property) gradients, etc., all affect the detailed behavior of practical metals and produce effects not predicted by continuum analyses.

Orange-peel in sheet forming operations is an obvious example where, with a limited number of grains through the thickness of the sheet, individual crystals deform irregularly giving a rough surface appearance. The hexagonal metals such as zinc and magnesium alloys have few available slip planes and consequently display this effect. Under unfavorable conditions, the presence of large grains encourages cracking along grain boundaries during cold work in tungsten and molybdenum, as well as in zinc and magnesium; all these metals show considerable increase in ductility when their grain size is small. In hot working, grains can coarsen and this may set limits of workability, even though greater reductions are possible within the capacity of the forming equipment and without causing fracture.

Backofen (1972) explains that the *fracture centers* (particles, pores, and weak interfaces) produced by previous processing are broken, flattened, and strung together in sheets or fibers by most plastic-working operations (c.f. slag stringers in wrought iron, pearlite banding in steel, and flow-lines revealed by sectioning and deep-etching). Such *mechanical fibering* is the structural prerequisite for *crocodiling* in rolling and fractures nucleated at the centerline of wire drawn with high Δ reductions. The anisotropy produced by fibering after cold rolling of sheet affects bendability, since ductilities along and across the sheet are different, cracks occurring when the sheet is bent one way, but not the other. Anisotropy caused by mechanical fibering is remarkably resistant to being "annealed out," although at high temperatures its effects can eventually be diminished as particles are spheroidized, interfaces are sintered, and concentration gradients are dispersed.

Metallurgical composition can have a strong effect on formability. For example, a small amount of sulfur in low-carbon steels has a drastic effect for the results of the *hot twist* test employed to assess hot forgeability. The cause is *hot shortness* and is counteracted by increasing the amount of manganese in the steel to form relatively harmless MnS inclusions in place of brittle FeS.

Longitudinal streaks on the surface of extruded aluminum alloy bars resulting in low circumferential strength and laminated fractures are due to a string of particles of intermetallic compounds, and are frequently found in extrusion from multi-hole dies. They arise from central billet *segregation* and have their origin in oxide films or coarse grain regions.

Surface and subsurface intrusion defects are very marked in extrudates when the billet surface layer is harder than the billet body. This occurs in hot extrusion where chilling of a billet by the cooler container creates the harder surface coating; α/β brasses are prone to this problem. Finite element methods are now being used to investigate the trapping of oxide scale and inclusions in skin layers in metal forming.

3. *Buckling, Laps, Laminations, etc*

Figure 2 shows a closed-die forging where there is excess material in the web. It buckles during forging and develops laps. Die cavity radii have a significant effect on the formation laps called *cold shuts*. When flash is formed in closed-die forming producing fins, it may be pressed into the metal during a subsequent pass.

In open-die forging, flat tools with sharp edges produce sharp-edge indentations which fold over during subsequent bites. Similarly, high Δ reductions in rolling can form burr-like protrusions at the edges, which are folded and rolled back into the material on subsequent passes. Laps may also occur in the edging passes of a hot rolling sequence when the rolls are worn and require dressing.

Laps and folds would not matter if the metal healed fully when the folds are pressed back into the workpiece (as in the rehealing which occurs in extrusion with porthole and bridge dies). But unhealed pressed-back laps and folds are weak interfaces which become potential cracks. Such defects can open up during subsequent forming operations; even if they

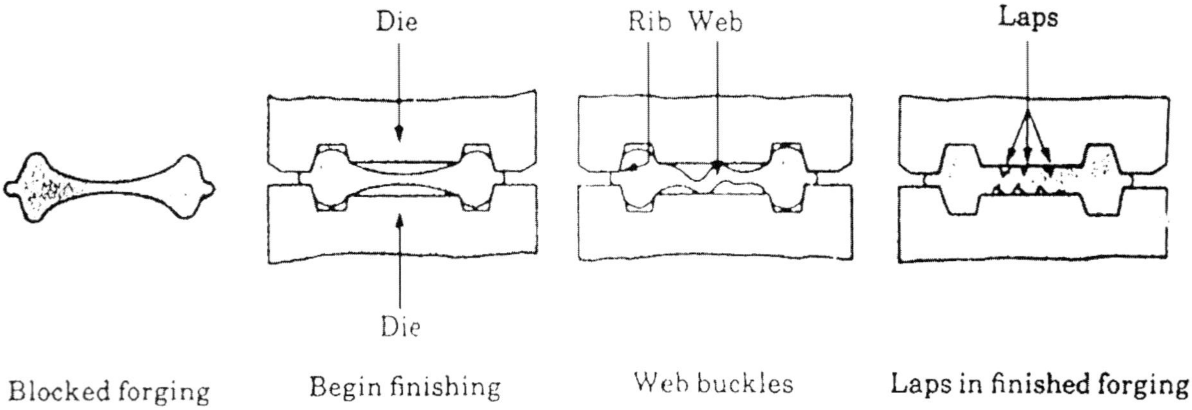

Figure 2
Laps produced by buckling of web in closed-die forging (after Kalpakjian 1984).

do not, they can be troublesome in service causing fatigue failures and/or leading to other problems such as corrosion and wear. Of course, the laps and folds themselves may crack before being folded over, leading to more defects, such as flash cracks in steel forgings.

Wrinkling in deep-drawing is due to multiple buckling of the undrawn part of the blank under the compressive stresses generated in the flange. *Puckering* refers to corrugations in the wall of the cup; similar axial defects in tube sinking are called *seams*. Wrinkling is sometimes seen in bending and contour forming on the inside of the bend. The tendency for wrinkling increases with increase of unsupported/unconstrained surface area of the sheet metal, decreasing thickness, and nonuniformity in sheet thickness. Wrinkles may also be initiated by trapped lubricants distributed nonuniformly between the surface of the sheet and the dies. For a limited range of products such as thin aluminum food trays, the wrinkles are left in place. In most situations they are unacceptable.

Blankholder pressure is important in controlling wrinkling: too high a blankholder force increases the punch load (because of friction) and leads to tearing of the cup wall. Conversely, if it is too low, wrinkling will occur in the flange. Deep drawing may be carried out without a blankholder provided that the sheet metal is relatively thick to prevent wrinkling.

In rolling, the ideal parallel gap is altered by (i) elastic deflexion of the rolls under load and (ii) thermal expansion of the rolls owing to dissipation of the work of deformation. Rolls are cambered (made barrel-shaped) to counteract elastic deflexion. When the roll gap is not parallel, the middle or the edges of the sheet will be 'over-rolled'. The compressive stresses induced cause buckling, giving so-called *wavy edge* or *wavy middle* defects (*shape defects*). Thermal camber takes time to develop, so that shape is a problem in the start-up of mills. Lubricant, cooling, and roll camber all interact in this complicated problem. The front and back mill tensions with which the sheet is rolled often hide shape defects, which do not become evident until the tensions are relaxed on coiling. *Shape-meters* have been designed to indicate flatness problems while the mill is operating. One measures the variation of radial contact stress against a bridle roll on exit from a millstand, loose regions not bearing down on the bridle roll as much as tight sections. Corrections for shape may be attempted by differential cooling of the rolls, or by roll-bending jacks acting on the roll bearings.

4. *Ductile Fracture in Metal Forming*

The favorable influence of high compressive hydrostatic pressure on increasing ductility of all types of material has long been known. Increasing compressive mean stress suppresses void nucleation and growth and delays fracture. This undoubtedly influenced the development of metalworking processes. Hydrostatic extrusion with substantial back pressure takes the idea to its limit. Practical forming methods which are predominantly tensile in nature, are rare.

Nevertheless, cracking can occur during metal forming. Well-known examples are peripheral cracking around the barreled side surfaces of upset (compressed) cylinders of ductile materials; surface cracking ("fir tree" cracking) of extrusions in hot working; (hidden) "chevron" cracking along the axis of extrusions and drawn wire; edge cracking in rolling; and (hidden) central cracks in the drawing of wire clad with a softer material; etc. (Fig. 3). Many empirical stress and strain criteria have been proposed over the years for fracture in metal forming. It is a feature of many that they give wrong predictions when applied to entirely different operations from those for which they were derived (Fig. 4).

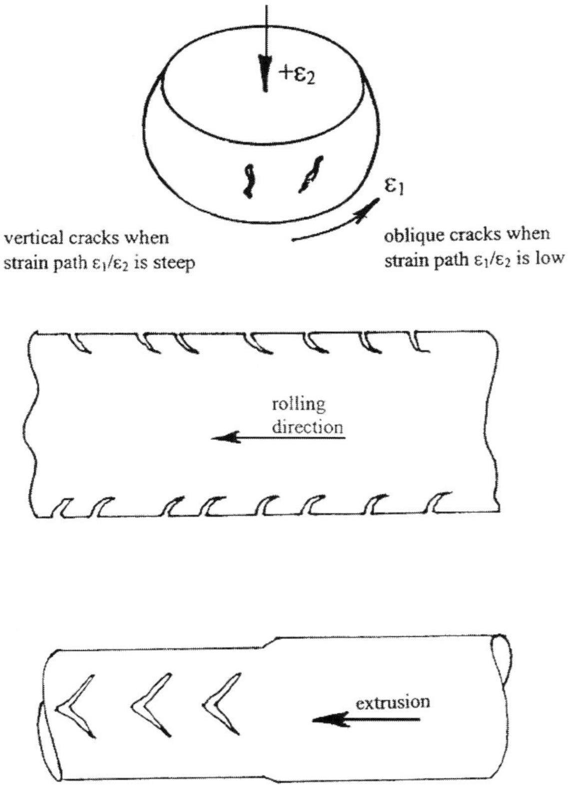

Figure 3
Crack formation in various forming processes.

It is well known that the hoop fracture tensile strain ε_{1f} and axial fracture compressive strain ε_{2f} local to the sites of fracture in upset cylinders falls on an approximately straight line failure locus of slope $\tan^{-1}(-\frac{1}{2})$ in diagrams like Fig. 4. The ε_{1f} intercept for different materials in different directions, or for the same material in different heat-treated or work-hardened states, indicates *workability*. The diagram has been employed extensively in solid forging (bolt heading, ring forging; bending, edge cracking in rolling, gear forging, and internal cracking in extrusion and hubbed-disk forging) and with powder preforms. Providing the strain path during forming does not cross the failure line (whatever the loading prehistory) it is said that no cracking will occur.

In sheet-forming, orthogonal in-plane plastic strains at fracture also lie on a line falling from left to right in diagrams like Fig. 4, but with a steeper slope, often at about $\tan^{-1}(-1)$, i.e., $\varepsilon_{1f}+\varepsilon_{2f}=$ constant, or constant true thickness strain ε_{3f} at fracture from volume constancy. Sheet forming differs from upsetting as it includes biaxial tensions. Furthermore, plane-strain necks form prior to fracture with through-thickness straining between neck formation and cracking.

Inspection of those empirical formulas which can be applied to more than one process, together with the knowledge that fewer cracks occur in fields with low Δ, and in cleaner microstructures, all lead to void growth mechanics as being the determining factor in crack formation in forming processes. A versatile criterion for ductile fracture, relates normalized hydrostatic stress $(\sigma_H/\bar{\sigma})$ and the von Mises strain at fracture $\bar{\varepsilon}_f$, as follows

$$(\bar{\varepsilon}_f - \bar{\varepsilon}_o) = 1.65(\bar{\varepsilon}_{fst} - \bar{\varepsilon}_o)\exp(-k\sigma_H/\bar{\sigma})$$

where k is a parameter to be found from calibration (see later), ε_{fst} is the true strain at fracture in simple tension, and $\bar{\varepsilon}_o$ is the threshold strain at which the accumulation of damage begins. The variation of this damage function throughout the deformation zone of a metal forming process can be evaluated by FEM as loading increases, thus identifying both the site of fracture and the strain at which cracking begins. Before FEM, candidate criteria for crack formation could only be assessed for simple states of stress, with slip line field analysis giving insight into more complicated fields. Alternatively, the use of gridded samples in experiments permits strain to be determined from which stresses are given by a constitutive law, thus permitting assessment of a fracture criterion.

A development of continuum plasticity hole growth modeling to determine ductile fracture is the employment of *porous plasticity* modeling. It has been used mostly for ductile fracture mechanics calculations but it has great potential for application to metal forming.

Hole growth models give sensible relations between hydrostatic stress and fracture strain, but predict microstructural particle spacing-to-size $(l/2r)$ values that are too low, i.e., the particles are predicted to be too close together (have too high a volume fraction) and/or the particles themselves appear to be bigger than they really are. This is because practical microstructures do not conform to the assumptions of the models. In consequence, the critical damage must be viewed as being attained in a "representative volume" of microstructure recently termed *damage cells* in porous plasticity calculations. Furthermore there may be distributions of different size particles all capable of voiding. The threshold $\bar{\varepsilon}_o$, below which damage does not accumulate is also important. Taking $\bar{\varepsilon}_o=0$ gives algebraic simplicity, but debonding around inclusions must first occur to form voids initially, and matrix-inclusion "strengths" vary with alloy type. In aluminum alloys with strong matrix-inclusion bonding, voids form only just prior to fracture. In spheroidized carbon steels with different volume fractions of cementite particles, ε_o was quite large (~ 0.5) so that nucleation contributed significantly to total ductility. In contrast, oxide or sulfide inclusions in steels and

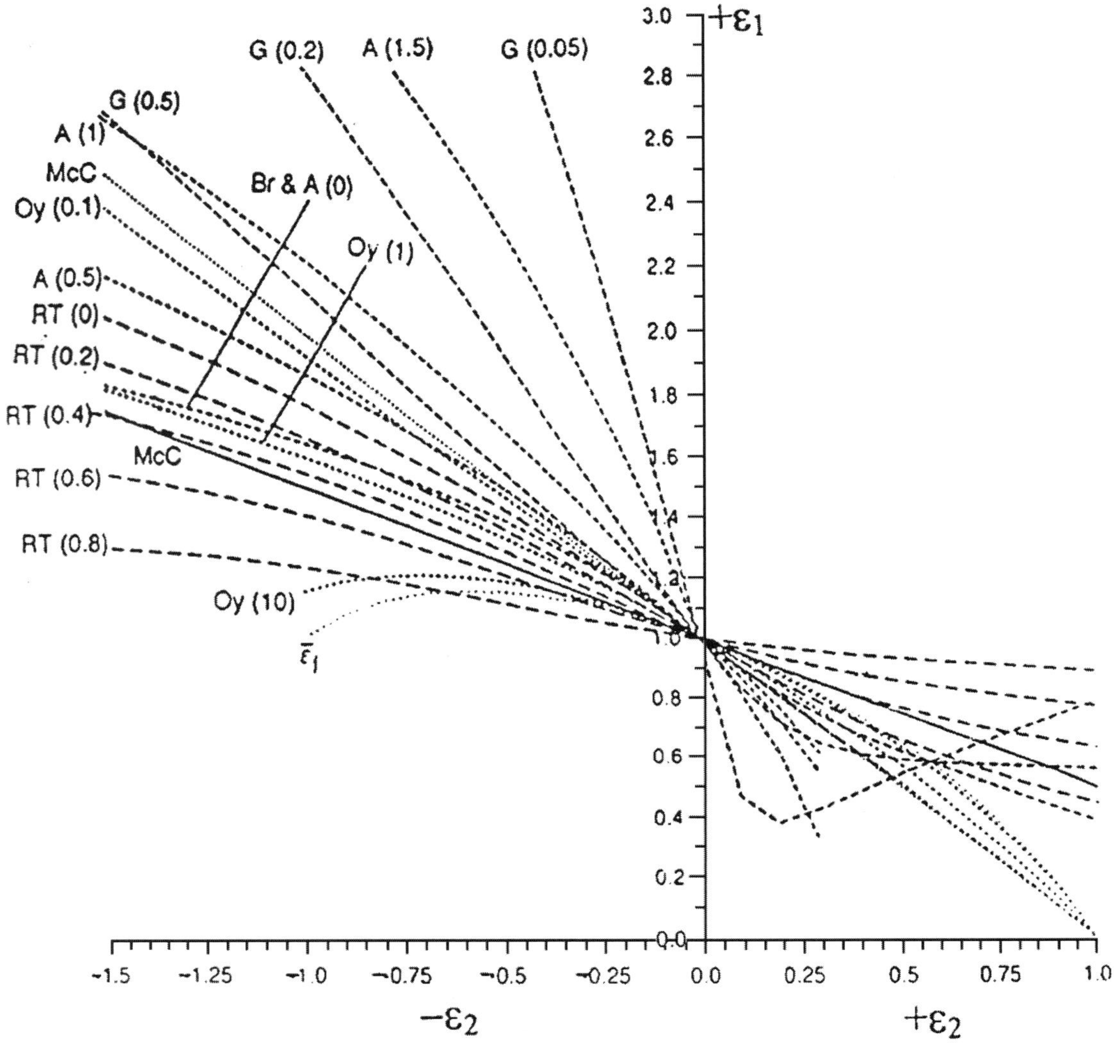

Figure 4
Predictions of various empirical and hole growth models for ductile crack initiation in plane stress (after Atkins 1997).

oxide particles in copper have much lower interfacial bonding. An implication of these studies is that $\bar{\varepsilon}_o$ is dependent upon hydrostatic stress, decreasing as σ_H becomes more tensile.

This raises the question of calibration of damage models. Experimental loci of fracture strains for particular metalworking processes may be used directly to design manufacturing routes to avoid fracture. In cases such as cracking in extrusion, those die geometry/friction conditions (large Δ factor) that induce local tensile stresses are simply avoided. Microstructural details and void growth mechanics are, perhaps, inconsequential. Problems arise when fracture is to be predicted in a metal forming operation on the basis of information in an entirely different process with an entirely different load history. Use of a single type of specimen to calibrate damage mechanics models, which are to be applied to conditions with greatly different $\sigma_H/\bar{\sigma}$, may very well be an *unsafe* procedure. It is clear also that incorporation of changing $\bar{\varepsilon}_o$ with load path will give different predictions for fracture along other load paths compared with predictions where $\bar{\varepsilon}_o$ is constant or zero. Calibration with only the one type of test piece would not identify that $\bar{\varepsilon}_o$ changes with $\sigma_H/\bar{\sigma}$, resulting in erroneous predictions for other loading paths. A further complication arises when necking occurs prior to fracture. Plane-strain necks accumulate damage at different rate from previous loading histories. As seen from the shape of the "forming limit diagram" for sheet materials, (see Sect. 5) necking strains do not change monotonically with strain ratio, being least for plane-strain tension, and greater for all other strain paths. The threshold strain may occur before necking for some paths but after necking for others. This will produce different shapes for failure loci. There is little, if any, experimental data on this important topic.

5. *Forming Limit Diagrams*

In the widest meaning, a forming limit diagram plots important variables, or groups of variables, to show combinations of conditions for which a particular method of forming is possible and those for which it is not possible. In tube bending, failure is either tensile at the outer sections of bends or compressive buckling at the inner radius. A forming limit diagram for bending thus has axes of (h/r) (bending strains) and (h/t) (propensity for buckling) where h is the distance from the neutral axis to the outermost sections, r is the radius of bend and t is the thickness. Formability envelopes depend on the section shape and material. Tension superimposed on bending (*draw bending, stretch forming*) shifts the neutral axis towards the inside of the bend, (it may even leave the section altogether), and elastic springback decreases. Achievable (h/t) increases, but achievable (h/r) decreases. (Elastic recovery is outside the scope of this article but might be thought of as a 'defect' in certain assembly problems.)

Another type of forming limit diagram is shown in Fig. 5, which relates the extrusion ratio to temperature for the extrusion of DTD 363 aluminum alloy. Since yield stress diminishes with increase of temperature, greater extrusion ratios are possible for the same extrusion pressure, as the temperature is increased. Similarly for a given extrusion ratio, lower pressures are required the higher the temperature. The maximum permissible extrusion temperature is in theory the alloy solidus (isothermal conditions, i.e., extremely slow ram speeds), but heat generated by the deformation heats up the billet, the limiting case for which would be adiabatic extrusion. Limiting constant pressure loci on the left of the diagram are given by plasticity theory, which also gives the irreversible work done, the temperature increase of the billet, and hence, the right hand limiting loci. Hot shortness can be incorporated in this sort of diagram.

The term "forming limit diagram" (FLD) is usually associated with the locus of the onset of localized necking (tensile instability) in the in-plane biaxial stretching of sheet material. FLDs are determined experimentally and are plots of the major and minor in-plane strain combinations at which localized necking occurs under different loading paths. They encompass necking under conditions where both strains are tensile (*stretching*) and where one is tensile

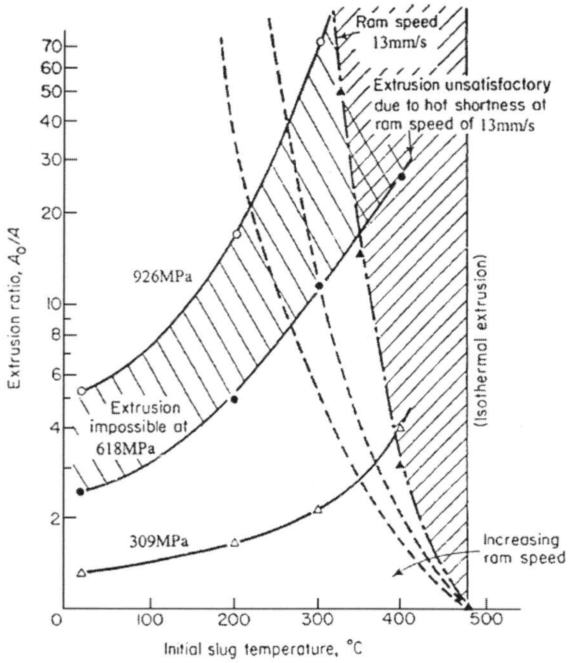

Figure 5
Limit curves for extrusion of an aluminum alloy (after Crane 1964).

and the other compressive (*drawing*). Strains are determined from the distortion of printed or etched small circles. A representative FLD for a low carbon steel is shown in Fig. 6.

Anisotropic materials displaying increased resistance to thinning are favored for pressing. The most important origin of anisotropy in metals is crystallographic, concerning preferred orientation of grains to give different "textures." In continuum terms, it is conveniently represented by the "r-factor" defined as the ratio of width strain to thickness strain in a strip tensile test ($r = 1$ for isotropy). If $r > 1$, the material thins at a lower rate than its isotropic equivalent. Under very special circumstances, the r-factor could have the same value in all directions in the plane of sheet material. Most often, however, r varies with direction, so an average r-factor is (arbitrarily) defined as:

$$\bar{r} = (r_0 + 2r_{45} + r_{90})/4$$

where the index 0 refers to the rolling direction of the sheet and the index 90 to across the sheet.

The r-factor is ultimately bound up with yield criteria for anisotropic solids and their associated flow rules. Since failure in cup drawing occurs when the punch load produces stresses in the cup wall exceeding its flow stress giving wall necking and

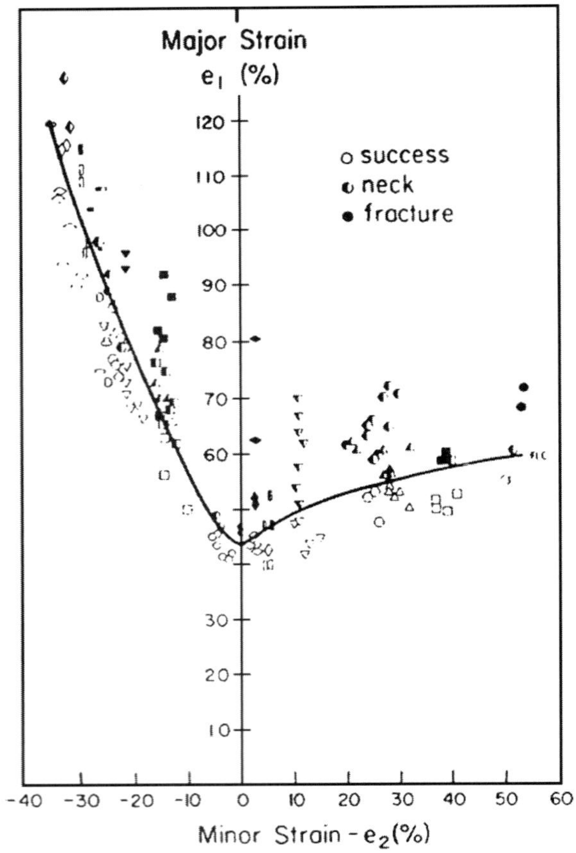

Figure 6
Forming limit diagram for a typical low carbon steel (after Hecker 1975).

fracture, the limiting draw ratio (LDR—the ratio of blank starting diameter to cup diameter) can be written in terms of $\bar{r}$, different relations applying for different yield criteria. It is well known that the greater $\bar{r}$, the greater the LDR. α-Ti alloys have the highest $\bar{r}$ (3 ~ 5) and very deep cups can be drawn. Hosford and Caddell (1983)gives $\bar{r}$ for various sheet metals; also listed are representative strain hardening exponents n and strain-rate sensitivities m in $\sigma = \sigma_0 \varepsilon^n \bar{\varepsilon}^m$, which also affect localized necking.

Bibliography

The following (where further references will be found) are only representative of an extremely large body of literature in different countries.

Atkins A G 1997 Fracture mechanics and metal forming. In: Rossmanith H P (ed.) *Fracture Research in Retrospect: G R Irwin Festschrift*. Balkema, Rotterdam, The Netherlands, pp. 327–50

Atkins A G, Mai Y-W 1985 *Elastic and Plastic Fracture*. Ellis Horwood, Chichester, UK

Backofen W A 1972 *Deformation Processing*. Addison-Wesley, Reading, MA, USA

Blazynski T Z 1989 Mode of deformation and material response. In: Blazynski T Z (ed.) *Plasticity and Modern Metal-Forming Technology*. Elsevier Applied Science, London, pp. 115–47

Clift S E 1992 Fracture in forming processes. In: Hartley P, Pillinger I, Sturgess C (eds.) *Fracture in Forming Processes in Numerical Modelling of Material Deformation Processes*. Springer, London, pp. 406–18

Crane F A A 1964 *Mechanical Working of Metals*. Macmillan, London

Dong Y, Tagavi K A, Lovell M E, Deng Z 2000 Analysis of stress in cross wedge rolling with application to failure. *Int. J. Mech. Sci.* **42**, 1233–53

Fischer F D, Kolednik O, Shan G X, Rammerstorfer F G 1995 A note on calibration of ductile failure damage indicators. *Int. J. Fract.* **73**, 345–57

Hecker S S 1975 *Sheet Metal Indust.* **52**, 671–5

Hosford W F, Caddell R M 1983 *Metal Forming: Mechanics and Metallurgy*. Prentice-Hall, Englewood Cliffs, NJ

Kalpakjian S 1984 *Manufacturing Processes for Engineering Materials*. Addison-Wesley, Reading, MA

Marciniak Z, Duncan J 1992 *Mechanics of Sheet Metal Forming*. Arnold, London

A. G. Atkins

Metal Forming: General and Economic Aspects

In a solidified state, metallic materials, which include steel, are plastically formed to a permanent set by forces applied externally.

The phenomenon of plasticity has its origin in the crystalline structure of metals. Its practical application in engineering and industry on a larger scale is to be found in the field of metal forming, or more specifically, rolling, open-die and closed-die forging, and extrusion technologies.

In the perfect crystal of a metal the atoms are arranged in periodic lattice structures. Plastic reshaping is produced where lattice layers being affected by shear stress drift off. Still, the lattice structure remains intact, there are merely slipping zones to be noted at the edges, and this is where the modification of shape proper takes place. Metals manufactured by conventional methods have lattice defects (vacancies, interstitial atoms, dislocations, grain boundaries, twin boundaries, phase boundary surfaces). Dislocations, for instance, cause the stress (and thus forces and energies also) required for forming to be two to three orders of magnitude smaller than in an ideal defectless crystal. While plastic forming is in progress, the number of dislocations is rising considerably. They are obstructing each other and/or piling up at obstacles, which results in an increase in resistance to deformation with progressive forming. Other lattice defects equally restrict ductility and increase strength—with the exception of the grain boundaries in creep which progresses slowly.

A basic requirement for any forming procedure is that stress conditions rendering the material ductile are available. Not only the yield stress of the material must be overcome but, frequently, also friction losses which, apart from the coefficient of friction in the contact surface between die and workpiece, are also affected by the geometry of the forming zone. If the geometry of the zone of deformation is not favorable, friction losses may amount to a multiple of the ideal energy rating calculated for the forming operation. This is the case, for example, when a very thin strip is rolled between relatively thick working rolls.

Forming of brittle materials, in particular, requires stress conditions with a high contribution of compressive stress in order to improve the formability of the material to be processed. This has led to the development of suitable metal-forming methods enabling, if possible, the application of pressure on all sides, such as extrusion or shape rolling (with grooved rolls).

1.1 Classification of Metal-forming Methods

(a) Classification by stress conditions. According to DIN 8580, forming is one of the principal kinds of manufacturing procedure, along with primary forming, cutting, joining, coating and changing of material properties. DIN 8582 (Fig. 1) shows the classification of metal-forming methods by their characteristic stress conditions:

- Compressive forming: rolling, open-die forming (open-die forging), closed-die forming (closed-die forging), extruding (cold extrusion);
- Tenso-compressive forming: e.g., drawing of wires, rods, and tubes (through tapered dies), deep-drawing of flat sheet-metal parts into hollow parts;
- Tensile forming: plates and bars;
- Forming by bending: plates and bars; and
- Shear forming: shifting, twisting.

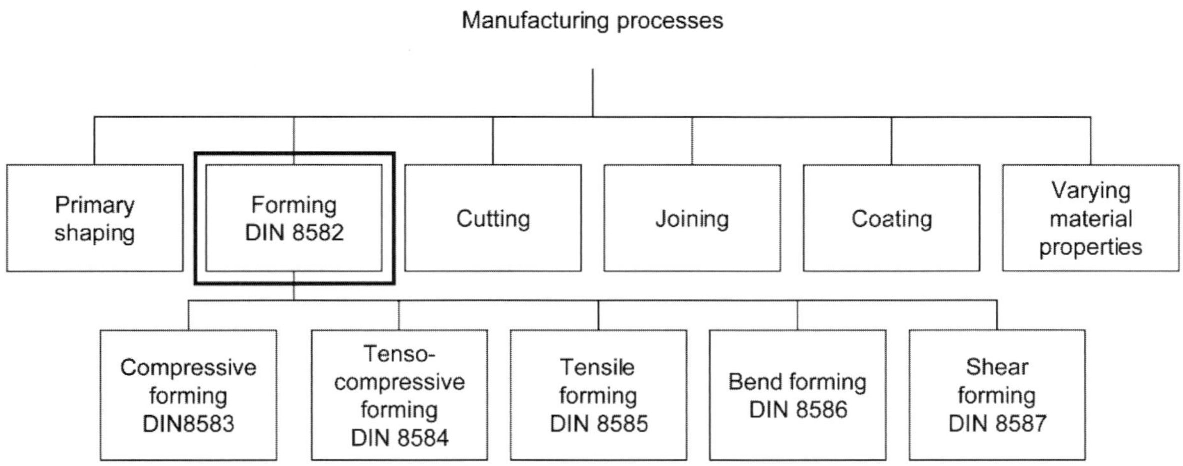

Figure 1
Manufacturing processes according to the DIN standard.

DIN 8583 to 8587 define and explain the individual forming methods.

(*b*) *Classification by product types.* Forming methods for the manufacture of shaped sheet-metal parts are referred to as sheet-metal forming methods, as against bulk-forming methods by which cast or forged solid semis are further processed into thick-walled, massive components. Figure 2 shows the most important forming principles.

(*c*) *Classification by initial temperatures.* Depending on whether the material is heated up prior to the forming process or not, one distinguishes between cold and hot forming processes. The increase in temperature caused by dissipative conversion of the energy of deformation into heat in the course of the process is not taken into account. In practice, the term warm forming is also used when reference is made to prior heating to a point below recrystallization temperature. Strictly speaking, hot forming would be the correct definition in this case, too.

In addition, it is important to know the number of workpieces to be produced (single piece manufacture or mass production) to enable selection of a forming method that is economical.

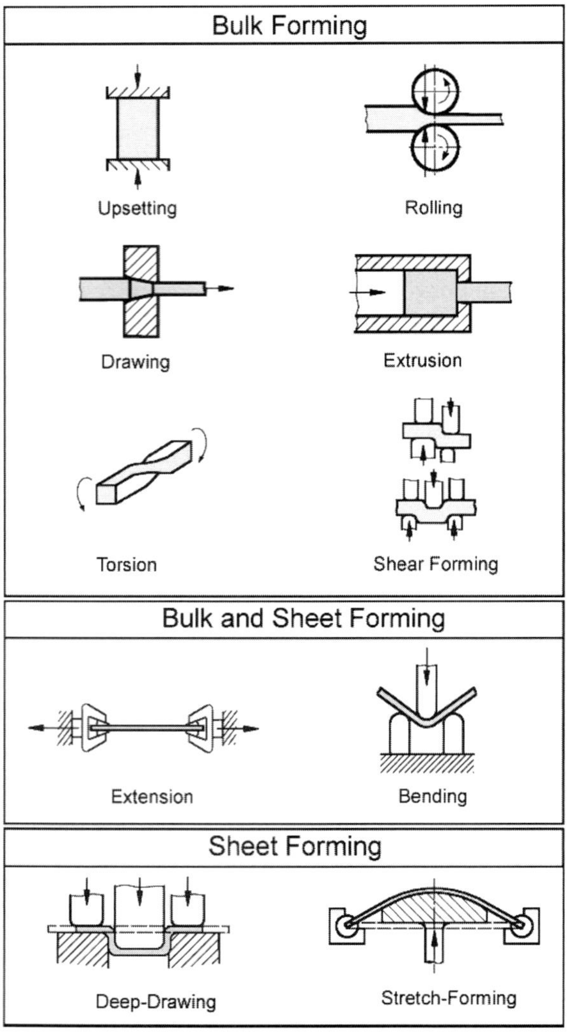

Figure 2
Metal forming principles.

1.2 Influencing Product Properties in the Course of Forming

Two basic tasks are handled by forming processes: setting of the desired geometry (shape designing) and obtaining a specific quality.

There are two fundamentally different strategies in shape designing: on the one hand, there is the possibility of die-related shaping, which means the final shape of the product is largely determined by the shape of the tool (e.g., closed-die forging). Minor deviations between the finished shape of the formed component and the shape of the die before its use in the process are caused by elastic deformation of the die and the workpieces during the forming process, as a result of mechanical load and of the change in the initial shape of the die produced by thermal effects and changes in the workpiece because of its cooling after the process. Thanks to precise numerical calculations, it is possible to consider such changes when the final geometry of a component is planned.

The other strategy of shaping is not die related. With this method, the material flow is controlled via simple controllable tools in such a way that the desired shape is produced (e.g., open-die forging).

The second task of forming processes is to influence the mechanical properties. Since real materials are, as a rule, multicomponent systems, it is advisable to take a look at those process parameters influencing quality at different levels (Fig. 3).

At the topmost level, the material in the whole workpiece is regarded as continuum, i.e., the material is averaged across all its phases. At this level variables such as force, energy, and power requirements of the process, average temperature in the workpiece, and global geometrical changes such as, for example, variations in length or width, are established. Normally, the elementary plasticity theory will be sufficient to calculate these parameters. Generally, this rough analysis serves to correlate global changes in shape, such as degree of deformation $\varphi_h = \ln(h_1/h_0)$ or the relative dimensional variations $\varepsilon_h = \Delta h/h_0$, with the properties of the finished product (e.g., strength and toughness).

Since nearly all forming processes cause inhomogeneous material flow, the local process history has to be taken into account for a more specific correlation

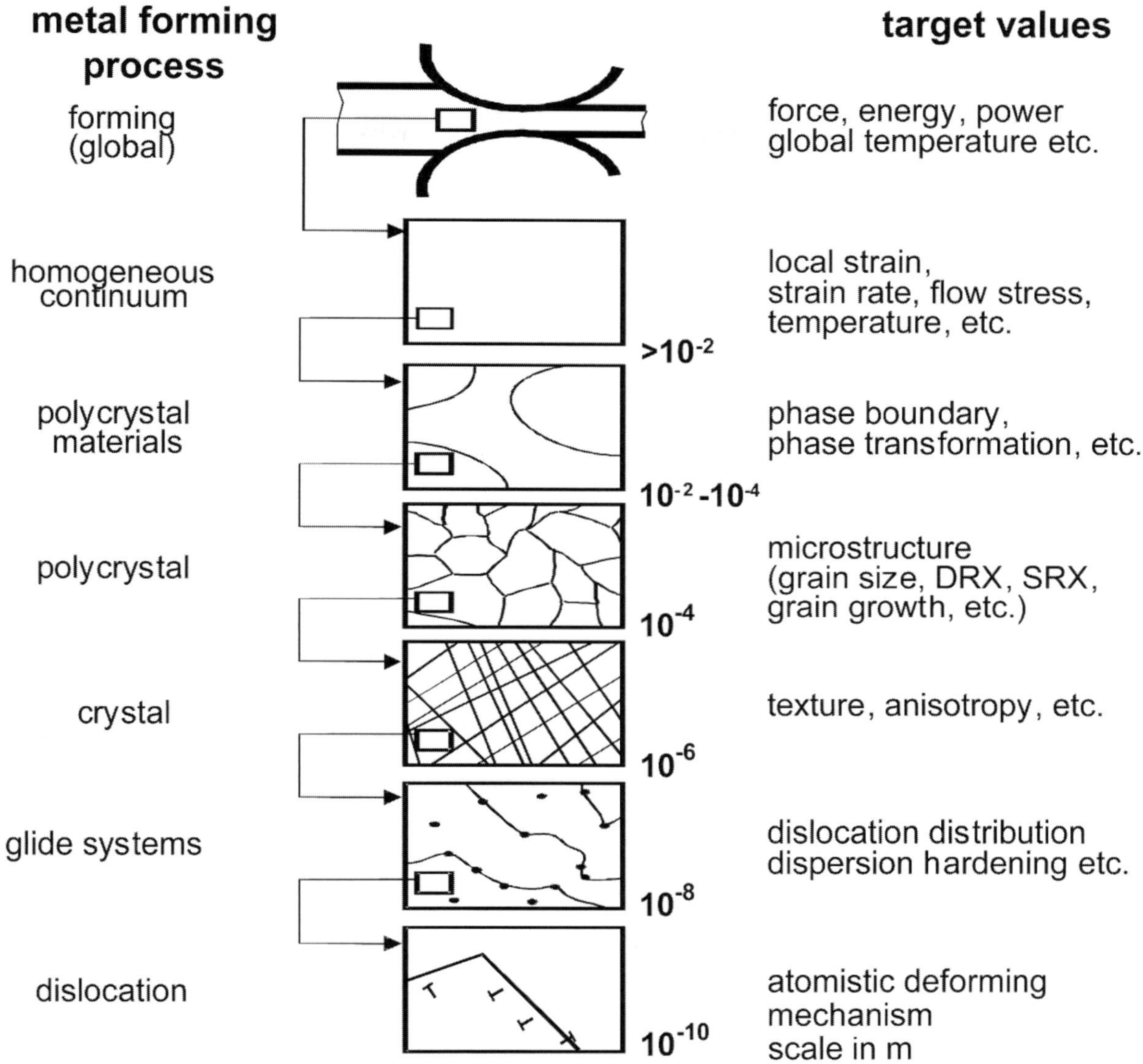

Figure 3
Parameters influencing quality at different levels.

of the process parameters with the product properties. This is accomplished by the next accuracy class, the advanced plasticity theory, which still describes the mostly multiphase material as continuum, locally, with a good approximation. The method now most frequently used to solve the differential equation system of the plasticity theory is the finite element method (FEM). It is used to calculate local shape changes, stresses, and temperatures and correlate them to the properties (e.g., strength, residual stresses) of the product measured in the relevant positions (depending in each case on priorities and computing time involved).

Both at the first and second level, the most commonly used terms to describe the material are the yield stress k_f as a function of logarithmic strain φ or degree of deformation ε, strain rate $\dot{\varphi}$ or rate of deformation $\dot{\varepsilon}$ and temperature θ. For a more precise description of the forming processes (third level) the material requires to be modeled in greater detail. For this purpose, the yield stress will ultimately have to be determined as a function of the concentration of alloy atoms in solid solution, dislocation density, grain size in polycrystals, stacking fault energy, frequency of twins, size, kind and occurrence of pores and inclusions, and texture. In view of the extremely extensive calculation work, in many cases, especially where results relevant for practical application are expected, use of statistical models would be an appropriate alternative.

These models average the processes in the course of forming, for example, in a grain. In this way different models for dynamic and static recrystallization and recovery as well as for transformation have been created. In addition, it is necessary to redefine in more detail the yield criterion on which the continuum mechanics are based and the flow condition law (e.g., Tresca and von Mises). The flow condition laws of the third level have to consider grain size, anisotropy, and pore shape in the case of powder–metallurgical material shape. Statements based on the microplasticity theory will become more significant with growing computer capacities.

The relationship between the process parameters and the properties finally obtained can also be described quantitatively. These correlations are mostly based on empirical or semiempirical functions.

The following sections describe the most important relationships in hot, cold, and warm forming.

2. *Hot Forming*

The resistance to deformation of metals is lower at high forming temperatures than at room temperature, so that force and energy requirements (other than purely thermal) are low when forming is performed at high temperatures. Casting defects can be minimized or even remedied by hot forming. Also, with high temperatures thermally activated modifications in the microstructure become possible, such as recrystallization and recovery. On the one hand, these processes counteract strain hardening of the material and thus reduce forming forces although forming continues, and on the other hand, properties of the metallurgical structure, e.g., grain sizes, can be additionally varied by controlled thermomechanical treatments (TMT).

The physical material properties, too, are essentially defined by hot forming in combination with the thermal treatment. The subsequent cold forming process only allows a fine adjustment.

2.1 *Influence of the Casting Structure on Hot Forming*

The as-cast structure depends on solidification and cooling conditions and therefore varies within a casting, which may cause an inhomogeneity of the mechanical properties. In addition, it is characterized by grain size, density, inclusions, networks of intermetallic phase parts, microsegregations, etc., and thus has a decisive influence on the material and component properties.

This means that the local conditions in the workpiece must be known to the manufacturing engineers to enable them to develop the forming strategy from this basis. What was said before equally applies here: it is true that global forming largely determines shaping. Foundry defects, however, can only be remedied by knowing the local conditions. Local modifications of the shape must either be so amply sized that pores, for instance, are closed again and welded during forging, or a maximum strain must not be exceeded in order to avoid bursting in subsequent rolling.

Thanks to recent casting methods such as continuous casting, thin strip casting, etc., the quality of the cast products has been considerably improved over the past years, which automatically also means a reduction in casting defects. Where large forgings, e.g., turbine shafts, are manufactured, for which cast ingots of up to some 600 t are required, the occurrence of foundry defects cannot be excluded. Any segregations, shrinkholes, pores, cavities, porous zones, and inclusions existing would have a lasting effect on downstream forming operations or, in the worst case, cause the part to fail in operation later on.

Such defects cannot be remedied before a precise analysis of the metal-physical processes and of the plasto-mechanical relationship has been made. It is not possible to draw any conclusions from shaping (globally) about local sizes and parameters. It is rather a matter of finding the right strategy for remedying the defects by precisely calculating local conditions within the workpiece to be formed; for instance, by using the slip line field theory or FEM simulation. In Fig. 4,for example, the local state of stress for cogging has been calculated as a function of

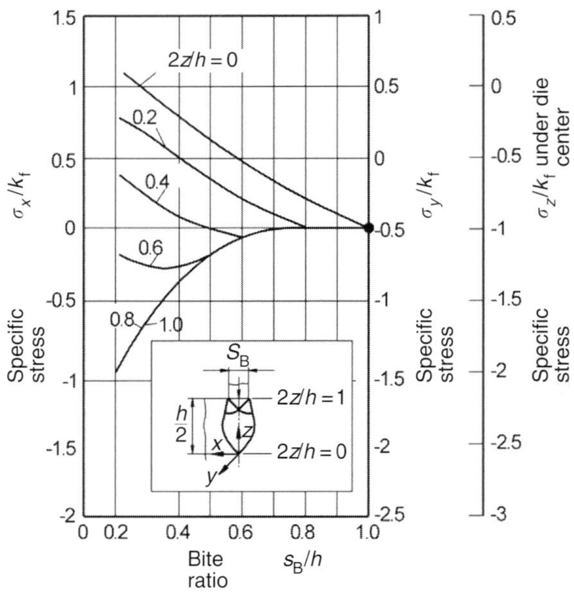

Figure 4
Stress state during cogging (below die center).

the bite ratio s_B/h. The result indicates the cause of the internal cracking (occurrence of tensile stresses in the core of the forging at a small bite ratio) (Nieschwitz 1988).

Process-related and metallurgical options, or a joint strategy of both, are indicated for the remedy of defects. Cavities can be closed again by upsetting, cogging, and also rolling, provided local forming is optimal. Controlled forming accompanied by an annealing treatment equally serves to remedy defects by initiating thermally activated processes, such as recrystallization or phase transformation, in order to accomplish restructuring.

2.2 *Influence of Microstructural Changes on Hot Forming*

Nearly all metals, except for some cast or metal powder parts, undergo changes in their microstructure in the course of hot forming. These changes decisively influence properties and quality of the final product. Process-engineered details such as forming force ratings, material flow and thus final dimensions, are determined by changes in the structure.

Consequently, forming behavior and any microstructural changes occurring in the material have to be considered when multistage forming processes are planned and engineered. Applying the so-called TMT (see also Sect. 2.5), this is taken into account.

Time–temperature control and time-controlled forming enable setting of a desired material state with controlled material properties. While hot forming is in progress, the material, initially of low-dislocation coarse-grained structure, is hardened because it is reshaped at a certain strain and strain rate. Hardening manifests itself in a clear rise in dislocation density along with an increase in yield stress. In the course of hot forming operations, thermally activated processes such as dynamic recovery or recrystallization may set in as a result of the high temperatures, counteracting hardening and thereby reducing the yield stress (as illustrated in Fig. 5).

These metal-physical processes lead to a rearrangement of dislocation structure, or even to a new formation of microstructure owing to grain boundary migration. Basically, the distinction is made between two processes responsible for removal of work hardening (see *Plastic Deformation: Role of Recovery and Recrystallization*, Doherty *et al.* 1997).

Recovery is characterized by migration and rearrangement of dislocations, the mobility of which is assisted by the high temperatures present during hot forming. Owing to the fact that dislocations tend to take energetically more favorable positions, destroy each other (annihilation), or leave the crystal, a spatially closed network of many small angle grain boundaries (polygonization) is formed. This new energetically more favorable substructure is created

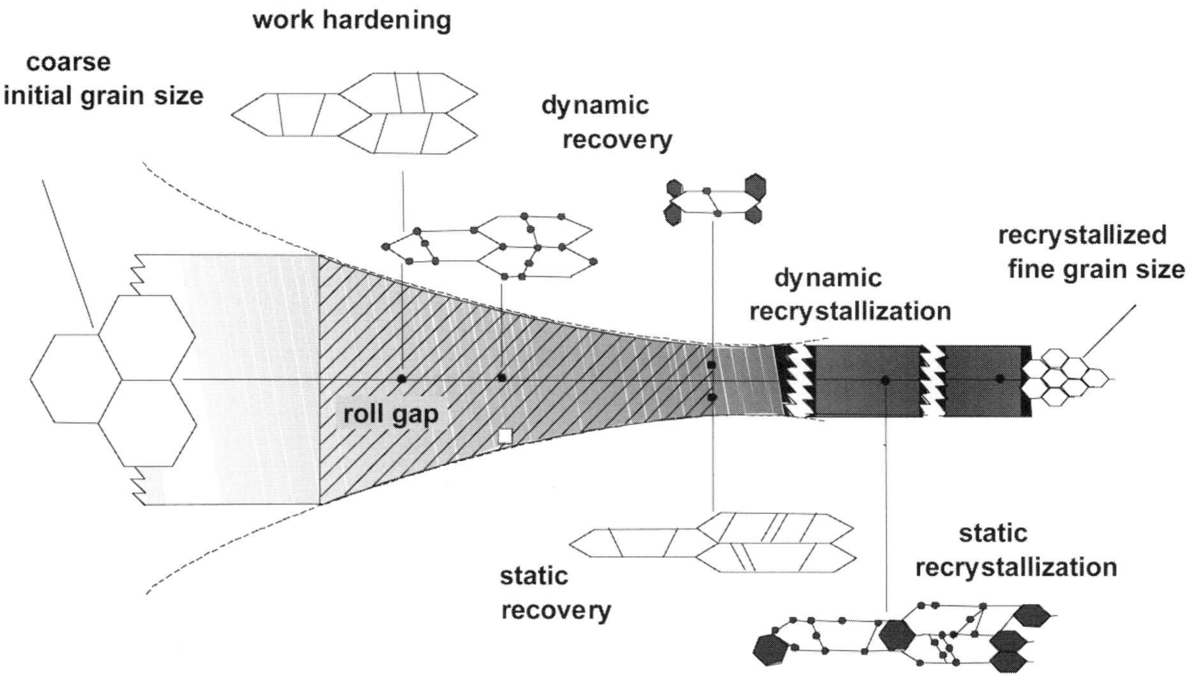

Figure 5
Structural change during and after hot forming.

solely by thermal activation. An incubation time is not required, so that the kinetics observed in the course of this process also differ fundamentally from that of the recrystallization.

Recrystallization, by comparison, describes the process of forming a new structure, by means of high angle grain boundaries emerging and migrating while annihilating the structure of deformation.

It is characterized by the processes of nucleation and growth, which are also summarized by the term primary recrystallization. In the course of an annealing treatment, primary recrystallization might still be followed by grain coarsening.

Forming technology makes use of these metallurgical phenomena in order to select the final properties of the product by a precisely defined structure. It has been verified by numerous tests that, for example, the tensile yield strength of unalloyed steel grades with low carbon contents rises in inverse proportion to the root of the ferrite grain size. This relation is, for the lower yield point, described by the Hall–Petch equation:

$$R_{eL} = \sigma_r + k_y d^{-1/2}$$

where σ_r is the stress to overcome the resistance, which must be available for initiation of the dislocation movement in the grain, k_y is a constant to describe the grain boundary resistance, and d is the average grain diameter.

2.3 *Toughness and Fracture Behavior*

The mechanical properties of modern structural material are essential for practical applications. A high tensile strength should be combined with, if possible, toughness over a wide temperature range and all kinds of loads must be accepted for reliable use. By optimal alloy selection in connection with perfectly matched forming and heat treatment, sophisticated high-tech materials are designed to suit requirements.

2.4 *Transformation*

Transformation rates among the most important procedures in the manufacture of components and is the basis for TMT of metallic materials. Mechanical properties such as strength, ductility, toughness, and fracture behavior are decisively determined by the kind and distribution of secondary phases in the structure.

In order to be able to benefit from the advantages of a transformation during heat treatment or TMT, it is absolutely necessary to know precisely the fraction of the precipitated phase as a function of annealing temperature and annealing period. This information is to be found in so-called continuous-cooling transformation diagrams (also referred to as CCT graphs). An extension to these graphs, showing the influence of a previous forming sequence of steels on austenite transformation, are the so-called forming-time continuous-cooling transformation diagrams (FTCCT). With strain being considered, they are significant for TMT as well as heat treatment by rolling and by forging heat.

2.5 *Thermomechanical Treatment*

The purpose of TMT is to provide, by selective control of temperature and forming conditions, a final product with material properties which this particular material would never be able to reach under conventional production methods. As already discussed above, mechanical and technological properties (strength, toughness, resistance to corrosion, and anisotropy) essentially depend on microstructure. Optimization of these properties by TMT requires precise knowledge of metal physical processes such as hardening, removal of work hardening (recovery, recrystallization), precipitation, and allotropic transformations. These four metal physical processes are directly connected with the use of microalloying elements which play an important part in TMT.

In the field of rolled products, both in the steel and nonferrous metal line (aluminum, copper), there are numerous TMT strategies, described in more detail in *Hot Rolling* (DeArdo 1997, Hirsch 1996).

3. *Cold Forming*

Cold forming is applied when either material hardening is to be exploited, or when the demand is for a minimal post-treatment of the finished part with respect to its shape. To control the shape and the final properties of the part, it is very important to know the yield curve. The hardening process, which is described by the yield curve, depends on the classical continuum mechanical theory on distortion. From the metallurgical point of view, the crystallographic texture brought into the material through different production steps has an important influence on hardening behavior. This will be described in the following section.

3.1 *Hardening*

(*a*) *Uniaxial stress state.* In the well-known tension test of a metallic bar at room temperature, the stress–strain curve is more or less linear at the beginning of the experiment and becomes nonlinear at a certain stress called the yield stress σ_y. For $\sigma < \sigma_y$ the material behavior is called linear-elastic (reversible distortion) and for $\sigma \geqslant \sigma_y$ inelastic or plastic (irreversible distortion). The elastic range is described by the Hookean law of uniaxial stress state. In the plastic range, the necessary stress for further distortion becomes disproportionately higher and higher in comparison to the elastic range. This effect is called hardening. It should be noted that the stress definition results in two different stress–strain curves: the nominal or Piola–Kirchhoff stress is related to the *initial* cross-sectional area of the bar, which results in a curve with a maximum. The sloping down of the curve after the maximum is called softening.

The true or Cauchy stress is related to the *actual* cross-sectional area of the bar and the curve steadily slopes upwards until the bar is broken. In practice, a distinction is made between three different kinds of hardening for different metals and metal alloys: isotropic, anisotropic, and kinematic hardening. In the case of isotropic hardening the stress–strain curve is mirrored against the stress axis and the strain axis of the graph when the load is reversed. In the case of kinematic hardening and reversing from tension to compression load, plastic yielding starts at a lower order of magnitude of the stress than in the previous tension test. This is called the Bauschinger effect. Anisotropic hardening is often found in rolled sheets when flat tension specimens lying in directions differing from the rolling direction are taken from the material. In subsequent tension tests, different yield stresses of the tension strips are measured.

(*b*) *Multiaxial stress states.* The definition of the yield stress in the tension test described above is no longer valid in multiaxial stress state. In the tension test one stress only is acting on the bar and therefore this is the only stress which is responsible for yielding. In a multiaxial stress state, all stresses are causing plastic yielding. Mathematically speaking, it can be said that yielding (under the assumption of invariant volume) is caused by a stress state which is expressed by a yield function of the deviatory stress tensor σ'_{ij}. This must be equal to the yield stress when yielding starts: $f(\sigma'_{ij}) = \sigma_y$. This function is independent of the coordinate system, and if the material is assumed to be isotropic, it is also independent of the load direction. Therefore, this function can be expressed in terms of the invariants of the deviatoric stress tensor: $f(I'_2, I'_3) = \sigma_y$ (the first one is zero). For isotropic material, von Mises assumed that the square of the second invariant is proportional to σ_y^2 which leads to the well-known function

$$\sigma_y = \sqrt{\frac{3}{2}\sigma'_{ij}\sigma'_{ij}}$$

In principal stress space, the implicit function

$$F(\sigma'_{ij}) = \frac{3}{2}\sigma'_{ij}\sigma'_{ij} - \sigma_y^2 = 0$$

represents a cylinder whose longitudinal axis is coaxial to the diagonal of the 3D deviatory stress space. An elastic stress state produces a point lying inside the yield locus, while a plastic stress state produces a point lying exactly on the yield locus. Points outside the yield locus are not valid. In the special case of a biaxial or plane stress state, one of the principal stresses is zero and the cylinder becomes an ellipse. The hardening types described above could also be found in the multiaxial stress state. If one imagines the yield locus as an ellipse in the case of plane stress, isotropic hardening would lead to homogenous expansion of the ellipse while anisotropic hardening would lead to a distorted ellipse. Kinematic hardening causes a translation of the ellipse through the 2D stress space. It should be noted that the yield locus has to be convex.

(*c*) *Physical reasons.* The reasons for hardening from the point of view of material science are briefly itemized below. The most important mechanism for plastic distortion is dislocation gliding. In any case, the reduction of dislocation mobility is the main reason for hardening in cold forming. The dislocation mobility is mainly reduced by the following mechanisms:

(i) Damming up of mobile dislocations in parallel planes at obstacles such as precipitates, and the evolution of dislocation forests.

(ii) Evolution of immobile dislocations (Lomer–Cottrell dislocations) caused by dislocation interaction.

(iii) Holding up of dislocations at high angle boundaries (grain boundary hardening). This hardening effect is the stronger the smaller the grains are.

3.2 Textures

Commonly, structural materials are aggregates of many grains (i.e., single crystals). Such aggregates are called polycrystals. The crystallographic orientation of the grains is generally nonrandom, which is attributable to many reasons, but mainly to the metal-forming process. All nonrandom orientation distributions are called preferred orientations or texture. During metal-forming processes the grain orientation is changed by lattice rotation, and the texture intensity increased. The crystallographic texture is of utmost importance for anisotropic mechanical behavior, and in a simple cup drawing process of rolled metal sheets the well-known "earing" phenomenon might show.

Each single crystal is strongly anisotropic in its behavior, but if the crystallographic orientation of the grains is nearly random, the behavior of the polycrystal is macroscopically isotropic, or quasi-isotropic. In metal forming, textures are mainly of interest in sheet manufacturing (because of the material behavior in deep drawing of car bodies or cans) in order to reduce production errors.

A modern possibility to represent textures is the so-called orientation distribution function (ODF) defined in a 3 D Euler space which, in comparison with pole figures, allows a mathematical description of textures. In the Euler space many typical textures can be identified as lying on certain fibers.

In order to reduce production errors and prototyping time for die design, an accurate prediction of the material flow and forming limitations with the aid of FEM is desirable. Deep drawing process simulations are often carried out using elastoplastic FEM codes which are based on the usage of the yield locus that separates the elastic and the plastic range. From the numerical point of view, knowing the yield locus is essential. In the isotropic case, the von Mises yield locus agrees well with experimental data and therefore it is mainly used for isotropic elastoplastic metal forming simulations. If, however, the material is strongly anisotropic in its mechanical behavior, the resulting yield locus cannot be described accurately by the (quadratic) von Mises yield surface, because it is only able to expand and not to undergo a deformation. Therefore, a lot of research work has been done and today some phenomenological anisotropic yield locus functions are available (e.g., see Barlat *et al.* 1999 and Hill 1993).

They contain material anisotropy coefficients, and one of the most important pieces of information of such functions is the number of these coefficients, how they can be obtained, and with what amount of experimental work (and cost). The most significant property of phenomenological anisotropic yield functions is their nonquadratic mathematical form. Barlat *et al.* (1999) have shown in FEM simulations of spherical cup deep drawing that using phenomenological anisotropic yield functions leads to a different number of wrinkles than the isotropic von Mises function. This is why imperfections in sheet metal forming are chiefly caused by anisotropy.

For the purpose of enabling prediction of crystallographic textures, several texture simulation models have been developed (see Van Houtte 1996). The best known one is the so-called Taylor model. The paper of Zhou *et al.* (1996) is very interesting where the influence of the presence of different texture components on FEM cup drawing simulations was investigated by calculating the yield locus from texture. It could clearly be seen which was the best mixture of texture components to obtain a minimal ear height in the cup. It should be noted that the prediction quality of present texture models differs for b.c.c. and f.c.c. metals (see *Texture*).

3.3 *Residual Stresses*

The development of residual stresses of a metallic material is caused by the manufacturing processes, which are, for example, casting, forming, cutting, joining vapor depositing, and annealing. In general, one has to distinguish between tensile and compressive residual stresses. Depending on the size of the material range in which residual stresses are homogenous, three different kinds of residual stresses are known (Preckel 1988, *Residual Stress: Definition*):

(i) Residual stresses of the first type are homogenous in wide material areas (several grains). A disturbance of the internal force and momentum equilibrium will cause macroscopic shape deviations.

(ii) Residual stresses of the second type are nearly homogenous in smaller material areas (e.g., one grain). Disturbance of the force and momentum equilibrium might cause macroscopic shape deviations.

(iii) Residual stresses of the third type are inhomogeneous across several atomic distances.

The most important influences of residual stresses on the material properties are the following:

(i) As an influence on fatigue strength, compression residual stresses increase the dynamical strength, caused by the superposition of tension and compression stresses.

(ii) In the presence of a corrosive medium some materials show a brittle behavior caused by inter-and/or intracrystalline cracks (stress corrosion cracking). This effect is mainly dependent on the tension surface stresses and cracks that are induced at the surface. In stainless steels, chloride ions are necessary for stress corrosion cracking. If such materials are used in a chlorine-containing environment, the presence of tensile stresses and residual stresses must be avoided.

(iii) The shape of parts is also influenced. Tension residual stresses lead to decreasing dimensions and compression stresses to increasing ones.

Since the experimental determination of residual stresses is very difficult, some researchers use the FEM based on elastoplastic constitutive laws to compute them. It should be noted that stresses calculated in displacement-based FEM codes are naturally not steady and jump from element to element. The use of very fine meshes will lead to a decreasing stress difference. Therefore, the FE meshes have to be very fine if the stress computation is a major issue, as in the case of residual stresses. Hoppe and Kopp (1991) investigated and validated different parameters influencing the residual stresses in wire drawing. With the aid of FEM simulations it was found that the optimal value for a cylindrical guidance of a drawing hole with respect to the elastic energy in the workpiece is

$$\frac{l}{d_1} = 0.92$$

where l is the length of guidance and d_1 the exit diameter. The elastic energy is regarded as a measure for the residual stresses brought into the material.

4. *Warm Forming*

When finished parts are to be manufactured, the surface quality and dimensional accuracy are, among other things, vital criteria for the product quality. The process typically used where demands concerning surface quality and dimensional accuracy are high is cold forming. This means that the workpieces are not especially heated immediately before manufacturing starts. Cold forming, however, meets technological limits because of the generally very high yield stresses. Metal-forming units cannot produce any force desired, and the formability of the material is limited. Since both these properties depend on the temperature to a considerable extent, the yield stress can generally be lowered and formability be improved by having a higher forming temperature. An increase in temperature to 900°C and even to 1200°C as normally used in steel hot forming results, however, in much scaling and consequently less dimensional accuracy and poorer surface quality so that further treatment is necessary. An alternative to this is the warm forming method. Usually, it takes place in a temperature range between room temperature and transformation temperature and so combines the advantages of hot forming with those of cold forming. For low-alloy carbon steels and case-hardening steels the range of blue shortness is a bottom limit. In a temperature range of about 200°C to 400°C the yield stress of these steel grades shows a relative maximum whose position is dependent on the speed: at higher rates it will be shifted towards higher temperatures. The yield stress may, in this case, reach values usually occurring at room temperature. With higher alloyed steels, the range of blue shortness does not become so apparent and is possibly only discernible by a slight drop of the yield stress. Above blue brittleness the yield stress is noted to drop until it rises again in the range of red shortness at approximately 800°C. Since, in the range of red shortness, scaling has already been considerably increasing, the range of red shortness limits the temperature range for warm forming. There is no blue shortness to be found in austenitic steels so that heating to 400°C would already be sufficient to obtain a considerable reduction of the yield stress (see also *Metal Forming (Warm): Comparison with Hot and Cold Forming*).

5. *Determination of Process-chain Efficiencies*

When assessing metal-forming processes by doing an economical balance, it is suggested to study the energy balance, since the amount of energy spent on the individual production steps is the dominant process parameter in most cases. A keyword in this context is cumulative energy input (CEI), which was defined in a concept issued by the Verein Deutscher Ingenieure 1995 as a characteristic value. Primary energies, by this definition, are all original sources of energy. These are, in particular, fuels such as oil and natural gas, but also regenerative ones such as water and wind.

The CEI is calculated from the sum of the primary energy input for the manufacture, use and waste removal of a product. In contrast, there is the so-called ideal energy input. Unlike the CEI, the ideal energy requirement is not related to any real production process or to a special technology. The ideal energy requirement is the energy input absolutely necessary to manufacture a product with specified properties. It is a theoretical quantity defined by a loss-free, chemically ideal model for the change of state aimed at. Material losses, for instance, caused by scaling during heat treatment have technological causes and are not necessarily the consequence of the physical change of state.

The relationship of ideal energy input and real CEI represents the efficiency of processes and can be used for their evaluation. To obtain to the efficiency of a process chain, one has to split up the entire process into individual increments which are first analyzed separately in order to calculate W_{id} for each step and the CEI. For the total efficiency, the relation is as follows:

$$\eta_{\text{process-chain}} = \frac{\min \sum W_{\mathrm{id}}}{\sum CEI} \tag{1}$$

The minimum ideal energy W_{id} is to be related to the overall process chain. It may well be more favorable to invest more energy into one process step in order to save energy in the next one by doing so (e.g., less heating of a workpiece although more forming force is required).

The ideal energy in heating a workpiece is related only to the mass that is actually being used for further processing. In particular, the scale incurred must be deducted from the starting mass. The ideal energy necessary to heat processes from a temperature θ_1 to θ_2 is, according to the first principle of thermomechanics for isobar processes, as follows:

$$W_{\mathrm{id}} = m_{\mathrm{w,used}} \int_{\theta_2}^{\theta_1} c_p(\theta)\mathrm{d}\theta \tag{2}$$

where $m_{\mathrm{w,used}}$ is the actually heated mass of the workpiece and c_p the temperature-dependent true specific heat capacity. When a round blank is heated in a gas furnace, the CEI is calculated as sum of the gas energy used, the energy consumption of the auxiliary equipment, the CEI of the auxiliary material and consumables used, the CEI for residual material recycling, and the CEI of the percentage of the mass lost as a result of scaling.

The ideal energy for a metal-forming process is calculated as product of the workpiece volume by work of deformation (integral of yield stress as a function of strain).

$$W_{id} = V \int k_f d\phi \qquad (3)$$

The relations for the calculation of the processes of upsetting, piercing, and shearing off are derived from Eqn. (3).

The process chain efficiency is obtained by calculating the minimum from the sum of the ideal individual energies—Eqn. (1) follows. Optimization parameters are, first, all internal state variables influencing the energy requirement. Essentially, these are the strain, the strain rate and the temperature. The strain is specified by the intended shaping, so it is not a variable. The lowest technologically feasible strain rate should be selected in order to design an energy-optimized process. The internal state variable left for energy minimizing is the temperature.

Determination of the process chain efficiency may reveal weaknesses in the production and show by which measures substantial energy savings are achievable. It is important to have realistic approaches to the calculation of the CEI and the ideal energy, as the present paper has demonstrated.

Bibliography

Barlat F, Kim J-B, Yang D-Y, Yoon J-W 1999 The effect of plastic anisotropy on compressive instability in sheet metal forming. *Int. J. Plast.* **16**, 649–76

DeArdo A J 1997 The physical metallurgy of thermomechanical processing of microalloyed steels. In: Chandra T, Sakai T (eds.) *Proc. THERMEC '97, Int. Conf. Thermomechanical Processing of Steels and Other Materials*. The Minerals, Metals and Materials Society, Warrendale, PA, pp. 13–29

Doherty R D, Hughes D A, Humphreys F J, Jonas J J, Juul Jensen D, Kassner M E, King W E, McNelley T R, McQueen H J, Rollet A D 1997 Current issues in recrystallization: a review. *Mater. Sci. Eng. A* **238**, 219–74

Hill R 1993 A user-friendly theory of orthotropic plasticity in sheet metals. *Int. J. Mech. Sci* **35**, 19–25

Hirsch J 1996 Thermomechanical processes in hot rolling of aluminium alloys. In: Hutchinson *et al.* (eds.) *Conf. Thermomechanical Processing—Theory, Modelling and Practice*. Swedish Society for Materials Technology, Stockholm, September 1996, pp. 78–92

Hoppe M, Kopp R 1991 Eigenspannungen in gezogenen und gerichteten Drähte; Stand der Erkenntnisse. *Draht* **42**, S885–94

Meyer L, Müschenborn W, Schriever U 1996 Principles, practice and results of thermomechanical processing of flat-rolled steel products. In: Hutchinson *et al.* (eds.) *Conf. Thermomechanical Processing—Theory, Modelling and Practice*. Swedish Society for Materials Technology, Stockholm, September 1996, pp. 93–119

Nieschwitz P, Kopp R 1988 Beitrag zur Qualitätsoptimierung beim Freiformschmieden. *Arch. Metall* **33**, 2

Preckel U 1988 *Eigenspannungen—Rechnergestützte Bestimmung in Umformteilen*. Verlag TÜV Rheinland, Köln, Germany

Van Houtte P 1996 Microscopic strain heterogeneity and deformation texture prediction. In: Liang Z, Zuo L, Chu Y (eds.) *Proc. ICOTOM11 Conf.*, Xi'an, Republic of China. International Academic Publishers, Beijing, pp. 236–47

Verein Deutscher Ingenieure (VDI) 1995 *Kumulierter Energie auf-wand—Begriffe, Definitionen, Berechnungsmethoden Entwurf zur VDI-Richtlinie 4600*. VDI, Düsseldorf. Germany

Zhou Y, Jonas J J, Sovoic J, Mahinde A, McEwan S R 1996 Effect of texture on earing in fcc metals: Finite element simulations. *Int. J. Plast* **14**, 117–38

R. Kopp

Metal Forming: Microparts

In metal shaping microparts are referred as structural parts which have dimensions below the millimeter range at least in two dimensions. Such parts are needed, for example, for systems like miniature hard disks with a capacity of more than 300 M byte and $\sim 2.5\,cm^2$. For the fabrication of such parts different methods might be used.

Investment casting is a method of primary shaping. It uses a thin ceramic shell, generated on the surface of a lost model. The geometry of the model, which might be complex, is reproduced, the minimum part size is limited to $100\,mm^3$.

Micro-electromechanical systems (MEMS) are manufactured using technologies which are commonly used in the production of microchips. These are deposition and etching methods or built-up methods (LIGA, LI tomography, GA laserforming reproduction). The difference from microelectronic chips is a larger thickness of the layers and the use of sacrificially released layers to allow the generation of movable elements. While the etching methods produce flat structures (height/diameter = aspect ratio < 1) preferably from silicon, LIGA can deliver metal parts (nickel, gold, and copper) with an aspect ratio of 10. The volume of these parts lies in the order of $\leqslant 1 \times 10^{-4}\,mm^3$.

Parts with a size between these dimensions, e.g., parts with a volume of $1\,mm^3$, can be manufactured by metal forming, i.e., controlled plastic deformation of a metal part without melting, removing, or adding material. Both sheet metal and bulk metal parts can be made as microparts. In the case of sheet metal parts the sheet thickness is in the micrometer range. This section is focussed on the metal forming methods.

1. *Forming Methods*

In principle, all methods of conventional metal forming techniques might also be used for micro metal forming. Due to the excellent capacity concerning form filling, superplastic forming of microparts was successfully demonstrated by Saotome *et al.* (1993; e.g., see Fig. 1). Even the details of the gear teeth of the small gear with a diameter of 1 mm are well developed after superplastic forging.

In addition, there are other methods which become of increasing interest when the part size is reduced, e.g., laser forming (see *Laser Forming of Metals*). Laser forming is based on the generation of thermal stresses induced by laser irradiation of the part. These stresses induce local plastic strains, which can be used for bending when a proper control of the beam intensity is applied. The bending angles are small and can be used for high precision aligning of mounted parts. As no solid tools are necessary the method can easily be applied in clean room conditions which are often a prerequisite for the production of the systems mentioned above.

Figure 1
Gear and double gear made by superplastic forging of $Pd_{40}Cd_{30}Ni_{10}P_{20}$ (after Saotome *et al.* 1993).

2. *Size Effects*

According to the theory of similarity the downsizing of a processes should be made by obtaining rules for similarity. For example, all dimensions of a part have to be reduced by the same factor, the length scale λ, to obtain similar results. So, all dimensions x, y, and z of a part should be reduced in a like manner according to $x' = \lambda x$. For the elastic, plastic, and thermal effects there are similar factors. The deviation of the measured values from the expected values, which occur in spite of the fact that the size reduction was done with a constant λ, are called *size effects*.

There are three groups of values where size effects occur when conventional metal forming processes are scaled down to the micro size: the mechanical behavior (especially strength), tribology, and adhesion. The effects can be categorized according to Geiger *et al.* (1996) as first and second order effects. First order effects can be accounted for by using analytical models for the process description or size independent finite element method (FEM) simulation. They are mainly due to the changes in the surface to volume fraction, which is increasing when the part size decreases. For example, the relative amount of the friction forces during deep drawing increases if the part size decreases, even if the theory of similarity is obtained.

Second order effects are due to the fact that the size of the workpiece and tools can be scaled down according to the rules of similarity, but the structure of the material (grain size), the surfaces (roughness), and the surface films remain nearly unchanged.

Therefore, the normalized values, e.g., the grain size divided by the sheet thickness, will increase while scaling down the dimensions of the part.

3. *Strength*

Most research is done on the second order size effects of mechanical strength. It has been shown that there is a transition from the polycrystalline to single crystal behavior, if the normalized thickness of the part, i.e., the fraction of part diameter and grain size, falls below a certain value (part diameter/grain size ~10, see Fig. 2). Above that value constant strength is detected if the grain size is held constant and the diameter of the samples is varied. As the sample size decreases to dimensions comparable to 10 times the grain size, the region with size effects is entered. The typical behavior of a polycrystal, which is determined by the interaction of dislocations with grain boundaries and described by the Hall–Petch equation, transforms to the behavior of a single crystal, as the restraints by the grain boundaries diminish. The strength of the material, extrapolated for an infinite grain size, decreases from $M\tau_0$ to $S^{-1}\tau_0$, where τ_0 is the shear stress for a single crystal, $M = 3.06$ the Taylor factor, and S the Schmid factor. Due to that the forming forces could fall at least a factor 1.53 (for $S = 0.5$, $M/S^{-1} = 3.06/2$) stronger than expected from the decrease of the geometry, if the forming force is linearly dependent on the flow stress.

It has been shown that the onset of the transition from polycrystalline to single crystalline material can be dependent on the grain size (Miyazaki *et al.* 1979), the material, and the stress state, i.e., the forming method (Geiger *et al.* 1999). In the case of bending the effect is more pronounced than in cold extrusion or shear cutting. It can be accounted for by a size dependent FEM calculation, defining three different zones:

(i) The core zone of the part, where the mechanical behavior is equal to that of a polycrystal.

(ii) The free surface zone, where the mechanical behavior approaches that of a single crystal.

(iii) The bond surface zone, where the mechanical constraints to the grains are stronger than in the core zone, as the tool surface will not deform like neighboring crystals in the core zone.

The thickness of these zones is correlated to a feature of the microstructure. This can be the grain size, but also the twin size can be relevant.

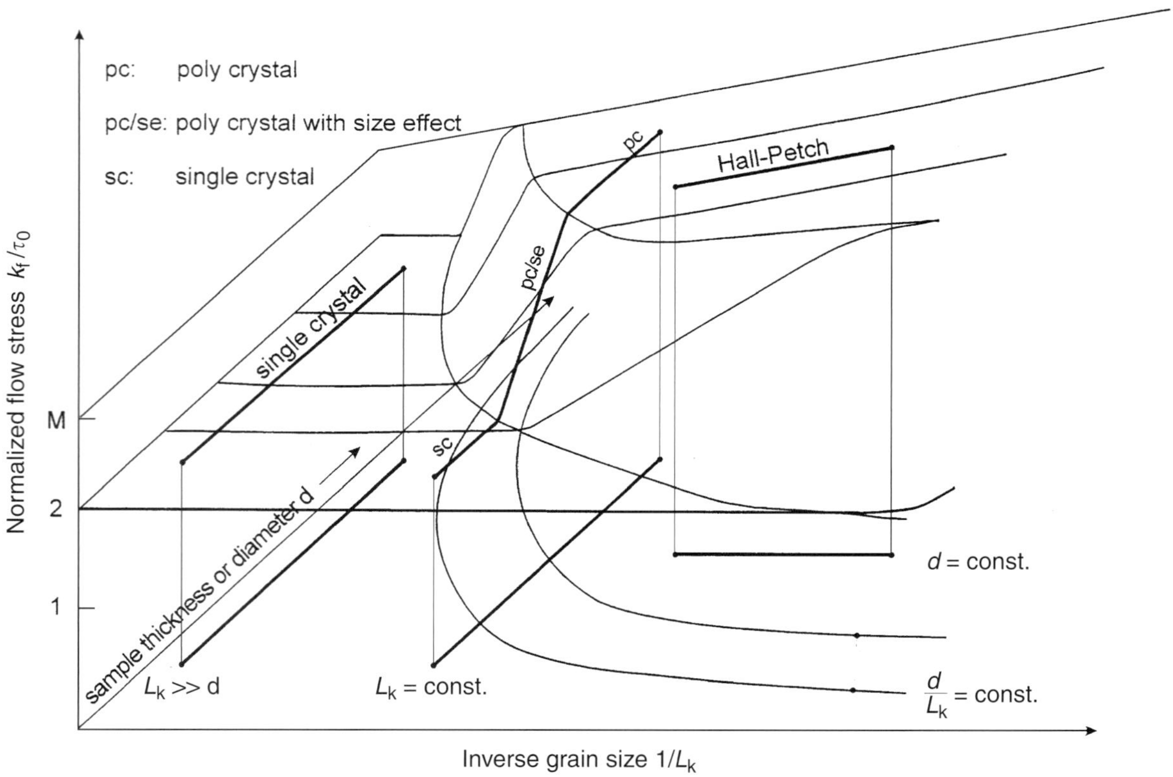

Figure 2
Influence of grain size and sample size on measured strength value.

4. *Accuracy*

While the changes in strength can be accounted for by using the size dependent FEM simulation, another feature gives rise to severe problems in forming micro parts. As shown in Fig. 3 the scatter of the properties (bending force, flow stress, and cutting force) increases linearly with the normalized grain size (grain size over part diameter). Due to this it becomes increasingly difficult to meet the strong demands in respect to accuracy, which are defined for micro parts (Jimma and Adachi 1993). The reason for the increase of scatter has its roots in the increasing fraction of grain size to part dimension. While in a polycrystal the anisotropy of the strength of single crystals is averaged out by the existence of a very large number of grains with different orientations, this is not the case for small parts. In that case there are only a small number of grains which determine the strength. As the orientation of the grains differs from sample to sample, so does the strength. The use of single crystals with defined grain orientation or nanocrystalline materials could reduce the scatter.

5. *Tribology*

The size effects in tribology were investigated in ring upsetting tests by Messner (1997) and combined forward–backward extrusion by Tiesler *et al.* (1999). The investigations were concentrated on the friction coefficient using liquid lubrication. A significant increase of the friction coefficient by a factor of five was noted when the sample size decreased by a factor four when the surface structure was held constant. This size effect was attributed to the changes in lubrication. The lubricant is entrapped in the roughness valleys of the sample. These valleys are interconnected but become isolated groups as the surface area, which is in direct contact to the tool, increases. The lubricant will be squeezed out of interconnected roughness valleys, while it will be able to transmit high normal pressure, if it is entrapped in closed valleys. With increasing sample size the portion of closed valleys increases. Therefore, the direct contact between tool and workpiece decreases, yielding a decreasing friction coefficient.

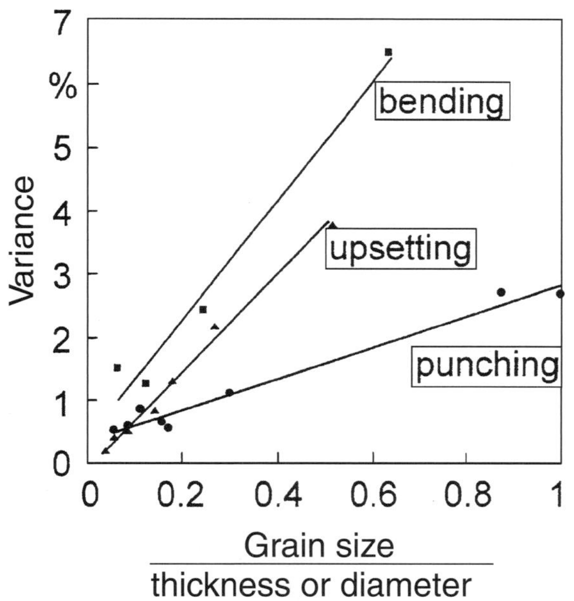

Figure 3
Increase of scatter for increasing normalized grain size. The maximum processing force was taken for bending and punching of $CuNi_{18}Zn_{20}$, the flow stress in upsetting for $CuZn_{15}$ after Geiger *et al.* 1999).

6. *Concluding Remarks*

In the metal forming of microparts, i.e., parts with at least two dimensions below 1 mm, size effects of first and second order occur. These effects must be accounted for in the process layout.

The effects on the mechanical strength and the tribological behavior have been already investigated, and essentials are settled. There is still a lack of research on the adhesion effects. It is possible to produce microparts using the metal forming technology. This is even true for gears which are very difficult to produce by metal forming in the normal size range.

Bibliography

Geiger M, Engel U, Eckstein R, Tiesler N 1999 In: Geiger M, Feldmann K (eds.) *Produktionssysteme in der Elektronik (Production Systems for Electronics)*. Meisenbach, Bamberg, Germany, pp. 85–123

Geiger M, Vollertsen F, Kals R 1996 Fundamentals on the manufacturing of sheet metal micro parts. *CIRP Ann.* **45**, 277–82

Jimma T Adachi T 1993 Recent trends in precise press working of electronic components. *Proc. 4th ICTP Advanced Technology of Plasticity*. Beijing, China, pp. 1547–1554

Messner A 1997 *Kaltmassivumformung metallischer Kleinstteile—Werkstoffverhalten, Wirkflächenreibung, Prozeßauslegung* [*Cold forging of Micro Parts—Material Behavior, Tribology, Process Design*]. Meisenbach, Bamberg, Germany

Miyazaki S, Fujita H, Hiraoka H 1979 Effect of specimen size on the flow stress of rod specimens of polycrystalline Cu–Al alloy. *Scr. Metall.* **13**, 447–9

Saotome Y, Itoh A, Amada S 1993 Superplastic Micro Forming of Double Gear for Milli-Machines. *Proc. 4th ICTP Advanced Technology of Plasticity*. Beijing, China, pp. 2000–5

Tiesler N, Engel U, Geiger M 1999 Forming of microparts—effects of miniaturization on friction. *Proc. 6th ICTP Advanced Technology of Plasticity*. Springer, Berlin, pp. 885–94

F. Vollertsen

Metal Forming: Specialized Procedures for the Aircraft Industry

Minimization of weight is still one of the main design features determining aircraft structures. It influences material selection as well as principles of design. The most widely used metals are high-strength aluminum and titanium alloys which are usually of very limited formability. At the same time, complex and highly integrated parts are used (e.g., NC-milled wing panels) which cannot be formed by conventional methods.

This article describes technologies developed specifically for aircraft manufacturing. "Shot peen forming" and "age creep forming" are the processes most suitable for the production of very large parts, e.g., wing skins, which are milled from sheets or plates. Very complex sheet parts of high-strength aluminum or titanium alloys can be produced by "superplastic forming."

The processes described in this article are suitable only for the small series production typical of the aerospace industry or for the manufacture of prototype parts, but they offer great flexibility as far as modifications or design changes of the parts are concerned.

1. *Shot Peen Forming*

1.1 Description of the Process Principle

Shot peen forming, as a variant of shot peening methods, is a cold forming process for all kinds of metal sheets. During the forming process the surface of the metal part is peened by small, round steel balls that are accelerated by air pressure or wheels. The balls hitting the surface act like a small forging hammer. This leads to a stretching of the surface zone and local plastic deformation. Furthermore, compressive stresses are introduced into the surface. The combination of these effects results in a convex curvature on the peened side of the metal part (see Fig. 1). This forming method was developed in the early 1950s and used by Lockheed (ICSP 1981) for forming of the wings of the Super Constellation.

Peen forming is especially suitable for large-surface parts with a large curvature and without sharp contour changes. The curvatures are normally in the elastic–plastic range of the metal (convex forming), but full plastic forming (concave forming) using a greater peening pressure is also possible, as is discussed below.

Peen forming is a free forming method that requires neither tools nor dies, thus avoiding the expense of the development and fabrication of dies. The method is highly adaptable to design changes that may become necessary after prototype production. By varying the peening parameters, such as air pressure, peening mass, etc., it is possible to produce the desired changes in the curvature of the part.

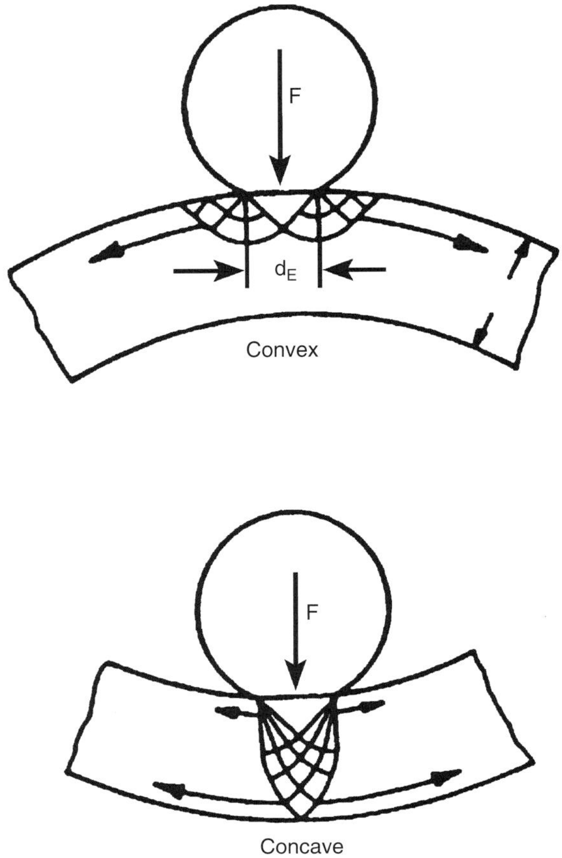

Figure 1
Curvature achievable with shot peen forming (convex: F = low; concave: F = high).

The formable wall thickness for aluminum is 1.2–51 mm and for steel is 0.4–24.4 mm. Modern peening machines permit the manufacture of parts with dimensions of up to 4.5 m in width and 35 m in length.

In addition to conventional peen forming it is also possible to peen form with elastic bending or stretching prestresses. When peen forming is combined with a superimposed bending stress, the part is elastically prestressed in the desired direction of curvature. The necessary prestressing forces are distinctly below those required for pure bending. The plasticized zone is greater than that of nonprestressed parts. The result is that the same curvature can be produced with lower peening intensity and without excessive increases in indentation diameters and/or surface roughness. This technique was used, for example, for forming a side shell of the Airbus A310 (Meyer *et al.* in ICSP 1987).

One example of a combination of different forming processes is the peen forming of a part very close to the end form in nonhardened temper followed by an autoclave process under pressure and temperature (similar to the creep forming process) during which final forming with high contour accuracy and precipitation hardening of the part occurs (EU Patent 0726106).

1.2 Peening Parameters

The main peening parameters are the shot size, the air pressure or velocity of the wheel, the shot mass per unit time, the velocity of the nozzle, the distance of the peening paths, and the coverage. The combination of these parameters results in the peening intensity, which can be measured by the so-called Almen-Test method (Burakowski in ICSP 1981). The Almen-Test, specified in MIL S-13165, contains three different kinds of test stripes (N, A, C) suitable for low, medium, and high peening intensities, and is based on the following principle. A metal strip, fixed in a block, is peened from one side to make it curved. The height of the curvature in millimeters serves as a measure of the peening intensity. Depending on the intensity, either a convex or a concave curvature is achieved. Convex peening occurs when the impact of the balls causes the part to deform against the direction of peening. If the kinetic energy of the balls is increased to a point where the part is highly plasticized, it will deform in the direction of peening (concave peening). This method was used, for example, for forming the tank bulkhead segments of Ariane 5 (Koehler *et al.* 1991).

1.3 Modeling of the Peen Forming Process

A simple computer model (Kopp *et al.* in ICSP 1987) has been established to permit prediction of peen forming parameters for uniaxially curved components. The input variables required for the model are sheet thickness, ball diameter, indentation diameter, adjusting factor, and final curvature. As a result, the model presents the relationship between the forming parameters in a graphic form. With the aid of the model the required coverage, peening time, and shot mass per unit area can be calculated for a given sheet thickness, final curvature, and selected indentation diameter. Technical feasibility and cost-effectiveness of the peen forming process can be predicted using this model. For multiaxially curved components the relationship may be very complex. Predictions of this kind are therefore not possible, and peening parameters have to be determined experimentally.

1.4 Applications

Shot peen forming is often used for the forming of large metal parts which are difficult to form by other methods such as stretch forming or age creep forming. Typical application examples are integrally stiffened wing skin parts for aircraft or large tank parts or segments. Peen forming is now used for series production of many aircraft, e.g., Grumman A-6, Boeing 747 and 727 (US Patent 114017), Lockheed C-130, McDonnell Douglas F-15 and DC-10 (Moore 1982), Fairchild A-10, De Havilland Canada Dash 7, and Airbus A310, A330, and A340.

Peen forming has also been used for the forming of spherical curved tank bulkhead segments of Ariane 5 (Wüstefeld *et al.* in ICSP 1996), Fig. 2.

Shot peen forming can also be used for reworking metal parts previously formed by means of another forming technology.

1.5 Influence on Material Properties

After peen forming, the two sides of a component show different distributions of residual stresses in the surface. A survey of this is given by Wohlfahrt in ICSP (1984). The value and depth of the compressive stresses depend on peening parameters such as intensity, shot size, and hardness (Fig. 3) (Koehler in ICSP 1984). These stresses normally improve the fatigue strength and corrosion fatigue properties. They also serve to prevent stress corrosion cracking (Meyer *et al.* in ICSP 1987). To intensify this effect, the parts are often treated by saturation peening with small shot before peen forming. Shot peening is also used to eliminate unwanted tensile stresses introduced into the parts during forming by other methods.

As peen forming is a cold forming method there are no changes in the microstructure of the metal parts. Depending on the peening parameters often a slight increase in tensile strength and a decrease of elongation is observed.

2. Age Creep Forming

2.1 Description of the Process Principles

Age creep forming was developed for the contour forming of large, integrally stiffened aircraft parts, such as wing panels. It is usually applied to parts made of precipitation heat treatable, high-strength aluminum alloys. The process combines a creep forming operation with the final artificial aging cycle of these alloys. The process was initially developed by Textron (Hambrick 1986) using the temperature and pressure capabilities of an autoclave. For thinner sheet parts, e.g., rocket fuel tank structures, mechanical clamping and heating in an oven are also used.

The principle of autoclave forming is shown in Fig. 4. A flat, machined part is placed on a female tool and sealed around the edges. The cross-section of the part does not have to be uniform; thickness variations in longitudinal and transversal direction

Figure 2
Ariane 5 tank bulkhead with segments formed by shot peen forming (courtesy of Dornier GmbH).

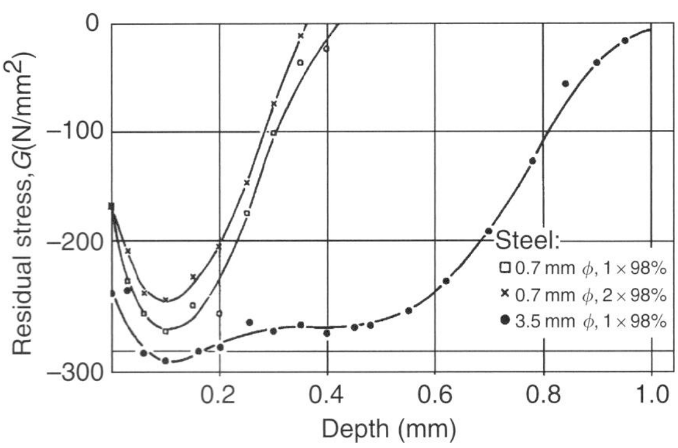

Figure 3
Residual stresses in an AlZnMg alloy peened by steel shot with different sizes and intensities: $I = 0.21$ mm A, 0.22 mm A, and 0.38 mm A.

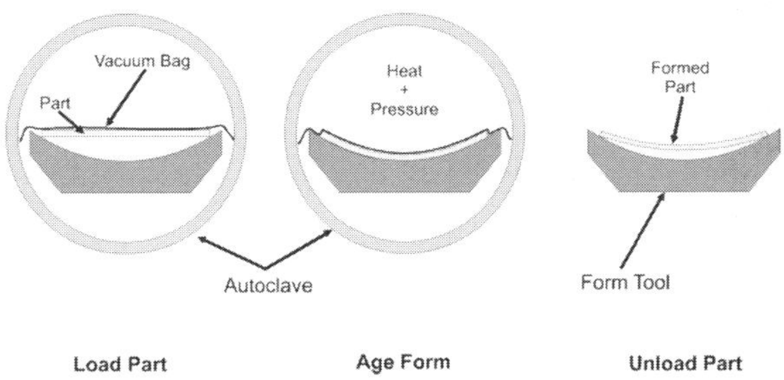

Figure 4
Principle of autoclave age creep forming.

are permissible. The tool normally has an overformed contour that allows for spring-back effects. The assembly is placed in an autoclave and a vacuum is created between the part and the tool. The autoclave is then closed and pressure and heat are applied. The required aging condition of the alloy determines the temperature–time cycle. On completion of the heating cycle the part is cooled and pressure and vacuum are released. As illustrated schematically in Fig. 5, the unloaded part springs back by a certain amount. A development phase is necessary for each new part in order to assess the exact spring-back effect. After optimization and freezing of the tool contour, the

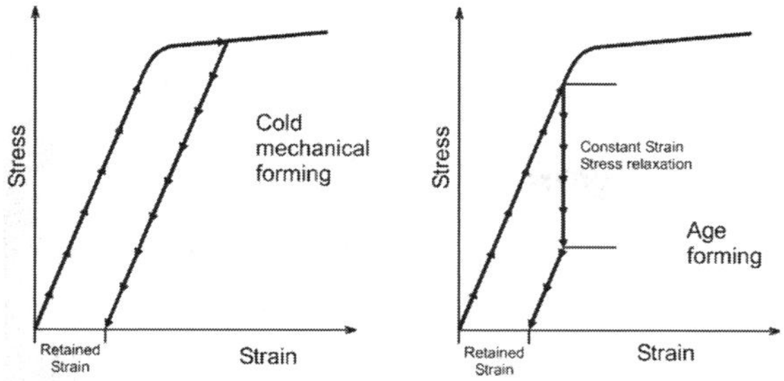

Figure 5
Stress–strain curves for age creep forming compared with conventional cold forming.

process is very reproducible. Induced internal stresses are very low compared with cold forming processes.

2.2 *Materials and Forming Mechanism*

In principle, the method is applicable to all alloys that can be artificially aged. The most commonly used aluminum alloys are of the types AlMgCu (2000 series), AlMgSi (6000 series), or AlZnMgCu (7000 series). Final temper is either the peak-aged or the over-aged condition. There are three possibilities for the starting temper condition of these materials:

(i) As quenched, i.e., room temperature aged (T3 or T4 temper).

(ii) Intermediate temper after a first aging step. In this condition most alloys are more stable than in T3; however, only the second aging step is usable for age creep forming.

(iii) Final temper. For most 7000 aluminum alloys an additional heating to about 150 °C is permitted for several hours. At this temperature stress relaxation is the preferred mechanism.

Once the part has been loaded and sealed, vacuum and pressure are applied. The part becomes elastically preformed; but normally complete contact with the tool is not achieved straight away. During heating the material becomes softer and the part creeps into the tool by plastic deformation. Subsequently, the part takes on the shape of the tool and the induced stress is relieved by stress relaxation. The material strength increases as the aging process continues. If an over-aged temper is required, material strength decreases again after passing over the peak-age condition. At the end of the aging cycle a certain elastic stress remains in the part which is less than the initial stress, as is illustrated in Fig. 5. After unloading, this stress is released by spring-back of the part. Figure 5 also demonstrates the difference between forming by conventional methods (e.g., rolling) and age creep forming. After spring-back the remaining internal stresses are very low. The remaining elastic stress, and therefore the amount of spring-back, depends on the material and the starting temper. Spring-back is lower with decreasing yield strength of the material. Geometry also affects the amount of spring-back. For thicker cross-sections and smaller radii, the relative spring-back decreases (higher effective maximum bending stress).

The microstructure and the hardening mechanism of the alloy and therefore also the mechanical properties of the alloy are not affected by the forming process.

2.3 *Modeling and Simulation*

Development of the final tool shape for any given shape of a part is normally an extensive, iterative process. In order to shorten this process or avoid it completely, various numerical simulation tools, mainly FEM models, have been developed to permit determination of the optimal shape of the die in order to achieve the specified final contour of the sheet. Even for parts with simple geometry simulation is quite complex, since different mechanical effects such as elastic deformation, plastic deformation, creep, stress relaxation, and residual stresses have to be taken into consideration.

2.4 *Equipment*

The process requires heat, pressure or clamping force, and time. To make full use of the ability of the process to form very large and thick parts, the autoclave technique is the best choice. Autoclaves of at least 10–15 m in length and of about 2–3 m in diameter are in use.

For smaller parts, a clamping technique can also be used instead of an autoclave. Using this technique the part is initially bent and mechanically fixed in the die.

2.5 *Applications*

Typical applications are very similar to those of peen forming. Age creep forming is mostly used for large, integrally stiffened aluminum parts, as used for pressure vessels in space technology or for fuselage or wing skin panels in the aircraft industry.

A very early application was the fabrication of fuel tank plates from stiffened sheets of 2219 alloy for the Saturn V space vehicle (Layton 1963). Some of the largest and most complex parts are for the wing skins of B-1B military aircraft using alloys 2124 and 2419. These skins are 15 m long and about 2.7 m wide and their thickness ranges from 2.5 mm to 63.5 mm (Hambrick 1986).

The most important application is the forming of wing skin panels for a wide range of civil and military aircraft, e.g., the wing skins of the Gulfstream IV and V and of the Airbus A330/A340. Ariane 5 panels are formed by stress relaxation in the fully over-aged condition of 7075-T7351 using a vacuum clamping technique.

3. Superplastic Forming

Superplastic forming is a unique, high-temperature forming process, characterized by extreme formability at very low flow stresses and strain rates. In the aircraft industry the term superplastic forming, or SPF, generally denotes a sheet forming process. A wider definition, however, would include bulk forming processes (e.g., isothermal forging) which may also be considered to be superplastic forming processes.

3.1 Superplasticity

The SPF process can only be applied to a few special materials. Generally these materials are characterized by very fine grain structures (grain sizes below 20 μm), which are stable against grain growth at elevated temperatures. At high temperatures ($>0.5\ T_m$) and low strain rates (10^{-4}–$10^{-3}\,s^{-1}$) an unusual deformation mechanism occurs in this type of microstructure, which can be interpreted as a continuous displacement of the grains to each other by grain boundary sliding, accompanied by diffusion-controlled shape accommodation of the grains. In this deformation mode the flow stress, σ_f, is low (<10 MPa) and largely depends on the strain rate, $d\varepsilon/dt$:

$$\sigma_f = k^*(d\varepsilon/dt)^m$$

The higher the value of the strain rate sensitivity coefficient, m, the more stable the material is against local strain rate increases and thus against early failure by local necking. For $m>0.3$, elongation values of several hundred percent up to more than 1000% without necking can be achieved in tensile testing—the material is superplastic. The value of m itself depends greatly on the strain rate, and the maximum value is only obtained in a small optimum strain rate range.

The small and stable grain sizes and high m values required for this process are inherent properties of only a very small group of standard materials, the most important example being the titanium alloy Ti6Al4V. In most other cases, and especially in the case of the technically important aluminum alloys, special thermomechanical processing at the sheet manufacturer is necessary in order to achieve the required microstructure. A broad overview of superplastic materials and the underlying deformation mechanisms is given in Nieh *et al.* (1997).

3.2 Processes and Equipment for Superplastic Sheet Forming

The most common process for superplastic forming of sheet material is a blow forming process similar to that used for thermoplastic polymer foils and sheets (Fig. 6). The superplastic sheet is placed between two heated tool halves and tightly clamped between them along its outer perimeter by some mechanical device, usually a hydraulic press. Then the gas pressure on the upper side of the sheet is slowly increased. The sheet starts to stretch and slowly forms down into the tool cavity on the pressure-free side. The geometry of this lower tool cavity defines the geometry of the final part. Maximum pressures below 10 bar are usually sufficient for the complete forming of the part.

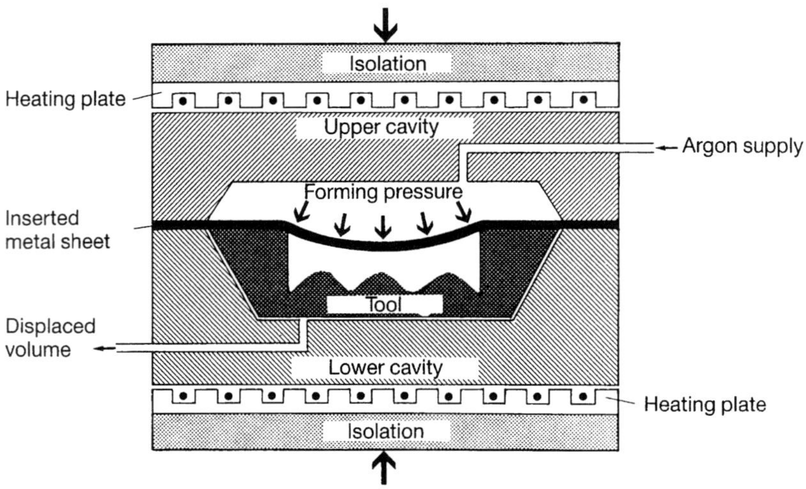

Figure 6
Superplastic forming process.

However, important limitations for part and tool design result from the stretch forming principle, which causes thinning of the sheet. Parts with low width-to-depth ratio and with small radii can easily be formed but will show pronounced thickness variations over the part, a fact that has to be taken into careful consideration during part design. Various modifications of the simple blow forming process have been developed in order to minimize the wall thickness variations in the final part (Plege and Schleinzer in Hori *et al.* 1991).

It is important to keep tight control of temperatures and of the forming gas pressure during the forming process. The gas pressure vs. time cycle must be optimized in such a way as to ensure that at all times and at each point of the forming sheet the strain rate is as close as possible to the optimum strain rate. Maximum forming gas pressures of about 10 bar are normally sufficient to form out even the finest details. A number of modeling and simulation tools for predicting thickness profiles and for optimizing the pressure–time cycle have been developed (Bonet *et al.* in Hamilton and Paton 1988).

Owing to the low strain rate, forming cycles are long compared to more conventional forming processes and range from 20 min up to 120 min for one forming cycle. SPF presses are available for small as well as for large parts up to about 2 m × 1 m.

3.3 Materials and Forming Parameters

The most important superplastic materials for structural aerospace applications are titanium alloys and a few high-strength aluminum alloys. The widely used titanium alloy Ti6Al4V is formed at temperatures around 910°C. Typical flow stresses are 8 MPa at optimum strain rates of about $0.0002\,s^{-1}$. High-temperature-resistant steel castings have proved to be successful for the dies. Owing to the high reactivity of hot titanium, high-purity argon must be used as the forming gas. However, an oxygen-enriched surface layer of 0.1–0.2 mm still remains after forming which has to be removed by chemical milling.

The most important of the high-strength aluminum alloys is the age-hardenable AlZnMgCu alloy

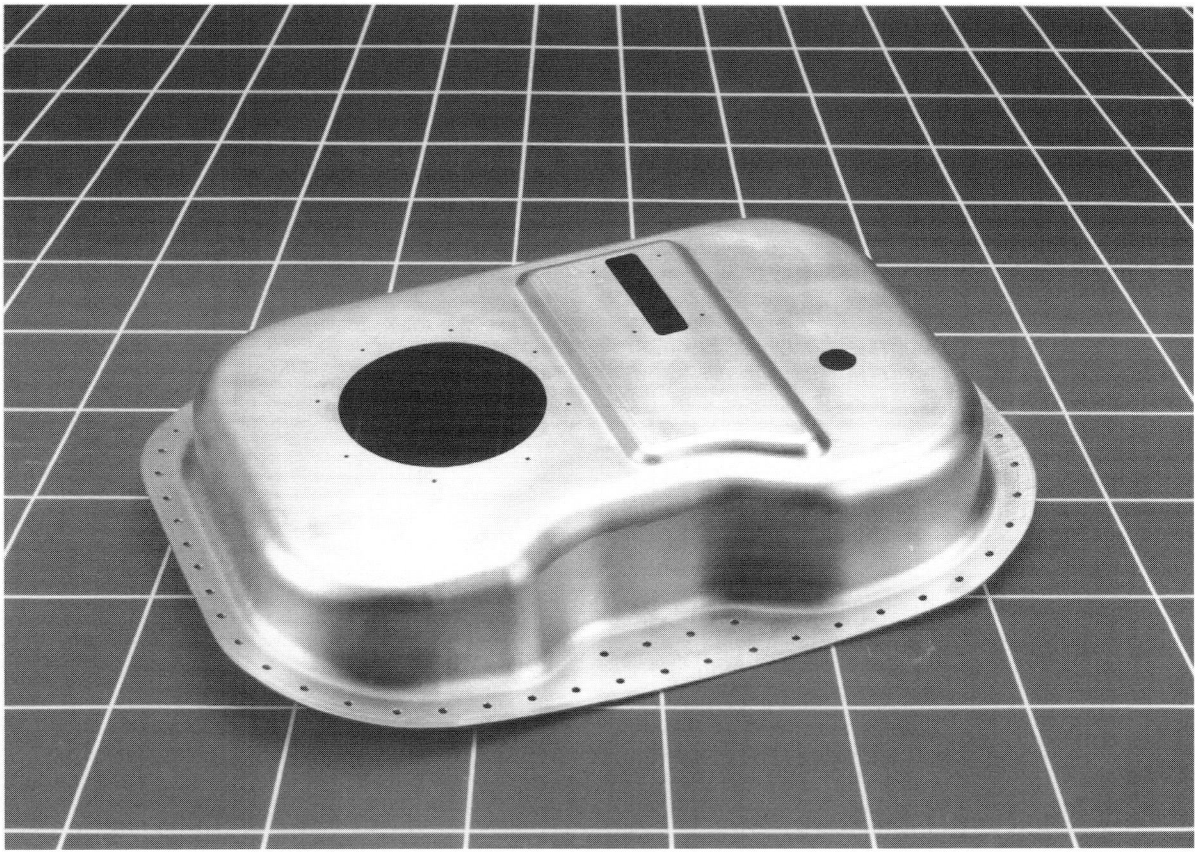

Figure 7
Service panel for Dornier 328 aircraft (superplastically formed from TiAl6V4 alloy).

AA7475. Typical processing parameters are a temperature of 520°C, a flow stress of 2 MPa, and a strain rate of 0.0002 s^{-1}. In order to avoid cavitation (i.e., the formation of small cavities in the sheet microstructure) the "back pressure technique" is recommended. Instead of starting the forming process with atmospheric pressure above and below the sheet, this technique entails first pressurizing both upper and lower tool cavities to at least 10 bar before initiating the increase of controlled forming pressure on the upper side. Steel tools are commonly used for series production; tools from aluminum or ceramic casting materials are sometimes used for prototypes and tests. Argon or dry and clean nitrogen are used as forming gases. Some lower strength aluminum alloys are also available in superplastic quality (Grimes and Butler in Hamilton and Paton 1988) for secondary applications.

The superplastic forming process by grain growth, softening, and cavitation effects influences the final mechanical properties of the materials used. Design allowables are up to 10% lower than for the same materials after conventional forming (McDarmaid *et al.* in Hamilton and Paton 1988).

3.4 Applications

Although the tooling required for SPF is simple, the process itself is by no means a cheap one. This is because of the high forming temperature and the long forming cycle. Application of SPF technology is therefore limited to parts with complex geometry, which make full use of the excessive formability of superplastic alloys thus avoiding the various designs using riveting, welding, etc. A typical example is shown in Fig. 7. This part has to be made from titanium for reasons of strength and corrosion. It would be impossible to form this part "in one shot" using conventional cold forming processes. On the contrary, an expensive and heavy structure riveted or welded from several smaller and simply shaped pieces of titanium would result. Similar considerations lead to the use of SPF titanium and aluminum parts in flap and door stiffeners (Hughes *et al.* in Hori *et al.* 1991), hoods, firewalls, spars (Plege and Schleinzer in Hamilton and Paton 1988), etc.

In some cases the SPF of titanium can be combined to great advantage with diffusion bonding between two or more sheets in one process (SPF/DB), resulting in highly integrated, multiwall structures with internal stiffeners, channels, or similar features (Winkler in Hori *et al.* 1991).

4. Summary

Large, complex structures, such as components for aircraft or space vehicles that cannot be formed by conventional methods such as stretch forming, often necessitate the use of special forming processes. Suitable forming processes are "peen forming" and "age creep forming," which are used, for example, for the fabrication of large wing skin panels for various aircraft.

Another special forming process, most suited to smaller parts requiring a high degree of forming, is "superplastic forming." This process can be used particularly for certain special alloys such as Ti6AlV4 and AA 7475, which show superplastic behavior in a limited temperature range. In this range, elongation values of more than 1000% without necking are possible.

The principles, advantages, and applications of these special forming processes are described in this article.

Bibliography

Hambrick D M 1986 Age forming aluminium in an autoclave *Proc. Int. Conf. Charlottesville*, VA, Vol. 2, pp. 883–900

Hamilton C H, Paton N E (eds.) 1988 *Proc. Int. Conf. on Superplasticity and Superplastic Forming*. Minerals, Metals and Materials Society, Warrendale, PA

Hori S, Tokizane M, Furushiro N (eds.) 1991 *Proc. Int. Conf. on Superplasticity in Advanced Materials*. Japan Society for Research on Superplasticity, Osaka, Japan

ICSP 1981 *Int. Conf. on Shot Peening*

ICSP 1984 *Int. Conf. on Shot Peening*

ICSP 1984 *Int. Conf. on Shot Peening*

ICSP 1996 *Int. Conf. on Shot Peening*

Koehler W, Hornauer K P, Bornschlegl H 1991 Manufacture of spherical Al-components by shot peening with regard to particular material properties. *Metall.* **45**, 1220–4

Layton R E 1963 Fabrication skin segments for the Saturn buster—simultaneous aging and forming. *Am. Machinist/ Metalworking. Manufact.* **5**, 67–70

Moore D 1982 The application of shot peen forming technology to commercial aircraft wing skins. In: *Shot Peening for Advanced Aerospace Design*, SP-528, SAE, p. 2

Nieh T G, Wadsworth J, Sherby O D 1997 *Superplasticity in Metals and Ceramics*. Cambridge University Press, Cambridge

W. Koehler, B. Plege, and K. F. Sahm

Metal Injection Molding

Metal injection molding (MIM) is a technology for manufacturing complex, precision, net-shape components for use in medical, automotive, industrial, firearms, and consumer industries. The potential of MIM lies in its ability to capture the shaping advantage offered by injection molding by using a starting mixture of fine metal powder and organic binder. The MIM technology provides advantages of cost reduction at volumes ranging from a few thousand to over millions of parts per year due to its ability to hold close tolerances. Typical tolerances achievable range from ±0.3% to ±0.5%, making the process highly cost competitive as compared to machining and investment casting. Further, MIM overcomes the dimensional and productivity limits of isostatic pressing and slip casting, the defects and tolerance limitations of investment casting, the mechanical strength of die-cast parts, and the shape limitation of traditional powder compacts.

The evolution of MIM technology has spanned over two decades and is an offshoot of ceramic injection molding (German and Bose 1997, Mutsuddy and Ford 1995). Along its evolution path are rapidly changing variants, reflecting different combinations of powders, binders, molding techniques, debinding routes, and sintering hardware. It is important to understand the processing steps, design guidelines, and limitations to fully appreciate the potential of this technology.

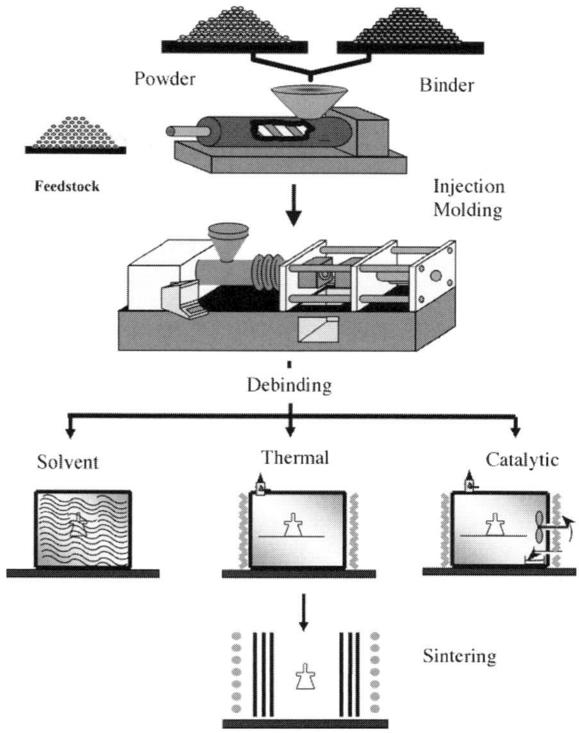

Figure 1
Process schematic of metal injection molding.

1. Processing Steps

There are four primary steps to producing metal injection molded parts as shown in Fig. 1.

1.1 Feedstock Formulation

The starting material for metal injection molding is a homogeneous pelletized mixture of metal powder and organic binder termed feedstock. The binder is simply a carrier medium for the powder and once a part is molded, the binder is removed in a subsequent step (termed debinding). A binder formulation is based on its ability to provide for a rigid molded part, ease of removal from a molded part, recyclability, and nontoxicity. Process considerations usually dictate three components to a binder system, a backbone that provides strength, a filler phase that is easily extracted during initial stages of debinding, and surfactant to tailor feedstock rheology. Commonly used binder systems, as listed in Table 1, include thermoplastics, thermosets, and gelation systems.

Powder selection for the MIM process involves a combination of tailored particle size distribution to maximize packing densities, with a mean particle size 5–15 μm to achieve a high sintered density. Examples of MIM-grade powder morphology using water atomization, gas atomization and carbonyl-decomposition are shown in Fig. 2. The metal powder and binder are hot mixed above the softening point of the binder constituents to provide for a uniform coating of the binder on the powder surface. A shearing action using a continuous extruder or batch mixer aids in mixture homogeneity. The mixture is pelletized to an appropriate shape for feeding into the molding machine. Typical binder volume fraction in feedstock range from 0.3 to 0.45.

1.2 Molding

The injection molding process is principally identical to conventional plastic injection molding. However, some machine hardware changes are usually required to process specific feedstocks based on compressibility and viscosity. The relatively high viscosity of feedstocks (from 10^2 to 10^3 Pa s at shear rates of 10 to $10^3\,s^{-1}$), coupled with the high thermal conductivity of a metal filled binder, and a tendency for the powder–binder mixture to separate at high shear rates defines the constraints for the molding process. Control of the molding process is vital for maintaining tight tolerances in subsequent steps. Most design advantages of MIM technology are captured

Table 1
Commonly used binder systems for MIM.

Thermoplastic binders
Wiech-type binders: polyolefin (PE/PP/PS) + waxes, oil + dispersant (fatty acids/ esters)
Water soluble binders: PEG/PVA + polyacrylates/polyamides/PVB
Polyacetal binder: POM + polyolefin + additives
Thermoset binders
Cross-linking of polymers: epoxy, phenoxy-resins
Gelcasting: methacrylamide + cross-linkers
Gelling/freezing binders
Hot gelling: methylcellulose/boric acid + water
Cold gelling: polysaccharides + water

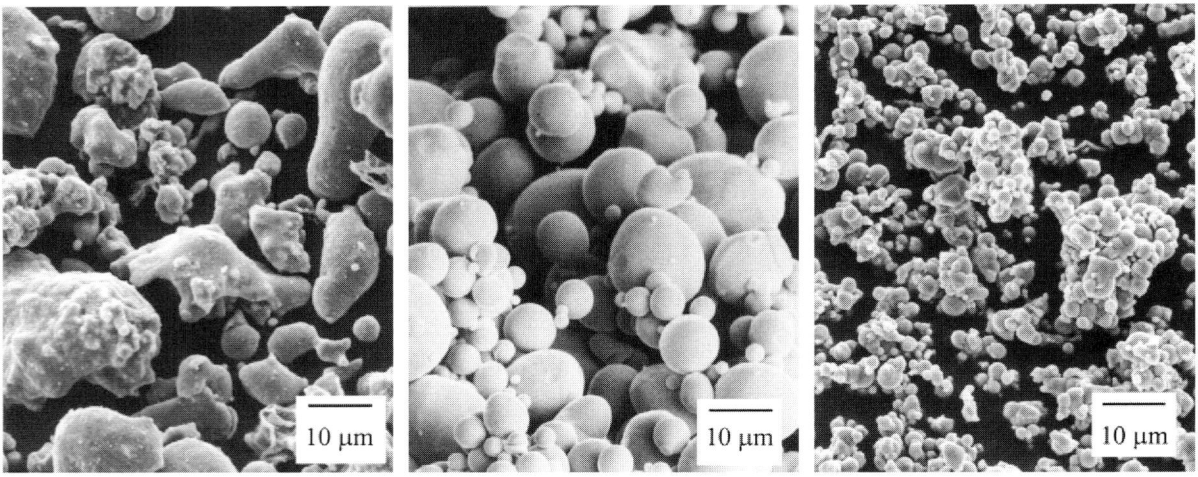

Figure 2
Scanning electron micrographs showing the morphology of MIM-grade powder manufactured using (a) water atomization, (b) gas atomization, and (c) carbonyl decomposition.

during molding by relying on the flexibility of incorporating complexities in the tool. A molded part is called a "green" part and is oversized to allow for part shrinkage during sintering.

1.3 Debinding

The debinding step involves the removal of the organic binder from a molded part using one of three different methods: thermal, solvent, or catalytic, depending on the composition and constituent of the binder. Thermal debinding involves application of heat to remove the binder by degradation, evaporation, or liquid extraction (wicking) (Angermann and Van Der Biest 1995). Temperatures for thermal debinding can range from 60°C to 600°C with the incorporation of dwell times at selected temperatures specific to the thermal decomposition of its binder constituent. The relatively long debind time associated with thermal debinding is greatly reduced using a solvent such as heptane or trichloroethane which can dissolve the soluble binder, e.g., wax in a wax-polyethylene-polypropylene-based binder. Some glycol- and polyvinyl alcohol-based water-soluble feedstocks use water as a solvent, while gelation-type binders, such as those based on polysaccharides plus water, rely on air drying to remove the water. With acetal-polyolefin-based feedstocks, a hybrid of thermal and solvent processes, termed catalytic debinding, uses nitric or oxalic acid vapor at temperatures between 110°C and 150°C to depolymerize polyacetal into formaldehyde gas (Krueger *et al.* 1993). The solid (acetal) to vapor (formaldehyde) catalytic degradation results in faster debinding (as compared to thermal debinding) while offering superior handling strength as compared to thermal or solvent-based feedstocks.

In all debinding methods, a skeleton of backbone binder remains to impart adequate strength and shape retention up to the onset of sintering. This remaining backbone is thermally removed between 200°C and 600°C in a presintering step.

1.4 Sintering

Depending on the alloy, debound parts are sintered at temperatures ranging from 1200°C to 1350°C. The fine particle size used in the MIM process results in high sintered density, which is concomitant with shrinkage (ranging from 14% to 20%). Therefore, molded parts are oversized to compensate for the sintering shrinkage. Sintered density ranges from 95% to 99% of theoretical, thus providing superior mechanical and corrosion properties as compared to press and sinter technology. Both batch and continuous sintering is employed for production. The underlying principles of sintering are discussed in and . The sintering time and temperature uniformity are of paramount importance in achieving a combination of desired mechanical properties and dimensional tolerance. Common sintering atmospheres employed for MIM alloys are reducing (H_2-based), inert (N_2- and Ar-based) or vacuum.

2. *MIM Alloys*

Common MIM alloys include austenitic stainless steels (304L and 316L), precipitation-hardened martensitic stainless steel (17-4PH), martensitic steels (420, 440C), alloy steels based on Fe–2%Ni and Fe–8%Ni (containing up to 0.5% carbon), 4140, 8620, tool steels including M2, soft magnetic alloys including Fe–50%Ni, Fe–3%Si, 430L, and alloys for glass-to-metal sealing applications such as kovar. The mechanical properties of MIM alloys show a broader range as compared to the wrought values. These variations result primarily from the differences in commercial production practices using different binders, debinding methods, sintering technology, and powders. However, optimal processing results in mechanical and corrosion properties that are comparable (in some case superior) to its wrought counterpart.

3. *Design Considerations and Limitations*

Parts for metal injection molding are designed based on shape, size, and geometry, for optimum performance. A summary of desirable design features is listed in Table 2. From a tooling perspective, the designed part must incorporate a parting line, gate, and ejector system. The avoidance of sharp corners, thick-to-thin transitions, and the use of draft angles, and corner radii provide for optimum moldability. Additional design restrictions such as the maximum part thickness are determined by the debinding limitations imposed by a specific feedstock. Generally, most MIM parts have a maximum cross-section thickness <7 mm. In this respect, acetal-based feedstocks (using catalytic debinding) have proven processing and handling advantages, and are capable of manufacturing parts with thicker cross-sections with virtually no debind constraints. The polysaccharide–water (gelation-type) system with its air-drying step replacing conventional debinding also has potential for manufacturing components with thick cross-sections.

The relatively high raw-material cost as dictated by unique powder specification also constraints the

Table 2
Typical design guidelines for MIM.

	Specifics	Reason
Restrictions	No undercuts on internal bores	Difficult to tool
	No inside close cavities	Impossible to tool
	Corner radius >0.1 mm	Avoid sharp edges/reduce stress
	0.5 to 2° draft	Facilitate ejection
	Smallest hole diameter $=0.1$ mm	Poor steel condition
	Minimum wall thickness $=0.5$ mm	Difficult to fill thinner sections
Desirable features	Gradual section thickness change	Ease of fill/reduce stress
	Largest dimension <100 mm	Processing difficulty
	Weight <100 g	Material cost/debinding constraints
	Wall thickness <6–10 mm	Facilitate debinding
	Incorporation of flat surfaces	Provide sintering support
Allowed design features	Holes at angles to one another Hexagonal, square, blind, D-shaped, keyed, and flat-bottom holes	All these features can be easily incorporated into the part design and can be molded
	Stiffening ribs and knurled features	
	Protrusions and studs	
	External threads	
	Part number, identification logo in die	

MIM technology to parts weighing <100 g. However, many heavier components whose complexity overrides the cost constraints have been produced using MIM. The best examples for this technology are parts that have been designed specifically to take advantage of the ability of the MIM process to capture design complexity as offered during the molding step.

Generally, the MIM process is not competitive as compared to stamping or screw machining. It competes with investment casting by providing better surface finish, fine-scale homogeneous microstructure, thinner and more complex wall sections, and better dimensional tolerances. With respect to machining, the MIM process results in cost reduction at medium to high volumes especially if the parts require multiple machining operations. In cases where superior mechanical performance is required, MIM also competes with press and sinter by providing parts with a higher density, but at a higher cost.

4. Concluding Remarks

The MIM technology has seen rapid advances in the 1990s and is a strong competitor to machining and investment casting given the right combination of complexity and production quantity. Annual worldwide growth rates for MIM is projected to be between 20% and 30% (German 1999). The ability of the process to maintain tight tolerance is strongly dependent on several factors ranging from raw material selection to process control. Commercially available feedstocks (with rapid debinding) have resulted in distinct processing advantages such as fabrication of thicker components, while allowing for standard raw materials to be available in the market. The MIM technology continues to penetrate new markets with the introduction of alloys, including cobalt-based alloys for wear resistant applications, and titanium- and nickel-based alloys for medical and aerospace applications.

Bibliography

Angermann H H, Van Der Biest O 1995 In: Bose A, German R M, Lawley A (eds.) *Reviews in Particulate Materials*. Metal Powder Industries Federation, Princeton, NJ

German R M 1999 Wear applications offer further growth for powder injection molding. *Met. Powd. Rep.* **54**(6), 24–8

German R M, Bose A 1997 *Injection Molding of Metals and Ceramics*. Metal Powder Industries Federation, Princeton, NJ

Krueger D, Blomacher M, Weinand D 1993 Rapid catalytic debinding MIM feedstock: a new technology grows into a manufacturing process. In: Gaspervich T, German R M (eds.) *Advances in Powder Metallurgy and Particulate Materials*. PMTech93, Tennessee, pp. 165–80

Mutsuddy B C, Ford R G 1995 *Ceramic Injection Molding*. Chapman and Hall, London

R. Tandon

Metal Working: Cold Extrusion

Extrusion processes can be classified into two groups: the hot processes, which are mostly used for manufacturing semiproducts (see *Metal Working: Extrusion*), and the cold processes, which are used for the extrusion of components. Extrusion processes are further classified according to the material flow in relation to the direction of the tool movement (forward, backward, radial) and according to the initial shape of the component (solid, hollow). Figure 1 shows some examples of cold extrusion processes with the material flow in one direction (Figs. 1(a) and (b)) and with a divided material flow combined in one tool (Figs. 1(c) and (d)). It should be noted that the term "extrusion" is also used in the context of the manufacture of polymers (see *Metal Working: Extrusion*).

1. History

Industrial implementation of these processes, especially of the cold extrusion of steels, which was invented in Germany before World War II, was limited owing to very high tool loads. Mass production started in the 1960s, primarily in the automotive industry. Toyota became interested in cold extrusion in 1958, and by the late 1980s every Toyota automobile (1800–2000 c.c. engine) had more than 40 kg of cold extruded components, not including bolts and nuts.

The reasons why the global production of cold extruded steel components amounts to more than one million tons per year are the improved mechanical properties resulting from strain hardening and the possibility of optimizing the material texture (Fig. 2), and high productivity of parts with very high shape accuracy (some tens of components per minute per machine). This leads to near net shape forming (NNSF) or net shape forming (NSF), where only some additional machining operations are needed before the parts are ready for assembly (Fig. 3).

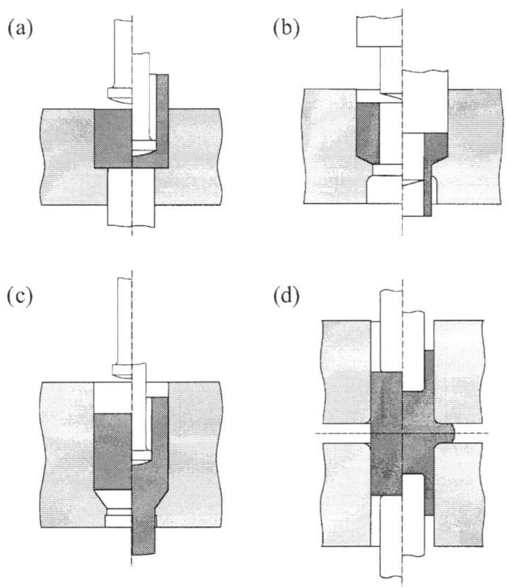

Figure 1
Some examples of cold-extrusion processes: (a) backward can; (b) forward tube; (c) forward rod and backward can; (d) radial and backward can and forward can.

2. Cold Extrusion Process Planning

The cold extrusion of metals is a manufacturing process that is part of an integral production system. It starts with the production of a slug.

Before entering the first extrusion operation (Fig. 4) slugs must be heat treated to lower their flow resistance and to improve their formability. In many cases this is done by soft annealing, i.e., by keeping the slugs at a defined temperature (for steels between 680°C and 730°C) for several hours, and then slowly cooling them.

As forming pressures and contact surface expansions are very high, the control of the tribological system (friction, lubrication, and tool wear) is critical for the success of the extrusion process. To keep the lubricant inside the deformation zone during the whole process, an appropriate surface roughness is needed. In difficult conditions a special lubricant carrier is needed which adheres to the surface of the specimen and also forms a surface profile capable of keeping the lubricant inside the zone. The most frequently used surface carriers are zinc phosphate and iron oxalate, in some cases also copper. As lubricants, mineral oils with additives are used; in more difficult conditions soaps and molybdenum disulfide are needed.

Heat- and surface-treated slugs can then enter the extrusion operation. Their number depends on several factors: the shape of the component, the material, available equipment, the final shape accuracy, the mechanical properties of the final parts, tribology, etc. In cases when the formability of the component is exhausted, or the tool loads have reached the highest permissible level, the heat treatment and consequently also the surface treatment of the slugs must be repeated.

To eliminate strain hardening, a new grain structure is needed. For better mechanical and forming conditions, a fine-grained structure is preferred. The recrystallization process (temperature and holding time) strongly depends on the degree of plastic deformation during the previous extrusion. In general, the grains are finer if the temperature is lower and the previous cold deformation is greater (see *Primary Recrystallization*). The process loops generally consist of heat treatment, surface treatment, and extrusion. The number of loops depends on how difficult are the conditions.

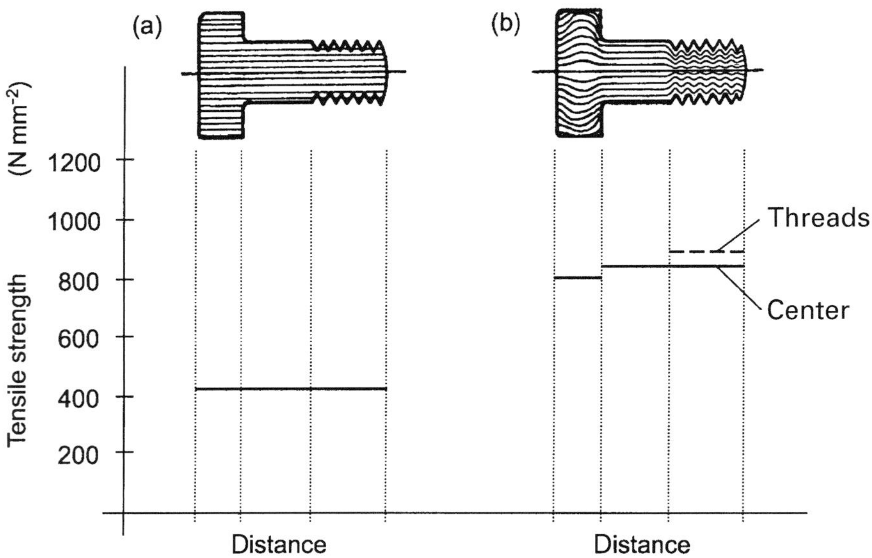

Figure 2
Comparison of average tensile strengths of (a) a machined and (b) a cold extruded screw with cold rolled threads for a steel containing 0.15% carbon.

Figure 3
Examples of NNSF and NSF parts for the automotive industry (courtesy of Krupp Presta AG, Liechtenstein, and Iskra Avtoelektrika d.d, Slovenia).

There are several methods of determining the necessary extrusion forces and tool loads. With the advent of powerful computers, methods such as the finite element method (FEM) offer the possibility of calculating the stress and strain distributions inside the extruded part and the tool much more precisely and thereby of optimizing the integral extrusion process.

3. *Materials for Cold Extrusion*

Materials suitable for cold extrusion should have good formability, which means they should have low yield and flow stresses and they should allow substantial strains without any cracks or other defects (see *Metal Forming: Formability*). Formability depends mainly on microstructure.

3.1 *Steels*

Unalloyed and low-alloyed steels can be efficiently cold extruded if the alloying elements have the following values: $C<0.35\%$ for backward, $C<0.60\%$ for forward extrusion; Si 0.15–0.35%; $Mn<1\%$ with slight increase in flow stress, possibly $<2\%$; $Mn<0.3\%$ if $C<0.3\%$; $Mn<1.35\%$ for $C<0.12\%$; S, $P<0.035$–0.040%; $Ni<1\%$; $Cr<0.9\%$; $Mo<0.4\%$.

Stainless steels are known to be less suitable for cold extrusion for two reasons: (i) very intensive strain hardening causes an increase of tool loads and (ii) their chemical composition demands a special treatment with lubricant carriers (oxalating).

3.2 *Nonferrous Metals*

Aluminum, copper, and aluminum and copper alloys were the first metals used in cold extrusion. Their formability is very high and tool loads are therefore not critical. Pure aluminum has the best formability. If it is alloyed with magnesium (up to 2%), copper (up to 1%), and especially with silicon (up to 1%) the formability of the alloys decreases.

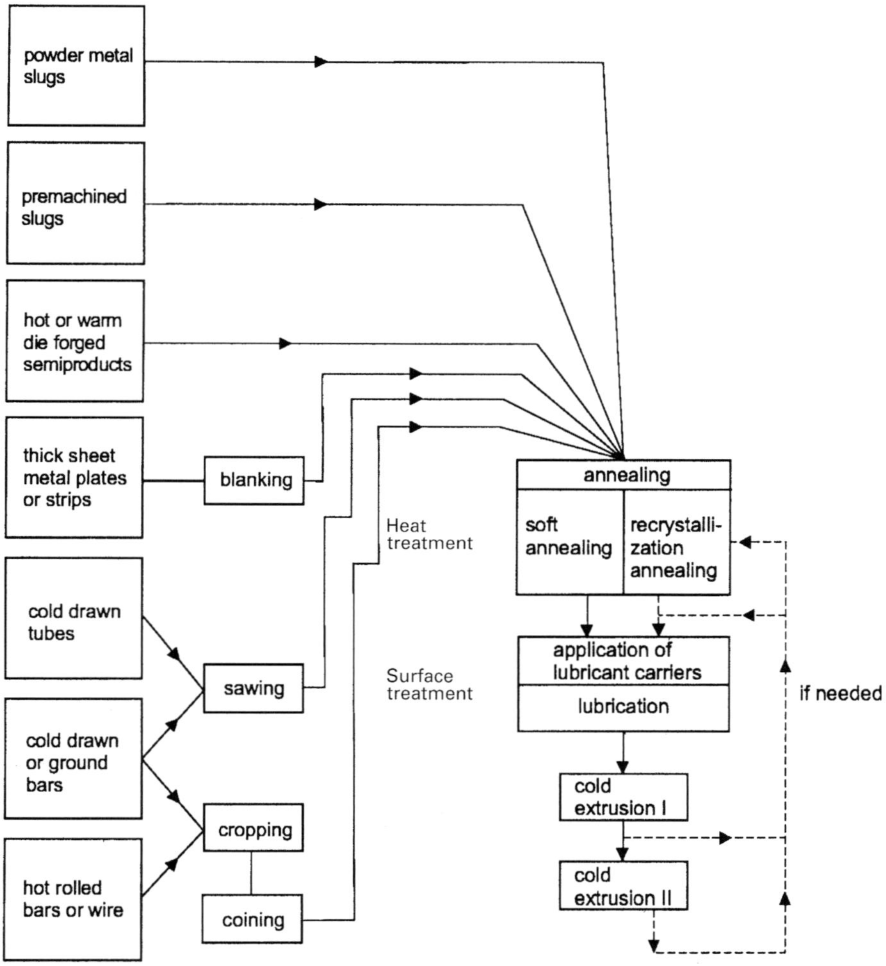

Figure 4
General overview of the cold-extrusion process.

Pure copper also has a very good formability. The formability of different types of brasses decreases by increasing the zinc content (37% is the limit value for successful cold extrusion).

3.3 *Cold Extrusion of Metals with Low Formability*

It is known that in alloying metals in order to achieve better machinability their formability is often decreased. However, it has been found that the cold extrusion of metallic alloys prepared for machining is possible if a hydrostatic pressure is added to the process. This approach demands higher tool loads and adequate tribological solutions.

4. *Tools for Cold Extrusion*

A successful industrial implementation of cold extrusion depends on reliable tooling support. Tools must sustain very high cyclic loads, have good toughness, and good wear resistance. In order to achieve a reasonable tool life, an axisymmetrical shape of cold-extruded components is preferred (Figs. 1–3). If this is not possible, tools should be designed to compensate for bending and eccentric loads.

The tool assembly for cold extrusion (Fig. 5) consists of active parts being in contact with the workpiece (punch, counter punch or ejector, and die), and supporting, clamping, and guiding elements.

When designing a tool and selecting tool materials there are many factors to be considered. The most important ones are (Kuzman and Kruăič 1998):

- the shape and the geometrical accuracy of the part;
- punch loads for extrusion and drawing out, counter punch loads, die loads;
- the number of components to be produced;

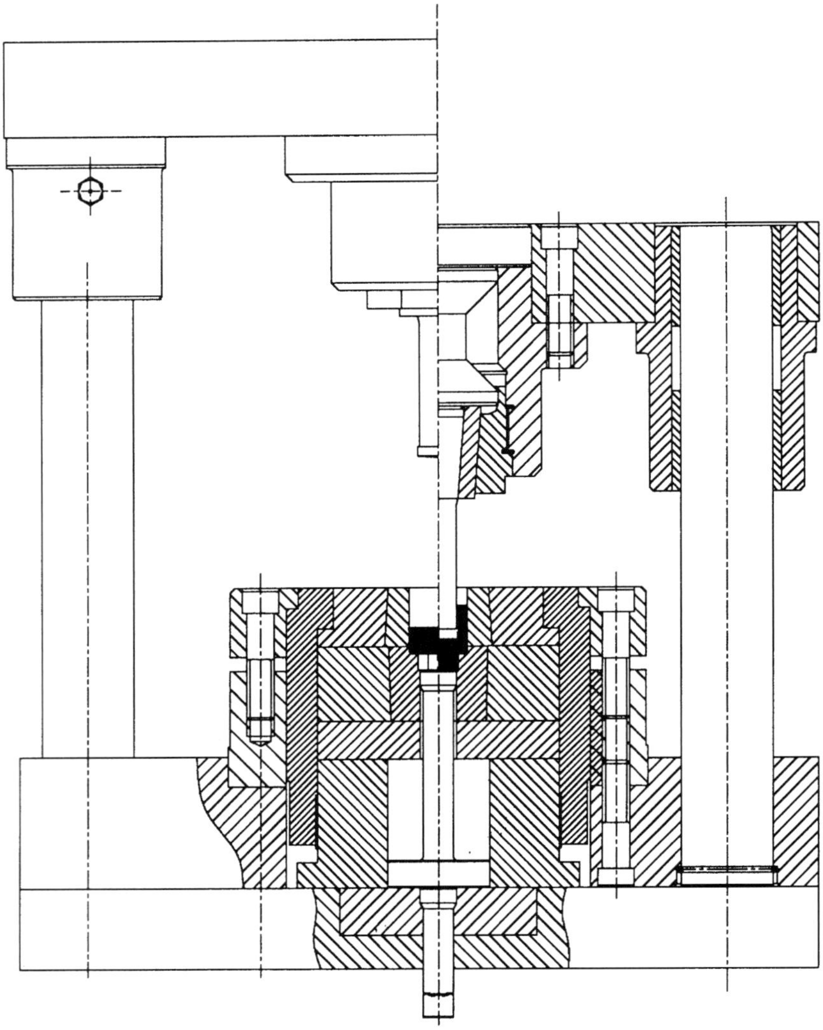

Figure 5
Tool assembly for cold extrusion of a special nut with a radially and axially prestressed die.

- available machinery, systems for handling slugs and components;
- control of the heat generation and transfer;
- control of the elastic response of the system part/tool/machine;
- distribution of lubricants, avoiding the accumulation of used lubricants;
- the forming sequences must be defined in such a way that the material can follow natural flow lines, i.e., it should have a laminated, equalized, and symmetrical flow; and
- the stress distribution in a tool should be as homogeneous as possible—drastic shape changes, small radii, acute angles, and marks can cause stress concentrations and initiate cracks.

Design and manufacture of extrusion dies demands knowledge and experience. Owing to the high radial pressures in die cavities, dies are reinforced (radially prestressed) by one or more stress rings. For very high pressures strip wound dies are used. In cases where high stress concentrations are expected in the axial direction, the dies are axially split. Transversally split dies can also be axially prestressed (Fig. 5).

The selection of materials for punches and die inserts depends on various parameters. High-speed steels are used the most because of their high working hardness, wear resistance, and adequate toughness. Very often steel with 0.9% C, 6.5% W, 4% Cr, 5% Mo, and 1.9% V (M2 after AISI) is used.

Tools made of cemented carbides generally have better wear resistance than those made of high-speed steel. Even though cemented carbides are much more brittle than tool steels, they have often replaced tool steels. Because they are much more sensitive to tensile stresses and stress concentrations, they should be implemented only after careful considerations. For extrusion punches low cobalt grades (5–15%) with medium grain size are recommended, whereas for die inserts medium grades (15–30%) with coarse to medium grain size are recommended.

In 1967 a group of international experts in the field of cold extrusion (the International Cold Forging Group, ICFG) was established. One of their aims was to disseminate knowledge of relevant processes. The group has published a lot of documents and data sheets. In a booklet from 1992 useful information concerning tools for cold extrusion can be found.

5. *Machining of Cold-extruded Components*

Modern cold-extrusion technology produces NNSF or NSF components, where finish machining is performed only on some surfaces with a few thin cuts. To control such machining processes it is necessary to know the technological history of the part being previously cold extruded. Problems connected with precise machining can be divided into three groups, as discussed below.

5.1 Effect of Strain Hardening

As the shape of the components is complex, the strain hardening distribution, which results in residual stresses, is highly inhomogenous. If some part or layer of the workpiece is removed by machining, the state of equilibrium is disturbed. In such cases the workpiece is usually deformed because the residual stresses are relaxed and are trying to attain a new state of equilibrium.

5.2 Effect of Recrystallization Annealing

Recrystallization annealing is frequently used after cold extrusion with large plastic deformations because it re-establishes good formability to the material, which can then be again cold deformed. If the strain distribution after the extrusion operation is not homogeneous, which is almost always the case, an inhomogeneous grain size structure can be expected after annealing. The new structure will retain its characteristics, even if a light finish cold-extrusion operation is performed (coining or sizing). During the machining of such components, the cutting process and especially the surface quality obtained depend on different grain size distribution. To control the finish machining and to obtain uniform surface quality, the optimal machining parameters should not be kept fixed throughout the process—they must be adapted to the volumetric grain size distribution in the component.

5.3 Dimensional Accuracy and Surface Quality in Turning Cold-extruded Components

In cold extrusion, materials with relatively low flow stresses and good formability are most commonly used. The applied heat treatments are oriented towards obtaining optimal extrusion conditions which are very often not suitable for later machining operations. At turning such components, chip formation is continuous. A long ribbon chip can therefore wind around the tool and the workpiece, causing damage to the machined surfaces. The problem can be solved by proper cutting tool geometry and adequate turning process parameters.

Dimensional inaccuracy of cold-extruded components causes clamping problems for later machining, where form errors must be considered very carefully.

6. *Conclusions*

Cold extrusion of metals, especially steels, is one of the fastest growing manufacturing technologies. The reasons for its rapid development are material and energy savings, high productivity, cost efficiency, and mechanical properties adapted to demands. Because of very high initial costs (technology and tool development, special machinery) cold extrusion was first implemented in mass production. A lot of examples can also be found where the quantities of the products do not exceed a few thousands. This is possible because of the standardization of tools, the use of experience together with similarity laws, and integral computer support from the part to the tool design and manufacture.

Bibliography

ICFG International Cold Forging Group 1967–1992: Objectives, History, Published Documents. Meisenbach Verlag, Bamberg, Germany, pp. 33–82

ISO 1993 Steels for cold heading and cold extruding. ISO 4954, 2nd edn., p. 38

Kuzman K, Kruăič V 1998 Tools—one of the mayor attributes for successful cold forging. In: *Proc. Int. Conf. on Forging and Related Technology, IMechE Conf. Trans. 1998-3.* Professional Engineering Publishing, pp. 333–40

Lange K 1985 *Handbook of Metal Forming.* McGraw-Hill, New York, Chaps.13, 15

Spur G, Stöferle Th 1984 *Handbook on Manufacturing.* Carl Hanser Verlag, Munich, Germany, Vol. 2/2, Chap. 3.10.3 (in German)

K. Kuzman

Metal Working: Cold Rolling

Although some small profiles are cold-section rolled (or, more often, drawn between idle rolls), and although some seamless tubes are cold-pilgered, the vast majority of cold-rolled metal is in the form of flat semi-products, i.e., such that the thickness is much smaller than the width and length. Such products are named sheets and strips, and are mostly sold as coils; they are the subject of this article. "Rolling" means that the product (here, the strip) is reduced in cross-sectional area (here in thickness) by two axisymmetric tools named rolls, counter-rotating around their respective symmetry axis; friction is the driving force that entrains the product through the *bite*; the *gap* between the rolls defines the thickness of the product.

1. *Position of Cold Rolling in Metal Forming*

Probably more than 90% of all metal production is hot rolled; cold rolling is performed on 30% of carbon steel, 60% of stainless steel, and 20% of copper alloys: overall, about 30% of the world metal production. The advantage of rolling over other possible thickness-reducing processes such as strip drawing (in fixed dies or idle rolls) is that normally it imposes a triaxially compressive state of stress favorable to ductility, so that larger reductions per pass may be taken (60% is common in aluminum rolling). However, in view of the large cold reduction to be performed, cold rolling is almost always a multipass operation.

Before cold rolling, the metal has almost always undergone a hot-rolling stage, although thin strip casting, which short-cuts hot working, tends to develop (Steffen and Tacke 1999). Some properties or defects are inherited from the hot-rolling stage and must be accounted for in the design of the cold-rolling schedule (e.g., pickling to eliminate oxides prior to steel cold rolling). *After* cold rolling, it will most of the time go through deep drawing or stamping, other sheet forming processes (spinning, bending, roll forming) or cutting processes (piercing, fine blanking). The necessary properties (geometric tolerances, surface finish, mechanical resistance) must therefore be imparted at the cold-rolling stage: cold rolling, as any forming process, cannot be isolated from the previous and next processing stages. Thus, owing to stringent requirements on ductility and mechanical properties in sheet forming operations, the metal often has to be annealed at the end of the cold-rolling process, or before the end if a degree of work-hardening is required from the finished strip.

Cold rolling means that no heating device is used just before or during rolling. However, self-heating of the product by its plastic deformation is significant. Working temperatures of 100°C to 150°C are common, 200°C is sometimes reached. Certain metals are indeed rolled, from the metallurgical and microstructural point of view, in their hot deformation range $T(\mathrm{K}) > 0.5\ T_m$ (T_m is the melting temperature). Thus, lead, tin or zinc recrystallize during cold rolling; aluminum alloys reach the limit between the cold (non-thermally activated) and hot (thermally activated) range.

A number of applications of rolled strips require coatings. This is particularly true for corrosion protection of steel (the majority of car body panels are now galvanized, mainly by hot dip-coating; steel strips for food packaging are tinned). Coating is often performed before the last rolling operations.

Rolling is also used to cold-bond strips of different natures: different aluminum alloys, aluminum or stainless steel on steel, etc. (Bay 1986) (see *Joining of Metals*).

2. *Requirements of Cold-rolled Strips*

2.1 *Geometry, Tolerances*

A strip should have as constant as possible a thickness (or gauge), in the rolling direction (RD) and in the transverse direction TD ("thickness profile"). Furthermore, it should be flat (i.e., the mean surface should be a plane, devoid of undulation or waviness). Thickness variations in RD may be caused by imperfect automatic gauge control at transients (strip heads, or welds in continuous rolling); small periodic variations may also be owing to mill vibrations or "chatter" (Yun et al. 1998). Thickness profile defects (variations along TD) and flatness defects are connected with mill stretching and roll elastic deformation (Fig. 1). Owing to the applied contact stresses with the strip, the work roll (WR) bends as a beam over all its length, giving a roughly parabolic strip thickness variation along TD. Furthermore, in the contact area, compression of the roll metal results in the elastic "indentation" of the roll by the strip, called roll flattening. It results in a sudden drop of thickness by a few tens of microns near the strip edge (the edge-drop defect), at the transition between the flattened (in contact with the strip) and non-flattened areas of the roll. When rolling thin gauges, the flattening indentation of the roll may be so large that the upper and lower WR touch on either side of the strip, a situation called "roll kiss," systematically found in the final passes of aluminum thin-foil rolling.

Thus a strip incoming in a bite has a varying thickness along TD owing to previous rolling passes ("strip crown"). The deformed, bent rolls leave a nonconstant gap as well. The difference between these two profiles will bring about differential reduction, thus differential elongation of the strip, depending on position in TD. Longitudinal stresses result in the inter-stands or on the finished strip (residual stresses); if large enough, they induce a buckling phenomenon giving rise to flatness defects.

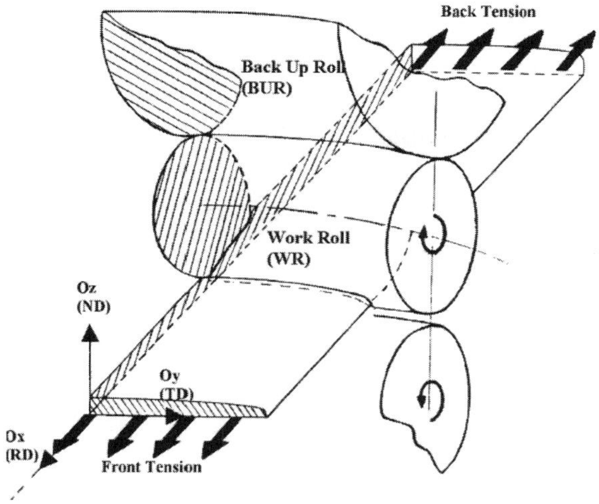

Figure 1
Schematic representation of the strip/roll contact (here the left upper part of a four-high mill). Note the roll deformations (bending and flattening, see text).

These may occur at different TD positions depending on the respective strip and roll profiles: long edges consist in a waviness of the edges, long center is a waviness in the middle of the strip; quarter-buckles are in between (Roberts 1979).

It must be noted that strip crowns are inherited from hot rolling. Trying to eliminate them at cold rolling will indeed result in differential elongation, residual stresses, and flatness defects: it may be said that the final profile and flatness of a cold-rolled strip is largely dependent on the geometrical quality of the hot coil, and the corrective capabilities of cold rolling are limited.

Full details of the mill designs, devices used to control the thickness, its profile and the strip flatness (profile and flatness actuators) along with the corresponding automatic control systems can be found in, for example, the publication by Roberts (1979): elaborated roll stack systems where smaller work rolls (300–700 mm in diameter, except for cluster mills for harder metals, with diameters 50–100 mm) are supported by bulkier and more rigid back-up rolls (1000–1800 mm in diameter); "roll crown" (variation of diameter along TD, by a few tens of microns); hydraulic roll counter-bending systems; TD-shiftable crowned or tapered rolls; roll crossing; TD-differential roll cooling (Berger *et al.* 1987).

2.2 *Mechanical Properties and Microstructures*

The most stringent requirements are put on cold-rolled strip if it is to be deep drawn. Stamping or deep drawing apply stresses tensile in nature, which may promote necking and fracture. Three characteristics are therefore desirable:

(i) A high strain hardening exponent (and if possible a high strain rate sensitivity) so as to oppose strain localization and necking: hence the importance of annealing.

(ii) A high Lankford coefficient, i.e., the metal should elongate more easily in the in-plane directions as compared to the normal direction ND, so as to resist thinning in tension and thickening in compression (see *Metal Forming (Cold): Dislocation Mechanisms and Microstructural Changes*). Typical values of 2 or above are achieved for carbon steel, whereas values around 1 are obtained for aluminum alloys.

(iii) Furthermore, in-plane anisotropy should be kept as low as possible to avoid the earing defects.

These characteristics must be achieved in the cold-rolled coil, along with sufficient ductility for further processing. This is why annealing (batch or, more and more, continuous) is often performed at the end, or shortly before the end, of the cold rolling process.

2.3 *Surface Finish and Surface Properties*

Specifications on surface finish are motivated either by aesthetic reasons (e.g., bright-annealed stainless steel for domestic applications or in the building industry) or by technical needs connected with further processing: avoiding seizure and galling in deep drawing requires a sufficient roughness, so that large "valleys" may store lubricant and receive wear debris. Requirements on roughness often go beyond the traditional *Ra* value (e.g., skewness *Sk* and/or peak density are specified). The cold rolling process must therefore be able to build a well-controlled roughness pattern on the strip, which requires:

(i) a careful preparation and maintenance of roll surface using mostly grinding and sandblasting, but also electro-discharge texturing, laser or electron beam texturing. The latter three are mainly used in steel strip rolling for the automobile industry;

(ii) a good knowledge of interfacial phenomena ("roughness transfer") and lubrication mechanisms.

A major consideration is strip cleanliness. Annealing may crack lubricants and deposit carbonaceous residues, which stain the product. Lubricants must be carefully optimized to avoid this, using only small molecules (aluminum or copper alloys), replacing tallow and vegetable oils by synthetic esters for carbon steel, etc. Many details on tribological mechanisms relevant to rolling processes, along with lubricant composition and performance, may be found in the publication by Schey (1987).

In the case of carbon steel, a corrosion protection is also needed between the end of rolling and the delivery at the customer's. A protective oil is thus coated on the strip; it is more and more used as the lubricant for deep drawing.

2.4 *Automatic Control and Modeling*

Modern rolling mills, and specially cold tandem mills with high productivity, are equipped with a number of transducers participating in automatic control systems: rolling force (on exception rolling torques), strip tensions between stands, average gauge and thickness profile at exit, strip tension profile (shape measuring systems) between certain stands, sometimes temperature, parameters pertaining to profile and flatness actuators (forces, oil pressures, linear or angular position), lubricant outputs and temperature, coil length and weight, etc. Data logging from and process control through these transducers require a large computing capacity.

Control systems (e.g., control of thickness by the force, or of the thickness profile via the roll bending forces) require models linking the input and output parameters: for instance how much more load δF is needed to change the exit thickness by δh; this of course depends on all other rolling conditions and parameters, the "working point." Simplified "on line" models are used to do this: they linearize the relationship (i.e., estimate $\partial F/\partial h$) around a certain working point. Artificial intelligence techniques are more and more used for this purpose. They may also be derived from more complex "off-line" models solving the thermo-mechanical equations of the system. Full account must then be taken of the coupling between the elastic deformation of the tools, the settings of the profile and flatness actuators, and the plastic deformation of the strip. The intensity of the coupling is all the stronger as thin, hard strip and high friction are considered. Such knowledge models may be complemented by a thermal analysis of the system, by lubrication and friction models, and by microstructure evolution models, all of which may also be coupled to the main thermo-mechanical analysis.

3. *Basic Results on the Mechanics of Cold Rolling*

3.1 *Neutral Point and Non-skidding Condition*

Experience shows that safe rolling conditions consist of a strip entering the bite at a speed V_{in} smaller than the roll speed V_r, and exiting at V_{out} which is slightly larger than V_r. If $V_{out} < V_r$, skidding occurs: although it remains possible to roll smoothly, it is generally considered unsafe, as rolling may become unstable, jerky, chatter may follow, etc. The forward slip $S = (V_{out} - V_r)/V_r$ is introduced to quantify these tendencies; normally, it is positive but skidding means that it is negative.

Since $V_{in} < V_r$ and $V_{out} > V_r$, by continuity, there exists a plane or a zone in the bite where the strip speed $V(x) = V_r$: this is the neutral point, at position x_n. Before x_n (on the entry side), the roll moves faster, thus friction (counted positive) is driving the strip; after x_n, the roll is slower than the strip and brakes it (friction stress is counted as negative). This reversal of the direction of friction is a very important aspect of rolling processes, with consequences for thermal transfer, lubrication, stress states, surface properties, etc.

Skidding is also defined by the fact that the neutral point becomes virtual, out of the bite on the exit side (since $V(x) < V_r$ everywhere). The position of the neutral point can be found by averaging the normal and tangential stresses and writing the force balance on the strip. In the absence of unbalanced strip tensions:

$$\alpha_n = \frac{\alpha}{2} - \frac{\alpha^2}{4\mu} \qquad (1)$$

where α_n is the angular position (see Fig. 2) of the neutral point, α is the bite angle:

$$\alpha \approx \sqrt{\frac{h_{in} - h_{out}}{R}} \qquad (2)$$

and μ is the Coulomb friction coefficient. This shows that decreasing friction decreases α_n, thus driving the neutral point towards the exit side. Skidding occurs when $\alpha_n = 0$, i.e.:

$$\mu_{min} = \frac{\alpha}{2} \qquad (3)$$

where $\mu \geqslant \alpha/2$ is the non-skidding condition. It states that a certain level of friction is necessary to roll safely. The friction needed is all the higher as larger reductions are desired, and as the roll radius is smaller. In practice, μ_{min} ranges from 0.08 for the first pass on a Sendzimir mill with 100 mm diameter rolls (stainless steel) to 0.03 in the first stand of a steel or aluminum tandem mill, 0.01 in the last stand on a tinplate tandem mill, and 0.005 in the last passes of aluminum foil rolling.

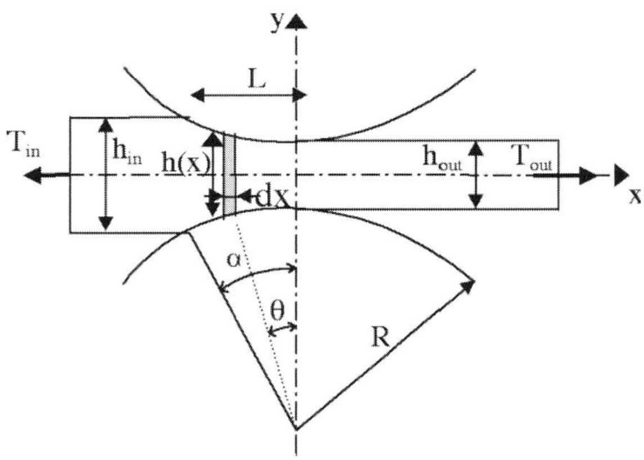

Figure 2
Slab method modeling of rolling: geometry and notations.

However, too large friction is detrimental to strip surface quality, to profile and flatness, because it increases the strip-roll contact stresses; and to the microstructure owing to temperature increase. Thus for each rolling operation, an optimal friction range has to be determined, then obtained and maintained on the mill.

3.2 Mechanical Equations, Stress State

Owing to the very large width/thickness (w/h) ratio, spread is practically negligible in cold rolling, which is therefore a quasi-plane strain process *if roll bending is neglected.* Furthermore, in most cold rolling operations, the contact arc length/thickness ratio is also large ($L/h > 3$), so that the principal axes of strain rate ε and stress tensor σ almost coincide with RD and ND (normal direction), and the variations along ND are small compared with those along RD: these are conditions under which the slab method (SM) is known to perform well. It can be used to explain many aspects of the process. The equilibrium equations along RD and ND are grouped with the plastic yield criterion in plane strain, giving (see notations in Fig. 2):

$$\frac{\mathrm{d}(h\sigma_{xx})}{\mathrm{d}x} = 2\left(\sigma_{xx} - \frac{2\sigma_0}{\sqrt{3}}\right)\frac{\frac{1}{2}\frac{\mathrm{d}h}{\mathrm{d}x} \pm \mu}{1 \mp \mu\frac{1}{2}\frac{\mathrm{d}h}{\mathrm{d}x}}, \quad (4a)$$

$$\sigma_{yy} = \sigma_{xx} - \frac{2\sigma_0(\bar{\varepsilon}, \dot{\bar{\varepsilon}})}{\sqrt{3}} \quad (4b)$$

$$\text{Strain } \bar{\varepsilon} = \frac{2}{\sqrt{3}} \ln \frac{h_{\mathrm{in}}}{h(x)}, \quad (5a)$$

$$\text{strain rate } \dot{\bar{\varepsilon}} = \frac{2}{\sqrt{3}} \frac{V_{\mathrm{r}}\frac{1}{2}\frac{\mathrm{d}h}{\mathrm{d}x}}{h(x)} \quad (5b)$$

(Note that $\mathrm{d}h/\mathrm{d}x = -2\tan\theta$ if roll flattening is neglected).

In the last fraction of Eqn. (4a), the $\pm$ signs represent the reversal of the friction stress at the neutral point x_n; the sign above is valid after x_n, the sign below is to be used before x_n. Note that the position x_n is an unknown of the system. σ_0 is the yield stress, μ is the Coulomb friction coefficient, θ is the current angle, with its origin in the plane of the roll centerlines, where $x = 0$ (Fig. 2), so that $x = -R\sin\theta$ (R is the roll radius), $\theta = 0$ at exit, and $\theta = \alpha \approx \sqrt{(h_{\mathrm{in}} - h_{\mathrm{out}})/R}$ at entry.

Integration of this equation with the boundary conditions shown in Fig. 2 leads to the "friction hill" normal stress profiles of Fig. 3. The normal stress $|\sigma_n|$ increases, first very fast from 0 at entry to $(2\sigma_0/\sqrt{3})(1 - T_{\mathrm{in}})$ at the end of the entry elastic zone, then to an apex at the neutral point, and decreases down to $(2\sigma_0/\sqrt{3})(1 - T_{\mathrm{out}})$ at the end of the plastic zone and to 0 at exit. This particular shape, linked to the reversal of friction at x_n, is called the "friction hill." Note that the rolling force is practically equal to the area under the friction hill, whereas the torque is deduced by subtracting the area under the curve after the neutral point (resisting friction) from the area under the curve before the neutral point (driving friction).

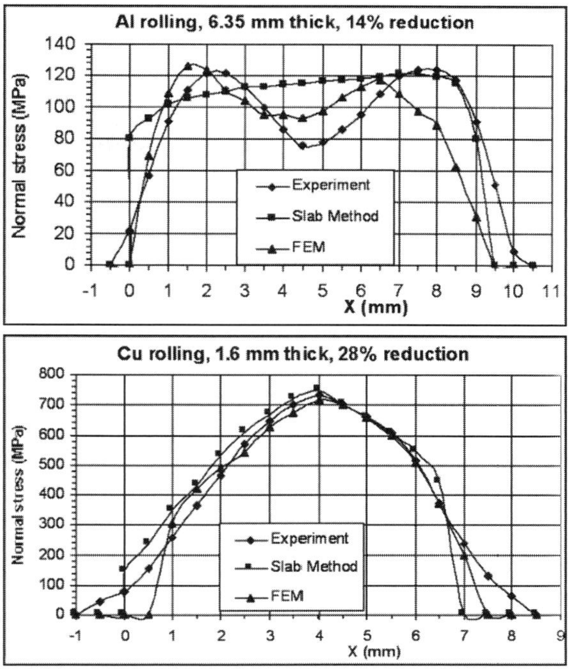

Figure 3
Comparison of experiments by Al-Salehi *et al.* (1973) with results from the slab method and the finite element method (FEM). Roll diameter 160 mm. Aluminum rolling: $L/h = 1.45$; copper rolling: $L/h = 3.7$.

The normal stress at x_n (and similarly the "slope" of the friction hill) is a fast increasing function of $\mu L/h_{\mathrm{out}}$ where the contact length L is given by $L \approx \sqrt{R.(h_{\mathrm{in}} - h_{\mathrm{out}})}$ the magnitude of normal stresses (which in turn govern the rolling force and torque, the roll deformation , etc.) increases with the friction coefficient μ, *all the more as the bite aspect ratio L/h increases.* Rolling of "thin strips" (in the sense of large L/h) is thus very sensitive to friction (this is true for all forming processes), which means that large reductions will lead to larger stresses, as expected, but that large rolls radii may considerably increase the stresses as well. The rolling force increases still faster with roll radius, since higher stresses have to be integrated through a longer bite.

3.3 Extension to Thicker Strips

The previous results, obtained by the slab method, are reliable provided that the hypotheses of the method are well verified: in particular, the identity of RD, TD, and ND and principal directions of strain-rate and strain is valid only for "thin strips," i.e., $L/h > 3$ (roughly). In other situations, results will deviate significantly, so that another technique is required: the Finite Element Method (FEM) must now be used. Figure 3 compares, for two cases:

(i) experimental results (Al Salehi *et al.* 1973) showing the "friction hill";

(ii) the same curve computed by the slab method (fitting of friction coefficient on the rolling load);

(iii) the same curve obtained by the FEM.

Obviously, both theoretical analyses agree for the thin (copper) strip such that $L/h = 3.7$. They differ greatly for the thicker (aluminum) strip, where $L/h = 1.2$: in particular, the SM is unable to capture the initial peak, due to strain rate concentration at the entry point.

4. Roll Elastic Deformation and its Consequences

4.1 Roll Flattening

Apart from its contribution to the edge-drop defect, which is a fully three-dimensional effect, roll flattening has another, sometimes strong, impact on the mechanics of the rolling process through the increased contact arc length. This impact is even essential in some cases such as thin foil rolling. The problem becomes a coupled one. The strip deformation may be modeled by the SM, or FEM if $L/h < 3$; the roll deformation may be analyzed by an elastic FEM, or influence function techniques. All the difficulty is in the coupling Gratacos *et al.* (1992). Figure 4 Fleck *et al.* (1992) shows the evolution of the contact arc shape and length and of the friction hill as thinner and thinner strips are rolled. A flat area appears on the contact arc, in which the strip metal may become locally elastic; sliding is very small in this flat area; the neutral point lies near its end. The reduction is performed by halves at the entry and at the exit. The contact pressure evolves from a sharp friction hill type to an elliptic (hertzian contact) profile, with a sharp peak on the exit side.

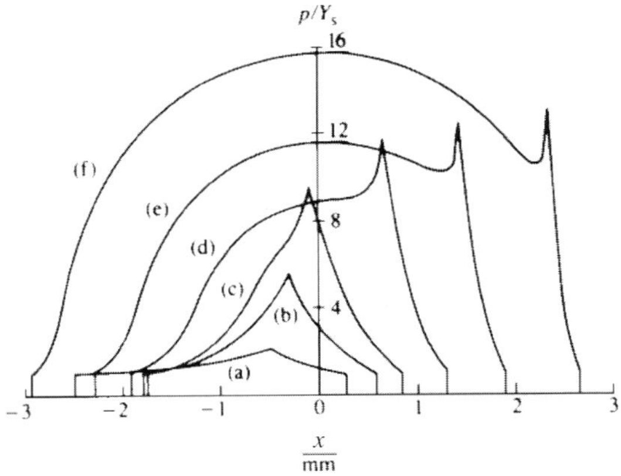

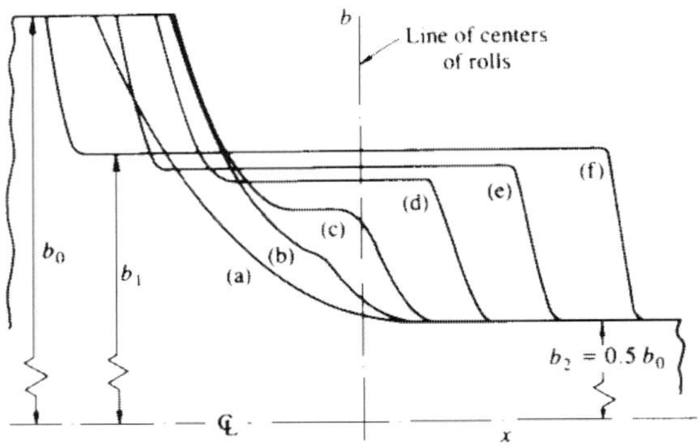

Figure 4
Aluminum foil rolling. Reduction 50% in all cases. The initial thickness varies between 140 µm, case (a), and 13 µm, case (f) Fleck *et al.* (1992). Upper diagram: normal stress divided by yield stress Y_s. Note the transition between a "friction hill" pattern towards a "Hertzian pressure profile" when thickness is reduced. Lower diagram: roll profile. Note the occurrence of a flattened zone and its lengthening as the thickness is reduced.

4.2 Heterogeneous Flow and Residual Stresses

Even in the ideal case where a strictly rectangular strip enters a strictly parallel bite, deformation will be slightly heterogeneous:

(i) Shear is more intense at the contact of the rolls, so that a slightly larger strain is found at and beneath the surface as compared to the core of the strip; elongation tends to be larger at the surface as well.

(ii) Elongation tends to be smaller near the edges of the strip because of a competition with lateral flow (spread).

Of course, apart from transient effects, the strip moves away from the bite as a rigid body, with a length independent on ND or TD. This means that, at the exit of the bite, differential elongation tendencies are compensated by two phenomena:

(i) Transverse flow tends to redistribute the metal from the most elongated "fibers" to the least elongated ones; this effect is weaker in cold than in hot rolling owing to higher w/h ratio.

(ii) Compressive, *residual* longitudinal stresses shorten the longest fibers, while tensile stresses elongate the short ones (sometimes plastically in hot rolling, always elastically in cold rolling).

Thus, the surface is in tension, while the core is in compression (the reverse may be true for very small reductions, i.e., temper rolling). Similarly, the edges are in tension, and the center is in compression.

Furthermore, as the temperature is also heterogeneous in the bite, differential cooling down to room temperature would add another residual stress field: the warmest points would retract more, but are restricted by the cooler ones; the former points are in tension, the latter in compression. As the surface is generally the warmest as well as the most strained area, tensile elastic and thermal residual stresses mount up there.

In general, the strip is maintained under global tension; during rolling, the average tension stress corresponding to this tension force is added to the residual stress pattern described above.

4.3 Roll Deformation and Strip Profile, Stresses, and Flatness

Things get much more complicated, and residual stresses much larger, when the non-affinity of strip crown and WR profile, owing to roll deformation, causes differential reduction, which becomes the dominant term of the differential elongation.

Generally, the transverse flow is far from compensating differential elongation coming from edge effects, thermal dilation of rolls, and strip/roll crown difference. Residual stresses may thus be so large as to induce buckling in the compressed areas of the inter-stand: an *immediate* flatness defect has occurred. In less severe cases, the strip is flat on the mill and in the coil, but *delayed* flatness defects appear when a blank is cut from it before deep-drawing. Immediate flatness defects and dangerous residual stresses, which might lead to delayed flatness defects, can be significantly alleviated by straightening, e.g., a succession of small, alternate and decreasing plastic strains applied by bending under tension (see *Metal Working: Straightening*).

In fact, transverse flow and outlet profile result from a strong coupling between strip deformation, which imposes the contact stresses and lateral flow, and roll elastic deformation, which imposes the tool geometry as a boundary condition. Three-dimensional coupled modeling of the strip and roll stack deformation and of temperature field is therefore the basis for the prediction of the strip profile and flatness. Flatness and profile actuators must be included in such models, which tend to control and optimize their action. Two kinds of strip metal plastic flow modeling are available:

(i) Those based on a three-dimensional modeling, now by the FEM.

(ii) Those using a set of two-dimensional (SM) models at various locations along TD Pawelski *et al.* (1989).

Both give a distribution of rolling load $f(y)$ or contact stresses $\sigma_n(x, y)$ which are used as boundary conditions for roll stack models. The latter are sometimes based on elastic FEM, but less costly approaches may be derived by adequately merging known solutions of elementary problems Hacquin *et al.* (1998). Special attention must be paid to the algorithm coupling roll and strip deformation, and thermal effects both in strip and roll may not be overlooked.

Bibliography

Al-Salehi F A R, Firbank T C, Lancaster P R 1973 An experimental determination of the roll pressure distributions in cold rolling. *Int. J. Mech. Sci.* **15**, 693–710

Bay N 1986 Cold welding. Parts I–III. *Met. Construct.* **6**, 369–72; **8**, 486–90; **10**, 625–9

Berger B, Neuschütz E, Mücke G 1987 Experiences with a flatness control system including axial roll shifting In: *Proc. 4th Int. Steel Rolling Conf.* Association Technique de la Sidérurgie (ATS), Paris, pp. E13.1–6

Fleck N A, Johnson K L, Mear M E, Zhang L C 1992 Cold rolling of foil. *Proc. Inst. Mech. Eng.* **206**, 119–31

Gratacos P, Montmitonnet P, Fromholz C, Chenot J-L 1992 A plane strain elastoplastic FEM for cold rolling of thin strip. *Int. J. Mech. Sci.* **34**(3), 195–210

Hacquin A, Montmitonnet P, Guillerault J-Ph 1998 A 3D semi-analytical model of rolling stand deformation with finite element validation. *Eur. J. Mech. A (Solids)* **17**(1), 79–106

Pawelski O, Rasp W, Rieckmann J 1989 Description of elastic and plastic deformations in cold rolling mills. *Stahl und Eisen.* **20**, 49–57 (in German)

Roberts W L 1979 *Cold Rolling of Steel.* Dekker, New York

Schey J A 1987 *Tribology in Metalworking.* ASM, Metals Park, OH

Steffen R, Tacke K H 1999 State of steel strip casting. *Stahl Eisen* **119**(6–7), 129 (in German)

Yun I S, Wilson W R D, Ehmann K F 1998 Review of chatter studies in cold rolling. *Int. J. Mach. Tools Man.* **38**, 1499–530

P. Montmitonnet

Metal Working: Drawing of Bars and Wires

The purpose of the process of wire drawing is to reduce the diameter (D_o) of a rod or wire to a smaller size (D_f), by drawing the incoming raw material through a converging (usually conical) die (see Fig. 1). Note that the radius $R_f = (1/2)D_f$. The percentage reduction in area of the wire becomes $r\% = 100[1 - (D_f/D_o)^2]$. The required drawing force ($F = R^2{}_f\sigma_{xf}$) is mainly dependent on the reduction ($r\%$), the semi die angle (α), and the friction resistance to sliding (m) between the wire and the die. It is also linearly proportional to the flow strength (σ_0) of the drawn wire. The characteristics of the relative drawing stress over the flow strength (σ_{xf}/σ_0) as a function of the three input-independent parameters can be expressed symbolically as $(\sigma_{xf}/\sigma_0) = f(r\%, \alpha, m)$. This relationship (Fig. 2) has been studied experimentally and formulated by Siebel (1947) in his pioneering work. Avitzur (1979) derived this relationship analytically as follows:

$$\frac{\sigma_{xf}}{\sigma_0} = \frac{\sigma_{xb}}{\sigma_0} + 2f(a)\ln\left(\frac{R_0}{R_f}\right) + \frac{2}{\sqrt{(3)}} \times \left[\frac{\alpha}{\sin^2\alpha} - \cot\alpha + m\cot\alpha\,\ln\left(\frac{R_0}{R_f}\right) + m\,\frac{L}{R_f}\right] \quad (1)$$

where

$$f(a) = \frac{1}{sin^2\alpha}\left(1 - (cos\ a)\sqrt{1 - \frac{11}{12}\,sin^2\alpha} + \frac{1}{\sqrt{11 \times 12}}\,ln\,\frac{1 + \sqrt{\frac{11}{12}}}{\sqrt{\frac{11}{12}}\,cos\,\alpha + \sqrt{1 - \frac{11}{12}\,sin^2\alpha}}\right)$$

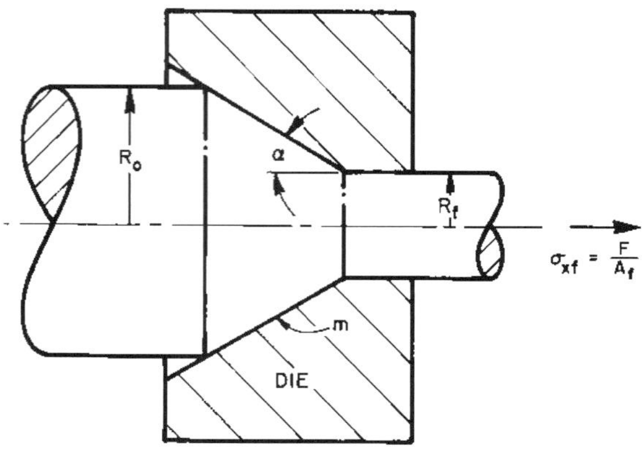

Figure 1
Drawing force.

1. *Limit Analysis*

Exact analytical solutions are not available for such problems in metal forming as flow through conical converging dies (see Fig. 1). Approximations and simplifying assumptions are inevitable and many approaches, such as slug equilibrium, slip line techniques and others, have been partially successful. With advances in computer science and technology, numerical approaches are emerging.

Limit analysis, as an analytical tool, is a promising approach that is being used with increasing frequency. In this approach, as applied to the study of drawing or extrusion force, two approximate solutions are developed. One, the upper-bound solution, provides a value that is known to be higher than or equal to the actual force; the other, the lower-bound solution, provides a value that is known to be equal to or lower than the actual force. The actual force thus lies between the two solutions. For example, in Fig. 2 with drawing stress as ordinate and semicone angle of the die as abscissa, upper- and lower-bound solutions are plotted for several reductions together with corresponding measured values of actual stress. Even when experimental results are not available, it is

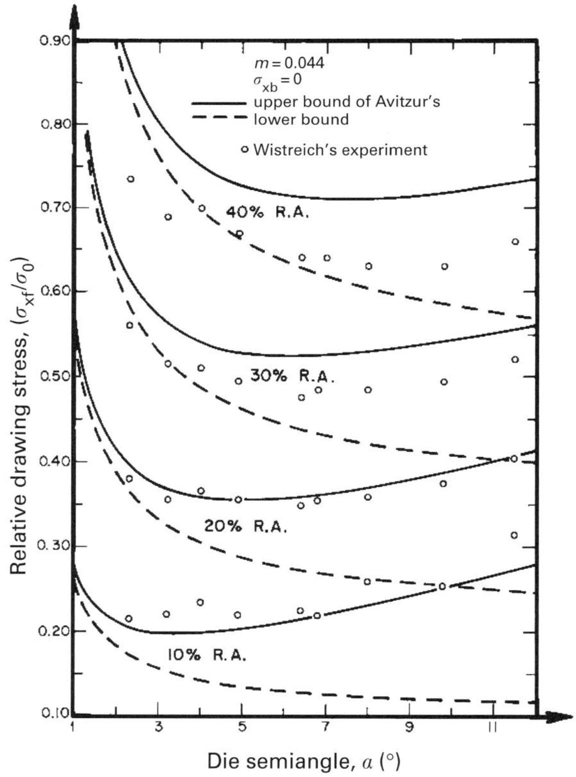

Figure 2
Relative drawing stress.

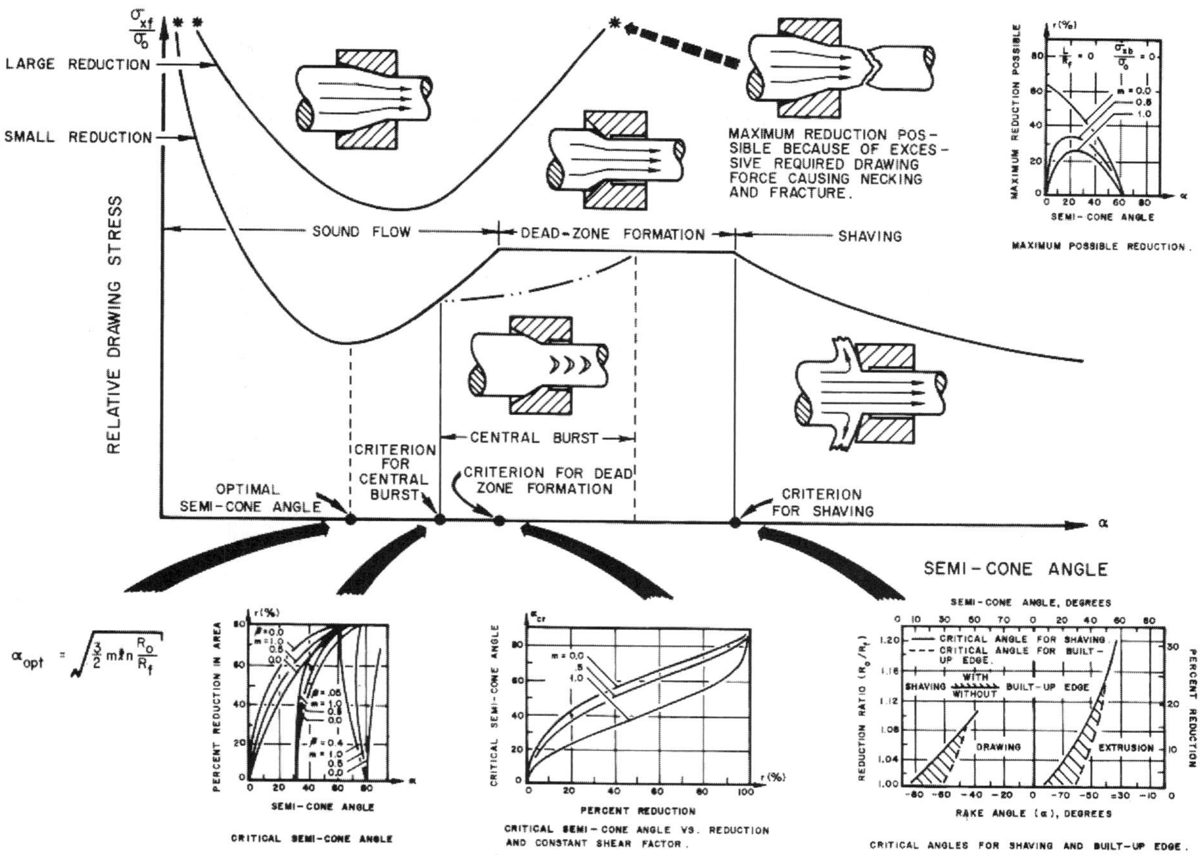

Figure 3
Effect of cone angle on metal flow.

expected that the actual stress and the exact solution, if these were available, would lie between the upper and lower bounds as obtained analytically. Thus, by limit analysis, an approximate solution is given with an estimate of the maximum possible error. The gap between upper- and lower-bound solutions may be narrowed by providing several upper bounds, and choosing the lowest upper bound, and by providing several lower bounds, and choosing the highest lower bound. Upper- and lower-bound solutions are obtained by following strict rules (including the requirement of proper description of friction behavior and material characteristics), thereby rendering them to become upper- and lower-bound solutions. We observe the effect of the main processing parameters on the required drawing force when a sound product is produced. A major undertaking of a study of the process goes much further. In the rest of this article we study deviations of the material flow that may cause the process to fail and discuss how to prevent failure.

2. *Flow Patterns*

A collective display of flow patterns is given in Fig. 3. The basic graph shows the relative drawing stress as a function of the semicone angle of the die. The insets represent the various modes of deformation and several associated criteria. This display emphasizes the broad scope of possibilities afforded by application to metal forming of the principle of minimum energy. Computation of the required forces and determination of the optimal die angle or maximum possible reduction without tear of the rod at the exit is an impressive achievement by itself; steps beyond that have been taken when criteria for fracture and other flow patterns are developed.

It is worth noting that while sound flow is considered to belong to the field of plasticity, shaving to metal cutting, and central burst to fracture mechanics, in this discussion all these phenomena have been treated uniformly by the upper-bound approach. Identical analytical tools have been used under the principle of minimum energy. Fracture, for example, has not been accounted for by a separate metal property. One does not need a separate property, and indeed it appears that none exists. Fracture is the outcome of processing and the appropriate stress–strain relationships and yield criteria alone. Yield criteria indicate when plastic flow will commence. A material can undergo any amount of deformation: it is important only that the deformations are applied in the right fashion, without exhausting the ability of the material to undergo further deformation. For all these studies, whether of fracture, metal cutting, or plastic flow, a single characteristic approach yields the answers: the upper-bound approach with the associated concept of minimum energy.

Other phenomena can be studied likewise. Suppose that one observes a new mode of fracture, which has not yet been studied. Let Fig. 4 serve as an example: no criterion for the prevention of this fir tree (called also "fishskin" or "snakeskin") fracture is yet available. All that one has to do—and it may require some doing—is to propose a velocity field that leads to this defect; to compute the internal power, shear, and friction losses associated with the flaw; and to determine the range of parameters in which the total drawing or extrusion stress required is least among the power requirements of all previously known flow patterns. There are many more patterns that occur and which may be studied in this manner.

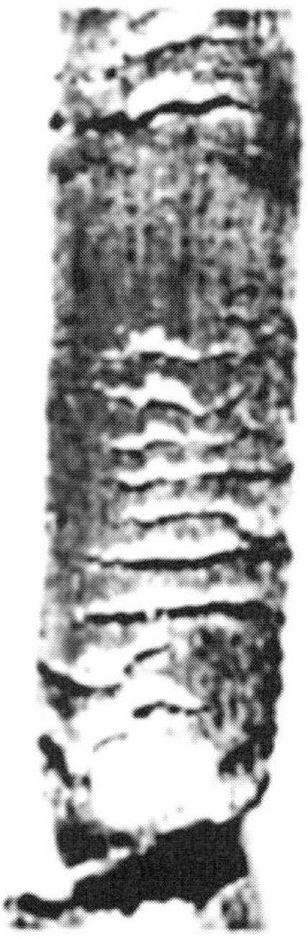

Figure 4
Fir tree defect.

3. *Friction*

Friction is a predominant factor in studies of wire drawing. Quantitative evaluation of friction existing during forming is a necessity. The work of Wistreich (1955) provides also a means to measure friction independently.

4. *Texture*

Significant progress in wire quality has been achieved through the control of texture. Texture is defined as the preferred orientation of heavily deformed grains in correlation with the direction of maximum strain. A severe amount of deformation in the wire drawing or rolling process will develop this preferred orientation, also known as fiber texture, owing to grain alignment similar to fibrous materials. An ideal wire texture represents axisymmetric grain orientation in the direction of the wire axis. Depending on the crystallographic structure of deformed metal different crystallographic orientations will dominate. In b.c.c. metals the fiber texture is in the $\langle 110 \rangle$ direction, while f.c.c. metals can have double orientation in the $\langle 111 \rangle$ and $\langle 100 \rangle$ directions. In aluminum, which is a high stacking fault metal, the $\langle 111 \rangle$ texture dominates while low stacking fault metals and alloys such as brass or silver have mainly $\langle 100 \rangle$ texture. There are two distinct types of texture, a crystallographic and mechanical one. A crystallographic texture is as a result of grain orientation. A mechanical texture describes alignment of inclusions and second-phase constituents in the main direction of mechanical working. Both types of texture influence mechanical properties of deformed wires and rods.

When cold-drawn wire undergoes recrystallization, its microstructure will form a new and stronger texture called the annealing texture or recrystallization texture. However, there is a correlation between deformation and annealing textures. Strong $\langle 111 \rangle$ texture results in lower recrystallization temperature in copper wire. The new annealed texture will result in new physical properties of the wire. Some processing parameters such as total reduction, die angle, and strain uniformity measured by redundant work influence fiber texture and therefore final physical properties.

5. *Conclusions*

Wires made of ferrous alloys are used in many structural applications such as bridge cables, tire wire, reinforced concrete, etc. Copper-based alloys are used in electric and magnetic devices. The miniaturization in the microelectronic industry imposes severe demands to produce ultrafine electronic wire. Progress in making more sophisticated smaller devices (electronic chips) depends to a great degree on the availability of finer wire.

See also: Metal Forming (Cold): Dislocation Mechanisms and Microstructural Changes, Metal Forming: General and Economic Aspects

Bibliography

A comprehensive list of references on wire drawing is available at the website http://www.metalforminginc.com/Publications/publications.shtml

Avitzur B 1979 *Metal Forming: Processes and Analysis*. Krieger, New York

Kraft F F, Chakkingal U, Baker G, Wright R N 1996 The effects of die angle on texture and annealing response of ETP copper wire. *J. Mater. Process. Technol.* **60**, 171–8

Pops H 1996 Metallurgy and technology of commercial copper electrical conductor wires. In: Paris H G, Kim D K (eds.) *Metallurgy, Processing and Applications of Metal Wires*. TMS, Warrendale, PA

Siebel E 1947 Der derzeitige Stand der Erkenntisse uber die Mechanischen Vorgange beim Dratziehen. *Stuhl Eisen* **11/12**, 171–80, (in German)

Wistreich J G 1955 Investigation of the mechanics of wire drawing. *Proc. Inst. Mech. Engrs.* **169**, 654–65

B. Avitzur and W. Z. Misiolek

Metal Working: Extrusion

Extrusion as a deformation process for the production of bars, tubes, and profiles from light and heavy metals, steel, as well as metallic composite materials, is subject to many influences with regard to product quality such as initial material structure, extrusion velocity, local process parameters, temperature, velocity, strain, strain rate, and stresses. With regard to high billet temperatures in the case of heavy metals and high-temperature materials such as intermetallics additional attention must be given to heat losses during material transfer from the preheat furnace to the press equipment as well as heat losses occurring during deformation, when the hot workpiece undergoes chilling due to the much colder toolings. To minimize these heat losses usually the billets are enclosed in a sacrificial can (e.g., for TiAl) and powdered with glass lubrication. Special die design and die materials (e.g., ceramic die inserts) are necessary.

This article describes in more detail the interrelations for the extrusion of light metals. In this case the initial material structure will be influenced by various production steps during billet preparation, whereas the type of extrusion has an effect on local process parameters.

1. *Billet Preparation*

1.1 *Casting and Homogenization*

Aluminum billet can be produced by casting processes with permanent mold, both fully continuous or semicontinuous direct-chill (DC), horizontal, or vertical routes. The vertical DC route is the most widespread. The microstructure developed by DC casting is not optimal for extrusion. In comparison with the equilibrium status, the distribution of the soluble alloying elements is not uniform. The percentage by volume of the eutectic precipitate is increased, the solidus temperature is decreased, and the arrangement of the particles for a deformation process is unfavorable. The as-cast billet should be further heat treated by means of a homogenization process.

The microstructure of a DC cast billet of AA2024 is shown in Fig. 1. Figure 1(a) and (b) clearly indicate the dendritic form of the solidified billet. The grain structure revealed in Fig. 1(b) has an average grain size of 90μm. Cored dendrites are surrounded by the interdendritic eutectic, which appears darker in the optical micrographs. The scanning electron microscope (Figs. 1(c) and (d)) reveals the darker interdendritic eutectic to consist of either $\alpha+(FeMnSiCu)Al$ or $\alpha+\theta+(Cu_2Mg_8Si_6Al_5)$, while the lighter eutectic is $CuAl_2$, $CuMgAl_2$, or $CuFeMnAl_6$.

The cored dendrite structure obtained in the as-cast alloy is completely eliminated after the 24 h

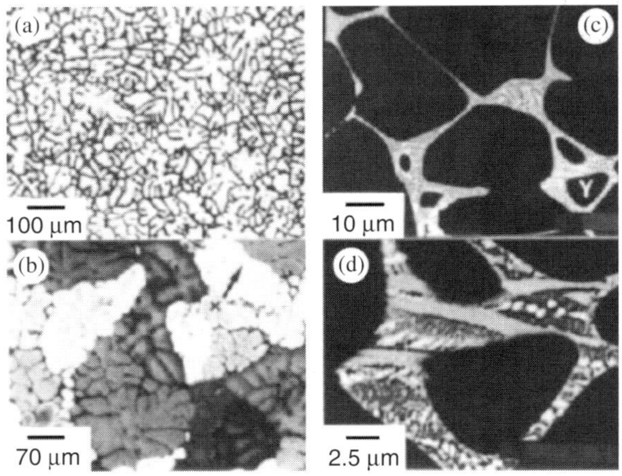

Figure 1
As-cast microstructure of AA2024 (Sheppard 1996).

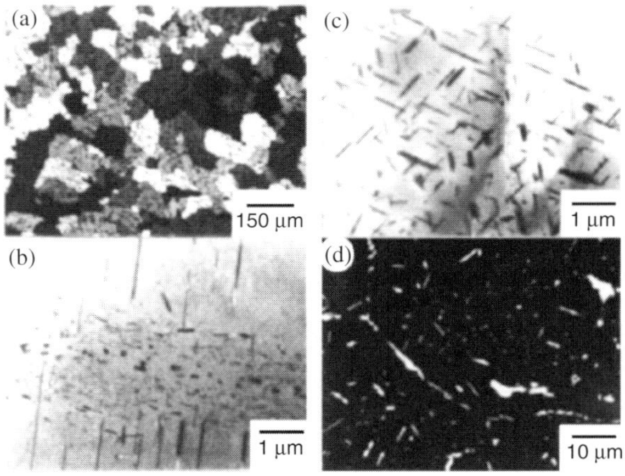

Figure 2
Microstructure of homogenized billets (Sheppard 1996).

homogenization process at 500°C. The homogenized structure is shown in Fig. 2. Figure 2(a) indicates that most of the copper and magnesium have been dissolved from the interdendritic eutectic phase. This structural change is represented in Figs. 2(b) and (c). Segregation is also eliminated and the soluble constituents have been distributed throughout the cell by diffusion. Homogenization leaves some $CuMgAl_2$ and $CuAl_2$ out of solution. All the iron-containing phases undergo transformation to Al_7Cu_2Fe and $FeMnAl_6$ and remain in the dendrite core. Manganese is precipitated from solid solution as $Cu_2Mn_3Al_{20}$. The needle-like structure shown in Fig. 2(c) is identified as $CuAl_2$ or $CuMgAl_2$. Figure 2(d) reveals the transition metal particles along the grain boundaries and prior to dendrite boundaries.

1.2 Cooling after Homogenization

The influence of a different billet cooling rate after homogenization on extrusion force and the mechanical product properties can be seen in Figs. 3 and 4, respectively. Four conditions of the billet heat treatment have been investigated as indicated in the figures.

The step-cooled billet requires the least and the quenched billet the highest extrusion force. Conventionally treated billets and solutionized billets are in between. The step-cooling treatment was developed to increase the extrudability of AA6063. If a step-cooled billet is warmed up slowly during the preheating process prior to extrusion, the positive influence of step-cooling treatment is lost, for the precipitate is dissolved during the warming up time. The profile produced from a step-cooled billet is of a lower mechanical stiffness than the other profiles (Fig. 4). This is also a disadvantage of this manner of billet treatment.

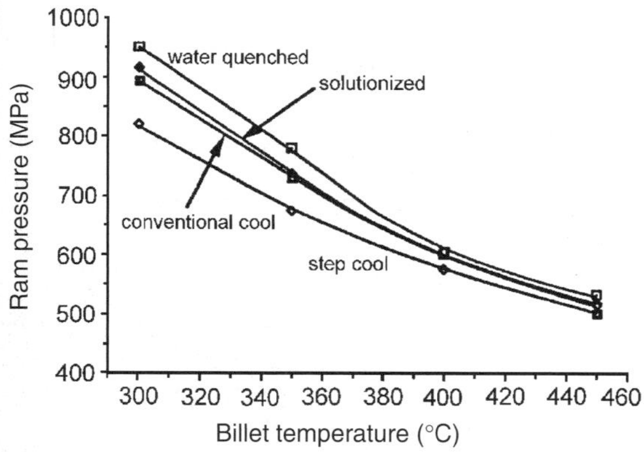

Figure 3
Ram pressure as a function of billet temperature and cooling procedure Ricks *et al.* (1992).

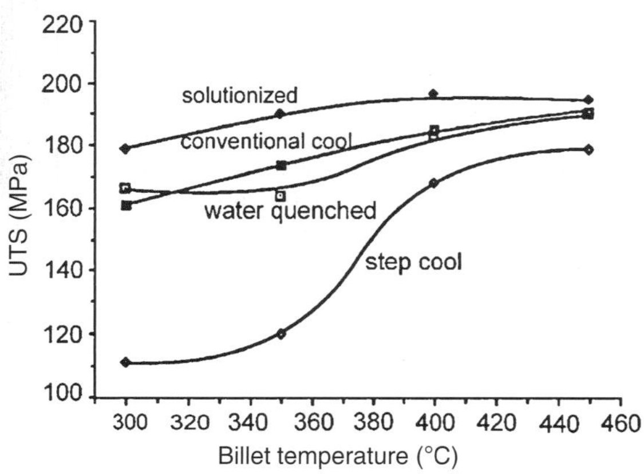

Figure 4
Tensile strength as a function of initial billet temperature and cooling procedure (Ricks *et al.* 1992).

Figure 3 shows a slight increase of extrusion pressure with solutionized billets. This is caused by the larger amount of magnesium and silicon dissolved in the solid-solution with aluminum.

1.3 Preheating

The process limitations concerning extrusion speed as a function of billet temperature are illustrated in so-called limiting diagrams. Due to the presence of precipitates in conventional aluminum alloys, billet preheating to press temperature has an influence on the attainable extrusion speed. A modified theoretical limiting diagram, considering the phase change of the material, can be constructed by combining the proposed limiting diagram with the phase diagram (Lefstad *et al.* 1996). Such a diagram is represented in Fig. 5 for AlMgSi alloy. The extrusion speed limitation is the reason for tearing. This limiting line is divided in two different lines with a sudden increase in speed at a certain temperature T_a (solvus temperature). At billet temperatures below T_a, the material still contains Mg_2Si particles, precipitated during cooling after homogenization and/or during preheating before extrusion (when the billet is heated directly from room temperature). At billet temperatures above T_a, the material no longer contains Mg_2Si particles. A sudden increase in maximum extrusion speed occurs. If the billet is overheated, the speed increases in comparison with that of the directly heated material (the dashed line in the limiting diagram). Overheating above T_a for a sufficient time dissolves all Mg_2Si particles. Then the material is cooled down rapidly to a temperature below T_a, preventing any precipitation. As shown in Fig. 6, this

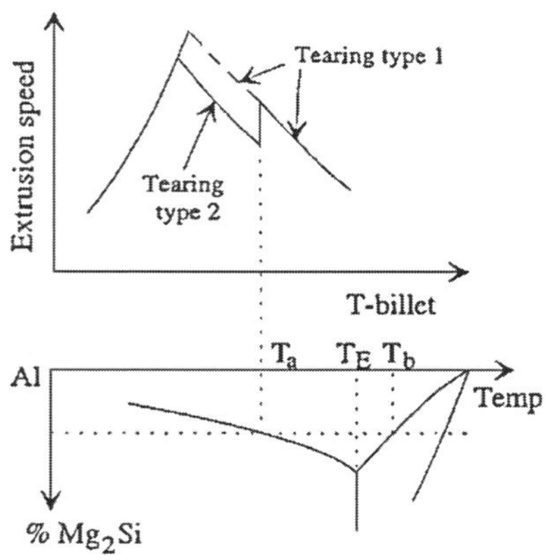

Figure 5
Limiting diagram for 6000-alloys combined with phase diagram (Lefstad *et al.* 1996).

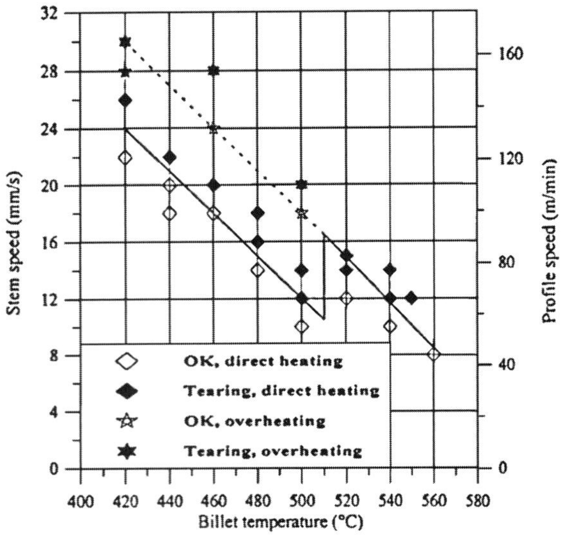

Figure 6
Extrusion speed limit curve of AA6082 regarding tearing as function of billet temperature (Lefstad *et al.* 1996).

procedure allows for an increase of the maximum allowable speed limit by 30–40%.

1.4 Grain Refining

In addition to the above-mentioned procedures, the extrudability of metallic material can be improved by refining the crystal grains. Due to grain refinement, the deformation stiffness will be covered. This does not mean, however, that the tensile strength of the product will decrease, as the microstructural development during the extrusion process determines substantially the mechanical properties. Small amounts of manganese or chromium act as a grain refinement agent. In AA6005 the form and distribution of the secondary phase AlFeSi change with the addition of manganese. $(FeMn)_3SiAl_{15}$ forms α-phase while Al_5FeSi forms β-phase. These phases reside in the cell boundaries and form an intermetallic distribution. If no manganese is added, the AlFeSi phase has a hexagonal structure; with manganese added, a cubic structure forms. The transformation of the plate-like β-phase into the more rounded α-phase by the addition of a small amount of manganese involves decreasing the Si/Fe ratio and coincides with particle refinement. This results in lower flow stress and improved hot ductility.

2. *Extrusion Process*

2.1 Material Flow

An influence on material flow, and therefore on product quality and mechanical properties, is given by the kind of extrusion itself. Basically three different extrusion processes, i.e., direct, indirect, and hydrostatic extrusion, are known with each involving different mechanisms and each producing different and characteristic patterns of flow. Figure 7 shows four basic flow patterns (S, A, B, and C) and two extended flow patterns A_1 and B_1, which are developed for the aluminum extrusion. Flow patterns of type S occur in the extrusion with glass lubrication as well as in indirect extrusion with die lubrication. The friction between the billet and the container wall including the die is almost zero. If the friction between billet and die can no longer be neglected yet the friction with the container wall is almost zero, then type A flow occurs. Flow patterns of this type are frequently found in the direct extrusion of soft materials. Under these boundary conditions, the material in the corner (as in a dead zone) between the die and the container will remain unextruded. A type B flow pattern occurs if the friction with the container wall, as well as with the die, are too high to be neglected. Such a type of flow diagram can be found in the direct extrusion of copper alloys. Type C applies for the hot extrusion of α/β brasses, which are extruded in a container of lower temperature than the billet. The friction is very large and the deformation stiffness of the material in the outer layer of the cooling billet becomes higher than in the bulk, so that the outer surface forms a relatively rigid shell. A large dead zone is developed extending from the die to the rear end of the billet.

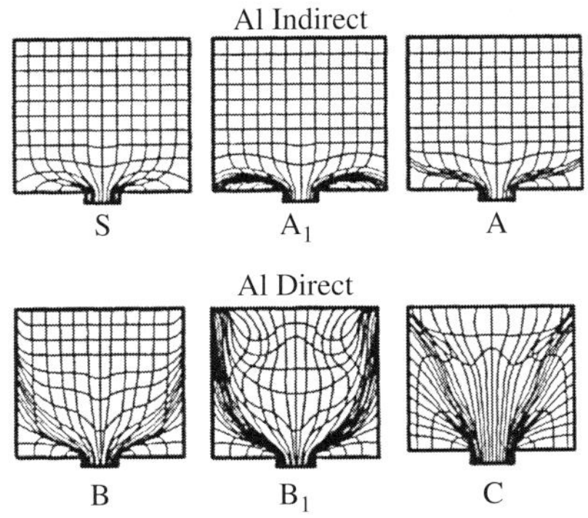

Figure 7
Extrusion flow patterns (Valberg 1996).

A special variant of indirect extrusion is indirect extrusion with active friction force (IEAFF). In IEAFF, the active friction is obtained by moving the container towards the product exit with a certain speed. As a result, a radial flow occurs in front of the dummy block (from the axis to the periphery, labeled 3 in Fig. 8) and in front of the die (from the periphery to the die orifice, labeled number 5 in Fig. 8). The flow of material at the peripheral layer of the billet (labeled 4) causes an acceleration of the material flow

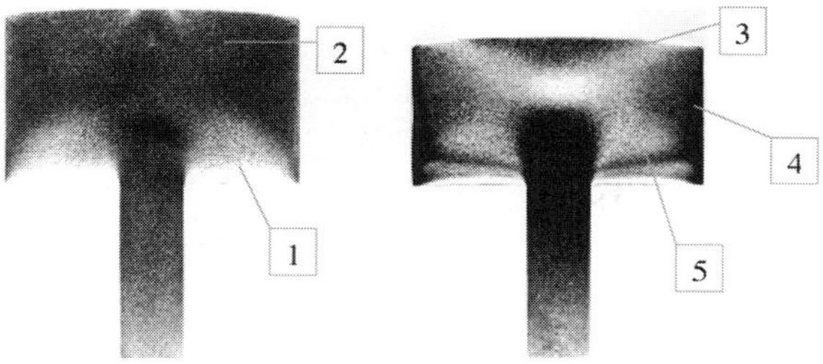

Figure 8
Comparison of the macrostructure from an indirectly extruded billet (left) and an indirectly extruded one with the additional use of active friction force (right).

at the periphery of the profile within the area of the die orifice and a deceleration at the profile axis so that the velocity becomes uniformly distributed cross-wise to the profile. Figure 8 represents the comparison of IEAFF and indirect extrusion. The macrostructure of the indirectly extruded billet reveals only two zones (labeled 1 and 2). The change of the local process parameters in the deformation zone as a result of the intensification of the radial flow causes a minimized residual stress at the profile surface, which means a reduction of tearing (Moroz and Müller 1995).

The different flow behavior in the container and in front of the die inevitably influence the extrusion force, heat development, and surface quality as well as the structure of the extrudate.

In the case of direct extrusion of aluminum alloys, the so-called isothermal extrusion allows for producing extrudates with a constant profile temperature, which leads to uniform mechanical properties. To compensate the heat development caused by the friction between the container and the billet, both tapered heated billets or ramp speed profiles are used.

In the case of hydrostatic extrusion it is possible to produce extrudates with different properties just by choosing extrusion speed, extrusion ratio, and extrusion temperature, because this process is nearly adiabatic and the deformation heat remaining in the extrudate is directly given by the applied extrusion pressure. For instance, a total extrusion pressure of about 1000 MPa leads in the case of copper extrusion to a 300°C temperature rise.

2.2 *Microstructure*

Subgrain size is dependent on processing conditions. It increases with rising temperature and falling strain rate. The subgrain diameter relates to the temperature compensated strain rate (Z), with $Z = \dot{\varphi}\exp(Q_W/RT)$. For a number of aluminum alloys it has been shown that a relationship of the form $d^{-1} = a + b\ln Z$ is in good agreement with measured subgrain sizes (Fig. 9).

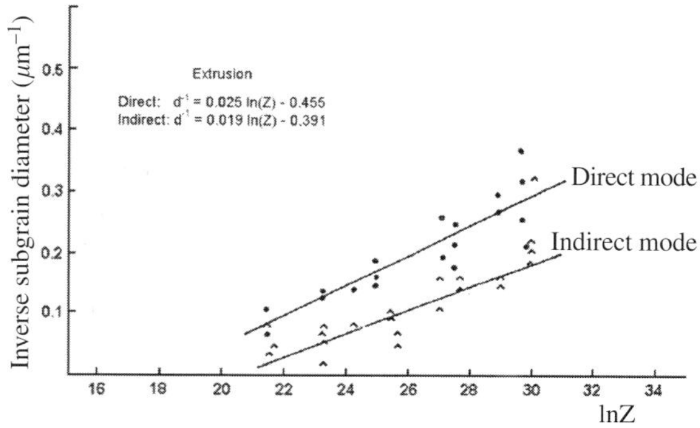

Figure 9
Subgrain variation with ln Z for AA6063 (Clode 1992).

Figure 10 reveals the difference of the microstructures between directly and indirectly extruded profiles as a function of temperature with constant extrusion speed. The structure of the profile directly extruded at 350°C shows a fibrous appearance speckled with statically recrystallized grains, whereas indirect extrusion displayed a more highly recrystallized structure with a slight aspect towards the extrusion direction (upper panels of Fig. 10). This difference is caused by the total stress and strain rate during the direct extrusion process, which stimulate the generation of a great number of nucleation sites. These will impede grain growth so that the microstructure of the directly extruded product is fine-grained. During the indirect extrusion the grains undergo less distortion and elongation in the extruding direction.

Another factor governing the structural differences is the quench delay, due to the tooling arrangement for indirect extrusion. As the initial billet temperature rises to 450°C, greater static recrystallization is obtained for both extrusion modes. The directly extruded structure continues to exhibit a directionality in its recrystallized grains. The grain size is smaller than the equivalent indirect structure as a result of a combination of increased nucleation sites and reduced time due to the quench delay.

2.3 *Texture*

The influence of temperature and extrusion speed on the formation of textures has also been examined, e.g., with the aluminum alloy $AlSi_{25}Cu_4Mg$ containing a high level of silicon after indirect extrusion. The variation of extrusion temperatures did not result in any noticeable microstructural difference of the extrudates. Texture analysis reveals that the silicon particles, due to their high strength and their globular shape, do not show any preferred orientation. In the center of the samples, the aluminum matrix develops, however, a $\langle 111 \rangle$–$\langle 100 \rangle$ double

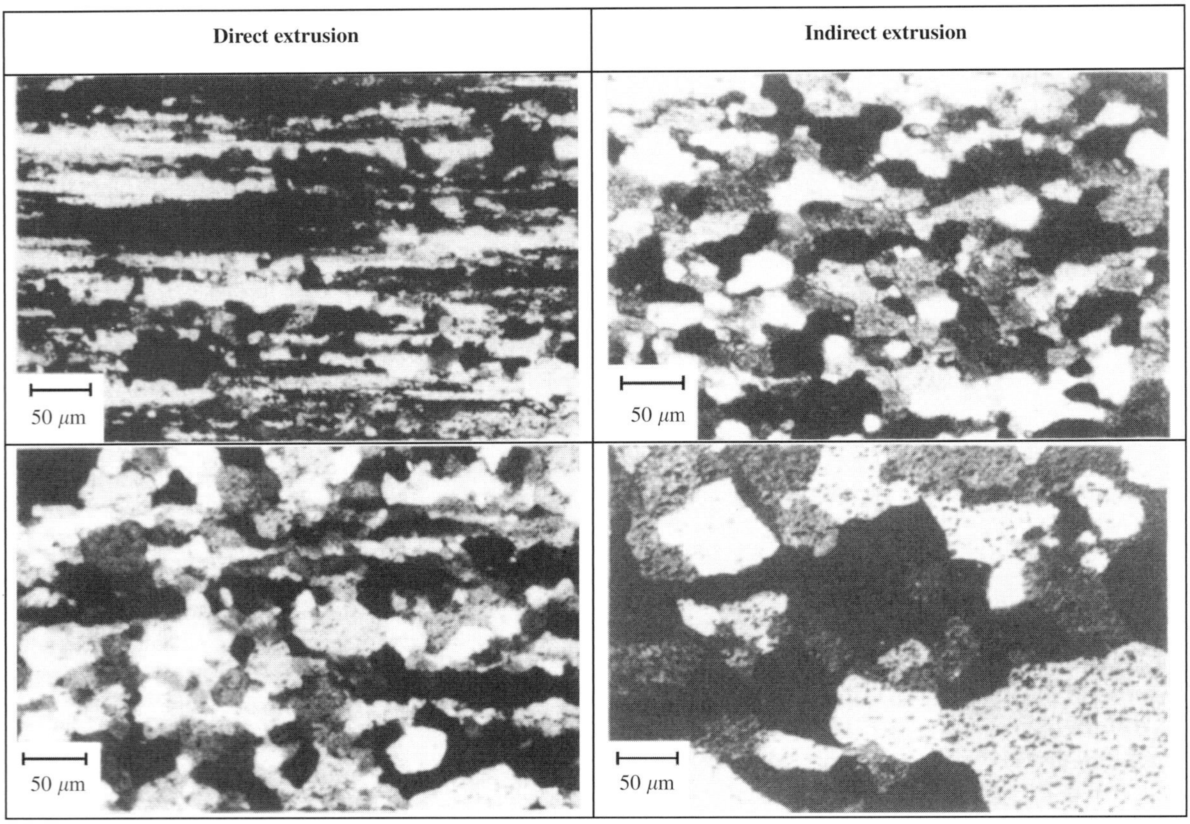

Figure 10
Influence of billet initial temperature and extrusion mode on microstructure development. Billet temperature: 350°C (above), 450°C (below) (Clode 1992).

fiber texture with a very weak ⟨111⟩ component. Increasing the initial billet temperature causes an increasing strength of the ⟨100⟩ fiber component. The ⟨111⟩ fiber component is, however, independent of the deformation temperature. Regarding the extrusion speed, experimental data reveal that the strength of ⟨111⟩ fibers does not depend on the extrusion speed, while the strength of the ⟨100⟩ fiber component increases with the increasing extrudate speed. Therefore, the ratio of the ⟨100⟩–⟨111⟩ fiber components increases according to the extrusion speed.

3. Concluding Remarks

Aspects of extrusion with regard to the product quality of extrudates are described. Two main influences for the optimization of product quality are considered: billet preparation and extrusion processing. Casting and homogenization, cooling, preheating, and grain refining of the billets affect the extrudability and mechanical properties of extrudates. The influences of local process parameters on the formation of flow patterns, microstructure, and texture are discussed.

Bibliography

Bryant A J B, Fielding R A P 1998 Developments in billet and extrusion metallurgy on the operation of the extrusion process. *Light Metal Age* **(April)**, 6–34

Clode M P 1992 Material flow and microstructural development during extrusion of AA6063. *Aluminium Association, 5th Extrusion Technology Seminar* **2**, 79–99

Heegn H, Müller K, Koch M 1999 Strangpressen von mechanisch legiertem Titan-Aluminid-Pulver. *Aluminium* **75**(4), 304–8

Lefstad M, Reiso O 1996 Metallurgical speed limitations during the extrusion of AlMgSi-alloys. *Aluminium Association, 6th Extrusion Technology Seminar* **1**, 11–21

Moroz B, Müller K 1995 Indirektes Strangpressen mit aktiver Reibung bei der Herstellung schwer preßbarer Aluminiumwerkstoffe. *Aluminium* **71**(1), 80–8

Pyzalla A, Wegener J, Müller K, Liss K-D 2000 Influence of extrusion parameters on microstructure and texture of

$AlSi_{25}Cu_4Mg_1$. *Proc. ICAA-7*, 9–14 April 2000, Charlottesville, VA **Part 2**, 781–6
Reiso O, Hafsas J E, Sjothum O, Tundal U 1996 The effect of cooling rate after homogenisation and billet preheating practice on extrudability and section properties. *Aluminium Association, 6th Extrusion Technology Seminar* **1**, 1–10
Ricks R A, Parson R C, Yiu H L, Court C A 1992 Microstructural optimisation for extrusioin of 6063 alloys. *Aluminium Association, 5th Extrusion Technology Seminar* **2**, 57–69
Sheppard T 1996 Recrystallization kinetics and prediction of recrystallized layer thickness in some Al-alloys. *Aluminium Association, 5th Extrusion Technology Seminar* **1**, 163–70
Werner R I (ed.) 5th Extrusion Technology Seminar. Vol. 1 and Vol. 2. Aluminum Association, Chicago, 1992
Werner R I (ed.) 6th Extrusion Technology Seminar. Vol. 1 and Vol. 2. Aluminum Association Chicago 1996
Valberg H 1996 A modified classification system for metal flow adapted to unlubricated hot extrusion of aluminium and aluminium alloys. *Aluminium Association, 6th Extrusion Technology Seminar* **2**, 95–100

K. Müller

Metal Working: Straightening

In the course of a continuous straightening process the material is generally bent several times in opposing directions. This bending action produces axial normal stresses partly reaching yield strength. They bring about small plastic strains, normally below $\varphi = 0.01$. Very often there is an additional load superimposed on bending, e.g. oval bending when tubes are straightened, or a transverse rolling operation at high pressure when bars are straightened to get better straightness and a smoother surface.

The pattern of residual stresses existing before straightening, and resulting from earlier deformation or thermal treatment, is transformed into a different typical pattern by straightening. This also involves characteristic changes in dimensions.

Since straightening causes plastic deformation, it is often accompanied by cold work hardening. If straightening is preceded by cold deformation, also in an axial direction, work hardening may not take place because of the Bauschinger Effect. Frequently, a considerable *decrease* in tensile strength, and especially in yield strength, is in fact measured.

1. *Straightening of Round Bars on a Two-roll Straightener*

When round cold finished bars are produced, the cold drawn or centerless turned bars are, generally, straightened and polished in the final operation. Nowadays, this operation is usually performed by a two-roll straightener. A pair of rolls is arranged at an angle to the direction of bar passage through the machine, as shown in Fig. 1, with the axis of roll 1 crossing that of roll 2 on different horizontal planes, creating in this way a screw motion for the bar running through the curved straightening gap. The bar is plastically bent by the bottom roll (with a cylindrical or convex shape) into the hollow of the concave top roll. Each section of the bar is repeatedly bent in all cross directions. When a bar section passes the exit half of the straightening gap, its curvature decreases from one bending step to the next one, until it is (almost) zero. In order to improve straightness and to ensure proper polishing of the bar surface, a pressing force in the middle of the straightening gap is usually superimposed onto the pure bending work (see Fig. 1(b)).

2. *Cold Work Hardening and Decrease of Strength by Bauschinger Effect*

When non-predeformed (e.g., centerless turned) bars with small diameters are straightened (in the usual way, i.e. by bending with superimposed pressure), a considerable increase in tensile and yield strength is generally noted. The selection of different straightening pressure rates enables adjustment of a tensile strength specified by the customer up to a bar diameter of about 40 mm.

When predeformed (drawn) bars are straightened, just a very small amount of work hardening, or in fact a *decrease* in strength, will be involved. This is caused by the Bauschinger Effect and by the fact that, because of predeformation, straightening takes place in an essentially flatter range of the stress-strain curve.

The Bauschinger Effect, which appears when the direction of deformation is reversed, causes a decrease in strength near the bar surface, when straightening is carried out by pure bending without any additional pressure being applied. Plastic deformation by bending is only noted close to the bar surface. When pressure is applied, which will mainly produce radial deformation, the decrease in strength is occasionally compensated near the surface.

Following are some typical values taken from a US plant, where hot rolled round steel wire from the

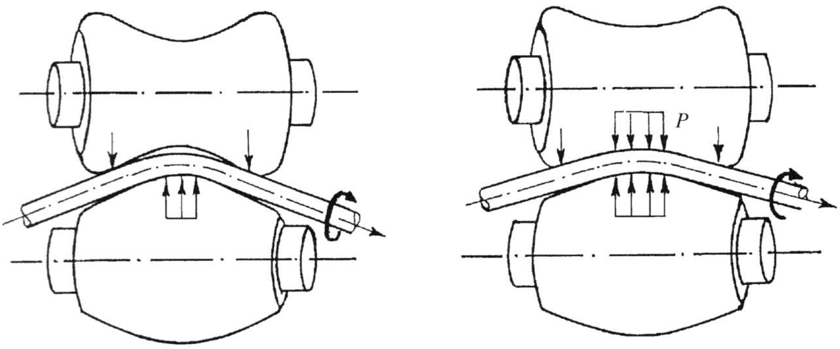

(a) (b)

Figure 1
Different modes of operation of a two-roll straightener for round bars: (a) straightening by pure bending; (b) straightening by bending with superimposed pressure P.

coil is prestraightened, shot-blasted, drawn, cut and straightened on a two-roll straightener integrated into a combined drawing line.

The example is steel C 45 of 32 mm hot rolled diameter drawn to 29 mm, corresponding to a reduction of 18%. The increased yield strength of 765 N mm^{-2} obtained by work hardening in the course of drawing was reduced by about 14% to 655 N mm^{-2} in the course of the two-roll straightening operation. The tensile strength dropped by about 6%, from 880 N mm^{-2} before straightening to 820 N mm^{-2} after. The Bauschinger effect resulted in a decrease in tensile strength and an even greater decrease in yield strength.

When the degree of bending is reduced and the pressure is increased on the two-roll straightener, the decrease in strength (of drawn bars) is smaller. When bending is increased and pressure reduced, the decrease in strength is larger.

3. *Residual Stresses and Dimensional Changes*

Two-roll straightening produces characteristic changes in bar diameter and length. The behavior is indeed different, depending on whether a bar has first been cold-worked (i.e. drawn) before or not (i.e. centerless turned).

Figure 2(a) (left-hand side) shows that the diameter of a turned bar hardly changes if it is straightened

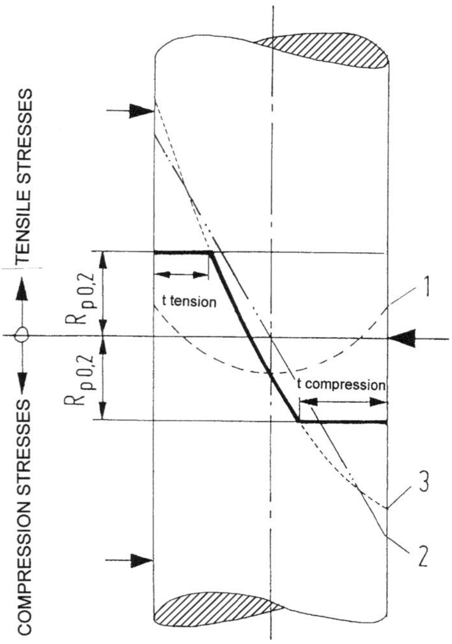

Figure 3
Residual stresses from drawing and stresses under load during pure bending. 1 = axial residual stresses after drawing; 2 = axial stresses during bending (ideally elastic); 3 = resulting axial stresses.

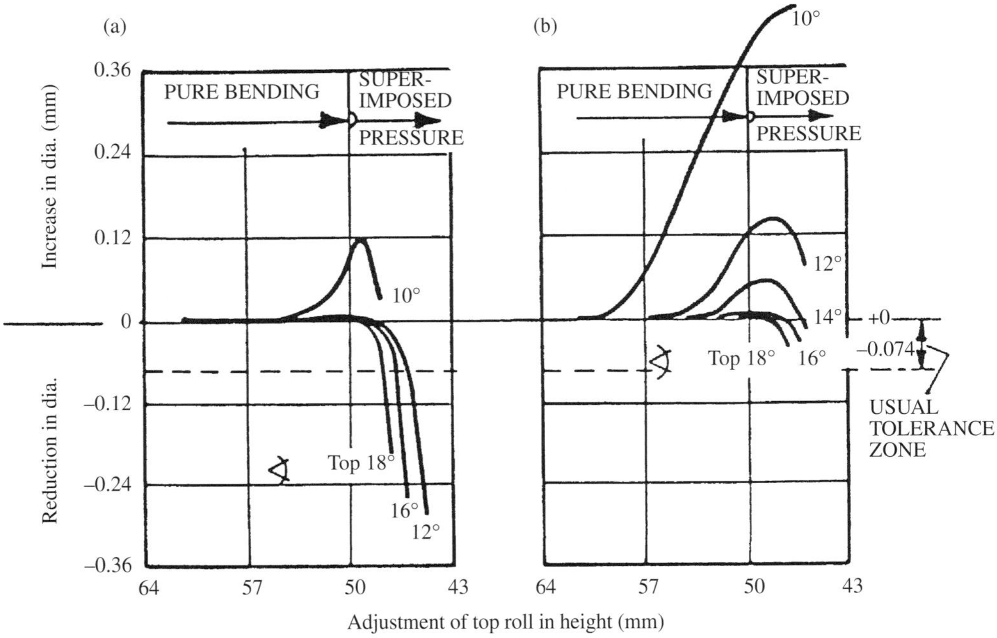

Figure 2
Bar diameter changed by straightening: (a) turned bars, (b) drawn bars. Steel C45 (1045), diameter 50 mm.

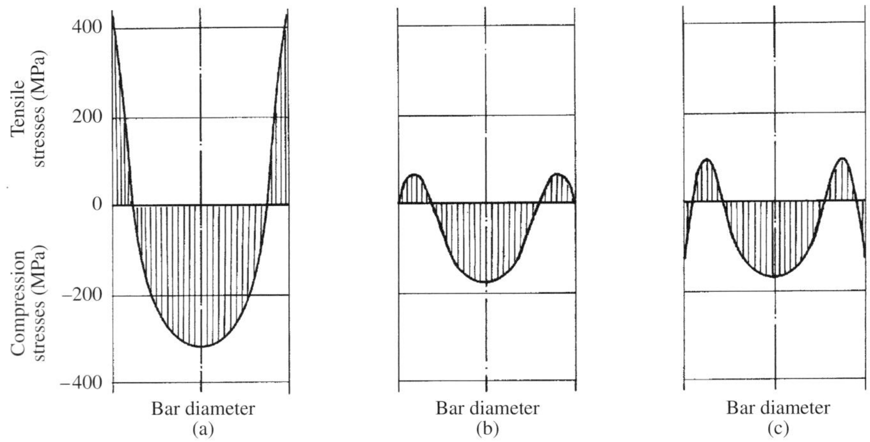

Figure 4
Axial residual stresses after drawing and after downstream two-roll straightening: steel 9 SPb 23, drawn from 24 mm dia. to 20 mm dia: (a) drawn; (b) drawn and straightened (by pure bending); (c) drawn and straightened (by bending with superimposed presssure).

by pure bending (Fig. 1(a)). If, however, along with an increase in roll adjustment the straightening gap in the middle of the rolls comes close to the diameter (here 50 mm), and equals it in the end, and adjustment is continued (see Fig. 1(b)), the radial pressure on the bar being built up will become higher and higher. The effect to be noted is a clear *decrease* in bar diameter (and an increase in bar length, because the volume is always constant).

When cold drawn bars are straightened by pure bending (Fig. 2(b), right-hand side), one is surprised to note an unmistakable *increase* in diameter, proportional to the amount of bending. The length becomes shorter because the volume remains constant. As soon as pressure is added (by adjusting the top roll to a dimension of less than 50 mm), there is a tendency towards a decrease in diameter, resulting from radial compressive stresses produced.

The strange *increase*in diameter produced by pure bending results from the pattern of residual stresses in the bar originating from predeformation (drawing). Figure 3 shows the characteristic pattern of residual stresses present after drawing, with the usual reductions in cross section (10% to 30%). In this graph, residual stresses already existing before were added to the stresses resulting from bending on the straightener. It is obvious that penetration of plastic deformation on the side under tension ("t tension" on the left) has less depth compared to that on the side of compression stress ("t compression" on the right). This means that upsetting in the axial direction, applied to the bar on the compression side, is stronger than stretching on the tension side. With upsetting by compression not being fully compensated, the bar length is reduced.

With turned as well as with drawn bars, the operator of the straightener is able to influence the usual close-toleranced diameter by skillfully combining bending and pressure. This is an essential advantage of the two-roll straightening process in comparison to other straightening processes.

Figure 4 shows how the characteristic pattern of axial residual stresses produced by drawing (Fig. 4(a)) is minimized by pure bending (Fig. 4(b)) on a two-roll straightener. The risky tensions near the surface, especially, are reduced to low values. If pressure is superimposed (Fig. 4(c)) compressive residual stresses in the axial direction are produced near the surface. This may considerably increase the fatigue strength of a highly loaded structural member manufactured from such a bar.

4. Conclusion

Nowadays, the two-roll straightening process has become the straightening method most frequently used for round bars. There are many reasons for this development. Straightness and surface quality are excellent, straightening speed is high, and the straightening effect comprises both ends of the bar. By combining pure bending with pressure (by transverse rolling) at different rates, yield and tensile strength, residual stresses (and thereby fatigue strength), and last but not least dimensional changes of the material, can all be influenced within a notably wide range.

Bibliography

Fangmeier R 1984 Grundlagen für das Richten von Flachmaterial, Profilen, Rundstäben und Rohren. *Arch. Eisenhüttenwesen* **55**, 251–5 [in German]

R. Fangmeier

Metal Working: Stretching of Sheets

The main processes of sheet metal forming are the stretch forming, deep drawing, or in general a combination of both. During stretch forming, the sheet material will be shaped by the punch but squeezed between the drawing die and the blank holder. The material flows under a biaxial stress and strain condition. In contrast to the deep drawing process the cup will shaped with a reduction in sheet thickness.

1. Stretch Forming and Deep Drawing

The deep drawing process is often a combination of stretch forming and deep drawing. At the beginning of the forming process, when the punch moves downwards and form out the bottom, a general stretch forming process results. When the bottom is formed, the deep drawing process follows, characterized by transfer of the drawing force from the punch through the cup wall into the flange. Within the flange results the main forming process with radial and tangential compression loads. Figure 1 shows the whole process of deep drawing, including the stretch forming and deep drawing processes.

During the deep drawing process, different stress conditions result in the material. Figure 2 shows a drawn cup with the different loading zones. The zones of interest are the flange, the side wall (cup wall), and its bottom. The stress condition in the flange area is a tensile load in radial direction and compression load in tangential direction. The stress conditions determine the forming process at deep drawing. The important stress condition in the cup wall is the point of transition of the punch radius to the cup wall area. This loading zone is characterized by a plane strain load, whereas the cup bottom has a biaxial stress load. The cup bottom and its biaxial stress load determine the stretch forming process.

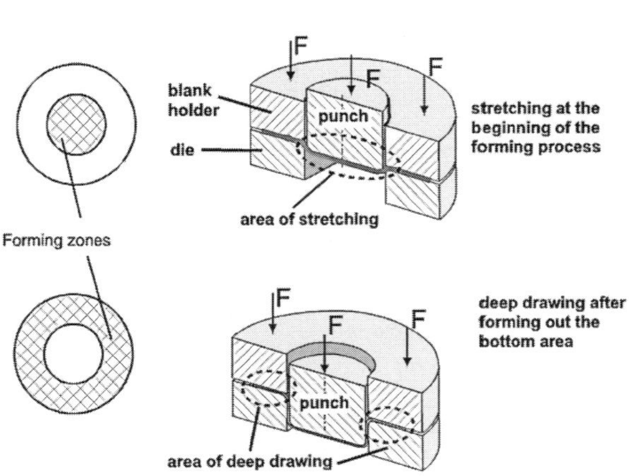

Figure 1
Stretching and deep drawing.

2. Formability of Materials

The formability of different materials varies depending on the stress conditions. The stress conditions lead to strain forming, which can be detected on the surface with electrochemically marked or silk-screen printed grids. The determination of the forming limit can be measured by different methods. In dependence on the stress conditions and the material, a forming limit curve (FLC) is determined as shown in Fig. 3. The forming limit diagram shows the forming zones of deep drawing, uniaxial tensile, plane strain, and stretch forming. The lowest formability for sheet metal takes place under plane strain conditions.

3. Significant Mechanical Values as Indicators of Formability

To get significant values for the forming process, different test methods have been developed. These

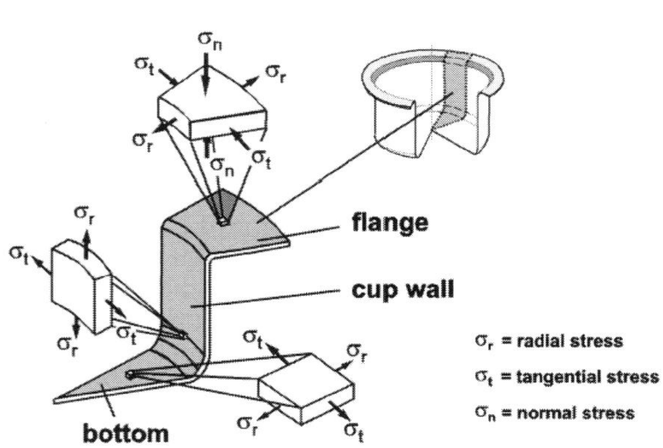

Figure 2
Stress-conditions in a drawn cup.

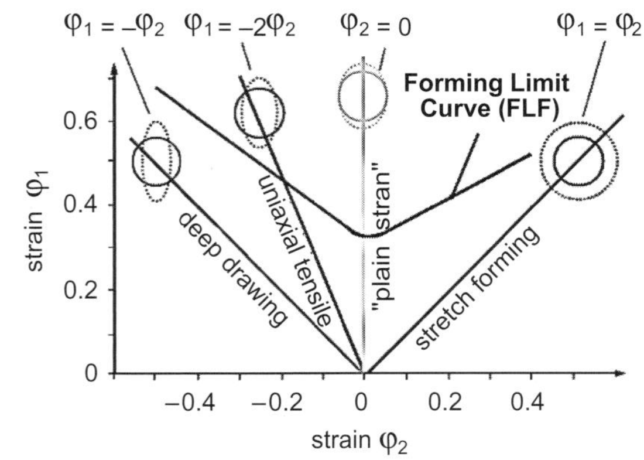

Figure 3
Forming limit curve.

tests show the suitability of sheet materials for stretch forming, deep drawing, or a combination of both Doege *et al.* (1997). However, the uniaxial tensile test is the only one which is internationally standardized and consequently is the most accepted.

In general, the mechanical values are determined by the tensile test according to the European Standards EN 10002-1 (1991) and EN 10130 (1991). This is due to the simple realization and analysis of this test and the good reproducibility of the key values. However, the values are needed for a basic statement about the deformation behavior of the materials. The parameters which characterize anisotropy and strain hardening are especially important for sheet metal forming, although they are only defined for low-carbon steels. At present, these values are used to get information about yielding and deformability of materials and consequently about the behavior during deep drawing and stretch forming. Concerning high-strength steels, the validity of the relationship between key values and deformity must be viewed critically.

3.1 *Anisotropy*

During deep drawing or stretch forming, there is a relationship between the material flow in the sheet plane and the sheet thickness. This behavior is described mathematically by the definition of the vertical anisotropy r, which is the ratio of true strain in the width (φ_b) to true strain in the thickness (φ_s) in a tensile test of sheet metal (Fig. 4),

$$r = \frac{\ln(b/b_0)}{\ln(s/s_0)} = \frac{\varphi_b}{\varphi_s} \qquad (1)$$

where b_0 and s_0 are the width and thickness of the tensile specimen and b and s the width and thickness after elongation. φ_b and φ_s are the logarithmic changes of the width and of the thickness.

The ability of the material to flow within the sheet plane increases with higher vertical anisotropy. Therefore, r values of approximately $r = 1$ mean that the material flows within the sheet plane as well as from the thickness. Transferring this to the deep drawing process, a higher r value corresponds to a better deep drawability.

During the deep drawing process, material flowing out of the width leads to increased tangential stresses in the area of the flange. A decrease of the thickness over the whole cup wall area occurs. In contrast to this the material flowing out of the thickness causes an impairment of the material. According to this theory, the deep drawing quality of sheet metal increases with higher r values. However, the significance of the r values should be regarded critically, because isotropic steels which are approximately at $r = 1$ have good deep drawing qualities.

Furthermore, the vertical anisotropy r varies with the angle of the rolling direction (RD). A characteristic value for this is the planar anisotropy, which is calculated by different r values obtained on specimens at different angles relatively to the rolling direction (0°, 45°, and 90°). The following equation defines the planar anisotropy Δr;

$$\Delta r = \frac{r_0 + r_{90} - 2r_{45}}{2} \qquad (2)$$

At deep drawing of cups, the anisotropy of the material leads to earings in the flange area. The planar anisotropy Δr describes the tendency of the material for earings. In contrast to anisotropic materials, isotropic materials do not have this effect and the planar anisotropy is approximately $\Delta r = 0$.

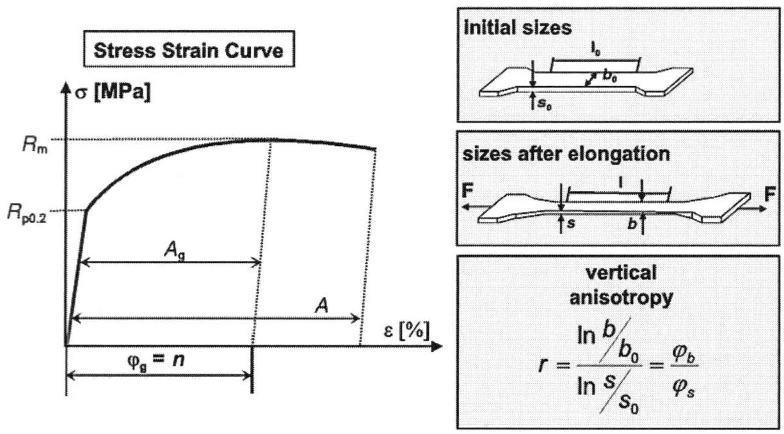

Figure 4
Anisotropy value r.

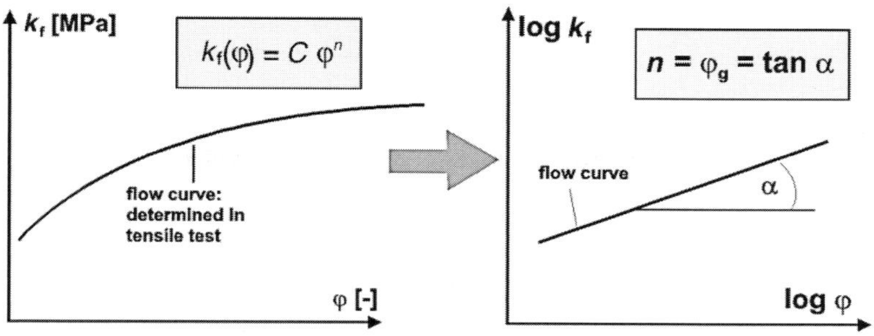

Figure 5
Hardening exponent n.

Morever, the average vertical anisotropy is defined as the arithmetic median of the anisotropy values relatively to the different rolling directions;

$$r_m = \frac{r_0 + r_{90} + 2r_{45}}{4} \tag{3}$$

3.2 Hardening Exponent

The hardening exponent n characterizes the strain hardening of a material during the forming process. It is defined by the gradient of the flow curve in a double-logarithmic form (Fig. 5). This definition can be applied with sufficient accuracy to the most low-carbon and low-alloyed steels. The strain hardening exponent of these materials can be supposed to be linear in the range of 10–20% plastic elongation. To describe the flow curve, the early Ludwik equation is often used, which is shown here in the form due to Hollomon;

$$k_f(\varphi) = C\varphi^n \tag{4}$$

where k_f is the flow stress, φ the logarithmic strain, C the stress at $\varphi = 1$, and n the hardening exponent. Furthermore, Reihle (1961) defines the relationships;

$$n = \varphi_g \tag{5}$$

$$C = R_m\,(e/n)^n \tag{6}$$

where R_m is the tensile strength, e is the basis of the natural logarithm, n is the hardening exponent, and φ_g is the effective strain at uniform elongation.

During the deep drawing process, materials with low n values show cracks in the point of transition of the punch radius end to the cup wall area before the bottom has finally shaped out. In contrast, materials with high n values transfer the local work hardening to large cross-sections, even after a small drawing depth. Consequently, materials with high hardening exponents should be preferred for stretch forming operations. However, an increasing n value at the beginning of the stretch forming period would be ideal concerning deep drawing. A decreasing n value seems to be ideal for the whole forming process.

4. Concluding Remarks

Concerning the processing of metal sheets during deep drawing and stretch forming, the simple tensile test can be used to determine basic mechanical values. These values guarantee a sufficient transferability to the stretching or deep drawing process. In spite of many existing testing methods for the characterization of mechanical properties of metal sheets, the tensile test is internationally standardized. This is possible, because of the simple realization and the good significance of the values for characterizing metals in deep drawing and stretch forming processes. Especially the determination of the vertical anisotropy r and the strain hardening exponent n is very important for forming processes. However, in spite of the relatively easy way to determine these values and the good transferability to practical forming processes, these values are use for characterized non-low-carbon steel sheet. In particular for the evaluation of high-strength steel and light-weight alloys, the description using r and n values, allow only restricted conclusions regarding formability.

See also: Metal Forming: Formability

Bibliography

Doege E, Dröder K, Griesbach B 1997 On the development of new characteristic values for the evaluation of sheet metal formability. *J. Mater. Proc. Technol.* **71**, 152–9

European Standard EN 10002-1 Metallic materials; tensile testing; part 1: method of testing at ambient temperature April 1991

European Standard EN 10130 Cold rolled low carbon steel flat products for cold forming; technical delivery conditions October 1991

Goodwin G M 1968 Application of strain analysis to sheet metal forming problems in the press shop. *Soc. Automotive Engineers* 680 093, 380–7
Hollomon J H 1945 Tensile deformation. *Trans. Met. Soc. AIME* **162**, 268–90
Keeler St P 1965 Determination of forming limits in automotive stampings. *Soc. Automotive Engineers* 650 535, 1–9
Lange K 1990 *Umformtechnik—Handbuch für Industrie und Wissenschaft, Band 3: Blechbearbeitung*. Springer, Berlin
Reihle M 1961 Ein einfaches Verfahren zur Aufnahme der Fließkurve von Stahl bei Raumtemperatur [A simple method for the determination of flow curves of steel at room temperature]. *Arch. Eisenhüttenwesen* **32**, 331–6

E. Doege and T. Hallfeld

Modification of Wood: Mechanical Properties and Application

Wood modification is one of the possible ways for the utilization of less valuable wood species. Smart modification opens a prospect not only for the most complete utilization of wood but also for the production of new, high-quality wood-based materials. Wood composites with specific properties can be tailored under various processing routes. They can be used as substitutes for nonferrous metals and other valuable materials. The behavior of modified wood under load and exposed to various factors depends on the method of modification.

1. *Methods of Modification*

A wooden cell wall is a polymeric structure mainly comprised of cellulose, hemicellulose, and lignin. All these components contain hydroxyl groups that play a key role in the interaction between water and wood. As wood takes up water in moist conditions, water molecules settle within the wood polymers forming hydrogen bonds with the hydroxyl groups. Water is thus inserted throughout the cell wall components, and the wood swells. Modified wood is generally obtained by making use of physical, chemical, and mechanical methods at different structural levels.

1.1 *Chemical and Physical Methods*

We first consider wood on a gross structural level as a composite consisting on the one hand of cell wall material and on the other of empty wall volume with definite shape and orientation. Methods for wood modification can be roughly classified into physical and chemical routes, which provide:

(i) a decrease of the wood wall to void volume fraction and a change of the mutual orientation of various cell groups (these transformations are achieved by physical methods) and

(ii) changes in the composition and properties of the cell wall material, essentially by chemicals.

The physical and chemical methods can be combined in the greatest majority of instances. The chemical modification is designed to substitute hydrophobic chemical groups for the OH^- groups by etherification, esterification, oxidation, silylation, or other chemical reactions (for an overview of methods, see Militz *et al.* 1997). An example is provided by the acetylation, a well-known technique in many industrial and scientific areas. The acetylation of wood is utilized to increase the quantity of acetyl groups that are already present in small amounts in natural wood. Formaldehyde has been employed to treat wood composites in order to reduce swelling and to increase moisture resistance and mechanical properties.

The goal of chemical treatments is, in addition, to extend the lifetime of wood products whose protection against biotic and abiotic agents of deterioration may be required too (see Wood: Chemical Treatment Processes). Hence, whereas the purpose of wood treatment is to prevent biological attack, it can also impart resistance to fire, chemical, mechanical, or UV light damage as well as the risk of water sorption.

Heat treatment has been known long ago as a beneficial method to modify wood properties (Kollmann and Schneider 1963, Kollmann and Fengel 1965, Hillis 1984). Temperatures over 150°C alter the physical and chemical properties of wood permanently. Heat treatment reduces wood shrinkage and swelling and improves the equilibrium moisture content (Homan and Jorissen 2004). High temperatures improve rot resistance and reduce the susceptibility to wood decay by fungus. Undesired side effects are a loss of strength and an increased brittleness. On average, woods exhibit a lower strength after thermal modification, and they are very brittle in tension and bending. Giebeler (1983), for instance, reports a reduction of the strength of wood of 20% to 50% after thermal treatment between 180°C and 200°C.

The extent to which wood properties are changed during heat treatment depends on (i) the maximum temperature and the maximum duration of the heat treatment at the nominal temperature; (ii) the gradient of temperature; (iii) the maximum length of the entire heat treatment; (iv) the presence and the amount of water vapor; (v) the nature of the kiln drying process before the actual heat treatment; and (vi) the characteristic properties of the wood species.

Compared to untreated wood, the fracture properties of chemically and thermally treated wood deteriorate. Tension tests on modified veneer show that the above-mentioned types of treatment do not affect the modulus of elasticity (MOE) significantly, but the tension strength is reduced down to 60% after thermal treatment. Bending tests show that short heat treatments improve both the bending strength and MOE.

1.2 *Mechanical Methods*

The mechanical methods aimed at changing the wood's natural gross structure can be divided into several groups. Two groups, related to the change in the shape and specific volume of the constituents, are:

(i) compression as a result of the flattening of wood cavities or of the replacement of the volume of cavities with wall material; and

(ii) filling the cavities with other materials.

Two other groups deal with the transformation of the natural wood structure by mechanical division into layers or particles for the production of a new material, namely:

(iii) the production of composites from wooden flakes and chips; and
(iv) the production of reinforced and laminar wooden composites (Rocens 1975).

The wood cellular structure is significantly and durably changed during compression at high temperature and under steam pressure and results in a material with new properties. The strength and the dimensional stability of a compressed wood sample can be highly influenced by structural modifications of the cell wall under compressive strains. One of the major factors influencing the mechanical and physical behavior of densified wood is how and to which amount cellular collapse takes place (Bodig 1966). Depending on the environmental conditions and on the nature of the cell wall material, cellular collapse occurs by elastic buckling, plastic yielding, or brittle crushing (Wolcott *et al.* 1990). Changes in the cellular structure of wood can be observed with electronic or light microscopy and quantified using image analysis techniques.

In the case of a laminar composite, rotary-cut veneers are cross-laminated to form plywood, and the negligible longitudinal shrinkage of wood restrains much of the in-plane transverse shrinkage and swelling. As a result, thickness swelling increases but the volumetric swelling of plywood is less than the free swelling of wood (Kollmann and Côté 1968, Bodig and Jayne 1982). As external swelling is restrained, internal expansion occurs within cell voids.

The in-plane stability increases with the number of plies, approaching isotropy. In parallel laminated wood, grain orientation is randomized, minimizing the shape distortion of the components submitted to changes of moisture content. Because of added compaction during the pressing, flakeboards made of lower-density species show a greater degree of thickness swelling. The positioning of flakes relative to the plane of the board and resin content is important in both linear and thickness swelling.

1.3 *Modification by Compression*

Wood properties can be tailored by increasing the volume of cell wall material per unit volume of wood. This is realized by compressing wood perpendicularly to the grain (Bodig 1966). Plasticization is necessary in order to accomplish nondestructive compression of the wood cell wall material (Rocens 1976a). Under appropriate control methods, the process of plasticization can be brought to the necessary compliance. It is established that two parameters, the velocity of transverse ultrasonic (US) waves and logarithmic damping decrement, adequately reflect the kinetics of the physical and chemical processes involved in wood plasticization with ammonia. The stabilization of these parameters indicates that a steady state of mass transfer and the equilibrium of physical and chemical processes in wood have been reached. These methods are well suited to the control of the active plasticization process.

Compression affects wood elasticity and it increases the volume fraction of cell wall material. The resulting changes of the propagation rates of US waves (Rocens 1983) are able to reflect different stages in the compression process. The nature of strain variation under stress depends on the particularities of wood structure. For instance, compressing birch—a typical diffuse-porous wood—flattens the cavities of libriform vessels and microcapillaries (Fig. 1). The shape of the cavities and the void volume change simultaneously. It is established that in the course of compression, the cross section of vessels transforms itself from circular into elliptic with a long axis oriented normal to the direction of force application.

In the first stage of compression, the elastic part of the strain depends on the stress level, while in the next stages, which actually coincide with the gross

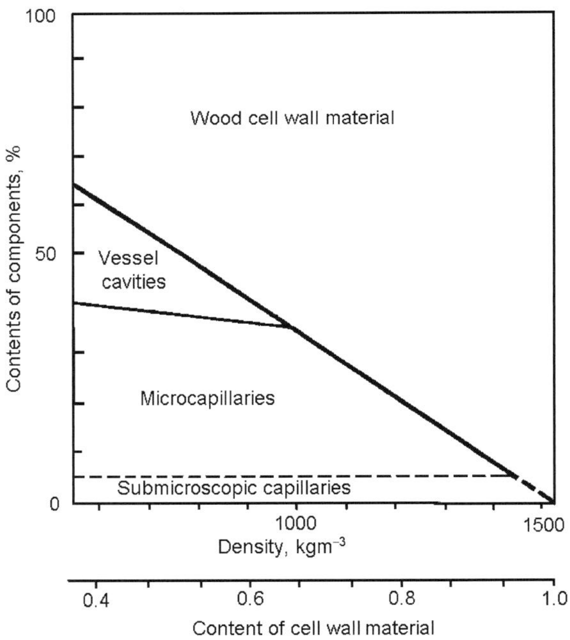

Figure 1
Volumetric content of the components of densified birch wood vs. the density or relative content of material in cell walls. Modified from Rocens K 1983 Macrostructure theory of modification of wood properties. *J. Appl. Polym. Sci., Appl. Polym. Symp.* **37**, 923–41, with permission from John Wiley & Sons, Inc.

structural variation, the elastic strain depends on both the stress level and the duration of stress action (Rocens 1976b, Wolcott *et al.* 1990). At fixed temperature, moisture content, and loading, a transformation of elastic and elastic-delayed strains into residual strains takes place gradually. The rate of the transformation increases with increasing stress. Neither the residual nor the elastic-delayed strains follow a linear dependence of strains on applied stress.

2. *Mechanical Properties of Modified Wood*

2.1 *Modeling of Wood Structure*

A composite material with mutually parallel rectangular cavities is considered to model solid wood with artificially changed gross structure (Fig. 2). The axes of Cartesian co-ordinates are oriented in directions corresponding to those of wood elasticity as follows: x_1—parallel to the grain; x_2 and x_3—perpendicular to the grain, that is, tangential and radial with respect to the annual rings, respectively. The filling material covers all the wall cavities of the tetragonal matrix with layers of equal thickness, and there is total adhesion between the matrix and the filler. Based on the general Hook's law relating stresses and strains, the components of the elastic compliances of modified wood can be determined using the assumptions of Reuss or Voigt (Reuss 1929, Voigt 1910).

The principal elastic compliances of solid wood with a natural gross structure whose cavities are filled with wall material or other materials (polystyrene, epoxy resin, metal alloys, etc.) have been determined based on a composite model. In the case of cavity filling with wall material, Rocens (1983) has shown that the analytical estimates of the compliances in the main directions of elasticity, S_{11}, S_{22}, S_{33}, together with the shear compliances, S_{55} and S_{66}, of modified wood exhibit a fair consistency with experimental results. Resulting from modification by compression,

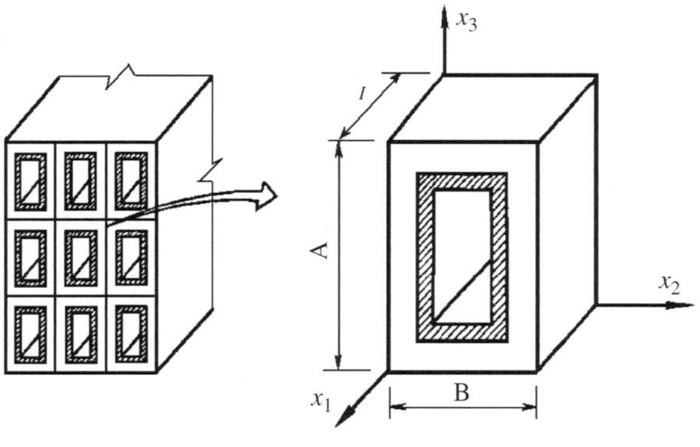

Figure 2
Composite model with parallel tube-shaped cavities.

the compliances S_{33}, S_{44}, and S_{55}, which characterize the deformability in transversal directions, appear to be higher in comparison with natural wood. The resistance of cell wall material to the action of stresses σ_1, σ_2, and σ_6 increases, resulting in a decrease in compliances S_{11}, S_{22}, and S_{66}.

Figure 3 compares the data obtained from the testing of birch wood–polystyrene composite, produced by a radiation-catalytic method, to analytically determined values of the shear compliances S_{44}, S_{55}, and S_{66} in relation to the volume of filler material. The analysis of the elastic compliances of modified wood as a function of the volume of the cavities shows that replacing the cavities with cell wall material is the most beneficial process route to compliances S_{11}, S_{22}, and S_{66}, while for compliances S_{33}, S_{44}, and S_{55} it is the filling of cavities with other materials.

2.2 *Experimental Investigation of Densified Wood*

The densification of plasticized wood has been performed by compression under uniform strain rate resulting in a wood structure in the density interval 450–1200 kg m^{-3} (Rocens 1983). Changes of cellular structure have been examined under reflection microscopy on sections of wood samples. The densification occurs first when vessel elements and libriform fibers become flat, and after cell walls are densified. The elastic properties of densified wood were determined by mechanical loading (in tension, compression, and bending) and US methods (Brauns and Rocens 2003). For a densification in the radial direction, the relationships between the elastic

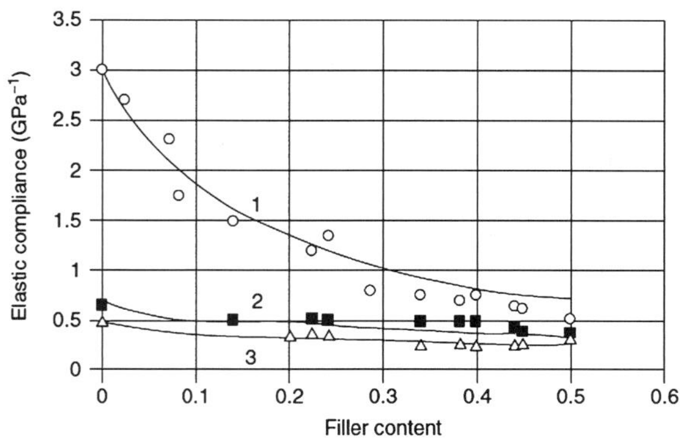

Figure 3
Elastic compliances of wood filled with polystyrene as function of filler content: 1, S_{44}; 2, S_{66}; 3, S_{55}. Experimental points determined by US method: ○, S_{44}; ■, S_{66}; Δ, S_{55}. Modified from Rocens K 1983 Macrostructure theory of modification of wood properties. *J. Appl. Polym. Sci., Appl. Polym. Symp.* **37**, 923–41, with permission from John Wiley & Sons, Inc.

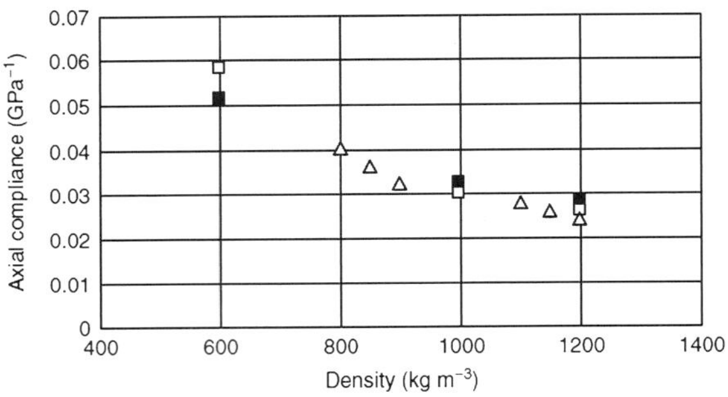

Figure 4
Variation of wood axial compliance with density: Δ, compression; ■, tension; □, bending. After Brauns J, Rocens K 2003 Estimation of technological stresses in wood composites. In: Žiliukas A, Schmidt F, Pritykin I, Januteniene J, Leišis V (eds.) *Proceedings of 3rd International Conference on Strength, Durability and Stability of Materials and Structures*. Kaunas University of Technology, Kaunas, Lithuania, pp. 11–23.

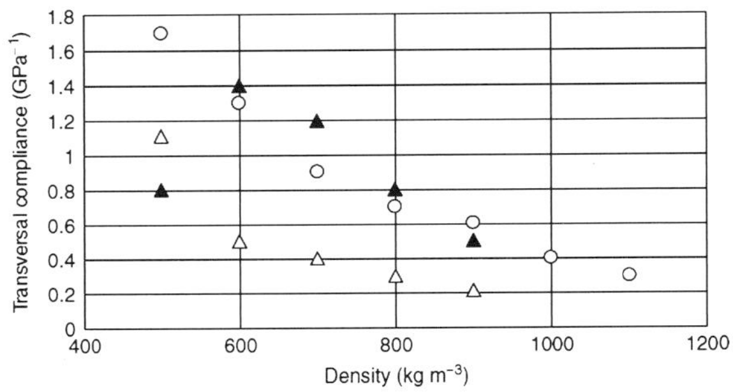

Figure 5
Variation of wood transversal compliance with density: tangential direction (○, compression; Δ, US method); radial direction (▲, US method). After Brauns J, Rocens K 2003 Estimation of technological stresses in wood composites. In: Žiliukas A, Schmidt F, Pritykin I, Januteniene J, Leišis V (eds.) *Proceedings of 3rd International Conference on Strength, Durability and Stability of Materials and Structures*. Kaunas University of Technology, Kaunas, Lithuania, pp. 11–23.

compliances and the density of solid wood are nonlinear, exhibiting differences in the longitudinal, tangential, and radial directions (Figs. 4 and 5). Together with elastic properties, the strength characteristics of densified wood increases in all three—longitudinal, tangential, and radial—directions. The effect varies of course according to the wood species investigated and to the technology utilized. Depending on the loading case, strength can be increased by a factor of 1.5–3 and even more.

3. *Modification of Wood in Wooden Composites*

Hot pressing is one of the important stages during the production of wood-based composites when mats of resinated wood fibers, particles, flakes, or veneers are consolidated under heat and pressure to create close contact and form bonds between the wood constituents. Mat deformation is, however, not uniform across the material thickness (Xu and Winistorfer 1995), resulting in a density profile in pressed composite boards. This in turn has significant effects on the physical and mechanical properties of the final product (Geimer *et al.* 1975). The initial mat structure and the mechanical properties of its components govern the overall mat stress–strain relationship (Zambori *et al.* 2001). Due to the viscoelasticity of wood (Schniewind and Barrett 1972, Wolcott *et al.* 1990), some elements exhibit time-dependent creep behavior and they may undergo stress relaxation.

As resin in wooden composites is employed in limited amount, bond efficiency relies on the degree of mat densification, which should be high. Increasing mat density, however, causes detrimental effects such as increased weight and wood consumption, and excessive thickness swelling in service when the product is subjected to high humidity conditions (McNatt 1974).

4. *Structural Changes in Technological Pressing of Wooden Composites*

Studies on wooden composites have focused on the relationships between processing parameters and material properties caused by densification (Brauns and Rocens 2003). In manufacturing of a composite board, the basic function of the pressing operation is the development of an adhesive bond between individual flakes or veneer sheets while minimal glue spread is utilized.

The structure of a mat made of fibers exhibits a double cellularity. Void space among the strands results from the inherent randomness of the deposition process, and the flakes include the cell lumen. It is thus possible to distinguish not only two different voids, but also two different densities: the density of the mat (γ_m), and the average density of the fibers (γ_f). During the first stage of pressing, void space (V_s) between flakes is eliminated from the mat structure and the flakes get into contact. Further pressurization enhances the collapse of cell walls and the lumen volume is decreased. The volume of the mat is the sum of the volume of the flakes (V_f) and of the void space:

$$V_m = V_f + V_s \tag{1}$$

At this stage, both the density of the mat and that of constituent flakes increase.

The densities of the mat and of the flakes are given by

$$\rho_m = \frac{m_f}{V_m} \quad (2)$$

and

$$\rho_f = \frac{m_f}{V_f} \quad (3)$$

respectively, where m_f is the mass of flakes. Figure 6 shows the change of these densities during compression. At the beginning of the pressing, the density of the mat is lower than the density of the flakes because of the large amount of space in the structure. As the mat is compressed, the mat density soon increases to near the flake density obtained at the end of the compression.

4.1 Viscoelastic Behavior of Wooden Composite

Viscoelasticity influences wood modification when wood flakes or veneers are under static pressure and subjected to changes in temperature and moisture content. In these situations, wood can be adequately treated as a linear viscoelastic material (Schniewind and Barrett 1972).

The high pressure applied during the manufacture of wood composites, coupled with the random mat structure, results in a nonlinear and nonuniform mechanical response of the wood constituents. Time-dependent stress–strain relationships, instantaneous and delayed strain recovery, permanent deformation, and temperature- and moisture-dependent stress relaxation may intervene during the pressing process (Heebink 1972).

The beginning of the pressing is aimed at achieving the designed mat thickness. In the next stage, the thickness is fixed and stress relaxation takes place. In the final stage, the rate of mat deformation is small and the stress–strain relationship may be written as follows:

$$\frac{\partial \sigma}{\partial t} + C\frac{E}{H}\sigma = C\sigma_0 \quad (4)$$

where E and H are elastic and long time modulus, respectively. The long time modulus can be expressed as the ratio $\sigma/\varepsilon(t \to \infty)$. The coefficient C is determined from relaxation curves in terms of the relaxation time constant, τ_r, and the viscoelastic characteristic, K:

$$C = \frac{H}{E\tau_r}\,e^{-(t-\tau_r)/K} \quad (5)$$

From Eqn. (4), the time dependence of stress relaxation during the formation of a wood board is written as

$$\sigma_r(t) = \sigma_0\left[\left(1 - \frac{H}{E}\right)\exp\left(-Ce^{-(t-\tau_r)/K}\,\frac{E}{H}\,t\right) + \frac{H}{E}\right] \quad (6)$$

where $\sigma_0 = \sigma(t = 0)$. A comparison with experimental results is provided in Fig. 7.

The elastic resistance of wooden mats also depends on drying during the pressing of the board. In the case of fixed mat thickness, the drying stresses are

$$\sigma_{sh}(t) = \frac{E\eta(W_0 - W_*)}{t_*}\,t \quad (7)$$

where η is the shrinkage coefficient and W_0 is the initial moisture of wood flakes; t_* is the pressing time and W_* is mat moisture at time t_*. The total stresses

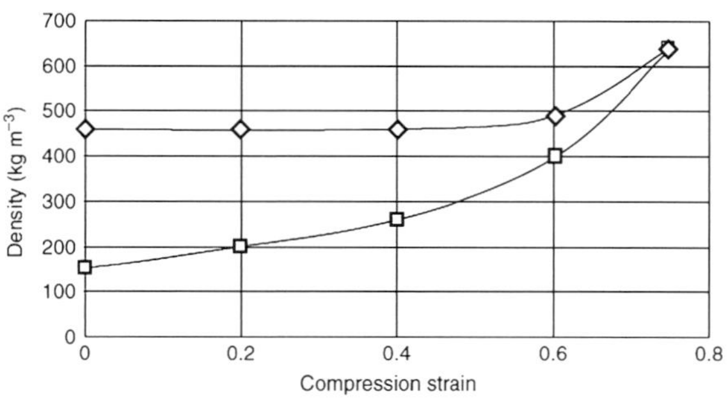

Figure 6
Variation of mat density (□) and flake density (◇) during compression. After Brauns J, Rocens K 2003 Estimation of technological stresses in wood composites. In: Žiliukas A, Schmidt F, Pritykin I, Januteniene J, Leišis V (eds.) *Proceedings of 3rd International Conference on Strength, Durability and Stability of Materials and Structures*. Kaunas University of Technology, Kaunas, Lithuania, pp. 11–23.

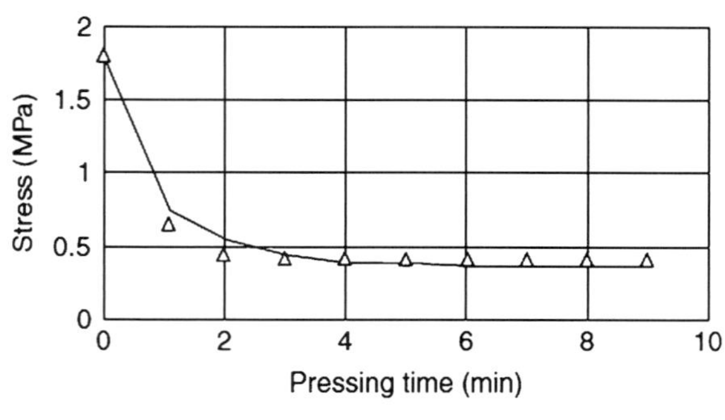

Figure 7
Stress relaxation with densification. After Brauns J, Rocens K 2003 Estimation of technological stresses in wood composites. In: Žiliukas A, Schmidt F, Pritykin I, Januteniene J, Leišis V (eds.) *Proceedings of 3rd International Conference on Strength, Durability and Stability of Materials and Structures*. Kaunas University of Technology, Kaunas, Lithuania, pp. 11–23.

in a wooden mat during the drying stage of a pressing process is the sum of the above expressions

$$\sigma(t) = \sigma_r(t) + \sigma_{sh}(t) \quad (8)$$

4.2 Influence of Composite Structure on Hygromechanical Behavior

The distribution and properties of densified flakes determine the load-carrying capacity of the board. Oriented strand boards can increase the strength and the stiffness of wooden composites. Nevertheless, the warping of such panels is very high (McNatt 1973). As a result, the quality of wood composite boards depends on the properties and alignment of flakes (Geimer 1982).

In general, wooden composites do not display symmetry with respect to the midplane; their properties can be either asymmetric or antisymmetric (Brauns and Rocens 1994a). The composite is analyzed as a system of parallel elementary layers. The direction of layer k is $\varphi^{(k)}$ (Fig. 8). The local coordinate system $_i$ is associated with the principal directions of the elementary layers. In order to approximate the stress state, stress gradient-dependent force-stresses and couple-stresses are used in the model.

The dependence of the strains and stresses on moisture content changes, ΔW, can be written as follows in the principal axes of an elementary layer:

$$\varepsilon'_{ii}(W) = \beta'_{ii}(W)\Delta W \quad (9)$$

$$\sigma'_{jj}(W) = r'_{jj}(W)\Delta W \quad (10)$$

where $i, j = 1,2$.

The specific moisture strains β'_{ii} and constraining stresses r'_{jj} in Eqns. (9) and (10) are related to the composite material stiffness (Tsai and Hahn 1980, Brauns and Rocens 2004).

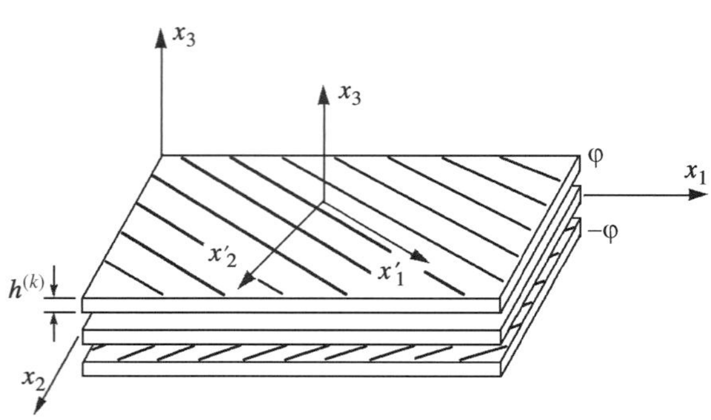

Figure 8
Multilayer model of wood composite.

The multilayer model and the model based on the laminate analogy (Halpin *et al.* 1971) can be used to determine the behavior of layered and fiber composites. Nonsymmetric moisture distribution causes linear expansion and hygromechanical warping. Curvatures resulting from hygrodeformation of wood were computed using the method developed in Brauns and Rocens (1994a, 2004) and (see *Elasticity in Wood and Concrete: Hygromechanical Effects*).

To illustrate the effect of particle alignment on hygromechanical and mechanical properties, laminates with elementary layers oriented differently with respect to, for example, machine direction (x_1) can be considered (Brauns and Rocens 1997). For any laminate, the average angle (disregarding the sign) is defined as $\bar{\varphi}$, and the relative alignment ξ may be written as follows:

$$\xi = [(\pi/4 - \bar{\varphi})/\pi/4] \cdot 100\% \quad (11)$$

defining a random board with zero alignment.

This model has been applied to wooden composites consisting of veneers, finite-length flakes, and glue (Brauns and Rocens 1991). The effects of densification during material fabrication are taken into account (Brauns and Rocens 1994b). The technical characteristics of densified birch veneer used in the analysis are: modulii of elasticity $E_1 = 16\,500$ and

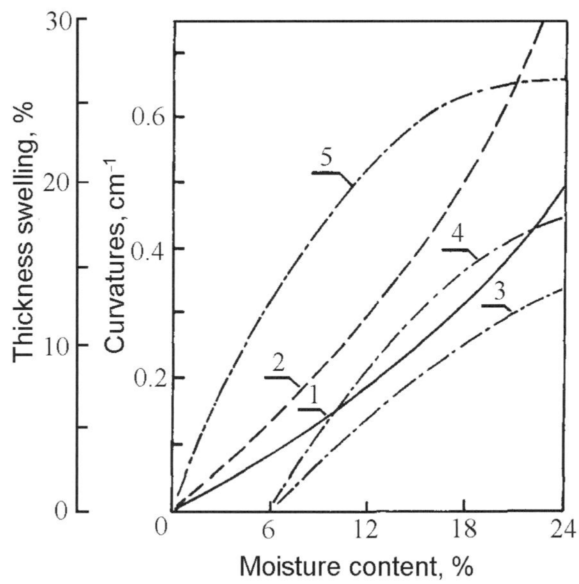

Figure 9
Thickness swelling (ε_3) of plywood: 1, customary wood; 2, densified wood; curvatures: 3, κ_1; 4, κ_2 (nonsymmetric moisture distribution); warping: 5, $\kappa_6 \times 10$ (antisymmetric structure). After Brauns J, Rocens K 1997 Hygromechanical behaviour of wooden composites. *Wood Sci. Technol.* **31**, 193–204.

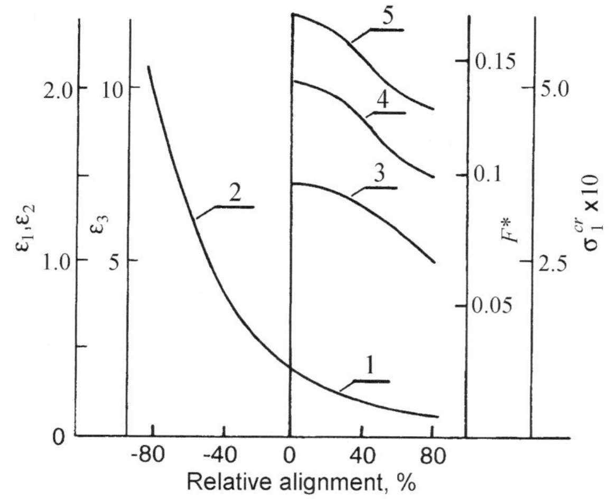

Figure 10
Effect of flake alignment on the swelling and mechanical properties of flakeboard for 12% moisture content: 1, ε_1; 2, ε_2; 3, ε_3 (5); 4, $F*$ (kN); 5, σ_1^{cr} (MPa). After Brauns J, Rocens K 1997 Hygromechanical behaviour of wooden composites. *Wood Sci. Technol.* **31**, 193–204.

$E_2 = 700$ (MPa); shear modulii $G_{12} = 900$, $G_{13} = 1500$, and $G_{23} = 300$ (MPa); Poisson's coefficients $\nu_{21} = 0.45$, $\nu_{31} = 0.34$, and $\nu_{32} = 0.30$. The corrections for moisture change of the elasticity and shear modulii, respectively, are: $\alpha_1 = 250$, $\alpha_2 = 25$, $\alpha_{12} = 25$, $\alpha_{13} = 30$, and $\alpha_{23} = 20$ (MPa). The specific moisture strains at given initial conditions are: $\beta_1 = 0.00003$, $\beta_2 = 0.002$, $\beta_3 = 0.003\,(\%)^{-1}$.

The midplane strains and curvatures of wooden composites depend on the content and distribution of moisture. The hygroscopic deformation of plywood is presented in Fig. 9. The model contains seven veneers with equal thickness of 1.4 mm. The partial density of plywood (birch) is 660 kg m^{-3}. The results, which compare the properties of densified and customary wood, illustrate the effect on curvature of an asymmetric nonlinear distribution of moisture and antisymmetric structure at uniform moisture content.

The relationship between linear expansion, thickness swelling, and flake alignment for an increase of moisture content of 12% is shown in Fig. 10. To improve the board performance, it is preferable to form boards with relative flake alignment $\xi = 0-40\%$, for which the effect of flake alignment on longitudinal expansion strains ε_1 and ε_2 is negligible while thickness-swelling ε_3 reaches a maximum. Figure 10 also shows relationships between flake alignment and specific concentrated reaction $F*$ in bending (for a deflection of 1 cm) and critical stress σ_1^{cr} in compression of a hinged square flakeboard.

5. *Concluding Remarks*

The constituents of natural wood can be modified by physical, chemical, and mechanical methods. The purposes of chemical and thermal treatment are to hamper biological attack, to impart resistance to fire, chemical, mechanical, or UV light damage, as well as to prevent the risk of water sorption. Wood can be mechanically modified by compression, to flatten its cavities, and by cavity-filling with other materials. Making use of oriented strand boards and plywood can improve wood strength and stiffness. The mechanical and hygromechanical quality of wood composite boards depend on the properties of layers and on the alignment of these.

See also: Wood: Chemical Treatment Processes

Bibliography

Bodig J 1966 Stress–strain relationship for wood in transverse compression. *J. Mater.* **1**, 645–65

Bodig J, Jayne B A 1982 Mechanics of Wood and Wood Composites. . Van Nostrand Reinhold Co, New York

Brauns J, Rocens K 1991 Applied model of composite with incomplete bonding and finite length of reinforcing elements. *Mech. Compos. Mater.* **27**, 567–73

Brauns J, Rocens K 1994a Hygromechanics of composites with unsymmetric structure. *Mech. Compos. Mater.* **30**, 601–7

Brauns J, Rocens K 1994b Nonlinear moisture deformation of a composite with a densified reinforcement. *Mech. Compos. Mater.* **30**, 222–6

Brauns J, Rocens K 1997 Hygromechanical behaviour of wooden composites. *Wood Sci. Technol.* **31**, 193–204

Brauns J, Rocens K 2003 Estimation of technological stresses in wood composites. In: Žiliukas A, Schmidt F, Pritykin I, Januteniene J, Leišis V (eds.) *Proceedings of 3rd International Conference on Strength, Durability and Stability of Materials and Structures.* Kaunas University of Technology, Kaunas, Lithuania, pp. 11–23

Brauns J, Rocens K 2004 Design of humidity sensitive wooden materials for multi-objective application. *Wood Sci. Technol.* **38**, 311–21

Geimer R L 1982 Dimensional stability of flakeboards as affected by board specific gravity and flake alignment. *Forest Prod. J.* **32**, 44–52

Geimer R L, Montrey H M, Lehman W F 1975 Effects of layer characteristics on the properties of three-layer particleboards. *Forest Prod. J.* **25**, 19–29

Giebeler E 1983 Dimensionsstabilisierung von Holz durch eine Feuchte/Wärme/Druck-Behandlung. *Holz Roh- Werkstoff* **41**, 87–94

Halpin J C, Jerine K, Whitney J M 1971 The laminate analogy for 2 and 3 dimensional composite materials. *J. Comp. Mater.* **5**, 36–49

Heebink B G 1972 Irreversible dimensional changes in panel materials. *Forest Prod. J.* **22**, 44–8

Hillis W E 1984 High temperature and chemical effects on wood stability. *Wood Sci. Technol.* **18**, 281–93

Homan W J, Jorissen A J M 2004 Wood modification developments. *HERON* **49**(4), 361–86

Kollmann F, Côté W A 1968 Principles of Wood Science and Technology. Vol. 1: Solid Wood. Springer, Berlin, Heidelberg, New York
Kollmann F, Fengel D 1965 Änderungen der chemischen Zusamensetzung von Holz durch thermische Behandlung. *Holz Roh-Werkstoff* **23**, 461–8
Kollmann F, Schneider A 1963 Über das Sorptionsverhalten wärmebehandelter Hölzer. *Holz Roh- Werkstoff* **21**, 77–85
McNatt J D 1973 Buckling due to linear expansion of hardboard siding. *Forest Prod. J.* **23**, 37–43
McNatt J D 1974 Effects of equilibrium moisture content changes on hardboard properties. *Forest Prod. J.* **24**, 29–35
Militz H, Beckers E P J, Homan W J 1997 *Modification of Solid Wood: Research and Practical Potential.* International Research Group on Wood Preservation, IRG/WP 97-40098
Reuss A 1929 Berechnung der Fliessgrenzen von Mischkristallen und Grund des Plastizitätsbedingung für Eiskristall. *ZAMM* **9**, 49–58
Rocens K 1975 Grundlagen von Berechnungen an armirten Lagenholzwerkstoffen. *Holztechnologie* **16**(2), 112–6
Rocens K 1976a Rheological features of wood plasticized with ammonia. *J. Appl. Polym. Sci., Appl. Polym. Symp.* **28**, 1109–16
Rocens K 1976b Die rheologische Eigenschsften von mit Ammoniak plastiziertem Holz. *Holztechnologie* **17**(1), 40–45
Rocens K 1983 Macrostructure theory of modification of wood properties. *J. Appl. Polym. Sci., Appl. Polym. Symp.* **37**, 923–41
Schniewind A P, Barrett J D 1972 Wood as a linear orthotropic viscoelastic material. *Wood Sci. Technol.* **6**, 43–57
Tsai S W, Hahn H T 1980 Introduction to Composite Materials. Technomic Publ. Co, Westport
Voigt W 1910 Lehrbuch der Kristallphysik. Teubner, Berlin
Wolcott M P, Kamke F A, Dillard D A 1990 Fundamentals of flakeboard manufacture: viscoelastic behaviour of the wood component. *Wood Fiber Sci.* **22**(4), 345–61
Xu W, Winistorfer P M 1995 A procedure to determine thickness swell distribution in wood composite panels. *Wood Fiber Sci.* **27**, 119–25
Zambori B G, Kamke F A, Watson L T 2001 Simulation of the mat formation process. *Wood Fiber Sci.* **33**, 564–79

J. Brauns and K. Rocens

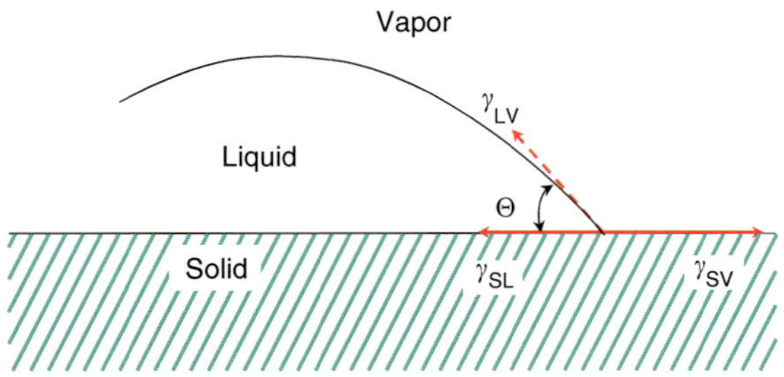

Plate 1
Surface-tension forces determine the contact angle Θ for a liquid droplet on a solid surface (For B/W version, see page 30).

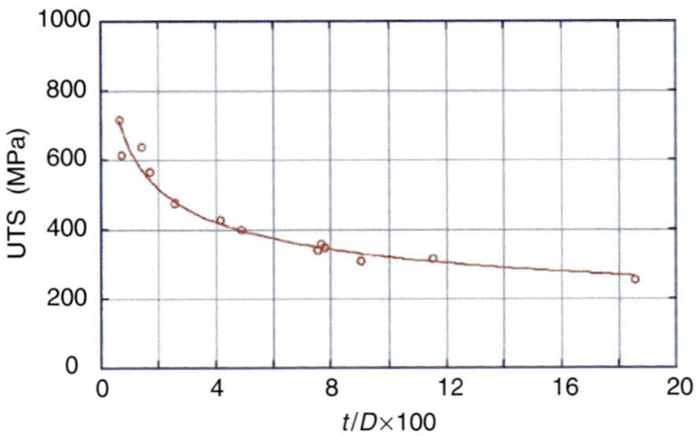

Plate 2
The effect of interlayer thickness on the ultimate tensile strength of maraging steel specimens brazed with Ag–4% Pd filler metal. t = interlayer thickness, D = specimen diameter. Reproduced from Saxton H J, West A J, Barrett C R Deformation and failure of brazed joints – macroscopic considerations, from *Metall. Trans.* volume **2**, 1971, page 999, with permission from The Minerals, Metals & Materials Society (For B/W version, see page 31).

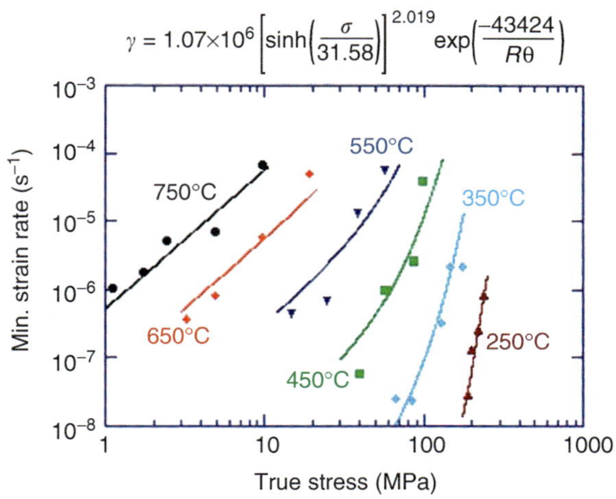

Plate 3
Steady-state creep vs. applied true stress for monolithic Ag–28Cu specimens tested at various temperatures. The data can be described by the expression presented above the graph, where γ = steady-state creep rate, σ = applied true stress, and Θ = temperature (For B/W version, see page 35).

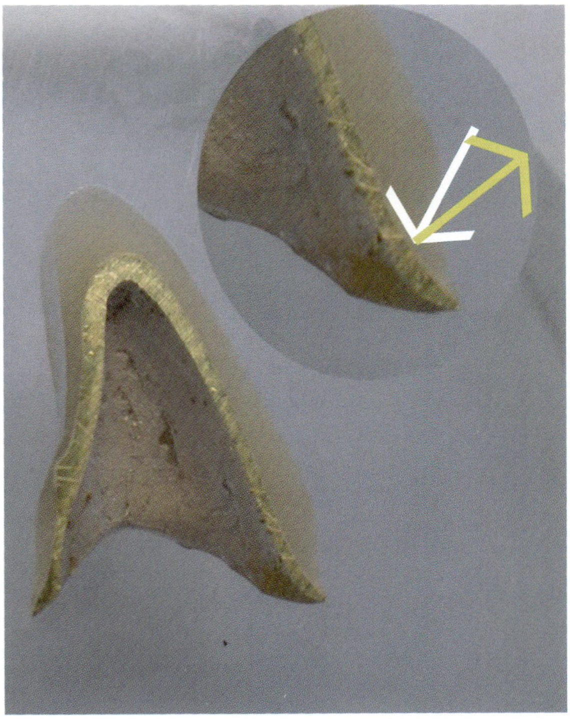

Plate 4
An example of a porcelain-fused-to-metal crown made from the developed alloy. The bottom of the crown sits below the gum line. The enlarged inset shows how the light reflected off the alloy surface will be yellow thus complementing the natural gum tissue (For B/W version, see page 199).

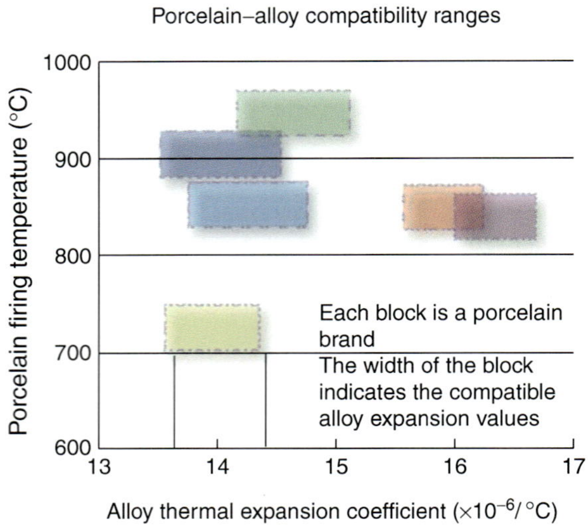

Plate 5
Porcelain–alloy compatibility ranges. Each block represents a porcelain brand and is positioned vertically dependent upon its fusion temperature. The width of the block indicates the alloy thermal expansion values that are compatible with the porcelain (For B/W version, see page 201).

Plate 6
Alloy oxide colors resulting from varying the nature and content of base elements in a 99% gold alloy. The final alloy developed is shown on the top left (For B/W version, see page 202).

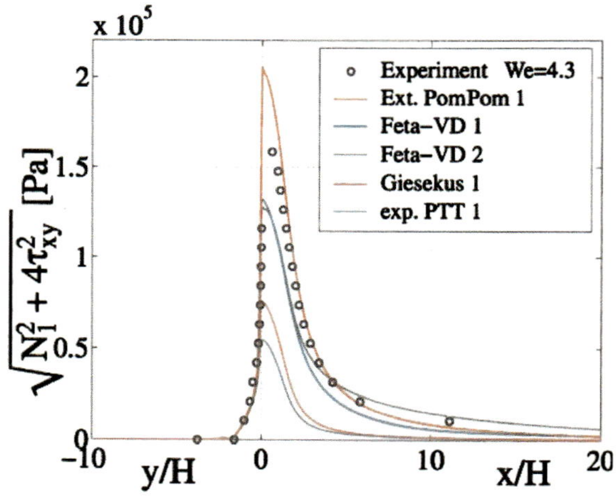

Plate 7
Normal stress along the symmetry line (For B/W version, see page 545).

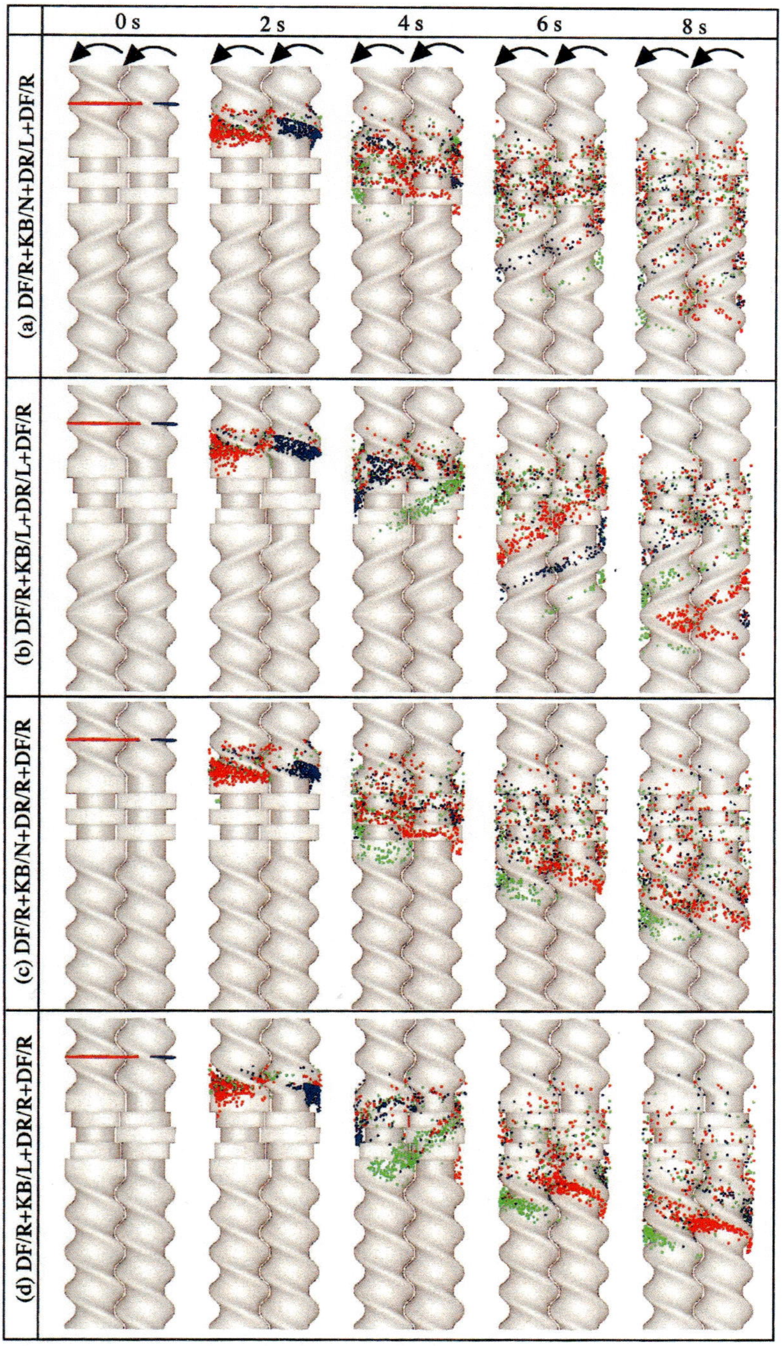

Plate 8

Comparisons of snapshot marker dispersions in four kinds of screw assemblies. The number of markers is 1338. Initial markers are set at X–Y plane of center of first DF/R, with each marker distance being 0.80 mm. Rotational speed is 60 rpm, flow rate is 10 $cm^3\ s^{-1}$. Fluid is pure viscous non-Newtonian. The DF/R and DF/L are right- and left-handed double flighted screws; KB/N and KB/L are both 90° and 135° stagger angles kneading blocks. Barrels are removed (For B/W version, see page 575).

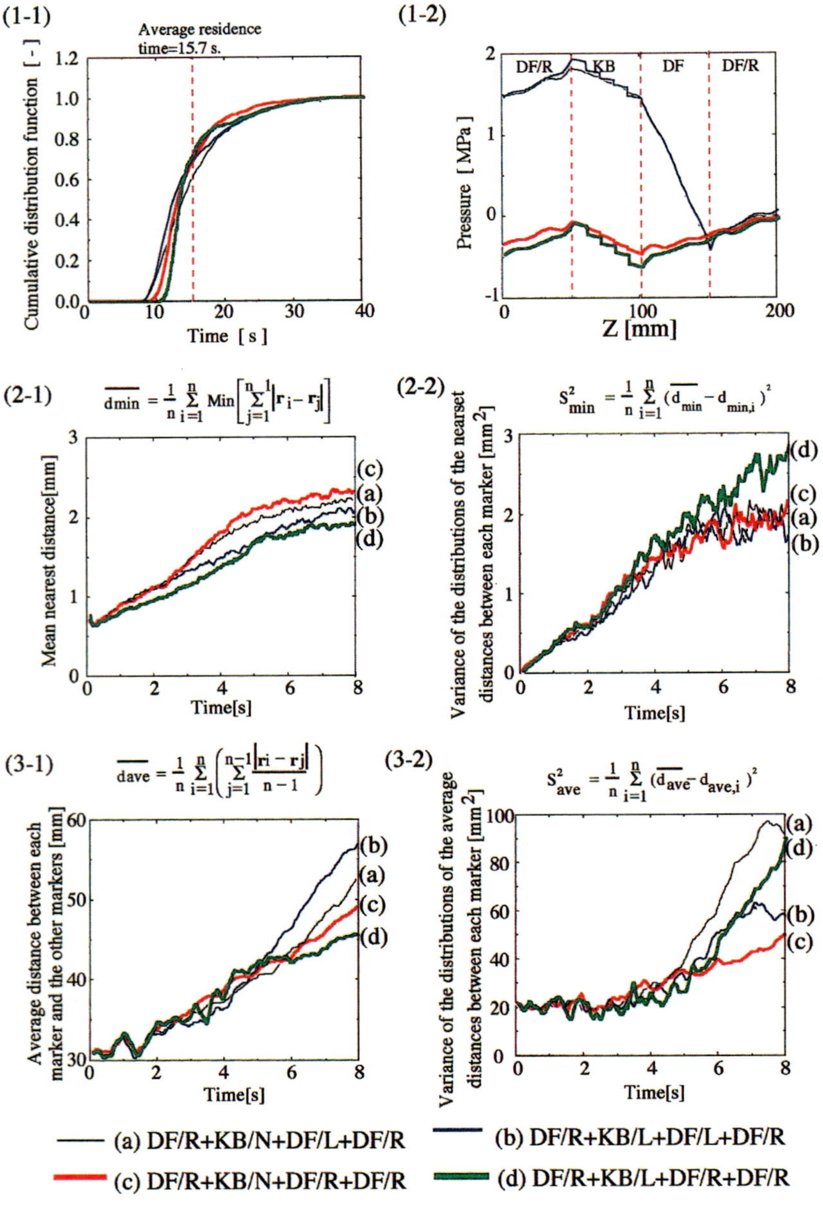

Plate 9

Comparison with (1-1) cumulative residence time distribution, (1-2) cross area average pressure, (2-1) mean nearest distance, (2-2) variance of the mean nearest distance, (3-1) mean value of the average distances and (3-2) variance of the mean value of the average distances for four kinds of screw assemblies. Rotational speed is 60 rpm, flow rate is $10\,cm^3\,s^{-1}$. Fluid is pure viscous non-Newtonian. The DF/R and DF/L are right- and left-handed double flighted screws; KB/N and KB/L are both 90° and 135° stagger angles kneading blocks (For B/W version, see page 576).

N

Nanomaterials

Much of the technological progress since the beginning of the twentieth century has been in the area of information technology. This progress is a result of achieving higher and higher densities of electronic components and bits in storage media. Figure 1 illustrates this progress in the area of magnetic recording media. The advantages of being able to fabricate structures atom by atom were emphasized by Feynman (1960) in his article "There's plenty of room at the bottom," where he pointed out that if a bit of information requires only 100 atoms, then all the books ever written could be stored in a cube 0.02 inches (0.05 cm) on each side.

Nanomaterials are important in many areas other than information technology. This article discusses the fabrication, characterization, and properties of nanomaterials, defined as materials having a characteristic length less than 100 nm. Several books review the subject of nanomaterials (Edelstein and Cammarata 1996, Garcia *et al.* 1997).

1. Synthesis

It should be noted that methods that produce the most desirable materials usually also only produce a small quantity of material. It is easier generally to make nanoparticles and very thin films than to make ordered arrays of nanoparticles or nanowires. Making nanowires requires breaking the symmetry.

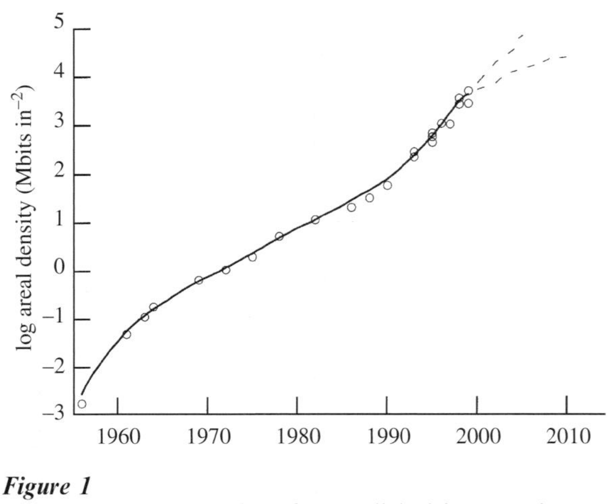

Figure 1
Increase in areal density of IBM disk drives vs. time (data taken from the IBM website).

1.1 Synthesis of Nanomaterials in a Vapor

One of the early methods for producing nanomaterials was to achieve supersaturation by using an inert gas. The inert gas limits the diffusion of atoms away from the region around the source. The sources used include evaporation sources, sputtering sources, laser ablation sources, and electric arcs. The choice of source is often dictated by the material to be made as nanoparticles. For example, evaporating refractory materials is very difficult. Laser ablation and sputtering have the advantage that they can be used for most materials. The reversible work (free energy), $W(n)$, done in forming a cluster containing n atoms or molecules is given by:

$$W(n) = -nk_B T \ln\left(\frac{P}{P_e}\right) + 4\pi\sigma\left(\frac{3vn}{4\pi}\right)^{2/3} \quad (1)$$

where k_B is the Boltzmann constant, T is the temperature, P is the vapor pressure, P_e is the equilibrium or "saturation" vapor pressure at the temperature of the vapor, v is the volume per molecule in the bulk liquid, and σ is the interfacial tension or surface energy per unit area.

Equation (1) expresses the competition between the "bulk" and "surface" free energy in determining the cluster stability. Because of the positive interface contribution in Eqn. (1), there is a free-energy barrier which impedes nucleation. Thus, if the cluster is not larger than some critical size that depends on the temperature, it will be unstable.

(*a*) *Inert gas condensation.* Nanomaterials can be produced in an inert gas in several different ways. For example, one can collect the nanoparticles from a place in the chamber near where they are produced. The apparatus for using this approach consists of a vapor source inside a vacuum chamber containing an inert gas, usually argon or helium. The vapor source can be an evaporation boat or a sputtering target. Supersaturation is achieved above the vapor source and nanoparticles are formed. There is a collection surface above the source that is often cooled to liquid nitrogen temperatures. A convective flow of the inert gas is set up between the warm region near the vapor source and the cold

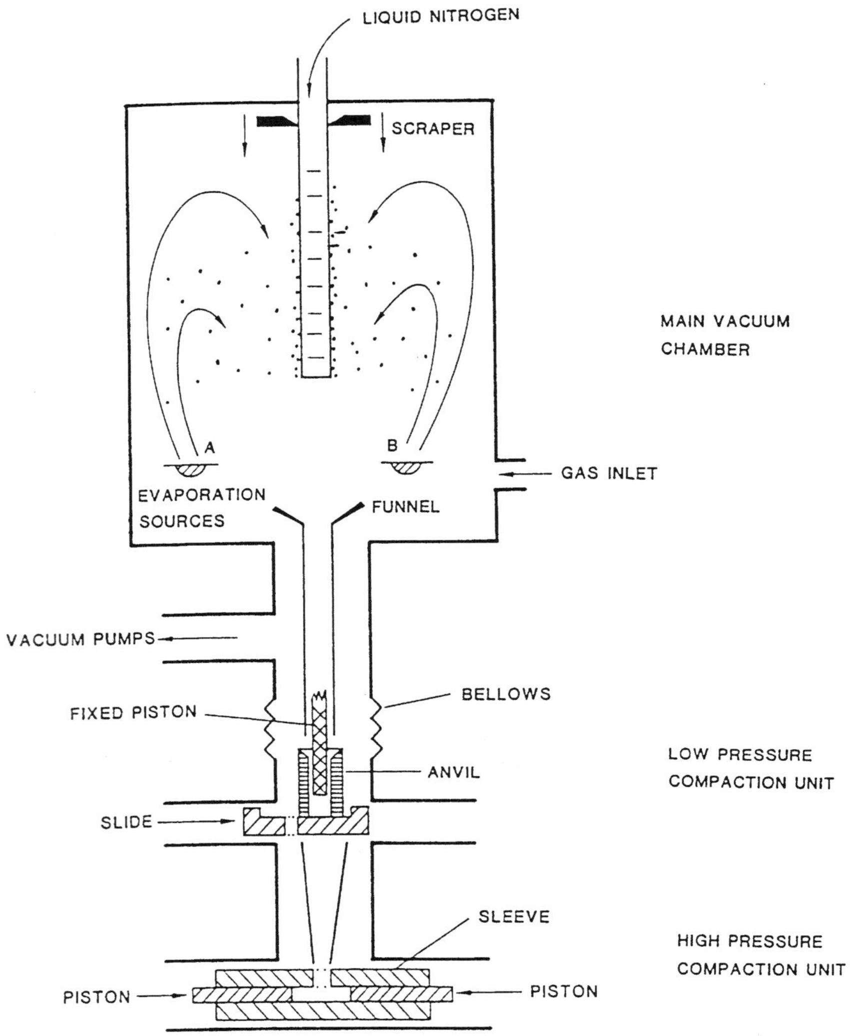

Figure 2
Schematic diagram of apparatus for making, collecting, and compacting nanoparticles in an inert atmosphere.

surface. This flow carries the nanoparticles to the cold surface where they are collected. A good description of the process was published by Granqvist and Buhrman (1976). Figure 2 is a schematic representation of this type of apparatus which makes compacted pellets of nanoparticles. Using such an apparatus, it is not uncommon to produce gram quantities of material during a single run. Nanoparticles of a large variety of materials, including oxides, can be made by this method. The nanoparticles produced range in size from 1nm to 100 nm. This approach permits one to make large quantities of material but suffers from the drawback that there is usually a large distribution of particle sizes.

Alternatively, one can set up a beam of the nanoparticles by including a forced gas flow. If there is a forced gas flow, the particles can be collected at a considerable horizontal distance away from where they are formed. This method has the advantages that one can use a mass spectrometer to narrow the size distribution and further processing of the nanoparticles can be done in the beam. For example, the particles can be coated with another phase or reactions can occur if gases such as oxygen or nitrogen are added downstream to the vapor.

(*b*) *Clusters and cluster beams.* Clusters can be produced by the expansion of atoms from a high-pressure gas source into a low-pressure ambient background. The clusters can be selected by size using a mass spectrometer. One of the interesting findings of research on clusters was the discovery that clusters of some sizes are more abundant than other sizes (Knight *et al.* 1984). The numbers of atoms in the more abundant clusters are called magic numbers. For clusters containing alkali metal atoms, one can treat the cluster using a free electron model. In this model, the valence electrons occupy spatially delocalized orbitals. There are nZ such electrons in a cluster of n atoms, each with a valence of Z ($Z=1$ for sodium, 2 for magnesium, 3 for aluminum). Similar to the stability of atoms with filled atomic shells, clusters whose total number of valence electrons fills an electronic shell are especially stable. Therefore, the magic numbers observed as peaks in the abundance spectra (e.g., for sodium at $n=8$, 20, 40, 58, 92) correspond to major electronic shell closings.

Cluster beams can be used to deposit material on substrates. Cluster beam deposition falls into two classes: (a) depositions in which the clusters are ionized and accelerated so that they have energies of the order of kiloelectron volts before reaching the substrate and (b) deposition of low-energy neutral cluster beams. If high-energy cluster beams are used, the clusters fragment on the substrate. High-energy cluster beams have been used to produce epitaxial coatings. If low-energy cluster beams are used, then the clusters do not fragment on striking the substrate, and nanostructured thin films are produced. The subject of cluster-assembled materials has been reviewed by Melinon *et al.* (1995).

(*c*) *Electric arc discharges.* Supersaturation of a material in an inert atmosphere to produce nanomaterials can be achieved if one strikes an electric arc between two electrodes containing the material. This method is frequently used to produce fullerenes like C_{60} and carbon nanotubes (Iijima 1991). The high temperatures in the arc lead to the sublimation of the material.

1.2 Synthesis of Nanomaterials on Surfaces

(*a*) *Nucleation and growth.* The nucleation of nanoparticles on a surface differs from nanoparticle formation in a vapor in that one must consider the interface between the particle and the substrate. Another difference is that atoms can diffuse on the substrate. Because of this diffusion, the larger particles can increase in size as atoms diffuse away from the surface of the smaller particles. This occurs because the atoms are less tightly bound to the surface of the smaller particles. This process is called Ostwald ripening (see Sect. 1.6).

(*b*) *Molecular beam epitaxy* (*MBE*). Single-crystal films can be deposited by MBE, which usually involves using an ultrahigh vacuum, a low deposition rate, and a substrate whose lattice constant nearly matches the lattice constant of the crystal. Nonequilibrium structures, such as b.c.c. cobalt and f.c.c. iron, can be grown by MBE.

(*c*) *Interfaces.* Desirable properties sometimes result from the segregation of atoms or nanocrystals at interfaces. For example, the chromium atoms that segregate at grain boundaries weaken the magnetic coupling between the cobalt particles in magnetic recording media. Precipitate hardening improves the properties of aluminum alloys. These precipitates often are nanocrystalline (Cahn 1965).

(*d*) *Coatings.* Hard protective coatings are needed for cutting tools and other applications. The properties of these coatings depend on the microstructure. For example, titanium nitride films with micrometer-sized grains used for these coatings contain titanium inclusions, which decrease the hardness. Alternative processing can sometimes improve the properties. It was found that nanocrystalline TiN films produced by filtered arc evaporation (Konyashin and Fox-Rabinovich 1998) were free of inclusions and had Vickers hardness values as high as 40 GPa. Generally, one can increase the hardness of a coating by decreasing the particle size. Nanocrystalline boron nitride films, prepared by sputtering, have hardnesses of 1.5 to 2 times those of films of large-grained polycrystalline boron nitride.

(*e*) *Lithography.* Lithography is the process of forming a pattern on an electronic material (see *Optical Lithography*). The pattern is formed in an imaging layer (resist) on top of a semiconductor by selected area irradiation. The resist can act as either a template or as an etch mask. The radiation can be ultraviolet photons, electrons, ions, or x rays. In production, photons are used, and the feature size has been reduced to 0.3–0.5 μm. Because of diffraction effects, the feature size cannot be reduced much below the wavelength of the photons unless a near-field optical microscope is used.

Methods that can be used to fabricate patterns on a nanoscale are too costly for commercial application. For example, serial writing with a scanning tunneling microscope (STM) or with an atomic force microscope (AFM) is a very slow process. There are efforts underway to use arrays of STM tips to perform parallel lithography.

1.3 Chemical Synthesis Methods

Chemical synthesis offers a route for producing large quantities of nanomaterials economically. One can start with an aqueous or nonaqueous solution containing salts and add a reagent, e.g., a hydroxide. Once the solution becomes supersaturated with the product, precipitation occurs by either homogeneous or heterogeneous nucleation (see *Nucleation*). To form particles with a narrow size distribution nearly all the precipitation must occur at the same time and there must be no further nucleation as the particles grow. The properties of the particles are largely determined by the reaction rates, which in turn are affected by the concentration of reactants, the temperature, the pH, and the order in which the reagents are added to the solution. Multiphase nanomaterials are more difficult to produce by chemical means since each phase will usually require different conditions to precipitate. The particle size can be limited by having a very large number of nucleation sites, using reverse micelles, or by capping the surface.

(*a*) *Reverse micelles.* Many surfactants are rod-like molecules having a hydrophobic and a hydrophilic end. When oil, water, and a surfactant are mixed in the proper proportions, the surfactant molecules arrange themselves to form spherical shells with water pools filling the space inside the shell. This geometric arrangement of the surfactant and water, which is called a reverse micelle, occurs to minimize the energy. The size of the reverse micelle can be controlled because the reverse micelle radius is linearly dependent on the ratio of the amount of water to the amount of surfactant (Pileni 1993). Most reactions that can be performed in water can also be performed in the water pools inside the reverse micelles. Thus, one can precipitate nanoparticles inside reverse micelles. The size of the nanoparticle is limited by the size of the reverse micelle. There is a weak interaction between the nanoparticle and the inside surface of the micelle.

Agglomeration is a common problem in processing nanomaterials. Because of their large surface area, nanoparticles often agglomerate to decrease their surface energy. Agglomeration hinders the consolidation of nanoparticles and is generally undesirable (see Sect. 2). Surfactants can be used to prevent agglomeration. Nanoparticles prepared inside reverse micelles do not agglomerate.

(*b*) *Capping.* Capping agents, which covalently bind to the surface of growing clusters, can be added to solutions to prevent further growth of the clusters. Thiolates are the most common capping agents. Capping also limits agglomeration.

(*c*) *Reduction.* The precipitated nanoparticles are often oxides or hydroxides. If metallic nanoparticles are needed, the oxides or hydroxides can be reduced by heating them in flowing hydrogen. Thermal reduction in a polyol such as ethylene glycol can also be used.

(*d*) *Macromolecules.* Metal cluster compounds can be produced which are macromolecules. They consist of a metallic core surrounded by a ligand shell composed of ligand atoms (chlorine, iodine, oxygen, etc.) or ligand molecules (de Jongh 1994). The ligand shell stabilizes the molecules against oxidation. Although the cores are usually small, containing 20–30 atoms, macromolecules have been made with cores containing 2057 palladium atoms.

1.4 *Mechanical Synthesis Methods*

High-energy ball milling, also called mechanical attrition, can be used to reduce the grain size of materials from many micrometers to 2–20 nm (see *Mechanical Alloying*). This is a result of the cold-working process creating large-angle grain boundaries. Most of the reduction in grain size occurs rapidly, but the process slows, and long times are required to reach the smallest sizes. This process has the advantage of being relatively inexpensive and can be easily scaled up to produce large quantities of material. Usually, to maximize the energy of collision, high-mass hard-steel or WC balls are used. Contamination by materials removed by the balls is a major concern. Severe mechanical deformation and plastic deformation at high strain rates (10^3–$10^4\,s^{-1}$) occurs during the process. Initially, shear bands are formed consisting of a high density of dislocations. Later these dislocations annihilate and recombine as small-angle grain boundaries forming nanometer-sized grains. Finally, the orientation of these nanometer-sized grains is randomized.

The range of solubility of multicomponent systems is greatly increased by mechanical attrition. Mechanical attrition can also produce metastable materials. If the milling is done in the presence of O_2 or N_2, oxides or nitrides can be formed.

1.5 *Other Synthesis Methods*

Nanocrystalline oxides can by prepared using the sol–gel process. Nanocrystalline cylinders can be electrodeposited inside the pores of membranes. The size of the nanoparticles can be controlled by synthesizing them inside zeolite hosts.

1.6 *Effects of Heat Treatment*

Heat treatment can be used to create nanomaterials and to modify their size distribution. One can produce amorphous, metastable mixtures of immiscible elements by sputtering or other rapid cooling techniques. Heating causes these metastable mixtures to phase-separate, forming multiphase mixtures of nanosized crystals. Further heating causes the larger crystals to grow and the smaller crystals to disappear. This diffusion process is an example of Ostwald ripening (see Sect. 1.2.1).

1.7 Self-assembly

Under some conditions, unusually regular nanomaterials form by self-assembly. For example, stress due to lattice mismatch drives islands to form when germanium is deposited on the (001) surface of silicon. These islands can have shapes which include square-based pyramids and domes. Using advanced synthesis techniques, nearly perfect monodispersed nanocrystals of CdSe can be formed that self-assemble into three-dimensional semiconductor superlattices of quantum dots (Murray *et al.* 1995). Quantum dots are nanosized particles in which the confinement of the electrons and holes shifts the quantum mechanical energy levels and, hence, modifies the optical properties (see Sect. 5.8). Figure 3 shows a CdSe superlattice. Another example of self-assembly was found by sputtering molybdenum at an argon pressure that was just above the threshold pressure for particle formation. In this case, the molybdenum particles that form are cubes that self-assemble to form larger $n \times n \times n$ ($n = 2, 3, 4$) cubes (Kaatz *et al.* 1993, Edelstein *et al.* 1991). Examples of $3 \times 3 \times 3$ molybdenum cubes are shown in Fig. 4. Fullerene clusters of C_{60} were also found to self-assemble into clusters with magic numbers. The evidence for this is the intensity anomalies observed (Martin *et al.* 1993) in the mass spectra of clusters made of fullerenes. These anomalies occur at fullerene cluster sizes that are similar to the magic numbers for argon clusters with increased stability. For the argon and C_{60} clusters, the stability is a result of geometric effects and not the electronic effects discussed in Sect. *1.1.2*. For self-organization to occur, it is important that the entities have a narrow size distribution.

2. Consolidation

To obtain nanomaterials in bulk form the nanoparticles must be consolidated while maintaining the nanosizes of the crystals (see *Sintering*). The usual goal is to form a high-density solid that is free of voids. Sintering is a common technique for consolidating metallic and ceramic materials. The material is first compacted into a low-density solid which may contain binders. Next, higher temperatures and sometimes pressures are used to increase the density by the diffusion of vacancies out of the pores. Agglomeration of the starting material creates large voids that are difficult to eliminate. Wet-processing techniques are often used to compact ceramics but are less desirable to use on metals because they tend to form oxide surface layers.

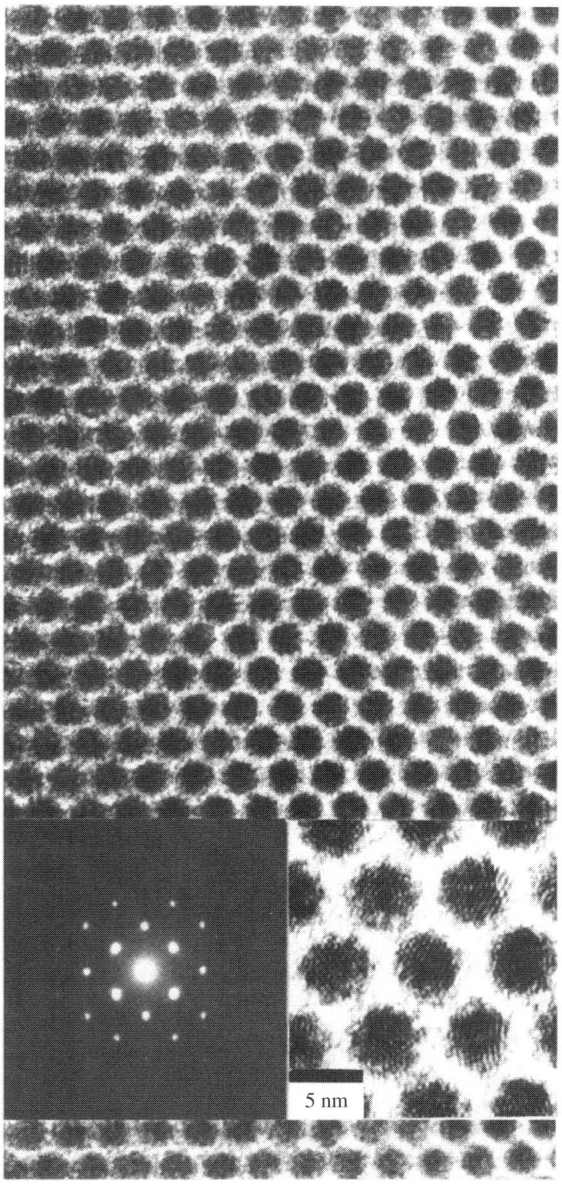

Figure 3
Bright field TEM micrograph of self-assembled structure of CdSe nanoparticles. Lower left insert is TEM micrograph taken in the diffraction mode showing that the particles are single crystals (courtesy of C. Murray).

When pressure is applied to nanocrystalline metals during cold compaction, the metal particles deform plastically and are extruded into the pores. Compacted nanocrystalline metals can have densities

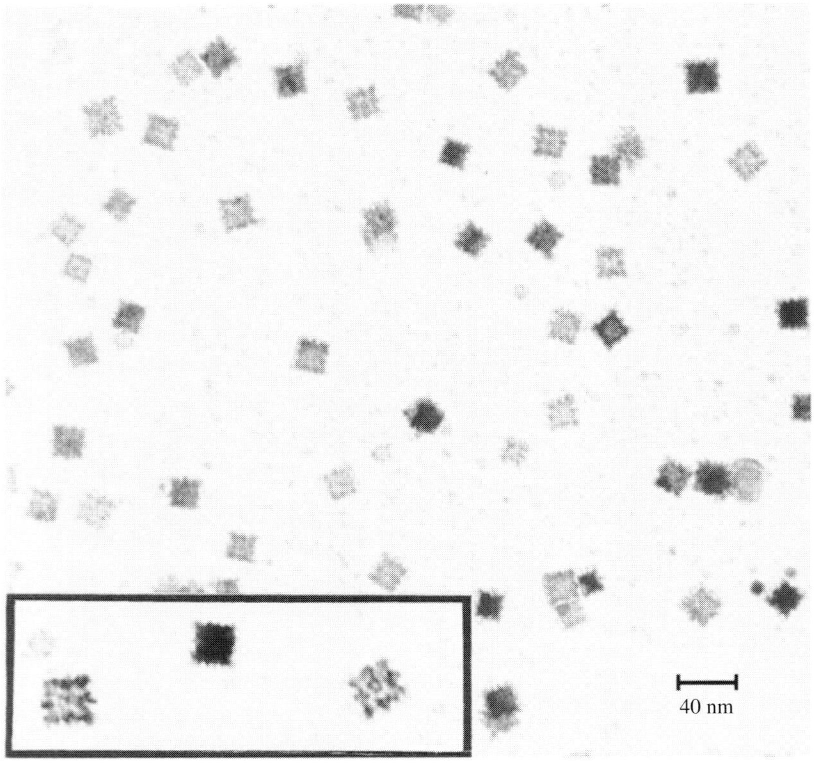

Figure 4
TEM image of self-assembled molybdenum cubes made by high-pressure sputtering (after Edelstein *et al.* 1991).

which are 96–98% of the bulk. In contrast, ceramic particles do not plastically deform during cold compaction and sintering must be used. Nanocrystalline ceramics are more difficult to consolidate by compaction than nanocrystalline metals.

During sintering of ceramics, densification occurs first by neck growth at the contact point between adjacent particles. Then the tubular pores open to the outside surface, and these tubular pores shrink in diameter. Next, the narrow pores become unstable and pinch off to form spherical pores. Finally, these spherical pores are eliminated. These processes are illustrated in Fig. 5.

3. Fullerenes

3.1 Fullerene Clusters: Primarily C_{60}

The mass spectra of laser-ablated carbon shows that clusters of 60 carbon atoms are unusually stable (Kroto *et al.* 1985). It was found that clusters containing 70 carbon atoms, though not as stable as the C_{60} clusters, were also more stable than other sized clusters. The C_{60} clusters turned out to be soccer-ball-shaped molecules with icosahedral symmetry. The spherical structure consists of 12 pentagons and 20 hexagons. Fullerenes were first produced in large quantities by striking an arc between carbon rods in an inert atmosphere. It is possible to grow single crystals of C_{60} molecules. Crystalline C_{60} has an f.c.c. structure above about 255K, but transforms below this temperature to a simple cubic structure. Crystalline C_{60} has a bulk modulus of 18GPa. Other atoms can be attached to a C_{60} molecule or incorporated inside its hollow interior. The molecules based on C_{60} have been reviewed by Taylor (1995).

3.2 Carbon Nanotubes

Carbon nanotubes (Ebbesen 1997) were first observed by Iijima (1991) and were subsequently produced in larger quantities. The tubes, either single- or multiple-walled nanotubes, consist of concentric, cylindrical, graphitic carbon layers capped on the ends with fullerene-like domes. Carbon nanotubes can be synthesized by carbon arcs, laser vaporization, catalytic combustion, chemical vapor deposition, and ion bombardment. Atomic force microscopy (Wong *et al.* 1997) was used to investigate the mechanical properties of multiwalled nanotubes. The Young's

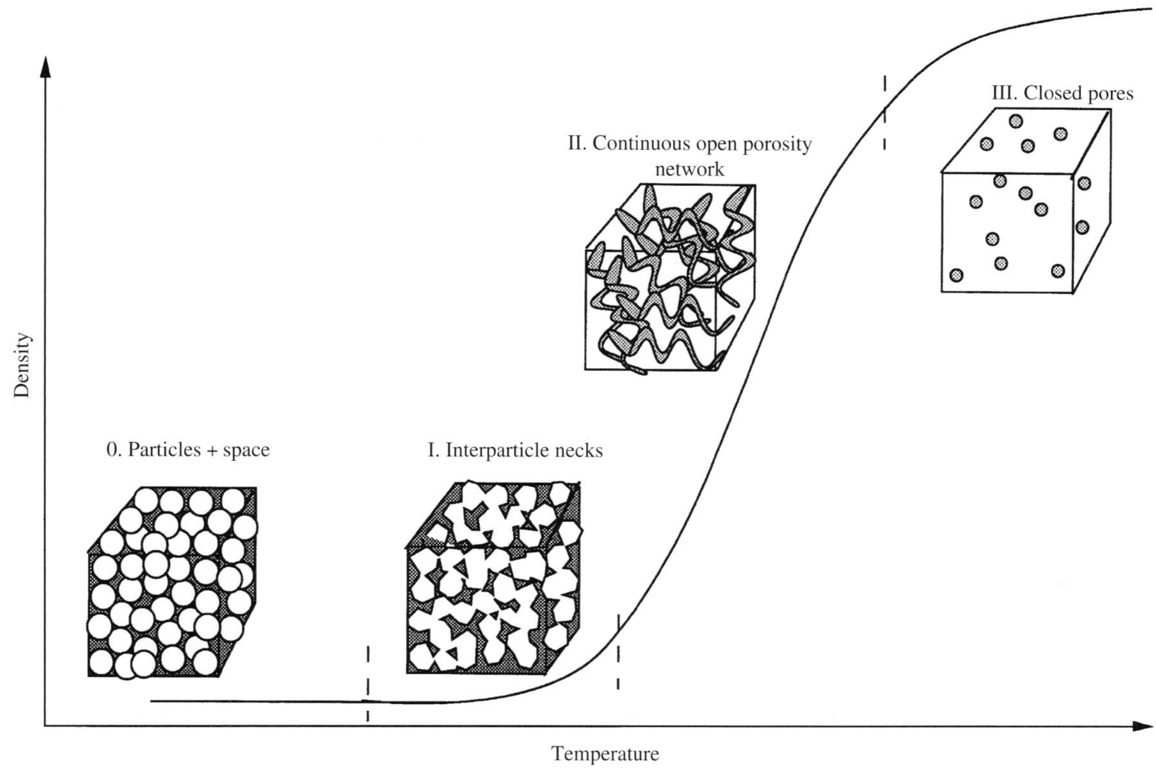

Figure 5
Schematic diagram of the different stages of sintering a powder compact.

modulus was measured to be 1.26TPa. Bernholc *et al.* (1998) discuss the growth and mechanical properties of carbon nanotubes.

There is great interest in single-walled carbon nanotubes because of their electronic and mechanical properties. Single-walled carbon nanotubes act as quantum wires in that the electrical conduction seems to occur through discrete electron states that are coherent over long distances (Tans *et al.* 1997). Rinzler *et al.* (1998) describe a process for large-scale purification of single-walled carbon nanotubes. Electromechanical actuators based on sheets of single-walled carbon nanotubes can generate higher stress than natural muscle and higher strains than high-modulus ferroelectrics (Baughman *et al.* 1999).

4. Characterization

4.1 Microscopy Techniques

Although nanomaterials are too small to be investigated using optical microscopy, several other microscopic techniques do allow a direct examination. Of these, transmission electron microscopy (TEM) is probably the most useful. Using TEM one can observe the shape of the nanoparticles, determine the crystal structure and orientation, and measure the particle size distribution. Modeling is required to understand the image if a high-resolution electron microscope is used.

STMs and AFMs can be used to measure the surface topography of nanomaterials. Although atom resolution is possible with an STM, a limitation is that only conducting material can be examined. Considerable effort is being made to improve STM and AFM tips. These efforts include trying to make sharper tips by attaching carbon nanotubes to the ends of the tips and treating them so that they are chemically specific. An AFM can be converted to a magnetic force microscope (MFM) if the AFM tip is replaced by a ferromagnetic one; MFM resolution is 50 nm.

4.2 Scattering Techniques

Nanomaterials can be characterized by a variety of scattering techniques. The simplest, most direct method is x-ray scattering, which can be used to investigate the phase and particle size. The particle size can be estimated from the width of the diffraction peaks using the Scherrer equation:

$$D = \frac{\kappa\lambda}{\beta\cos\theta} \quad (2)$$

where D is the average crystallite size perpendicular to the reflecting planes, λ is the x-ray wavelength, θ is the Bragg angle, β is the width of the peak in radians due to the finite size of the crystal, and κ is a constant close to unity. Because strain also broadens the x-ray lines, more complicated methods are needed to make a precise measurement of the average particle size. Klug and Alexander (1974) provide a method for separating size effects from strain.

Small-angle x-ray scattering (SAXS) can be used to investigate particles and the quality of multilayers. In the case of multilayers, SAXS can determine the roughness and the width of the interfaces. Small-angle neutron scattering (SANS) can be used to study porosity. If there is a broad particle size distribution, the SAXS and SANS spectra are broadened and more difficult to interpret.

Extended x-ray absorption fine-structure spectroscopy (EXAFS) provides atom-specific structural information. Specifically, it provides information on the radial distance and number of atoms of a given type which are neighbors of a given atom type. EXAFS measurements can only probe local short-range order because of the short lifetime of the photoelectron.

Raman spectroscopy measures the energy shifts of scattered light due to creating optic mode excitations. In some cases, Raman spectroscopy can be used to determine the particle size and oxidation state of a nanomaterial.

4.3 *Indentation Measurements*

The hardness of nanomaterials, which is the easiest and most frequently applied measurement of mechanical properties, can be investigated by standard indentation measurements or nanoindentation measurements. In nanoindentation measurements, the force and displacement are continuously monitored.

4.4 *Other Characterization Techniques*

Positron annihilation spectroscopy (PAS) is a sensitive technique for the detection of free volumes. In PAS, positrons are implanted into the solid from a radioactive source. After implantation, the positrons are slowed to thermal energies before they decay. The rate of their decay is determined by the overlap of the positron wave function and local electron density. If the positron is trapped at a free volume such as a pore, its decay rate is decreased because the local electron energy density is reduced. Positron lifetimes are sensitive to very small free volumes.

The density can be investigated by the Archimedes method. The structure of pores that are open to the surface and one another can be investigated by the BET (Brunauer–Emmett–Teller) N_2 absorption method. NMR can be used to estimate the strains in nanomaterials.

5. *Properties*

5.1 *Structural Properties*

(*a*) *Clusters.* The crystal structure of very small clusters, containing less than about 20 atoms, is different from the bulk. Somewhat larger clusters either have metastable structures or the bulk structure but with a reduced lattice constant. The reduced lattice constant results from a competition between the surface energy and the elastic energy. The minimization of the free energy gives rise to a Laplace pressure given by:

$$P = \frac{2f}{r} \quad (3)$$

where f is the surface stress and r is the radius. The reduction of the lattice constant of aluminum is shown in Fig. 6. Large clusters usually have the bulk values for their lattice constant and the bulk structure.

(*b*) *Nanocrystalline solids.* In nanocrystalline solids, a relatively large fraction of the atoms are interface atoms and the properties of the material depend on the interface structure. Two different

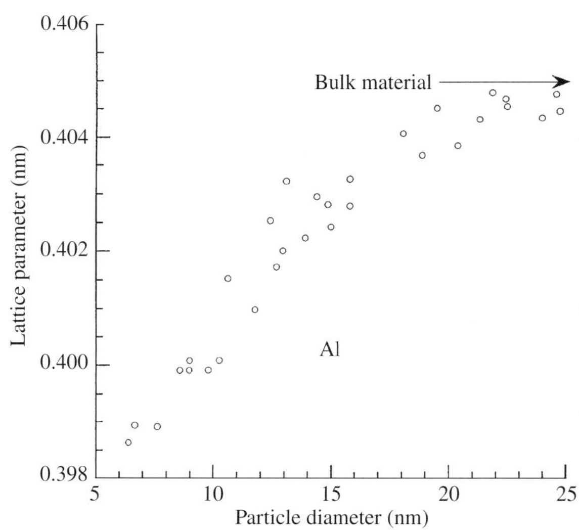

Figure 6
Reduction of the lattice constant of aluminum nanoparticles as a function of particle size (after Woltersdorf *et al.* 1981).

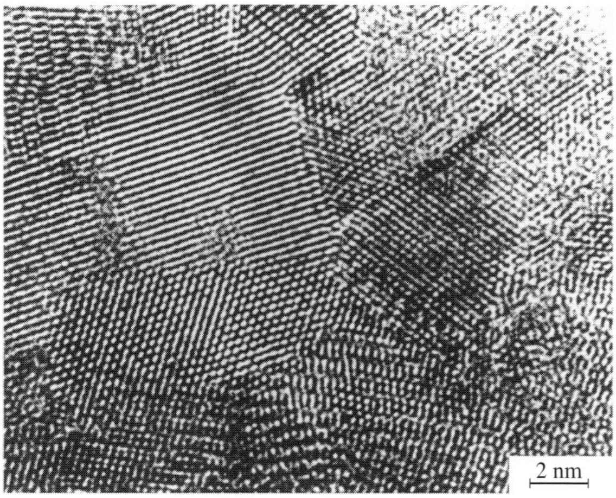

Figure 7
High-resolution TEM micrograph of nanocrystalline palladium.

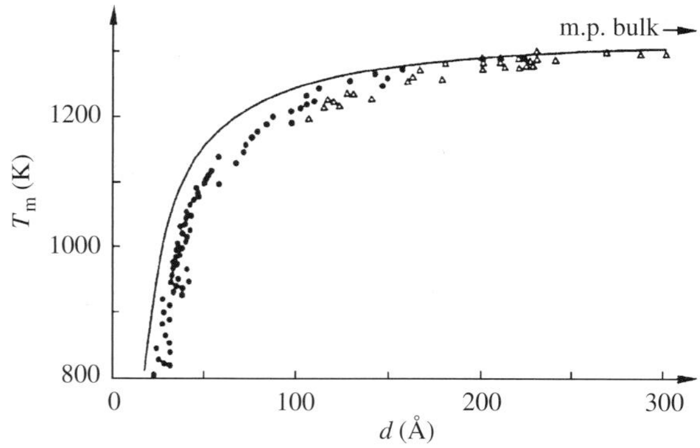

Figure 8
Reduction of the melting temperature of gold nanoparticles as a function of particle size; the solid line is the theoretical curve (after Borel 1981).

views have been presented to describe the structure of nanocrystalline interfaces. One early view was that interface atoms are "gas-like" in that they are more disordered and have a lower density than interface atoms in micrometer-sized polycrystalline materials. The second view was that the grain boundaries are similar to those of coarser-grained material. More recent measurements indicate that the grain boundaries in nanomaterials are similar to grain boundaries in coarse-grained materials but that there are subtle differences which depend on the processing of the material. Figure 7 shows a high-resolution TEM micrograph of nanophase palladium. The sample was produced by consolidating gas-condensed particles. The grains are equiaxed and have a lognormal size distribution.

5.2 *Phase Transitions*

The thermodynamic properties of a nanomaterial can be very different from its bulk properties. For example, the melting temperatures of gold (Borel 1981) and aluminum (Lai *et al.* 1998) are depressed. Figure 8 shows the melting temperature of aluminum as a function of particle size. Simulations indicate that the melting process initiates at the surface. Since microelectronics have tolerances approaching only a few nanometers, changes in the physical properties at small dimensions are increasingly important.

5.3 *Thermal Diffusion*

Thermal diffusion in nanostructured materials can be orders of magnitude faster than in coarse-grained materials. The large number of void and grain boundaries enhances the diffusivity. Sintering to get rid of the voids significantly decreases diffusion.

5.4 *Mechanical Properties*

The surface and interface stress are important in nanostructured materials. For coarse-grained materials the yield stress, σ, is empirically found to satisfy the Hall–Petch relation:

$$\sigma = \sigma_0 + \frac{k}{\sqrt{d}} \tag{4}$$

where σ_0 is a friction stress, k is a constant, and d is the grain size. A similar relationship exists relating the hardness and d. The basis for the Hall–Petch relation in coarse-grained materials is the interaction of dislocations at grain boundaries. In the Cottrell model, yielding occurs when the concentration of stress ahead of a pile-up of dislocations at a grain boundary is large enough to generate a Frank–Read source. If the Hall–Petch equation were still valid when d is reduced to 10 nm, nanomaterials would have remarkable strength. Unfortunately, the ideas providing the basis for the Hall–Petch relationship are not applicable at nanometer dimensions.

Because of limitations on the amount of material, most of the measurements of the strength of nanomaterials are microhardness or nanoindentation measurements. Another difficulty is that the measurements are affected by flaws and porosity. Nevertheless, the measurements indicate that nanocrystalline metals and ceramics do have improved strength due to suppressing dislocation motion. The scatter in the data is too large to determine whether the increase in hardness is proportional to $1/d$ or $1/\sqrt{d}$. In any case, the improvements in strength are not as large as predicted by Eqn. (4), and for the smallest size grains the strength appears to decrease with decreasing grain size. One limitation on the strength may be softening due to enhanced diffusion.

Despite not fulfilling the strength expectations based on Eqn. (4), nanocomposites do have advantageous mechanical properties. For example, intragranular bonding and thermal expansion mismatch can result in toughening. The presence of another phase also tends to inhibit coarsening. Three-point bending tests have shown that nanocomposites can have strengths greater than 1 GPa. Nanocrystalline Co–WC exhibits a hardness which makes it attractive for machine cutting tools (Kear and McCandlish 1993).

Superplasticity is the property that allows some materials to undergo large tensile deformation without necking or fracture. Many metals are superplastic. The common features of superplastic materials are small equiaxed grains and high-energy grain boundaries. The phenomenological relation for the superplastic strain rate $\dot{\varepsilon}$, is:

$$\dot{\varepsilon} = A\sigma^n \exp(-E_a/kT)/d^3 \quad (5)$$

where σ is the applied stress, A is a constant, E_a is the activation energy, d is the grain size and T is the temperature. For superplastic materials $n \approx 2$. Because d is small in nanocrystalline materials, one expects that strain rates should be higher in nanomaterials.

The strain-rate sensitivity is the inverse of the value of n in Eqn. (5) and is obtained by measuring $\dot{\varepsilon}$ as a function of σ. Although the room-temperature values for the strain-rate sensitivity are considerably larger for nanocrystalline material than for coarse-grained materials, they are more than an order of magnitude smaller than the characteristic value of 0.5 for superplastic materials. Nanocrystalline materials are not superplastic at room temperature. Nevertheless, nanocrystalline materials do exhibit superplasticity at higher rates and lower temperatures than coarse-grained materials. The improvements would be greater if grain growth did not occur during the deformation. Because of the limited availability of nanomaterials and competition from other superplastic materials, applications like diffusion bonding may be more important than near net shape forming.

5.5 *Transport*

Grain-boundary scattering and impurities at grain boundaries increase the electrical resistivity of nanocrystalline materials. In addition to this general property, mixtures of a nanocrystalline metal and a nanocrystalline insulator exhibit interesting transport behavior as a function of the concentration, x, of the metal species. Below a critical concentration, x_c, called the-percolation threshold, there is no continuous conducting path of metallic particles in contact with one another, and the resistivity is high. For $x > x_c$, the resistivity rapidly decreases with increasing x to a value that is limited by grain boundary scattering. For a random mixture, x_c is 16 vol.%. It is interesting that x_c for nanomaterials produced by phase separation is 40–50 vol.%. This higher value is a result of the fact that phase separation is not a random process. In phase separation, like atoms diffuse toward one another.

The electrical conductivity for $x < x_c$ is strongly temperature dependent because it involves both thermal activation and tunneling between metallic grains. The electrical charging energy required to add or subtract a single electron to a grain is large because the capacitance of the grain is proportional to its diameter. Coulomb blockade devices are based on the discrete changes that occur in the conductance when a single electron tunnels to or from a nanoparticle. In the limit of low temperatures, the resistivity is proportional to $\exp(B/T^{1/2})$ where B is a constant.

5.6 *Magnetic Properties*

The magnetic properties of small clusters may be different from those of the bulk material. For example, clusters of sodium, potassium, and rhodium are ferromagnetic, whereas these elements are paramagnetic in their bulk form. Another difference is that the magnetic moment of atoms on the surface of a nanoparticle may either be larger or smaller than in the interior.

Ferromagnetic materials form domains to decrease the magnetostatic energy. The additional energy required is the domain wall energy. Most magnetic nanoparticles are single domain, because the decrease in magnetostatic energy in adding a domain wall is smaller than the domain wall energy.

(*a*) *Superparamagnetism.* The remanence (M_r) is the residual magnetization that remains after the material is first saturated in a magnetic field and then the field is reduced to zero. For weakly interacting particles the total magnetic anisotropy per particle (E_A), which increases with the volume of the particle, determines the remanence. If $E_A \gg kT$, the particles will have a value for M_r which is a significant fraction of the saturation magnetization. If $E_A \approx kT$ then $M_r = 0$ because the anisotropy is not large enough to prevent the magnetization of the particles from changing its direction. In this case, the particles are said to be acting superparamagnetically. Such particles are useless for magnetic recording. There is a temperature, called the "blocking" temperature, below which the magnetization becomes stable against thermal fluctuations and $M_r \neq 0$. The blocking temperature depends on the particle size, the applied field, and the method of measuring the magnetization. The progress in increasing the density of magnetic recording media, shown in Fig. 1, has relied on using ever-smaller particles. A way is needed to avoid

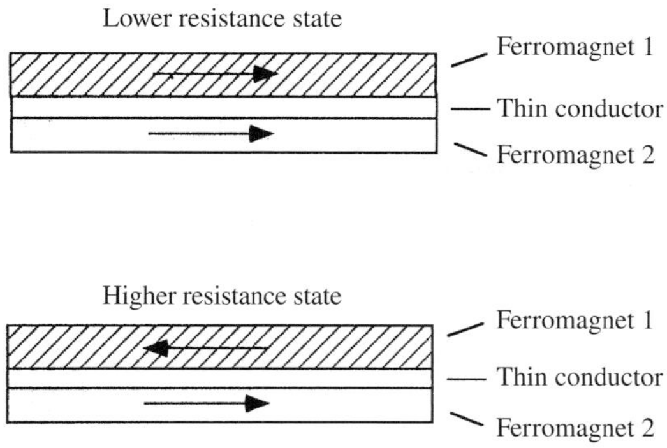

Figure 9
Schematic diagram of a giant magnetoresistance spin valve. The individual layer thicknesses are a few nanometers. The arrows indicate the direction of the magnetizations of the two ferromagnets.

superparamagnetic behavior if this progress is to be maintained.

(*b*) *Giant magnetic resistance* (*GMR*). There is great interest in magnetic field sensors based on the phenomenon of GMR. Magnetic recording read heads based on GMR are being used to read high-density magnetic recording media. An example of a GMR magnetic field sensor called a spin valve is shown in Fig. 9. The low (high) resistance state occurs when the magnetizations of the two ferromagnets are parallel (antiparallel). These sensors exhibit magnetoresistances of 15% in fields of a few tens of oersteds. They are also being developed for use as memory elements.

(*c*) *Magnetotunneling*. Magnetotunneling devices have advantages over GMR devices that include higher magnetoresistance, higher overall impedance, and increased thermal stability.

5.7 *Chemical Properties*

The large surface area of nanomaterials makes them useful as catalysts. Metal or metal oxide particles on nonmetallic supports are often used to perform heterogeneous catalytic reactions. It is important that the particles are highly dispersed to maximize the contact area of the catalyst and reactants. The chemical activity depends on the size of the particles. For example, the activity of oxygen reduction by platinum nanocrystallites on carbon is a maximum when the platinum particles are about 3 nm in size. Platinum alloys on carbon blocks are used in fuel cells. Agglomeration of the catalytic particles is minimized by using microporous supports such as aluminosilicate zeolites. If the reaction is too vigorous, atoms and/or molecules will move and the size of the particles will increase and they will tend to coalesce. Electrochemical nucleation can be used to control the rate of nucleation and growth.

5.8 *Optical Properties: Quantum Dots*

In semiconductors, electrons and holes form hydrogen-like quantum systems called excitons which play an important role in the optical properties of semiconductors. Novel optical properties result when the size of semiconductor particles, R, becomes comparable to the bulk exciton Bohr radius which is typically a few nanometers. These nanoparticles are called quantum dots. Quantum dots of CdS, CdSe, and CdTe have been prepared in glass. The exciton energy levels are given by:

$$\varepsilon_i = \frac{\hbar^2}{2m_{ex}}\left(\frac{\alpha_{nl}}{R}\right)^2; \quad \frac{1}{m_{ex}} = \frac{1}{m_e} + \frac{1}{m_h} \qquad (6)$$

where nl are the eigenstates, and m_e and m_h are the masses of the electrons and holes, respectively. Reducing R increases the energy level spacing, i.e., causes a blue shift. Quantum dots also have important nonlinear optical properties.

See also: Sintering

Bibliography

Baughman R H et al. 1999 Carbon nanotube actuators. *Science* **284**, 1340–4

Bernholc J, Brabec C, Buongiorno Nardelli M, Maiti A, Roland C, Yakobson B I 1998 Theory and growth and mechanical properties of nanotubes. *Appl. Phys. A* **67**, 39–46

Borel J-P 1981 Thermodynamical size effect and the structure of metallic clusters. *Surf. Sci.* **106**, 1–9

Cahn R W 1965 *Physical Metallurgy*. North-Holland, Amsterdam

Ebbesen T W 1997 *Carbon Nanotubes: Preparation and Properties*. CRC Press, Boca Raton, FL

Edelstein A S, Cammarata R C (eds.) 1996 *Nanomaterials: Synthesis, Properties, and Applications*. Institute of Physics, Bristol, UK

Edelstein A S, Chow G M, Altman E I, Colton R J, Hwang D M 1991 Self-arrangement of molybdenum particles into cubes. *Science* **251**, 1590–2

Feynman R P 1960 *Eng. Sci.* **23**, 22; reprinted 1992 There's plenty of room at the bottom. *J. Micromech. Syst.* **1**, 60–6

Garcia N, Nieto-Vesperinas M, Rohrer H (eds.) 1997 *Nanoscale Science and Technology*. Kluwer, Dordrecht, The Netherlands

Granqvist C G, Buhrman R A 1976 Ultrafine metal particles. *J. Appl. Phys.* **47**, 2200

Iijima S 1991 Helical microtubules of graphitic carbon. *Nature* **354**, 56–8

de Jongh L J (ed.) 1994 *Physics and Chemistry of Metal Cluster Compounds*. Kluwer, Dordrecht, The Netherlands

Kaatz F H, Chow G M, Edelstein A S 1993 Narrowing sputtered nanoparticle size distributions. *J. Mater. Res.* **8**, 995–1000
Kear B H, McCandlish L E 1993 Chemical processing and properties of nanostructured WC–Co materials. *Nano-Structured Mater.* **3**, 19–30
Klug H P, Alexander L E 1974 *X-ray Diffraction Procedures for Polycrystalline and Amorphous Materials*, 2nd edn. Wiley, New York
Knight W D, Clemenger K, de Heer W A, Saunders W A, Chou M Y, Cohen M L 1984 Electronic shell structure and abundance of sodium clusters. *Phys. Rev. Lett.* **52**, 2141–3
Konyashin I, Fox-Rabinovich G 1998 Nanograined titanium nitrided thin films. *Adv. Mater.* **10**, 952–5
Kroto H W, Heath J R, O'Brien S C, Curl R F, Smalley R E 1985 C_{60}: Buckminsterfullerene. *Nature* **318**, 162–3
Lai S L, Carlsson J R A, Allen L H 1998 Melting point depression of Al clusters generated during the early stages of film growth: nanocalorimetry measurements. *Appl. Phys. Lett.* **72**, 1098–100
Martin T P, Näher U, Schaber H, Zimmermann U 1993 Clusters of fullerene molecules. *Phys. Rev. Lett.* **70**, 3079–82
Melinon P et al. 1995 From free clusters to cluster-assembled materials. *Int. J. Mod. Phys. B* **9**, 339–97
Murray C B, Kagan C R, Bawendi M G, Kagan C R 1995 Self-organization of CdSe nanocrystallites into three-dimensional quantum dot superlattices. *Science* **270**, 1335–8
Pileni M P 1993 Reverse micelles as microreactors. *J. Phys. Chem.* **97**, 6961–73
Rinzler A G et al. 1998 Large-scale purification of single-wall carbon nanotubes: process, product, and characterization. *Appl. Phys. A* **67**, 29–37
Tans S J et al. 1997 Individual single-wall carbon nanotubes as quantum wires. *Nature* **386**, 474–7
Taylor R (ed.) 1995 *The Chemistry of Fullerenes*. World Scientific, Singapore
Woltersdorf J, Nepijko A S, Pippel E 1981 Dependence of lattice parameters of small particles on the size of the nuclei. *Surf. Sci.* **106**, 64–9
Wong E W, Sheehan P E, Lieber C M 1997 Nanobeam mechanics: elasticity, strength and toughness of nanorods and nanotubes. *Science* **277**, 1971–5

A. S. Edelstein

Nickel Alloys: Thermal Treatment and Thermomechanical Processing

The availability of useful product forms is the primary goal of the production and thermomechanical processing of nickel alloys. Many product shapes are manufactured, including flat products such as plate, strip and sheet, solid sections such as rounds, octagons, and squares, and hollow sections such as seamless tubing. These shapes are useful only if the application requirements including dimensional specifications and physical and mechanical properties are met. The chemistry and microstructure of the alloys control these properties.

Production of nickel alloys involves a sequence of manufacturing processes that convert raw material into the several product forms. A schematic diagram showing the overall material flow and production steps is portrayed in Fig. 1. Manufacturing starts with primary melting of raw materials and scrap.

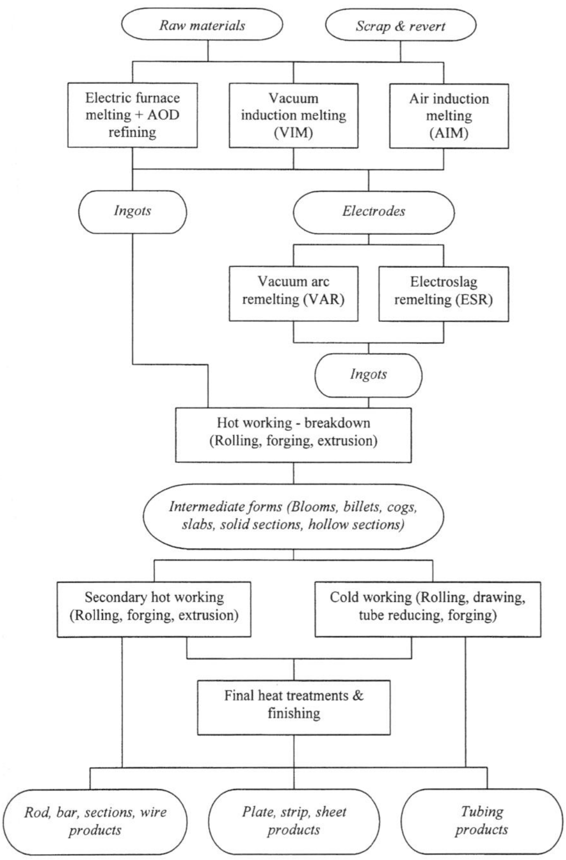

Figure 1
General process flow for manufacture of nickel alloy product forms.

This may be followed by a remelting process to produce improved ingots for the start of the mechanical working processes. The ingots are then hot-worked to yield intermediate products more suitable to subsequent working. Secondary hot-working or cold-working is used to shape the alloys into their final form. Heat treatments and finishing operations, such as straightening and surface conditioning, complete the production cycle.

1. Primary Melting

The primary melting and casting operations produce ingots of the required chemical composition and alloy microstructure. Chemistry control is paramount, not only for the major alloying elements, but also for active elements such as aluminum, titanium, and contaminants such as lead, bismuth, arsenic, and sulfur. The ingot structure should be uniform and clean. Unfortunately, segregation of alloying elements due to gravity and solidification parameters can occur on both macro and micro scales. Casting practices aim to minimize the presence of inclusions, internal stress, and ingot cracking.

Three melting and refining methods are used in primary melting: (i) the combination of electric furnace (EF) melting followed by argon–oxygen–decarburization (AOD) refining; (ii) air induction melting (AIM), and (iii) vacuum induction melting (VIM).

The EF–AOD method has several advantages for alloy melting. The electric furnace provides a molten bath of an approximate alloy composition in preparation for further chemistry refining and tailoring in the AOD process. Because the two operations are performed in sequence and the electric furnace can achieve very rapid melting, the EF–AOD combination can produce large batches of alloy efficiently. The presence of a slag cover and the injection of gases such as oxygen allow significant refining of the melt bath to occur during AOD. Thus, large amounts of scrap material can be used as input to the process. Another result of the refining action is that the EF–AOD process is not well suited to producing alloys containing large amounts of active elements such as aluminum and titanium. These elements tend to be lost to the slag during the refining action. Nickel–chromium–iron alloys are typically melted using the EF–AOD combination.

The AIM process can produce relatively high-purity alloys. This is a melting rather than a refining process and it requires cleaner raw material input than for the EF–AOD. Therefore, scrap use is limited. Melting is performed open to air and there is essentially no slag present. Composition adjustments must be made by raw material additions to the charge as only minimal change is possible using this technique. Nickel–copper alloys are typically produced by

AIM. Alloys containing active or volatile elements can be made by AIM if these elements are added at the end of the process just prior to casting. This minimizes the time for oxidation and/or volatilization loss.

The primary advantage of VIM is that the process is carried out in a vacuum chamber. The batch size is limited to about 20–25 tons because the entire operation must be carried out in the chamber. Major chemistry adjustments are achieved by additions to the charge. Due to the vacuum environment, degassing and removal of volatile contaminants occurs. High-purity alloys containing large levels of active elements are produced using this technique. Nickel–chromium alloys and nickel-based superalloys use VIM as the primary melting process.

2. *Remelting*

Remelting processes are used to improve the quality of ingots cast from primary melting. This benefits alloy chemistry as well as facilitating the hot-working processes. Since the remelting processes involve only a small molten pool at any given time, chemical segregation is minimized and the alloy grain size is controlled. Minor changes in chemical composition can also be effected. Nickel alloys with more complex chemistries are usually remelted.

The most common remelting techniques for nickel alloys are vacuum arc remelting (VAR) and electroslag remelting (ESR). Both techniques are consumable electrode melting processes in which the heat is generated by an electric arc between the ingot electrode cast from the primary melting process and some type of base plate located at the bottom of the mold. VAR involves sequential melting of the electrode and is performed in a vacuum. Exposure of the molten droplets to the reduced pressure decreases the amount of dissolved gases and volatilizes a portion of the tramp elements in the alloy. ESR also involves sequential melting of the electrode. It is performed in air, but a conductive slag is present between the melting electrode and the solidifying ingot. The molten droplets are thus exposed only to the slag. Refining of the composition and minor chemical composition adjustments occur as a result.

3. *Hot Working*

Thermomechanical processing of nickel alloys is used to change the shape of the cast ingots into the final product form. Such processing includes hot working, cold working, and various types of thermal and heat treating techniques. Hot working is deformation performed at temperatures and strain rates where simultaneous recrystallization and deformation occur and strain hardening is avoided. All ingots are subjected to hot working for the initial deformation after melting and/or remelting. This is called ingot breakdown. One purpose of this process is to reduce the ingot to a size and section that is appropriate for later hot or cold working. Breakdown also decreases chemical segregation and refines the alloy microstructure.

Three types of hot working are used for breakdown—rolling, forging, and extrusion. The process selection depends on the facilities available, alloy chemistry, and the shape of the finished form. Facilities include not only the mechanical working equipment itself, but also the type and capacity of the furnaces needed to heat the ingots before and intermittently between working sessions. Alloy chemistry affects the selection by defining the incipient melting temperature, hot working temperature range, microstructural stability, and deformation flow stress. Product shape is critical when the form is hollow, such as tubing.

Hot rolling reduces the section of the ingot by passing it through driven rotating rolls. Typical intermediate shapes produced by rolling include blooms, billets, rods, and plates. Breakdown forging reduces the section of the workpiece by compressive force between dies. This is done by sequential deformation along the length of the workpiece. Because forging requires significant time, one or more reheats are required to maintain the alloy within the desired temperature range during working. Computer-controlled mechanical manipulators are commonly used to rotate the material during forging to minimize workpiece chill due to die contact during working. Extrusion converts the workpiece into an intermediate length with uniform cross section by forcing the alloy through a die. If a hollow section is desired, a mandrel is used to form the interior hole. Extrusion is usually the preferred process when the alloy has a limited hot working temperature range and/or a high resistance to deformation.

Secondary hot working produces some finished product forms. Where hot working is the final deformation process, the dimensional tolerances of the product are relatively broad, or a subsequent finishing operation such as machining is used to produce the required dimensions. The most common of these forms are hot-rolled plate and hot-rolled or forged rod. Hot extrusion is used to produce larger-sized tubing where dimensions are not critical.

4. *Cold Working*

Cold working is deformation in an alloy accompanied by work hardening—that is, increased strength. Cold working is used as the final working step when tight dimensional tolerances and manipulation of alloy properties are required. Adjusting the amount of deformation and the rate at which the deformation is performed can control both strength and ductility. Cold working also produces a good surface finish.

Common cold-working processes used for nickel alloys are rolling, drawing, tube reducing, and forging. Factors affecting the choice of process include alloy characteristics, equipment availability, and shape of the finished form. Alloy characteristics include work hardening rate, strain rate sensitivity, and deformation flow stress. Equipment considerations are similar to those for hot working with the added concern for lubrication between the equipment and workpiece. Rolling is used to produce flat products such as sheet and strip. It can also be used to produce rods. Drawing can produce sections, i.e. rounds and hexagons, as well as hollows. Tube reducing is normally used only for hollow shapes and tubing.

5. *Heat Treatment and Finishing*

Several types of heat treatments are used to process nickel alloys. These include homogenization, stress relieving, annealing, and aging. Homogenization is typically used to treat ingots after melting to decrease or eliminate chemical segregation. These high-temperature heat treatments are often quite long, e.g., greater than 24 hours, due to the size of the ingots and the extent of segregation. Stress relief treatments are used to reduce casting strains or to decrease residual stress resulting from cold work. Annealing treatments involve elevated temperature exposures to remove the effects of cold work or promote the dissolution of precipitates in the microstructure to form a solid solution. Alloys are usually softened by the annealing treatments. Sheet and strip are often given bright anneals as the final process. This annealing is performed in a protective atmosphere to prevent discoloration of the smooth rolled surface. Aging, also called precipitation-hardening, is used to precipitate beneficial constituents in alloy microstructures. Aging treatments are usually performed after annealing or cold working to obtain the desired combinations of strength and ductility in the product.

Special conditioning operations are used to remove surface oxides, achieve a required surface finish, or obtain the final shape dimensions. Pickling is a chemical process that is used to remove oxides from hot-worked material. When the nickel alloy contains a significant level of chromium, a fused salt pretreatment must often be used prior to pickling to facilitate the oxide removal. Surface blasting is also used to remove hot-work oxides. The blast medium for nickel alloys is usually a fine oxide grit or stainless steel shot. Machining is the operation of choice for obtaining the required final shape dimensions. Machining operations include turning, grinding, and buffing.

See also: Nickel-based Superalloys: Alloying Methods and Thermomechanical Processing

Bibliography

Betteridge W, Heslop J 1974 *The Nimonic Alloys*, 2nd edn. Crane, Russak & Co, New York, pp. 115–56

Mankins W R, Lamb S 1990 Nickel and nickel alloys. In: *Metals Handbook*, 10th edn. ASM International, Metals Park, OH, Vol. 2, pp. 428–45

Tundermann J H, Tien J K, Howson T E 1996 *Kirk–Othmer Encyclopedia of Chemical Technology*, 4th edn. Wiley, New York, Vol. 17, pp. 1–17

J. H. Weber

Nickel-based Superalloys: Alloying Methods and Thermomechanical Processing

Nickel-based alloys are multicomponent alloys with carefully balanced compositions. These alloys are used under severe operating conditions and must function at temperatures that are high percentages of their melting temperatures. Such compositions with high levels of active and hardener elements are inherently more difficult to produce. Composition control is critical and mechanical working capabilities are often very limited. In addition to the processing limitations, achievement of the desired properties requires unique alloy microstructures that are carefully designed and require sophisticated manufacturing technologies. The goal of alloy processing and manufacture is to produce parts that incorporate the appropriate combinations of shape configuration with the required composition and designed microstructures to meet the needs of the selected application.

The production of nickel-based superalloys includes several production paths. These sequences of processes are dependent on the alloy composition and the part or product form being produced. Three general methodologies are used: investment casting, powder metallurgy, and the more conventional wrought processes. Investment casting is used for complex shape parts made from alloys with high levels of active elements. The alloys are directly cast into near-net or net shapes. With variation of the solidification practices, the microstructural grain morphology can be equiaxed, directional, or even single crystal. Powder metallurgy is used to produce alloy compositions that are either difficult to cast or cannot be made by any type of conventional melt + cast practice. An additional complication when using powder metallurgy is the need to consolidate the powder produced into useful forms with acceptable properties. Conventional wrought manufacturing practices are used to produce the simplest, that is, lean, superalloy compositions. Wrought manufacturing processes for nickel-based superalloys must be more carefully controlled than those for nickel engineering alloys. However, the range of products for superalloys includes all of the generic forms—plate, sheet, strip, and sections such as rounds and octagons. Figure 1 shows a schematic for the

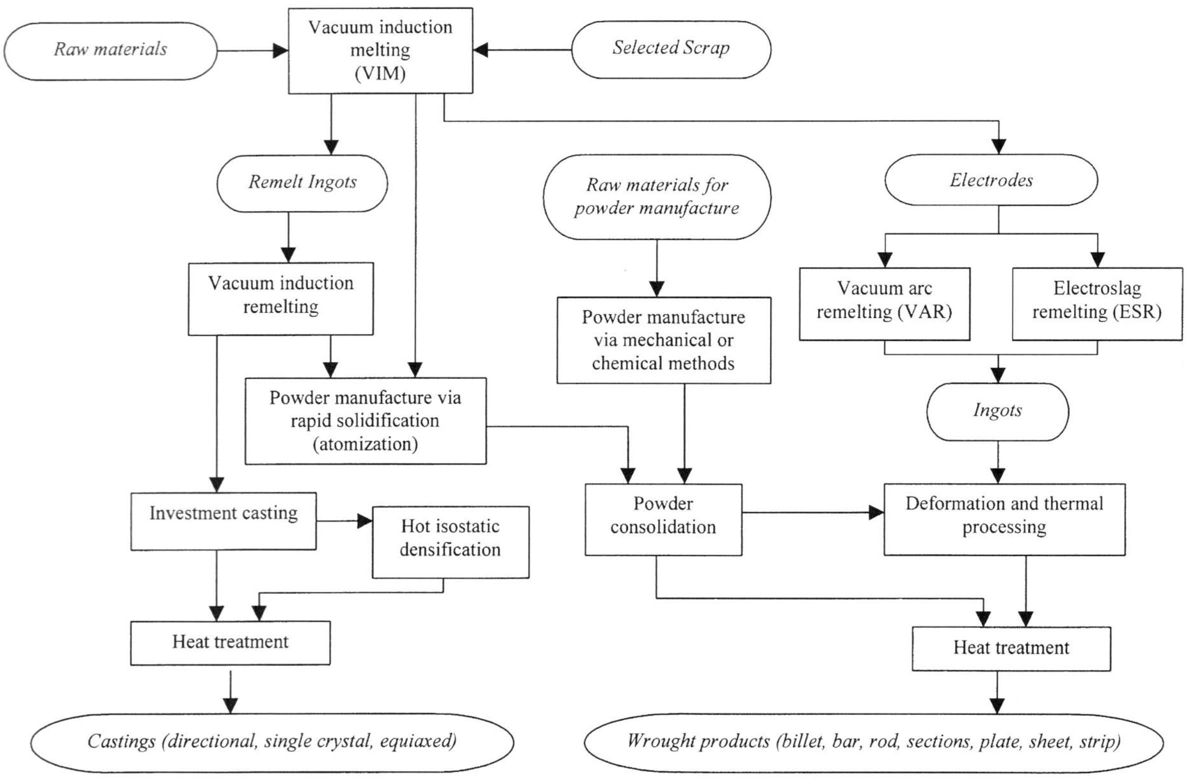

Figure 1
General process flow for manufacture of nickel-based superalloy product forms.

overall material flow and production steps for the manufacture of cast, powder metallurgy, and wrought nickel-based superalloys.

1. *Cast Alloy Processing*

Alloys that are used as complex shapes such as gas turbine engine blades and vanes are normally produced by direct casting to finished shape. These applications also have severe service requirements. This combination of shape and performance severity can only be met by the highly alloyed compositions. These alloys have limited ductility at both ambient and elevated temperatures and, thus, are not amenable to mechanical working processes to form the desired part geometries. The general processing sequence to manufacture these cast parts includes the following steps: primary melting, remelting, investment casting, heat treatment, and final surface conditioning or coating.

1.1 Melting

The highly alloyed nickel-based superalloys are almost always produced by vacuum induction melting (VIM). Melting, alloying, purification, and solidification are all carried out under vacuum. Metallurgical considerations and the required alloy cleanliness dictate this method. Chemistry control is required for major alloying elements, for active elements such as aluminum, titanium, niobium, and vanadium, and for tramp contaminants including lead, bismuth, tin, and others. Excellent alloy cleanliness and deoxidation are required to achieve acceptable performance in use.

Vacuum induction melting is a highly flexible process that allows independent control of bath temperature, chamber pressure, melt bath stirring, and cycle time. The VIM practice begins with melting of the prepared charge. This initial charge typically contains the major alloying elements that are not reactive. After meltdown, the bath temperature is adjusted. The deoxidation reaction, sometimes called the "carbon boil," begins. This gas evolution, refining reaction is the result of the combination of carbon with oxygen from the molten bath to form carbon monoxide gas. It is a heterogeneous reaction that occurs preferentially along the walls of the vessel and is promoted by electromagnetic stirring of the bath. The nitrogen content of the bath may also be decreased at the same time due to scavenging by the CO gas as it passes through the bath. Volatilization of certain tramp elements such as lead, silver, bismuth, selenium, and tellurium occurs due to the vacuum environment above the molten bath. The final stage of the melting process is the addition of the active elements such as aluminum, titanium, zirconium, and hafnium, followed by temperature adjustment in preparation for pouring and casting.

Pouring and casting are performed under vacuum. Casting is usually performed in a second chamber adjacent to the melting one. The molten alloy is conveyed to the molds using a ceramic-lined trough. The pouring rate is restricted to allow heterogeneities to rise to the surface in the molds before solidification. When higher pouring rates are desired or the bath exhibits gross cleanliness problems, ceramic foam filters are sometimes used to clean the molten stream before casting. Remelt bars for investment casting are the most common form for the VIM cast product. In a few instances, the VIM bath is directly converted into castings or atomized powder.

Separate VIM remelting furnaces are commonly used for the investment casting process and production of atomized alloy powders. This separation facilitates better control of the subsequent secondary casting and atomization processes. The casters and atomizers are sometimes at separate facilities or even different companies.

1.2 Investment Casting

Production of complex shapes with good dimensional control is achieved using the investment casting process. Investment casting is the casting of parts into ceramic shell molds. Preparation of the molds for investment casting includes: manufacture of dies having negative images of the part shapes; wax injection to form positive images of the parts; assembling wax forms with casting gates and risers; multiple sequences of slurry coating and addition of dry ceramic powder to the wax assemblies; heating to remove the wax forms; and firing the ceramic to produce a rigid shell mold. These molds are then placed in a vacuum chamber where the alloys are melted and poured into the molds. After casting the parts are separated, cleaned, straightened, and inspected.

Cast parts with microstructures comprising equiaxed grains, directional grains, or single crystals can be produced using variations in the process parameters. Equiaxed parts are produced by using mold face coats to enhance grain nucleation, low pouring temperatures, and control of the mold preheat temperature, mold thickness and assembly configuration, and cooling rate after casting. Directionally solidified structures are produced by incorporating a secondary withdrawal furnace during casting and using chill plates at the ends of the molds. These allow careful control of the thermal gradient in the mold and crystal growth rate. Production of single crystal cast parts is the most complex. Here all of the controls for producing directional structures are augmented by seed crystals or helical segments in the mold to promote the crystal growth orientation.

Hot isostatic press (HIP) densification is an optional step in the production of equiaxed cast parts. Solidification of these parts can allow the formation of porosity due to shrinkage in the mold. Such porosity can be deleterious to part performance in service. Closure of this porosity is achieved by undergoing HIP at controlled combinations of temperature, pressure, and time.

1.3 Heat Treatment

Multistage heat treatments are given to superalloy products and parts to develop alloy microstructures appropriate for the intended service application. Typically, this involves a sequence of a solution anneal followed by one or more precipitation aging treatments. The solution anneal is performed to allow some or all of the carbides, and other constituents such as γ' and γ'' phases, to go into solid solution. This anneal is usually followed by some type of rapid cooling to prevent dissolution and reprecipitation. This prepares the alloy for later heat treatments where control of temperature, time, and cooling rate promotes selective precipitation of the various phases to the desired locations in the microstructure. For example, precipitation of carbides is often directed towards grain boundaries, whereas precipitation of the intermetallic phases γ' and γ'' is primarily within the grain interior. Duplex aging treatments can be used to develop hierarchies of precipitates with different size and size distributions within the grains.

2. *Powder Metallurgy Processing*

Powder metallurgy methods are used to manufacture nickel-based superalloys that are very highly alloyed and are difficult to cast or cannot be made by conventional casting technologies. Powder metallurgy can overcome casting and processing limitations and can produce a finer controlled grain size. Disadvantages attendant with using powder metallurgy are concerns regarding gas content of the powders produced via atomization, the potential for nonmetallic inclusions, and the existence of prior particle boundary (PPB) oxides and carbides in the consolidated product. The general processing sequence for the manufacture of powder metallurgy components includes powder production, powder screening and cleaning, consolidation, thermomechanical processing, and heat treatment.

2.1 Powder Production

Atomization is powder production using melting techniques. The most common form of atomization is inert gas atomization. Remelt bars from primary VIM melting are remelted in a secondary VIM furnace. The molten alloy is poured into a ceramic lined tundish with a small orifice in the bottom. Alloy flowing through the orifice at a controlled rate is disintegrated into fine droplets by impingement of high-pressure inert gas, typically argon. The alloy droplets freeze during their fall to the bottom of the atomizing chamber. Powder size and size distribution are controlled by varying the chamber pressure, gas pressure, gas flow rate, and gas nozzle configuration.

Atomized powder is also produced using centrifugal methods. An alloy feed electrode is melted on the end by an electric arc or plasma. Centrifugal force is created either by high rotational speed of the electrode being melted, or by high rotational speed of a chill plate under the electrode. In either situation, powder is formed by the controlled melt of the electrode and the disintegrating force of the rotating component.

Mechanical and chemical methods of producing powder are the only practical means to produce oxide dispersion-strengthened (ODS) nickel-based superalloys. The only mechanical powder production method in commercial production is mechanical alloying (MA). MA is used to incorporate a controlled fine dispersion of oxides into complex, highly alloyed alloys. This process can yield essentially any composition as long as one of the input components is slightly compressively deformable. It is a dry milling process wherein the input raw material powder components are subjected to repetitive welding and fracture. The resultant composite powder has the nominal composition of the raw material charge containing the uniformly dispersed oxides. To preserve cleanliness, the milling is performed in a controlled atmosphere. Very close chemistry control can be obtained using MA. Chemical processing involves precipitation of fine alloy powder from aqueous solutions. The compositions of these powders are limited to simple matrices such as nickel–chromium and a single oxide such as thoria. Complex compositions with high levels of active elements are not possible via chemical processing.

Sizing and cleaning are performed on the powders after manufacture. Sizing is done by sieving to remove the coarsest and/or finest powder fractions. This process optimizes the size distribution of the powder materials for subsequent consolidation. Selected sieving is sometimes used to remove contaminated powder particles. Chemical cleaning of powders may be used to remove contaminated powder surface layers and minimize later PPB effects.

2.2 Consolidation

Compacting the powder particles into solid forms involves three steps: encapsulation of the powder in a container; evacuation of the sealed container to remove contained gas; and the compaction process itself. The container for the powders is normally made of mild steel, but glass and ceramic containers are also used. After sealing, the containers are evacuated to remove interparticle gases and reduce particle surface contamination. Evacuation made be done at either ambient or elevated temperatures.

The most common consolidation practice for superalloy powders is hot isostatic pressing (HIP). This is performed in an autoclave, a large pressure vessel, using isostatic gas pressure to densify the powder into a solid. The control parameters during HIP are the temperature, gas pressure, and process time. The consolidated form is dependent upon the shape and design of the powder container. Shapes range from simple round cylinders to relatively complex three-dimensional disk configurations. Canned powders are also consolidated by hot extrusion and forging. The advantage of these mechanical working–consolidation practices is better interparticle bonding due to particle deformation and surface shearing. Extrusion and forge consolidation is typically limited to relatively simple shapes.

2.3 *Thermomechanical Processing*

The fine grain structure and the lack of chemical heterogeneity allow consolidated powder metallurgy forms to exhibit higher hot ductility and lower deformation stresses than their cast counterparts. Limited mechanical working is thus possible. Conventional metalworking processes under controlled conditions are used to produce the desired shapes. A more detailed discussion of metalworking is presented below in Section 3.

2.4 *Heat Treatment*

Multistage heat treatments are given to powder metallurgy superalloy products to develop alloy microstructures appropriate for the application. Heat treatment involves a sequence of a solution anneal followed by one or more precipitation aging treatments. The solution anneal is performed to allow carbides and precipitated constituents to go into solid solution. This anneal is usually followed by some type of rapid cooling to prevent dissolution and reprecipitation. Directional heat treatments can be used for ODS alloys to promote the formation of very coarse, elongated grains. These anneals and treatments prepare the alloy for later heat treatments where control of temperature, time, and cooling rate promotes selective precipitation of the various phases to the desired locations in the microstructure. Single or multistage aging treatments are then used to develop the desired precipitate size and size distributions.

3. *Wrought Superalloys*

Wrought superalloys are produced using the same conventional processing techniques as the engineering nickel alloys. The general process sequence includes melting, remelting, hot working, some cold working for sheet and strip products, and final heat treatments. A detailed presentation of these techniques is given in *Nickel Alloys: Thermal Treatment and Thermomechanical Processing*. The discussion below will focus on the process differences compared with those for nickel alloys.

3.1 *Melting and Remelting*

Vacuum induction melting is the preferred melting method for superalloys due to the specifications for alloy chemistry and purity. Commercial grade superalloys are melted using select raw materials, some foundry or other revert, and lower purity elemental raw materials. Aerospace grade superalloys also use select materials, but the revert material, if used at all, is purposely cleaned prior to melting. Premium quality grade alloys are produced only from the highest quality raw materials. In addition, premium quality methods require very tight control of both the VIM melting and remelting procedures.

Superalloys may also be produced by electron beam melting. This specialized melt process involves melting the charge under high vacuum using an electron beam. The charge is sequentially melted on a water-cooled hearth with no refractories. Very high purities are achievable due to the vacuum, sequential melting, and slag flotation during melting. This newer melting process is costly and its use to date is limited.

3.2 *Hot and Cold Working*

Hot and cold working of nickel-based superalloys is used to change the shape of the cast ingots into the final product form. All ingots are subjected to hot working. Three types of hot working are used: forging, rolling, and extrusion. Hot working destroys the ingot structure and replaces it with an improved one having better grain structure and chemical homogeneity. The highly alloyed compositions of superalloys limit the hot working temperature range and several reheats are typically used during the working process. In addition to uniform section rod, hot forging is used to produce shapes having varying section size along the workpiece length. Extrusion of nonround or rectangular sections is quite common. These sections are sometimes cut into short lengths and forged into complex shapes.

Cold rolling of superalloys produces sheet and strip with good dimensional control and a smooth surface finish. These product forms are only obtained with the leaner superalloy compositions. Even with these alloys, the amount of cold work between softening anneals is limited and numerous anneals are required.

3.3 *Heat Treatment*

Wrought superalloys are given the same multistage heat treatment sequences as are cast and powder metallurgy superalloys. The overall heat treatment sequence involves a solution anneal followed by one or more precipitation aging treatments. Carbides and precipitated constituents are dissolved into the solid solution during annealing. Anneals are followed by some type of rapid cooling to prevent reprecipitation. Single or multistage aging treatments with controlled temperature, time, and cooling rate cycles promote precipitation of the various phases in the microstructure.

3.4 *Finishing Processes*

Finishing operations for all types of nickel-based superalloys include not only the typical pickling, machining, and blasting but also sophisticated coatings. The severe conditions these materials must endure often require protection of the parts with even more highly alloyed or ceramic coatings to prevent environmental surface degradation.

See also: Nickel Alloys: Thermal Treatment and Thermomechanical Processing

Bibliography

Benjamin J S 1986 *Mechanical Alloy for Preparation of Superalloys, Advanced High Temperature Alloys: Processing and Properties*. American Society for Metals, Metals Park, OH, pp. 85–91

Betteridge W, Heslop J 1974 *The Nimonic Alloys*, 2nd edn. Crane, Russak & Co, New York, pp. 115–56

Erickson G L 1990 Polycrystalline cast superalloys. In: *Metals Handbook*, 10th edn. ASM International, Materials Park, OH, Vol. 1. pp. 981–94

Gell M, Kortovich C S, Bricknell R H, Kent W B, Radavich J F (eds.) 1985 *Superalloys 1984*. The Metallurgical Society of AIME, Warrendale, PA

Harris K, Erickson G L, Schwer R E 1990 Directionally solidified and single-crystal superalloy. In: *Metals Handbook*, 10th edn. ASM International, Materials Park, OH, Vol. 1, pp. 995–1006

Stoloff N S 1990 Wrought and P/M superalloys. In: *Metals Handbook*, 10th edn. ASM International, Materials Park, OH, Vol. 1, pp. 950–80

J. H. Weber

Nuclear Reactor Fuel Fabrication (Including Quality Control)

The world's first self-sustaining nuclear fission chain reaction was demonstrated in December 1942 by Fermi and his co-workers in the Chicago "pile" by using some 4 tons (4.1×10^3 kg) of natural uranium fuel in the crude form of lumps of metal and oxide distributed throughout a matrix of 400 tons (4.1×10^5 kg) of graphite. Thereafter, a number of metallic, ceramic, and dispersion fuels, in both pin and plate geometries were designed, fabricated, and irradiated in the 1950s and 1960s in a variety of reactor systems with different neutron energies and with different combinations of coolant, moderator, and cladding materials from which viable fuel systems have emerged for power reactors and research reactors. The nuclear fuels industry is now established in several parts of the world catering for the requirements of more than 300 research reactors that are used for production of radioisotopes, materials irradiation, basic research, training, and education in 65 countries, and some 430 nuclear power reactors with a total installed capacity of 350 GWe in 31 countries.

This article reviews the fabrication of conventional and advanced fuels for the commonly used nuclear research reactors, namely the materials test reactor (MTR) and TRIGA (training, research, isotopes, general atomic); and the popular nuclear power reactors, namely the light water-cooled reactor (LWR), the pressurized heavy water reactor (PHWR), and the liquid metal-cooled fast breeder reactor (LMFBR). LWRs are of two main types: the pressurized water reactor (PWR) and the boiling water reactor (BWR). These are used extensively all over the world and are in operation in the USA, France, Japan, Germany, Sweden, Belgium, Switzerland, Russia, Ukraine, and South Korea. The Russian version of the PWR is known as VVER. The PHWR is the backbone of the nuclear power programs in Canada (where it is known as CANDU) and India. There are very few LMFBRs, but they are likely to play a major role in the coming years.

Uranium and thorium, the basic materials for "nuclear fuels," are radioactive and among the heaviest elements occurring in nature. Uranium occurs natural mainly in two isotopic forms, the "fertile" ^{238}U (99.3%) and the "fissile" ^{235}U (0.7%). Thorium occurs in nature almost entirely as the "fertile" ^{232}Th isotope. ^{235}U is the only fissile isotope occurring in nature. The two synthetic fissile isotopes, ^{239}Pu and ^{233}U, are usually produced by neutron capture reactions of ^{238}U and ^{232}Th, respectively, in nuclear reactors. ^{239}Pu is often associated with ^{241}Pu, which is also a fissile isotope.

Nuclear fuels are made up of one or more fissile isotopes, namely ^{235}U, ^{239}Pu, and ^{233}U, often in combination with a fertile isotope, ^{238}U or ^{232}Th, and are usually in the solid form of metals, alloys, oxides, carbides, nitrides, silicides, etc., of uranium, plutonium, and thorium. The term "fuel" defined as a material for producing heat is in the true sense not applicable to nuclear fuel, since the heat in this case is generated not by burning but by fission or disintegration of the fissile isotopes. Since uranium, plutonium, thorium, and their fission products are radioactive and hazardous to health to varying degrees, nuclear fuels are hermetically sealed inside a nonradioactive structural material, known as cladding. The cladding is an integral part of the nuclear fuel element. It serves as primary containment for radioactive fission products, acts as a barrier avoiding direct contact of the fuel with coolant, and transfers fission heat energy from fuel to coolant. Nuclear fuel elements are usually manufactured in the shape of plates, pins, or rods and are assembled in specific geometric configurations using spacers, end fittings, and other supporting hardware. The package of fuel elements is called a fuel assembly.

1. *Fuels for Research and Power Reactors*

Nuclear research and power reactors operate mainly by the fission of ^{235}U by slow or thermal neutrons. The ^{239}Pu by-product from these thermal reactors is the best fissile material for LMFBRs. However, since the LMFBRs are yet to be deployed commercially, the plutonium is recycled in the interim in several LWRs in France, Germany, Belgium, Switzerland, and Japan. ^{233}U is the ideal fissile material for thermal reactors and it is possible to have a self-sustaining ^{232}Th–^{233}U fuel cycle. ^{233}U fuels have so far been manufactured only on a laboratory scale, primarily for irradiation testing experiments. Table 1 summarizes the conventional and advanced fuels and cladding materials for research and power reactors.

The conventional fuels for LWRs and PHWRs are slightly enriched uranium (1.5–5% ^{235}U) and natural uranium, respectively, in the form of high-density cylindrical UO_2 "pellets," which are stacked inside zirconium alloy cladding tubes and hermetically sealed. Many aspects of uranium oxide fuel are reviewed in Belle and Lustman (1958). In VVER fuel, the uranium oxide pellets have a central hole. For BWRs and PWRs, zircaloy 2 and zircaloy 4 are used as cladding materials, respectively. The Russian VVER and RBMK reactors use Zr–1% Nb as cladding material. The quaternary Zr–Nb–Sn–Fe alloys are being developed as advanced cladding materials for very high burn-up LWR fuels. Zircaloy 4-clad very high-density ($\geqslant$96% theoretical density (TD)) natural uranium oxide is the reference fuel for PHWRs. The inner surface of the PHWR cladding tube is coated with graphite. The advanced PHWR

Table 1
Conventional and advanced fuels for MTRs, LWRs, PHWRs, and LMFBRs.

Reactor	Conventional	Advanced/alternative fuel
Research reactors		
MTR	HEU (>93% ^{235}U): (i) Al–UAl_x (density ≤ 1.7 gU cm^{-3}); (ii) Al–U_3O_8 (density ≤ 1.3 gUcm^{-3})	LEU (<20% ^{235}U): (i) Al–UAl_x (density = 2.3 g U cm^{-3}); (ii) Al–U_3O_8 (density = 3.2 g U cm^{-3}); (iii) Al–U_3Si_2 (density = 4.8 g U cm^{-3}); (iv) Al–U_3Si (density = 6.0 g U cm^{-3}); (v) Al–U–Mo and Al–U–Zr–Nb (density = 7–8 g U cm^{-3})
TRIGA	Aluminum- or stainless steel-clad $UZrH_x$ pin assembly (density = 0.5 gU cm^{-3})	Aluminum- or stainless steel-clad $UZrH_x$ pin assembly (density ≤ 3.7 g U cm^{-3})
LWR: BWR, PWR, and VVER		
Fuel	LEU (≤5% ^{235}U) as UO_2	LEU (5–10% ^{235}U) mixed uranium–plutonium oxide (≤10% PuO_2)
Cladding	Zircaloy 2 (BWR), Zircaloy 4 (PWR)	Zr–Sn–Nb–Fe and Zr– Nb–O alloys
Burn-up	$2 \times 10^4 - 3 \times 10^4$ MWD/t	High: ≤6×10^4 MWD/t; ultrahigh: ≤9×10^4 MWD/t
PHWR		
Fuel	Natural UO_2	REU, SEU in the form of UO_2, $(Th,Pu)O_2$, and $(Th,^{233}U)O_2$, containing ≤2% fissile material
Cladding	Zircaloy 4	Zircaloy 4
Burn-up	6700 MWD/t	$1.5 \times 10^4 - 2 \times 10^4$ MWD/t
LMFBR		
Fuel	HEU in the form of UO_2 and $(U,Pu)O_2$ (≤25% Pu)	(U,Pu)C, (U,Pu)N, and U–Pu–Zr, (≤25% Pu)
Cladding	Stainless steel D-9	Stainless steel (HT-9 or oxide dispersed)
Burn-up	10^5 MWD/t	≤2×10^5 MWD/t
Breeding ratio	≤1.2	≤1.5

fuels aim at improved burn-up. Accordingly, recycled uranium (REU, 0.8–1.1% ^{235}U) oxide and slightly enriched uranium oxide, $(U,Pu)O_2$, $(Th,Pu)O_2$, and $(Th,^{233}U)O_2$, are being developed. The carbides and nitrides of uranium, plutonium, and thorium are not compatible with water and are hence not used as water-cooled reactor fuels.

Austenitic stainless steel (mostly type D-9)-clad mixed uranium–plutonium oxide (MOX) pellets containing between 15% and 25% PuO_2 have been the driver fuel for the first generation of demonstration and prototype LMFBRs in France, the UK, and Japan. The Russian LMFBRs use highly enriched uranium oxide pellets as fuel. Mixed uranium–plutonium monocarbide (MC), mononitride (MN), and U–Pu–Zr are considered as advanced LMFBR fuels on the basis of their superior breeding ratio, higher thermal conductivity, and better chemical compatibility with sodium coolant and stainless steel cladding compared to MOX fuel.

The HT-9 type and oxide-dispersed ferritic steels have emerged as advanced cladding materials for LMFBRs. India has been able to use mixed uranium–plutonium monocarbide driver fuel in their fast breeder test reactor (FBTR) with a hitherto untried plutonium-rich fuel composition (Ganguly and Rodriguez 2000). Metallic fuels, U–Zr and U–Pu–Zr, have been used mostly in the experimental breeder reactor (EBR-II) in the USA and are considered to be the reference fuels for the "inherently safe, integral fast reactor" operating on a proliferation-resistant fuel cycle with co-located facilities for fuel reprocessing and refabrication (Chang 1989, Stevenson 1987).

"Vi-pac" or "sphere-pac" fuel elements containing high-density microspheres of mixed uranium–plutonium oxide, monocarbides, and mononitrides are also considered as advanced LMFBR fuels. In vi-pac pins, the fuel microspheres of two or three size fractions are vibrofilled in a metallic cladding tube and encapsulated.

2. *Challenges in Manufacturing Nuclear Fuels*

The following are the unique features of nuclear fuels, which are taken into consideration during their fabrication.

2.1 *Radioactivity*

Uranium, plutonium, and thorium emit alpha particles and are radioactive to varying degrees. Hence, for radiation safety, materials bearing these elements are handled in a well-ventilated special operating area such that the radioactive aerosols are contained within the plant. ^{239}Pu and ^{233}U because of their high radiotoxicity and specific radioactivity are handled inside leak-tight glove boxes or hot cells, maintained at slightly negative pressure. Beta–gamma and neutron shielding are not needed for handling pure ^{239}Pu and ^{233}U isotopes. However, the ^{239}Pu obtained by reprocessing spent uranium fuels from power reactors contains significant quantities of other plutonium isotopes, namely ^{238}Pu, ^{240}Pu, ^{241}Pu, and ^{242}Pu, which release spontaneous neutrons (^{238}Pu, ^{240}Pu, and ^{242}Pu) and gamma rays (^{241}Pu decays to ^{241}Am, which is a strong gamma emitter). Likewise, the ^{233}U obtained by reprocessing spent thoria-bearing fuels or blanket materials always contains ^{232}U. Hence, beta–gamma and neutron shielding is necessary for handling plutonium and ^{233}U-bearing materials. In order to keep the radiation exposure of the plant personnel well below the permissible limits, great emphasis is given to remote working, automation, and minimization of radiotoxic dust hazards.

2.2 *Pyrophoricity*

Uranium and plutonium metal, monocarbide, and mononitride, and uranium silicide are highly susceptible to oxidation and hydrolysis and are pyrophoric in powder form. Hence, these materials are required to be handled in leak-tight enclosures under a high-purity inert cover gas (argon or nitrogen containing <25 ppm each of oxygen and moisture).

2.3 *Nonstoichiometry*

The dioxide, monocarbide, mononitride, and silicides of uranium and plutonium are nonstoichiometric compounds. Strict control of composition is necessary in order to ensure stability and to avoid chemical interaction between fuel and cladding and fuel and coolant. Hyperstoichiometric $(U,Pu)O_2$ fuel containing a higher percentage of plutonium oxide may cause fuel–clad and fuel–coolant chemical interactions. In addition, in the case of oxide or mixed oxide fuels, there is significant reduction in thermal conductivity even with slight deviation of the oxygen to metal ratio from 2.00. Monocarbides and mononitrides of uranium and plutonium have line compositions. A slight deviation from the stoichiometric ratio leads to an unwanted metallic second phase for hypostoichiometric compounds and higher carbide or higher nitride for hyperstoichiometric composition. The metallic second phase forms a low-melting eutectic with the iron and nickel of stainless steel cladding and the higher carbides and higher nitrides cause clad carburization or nitridation. Hence, only a controlled amount of the higher carbide second phase is acceptable in monocarbide fuel.

2.4 *Criticality*

Fissile materials are permitted to be handled only in limited quantities (a few kilograms) in order to avoid an uncontrolled nuclear fission chain reaction or what is known as a "criticality accident."

2.5 *Accountability*

Precise and real-time accounting of special nuclear materials (SNM: fissile and fertile) is necessary to prevent criticality accidents and also to avoid proliferation of SNM for nonpeaceful purposes. Hence, high security arrangements are needed for SNM.

3. *Fabrication of Research Reactor Fuel*

The common fuels for water-cooled nuclear research reactors are as follows:

- aluminum-clad, natural uranium metal in the form of rods or clusters of pins;
- aluminum alloy-clad, low-enriched uranium (LEU; 19.75%) as $Al–UAl_x$, $Al–U_3O_8$, or $Al–U_3Si_2$ in the form of plate fuel assemblies for MTRs; and
- aluminum- or stainless steel-clad, low-enriched uranium in the form of uranium zirconium hydride rods for TRIGAs.

3.1 *Uranium Metal Fuel*

The natural uranium metal fuels used in NRX (Canada) and CIRUS and Dhruva (India) reactors are typical of metallic uranium fuel and are fabricated in two steps (Vijayaraghavan 1989). First, a uranium metal button is prepared by either magnesiothermic or calciothermic reduction of uranium tetrafluoride in a steel mold or bomb.

In the second step, the uranium metal button is subjected to vacuum melting and casting at around 1300°C in an induction furnace using an yttria-coated graphite crucible and mold. The cast uranium metal ingot is evaluated by ultrasonic testing for intercepting any casting defects and subjected to conventional shaping treatments including extrusion, forging, rolling, swaging, drawing, machining, etc., to obtain the uranium fuel in the desired geometric shape and dimension. Usually, the uranium metal is hot rolled in the alpha phase (orthorhombic structure stable up to 665°C) and protected from oxidation by using fused salt or inert gas atmosphere. However, since alpha uranium has anisotropic properties, it is susceptible to irradiation growth.

In order to avoid this, the uranium metal is often alloyed with molybdenum or zirconium to retain the stable b.c.c. gamma phase structure at ambient temperature. The gamma phase is relatively soft and amenable to cold working.

3.2 Low-enriched Uranium Dispersion Fuels

The MTR was initially developed in the USA in the 1950s and was later used all over the world. Until 1977 high-enriched uranium (HEU; $\geqslant 85\%$ ^{235}U) was used as the fuel in these reactors in the form of aluminum alloy-clad Al–UAl_x and Al–U_3O_8 dispersion fuels with uranium densities of 1.7 g cm^{-3} and 1.3 g cm^{-3}, respectively, in both pin and plate geometries. In 1978 Argonne National Laboratory (ANL) in the USA initiated the program on reduced enrichment for research and test reactors (RERTR) in order to avoid diversion of HEU for nonpeaceful purposes. The primary objective of the program was to substitute HEU with low-enriched uranium (LEU; $\leqslant 20\%$ ^{235}U) fuel with high uranium density. Accordingly, several LEU-based high uranium density fuels have emerged as driver fuels for MTRs (RERTR 1996, 1998). The qualified uranium densities in LEU-based fuels extend up to 2.3 g cm^{-3} for Al–UAl_x, 3.2 g cm^{-3} for Al–U_3O_8, and 4.8 g cm^{-3} for Al–U_3Si_2 fuel.

The advanced LEU-based MTR fuels consist of Al–U_3Si with a uranium density in the region of 6.0 g cm^{-3} and Al–U–Mo and Al–U–Zr–Nb with a uranium density in the range 7–8 g cm^{-3}. The alloy or dispersion fuel elements are usually manufactured in the form of plates in which the fuel "meat" is sandwiched between aluminum alloy cover plates. The fuel meat consists of UAl_x, U_3O_8, U_3Si_2, U_3Si, U–Mo, or U–Zr–Nb dispersoids, large (44–150 μm) in size compared with the recoiled range of fission products (6–12 μm), which are uniformly dispersed in an aluminum or aluminum alloy matrix. An ideal dispersion "is an array of uniform, spherical fuel particles, sized to permit a reasonable amount of matrix metal between particles to remain undamaged by fission fragments." In order to avoid any overlapping of damaged fission recoil regions, the volume fraction of the fuel particles should be within 70% and the average size of the fuel particles should not be less than 44 μm. The principal steps in the fabrication of aluminum-clad, aluminum matrix plate-type dispersion fuels are the following:

- preparation of aluminum matrix dispersion-type fuel in the form of rectangular plates;
- picture framing and encasing of fuel meat in aluminum or Al–Mg alloy plates to form a sandwich between the top and bottom cover plates of aluminum or Al–Mg alloy;
- hot rolling of the sandwich for metallurgical bonding between mating surfaces of the core and cover plates;
- x-ray radiography for locating the outline of the fuel meat in the roll-bonded sandwich followed by shearing in the region outside the fuel outline to obtain the fuel elements of the desired dimension; and
- assembling several fuel elements in certain geometric configuration to form the fuel assembly.

Aluminum-clad, Al–23% ^{233}U alloy plate fuel has been manufactured in India by the classical melting and casting route (Ganguly *et al.* 1991a). The increasing divergence between the solidus and the liquidus of the hypereutectic Al–U alloys imparts a pronounced tendency for macrosegregation especially during slow cooling of the ingots and leads to the formation of coarse, columnar grains of UAl_4, which are very brittle and are responsible for severe cracking of the ingots during subsequent rolling. This problem is avoided by the addition of a small quantity (1–2%) of zirconium during the Al–U master alloy preparation and by rapid solidification of the cast alloy with the use of a water-cooled metallic mold. Zirconium addition is found to stabilize the relatively softer and denser UAl_3 phase and eliminate the dendritic morphology of UAl_4 platelets in the cast Al–U alloy.

Fuel particles of UAl_x, U_xSi_y and U–Mo alloy in the size range 44–150 μm are fabricated by vacuum induction or arc melting of aluminum and uranium, silicon and uranium, or molybdenum and uranium, followed by casting in the form of buttons, crushing, and sieving. The U_3Si is tough and difficult to grind and is more susceptible to oxidation than U_3Si_2. In addition, the former has higher irradiation swelling. Al–U_3Si_2 fuel with a uranium density ~4.8 g cm^{-3} has evolved as a standard LEU-based fuel for research reactors. With Al–U–Mo or Al–U–Zr–Nb fuel, it is possible to obtain uranium densities in the range 7–8 g cm^{-3}.

3.3 Uranium–Zirconium Hydride Fuel

Aluminum- or stainless steel-clad uranium–zirconium hydride ($UZrH_{1.0}$) fuel in the configuration of cylindrical pins is used in some 65 TRIGA reactors worldwide (Razvi 1998). Initially, HEU fuel was used in TRIGA with a uranium density ~0.5 g cm^{-3}. After the initiation of the RERTR program, the HEU was replaced by LEU with a uranium density as high as 3.7 g cm^{-3} in $UZrH_x$ fuel (RERTR 1996, 1998). The uranium enrichment is ~20% and zirconium hydride serves as a moderator. Since the hydrogen is intimately mixed with uranium, the U–Zr hydride fuel ensures inherent safety of TRIGA reactors. The TRIGA fuel rods are manufactured in several steps. First, homogenous ingots of uranium–zirconium alloy are prepared by double arc melting zirconium sponge and uranium metal piece. Next, the ingots are hot extruded and machined to the required size. Finally, the alloy is hydrided in a tube furnace under controlled hydrogen atmosphere at around 750°C to form uranium–zirconium hydride rods with hydrogen to zirconium atom ratio of 1:1. The microstructure of the uranium–zirconium hydride fuel thus prepared consists of alpha zirconium, zirconium hydride, and the epsilon phase of zirconium–uranium alloy. The uranium–zirconium hydride rods are assembled into aluminum or stainless steel tubes together with graphite rods and burnable poison disks at both ends and encapsulated to form cylindrical fuel rods.

4. *Fabrication of Ceramic Nuclear Fuels for Power Reactors*

Figure 1 summarizes the major steps followed in the fabrication of oxide and mixed oxides of uranium, plutonium, and thorium and mixed uranium–plutonium monocarbide and mononitride fuels by powder pellet, vibratory compaction, and sol–gel microsphere pelletization (SGMP) processes. UO_2, PuO_2, and ThO_2 are isostructural (f.c.c., CaF_2 type), completely solid soluble in each other and have very similar thermodynamic and thermophysical properties. Hence, the manufacturing processes for oxide and mixed oxide fuels are similar. The manufacturing processes for mixed uranium–plutonium monocarbide and mononitride fuels are likewise the same as for carbide or nitride, respectively, since UC, PuC, UN, and PuN have the same f.c.c., NaCl-type crystal structure and are completely solid soluble in each other.

The powder pellet route is followed in most countries for fabrication of both low ($\leqslant$85% TD) and high ($\geqslant$94% TD) density oxide, carbide, and nitride fuel pellets, starting with UO_2, PuO_2, and ThO_2 powders. The mixed uranium–plutonium monocarbide and mononitride powders are usually prepared by carbothermic synthesis of tableted, or granulated oxide plus carbon mixture in vacuum and flowing nitrogen, respectively. The major drawback of the powder pellet process is generation and handling of fine powder or dust particles of the fuel and the problem of radiotoxic dust hazard associated with it. Further, fine powders have poor flowability, which makes remote and automated processing somewhat difficult. In addition, fine powder of carbide and nitride are highly pyrophoric and require a very high purity inert cover gas for handling. The objectives of advanced methods

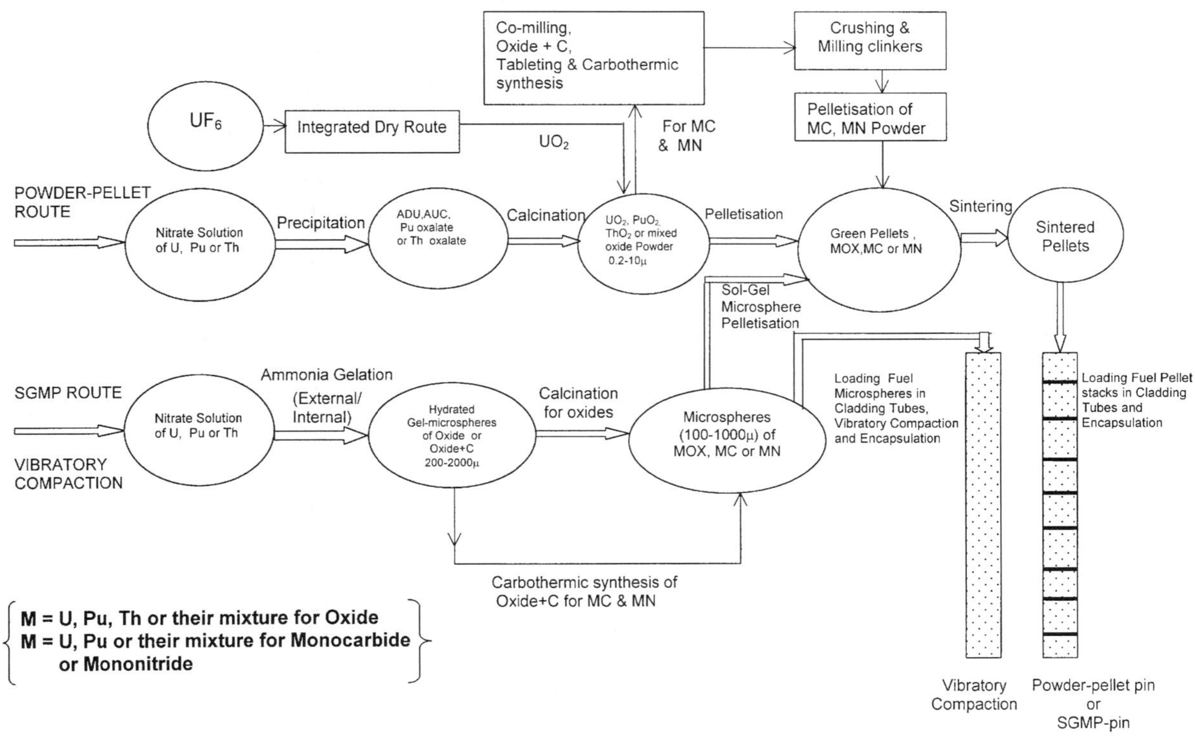

Figure 1
Powder pellet, vibratory compaction, and SGMP processes for fabrication of ceramic nuclear fuels.

of fabrication of ceramic nuclear fuels are safety of plant personnel, minimization of cost, and improved microstructure for high burn-up. The major advanced processes developed so far are as follows:

- advanced powder pellet route: low-temperature oxidative sintering (LTS) of UO_2 and $(U,Pu)O_2$ pellets; admixing TiO_2 dopant for obtaining UO_2 and $(U,Pu)O_2$ pellets of large grain size; low-temperature sintering of ThO_2, $(Th,U)O_2$, and $(Th,Pu)O_2$ pellets by MgO, CaO, or Nb_2O_5 "sintering aid"; pellet impregnation technique for fabrication of high-density $(Th,^{233}U)O_2$ pellets;
- vibro–sol or vibratory compaction process for oxide, carbide, and nitride fuels;
- SGMP of oxide, carbide, and nitride pellets; and
- combined SGMP–LTS process for UO_2 and $(U,Pu)O_2$ pellets.

4.1 Preparation of UO_2 Powder

Sinterable-grade uranium oxide powder is prepared either by the integrated dry route (IDR) starting with UF_6 or by employing the wet chemical routes starting with magnesium diuranate (MDU) or UF_6. The most common wet routes are based on ammonium diuranate (ADU) and ammonium uranyl carbonate (AUC) processes.

The IDR is extensively followed in France, Russia, the UK, and the USA for manufacturing enriched UO_2 powders (1.5–5% ^{235}U) (Chotard *et al.* 1991). The IDR is carried out inside a rotary kiln involving single-stage gas–gas and gas–solid reactions. The vaporized UF_6 is first reacted with superheated dry steam to form uranyl fluoride (UO_2F_2) by hydrolysis. Next, the product is routed through the second part of the kiln, consisting of a rotating tube at 600–700°C, where it is reduced to UO_2 by a countercurrent flow of hydrogen and steam. The overall reaction is as follows:

$$UF_6 + 2H_2O + H_2 = UO_2 + 6HF$$

The process does not generate any liquid effluent. The only by-product is high-purity HF, which could be recovered and sold. The specific surface area of IDR-derived UO_2 powder is generally low, about $2\,m^2g^{-1}$, compared with the powder produced by the wet route ($2.5–6\,m^2g^{-1}$). The powder consists of spherical particles or crystallites with diameters of $\sim 0.2\,\mu m$ aggregated to form brittle particles of $\sim 1\,\mu m$. Despite a low specific surface area, the IDR-derived UO_2 powder has excellent sinterability. It is usually transferred directly into orbital screw blenders for homogenization. If necessary, pore formers could be introduced at the blending stage.

The ADU route is most popular in Canada, India, and Brazil (Hastings 1989, Sejnoha 1999). To prepare natural uranium oxide powder for manufacturing PHWR fuel, it is essential to purify the MDU feed material in order to keep rare earths, boron, and cadmium, which are high neutron absorbers, to less than 1 ppm. The MDU cake is dissolved in nitric acid and purified by solvent extraction with an organic solvent, namely tributyl phosphate (TBP) in kerosene. Uranium is preferentially extracted into the organic phase, leaving the impurities in the aqueous phase. The uranium-laden organic, after a scrubbing treatment, is stripped with water to recover uranium as pure uranyl nitrate aqueous solution. By further steps of ammonia precipitation, filtration, calcination, hydrogen reduction, and stabilization, pure uranium and sinterable UO_2 powder is produced. If UF_6 is used as a starting material, it is subjected to hydrolysis followed by ammonia treatment to obtain ADU precipitate, which after washing and drying is pyrohydrolyzed to remove traces of fluoride and to yield U_3O_8, which after hydrogen reduction and stabilization leads to the formation of sinterable-grade UO_2 powder. The ex-ADU, UO_2 powder is extremely fine ($<1\,\mu m$), and has a specific surface area in the range $3–5\,m^2g^{-1}$, oxygen to metal ratio between 2.07 and 2.15, and bulk density of $1.8–2.0\,g\,cm^{-3}$. The powder is not free flowing and requires additional process steps, namely precompaction and granulation or roll compaction and granulation for producing free-flowing UO_2 granules that are suitable as press-feed material for pelletization in high-speed automatic hydraulic or mechanical presses.

The AUC process was developed in Germany (Assmann and Becker 1979) and is used in South Korea and Argentina. It can use both UF_6 and uranyl nitrate as feed materials and produces free-flowing UO_2 powder suitable for direct pelletization. Unlike ADU, AUC is crystalline, free flowing, and relatively coarse ($40\,\mu m$). Calcination of AUC at around 900 K produces oxide powder that is relatively coarse ($10–20\,\mu m$), free flowing, and with relatively higher specific surface area ($5\,m^2g^{-1}$), suitable for direct pelletization. The AUC process has also been extended to the fabrication of $UO_2–Gd_2O_3$ fuel pellets for LWRs. The coprecipitated uranium–gadolinium mixed oxide thus produced avoids free Gd_2O_3 in sintered pellets, ensuring excellent microhomogeneity of the gadolinium burnable poison in fuel pellets. The ammonium uranium plutonium carbonate (AuPuC) route is an extension of the AUC process and has been extensively used for manufacturing free-flowing $(U,Pu)O_2$ powders in Germany (Assmann *et al.* 1988, Roepenack *et al.* 1984).

4.2 Preparation of ThO_2 and PuO_2 Powders

ThO_2 and PuO_2 powders are prepared by air calcination of the oxalate precipitates obtained by the addition of oxalic acid to thorium nitrate or plutonium nitrate solution. The calcination temperatures are approximately 850 K and 1050 K, respectively, for plutonium oxalate and thorium oxalate. In order to dope CaO or MgO a sintering aid (0.2–0.4 w/o) in the ThO_2 powder, $Ca(NO_3)_2$ or $Mg(NO_3)_2$, is usually added to the thorium nitrate solution prior to precipitation. The oxalate-derived ThO_2 and PuO_2 powders are extremely fine (<1 μm) and have poor flowability and need to be granulated to produce free-flowing press-feed materials. In addition, the oxalate-derived ThO_2 powder has platelet morphology and requires premilling to break the platelets.

4.3 Preparation of Carbide and Nitride Powders

Figure 2 summarizes the different techniques for the preparation of mixed uranium plutonium monocarbide (MC) and mononitride (MN) in the form of powders and microspheres, starting with metals, oxides, or nitrate solutions of uranium and plutonium as feed materials (Sowden *et al.* 1964, Leary and Kittle 1977, IAEA 1984, 1988). Uranium forms three different carbides, namely UC, U_2C_3, and UC_2. Similarly, carbides of plutonium include Pu_3C_2, PuC, PuC_2, and Pu_2C_3. The U–N system is characterized by four compounds, namely UN, α-U_2N_3, β-U_2N_3, and a nitrogen-deficient UN_2. The Pu–N system has only one compound, the mononitride PuN.

The process involving hydriding and dehydriding of massive metal at 450–525 K and 800–1000 K, respectively, for obtaining fine metallic powder of high surface area followed by carburization with methane or propane at 1000–1100 °C for synthesis of monocarbide powder and nitridation with N_2 or N_2+H_2 mixture for preparation of mononitride powder was used in the initial years of carbide and nitride fuel development (Matzke 1986). This method of synthesis of MC and MN has two main advantages. First, the reaction temperatures are low, which, in turn, avoids plutonium volatilization losses. Second, the end products are fine and highly reactive MC and MN powders, which can be pelletized and sintered directly.

For preparation of MC and MN, all techniques using uranium and plutonium metals as starting materials are uneconomical because these metals

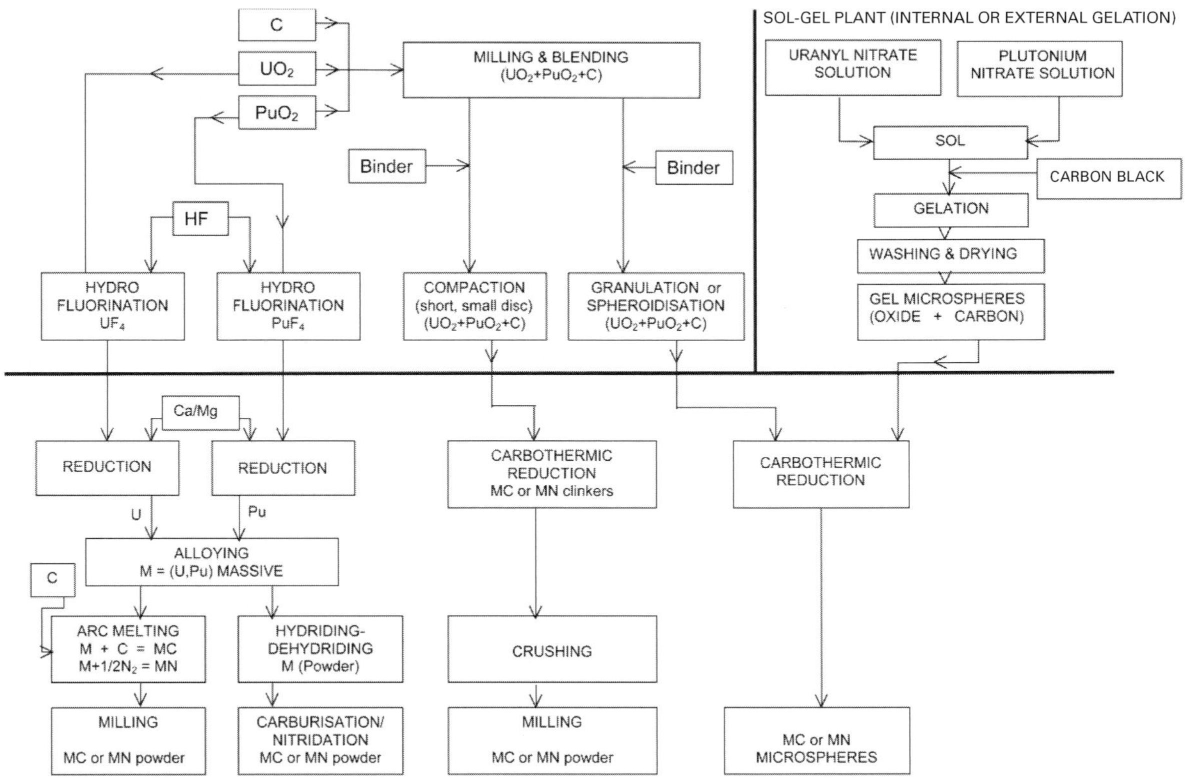

Figure 2
Summary of different methods for preparation of (U,Pu)C and (U,Pu)N powders and microspheres starting with metals, oxides, or nitrate solutions of uranium and plutonium.

are prepared from their oxides via expensive steps such as hydrofluorination and calcio- or magnesiothermic reduction. Moreover, storing uranium and plutonium in bulk quantities is associated with criticality hazards. Uranium and plutonium powders are pyrophoric and chemically unstable. Oxides have advantages as starting materials mainly because of economic incentives. Further, oxides are chemically stable and are the easiest and safest to store.

Carbothermic reduction of mechanically mixed or coprecipitated oxide is most attractive for large-scale production of MC and MN powders and has, therefore, been studied intensively in all the laboratories associated with the development of MC and MN fuels (IAEA 1988). In this process, a high degree of microhomogeneity of the starting oxide–carbon mixture is necessary. Otherwise, localized deficiencies and excesses of carbon lead to the formation of unwanted second phases. The requisite homogenization is achieved either by prolonged blending of the oxide–carbon powder mixture or alternatively by the wet chemical route, popularly known as the sol–gel process. The homogeneous oxide–carbon powder mixture is compacted into short (2–3 mm long) and small diameter ($\leqslant$25 mm) tablets in order to increase the oxide–carbon contact surface.

(*a*) *Carbothermic synthesis of monocarbide.* The overall simplified chemical equation for preparation of monocarbide from oxide is

$$MO_2 + 3C = MC + 2CO$$

Mixed uranium–plutonium monocarbide powder is usually prepared by single-step carbothermic reduction of a tableted powder mixture of oxide and carbon. Two-stage reduction (Richter 1976) and three-stage reaction sintering routes have also been developed. Thermodynamic calculations show that it is not possible to produce oxygen-free MC by the single-step direct solid-state carbothermic reduction of mechanically mixed oxide without incurring significant plutonium volatilization loss. During carbothermic reduction, control of the partial pressure of CO is very important since the evolution of this gas not only constitutes the primary reduction mechanism, but also controls the kinetics of this reaction. The two-step solid-state synthesis in a static bed aims at the preparation of a single-phase MC with very low oxygen and nitrogen impurities and with practically no loss of plutonium by volatilization.

In the two-step process, carbothermic reduction of the oxide is carried out with excess carbon in order to form M_2C_3 in the first step. M_2C_3, unlike MC, has very little oxygen and nitrogen solubility and can be prepared at relatively low temperatures ($\leqslant$1500 K). In the second step, the M_2C_3 is crushed, milled, and treated with hydrogen at 1125 K in order to reduce it to MC. Any free carbon present is removed as methane. A major drawback of the two-stage carbothermic reduction is that pure H_2 reduction is not recommended for plutonium-bearing material because it is associated with explosion hazards. In a nonexplosive Ar + 8% H_2 gas mixture, the partial pressure of hydrogen is inadequate for complete conversion of M_2C_3 to MC.

In three-stage synthesis (Wedemeyer and Gunther 1975) UC is prepared by carbothermic reduction of UO_2 at 2073 K. In the second stage, plutonium oxycarbide is produced by discontinuing the carbothermic reduction at 1473 K. And finally, in the third stage, uranium carbide and plutonium oxycarbide–carbon powders are blended, compacted, and subjected to reaction sintering. There are three advantages of this method. First, since the primary reduction of PuO_2 is at a low temperature (1473 K), the plutonium volatilization loss is low. Second, a carbon stoichiometric correction is possible before the sintering process because of the separate preparation of uranium carbide and plutonium oxycarbide. Third, the loading of plutonium is a minimum and the UC is produced separately without the penalty of plutonium hazards. A significant disadvantage of this method is the microstructural inhomogeneity of the sintered pellet. In the interior of the pellet a coarsely porous structure is observed while at the edges of the pellet a highly densified zone is seen. Further, because of the substantial release of carbon monoxide during reaction sintering, the maximum achievable sintered density is limited to 94% TD.

(*b*) *Carbothermic synthesis of mononitride.* The carbothermic synthesis of mononitride is represented by the overall reaction

$$MO_2 + 2C + \tfrac{1}{2}N_2 = MN + 2CO$$

Nitrogen plays the dual role of reactant and carrier for the removal of CO. The product of the reaction will have the general formula ($MN_{1-x-y}C_xO_y$) and the relative levels of oxygen and carbon retained in MN will depend on: (i) N_2 flow rate and, in turn, the partial pressure of N_2 and CO, (ii) the oxide to carbon mole ratio, and (iii) use of H_2. Around 10% excess carbon content in the oxide–carbon mixture, a gas atmosphere of N_2 followed by $N_2 + H_2$ followed by argon, a gas flow of around 1000 L h^{-1}, and a carbothermic synthesis temperature in the range 1825–1925 K are the optimum parameters for obtaining relatively high-purity mixed uranium–plutonium mononitride with minimum residual impurities of oxygen and carbon (Ganguly 1991a).

4.4 Sol–Gel Process for Preparation of Oxide, Carbide, and Nitride Microspheres

The sol–gel process is an alternative wet chemical route whereby "ammonia gelation," dust-free, free-flowing, coarse, and spherical (10–1000 μm in diameter) hydrated gel microspheres of oxide (UO_3, ThO_2, or PuO_2) or mixed oxide and oxide + carbon mixtures of very good microhomogeneity are produced from the heavy metal nitrate feed solutions. The ammonia gelation is achieved either externally via NH_3 gas and NH_4OH or "internally" via an added ammonia generator, namely hexamethylene tetramine (HMTA). The ThO_2 and ThO_2-based mixed oxide microspheres, namely $(Th,U)O_2$ and $(Th,Pu)O_2$, are most efficiently prepared by the external gelation of thorium (EGT) process of Germany. For UO_2 and urania-based mixed oxide, namely $(U,Pu)O_2$, the external gelation of uranium (EGU) process of Germany, the SNAM process of Italy, and the KEMA internal gelation of uranium (IGU) process followed in The Netherlands, Switzerland, and India are popular (IAEA 1974).

The major steps in the sol–gel process are as follows:

- Preparation of sol or broth from nitrate solutions of uranium, thorium, or plutonium.
- Ultrasonic dispersion of fine carbon black particles in the sol or broth for preparation of "porous" oxide microspheres after controlled calcination and carbide and nitride microspheres after carbothermic synthesis (for preparation of "nonporous" oxide microspheres this step is not needed).
- Droplet formation by a vibrating nozzle.
- Droplet gelation in NH_3 gas and NH_4OH bath (for external gelation) or silicon oil bath at 90°C (for internal gelation). In the internal gelation process, microwave heating is also used in place of the silicon oil bath.
- Washing of gel microspheres.
- Drying of gel microspheres.

In the EGU process, the broth is prepared by mixing uranium nitrate solution (1 mol L^{-1}) with urea (4 mol L^{-1}) and NH_4NO_3 (2.5 mol L^{-1}) and boiling the mixture for 30 min. A small amount of poly(vinyl alcohol) (5 g L^{-1}) is also added to the mixed solution for mechanical stability of the gel. The same procedure is followed when the EGU process is extended for preparation of mixed uranium–plutonium oxide microspheres starting from nitrate solutions of uranium and plutonium. In the EGT process, the thorium-bearing sols are prepared by passing a controlled amount of ammonia gas through the thorium nitrate solution at 80–100°C, such that 85–88% of preneutralization reactions are completed.

The uranium/thorium-bearing broth/sol is dispersed into droplets, which pass through a curtain of NH_3 gas and fall into a 10 w/o aqueous ammonia gelation bath. In the gelation bath, the droplets gel within seconds to rigid gel particles, which are then washed in 1% ammonia solution for removal of NH_4NO_3 and dried on a continuous belt drier at 200–250°C.

In the IGU process, the uranium and plutonium nitrate solutions are mixed with HMTA (ammonia generator) and urea in mole proportions of 1.25, 1.75, and 1.75, respectively, and cooled to 0°C (Vaidya *et al.* 1987). Urea prevents premature gelation of the solution. Droplets of the sol or broth fall into the silicon oil gelation bath at 90 ± 1°C that decomposes HMTA to release ammonia for conversion of the droplets into hydrous gel particles. The gel particles are washed with carbon tetrachloride to remove oil and in a 3 M NH_4OH solution to remove ammonium nitrate.

Ganguly and Basak (1991), Ganguly *et al.* (1986b, 1987), and Hastings (1989 pp. 398–413) made the following modifications to the external and internal gelation processes for production of soft and porous UO_2, $(U,Pu)O_2$ ($\leqslant 30\%$ PuO_2), ThO_2, ThO_2–2% UO_2, and ThO_2–5% PuO_2 microspheres suitable for pelletization and sintering to high-density pellets (the plutonium was simulated with cerium in most cases):

(i) use of a heavy metal nitrate feed solution of lower molarity ($\leqslant 1$);

(ii) addition of carbon black (30 g mol^{-1} heavy metal) pore former to the sol or broth prior to gelation and later removing the carbon black from the gel microspheres by controlled air calcination at 700°C to form porous oxide microspheres; and

(iii) addition of dopants to the heavy metal nitrate feed solutions, which enhance densification and/or grain growth during sintering of oxide pellets.

The hydrated gel microspheres of uranium oxide or mixed oxide thus obtained are subjected to air calcination followed by H_2 reduction and stabilization treatment in order to obtain porous, free-flowing, and soft microspheres with the desired oxygen to metal ratio and specific surface area in the range $10 \pm 5\, m^2 g^{-1}$. The porous microspheres crush easily during pelletization and have excellent sinterability. If carbon black pore former is not added, relatively hard and nonporous microspheres are produced.

Mixed uranium–plutonium monocarbide and mononitride microspheres are prepared in two stages. First, hydrated gel microspheres of $UO_3 + PuO_2 + C$ are prepared by the ammonia internal or external gelation process starting with uranyl nitrate and plutonium nitrate solution. Next, the gel microspheres are subjected to carbothermic synthesis at 1450°C in vacuum (1 Pa) for monocarbide and in flowing nitrogen atmosphere (flow rate of 1.2 $m^3 h^{-1}$) at 1550°C for mononitride. The sol–gel process is particularly attractive for the preparation of monocarbide and mononitride microspheres for the following reasons:

(i) the number of process steps is less than half compared to the conventional powder route;

(ii) generation and handling of fine carbide and nitride powders are avoided, which, in turn, reduces fire and radiotoxic dust hazards;

(iii) high specific surface area and chemical reactivity of gel microspheres containing oxide–carbon mixtures reduce significantly the carbothermic reduction temperature and the soaking period; and

(iv) a high degree of microhomogeneity is achieved in the fuel.

4.5 Cold Pelletization and Sintering

The oxide, carbide, or nitride fuel pellets are universally manufactured by the classical powder pellet route starting with fine powders. The alternative SGMP process is dust-free and has several advantages over the powder pellet route but has not been employed anywhere on an industrial scale.

In the powder pellet route, the fine powders of oxide, carbide, and nitride are granulated by precompaction and granulation, roll compaction and granulation, or extrusion in order to obtain free-flowing, coarse (~ 1 mm diameter) granules to ensure constant filling of the dies of the multipunch double-acting hydraulic press or rotary press. UO_2 and $(U,Pu)O_2$ powders obtained by the AUC and AUPuC process, respectively, are free flowing and can be directly pelletized. In the powder pellet route, the pore formers (e.g., poly(vinyl alcohol), methyl cellulose), binders (paraffin wax, naphthalene, etc.), and additives for sintering aid and grain growth are added to the powder during milling or blending prior to granulation.

In case of the SGMP process, the dust-free and free-flowing, oxide, carbide, or nitride microspheres are directly pelletized and sintered. The sintering aid and grain growth additives are added to the solution or sol prior to gelation. The lubricants, if any, are added to the granules or sol–gel particles prior to pelletization.

The sintering of oxide, carbide, or nitride pellets is usually carried out at a high temperature (1600–1750 °C) in a reducing atmosphere (H_2, cracked NH_3, N_2+H_2, $Ar+H_2$) in a molybdenum resistance furnace of the batch type or continuous pusher or walking beam type. In the case of carbide and nitride, the sintering is also carried out in vacuum or argon atmosphere. Fabrication of high-density UO_2 and $(U,Pu)O_2$ pellets is possible by LTS ($\leqslant 1200$ °C). In LTS and SGMP–LTS, there is significant conservation of electrical energy as a result of which the fuel fabrication cost is reduced and the requirements of molybdenum heating element and charge carriers are avoided.

(a) Sintering of UO_2 and $(U,Pu)O_2$ pellets for water-cooled reactors. Industrial production of UO_2 fuel pellets for water-cooled reactors is universally carried out by cold pelletization of UO_2 powder followed by sintering at 1700–1750 °C in a high-temperature resistance furnace using molybdenum heating elements and molybdenum charge carriers. Industrial-scale production of mixed uranium–plutonium oxide (MOX) fuel for LWRs has so far been carried out mainly in France, Germany, Belgium, and the UK. In addition, several MOX pilot plants are in operation in Japan, India, and Russia (IAEA 1997).

Figures 3(a) and (b) summarize the process flowcharts followed for production of MOX fuel pellets by the micronized master (MIMAS) process and the short binderless route (SBR), respectively. The MIMAS process was developed by Belgonucleaire in the mid-1980s and is being followed in the Dessel plant in Belgium and the MELOX plant in France. These two plants supply MOX reload fuel to several PWRs and BWRs. The MIMAS process produces a master blend of UO_2 and 30% PuO_2 by ball milling, which also ensures micronization of the UO_2 and PuO_2 particles. In the second blending operation, free-flowing UO_2 powders, either of the AUC or the ADU type, is added in order to have 2–8% PuO_2 in the final blend depending on the requirement. The double blending ensures very good isotopic homogeneity of the plutonium in the product, even with plutonium of various origins, forms, and sources. In addition, in the MIMAS process, the rejected pellets, grinding powder, and other scraps are subjected to dry recycling, which is reintroduced at the first or the second blending step. The sintered MOX pellets do not contain any undissolved PuO_2 particles and ensures easy solubility in nitric acid.

The SBR was developed by British Nuclear Fuels Ltd (BNFL) in the early 1990s, based on which the MOX Demonstration Facility (MDF) was set up at Sellafield, UK. The SBR uses a high-energy attritor mill to blend UO_2 and PuO_2 feed powders and a spheroidizer to condition the powder before it is used as a feed to a rotary compaction press. At the milling stage, zinc stearate lubricant and Conpor pore former are added. The MOX pellets are sintered at 1650 °C over a cycle time of 24 h. Based on the successful operation of MDF, the Sellafield MOX plant (SMP) has been set up. The use of attritor milling ensures good microhomogeneity of the co-milled powder, which in turn leads to complete solid-solution formation of UO_2 and PuO_2 during sintering. The MOX plant in India also uses an attritor for co-milling ex-ADU UO_2 and oxalate-derived PuO_2 powders.

The lubricant is added at the attrition stage after which the co-milled powder is subjected to precompaction and granulation leading to free-flowing oxide powder that is pelletized and sintered. The MOX fuel thus produced would have excellent

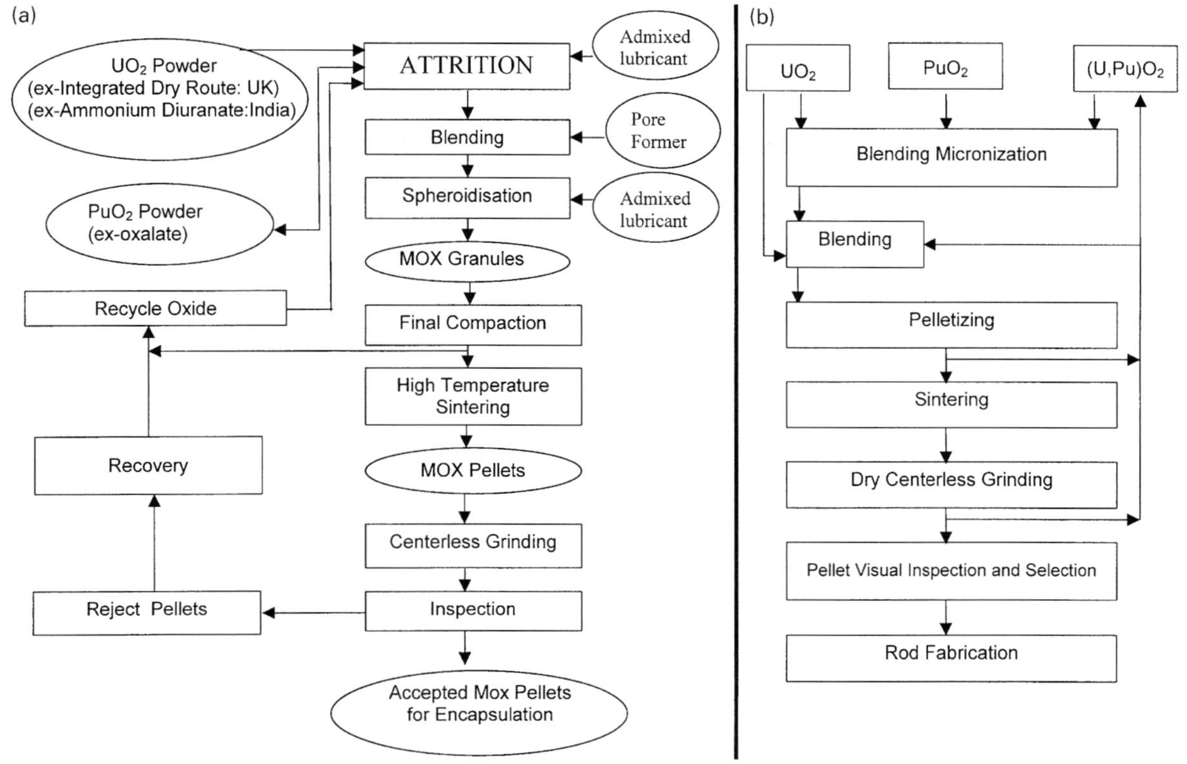

Figure 3
(a) Process flowchart based on the SBR followed in the UK and India for fabrication of mixed uranium–plutonium oxide (MOX) fuel pellets involving attrition of UO_2 and PuO_2 powders. (b) MIMAS blend process followed in Belgium and France for fabrication of mixed uranium–plutonium oxide (MOX) fuel pellets for LWRs and FBRs.

microhomogeneity and is expected to perform better than UO_2. In addition, such MOX fuel would be easy to reprocess since undissolved PuO_2 particles would be absent. The microwave denitration of mixed nitrate solutions of uranium and plutonium followed in Japan and the AUPuC process of Germany lead to homogeneous MOX powder, which after pelletization and sintering ensures complete solid-solution formation between UO_2 and PuO_2, leaving no undissolved PuO_2 particles.

For high burn-up PHWR and LWR fuels, it is essential to manufacture UO_2 and $(U,Pu)O_2$ pellets of large grain size (from 10–15 μm to ~50 μm) in order to minimize fission gas release from the fuel during operation in the reactor and to avoid pellet–cladding interaction (PCI) failures. This is also true for mixed urania–gadolinia (gadolinium as burnable poison) fuel pins for LWRs. Pellets of large grain size are mostly obtained by addition of dopants such as TiO_2 and Nb_2O_5, which enhance cation diffusion in the UO_2 or $(U,Pu)O_2$ lattice. Large-grained pellets can also be obtained at relatively low temperature (1200°C) by starting with oxide powder having an oxygen to metal ratio in the range 2.25–2.27 by sintering in controlled oxidizing and reducing atmospheres.

The major breakthroughs in the fabrication of UO_2 and $(U,Pu)O_2$ pellets are: (i) LTS, also known as the NIKUSI process developed in Germany (Doerr *et al.* 1986), and (ii) the subsequent SGMP–LTS process developed in India (Ganguly 1993, Ganguly and Basak 1991).

In the LTS process, the sintering of UO_2 or $(U,Pu)O_2$ is carried out at 1100–1200°C for a very short duration (soaking at 1100–1200°C for less than 1 h) in a three-zone continuous sintering furnace maintained at three different atmospheres. The diffusion of uranium cation (D^U) in UO_{2+x} is proportional to x^2. Accordingly, in the LTS furnace, the sintering of hyperstoichiometric UO_2 or $(U,Pu)O_2$ pellets is carried out in the oxidative zone containing a mixture of air and nitrogen with 500–1000 ppm oxygen in order to facilitate the densification process. After densification, the hyperstoichiometric oxide or MOX pellets are reduced to stoichiometric oxide in the reducing zone of pure hydrogen or nitrogen plus hydrogen atmosphere. A nitrogen curtain is used between the oxidizing and reducing zones.

In the SGMP–LTS process, sol–gel-derived dust-free and free-flowing oxide or mixed oxide microspheres are used for pelletization, thus minimizing the radiotoxic dust hazard associated in the powder route and facilitating remote and automated compaction of pellets. High-density UO_2 and $(U,Pu)O_2$ pellets have been manufactured by the SGMP–LTS process using the IGU and EGU processes for preparation of hydrated gel microspheres. In most cases, for preparation of porous microspheres, carbon black pore former is added to the solution prior to gelation and later removed by controlled calcination. The porous microspheres thus prepared are found to have a low crushing strength (2.5 N per microsphere); as a result the microspheres disintegrate easily during pelletization at 350 MPa and yield sintered pellets without microsphere boundaries and with uniformly distributed "closed" spherical pores in the ideal diameter range of 2–5 μm. Such a pore structure contributes to improved thermal conductivity and better retention of fission gas and accommodation of fuel swelling. The high-density oxide and mixed oxide fuel pellets are found to have a single-phase, equiaxed microstructure. The non-porous mixed oxide microspheres retain their individual identity even after pelletization and sintering and lead to a "blackberry" structure with a lot of open porosity (Ganguly and Hegde 1997). Such low-density pellets with open porosity are recommended for LMFBR fuel to facilitate fission gas release and in turn reduce fuel swelling. The mixed oxide pellets manufactured by the SGMP process ensure complete solid-solution formation between UO_2 and PuO_2 and the absence of any undissolved PuO_2. High-density UO_2–5% Gd_2O_3 pellets have also been prepared by the SGMP process, using the EGU process for the preparation of hydrated gel microspheres.

(*b*) *Sintering of ThO_2, $(Th,U)O_2$, and $(Th,Pu)O_2$.* High-density ThO_2, $(Th,U)O_2$ (<4% UO_2), and $(Th,Pu)O_2$ (⩽4% PuO_2) pellets have been fabricated successfully by the classical powder metallurgy route using approximately 0.2–0.4% MgO as a sintering aid (Hastings 1989, pp. 398–413). The MgO is usually added in the form of $Mg(NO_3)_2$ in thorium nitrate solution. The sintering is carried out at 1923 K for 4 h in argon + 8% hydrogen atmosphere. High-density ThO_2, $(Th,U)O_2$ (2% UO_2), and $(Th,Ce)O_2$ (4% CeO_2 with cerium as simulator for plutonium) pellets have been successfully fabricated by the SGMP process, using the ammonia EGT process for the preparation of the hydrated gel microspheres (Ganguly *et al.* 1986b). Calcium nitrate (1%) was added to the thorium nitrate solution in order to have around 0.4% CaO sintering aid in the ThO_2 and ThO_2-based mixed oxide pellets. In order to obtain soft and porous microspheres, thorium nitrate solution of low molarity (1–1.2) was used and carbon black was added as a pore former to the sol prior to gelation and later removed by controlled calcination in air.

(*c*) *Pellet impregnation technique for preparation of $(Th,U)O_2$ pellets.* The pellet impregnation technique (Hastings 1989 pp. 398–413) is ideally suitable for remote and automated fabrication of highly radiotoxic $(Th,{}^{233}U)O_2$ fuel containing up to 2% ^{233}U, which is an advanced fuel for PHWRs. In this process, ThO_2 pellets of relatively low density (75–80% TD) are fabricated in an unshielded conventional plant by cold pelletization of ThO_2 powder followed by sintering at relatively low temperature (1373–1573 K). The ThO_2 pellets are then transferred inside a shielded alpha-tight cell and impregnated with acid-deficient uranyl nitrate solution for half an hour in vacuum. After impregnation, the pellets are vacuum dried, calcined, and sintered in vacuum or hydrogen atmosphere at 1973 K to obtain high-density $(Th,{}^{233}U)O_2$ pellets with single-phase microstructure. Thus, the pellet impregnation route avoids the handling of fine ^{233}U-bearing powder and involves fewer process steps. In addition, only a few parts of the process equipment are required to be exposed to ^{233}U and to be confined in shielded facilities, thereby making plant maintenance easier.

(*d*) *(U,Pu)C and (U,Pu)N pellets for LMFBRs.* Among all the existing methods of consolidation of (U,Pu)C and (U,Pu)N fuels for LMFBRs, the cold pelletization and sintering is the simplest, is well established, and the most widely accepted. MC and MN fuels have been produced on a small scale (batch size ⩽5 kg) mainly for out-of-pile property evaluation and irradiation testing. The cumulative production of MC and MN fuel worldwide will not exceed 1000 kg and 200 kg, respectively. The starting materials consist of either fine powders or sol–gel-derived synthesized microspheres of MC and MN. The MC and MN products synthesized by different methods are subjected to the following treatment for obtaining fine powder:

- crushing and ball milling arc-melted MC and MN buttons;
- crushing and ball milling clinkers, after carbothermic reduction of tableted oxide–carbon powder mixture; and
- milling oxide–carbon spheroids, granules, or microspheres after reaction sintering.

Suitable binders and sintering aids (if any) are added to the milled powder and the milling is continued for proper homogenization. The powders are then compacted into pellets (length to diameter ratio of 1.0–1.5), preferably in a double-action press at 140–350 MPa, followed by sintering in vacuum, argon, or argon–hydrogen mixture in the temperature range 1673–2173 K.

The degree of densification and the microstructural features of sintered MC and MN pellets depend, besides sintering temperature and time, on the average particle size or surface area of the powder, the binding agent (i.e., binder), the sintering aids (if any), and the plutonium and the oxygen contents of the powder. Most of the studies reported on MC and MN pellets pertain to uranium-rich composition with maximum PuC and PuN contents of 20%. Extensive investigations have been carried out in India on the densification of plutonium-rich mixed uranium–plutonium monocarbide and mononitride keeping their potential use as driver fuel in FBTRs (Ganguly *et al.* 1986a, 1991a, 1994). The densification behavior of MC and MN pellets of both uranium- and plutonium-rich compositions are found to be similar.

The powder compaction pressure in the range 140–350 MPa and the sintering atmosphere play only minor roles in controlling the sintered density of MC and MN. For MN pellets, although vacuum sintering is marginally more effective than sintering in argon or nitrogen atmosphere, the decomposition of MN in vacuum limits the maximum permissible sintering temperature to 1873 K. For argon and nitrogen sintering, the MN pellet density increases from about 75% TD at 1773 K to around 95% TD at 2173 K.

The general conclusion regarding the effect of the surface area of UC or (U,Pu)C powders on the sintered density is that powders of surface area of $1.5–2.0\,m^2g^{-1}$ are adequate for achieving high sintered densities (~96% TD) at temperatures of ~1873 K. UC and (U,Pu)C powders of these surface area values can be obtained by ball milling in a vibratory mill with tungsten carbide balls for 12–16 h.

A small quantity of nickel powder (0.1–0.2 wt.%, freshly hydrogen reduced) when admixed with UC, PuC, or (U,Pu)C powder is found to act as a very effective sintering aid. High-density (≥95% TD) pellets are consistently obtained at relatively low temperatures (1723–1823 K) even with considerable variations in the specific surface area, oxygen contents, and sintering conditions of the carbide pellets. However, nickel promotes grain growth and the formation of M_2C_3. Monocarbide pellets with nickel always contain 5–15% M_2C_3 and have a microstructure with angular or rectangular grains and banded structure within the grains.

The densification of UN and (U,Pu)N powder compacts is extremely sensitive to the average particle size (or the surface area) of the nitride powders. A small amount (1–5%) of oxide phase (MO_2) has been found to act as a sintering aid for UN and (U,Pu)N. The effect of oxygen on the densification of MC and MN depends on the form in which the oxygen is present. For MC, the sintered density is not at all affected if the oxygen is dissolved in the monocarbide grains. However, if the oxygen is present as an oxide phase, it acts as a barrier to the interdiffusion of uranium, plutonium, and carbon and inhibits sintering. The sinterability of mononitride is enhanced when a small amount of oxide is present as a second phase.

The mixed carbide and mixed nitride microspheres prepared by the sol–gel process are nonporous and have a relatively high crushing strength (20 N per microsphere). As a result, the microspheres retain their individual identity even after pelletization at high pressure (840 MPa) and sintering at high temperature (1973 K), resulting in a "blackberry structure" with microsphere boundaries and open pores. The blackberry microstructure in sintered pellets is desirable for LMFBR fuel as it facilitates fission gas release, which, in turn, minimizes fuel swelling.

(e) Vibratory compaction. In this process, high-density microspheres (≥98% TD) of oxide, carbide, or nitride fuels of 1, 2, or 3 size fractions in the range 10–1000 μm are directly loaded into one end of a closed fuel cladding tube and subjected to vibratory compaction in order to obtain fuel pins of controlled smear density (weight of the fuel pin divided by internal volume of cladding tube) in the range 60–90% TD. The microspheres are generally prepared by the sol–gel process. In some cases, the microspheres are prepared by the dry route involving spheroidization of powder. For vibratory compaction, various types of pneumatic and electro dynamic vibrators are used to obtain high compaction efficiencies. For smear densities typical of LMFBR fuels (75–80% TD), particles of 2 size fractions, typically 750 μm and 70 μm, are used in the ratio of 3:1. The highest achievable smear density of 90% TD is obtained by mixing the coarse (750–1000 μm), medium (70–100 μm), and fine (7–10 μm) fractions in the ratio 7:2:1. The rates of infiltration and settling of the fine particles in the voids formed by the packing of the coarse particles depend on the frequency, the waveform, and the acceleration of vibration, as well as on the particle shape and size fractions and the internal geometry of the fuel cladding tubes.

The vibratory compaction process avoids generation and handling of fine radioactive powder of uranium-, plutonium-, or thorium-bearing oxide, carbide, or nitrides. The dust-free and free-flowing microspheres minimize radiotoxic dust hazard and the process is amenable to remote and automated fuel fabrication. The process is particularly attractive for carbide and nitride fuels because the number of process steps is less, and it is recommended for fabrication of highly radioactive plutonium- and ^{233}U-bearing fuels. Apart from this, the vibratory compaction route has the maximum flexibility of operation because fuel cladding tubes of any internal dimensions can be easily vibro-filled to a wide range of smeared density (60–90% TD) by packing microspheres of 1, 2, or 3 sizes. However, the maximum achievable smeared density of a vibro-filled fuel pin is 90% TD. Hence, the process is not suitable for manufacturing LWR or PHWR fuels, which have very high (⩾94% TD) fuel pellet density and high smeared density. The vibro-filled fuel pins are suitable only for use as helium-bonded fuel pins for LMFBRs, which require low smeared density (75–80% TD).

5. *Fabrication of Metallic Fuels for Fast Reactors*

U–Pu–Zr alloys have emerged as safe and commercially viable advanced LMFBR fuels. They are fabricated by vacuum induction melting of uranium, plutonium, and zirconium metals in a yttria/zirconia-coated graphite crucible at around 1450°C, followed by injection casting (by furnace pressurization with 0.05 MPa argon or helium) in a closed-end Vycor (SiO_2) mold coated with yttria, zirconia, or boron nitride (Chang 1989). The zirconium is added to increase the solidus and fuel–clad eutectic temperatures. Further, oxygen and nitrogen impurities in the fuel form ZrO_x and ZrN_x diffusion barriers on the fuel–clad interface and delay eutectic formation with nickel in the stainless steel cladding. The cast pins require no further surface finishing since the metallic fuels use sodium bonding with large gap in the annular space between pellet and cladding.

6. *Conclusion*

The fabrication of nuclear fuels is unique in several ways since uranium, plutonium, and thorium are radioactive and are associated with the problems of criticality (uncontrolled nuclear fission chain reaction), health hazard, security, and pyrophoricity (in some cases). The fuel has to meet stringent dimensional and chemical specifications in order to ensure satisfactory performance in the reactor. The manufacturing technology of both metallic and ceramic fuels for research and power reactors has been well established over the years. The aluminum alloy-clad aluminum matrix dispersion fuel containing U_3Si_2 particles, the reference fuel for MTRs, is fabricated by the picture-frame technique followed by roll bonding. The manufacturing process of aluminum alloy- or stainless steel-clad U–ZrH fuel for TRIGA reactors consists of controlled hydriding of uranium–zirconium alloy rods. For water-cooled nuclear power reactors, zirconium alloy-clad high-density uranium oxide pellets are the reference fuel. Mixed uranium–plutonium oxide containing up to 5% PuO_2 and 25% PuO_2 have been used in PWRs and LMFBRs, respectively. $(Th,U)O_2$ and $(Th,Pu)O_2$ are advanced fuels for PHWRs. The oxide and mixed oxide fuels are mostly manufactured in the form of small cylindrical pellets of controlled density and microstructure by the classical powder pellet route. The mixed uranium–plutonium monocarbide and mononitride advanced fuels for LMFBRs are also fabricated by the powder route starting with UO_2 and PuO_2 powder. The SGMP process is a dust-free advanced method for manufacturing oxide, carbide, and nitride fuels and is best suited for remote and automated fabrication of highly radiotoxic plutonium- and ^{233}U-bearing fuels. The vibratory compacted oxide, carbide, and nitride fuels are suitable for manufacturing LMFBR fuel pins with relatively low packing density (⩽85% TD). The U–Zr and U–Pu–Zr advanced metallic fuels for fast reactors are fabricated by vacuum induction melting followed by injection casting.

Bibliography

Assmann H, Becker M 1979 Technology of UO_2 fuel fabrication by the AUC process. *Proc. European Nuclear Conf.* European Nuclear Society, Hamburg, Germany

Assmann H, Peehs M, Roepenack H 1988 Survey of binary oxide fuel manufacture and quality control. *J. Nucl. Mater.* **153**, 115

Belle J, Lustman B 1958 Preparation of UO_2, properties of UO_2 and sintering characteristics. *USAEC Report TID 7546*

Chang Y I 1989 The integral fast reactor. *Nucl. Technol.* **88**, 129

Chotard A, Ledac A, Bernardin M 1991 Fabrication characteristics and in-pile performance of UO_2 pellets prepared from dry route powder. In: Ganguly C, Roy S K, Roy P R (eds.) *Advanced Ceramics, Key Engineering Materials.* Trans Tech, Switzerland, Vols. 56, 57, pp. 471–88

Doerr W, Peehs M, Assmann H 1986 Advantages of the oxidative UO_2 sintering process NIKUSI. *Proc. 4th Int. ENS/ANS Conf.* European Nuclear Society, Geneva, Switzerland, Vol. 4, p. 233

Ganguly C 1989 *Nuclear Fuel Fabrication*, Materials Science Forum. Trans Tech, Switzerland, Vols. 48, 49

Ganguly C 1993 Sol–gel microsphere pelletization—a "powder-free" advanced process for fabrication of ceramic nuclear fuel pellets. *Bull. Mater. Sci.* **16**, 509–22

Ganguly C, Basak U 1991 Fabrication of high-density UO_2 fuel pellets involving sol–gel microsphere pelletization and low temperature sintering. *J. Nucl. Mater.* **178**, 179–83

Ganguly C, Basak U, Vaidya V N, Sood D D, Roy P R 1989 Sol–gel microsphere pelletization of UO_2 and UO_2–PuO_2 pellets of PHWR fuel specifications using internal gelation process. In: Hastings I J (ed.) *Proc. 2nd Int. Conf. CANDU Fuel.* Canadian Nuclear Society, Pembroke, Canada, pp. 108–24

Ganguly C, Hegde P V 1997 Sol–gel microsphere pelletization process for fabrication of $(U,Pu)O_2$, (U,Pu)C and (U,Pu)N fuel pellets for the prototype fast breeder reactor in India. *J. Sol–Gel Sci. Technol.* **9**, 285

Ganguly C, Hegde P V, Jain G C 1994 Fabrication of $(Pu_{0.55}U_{0.55})C$ fuel pellets for the second core of FBTR. *Nucl. Technol.* **105**(3), 346–54

Ganguly C, Hegde P V, Sengupta A K 1991a Preparation, characterization and out-of-pile property evaluation of (U,Pu)N fuel pellets. *J. Nucl. Mater.* **178**, 234–41

Ganguly C, Jain G C, Hegde P V, Basak U, Mehrotra R S, Majumdar S, Roy P R 1986a Development and fabrication of 70% PuC–30% UC fuel for the Fast Breeder Test Reactor in India. *Nucl. Technol.* **72**, 59–69

Ganguly C, Langen H, Zimmer E, Merz E 1986b Sol–gel microsphere pelletization process for fabrication of high density ThO_2–2% UO_2 fuel for advanced pressurized heavy water reactors. *Nucl. Technol.* **73**, 84–95

Ganguly C, Linke U, Kaiser E 1987 Characterization of $(U,Ce)O_2$ pellets prepared by the sol–gel microsphere pelletization. *Metallography* **20**, 1–14

Ganguly C, Prasad G J, Mahule K N, Ghosh J K, Asari K B J, Chandrasekharan K N, Muralidhar S, Balan T S, Roy P R 1991b Fabrication experience of Al–U^{233} and Al–Pu plate fuel for PURNIMA-III and KAMINI research reactors. *Nucl. Technol.* **96**, 72–83

Ganguly C, Rodriguez P 2000 Indian nuclear fuel cycle: options and strategies. *Proc. 3rd Meeting Int. Working Group on Nuclear Fuel Cycle Options (IWGNFCO)*. International Atomic Energy Agency, Vienna

Ganguly C, Roy S K, Roy P R 1991c *Advanced Ceramics, Key Engineering Materials*. Trans Tech, Switzerland, Vols. 56, 57

Hastings I J (ed.) 1989 *Proc. 2nd Int. Conf. CANDU Fuel*. Canadian Nuclear Society, Pembroke, Canada

Hastings I J (ed.) 1989 *Proc. 2nd Int. Conf. CANDU Fuel*. Canadian Nuclear Society, Pembroke, Canada

IAEA 1974 IAEA technical report. *Proc. Int. Atomic Energy Agency Panel Meeting on Sol–Gel Processes for Fuel Fabrication*. International Atomic Energy Agency, Vienna

IAEA 1984 IAEA-TECDOC-352. *Proc. IAEA Advisory Group Meeting on Advanced Fuel Technology and Performance*. International Atomic Energy Agency, Vienna

IAEA 1988 IAEA-TECDOC-466. *Proc. IAEA Technical Committee Meeting on Advanced Fuels for Fast Breeder Reactors: Fabrication and Properties and their Optimization*. International Atomic Energy Agency, Vienna

IAEA 1997 IAEA-TECDOC-941. *Proc. IAEA Technical Committee Meeting on Recycling of Plutonium and Uranium in Water Reactor Fuel*. International Atomic Energy Agency, Vienna

Leary J, Kittle H 1977 *Proc. Topical Meeting on Advanced LMFBR Fuels*, ERDA 4455. American Nuclear Society

Mansfield R G 1976 Preparation of uranium–plutonium mixed carbide, ORNL–tr–4180, USA, English translation of Wedemeyer H, Guenther E 1975 Report KfK 2238, Karlsruhe, Germany

Matzke Hj 1986 *Science of Advanced LMFBR Fuels*. North Holland, Amsterdam

Razvi J 1998 TRIGA: 40 years of success. *General Atomics Update*, August

RERTR 1996 *Proc. 19th Int. Meeting on Reduced Enrichment for Research and Test Reactors (RERTR)*, RERTR 19. Korea Atomic Energy Research Institute, Seoul, Korea

RERTR 1998 *Proc. 21st Int. Meeting on Reduced Enrichment for Research and Test Reactors (RERTR)*, RERTR 21. Sao Paulo, Brazil

Richter K 1976 Preparation methods for uranium plutonium non-oxide fuels *Proc. European Nuclear Conf.* Pergamon Oxford, Vol. 7.

Roepenack H, Schneider V, Druckenbrodt W G 1984 Experience with AuPuC—Co-conversion process for mixed oxide fuel fabrication. *Am. Ceram. Soc. Bull.*, 63

Sejnoha R 1999 *Proc. 6th Int. Conf. on CANDU Fuel*. Canadian Nuclear Society, Niagara Falls, Canada, Vols. 1, 2

Sowden R G, Hodge N, Morten-Smith M S, White D V 1964 *Carbides in Nuclear Energy*, Macmillan, London, Vols. 1, 2

Stevenson C E 1987 *The EBR II Fuel Cycle Story*. American Nuclear Society

Vaidya V N, Mukherjee S K, Joshi J K, Kamat R V, Sood D D 1987 A study of chemical parameters of the internal gelation based sol–gel process for uranium dioxide. *J. Nucl. Mater.* **148**, 324

Vijayaraghavan R 1989 Experiences in the fabrication of aluminium clad metallic uranium fuel. In: *Materials Science Forum*. Trans Tech, Switzerland, Vols. 48, 49, p. 329

Wedemeyer H, Gunther E 1975 *Germon Report Kfk 2238*

Zimmer E, Ganguly C, Borchardt J, Langen H 1988 SGMP—an advanced method for fabrication of UO_2 and MOX fuel pellets. *J. Nucl. Mater.* **152**, 169–77

C. Ganguly

P

Particulate Processing (Powder Metallurgy)

Most writers date the beginning of powder metallurgy technology to the production of compacted platinum by Wollaston in 1813. However, there were many uses of metallic powders long before this. One of the first "recorded" mentions of metal powders comes from the tale of Jason and the Golden Fleece. While some will dismiss this as only fable, historical data show that people of Colchis used the fleece of sheep to consolidate the particles of alluvial gold.

Powders were also used for decorative purposes on ceramic objects as well as in paints and inks as early as 5000 BC.

Some of the basic principles of powder metallurgy were also used as early as 5000 years ago to prepare iron implements. Certainly the methods were crude and the objects that were formed varied considerably in quality. The famous "Damascus Steels" may be also considered as a product of powder metallurgy. The earliest description of the "Damascus Blades" dates back to AD 540, yet they may have been used by Alexander the Great as early as 300 BC. Actually the process did not originate in Damascus, but rather in India, where it was known as "Wootz." Like any steel-making process, the manufacture required the removal of oxygen from the iron ore and the addition of carbon for strength. The iron ore was mixed with charcoal and heated in a ceramic-lined stone hearth with air from a bellows to increase the temperature for quicker reaction. The reduced iron had a rather spongy texture and was then hammered to expel impurities. The pieces of wrought iron were then packed into a crucible with more charcoal and then reheated. When a certain amount of liquid metal was formed, the mass was slowly cooled in the furnace. The resulting cake, or "Wootz," was about the size of a hockey puck and could be easily traded or shipped in this form. The smiths then reheated the hammered cake into swords whose characteristic pattern was a result of carbon distribution.

Even as late as 1957 a similar process was still being practiced by the Matakam tribe in Africa to produce tools and jewelry.

Another product of this type of manufacture was the "Delhi Pillar" of India made in about AD 300, and at an estimated weight of six tons this even supersedes the largest powder metallurgy part made today. From this time the production of powders and/or objects seems to have ceased until about the eleventh century when a monk named Theophilus described a method of making powders of silver, copper, and brass by grinding in a mortar and pestle.

Around AD 1000 the Vikings established settlements in Greenland but they had no native metals to supply their needs for weapons and tools. They, however, found iron nodules in their exploration of the northern tip of Labrador. These nodules of "bog iron" are formed when certain bacteria precipitate iron from the water. It was from these spongy nodules that the Vikings forged some of their tools and weapons. While it was an inferior iron, it was something that the local natives in their Stone Age culture did not do.

The Medieval German writer Georgius Agricola in his *De Re Metallica*, published in 1556, describes a method of granulating lead and gold by pouring the molten metals in water agitated by brooms. It could be said that this was an early practice of metal powder atomization.

Either powder metallurgy lapsed into the realm of the forgotten in the seventeenth century or all of the records of that period have disappeared. In the years between 1750 and 1850 over 20 metals were produced for the first time and, almost without exception, they were made in the powder form.

Wollaston's work on platinum occurred in the early part of the nineteenth century, around 1829, which has been described in many publications dealing with powder metallurgy. A similar method was developed by Sobolewskoy in Russia, who claimed the method to be in use as early as 1826. The Russians also pressed platinum in a thick iron ring, heated it, and then restruck it. The resulting disks were then rolled and minted into coins. This might be considered as the forerunner to modern powder forging. In the years between 1830 and 1841 Osann developed a method for producing pressed and sintered copper powder, which is probably the basis of modern practices.

Around 1840 Henry Bessemer began the manufacture of flake powders from copper and copper alloys. It is interesting to note that Bessemer did not patent his process for fear of infringement and he kept the process secret for over 40 years. It is said that the money he made from the process (since he sold his powders at a price higher than that for silver) paid the costs for his experiments leading to the discovery of the Bessemer process for making steel.

Towards the middle of the nineteenth century powder metallurgy as a means of producing useful articles ceased completely. It is usually easier to trace growth than a decline: growth is usually apparent, while decline is not apparent until long after the trend has been underway. Why then did powder metallurgy not take its place as a productive source for articles people needed? Except for some sporadic experimentation, an entire generation completely forgot all the experiences and practices. The only explanation is that the fusion methods were far superior, and that the temperatures needed for melting and casting could be obtained and controlled with better refractories. Added to these factors was the fact that no simple machines existed that were capable of powder metal processing. Also about this time the modern steel age was ushered in by the inventions of Kelly and Bessemer.

In 1889 Mond discovered the carbonyl process for making nickel powders, although actual production of nickel by this process did not occur until 1895. We might consider this development as the beginning of the nanoscale powder era, which is the topic of many publications.

The lack of interest in powder metallurgy continued until 1908 when a patent was granted to Lowendahl for a bearing of sintered iron.

The refractory carbide industry first emerged in 1909 when Connell described a process for fusing tungsten beds with carbon to make bearings for electric meters. The first commercial production of sintered carbides was made possible by patents granted to Lohmann in 1914. The modern methods of producing cemented carbides can be traced to the work of Schroter who was granted a patent in 1925. Both the Lohmann and Schroter patents were later invalidated, although the methods described by Schroter are still in use. Later the Schroter patents were assigned to the General Electric Company in the USA whose Carboloy Division started production in 1928, at the same time as did the Firth Sterling Steel Company.

Commercial sponge iron powder production dates back to 1903 with the invention of the Sieurin process in Sweden. Since 1911 the Hoganas Corporation has continuously manufactured sponge iron, although the early use for this product was as a feedstock for steel making.

The birth of the porous bearing industry is generally accepted by many to be the development of the "Genelite" bearing material by General Electric in 1916. However, fundamentals of self-lubricating bearings were developed by Gwinn in 1870. Gwinn used a mixture of 99 parts of tin prepared by rasping or filing, and one part of petroleum still residue. The mixture was stirred while being heated and then pressed into a mold to shape it. The original use for bearings was as journal bearings in the railroad industry.

Apparently the concepts and methods of Gwinn were forgotten until the development of the "Genelite" bearing in 1916. The success of the Genelite bearing or the interest in self-lubricating bearings sparked a revival in the early 1920s. A number of patents were granted to Bound Brook Bearings, General Motors, and Chrysler Motors. In 1940 Roland Koehring of General Motors patented a process of bonding a powder metal layer with a lubricant or lead. The development of the porous bearings in the 1920s stimulated the production of copper powders. The first methods were by electrolysis and the reduction of copper oxides. Later such processes as cementation and hydrometallurgy were developed. Additional growth of the copper powder industry was encouraged by the expansion of the iron–copper structural parts industry.

From 1908, when the first use of iron powder was patented by Lowendahl, until the middle 1930s, little use was made of iron powder. The manufacture of oil pump gears by General Motors and door wedges by Chrysler encouraged a further increase in the use of iron powder, particularly in automotive applications. This continues to be the case today as automotive application estimates exceeded 30 lb (13.5 kg) per vehicle in 1999.

The advent of World War II and the demands of Germany for iron powder to make artillery shell driving bands effectively cut off the supply to the USA. Germany had found that iron could be substituted in these applications where copper was not available. The loss of Swedish iron powder imports was balanced by several US firms entering the field. At the end of the war, Swedish iron again became available and the use of iron rapidly accelerated.

Some early work by General Motors in 1942 was toward the hot forging of iron powder briquettes. No mention is made of any useful articles resulting from these experiments. The Amplex Division of Chrysler hot forged a machine-gun pawl from iron powder at about this time. Thus was born the modern process known as powder forging, a major aspect of the powder metallurgy industry.

The various stages through which powder metallurgy has progressed are somewhat analogous to archaeological ages. The "Stone Age" of powder metallurgy might be considered as the period in which Coolidge did his work on tungsten and the start of the work on the refractory metal carbides. As in the archaeological Stone Age, iron and copper powders were known but little used. The "Bronze Age" of powder metallurgy came in about in the early 1920s when the emphasis was on the manufacture of porous bronze bearings. Then in the early 1930s the powder metallurgy "Iron Age" began. This paralleled the historic Iron Age starting with low-density, simple iron and iron–copper parts. As the industry began to grow with the acceptance of powder metallurgy as a sound manufacturing technique, the size of the parts increased and the mechanical properties were upgraded. The property improvements continue and every year at all the major powder metallurgy industry conferences there are reports about how the processing has been improved and the properties are now approaching those of wrought metals.

Warm compaction is now becoming general and numerous parts are now being used in automotive applications. Metal injection molding (MIM) is now used for the complex parts that are beyond the capabilities of conventional compaction processes. The fine powders used in MIM can sinter at a much faster rate that the coarser counterparts resulting in finished products that approach the near full density of the alloy composition.

Powder metallurgy has even entered the "space age." Some of the first applications were pressed beryllium bodies for re-entry missiles. The heat shields for the Mercury capsules were forged from a 58 in (148 cm) diameter by 8 in (20 cm) thick billet of beryllium. Other billets up to 75 in (190 cm) in diameter and weighing 1.1×10^4 lb (4990 kg) have been produced at 99% of the pore-free density in the hot pressed state and machined without forging. As the demands continue to grow, powder metallurgy will be used to solve many of the outstanding challenges.

1. Powder Metallurgy

The manufacturing of engineered products using particulate processing (powder metallurgy) is one of the oldest and one of the fastest growing technologies of the metalworking industries. The powder metallurgy (P/M) process is widely accepted for many components because of its ability to fabricate high-quality, complex parts with close tolerances in a cost-effective manner. The materials are tailored to meet the properties of the end product application. The materials cover almost the entire spectrum of metals. It has been said that all metals can be processed in particulate form, but some metals have attributes that make them more readily processed by these techniques.

The interest in and growth of P/M are brought about by the fact that not only can a wide spectrum of metals and alloy compositions be used, one can also control whether there should be porosity or not. In addition, the process can be engineered to produce materials for specific applications with properties that are not possible by other metal-processing techniques, an example being a controlled porosity material for filtration or as a carrier for lubricants in bearing applications. Another application is for the use of composite compositions, such as tungsten and silver, used in contacts where it is desired to have the spark resistance of tungsten and the current-carrying capabilities of silver. There are numerous applications where it is the goal of processing to eliminate most or if not all the porosity. This is achieved through the hot working of powder preforms by either hot (or warm) pressing or by upset forging of preshaped preforms. An example of the use of this technique is in the powder forging of connecting rods for automotive engines. In addition, almost pore-free components can be produced through the use of MIM and the use of sintering activators. This technique eliminates the restrictions that the part must have straight-walled characteristics to be able to be ejected from the die.

One consideration in the manufacture of components by the P/M process is that the final performance of the part is directly related to the density, composition, and microstructure. Increasing the density will result in an overall improvement of the mechanical properties. This includes tensile properties, impact, and fatigue as well as the apparent hardness. Selection of the composition plays a major role also in determining the performance characteristics. Because P/M compositions are tailor-made for the application, one can be selective in the choice of elements. Furthermore, selection of composition can be made based on the least expensive elements that still meet the performance requirements. The microstructure of the P/M part, as with conventional wrought metallurgy, is quite important in determining part performance. One can adjust the composition to give the best properties, whether it be as-sintered or heat-treated parts.

2. Powder Fabrication Techniques

The starting particulates are prepared by a number of different processing techniques, discussed in greater detail in this article. These particles can be produced by various fabrication techniques that include mechanical, electrolytic, chemical, and atomization.

2.1 Mechanical Methods

The most common methods of mechanical means for preparation of particulates are through impaction, attrition, shearing, and compression. Impact formation of particles results from the delivery of a high and rapid load to a metal causing fracture and reducing the size of the metal. If this impact load is done in multiple blows, then the particle size is continually reduced. The process is generally time controlled to obtain the desired particle size or distribution. The attrition process is the rubbing of powder particles against each other causing the reduction of the particle(s). Again this process is time controlled. The shearing process is where a material is fractured along shear planes and thus forming particles. This process works especially well where the base material is hard and brittle. The process does not lend itself to the production of fine powders. The compression process is where a compressive load is applied to the base metal causing fracturing. Like the shearing process, this process works best on materials that are hard and brittle. Particles produced by this process tend to be coarse and are not generally used in conventional P/M compaction. Frequently, powders are prepared by a combination of two or more of these preparation techniques. For example, machine turnings can be further reduced in size by using ball milling in order to obtain a fine powder. Under normal circumstances because of the high amount of cold work induced in the particles, they have to be annealed to soften them in order to allow for compaction and densification of the part prior to sintering. The annealing step has to be done in a protective atmosphere and the temperature must be sufficiently high to soften the particles, but low enough so that particle sintering does not occur.

2.2 *Electrolytic Methods*

Powders can be precipitated at the cathode of an electrolytic cell under controlled processing conditions. Powders thus formed have a high chemical purity, which is associated with this processing. The metal is dissolved at the anode and deposited at the cathode, with the reaction being driven by an electrical current passing through the electrolyte. The cathode deposit is generally a dendritic structure and porous in nature and is easily removed. The removed metal is then washed and ground to control the desired particle size for further processing. The powder is screened to control the mesh size distribution and finally annealed to reduce any surface oxides and to soften the particles so that they might have good compression characteristics. The powders most commonly produced by this method are iron, copper, zinc, manganese, palladium, and silver.

2.3 *Chemical Methods*

This method can be employed to produce almost all powders and includes decomposition of a solid by a gas, thermal decomposition, and precipitation from a liquid or gas.

In the gaseous reduction process, a purified ore is commonly used as the starting material. An example of this technique is the reduction of magnetite iron ore to produce sponge iron powder. The sponge cake is crushed, ground, screened, and annealed resulting in a powder that is widely used for low- to medium-density part applications. In addition to iron some grades of copper, molybdenum, and tungsten are produced in this manner.

Powders can be produced by a combination of vapor decomposition and condensation. The carbonyl processes for iron and nickel are the most common examples of this method. The metals are reacted with carbon monoxide to form the carbonyls. The formation requires both pressure and heating. The powders thus produced are characterized as being quite fine and somewhat spherical in shape. The metal powders that are candidates for this process are nickel and iron. Other metals are not considered because of the high energy costs and health hazards associated with their processing.

Powders can be produced by precipitation from a liquid. In this process a dissolved metal salt can be treated to produce either a metallic or metal precipitate. Chemically precipitated powders are characterized as being very fine with high degrees of purity. These powders tend to agglomerate. The candidate metals for this type powder forming are copper, cobalt, and nickel. The powder can also be precipitated from a gas. Reactive metals are candidates for this of processing because the particles are formed without melting and contact with a crucible, which could be the source of contamination. Candidates for this type processing include nickel, cobalt, zirconium, titanium, niobium, and molybdenum.

2.4 *Atomization Methods*

Atomization involves the formation of powder from molten metal using a spray of either a gas or liquid. This is the most significant method of producing metal powder particles. In gas atomization, generally inert gases like nitrogen, helium, or argon are used for breaking up the stream of molten metal. The liquid metal is disintegrated by rapid expansion of the gas out of a nozzle. Super alloys are ideal candidates for this type of preparation. During the flight of the particles in the collection chamber, the particles cool and solidify. The particle size can be controlled by adjusting the melt temperature, the nozzle orifice opening, and the volume and pressure of the gas. Powders made by gas atomization tend to somewhat spherical in shape. Many metals are made by atomization using a liquid as the medium and water is quite often the choice because it is inexpensive. As with gas atomization the process is controlled through the metal melting temperature, the nozzle design, and the liquid volume and pressure. The process is generally designed so that the particles are somewhat irregular in shape. The shape might be classified as being "potato like". This is done so that the particles have a certain amount of interlocking during the compression cycle, and the compacts have good green strength so that they resist fracture while being transported to the sintering furnace. After atomization, the particles thus formed are screened and annealed. During the annealing the oxides that might have formed during the atomization are reduced and the particles softened should they be capable of being hardened by quenching. The annealing is done in a protective reducing atmosphere and the temperature is high enough to be effective for the reduction process, but not sufficiently high to get particles to sinter. After the annealing there is at times a light cake formed which must be broken up. Care is taken so as not to induce any cold work into the particles.

3. Part Fabrication

3.1 Powder Mixing

Base metal powders do not, in many instances, have the desired chemical composition required in the finished part. Consequently, additions are made in the form powders to achieve this goal. An example of this is with ferrous P/M powders where many of the desired elemental additions are made as fine powder particles rather than by alloying them with the base iron during the powder manufacture stage. Should these elements be ferrite strengtheners they would have an influence on the compression characteristics of the powder and thus the powder would not be able to be compacted to the desired density range to achieve the requirements of the finished component. Good examples of this are with the commonly used P/M steels where copper, nickel, and manganese are the primary additives. Carbon also falls into this category and to achieve the finished carbon level necessary to achieve the required properties, a graphite addition is made to the premix. The graphite reacts with the base iron, thus forming a steel. This level of carbon can be rather precisely controlled, to much the same levels achieved with conventional wrought steel compositions.

Some base powders are produced with some alloying elements already in solution within the metal. An example of this is molybdenum in iron. It has been found that molybdenum up to several percent can be prealloyed without having any appreciably effect on the compressibility of the powder. This has been quite beneficial in producing parts that require additional strength properties over straight iron or even iron–copper and iron–nickel P/M materials. Furthermore, molybdenum can improve the hardenability of the product; therefore molybdenum steels are ideal candidates for hybrid heat-treatable alloy compositions.

This premixing is accomplished in mixers which when charged with the additives are rotated allowing the powders to tumble and thus mix thoroughly to a uniform composition. This is necessary because the final mix composition has a profound influence not only the finished part properties but also on the dimensional characteristics as the parts are passed through the sintering furnace. Controlling the dimensional characteristics is one of the fundamental reasons for the success of the P/M process, namely the ability to produce high volumes of duplicate parts with good properties and dimensional control, and of course at a cost which is less than other metal-fabrication processes. It is for this reason that the automotive industry has converted a large number of parts to P/M.

Generally speaking, a dry lubricant is added to the metal powder premix. The purpose of this lubricant is to reduce the friction between the powder particles during compact, as well as to reduce the friction between the particles and the die wall. It also serves to help slide the part out of the die on the ejection cycle of the press. Common lubricants are metallic soaps such as stearates of zinc, iron, and barium. Waxes like ethylenebisstearamide are becoming more widely accepted. The amount and type of lubricant are held to a minimum in order to achieve the desired compression. The lubricants, being of a low specific gravity, can take up a considerable volume in the premix. The more lubricant, the lower the pore-free density of the premix. Naturally one wants to be able to achieve high densities during part compaction, so a compromise has to be made between overcoming the frictional factors and achieving high part density. In addition, the lubricant must be completely "burned off" prior to reaching the sintering temperature. Any residue or ash left by the lubricant impedes the particle sintering, thus resulting in parts with poor mechanical properties.

Developments have shown that the properties of the premix can be enhanced through the addition of binders. This binder treatment technology is being offered by several powder producers and is becoming more widely accepted for part production. These premixes generally have better flow characteristics, reduced dusting potential, and a more uniform chemical distribution of the additives. In this development the binder must not in any way decrease the compressibility of the premix and must not leave an ash or residue to interfere with the sintering process. These developments have resulted in proprietary material compositions of powder suppliers.

Another advantage of the P/M process is that it is essentially a scrap-free process. Except for a small amount of powder lost when filling the die cavity, all the powder is formed into usable parts. During the set up of the compaction press, any parts which are formed and do not conform to the dimensional tolerances that are set as the target, can be ground and the powder reclaimed by mixing back with the original mixes.

3.2 Compaction

The premixed powders can be consolidated by compaction in steel or ceramic tooling to form the part shape, which can have multiple levels with a complex shape. This technique allows the parts to be compressed at very fast rates. Mechanical presses are the fastest; hydraulic presses can be used but they normally are significantly slower. The compaction cycle is generally made up of die filling, entrapment of the powder within the tooling, compression of the powder through the application of pressure from both the upper and lower moving rams, and finally ejection of the part from the die. This total cycle takes a fraction of a minute and the controlling parameter is how fast the powder fills the die cavity. That is why the flow rate of the powder premix is so important. Another controlling factor is the height of the finished part. Normally the loose powder height is compressed by a factor of 2–3 times. A fill height, for example, of 3 in (7.6 cm) would end up as a part with a height of 1.5–2.0 in (3.8–5.1 cm). The fill depth of a compaction press normally does not exceed about 8 in (20 cm). Therefore, the maximum height of a part in the pressing direction is about 3–4 in (7.6–10 cm). Press capacity is not a limiting factor, the only limit being the expense of building large-capacity presses with multiple tool movements. Some of the largest presses have capacities in the region of 2000 tons (2030 t).

The compaction pressure can be in the range 20–60 tsi (3×10^8–9×10^8 Pa). This limits the maximum size of pat that can be compacted on a given press. For example, if one wants to compact a part with a projected surface area of 5 in^2 (160 cm^2) and the compaction pressure to achieve the part green density (density of the as-pressed part) is 50 tsi (7.7×10^8 Pa), a compaction press is required with a capacity of at least 250 tons (5×50) (254 t). Considering that one of the largest presses available has a capacity of about 2000 tons (2030 t), then using a 50 tsi (7.7×10^8 Pa) compaction pressure would limited the part to a projected surface of 20 in^2 (260 cm^2). This is not very large in comparison to parts processed by some other metalworking techniques such as forging or casting. However, the precision of parts made by the P/M process is more than likely to be considerably better than can be achieved with other methods.

Some powders are compacted by means other than conventional die methods. Some reactive metal powders must be containerized either in a can or wrapped in foil to prevent the powder from reacting with the atmosphere. The containerized metals are then consolidated by means of isostatic gas pressurization or use of a liquid for transmitting the pressure. Hot working of the container is also used. This includes hot pressing and hot forging. Naturally the shape of the part formed is rather simple and is most likely a billet that has to be formed by some other working technique or perhaps has to be completely machined to obtain the desired part shape and tolerances.

Cold isostatic pressing is used to form some rather complex-shaped parts. The powder is placed in a pliable mold, which is placed in a chamber where a fluid is pressurized around the mold. The part formed in this manner takes the shape of the mold. Obviously the compressed part is smaller in dimensions than the mold because the apparent density of the powder mixture is less than that of the compressed part. The part can be removed rather easily from the mold. The process has some distinct advantages over conventional die pressing in that the pressure is applied uniformly around the compact rather than just applied from the top and bottom punches. The pressed part has a more uniform density and is not restricted by the frictional forces of the tool members. The parts can be pressed to lengths that are impractical for conventional compaction. Generally in conventional compaction the maximum part height is about 4–5 in (10–13 cm). With isostatic pressing, cylinder liners for automotive engines can be formed with lengths of 15–18 in (38–46 cm). Another advantage of this processing technique is that the dimensional change during sintering is much more uniform and predictable because of the uniform density distribution within the part. The disadvantage of the technique is that the rate of the cycle is much slower than for conventional compaction. Therefore, to be economical the parts have to have some unique characteristic like the cylinder liner mentioned above.

MIM is another method of consolidating the powder mass into a component part. The process begins by mixing fine base powder particles (less than 20 μm) with a binder. The mixture is heated, granulated, and pelletized. These pellets are injection molded into the desired shape. The binder is then removed and the remaining powder shape is sintered. The parts can, in many instances, be used in the as-sintered condition because the part densities are close to the pore-free density of the base alloy. Should it be necessary, the parts can be further processed in secondary operations to further increase the density, heat treatment, or machining of contours not possible in the molding operation. The advantage of this process is that extremely complex shapes are feasible. The shapes are typical for conventional injection molded parts, except that with the powder approach alloy compositions can be selected which are not practical by conventional practices.

3.3 Sintering

Sintering remains an operation of the P/M process that is not widely understood, perhaps because one cannot see it happen. It is generally carried out in a furnace operating at an elevated temperature. This temperature is below the melting temperature of the major base powder. So, in effect, the process is a solid-phase bonding process where the particles are fused together resulting in a component that has controllable properties. Because the temperatures are high enough that rapid oxidation occurs during processing, a protective and reducing atmosphere is supplied to the furnace.

Some metal powder mixtures for making structural parts contain a second additive that has a lower melting temperature than the sintering temperature of the base metal powder. In this case, the low-melting liquid phase provides a rapid acceleration of the sintering process. Wetting is a requirement for the liquid phase. As the sintering progresses, the liquid can diffuse into the microstructure and form a new alloy system. An example of this is the iron–copper system, where copper powder is mixed with iron powder. A sintering temperature is selected that is above the melting temperature of copper but below that of iron. The copper wets the iron and is uniformly distributed throughout the part. Copper is soluble in iron so with further time at the sintering temperature, the copper is diffused. The copper imparts higher mechanical properties than the iron has by itself. This type of processing is called "liquid-phase sintering."

The sintering process has four basic stages when the parts are passed through the furnace. These are heat-up, lubricant burn-off, sintering, and part cooling. The heat-up stage is necessary so that the parts are not thermally shocked by a rapid temperature change. The lubricant added to aid in the compaction cycle must be removed prior to the part reaching the sintering zone of the furnace. The lubricant can be either melted out or burned off. The decomposition products from the lubricant must be flushed from the furnace. The sintering atmosphere accomplishes this in its movement from the entry between the vestibule of the furnace and the high-temperature zone. A positive pressure is required for the flushing. The lubricant removal is at a substantially lower temperature than the sintering temperature and varies depending on the type of lubricant used. Generally the temperature is about 650°C (1200°F). The lubricant removal can be done in a separate operation, but for economical reasons it is generally part of the continuous operation of the furnace. The sintering or high-temperature zone of the furnace is where the powder metal compact is consolidated into a usable mass. The part is slowly moved from the burn-off zone into the high-heat zone. Again this is done so the part is heated slowly to prevent thermal shock. After the appropriate time at a temperature to achieve the sintering, the part then is transported to the cooling zone. This zone is generally water jacketed to enable the part to cool in a reasonable period of time while still under the protective atmosphere. The part is then removed from the furnace at the exit.

The transport of the part through the various zones of the furnace can be done through pushers or through a continuous metal mesh belt that runs through the total length of the furnace. In the case of pusher furnaces, the parts must be loaded onto plates or placed in boats. The loaded parts are then transported with a pusher rod through each of the furnace zones. Some sintering furnaces are roller hearth types as higher furnace loads are possible. If a belt is used it must be made from an alloy composition that has a good lifetime through this thermal processing. It also must have a good resistance to stretching at elevated temperatures because the furnace charge of parts can be quite heavy at times. In many instances the parts are placed directly on the belt, but if flatness is a major concern the parts can be loaded on ceramic trays. As one might expect, when trays are used this increases the furnace charge and adequate time must be added to compensate for this additional mass. For the process to be economical, sintering furnaces are generally run continuously with little downtime.

The furnaces are heated by gas or electrical means. The selection depends on the least expensive method in the area where the furnace is being installed. The concern is that the temperature range can be controlled to relatively tight tolerances. This is necessary because the dimensional change characteristics of the part are directly related to the sintering temperature, time, and atmosphere.

Generally the furnaces have hearths which can be either refractory brick linings or metallic walled that can be either fabricated or cast. Metallic hearths are preferred because the properties of the sintering atmosphere can be controlled to a greater degree.

The protective atmospheres used most likely to be reducing in nature to ensure low oxygen or oxide content in the finished parts. A lower oxygen content in the finished part results in generally better dynamic properties.

Hydrogen is an ideal atmosphere for sintering because of its reducing potential for oxides. However, it is rarely used in the pure form because it is somewhat expensive. Dissociated ammonia (75 vol.% nitrogen–25 vol.% hydrogen) is an alternative. Nitrogen–hydrogen mixtures are quite attractive for use in sintering. Forming gases with less than 10 vol.% hydrogen are adequate in most sintering applications. Pure nitrogen is used with some nonferrous metals where there is no concern about nitriding.

When possible, endothermic or exothermic gas atmospheres are used. In these cases, the carbon potential of the sintering gas atmosphere can be controlled to precise ranges and thus prevent decarburization of steels. They can also be used where a carburized case is desired for the finished part. P/M parts generally respond to case hardening at much faster rates than conventional steels because the interconnected porosity provides easy access for the gas to penetrate the structure. In essence, the exposed surface-to-volume ratio of P/M parts is greater than that of a similar shaped wrought metal part. Actually, the diffusion rates of metals of the same composition are identical.

A vacuum is used for sintering of reactive metals. Partial vacuum levels are in many instances adequate for the processing of parts. Continuous vacuum furnaces have been developed so the restriction of operation in batch modes has been overcome. The operation is, however, still slower than continuous belt-type furnaces.

There has been a concerted effort to develop faster cooling rates for furnaces. These efforts have been fruitful and accelerated cooling is possible. This is important because some commercial P/M compositions are capable of being sinter hardened. Here the part is above the austenitizing temperature during sintering, so increased cooling can be used with selective compositions to obtain a transformation to either a bainitic or martensitic microstructure. This has proved to be a cost-effective alternative and is now in widespread use. The added benefit of this processing is that liquid quenching is avoided and part distortion is minimized. Naturally the sinter-hardened parts should be tempered as is appropriate for heat-treated parts. Tempering is particularly important where dynamic properties are an important criterion of the application.

In most P/M part fabrication operations, the sintering operation seems to be the rate-limiting step. The compaction operation is done at a faster rate than sintering. In a three-shift operation, it is not unusual to find a two-shift operation for compaction and a three-shift operation for sintering. Sintering the most costly of the processing steps.

3.4 Secondary Operations

After the parts are sintered, it is often necessary to perform secondary operations. These operations can be simple sizing of the part to achieve better qualification of the dimensions or repressing to achieve increased densification and thus improve the mechanical properties.

Parts can be machined to obtain contours that are not possible to compact, e.g., reverse tapers, side holes, threads, and undercuts. Because the P/M part contains porosity, its machining characteristics are not the same as a similar wrought composition. The cuts tend to be interrupted. Many of these differences are overcome by the adjustment of the parameters of speed, feed, and depth of cut. In addition, different tool steels or carbides are used for the processing. Likewise, milling and honing require some special attention to the processing parameters. Grinding of P/M parts is seldom any problem.

Parts can be joined or welded to form assemblies. This joining can be done during the sintering operation by the use of special brazing alloys specifically designed for furnace processing. Parts can be assembled in the green condition and joined by the use of brazing alloys or by selective copper infiltration as is possible with ferrous-based metal powder parts. Welding practices like friction, laser and electron beam, and MIG and TIG have been successfully used with P/M parts.

Most common heat treatments can be used for P/M components. Generally speaking, it is immaterial whether the part is formed by P/M processing or some other metal-processing technique. It is the alloy composition that is the important factor. So, P/M parts follow all the traditional metallurgical principles. Parts can be annealed, carbo-nitrided, carburized, case hardened, nitro-carburized, precipitation hardened, steam treated, stress relieved, tempered, and through hardened.

Since the range of P/M alloys available is too large to be cover in a short paragraph on heat treatment, one can generalize by saying that the practices are tailored to meet the requirement of the finished component. This might mean that one needs surface wear properties, or strength through the cross-section to meet performance requirements. With nonferrous alloys precipitation hardening can be used. The main concern is to control the microstructure. As discussed above, the final performance characteristics of a P/M part are related to the DCM, density, composition, and microstructure. The density is controlled during compaction, repressing, and forging.

P/M parts can be plated by vapor deposition or conventional electroplating. In either case the major concern is the porosity. The voids must be filled by some means to assure a continuous surface coating. Should any of the plating solution be absorbed into the porosity, it can bleed out at a later time and cause surface staining or corrosion. Filling the pores can be accomplished by infiltrating the part with a metal, or impregnating it with a plastic that does not react during the plating. If the density of the compacted and sintered part is above 90% of the pore-free density of the base powder alloy, there is little, if any, porosity interconnected with the surface. Under these conditions, filling the pores can be omitted from the processing cycle.

4. Summary

The purpose of this article is to set the stage for other contributions on the various stages of P/M processing of parts. It is certainly meant to be nothing more than an introduction. The details of the P/M process are covered in other articles in the areas of atomization, powder characterization, compaction, sintering (both solid state and liquid phase), MIM, metal joining processes, and heat treatment. It is strongly recommended that the reader consult these discussions for a clearer picture of the powder metallurgy process.

D. Apelian

Plastic Forming

Plastic deformation of ceramic raw materials is used to manufacture a variety of products including bricks, tubing, electrical insulators, microelectronic sensors, and refractory components. Clay-based mixtures are rendered plastic by the simple addition of water and are used extensively to form traditional ceramics such as pottery, bricks, and tableware. Nonclay ceramic powders, such as oxides, carbides, and nitrides, are generally used to produce advanced engineering components and are made workable by the addition of organic plasticizers and an organic or aqueous liquid.

This article begins with a review of standard and emerging industrial methods for the plastic forming of ceramics. Methods of measuring the plasticity of ceramic raw materials are then summarized before a review of the critical factors involved with raw material formulation. Plastic molding of art pottery will not be discussed, as this topic is reviewed in the ceramic art literature. Ceramic injection molding is an important plastic forming operation and is reviewed elsewhere (see *Injection Molding*).

1. Plastic Forming Methods

1.1 Extrusion

Extrusion is a common plastic forming method for industrial ceramics. Extrusion is a process of shaping an object by compacting a plastic mass in a high-pressure cylindrical chamber (barrel) and pushing it through the barrel towards a specially shaped die orifice. The cross-section of the extruded column (the pug) takes the shape of the die orifice. As the plastic mass exits the die, it is supported on a flat or shaped surface to prevent distortion. Sections of the desired length are formed by cutting through the pug with stiff wires or rotating knives. Ceramic extrusion has been comprehensively reviewed (Norton 1970, Jones and Berard 1972, Richerson 1982, Reed 1995, Janney *et al.* 1987).

Extruded products can be very small, such as tubes of under 1.5 mm in diameter, or large pipes up to 1 m in diameter. Electronic substrates and sensors are made by extruding flexible ceramic sheets (Wu and Donato 1995, Takebe *et al.* 1992, Thompson 1963). Complex cross-sections can be extruded, such as thin-walled honeycomb structures that are used for heat exchangers, woodstove combustors, molten metal filters, and catalyst supports (Lachman 1984).

There are two main types of extrusion equipment: (i) auger extruders and (ii) piston extruders. In both cases, the barrel and the die are usually made of a tough, wear- and corrosion-resistant alloy steel. In auger extrusion, an auger is located inside the cylindrical barrel and forces the plastic material towards the die. Piston extruders have essentially the same design, except that the auger is replaced by a piston fitted to the inside of the barrel and forces the plastic material toward the die. Piston extrusion is a batch process, whereas auger extrusion is continuous. Extrusion presses capable of exerting 900 kN or more are commercially available; however, presses of up to 250 kN capacity are more common.

Before extrusion begins, a plastic mix is fed through a vacuum pug mill where it is kneaded to remove air and to achieve uniform consistency. This de-aired material can then be continuously fed into an auger extruder, or it may be formed into billets that subsequently feed into a piston extruder.

Clay-based extrusion is well known, and body formulations are made workable by the simple addition of water and small amounts of a deflocculant, a wetting agent, and a lubricant. Janney *et al.* (1987) compiled an extensive bibliography on clay-based extrusion. Nonclay ceramics are rendered plastic either by the addition of clay or organic additives (Sarkar and Greminger 1988, Zheng *et al.* 1994, Shanefield 1995, Onoda 1978, Chen and Cawley 1989).

A variety of defects can arise during extrusion, such as lamination, surface and edge tearing, bridge and core cracking, column splitting, segregation, and preferred orientation of particles. Robinson (1978) conducted an extensive review of the causes and potential cures for these defects. Funk (1984) also reported a series of case studies identifying typical problems with porcelain extrusion.

Extrusion is often used as a preliminary step to other forming methods. For example, long, single-piece, high-voltage electrical insulators are formed by lathe-turning porcelain cylinders that have been extruded and dried (Jones and Berard 1972).

Brady *et al.* (1995) fabricated unique textured microstructures by coextrusion. This method entails preparation of a composite feed rod by wrapping a core rod of plastic material (e.g., a ceramic powder slurry) with a second plastic material (e.g., a metal powder slurry). This composite feed rod is then extruded into a thin filament. Many such filaments are molded into a composite rod that is subsequently re-extruded to produce an extrudate with a textured, multifilament microstructure. Examples include Si_3N_4/BN, Al_2O_3/graphite, and ZrO_2/Ni. Slurry formulations generally entail use of a high molecular weight thermoplastic polymer (e.g., ethylene vinyl acetate) and a processing aid such as mineral oil or polyethylene glycol. High-temperature extrusion equipment is needed. Coextruded composite rods can also be pressed into thin sheets between heated lamination dies in a hydraulic press.

Tarpley *et al.* (1961) evaluated the extrusion behavior of several clay-based and nonclay-based ceramic suspensions. In all cases the rate of extrusion can be increased by ultrasonically vibrating the extruder die. Ultrasonic vibration of the die enables extrusion of materials that are normally not extrudable because of their low plasticizer or water content.

The effects of different slurry formulation variables on extrusion flow behavior can be analyzed using the Benbow equation, which describes the extrusion pressure, P, needed to produce a given extrudate velocity, V (Benbow *et al.* 1987, 1989):

$$P = (4L/D)(\tau_f + k_f V^m) + 2\ln[(D_o/D)(\tau_b + k_b V^n)] \quad (1)$$

where D_o is the barrel inside diameter, D is the inside diameter of the die orifice, L is the die land length, τ is the yield stress, k is a velocity factor, the subscripts b and f refer to the plastic body and the die land slip film, respectively, and n and m are the shear thinning exponents for the plastic body and slip film, respectively. Benbow *et al.* (1989) reported that $m = n = 1.0$ when fine, rigid particles are mixed with a low-viscosity liquid and extruded through square entry dies of circular cross-section.

The parameters τ_b, τ_f, k_b, and k_f can be evaluated for a given batch formulation using a mechanical testing machine that is set up to operate a laboratory-scale piston extruder. The mechanical testing machine pushes the piston into the filled barrel at a constant speed while measuring the force transmitted to the piston. Extrusion tests are then conducted at different piston speeds with interchangeable dies having different aspect ratios (L/D). Steady-state forces and velocities are then compiled and fitted to Eqn. (1) in order to calculate τ_b, τ_f, k_b, and k_f.

Table 1 gives an example where this process is used to study the ability of a maltodextrin sugar to improve extrusion of aqueous Al_2O_3 suspensions (Schilling *et al.* 1999). Filter pressing is first used to make a slurry paste containing deionized water and 49 vol.% submicrometer alumina powder. This paste exhibits "clay-like" plasticity from the standpoint that the parameters τ_b, τ_f, k_b, and k_f are similar to those of the kaolin formulation listed in Table 1. The solids concentration of the same alumina plus water suspension is subsequently raised to 57 vol.% by filter pressing, and the resulting filter cake is too stiff and brittle to be extruded. The subsequent addition of 3% maltodextrin sugar makes this same 57 vol.% alumina plus water system easy to extrude, although it has higher yield stresses and higher velocity factors than does kaolin.

1.2 Jiggering and Roller Forming

Jiggering is used to manufacture round, flat pieces such as tableware (Norton 1970, Jones and Berard 1972). A cylindrical column of a plastic, clay-based material is first de-aired and extruded from a vacuum auger. The extruded column is then sheared into disk-shaped bats. A single bat is then lowered on to the upper surface of a rotating plaster mold. A formed tool (template or knife) is then brought down on the upper surface of the rotating clay, and water is sprayed on this upper surface for lubrication. The template forms the upper surface of the clay while pressing the bat onto the contours of the rotating plaster mold. Templates can be made of hardened steel; however, template wear can cause discoloration of the kiln-fired ceramic. Templates can be made of a biscuit-ware ceramic in order to avoid this problem. Jiggering can be automated or performed manually.

The forming template is often a shaped roller, in which case the forming operation is called roller forming (Clough 1964). In this case, a polished steel roller swings down on the revolving mold and rolls the clay body out to the proper shape. A trimming knife slices off the excess clay at the edge during rolling. Cups and flatware can be made by this process.

1.3 Miscellaneous Forming Methods

Wet pressing is used to produce electrical insulators and special refractory shapes (Norton 1970, Jones and Berard 1972). In wet pressing, a plastic mass is formed by pressing a de-aired and extruded slug into a steel or plaster die. As pressure is applied, the plastic mass flows into the contours of the die. Wet pressing is largely confined to the manufacturing of clay-based products. In one version of this process, a steel die is heated, causing a steam cushion to develop between the plastic mass and the die surface.

Table 1
Packing density and binder effects on extrusion of Al_2O_3.

Material	Benbow parameters			
	τ_b (MPa)	k_b (MPa s m^{-1})	τ_f (MPa)	k_f (MPa s m^{-1})
49 vol.% Al_2O_3 + water	0.34	2.8	0.05	1.0
57 vol.% Al_2O_3 + water	a	a	a	a
57 vol.% Al_2O_3 + water + 3 wt.% maltodextrin	1.72	16.4	0.13	4.26
60 vol.% kaolin + water	0.42	3.7	0.03	0.69

[a]Samples too stiff to be extruded.

The steam cushion acts as a lubricant to promote easy removal of the formed part. If the die is not heated, it must be lubricated. In some operations, the die rotates in order to produce a smooth surface finish to the molded ware.

The ram process entails squeezing a lump of plastic slurry between two damp molds that are mounted in a hydraulic press. The molds are normally made of porous plaster, although porous ceramic molds can also be used that are more wear resistant. The pores in each mold are continuous and a few micrometers in size. Embedded in each mold is plastic or metal tubing that is used to connect a vacuum or compressed air to the pores in a mold. As the mold halves close, the plastic mass fills the mold cavity and the excess material extrudes out forming a flashing. At this stage, compressed air is forced through the bottom mold and vacuum is simultaneously applied to the upper mold. The mold halves subsequently open and the vacuum in the upper mold lifts the pressed ware away from the lower mold. The ware is then released onto a platen. The ram process is largely used to produce clay-based ceramics. Schilling *et al.* (1999) have developed aqueous, polysaccharide-based formulations that allow ram pressing of non-clay ceramics.

Compression molding, transfer molding, vacuum forming, and blow molding originated in the plastics manufacturing industry and are used to produce ceramic components (Haunton *et al.* 1990, Hammond and Evans 1991). In each case, a mixture of ceramic powder and a thermoplastic resin is first heated, vacuum de-aired, and compounded in a plastics extruder. Compression molding entails feeding a bat of this extruded material into a heated die cavity where the mix plastically deforms under pressure to fill the cavity. Blow molding and vacuum forming, respectively, entail the use of compressed air or vacuum to force a bat of plastic material into a heated die cavity.

Roll compaction (also called roll forming) produces a flexible, thick ceramic sheet by squeezing a plastic mass between two rotating metal cylinders (Richerson 1982). Applications include microelectronic substrates and sensors. High differential shrinkage and inadequate surface finish are common problems with roll compaction. It is possible to alleviate these problems by the use of sheet extrusion instead of roll compaction (Wu and Donato 1995, Takebe *et al.* 1992, Thompson 1963). Calendaring is similar to roll compaction, except that grooves are machined along the length of a roller surface. Corrugated tapes are produced that are used for ceramic heat exchangers and catalyst supports (Richerson 1982).

Intricate ceramic parts can be machine stamped from predried sheets of plasticized ceramic materials. The sheets can be made by roll forming or tape casting. Forming the material initially into a sheet can eliminate some of the problems associated with conventional dry pressing of small, thin parts (e.g., porosity gradients, costly high-pressure dies).

Griffith *et al.* (1995) reported on the microfabrication of ceria–zirconia composites by making simple surface impressions onto soft ceramic green tapes using a plasma-etched polyimide pattern as a micromold. Features as small as ten micrometers can be replicated.

2. *Analyzing Plasticity*

The term plasticity refers to permanent deformation without rupture when the applied load produces a shear stress exceeding the shear strength of a material. A variety of empirical tests to determine indices of plasticity exist in the literature. These tests can be categorized as torque of mixing, indentation, torsion, unconfined compression, vane rheometry, capillary rheometry, and simple shear tests (Phelps *et al.* 1989, Reed 1995, Norton 1970, Steffe 1996). Each test involves the application of a static or sinusoidally varying load while simultaneously measuring the deformation of the plastic material.

Indices of plasticity vary according to the test method used. Workability is a common index of plasticity that is used in clay-based ceramics. It is measured during axial compression of a billet of plastic material and is defined as the product of the maximum stress and maximum compressive displacement before cracking occurs (Norton 1970). Steffe (1996) performed a comprehensive review of standard methods for measuring the rheological properties of liquids and viscoelastic media.

Computer simulations have been developed to model the plastic flow behavior of ceramic powder suspensions. Two generic types of simulations exist: (i) finite element methods that predict localized plastic flow based on predetermined measurements of the elastic and plastic properties of a given suspension (Zheng *et al.* 1992), and (ii) discrete element simulations, where the macroscopic deformation of a granular medium is calculated based on the movements of individual particles and the friction and adhesion of particle–particle contacts (Aizawa 1995, Cundall and Hart 1992, Hong and Greil 1995). These methods have great potential for the computer-aided design of plastic forming operations as computers become faster and the simulation software continues to advance rapidly.

3. *Raw Material Preparation*

Plastic feed materials are commonly prepared by one of two methods: (i) batching and mixing raw materials and additives in a high-shear mixer, or (ii) mixing a low-viscosity, dilute slurry and then filter pressing the slurry to form a plastic filter cake.

In applications where de-aired plastic materials are needed, the media produced by methods (i) or (ii) subsequently undergo further processing in a pug mill. In this case, paddles or augers mix the plastic feed material and force it through a shredder into a vacuum chamber, which de-airs the shredded material. Shreds that are smaller in cross-section are more efficiently de-aired without excessive surface drying. De-aired shreds are then continuously fed into an auger, which mixes, consolidates, and extrudes the plastic material through a die. The extrudate can be used as a formed product or as a controlled feed material for a subsequent forming operation.

Compositions containing clay are made workable by the simple addition of water and small amounts of a deflocculant, a wetting agent, and a lubricant. Mixtures of nonclay ceramic powders can be rendered plastic by two possible methods: (i) the addition of clay and (ii) the addition of organic additives. Typically, 50–65 vol.% of a liquid medium is added to nonclay ceramic powder in order to achieve adequate plasticity for forming.

Organic additives are generally classified as binders, plasticizers, and lubricants. Binders provide wet strength during forming and dry strength after drying. Plasticizers and lubricants lower the frictional forces present during forming and permit forming at lower pressures. Methyl cellulose, poly(vinyl alcohol), poly(ethylene oxide), polyacrylimides, water-soluble starches, dextrins, maltodextrins, gums, proteins, and other organic materials are common binders that can be used in the presence of water. Common plasticizers include water, glycerin, ethylene glycol, di-, tri-, tetraethylene glycol, poly(ethylene glycol), propylene glycol, poly(propylene glycol), dibutyl phthalate, and dimethyl phthalate. Waxes, stearates, and water-soluble oils are examples of lubricants that are used in the presence of water.

Shanefield (1995), Onoda (1978), and Reed (1995) have reviewed the effects of different organic additives on the plastic flow behavior of nonclay ceramics. Bloor (1957), Janney *et al.* (1987), and Phelps *et al.* (1989) have reviewed the effects of formulation variables on the plasticity of clay-based ceramics.

Temperature strongly affects binder moldability. For example, poly(vinyl alcohol) is brittle and elastic at 20°C. As the temperature increases above this, a critical temperature is reached (T_g, the glass transition temperature) at which poly(vinyl alcohol) undergoes a transition from elastic behavior (fracture at a strain greater than about 5%) to time-dependent viscoelastic deformation (no fracture at a strain exceeding 100%). T_g is a characteristic temperature that defines where a given polymer exhibits a dramatic change in not only the resistance to mechanical deformation, but also the thermal expansion and specific heat. T_g is commonly evaluated in the laboratory by mechanical property measurements, dilatometry, or calorimetry. Plastic forming of a given ceramic mixture typically occurs above T_g. Glass transition temperatures are higher for binders with stiff side groups that resist the movement of polymer molecules and also for binders containing polar side groups that bond polymer chains strongly together. T_g is generally higher for a binder of higher molecular weight and when the amount of cross-linking between molecules is greater.

Plasticizers are small molecules that are distributed among the larger binder molecules to reduce T_g and intermolecular hydrogen bonding. Plasticizers increase the distance of separation between binder molecules, thus reducing van der Waals forces of attraction between binder molecules. Plasticizers increase binder flexibility but also decrease binder strength.

A liquid with a low vapor pressure at the molding temperature is usually selected as a plasticizer. A platicizer with a relatively high boiling point, such as glycerol or polymerized ethylene glycol, is important when the system is exposed to elevated temperatures. Plasticizers for thermoplastic polymers, such as polyethylene and polystyrene, include oils and waxes that are molten at the molding temperature. Additives such as stearic acid and oleic acid are plasticizers for waxes.

Since organic additives must be removed prior to firing the ware, only enough is used to develop maximum plasticity required for the particular forming process. Too much water or organic material may result in warping and excessive shrinkage during drying. Too little will result in a tendency to crumble and the need for excessively high forming pressures. If organic additives are extracted too rapidly during debinding, cracking, bloating, or distortion of the ware can result. Inadequate extraction results in cracking, bloating, or contamination during subsequent high-temperature firing.

Bibliography

Aizawa T 1995 Granular modeling for computer aided design of colloidal processing. *Ceram. Trans.* **54**, 251–69

Benbow J J, Lawson T A, Oxley E W, Bridgwater J 1989 Prediction of paste extrusion pressure. *Am. Ceram. Soc. Bull.* **68**, 1821–4

Benbow J J, Oxley E W, Bridgwater J 1987 The extrusion mechanics of pastes—the influence of paste formulation on extrusion parameters. *Chem. Eng. Sci.* **42**, 2151–62

Bloor E C 1957 Plasticity: a critical survey. *Trans. Br. Ceram. Soc.* **56**, 423–81

Brady G A, Hilmas G E, Halloran J W 1995 Forming textured ceramics by multiple coextrusion. *Ceram. Trans.* **51**, 297–301

Chen A Y, Cawley J C 1989 Extrusion of Al_2O_3 ceramics with low organic content. *Ceram. Eng. Sci. Proc.* **10**, 773–83

Clough A R 1964 Plastic forming by roller machine. *J. Br. Ceram. Soc.* **1**, 391

Cundall P A, Hart R 1992 Numerical modeling of discontinua. *Int. J. Eng. Comp.* **9**, 101–13
Funk J E 1984 Extrusion of electrical porcelain bodies. *Advances in Ceramics*. American Ceramic Society, Westerville, OH, Vol. 9, pp. 184–92
Griffith M L, Barda A R, Taylor N, Halloran J W 1995 Micromolding ceramics with photolithographic polyimide patterns. *Ceram. Trans.* **51**, 321–5
Hammond P, Evans J R G 1991 On the blow molding of ceramics. *J. Mater. Sci. Lett.* **10**, 294–6
Haunton K M, Wright J K, Evans J R G 1990 The vacuum forming of ceramics. *Br. Ceram. Trans. J.* **89**, 53–6
Hong C W, Greil P 1995 Discrete element modeling of colloidal powder processing. *Ceram. Trans.* **54**, 235–49
Janney M A, Vance M C, Jordan A C, Kertesz M P 1987 Bibliography of ceramic extrusion and plasticity, Oak Ridge National Laboratory report ORNL-6363. Oak Ridge National Laboratory, Oak Ridge, TN
Jones J T, Berard M F 1972 *Ceramics Industrial Processing and Testing*. Iowa State University Press, Ames, IA
Lachman I W 1984 Porosity in extruded cellular ceramics. In: Mangels J A, Messing G L (eds.) *Advances in Ceramics: Vol. 9. Forming of Ceramics*. American Ceramic Society, Westerville, OH, pp. 201–11
Norton F H 1970 *Fine Ceramics, Technology and Applications*. Krieger, Malabar, FL, pp. 130–47
Onoda G Y 1978 The rheology of organic binder solutions. In: Onoda G Y, Hench L L (eds.) *Ceramic Processing Before Firing*. Wiley, New York, pp. 235–51
Phelps G W, Maguire S G, Kelly W J, Wood R K 1989 *Rheology and Rheometry of Clay–Water Systems*. Cyprus Industrial Minerals, Sandersville, GA
Reed J S 1995 *Principles of Ceramic Processing*, 2nd edn. Wiley, New York
Richerson D W 1982 *Modern Ceramic Engineering, Properties, Processing, and Use in Design*. Marcel Dekker, New York, pp. 198–211
Robinson G C 1978 Extrusion defects. In: Onoda G Y, Hench L L (eds.) *Ceramic Processing Before Firing*. Wiley, New York, pp. 391–407
Sarkar N, Greminger G K 1988 Methylcellulose polymers as multifunctional processing aids in ceramics. *Bull. Am. Ceram. Soc.* **62**, 1280–4
Schilling C H, Bellman R, Smith R M, Goel H, Giesche H 1999 Plasticizing aqueous suspensions of concentrated alumina with maltodextrin sugar. *J. Am. Ceram. Soc.* **82**, 57–66
Shanefield D J 1995 *Organic Additives and Ceramic Processing*. Kluwer, Boston
Steffe J F 1996 *Rheological Methods in Food Process Engineering*. Freeman Press, East Lansing, MI
Takebe H, Yoshida M, Hayashi K, Morinaga K 1992 Fabrication of alumina sheets by extrusion. *J. Ceram. Soc. Jpn.* **100**, 750–4
Tarpley W B, Yocoum K H, Phesant R 1961 Ultrasonic extrusion: reduction in vehicle and plasticizer requirements for non-clay ceramics. US Atomic Energy Commission report NYO-10006
Thompson J J 1963 Forming thin ceramics. *Am. Ceram. Soc. Bull.* **42**, 480–1
Wu X L K, Donato T 1995 Binders for flexible ceramic sheet extrusion. *Ceram. Trans.* **62**, 289–96
Zheng J, Carlson W B, Reed J S 1992 Flow mechanics on extrusion through a square entry die. *J. Am. Ceram. Soc.* **75**, 3011–6
Zheng J, Reed J S, Verma S K 1994 Ceramic extrusion with PEG. *Am. Ceram. Soc. Bull.* **73**, 61–6

C. H. Schilling

Polymer Fiber Processing: High-performance Fibers

Many studies have aimed to improve properties such as strength and modulus of polymer fibers to those of steel in strength and modulus Jiang *et al.* (1993). After the invention of nylon as the first man-made fiber, many fibers, including polyethyleneterephthalate and polyethylene (PE), have been developed and commercialized. Even these generic polymers can, however, achieve high strength levels by drawing. In order to make further progress it is necessary to design the molecular structure and fiber morphology from basic principles. Conditions to fulfill the above mentioned progress are as follows: (i) selecting materials with a molecular backbone linkage that is strong enough (double bond, heterocyclic ring, etc.) and that have a linear conformation and an occupied volume in crystal state that is small and compact; and (ii) drawing the molecules into a fiber such that they are made parallel to the fiber axis, to form a so-called *extended chain structure*.

1. *Molecular Design and Morphological Picture*

First we illustrate the design of molecule for achieving a high modulus. Relationships between chemical molecular structures and mechanical properties have been discussed since the 1960s (e.g., Tashiro *et al.* (1996) who performed theoretical calculations). In general, a polymer chain can take various forms, such as planar zigzag, helical, linear, and heterocyclic conformations. Theoretical predictions suggest that a strong linkage backbone with linear conformation is one of characteristics for molecules that show high mechanical properties.

Fiber morphological design is discussed next. To explain the relationship between fiber properties and morphology schematically, Hosemann (1962) and Peterlin (1969) derived some models. If a disordered structure exists in the fiber, such as a long-period structure formed by lamella blocks (crystal and amorphous alternating structure), molecular ends, voids, and/or a skin-core structure, they will become sources of strain and stress concentrations once the fiber is deformed. Molecular orientation is one of factors to control the fiber modulus (Northolt and Sikkema 1990). Even chain ends give modulus reduction (Jones and Martin 1995) and actual fiber strength and modulus fail to reach their potential limit values. In the ideal structure of a high-performance fiber, infinitely long molecules make up the crystalline state and are aligned perfectly to the fiber direction.

As examples of fibers that possess the above ideal structure, ultra-high molecular weight PE (UHMW-PE) fibers (Dyneema and Spectra), poly-p-phenyleneterephthalamide (p-aramid) fiber, (Kevlar, Twaron), polypyridobisimidazole (PIDP) fibers, and poly-p-phenylenebenzobisoxazole (PBO) fibers (Zylon) have been developed and commercialized. Their properties are summarized in Table 1 and their repeat units are shown in Fig. 1. It is possible to estimate how much a developed fiber approaches the ideal structure by comparing its mechanical properties with the theoretical values. Theoretical and measured crystal modulus values have been obtained and accumulated systematically on many polymers. Various methods such as lattice mechanics (Tashiro *et al.* 1978, Tashiro and Kobayashi 1996) and a method solving a quantum mechanical equation with approximation Shoemaker *et al.* (1992) are used for this purpose. The crystal modulus can be measured by detecting a change of diffraction spots of fibers under deformation as a function of the

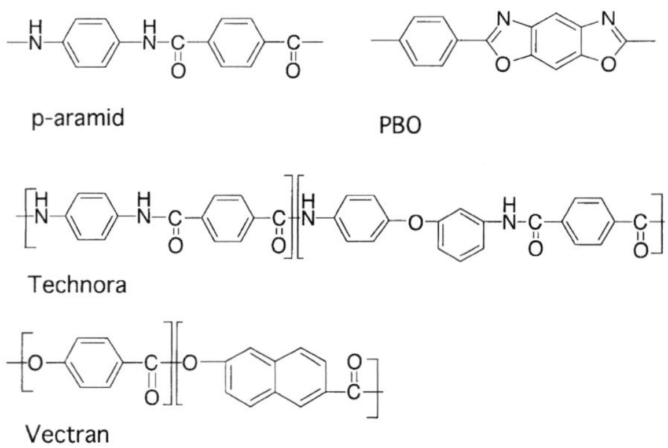

Figure 1
Molecular repeat units.

Table 1
Mechanical properties of typical high-performance fibers.

Fibers	PE	p-Aramid	PBO	Technora	Vectran	PET	Steel
Fiber strength (GPa)	3.0	3.5	5.8	3.1	2.9	1.2	2.8
Fiber modulus (GPa)	172	186	280	70	69	12	200
Measured crystal modulus (GPa)	235	156	480	91	126	95	—
Theoretical modulus (GPa)	295	182	460	—	—	108	—

applied stress Sakurada *et al.* (1962). It has been confirmed that the fiber modulus of p-aramid realizes its theoretical value, while there is still a difference between theory (higher) and experimental values for UHMW-PE and PBO fibers. There are some ideas to explain this difference in terms of structural defects such as molecular misorientation and structural inhomogeneity along the fiber axis, which will be discussed later.

2. *Production*

In order to make molecules aligned to the fiber axis two well-known methods have been developed. In the gel-spinning method flexible chain molecules such as UHMW-PE are spun and drawn with reduction of entanglement densities. The liquid-crystal spinning method makes use of the rigid-rod molecular nature, such as in p-aramid and PBO, that shows liquid crystal state (domains with an alignment of molecules) spontaneously in a certain thermodynamic condition of solution.

2.1 Gel-spinning Method

The development of the gel-spinning method dates back to the invention of ultra-drawing of gel-like fibers by Smith and Lemstra (see *Processing Flexible Polymers to High Performance Fibers*). The fiber is spun from a UHMW-PE solution with a dry-wet, or airgap spinning technology, followed by a drawing process. One of key points of this method is that UHMW-PE (with a molecular weight in excess of one million) is dissolved in a solvent (e.g., xylene) and a quasi-dilute solution is prepared. Being different from the melt state of UHMW-PE, the molecules in the quasi-dilute solution are isolated relative to each other, while only some entanglement between the molecules is present. A gel fiber is then made from this solution by cooling. Molecules in the fiber take on a swollen network structure and the density of the entanglement network is kept low. In addition, the interaction force between PE molecules is so low that an increase in drawing stress of the gel fiber is suppressed even when orientation-induced crystallization occurs. Due to this the gel fiber can be drawn up to by more than 30-fold and the process is called *ultra-drawing*. After drawing the fiber shows high molecular orientation in x-ray diffraction and an absence of a long period in small-angle x-ray scattering (SAXS, Ohta *et al.* 1988) which is characteristic to a melt spun PE fibers.

2.2 Liquid-crystal Spinning Method

As the molecular conformation becomes more rigid, the molecule behaves more like a rod than a coil when in solution; therefore, the excluded volume increases. The thermodynamic state of the solution becomes unstable beyond a certain concentration, and the liquid-crystal domains in which molecules are aligned in a certain direction starts to coexist with the isotropic solution, spontaneously. Molecules in p-aramid and PBO fibers can be extended by utilizing the above feature. The fibers are spun from strong acid solutions with a dry-jet/wet (air gap) spinning method. After extrusion of the spinning solution from a spinneret, the fiber is drawn in the air gap (between the spinneret and the coagulation bath) in which the molecular orientation is increased by the extension flow. Subsequently the drawn molecules are fixed in a coagulation bath with a nonsolvent, followed by a washing process to wash out residual solvent and nonsolvent in the fiber, and dried. The rigid-rod molecules have such a long relaxation time that the extended structure obtained in the air gap can be retained without molecular relaxation. Being different from UHMW-PE fibers, these as spun fibers show a high modulus. To raise the fiber modulus even more, fibers are heat-treated under tension in some cases. Better orientation makes the modulus higher (Northolt and Sikkema 1990).

2.3 Other Methods

One of disadvantages in spinning p-aramid and PBO is that strong acids are required to prepare the spinning solutions. This makes the production equipment longer and more complicated (e.g., the extruder needed to obtain the spinning solution and the coagulation baths) than in a melt spinning process. By reducing the molecular rigidity, by copolymerizing with a flexible spacer into the molecular backbone, the above disadvantage can be moderated. Technora is spun from an isotropic (organic) solution. To raise the molecular orientation of the as-spun fiber, it is drawn up to $\times 10$ with a special drawing technique (Ozawa 1987). The resulting fiber shows a high molecular orientation by x-ray diffraction pattern Nakamae *et al.* (1987). Vectran is spun with a melt spinning method (Chung 1986). Because of the semi-flexible nature in the molecular structure, a quasi-liquid crystal state can be realized in the extension flow in the drawing process and the resulting fiber shows a high modulus and strength.

3. *The Relationship Between Fiber Morphology and Properties*

The influences of molecular misorientation and of inhomogeneity along the fiber axis to the fiber modulus are surveyed in this part. Figure 2 shows an x-ray diffraction pattern from a PBO (Zylon HM+) fiber. Sharp reflection spots exist on the equator, which demonstrate a high molecular orientation along the fiber axis. The Herman's orientation

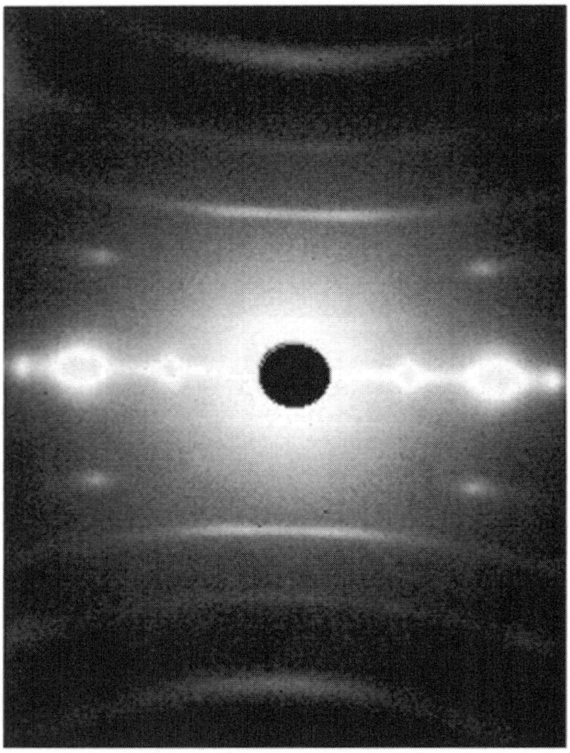

Figure 2
X-ray diffraction pattern of PBO HM + fiber.

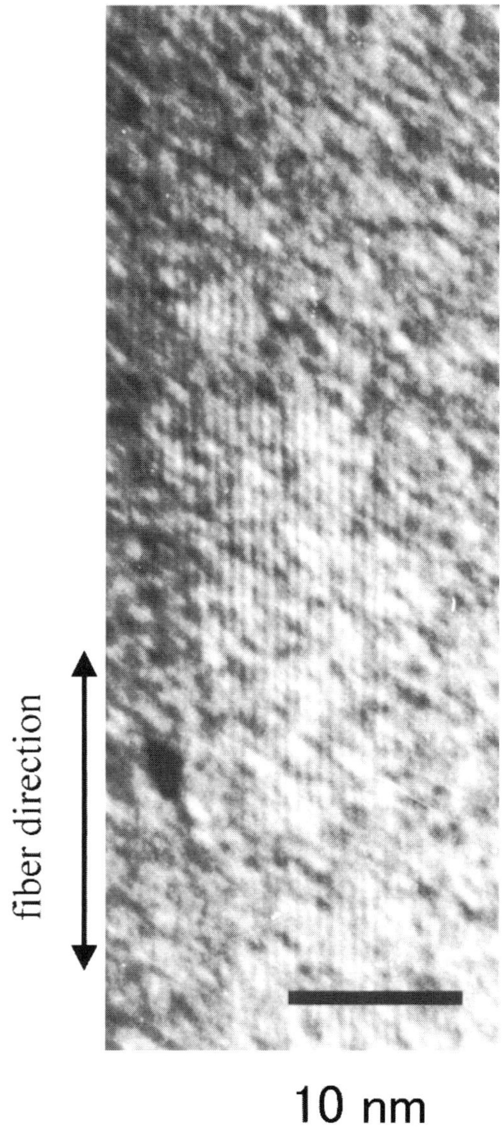

Figure 3
Lattice image of PBO HM + fiber.

coefficient has been introduced to express the degree of molecular orientation quantitatively, and measured values of more than 0.99 are reported to be common for UHMW-PE (Busing 1990), p-aramid (Northolt and Sikkema 1990) and PBO Kitagawa *et al.* (1998) fibers. Northolt and Sikkema (1990) treated the relationship between molecular orientation and fiber modulus with a simple model. The method can predict the crystal modulus from measured fiber modulus values in a certain range of molecular orientation by removing the orientation-dependent factor by extrapolation. There exists an agreement between the extrapolated crystal modulus based on this estimation and the crystal modulus directly measured by x-ray diffraction in the p-aramid fiber, while PBO fiber shows a difference. It means that there is still a missing factor, other than the orientation, to explain this gap in PBO fibers having an extended structure.

Next, structural inhomogeneity along the fiber axis is discussed. It was reported that the intensity of the SAXS pattern has a strong relationship with the presence of a long-period structure and modulus in PE fibers Wu *et al.* (1980). UHMW-PE and p-aramid fibers do not give such SAXS scattering, while PBO HM fiber (heat-treatment under tension) show a four-point SAXS pattern. The PBO molecule is so rigid that there is much less chance to form lamella blocks made of folding molecules, which normally can be detected as the long-period structure. There is still a periodical inhomogeneity along the fiber axis, which gives the SAXS four-point pattern in the PBO case and that also may be the source of hindrance for the fiber modulus to reach the value of its crystal.

As other methods to inspect fibers, such as scanning and transmission electron microscopes are often used. As a method to estimate molecular alignment directly, high-resolution transmission electron microscopy (lattice imaging) has been made available. The fact that p-aramid fiber takes the extended chain structure was first proved by this method Dobb *et al.* (1977). In Fig. 3 a lattice image of a PBO fiber (Zylon HM+) is shown. It can be confirmed that there are fringes parallel to the fiber axis, which reflect the alignment of molecules and persist over 30 nm. It can be also seen in the figure that some part showing no fringes coexists with crystal regions. This indicates that there still remain defects, such as molecular ends, and conformational disorder. Some models that express the fiber structure features have been proposed and they include all results of inspection of fibers with analytical equipment, for high-performance fibers (Dobb *et al.* 1977, Northolt and Sikkema 1990, Yabuki *et al.* 1975). In the main, attention is paid to the molecular orientation and inhomogeneity along the fiber axis to understand the fiber modulus. Fiber strength is also governed by structural defects. There are some theoretical studies to relate these with the strength Termonia *et al.* (1985). The compressive strength of fibers containing an extended structure is also one of important properties to be improved. It is noteworthy that PIDP fibers give a higher value (1.6 GPa) than any of the other super fibers ($\sim$0.6 GPa) (Sikkema 1998). It is plausible that hydrogen bonding in PIDP fiber forms interactions between neighboring molecules and reinforces the compressive strength, while p-aramid fiber (also having hydrogen bonding) shows a lower value than PIDP. Further study is needed to explain the above discrepancy quantitatively from the viewpoints of morphology and theory.

4. Concluding Remarks

Selecting a suitable molecule is the first step to start developing a high performance fiber. The fiber should take an extended structure that shows high molecular orientation and homogeneity along the fiber axis. As far as the Young's modulus is concerned, a high crystal modulus is the target to be achieved in developing high performance fibers and it is realized as the actual modulus in some fibers. Existing theories can estimate the theoretical modulus with fair accuracy. But concerning other properties, they need to be further refined to estimate, for example, compressive strength.

See also: Processing Flexible Polymers to High Performance Fibers; Processing Rigid Polymers to High Performance Fibers; Solid-state Extrusion to High Performance Fibers

Bibliography

Blackwell J, Biswas A 1987 Structure and properties of thermotropic liquid crystalline copolyesters. In: Ward I M (ed.) *Development in Oriented Polymers 2*. Elsevier, London, Chap. 5, pp. 153–97

Busing W R 1990 X-ray diffraction study of disorder in Allied Spectra-1000 polyethylene fibers. *Macromolecules* **23**, 4608–10

Chung T 1986 The recent developments of thermotropic liquid crystalline polymers. *Polym. Eng. Sci.* **26**, 901

Dobb M G, Johnson D J, Saville B P 1977 Supramolecular structure of a high-modulus polyaromatic fiber (Kevlar 49). *J. Polym. Sci. Polym. Phys.* **15**, 2201–11

Flory P J 1953 *Principles of Polymer Chemistry*. Cornell University Press, New York

Hosemann R 1962 Crystallinity in high polymers, especially fibers. *Polymer* **3**, 349–92

Jiang H, Adams W W, Eby R K 1993 High performance polymer fibers. In: Cahn R W, Haasen P, Kramer E J (eds.) *Materials Science and Technology*. VCH, New York, Chap. 13, pp. 596–652

Jones M-C G, Martin D C 1995 Molecular stress and strain in an oriented extended-chain polymer of finite molecular length. *Macromolecules* **28**, 6161–74

Kitagawa T, Ishitobi M, Yabuki K 2000 An analysis of deformation process on poly-p-phenylenebenzobisoxazole (PBO) fiber and a structural study of the new high-modulus type PBO HM+ fiber. *J. Polym. Sci., Polym. Phys.* **38**, 1605

Kitagawa T, Murase H, Yabuki K 1998 Morphological study on poly-p-phenylenebenzobisoxazole (PBO) fiber. *J. Polym. Sci. Polym. Phys.* **36**, 39–48

Lemstra P J, Kirschbaum R 1985 Speciality products based on commodity polymers. *Polymer* **26**, 1372–84

Lemstra P J, van Aerle N A J M, Bastiaansen C W M 1987 Chain-extended polyethylene. *Polymer* **19**, 85–98

Nakamae K, Nishino T, Shimizu Y, Matsumoto T 1987 Experimental determination of the elastic modulus of crystalline regions of some aromatic polyamides, aromatic polyester, and aromatic polyester ketone. *Polymer* **19**, 451–459

Northolt M G, Sikkema D J 1990 Lyotropic main chain liquid crystal polymers. *Adv. Polym. Sci.* **98**, 115–77

Ohta T, Kunugi T, Yabuki K 1988 *Koukyoudo Koudanseiritsu Seni*. Kyoritsu, Tokyo

Ozawa S 1987 A new approach to high modulus, high tenacity fibers. *Polymer J.* **19**, 119

Peterlin A 1969 Bond rupture in highly oriented crystalline polymers. *J. Polymer Sci. Polym. Phys.* **7**, 1151

Sakurada I, Nukushina Y, Ito T 1962 Experimental determination of the elastic modulus of crystalline regions in oriented polymers. *J. Polym. Sci.* **57**, 651–60

Shoemaker J R, Horn T, Haaland P D, Pachter R P, Adams W W 1992 Electronic structure and properties of strained polymers: 1. Methods for polyethylene. *Polymer* **33**, 3351–3356

Sikkema D J 1998 Design, synthesis and properties of a novel rigid polymer, PIDP or 'M5': high modulus and tenacity fibers with substantial compressive strength. *Polymer* **39**, 5981–6

Tadokoro H 1979 *Structure of Polymers*. Wiley-Interscience, New York

Tashiro K, Kobayashi M 1996 Molecular theoretical study of the intimate relationship between structure and mechanical properties of polymer crystals. *Polymer* **37**, 1775–86

Tashiro K, Kobayashi M, Tadokoro H 1978 Calculation of three-dimensional elastic constants of polymer crystals. *Macromolecules* **11**, 908–20

Termonia Y, Meakin P, Smith P 1985 Theoretical study of the influence of the molecular weight on the maximum tensile strength of polymer fibers. *Macromolecules* **18**, 2246–52
Wu W, Simpson P G, Black W B 1980 Morphology and tensile property relations of high-strength/high-modulus polyethylene fiber. *J. Polym Sci. Polym. Phys.* **18**, 751
Yabuki K, Ito H, Ota T 1975 Studies on the fine structure and the physical properties of poly (p-phenylenetrephthalamide) (Part 1). *Sen-i Gakkaishi* **11**, 20

T. Kitagawa

Polymer Fiber Processing: Modeling

In an ordinary melt spinning process, molten polymer is extruded from a circular nozzle with a diameter of 0.2–0.6 mm and pulled down using godet rolls and a winding system, which is placed 3–5 m below the nozzle. Thinning, cooling, and structure development of the polymer melt proceed while the extruded polymer, which is usually referred to as the spinline, runs through the quenching air-flow. A schematic of the melt spinning process is shown in Fig. 1.

The concept for the modeling of the melt spinning process was first given by Ziabicki (1957) and Andrews (1959), who separately proposed equations on the thinning and cooling of the spinline. Kase and Matsuo (1965) generalized this problem by introducing simultaneous differential equations for three variables—diameter, tension, and temperature—which vary with the axial distance from the spinning nozzle. Empirical equations relating to cooling and elongational viscosity of the spinline, which are indispensable to solve the spinning process, were also proposed by the same authors, and the results of numerical simulation were reported. This was the first successful attempt in the field of the modeling of polymer processing.

Realizing the importance of the transient state of the spinning process, Kase and Matsuo (1965, 1967), proposed the partial differential equations for non-steady-state spinning by introducing time as the second independent variable. To minimize the CPU time for numerical calculations, normalized linear perturbation variables were introduced.

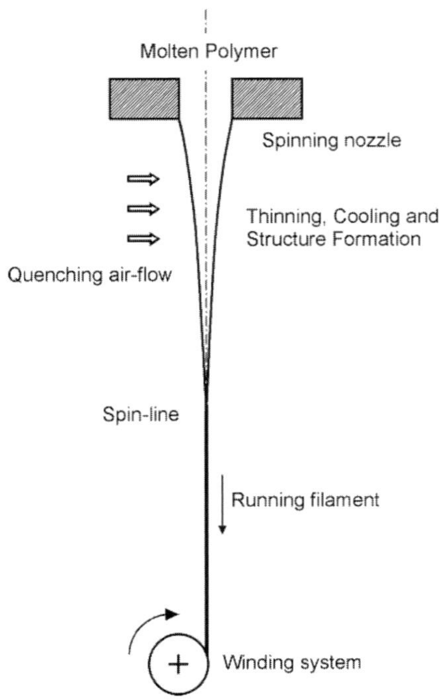

Figure 1
Schematic representation of the melt spinning process.

1. Fundamental Governing Equations

A set of partial differential equations, based on the assumptions of (i) circular cross-section; (ii) thin thread line (temperature, stress, structure, etc., in the cross-section of the spinline are uniform); (iii) negligible die-swell; (iv) the molten polymer being Newtonian fluid; and (v) negligible heat conduction along the spinline, are as follows.

Equation of continuity:

$$\frac{\partial A}{\partial t} + \frac{\partial (AV)}{\partial x} = 0 \tag{1}$$

Equation of momentum balance:

$$\frac{\partial F}{\partial x} = \rho A\left(\frac{\partial V}{\partial t} + V\frac{\partial V}{\partial x}\right) - \rho A g + 2\pi R \tau_{\mathrm{f}} - 2\pi\gamma\frac{\partial R}{\partial x} \tag{2}$$

Equation of energy balance:

$$\frac{\partial T}{\partial t} + V\frac{\partial T}{\partial x} = -\frac{2h}{\rho C_p R}(T - T_{\mathrm{a}}) \tag{3}$$

Constitutive equation:

$$\frac{\partial V}{\partial x} = \frac{F}{A\eta_{\mathrm{e}}} \tag{4}$$

where t and x are the time and the axial distance along the spinline; F, T, V, A, and R are the tension, temperature, velocity, cross-sectional area, and radius of the spinline; γ, C_p, and η_{e} are the surface tension, heat capacity, and elongational viscosity of the polymer melt; and g and T_{a} are the acceleration of gravity and the ambient temperature. Empirical equations for the coefficient of air-friction and the Nusselt number can be developed for estimating the air-friction stress τ_{f} and the heat transfer coefficient h, respectively.

In the case of steady-state spinning, terms differentiated by ∂t can be omitted and the relation between V and A can be obtained from Eqn. (1) as $W = \rho \mathrm{AV}$, where W is the throughput rate. Consequently, Eqns. (2)–(4) can be solved numerically as a set of simultaneous differential equations. When calculation is based on the finite difference method, F, T, and V at the nozzle exit are needed as a boundary condition. Because of the lack of information on F at $x = 0$, reiteration of the calculation is necessary to satisfy another boundary condition, i.e., V at the take-up. It should be noted that in steady-state spinning, position x can be converted to time t using the equation

$$t = \int_0^x \frac{dx}{V} \tag{5}$$

2. *Typical Results of Simulations of the Steady-state Spinning of a Newtonian Fluid*

Examples of the results of calculation for the melt spinning of poly(ethylene terephthalate) (PET) at take-up velocities of 1 and 4 km min^{-1} and a throughput rate of 2 g min^{-1} are shown in Fig. 2. In early attempts to analyze the melt spinning process, tension was assumed to be constant along the spinline. This assumption is valid only when the melt spinning of highly viscous material at low take-up velocity is considered. The effects of inertia and air-friction forces (the first and third terms in the right hand side of Eqn. (2)) become dominant with an increase in the take-up velocity, and the spinline tension increases steeply with increasing distance from the spinning nozzle. On the other hand, the effect of surface tension is negligible, except for materials with extremely low viscosity. Under the condition of constant throughput rate, the cooling behavior of the spinline with respect to x is not much affected by the change in the take-up velocity. If the cooling behavior is compared after converting the independent variable x to the other independent variable t, using Eqn. (5), rapid cooling of the material at a high-take-up velocity can be recognized clearly, as shown in Fig. 3.

3. *Improvement of Governing Equations*

3.1 *Constitutive Equation*

For melt spinning of typical polycondensates, such as polyesters and polyamides at relatively low take-up velocities, the use of the constitutive equation for a Newtonian fluid would yield reasonably accurate results. On the other hand, for the melt spinning of polyolefins such as PE and PP, and especially for the polymers with branching or a wide molecular weight distribution, the use of a viscoelastic model is recommended, even for extremely low take-up velocities. Various differential-type constitutive equations have been incorporated, such as the upper-convected Maxwell model, the Phan Thien–Tanner model, and

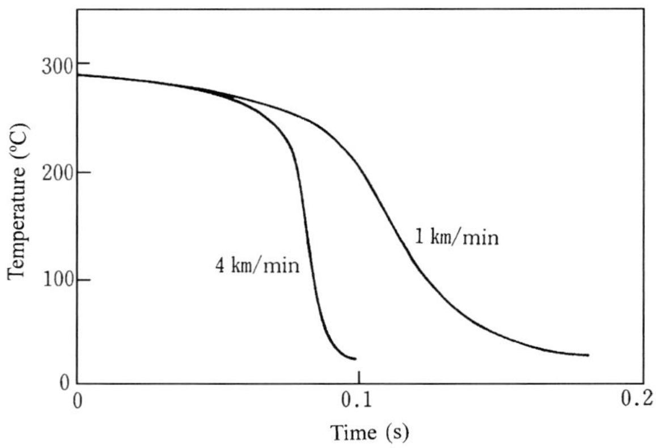

Figure 3
Time vs. temperature relation of the spinline for two different take-up velocities.

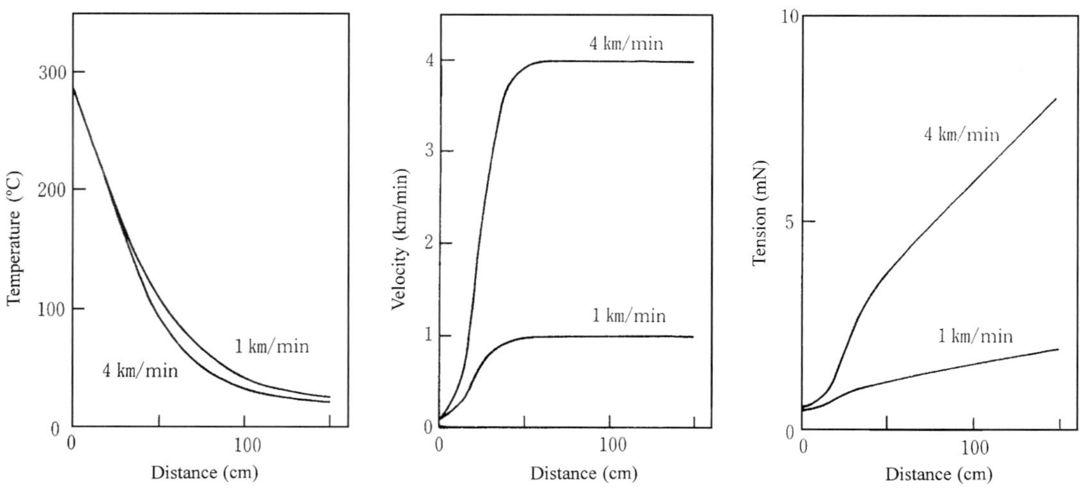

Figure 2
Variation of temperature, velocity, and tension along the spinline for two different take-up velocities.

the White–Metzner model, and integration-type constitutive equations such as the Wagner model and the K-BKZ model have been reported.

3.2 Structure Development

One of the objectives of modeling melt-spinning processes is the prediction of structure and properties of as-spun fibers, which vary significantly depending on the spinning conditions. Theoretical considerations of the molecular orientation and crystallization in the spinline were first given by Ziabicki (1967). Development of the molecular orientation of PET under ordinary spinning conditions was also reported to be predictable by simply applying the stress-optical law, i.e., the linear relation of birefringence with the spinline stress. Crystallization behavior in the spinline was calculated by incorporating an equation for the non-isothermal crystallization kinetics into the simulation program (Shimizu *et al.* 1985, Katayama and Yoon 1985). A temperature- and stress-dependent crystallization rate function was used for the calculation. The effects of the evolution of latent heat and the hardening of the polymer melt, accompanying crystallization, were also incorporated.

3.3 Variations in the Fiber Cross-section

Evolution of the temperature distribution in the cross-section of the spinline was predicted by considering the heat conduction in the direction perpendicular to the fiber axis. From the temperature distribution, stress and structural variations were also estimated. The analyses on the cross-sectional temperature distribution can be used to predict the shrinkage of PP fibers, caused by the cross-flow quenching, and the variation of structure across the cross-section of high-speed spun fibers. It was also pointed out that the empirical equation for the cooling of the spinline should be modified when the cross-sectional temperature variation is incorporated into the simulation (Vassilatos *et al.* 1992).

3.4 Non-circular Fiber Spinning

Various types of fibers with non-circular cross-sections are produced in the fiber industry. For the modeling of the spinning process of such fibers, the difference in the cross-sectional shape of the spinning nozzle and the as-spun fibers must be taken into account. A differential equation for the change of cross-sectional shape along the spinline was proposed, assuming that the surface tension is the major driving force for viscous flow in the plane perpendicular to the spinline axis. The equation was incorporated into the numerical simulation of the melt spinning process to predict the change in cross-sectional shape of rectangular and hollow fibers (Takarada *et al.* 2001). The finite element method was also applied for the prediction of the change in cross-sectional shape of non-circular fibers (Oh *et al.* 1997). It must be noted that a modification of the empirical equations for the cooling and air-friction of a circular cross-section spinline is necessary for these calculations.

3.5 Multicomponent Spinning

Multicomponent fibers in which more than two kinds of polymers are arranged in the cross-section of a single fiber are widely produced in the fiber industry. Typical examples of this type of fiber are the sheath-core type and the side-by-side type fibers. Ultrafine fibers for synthetic suede, highly crimped fibers, thermal bonding fibers, etc., are produced utilizing this technology. The melt-spinning behavior of bicomponent fibers was simulated, incorporating a viscoelastic model and the effect of the evolution of radial stress (Lipscomb 1994). It should be noted that, although the thinning and cooling behavior of the spinline is governed by the averaged properties of respective components, the tensile stress applied to each component in multicomponent spinning becomes significantly different from that in single-component spinning of each polymer. This leads to the development of a unique fiber structure, which was found to be predictable throughout the modeling of the multicomponent spinning process (Kikutani *et al.* 1996).

3.6 Multifiber Spinning

For the production of filaments and staples, several tens to several thousands of fibers are melt-spun simultaneously from a single spinning head. The cooling behavior of each spinline in the fiber bundle varies, depending on the position of each fiber in the bundle and the type of air flow of the quenching air. A numerical simulation program, coupling the computations of the air flow in the fiber bundle and the thinning and cooling of each spinline, was developed so that the variation of the fiber structure in the fiber bundle could be predicted (Ishihara *et al.* 1989, Beyreuther *et al.* 1997).

3.7 Spun-bond Process

For the production of spun-bond fabrics, polymer melts extruded from a spinning nozzle are pulled down, applying pneumatic force, and distributed on a moving wire netting. One of the important characteristics of this process is the fact that the take-up velocity cannot be set as a spinning condition, as it can in the case of the spinning process with a winder. The numerical simulation of the spun-bond process is performed considering the balance of the spinline tension and the friction force applied in the air-ejector (Ishihara *et al.* 1989, Beyreuther *et al.* 1997).

3.8 Non-steady-state Spinning

The effect of the fluctuation of spinning conditions, such as the quenching airflow, on the longitudinal inhomogeneity of as-spun fibers can be analyzed by solving the partial differential equations, Eqns. (1)–(4), under non-steady-state conditions. This can be applied for the analysis of the stability of the spinline, and also for the prediction of the draw resonance behavior, a periodic fluctuation of the fiber diameter.

4. Concluding Remarks

The fundamentals of the modeling of melt spinning processes for both steady and non-steady states were already established in the very first paper on this subject (Kase and Matsuo, 1965). Since then, various modifications have been attempted, and the applicability of the model for the prediction of industrial processes has been improved. One important subject requiring further improvement in this field may be the quantitative estimation of structure development, especially the orientation-induced crystallization in the process. Difficulties in establishing a practical model for structure formation arise because of (i) the extremely high crystallization rate in the spinline, in comparison with the quiescent state; (ii) the extremely high strain rate, accompanied by neck-like deformation (Kikutani *et al.* 1999); and (iii) the uncertain mechanisms of solidification, etc. Inclusion of the effect of structure formation into modeling is a common and important subject in various types of polymer processing, and research in the fields of both modeling and on-line measurement needs to be continued intensively for further clarification of the processes.

Bibliography

Andrews E H 1959 Cooling of a spinning thread-line. *Brit. J. Appl. Phys.* **10**, 39

Beyreuther R, Brunig H, Blechschmidt D 1997 Nonwoven process–realized with the help of a fiber spinning model. *Intern. Fiber J.* **12**, 129

Ishihara H, Hayashi S, Ikeuchi H 1989 Computer simulation of multi filament air jet melt spinning. *Intern. Polymer Processing* **VI**, 91

Kase S, Matsuo T 1965 Studies on melt spinning. 1. Fundamental equations on the dynamics of melt spinning. *J. Polym. Sci.* **A3**, 2541

Kase S, Matsuo T 1967 Studies on melt spinning. 2. Stead state and transient solutions of fundamental equations compared with experimental results. *J. Appl. Polym. Sci.* **11**, 251

Katayama K, Yoon G 1985 Polymer crystallization in melt spinning: mathematical simulation. In: Ziabicki A, Kawai H (eds.) *High-speed Fiber Spinning*. Wiley, New York, p. 207

Kikutani T, Radhakrishnan J, Arikawa S, Takaku A, Okui N, Jin X, Niwa F, Kudo Y 1996 High-speed melt spinning of bicomponent fibers: mechanism of fiber structure development in poly(ethylene terephthalate)/polypropylene system. *J. Appl. Polym. Sci.* **62**, 1913

Kikutani T, Nakao K, Takarada W, Ito H 1999 On-line measurement of orientation development in the high-speed melt spinning process. *J. Polym. Eng.* **39**, 2349

Lipscomb G G 1994 The melt hollow fiber spinning process: steady-state behavior, sensitivity and stability. *Polym. Adv. Tech.* **5**, 745

Oh T H, Lee M S, Kim S Y, Shim J 1997 Studies on melt-spinning process of hollow fibers. *J. Appl. Polym. Sci.* **68**, 1209

Shimizu J, Okui N, Kikutani T 1985 Simulation of dynamics and structure formation in high-speed melt spinning. In: Ziabicki A, Kawai H (eds.) *High-Speed Fiber Spinning*. Wiley, New York, p. 173

Takarada W, Ito H, Kikutani T, Okui N 2001 Studies on high-speed melt spinning of non-circular cross-section fibers. III. Modeling of melt spinning process incorporating change in cross-sectional shape. *J. Appl. Polymer Sci.* **80**, 1589–600

Vassilatos G, Schmelzer E R, Denn M M 1992 Issues concerning the rate of heat transfer from a spinline. *Int. Polymer Processing* **VII**, 144

Ziabicki A 1957 Uber einige Regelmassigkeiten beim Spinnprozess Syntheticher Fasern. *Faserforsch. Textil* **8**, 467

Ziabicki A 1967 Kinetics of polymer crystallization and molecular orientation in the course of melt spinning. *Appl. Polym. Symp.* **6**, 1

T. Kikutani

Polymer Fibers: Formation and Structure

Conventional fiber formation starting from semi-crystalline polymers consists of three processes: melt-spinning, drawing, and annealing. The drawing and annealing processes are usually combined, as shown schematically in Fig. 1. The spinning and the drawing and annealing processes also can be conducted continuously, and are then known as a spin-draw process.

In the spinning process, either melt- or solution spinning, polymers are extruded through tiny orifices to form thin threads and taken up using a winder. Since polymer molecules are not well aligned along the fiber axis at this stage, as-spun fibers usually exhibit poor mechanical properties such as low modulus, low tenacity, and high elongation. Therefore, the drawing and annealing processes are necessary to fabricate fibers with well-developed structure.

In the drawing and annealing processes, as-spun fibers pass through several sets of rolls with different surface velocities. Fiber temperature is controlled either by heated rolls, plate heaters, or some other heating device. Drawing brings about molecular orientation and mechanical anisotropy, whereas annealing enhances the crystallization and thermal stability. Both properties are needed as fundamental characteristics of the finished fibers.

If the take-up velocity is increased significantly in the melt spinning process, the extensional stress applied to the spinning line increases and the as-spun fibers start to have high molecular orientation. At the same time, since the crystallization rate increases with an increase in molecular orientation, crystallization can be promoted in the spinning line even though the cooling of the molten polymer is extremely fast. This type of crystallization is defined as orientation-induced crystallization. Because of significant fiber structure development, a high-speed melt-spinning process can provide fibers directly applicable for their end use (Ziabicki and Kawai 1985).

1. History of the Fiber Structure Model

Fiber structure models were initially developed for natural fibers. A model proposed for cellulose fibers was the 'fringed micelle model'. The most important structural difference between synthetic and natural fibers is in the existence or nonexistence of periodic structure along the fiber axis (Kaji 1997).

The 'fringed-micelle micro-fibril model,' however, cannot explain (i) the SAXS intensity on the equator, and (ii) the mechanical properties of fibers. The 'paracrystalline layer structure model,' in which the concept of folded-chain crystal is introduced, was for the crystalline region.

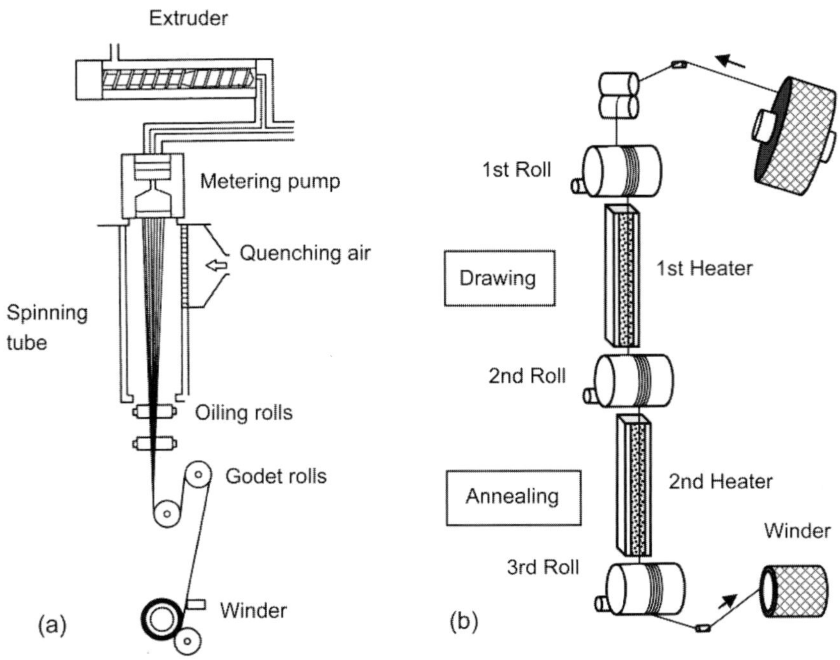

Figure 1
Schematic illustrations of (a) melt spinning and (b) drawing and annealing processes.

'Shish-kebab structures,' in which extended chains form the 'shish,' and crystalline lamellae epitaxially grown from the side surface of the shish form the 'kebab,' have become widely accepted. This structure was originally called the 'row structure,' as all the shish are oriented along the flow direction. Considering the overlapping of lamellae grown from adjacent shish, the 'shish-kebab model' is also referred to as the 'interlocking shish-kebab model' (Bashir *et al.* 1986).

Although the shish-kebab structure has been widely observed in highly oriented polyethylene (PE), polypropylene (PP), and polyamide 66 (PA66), further investigation may be necessary concerning the applicability of this model for polymers with lower crystallizability, such as aromatic polyesters.

2. Aromatic Polyesters

2.1 Fiber Processing and Structure Development

Melt-spinning processes can be applied to the production of fibers from aromatic polyesters such as poly(ethylene terephthalate) (PET), poly(trimethylene terephthalate) (PTT), poly(butylene terephthalate) (PBT), and poly(ethylene naphthalate) (PEN). These polymers have relatively low crystallization rates. The glass transition temperatures of these polymers are higher than room temperature except for PBT, which has a glass transition temperature close to room temperature. Therefore, the ordinary melt-spinning process of aromatic polyesters provides amorphous fibers with low molecular orientation. Development of the fiber structure can be achieved by the drawing of as-spun fibers at a temperature near the glass transition temperature, followed by the annealing of the drawn fibers at a temperature between the glass transition and melting temperatures. A temperature close to the maximum-crystallization-rate temperature may be preferable for the annealing process.

In high-speed melt spinning of these polymers, molecular orientation increases with an increase in the take-up velocity. Changes in the birefringence of various aromatic polyester fibers with an increase in the take-up velocity are summarized in Fig. 2. The difference in birefringence development behavior originates from the differences in molecular rigidity, intrinsic birefringence, etc. In these polymers, orientation-induced crystallization starts to occur at the take-up velocities of around 2–5 km min^{-1}. In each polymer, the starting point of crystallization varies slightly, depending on the extrusion and cooling conditions and the molecular weight of the polymers. After the onset of orientation-induced crystallization, a higher take-up velocity generally yields higher molecular orientation, higher crystallinity, and higher ordering in the crystalline region. Conversely, molecular orientation in the amorphous region is relatively low in comparison with that in the fibers produced through conventional drawing and annealing processes.

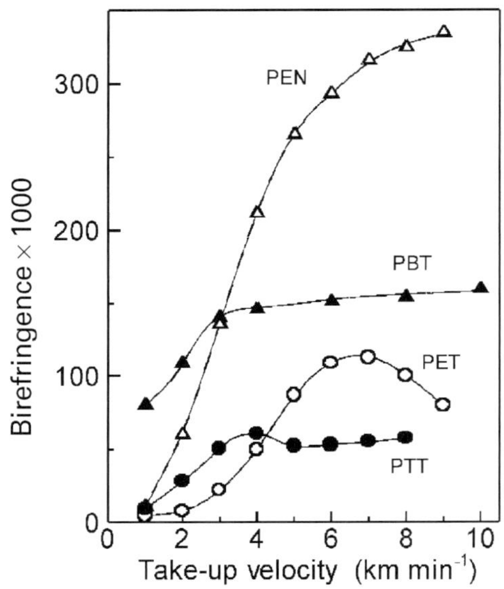

Figure 2
Birefringence development of various aromatic polyester fibers with increasing take-up velocity in the high-speed melt spinning process.

As a typical example, structural changes in PET fibers from the high-speed melt-spinning process are summarized in Fig. 3. Initiation of orientation-induced crystallization is represented by the change in the WAXS pattern and a sudden increase of density at around 4 km min^{-1}. Birefringence of as-spun PET fibers is roughly predictable from the stress–optical coefficient and the tensile stress acting on the spinning line near the solidification point. When the take-up velocity exceeds 7 km min^{-1}, however, molecular orientation, crystallinity, and mechanical properties start to decrease. The development of significant structural variation was found in the cross-section of PET fibers prepared in this take-up velocity region.

2.2 Crystal Modification

The crystal unit cell of all the aromatic polyesters discussed above is triclinic. No polymorphism is reported for PET and PTT, while PBT and PEN are known to have α- and β-form crystals. In comparison with PET, PTT has an extremely low crystalline modulus in the chain direction. This is because the aliphatic part of the crystalline structure of PET has a trans conformation, whereas that of PTT has a highly coiled structure of gauche–gauche conformation (Desborough 1979). Therefore, the unit cell

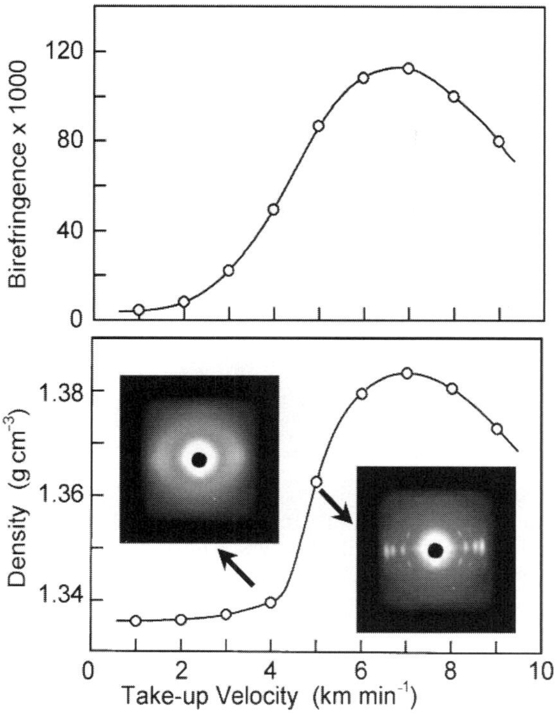

Figure 3
Change in fiber structure of PET in high-speed melt spinning process.

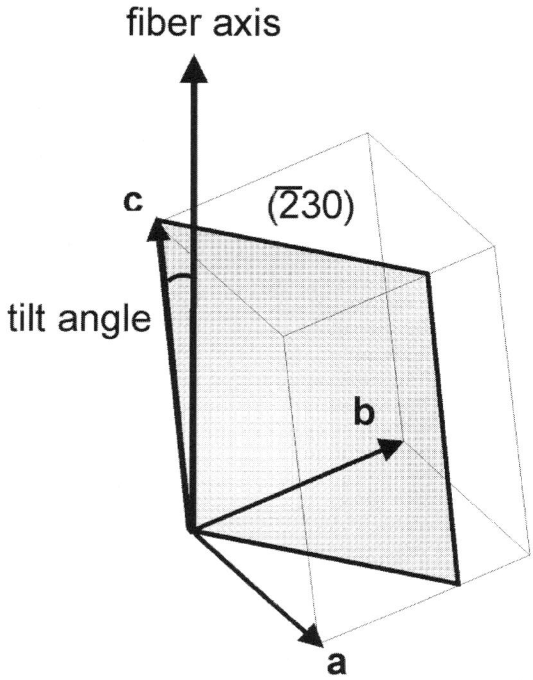

Figure 4
Schematic illustration explaining the tilting of c-axis along ($\bar{2}$30) plane in PET fibers.

parameters of PTT show a distinct mechanical response upon the application of tensile stress to the fibers (Jakeways 1975).

PBT fibers prepared both through the conventional process and the high-speed melt-spinning process have α-form structure (Fujimoto *et al.* 1988). The crystal form changes from α to β when a tensile strain of about 12% is applied, but reverts to α form upon the removal of strain (Yokouchi *et al.* 1976). As the c-axis length of the β-form unit cell is about 8% longer than that of the α form, the stress–strain curves of PBT fibers exhibit a remarkable yielding behavior showing a distinct plateau region. The yielding stress corresponds to the critical transition stress from α to β form (Tashiro *et al.* 1980).

In the case of PEN, α-form crystals are formed in the conventional spinning and drawing processes, whereas the coexistence of α and β form crystals is confirmed in the high-speed spun fibers. Even though the β-form unit cell of PEN has a shorter c-axis length and higher crystalline density than the α form, the amount of β-form crystals increases with an increase in the take-up velocity, that is, with an increase in the tensile stress applied during the orientation-induced crystallization. In the quiescent state, β-form crystals can be developed only when the crystallization temperature is higher than 220–245°C, whereas the formation of the β form occurs at much lower temperatures in the high-speed spinning line (Miyata *et al.* 1997).

3. *Tilting*

Another interesting feature in the structure of aromatic polyester fibers is the tilting of the c-axes of the crystalline unit cell away from the fiber axis. This means that the molecular chains on average are not aligned parallel to the fiber axis. The direction and degree of tilting can be analyzed from the shifts of crystalline reflections from layer lines in the WAXS pattern. In PET fibers, tilting of the c-axis occurs along the ($\bar{2}$30) plane as shown schematically in Fig. 4. The tilt-angle of conventional PET fibers is about 5°, whereas that of high-speed spun fibers is reported to decrease from 10° to 2° with an increase in the take-up velocity (Shimizu *et al.* 1984). In PBT fibers, tilting of about one degree was found both in α- and β-form crystals. The direction of tilting is along the ($\bar{2}$10) and (130) planes, respectively. The tilt-angle of high-speed spun PBT fibers also decreases with an increase in the take-up velocity (Fujimoto *et al.* 1988). In PET and α-form PBT fibers, tilting is essentially parallel to the crystallographic plane containing the benzene ring. On the other hand, tilting of about 3–5° occurs along the ($\bar{1}20$) plane in PTT fibers, although the molecular chain in the crystal is in 2_1 helix conformation.

4. *Aliphatic Polyesters*

Some of the aliphatic polyesters, which generally have lower glass transition and melting temperatures in comparison with aromatic polyesters, attract interest because of their biodegradability. Melt spinning of these polymers, however, is difficult to control, as the extruded fibers tend to reach the take-up device in a rubbery state. This is caused by a combination of two factors: the low crystallization rate, and the glass transition temperature being lower than the room temperature. Crystallization of these polymers may be controlled through the installation of a hot tube or hot water-bath into the spinning line.

4.1 *Poly(3-hydroxybutylate)*

Poly(3-hydroxybutyrate) (PHB), and copolymers of PHB with 3-hydroxyvalerate (P(HB/HV)), are produced by certain bacteria. The fiber structure of PHB and P(HB/HV) is mainly composed of α-form crystals of PHB, but small amounts of β-form crystals are known to coexist. The molecular chain in the α-form crystal is in 3_1 helix conformation, whereas that in the β-form crystal is almost in the planar zigzag conformation (Orts *et al.* 1990). The α- and β-form crystals usually exhibit ordinary c-axis orientation, and the amount of β form varies depending on the drawing and annealing conditions (Furuhashi *et al.* 1998) or the take-up velocity in the high-speed spinning process (Schmack *et al.* 1999).

In some cases, the α-form crystals in drawn and annealed fibers exhibit peculiar bimodal orientation, which is represented by the coexistence of diffraction spots and rings in the WAXS pattern. Some rings are complete, with a higher intensity near the meridian, while other rings are incomplete, with a lack of intensity near the meridian, as shown in Fig. 5. Detailed analyses of the ring pattern reveal that the c-axis is preferentially oriented perpendicular to the fiber axis, whereas the a- and b-axes are randomly distributed around the c-axis (Furuhashi *et al.* 1998).

4.2 *Polylactide*

Poly(L-lactide) (PLLA) has a glass transition temperature higher than room temperature. PLLA fibers prepared through conventional fiber formation processes and high-speed melt-spinning processes are known to exhibit α-form crystals, whereas the β-form crystals can be developed in the fibers drawn to extremely high draw ratios (Hoggsteen *et al.* 1990).

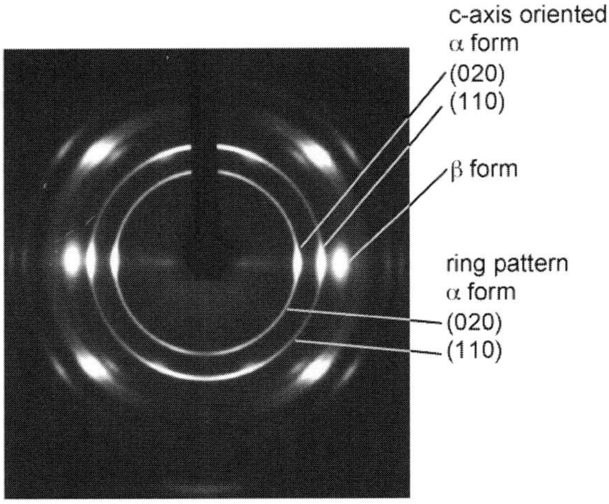

Figure 5
Bimodal orientation of α form crystals of PHB in drawn and annealed P(HB/HV) fibers. Coexistence of β form crystals is also indicated.

The molecular chain in the α- and β-form crystals is in 10_3 and 3_1 helix conformation, respectively.

5. *Aliphatic Polyamides*

5.1 *Fiber Processing and Structure Development*

Fiber formation behavior in aliphatic polyamides (PA) is sensitive to the ambient conditions. This is due to the highly hygroscopic nature of the polymers. Structural parameters and mechanical properties are also sensitive to humidity because of the change in water content in the fibers.

In ordinary melt-spinning processes with polyamide 6 (PA6) and polyamide 66 (PA66), amorphous as-spun fibers can be obtained, as the glass transition temperature of these polymers is around 40 °C, which is higher than room temperature. (In the case of PA66, which has a significantly higher crystallization rate than PA6 in the quiescent condition, spherulitic structures may be developed in the spinning line if the spinning conditions result in a relatively low cooling rate.) On exposure to moisture after spinning, however, diffusion of water into the amorphous as-spun fibers results in crystallization. This process is referred to as the conditioning process.

When as-spun fibers of PA6 and PA66 are prepared at take-up velocities from 1 to 3 km min^{-1}, birefringence increases significantly and fibers elongate spontaneously during the conditioning process. The amount of spontaneous elongation may be suppressed through the enhancement of crystallization in the spinning process, by applying water to the surface of running fibers before winding. Drawing of the as-spun fibers of PA6 and PA66 usually is conducted at room temperature. Therefore, the drawing process is called cold drawing. The reduction of the glass transition temperature by moisture uptake and heat generation during the drawing process provides enough mobility to molecules for the formation of crystalline fiber structure. Higher annealing temperatures are necessary for the production of high-strength polyamide fibers, and highly oriented and crystallized fibers can be obtained in a similar manner to the aromatic polyesters.

5.2 *Crystal Modification*

PA66 has crystal modifications of α and β forms. Both the drawn and annealed fibers and high-speed spun fibers (Shimizu *et al.* 1985) exhibit α-form structure. Very weak (002) reflections of the β form were found to coexist in the WAXS pattern of high-speed spun fibers.

Various crystal modifications such as α, β, γ, and smectic (pseudo-hexagonal) forms have been reported for PA6. A reasonable rationalization is that there are only two basic crystallographic forms, α and γ. Although there are some differences in reflection intensities, the unit cell parameters of α and β forms are identical. Although γ and smectic form crystals have different thermal stabilities, these two forms exhibit similar WAXS patterns. In drawn and annealed PA6 fibers, α and β forms coexist, and the proportion of α form increases with increases in the crystallization temperature and applied tension.

In high-speed melt-spinning processes, orientation-induced crystallization starts to occur at take-up velocities of around 3–4 km min^{-1}. In this region, the γ form dominates the crystalline structure of as-spun fibers, and the proportion of α form increases with an increase in take-up velocity (Shimizu *et al.* 1985). Higher molecular weight also yields greater proportions of α-form crystals (Koyama *et al.* 1986). These results suggest that the higher tensile stress and higher crystallization temperature lead to the preferred formation of the α form. Heuvel *et al.* (1976) confirmed that orientation distribution of PA6 α-form crystals is axis-symmetric with respect to the fiber axis, and there is no predominant direction in the rotation of the unit cell.

6. *Polyolefins*

Polyolefins such as polyethylene (PE) and polypropylene (PP) have high crystallization rates and their glass transition temperatures are below room temperature. Therefore, these polymers need to be crystallized in the spinning line to obtain solid fibers. Although the usage of PE fibers is limited, PP fibers are widely used as nonwoven fabrics and other applications such as carpets. In the spun-bond process, which is one of the important technologies in producing nonwoven fabrics, melt-spun fibers are directly spread on a wire mesh to obtain fabrics.

Although a wide range of drawing temperatures can be applied for the drawing of as-spun PE and PP fibers, drawing temperatures near the crystalline dispersion temperature may be preferable because the starting fibers already have high crystallinity.

6.1 *Polyethylene*

When the tensile stress applied to the spinning line is low or the take-up velocity is low, the crystallographic b-axis tends to orient perpendicular to the fiber axis, whereas the a-axis tends to orient along the fiber axis. The orthorhombic unit cell rotates around the b-axis gradually with an increase in the tensile stress or the take-up velocity, and eventually ordinary c-axis oriented crystals can be formed, as shown in Fig. 6. It should be noted that the density or crystallinity of PE fibers hardly changes, irrespective of the extensive variation in fiber formation conditions. Drawn PE fibers usually exhibit high molecular orientation with high c-axis orientation of the crystals.

Recent improvements in polymer synthesis technologies have brought high spinnability to PE polymers. Structure development in PE high-speed spinning processes varies slightly depending on the type of PE polymers: high density PE, linear-low- density PE, or

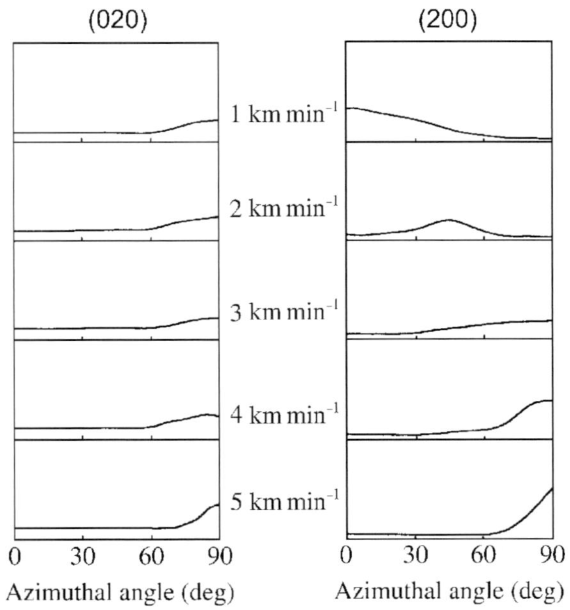

Figure 6
Changes in azimuthal intensity distribution of (020) and (200) reflections of PE fibers in wide-angle x-ray diffraction measurements. Take-up velocities are indicated.

metallocene-catalyzed linear low-density PE. In the case of linear low-density PE, formation of a-axis oriented crystals in the low speed region was not detectable (Cho *et al.* 2000).

6.2 *Polypropylene*

PP generally has a lower crystallization rate than PE. More importantly, the crystallization rate of PP is comparable to the cooling rate the polymer experiences in the spinning line. Therefore, the development of crystalline structure in the melt-spinning process depends strongly on the spinning conditions and polymer characteristics.

When the take-up velocity is low, and therefore the cooling rate and the applied tensile stress are low, monoclinic crystals can be formed in the as-spun fibers. The crystalline orientation is nearly isotropic in these fibers. Formation of monoclinic crystals suggests that the crystallization proceeds at relatively high temperatures. When the take-up velocity is increased, the crystalline structure changes to the smectic (pseudo-hexagonal) form, because of the enhancement of cooling and a reduction in the crystallization temperature. There can be additional development of smectic crystals even after the spinning process, because the glass transition temperature of PP is below room temperature. Further increase of the take-up velocity leads to a significant development of molecular orientation. The molecular orientation enhances the crystallization, and because of the increase in the crystallization temperature, the crystal structure changes from isotropic smectic crystals to highly oriented monoclinic crystals (Cao *et al.* 1989).

The full series of structural changes described above can be observed when the molecular weight is low, and the extrusion temperature and throughput rate are high. Under these conditions, the mechanism of orientation-induced crystallization becomes dominant only at high take-up velocities. If molecular weight is high, the extrusion temperature and throughput rate are low; then the crystalline form simply remains monoclinic, and the crystalline orientation changes from nearly isotropic to a highly oriented state with an increase in the take-up velocity.

Another important feature of PP fibers is the development of bimodal orientation, i.e., the coexistence of c- and a*-axis oriented monoclinic crystals. Bimodal orientation can be confirmed from WAXS patterns in which the (110) reflection splits to two peaks, one on the equator and the other near the meridian, whereas (040) reflection remains on the equator as shown in Fig. 7. This structure resembles the so-called crosshatch structure, in which epitaxial growth of daughter lamellae occurs on the side-surface of oriented parent lamellae.

On-line measurements of the development of crystalline structure were conducted utilizing a synchrotron radiation source (Kolb *et al.* 2000). The experimental results suggested that the development of c- and a*-axis-oriented crystals proceeds almost simultaneously. The amount of a*-axis-oriented crystals varies depending on spinning conditions, whereas the highly drawn and annealed fibers usually exhibit only the c-axis-oriented crystals.

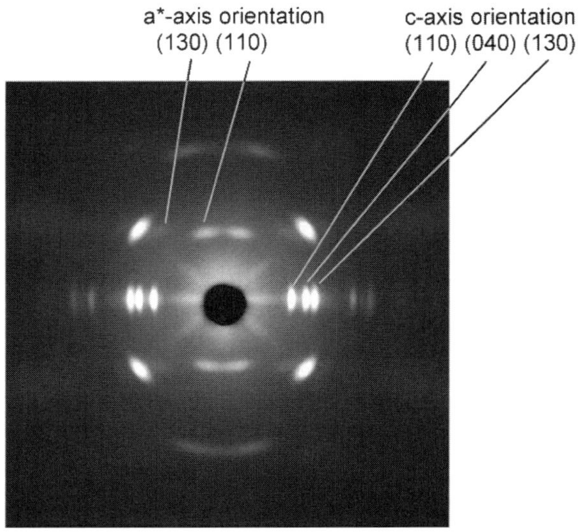

Figure 7
Bimodal orientation in PP fibers.

Although the molecular orientation of as-spun fibers strongly depends on the tensile stress applied to the spinning line, birefringence of as-spun fibers is significantly larger than that predicted from the applied stress and stress-optical coefficient of the polymer melt (Shimizu *et al.* 1985). The nonvalidity of the stress-optical rule suggests that crystallization is accompanied by a significant development of molecular orientation.

7. *Concluding Remarks*

Fiber structure in semicrystalline polymers has been explained, based on the degree of molecular orientation and differences in crystalline structure. Crystal modifications appearing in the fiber structure are summarized in Table 1. It is still difficult to find a general rule between the unit cell and the type of crystal modifications appearing in the fiber processing. It appears, however, that the crystal form with higher density is preferred in high-speed melt-spinning, especially in the extremely high-speed region. The only exception in Table 1 is PA66. It should be noted here that the unit cells with longer fiber repeat distances are not necessarily preferred, although the crystallization is strongly accelerated by high tensile stress.

Table 1
Crystal modifications appearing in fiber structure.

		Unit-cell parameters			
Polymer	Crystal modification crystal system	a (nm) α	b (nm) β	c (nm) γ	L_r (nm) ρc (g cm^{-3})
Aromatic polyesters					
PET	#	0.456	0.594	1.075	1.075
	Triclinic	98.5	118	112	1.455
PTT	#	0.46	0.62	1.83	1.83
	Triclinic	98	90	112	1.43
PBT	α (*)	0.483	0.594	1.159	1.159
	Triclinic	99.7	115.2	110.8	1.404
	β	0.495	0.567	1.295	1.295
	Triclinic	101.7	121.8	99.9	1.283
PEN	α	0.651	0.575	1.32	1.32
	Triclinic	81.3	144	100	1.407
	β (*)	0.926	1.559	1.273	1.273
	Triclinic	121.6	95.6	122.5	1.439
Aliphatic polyesters					
PHB	α (*)	0.576	1.32	0.596	0.278
	Orthorhombic	90	90	90	1.26
	β	0.522	0.522	#	#
	Hexagonal	90	90	120	#
PLA	α (*)	1.06	0.61	2.88	0.288
	Orthorhombic	90	90	90	1.2845
	β	1.03	1.82	0.9	0.3
	Orthorhombic	90	90	90	1.286
Aliphatic polyamides					
PA66	α (*)	0.49	0.54	1.72	1.72
	Triclinic	48.5	77	63.5	1.24
	β	0.49	0.8	1.72	1.72
	Triclinic	90	77	67	1.25
PA6	α (*)	0.956	1.724	0.801	1.724
	Triclinic	90	67.5	90	1.23
	γ	4.79	4.79	1.67	1.67
	Hexagonal	90	90	60	1.17
Polyolefins					
PE	#	0.74	0.493	0.253	0.253
	Orthorhombic	90	90	90	1
PP	α (*)	0.665	2.096	0.65	0.65
	Monoclinic	90	99.3	90	0.938
	γ	0.65	0.65	#	#
	Hexagonal	90	90	120	#

* Preferred crystalline modification in high-speed melt spinning.
L_r: chain length /repeating unit.

There are more complicated and hierarchical aspects in fiber structure. Important features that could not be included in the previous explanations are the concepts of entanglement and network structure.

Long and Ward (1991) found that the true-stress vs. true-strain curves of PET fibers prepared under a wide range of processing conditions can be superposed by simply shifting the curves along the true-strain axis, as shown in Fig. 8. The entanglement density estimated from the network draw ratio, which can be estimated from the amount of shift for curve matching, and thermal shrinkage stress are roughly

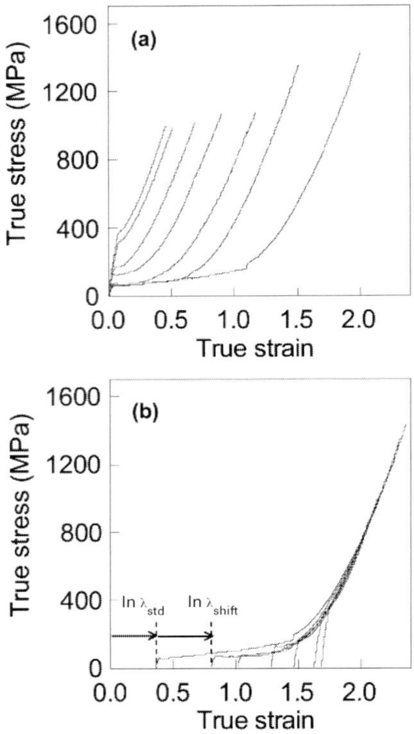

Figure 8
(a) True stress versus true strain curves of PET fibers prepared by various processing conditions. (b) Matching of true stress versus true strain curves of PET fibers. Superposed curve corresponds to the network draw ratio of unity. λ_{std}, network draw ratio of standard fibers; λ_{shift}, draw ratio corresponding to axial shift along the true strain axis needed for curve matching.

comparable to those obtained from the plateau modulus measured through the rheological measurements of the polymer melt. These results indicate that PET fibers retain a level of high-order structure that is not affected by processing.

Fundamental studies of orientation-induced crystallization behavior suggest that the starting of crystallization is not controlled simply by the degree of molecular orientation, but that the nonlinear characteristics of elongational viscosity and the state of molecular entanglement also play an important role (Okamoto *et al.* 1998). In this context, the 'shish' in the 'shish-kebab' structure corresponds to the tightened molecules between two adjacent entanglements, that is, the starting point of the crystallization. Therefore, the role of entanglement should be considered more extensively for the better understanding of the mechanism of fiber structure development. Several types of mechanical measurements have been proposed for the analysis of entangled structures. However, at present there are no structural measurement techniques applicable for the direct analysis of the state of entanglement.

Bibliography

Bashir Z, Hill M J, Keller A 1986 Comparative study of etching techniques for electron microscopy using melt processed polyethylene. *J. Mater. Sci. Lett.* **5**, 876–8

Cao J, Kikutani T, Takaku A, Shimizu J 1989 Nonisothermal orientation-induced crystallization in melt spinning of polypropylene. *J. Appl. Polym. Sci.* **37**, 2683–97

Cho H H, Kim K H, Ito H, Kikutani T 2000 Fine structure and properties of polyethylene fibers in high-speed spinning. *J. Appl. Polym. Sci.* **77**, 1195–206

Desborough I J, Hall I H, Neisser J Z 1979 The structure of poly(trimethylene terephthalate). *Polymer* **20**, 545–52

Fujimoto K, Iohara K, Owaki S, Murase Y 1988 Structure and properties in high-speed melt spinning of poly(butylene terephthalate). *Sen'i Gakkaishi* **44**, 394–401

Furuhashi Y, Ito H, Kikutani T, Yamamoto T, Kimizu M, Cakmak M 1998 Structural analysis of poly(3-hydroxybutyrate-co-3-hydroxyvalerate) fibers prepared by drawing and annealing processes. *J. Polym. Sci.* **36**, 2471–82

Heuvel H M, Huisman R, Lind K C J B 1976 Quantitative information from x-ray diffraction of nylon-6 yarns. *J. Polym. Sci., Polym. Phys. Ed.* **14**, 921–54

Hoggsteen W, Postema A R, Pennings A J, Brinke G T 1990 Crystal structure, conformation, and morphology of solution-spun poly(L-lactide) fibers. *Macromolecules* **23**, 634–42

Jakeways R, Ward I M, Wilding M A, Hall I H, Desborough I J, Pass M G 1975 Crystal deformation in aromatic polyesters. *J. Polym. Sci.* **13**, 799–813

Kaji K 1997 From the fringed-micelle microfibril model to the shish-kebab model. *Sen'i Gakkaishi* **53**, P132–6

Kolb R, Seifert S, Strbeck N, Zachmann H G 2000 Simultaneous measurements of small- and wide-angle x-ray scattering during low speed spinning of poly(propylene) using synchrotron radiation. *Polymer* **41**, 1497–505

Koyama K, Suryadevara J, Spuriell J E 1986 Effect of molecular weight on high-speed melt spinning of nylon-6. *J. Appl. Polym. Sci.* **31**, 2203–29

Long S D, Ward I M 1991 Tensile drawing behaviour of polyethylene terephthalate. *J. Appl. Polym. Sci.* **42**, 1911–20

Mezghani K, Spruiell J E 1998 High-speed melt spinning of PLLA filaments. *J. Polym. Sci. Pt. B—Polym. Phys.* **36**, 1005–12

Miyata K, Kikutani T, Okui N 1997 Fiber structure formation in ultra-high-speed melt spinning of poly(ethylene-2,6-naphthalene dicarboxylate). *J. Appl. Polym. Sci.* **65**, 1415–27

Okamoto M, Kubo H, Kotaka T 1998 Elongational flow-induced crystallization of poly(ethylene terephthalate) under supercooled state. *Polymer* **39**, 3135–42

Orts W J, Marchessault R H, Bluhm T L, Hamer G K 1990 Observation of strain-induced β form in poly(β-hydroxyalkanoates). *Macromolecules* **23**, 5368–70

Schmack G, Tändler B, Vogel R, Beyreuther R, Jacobsen S, Fritz H G 1999 Biodegradable fibers of poly(L-lactide) produced by high-speed melt spinning and spin drawing. *J. Appl. Polym. Sci.* **73**, 2785–92

Shimizu J, Kikutani T, Takaku A, Okui N 1984 Fiber structure formation in high-speed melt spinning of poly(ethylene terephthalate): characterization of fiber structure by considering oriented mesophase. *Sen'i Gakkaishi* **40**, T177–84

Shimizu J, Okui N, Kikutani T 1985 Fine structure and physical properties of fibers melt-spun at high speeds from various polymers. In: Ziabicki A, Kawai H (eds.) *High-Speed Fiber Spinning*. Wiley, New York, pp. 429–83
Tashiro K, Nakai Y, Kobayashi M, Tadokoro H 1980 Solid-state transition of poly(butylene terephthalate) induced by mechanical deformation. *Macromolecules* **13**, 137–45
Yokouchi M, Sakakibara Y, Chatani Y, Tadokoro H, Tanaka T, Yoda K 1976 Structure of two crystalline forms of poly(butylene terephthalate) and reversible transition between them by mechanical deformation. *Macromolecules* **9**, 266–73
Ziabicki A and Kawai H (eds.) 1985 *High-Speed Fiber Spinning*. Wiley, New York

T. Kikutani

Polymer Film Casting: Modeling

The film-casting process is used industrially to manufacture thin flat sheets or films. The process itself is portrayed in Fig. 1, along with the respective Cartesian co-ordinate system used in the analysis. In this process, a polymeric material is extruded through a rectangular die or slit and taken up at the drum or chill roll. Typical thicknesses can vary from 10 μm to 2500 μm, whereas lateral dimensions range from 40 cm to 320 cm. For untreated films, the film-casting process has operating speeds in the range of 120–400 m min^{-1}, whereas if the film undergoes biaxial orientation, speeds are 280–350 m min^{-1} (Baird and Collias 1995). Cast films are primarily used in the packaging industry for foodstuffs and other consumer products. Other uses include magnetic strips for audio and video tapes. Film casting is very similar to the fiber spinning, except where a fiber is drawn uniaxially (stretching of a cylindrical rod), the film or sheet is drawn planarly (stretching of a flat surface). After the film reaches the drum or chill roll, it is usually trimmed at the sides to remove the thick edges (*edge beads*).

Early experimental work was conducted by Kase (1974), who presented results on the film casting of a polypropylene (PP) melt. Dobroth and Erwin (1986) examined the causes of thick edges or the edge-bead effect (also called the *bone effect*), and they attributed it to the change from planar extension in the center of the film to uniaxial extension at the edges. Studies have been conducted by Pis-Lopez *et al.* (1996a, 1996b) who analyzed the stability of the process with numerous constitutive models (upper-convected Maxwell (UCM), Carreau, and Giesekus). Another major source of studies has emerged from Agassant's group (d'Halewyn *et al.* 1990, Barq *et al.* 1992, Silagy *et al.* 1996), who have analyzed stability and performed simulations with the Newtonian and UCM models. An attempt was made by Alaie and Papanastasiou (1991) to simulate the data of Kase (1974) employing a 1D approximation with no *neck-in effect* (thinning in the third direction) and using a modified K-BKZ integral constitutive model. Debbaut *et al.* (1995) performed a 2D analysis with the use of the Newtonian and UCM models. Sakaki *et al.* (1996) performed a full 3D analysis using only the Newtonian model. These approaches have been compared to the quasi 1D approach by Beaulne and Mitsoulis (1999) given here.

1. *Mathematical Modeling*

The equations of mass and momentum are averaged over the cross-section of the film to produce the 1D equations. The equation of mass (continuity) becomes

$$\rho(4wh)u = \rho Q = \text{constant} \tag{1}$$

Here, ρ is the polymer density, w is the film half-width, h is the film half-thickness, u is the film velocity, and Q is the volumetric flow rate. The corresponding equation of momentum is

$$\frac{d}{dx}[4wh(\tau_{xx} - \tau_{zz})] = \frac{dF}{dx} = 0 \tag{2}$$

where x is the downstream direction. It should be noted that the effects of air drag, inertia, and gravity have been neglected in Eqn. (2). These effects are usually negligible due to the high viscosity of the materials used in the film-casting process. Eqns. (1)

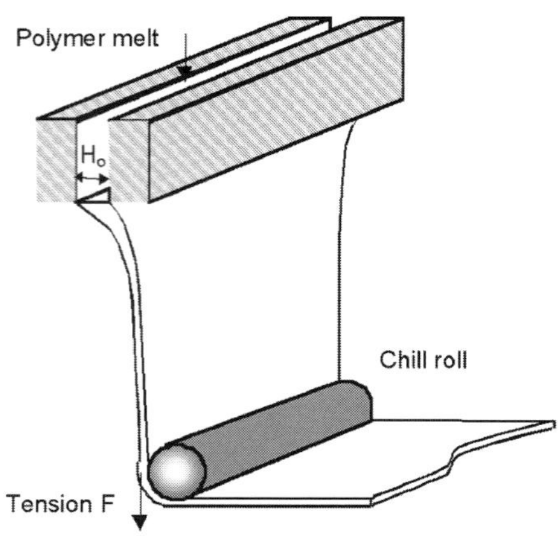

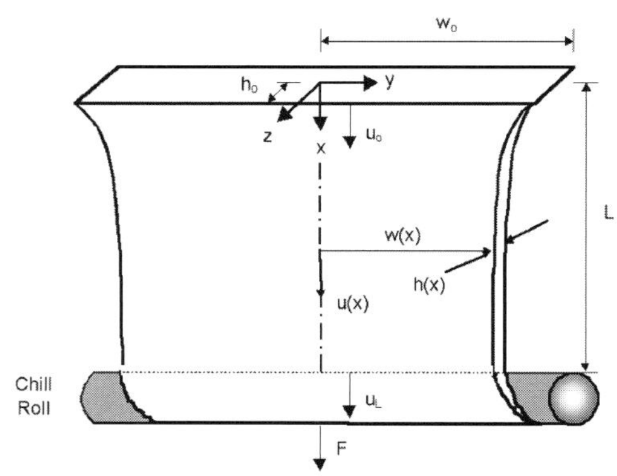

Figure 1
Sketch of the film-casting process. The melt is extruded through a slit die and is taken up at the chill rolls. Planar extension is effected by the drawing force at the chill rolls.

and (2) can be combined to give

$$\frac{d}{dx}\left[\frac{\tau_{xx} - \tau_{zz}}{\rho u}\right] = 0 \quad (3)$$

At the free surface, no fluid crosses the boundary and there is a no stress condition at the free surface, i.e.,

$$\bar{\bar{\sigma}} \cdot \bar{n} = 0 \quad (4)$$

When $\bar{n}$ is the normal vector to the free surface on the xy free-surface plane, Eqn. (4) gives

$$\frac{\tau_{yy} - \tau_{zz}}{\tau_{xx} - \tau_{zz}} = \left(\frac{dw}{dx}\right)^2 \quad (5)$$

The equation of energy in 1D form is given as

$$\rho c_p Q \frac{dT}{dx} + (4w + 4h)h_c(T - T_a) + (4w + 4h)\sigma_{SB}\varepsilon(T^4 - T_a^4) = 0 \quad (6)$$

where c_p is the heat capacity, T is the temperature of the film, h_c is the heat transfer coefficient, σ_{SB} is the Stefan–Boltzmann constant, and ε is the emissivity. The subscript a refers to ambient conditions. Eqns. (6) and (1) can also be combined to give

$$\rho c_p Q \frac{dT}{dx} + \left(4w + \frac{Q}{wu}\right)h_c(T - T_a) + \left(4w + \frac{Q}{wu}\right)\sigma_{SB}\epsilon(T^4 - T_a^4) = 0 \quad (7)$$

Single boundary conditions are required for Eqns. (3), (5), and (7) since they are first-order differential equations. These are:

(i) The velocity at $x = 0$ is the velocity at the die exit, u_o

$$u = u_o \quad \text{at } x = 0 \quad (8a)$$

(ii) The width at $x = 0$ is the width at the die exit, w_o

$$w = w_o \quad \text{at } x = 0 \quad (8b)$$

(iii) The temperature at $x = 0$ is the die temperature, T_o

$$T = T_o \quad \text{at } x = 0 \quad (8c)$$

To make the solution unique, an additional boundary condition is required. This is done by specifying the velocity at the take-up position (chill roll).

(iv) The velocity at $x = L$ is the take-up velocity, u_L

$$u = u_L \quad \text{at } x = L \quad (8d)$$

To proceed with the analysis of a viscoelastic fluid, the stresses τ_{xx}, τ_{yy}, and τ_{zz} in Eqns. (2), (3), and (5) must be expressed in terms of the velocity by means of a constitutive equation. The one chosen by Beaulne and Mitsoulis (1999) is the K-BKZ integral constitutive equation with multiple relaxation times, proposed by Papanastasiou *et al.* (1983) and further modified by Luo and Tanner (1988). The modified K-BKZ model reduces to the Newtonian model when a single relaxation time $\lambda_1 = 0$ is used, or, respectively, a small number, say $\lambda_1 = 0.001$. The K-BKZ model also reduces to the upper-convected Maxwell (UCM) model when a non-zero single relaxation time is used and the strain-memory function of the model equals 1. The UCM model is routinely used for validation purposes of a viscoelastic scheme. In viscoelastic simulations, the dimensionless quantity of choice is the Weissenberg number $Ws = \lambda u_o/L$.

2. *Results and Discussion*

2.1 Newtonian Case

At a very small Weissenberg number ($Ws = 0.001$), the viscoelastic case reduces to the Newtonian one. Figure 2(a) shows the simulation results with the modified K-BKZ model (with $\lambda_1 = 0.001$) along with the numerical solution for the Newtonian model with an aspect ratio $A = L/w_0 = 0.8$ and a draw ratio $D_R = u_L/u_0 = 10$. Also shown in Fig. 2(a) are the results from Silagy *et al.* (1996) for the same conditions. There is excellent agreement between different simulations and approaches for the Newtonian case.

2.2 Upper-convected Maxwell (UCM) Case—Comparison with 2D Simulations

Isothermal 2D simulations for the film-casting process have been conducted by Debbaut *et al.* (1995) with the Newtonian model and the upper-convected Maxwell (UCM) model. The 2D analysis also accounts for the edge-bead effect with an additional equation, which allows variance in the film thickness in the y-direction as well as the x-direction. The 1D analysis cannot predict the edge-bead effect; only 2D and 3D approximations can account for this. Figure 2(b) shows the simulation results for the neck-in effect (w_L/w_o) as a function of the draw ratio (D_R) for both the Newtonian model and the UCM model. The operating conditions used in the simulations are: $w_o = 50$ cm, $h_o = 0.05$ cm, $u_o = 10$ cm s^{-1}, and $L = 50$ cm. Agreement between the two approaches

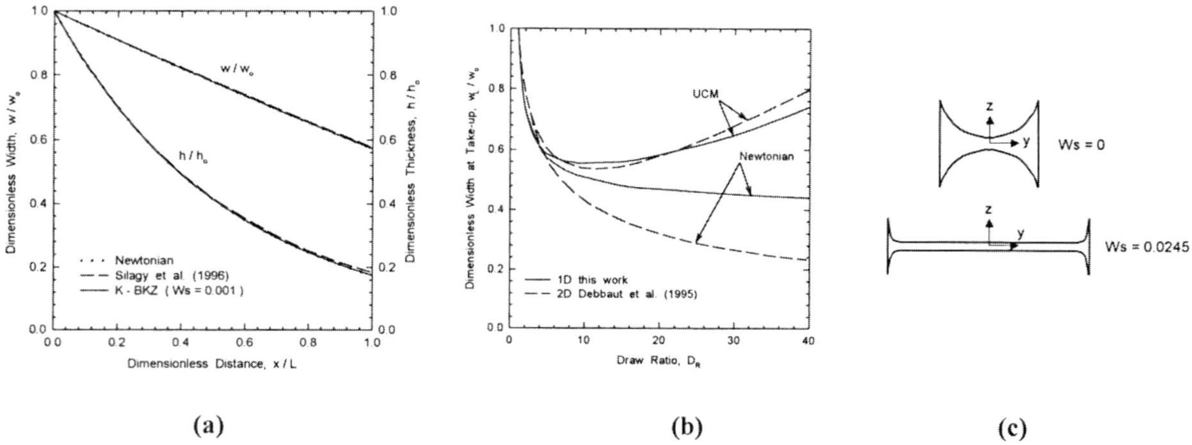

Figure 2
(a) Predictions of the viscoelastic simulation (solid line) in the limiting case of low Weissenberg number (Ws = 0.001) compared with the Newtonian solution (open symbols). The aspect ratio $A = L/w_0 = 0.8$ and the draw ratio $D_R = u_L/u_0 = 10$. (b) Predictions of 1D versus 2D approach in simulations of film casting of a Newtonian and a UCM fluid. (c) Edge-bead effect from the 2D simulations by Debbaut *et al.* (1995) showing that viscoelasticity reduces the curvature of the cast sheet.

for the Newtonian model is good for small draw ratios ($D_R < 3$) but poor at higher draw ratios ($3 < D_R < 40$). This is due to the ability of the 2D analysis to predict the edge-bead effect, whereas a 1D analysis is incapable of predicting this.

With respect to the UCM model, from Fig. 2(b), the 1D results agree quite well with the 2D results. The agreement is due to the *sharp* edge-bead effect resulting from the high elasticity of the UCM model, in contrast to the behavior of Newtonian model. To illustrate this Fig. 2(c) depicts the predicted edge-bead effect from the 2D analysis of Debbaut *et al.* (1995) for both the Newtonian model and the UCM model for a viscoelasticity level defined by the Weissenberg number $Ws = \lambda u_o/L = 0.0245$. The predicted edge-bead effect from the Newtonian model is greater in cross-sectional area and thus greatly alters the other dimensions (i.e., thickness at the centerline and width). The predicted edge-bead effect from the UCM model is *sharper* (i.e., less cross-sectional area) and lessens the significance on the thickness and width. In industrial operations, the edge-bead is simply trimmed to produce a film of uniform thickness. Therefore, the 1D and 2D analyses give similar viscoelastic predictions for the neck-in effect and the film thickness at the centerline, which are industrially important.

2.3 Non-isothermal K-BKZ Case—Comparison with Experiments

Nonisothermal film casting experiments were conducted for a PP melt at 215 °C by Kase (1974).

Table 1
Film-casting operating conditions for simulation based on experiments with PP conducted by Kase (1974).

Property	Value
Film width at die, w_o (cm)	1000
Melt velocity at die exit, u_o (cm s^{-1})	1.04
Film thickness at die, h_o (cm)	0.10
Length of film, L (cm)	80.0
Draw ratio, D_R	33.0
Inlet temperature, T_o (°C (K))	215 (488)
Air temperature, T_{air} (°C (K))	25 (298)

Measurements of the film thickness and the film temperature were taken with respect to distance for different draw ratios (D_R). The operating conditions used in the simulation and the experiment are given in Table 1. These data have been previously simulated by Alaie and Papanastasiou (1991) with a 1D approach with no variation of the film width.

Figure 3 shows the K-BKZ simulation results for the thickness and temperature profiles along with the experimental measurements by Kase (1974). Good overall agreement is obtained between simulations and experimental data with subtle differences. The predictions for the final thickness of the film (3.16 μm) is only slightly higher than the experimentally observed final thickness of 3.0 μm (5.7% error). As mentioned earlier, the final uniform thickness of the film is industrially important. The numerical scheme developed by Beaulne and Mitsoulis

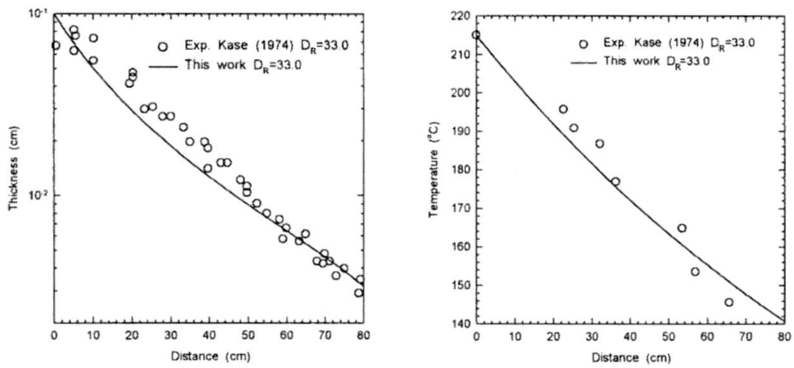

(a) (b)

Figure 3
Comparison of K-BKZ simulations by Beaulne and Mitsoulis (1999) and experiments conducted by Kase (1974) for a PP melt: (a) film thickness and (b) film temperature. Operating conditions are given in Table 1.

(1999) for integral constitutive equations demonstrates itself as a useful tool for quantitative predictions of film thickness and temperature for the film-casting process. However, a fully 3D simulation is necessary for cases where the neck-in and edge-bead effects have to be fully accounted for. This is the current state of affairs in the numerical simulation of the film-casting process.

See also: Polymer Rheology for Melt Processing: Complex Flows; Polymer Fiber Processing: Modeling

Bibliography

Alaie S M, Papanastasiou T C 1991 Film casting of viscoelastic liquid. *Polym. Eng. Sci.* **31**, 67–75
Baird D G, Collias D I 1995 *Polymer Processing: Principles and Design*. Butterworth-Heinemann, Boston
Barq P, Haudin J M, Agassant J F 1992 Isothermal and anisothermal models for cast film extrusion. *Int. Polym. Process.* **7**, 334–49
Beaulne M, Mitsoulis E 1999 Numerical simulation of the film-casting process. *Int. Polym. Proc.* **14**, 261–75
Debbaut B, Marchal J M, Crochet M J 1995 Viscoelastic effects in film casting. *Z. Angew. Math. Phys.* **46**, S679–98
d'Halewyn S, Agassant J F, Demay Y 1990 Numerical simulation of the cast film process. *Polym. Eng. Sci.* **30**, 335–40
Dobroth T, Erwin L 1986 Causes of edge beads in cast films. *Polym. Eng. Sci.* **26**, 462–7
Kase S 1974 Studies on melt spinning IV. On the stability of melt spinning. *J. Appl. Polym. Sci.* **18**, 3279–304
Luo X-L, Tanner R I 1988 Finite element simulation of long and short circular die extrusion experiments using integral models. *Int. J. Numer. Methods Eng.* **25**, 9–22
Papanastasiou A C, Scriven L E, Macosko C W 1983 An integral constitutive equation for mixed flows: viscoelastic characterization. *J. Rheol.* **27**, 387–410
Pis-Lopez M E, Co A 1996a Multilayer film casting of modified Giesekus fluids. Part 1. Steady-state analysis. *J. Non-Newton. Fluids* **66**, 71–93
Pis-Lopez M E, Co A 1996b Multilayer film casting of modified Giesekus fluids. Part 2. Linear stability analysis. *J. Non-Newton. Fluids* **66**, 95–114
Sakaki K, Katsumoto R, Kajiwara T, Funatsu K 1996 Three-dimensional flow simulation of a film-casting process. *Polym. Eng. Sci.* **36**, 1821–31
Silagy D, Demay Y, Agassant J F 1996 Study of the stability of the film casting process. *Polym. Eng. Sci.* **36**, 2614–25
Tanner R I 1985 *Engineering Rheology*. Clarendon, Oxford

E. Mitsoulis

Polymer Glasses, Welding of: Processes and Mechanisms

The welding of polymer glasses is important in applications where items or components cannot be manufactured solely using conventional polymer processing techniques such as injection molding, blow molding, and extrusion. Welding may also be selected in cases where economic considerations make the manufacturing of items in one operation too costly. Sometimes welds are formed during the manufacture of items using conventional polymer processing techniques such as injection molding, where weldlines can be formed, or extrusion of polymer pipe, where knitlines are formed.

For all of the above reasons, the welding of items made from polymer glasses is important in their manufacture and successful operation. Welding is only one way to join polymer glasses—the use of adhesive bonding and mechanical fastening being two other possibilities. However, welding is generally selected in cases where joints with good mechanical integrity over long periods are required and where the joint has to provide a seal for fluids, for example, the joints in uPVC (poly(vinyl chloride)) window frames.

This article describes the physical mechanisms involved when polymer glasses are welded. The key variables involved in weld formation are introduced, namely temperature, time, and pressure, and their significance is explained in terms of the development of contact area and polymer chain interdiffusion. Theories for welding are described in chronological order. The technology of welding polymer glasses is then reviewed, with each commercially applied process being explained and illustrated with a typical example.

1. *Welding Similar and Dissimilar Polymer Glasses*

The welding of components made from chemically similar polymer glasses occurs when they are in intimate contact at a temperature above the glass transition temperature. Under these conditions, polymer chains have mobility and will interdiffuse. If, after a certain period of time the temperature at the interface falls below the glass transition, the polymer chains lose their mobility and freeze. Some polymer chains will freeze in a position such that they straddle the weld interface and are responsible for the weld strength.

If a weld is attempted between dissimilar polymer glasses, the physical situation is different. For a good weld to be formed, miscibility of one polymer with the other is needed. The theory required to determine dissimilar polymer miscibility has been developed using thermodynamics and statistical physics (Flory 1953). However, in its simplest form, miscibility can be predicted from the equation of the free energy of mixing:

$$\Delta G_{\mathrm{m}} = \Delta H_{\mathrm{m}} - T\Delta S_{\mathrm{m}} \tag{1}$$

where ΔG_{m} is the change in the Gibbs' free energy of mixing, ΔH_{m} is the change in enthalpy, T is the absolute temperature, and ΔS_{m} is the change in entropy on mixing. If ΔG_{m} is negative then the polymers are miscible but if it is positive then they are not miscible. ΔS_{m} is always positive because mixing involves an increase in disorder and hence an increase in entropy. Therefore the miscibility of two polymers generally depends on the change in enthalpy on mixing.

The enthalpy of mixing is usually positive, i.e., opposing mixing, and this is why most polymers are immiscible. For a rigorous treatment of dissimilar polymer miscibility see Flory (1953), Olabisi *et al.* (1979), and Wool (1995). Tables containing lists of miscible polymers are also available in Watson (1988) and Brandrup and Immergut (1989).

2. *Theories for Welding*

If two pieces of uncross-linked rubber at room temperature are brought together under sufficient pressure to guarantee intimate contact at the interface, a joint will be formed between them after a few seconds. The glass transition temperature of uncross-linked natural rubber is well below room temperature and so it is straightforward to see that this situation is analogous to joining two pieces of polystyrene above their glass transition: i.e., a weld is formed.

Many of the early experiments on welding of thermoplastic polymers, based on rubber (Voyutskii 1963), were carried out in Russia in the 1950s (Vasenin 1961). The researchers undertook a number of experiments on "tack" in the rubber, that is, the build-up of mechanical strength between rubber components in intimate contact. A set of experiments was performed where weld strength or tack was measured between two pieces of polyisobutylene as a function of temperature while keeping time and pressure constant. The results of these experiments showed an exponential form when plotted on a graph. From these results, the researchers deduced that the development of weld strength was due to the interdiffusion of polymer chains as shown schematically in Fig. 1.

Mathematically, the theory predicted that the weld strength, A, should be given by:

$$A = A_0\exp(-E/RT) \tag{2}$$

where A_0 is a constant, E is the activation energy for the process, R is the gas constant, and T the absolute temperature.

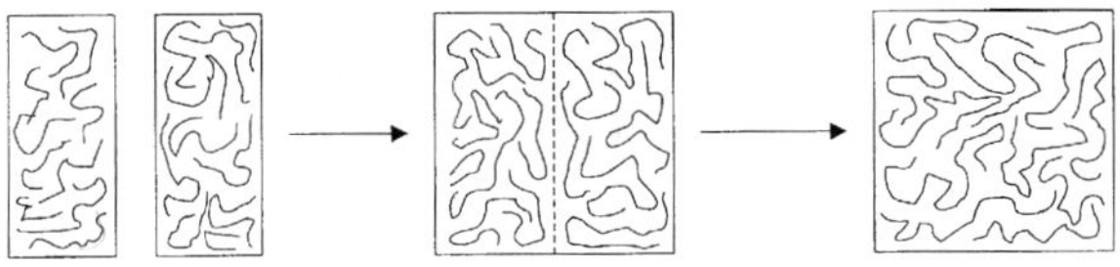

Figure 1
Scheme of coalescence of high polymers. (after Voyutskii 1963)

In the 1970s a theory to describe the motion of polymer chains above their glass transition temperature was developed by De Gennes (1971). The theory, called reptation after the Latin *reptare*—to creep, was used to make predictions about polymer self-diffusion, such as the self-diffusion coefficient D,

$$D \propto M^{-2}$$

where M is the molecular weight of the polymer. This was confirmed experimentally for a number of polymers.

In the reptation theory, polymer chains are considered to lie in a tube whose walls are defined by the positions of neighboring chains. As the chain moves backwards and forwards in the tube, its ends can choose different paths in the surrounding entangled polymer chains (Fig. 2). In this way, after a certain time T_r (the tube renewal time), it can escape the tube defining its original position. De Gennes showed that $T_r \propto M^3$ (see also *Polymer Rheology for Melt Processing: Molecular Modeling*).

The reptation theory was applied successfully to the welding of glassy thermoplastic polymers such as polystyrene (Wool 1995). From this work, Wool was able to predict the strength of welds made in glassy thermoplastic polymers as a function of temperature and time.

More recently the welding process has been modeled using Monte Carlo techniques (Haire and Windle 2001) to produce algorithms that can accommodate polydisperse systems under any combinations of temperature and time.

3. *Physical Mechanisms Affecting Welding*

Welds are made in thermoplastic polymers by first raising the temperature above their glass transition temperature T_g. The parts to be welded are pressed together so that they are in intimate contact. Sufficient time at temperature is allowed for welding to occur and then the joint is cooled under pressure to a temperature below the glass transition. The main parameters affecting welding are therefore temperature, pressure, and time. A theory to explain the significance of these parameters and how they affect weld strength has been developed over the last 50 years, as explained above. However, this only explains one aspect of welding, i.e., the interdiffusion of polymer chains at the weld interface. Although this mechanism must be present for weld strength to develop, other factors also greatly affect the strength of welds.

In commercial practice, welds in thermoplastic polymers can be made using any one of 20 techniques. In each case, generally there will be some flow of molten polymer and a thermal gradient in space and time centered on the weld interface. These two factors will create some local orientation of polymer chains, a distribution of residual mechanical stress, and a change in geometry in the region of the weld. It is the concentration and distribution of these effects that result in welds having a markedly different load-bearing capability. A diagram showing the occurrence of the factors contributing to the mechanical properties of welds in polymer glasses is shown in Fig. 3.

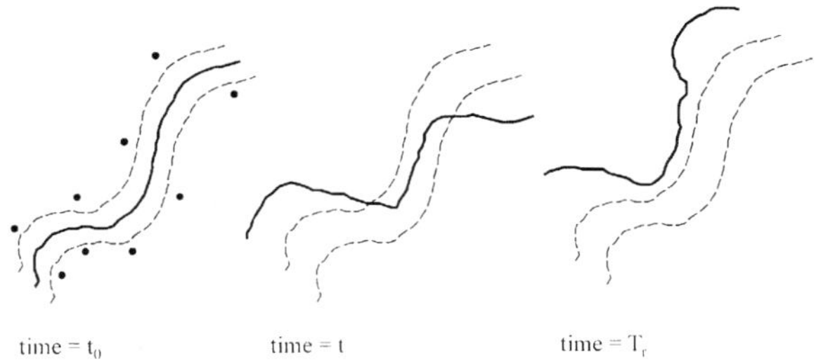

Figure 2
Position of a chain in a melt defined by a tube at time t_0, at some later time t, when the chain has just started to leave the tube, and at a time T_r (later again), after completely leaving the tube (after De Gennes 1971).

4. *Welding Techniques*

There are approximately 20 different welding techniques in commercial use. These can be divided into three groups (Wise 1999): those in which heat is generated by mechanical movement; those in which heat is generated by an external source and then introduced using thermal conduction; and those using electromagnetism directly.

4.1 Techniques in Which Heat is Generated by Mechanical Movement

(*a*) *Spin welding.* One component is rotated relative to the other and pressed against it such that heat is generated between them by friction. The main welding parameters are rotational speed, friction pressure, forge pressure, displacement, and weld duration. Applications include fittings welded to PVC pipes using spin welding. This technique is restricted to parts having cylindrical symmetry.

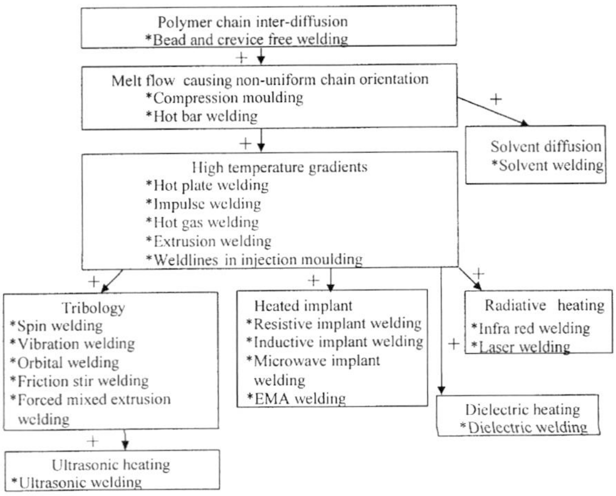

Figure 3
Additive hierarchical representation of 20 welding techniques and the physical factors that affect their mechanical properties. The diagram should be read from top to bottom and the factors (each represented by one box) are added. For example, laser welding includes effects from the following factors; polymer chain inter-diffusion, melt flow causing nonuniform polymer chain orientation, high temperature gradients, and radiative heating (after Wise 1999).

(*b*) *Vibration welding*. Components are brought into contact, under pressure, before being moved so that the joint area rubs together with a linear reciprocating motion. Heat is generated by friction and, as soon as the thermoplastic at the joint melts, the vibration stops, the parts are aligned, and the joint cools under pressure. The main processing parameters are weld time, weld pressure, vibration amplitude, and frequency of vibration. This technique is used in the welding of car bumpers where the joint is approximately 1.5 meters long and parts are welded at the rate of 70 per hour.

(*c*) *Ultrasonic welding*. Components are held together under pressure and are then subjected to ultrasonic vibrations such that the joint is compressed and then released at the frequency of the applied ultrasound. The process generally occurs at 20 kHz or 40 kHz and the heat is generated in the material because of viscoelastic loss and plastic deformation. Features that concentrate the dynamic mechanical stress introduced by the ultrasound (called energy directors) are usually molded into the opposing surfaces of the plastic components to be welded. Weld times are typically one second or less. Main welding parameters are frequency, weld time, weld pressure, trigger pressure (the pressure at which the ultrasound is initiated), and hold time. Ultrasonic welding is widely used in many industrial sectors, including automotive, domestic appliance, and biomedical, e.g., welding the sheath for catheter applicators.

(*d*) *Orbital welding*. Components are held together under pressure and one describes an orbital motion relative to the other such that each point on the surface of one component orbits a different point on the surface of the other component. The orbital motion, which may be of the order of a few millimeters in diameter, ceases after sufficient material has melted and the thermoplastic polymer then solidifies to form a weld. Orbital welding is a relatively recently developed welding technique but applications in the area of welding lids to fluid reservoirs have already been exploited.

(*e*) *Friction-stir welding*. Components are clamped together and a rotating metallic tool is drawn through the joint in a similar manner to a milling operation. The tool has a relatively smooth surface and the material is heated by this action producing a weld as the tool moves. In another variation, welds have also been produced by a tool moving with a reciprocally oscillating motion, in a similar fashion to a jig-saw. Friction-stir welding of thermoplastic polymers has yet to be exploited commercially although applications are anticipated in the fabrication of tanks and vessels where butt welds possessing good mechanical properties are required.

4.2 *Techniques Employing an External Heat Source*

(*a*) *Hot-plate welding*. Components to be welded are held in fixtures that press them against a heated tool. Heating takes place in two stages; first, the heated tool heats the surfaces of the materials so that full contact with the heated tool is obtained but mechanical stops prevent further displacement of the parts; the components continue to be heated until they are softened some distance from the heated tool. The fixtures then open, the heated tool is withdrawn, and the fixtures force the softened surfaces of the parts together. The parts are held together at a set distance or pressure until the polymer has cooled and a weld is formed. Main welding parameters include heating time, heating pressure or displacement, temperature of the heated tool, welding pressure or displacement, and welding time. Applications of hot-plate welding include the joining of extruded profiles to make window frames and the joining of pipes for the distribution of gas and water.

(*b*) *Hot-bar and impulse welding*. Hot-bar welding is mainly applicable to components in film form, that is, having a thickness of less than 0.5 mm. The technique involves the pressing of two thermoplastic polymer films against a hot bar where they melt together to form the weld. Since this technique relies on the conduction of heat through the material, there is a practical limit to the thickness of films which can be welded. Welding parameters important to this process are bar temperature, weld pressure, and weld time. Hot-bar welding has found wide application in the packaging industry where it is used for the sealing of films and bags.

Impulse welding is a variation of hot-bar welding where one or both bars are heated and then allowed to cool rapidly, thus allowing components being welded to experience a well controlled heating and cooling regime while under pressure.

(*c*) *Hot-gas welding.* Hot-gas welding involves a stream of gas which is heated electrically in a tool before being applied to the joint between two thermoplastic polymer components where it softens the polymer. A filler rod, made of the same type of polymer as the components being welded, is consumed by the process by being melted and added to the joint. Important welding parameters are gas type, gas flow rate and temperature, and the angle of the filler rod to the parts being welded. Hot-gas welding is widely used to make tanks, vessels, and liners to contain fluids, particularly where the finished product is too large for other welding techniques to be easily applied.

(*d*) *Extrusion welding.* Extrusion welding is a near relative of hot-gas welding and is based on the extrusion of thermoplastic through a heated die. As it emerges from the die in a molten form it is forced into the joint between two thermoplastic polymer components, which themselves have been softened by a hot-gas gun which forms part of the extrusion welding machine. Important welding parameters include the temperature of the hot gas, hot-gas flow rate, extrusion speed, and barrel temperature. Applications for the technique include the fabrication of tanks, vessels, and pipes.

(*e*) *Flash free welding.* Flash free welding is a technique for constraining the molten polymer produced during welding. This prevents any change in the geometry of the parts being welded due to the flow of molten polymer. The molten polymer is usually constrained using mechanical supports. The best example of this is the BCF (bead and crevice-free) welding technique developed for producing welds in thermoplastic polymer pipes with a completely smooth geometry. This is particularly important when producing very clean water systems for industrial sectors such as microelectronics.

(*f*) *Solvent welding.* Solvent welding involves the application of a polymer solution to both components to be joined such that the solvent causes swelling and plasticization of the polymer. The components are then pressed together and the polymer chains which have been mobilized by the action of the solvent are free to diffuse across the joint interface and entangle with polymer chains in the other component. A weld forms as the solvent diffuses or evaporates from the joint. Another way to view the formation of a solvent weld is to consider the solvent-impregnated region of the components as having a glass transition (T_g) temperature that has been lowered by the presence of the solvent to a temperature below ambient. It is then straightforward to see how a weld can be formed at room temperature and how this is analogous to a hot-plate weld, for example, where there is no solvent present but the required polymer chain mobility is generated by the increase in temperature. Solvent welding is used for welding pipes made from thermoplastic polymers in industries involved with fluid handling, e.g., domestic water systems.

4.3 Techniques that Directly Employ Electromagnetism

(*a*) *Resistive implant welding.* Resistive implant welding involves trapping an electrically conducting insert between the two parts to be joined and then resistively heating the insert by passing a high electric current through it. As the insert heats due to resistive losses the surrounding thermoplastic softens and, with the aid of applied pressure, causes the parts to come into intimate contact. A weld forms as the thermoplastic cools. Resistive implant welding has been applied to the welding of automotive bumpers and the joining of thermoplastic polymer pipes.

(*b*) *Induction welding.* Induction welding strictly describes welding techniques where heat is generated by an induction field. The two most common mechanisms by which heat can be generated in this way are eddy current heating and heating due to hysteresis loss. Hysteresis loss is described more fully under electromagnetic (EMA) welding below.

Induction heating by eddy current generation is similar to resistive implant welding in that an electrically conducting insert is required. A work coil is then placed in close proximity to the joint. As an electric current at high frequency passes through the work coil a dynamic magnetic field is generated whose flux links the implant. Electric currents are induced in the implant and, when these are sufficiently high to heat the conducting material, the surrounding thermoplastic polymer parts soften. Pressure is required to bring the parts into intimate contact and a weld forms as the joint cools.

Applications have included sealing plastic-coated metal caps to plastic bottles and welding grilles to the front of loudspeaker units.

(*c*) *EMA welding.* This technique harnesses hysteresis loss in ferromagnetic materials as the main heating mechanism. All ferromagnetic materials exhibit some hysteresis loss, the magnitude of which is generally a function of the applied magnetic field. In the EMA welding process an insert comprising particles of ferromagnetic material in a matrix of polymer is trapped between the parts to be welded. Energy is applied by a work coil placed in close proximity to the implant. Applications have included the welding of automotive and domestic appliance components.

(*d*) *Dielectric welding.* Dielectric welding (also called high-frequency or radio-frequency welding) relies on certain properties of the material in the parts being welded to cause the generation of heat in a rapidly alternating electric field. This means that only certain materials can be dielectrically welded, including poly(ethylene terephthalate) (PET) and PVC. The technique involves subjecting the parts to be welded to a high-frequency electric field, which is normally applied between two metal bars. These bars act as pressure applicators during heating and cooling. The dynamic electric field causes the molecules in some thermoplastic polymers to oscillate (depending on the presence of appropriate dipoles), thereby raising the thermal energy in the material. A measure of a material's ability to do this is the quantity called the dielectric loss factor which is a function of temperature and frequency. Applications for dielectric welding have included garment manufacture, upholstery, and packaging.

(*e*) *Microwave welding.* The use of microwaves for welding polymers has been technologically feasible for many years but has yet to be practiced commercially on any considerable scale. Developments have been reported where microwave radiation has been applied to welds in thermoplastic polymers by a coaxial cable feed, by wave-guide feed, by single-mode resonant cavities, and by multi-mode resonant cavities. Microwave welding will generally involve an insert placed at the joint interface to absorb the microwave energy preferentially from the thermoplastic parts to be welded. This is because most thermoplastic polymers will not absorb significant amounts of microwave energy at frequencies around 2.45 GHz. Exceptions to this are polymers that are filled with microwave absorptive particles such as carbon, ferromagnetics, and ferrimagnetics. The inserts should be absorptive to microwave radiation and could be made from materials including carbon, metal, electrically conducting polymers, and ferrites. There have been several applications for microwaves in materials processing, such as the curing of rubber and curing of polyurethanes in the automotive industry.

(*f*) *Infrared welding.* Infrared welding has been used to weld thermoplastic polymers in a manner similar to hot-plate welding but with the exception that there is no physical contact with the parts to be welded (reducing the chance of weld contamination and also decreasing the weld time). The parts are attached to fixtures on the machine and the infrared heat source applies energy to the surfaces to be joined. After the heat source has been removed the parts are forced together for a certain time and, at a certain pressure, a weld is formed as the polymer cools. Infrared heat sources employed for welding have included tungsten filament lamps and metal plates coated with special ceramics that emit strongly in the infrared. Infrared welding has been successfully applied to the welding of thermoplastic polymer pipes.

(*g*) *Laser welding.* Laser welding was first demonstrated in the 1970s when lasers were first used industrially. The technique involves the application of intense, coherent electromagnetic energy, usually in the infrared or visible region of the spectrum. Usually at these wavelengths the energy is absorbed at the surface of the thermoplastic polymers and welds can be made very quickly and effectively between thin sheets of material or films. Welding speeds can be many tens of meters per minute making laser welding an ideal technique for high-volume production in areas such as the packaging industry. Subsequently it has been shown that laser energy at certain wavelengths can be made to penetrate significant thicknesses of most thermoplastic polymers. This has led to the development of transmission laser welding, where parts several millimeters in thickness have been welded using the energy deposited at the joint interface. This is achieved by the inclusion of an absorptive material at the joint interface, such as carbon black or certain dyes that absorb specific wavelengths but are otherwise transparent. Transmission laser welding has been applied to the manufacture of the plastic cases of electronic keys for motor vehicles.

Bibliography

Brandrup J, Immergut E H (eds.) 1989 *Polymer Handbook*. 3rd edn. Wiley, New York

De Gennes P G 1971 Reptation of a polymer chain in the presence of fixed obstacles. *J. Chem. Phys.* **55**, 572

Flory P J 1953 *Principles of Polymer Chemistry*. Cornell University Press, New York
Haire K R, Windle A H 2001 Monte Carlo simulation of polymer welding. *J. Comput. Theor. Polymer Sci.* **11**, 227–40
Olabisi O, Robeson L M, Shaw M T 1979 *Polymer/Polymer Miscibility*. Academic Press, New York
Vasenin R M 1962 *Vys. Soed.* 1961 **3**, 679–85; RAPRA Translation 1010 by R Moseley
Voyutskii S S 1963 *Autohesion and Adhesion of High Polymers*. Wiley, New York
Watson M N (ed.) 1988 *Joining Plastics in Production*. The Welding Institute, Cambridge
Wise R J 1999 *Thermal Welding of Polymers*. Abington Publishing, Cambridge
Wool R P 1995 *Polymer Interfaces—Structure and Strength*. Hanser Verlag, Munich, Germany

R. J. Wise

Polymer Injection Molding: 3D Modeling

Injection molding is a popular process in producing plastic parts of a complex shape at low cost and is frequently used for plastic manufacturing because of its versatility and efficiency. In injection molding, computer aided engineering (CAE) is a useful tool for developing new high-quality molds and products. Problems with a new product are detected and eliminated by computer simulation at the early stage instead of during pre-production. Since the CAE software used at present is based on the Hele–Shaw flow assumption, which represents a 2D planar flow, it is called two and one-half dimensional (2.5D) simulation. The 2.5D simulation is practically reasonable for large and thin-walled products. However, a more detailed and accurate prediction is demanded to understand the physical phenomena for the optimal design of small and thick-walled products. Full 3D simulation is an aggressive approach to account for this issue.

1. *3D Effects*

The small and thick-walled products have 3D effects that cannot be analyzed by 2.5D software. Batch (1994) described the 3D effects in detail. The important effects are (i) edge effects through narrow channels on flow, (ii) transverse flow in corners and cavity thickness change effects on flow, (iii) heat transfer at edges of runners and narrow cavities on cooling, (iv) local shrinkage at inside corners and its influence on warpage, and (v) shape effects at thickness changes and their influence on warpage. The objective of the 3D simulation is to account for these effects and to provide a good understanding of the relation between molds, machines, materials, and final products.

2. *Modeling*

Polymer melt flow during the filling process is assumed to be a nonisothermal flow of an incompressible, highly viscous, and non-Newtonian fluid with free surfaces. The inertia force is neglected in the momentum equation because the Reynolds number of the polymer melt flow is low. The governing equations are given below (Inoue and Matsuoka 1994):

Continuity equation

$$\nabla \cdot \mathbf{u} = 0 \tag{1}$$

Momentum equation

$$\rho(\delta \mathbf{u}/\delta t) + \nabla P - \nabla \cdot 2\eta \mathbf{D} - \rho \mathbf{g} = 0 \tag{2}$$

Energy equation

$$\rho C_P[(\delta T/\delta t) + (\mathbf{u} \cdot \nabla)T] = k\nabla^2 T + 2\nabla\eta\mathbf{D} : \nabla\mathbf{u} \tag{3}$$

where $\mathbf{u}$ is the velocity vector, ρ is the density, t is the time, P is the pressure, η is the viscosity, $\mathbf{D}$ is the strain rate tensor, $\mathbf{g}$ is the gravitational acceleration, C_P is the specific heat, T is the temperature, and k is the thermal conductivity.

The viscosity is approximated by using a power-law and Arrhenius model as a constitutive equation:

$$\log \eta = c_1 + c_2 \log \dot{\gamma} + c_3/T \tag{4}$$

$$\dot{\gamma} = (2\mathbf{D} : \mathbf{D})^{1/2} \tag{5}$$

where c_1, c_2, and c_3 are constants experimentally determined for each polymer and $\dot{\gamma}$ is the shear rate.

For tracking the free surfaces, the following convective equation is described by the pseudo concentration method or volume of fluid method.

$$\delta F/\delta t + (\mathbf{u} \cdot \nabla)F = 0 \tag{6}$$

F is the pseudo concentration defined as the concentration of a pseudo fluid, which flows with the actual fluid. F varies from zero to one. The area where F is equal to one is fully filled by the fluid. If F is zero, the area is empty. Then, the location of the melt front is determined where $0 < F < 1$.

Lo *et al.* (1994) described a 3D model to simulate both the filling and curing stages of the reaction injection molding process. Haagh *et al.* (1997) reported a 3D numerical simulation for the gas-assisted injection molding to predict the gas distribution inside products.

3. *Numerical Methods*

3D simulations require high speed and large memory computers to solve the governing equations. Therefore, the most intensive work in 3D modeling has been directed towards how to reduce the calculation time and memory to make computation practical. Inoue and Matsuoka (1993) and Inoue *et al.* (1996) have developed a fast and low memory algorithm to compute 3D mold filling by combining several effective methods. The spatial derivative terms of the governing equations are discretized by using a finite volume method. The time integration in the momentum equation is calculated by a two-stage rational Runge–Kutta scheme. To correct the velocity and pressure fields, the simplified marker and cell method is used. The donor–acceptor method is applied to discretize the advective term in the convective equation. Then, the computing time is dramatically reduced and a 3D simulation can be run

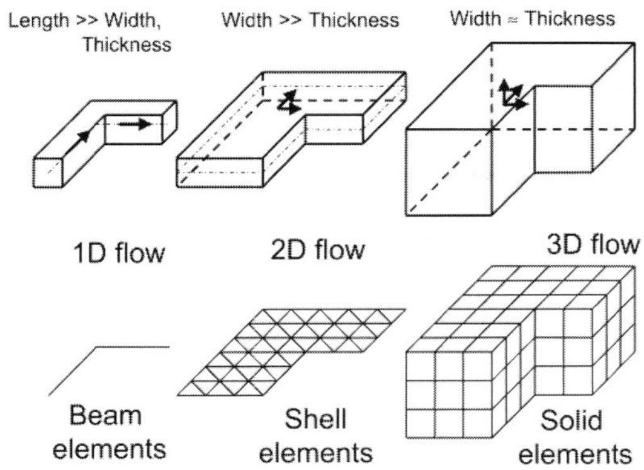

Figure 1
Elements for constructing meshed models.

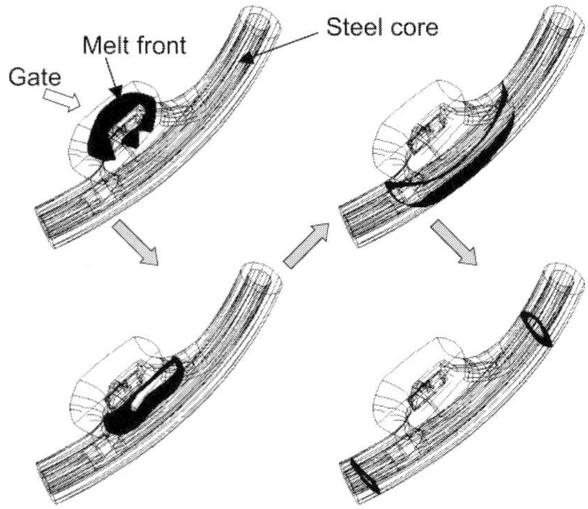

Figure 2
Snapshot of melt fronts during the mold filling of an automotive steering wheel calculated by the 3D simulation.

not only on workstations but also on personal computers.

The governing equations are solved by the finite element method (Haagh *et al.* 1997, Rajupalem *et al.* 1997, Hétu *et al.* 1998) and the finite difference method (Lo *et al.* 1994, Zachert and Michaeli 1997).

The shape of a product must be divided into many small elements (Fig. 1). Beam elements consider only a longitudinal flow through a channel with a length greater than the width and thickness such that a cross-sectional flow can be neglected. Shell elements are defined in the mid-plane of products for 2.5D simulation. The 3D simulation requires solid elements to predict flow in every direction. Shapes of the solid elements used are tetrahedral, pentagonal and hexahedral. The meshing work is the most time consuming process in CAE. Using a voxel method, 3D meshed models are automatically generated from 3D CAD data exported in a stereolithography file format (Nakano and Tanaka 1998). For a large product, it is difficult to make a mesh only with solid elements because of the huge number of elements required. In this case, a hybrid meshing technique has been developed by combining beam, shell and solid elements (Inoue *et al.* 1998).

4. *Industrial Applications*

3D mold filling simulations have been used already in practice. The flow prediction of an automotive steering wheel was carried out by Inoue *et al.* (1996). Steel cores are inserted into the steering wheel. Then the 3D complex flows at joint portions between the wheel and spokes are simulated. Figure 2 shows a snapshot of melt fronts in the area of a joint. The melt fronts are predicted as a surface. The results give us a good understanding of the flow behavior and allow us to prevent defects in the flow, like weldlines, voids, and air traps. Other industrial applications were reported on the mold filling of a door handle (Hétu *et al.* 1998) and an encapsulate molding of semiconductor devices by Shintai and Takenaka (1999).

Computer programs, MF/FLOW3D (Moldflow), SUNDYMOLD (Plamedia), TIMON3D (Toray), and Remylop (Toyota CRDL), are now commercially available for 3D CAE of injection molding processes.

The 3D technology provides high accurate predictions of pressure, temperature, and velocity because flow in all directions is considered. Therefore, it will be a useful tool for understanding complex flow phenomena in detail and developing novel injection molding techniques and products. The 3D simulation can be extended to predictions of post-filling, mold cooling, fiber orientation, and warpage, and is expected to be part of the next generation of CAE tools for injection molding.

See also: Polymer Injection Molding: Multimaterial Molding; Polymer Injection Molding: Gas-assisted Molding

Bibliography

Batch G 1994 3D effects in injection molding simulation. *Soc. Plast. Eng. Tech. Papers* **40**, 547–53

Haagh G A A V, Zuidema H, van de Vosse F N, Peters G W M, Meijer H E H 1997 Towards a 3-D finite element model for the gas-assisted injection moulding process. *Int. Polym. Process.* **3**, 207–15

Hétu J F, Gao D M, Garcia-Rejon A, Salloum G 1998 3D finite element method for the simulation of the filling stage in injection molding. *Polym. Eng. Sci.* **38**, 223–36

Inoue Y, Higashi T, Matsuoka T 1996 Numerical simulation of 3-dimensional flow in injection molding. *Soc. Plast. Eng. Tech. Papers* **42**, 744–8

Inoue Y, Matsuoka T 1998 3-D flow analysis for injection molding. *Ext. Abs. Polym. Process. Soc.* **14**, 319–20

Inoue Y, Matsuoka T 1993 Numerical simulation of 3-dimensional flow for injection molding process. *J. Soc. Rheology Jpn.* **21**, 175–80

Inoue Y, Matsuoka T 1994 Numerical simulation of 3-dimensional flow in an injection molding runner. *Abs. Pacific Conf. Rheology Polym. Process* **1**, 253–4

Lo Y, Reible D D, Collier J R, Chen C 1994 Three-dimensional modeling of reaction injection molding. *Polym. Eng. Sci.* **34**, 1393–400

Nakano R, Tanaka T 1998 Example of 3D injection molding CAE: directly connected with 3D CAD. *Ext. Abs. Polym. Process. Soc.* **14**, 321–2

Rajupalem V, Talwar K, Friedl C 1997 Three-dimensional simulation of the injection molding process. *Proc. Soc. Plast. Eng. Tech. Papers* **43**, 670–3

Shintai A, Takenaka O 1999 Integrated mold technology for semiconductor device. *SAE Technical Paper Series*, 1999-01-0161

Zachert J, Michaeli W 1997 Simulation and analysis of three-dimensional polymer flow in injection moulding. *Proc. Soc. Plast. Eng. Tech. Papers* **43**, 649–53

T. Matsuoka and Y. Inoue

Polymer Injection Molding: C-Mold

The advent of computer-aided engineering (CAE) technology for injection molding and related processes is having a rapidly growing impact on the plastic molding industry. Proper use of this technology has enhanced engineers' ability to handle all aspects of the molding process and has proven to benefit productivity, product quality and time-to-market.

C-MOLD is a suite of commercial CAE software products and knowledge management technology for polymer injection molding. Developed by Advanced CAE Technology, Inc. (under the name of C-MOLD), a US company founded in 1986 and purchased by Moldflow Pty. Ltd in 2000, the technology has evolved from a focused research effort by the Cornell Injection Molding Program (CIMP) at Cornell University, USA. Since its establishment in 1974, CIMP has advanced the scientific understanding of injection molding, as detailed in its progress reports (Wang *et al.* 1975–2000) and various publications listed in the bibliography. This article provides a brief description of C-MOLD, the required input and typical output data, different levels of in-depth analysis and design optimization, as well as the underlying theoretical bases for the process simulation and material modeling.

1. C-MOLD Products

C-MOLD has three product lines, Project Engineer, 3D QuickFill, and Advanced Solutions, which meet the needs of people involved in the various stages of design and manufacture of injection-molded parts. Project Engineer and 3D QuickFill provide intuitive implementation and fast turnaround time, whereas Advanced Solutions serve simulation analysts who need more user control and in-depth predictions. All three of the product lines share the same underlying theoretical bases, file format, databases, and user interface.

C-MOLD Project Engineer requires only the general part shape and thus is a tool suitable for the entire product development cycle, especially when the detailed design parameters have yet to be decided. Project Engineer's distinctive feature is a division of the design strategy into well-defined steps. Each step has a design objective with design parameters that may be optimized. By completing these steps, all of the important design parameters can be determined, including material and machine selection, runner layout and balancing, part thickness, process window, and cost.

C-MOLD 3D QuickFill performs filling simulations on solid CAD models in stereolithography (STL) format. No finite-element meshing procedure is required, and simulation results are displayed directly on the solid model. It helps part designers, mold designers, and molders incorporate engineering and manufacturing considerations into the early stages of product development. Output data include melt front advancement, pressure and temperature at the end of fill, injection pressure trace, cooling time, orientation, sink marks, air traps, weld and meld lines, design advice and specifications, and cost. C-MOLD QuickDesign, the successor to C-MOLD 3D QuickFill at the time of writing, performs additional cooling simulations and ejector-pin analysis.

C-MOLD Advanced Solutions encompass a wide range of molding processes. As shown in Fig. 1, these include thermoplastics injection molding, gas-assisted injection (Turng 1996, Wang *et al.* 1998), microchip encapsulation (Han *et al.* 1997), injection/compression molding (Wang 1999), co-injection (Turng *et al.* 1993), reaction injection molding, RTM, and SRIM (Chen and Wang 1993), as well as blow molding and thermoforming (Santhanam and Himasekhar 1995). C-MOLD Advanced Solutions provide rigorous analysis for advanced moldability considerations using variable ram speed and valve gate control, productivity concerns regarding cycle time and cooling system design, and performance issues such as fiber orientation, residual stresses, and, as shown in Fig. 2, dimensional stability. C-MOLD's built-in Design Optimization functions perform automatic optimization for gate location, runner sizing, and process conditions.

C-MOLD KnowHow!, a Web-based knowledge management system (KMS) for injection molding (see Turng and DeAugistine 1999), is aimed at helping the molding industry to create, retain, share, and leverage engineering data and know-how related to plastics part design and manufacturing. It includes an electronic library for injection molding and it serves as the gateway to C-MOLD CAE tools. KnowHow! enables users to contribute their own documents, including Web-based C-MOLD simulation reports, into the system. Implemented within the framework of the Internet and Web technologies, KnowHow! facilitates development of corporate memory and

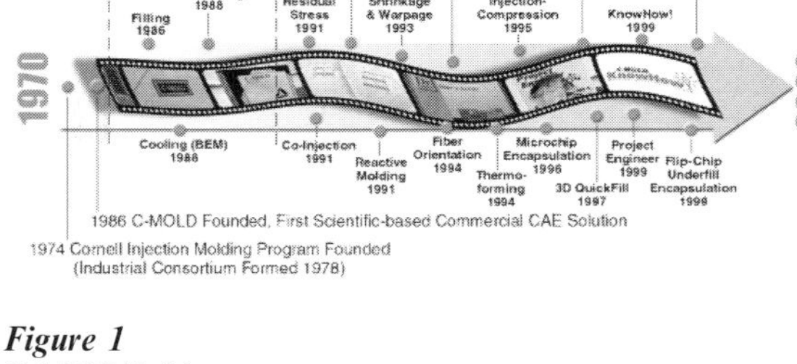

Figure 1
C-MOLD history.

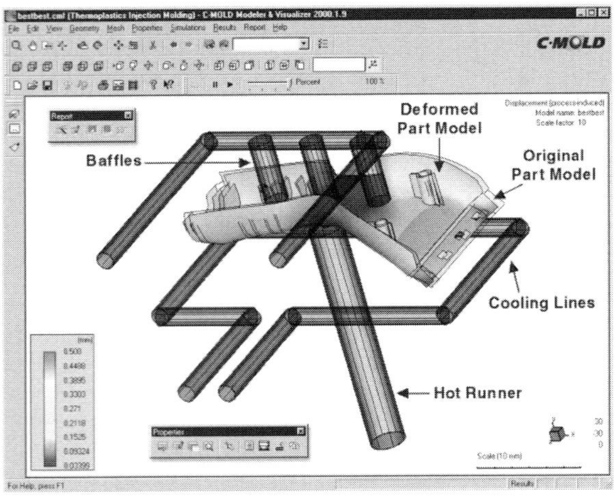

Figure 2
C-MOLD output data. This displacement plot shows predicted shrinkage and warpage (with a scale factor of 10).

collaboration among geographically dispersed team members.

2. *Modeling for Thermoplastics Injection Molding*

C-MOLD performs an integrated analysis of fluid flow and heat transfer in injection molding. The analysis is based on a hybrid finite-element/finite-difference/control-volume numerical solution of a generalized Hele-Shaw flow for a viscous fluid of variable density under asymmetric mold-cooling conditions. Mathematical formulations, material and process modeling, numerical implementation, and experimental verifications are detailed in Hieber and Shen (1980), Wang and Wang (1986), Chiang *et al.* (1991, 1993) and the references therein. The thermal analysis of the mold employs a modified 3D boundary-element numerical method (Himasekhar *et al.* 1992). The flow and thermal histories of the polymer is then used to calculate the fiber orientation distribution (Gupta and Wang 1993, Chiang *et al.* 1994), the resulting material properties, the thermal-induced residual stresses, and the shrinkage and warpage (Santhanam *et al.* 1992, Jin and Zhao 1998).

For material modeling, the Cross-WLF viscosity model (see Hieber 1987) has been employed to describe the rheological behavior of the polymer melt during the filling and postfilling stages. For residual stress prediction, the polymer is assumed to be a transversely isotropic viscoelastic material with thermal response modeled through the assumption of thermorheological simplicity (Santhanam 1992). For resins that lack the relaxation moduli for the viscoelastic model, a more numerically efficient instant-freeze model is used to calculate the molded-in stresses (Jin and Zhao 1998). To account for the effect of high cooling rates during molding on the crystallization behavior of semicrystalline polymers, fast-cooling pressure– volume–temperature (pvT) data are used in the simulation (Han and Wang 1997).

3. *Concluding Remarks*

The viability of CAE software (including C-MOLD) depends not only on the scientific principles, mathematical modeling, and numerical implementation used to develop the software, but also on the reliability of the input data, including the geometry description and finite-element mesh configuration, material properties, and process conditions. The best practice for applying CAE is to routinely incorporate it into the process of designing and manufacturing plastic parts to enhance molding efficiency and cost effectiveness. The C-MOLD products described in this article respond to the various levels of complexity and flexibility required to select the most suitable molding process and material and to design and manufacture plastic parts quickly and economically.

See also: Polymer Injection Molding: Modeling for Properties; Polymer Injection Molding: Moldflow; Polymer Injection Molding: 3D Modeling; Polymer Injection Molding: Multimaterial Molding; Polymer Injection Molding: Gas-assisted Molding

Bibliography

Chen T Y, Wang V W 1993 A unified simulation for reactive polymer processing: RIM, SRIM and RTM. *Annu. Tech. Conf. Soc. Plast. Eng.* **39**, 2421–5

Chiang H H, Hieber C A, Wang K K 1991 A unified simulation of the filling and post-filling stages in injection molding. *Polym. Eng. Sci.* **31**, 116–39

Chiang H H, Himasekhar K, Santhanam N, Wang K K 1993 Integrated simulation of fluid flow and heat transfer in injection molding for the prediction of shrinkage and warpage. *J. Eng. Mater. Technol.* **115**, 37–47

Chiang H H, Santhanam N, Himasekhar K, Wang K K 1994 Integrated CAE analysis for fiber-filled plastics in injection-molding. In: Himasekhar K, Prasad V (eds.) *Advances in Computer-Aided Engineering (CAE) of Polymer Processing*. ASME, New York

Gupta M, Wang K K 1993 Effect of processing conditions on fiber orientation in injection-molded composites. *Annu. Tech. Conf. Soc. Plast. Eng.* **39**, 2290–5

Han S, Wang K K 1997 Shrinkage prediction for slowly-crystallizing thermoplastic polymers in injection molding. *Intern. Polym. Proc.* **12**, 228–37

Han S, Wang K K, Crouthamel D L 1997 Wire-sweep study using an industrial semiconductor-chip-encapsulation operation. *J. Electron Packaging* **119**, 247–54

Hieber C A 1987 Melt-viscosity characterization and its application to injection molding. In: Isayev A I (ed.) *Injection and Compression Molding Fundamentals*. Marcel Dekker, New York, Chap. 1
Hieber C A, Shen S F 1980 A finite-element/finite-difference simulation of the injection molding filling process. *J. Non-Newton. Fluid* **7**, 1–32
Himasekhar K, Lottey L, Wang K K 1992 CAE of mold cooling in injection molding using a three-dimensional numerical simulation. *ASME J. Eng. Ind.* **114**, 213–21
Jin X, Zhao Z 1998 A shrinkage-warpage prediction method for injection-molded fiber-filled polymer composite parts. *Annu. Tech. Conf. Soc. Plast. Eng.* **44**, 614–8
Santhanam N 1992 Analysis of residual stresses and post-molding deformation in injection-molded components. Ph.D. thesis, Cornell University, Ithaca, New York
Santhanam N, Chiang H H, Himasekhar K, Tuschak P, Wang K K 1992 Postmolding and load-induced deformation analysis of plastic parts in the injection molding process. *Adv. Polym. Technol.* **11**, 77–89
Santhanam N, Himasekhar K 1995 A design tool for blow molding and thermoforming processes. In: Shen S F, Dawson P (eds.) *Simulation of Materials Processing: Theory, Methods and Applications*. Balkema, Rotterdam, pp. 1059–63
Turng L S 1996 Computer-aided engineering for the gas-assisted injection molding process. In: Stevenson J F (ed.) *Innovation in Polymer Processing: Molding*. Hanser Publishers, Munich, Chap. 2
Turng L S, DeAugustine D 1999 A web-based knowledge management system for the injection molding process. *Plast. Eng.* **55**, 47–50
Turng L S, Wang V W, Wang K K 1993 Numerical simulation of the coinjection molding process. *Trans. ASME* **115**, 48–53
Wang K K *et al.* 1975–2000 Cornell Injection Molding Program (CIMP) Progress Reports Vols. 1–21, Cornell University, Ithaca, New York
Wang T J 1999 Numerical simulation and process window design of injection/compression molding. *Annu. Tech. Conf. Soc. Plast. Eng.* **45**, 658–64
Wang T J, Chiang H H, Lu X, Fong L 1998 Computer simulation and experimental verification of gas-assisted injection molding. *J. Inj. Mold. Technol.* **2**, 120–7
Wang V W, Hieber C A, Wang K K 1986 Dynamic simulation and graphics for the injection molding of three-dimensional thin parts. *J. Polym. Eng.* **7**, 21–45

K. K. Wang and L.-S. Turng

Polymer Injection Molding: Gas-assisted Molding

In gas-assisted injection molding (GAIM), gas is injected into a mold that has been partially filled with polymer (see Fig. 1). The gas drives the molten polymer core further into the mold until it is filled completely. The penetrating gas leaves behind a polymer layer at the mold walls, yielding a product with a polymer skin and an inner gas channel. The gas typically penetrates along the path of least flow resistance.

After the mold has been entirely filled, gas is used to transmit the packing pressure to the polymer that is being cooled. Any shrinkage of the polymer material near the gas channel is compensated for by an enlargement of the gas core. Once all polymer material has solidified, the gas pressure is released. The product is then further cooled until it has retained sufficient rigidity to eject it from the mold.

The gas can be supplied in a volume-controlled manner (commonly called the *Cinpres* process) or in a pressure-controlled manner (the *Airmould* process). It can either be injected through a needle in the extruder nozzle, or directly into the mold through separate gas injection needles. In the blow-out technique, the gas is injected after the mold has been filled *completely* with polymer. The gas then forces part of the molten polymer out of the mold, either back into the injection unit or into a separate cavity. Finally, in the retracting core process, a mold containing a movable core part is completely filled with polymer. Then, the core part is withdrawn while gas is injected, filling up the space that is left by the retracting core or a moving mold wall (Rennefeld 1996).

All these techniques require a GAIM unit, consisting of a gas pressure generator, a gas injection unit, and an injection control unit, to be attached to a conventional injection molding machine. Usually, nitrogen gas is used to avoid burning effects. Up to 95% of the pressurized gas in the gas core can be regained when the gas pressure is released after the packing stage. To reuse this gas, a gas filtering system is needed as well.

1. Advantages

The main characteristic of GAIM is the fact that the pressure drop in the gas core is negligibly small compared to the pressure drop in an equivalent molten polymer core. Consequently, the gas pressure can be considered constant throughout the gas core, and this accounts for most of the advantages of GAIM (Turng 1995):

(i) *Reduced clamp force.* For a given product geometry, the machine clamp force is determined by the maximum pressure, which is usually the packing pressure. In conventional injection molding, this pressure has to be very high to overcome the large pressure drop between the injection point and the mold extremities, because of the polymer's high viscosity (see Fig. 2). In GAIM, a major part of the distance between injection point and mold extremities is covered by the gas channel, in which the pressure drop is considerably smaller than the pressure drop in conventional injection molding. Consequently, the required machine clamp force can be decreased accordingly.

(ii) *Reduced sink marks.* Sink marks are caused by shrinkage and are therefore most likely to occur at relatively thick mold parts, such as ribs and bosses, which are typically the parts along which the gas penetrates. The shrinkage in such parts manifests itself as a local enlargement of the gas core, instead of a sink mark at the outer surface (see Fig. 3).

(iii) *Reduced warpage.* Warpage is caused by residual stresses, which fall into two categories: thermally and pressure-induced stresses, and flow-induced stresses. The first are caused by inhomogeneous cooling

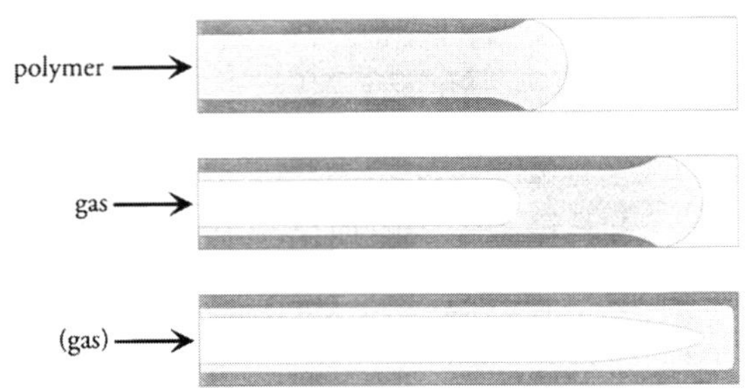

Figure 1
The principle of gas-assisted injection molding: polymer injection (top), gas injection (middle), and packing (bottom). Light gray = polymer melt, dark gray = solidified polymer, white = gas/air.

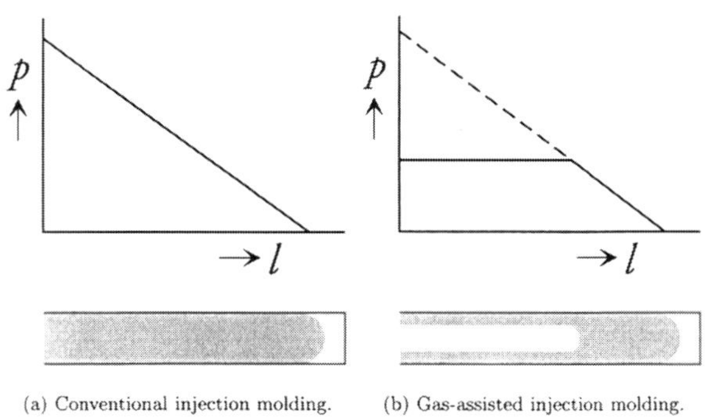

Figure 2
The principle of injection pressure reduction (for a thick-walled part) by gas-assisted injection molding compared to conventional injection molding.

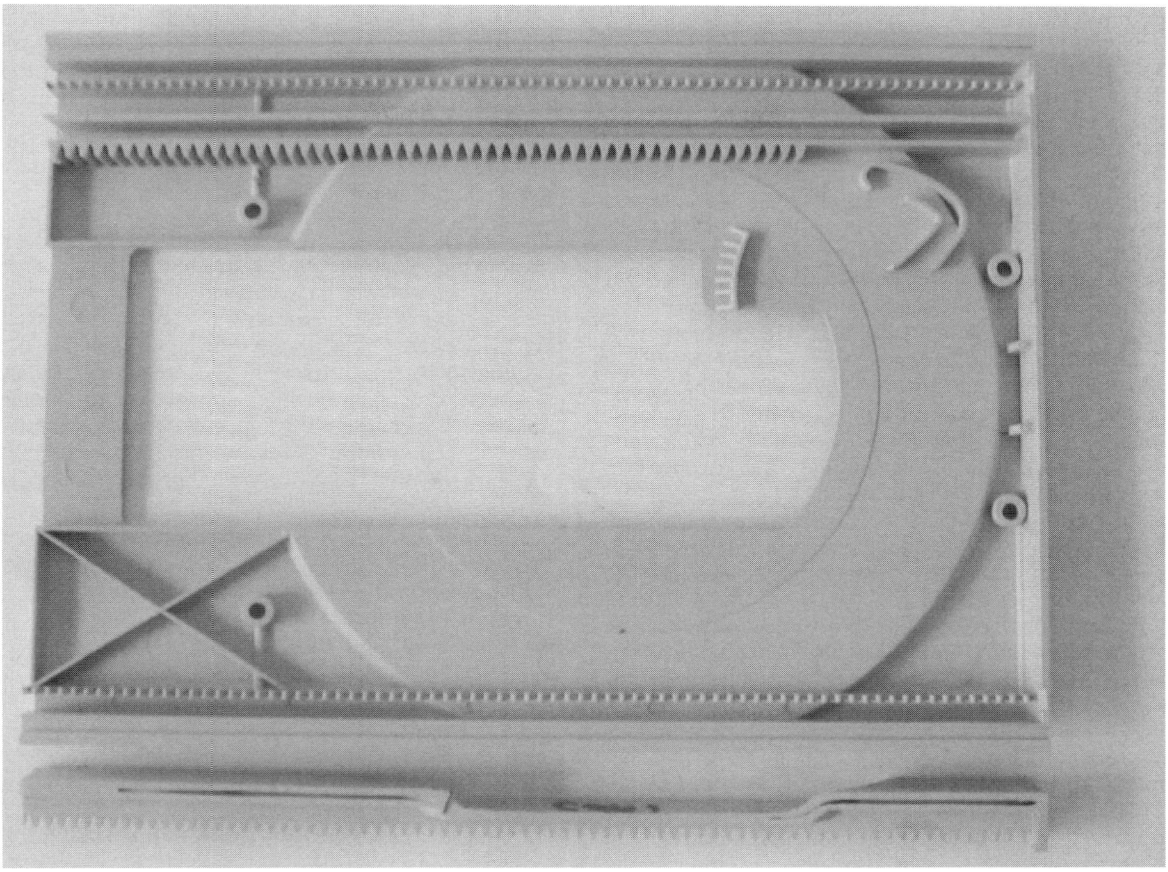

Figure 3
Gas-assisted injection molding was successfully applied to reduce sink marks and warpage in this CD-tray from a CD-player (photograph courtesy of the Philips Centre for Manufacturing Technology, Eindhoven).

under pressure: different material elements solidify at different pressure conditions and different cooling rates. This leads to density differences in the final product, which give rise to residual stresses. The packing pressure in GAIM is both lower and more uniform than in conventional injection molding, thus reducing the density differences and, ultimately, the thermally and pressure-induced residual stresses.

Flow-induced stresses are caused by orientation of polymer molecular segments during flow, and characterized by a competition between stretching (orientation) and relaxation ('deorientation'). In conventional injection molding, the largest flow-induced stresses develop during the packing stage, when the polymer in the product core is still stretched, while relaxation is suppressed due to the relatively low temperatures. In GAIM, however, the product core is not a polymer melt, but (primarily) gas. Hence, hardly any orientation takes place and flow-induced residual stresses can be significantly reduced.

The reduction of residual stresses in GAIM products may well be used to avoid warpage problems. However, one should be aware that if the injection of gas leads to large differences in the polymer layer thickness, warpage may even be increased.

The demand for thick mold parts—which are regarded as potential sources for product imperfections in conventional injection molding—along which the gas is to penetrate, may be turned into an advantage as well. For instance, hollow ribs in GAIM products provide structural rigidity without adding much weight and without causing sink marks (Yang *et al.* 1996). Thus, with GAIM, new product designs become possible, and are generally necessary. Furthermore, GAIM has been reported to reduce the molding cycle times in some (but definitely not all) cases. It is also versatile in the sense that GAIM can be done with virtually any polymer, including reinforced, thermosetting, and semicrystalline plastics (Eyerer *et al.* 1993; Rennefeld 1996).

2. *Pitfalls*

On the other hand, however, there are some pitfalls in the practical application of the GAIM process (Barton and Turng 1994). First, the ratio of polymer and gas to be injected has to be correct: if too much polymer is present in the mold, one will not benefit optimally from the advantages GAIM has to offer; if not enough polymer is present, the penetrating gas front will overtake the polymer melt front before the mold is filled, resulting in what is know as *gas breakthrough*. At the switchover from polymer to gas injection, the polymer front will halt momentarily. This usually becomes visible as so-called *hesitation marks* at the product surface, which should be hidden by, for example, painting or by having these marks occur at locations that are not externally visible. The blow-out technique mentioned in the introduction avoids these hesitation marks.

The thick-walled part into which the gas is supposed to flow may well have a significant lower flow resistance, giving rise to the *race-track effect*: the polymer front in the thick-walled part runs far ahead of the fronts in the adjacent thin-walled sections. The gas-leading channel may thus be filled prematurely, causing the gas to flow into the thin-walled sections, where it will proceed as irregularly shaped fingers. This is known as the *fingering effect*, and should be avoided in GAIM. Furthermore, as the gas penetrates along the path of least resistance by definition, gas flow bifurcations are destined to lead to uncontrolled gas flow. Moreover, in curved gas-leading channels, the path of least resistance lies closer to the inner wall of the bend than to the outer wall, causing the gas front to penetrate accordingly. This can lead to weak spots and, therefore, sharp corners and transitions in gas-leading channels are to be avoided.

3. *Modeling*

Commercial injection molding simulation packages, such as Moldflow and C-MOLD, have been incorporating GAIM options to provide computerized design and optimization tools for GAIM processes and products. Most of these packages are based on the so-called *thin-film* or *Hele-Shaw approximation*, in which the fact that most injection parts are thin-walled is used to decouple flow and thermal effects in mid-plane and thickness directions; this is known as the $2^{1/2}$-*D approach*.

However, analyses of the gas flow in GAIM have revealed that the gas penetration mechanism is governed by three-dimensional flow phenomena, which the $2^{1/2}$-*D* approach fails to take into account (Poslinski *et al.* 1995, Costa *et al.* 1995, Haagh *et al.* 1998). Three-dimensional GAIM modeling has been helpful in increasing the understanding of the GAIM process in terms of the effects of process parameters and product design on the gas distribution in GAIM products. The current challenge is to extend these models towards the prediction of product properties, such as (the reduction of) residual stresses and sink marks.

4. *Conclusion*

The GAIM process offers the possibility to reduce or even avoid common plastic products faults such as sink marks and warpage, provided that the product designer takes into account a number of specific GAIM design guidelines. As such, GAIM has been gaining increasing interest in the polymer-processing industry. Process modeling is expected to improve GAIM product design through software for computer-aided design. Future developments in this area will focus on the prediction of the final product properties.

See also: Polymer Injection Molding: Modeling for Properties; Polymer Injection Molding: Moldflow; Polymer Injection Molding: C-Mold; Polymer Injection Molding: 3D Modeling; Polymer Injection Molding: Multimaterial Molding

Bibliography

Barton K S, Turng L S 1994 General design guidelines for gas-assisted injection molding using a CAE tool. *ANTEC Proceedings*. Society of Plastic Engineers, Brookfield, CT, pp. 421–5

Costa F S, Thompson W, Friedl C 1995 An adaptation of the boundary element method for gas injection molding. In: Shen S-F, Dawson P (eds.) *Simulation of Materials Processing: Theory, Methods and Applications* (*NUMIFORM 95*). Balkema, Rotterdam, The Netherlands, pp. 1113–8

Eyerer P, Märtins R, Bürkle E 1993 Spritzgiessen mit Gasinnendruck. [Injection molding with internal gas pressure.]. *Kunststoffe* **83**(7), 505–17

Haagh G A A V, Zuidema H, Van de Vosse F N, Peters G W M, Meijer H E H 1998 Towards a 3-D finite element method for the gas-assisted injection molding process. *Int. Polym. Process.* **12**(3), 207–15

Poslinski A J, Oehler P R, Stokes V K 1995 Isothermal gas-assisted displacement of viscoplastic liquids in tubes. *Polym. Eng. Sci.* **35**(11), 877–92

Rennefeld C 1996 *Konstruktive Optimierung von Thermoplastformteilen und Spritzgiesswerkzeugen für die Gasinnendrucktechnik*. [Constructive optimization of thermoplastic products and molds for the internal gas pressure technique.]. Shaker Verlag, Aachen, Germany

Turng L S 1995 Development and application of CAE technology for the gas-assisted injection molding process. *Adv. Polym. Technol.* **14**(1), 1–13

Yang S Y, Huang F Z, Liau W N 1996 A study of rib geometry for gas-assisted injection molding. *Polym. Eng. Sci.* **36**(23), 2824–31

G. A. A. V. Haagh

Polymer Injection Molding: Modeling for Properties

In an ever-growing number of applications, injection molded products must meet tight demands concerning their mechanical and optical properties and dimensional stability. As a consequence, the ultimate aim of numerical simulations of the injection molding process is not only to analyze the processing stages, but also to calculate the mechanical, and frequently also optical, properties of the product, starting from the intrinsic rheological properties and the prevalent processing conditions. To verify the results of such simulations, measurement techniques that are able to assess the molecular orientation, residual stresses, and density distributions must be available.

1. General Remarks

The typical approach to model the injection molding process is the so-called $2\frac{1}{2}$D approach. In that case, the ratio of the product wall thickness h and a characteristic length l in the mid-plane of the product must be much less than unity. In this paper, an overview is given of the modeling of all stages of the production process using this $2\frac{1}{2}$D approach. The analysis is restricted to amorphous thermoplastic polymers.

Residual stresses in injection molded products stem from two main sources. The first is *flow induced stresses* caused by viscoelastic flow of the polymer during the filling and post-filling stage of the injection molding process. These stresses correspond with the orientation of macromolecules. This orientation is partly frozen-in by the rapid cooling of the product in the mold. The second is *thermally induced* and *pressure induced stresses*, which are caused by differential shrinkage and compression during solidification, respectively. In absolute values, the thermally induced stresses are usually substantially larger than the frozen-in flow induced stresses. However, the molecular orientation, as reflected in the frozen-in flow induced stresses, determines the anisotropy of mechanical, thermal, and optical properties and influences the long-term dimensional stability of an injection molded product.

A decoupled method is proposed to calculate flow induced stresses. Firstly, the kinematics of the flow field are determined, employing a viscous, generalized Newtonian constitutive law for the Cauchy stress tensor, in combination with the balance laws. This is realized for all stages of the process: injection, packing, holding, and cooling. The flow kinematics are subsequently substituted in a viscoelastic constitutive equation to calculate the transient stresses. One constitutive model is used: a compressible version of the *Leonov* model, which falls into the category of differential models. In the decoupled method, the flow kinematics are consequently supposed not to be influenced by the viscoelastic character of the flowing polymer melt. This decoupled method has a number of advantages compared to a coupled viscoelastic computation: the computation time is reduced considerably, an arbitrary viscoelastic constitutive equation can be employed easily, and no restrictions on the complexity of the flow field are imposed. In the case of 2D geometries, the validity of this approach is investigated by comparison of the results with those of a fully coupled viscoelastic calculation. These calculations show that the results obtained by the decoupled method are in acceptable agreement with the results of a fully coupled viscoelastic calculation.

For the calculation of thermally induced stresses, a thermo-viscoelastic constitutive law, a linearized form of both viscoelastic constitutive models mentioned above, is employed. In order to attain realistic results, special attention must be paid to the boundary conditions. In particular, it is shown that not only the temperature history, but also the pressure history, has a marked influence on the residual stress state of an injection molded product.

The above approach is applied to one representative example. Computed results are compared with experimental results.

2. Governing Equations

2.1 Fluid State

Assuming a generalized Newtonian behavior of the polymer melt, and neglecting inertia effects, neglecting elastic effects in the energy equation, assuming a rigid mold, and adopting the classical thin film (lubrication) approximation, the resulting set of equations that describe the injection molding process may be written as (Douven *et al.* 1995):

$$\vec{\nabla}\cdot(S\vec{\nabla}p) - \int_{-h/2}^{h/2}\kappa\dot{p}\,\mathrm{d}z = -\int_{-h/2}^{h/2}\alpha\dot{T}\,\mathrm{d}z \qquad (1)$$

$$\rho c_p\dot{T} = \frac{\partial}{\partial z}\left(\lambda\frac{\partial T}{\partial z}\right) + \eta\dot{\gamma}^2 - \frac{T}{\rho}\left(\frac{\partial\rho}{\partial T}\right)\dot{p} \qquad (2)$$

where the parameter S obeys

$$S = J_2 - \frac{J_1^2}{J_0},\ J_i = \int_{s^-}^{s^+}\frac{z^i}{\eta}\mathrm{d}z,\ i = 0,\ 1,\ 2 \qquad (3)$$

To obtain these equations, a coordinate system has been defined in the mid-plane of the cavity. The x and y coordinates are chosen within this mid-plane, and the z coordinate is perpendicular to the mid-plane.

The gradient operator $\vec{\nabla}$ is chosen with respect to the mid-plane, while h denotes the cavity height and s^+ and s^- denote the position of the solidified layers with respect to the mid-plane. The pressure p in the cavity is assumed constant across the thickness of the cavity. Compressibility is accounted for by the inverse bulk modulus κ, while thermal expansion is represented by α. The viscosity η is a function of the shear rate $\dot{\gamma}$ and temperature T, and is frequently described using a Carreau–Yasuda or Cross model. In the above set of equations, the density ρ is chosen to be pressure- and temperature-dependent and modeled using the so-called Tait equation (Douven *et al.* 1995). An improved description of the density is obtained using the KAHR model (Caspers 1995), which includes the history dependency of the density.

Once the pressure equation (Eqn.(1)) and the temperature equation (Eqn.(2)) have been solved, the velocity field in the cavity can be calculated. The resulting velocity and temperature field is used to compute flow-induced stresses. For this purpose, the compressible Leonov model has been used in the past (Baaijens 1991). It must be emphasized however, that at present more appropriate models are available to describe the complicated non-linear rheology of polymer melts. In particular, extensional data are poorly described by the Leonov model. Nevertheless, since the above modeling of the injection molding process includes shear phenomena only, the Leonov model is a suitable choice. It is also attractive from practical point of view since it only requires linear viscoelastic data to identify all the model parameters (i.e., the relaxation times and associated moduli).

2.2 *Solidified State*

The stresses that result in the solidified state are a combination of several effects. First of all, the relaxation times associated with flow-induced stresses increase enormously when the temperature drops below the glass transition temperature. Therefore these stresses can no longer relax in the solidified state (they are "frozen-in"). Secondly, if a material volume enters the solidified state from the molten

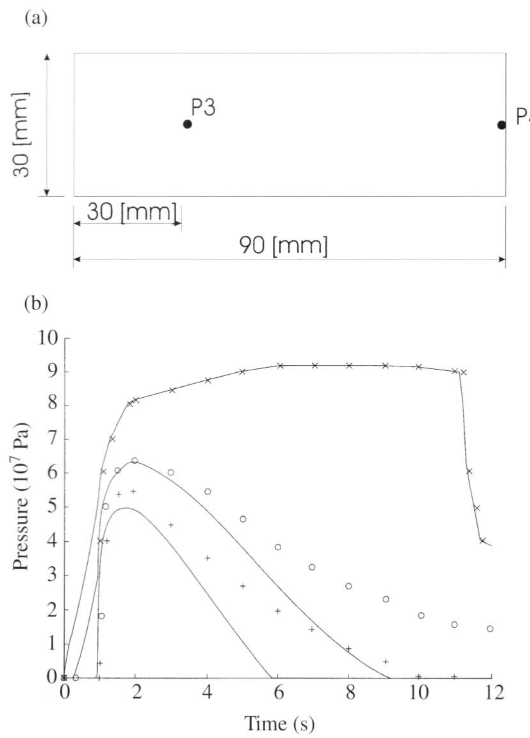

Figure 1
(a) Dimensions of the flat plate and location of pressure transducers; (b) pressure trace at three locations in the mould. The symbols represent experimental results at the gate ($\times$), P3 (○) and P4 ($+$).

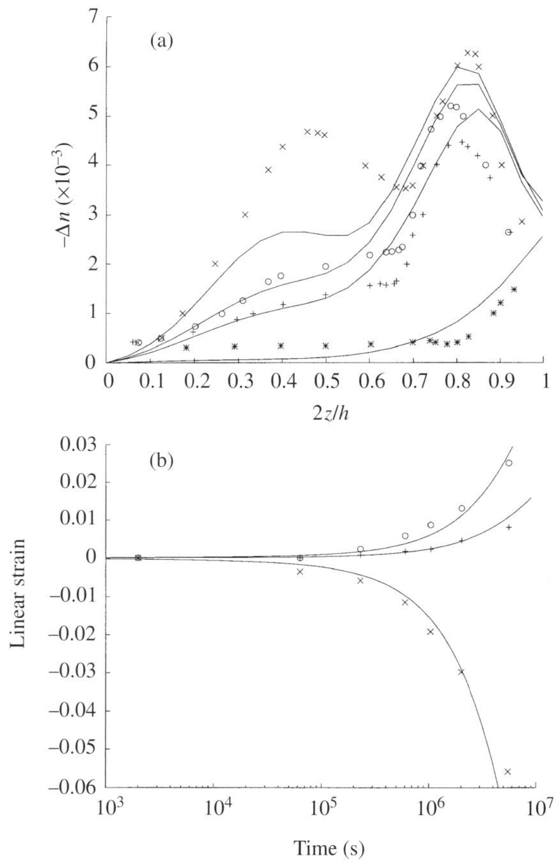

Figure 2
(a) Computed and experimental birefringence across the thickness of the product; (b) dimensional stability.

state, it is generally compressed due to the high pressure present during the injection and packing stage. Finally, differential shrinkage due to the inhomogeneous temperature distribution results in thermally induced residual stresses.

In the solidified state, the deformations are sufficiently small to allow a linearization of the Leonov model to compute the so-called thermal- or pressure-induced stresses. The equations that result are those of a linear thermo-viscoelastic model (Douven *et al.* 1995) that accounts for both pressure and temperature effects. As in the fluid state, the history dependency of the density may be included by adopting the KAHR model (Caspers 1995).

3. Application

The dimensional stability of an injection molded, rectangular, flat plate, as depicted in Fig. 1(a) is investigated. The computed pressure trace at three locations in the mold is compared to experimentally obtained values in Fig. 1(b). The locations at which the pressure is obtained are indicated in Fig. 1(a). Further details of this analysis may be found in Caspers (1995).

Based on the resulting velocity and temperature distribution, the flow induced residual stresses have been computed and transformed to experimentally observable birefringence. Figure 2(a) compares the computed birefringence using six relaxation times in the compressible Leonov model at 25 (×), 41 (○), 60 (+) and 110 (*) mm from the gate. The primary peak in the birefringence profile is located close to the solidified layer and is formed during the injection stage. The secondary peak, particularly noticeable close to the gate, is formed during the packing stage. Based upon these flow induced residual stresses, the dimensional stability of the plate is investigated. The storage temperature is set to 343 K. At this temperature, the impact of density relaxation effects on dimensional changes is negligible compared to elastic recovery effects due to flow induced stresses. Figure 2(b) shows the length change of the molded part (×), and the width change at two locations from the gate, (○) about half way and (+) at the end of the plate, all expressed in terms of linear strain. Due to elastic recovery, the plate shortens and widens inhomogeneously. The solid line corresponds to the computed results, while the symbols represent experimental observations.

Bibliography

Baaijens F P T 1991 Calculation of residual stresses in injection moulded products. *Rheologica Acta* **30**, 84–299

Caspers L 1995 Vlp: an integral approach to the simulation of injection moulding. Ph.D. thesis, Eindhoven University of Technology

Douven L F A, Baaijens F P T, Meijer H E H 1995 The computation of properties of injection-moulded products. *Prog. Polym. Sci.* **20**, 403–57

F. P. T. Baaijens

Polymer Injection Molding: Moldflow

Injection molding is a widely used technique for mass-producing articles with a high degree of geometrical complexity. A typical injection molding process consists of four stages: (i) filling of the molten polymer into the mold, (ii) packing of more material into the mold under high pressure to compensate for volumetric shrinkage of the material as it cools, (iii) cooling during which the material solidifies while in the mold until it is sufficiently solid, and (iv) ejection of the solidified product from the mold. During filling, packing, and cooling, the material experiences a complex thermomechanical history which leads to changes in local specific volume. While the part is in the mold, it is constrained within the plane of the part and so stresses develop in the part during solidification. Upon ejection, the relaxation of these stresses causes instantaneous shrinkage that is usually anisotropic and nonuniform throughout the molded part. Further shrinkage may also occur during cooling after ejection. The anisotropic and nonuniform shrinkage behavior will result in a degree of warpage.

The final properties of an injection molded part, produced from fiber-reinforced composites, are to a large extent determined by the fiber orientation distribution. Unless the embedded fibers are randomly oriented, they introduce anisotropy in the thermomechanical properties of the material. The anisotropic behavior has a great influence on shrinkage and warpage. The fiber orientation distribution is induced by kinematics of the flow during filling and, to a lesser extent, packing. An extensive literature deals with flow-induced fiber orientation while much other work has been devoted to micromechanical models, which estimate anisotropic elastic and thermal properties of the fiber–matrix system from the properties of the constituent fiber and matrix materials based on given microstructures. Comprehensive reviews of both research areas have been given in two recent books edited by Advani (1994) and Papathanasiou and Guell (1997) where many references can be found.

However, little has been done to relate results from the two research areas and to establish quantitative relationships between the flow-induced fiber orientation and the ultimate dimensional stability of the molded part. Osswald and Tseng (1994) applied micromechanical models to fiber-reinforced thermosets and were able to predict shrinkage and warpage of the molded part. A similar approach was used by Zheng *et al.* (1996) for injection molding of fiber-reinforced thermoplastics. More sophisticated analyses including the use of anisotropic thermoviscoelastic formulations for residual thermal stresses have also been reported (e.g., see Rezayat and Stafford 1991, Zheng *et al.* 1999).

Analysis of the final properties of injection molded short-fiber composite parts requires accurate prediction of flow-induced fiber orientation. Several different fiber suspension theories and numerical methods are available for the calculation of the motion of fibers during flow. Only the work of Folgar and Tucker (1984) is briefly reviewed here since it is the most relevant. The reader is referred to Phan-Thien and Zheng (1998) for additional reviews and discussions on the constitutive theories of fiber suspensions.

1. Injection Molding Simulation

All fluid flow problems involve solution of the equations that express conservation of mass, momentum and energy. In what follows a brief summary of the governing equations and their simplification is given. Details may be found in Kennedy (1995).

The conservation equations take the form:
Mass

$$\frac{\delta\rho}{\delta t}+\vec{\Delta}(\rho\vec{v})=0 \tag{1}$$

Momentum

$$\rho\frac{\delta\vec{v}}{\delta t}=-\vec{\nabla}p+\vec{\nabla}\cdot(\eta\underline{\dot{\gamma}})-\rho(\vec{v}\cdot\vec{\nabla}\vec{v}) \tag{2}$$

Energy

$$\rho c_p\left(\frac{\delta T}{\delta t}+\vec{v}\cdot\vec{\nabla}T\right)=\beta T\left(\frac{\delta p}{\delta t}+\vec{v}\cdot\vec{\nabla}p\right)+\eta\dot{\gamma}^2+\vec{\nabla}\cdot(k\vec{\nabla}T) \tag{3}$$

where ρ is the material density, t is time, $v^{\rightarrow}$ is the velocity of the melt, p is pressure, η is the melt viscosity, $\underline{\dot{\gamma}}$ is the rate of deformation tensor, c_p is the specific heat of the material, T is temperature, β the coefficient of thermal expansion, $\dot{\gamma}$ is the shear rate, and k the thermal conductivity. It should be noted that the viscosity of polymers depends on pressure, temperature and shear rate and this dependency must be incorporated in the simulation.

Adopting a Cartesian coordinate system, in which z is the gap direction of the cavity, and assuming the cavity thickness is small compared to the other dimensions, the mass and momentum equations may be reduced to the single equation:

$$\frac{\delta}{\delta x}\left(S\frac{\delta p}{\delta x}\right)+\frac{\delta}{\delta y}\left(S\frac{\delta p}{\delta y}\right)=\frac{1}{2}\int_{-H}^{H}\left\{\kappa\frac{\delta p}{\delta t}-\frac{\beta}{\rho c_p}\left[\eta\dot{\gamma}^2+\frac{\delta}{\delta z}\left(k\frac{\delta T}{\delta z}\right)\right]\right\}\mathrm{d}z \tag{4}$$

where

$$S = \frac{1}{2}\left\{\int_{h-}^{h+} \frac{z^2}{\eta} dz - \frac{\left(\int_{h-}^{h+} \frac{z}{\eta} dz\right)^2}{\int_{h-}^{h+} \frac{1}{\eta} dz}\right\}$$

in which h^+ and h^- are, respectively, the upper and lower z coordinates of the frozen layer position, κ is the coefficient of compressibility and H is half the wall thickness.

With the additional assumption that convection in the z-direction may be ignored, the energy equation takes the form:

$$\rho c_p\left(\frac{\delta T}{\delta t} + v_x\frac{\delta T}{\delta x} + v_x\frac{\delta T}{\delta x}\right) = \beta T\frac{\delta p}{\delta t} + \eta\dot{\gamma}^2 + \frac{\delta}{\delta z}\left(k\frac{\delta T}{\delta z}\right) \quad (5)$$

The left hand side of the energy equation represents the rate of change of temperature and convection while the terms on the right side account for heat of expansion/compression, viscous dissipation and conduction to the mold, respectively.

These equations, and their respective boundary conditions, are generally solved with a hybrid approach (Hieber and Shen 1980). The pressure solution is found from Eqn. (4) using finite elements and the temperature field is obtained from Eqn. (5) using the finite difference method (Güçeri 1989). The result of the simulation is the history of the pressure, velocity, and temperature at any position in the cavity and at any time during the filling and packing phases.

2. *Fiber Orientation Analysis*

In simplest terms, fibers tend to become aligned in the flow direction when the flow is converging and align transverse to flow where the flow diverges (Fig. 1). With multiple injection points and complicated geometry, the final orientation distribution can be extremely complex. Simulation can assist here by first calculating the final orientation state and then using this information to derive the anisotropic thermomechanical properties of the material such as elastic moduli, Poisson's ratio, and linear coefficients of expansion. These derived properties may be used for subsequent structural analysis of the part. They may also be used for determining shrinkage and warpage of the component.

Usually, fiber suspensions are classified into three concentration regimes according to the fiber volume fraction, φ, and the fiber aspect ratio α_R (defined by the length-to-diameter ratio, L/d). The volume fraction satisfies $\varphi = n\pi d^2L/4$ for rod-like fibers where n is the number density of the fibers. A suspension is called dilute if the volume fraction satisfies $\varphi\alpha_R^2 < 1$. In dilute suspensions, each fiber can freely rotate. The region in which $1 < \varphi\alpha_R^2 < \alpha_R$ is called semi-concentrated, where each fiber has only two rotating degrees of freedom. Finally, the suspension with $\varphi\alpha_R > 1$ is called concentrated, where the average distance between fibers is less than a fiber diameter, and, therefore, fibers cannot rotate independently except around their symmetry axes. Any motion of the fiber must necessarily involve a co-operative motion of surrounding fibers. Most commercial composites commonly used in injection molding fall into the semi- or highly concentrated regimes.

The orientation of a fiber can be described by a unit vector $\vec{p}$ along the axis of the fiber or the angle (θ, φ) as shown in Fig. 2. There are several choices

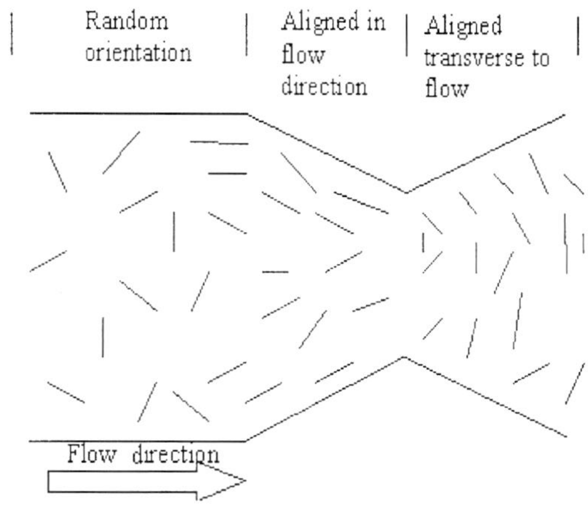

Figure 1
Fiber orientation distribution in converging and diverging flow regions.

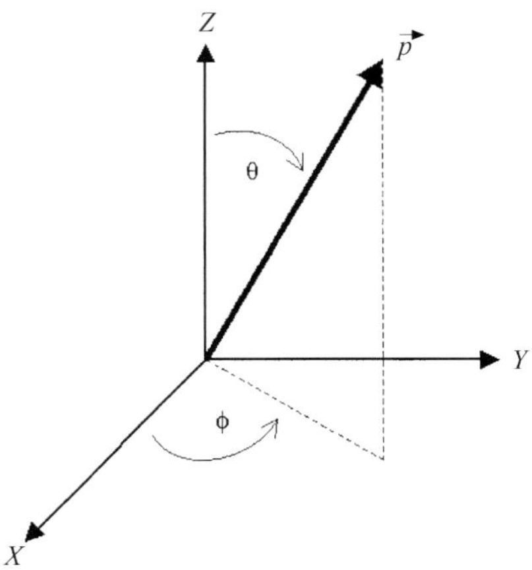

Figure 2
Definition of θ, φ, and $\vec{p}$.

available to model the motion of the fibers. One is to track the motion of each fiber, including explicitly determining the fiber–fiber interaction. Whilst possible, this method is not really practical for a large number of fibers in nonhomogeneous flows. An alternative is to use a probability distribution function, $\Psi(\vec{p})=\Psi(\theta,\phi)$, whose value represents the probability of having a fiber between the angles φ and $\varphi+\mathrm{d}\varphi$ and θ and $\theta+\mathrm{d}\theta$.

Given such a distribution, it is assumed that one end of the fiber is indistinguishable from the other and so, $\Psi(\theta,\varphi)=\Psi(\pi-\theta,\varphi+\pi)$. Also, the integral of the distribution over all possible directions must be one. That is,

$$\int_0^{2\pi}\int_0^{\pi}\Psi(\theta,\ \phi)\sin\theta\,\mathrm{d}\theta\ \mathrm{d}\phi=1$$

Although the distribution function completely describes the state of orientation, it is still too cumbersome to implement in simulation.

A far more compact method involves the use of orientation tensors (Advani and Tucker 1987). Orientation tensors are defined as even ordered tensors by relationships such as

$$a_{ij}=\int\Psi(\vec{p})p_ip_j\mathrm{d}\vec{p}$$

Similarly it is possible to define a fourth order tensor

$$a_{ij}=\int\Psi(\vec{p})p_ip_jp_kp_l\mathrm{d}\vec{p}\text{, and so on.}$$

The starting point for using orientation tensor to predict fiber orientation is the evolution equation of Jeffery (1922), derived by modeling the fiber as a rigid ellipsoid suspended in a Newtonian fluid. The microstructure of the suspension, characterized by the unit vector field $\vec{p}$, evolves in time as follows:

$$\dot{p}_i=-\frac{1}{2}\omega_{ij}p_j+\frac{1}{2}\lambda(\dot{\gamma}_{ij}p_j-\dot{\gamma}_{kl}p_kp_lp_i)$$

where

$$\dot{\gamma}_{ij}=\frac{\partial v_j}{\partial x_i}+\frac{\partial v_i}{\partial x_j},\quad \omega_{ij}=\frac{\partial v_j}{\partial x_i}-\frac{\partial v_i}{\partial x_j},$$

v_i is the fluid velocity and $\lambda=(a_{\mathrm{R}}^2-1)/(a_{\mathrm{R}}^2+1)$ is a constant that depends on the shape of the particle and is approximately one for slender rods.

Jeffery's equation is valid for dilute fiber suspensions where there are no fiber–fiber interactions. It was extended to concentrated solutions by Folgar and Tucker (1984) who added a diffusion term to account for fiber–fiber interaction. In terms of the orientation tensor, the Folgar–Tucker equation has the form

$$\frac{Da_{ij}}{Dt}-\frac{1}{2}\lambda(\dot{\gamma}_{ik}a_{kj}+a_{ik}\dot{\gamma}_j-2\dot{\gamma}_la_{ijkl})$$
$$-\frac{1}{2}(\omega_{ik}a_{kj}-a_{ik}\omega_{kj})+2C_I\dot{\gamma}(\delta_{ij}-3a_{ij})\qquad(6)$$

This model is closely analogous to a rigid dumbbell model with rotary Brownian motion (Bird *et al.* 1987). The term $C_{\mathrm{I}}\dot{\gamma}$ plays a role like the diffusivity in rotary Brownian motion. The constant C_{I} for a given suspension is assumed to be isotropic and independent of the orientation state, as a first approximation. The Folgar–Tucker has been widely used to determine the orientation of fibers in injection molding. The evolution of the fiber in a flow is obtained by solving this equation using the kinematics determined by solution of Eqns. (4) and (5).

The major problem with the model is the choice of an appropriate value for the coefficient C_{I}. Some progress has been made in this area recently. Fan *et al.* (1988) have recently presented a direct numerical simulation of fiber–fiber interactions. Short-range interaction is modeled by lubrication forces. Long- range interaction was calculated using a boundary element method. The hydrodynamic force and torque on each fiber were calculated to determine the motion of the fiber. Although the direct simulation method is currently limited to a simple shear flow, the numerical results can be used to produce macroscopic properties of the suspension, including the Folgar–Tucker constant C_{I}. This may then be used in finite element simulations for more complex flows. This represents the current state of the art with regard to fiber orientation and details of the application to injection molding problems may be found in Zheng *et al.* (1999).

After determining the resulting distribution of fiber orientation in an injection-molded part; it is possible to predict mechanical properties for the composite. Moduli and Poisson's ratios may be determined using a variety of mechanical theories for composite materials. These properties may then be used in structural analysis to determine the behavior of the component under load. Similarly the calculated mechanical properties may be used in warpage analysis to predict the deformation due to residual stresses developed in processing.

See also: Polymer Injection Molding: Modeling for Properties; Polymer Injection Molding: C-Mold; Polymer Injection Molding: 3D Modeling

Bibliography

Advani S G (ed.) 1994 *Flow and Rheology in Polymer Composites Manufacturing*. Elsevier, Amsterdam

Advani S G, Tucker C L III 1987 The use of tensors to describe and predict fiber orientation in short fiber composites. *J. Rheol.* **31**, 751–84
Bird R B, Curtiss C F, Armstrong R C, Hassager O 1987 *Dynamics of Polymer Liquids: Vol. II. Kinetic Theory*. 2nd edn. Wiley, New York
Fan X J, Phan-Thien N, Zheng R 1988 A direct simulation of fiber suspensions. *J. Non-Newton. Fluid* **74**, 113–36
Folgar F P, Tucker C L III 1984 Orientation behavior of fibers in concentrated suspensions. *J. Reinf. Plast. Compos.* **3**, 98–119
Güçeri S 1989 Finite Difference Solution of Field Problems. In: Tucker III C L III (ed.) *Computer Modeling for Polymer Processing*. Hanser, New York
Hieber C A, Shen S F 1980 A finite element/finite difference simulation of the injection molding filling process. *J. Non-Newton. Fluid* **7**, 1–32
Jeffery G B 1922 The motion of ellipsoidal particles immersed in viscous fluid. *Proc. Roy. Soc.* **102**, 161–79
Kennedy P K 1995 *Flow Analysis of Injection Molds*. Hanser, New York
Osswald T A, Tseng S-C 1994 Compression Molding. In: Advani S G (ed.) *Flow and Rheology in Polymer Composites Manufacturing*. Elsevier, Amsterdam
Papathanasiou T D, Guell D C (eds.) 1997 *Flow-induced Alignment in Composite Materials*. Woodhead, Cambridge
Phan-Thien N, Zheng R 1998 Macroscopic modeling of the evolution of fiber orientation during flow. In: Papathanasiou T D, Guell D C (eds.) *Flow-induced Alignment in Composite Materials*. Woodhead, Cambridge
Rezayat M, Stafford R O 1991 A thermoviscoelastic model for residual stress in injection molded thermoplastics. *Polym. Eng. Sci.* **31**, 393–8
Zheng R, Kennedy P, Phan-Thien N, Fan X J 1999 Thermoviscoelastic simulation of thermally and pressure induced stresses in injection molding for the prediction of shrinkage and warpage for fiber-reinforced thermoplastics. *J. Non-Newton. Fluid* **84**, 159–90
Zheng R, McCaffrey N, Winch K, Yu H, Kennedy P 1996 Predicting warpage of injection molded fiber-reinforced plastics. *J. Thermopl. Compos. Mater.* **9**, 90–106

P. Kennedy and R. Zheng

Polymer Injection Molding: Multimaterial Molding

The versatility of the injection molding process can be increased by combining multiple polymers in one product without assembly. The combination of materials can be chosen such that a higher level of functionality of a product can be reached compared to the conventional molding process. In the multimaterial injection-molding process (MMIM) as it is considered here, two or more polymers are injected sequentially or simultaneously into a mould to make products that consist of, for example, a layered structure. Due to the complex flow patterns during filling of a mold, it is not always known what the exact material distribution throughout the product will be in the case of geometrically complex products or multiple materials. For a successful application of this technique, it is desirable to predict the material distribution or to solve the inverse problem of how to inject the multiple material so that the desired (layered) structure will be reached.

1. *Multimaterial Injection Molding*

Two different techniques are frequently used: the multishot technique, in which a (semi-)product after solidification is placed into a larger cavity followed by the injection of the second material against or around the previous shot (also known as overmolding). In the technique considered here, which is also known as sandwich molding or coinjection, the different materials will flow together during mold filling, so there is only one cooling step involved. The distribution of the materials in the product depends on the position of the gate, the geometry of the injection nozzle, and the injection sequence (simultaneous and/or sequential). By changing this sequence (A,B; A,B,A; A/B; A,A/B,A; etc.) the thickness and position (centric or eccentric) of the resulting layers can be controlled.

The MMIM technique was originally developed by ICI (Garner and Oxley 1969). The process aimed at the production of large, lightweight, and stiff products by sequential injection of two components (A,B; A,B,A), where the component B (the core layer) typically consisted of the A material with a physical or chemical foaming agent. During the 1990s, the MMIM technology has been evaluated also for thin-walled products using the Battenfeld two-channel technique (Eckardt 1987). With this technique, not only sequential injection is possible, but also simultaneous injection of two components (A,A/B,A; A,B/A,A). In this way, thin, eccentric layers could be realized in a product. Examples of applications are: shielding against electromagnetic interference (EMI) by a conductive layer inside the product, special barrier material in the core, recycling of thermoplastic materials with scrap material in the core (Eckardt 1987, Tung and Kudert 1990, Donovan *et al.* 1975)

The sandwich molding technique has been used in the past also to acquire insight into the mold filling process (White and Lee 1975, Young *et al.* 1980) for isothermal conditions. However, in nonisothermal conditions it is not always straightforward to understand how the material distribution will look. Therefore, numerical simulation tools are developed (e.g., Turng *et al.* 1993, Schlattter *et al.* 1995, Vos *et al.* 1991, Peters *et al.* 1994, Zoetelief *et al.* 1997) which can help one to obtain the optimal core/skin ratio in a product. In all cases, the "classical" codes are extended with tracking techniques.

2. *Particle Tracking*

In order to keep track of the multiple materials injected, abstract distinct points in the flow are defined that have to be followed in time and space; *particle tracking*. Particle paths can be constructed by integrating the reversed velocity field starting from every nodal point in the mesh (Crochet *et al.* 1984). When applying this method to injection molding simulation and solving the initial value problem with, for example, a Runga-Kutta method, this requires storage of the complete velocity field for every time step. To get around this problem, Peters *et al.* (1994), and Zoetelief *et al.* (1997) used a conservation-of-identity method for MMIM by defining material particles by their unique identity. The transport of the identity during the flow can be described by a transport equation (pure convection):

$$\dot{\underset{\sim}{\xi}} = \frac{D\underset{\sim}{\xi}}{Dt} = \underset{\sim}{0} \quad \text{in } \Omega_m, \quad \text{with } \underset{\sim}{\xi}(\vec{x},t) = \underset{\sim}{f}(\vec{x},t) \text{ on } \Gamma_e \qquad (1)$$

where $\dot{\underset{\sim}{\xi}}$ is the material derivative and $\underset{\sim}{\xi}$ is, for instance, a column containing those properties of the particles that are considered important, like the starting position and time (x_0,y_0,z_0,t_0), the viscosity, color, etc. Γ_e is the injection area and Ω_m the cavity geometry. The problem of particle tracking is solved here with a Eulerian approach (fixed frame of reference). Then the conservation of identity can be expressed as:

$$\frac{\partial \underset{\sim}{\xi}}{\partial t} + \vec{v} \cdot \vec{\nabla} \underset{\sim}{\xi} = \underset{\sim}{0} \qquad (2)$$

where $\partial\underset{\sim}{\xi}/\partial t$ is the spatial derivative.

In this representation, the change of label values in space is calculated instead of tracking predefined particles.

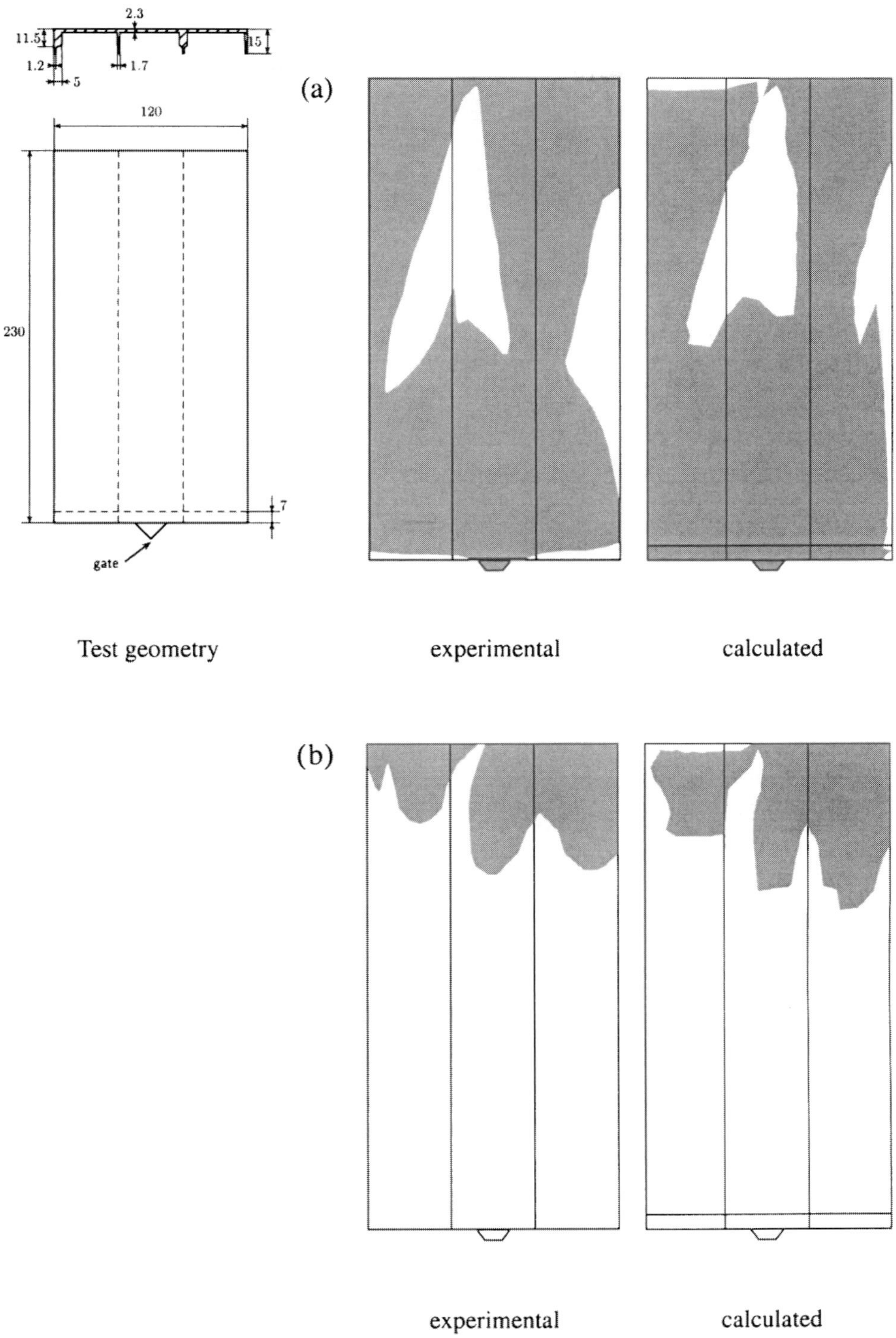

Figure 1
Color distribution in the midplane (a) and breakthrough areas (b). Switch point is 50% ($T_{inj1,2} = 571$, 556 K).

3. Front Flow

In the co- and sandwich-molding technique, the layered structure in the product is mostly obtained by relying on the so-called fountain flow effect at the melt front (Dupret and Vanderschuren 1988, Manas-Zloczower *et al.* 1987, Castro and Macosko 1982). The *fountain flow* occurs in flows with one or more free boundaries and a no-slip condition at the walls causing fluid elements adjacent to the moving front to spread towards the walls since fluid elements near the center move at a higher speed than the local average speed across the channel. So it is important to include this effect in the simulations.

4. Sandwich Molding

As an example of the predictive capabilities of the code, the following product is chosen: a strip of $230 \times 120 \times 2.3$ mm with four stiffener ribs and a beam-like edge of $7 \times 9 \times 120$ mm (see Fig. 1). The ribs are 12.7 mm in height and have different cross-sections. The product is filled by sequential injection of two polymer melts only differing in color (white, then dark), which are referred to as skin and core material, respectively. The strips were injected in a mold with a constant temperature of 348 K. The core material is injected after the product has been filled up to the volume percentage of 50%.

Since, in sequential injection, the material, or its color, only changes in time, the labels containing the entrance time into the mold provide all necessary information for visualization of the final distribution. Then the core material distribution is visualized by constructing contour plots of the labels in the mid-plane of the product, while the breakthrough areas are easily found by plotting the label field at the surface.

Figures 1(a) and 1(b) show the color patterns in the midplane and at the surface, respectively. It may be obvious that, for such a relatively simple geometry, the color distribution can become rather complex.

5. Inverse Problem Solutions

The ultimate goal of simulation of MMIM processes is to predict the process conditions given the required product properties. The task to be completed can be summarized as:

Given the particle distribution at the end of the filling stage and the definition of a required layer distribution, determine the injection configuration and sequence necessary in order to attain that preset definition.

By defining a specific, desired shape and position of a component inside a product, only the label values of these particles have to be determined to know what configuration should have been injected to realize this desired multicomponent product. The main advantage of this method is the use of field information rather than tracking a set of particles. For one-component flow, as occurs in multicolor injection, the inverse mapping can be carried out straightforwardly (as a postprocessing task). For multicomponent flows, however, the material distribution influences the flow kinematics and vice versa. In that case, the injection configuration has to be obtained iteratively, e.g., by following the steps:

(i) Perform a filling simulation with one component.

(ii) Determine the injection sequence given the required material distribution.

(iii) Perform a multicomponent filling simulation with the injection sequence determined under step (ii) as input.

(iv) Repeat steps (ii) and (iii) until the injected and calculated injection sequences correspond to the desired accuracy.

It is important to note that there is no guarantee that this iterative process will converge. However, it is our experience that for low Reynolds number flows, as is the case in MMIM, convergence is established to a sufficient degree.

See also: Polymer Injection Molding: Modeling for Properties; Polymer Injection Molding: Moldflow; Polymer Injection Molding: C-Mold; Polymer Injection Molding: Gas-assisted Molding

Bibliography

Castro J M, Macosko C W 1982 Studies of mold filling and curing in the reaction injection molding process. *AIChE. J.* **28**, 250–60

Crochet M J, Davies A R, Walters K 1984 *Numerical Simulation of Non-Newtonian Flow*. Elsevier, Amsterdam

Donovan R C, Rabe K S, Mammel W K, Lord H A 1975 Recycling plastics by two-shot molding. *Polym. Eng. Sci.* **15**, 774–80

Dupret F, Vanderschuren L 1988 Calculation of the temperature field in injection molding. *AIChE J.* **34**, 1959–72

Eckardt H 1987 Co-injection charting new territory and opening new markets. *J. Cell. Plast* **23**, 555–92

Garner P J, Oxley D F 1969 British patent No.1, 156, p. 217, Assigned to ICI

Manas-Zloczower I, Blake J W, Macosko C W 1987 Space–time distribution in filling a mold. *Polym. Eng. Sci* **27**, 1229–35

Peters G W M, van der Velden P J L, Meijer H E H, Schoone P 1994 Multilayer injection molding. Part 2: Particle tracking in reactive moulding. *Int. Polym. Process.* **9**, 258–65

Schlatter G, Davidoff A, Agassant J F 1995 Numerical simulation of the sandwich injection-molding process. *Proc. SPE ANTEC-95*. pp. 456–60

Tung T T, Kudert F G 1990 An injection molding process for producing encapsulated multilayer articles. In: *Proc. SPE ANTEC-90*. pp. 760–2

Turng L S, Wang V W, Wang K K 1993 Numerical simulation of coinjection molding filling process. *J. Eng. Mater. Technol.* **115**, 48–53
Vos E, Meijer H E H, Peters G W M 1991 Multilayer injection molding. *Int. Polym. Process.* **6**, 42–50
White J J, Lee B L 1975 An experimental study of sandwich injection molding of two polymer melts using simultaneous injection. *Polym. Eng. Sci.* **15**, 481–5
Young S S, White J J, Clark E S, Oyanagi Y 1980 A basic experimental study of sandwich injection molding with sequential injection. *Polym. Eng. Sci* **20**, 798–804
Zoetelief W F, Peters G W M, Meijer H E H 1997 Numerical simulation of the multicomponent injection molding process. *Int. Polym. Process.* **12**, 216–27

W. F. Zoetelief

Polymer Matrix Composites: Matrices and Processing

Polymer matrix composites, and fiber-reinforced plastics (FRPs) in particular, are commonplace in the industrialized world; their applications range from primary structural aircraft components to tennis racquets, and many household appliances incorporate some form of fiber reinforcement. The development of engineering FRPs began once significant quantities of glass fiber became available in the 1940s, and was initially performance driven (e.g., for aerospace). Since then, reduction of materials costs has greatly widened their range of applications, and the impetus for continued development stems mainly from economic factors such as efficient fabrication and transformation, and a need for the energy savings possible with low-weight, high-strength, and high-modulus materials. However, FRPs have the major disadvantage with respect to metal- and ceramic-based structural materials that their service temperatures are limited to below 400°C, even for the most advanced matrices.

An overview is given in this article of the most widely used FRP matrix materials and their relative merits. The various stages of processing to form composite parts are discussed, with reference to cost forecasting, life cycle engineering, and developing technologies. Finally, the important question of process–structure–property relationships in thermoset- and thermoplastic-based FRPs is addressed.

1. Matrix Materials

Apart from injection molding thermoplastics, which typically contain short (< 1 mm) fibers, most FRPs are based on thermoset resins. Thermosets can be cured catalytically, thermolytically, or photolytically over a wide range of temperatures to give highly cross-linked solids. One of their principal advantages for FRPs is their low initial viscosity, which facilitates fiber impregnation, particularly where the base "A-stage" resin consists of monomeric or oligomeric components. Partial cure may then be used to produce solid "B-stage" prepregs used as precursors for a variety of processing routes. There is nevertheless great interest in thermoplastic polymers; although the high molar mass of thermoplastics implies that higher processing temperatures or specialized techniques are needed for impregnation, this may be outweighed by other factors: (i) they are generally nonreactive, implying an indefinite shelf-life for precursors; (ii) forming processes are reversible, facilitating recycling and repair; and (iii) cycle times are not limited by reaction kinetics, and mixing steps, which may involve toxic precursors and need careful control to ensure reproducibility, are eliminated. Table 1 summarizes the main trends, although it should be stressed that there is considerable overlap, the toughness of FRPs based on both classes of resins being highly sensitive to, for example, fiber morphology, interfacial adhesion, and the presence of additives. Moreover, each class of resins encompasses a wide variety of materials and materials properties, which will now be presented in more detail along with typical applications. A more systematic discussion of the applications of polymer matrix composites is to be found.

1.1 Thermoset Resins

There are several main types of thermosets, distinguished by their chemistry, curing mechanisms, and final properties (Charrier 1990). Those discussed here are the most important for FRPs, namely polyesters, epoxies, and high-performance thermosets, although other thermosets, including polyurethanes (PUs), silicone rubbers, and phenol–formaldehydes (PFs), are also sometimes used as FRP matrices (e.g., PF–aramid composites for aircraft interiors).

(*a*) *Unsaturated polyesters.* Unsaturated polyesters (UPs) are the most widely used thermosets for FRPs. They consist of prepolymers cross-linkable by addition, initiated photolytically or by thermal decomposition of peroxides. Processing involves most of the methods described in the next section, and indeed UPs were originally developed for use in FRPs. However, to ensure adequate flow behavior, up to 50 wt.% styrene co-reagent is often added to give a so-called UPE resin. Environmental concerns have since led to the use of low styrene emission additives in UPEs. These segregate to, and seal, the surface during the final stages of cure, compromising adhesion in some cases. Another potential difficulty is excessive shrinkage, but this is largely avoided by the use of thermoplastic low profile additives. Once cured, unsaturated polyesters have heat deflection temperatures (HDTs) ranging from 80°C to 150°C in certain vinylester resins (the HDT is a softening temperature measured according to industrial norms and marks the approximate upper limit to the range of service temperatures). Their main composite applications include the construction industry, the transport

Table 1
Property and process characteristics of thermosets and thermoplastics.

Property	Thermoset	Thermoplastic
Modulus	High	Medium
Service temperature	High	Medium
Toughness	Medium	High
Viscosity	Low	High
Processing temperature	Low	High
Recyclability	Limited	Good

industry (boats, car body panels), furniture, and chemical storage tanks.

(*b*) *Epoxy resins.* Epoxy resins may also be used in conjunction with most composite processing techniques, and are particularly suited to B-stage processing. Cure generally involves polyaddition with a multifunctional co-reagent, referred to as the "hardener," to produce a three-dimensional network with low shrinkage, and excellent adhesion and environmental resistance. Drawbacks of conventional resins include somewhat higher viscosities than in UPEs, long cure times, and moisture uptake. A wide range of resins and co-reagents is nevertheless available, products ranging from tough low-temperature epoxies to more brittle, high-temperature epoxies for advanced FRPs (prepregs for aircraft components account for around 30% of total production). Commonly used aliphatic amine hardeners permit room temperature cure, but the reaction is strongly exothermic, causing problems in thick moldings. Aromatic amine hardeners require higher curing temperatures but are better suited to large parts and give relatively high HDTs (up to 230°C). Anhydride hardeners are less toxic and less polar than amines, but need an accelerator. With suitable catalysis, epoxies may also be cured thermally or photolytically without a co-reagent.

(*c*) *High-performance thermosets.* High-performance thermosets include allyls, cyanate esters, and liquid crystalline thermosets. Among the most successful are addition polyimides, developed mainly for the aerospace and electronics industries. Their HDT may reach 400°C after cure, and they have excellent stiffness and environmental resistance. The cure mechanism is thermally induced free radical addition of a polyimide precursor via reactive end groups in the presence of peroxide, avoiding problems with the volatile by-products of conventional condensation reactions. Applications include jet and automobile engine components, heat-resistant panels, etc. However, some potential future applications of high-performance FRPs, such as reusable launch vehicles, will require HDTs in excess of 500°C, setting the standard for the next generation of thermoset matrix materials.

1.2 Thermoplastic Resins

Both semicrystalline and amorphous glassy polymers form viscous melts that need to be processed at temperatures well above the glass transition temperature, T_g. However, since semicrystalline polymers melt well above T_g, it is the melting point, T_m, that defines the processing window. Whereas thermoset designations tend to be generic, most thermoplastics have well-defined linear architectures that undergo little change during processing. Their property profiles are generally adjusted for different processing routes by varying the molar mass or additive package (Charrier 1990); however, it should be borne in mind that other important properties such as toughness and ultimate tensile strength are also highly sensitive to changes in molar mass.

(*a*) *Semicrystalline thermoplastics.* Semicrystalline thermoplastics are tough and resistant to organic solvents, but many inexpensive and easy-to-process low T_g polymers, such as polyethylene, also show poor adhesion owing to their lack of functionality, and low softening temperatures, often associated with excessive creep. A good compromise between cost and performance is provided by polypropylene (PP, $T_m \approx 170°C$), for which functionalized additives are available to improve adhesion. PP is widely used in FRP applications in the automotive industry, for example, owing notably to glass mat thermoplastic technology. Polyamides (PAs or nylons), such as PA6 and PA6,6, are also popular, having higher HDTs than most thermosets, although they are moisture sensitive and more than twice as expensive as PP. Another polyamide of interest for FRPs is PA12, the versatile synthesis and functionality of which make it especially attractive for reactive FRP processing.

High T_g semicrystalline polymers such as poly (phenylene sulfide) (PPS) and poly(ether ether ketone) (PEEK) require correspondingly high processing temperatures, PEEK melting at about 345°C, for example. However, these polymers have outstanding properties, including exceptional toughness and fatigue resistance. Particular effort has gone into PEEK–carbon fiber composite technology, leading to the so-called aromatic polymer composite (APC) prepregs, considered to be a major step in the development of advanced thermoplastic matrix FRPs (Cogswell 1992). However, PEEK is nearly two orders of magnitude more expensive than PP (Charrier 1990).

(*b*) *Amorphous glassy thermoplastics.* Aromatic high-performance amorphous thermoplastics such as poly(ether sulfone) (PES, $T_g \approx 230°C$) or poly (ether imide) (PEI, $T_g \approx 215°C$) have also attracted interest as advanced FRP matrix materials. They are less expensive than PEEK and tougher than thermosets, but there remain (arguably unjustified) concerns over their solvent resistance.

2. *Processing*

Figure 1 summarizes the various routes by which raw materials (resin plus fiber) are transformed into the final part (Månson 1994). A description of the most important of these now follows. Further details concerning specific types of reinforcement, and glass fibers in particular, are to be found.

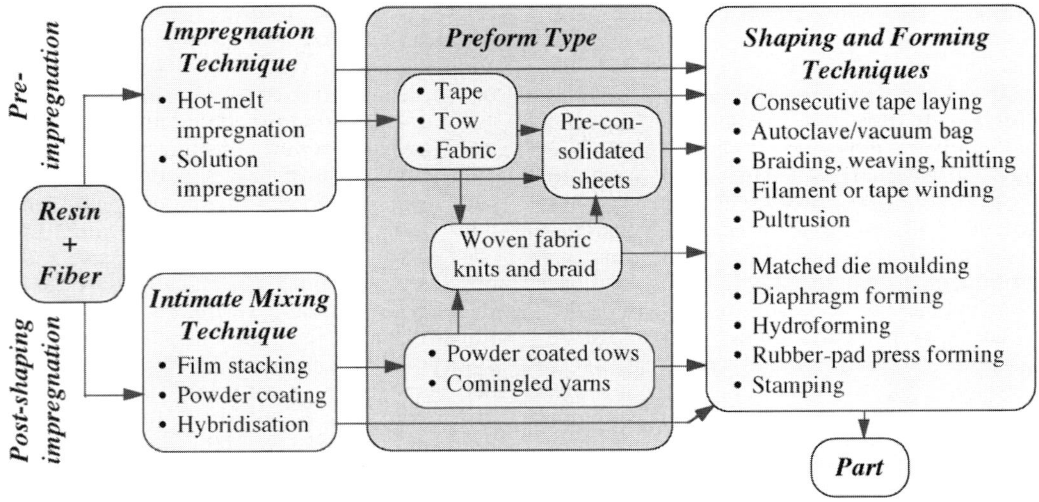

Figure 1
Alternative manufacturing routes: from resin and fiber to finished part (after Månson 1994).

2.1 Fibers, Prepregs, Feedstocks

Fibers are usually supplied in the form of rovings (glass fibers) or tows (carbon fibers). Rovings consist of straight continuous glass-fiber strands or bundles of about 200 filaments; the number of strands depends on the end use, and these may be several kilometers long. Tows are likewise available in various configurations. The fibers are typically sized (coated) during production to promote wetting and adhesion, silane coupling agents being widely used for glass fibers. They may be woven into fabrics if required, including *ad hoc* fiber preforms for specific applications. Chopped fiber lengths range from less than 1 mm for injection, to around 50 mm in randomly oriented mats for laminates.

Of major concern in FRP processing are trapped air and volatiles, since voids substantially reduce the final strength. Low-viscosity resins, vacuum outgassing, and consolidation all help reduce porosity. Use of preconsolidated sheets or prepregs also facilitates certain processes (see Fig. 1). These are available as impregnated fabrics or continuous warp sheet typically containing 60–65 vol.% uniaxially oriented fibers. Impregnation with thermosets may be achieved by dipping or hot melt processing (rolling of the fibers together with a resin film). When the viscosity is high, flow distances must be kept to a minimum, and intimate mixing of fine polymer powders or fibers with the fiber reinforcement often precedes melt processing of thermoplastic composites. Solvent processing is one alternative; it may lead to problems with residual solvent in the later stages of fabrication but has been used with success for PEI prepregs, for example. Another possibility is to chain extend relatively low molar mass precursors *in situ* by reaction of suitable end groups, an approach frequently used for high-performance thermoplastic FRPs.

Sheet and dough molding compounds are mostly based on unsaturated polyester resins containing filler and chopped fibers. Sheet molding compounds (SMCs) are produced from chopped strand mat placed between layers of filled resin containing thickening agents to promote moldability. Impregnation involves rolling of the resulting sandwich structure. SMCs usually contain 25–35 wt.% fibers but the fiber content can be up to 65 wt.%, in which case they are called HMCs. XMCs are similar to SMCs, but incorporate surface layers containing continuous fiber rovings. Dough molding compounds (DMCs), or bulk molding compounds (BMCs), contain shorter fibers than SMCs and are produced by mechanical mixing, which results in a more markedly three-dimensional fiber orientation distribution. Glass mat thermoplastics (GMTs) can be produced in a similar manner to SMCs, but the high matrix viscosity means that success is very dependent on the choice of fiber mat. Powder processing routes are an interesting alternative in this case.

It is sometimes advantageous to preassemble reinforcing fibers to fit a given shape, rather than rely on flow to distribute them throughout a mold. If displacement of the fibers during subsequent introduction of the resin is a problem, they can be anchored in place by spraying with a binder, or substituted by a preimpregnated fabric shaped into a suitable perform.

2.2 Shaping and Forming Techniques

The simplest forming technique is hand placement, where woven fabric or chopped strand mat is placed on a nonstick former and premixed liquid thermoset resin added with a brush, say, or by spraying. It is versatile and requires low investment, but the resulting mechanical properties are limited and the manufacturer may fall foul of legislation against volatiles. It is used for one-off or short production runs, including large objects, such as boat hulls. However, for reproducible, high-quality, cost-effective production other techniques must be considered.

(*a*) *Press molding.* Press molding using matched male and female tools is the most widely used technique for volume production of FRPs, being suitable for most types of precursor. With SMCs or DMCs, complex shapes can readily be formed at pressures of 3–15 MPa with cycle times of the order of minutes depending on the cure rate and the part size. It is also used for continuous fiber FRPs, but the limited flow restricts the practically attainable range of wall thickness and curvature. For mass production, steel tools and hence additional capital investment may be necessary, but for short runs, epoxy composite tools are often adequate. Variations include rubber-pad press molding, hydroforming, diaphragm forming, and cold stamping.

(*b*) *Vacuum and autoclave molding.* In vacuum molding, atmospheric pressure consolidates the material during cure, avoiding the need for a press. A prepreg or fabric laminate is placed on a mold, covered with a flexible membrane and the assembly sealed and evacuated. It provides a low-cost alternative to press molding for large, thin shells. If further consolidation is needed, cure may be done in an autoclave. Autoclaves usually operate at up to 1.5 MPa pressure and are widely used in the aerospace industry, having the advantage over press molding of cheap tooling for large parts and being ideal for complex structures such as honeycomb core sandwich composites. Indeed almost all prepreg material used in the aerospace industry is autoclaved. It is less suited to high-volume manufacture, although progress in automatic cutting and lay-up operations has led to considerable cost reductions and new market opportunities.

(*c*) *Resin transfer molding, reaction injection molding.* In resin transfer molding (RTM), the fiber reinforcement (either dry fibers or a preform) is placed in the mold cavity, and the resin then introduced by injection. RTM generally requires low-viscosity resins, but impregnation can also be improved by vacuum-assisted injection. It is used to produce large moldings such as car body shells, and the low injection pressures mean the molds themselves can be made from FRP, with considerable cost savings. However, although RTM is cheaper than prepreg processing, it is more limited in its range of application. Reaction injection molding (RIM) makes use of rapidly reacting components mixed just prior to injection, permitting very short cycle times, and short fiber reinforcement can also be introduced in one or both reagents (this is called reinforced reaction injection molding (RRIM)). It is competitive with press molding for the manufacture of a variety of automotive and domestic components.

(*d*) *Pultrusion.* Pultrusion is the continuous fiber FRP equivalent of extrusion, permitting continuous production of constant profile parts such as fishing rods, antennas, and channel profiles (or constant section parts, if the profile is modified *in situ*). Unlike in extrusion, however, the reinforcing fibers themselves are used to *pull* the material through a heated forming die, as shown schematically in Fig. 2. Rovings or tows are passed through a resin bath followed by a series of carding plates or "preformers," which force the resin into the fibers. The impregnated fibers are then pulled through the die, where the resin gels and cures. Preheating by radio-frequency methods may be used to increase throughput, resulting in production rates of up to several meters a minute. In some cases the resin is also injected directly into the die, eliminating the need for a preimpregnation step.

(*e*) *Filament winding.* Strong, stiff shell structures can be made by winding bands of continuous fibers impregnated with resin onto mandrels, which are removed after cure. Filament winding is widely used for the manufacture of reinforced plastic pipe and other cylindrical objects. However, modern filament winding techniques can also be used to produce complex structures such as aircraft fuselages. In such cases, the fibers must be wound along constantly changing nonrepetitive geodesic paths to prevent slippage. Automation of this technique has led to the development of advanced computer-based integrated design and manufacturing methods.

(*f*) *Developing technologies.* Some representative examples are now be given of developing technologies aimed at improving, or providing alternatives to established processing routes. The resin infusion process (RIFT) and its variants, in which a vacuum is used to draw resin into a preform or fiber bed (cf. RTM), has been successfully applied to large objects such as boat hulls. Large injection molding body technology (LIMBT), offers a viable alternative to press molding of SMCs and GMTs, contrary to the widely held notion that injection molding facilities become prohibitively expensive as part sizes increase; for example, it has been used to

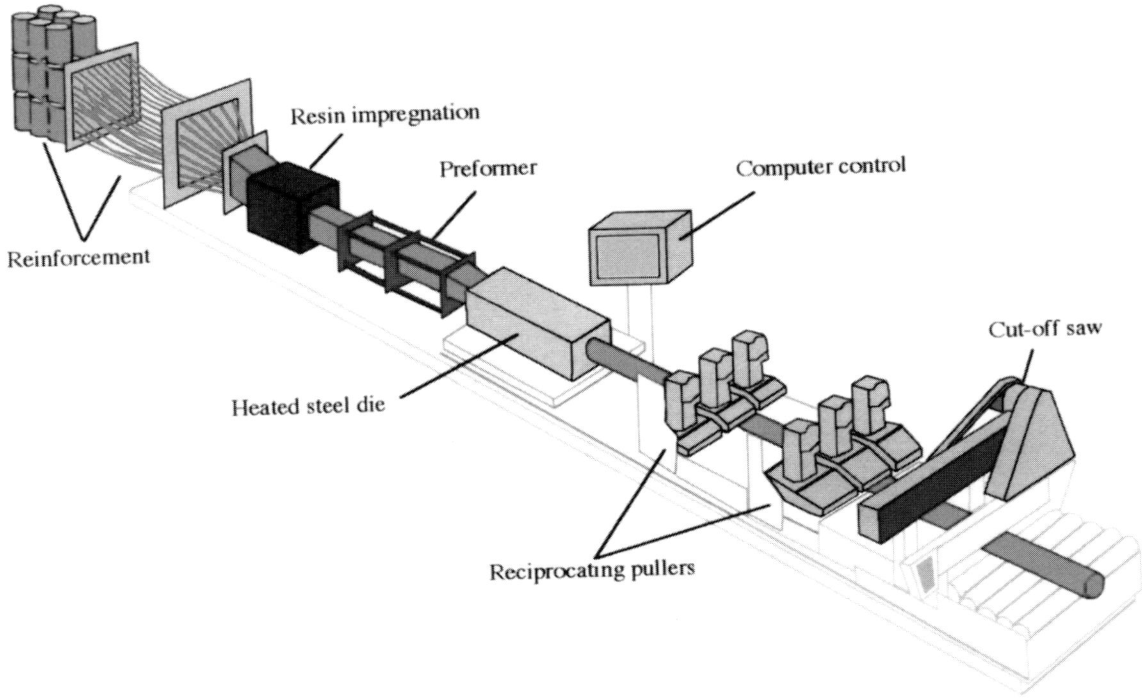

Figure 2
Schematic of the pultrusion process.

make large automotive body components from various thermoplastic FRPs (Argeropoulos *et al.* 1999). Integrated processing (IP) is aimed at improving manufacturing flexibility and adding value to composite products by combining different processes and materials; selective placement of reinforcement and integration of functions are thus achieved in short cycle times (Månson *et al.* 1995, Bourban *et al.* 1998). Reactive thermoplastics processing, developed for PA12-based composites, uses monomeric precursors for impregnation followed by polymerization *in situ*, avoiding the problems associated with the use of melt or solution impregnation techniques for thermoplastics (Schmid and Eder 1997, Bourban 2000).

2.3 Choice of Materials and Processing Routes: Composite Processing Maps

Figure 3 shows a map of various composite processing methods arranged according to the combination of resin viscosity and reinforcement aspect ratio to which they are most suited. If viscosity is an overriding consideration it will generally dictate the choice between thermoplastics and thermosets, but it is immediately clear from Fig. 3 that the range of processes is broad enough for solutions to be available for thermoplastics or thermosets for any required reinforcement aspect ratio. Such "composite processing maps" may be based on other parameters, such as the potential for control of reinforcement orientation and aspect ratio; injection molding, for example, gives little orientation control and is limited to low aspect ratios, whereas three-dimensional control and long fiber orientation require techniques such as filament winding, hand lay-up and RTM/preform technologies. Figure 4 gives a further example that is particularly important for costing (see below), namely the relationship between the suitability of a given process route for complex shapes and its suitability for high-volume production, as defined by the cycle time.

2.4 Cost Modeling, Life Cycle Engineering

Since material costs are now substantially irreducible, and given that manufacturing often represents close to 60% of expenditure on FRP composite parts, the processing route is generally the most critical factor in cost-effective production. Assessing the economic viability of a processing route requires comparative cost analysis and hence reliable data and cost estimation methods. A variety of approaches for evaluating manufacturing costs have been proposed

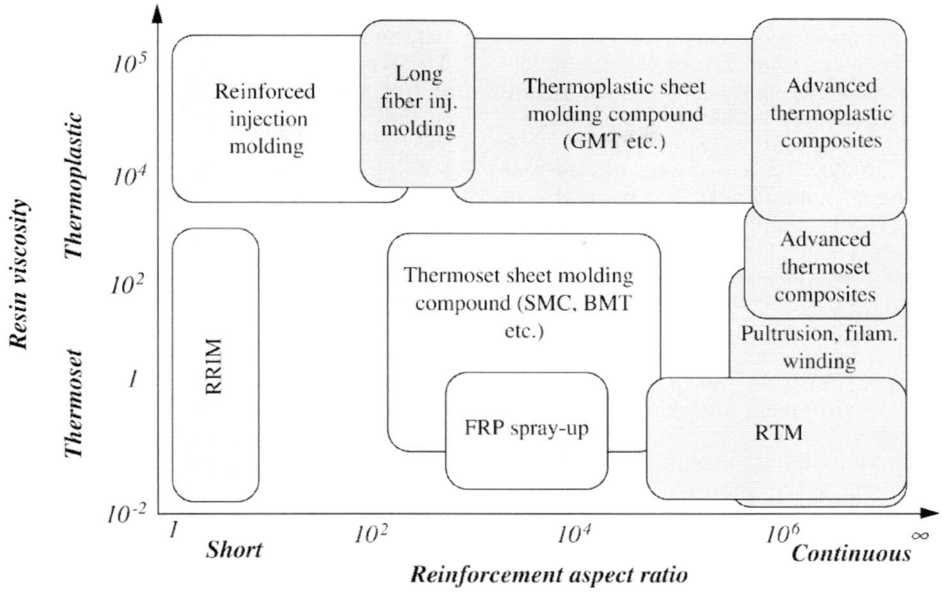

Figure 3
Composite processing map: viscosity and reinforcement aspect ratio (after Månson 1994).

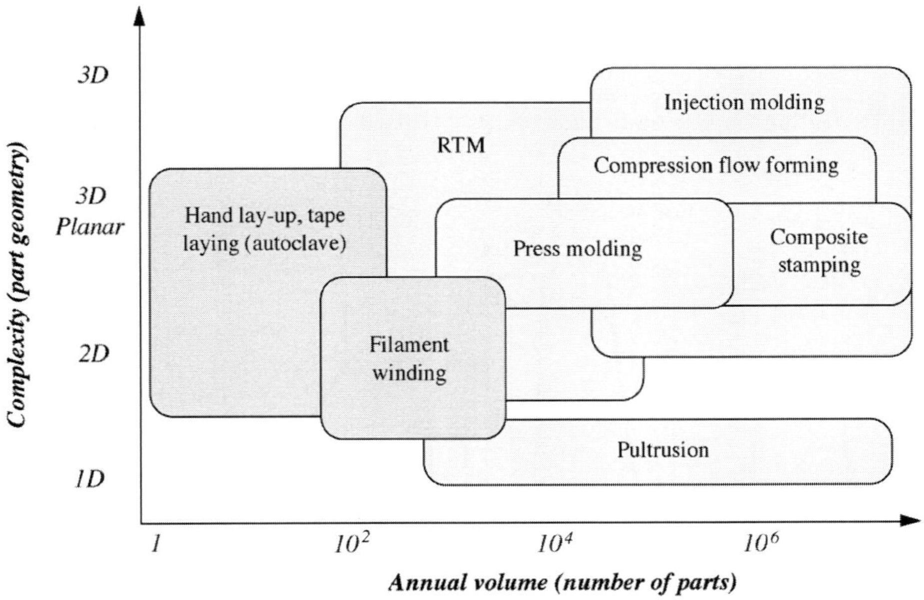

Figure 4
Composite processing map: complex shaping and volume production (after Månson 1994).

(Månson *et al.* 2000): comparative studies, process-oriented cost models, parametric cost models, and process flow simulation. Comparative studies are based on historical data and may suffice when envisaging the use of a standard technique for a component similar to those already in production. Process-oriented cost models are better adapted to new processes, and incorporate identification and quantification of cost drivers, but require detailed process information and a precise definition of the components. Parametric models are more versatile and allow easy manipulation of process and economic factors; however, they are based on the assumption of independent manufacturing steps, and may consequently underestimate costs. Such difficulties may be avoided by using process flow simulations, which take into account interactions between the different manufacturing steps, and provide forecasting tools that are also suitable for improvement studies, identifying bottlenecks and estimating repair costs. Given the growing concern for environmental issues, the successful introduction of a new product or process may also require establishment of a life cycle plan, particularly for the large-volume applications. As well as the impact of material on the environment during manufacture, service, and disposal, it is important to consider conservation of raw material value. This is a potentially important issue when choosing between thermosets and thermoplastics, since recycling of the former is severely limited. Furthermore, recovery from continuous fiber FRPs is difficult even for thermoplastic matrices. One alternative under consideration is recovery of the monomer via tertiary recycling.

3. Process–Structure–Property Relationships

The thermal cure history shown in Fig. 5(a) is typical of prepreg and autoclave processing sequences for a thermoset FRP. During impregnation, the resin viscosity first decreases owing to the rise in temperature, then increases as cross-linking proceeds. The reaction is interrupted by cooling, and the resulting B-stage material is stored at low temperature to inhibit further reaction. During final processing, the viscosity again passes through a minimum, tending to infinity as the gel point is approached. A two-step final cure is often used to improve consolidation and dimensional stability, after which the part is cooled and removed from the mold or former. The final properties of thermoset FRPs are strongly dependent on the details of such time–temperature trajectories. Figure 5(b) shows a schematic time–temperature–transformation (TTT) diagram illustrating the main trends (Pang and Gillham 1990). During isothermal cure at temperatures between T_g^{gel} and T_g^{∞}, the polymer gels and then vitrifies, defining optimum conditions for final processing. Isothermal cure between T_g^{gel} and T_g^0, the initial glass transition temperature (or melting point), would cause the polymer to vitrify prior to gelation, hindering further reaction.

In a typical processing cycle for a thermoplastic FRP, the components are first heated to above T_m of the resin (or T_g for a fully amorphous resin). Pressure is applied for long enough to give adequate consolidation, and the resulting composite is cooled with the pressure maintained. Optimum processing conditions may be defined with the aid of processing windows, such as that shown schematically in Fig. 6. The lower bound for the cooling (and heating) rate is fixed by economic constraints on the cycle time.

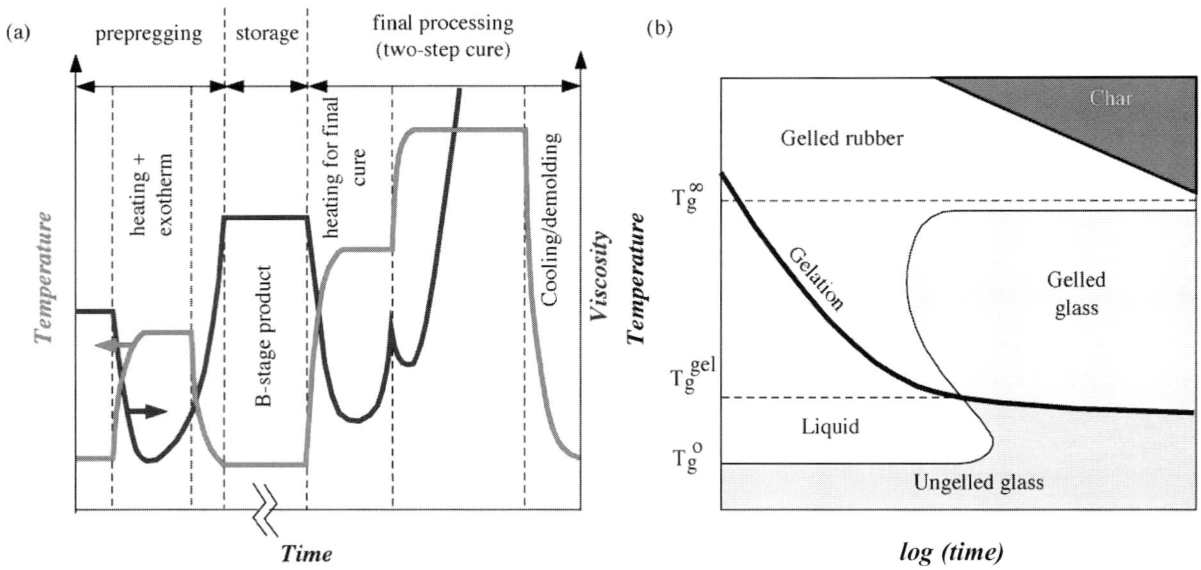

Figure 5
(a) Typical thermal cure history for a thermoset composite. (b) Isothermal TTT diagram associated with the cure of a thermoset polymer.

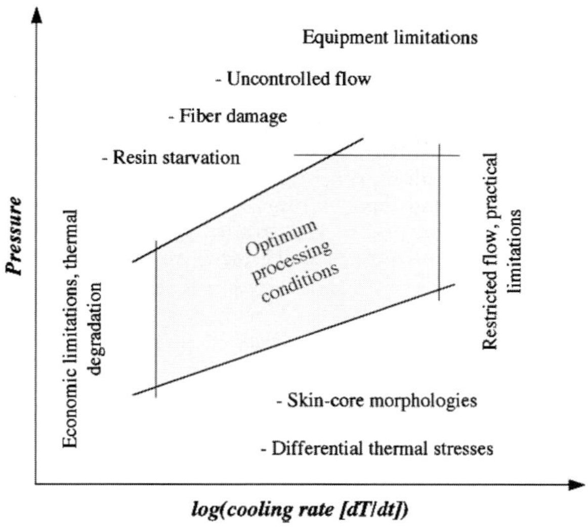

Figure 6
Generic processing window for a thermoplastic composite (after Månson 1994).

Slow cooling and heating may also result in prolonged exposure of the resin to high temperatures, leading to thermal degradation. At very high cooling rates, however, considerable thermal gradients may develop through the thickness of the component, leading to pronounced skin–core morphologies in semicrystalline resins, and excessive thermal stresses, as well as changes in primary morphological parameters such as spherulite size, lamellar thickness, and the degree of crystallinity. Increasing the hold pressure may help suppress defects resulting from internal stresses, but a maximum pressure will be imposed either by the performance of the press, or by the cost of high-pressure molds. High pressure may also damage the fiber bed and/or cause resin starvation.

4. Conclusion

Although the initial driving force for the development of FRP composites was performance, economics is now the overriding consideration. Raw material costs are generally higher than for metals, but this is offset by weight reduction, increased design freedom, and reduced "part count" in a given application; even image can be a strong selling point for sports goods. Nowadays, it is innovation in processing that offers the greatest scope for increasing the engineering viability of polymer matrix composite materials, and for paving the way to new applications of these materials. Doubtless some of the techniques under development will shortly join or supplant the "standard methods" referred to here.

Bibliography

Argeropoulos J G, Fielding C R, Oswald L J 1999 *SAC 1999-01-3244*
Bourban P-E 2000 Liquid molding of thermoplastic composites. In: Kelly A, Zweben C (eds.) *Comprehensive Composite Materials*. Pergamon, Oxford, Vol. 2, pp. 965–78
Bourban P-E, Bögli A, Bonjour F, Månson J-A E 1998 *Compos. Sci. Technol.* **85**(5), 633–7
Charrier J-M 1990 *Polymer Materials and Processing*. Hanser Verlag, Munich, Germany
Cogswell F N (ed.) 1992 *Thermoplastic Aromatic Polymer Composites*. Butterworth-Heineman, Oxford
Månson J-A E 1994 In: Chawla K K (ed.) *New Demands on Composite Materials*. TMS, Warrendale, PA
Månson J-A E, Bourban P E, Bonjour F 1995 *Pat. PCT*, WO 96/36477
Månson J-A E, Wakeman M D, Bernet N 2000 Composite processing and manufacturing—an overview. In: Kelly A, Zweben C (eds.) *Comprehensive Composite Materials*. Pergamon, Oxford, Vol. 2, pp. 577–608
Pang K P, Gillham J K 1990 *J. Appl. Polym. Sci.* **39**, 909–33
Schmid E, Eder R 1997 *German Pat.* 19 602 684 C1

C. J. G. Plummer, P. -E. Bourban,
and J. -A. E. Månson

Polymer Modified Wood

Polymer modified woods (PMWs) are kiln-dried, permeable lumber (such as birch, beech, red oak, maple, pine, poplar, rubberwood, and alder) or composites (such as plywood, oriented strandboard, particleboard, and fiberboard) which contain polymer (plastic) in their cell cavities or cell walls. PMWs are different from wood–plastic composites, which are largely a blend of thermoplastic resins and wood particles.

1. *How PMWs Are Made*

PMWs are made by impregnating the wood with liquid monomer or prepolymer mixtures and then polymerizing the liquid within the wood. Many monomers that contain vinyl groups (a carbon–carbon double bond) have a viscosity useful for wood impregnation at room temperature and also react well to form the polymer within the wood. The monomer most widely used for PMWs is methyl methacrylate.

Impregnation is carried out in a pressure vessel. First, a vacuum is drawn to remove air from the cells of the wood. Next, the liquid monomer is admitted to the pressure vessel while maintaining the vacuum. Atmospheric pressure is then restored to the vessel and additional pressure may be applied to help move the liquid uniformly into the wood. After impregnation the wood is removed from the vessel and placed in a curing chamber.

Curing (polymerization) of vinyl-type monomers starts with the generation of reactive groups using either temperature-sensitive catalysts or penetrating radiation such as gamma rays. The reaction proceeds with monomer molecules bonding together to form large polymer molecules and releasing heat as a by-product.

2. *Types*

There are three major types of PMWs. The cell-lumen type contains polymer, which fills cell cavities but has not reacted with the cell wall. Most vinyl monomers produce this type of PMW. Cell-wall PMWs are made using monomers, such as furfuryl alcohol monomer and phenol–formaldehyde prepolymer, which modify the cell wall. Phenol–formaldehyde prepolymer can be impregnated into thin veneers because of its high viscosity, and the impregnated veneers are glued together to make solid boards called "Impreg" and "Compreg." Cell-wall PMWs have high dimensional stability, weatherability, and resistance to chemicals but are more brittle than the wood from which they are made. Combination monomer formulations, the third type, are designed to make PMWs with properties intermediate between those of cell-lumen and cell-wall PMWs.

PMWs are typically made by forcing as much liquid monomer into wood as possible, using what is called in the wood preservation industry the "full cell" process. Made this way, cell-lumen PMWs have a density very close to $1.1\,g\,cm^{-3}$ and cell-wall PMWs $\sim 1\,g\,cm^{-3}$ regardless of the wood from which they were made. Because of this uniform density, PMWs have more uniform piece-to-piece properties than wood.

2.1 Cell-lumen PMWs

The hollow cavities in wood cells makes them susceptible to crushing in side-loading and buckling in longitudinal compressive loading. In cell-lumen PMWs the cavities are filled or nearly filled with polymer. The individual cells are thus reinforced by the polymer in their cavities, dramatically increasing hardness and compressive strength of the wood perpendicular to the grain. Since buckling of cells is the first failure mode in bending, the reinforcing effect of filled cell lumens in these PMWs is noticeable in bending-mode mechanical values. Toughness is also increased because the polymer adds more material to the wood which must be fractured.

The blocking of cell cavities in cell-lumen PMWs greatly retards moisture movement, lowering water vapor diffusion coefficient values, and causes longitudinal and transverse movements to be about equal. Blockage of cell cavities should have a retarding effect on wood-boring insects and marine organisms. Overall swelling in room-temperature water is lowered. Since moisture plays a strong role in weathering, PMWs are more resistant to weathering than wood. As cell cavities are the main route of fungal invasion, filling them with polymer greatly retards decay.

Dyes can be added to the monomer mixture to make colored PMWs, which have the same physical properties, except for color, as those made with clear polymer.

2.2 Cell-wall PMWs

The wood cell walls in cell-wall PMWs are chemically altered. Since the polymer in cell-wall PMWs also partly fills cell lumens, cell-wall PMWs have some of the characteristics of cell-lumen PMWs. The most noticeably changed properties of cell-wall PMWs are those which depend on moisture movement and uptake, including swelling, water vapor diffusion coefficient, and weathering. Wood-destroying organisms do not degrade them and they are highly resistant to acids and bases. However, cell-wall PMWs are more brittle than the parent wood cell wall, which is their chief disadvantage. The dark-brown color of furfuryl alcohol-based PMWs is useful for dark-colored products, but the darkness sometimes limits their usefulness.

2.3 Combination PMWs

Combination PMWs attempt to obtain some of the better properties of both cell-lumen and cell-wall PMW with less of their respective disadvantages. In moisture effects, toughness, and decay resistance they are intermediate.

2.4 Low Polymer Content PMW

There is a class of wood treatment that causes a chemical reaction to occur in the cell wall. These processes use either a small amount of chemical or no chemical and heat. Such material is called "modified wood." Examples that have reached the commercial stage are acetylation and heat treatment.

It has been recently shown that cell-wall PMW with low polymer content has reduced shrinkage and swelling and is resistant to microorganisms. A small amount of monomer or prepolymer reacts with the cell wall to provide protection. Weight gains as low as 15% have been shown to provide increased lifetime in adverse environments. Change in mechanical properties is small because there is so little polymer added. This class of modified wood falls between PMW produced by a full cell monomer impregnation process and modified wood that uses small amounts of chemical or heat. One such process that is being commercialized is called "furfurylation" because it is based on polymerized furfuryl alcohol. Another process under development uses phenol-formaldehyde prepolymers.

Low polymer contents are obtained by using a solution of the monomer or prepolymer in a solvent to impregnate wood with a full cell impregnation process and then causing the solvent to evaporate. Water is the preferred solvent. For furfurylated wood, polymerization can be accomplished by heating before or during drying.

3. Uses of PMWs

Cell-lumen PMWs made from vinyl monomers have been used for many years in flooring, stair parts, and sporting goods. It is called "acrylic-impregnated wood flooring" in industry because it is produced using methyl methacrylate monomer. There is interest in using alternative monomer formulations and curing methods in other products, particularly where hardness, wearability, corrosion resistance, weatherability, decay resistance, moisture resistance, stiffness, color choice, and nontoxicity are desired. Examples include railway ties, highway posts, outdoor furniture, and counter-tops. When sanded and polished, PMWs have a deep luster superior in performance and appearance to most surface coatings. Adding color increases style options. The attractiveness of PMW products has also generated interest among craftspeople who make musical instruments, jewelry, and kitchenware.

"Impreg" and "Compreg" have been used for many years in knife handles, pistol grips, and other small items. With dyes added to the veneer layers, an attractive color pattern can be produced. Furfuryl alcohol cell-wall PMWs have been used for knife handles and kitchenware, which are exposed to frequent mechanical dishwashing. It has potential as a substitute for some dark-colored tropical hardwoods.

Low polymer content PMW is similar in density, mechanical, and handling properties to untreated wood. Therefore, its potential is uses in which wood would be susceptible to decay and attack by insects

Table 1
Examples of Brinell hardness values for wood and polymer-modified wood (PMW).

	Wood	Cell-lumen PMW	Cell-wall PMW
Sugar maple (*Acer saccharum*)	6	18	
Yellow birch (*Betula alleghaniensis*)	4	18	10
Scots pine sapwood (*Pinus sylvestris*)	2	12	

Table 2
Examples of physical properties of maple wood and PMW.

	Wood	Cell-lumen PMW	Cell-wall PMW
Percent weight loss from fungal decay in soil block test	93	13	0
24 h, 21 °C percent tangential swell in liquid water	14	4	1
Longitudinal water vapor diffusion coefficient, $10^{-12} m^2 s^{-1}$	25	5	2
Transverse water vapor diffusion coefficient, $10^{-12} m^2 s^{-1}$	7	5	0.5

Table 3
Examples of mechanical properties (in MPa) of wood and cell-lumen PMW at oven-dry and 12% moisture content (MC).

	% MC	Wood	PMW
Sugar maple bending elastic modulus	0	14800	17200
	12	13100	15700
White pine bending elastic modulus	0	10300	14100
	12	6700	8900
Sugar maple tension parallel elastic modulus	0	16500	18000
	12	14600	16500
Sugar maple compression parallel elastic modulus	0	13300	16400
	12	11400	15100
Yellow birch compression perpendicular elastic modulus	0	1700	2900
Yellow birch compression perpendicular yield stress (elastic limit)	0	20	70
Yellow birch compression perpendicular ultimate (maximum) stress	0		120

Table 4
Property changes for low polymer content furfurylated wood (~30% weight gain based on wood). The variability in the furfurylated wood results from variations in original wood density and polymer content.

Properties	Untreated wood	Furfurylated wood
Physical		
Density (g/cc)	0.4–0.7	15–35% increase
Color		Slightly browner
Shrinkage		50–75% less
Hygroscopicity		50–75% less
Durability		
For use	Indoors and dry outdoors	Exposed outdoors and ground contact
Mechanical		
Compression parallel to grain		Same
Impact bending		20% less
Hardness (Brinell)	2–6	3–7

and marine borers but a high polymer loading would be undesirable or unneeded.

4. Example Properties

Tables 1–4 give the properties of some PMWs and the woods from which they were made.

5. Conclusions

Filling wood with polymers produces a denser material with some properties noticeably enhanced over those of the parent wood. Therefore, making PMWs increases utility for some products for which wood is already used and also increases the applications in which wood can be used. Using low amounts of polymer to modify wood increases its moisture resistance and durability with little density increase or change to other properties.

Bibliography

Furuno T, Imamura H, Kajita H 2004 The modification of wood by treatment with low molecular weight phenol-formaldehyde resin. *Wood Sci Technol* **37**, 349–61

Kumar S 1994 Chemical modification of wood. *Wood Fiber Sci.* **26**(2), 270–80

Meyer J W 1984 Wood–polymer materials. In: Rowell R M (ed.) *Chemistry of Solid Wood*, Advances in Chemistry Series. 207. American Chemical Society, Washington, DC, pp. 257–90

Rowell R M, Konkol P 1987 *Treatments That Enhance Physical Properties of Wood*, Report FPL-GTR-55, US Department of Agriculture, Forest Service, Forest Products Laboratory

Schneider M H 1994 Wood–polymer composites. *Wood Fiber Sci.* **26**(1), 142–51

Westin M, Lande S, Schneider M 2003 Furfurylation of wood—process, properties and commercial production. *Proceedings of European Conference on Wood Modification*

M. H. Schneider

Polymer Processing: A Modeling Approach

The development of polymer forming processes starts with the development of the main thermoplastic polymers, even if several of these techniques were already present in the rubber and food industry. One may imagine today that these technologies are fixed, based on a consolidated know-how. In fact this is not true:

- Traditional processes, such as extrusion, injection molding, and film blowing have been, and continue to be, subjected to significant technological progress. There is, for example, almost nothing in common between the single screw extruders just after the Second World War (e.g., short screws, smooth barrel, and uniform heating) and the sophisticated machines which are proposed today (e.g., long screws, grooved and cooled barrel in the solid conveying zone, barrier screws, double flight plasticating geometry, two-stage screws with intermediate venting, and mixing zones).
- New processes appear continuously, e.g., injection stretch blow molding and gas-assisted injection molding.
- Old processes come back: torpedo extrusion system for polymer recycling, compression molding of reinforced thermoplastics.

All these developments have been done originally by engineers with extensive knowledge of mechanics and polymer science. Physical and numerical modeling of the polymer forming processes was not really considered at that time except to confirm intuitive approaches. Nevertheless, Tadmor and Klein (1970), by looking carefully at the plastification mechanism in a single screw extruder, induced a real revolution in the screw design during the early 1970s. In the same way, Pearson and Matovich (1969) pointed out that the periodic instability observed in fiber spinning was not the result of a poor process control but was due to a hydrodynamic instability. This promoted significant contributions in the manufacturing of polymer fibers.

Today, process modeling is well recognized and an increasing number of polymer converters currently use these methods to design extrusion dies and mold geometries or to optimize processing conditions. The next step, and this is a revolution, consists in using numerical modeling to predict the properties of the produced part, and then, to tailor the physical properties of the polymer which will be used to produce this part (e.g., molecular weight distribution, crystallinity, and additives).

In that sense, numerical modeling will become a tool, not only for polymer converters and machine designers, but also for polymer producers and for end-users of polymer parts.

1. *The Challenges for Polymer Process Modeling*

Polymer process modeling is not an easy task for several reasons:

- The part geometry and flow kinematics are complex. Therefore, people often use approximations (e.g., lubrication approximations and heat transfer coefficients) which are not always valid.
- Phase transitions (solid to liquid during the plastification process and then liquid to solid during solidification) are not always well understood.
- The rheological behavior is also complex, especially in the liquid state (shear thinning, temperature dependence, and viscoelasticity) which leads to highly nonlinear constitutive equations.
- There is a strong coupling between mechanical and temperature balance equations: high viscosity induces viscous dissipation which in turn reduces the viscosity.

In the following, several examples concerning die or mold design, cast film, reactive extrusion, or reinforced polymers extrusion will be presented.

1.1 Coat Hanger Die Design

The problem is to design the die (Fig. 1(a)) in order to achieve a more regular polymer flow distribution at the die exit. The geometry is 3D but one of the dimensions (the thickness) is small. The Hele Shaw (lubrication) approximations, as well as a mean temperature approximation throughout the thickness, can be used. This leads to a thermomechanical problem with two unknowns (the pressure and mean temperature) which can be solved using either finite element or finite difference methods (Vergnes *et al.* 1984).

Pressure and flow rate distributions are shown in Fig. 1(a). The flow rate distribution at the die exit is highly influenced by the power law index n ($\eta = k\dot{\gamma}^{n-1}$). As a consequence it is dangerous to optimize die design for one polymer and to generalize it to another polymer (Fig. 1(b)). This flow rate distribution is also influenced by the thermal regulation of the die, which points out strong coupling between mechanical and thermal phenomena (Fig. 1(c)).

1.2 Cast Film Process

The polymer film or sheet which has been extruded in Sect 1.1 may be stretched over a short distance between the coat hanger die and a chill roll (Fig. 2(a)). There is a marked necking phenomenon as well as the development of a heterogeneous film thickness. This depends on the draw ratio, the stretching distance, and the polymer rheology (Silagy *et al.* 1998).

As a first approach the film thickness is considered to be uniform throughout the width of the film which leads to a 1D problem with three unknowns (velocity

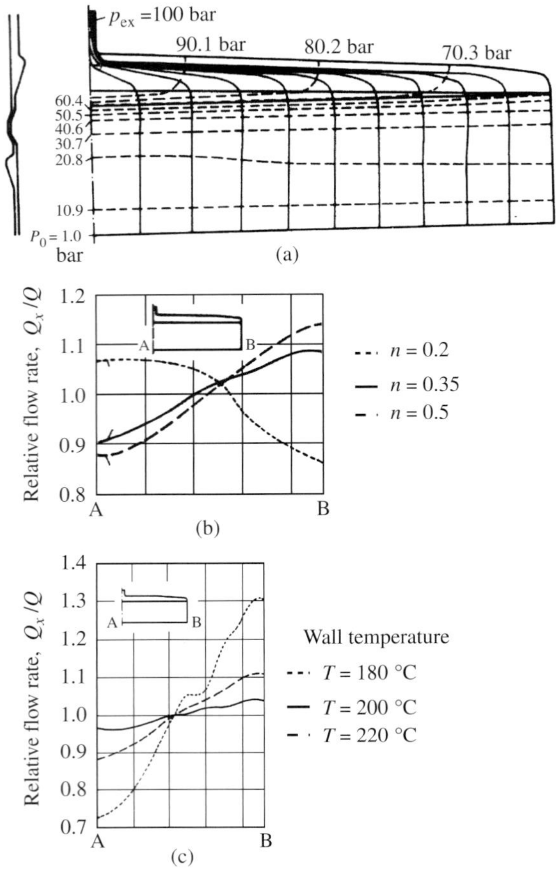

Figure 1
(a) Typical coat Fger die geometry, pressure and flow line distribution; (b) flow rate distribution at die exit as a function of the power law index n; and (c) flow rate distribution at die exit as a function of the die temperature (after Agassant *et al.* 1996).

$u(x)$, thickness $e(x)$ and width $W(x)$). The final dimensions of the film (width and thickness) are markedly influenced by the rheology (Fig. 2(b)).

In a second step the heterogeneous film thickness throughout the width is accounted for leading to a 2D model. Here again the shape of the film will depend on the rheology (Figs. 2(c) and 2(d)) but, for the same constitutive equation, 1D and 2D results are very different.

This example clearly points out the influence of the kinematic assumption (a 1D model is not appropriate) and of the constitutive equation (a Newtonian model gives erroneous results).

1.3 Injection Molding

There are numerous questions in injection molding: location of gates, position of weld lines, injection pressure, temperature distribution, weight, and deformation of the final part. A lot of numerical efforts have been devoted to mold filling modeling and there is strong competition between commercial software packages. Most of these software packages are based on Hele Shaw approximations, which are often but not always valid, especially when the parts present important thickness variations. In order to overcome these limitations a 3D injection molding software (REM3D) has been developed (Pichelin and Coupez 1998).

A 3D volume mesh of the part is automatically generated, starting from a surface mesh of the mold. At each time step in the problem systems of equations are successively solved. These include:

- the Stokes equations both in the filled area and in the empty area;
- the flow front propagation equation; and
- the temperature balance equations both in the cavity and in the surrounding mold.

Figure 3 shows successive flow front locations during the mold filling of a technical lens (Schneider Company).

1.4 Reactive Extrusion

Corotating twin screw extrusion is a widely used processing technique because it allows the processor to decouple the flow rate and the dissipated energy provided to the polymer material. It is a good mixing device. The residence time distribution can be controlled. Therefore, it is possible to control chemical reactions.

Here the influence of processing conditions and of screw design on the peroxydic degradation of polypropylene is demonstrated. Ludovic software (Berzin *et al.* 2000) allows the prediction of filled and unfilled zones along the screw and computation of pressure and temperature distributions (Figs. 4(a) and 4(b)).

Peroxydic degradation equations are strongly coupled to the temperature balance equation (exothermal reaction) and mechanical balance equation (viscosity is highly molecular mass dependent). Figure 4(c) shows the molecular weight decrease along the machine for different screw speeds.

This example shows that modeling is useful, not only to optimize processing conditions, but also to predict material properties (here, through the weight average molecular weight).

1.5 Short Fiber Orientation in Pipe Extrusion

Introducing short glass fibers into a thermoplastic polymer allows mechanical properties to be markedly increased. These materials are processed in classical tools (e.g., a single screw extruder and a tube die) (Ausias *et al.* 1994).

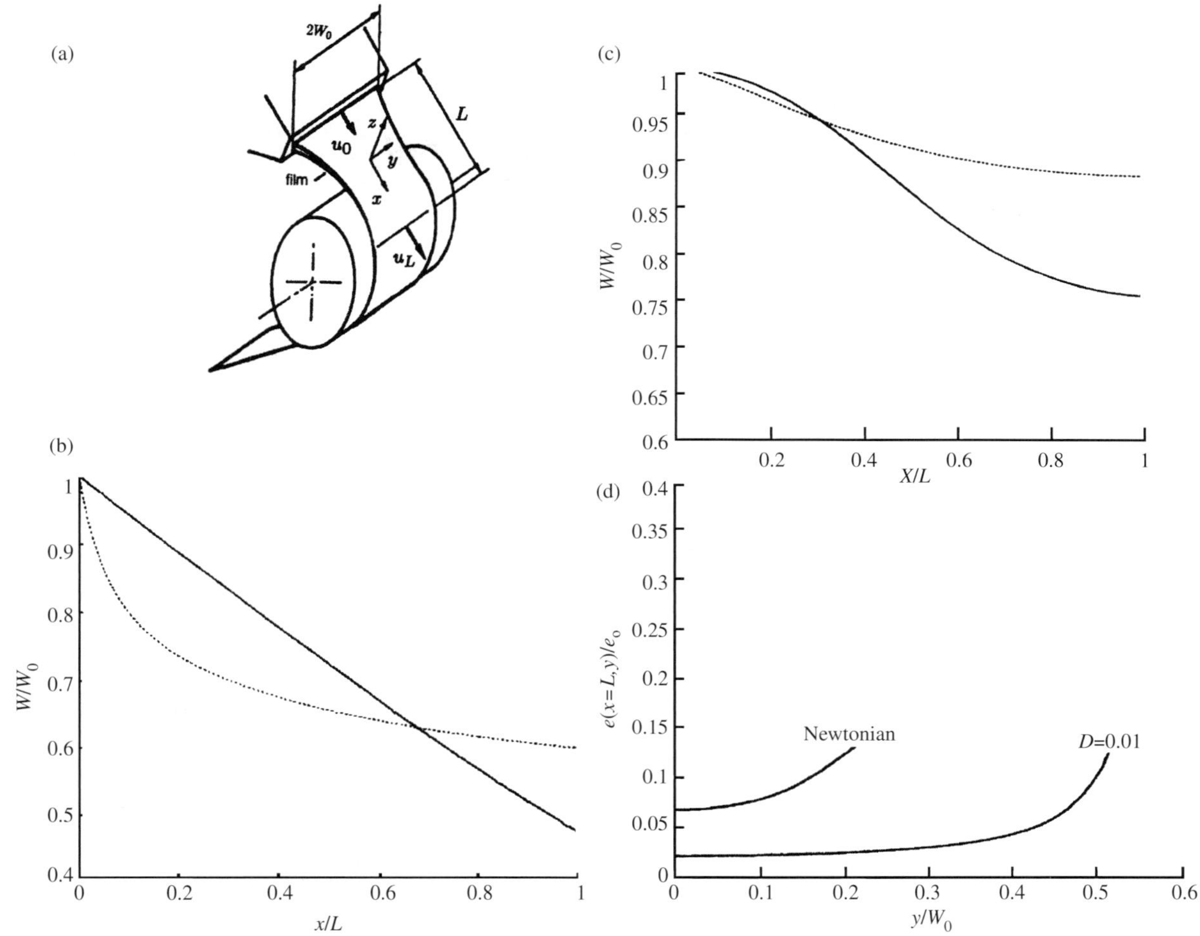

Figure 2
(a) Cast film process (after Silagy *et al.* 1998); (b) 1D model, Newtonian (solid line), viscoelastic (dashed line); (c) Newtonian (solid line), viscoelastic (dashed line); and (d) film thickness distribution on the chill roll.

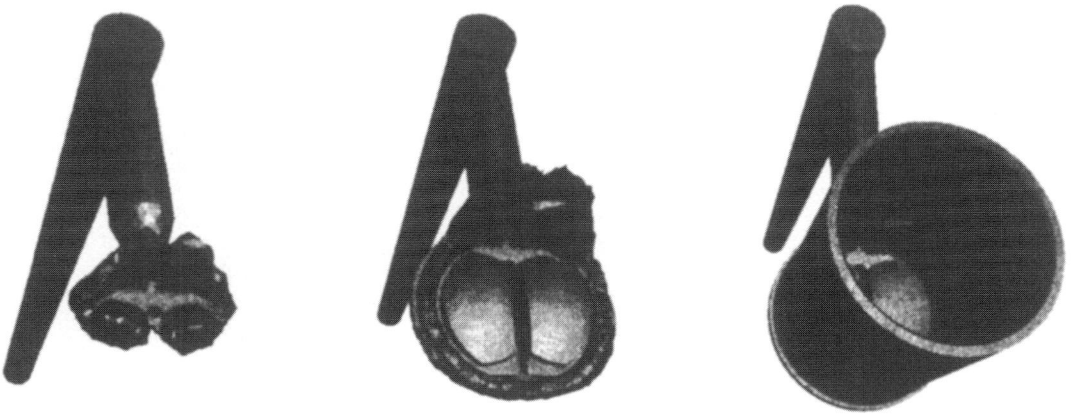

Figure 3
REM3D software. Flow front progression during the filling of a lens mold (courtesy of Schneider).

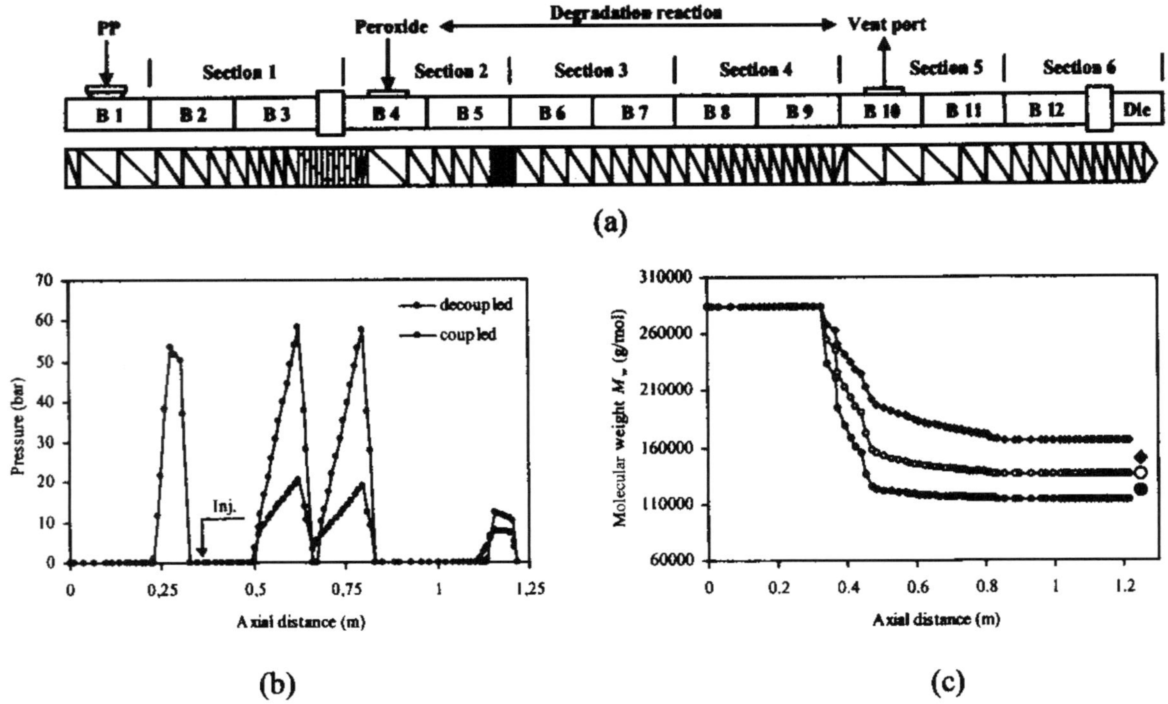

Figure 4
(a) Typical twin screw extrusion design; (b) pressure computation using Ludovic software (coupled mechanical/chemical and decoupled computation); and (c) weight average molecular weight along the machine for different screw rotation velocities.

But, with a classical tube die (Fig. 5(a)) all the fibers become oriented in the flow direction. There is no increase of the tube resistance to an internal pressure. Using a divergent die (Fig. 5(b)) leads to an orientation of a certain fraction of fibers in the hoop which will enhance mechanical properties. To account for these phenomena, it is necessary to introduce a constitutive equation for fiber orientation and an anisotropic rheology equation.

Figure 5(c) shows a fiber orientation calculation in the diverging die geometry: each ellipse represents locally both principal directions and eigen values of the orientation tensor (a circle corresponds to an isotropic orientation). In this case, numerical modeling allows the optimization of die geometry and also the polymer matrix rheology, the fiber content, and the fiber shape factor in order to achieve the required mechanical properties.

2. *Future Challenges*

Computation facilities now allow the calculation of polymer flow in complex geometries and with complex constitutive equations. Approximation methods (slab methods and lubrication approximation methods) remain useful but 3D calculation allows the determination of the degree of precision of these approximations for optimization purposes.

Nevertheless, it would not be reasonable to assume that numerical problems are now behind us. In order to produce useful numerical results it is of prime importance to know the precision and to develop pertinent error estimators. Decreasing computation time is a first order objective and this encourages parallel computation.

Moreover a numerical result is of interest only if it is based on a pertinent physical model. In that area, numerous questions remain open:

- What constitutive equation is to be used? Is it possible to postpone the development of more local viscoelastic constitutive equations?
- What are the real kinematic and temperature boundary conditions (e.g., accurate slip velocity conditions for reinforced polymers or elastomers, heat transfer coefficients, and thermal resistance)?
- What about the fusion and crystallization mechanisms? What kind of plastification mechanisms occur when there are only a few granules in the screw channel of a twin screw extruder? What is the

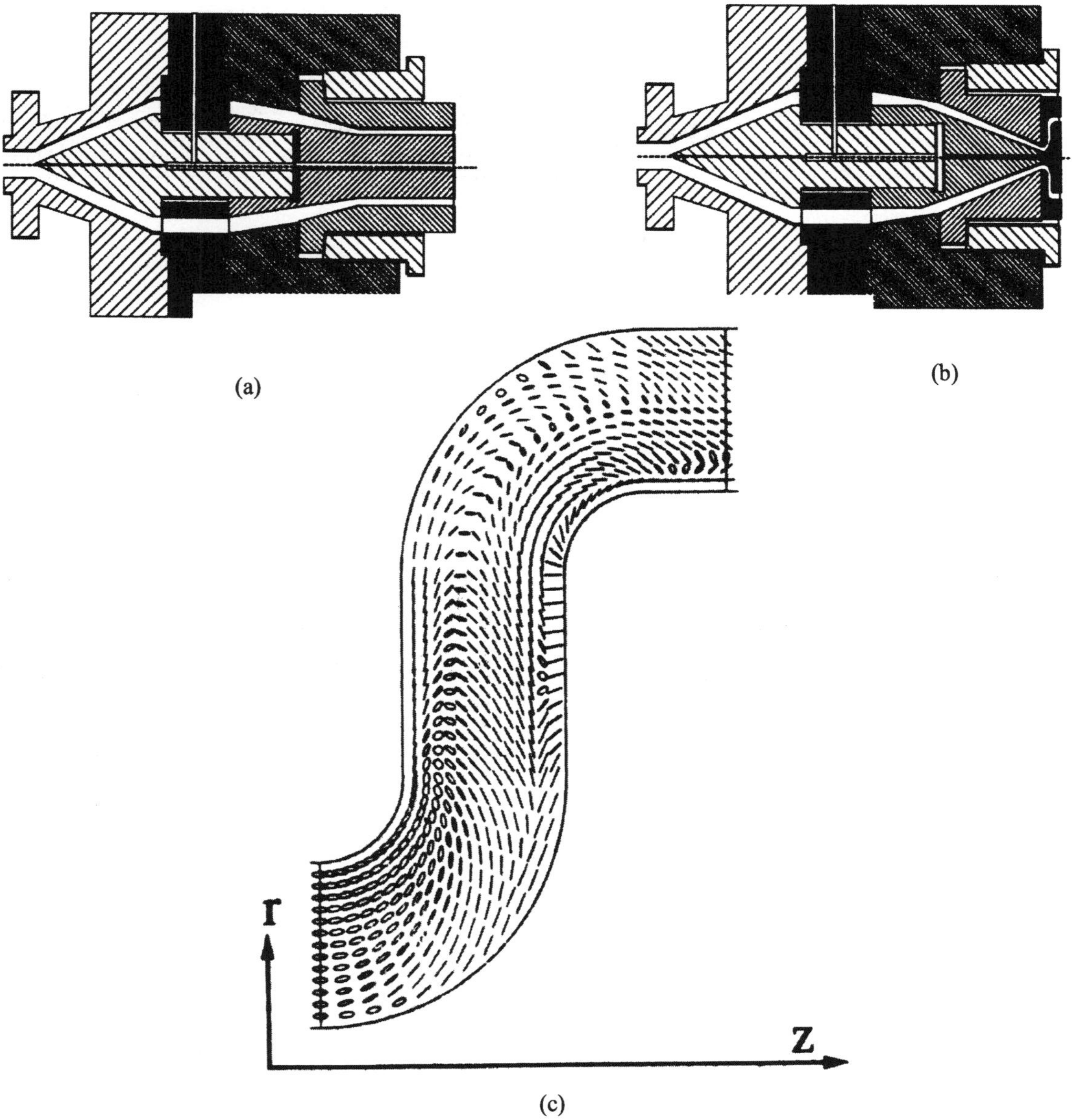

Figure 5
(a) Classical tube die geometry; (b) divergent tube die geometry; and (c) fiber orientation calculation in the divergent tube die geometry.

influence of high temperature gradients, of shear and tensile stresses, and of high pressures on crystallization mechanisms.

In addition, the links between thermomechanical flow parameters, structure and orientation development, and final properties of the produced part remain to be established.

As a conclusion, numerical modeling has proved to be a useful tool but there are numerous challenges to overcome. These can be met successfully by chemical engineers, physicists, mechanical engineers, mathematicians, and materials scientists working together in an interdisciplinary fashion.

See also: Polymer Single Screw Extrusion: Software Packages; Polymer Twin Screw Extrusion: 3D Modeling; Reactive Polymer Processing: Fundamentals of REX; Polymer Film Casting: Modeling

Bibliography

Agassant J F, Avenas P, Sergent J P, Vergnes B, Vincent M 1996 La mise en forme des matières plastiques, Technique et Documentation.

Ausias G, Agassant J F, Vincent M 1994 *Int. Polym. Proc* **9**, 51–60

Berzin F, Vergnes B, Dufossé P, Delamare L 2000 Modeling of peroxide initiated controlled degradation of polypropylene in a twin screw extruder. *Polym. Eng. Sci.* **40**, 344–56

Pearson J R A, Matovich M A 1969 *Eng. Chem. Fund* **8**, 606

Pichelin E, Coupez T 1998 Finite-element solution of the 3D mold filling problem for viscous incompressible fluid. *Comp. Methods Appl. Mech. Eng.* **163**, 359–71

Silagy D, Demay Y, Agassant J F 1998 Stationary and stability analysis of the film casting process. *J. Non-Newtonian Fluid Mech.* **79**, 563–83

Tadmor Z, Klein I 1970 *Engineering Principles of Plasticating Extrusion.* Van Nostrand Reinhold, New York

Vergnes B, Saillard P, Agassant J F 1984 Non-isothermal flow of a molten polymer in a coat-hanger die. *Polym. Eng. Sci.* **24**, 980–7

J. F. Agassant

Polymer Processing: On-line Monitoring

The use of process measurement techniques to characterize materials during processing is increasing with the development of analytical techniques from retrospective, laboratory-based methods to near-real-time process diagnostic and process control tools. The majority of polymers are processed in the melt phase, which will be the focus for this article.

There are three principal reasons for measurements on polymer melts (see Fig. 1 which uses rheological measurements as an example):

scientific understanding of particular features of the material, e.g., studies on how molecular topology influences bulk rheology in process-relevant flows, or the use of spectroscopy to assess molecular feature changes in a polymer during processing;

modeling of real process flows, e.g., development of rheological models and constitutive equations for use in computer modeling of die flows, cavity filling, and validation of process modeling;

process trend monitoring, to evaluate if a manufacturing process is under control, with potential for closed-loop control, by use of specific property- or process-variable measurements.

The range of measurement techniques that can be applied to polymer melts includes: rheometric, optical, ultrasonic, spectroscopic (IR, UV, dielectric, and NMR), which all aim to measure flow properties, polymer structure, morphology and composition, additive concentration, dimensions, appearance, color, and consistency of production. This article will emphasize in-process rheometry, which is the most developed technique (Coates *et al.* 1998a, 1998b, 1999) with brief comment on other techniques, many of which are emerging.

1. *Off-line, On-line, and In-line Measurements*

Off-line measurements are laboratory tests, such as rotational or capillary rheometry, fiber orientation and distribution, etc., which are undertaken on

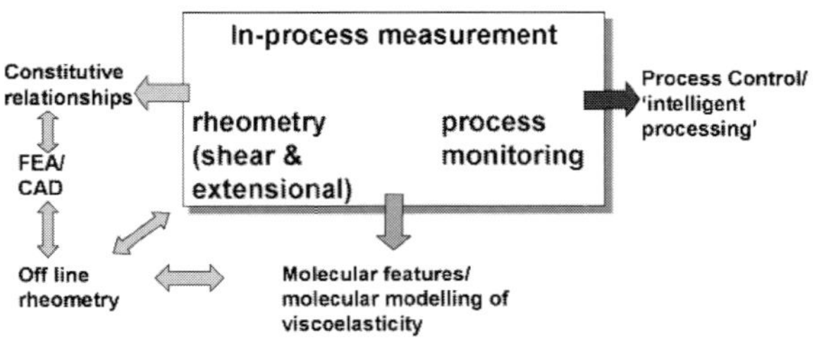

Figure 1
Schematic of the use of in-process rheometry and associated process monitoring; here the rheometry measurements simultaneously provide process monitoring.

samples of polymer away from the processing operation. *On-line* measurements are made on a sample of polymer abstracted from the process stream. These are sometimes also known as "at-process" measurements, and may include versions of standard "off-line" tests. A good example is an on-line capillary rheometer which has several dies Coates *et al.* 1998a). Melt flow is driven by a gear pump which takes melt from the process, typically in an adapter which allows a small "side-stream" flow to the main melt flow. By this means, flow rates can be separately controlled from the process flow, thus providing a melt flow characterization over a range of shear rates. *In-line* measurements are made directly on the process flow stream in appropriate locations in the processing machine (e.g., at the extrusion die or in the nozzle of an injection molding machine).

On-line and in-line measurements may be collectively described as "in-process" measurements. Advantages and disadvantages of in-process measurements are discussed below (see also Table 1).

1.1 The Need for In-process Measurements

Polymers melts are history-dependent materials—current responses depend on stress, strain, strain rate, and temperature history of the melt. Consequently, standard "off-line" (i.e., laboratory) assessments of polymer melt properties may be misleading, particularly if the off-line tests involve significantly different histories from those encountered in processing. For example, melt residence times in a standard capillary rheometer, used to characterize the shear rheology of a melt, may be several minutes (typically more than five minutes) whereas process melt residence times in an extruder are typically around one minute (but may be greater in complex processes). If the polymer is affected physically or chemically by dwell time at a given temperature, (e.g., poly(ethylene terephthalate), poly(vinyl chloride) (PVC), and many engineering polymers) it may be adversely affected during the off-line test so that the material tested is not the same as that processed. Similarly, a filled material may respond differently (e.g., in the distribution of filler with a consequent effect on bulk rheology) to the conditions imposed in off-line tests compared with those experienced in process flows. This does not mean that all off-line tests are not useful; it is, however, clearly important to interpret off-line measurements for a given polymer with care. Further disadvantages of off-line tests, of particular relevance to control of manufacturing processes, are: (i) the representative nature of the test sample—only very small amounts of a material can be measured in practice and this may not be typical of the bulk of the material being processed; and (ii) the total time taken to make measurements—this may typically be of the order of hours, by which time tonnes of material

Table 1
Off-line, on-line, and in-line rheometers.

Issue	Off-line	On-line	In-line
Test time scale (including sample preparation)	Minutes to hours	Minutes	Seconds
Real-time measurement with respect to processing	No to ~hours delay	Close, ~minutes	Possible
Flow curve?	Yes	Yes	Only with wedge type
Typical shear strain rates	10^{-3} to $10\,s^{-1}$ (rotational) 10^{-1} to $10^5\,s^{-1}$ (capillary)	1 to $10^3\,s^{-1}$	1 to $10^6\,s^{-1}$
Relative extent of material sampled	Very small amount	Small amount	Moderate to large amount

whose quality is being sampled may have been processed.

In-process monitoring attempts to address these concerns and to provide an accurate, preferably "real-time," assessment of specific polymer properties, such as shear or extensional rheology, or composition. In addition, in-process monitoring can provide a sensitive indicator of stability of the process, including batch-to-batch variation (Kelly *et al.* 2000), and as such can provide the basis for meaningful statistical process control or other statistical assessments of process variation.

(*a*) *On-line measurements.* On-line measurement techniques in current use include on-line rheometers (OLRs), but all essentially provide shear flow measurements. In addition, assessments of extensional behavior have been inferred from entry pressure measurements (see, e.g., Coates *et al.* 1998b), mid- and near-infrared spectroscopy (Hansen and Vedula 1997) (mainly concerned with molecular features and composition), and UV spectroscopy (mainly for additives) (Herman and Hope 1997).

On-line devices for polymer melts are all used on extruders. They take a sample of the process stream, of necessity from the barrel wall region, though this may consequently be untypical of other regions of the flow (e.g., in terms of additive concentration or other material composition features, and residence time/temperature history). Additional residence time and strain history are necessarily incurred in transferring the melt to the measuring section, normally along a temperature-controlled path via a gear pump, at volumetric flow rates which are typically small compared with those for the main process flow. However, the melt can be "conditioned" at a chosen temperature, and in the case of OLRs, viscometric flows can be obtained, for example, using capillaries. A range of shear rates can then be studied, normally by ramping through gear pump speeds and allowing pressure equilibration at each speed. A typical shear rheometry test with one capillary at extrusion-relevant shear rates (up to $\sim 1000\,s^{-1}$) takes several minutes. This also fixes the maximum rate at which the processing operation can be sampled. A key advantage of obtaining rheology data over a range of shear rates is that changes or checks on melt process stability or polymer batch variations will be more readily discerned. In "bypass" OLRs polymer is discarded; "recycle" types pump the tested melt back into the process. Early work by Rauwendaal and Fernandez (1985) was followed by the use of commercial devices, (see, e.g., Fritz and Stohrer 1988). Extensive OLR studies have now been reported (Coates *et al.* 1998a, 1998b, *Kelly et al.* 1997).

(*b*) *In-line measurements.* In-line measurements appear to be ideal in that sensors (preferably flush-mounted with flow boundaries) monitor melt properties in the actual process flow; essentially "real time" assessment is possible, depending on sampling, signal processing, and data processing requirements. Current in-line devices are used for rheometric, optical (including colour and composition), dielectric, ultrasonic, near-infrared, UV, and conductance measurements. These are typically "single-point" measurements, fixed by the process operating condition (e.g., an in-line rheometer (ILR) normally operates at a single wall shear rate Coates *et al.* 1998a, 1998b, 1999), rather than measurements over a range of conditions. However, a significant proportion of the main process flow passes the sensors, i.e., a far greater proportion of melt is sampled than by on-line techniques.

Melt temperature fields are not controlled—this can be a significant problem in ILRs at higher flow rates; also, a typical manufacturing process may not have a readily available process geometry section which is suitable for in-line rheometry. The volumetric flow rate required for rheometry calculations has to be inferred from melt density at certain process temperatures and measured mass flow rate in extrusion; in injection-molding it is obtained from the injection rate. Nozzle ILRs are now preferred for the determination of shear flow properties for use in computer modeling of injection molding.

Attempts to enhance in-line devices include wedge geometries to provide a range of shear rates over which measurements are taken, e.g., a static version (Pabedinskas *et al.* 1991) and a rotating cam with one or more wedges (McBrearty and Tadros 2000) which can offer a wider range of shear rates.

Figure 2 shows a comparison of off-line, on-line, and in-line shear flow data generated for: (c) a linear low-density polyethylene (LLDPE); (a) an unplasticized PVC; and (b) a polyamide-6 (PA-6). The LLDPE shows little effect of processing, exhibiting very good agreement between results for all three techniques, whereas both the UPVC and PA6 are clearly process-affected materials.

Exciting possibilities are currently being developed for optical (Migler and Bur 1998), dielectric (McBrearty *et al.* 2000), ultrasound (Piche 1998, Brown *et al.* 2000), and near-infrared (Khettry and Hansen 1996) in-process measurements. Other prospects being explored include Raman in-process assessment of orientation, NMR velocity field imaging, radio-tracers for mixing and flow-rate sensing, and process tomography, e.g., using ultrasound. Combinations of measuring techniques will often be required for for effective process monitoring. Also, as computer power increases and process models are better validated, a "soft sensor" approach, which combines one or more in-process measurements with computer modeling of the chosen process variables, will provide a fuller assessment of field variables.

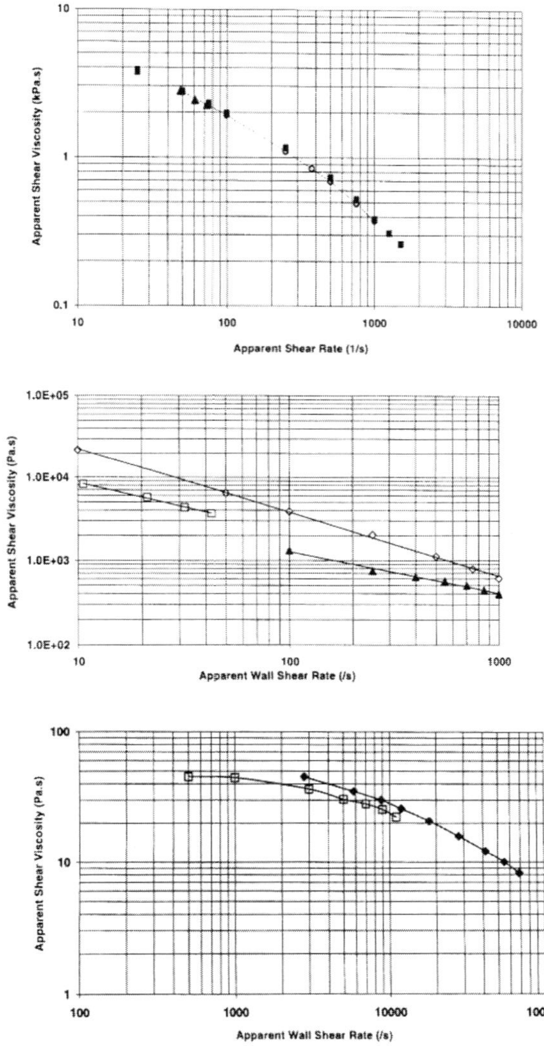

Figure 2
Shear viscosity of LLDPE at 190 °C using off-line capillary rheometry (■), on line capillary rheometry (○) and extrusion slit die in line (▲); shear viscosity of UPVC at 175 °C set temperature, from off-line capillary (◇), on-line capillary (▲), and extrusion in line slit (□) rheometry; shear viscosity of a polyamide-6 at 270 °C from off-line capillary rheometry (□) and injection nozzle ILR (♦).

Bibliography

Brown E C, Olley P, Coates P D 2000 In line melt temperature measurement during real time ultrasound monitoring of single screw extrusion. *Plast. Rubb. Composites* **29**, 3–13

Coates P D, Kelly A L, Martyn M T, Rose R M, Woodhead M 1999 *In-process Extensional Flow Measurements in Injection Moulding and Extrusion*. Bradford University, Bradford, UK

Coates P D, Kelly A L, Rose R M, Woodhead M 1998 *In-Process Shear Flow Measurements in Extrusion and Injection Moulding*. Bradford University, Bradford, UK

Coates P D, Kelly A L, Rose R M, Woodhead M 1998 *In-Process Entry Pressure Measurements in Extrusion and Injection Moulding*. Bradford University, Bradford, UK

Fritz H G, Stohrer B 1988 Polymer compounding process for controlled peroxide degradation of polypropylene. *Int. Polym. Process.* **1**, 31–40

Hansen M G, Vedula S 1997 In-line Measurement of Copolymer Composition and Melt Index. In: Coates P D (ed.) *Polymer Process Engineering '97*. Institute of Materials, London, pp. 89–102

Herman H, Hope P S 1997 Illuminating the Manufacturing Process—On line UV Spectroscopic Analysis. In: Coates P D (ed.) *Polymer Process Engineering '97*. Institute of Materials, London, pp. 65–77

Kelly A L, Coates P D, Dobbie T W, Fleming D J 1997 On-line rheometry: shear & extensional flows. *Plast. Rubb. Comp. Process. Applicn.* **25**, 313–8

Kelly A L, Woodhead M, Rose R M, Coates P D 2000 In process monitoring of polymer batch to batch variation in injection moulding. *Plast. Rubb. Comp.* **29**, 23–30

Khettry A, Hansen M G 1996 *Polym. Eng. Sci.* **36**, 1232
McBrearty M, Bur A J, Roth S 2000 Real time dielectric measurements during extrusion of filled polymers. *SPE Tech. Papers* **46**, 287–91
McBrearty M, Tadros S 2000 Novel in-line rheometer for polymer melts, compounds and solutions. *SPE Tech. Papers* **46**, 993–7
Migler K B, Bur A J 1998 Fluorescence based measurement of temperature profiles during polymer processing. *Polym. Eng. Sci.* **38**, 213–21
Pabedinskas A, Cluett W R, Balke S 1991 Development of an in-line rheometer suitable for reactive extrusion process. *Polym. Eng. Sci.* **31**, 305–75
Piche L 1998 Ultrasonic probe for on-line process monitoring. *SPE Tech. Papers* **44**, 3617
Rauwendaal C J, Fernandez F 1985 Experimental study and analysis of a slit die viscometer. *Polym. Eng. Sci.* **25**, 765–71

P. D. Coates

Polymer Processing, Process Control of

In this article, a survey of the possibilities of controlling the film thickness distribution, which is one of the most important quality characteristics of extruded films, is given as an example for process control of polymer processing.

1. Film Extrusion

The production of plastic film is one of the key processing methods employed for polymers. Major applications are found in packaging, information media and agriculture. Plastic film is essentially produced by flat and by blown film extrusion.

In the case of flat film extrusion, the polymer is formed into a flat molten web by a so-called slit die, which has the function of distributing the melt in such a way that it emerges from the die with an even velocity distribution. The melt is then deposited on a roll. In the chill-roll process this is a large-diameter cooling roll, followed by a number of smaller rolls. In the process with a calender, the melt is fed between two identical-diameter rolls with the roll gap set in such a way that a bank forms above it. Which of the processes is employed will depend primarily on the desired gauge of film (calender: films of 100 μm and above; chill-roll: films less than 100 μm thick).

With blown film extrusion the melt is formed into a horizontal tube, which is inflated with the aid of air. The melt is drawn upwards, laid flat by the appropriate guide mechanisms, fed through squeeze rolls, and finally rolled up. A volume of air is enclosed in the tube in the section between the take-off rolls and the extrusion die, so the tube diameter and hence the film width can be readily adjusted via the quantity of support air.

2. Thickness Profile Control in Flat Film Extrusion

2.1 Reasons for Thickness Fluctuations

The design of slit dies can generally only be optimized for an individual material or for materials which display similar viscosity behavior. If these dies are used for materials that differ greatly in rheological terms, a different distribution effect will result. Batch fluctuations, together with the accompanying fluctuations in viscosity, also make themselves felt here. Temperature inhomogenities and uneven heating inside the die have similar effects, influencing the viscosity and hence the melt distribution. While die layout should essentially be conducted independently of the operating point, it is not possible to completely eliminate differences in the distribution effect at different operating points.

A further feature of flat film production that needs to be taken into consideration is the fact that the melt is subject to a considerable amount of strain at right angles to the direction of extrusion while it is in the die. Plastic melts remember this deformation by virtue of their viscoelastic "memory" and, when they emerge from the die, experience a sudden reduction in width, the so-called neck-in, leading to a narrower film web. As a result, a pronounced thick point results in the edge zones, which has to be cut off once the film has cooled.

To illustrate the thickness distribution of flat film which has been produced without profile control, Fig. 1 shows the deviation from the mean film thickness over the width of the film. The influence of the neck-in at the edge of the film is clear to see while, in the center of the film, deviations of $\pm 4\%$ are to be found. An important factor to bear in mind when assessing this figure is that the thickness distribution results from all the influencing variables referred to above.

2.2 Measuring Film Thickness in Flat Film Extrusion

In order to smooth out the deviations from the mean value in the film thickness, it is first of all necessary to

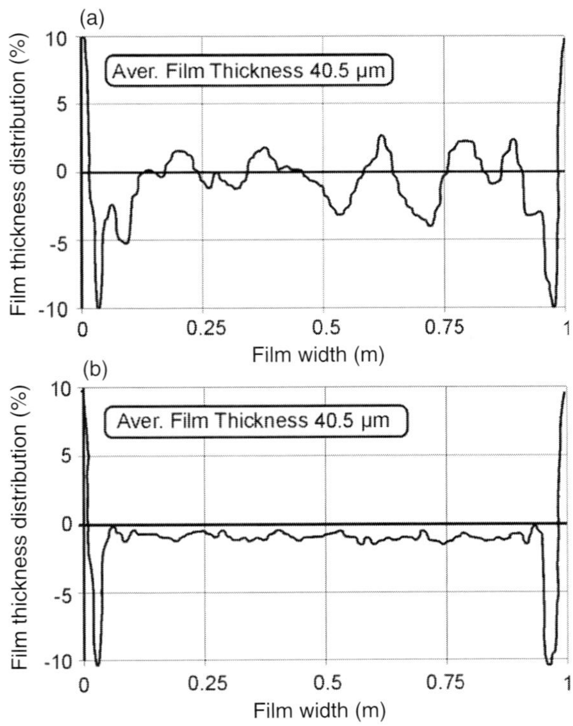

Figure 1
Thickness distribution of a flat film: (a) without thickness profile control; (b) with thickness profile control.

measure these deviations. Different measuring systems are available:

(*a*) *Distance measurement.* This contact measuring method uses inductive displacement sensing devices. The film passes over a roll and has a measuring wheel running over it from above. If the measuring wheel passes over a thick point it moves slightly upwards, and the sensor records the displacement. In doing this, the sensor travels over the entire width of the film. The main drawback to this system is that it is a contact system, which can cause surface defects on sensitive films.

(*b*) *Radiometry.* This involves the mass per unit area (the product of the density of the polymer and the thickness of the film) being measured by means of beta radiation, which is absorbed by the film as a function of the weight per unit area (see Fig. 2). The degree of transmission decreases exponentially with an increasing mass per unit area, while the amount of radiation absorbed by the measured product increases. It is clear that the radiation will be absorbed to a differing extent as a function of the emitter type, and hence the individual emitters can be used in different measuring ranges. If it is assumed that density of the film remains approximately constant during the measurement, the thickness of the film can be calculated directly. The particular advantages of the radiometric measurement are the robust design of the sensor, which will operate reliably under difficult environmental conditions, and the fact that this is a noncontact measurement. Resolutions of 0.1 μm are achieved with radiometric thickness analysis systems.

(*c*) *Infrared absorption.* This measurement principle also exploits the absorption properties of polymers, this time in the infrared light range. The top half of Fig. 3 shows the correlation between the degree of transmission of two polyethylene films of different thickness over the wavelength of light. A pronounced fall in the degree of transmission is seen at wavelengths of approximately 1.73 μm and 2.38 μm (with high degrees of absorption at the same time).

The lower half of Fig. 3 shows the measurement set-up for transmission through film. Infrared radiation is generated by a lamp, and this radiation is

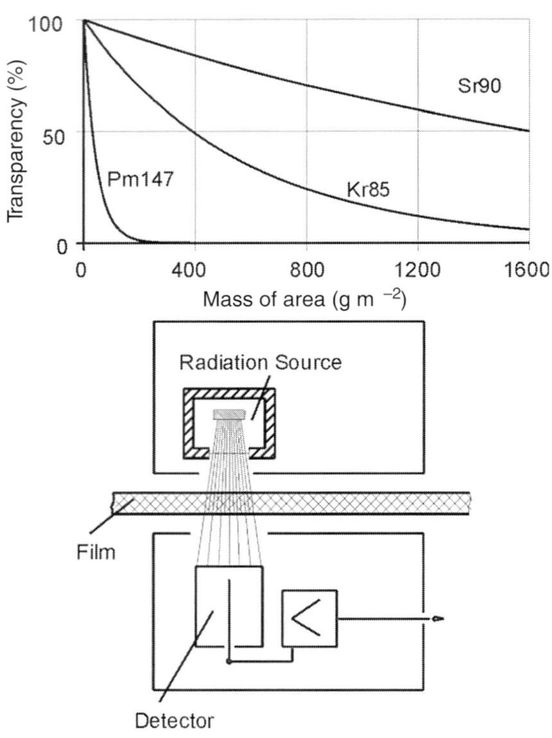

Figure 2
Radiometric measurement of film thickness.

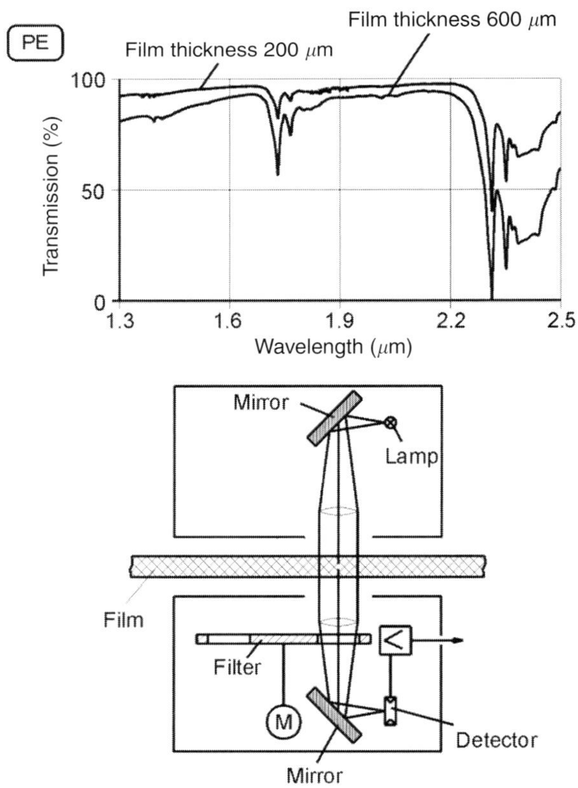

Figure 3
IR measurement of film thickness.

then reflected off a mirror and through the film to be measured. The radiation that is not absorbed is filtered, so that only the radiation components in a narrow wavelength range (e.g. 1.73 μm) can pass through. These are finally picked up by a detector which records the intensity of the radiation.

The intensity of the radiation recorded can be converted directly into a film thickness. Different filters can be employed for the measurement, permitting measurements to be conducted in different wavelength ranges. In this way, it is possible to register the influence of soiling and correct the measuring result. A further advantage is that this is a noncontact measurement. The measuring range of sensors of this type covers films in gauges of up to 10 μm, whereby a resolution of 0.5 μm can be achieved.

2.3 *Film Thickness Control in Flat Film Extrusion*

The thickness profiles measured with the aid of the methods referred to above are not only used for quality documentation, but also constitute the input variables for the thickness profile control system provided on a large number of plants. On flat film plants, the thickness profile is influenced via the flexible lip at the exit from the die, which can be adjusted locally by means of adjusting elements, thereby influencing the gap width. The manufacturers of dies of this type all offer the same concept, but with different forms of adjustment for the flexible lip.

A distinction is essentially drawn between three adjustment principles. Thermal expansion bolts occasion the lowest investment costs, but they consume a great deal of energy. The response time of thermal expansion bolts is in the range of 5 to 15 min, which has a corresponding impact on the control characteristics. Piezo translators display considerably more favorable properties, since their response time is in the millisecond range. The enormously high investment costs constitute a drawback; these are due primarily to the elaborate thermal separation required between the die and the piezo translator, since the translators may only be exposed to temperatures up to a maximum of 100°C. Adjusting elements with electric motors have recently been brought on to the market, which similarly have a short response time. Compared with the other two adjustment mechanisms, the considerably greater adjustment displacement constitutes an advantage here, making it possible to dispense with the elaborate pre-adjustment required for the other elements.

Figure 1, lower part, shows the way in which the thickness profile control system works. While the influence of the neck-in has not been reduced by comparison to the uncontrolled state, a largely balanced thickness profile is evident in the center of the web, with only slight fluctuations still. Leaving out the edge sections of the film, the deviations are approximately 0.5%.

3. *Thickness Profile Control in Blown Film Extrusion*

3.1 *Reasons for Thickness Fluctuations*

On tubular film plants too, the measurement and control of the film thickness constitutes the chief means of process monitoring and automation. The chief points that could be responsible for uneven thickness distribution around the circumference of the tubular film will thus be looked at below.

To begin with, mention must be made of inaccuracies in the distribution of melt in the die. Superimposed on this effect are production inaccuracies, such as non-circularity of the mandrel or the die, and also batch fluctuations, and hence viscosity fluctuations, in the mold. It is also possible for the melt to display an uneven temperature around its circumference if it has not been sufficiently homogenized in the extruder, or if the die is not uniformly heated, and this can lead to a changed distribution effect.

These inhomogenities influence the thickness of the film produced, since the melt has a different energy content here, and this energy has to be removed through air cooling. If it is first assumed that uniform cooling prevails over the circumference, a longer period of time is required to freeze the melt at points with a higher energy content. This means that the melt can be expanded for a longer period of time. It is now very much a function of the boundary conditions as to whether the film displays a thick point or a thin point in this case. A further key influencing factor is the cooling air flow. From the considerations set out above, it is clear that any unevenness in the cooling has a direct influence on the film thickness. A more pronounced cooling effect in certain areas will lead to more rapid melt freezing and hence to a thick point.

Figure 4 shows the thickness distribution of tubular film that has been produced without any profile control. It is clear to see that the die is not properly centered, leading to an excessively thin film between 0 and 180°, and an excessively thick film between 180 and 360°. The thickness fluctuations here are in the range of ±5%. Here too, it must be borne in mind that this thickness distribution is the result of all the influences set out above.

3.2 *Measuring the Film Thickness in Blown Film Extrusion*

In blown film extrusion it is not possible to measure the film thickness with the methods presented for flat film extrusion, since the film has to be accessible from both sides for these systems. For this reason, virtually

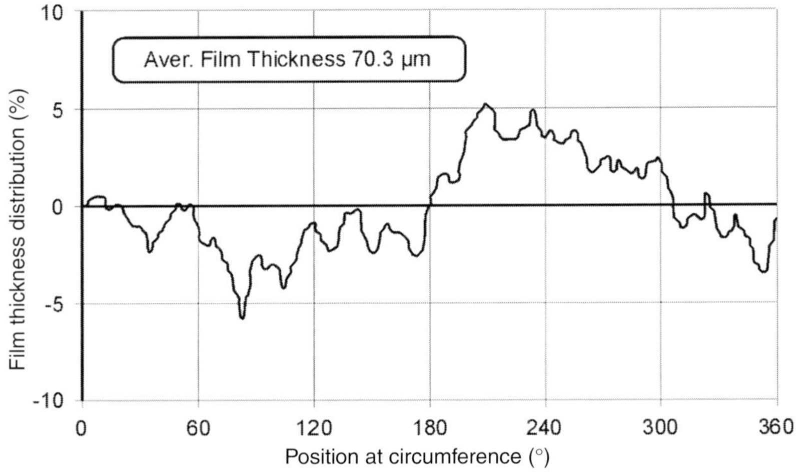

Figure 4
Thickness distribution of an uncontrolled blown film.

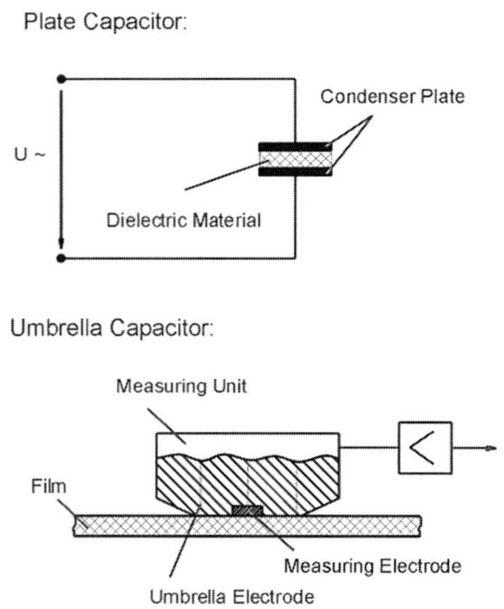

Figure 5
Capacitive measurement of film thickness.

exclusive use is made of capacitive thickness measuring systems on blown film plants, which rotate around the film tube.

Capacitive thickness measurement is based on the change in relative permittivity in the field of a measuring capacitor, which is brought about by thickness fluctuations in the film. Since the relative permittivity differs from one polymer to another, the measuring system has to be calibrated at the start of the measurement. Figure 5 shows the structure of a standard commercial guard ring capacitor, of the type generally employed for blown film extrusion, compared to a plate capacitor.

As with the plate capacitor, an electric voltage is applied between the shield and the measuring electrode, which forms a field into which the film is placed. In the case of homogeneous materials, the change in capacity in this capacitor constitutes a measure of the change in thickness, providing that the other test conditions are kept constant. Since the plastic material has a dielectric constant two to four times that of air, the capacity of the measuring capacitor changes with the thickness of the film.

The advantages of this measuring method are the very high response speed of the measuring sensor, and the possibility that exists for measuring very thin films. Changes in the dielectric constant of the polymer material have an influence on the measuring accuracy. These can be caused by temperature changes in the film, or through coloring agents and additives. Films up to some 5 mm thick can be measured with capacitive measuring sensors, with a resolution of approximately 0.1 µm.

3.3 Film Thickness Control in Blown Film Extrusion

To improve the thickness profile, profile control systems have been used on tubular film plants for monofilms since 1986, and on coextrusion plants since 1988. An essential distinction can be drawn between systems that have an influence on the melt flow in the die, and systems which are used to change the cooling of the tubular film on a localized basis. The processes employed today essentially involve increasing or lowering the local heat content of the film. If the heat content is increased, the film remains expandable for longer and a thinner film is produced at this point. In the case of the systems that have an influence inside the die, it is also necessary to take the modified distribution effect into account. A number of the processes available on the market are presented in greater detail below:

(*a*) *Mechanical principle with automatic dies.* Taking the findings from flat film extrusion, a concept has been developed for altering the gap width on a local basis by means of a flexible lip on the outer edge of the orifice. The force required to alter the shape of the outer lip was applied via geared motors with eccentric disks. One problem encountered with this system was that an adjustment at one point of the circumference had an impact at a different point. This system is thus no longer supplied.

(*b*) *Thermal principle with automatic dies.* In another concept, the temperature of the melt is influenced already in the die. The temperature and the local melt flow of the emerging melt are influenced directly beneath the melt output point via a local temperature control system for the orifice. The cartridge heaters used for this are in a position to heat the melt by a maximum of +3 K. The system cannot be retrofitted to existing plants, on account of the highly elaborate die design involved.

(*c*) *Thermal principle with segmented cooling rings.* In addition to this, systems with a cooling ring segmented by means of radial separator strips are available. These segments contain heating spirals with a maximum heating capacity of approximately 300 W, which lead to a temperature increase of 12 K maximum in the partial air flow. With this system, therefore, a local increase in the cooling air temperature causes the film to be cooled to a lesser extent, so that it freezes later and thus remains expandable for a longer period of time. If a thick point is detected by the thickness measuring system, the cooling air is heated at the appropriate point in order to reduce the film thickness there.

(*d*) *Local change in the main flow of cooling air.* On these systems, the main flow of cooling air is influenced on a segment-by-segment basis, so that more cooling air flows out of the segment in question, thereby ensuring that the melt freezes more rapidly. The period of time over which the film can be expanded is then shortened, leading to a direct increase in the thickness of the film. A segmentation of the internal cooling system is also implemented in a number of cases.

(*e*) *Segmented cooling ring with additional cooling air.* It is also possible to make use of additional cooling air. A twin-chamber cooling ring is employed in this case. The cooling air is fed to the upper main cooling lip and the lower control cooling lip in the correct proportions by means of a system of flaps. The lower cooling ring lip, which is of a segmented design, is equipped with adjusting elements that can be moved by a motor to change the channel cross-section, and hence influence the volume of cooling air in the segment. The upper cooling ring lip supplies the main cooling air; this rests on three adjusting elements over its circumference. These adjusting elements can be used to modify the basic air distribution in the main cooling air system, so as to balance out centering errors in the die and/or the cooling ring.

4. Conclusion

Film thickness control systems are available for flat film as well as for blown film extrusion lines. In spite of their different working principles, all systems are able to reduce the thickness differences in width direction down to ±1% of the average film thickness.

Bibliography

Feistkorn W 1988 Automatic blown film dies for high quality film. In: *Proc. 46th SPE ANTEC*. Society of Plastics Engineers. pp. 176–8

Feistkorn W 1989 Blasfolienextrusion—Heute besser und wirtschaftlicher denn je [Advances in blown film extrusion]. *Kunstst. Plast Eur.* **79**(11), 1178–81

Hensen F, Knappe W, Potente H 1986 *Kunststsstoff-Extrusionstechnik* [*Polymer Extrusion Technology*]. Hanser, Munich, Germany

Nentwig J 1994 *Kunststoff Folien* [*Plastic Films*]. Hanser, Munich, Germany

Spang A, Wüstenberg D 1993 On-line-Dickenmessung [On-line thickness measurement]. *Kunststoffe Plast Europe* **83**(11), 894–7

Spang A, Wüstenberg D 1994 Berührungslose On-line-Dickenmessung [Contactless on-line thickness measurement]. *Kunstst. Plast Eur.* **84**(4), 412–22

Wolf D 1996 Analysis of air cooling systems to increase cooling power and film quality of blown film extrusion lines. Ph.D. thesis, Universität Essen

E. W. Michaeli and J. Hauck

Polymer Rheology for Melt Processing: Complex Flows

The rheological behavior of polymer melts plays an important role during processing since it not only determines the extra stresses and pressure gradients and can reduce flow instabilities that yield process limits, but it also couples material, process, and production properties. Examples of (final) product properties determined by the process are frozen-in orientation, the structure of semicrystalline polymers, and long-term dimensional stability. For quantitative predictions of these qualities, the availability of reliable constitutive models is a prerequisite. Here, the performance of a new generation of constitutive differential equations (Verbeeten *et al.* 2000) based on the Pompom model (McLeish and Larson 1998) is presented.

1. Introduction

In recent years, significant progress has been made in the experimental and numerical analysis of viscoelastic polymer melt flow. From the numerical point of view, a number of well-established, mixed finite element formulations has been developed that are sufficiently stable, robust, and efficient to analyze a range of two- and three-dimensional flow problems. Typical examples are the so-called DEVSS/DG and DEVSS/SUPG methods. Experimentally, an increasing variety of flow geometries has become available, satisfying requirements such as near two-dimensionality of the flow (i.e., a depth-to-height aspect ratio of at least 8 in the whole flow domain). The measurement resolution of fringe patterns has improved significantly by the application of lasers (Schoonen 1998) such that accurate measurements can be made at critical locations in the geometry. Furthermore, the use of Mueller calculus allows the prediction of the observable fringe patterns in three-dimensional flow geometries (Bogaerds *et al.* 1999b).

Crucial to the quantitative prediction of velocity and stress fields in inhomogeneous flows is the availability of reliable constitutive models. Within the class of differential models, the multimode versions of the Giesekus and PTT model have been the most widely applied. A new generation of models (Verbeeten *et al.* 2000) has been developed based on the concepts of the Pompom model (McLeish and Larson 1998). These Pompom-based models are a major step forward in describing the rheology of polyethylene melts.

The objective of this contribution is to investigate the performance of an enhanced, single-equation version of the Pompom model in predicting the response to a reversed shear flow, and predicting the inhomogeneous flow of a low-density polyethylene (LDPE) melt in the cross-slot geometry. This particular version quantitatively predicts all available rheometrical shear and elongation data of a set of polymer melts (see Verbeeten *et al.* 2000).

2. Enhanced Pompom Model

To describe stresses of polymer melts, the Cauchy stress tensor σ is decomposed as:

$$\sigma = -pI + 2\eta_s D + \sum_{i=1}^{M} \tau_i \tag{1}$$

Here, p is the pressure term, I the unit tensor, η_s denotes the viscosity of a purely viscous component, $D=\frac{1}{2}(L+L^T)$ the rate of deformation tensor, in which $L=(\vec{\nabla}\vec{u})^T$ is the velocity gradient tensor, and $(\cdot)^T$ denotes the transpose of a tensor. The viscoelastic contribution of the ith relaxation mode is denoted by τ_i and M is the total number of different modes. A multimode approximation of the relaxation spectrum is usually necessary for a realistic description of the viscoelastic contributions.

Here, the constitutive behavior of the viscoelastic contribution is described with the differential Pompom model. This model is (mainly) developed for long-chain branched polymers. The multiple-branched molecule can be broken down to the several individual modes (Inkson *et al.* 1999). Every mode is represented by a backbone between two branch points, with on every end a number of dangling arms. The backbone is confined by a tube formed by other backbones. The original differential form by McLeish and Larson (1998), improved with local branch point displacement (Blackwell *et al.* 2000), is written in two decoupled equations and reads as follows:

$$\overset{\triangledown}{\mathbf{A}} + \frac{1}{\lambda_{0b}}\left[\mathbf{A} - \tfrac{1}{3}\mathbf{I}\right] = 0, \quad \mathbf{S} = \frac{\mathbf{A}}{I_A} \tag{2}$$

$$\dot{\Lambda} = \Lambda[\mathbf{D}:\mathbf{S}] - \frac{1}{\lambda_s}(\Lambda - 1) \tag{3}$$

$$\lambda_s = \lambda_{0s} e^{-\nu(\Lambda - 1)} \quad \forall \Lambda \leqslant q \tag{4}$$

$$\tau = S - G_0 I = G_0(3\Lambda^2 S - I) \tag{5}$$

The expression (5) for the extra stress differs (by a constant) from that proposed in McLeish and Larson (1998), but in Rubio and Wagner (1999), it has been shown that for the differential model, this is the correct form. Equations (2) and (3) are the evolution of orientation tensor $\mathbf{S}$ and backbone stretch Λ, respectively. $\mathbf{A}$ is an auxiliary tensor to get the backbone orientation tensor $\mathbf{S}$. The upper convected time derivative of the auxiliary tensor **A** is defined as:

$$\overset{\nabla}{\mathbf{A}} = \frac{\partial \mathbf{A}}{\partial t} + \vec{u} \cdot \vec{\nabla} \mathbf{A} - \mathbf{L} \cdot \mathbf{A} - \mathbf{A} \cdot \mathbf{L}^T \qquad (6)$$

The parameter λ_{0b} is the relaxation time of the backbone orientation. It is obtained from the linear relaxation spectrum determined by dynamic measurements. I_A is the first invariant of tensor **A**, defined as $I_A = \mathrm{tr}(\mathbf{A})$. The backbone stretch Λ is defined as the length of the backbone tube divided by the length at equilibrium. λ_{0s} is the relaxation time for the stretch and v a parameter which, based on the ideas of Blackwell *et al.* (2000), is taken to be $2/q$, where q is the amount of arms at the end of a backbone. Finally, G_0 is the plateau modulus, also obtained from the linear relaxation spectrum. Finite extensibility of the backbone is introduced by Eqn. (3), which only holds if the stretch Λ is smaller or equal to the number of dangling arms q.

The expression (2) has a similar structure as the upper convected Maxwell (UCM) model, and becomes unbounded if $\dot{\varepsilon}\lambda_{0b} \geqslant 1$. To overcome this problem and to reduce the set of equations, the orientation and stretch equations will be written as a single equation by using Eqn. (5). Therefore, Eqn. (2) is rewritten in terms of S:

$$\overset{\nabla}{\mathbf{S}} + 2[\mathbf{D} : \mathbf{S}]\mathbf{S} + \frac{1}{\lambda_{0b} I_A}\left[\mathbf{S} - \tfrac{1}{3} I\right] = 0 \qquad (7)$$

An expression is needed for I_A to get a full coupling between the stretch and orientation. Introduce the following definitions:

$$\mathbf{A} = \langle \vec{x}\,\vec{x} \rangle, \quad \mathbf{S} = \langle \vec{u}\,\vec{u} \rangle \qquad (8)$$

where $\vec{x}$ is a vector of unknown dimensionless length, representing a part of the backbone, and $\vec{u}$ is the unit vector. Assuming that the local vector $\vec{x}_i$ has a length of the averaged stretch $\Lambda = \langle \Lambda_i \rangle (\vec{x}_i = \Lambda \vec{u}_i)$, where Λ_i is the local stretch of a part of the backbone, I_A can be coupled with the stretch:

$$\mathbf{A} = \langle \vec{x}\,\vec{x} \rangle = \Lambda^2 \langle \vec{u}\,\vec{u} \rangle = \Lambda^2 \mathbf{S} = I_A \mathbf{S} \Rightarrow I_A = \Lambda^2 \qquad (9)$$

A second aspect of the Pompom model written in the form of equations (2) and (3) that needs improvement is zero second normal stress coefficient ($\Psi_2 = 0$) prediction by this form. To overcome this, the orientation equation (7) is modified to invoke a second normal stress coefficient. Instead of a UCM type, a Giesekus-like evolution equation for the orientation is proposed. Making use of \expression (9), the equation now reads:

$$\overset{\nabla}{\mathbf{S}} + 2[\mathbf{D} : \mathbf{S}]\mathbf{S} + \frac{1}{\lambda_{0b}\Lambda^2}\left[3\alpha\Lambda^4 \mathbf{S} \cdot \mathbf{S} + (1 - \alpha - 3\alpha\Lambda^4 \mathbf{I}_{\mathbf{S}\cdot\mathbf{S}})\mathbf{S} - \frac{1-\alpha}{3}\mathbf{I}\right] = 0$$

Here, α is a material parameter, defining the amount of anisotropy is molecular friction and Brownian motion. If $\alpha = 0$, the orientation equation simplifies to the UCM-type orientation equation (7). For nonzero α, also a nonzero second normal stress coefficient Ψ_2 is predicted. Moreover, Ψ_2 is proportional to α.

By taking the derivative of Eqn. (5):

$$\dot{\tau} = \dot{S} - G_0 \dot{I} = 6G_0 \Lambda \dot{\Lambda} S + 3G_0 \Lambda^2 \dot{S} \qquad (10)$$

and using Eqns. (3) and (2), it can all be written in a single equation for τ. This is the enhanced Pompom (EPP) model:

$$\overset{\nabla}{\tau} + \lambda(\tau)^{-1} \cdot \tau + \frac{G_0}{\lambda_{0b}}[f(\tau)^{-1} - 1]I = 2G_0 D \qquad (11)$$

with

$$\lambda(\tau)^{-1} = \frac{1}{\lambda_{0b}}\left[\frac{\alpha}{G_0}\tau + f(\tau)^{-1} I\right] \qquad (12)$$

$$\frac{1}{\lambda_{0b}} f(\tau)^{-1} = \frac{2}{\lambda_s}\left(1 - \frac{1}{\Lambda}\right) + \frac{1}{\lambda_{0b}}\left[\frac{1}{\Lambda^2} - \frac{\alpha I \tau \cdot \tau}{3G_0^2 \Lambda^2}\right] \qquad (13)$$

and

$$\Lambda = \sqrt{1 + \frac{I\tau}{3G_0}}, \quad \lambda_s = \lambda_{0s} e^{-v(\Lambda - 1)}, \quad v = \frac{2}{q} \qquad (14)$$

3. Computational Method

Numerous numerical formulations have been proposed for the analysis of viscoelastic fluid flow, see Baaijens (1998b) for a review. The two most frequently used mixed finite element methods are both based on the DEVSS (Discrete Elastic–Viscous Stress Splitting) formulation, first introduced by Guénette and Fortin (1995). If a continuous stress interpolation is applied, the SUPG method is generally applied to handle the convective terms in the constitutive model, giving the DEVSS/SUPG method, while the discontinuous Galerkin (DG) method is the method of choice for discontinuous stress interpolations, giving the DEVSS/DG method. If a separate interpolation of the velocity gradient tensor $L=(\nabla^{\rightarrow}u^{\rightarrow})^T$ is used in the constitutive equation, the so-called G versions (i.e., DEVSS-G/SUPG) appear. In this work, the DEVSS/DG method is applied because of its good convergence properties in both smooth and nonsmooth geometries (Baaijens 1998a), while a particularly efficient implementation for multiple relaxation time problems is selected (Bogaerds *et al.* 1999b)

$$\left(\mathbf{S}_i,\ \overset{\nabla}{\tau}+\lambda(\tau_i)^{-1}\cdot\tau_i+\frac{G_0}{\lambda_{0b}}\left[f(\tau_i)^{-1}-1\right]\mathbf{I}-2G_0D\right)-\sum_{e=1}^{K}\int_{\Gamma_i^e}\mathbf{S}_i:\vec{u}\cdot\vec{n}(\tau_i-\tau^{\text{ext}}{}_i)\mathrm{d}\Gamma=0 \quad (15)$$

$$\left(D_v,\ 2\overline{\eta}(D_u-E)+\sum_{i=1}^{M}\tau_i\right)-(\vec{\nabla}\cdot\vec{v},\ p)=0 \quad (16)$$

$$(G,\ E-D)=0 \quad (17)$$

$$(q,\ \vec{\nabla}\cdot\vec{u})=0 \quad (18)$$

where $(\cdot,\ \cdot)$ denotes the inner product in the domain Ω, τ_i^{ext} the extra stress tensor of the neighboring, upstream element, $\vec{n}$ the unit outward normal on the boundary of the element Ω^e, and $D_\phi=\frac{1}{2}(\vec{\nabla}\vec{\phi}+(\vec{\nabla}\vec{\phi})^T)$ with $\vec{\phi}=\vec{u},\vec{v}$. The stabilization term $2\overline{\eta}(D_u-E)$ has been added to the weak form of the momentum equation (15) introduced within the context of the DG formulation in Baaijens *et al.* (1997) following earlier work of Guénette and Fortin (1995) for the SU formulation.

The stabilization parameter $\overline{\eta}$ may, in principle, be chosen arbitrarily, but is set to $\overline{\eta}=\Sigma G_0\lambda_0$. It has been suggested to adapt $\overline{\eta}$ to the solution τ; however, it has been shown in Eqn. (13) that this may give rise to numerical instabilities for transient flows.

The governing equations are integrated in time by a backward Euler scheme. To obtain the solution of the nonlinear set of equations at each time step, a one-step Newton–Raphson iteration process is implemented. The resulting set of equations is solved in a decoupled fashion, as described in detail in Bogaerds *et al.* (1999a). This procedure allows an efficient solution of the subproblems using dedicated iterative solvers.

4. *Parameter Identification*

The parameters of the EPP (Extended PomPom) model are identified as follows. The linear relaxation spectrum is determined in the standard fashion based on linear viscoelastic data. The nonlinear parameter q and ratio $\lambda_{0b}/\lambda_{0s}$ are fitted on the uniaxial elongational data only, while α is chosen as $0.1/q$. Parameter sets were determined for two different LDPEs (Lupolen 1819H, BASF, and Stamylan LD2008 XC43, DSM) that were used in two different experiments, reversed shear flow and flow through a cross-slot device, respectively (see next section). Results are only given for the DSM LDPE melt. The steady η_u and the transient $\eta_u^+(t)$ uniaxial elongation viscocities are given in Fig. 1, all at a temperature of 150°C.

Other data, such as the shear viscosity and first normal stress coefficient, are *not* used in the parameter identification process. The predictions for these quantities, shown for the steady case only, are given in Fig. 2. In the current fit, only four relaxation times have been used, which has been found sufficient for the range of shear and elongation rates in the cross-slot flow experiment. The relative sudden drop

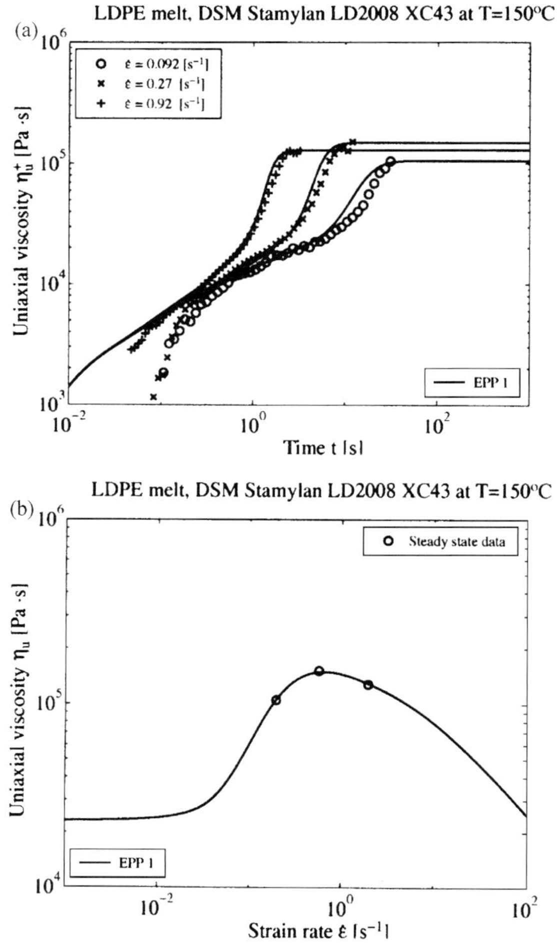

Figure 1
(a) Transient and (b) steady-state uniaxial elongation viscosity for the Stamylan LD2008 XC43 LDPE melt at 150°C. The solid line represents the fit of the EPP model.

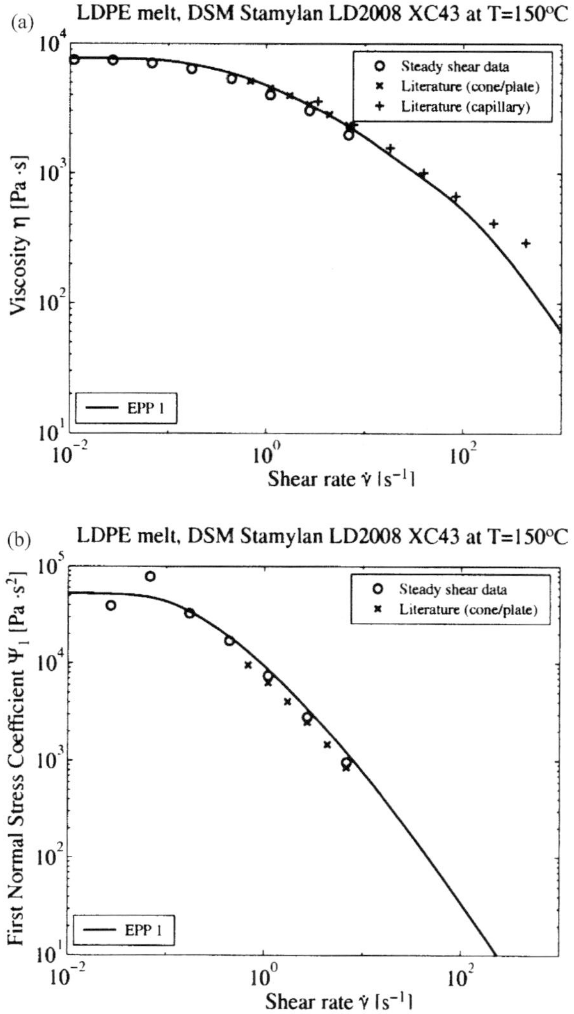

Figure 2
(a) Steady shear viscosity and (b) first normal stress coefficient of the Stamylan LD2008 XC43 LDPE melt at 150°C.

in predicted viscosity at shear rates beyond $10^2\,s^{-1}$, observed in Fig. 2(a), is due to the limited number of relaxation times used, and can easily be improved by increasing the number of relaxation times to describe the linear spectrum.

The Lupolen 1819 LDPE melt was fitted with six relaxation times (Verbeeten *et al.* 2000).

5. *Complex Flows*

The predictive performance of the EPP model is shown for two critical flows: the reversed shear flow and the cross-slot flow. Experimental data for the reversed shear flow is available from Kraft (1996) for

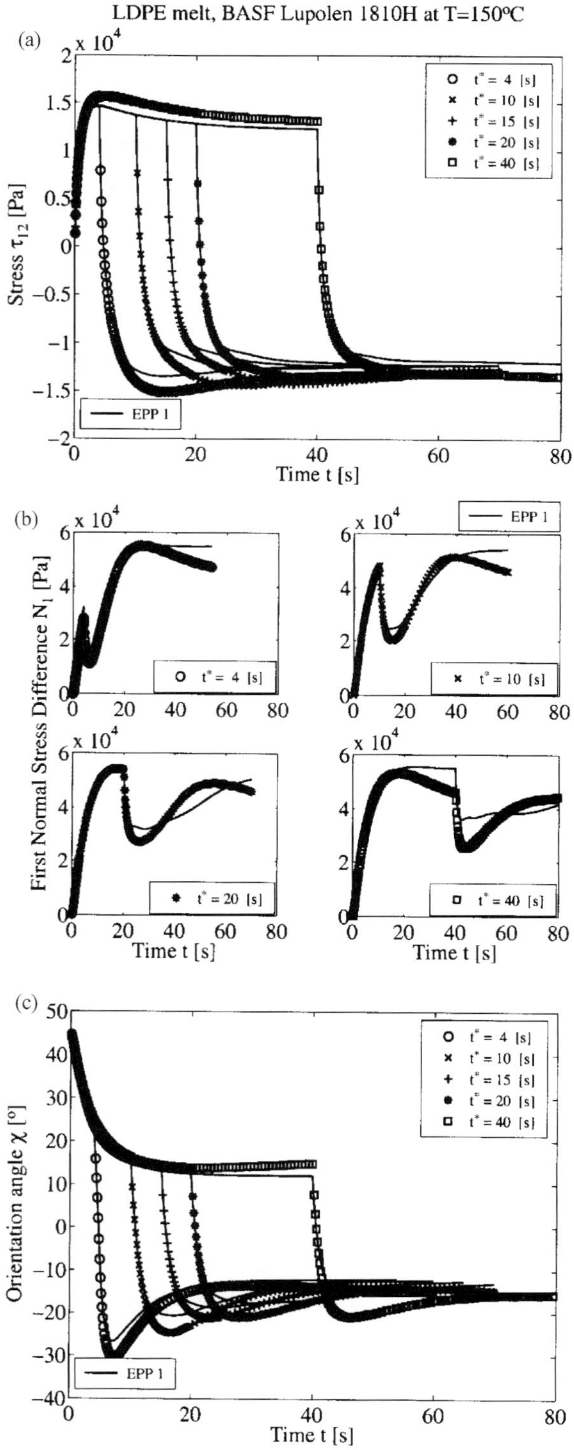

Figure 3
Reversed flow $\dot{\gamma} = 1(s^{-1})$. (a) Shear stress, (b) first normal stress difference N_1, (c) orientation angle for difference initial strains $\gamma = \dot{\gamma}\, t^* \cdot\ t^* = 4$, 10, 15, 20, 40 s.

a LDPE melt (Lupolen 1918H, BASF). This flow is known to be a very critical test for a constitutive model. Comparison (Fig. 4) shows that all experimentally observed features, the stresses and the orientation angle, are predicted rather well by the model. For the prediction, a six-mode fit was used.

Among the most frequently investigated flow geometries for polymer melts is the n:1 planar contraction, with, typically, $n=4$. The advantage of such a configuration is that it resembles a number of actual processing configurations, such as extrusion. It is, however, not an optimal test geometry (Schoonen 1998). A particularly challenging aspect of simulating inhomogeneous flows is the ability to predict both shear- and elongation-dominated flow regions simultaneously at finite strains. Although along the centerline of the 4:1 contraction, high elongational rates are achieved at the contraction region, the

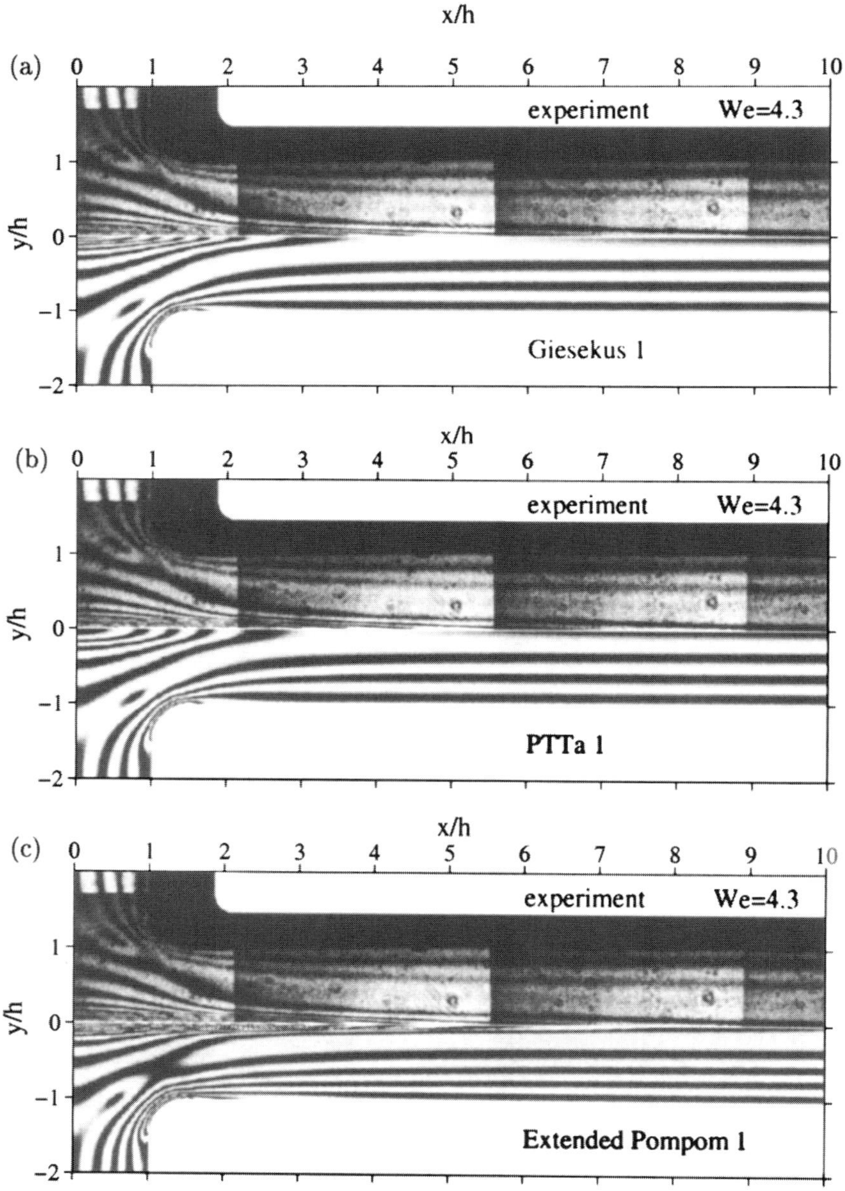

Figure 4
Predicted and measured birefringence pattern for (a) Giesekus, (b) PTT, and (c) EPP model.

residence time of fluid particles is low and only moderate strain levels are attained. A similar situation holds for the flow past a confined cylinder (Baaijens *et al.* 1997). Again, at the wake of the cylinder, high elongation rates are achieved, but the residence time is too short to achieve large strains. At the rear stagnation point, with infinite residence time, the elongation rate is zero.

The cross-slot flow geometry (Schoonen 1998, Bogaerds *et al.* 1999a, Peters *et al.* 1999) does not have these limitations. At the stagnation point, a finite elongational rate exists. Moreover, the elongational rate is fairly homogeneous in a region surrounding the stagnation point and the residence time in this region is sufficiently long to reach large strains. Sufficiently far upstream and downstream of the stagnation point, a fully developed shear flow exists, while elsewhere away from the stagnation point a combined shear/elongational flow is found. In conclusion: in contrast to many other geometries that have been investigated, the cross-slot geometry allows the investigation of steady elongation with large strains and combined shear/elongational flows.

To make a direct comparison of two-dimensional calculations with experimentally obtained birefringence data when using polymer melts, it is of critical importance that the flow geometry has a depth-to-height ratio of at least 8 throughout the domain of interest, otherwise three-dimensional effects have too large an influence on the experimental results. In fact, only the isochromatic fringes can be interpreted in a straightforward manner. Isoclinics can *not* be used to separate shear stresses from normal stresses due to end effects (see Schoonen 1998). If the aspect ratio is too small, or if the flow geometry is essentially three-dimensional, Mueller

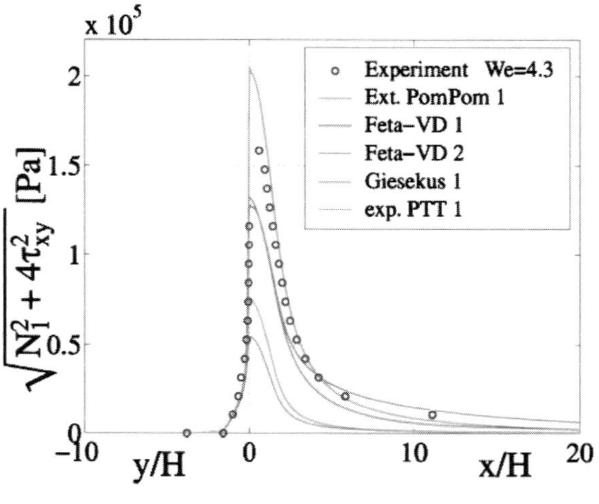

Figure 5
Normal stress along the symmetry line. (See colour plate Section)

calculus can be used to convert computed stress data to experimentally observable fringes (see Bogaerds *et al.* 1999a).

Figures 4(a) and 4(b) compare the predicted and the measured birefringence pattern using the Giesekus and the PTT model, respectively, at a Weissenberg number of 4.3. Figure 4(c) shows the results for the EPP model. Remarkably good agreement exists in the whole domain for the EPP model, white both the Giesekus and the PTT model are unable to predict the fringe patterns near and at the stagnation point. This is reflected more quantitatively in Fig. 5 where, for a number of models, the stress along the symmetry line is depicted. Of all models investigated (i.e., the Giesekus, PTT, Feta-VD (Peters *et al.* 1999), and the EPP model), the EPP model shows the best performance.

6. *Conclusions*

The enhanced Pompom model not only provides an excellent description of all available rheometrical data for a variety of polyethylene melts, as shown in Verbeeten *et al.* (2000), it also predicts the response of the reversed shear flow and the complex stress patterns in the inhomogeneous cross-slot flow geometry accurately. Even the stresses at the stagnation point, corresponding to a steady planar elongation flow, and the subsequent stress relaxation in the downstream section, are predicted with remarkable accuracy. Among the advantages of the enhanced version of the Pompom model—the EPP model—are that (i) the model only has a single differential equation, combining stretch and elongation, which is attractive from the computational point of view, (ii) the resulting equation does not show the unbounded behavior in elongation as the original Pompom model (due to the UCM-like expression for the evolution of the orientation tensor), (iii) a second normal stress difference is included, considered realistic for polymer melts, and (iv) the EPP model provides an improved description of the rheometrical data.

Numerical algorithms, such as the DEVSS/DG method, have been developed to the state where the analysis of a range of viscoelastic polymer melt flows is feasible with sufficient reliability, robustness, and efficiency. In fact, three-dimensional simulations using multiple relaxation times are possible and, moreover, birefringence data generated in such configurations can be interpreted using computational methods (Bogaerds *et al.* 1999a). Nevertheless, ample room for improvements persist since, for instance, unacceptably low limits in achievable Weissenberg numbers still exist, in particular for flows with geometrical singularities, while computational resource requirements (in terms of both memory and computing time) are quite demanding.

See also: Polymer Rheology for Melt Processing: Molecular Modeling; Polymer Film Casting: Modeling; Polymer Injection Molding: Modeling for Properties

Bibliography

Baaijens F P T 1998a An iterative solver for the DEVSS/DG method with application to smooth and nonsmooth flows of the upper convected Maxwell fluid. *J. Non-Newtonian Fluid Mech.* **75**, 119–38

Baaijens F P T 1998b Mixed finite element methods for viscoelastic flow analysis: a review. *J. Non-Newtonian Fluid Mech.* **79**, 361–85

Baaijens F P T, Selen S H A, Baaijens H P W, Peters G W M, Meijer H E H 1997 Viscoelastic flow past a confined cylinder of a LDPE melt. *J. Non-Newtonian Fluid Mech.* **68**, 173–203

Blackwell R J, McLeish T C B, Harlen O G 2000 Molecular drag–strain coupling in branched polymer melts. *J. Rheol.* **44**, 121–36

Bogaerds A C B, Verbeeten W M H, Baaijens F P T 1999a Success and failures of discontinuous Galerkin methods in viscoelastic fluid analysis. In: Cockburn B, Karniadakis G E, Shu C-W (eds.) *Discontinuous Galerkin Methods*. Springer, Berlin, pp. 264–70

Bogaerds A C B, Verbeeten W M H, Peters G W M, Baaijens F P T 1999b 3-D viscoelastic analysis of a polymer solution in a complex flow. *Comp. Methods Appl. Mech. Eng.* **180**, 413–30

Guénette R, Fortin M 1995 A new mixed finite-element method for computing viscoelastic flows. *J. Non-Newtonian Fluid Mech.* **60**, 27–52

Inkson N J, McLeish T C B, Harlen O G, Groves D J 1999 Modeling the rheology of low-density polyethylene in shear and extension with the multimodal Pompom constitutive equation. *J. Rheol.* **43**, 873–96

Kraft M 1996 *Untersuchungen zur Scherinduzierten Rheologischen Anisotropie von Verschieddenen Polyethylen-Schmeltzen*. PhD Thesis, ETH Zurich

McLeish T C B, Larson R G 1998 Molecular constitutive equations for a class of branched polymers: the Pompom polymer. *J. Rheol.* **42**, 81–110

Peters G W M, Schoonen J F M, Baaijens F P T, Meijer H E H 1999 On the performance of enhanced constitutive models for polymer melts in a cross-slot flow. *J. Non-Newtonian Fluid Mech.* **82**, 387–427

Rubio P, Wagner M H 1999 Letter to the editor: A note added to molecular constitutive equations for a class of branched polymers. The Pompom polymer (*J. Rheol.* **42**, 81). *J. Rheol.* **43**, 1709–10

Schoonen J 1998 *Determination of rheological constitutive equations using complex flows*. PhD Thesis, Eindhoven University of Technology

Verbeeten W M H, Peters G W M, Baaijens F P T 2000 Differential constitutive equations for polymer melts. *Proc. XIII Int Congress on Rheology*. British Society of Rheology, Glasgow, UK

G. W. M. Peters, W. M. H. Verbeeten, and F. P. T. Baaijens

Polymer Rheology for Melt Processing: Molecular Modeling

New catalysts for controlled industrial synthesis, such as the single-site metallocenes for the polyolefins, are allowing unprecedented control of manufactured polymer resins. The future lies in molecular engineering by design: optimization of molecular weight distribution (MWD), long chain branching (LCB), and co-monomer content for process and product. Great progress has been made in this field, driven also by the availability of anionically synthesized model materials as test cases, and the emergence of a quantitative model for polymer melt dynamics at a molecular level. Of great assistance is the dependence of key rheological properties on local (monomer) chemistry via only a few scaling parameters—much of the behavior is universal. Far greater variation is found within the structural parameters MWD and especially LCB. For each chemistry the key parameters are (i) the entanglement molecular weight M_e and (ii) the associated timescale for orientational relaxation of an entangled segment, τ_e. A subchain of M_e is just large enough to feel the topological constraints of the melt, and nearly universal behavior emerges when molecular weights are expressed in units of M_e. One consequence for theory is immediate: the task can be divided into treatments at two lengthscales. Atomistic modeling (principally by computer simulation) can attempt to compute the parameters M_e and τ_e from local chemical structure; then the connection to bulk rheology is made via the molecular structure, coarse grained at the level of entanglements. This article focuses on the latter stage. This is a highly interdisciplinary program: theoretical and experimental physics combine with synthetic chemistry as techniques brought to bear upon branched polymer melts. Not only careful rheological experiments, but also molecular probes such as neutron scattering, are providing further information for the remarkable theoretical models that have recently shed considerable light on this tangled tale.

1. Molecular Theories of Polymer Melts

The current molecular theories for melt rheology are the legacy of Rouse (1953), who calculated the rheological properties of a model assuming monodisperse Gaussian chains moving against a uniform frictional background. The central results of a shear stress proportional to shear rate and a relaxation modulus $G(t) \sim t^{-1/2}$ also apply qualitatively to melts with $M < M_e$, and to all melts under small strains and timescales $t < \tau_e$. The "Rouse time" τ_R marks the slowest mode of internal relaxation of such a chain, and grows with molecular weight M as $\tau_R \sim M^2$, inducing a growth of viscosity as $\eta \sim M$ (Rouse 1953).

The key challenge in melts is to deal with the effect of entanglement. At $M \sim 2M_e$ the molecular weight dependence of melt viscosity changes from $\eta \sim M^1$ to $\eta \sim M^{3.4}$. At the same time the frequency-dependent modulus begins to display a plateau in its elastic component of a value $G_0 = \rho RT/M_e$. Three approaches to extend the Rouse model into this regime, where chain *interactions* dominate can be identified; (i) transient network models, (ii) mode-coupling approaches, and (iii) tube models.

Network models are motivated from the similarity of the relaxation modulus in the plateau zone to that of a rubber, and from an early molecular picture of entanglements as spatially localized knots between two chains. Lodge (1964) showed that assuming that a melt possesses cross-links with finite lifetimes produces a Maxwell model. Variations on the model such as that of Giesekus (1982) endow the cross-link timescales with a dependence on the stress, essential to modify the original overprediction of strain-hardening in extension. No scattering experiment has ever detected evidence for localized cross-link structures in melts, however, and transient network models suffer from their requirement for modifications that cannot be derived molecularly. They do, however, apply to systems such as telechelic polymers that possess real temporary cross-links.

A phenomenological approach to glassy liquids due to Ngai has been applied to polymer melts (Yu *et al.* 1991). It incorporates the essential insight that dynamics of different structures in a melt are highly co-operative ("mode-coupling"). Incorporated into a series of *ansatzen* of the molecular weight dependence of the coupling, this insight can be made to produce predictions for $G(t)$ and mean monomer displacement (Schweizer 1997). The difficulty here lies in the lack of control of choice in the phenomenology.

The most successful and widely applied model begins with the assumption that melt topology confines the motion of chains to tube-like regions along their contours. Only chain ends are free of this constraint. Introduced dynamically by de Gennes (1971) to describe the diffusion of marker-chains in networks, the "tube model" was applied in approximate form to polymer melts by Doi and Edwards (1986). As well as requiring only two undetermined parameters for each chemistry, the model suggests natural extensions to strong flows, polydisperse and branched melts, and predicts not only rheology but also molecular configurations and motions. These have been checked in some cases by nuclear magnetic resonance (NMR, Komlosh and Callaghan 1998), neutron scattering (Muller *et al.* 1993) and numerical simulation (Putz *et al.* 2000). A recent revival of interest in self-consistency within non-equilibrium statistical mechanics has provided tools to check the validity of molecular (or any) rheological theories, sometimes suggesting refinements on thermodynamic grounds (Ottinger 2000).

2. *Linear Rheology of Linear Chain Melts*

If well-entangled linear chains are modeled as Gaussian springs with local friction confined to tubes, a series of dynamical modes emerge that determine the linear rheology. (i) At small times all local motion on scales larger than a statistical segment length is Rouse-like until monomer displacements are impeded by the tube. (ii) At longer times Rouse motion is confined to redistribution of monomers along the tube contour, except at the chain ends which progressively create new tube by "contour-length fluctuations." (iii) Beyond the Rouse time further motion depends on the curvilinear diffusion of the chain center of mass along its contour. This "reptation" motion allows complete reconfiguration of chain configuration after periods in which it repeatedly disengages from its tube (Fig. 1). The associated "reptation time" $\tau_{rept} \sim \tau_e M^3 / M_e$ marks the dominant stress relaxation time. (iv) Reptation of the neighboring chains causes small displacements of the tube at the timescale of τ_{rept}. This "constraint release" (CR) modifies the slowest relaxations of chain by giving the tube itself a slow Rouse-like dynamics. Stress relaxation depends both on the fraction of tube unvisited by a chain end and the reconfiguration of the tube by CR. A useful approximation of this effect, termed *double reptation*, calculates $G(t)$ by taking the *square* of the unrelaxed tube fraction (for a more detailed review see Rubinstein 1998).

The approximation in which only the dominant reptation mode is retained predicts a viscosity dependence of M^2 and a peak in $G''(\omega)$ sharper than observed. Inclusion of contour length fluctuations recovers the effective $M^{3.4}$ behavior and broadens the expected relaxation peak (Milner and McLeish 1998).

Considerable effort has developed the use of the relaxation spectrum as a rheological measure of MWD in linear polymers. The near-single exponential relaxation of monodisperse entangled polymers with a timescale strongly dependent on M allows the interpretation of broadening in the $G(t)$ signal as a quantitative measure of polydispersity. Careful subtraction of response due to Rouse motion and use of the double reptation approximation is beginning to yield reliable results (Thimm *et al.* 2000 and references therein).

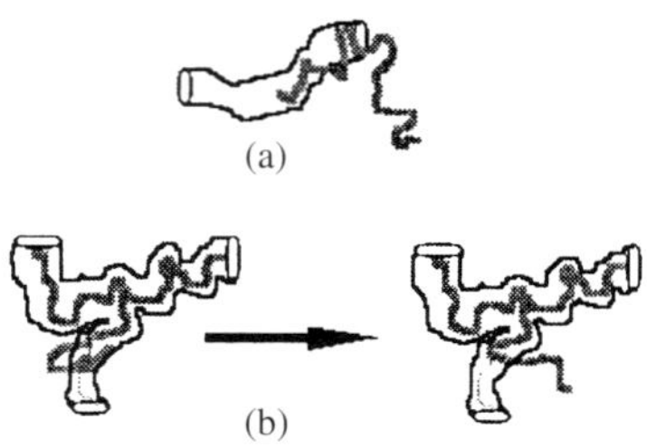

Figure 1
Reptation (a) and arm fluctuation (b) in linear and star molecules in their confining tubes.

3. *Linear Rheology of Branched Chain Melts*

Long chain branching of entangled melts radically changes the rheological response because the branch points suppress reptation. Now the only routes to tube reconfiguration are (ii) and (iv) above. Fluctuations of the contour length to very small values—so releasing entanglements close to a branch point—are *exponentially* rare in the branch molecular weight. In the well-studied case of "star polymers" this endows the viscosity, terminal relaxation time, and self-diffusion constant with a dominant dependence on the molecular weight of the star *arms* only, M_a, of $\exp(\nu M_a/M_e)$. To derive the value of $\nu = 8/24$ that agrees with experiments, CR must be accounted for. Entangled branched arms are affected by CR in an extreme limiting way such that all previously relaxed segments act as solvent for unrelaxed chain. This approach, called "dynamic dilution" has been applied successfully to a range of blends and architectures of model polymers qualitatively (McLeish and Milner 1999).

More complex architectures with multiple branch points can exhibit relaxations arising from local topological pieces of the molecule, but at widely separate timescales. For example, the H-shaped architecture shows a range of star-arm-like relaxations at intermediate frequencies, but a near-single relaxation at low frequencies arising from *reptation* of the "cross-bar" segments of the molecules in the effective entanglement (see Fig. 2). A current goal of molecular rheology is the implementation of structure-rheology relations derived for exact monodisperse architectures to the randomly branched molecules arising from industrial processing. Two important cases are the familiar low density polyethylene (LDPE) and long chain branched melts arising from vinyl incorporation on metallocene catalysts.

4. *Nonlinear Rheology—Approaches*

Any useful application of molecular rheology to melt-processing must naturally consider highly nonlinear response. Fortunately the tube picture is suggestive of nonlinear extension. The simplest assumption to make is that the tube constraints deform affinely in a strong flow, while permitting Rouse-like retraction of chains along the tube contours. In an approximation, this theory can be cast in to a

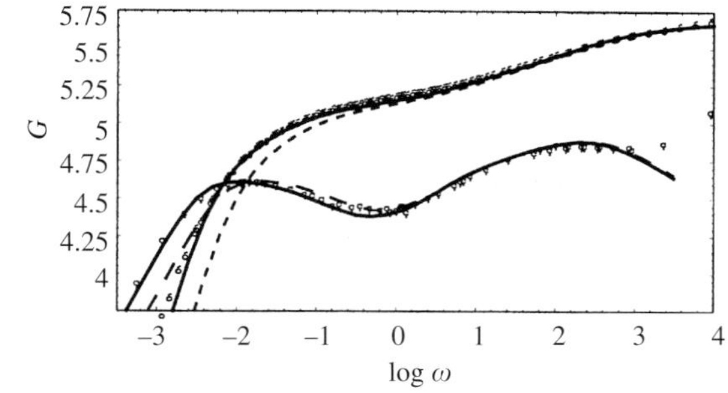

Figure 2
G' and G'' as a function of frequency ω for a polyisoprene melt of H-shaped polymers (log axes). Relaxations from the arms ($0<\log \omega<3$) and cross-bars ($\log \omega \sim -2$) are clearly separated, and predicted by a tube model (solid lines).

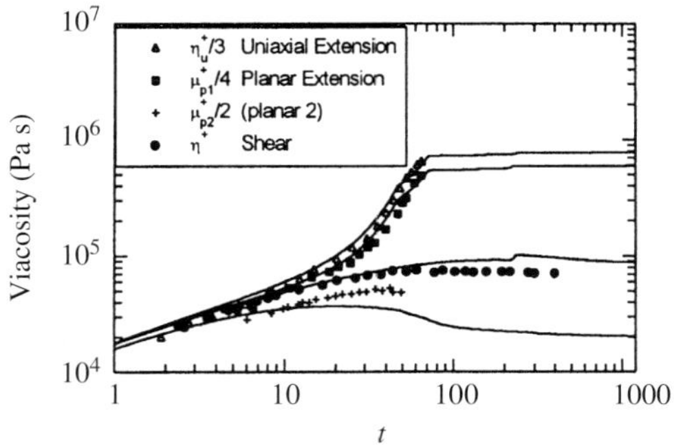

Figure 3
A "multimode pom-pom" model fitting simultaneously transients in uniaxial and planar extension as well as shear for the LDPE IUPACX.

familiar (KBKZ) integral constitutive form. Such a calculation predicts the stress-response of monodisperse linear chain melts after a step shear (the "damping function") very well, also correctly predicting absence of extension-hardening and a strong shear thinning. However, the latter is strongly *over-predicted*, expecting a maximum in shear stress as a function of shear stress, which would lead to unobserved instabilities in flow. Careful modifications of integral constitutive equations, motivated by molecular concepts, such as chain-stretch, have permitted closer modeling of data (Wagner *et al.* 1998).

Both rheological and neutron scattering experiments suggest that predicted stress-maximum is due to a process, missing from the theory, that served to mis-orient the chains locally in a fast flow. Current work suggests that retraction of neighboring chains can induce lateral diffusion of the deforming tube network, a process termed *convective constraint release* (CCR, Marrucci 1996). In order to turn this physics into a computational tool useful for processing, versions of CCR incorporating chain stretch are currently being formulated (Mead *et al.* 1998).

5. *Long Chain Branching—The Pom-pom Decomposition*

Paradoxically, more progress has been made in the nonlinear molecular rheology of long chain branched polymers. This is because of the dominance of segmental stretch of chain segments within tubes induced by the branch-points. For weakly nonlinear rapid deformations, the branch points move rather like cross-links, but on cessation of flow may diffuse back through the melt, driven by the slow retraction of dangling entangled arms. In stronger flows, a new process, termed *branch point withdrawal* collapses outer dangling arms into tube segments originally defined by deeper segments. This rapidly reduces the effective drag on the branch points, and limits the degree of extension hardening observed in the melt. The most extensive tests of these ideas have been made to date on the H-architecture (McLeish *et al.* 1999).

The essential consequences of tube theory for LCB polymers are (i) separation of the relaxation times of orientation (reptation or activated retraction) and stretch (curvilinear Rouse retraction in tubes), (ii) branch-point withdrawal, were encapsulated in a simple constitutive form termed the *pom-pom* equation (McLeish and Larson 1998). The form is a scalar stretch ($\lambda(t)$) equation

$$\frac{\partial}{\partial t}\lambda = \lambda(\mathbf{K}:\mathbf{S}) - \frac{1}{\tau_S}(\lambda - 1)$$

and another for the orientation tensor:

$$\mathbf{S}(t) = \int_{-\infty}^{t} \frac{dt'}{\tau_b(t')} \exp\left\{-\int_{t'}^{t} \frac{dt''}{\tau_b(t'')}\right\} \mathbf{Q}(t,\ t')$$

with the stress given by the product: $\sigma(t) = (15/4)G_0\varphi b^2\lambda^2(t)\mathrm{S}(t)$, and where τ_s may depend on λ.

This constitutive formulation illustrates the huge variety of forms that may arise when molecular-level thinking is employed. Their properties can differ markedly from standard phenomenology: for example, the pom-pom equations are highly extension-hardening in planar and uniaxial flows, but also very shear-thinning. Multi-mode variants of the pom-pom equations are currently achieving unparalleled success in representing quantitatively the rheology of LDPE in many different flow geometries and histories (Blackwell *et al.* 2000, see Fig. 3). In a differential approximation they carry the additional advantage of easy computation in flow solvers.

A challenge for the future is to derive equations of this and more faithful forms, and incorporate them in fully multiscale models of polymer processes.

See also: Polymer Rheology for Melt Processing: Complex Flows

Bibliography

Blackwell R J, McLeish T C B, Harlen O G 2000 Molecular drag-strain coupling in branched polymer melts. *J. Rheol.* **44**, 121–36

de Gennes P G 1971 Reptation of a polymer chain in the presence of fixed obstacles. *J. Chem. Phys.* **55**, 572–9

Doi M, Edwards S F 1986 *The Theory of Polymer Dynamics*. Oxford

Giesekus H 1982 A simple constitutive equation for polymer fluids based on the concept of deformation-dependent tensorial mobility. *J. Non-Newton. Fluid* **11**, 69–109

Komlosh M E, Callaghan P T 1998 High magnetic field gradient PGSE NMR in the presence of a large polarizing field. *J. Chem. Phys.* **109**, 10053–67

Lodge A S 1964 *Elastic Liquids*. Academic Press, New York

Marrucci G 1996 Dynamics of entanglements: a nonlinear model consistent with the Cox-Merz rule. *J. Non-Newton. Fluid* **62**, 279–89

McLeish T C B, Allgoner J, Bick D K, Bishko G, Biswas P, Blackwell R, Blottière B, Clarke N, Gibbs B, Groves D J, Hakiki A, Heenan R, Johson J M, Kant R, Read D J, Young R N 1999 Dynamics of entangled H-polymers. *Macromolecules* **32**, 6734–58

McLeish T C B, Larson R G 1998 Molecular constitutive equations for a class of branched polymers: the pom-pom polymer. *J. Rheol.* **42**, 81–110

McLeish T C B, Milner S T 1999 Entangled dynamics and melt flow of branched polymers. *Adv. Polym. Sci.* **143**, 195–256

Mead D W, Larson R G, Doi M 1998 A molecular theory for fast flow of entangled polymers. *Macromolecules* **31**, 7895

Milner S T, McLeish T C B 1998 Reptation and contour-length fluctuations in melts of linear polymers. *Phys. Rev. Lett.* **81**, 725–8

Muller R, Pesce J J, Picot C 1993 Chain conformation in sheared polymer melts as revealed by SANS. *Macromolecules* **26**, 4356–62

Putz M, Kremer K, Grest G S 2000 What is the entanglement length in a polymer melt. *Europhys. Lett.* **49**, 735–41

Rouse P E 1953 A theory of linear viscoelastic properties of dilute solutions of coiling polymers. *J. Chem. Phys* **21**, 1272

Rubenstein M 1998 Theoretical challenges in polymer dynamics. In: McLeish T C B (ed.) *Theoretical Challenges in the Dynamics of Complex Fluids*. Kluwer, Dordrecht, The Netherlands

Thimm W, Friedrich C, Marth M, Honerkamp 2000 On the Rouse spectrum and the determination of the molecular weight distribution. *J. Rheol.* **44**, 429–38

Yu H, Binder K, Schweizer K, McKenna C, Muthukum M, Sillescu H, Donth E J, Struik L, Lodge T P, Phillies G 1991 Polymer diffusion, dynamics and viscoelasticity. *J. Non-Cryst. Solids* **131**, 742–54

Ottinger H C 2000 Thermodynamically admissible reptation models with anisotropic tube cross-sections and corrective constraint release. *J. Non-Newtonian Fluid Mech.* **89**, 165–85

Schweizer K S, Fuchs M, Szamel G, Guenza M, Tang H 1997 Polymer mode-coupling theory of the slow dynamics of entangled macromolecular fluids. *Macromol. Theory Simul.* **6**, 1037–117

Wagner M H, Bastian H, Ehrecke P, Kraft M, Hachmann P, Meissner J 1998 Nonlinear viscoelastic characterization of a linear polyethylen (HDPE) melt in rotational and irrotational flows. *J. Non-Newtonian Fluid Mech.* **79**, 283–96

T. McLeish

Polymer Single Screw Extrusion: Modeling

The physical processes in plasticating extrusion are extremely complicated. In order to describe, simulate, and optimize these processes with mathematical/physical models, one has to simplify the problems to a level where calculations are possible with acceptable expenditure of mathematical formulation, programming, and time. This way of proceeding seems reasonable, since the boundary conditions of the process and the relevant material data functions are often not well known. However, the simplifications should not severely affect the accuracy of the calculated results. Running on a computer, the models allow the simulation of the extrusion process and the optimization of screw design.

1. *Basic Considerations*

Up until now, there has not existed a universal constitutive equation for polymer materials which is able to simultaneously describe the phenomena in the solids, melting, and melt zones. So it is necessary to divide the extruder into zones, in which specific physical models can be established.

1.1 The Functional Zones

The functional zones of a typical single screw extruder are presented in Fig. 1:

- feed hopper, with which the particulate solids are put into the extruder (gravity flow);
- solids zone (or feeding zone), which is in most cases cooled and in which the solid polymer is transported and more or less compressed;
- delay zone, in which Coulomb-type solids transport is changing to a viscous-type solids transport and melt pool formation starts;
- melting zone, where the polymer is melted due to viscous dissipation and heat transfer from the barrel;
- melt zone (or metering zone), where an additional pressure rise can be achieved, often accomplished with a melt-cooling (cooled barrel);
- mixing zone (often with a cooled barrel), with special shearing and/or mixing elements to improve melt homogeneity;
- forming zone (extrusion die), to give the extruded polymer a suitable preform for the subsequent calibration zone of an extrusion line, where the final product shape is defined.

The transition points between *functional* extruder zones and their lengths are often not well known and are highly dependent on the processing conditions (e.g., screw speed, pressure, temperature, polymer). For the most part, they also do not coincide with the *geometrical* zones of the screw (e.g., compression zone, metering zone). These facts make the modeling and calculating of the entire screw more difficult.

1.2 Polymer Data Functions

The material data of polymers are in principle dependent on temperature (T) and pressure (p). In some cases they are also dependent on surface roughness (Rt), sliding velocity (v), particle size of

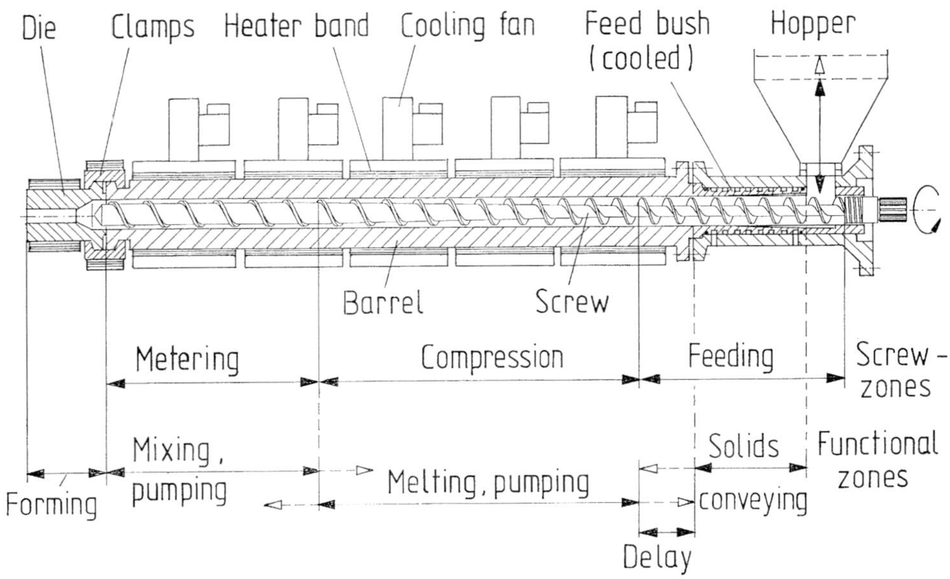

Figure 1
Functional zones and geometrical screw zones in a single-screw extruder

granules (d), the screw channel depth (h_s) in the feed hopper zone, shear rate ($\dot{\gamma}$), elongation rate ($\dot{\varepsilon}$), and time (t). Thermal material data of amorphous and semicrystalline polymers sometimes differ very much from each other.

An overview of the relevant material data, their dependence on physical conditions, and their importance for the different functional zones is given by Grünschloss (1997) and summarized in Table 1. Reference should also be made to VDMA (1981, 1982, 1983). For calculating purposes it is very suitable to approximate the material functions with mathematical equations. However, these functions are mostly nonlinear and sometimes difficult to formulate in closed analytical expressions.

1.3 Calculating Procedure for the Entire Screw

Following the material stream, at the interface cross section between the zones the relevant process data (e.g., pressure, polymer temperature, residence time, shear deformation) are transmitted from the previous zone as starting conditions to the following zone. The unknown mass throughput has accordingly to be calculated iteratively, such that a given pressure at the end of the screw is achieved.

1.4 Computation Methods

In many practical screw design problems, the exact material data are not known. In these cases one should use the material data of a similar and known polymer of the same family. Moreover, the exact boundary conditions, the temperature at the inner barrel wall and, particularly, the screw temperature are not well known either. Because of this situation, one should choose a suitable calculation method to avoid unnecessary effort in modeling and programming. The most comprehensive treatments of computer modeling of extrusion and polymer processing have been given by Tucker (1989).

(*a*) *Analytical methods.* Analytical methods should be used if the problem can be simplified to a one-dimensional problem. The required process quantities (e.g., throughput, pressure) can be calculated with closed mathematical equations. An iterative procedure could be established to account for the interdependence of material data and calculated physical properties. The main advantages of analytical methods are the good transparency between calculated results and input data, less effort in modeling and programming, and very short computation time. Depending on the degree of simplification, the results are sometimes not very accurate.

(*b*) *Numerical methods (FDM, FEM, BEM).* Numerical methods must be used if the problem is multidimensional (e.g., three-dimensional flow in mixing elements or complicated extrusion dies, temperature fields, streamlines) and/or if the geometry of the flow region is too complex. They need a high degree of mathematical formulation and programming. Even with commercial software packages on powerful computers, the computational times are rather long. Theoretically, the accuracy of the predictions could be very good, if the polymer data functions, the starting conditions, and the boundary conditions are controlled or well known. If this is not the case, numerical methods may produce no better results than good analytical methods.

Table 1
Overview of the material data functions for plasticating extrusion.

	Function	Zones
Bulk density of solids particles	$\rho_{Sch}(h, d)$	Hopper, solids
Density of compressed particles	$\rho_F(p, T, t)$	Solids, delay, melting
Polymer density	$\rho(p, T)$	Delay, melting, melt, mixing
Friction coefficient (outer)	$\mu_a(T, p, Rt, v)$	Hopper, solids, delay
Friction coefficient (internal)	$\mu_i(T, p, v)$	Hopper, solids, delay
Normal stress anisotropy of solids	$k(T, t)$	Solids, delay
Shear strength of solids	$k_f(T, p)$	Solids, delay, melting
Wall slip velocity	$v_{slip}(\tau_w)$	Delay, melting, melt, mixing
Shear viscosity	$\eta(\dot{\gamma}, T, p)$	Delay, melting, melt, mixing
Elongational viscosity	$\eta_e(\dot{\varepsilon}, T, p)$	Melting, melt, mixing
Enthalpy (or spec. heat capacity)	$h(T, p)$	Solids, delay, melting, melt, mixing
Heat conductivity	$\lambda(T, p)$	Solids, delay, melting, melt, mixing
Other physical conditions:		
Melt elasticity		Melting, melt, mixing
Wall slipping of polymer melt		Delay, melting, melt, mixing

1.5 Geometrical Data

In designing screw—and, if necessary, grooved—barrel sections one should have all the freedom possible. So the models should accept variable geometrical data (e.g., to calculate screws with variable pitch or channel depth). Geometries which are too complicated (e.g., bottom radius at the screw flight) can be approximated by suitable straight lines. It is useful to formulate the geometrical data as a function of the axial position (l) along the screw axis. These data are basically the following:

Inner barrel diameter (usually constant)	$D(l)$
Clearance barrel—screw flight	$\delta_b(l)$
Clearance barrel—barrier flight for barrier screws	$\delta_b(l)$
Channel depth of the screw	$h_s(l)$,
of the barrel groove	$h_z(l)$
Pitch of the screw	$s_s(l)$,
of the barrel groove	$s_z(l)$
Helix angle of the screw	$\varphi(l)$,
of the barrel groove	$\omega(l)$
Width of flight land of the screw	$e_s(l)$,
of the barrel groove	$e_z(l)$
Width of channel of the screw	$b_s(l)$,
of the barrel groove	$b_z(l)$
Number of channels of the screw	z_s,
of the barrel groove	z_z

Special geometries of mixing and shearing elements must be defined and modeled separately.

1.6 Boundary Conditions and Assumptions

The boundary conditions screw temperature $T_S(l)$ and barrel temperature $T_Z(l)$ are given as a function of the axial coordinate l. Further conditions are the average starting temperature of the solids (T_f) or the melt T_m (in case of a pure melt extruder), the screw speed n, and the melt pressure p_W at the end of the screw. The mass throughput $\dot{m}$ can be calculated iteratively or can also be given, if experimental results must be checked.

The basic assumptions in modeling are mostly simplifications as follows:

The process is steady-state (starting-up processes with time-dependent properties are very difficult to model);

the channels are completely filled with polymer (but: in vent extruders the screw channel is partially filled with melt);

the screw channel is unwound to a flat channel for modeling in Cartesian coordinates (in most cases this is tolerable, since the ratio of channel depth to screw diameter is small enough);

the polymer data functions are simplified or reduced to constant mean values (these simplifications are sometimes tolerable. To verify this hypothesis, the calculated results should be checked with experiments);

in melt-dominated extruder zones the melt is sticking to the wall (this is not the case in processing polymers which tend to wall slipping. So the possibility of wall slipping should be allowed for. It can have a very great effect in melting and melt mixing).

For a better explanation, the screw is taken to be stationary and the barrel rotates (a simple kinetic inversion which would not affect the process).

In modeling one should take into account—at least approximately—*all relevant conditions* of geometry, material, and process. So in the computer simulation it would be possible to study the influences of *all* conditions on the process.

2. Modeling the Feed Hopper

The static pressure P of a liquid with density ρ_L under a column of height h is isotropic proportional to the height, and is determined by $P = \rho_L g h$ (g = gravitational constant). In a column of particulate solids contained in a vertical bin, the pressure at the base is not proportional to the height of the column, because of the friction between the solids and the wall. Moreover, a complex stress distribution develops in the system, which depends on the properties of the particulate solids, the loading method, and the shape of the bin (prism, cone, or wedge). For analytical treatment the problem has to be simplified with the following assumptions (see also Tadmor and Gogos (1979) with many references):

the whole mass of particulate solids is moving like a plug towards the exit of the hopper (*mass flow* with no stagnation regions in the hopper);

the vertical compressive stress is constant over any horizontal plane;

the ratio of horizontal and vertical stresses is constant and independent of depth;

the bulk density is constant;

the wall friction is fully mobilized (slip condition at the wall), no adhesion of particles at the wall;

acceleration forces are neglected (but may be included in a more sophisticated analysis).

According to Fig. 2, a force balance at a volume element of height dh can now be set up. The weight F_g of the element plus the pressure force $(p + \mathrm{d}p)$ acting downwards is equal to the frictional force $q_0 B p$ at the surrounding walls supporting the element, plus the pressure force p acting upwards. For details of the derivation the reader is referred to Walker (1966).

2.1 Upper Part of the Hopper with Constant Cross Section

For this part the resulting vertical stress (or pressure) distribution is given as:

$$P = P_1 \exp[q_0 B_0 (h - h_1)] + \frac{\rho_{Sch} g}{q_0 B_0} \{1 - \exp[q_0 B_0 (h - h_1)]\} \quad (1)$$

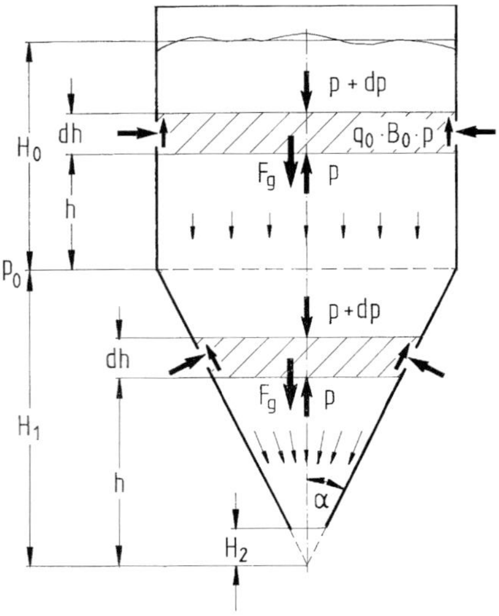

Figure 2
Force balance at a volume element in a hopper for particulate solids

where P_1 is the pressure at height h_1 and $q_0 = q_{cyl.} = 4/D$ (for cylindrical hoppers with diameter D), and $q_0 = q_{rec.} = (2a + 2b)/ab$ (for rectangular hoppers with the side lengths a and b), respectively, and with:

$$B_0 = \frac{\sin\delta \sin\kappa_0}{1 - \sin\delta \cos\kappa_0} \tag{2}$$

where

$$\kappa_0 = \beta_w + \arcsin\left(\frac{\sin\beta_w}{\sin\delta}\right), \quad \frac{\sin\beta_w}{\sin\delta} \leqslant 1 \tag{3}$$

and where β_w is the angle of external friction at the hopper wall

$$\beta_w = \arctan(\mu_a) \tag{4}$$

and δ is the angle of internal friction:

$$\delta = \arctan(\mu_i) \tag{5}$$

For the special case with $h_1 = H_0$ where $P_1 = 0$, Eqn. (1) reduces to the familiar Janssen equation:

$$P = \frac{\rho_{Sch} g}{q_0 B_0}\{1 - \exp[q_0 B_0 (h - H_0)]\}. \tag{6}$$

2.2 Lower Part of the Hopper with Convergent Cross-section

For convergent hoppers, the vertical stress (or pressure) distribution is given as:

$$P = P_0\left(\frac{h}{H_1}\right)^{q_1 B_1} + \frac{\rho_{Sch} g h}{q_1 B_1 - 1} \times \left[1 - \left(\frac{h}{H_1}\right)^{q_1 B_1 - 1}\right], \quad q_1 B_1 \neq 1 \tag{7}$$

with the coefficient $q_1 = q_{con.} = 2/\tan\alpha$ for conical and $q_1 = q_{wed.} = 1/\tan\alpha$ for wedge-shaped hoppers, respectively, and:

$$B_1 = \frac{\sin\delta \sin(2\alpha + \kappa_0)}{1 - \sin\delta \cos(2\alpha + \kappa_0)} \tag{8}$$

with κ_0 according to Eqn. (3). If the product of $q_1 B_1$ is unity, Eqn. (7) reduces to:

$$P = P_0\left(\frac{h}{H_1}\right) + \rho_{Sch} g h \ln\left(\frac{H_1}{h}\right), \quad q_1 B_1 = 1 \tag{9}$$

H_1 is the height where the vertical pressure is P_0. This can be the pressure at the base of a cylindrical hopper placed above the convergent hopper.

A typical calculated vertical pressure distribution of PE granules in a hopper with an upper cylindrical, a conical, and a lower cylindrical section is shown in Fig. 3.

3. *Modeling the Solids Conveying Zone*

The classical, one-dimensional theory of solids transport in single-screw extruders with a smooth inner barrel surface was derived by Darnell and Mol (1956). Tadmor *et al.* (1972) have further contributed to the theory by introducing pressure-dependent solids density and a non-isothermal model. A two-dimensional theory, which takes the transverse pressure gradient, gravity, and centrifugal forces into consideration, has been formulated by Lovegrove and Williams (1973, 1974). In principle the solids transport is determined by the geometry of the screw, the inlet pressure, the counterpressure at the end of the solids zone, and by the dynamic coefficient of external friction $\mu_a(T, p, v_g, R_t)$, the bulk density $\rho_{Sch}(h, d)$, the pressure-dependent density of the granules $\rho_F(p, T)$. In the case of a grooved feed section, the geometry of the grooves and the internal coefficient of friction $\mu_i(T, p, v_g)$ have to be considered, too.

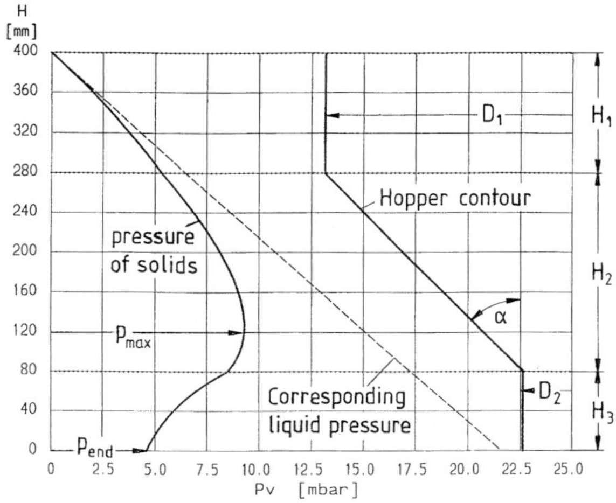

D_1 = 500 mm	D_2 = 100 mm	α = 45°
H_1 = 120 mm	H_2 = 200 mm	H_3 = 80 mm
ρ_{Sch} = 0.55 g/cm^3	μ_a = 0.2 -	μ_i = 0.5 -

Figure 3
Calculated vertical pressure distribution in a conical hopper

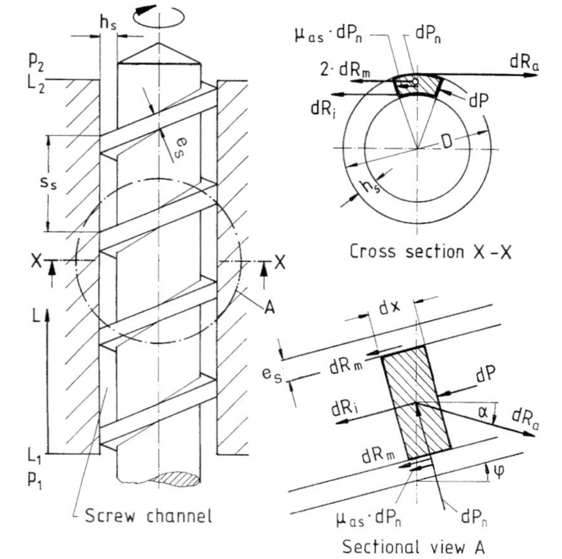

Figure 4
Force balance at a volume element in the smooth solids conveying zone

3.1 Solids Conveying Zone with a Smooth Inner Barrel Surface

As an example for an analytical treatment of the problem we consider the theory of Schneider, with the following basic simplifications and assumptions:

the screw channel is completely filled with particulate solids: melting does not occur;

the solids are moving like a plug in the screw channel;

the material data μ_a, ρ_{Sch}, and ρ_F are constant;

the pressure in the solids is not isotropic.

The most interesting quantity is the mass throughput which can be written as:

$$\dot{m} = \rho_{Sch} \dot{V} = \rho_{Sch} \pi (D - h) h \frac{s_s - (e_s z_s / \cos \varphi)}{s_s} v_{ax} \quad (10)$$

with the axial velocity v_{ax} of the solids

$$v_{ax} = \pi D n \frac{\tan \varphi \tan \alpha}{\tan \varphi + \tan \alpha} \quad (11)$$

where n is the screw speed, φ is the helix angle of the screw and α is the still-unknown conveying angle of the solids, which itself depends strongly on the pressure gradient dp/dx in the system. According to Fig. 4, a force balance at a volume element in the axial direction and a momentum balance in the peripheral direction leads after integration to the required relation between pressure increase p_2/p_1 and conveying angle α:

$$\cos \alpha = K \sin \alpha + C_1 + C_2 \frac{h_s}{L_2 - L_1} \ln \frac{p_2}{p_1} \quad (12)$$

and finally to

$$\alpha = \arcsin \left[\frac{\sqrt{1 + K^2 - \left(C_1 + C_2 \frac{h_s}{L_2 - L_1} \ln \frac{p_2}{p_1} \right)^2}}{1 + K^2} - \frac{K \left(C_1 + C_2 \frac{h_s}{L_2 - L_1} \ln \frac{p_2}{p_1} \right)}{1 + K^2} \right] \quad (13)$$

The dimensionless constants K, C_1, and C_2 are only dependent on the screw geometry and the frictional coefficients at the screw, μ_{as} and barrel surface, μ_{az},

$$K = \frac{E(\tan \varphi_m + \mu_{as})}{1 - \mu_{as} \tan \varphi_m} \quad (14)$$

$$C_1 = \frac{\mu_{as}}{\mu_{az}} \frac{k_2}{k_1} \frac{2 h_s E}{s_s - (e_s z_s / \sin \varphi_m)} (K \tan \varphi_m + E) + \frac{\mu_{as}}{\mu_{az}} \frac{k_3}{k_1} C \cos \varphi_i (K \tan \varphi_i + C) \quad (15)$$

$$C_2 = \frac{E \cos \varphi_m \sin \varphi_m}{\mu_{az} k_1} (K \tan \varphi_m + E) \qquad (16)$$

where $E=(D-h_s)/D$, $C=(D-2h_s)/D$, $\tan \varphi_m = s_s/[\pi(D-h_s)]$, and $\tan \varphi_i = s_s/[\pi(D-2h_s)]$. With the pressures (or normal stresses) p in the downchannel direction, p_z at the barrel, p_f at the screw flights and p_s at the screw bottom, the anisotropy coefficients of the pressures are defined as:

$$k_1 = \frac{p_s}{p};\ k_2 = \frac{p_f}{p};\ k_3 = \frac{p_z}{p} \qquad (17)$$

These coefficients could be calculated or set to $k_1 = 1$, $k_2 = 0.4 \ldots 0.5$ and $k_3 = 1$.

With Eqns. (13) and (11), the pressure-dependent mass throughput in Eqn. (10) can now be calculated.

Solving Eqn. (12) for p_2 leads to the exponential pressure distribution $p(L)$, as expected:

$$p(L) = p_1 \exp\left(\frac{L - L_1}{h_s} \frac{\cos\alpha - K \sin\alpha - C_1}{C_2}\right) \qquad (18)$$

If the material data should be considered as dependent on pressure, sliding velocity, or temperature, or if the screw geometry is varying, a classical analytical solution is no longer available. But, by simply modifying the dimensionless constants K, C_1, and C_2, one can calculate in each element the pressure increase (or drop) iteratively, which is then integrated numerically within the length $L_2 - L_1$ of the solids conveying zone. Temperature changes in the interface between solids and screw, or solids and barrel, can be calculated if the frictional heat generation is balanced with heat conduction and convection in the solids and—if necessary—with barrel and/or screw cooling.

3.2 Solids Conveying Zone with a Grooved Inner Barrel Surface

Grooves in the inner barrel surface increase the effective external friction μ_{az} and can improve the conveying capacity considerably. This is due to the fact that, in the area portion A_i of the grooves, the internal coefficient of friction μ_i is effective, which is always higher than μ_a.

(*a*) *Smooth barrel conveying behavior.* One approximation to estimate an average coefficient of friction $\bar{\mu}_z$ at the barrel:

$$\bar{\mu}_z = A_i \mu_i + (1 - A_i)\mu_a;\ 0 \leqslant A_i < 1 \qquad (19)$$

Grünschloss (1984) developed a more sophisticated method to calculate $\bar{\mu}_z$ with the aid of a power balance, including transverse flow in shallow tapered grooves. If the depth of the grooves and its surface roughness is high enough, the solids stagnate in the grooves and the calculation can be done in the same way as for the smooth zone, by simply using $\bar{\mu}_z$ instead of μ_{az}. Compared with a smooth barrel, the output will be increased and is still dependent on the counterpressure p_2, but less so.

(*b*) *Forced conveying behavior.* However, if the groove depth is on the same scale as the granule size, and if the grooves are polished to improve self-cleaning, a significant velocity v_z of the solids in the groove direction is observed, especially in helical grooves (Grünschloss 1979, 1987). In this case throughput and pressure increase in the grooves must be calculated, too, which can be done separately and in analogy to the screw channel. For this purpose, screw and barrel channel are subdivided into volume elements of equal axial length dL, in which material data are at first assumed to be constant. According to Fig. 5, a force and momentum balance can be set up at a volume element in the screw and in the opposite barrel channel (groove). The coupling forces between these elements are mainly the internal frictional forces dRSI and dRZI in the direction of α (Grünschloss 1979).

For a given throughput, and on condition that the axial pressure increases dp in the screw, dp_s, and groove, dp_z, are the same (dp = dp_s = dp_z), the unknown velocity v_z in the grooves as well as the angle α can be calculated iteratively and—if desired—with stepwise varying geometry and/or material data (inner loop). Thus all the vector diagrams of velocities shown in Fig. 6 are possible solutions in principle, whereas the "nut-on-screw" conveyance, "shear-off", and "stagnation" are the most likely ones. If the pressures at the beginning (p_1 and p_2) and end (L_1 and L_2) of the zone are known, the wanted mass throughput $\dot{m}$ can be calculated iteratively in an outer loop.

The "nut-on-screw" conveyance, where the solids move in the groove direction (helical or axial), like a nut on a rotating screw, is of great practical interest, because the throughput $\dot{m}$ is at a maximum and *completely independent* of counterpressure p_2 over a wide range, Grünschloß (1993).

4. Modeling the Delay Zone

A delay zone was first defined by Kacir and Tadmor (1972) to explain the observed delay in melting in single-screw extruders. According to this theoretical model the delay zone starts at the end of the solids zone, where the surface of the solid plug reaches the melting temperature of the polymer, and ends where the steady-state melting with the formation of the classical melt pool starts (see Fig. 1). The authors

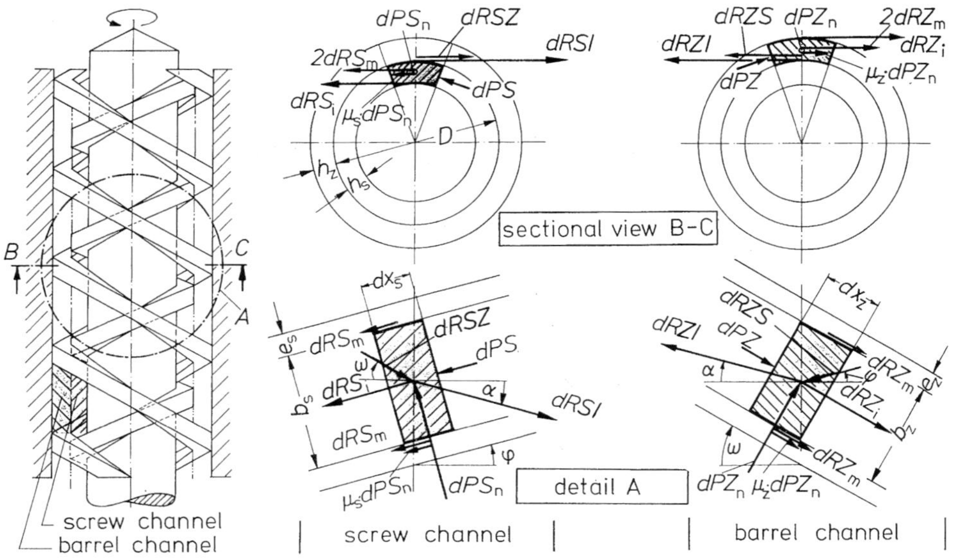

Figure 5
Force balances at the volume elements in the screw channel and barrel channel of a grooved solids conveying zone

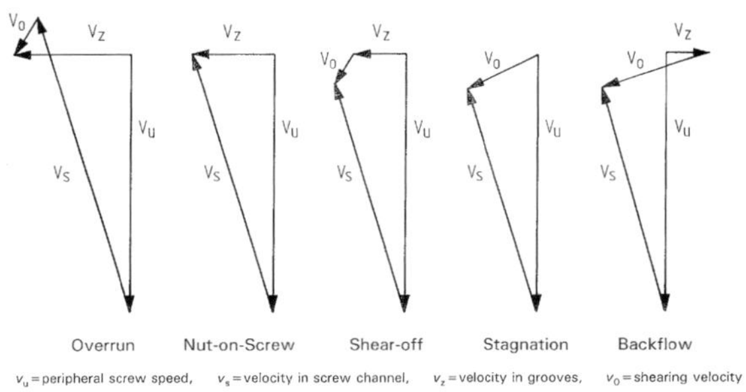

Figure 6
Vector diagrams of velocities in feed zones with axial grooves

assumed that there are at least two causes for this delay:

(i) the first traces of melt flow into the gaps between the particles rather than accumulating at the pushing flight to a melt pool; and

(ii) after formation of the melt film the melting mechanism can start to operate only after the film thickness δ_f exceeds the flight clearance δ_{Cl} several times. Only then will the cross-channel pressure drop developed in the melt film be high enough to segregate the solid bed at the trailing flight, allowing formation of a melt pool at the pushing flight.

Another very important phenomenon would be the wall slipping of melt (see Fig. 7). In the delay zone there must be a more or less gradual transition from the wall slipping of solids to the (assumed) wall sticking of melt. Even in the pure melt state, many polymers—and India rubber—show a more or less pronounced wall slip velocity v_{slip}, with or without a slip agent additive: see Ruoff *et al.* (1987), Joshi *et al.* (2000). It seems that v_{slip} can be approximated as a power law function of the wall shear stress τ_w. Many polymers show a sudden rise in wall slip velocity v_{crit} above a critical, temperature-dependent wall shear stress τ_{crit}

$$v_{slip} = v_w - v_f = v_0\left(\frac{\tau_w}{\tau_0}\right)^{n_1} \leqslant v_w;\ (\tau_w < \tau_{crit})$$
$$v_{slip} = v_{crit} + v_0\left(\frac{\tau_w}{\tau_0}\right)^{n_2};\ (\tau_w > \tau_{crit}) \qquad (20)$$

where v_w is the wall velocity, v_f is the maximum film velocity and v_0, τ_0, n_1 and n_2 are constants (Ruoff *et al.* 1987 for PE-LLD at 210°C: $v_0 = 5.5\,\mathrm{mm\,s^{-1}}$, $\tau_0 = 10^5\,\mathrm{Pa}$, $n_1 = 2.1$). With increasing melt temperature and increasing melt film thickness δ_f, v_{slip} will decrease (theoretically to zero). The slip velocity reduces the drag removal of melt in the film and is therefore an additional, if not the main, reason for the delay in melting. Besides this, with wall slipping at higher screw speeds the melting process is less efficient, because it is mainly dominated by heat conduction from the surrounding walls, rather than by viscous dissipation in the melt film (see also Maddock 1959).

The conveying mechanism in the delay zone is quite similar to that of the solids conveying zone, the only difference being that the forces at a volume element are no longer solid–solid friction forces but viscous drag forces. The magnitude of these forces

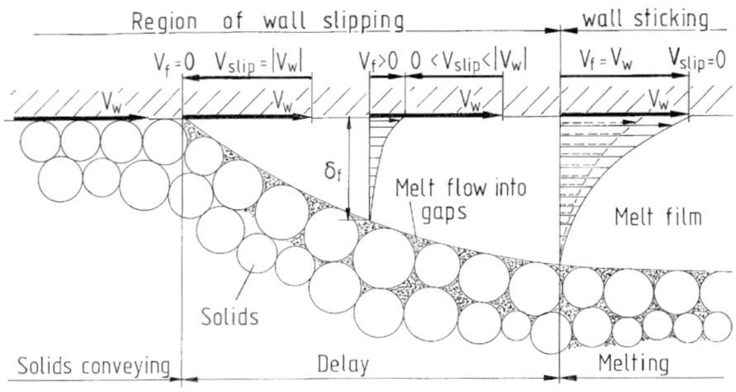

Figure 7
Principal processes in the delay zone: decreasing slip velocity vslip, increasing melt film velocity v_f, and melt penetration between particles

will thus depend on the shear stresses τ_w in the surrounding melt films. The calculations of the temperature profiles in the melt films and the solid bed, and the shear stresses, are quite similar to those for the melting zone (see later). For details the reader is referred to Kacir and Tadmor (1972).

5. *Modeling the Melting Zone*

The melting mechanism in a single-screw extruder was first experimentally analyzed by Maddock (1959) and verified by by numerous other authors. For some wall-slipping PVC-compounds, Klenk (1968) found that the melt pool forms at the trailing flight. No typical melt pool was found by Dekker (1976) processing polypropylene on an extruder with a screw diameter of 90 mm (see Fig. 8).

Maddock's melting mechanism was first analyzed mathematically by Tadmor (1966) and Tadmor and Klein (1970), whose work is at the basis of modifications and/or improvements in many further publications, the most important of which is Donovan (1971), who introduced solids bed acceleration. A critical review of these and other publications and their exceptional features is given in detail by Meijer (1980) and later by Lindt (1985).

Whereas Klenk's mechanism has not been further analyzed analytically, the Dekker melting model has found a certain acceptance. It seems that the Dekker mechanism is valid in larger extruders and in the end region of the melting zone (see Lindt 1976, Meijer 1980) where the melt films surrounding the solid bed are thick enough to allow melt circulation. In tapered screw channels, at the end of the melting zone the solid bed accelerates and tends to break up: see upper part of Fig. 9.

5.1 Modeling the Maddock Melting Mechanism

Halmos *et al.* (1978) have proposed one of the most reasonable models of Maddock's melting mechanism.

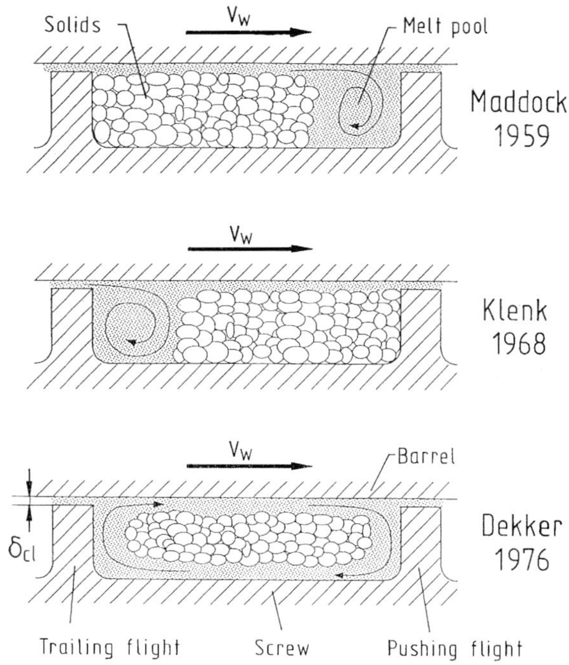

Figure 8
Different melting mechanisms in single-screw extruders

They consider additional melting from the screw root and trailing flight, allowing for solid bed deformation and acceleration in a tapered screw channel, and they make a force balance on the solid bed and the melt pool to calculate the pressure gradient. For this purpose, the cross section X–X of the screw in Fig. 9 is divided into six zones *A* to *F*, as shown in an unwound screw channel in Fig. 9. The deformable solids *A* are surrounded by the melt films at the barrel (*C*), at the screw root (*D*), at the trailing screw flight (*E*), and the melt pool *B*.

Melting occurs mainly at the interface between film *C* and solids *A* due to heat conduction and viscous heat dissipation in the melt film *B*, which is only some 0.1 mm thick. Due to the relative velocity v_w between solids and barrel, the melt produced (wall sticking) is immediately dragged away, whereby the melt film *C* is kept thin and heat conduction and dissipation are kept high. Afterwards, the melt is scraped at the pushing flight into the melt pool *B*. In zone *F* (screw clearance δ_{Cl}), a certain leakage flow leaves the screw channel and enters zone *C* again. The solids slide on the melt films *D* and *E* and are warmed up and molten from the hot screw and the melt pool *B*.

In the downstream direction, the melt pool *B* grows broader and pushes the deformable solids *A* to the trailing flight. For a high melting performance, the screw channel depth can be decreased to squeeze the solids and keep the melting area *C* at the barrel as large as possible.

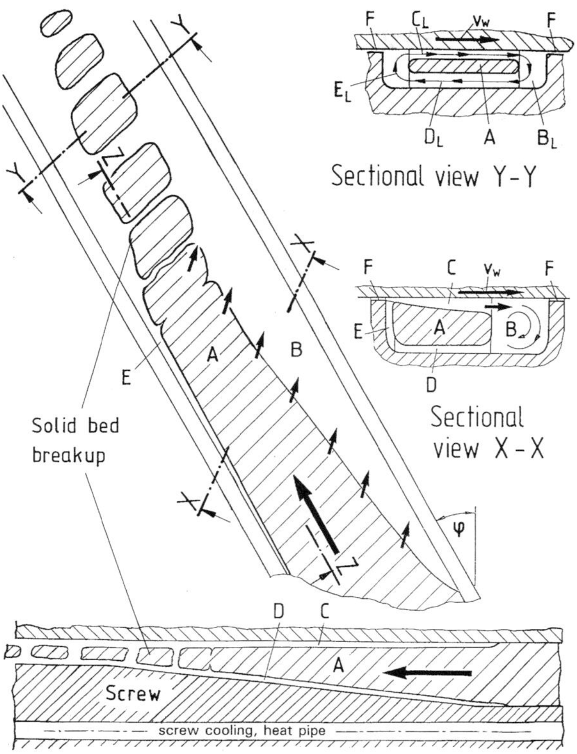

Figure 9
Melting mechanisms, solid bed break-up and 6-zone melting models in single-screw extruders

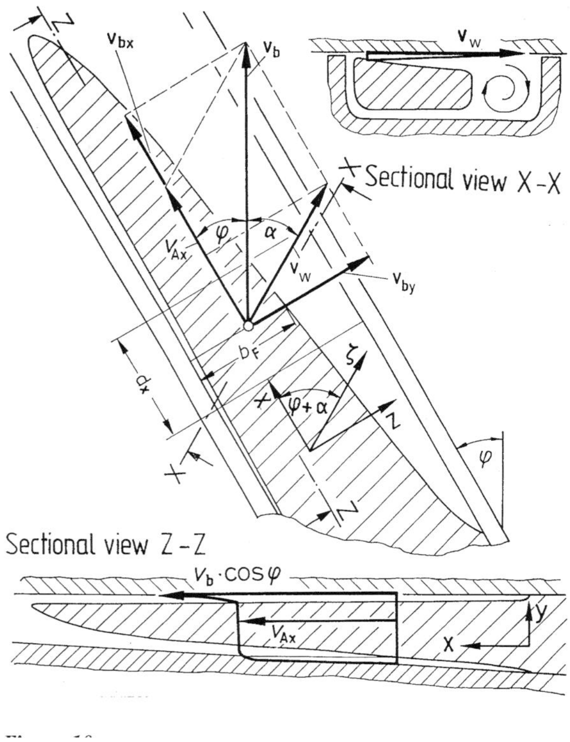

Figure 10
Velocities at a volume element in the solid bed of the melting zone

The screw channel is subdivided into volume elements in which the different velocities are defined (Fig. 10), and the energy balance (Fig. 11), mass balance (Fig. 12), and force balance (Fig. 13) at the solid bed and in the surrounding zones are formulated. The thermal processes in the zones can be treated independently of each other with acceptable accuracy to give analytical solutions. The pressure gradients in the melt pool *B* and in the solid bed must be equal, giving a condition for the width of the solid bed and its downstream acceleration, which must be calculated iteratively in each volume element. The features of the derivations are too complex to be given, so for details the reader is referred to the original paper.

However, they neglect the internal shear forces in the solid bed during deformation. A recently developed model and computer program (Grünschloss 1997) demonstrates the great influence of the solid bed shear strength on solid bed acceleration, melting performance, and downstream pressure gradient, particularly in tapered screw channels.

Maddock (1959) has already detected the large effect of lubricants on melting performance. Other authors confirm this finding. Therefore it seems

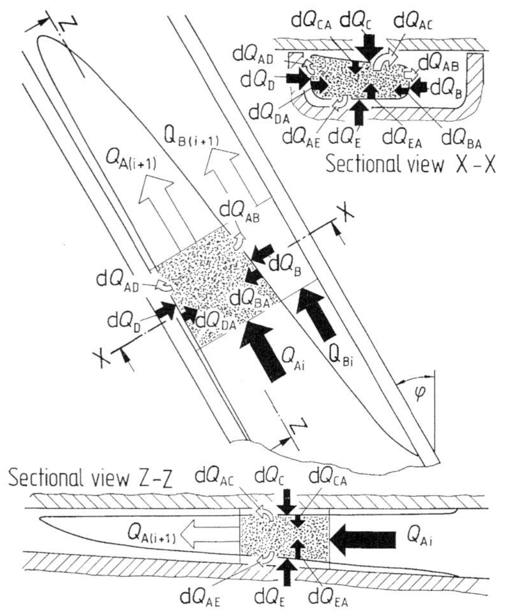

Figure 11
Energy balance at a volume element in the melting zone

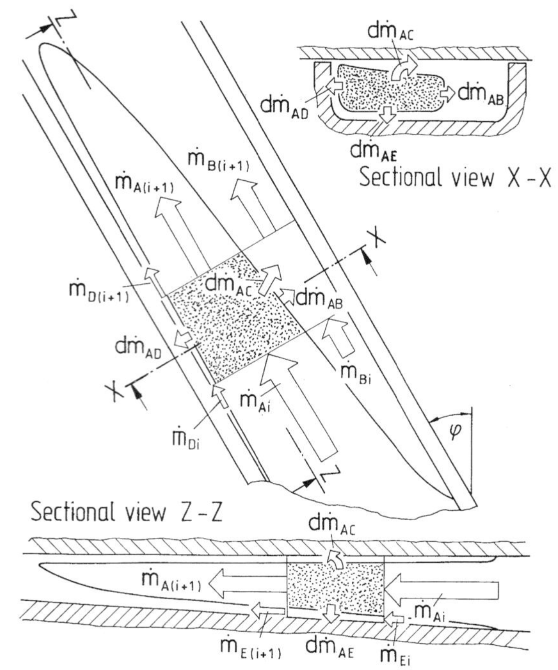

Figure 12
Mass balance at a volume element in the melting zone

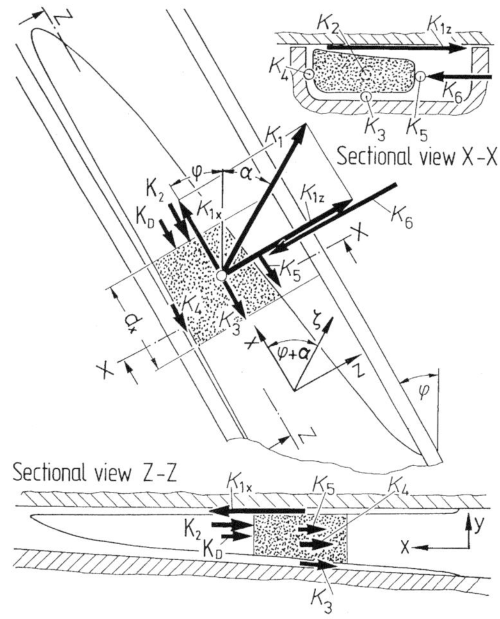

Figure 13
Force balance at a volume element in the solid bed of the melting zone

strictly necessary to introduce wall slipping in modeling the melting process (see also Sect. 4).

5.2 Modeling the Dekker Melting Mechanism

The Dekker melting mechanism was analysed by Lindt (1976), who introduced the calculation of the pressure gradient in the screw channel. He takes the solid bed to be rigid and considers only three zones, namely the regions A, C_L, and D_L (see Fig. 9: upper right), neglecting the zones B_L, E_L, and F. Similarly to the model described above, a simultaneously solved momentum and energy balance, together with a mass balance, gives the desired solution for melting rate, temperatures, and pressure gradient.

5.3 Modeling the Solids Breakup

The solids breakup mechanism was recently studied and modeled by Wong *et al.* (1997) and Zhu *et al.* (1997). Their numerical (finite difference) model, based on Lindt's approach but now including the zones B_L and E_L, allows the analysis of the dynamic behavior of the breakup of the (elastic/plastic) solid bed during melting. The physical properties at the interface of the different zones (see upper part of Fig. 9)—particularly the stresses at the solid bed—were analyzed. With their model, the location of solids bed breakup, pressure distribution, and complete melting can be estimated. Agreement with experiments is reported to be good.

5.4 Modeling Barrier Screws

Solids breakup, which is the main reason for surging in plasticating extrusion, should be and can be avoided: for example, with screw cooling or with barrier screws, where an additional screw flight with increased clearance δ_b separates the solids from the melt pool, allowing only melt or already workable material to pass through (Fig. 14). The melting mechanism is quite similar to that of Maddock (1959), and the mathematical treatment is analogous to that in Sect. 5.1. In the absence of solid bed breakup, modeling is more reliable and melt and solids channel can be treated separately, yet—as far as pressure development is concerned—not independently from each other. Comparative description and modeling of different barrier screw designs are given by Meijer (1980) and Ingen Housz and Meijer (1981).

6. ***Modeling the Melt Zone***

The melt conveying zone is the best understood extruder zone and has been treated by numerous authors. Reference should be made to the standard books of Schenkel (1963) for a comprehensive description of state-of-the-art polymer extrusion in

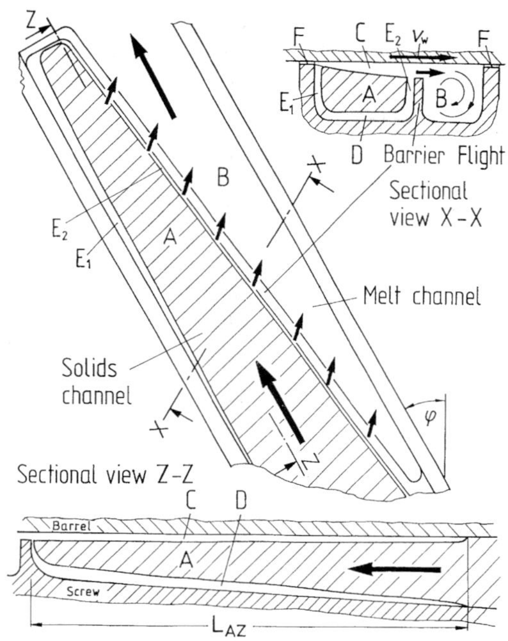

Figure 14
Modeling the melting mechanism in a barrier screw

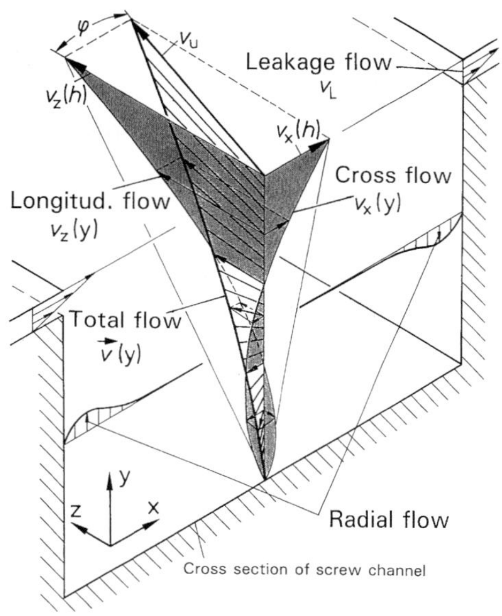

Figure 15
Different parts of flow in a melt-filled screw channel

1963 and the analytical solutions for Newtonian liquids; to Tadmor and Klein (1970) for a much more sophisticated treatment of the subject; and to Rauwendaal (1981). In these books many original papers are cited.

The fundamental flow of a wall-sticking melt in an unwound screw channel is shown in Fig. 15, with different velocity fields. The 2-dimensional velocity field of the total flow $\vec{v}(y)$ in the downchannel direction can be thought to be split up in a longitudinal flow $v_z(y)$ and a cross flow $v_x(y)$. The radial flow, which is created by the cross flow, is important for mixing and heat transfer between wall and melt, but is ignored in the following.

An analytical solution of the volume throughput G versus pressure gradient dp/dz relation is available, if the melt viscosity η is Newtonian and if—in a first approach—the influences of the screw flights and of channel curvature are neglected. The downchannel momentum balance reduces then to:

$$\frac{dp}{dz} = \eta \frac{d^2 v_z}{dy^2} \tag{21}$$

which after integration leeds to the velocity distribution $v_z(y)$

$$v_z(y) = v_u \cos\varphi \frac{y}{h_s} - \frac{h_s}{2}\frac{y}{\eta}\left(1 - \frac{y}{h_s}\right)\frac{dp}{dz} \tag{22}$$

with the peripheral screw speed v_u. Integration of $v_z(y)$ gives finally the volume throughput G:

$$G = G_d - G_p - G_l \tag{23}$$

with the drag flow G_d

$$G_d = z_s \tfrac{1}{2} b_s h_s v_u \cos\varphi F_d F_{Cd} \tag{24}$$

the pressure flow G_p

$$G_p = z_s \frac{b_s h_s^3}{12\eta}\frac{dp}{dz} F_p F_{Cp} \tag{25}$$

and the leakage flow G_l

$$G_l = z_s \frac{\pi^2 D^2 \delta \delta_{Cl}^3}{12\eta_{Cl} e_s}\frac{dp}{dz} \tag{26}$$

The corrections for the influence of the side walls are F_d and F_p and the corrections for the influence of the channel curvature are F_{Cd} and F_{Cp}. For flat channels with $h_s/b_s < 0.1$ and $h_s/D < 0.1$ these factors can taken to be unity.

For non-Newtonian polymer melts the viscosity η depends on the local shear rate

$$\dot{\gamma}(y) = \left[\left(\frac{dv_x}{dy}\right)^2 + \left(\frac{dv_z}{dy}\right)^2\right]^{\frac{1}{2}} \tag{27}$$

Therefore the above superposition of the drag and pressure flow is actually forbidden, and the problem should be solved numerically. Yet a good approximation can be achieved, if in the screw channel a "right" average viscosity η is defined to obtain the "right" value for the flow rate. For details see Tadmor and Klein (1970 p. 267 ff).

A less accurate but more practical possibility, which neglects the influence of different pressure gradients on the viscosity distribution, is to calculate the average viscosity from the viscosity shear rate function at an "average" shear rate $\dot{\gamma}_{av} = v_u/h_s$ in

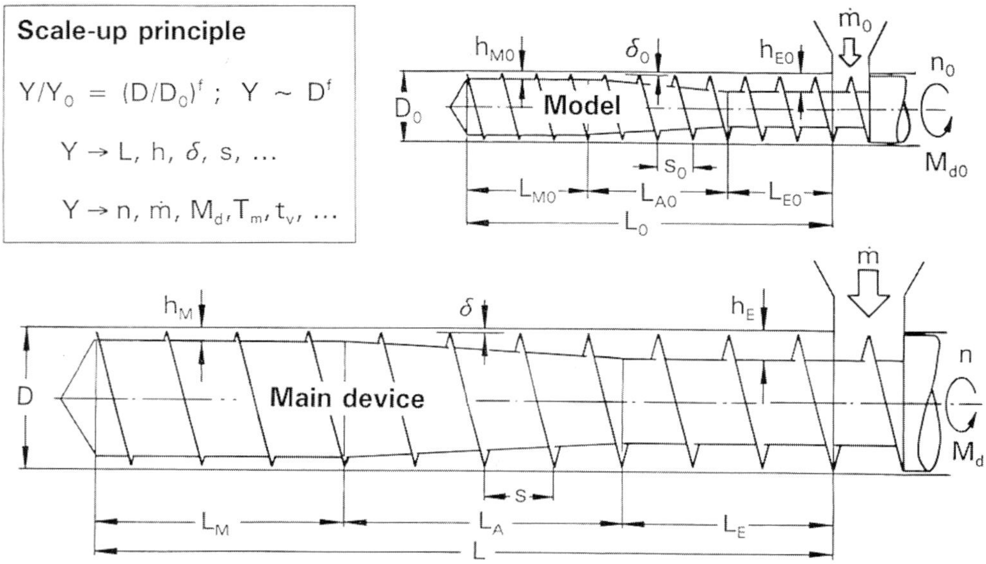

Figure 16
Model and main device in single-screw extruder scale-up

Table 2
Scale-up function f in Eqn. (28) according to Pearson and Grünschloss. The constant m is the index from the well-known power law $\dot{\gamma} = \varphi\tau^{m}$, and the constant σ holds for the change in bulk density ρ_{Sch}/ρ_{Sch0} with varying solids channel depth h_E/h_{E0}: $\sigma = \log(\rho_{Sch}/\rho_{Sch0})/\log(h_E/h_{E0})$.

Description	Variable	Scale-up function, f
Screw speed	n	$-2(m+1)/(m+3)$
Mass throughput	$\dot{m}$	$(m+5)/(m+3)$
Specific mass throughput	$\dot{m}/n$	$(3m+7)/(m+3)$
Torque	M_d	$(3m+7)/(m+3)$
Driving power	P	$(m+5)/(m+3)$
Heating/cooling power	H	$(m+5)/(m+3)$
Mean residence time	t_v	$2(m+1)/(m+3)$
Peripheral speed	v_u	$-(m-1)/(m+3)$
Mean shear rate	$\dot{\gamma}$	$-2m/(m+3)$
Mean shear deformation	γ	$2/(m+3)$
Length	L_E, L_A, L_M, L	1
Screw pitch and groove pitch	s_s, s_z	1
Screw clearance	δ	$(m+1)/(m+3)$
Channel depth (melt)	h_M	$(m+1)/(m+3)$
Channel depth (solids)	h_E	$(m+1)/[(m+3)(1+\sigma)]$
Initial groove depth	h_z	$(m+1)/[(m+3)(1+\sigma)]$
Groove width	b_z	$(m+1)/[(m+3)(1+\sigma)]$
Number of grooves	z_z	$1-(m+1)/[(m+3)(1+\sigma)]$

the screw channel and $\dot{\gamma}_{\text{Clav}} = v_u/\delta_s$ in the clearance, respectively.

By subdividing the melt zone into volume elements of length dz, variable geometry and changing melt temperatures can be accounted for. Due to the thermal feedback (melt temperature → viscosity → viscous heat dissipation → melt temperature), the resulting temperatures must be calculated iteratively in each element. For a given throughput and starting values for melt temperature and pressure, a stepwise numerical integration provides the final pressure and melt temperature at the end of the zone.

7. *Scale-up of Extruder Zones*

Scaling up extruder screws and grooved barrel sections, from a laboratory scale (D_0), the model, to a mostly larger production scale (D), the main device, is basically done on the assumption that polymer data, boundary conditions, pressure, and melt temperature remain unchanged. Yet this is not always the case, since polymer melts are non-Newtonian and the bulk density varies with channel depth. The main problem is that thermal effects related to a surface area (e.g., heat conduction) vary with $(D/D_0)^2$, and those related to the volume (e.g., heat convection, viscous dissipation) vary with $(D/D_0)^3$.

The most comprehensive work is probably the book of Potente (1981), who allowed for variation in melt temperature, viscosity, L/D relation, screw pitch, etc and extended his scale-up laws to twin-screw extruders. Grünschloss (1997) introduced later a variable bulk density in his scale-up laws of the solid conveying zone.

In all cases, the quantities Y of the main device are formulated as a power-law function of the model quantities: see Fig. 16.

$$\frac{Y}{Y_0} = \left(\frac{D}{D_0}\right)^f; \quad Y \sim D^f \tag{28}$$

where f could be a function of material data and/or geometrical variables. A list of the function f according to Grünschloß (1997) is given in Table 2.

See also: Polymer Processing: A Modeling Approach; Polymer Single Screw Extrusion: Software Packages

Bibliography

Darnell W H, Mol E A J 1956 Solids conveying in extruders. *SPE-J* **12**(4), 20–9

Dekker J 1976 Verbesserte Schneckenkonstruktion für das Extrudieren von Polypropylen. *Kunststoffe* **66**, 130–5

Donovan R C 1971 A theoretical melting model for plasticating extruders. *Polym. Eng. Sci.* **11**, 247–57

Grünschloss E 1979 Solids conveying in single-screw extruders with helical grooves in the barrel. *SPE-ANTEC* **37**, 160–5

Grünschloss E 1984 Zur Berechnung des mittleren Zylinderreibungskoeffizienten in genuteten Einzugszonen. *Kunststoffe* **74**, 405–9

Grünschloss E 1987 Feststoff-Förderung in Nutbuchsenextrudern. Video-Film, Institut für Kunststofftechnologie, Universität Stuttgart, Germany

Grünschloss E 1993 Schüttdichte und Massedurchsatz in Nutbuchsenextrudern. *Kunststoffe* **83**, 309–11

Grünschloss E 1997 Feststoff-Förderung im Einschneckenextruder (pp. 21–69), Aufschmelzen im Einschneckenextruder, (pp. 71–117). In: VDI-K (ed.) *Der Einschneckenextruder—Grundlagen und Systemoptimierung*. VDI, Düsseldorf, Germany

Halmos A L, Pearson J R A, Trottnow R 1978 Melting in single screw extruders: part III. *Polymer* **19**, 1199–216

Ingen Housz J F, Meijer H E H 1981 The melting performance of single-screw extruders. *Polym. Eng. Sci.* **21**, 352–9

Joshi Y M, Lele A K, Mashelkar R A 2000 A unified wall slip model. *J. Non-Newtonian Fluid Mech.* **94**, 135–49

Kacir L, Tadmor Z 1972 Solids conveying in screw extruders, part III: the delay zone. *Polym. Eng. Sci.* **12**, 387–95

Klenk P 1968 Plastifiziermodelle für die Verarbeitung benetzender und nichtbenetzender Thermoplaste auf Einschnecken-Extrudern. *Rheol. Acta* **7**, 75–8

Lindt J T 1976 A dynamic melting model for a single-screw extruder. *Polym. Eng. Sci.* **16**, 284–91

Lindt J T 1985 Mathematical modeling of melting of polymers in a single screw extruder: a critical review. *Polym. Eng. Sci.* **25**, 585–8

Lovegrove J G A, Williams J G 1973 Solids conveying in single screw extruders: the role of gravity forces. *J. Mech. Eng. Sci.* **15**, 114–22 and 195–199

Lovegrove J G A, Williams J G 1974 Pressure generation mechanisms in the feed section of screw extruders. *Polym. Eng. Sci.* **14**, 589–94

Maddock B H 1959 A visual analysis of flow and mixing in extruder screws. *SPE-J* **15**, 383–9

Meijer H E H 1980 Melting in single screw extruders: models, calculations, screw design. Dissertation, Technical University, Twente, The Netherlands

Potente H 1981 *Auslegen von Schneckenmaschinen Baureihen—Modellgesetze und ihre Anwendung*. Kunststoff-Fortschrittsberichte, Hanser, München, Germany

Rauwendaal C J 1981 *Polymer Extrusion*. Hanser, München, Germany

Ruoff M, Fritz H G, Geiger K 1987 Leistungssteigerung bei der LLDPE-Extrusion durch Verarbeitungshilfsstoffe. *Kunststoffe* **77**, 480–4

Schenkel G 1963 *Kunststoff-Extrudertechnik*. Hanser, München, Germany

Tadmor Z 1966 Fundamentals of plasticating extrusion I: theoretical model for melting. *Polym. Eng. Sci* **6**, 185–90

Tadmor Z, Klein I 1970 *Engineering Principles of Plasticating Extrusion*. Van Nostrand, New York

Tadmor Z, Broyer E 1972 Solids conveying in screw extruders, part II: non isothermal model. *Polym. Eng. Sci* **12**, 378–86

Tadmor Z, Gogos C G 1979 *Principles of Polymer Processing*. Wiley, New York

Tucker C L III 1989 *Computer Modeling for Polymer Processing—Fundamentals*. Hanser, München, Germany

VDMA 1981 *Kenndaten für die Verarbeitung thermoplastischer Kunststoffe, Teil 1: Thermodynamik*. Hanser, München, Germany

VDMA 1982 *Kenndaten für die Verarbeitung thermoplastischer Kunststoffe, Teil 2: Rheologie*. Hanser, München, Germany

VDMA 1983 *Kenndaten für die Verarbeitung thermoplastischer Kunststoffe, Teil 3: Tribologie*. Hanser, München, Germany

Walker D M 1966 An approximate theory for pressure and arching in hoppers. *Chem. Eng. Sci.* **21**, 975–97

Wong A C-Y, Zhu F, Liu R, Liu T 1997 Breakup of solid bed in melting zone of a single-screw extruder, part I: mathematical model. *Plast. Rubber Compos. Proc. Appl.* **26**, 336–42

Zhu F, Wong A C-Y, Liu R, Liu T 1997 Breakup of solid bed in melting zone of a single-screw extruder, part II: Strength of solid bed and experimental results. *Plast. Rubber Compos. Proc. Appl.* **26**, 343–50

E. Grünschloss

Polymer Single Screw Extrusion: Software Packages

Today, optimization of screw design is of great importance for economical plastics processing. The software REX (Rechnergestützte Extruderauslegung: computer-aided extruder layout) helps to gain important process information by simulating the process within a single screw extruder. Thus, necessary changes can be made before the actual screw is built without using the expensive and time-consuming *trial-and-error* method. The REX software is based on mathematical–physical models. These models incorporate the possibility of rapidly calculating important process values, e.g., the throughput and pressure profile along the screw length, while providing fast and convenient data entering.

1. Introduction

By reducing experimental investigations to a minimum the use of simulation software helps achieve both a fast and economical extruder layout as well as an optimization of an already existing extrusion process (Potente and Hanhart 1994, Potente and Jungemann 1996). One additional advantage of simulation software over experimental investigations is the fact that the extrusion process becomes more transparent. This is because one can look inside the machine. With simulations one can gain access to process values which are impossible or very difficult to measure. Thus process simulations can avoid a single screw extruder being seen as a black box. Contrary to hardware and time consuming simulation methods like, for example, the finite element method (FEM), REX incorporates mathematical–physical models which are very fast while running on small PC-like computers. In addition, REX is specialized for simulating single screw extruders. Thus data entering with REX is more convenient and much faster compared to most finite element programs because of their more general simulation approach leading to a time consuming preprocessing. Moreover, contrary to REX most of these simulation techniques do not incorporate models for solid conveying and melting within the extruder. Therefore, only melt flow can be observed by these programs.

2. Process Simulation

In order to set up a process it is first necessary to define the screw geometry. The user is requested to setup the screw by combining several screw sections like feed, compression, decompression, metering, and barrier sections as well as various shearing and mixing sections. For each individual screw section the user enters several geometric values, e.g., the channel height. Also the barrel has to be defined. A number of parameters, such as the diameter, are adopted from the screw data. Besides the heating zone data, the grooved bush geometry and cooling channel data are also part of the barrel if one simulates a grooved bush extruder. The material data describe the rheological, thermodynamic, and tribological behavior as well as the density. All the settings and boundary conditions which can additionally be adjusted on a real-life extruder, such as the screw speed, are entered under the heading of process parameters. Since REX offers many different simulation models the user has to determine which calculations the program should perform. Depending on the type of extruder the user can select between several throughput models. For smooth barrel extruders the throughput can be calculated assuming pure melt conveying. This is relevant for most extruders if the metering section dominates the throughput behavior. Another model (Jungemann 1998) which can be selected takes melt conveying, solids conveying, and plasticizing into account. For grooved bush extruders the throughput can be calculated assuming independency of back pressure or, alternatively, by using a combined solids conveying, plasticizing, and melt conveying model which takes a possible melting of the pellets in the grooves, as well as the melting performance, into account. Besides that, the pressure profile, temperature profile, power consumption, shear stress at the barrel wall, minimum and average residence time, and the overall melting performance along the screw length can be simulated. In addition, a number of curves can be calculated in order to gain information on the pressure dependency of the throughput.

The basis of the REX software are mathematical–physical models (Potente *et al.* 1992, Schöppner 1994) which have been developed by means of experimental investigations, finite element calculations, and theoretical analytical derivations. They describe various physical values with fairly simple mathematical equations. Figure 1 shows a simplified flow chart of the simulation algorithm. By dividing the complete screw into several intervals, and solving the equations for each interval, the complex extrusion process can be broken down. The complexity of the process, however, still incorporates some iterative numerical algorithms. The whole simulation runs in a certain order since some values can only be calculated if others are already known. Since the throughput is important for all other simulation results it is calculated first. Subsequently, the pressure and the melting profile are calculated, if selected. The temperature profile cannot be calculated until the melting performance is known. The simulation of the residence time and the power consumption require both the melting and temperature profile. Once the power consumption, melting, and temperature profiles are known, the wall shear stress profile can be calculated.

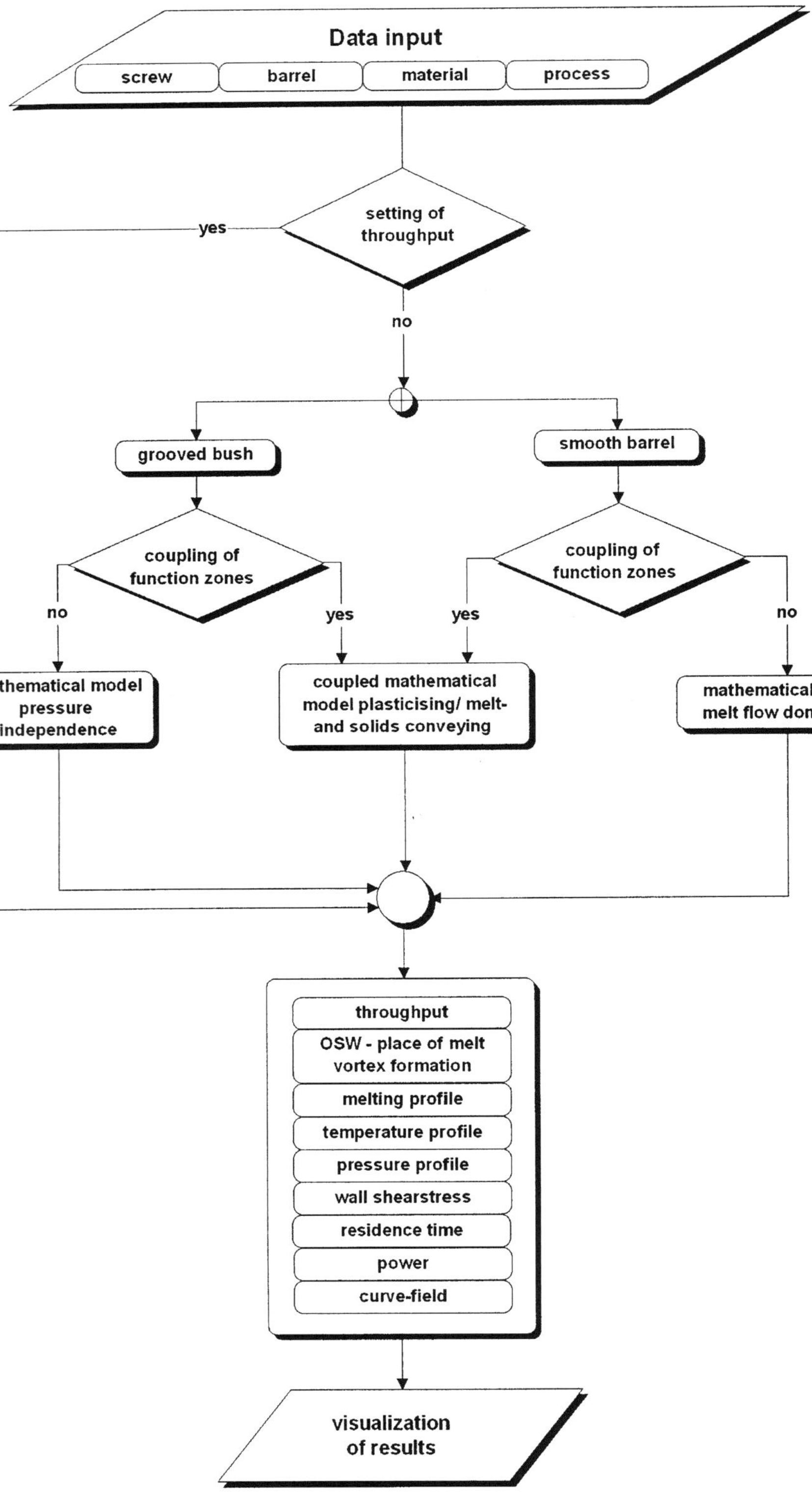

Figure 1
Flow chart of the simulation algorithm.

REX presents the simulation results as graphs or as a listing. Figure 2 shows an example of the results of pressure profile over screw length. Important parameter settings, i.e., the throughput and a schematic drawing of the screw, are also shown. Similarly the melting profile, temperature profile, residence time, and power consumption can be presented.

3. *Scale-up and Batch Processing*

Some additional features of the REX software simplify the data input and help the user to optimize the screw design. One is the scale-up feature. Based on an optimized screw, from which the geometrical data are used as input, the software generates a new screw geometry, with a user-defined diameter, by using certain scale-up rules. If modification of the scale-up equations is necessary, the parameters can be changed. Afterwards the screw generated can be optimized by performing some additional simulations. This feature helps to find a good screw design to start an optimization, and thus to design a series of extruder machines with differently sized screws.

A second feature is the batch processing module. This feature allows the user to run several simulations with a slight variation of one or two geometrical or process parameters. Thus, the influence of these parameters can easily be observed and the optimization of screw design is simplified. If REX runs in batch processing, all simulations are done consecutively. For one varying parameter REX presents the results with differently colored curves with each curve representing the result for one particular parameter setting. If two parameters vary, only single point data, e.g., the maximum pressure, can be shown in the graphs while each varying parameter belongs to one axis.

4. *Concluding Remarks*

The REX software incorporates the extrusion theory in a user friendly software package. With its easy-to-handle user interface and fast simulation algorithms,

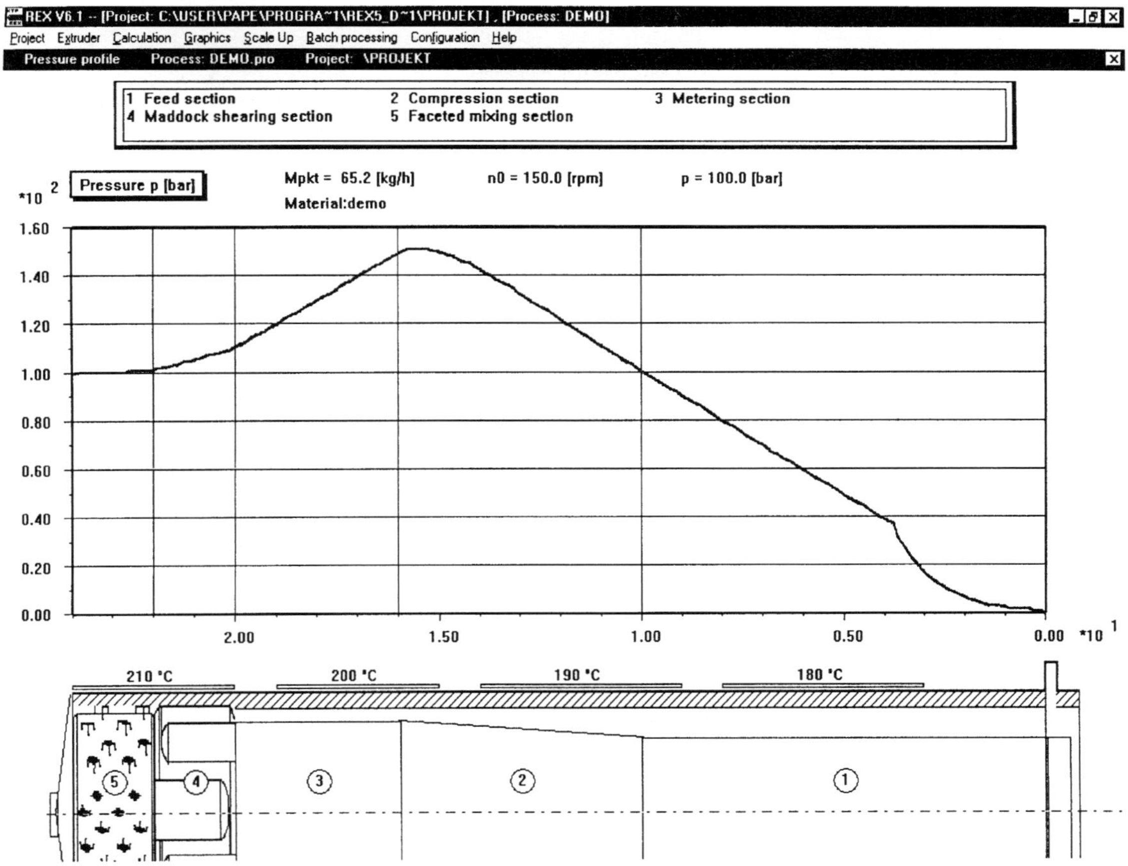

Figure 2
Pressure profile along the screw length.

it provides many features that help to optimize and lay-out an extruder design, as well as getting important information about the process itself. Therefore, it is a very useful tool for the plastics processing industry.

Bibliography

Jungemann J 1998 Verbesserte Modellbildung am Glattrohr-plastifizierextruder. Dissertation, Universität Paderborn, Germany

Potente H, Effen N, Hanhart W, Kessler H, Koch M, Michel P, Schöppner V, Schultheis M, Tappe P, Wortberg J 1992 *Rechnergestützte Extruderauslegung*. Handbuch des Kunststofftechnischen Seminars. Universität Paderborn, Germany

Potente H, Hanhart W 1994 Design and processing optimization of extruder screws. *Polym. Eng. Sci.* **11**, 937–45

Potente H, Jungemann J 1996 Auslegung und Optimierung mittels Simulation. *Plastics Special* **9**, 4–10

Schöppner V 1994 Simulation der plastifiziereinheit von einschneckenextrudern. Dissertation. Universität Paderborn, Germany

H. Potente and J. Pape

Polymer Twin Screw Extrusion: 2D Modeling

Fully intermeshing, corotating twin-screw extruders (TSEs) are frequently used in many operations that involve polymer blending, devolatilization, or the incorporation of solids such as fillers or reinforcing agents into a polymeric matrix. This is largely due to their flexible screw design, which can be fitted "to the job," good mixing capacity, and self-wiping action. Besides, the advantage of TSEs over conventional single screw extruders is their high solids intake rate and melting capacity.

This article will cover three aspects of twin-screw extrusion: (i) the *axial pressure profile* caused by the presence of subsequent zones which are partially filled, combined with locally pressurized zones; (ii) the mechanisms involved in the so-called *dissipative mix-melting process*; and (iii) the importance of screw geometry and typical sequence of screw elements required for optimal *mixing*.

1. Screw Geometry—Filled Lengths

The screw geometry of a TSE is modular, which means that screw elements are placed in a particular sequence (there are, of course, two identical screws) onto a shaft until the desired configuration for a given purpose is achieved. The sequence of screw elements corresponds to openings in the barrel, which is also segmented. This makes it possible to build separate zones, each with its own function, such as melting, additive or glass-fiber mixing, or devolatilization (Andersen 1998).

To make all of this possible, TSEs must be operated in a starve-fed mode, which means that the actual

Figure 1
Transition from partially filled to fully filled screw channel (indicated by the arrow), in a combination of a screw element with positive pitch (on the right) with a screw element with negative pitch (on the left). The image represents one frame of a high-speed camera film and shows the flow inside a Plexiglas-walled laboratory twin-screw extruder (screw diameter 25 mm). (Reproduced by permission of Hanser from "Mixing and Compounding—Theory and Practice".)

throughput is only a fraction of the maximum theoretical throughput. The material (which is metered by some gravimetric feeding device) only fills a fraction of the available cross-section of the extruder channel. The throughput of the extruder is fully determined by the input from the feeder and no longer by the feed from the hopper, as is the case with single-screw extruders. In the case of a twin-screw extruder, screw speed and throughput can therefore be controlled independently of each other within certain limits.

A typical screw geometry starts with an intake zone which consists of several conveying elements (uninterrupted flights) having a large pitch. The conveying screw elements that follow have a smaller pitch, in order to compact the granular or powdery polymer prior to melting. Since the cross-section (and hence the channel depth) is constant along the screw axis, the decreased pitch serves the same purpose as the gradual decrease in channel depth (the compression) that is common in single screw extruders.

The polymer is melted in a so-called kneading section. The screw elements in this section consist of a sequence of ellipsoidal discs of a certain width, with an axial stagger angle, e.g., 45° or 90°. These kneading elements—some of which even may have a negative pitch—have a lower pumping capacity than regular screw elements, and are therefore completely filled. This implies that at a certain distance upstream from the kneading elements there is a transition from partially filled to fully filled screw channels (see Fig. 1). Downstream of the pressurization/depressurization zone, the channel is again partially filled.

The kneading elements represent a flow restriction: with increasing throughput (at constant screw speed) the filled length increases. At constant throughput, the filled length decreases with increasing screw speed. The greater the filled length, the greater the polymer/cylinder contact area, and the greater the motor power that is required.

In cases where the TSE is fed with hot melt, e.g., when it serves as a finishing extruder, the filled length can be calculated, based on an analysis that involves local pressure gradients (Armstroff and Zettler 1973, Meijer and Elemans 1988, Wang *et al.* 1989, Vergnes *et al.* 1998).

2. *Melting*

One of the advantages of corotating twin-screw extruders is their large melting capacity. In a co-rotating twin-screw extruder, the melting of the polymer is not due to conduction melting with drag-induced melt removal (as in single screw extruders), but to the so-called dissipative mix-melting mechanism (Tadmor and Gogos 1979). In this mechanism, mixing and melting are closely linked. The energy needed for melting is introduced through the screw shafts and is converted into heat by constant deformation of the (mostly granular) material. Basically, there are three

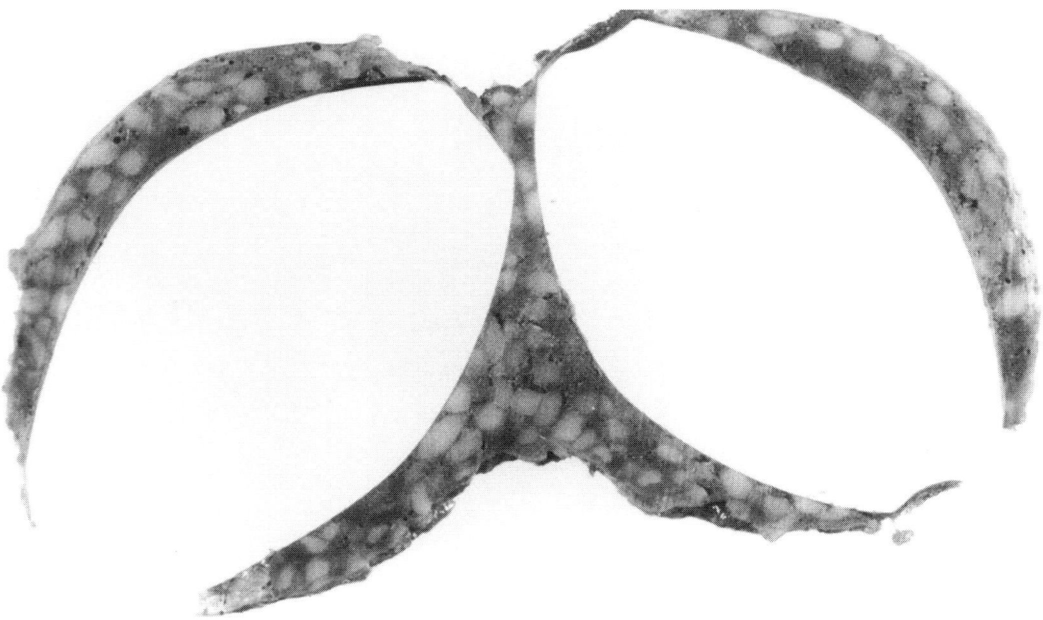

Figure 2
Mixture of melt with unmelted fragments, which is present at some cross-section along a kneading section in a corotating twin-screw extruder.

mechanisms going on: deformation of the polymer granules, dry friction between the granules, and viscous dissipation in the already molten melt films between the granules (Gogos *et al.* 1998, Todd 1992, Esseghir *et al.* 1997). As the melting progresses, the importance of the last mechanism increases. This melting process is not only efficient, but also uniform. At any cross-section of the extruder, there is a homogeneous mixture of melt and solids (see Fig. 2). The composition of this mixture changes from a solid-rich suspension to a melt-rich suspension within a short axial length (typically three diameters as compared to a melting length of about 10 diameters in single-screw extruders). Local "hot spots" are simply mixed into the bulk of the material. Since the energy has to be introduced by the shafts, the melting capacity and thereby the throughput are limited by the torque of the machine. Although a theoretical analysis of the melting process is still far from complete, recently some advancement has been made (Vergnes *et al.* 1999).

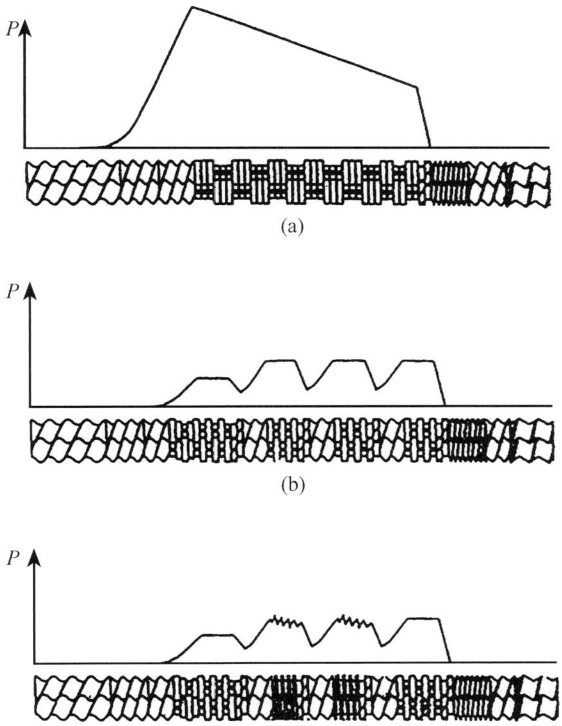

Figure 3
Possibilities of avoiding high pressures by using an optimized screw geometry for compounding of high loading of filler (powder) into a base polymer. Geometry "c" gives optimal dispersion, whereas screw "a" gives the worst dispersion. (Reproduced by permission of Giesel Verlag from *Kunststoffberater*, 1991, **10/91**, 24–8.)

3. Dispersive and Distributive Mixing

Co-rotating twin-screw extruders are often used in compounding operations that involve the dispersing of high concentrations of additives (mostly powders) in polymers. This is due to their good distributive as well as dispersive mixing performance, since all the material introduced into the extruder regularly proceeds through the various zones characterized by a high shear stress, followed by zones with a good distributive mixing (Manas-Zloczower *et al.* 1982).

The distributive mixing action in co-rotating twin-screw extruders is due to the frequent stretching and folding that takes place in the screw channel, especially in the so-called intermeshing region, where the material is passed from one screw onto the other (Teverovskiy and Manas-Zloczower 1999).

To enhance dispersive mixing, the screw can be composed in such a way that the base polymer is brought into contact with the additive powders virtually without any pressure. This can be advantageous for compounds that involve powders, which tend to form agglomerates under influence of hydrostatic pressure (Smith 1974, Mack 1991). To avoid pressure peaks (leading to agglomerates), the screw should contain a sequence of kneading sections (Colbert and Dillon 1991). This is illustrated in Fig. 3.

In recent years, considerable progress has been made with regard to the modeling of the mixing mechanisms (Avalosse and Crochet 1997) as well as the melting mechanisms (Gogos *et al.* 1998, Vergnes *et al.* 1999) in co-rotating twin-screw extruders. Continuation of the modeling along these lines will certainly lead to better screw design and thus to better performance of these extruders.

Bibliography

Andersen P 1998 In: Todd D B (ed.) *Plastics Compounding: Equipment and Processing*. Hanser, Munich, Germany

Armstroff O, Zettler H D 1973 Ansätze zur Berechnung von Zweischneckenextrudern. *Kunststofftechnik* **12**(9), 240–3

Avalosse T, Crochet M 1997 Finite-element simulation of mixing: 1. Two-dimensional flow in periodic geometry. *AIChE J* **43**(3), 577–87

Colbert J A, Dillon M 1991 Fortschritte im Bereich der Doppelschneckentechnik bei der Herstellung von Kunststoff-Compounds. *Kunststoffberäter* **10/91**, 24–8

Elemans P H M 1994 In: Manas-Zloczower I, Tadmor Z (eds.) *Mixing and Compounding—Theory and Practice*. Hanser, Munich, Germany

Esseghir M, Yu D W, Gogos C G, Todd D B, Tadmor Z 1997 Melting mechanisms of single-component polymers in co-rotating twin-screw kneading blocks through visual and microscopic analysis. *SPE ANTEC Tech. Papers* **55**, 3684–9

Gogos C G, Tadmor Z, Kim M H 1998 Melting phenomena and mechanisms in polymer processing equipment. *Adv. Polym. Technol.* **17**(4), 285–305

Mack M H 1991 Continuous compounding of color masterbatches. *Plast. Eng.* **12/ 91**, 39–42

Manas-Zloczower I, Nir A, Tadmor Z 1982 Dispersive mixing in internal mixers—a theoretical model based on agglomerate rupture. *Rubber Chem. Technol.* **55**, 1250–85
Meijer H E H, Elemans P H M 1988 The modeling of continuous mixers: 1. The corotating twin-screw extruder. *Polym. Eng. Sci.* **28**(5), 275–90
Smith M J 1974 Some physical properties of pigments and their importance in relation to dispersion in plastics. *J. Oil Col. Chem. Assoc.* **57**, 36–43
Tadmor Z, Gogos C G 1979 *Principles of Polymer Processing.* Wiley, New York
Teverovskiy M, Manas-Zloczower I 1999 http://www.osc.edu/abe/viz/Manas/dblcouette.html
Todd D B 1992 Melting of plastics in kneading blocks. *SPE ANTEC* 2528–36
Vergnes B, Delacour M L, Souveton G, Bouvier J M 1999 A study of polymer melting in a corotating twin-screw extruder. H E H Meijer (ed.) *Proc 15th Annu. Meeting Polymer Processing Society.* 's Hertogenbosch, The Netherlands (CD-ROM)
Vergnes B, Della Valle G, Delamare L 1998 A global computer software for polymer flows in corotating twin-screw extruders. *Polym. Eng. Sci.* **38**(11), 1781–92
Wang Y, White J L, Szydlowski W 1989 Flow in a modular intermeshing co-rotating twin-screw extruder. *Int. Polym. Proc* **IV**(4), 262–9

P. H. M. Elemans

Polymer Twin Screw Extrusion: 3D Modeling

Developments in computational technology have given rise to a new scientific field of study. In polymer processing, there is great demand for computational technology that aids in revealing complex and black-box phenomena as well as aiding in developing new process conditions. One of the important processes in polymer processing is mixing, compounding, and homogenizing of different polymers and additives, sometimes developing ordered structures of materials with enhanced properties. Twin screw extruders (TSEs) are widely used in mixing or compounding of highly viscous and immiscible polymeric materials. However, the flow and mixing phenomena inside TSEs are not well understood. The reasons are that the detailed 3D complex and time-dependent change of the flow geometry proves fundamentally important in the mixing process in the intermeshing zones. Furthermore, multiphase transition and interface-diffusion processes occur inside a wide range of the TSE. The 3D time-dependent numerical simulations require powerful computers, but they become quite effective tools in revealing these detailed phenomena and in designing optimal conditions, which could not be obtained from experimental studies.

Today, the finite element method (FEM) is widely used in applying 3D numerical simulations, because of its reliability in solving complex nonlinear partial differential equations and generality in applying unstructured meshes fitted to complex geometries. The filled zones of the typical geometries of TSEs, such as full-flighted screws, kneading and rotor blocks and their combinations, have been simulated by FEM and, by using thousands of markers, the mixing patterns have been visualized. Table 1 summarizes the applicable range of 3D numerical simulation results for TSEs at present. In the near future, unfilled and melting processes, including dynamic interface or surface motions with multiphase transitions, will be simulated by elaborate numerical methods along with theoretical developments.

1. Numerical Method

The basic assumptions of numerical simulations of internal TSE flow are as follows: (i) the fluid is incompressible and pure-viscous non-Newtonian (the

Table 1
The applicable range of 3D numerical simulation results for twin screw extrders at present.

Twin-screw element		Geometric configurations		Velocity distr.	Leakage flows	Shear stress distr.	Pressure drop	Residence time distr.	Marker tracking Animation	Temperature distr.	Reaction
1	Single-flighted (SF)	Rectangular channel		◎	◎	◎	◎	◎	○	○	○
		Self-cleaning		◎	◎	◎	◎	◎	○	○	○
2	Double-flighted (DF)	Self-cleaning		◎	◎	◎	◎	◎	◎	○	○
3	Rotor	Continuous		◎	◎	◎	◎	○	○	◎	○
		Discontinuous		◎	◎	◎	◎	○	○	○	○
4	Kneading blocks (KB)	Two-tip	Right-handed (R) stagger angle	◎	◎	◎	◎	◎	◎	◎	◎
			Neutral(N) stagger angle	◎	◎	◎	◎	◎	◎	◎	◎
			Left-handed(L) stagger angle	◎	◎	◎	◎	◎	◎	◎	◎
		Three-tip	Right-handed (R) stagger angle	◎	◎	◎	◎	◎	◎	○	○
			Neutral(N) stagger angle	◎	◎	◎	◎	◎	◎	○	○
			Left-handed (L) stagger angle	◎	◎	◎	◎	◎	◎	○	○
5	Assembly	3-pitch DF/R+KB/R+DF/R DF/R+KB/N+DF/R DF/R+KB/L+DF/R		◎	◎	◎	◎	◎	◎	◎	○
		4-pitch DF/R+KB/N+DF/L+DF/R DF/R+KB/L+DF/L+DF/R DF/R+KB/N+DF/R+DF/R DF/R+KB/L+DF/R+DF/R		◎	◎	◎	◎	◎	◎	○	○
		10-pitch 2DF/R+3KB/R+2KB/N +KB/L+2DF/R		◎	◎	◎	◎	◎	◎	○	○

◎ : Done ○ : Possible

Carreau model); (ii) inertia and gravitational forces are negligible and the fluid is in a quasi-steady state because of its high viscosity; (iii) the fluid sticks on solid surfaces; (iv) the flow channel is fully filled with a polymer melt; and (v) the flow is in an isothermal and quasi-steady state. These assumptions must be modified depending on the situation. For example, Ishikawa *et al.* (1999, 2000) calculated the simultaneous equations of energy and reaction in order to analyze the non-isothermal polypropylene peroxide degradation process in 3D kneading blocks. The conservation of mass and momentum, and the constitutive equations with the basic assumptions, are as follows:

$$\nabla \cdot \mathbf{v} = 0 \tag{1}$$

$$-\nabla p + \nabla \cdot \boldsymbol{\tau} = 0 \tag{2}$$

$$\boldsymbol{\tau} = 2\eta \mathbf{D} \tag{3}$$

$$\eta = \eta_0[1 + 2\lambda_c \mathrm{II_D}]^{n-1/2} \tag{4}$$

where $\mathbf{v}$ is the velocity vector, p is the hydrostatic pressure, $\boldsymbol{\tau}$ is the extra stress tensor, and $\mathbf{D}$ is the rate of deformation tensor. $\mathrm{II_D}$ is the second invariant of the $\mathbf{D}$, which is defined by $\mathrm{II_D} = \mathbf{D}{:}\mathbf{D}/2$. η_0, n, and λ_c are material constants that are determined by fitting the constitutive equation to experimental data of the shear viscosity (η) curve.

Funatsu *et al.* (1999) solved the equations by the conventional Galerkin FEM using 27 node tri-quadratic brick elements for the three velocity components, and the penalty function method for the pressure variable $p = -\lambda(\nabla \cdot \mathbf{v})$; λ is penalty number. Pressure values were obtained by the least square Galerkin FEM among the known velocities. Concerning the convection term, the streamline upwinding Petrov–Galerkin method is usually applied to get oscillation-free solutions. The finite element matrixes are decomposed with a direct method. The inlet and outlet boundary velocities are applied by using a periodical flow within the same geometry. The time-dependent flow caused by the screw rotations is constructed with a velocity distribution obtained for every 3 degrees of screw rotation.

The time-dependent properties, such as residence time distribution and Lyapunov exponent, in flowing systems, are obtained from marker tracking analysis. Either the Euler forward integration method or the fourth-order Runge–Kutta method is used for time integration in the consecutive velocity distribution. In some cases, the tetrahedral decomposition method is applied to search for the element where the marker exists.

2. Criteria of Mixing

Several kinds of mixing indices have been proposed to judge the mixing abilities of TSEs. Universal indices have not been established, because the criteria of mixing are dependent on its purpose. Most of the proposed indices relate to homogenized materials. One index expresses the local mixing performance, such as flow number, Lyapunov exponent, area stretching, etc. Others are related to the whole mixing system, including the residence time distribution for perfect mixing, pair correlation functions, Poincaré sections, etc. Cheng and Manas-Zloczower (1998) and Yao and Manas-Zloczower (1998) examined several mixing indices which dealt with distributive and dispersive mixings in TSEs. Funatsu *et al.* (1999) proposed two kinds of mixing indices, representing the time-dependent marker distribution from the viewpoint of the distance between markers, and accordingly applied them to 4-pitch assemblies of TSEs as described in the next section. One represents the mean nearest distance between markers; $\overline{d_{\min}}$ and its variance, $S^2_{\min}$. The other represents the mean value of the average distances between markers; $\overline{d_{\mathrm{ave}}}$ and its variance, S^2_{ave}. These mixing indices are defined by the following equations:

$$\overline{d_{\min}} = \frac{1}{n}\sum_{i=1}^{n} d_{\min,i} = \frac{1}{n}\sum_{i=1}^{n} Min\left[\bigcup_{j=1}^{n-1} |\mathbf{r}_i - \mathbf{r}_j|\right] \tag{5}$$

$$S^2_{\min} = \frac{1}{n}\sum_{i=1}^{n}(\overline{d_{\min}} - d_{\min,i})^2 \tag{6}$$

$$\overline{d_{\mathrm{ave}}} = \frac{1}{n}\sum_{i=1}^{n}(d_{\mathrm{ave},i}) = \frac{1}{n}\sum_{i=1}^{n}\left(\sum_{j=1}^{n-1}\frac{|\mathbf{r}_i - \mathbf{r}_j|}{n-1}\right) \tag{7}$$

$$S^2_{\mathrm{ave}} = \frac{1}{n}\sum_{i=1}^{n}(\overline{d_{\mathrm{ave}}} - d_{\mathrm{ave},i})^2 \tag{8}$$

where $\mathbf{r}_i$ is the position of marker i and n is the number of markers. These mixing indices indicate the local and the average distributive mixings in the TSE, respectively.

3. Examples of Simulation Results

One of the advantages of 3D direct numerical simulations is that we are able to examine and optimize the effects of changing screw dimensions, rotational speeds, flow rates, materials, etc., in further detail. Cheng and Manas-Zloczower (1998), Yao and Manas-Zloczower (1998), Funatsu *et al.* (1999), and Yoshinaga *et al.* (2000) compared the mixing performance of TSEs by marker tracking analysis using a non-Newtonian flow character. Funatsu *et al.* examined the influence of different screw combinations in mixing performance. The configurations are (i) DF/R + KB/N + DF/L + DF/R, (ii) DF/R + KB/L + DF/L + DF/R, (iii) DF/R + KB/N + DF/R + DF/R and (iv) DF/R + KB/L + DF/R + DF/R assemblies. Here, DF/R

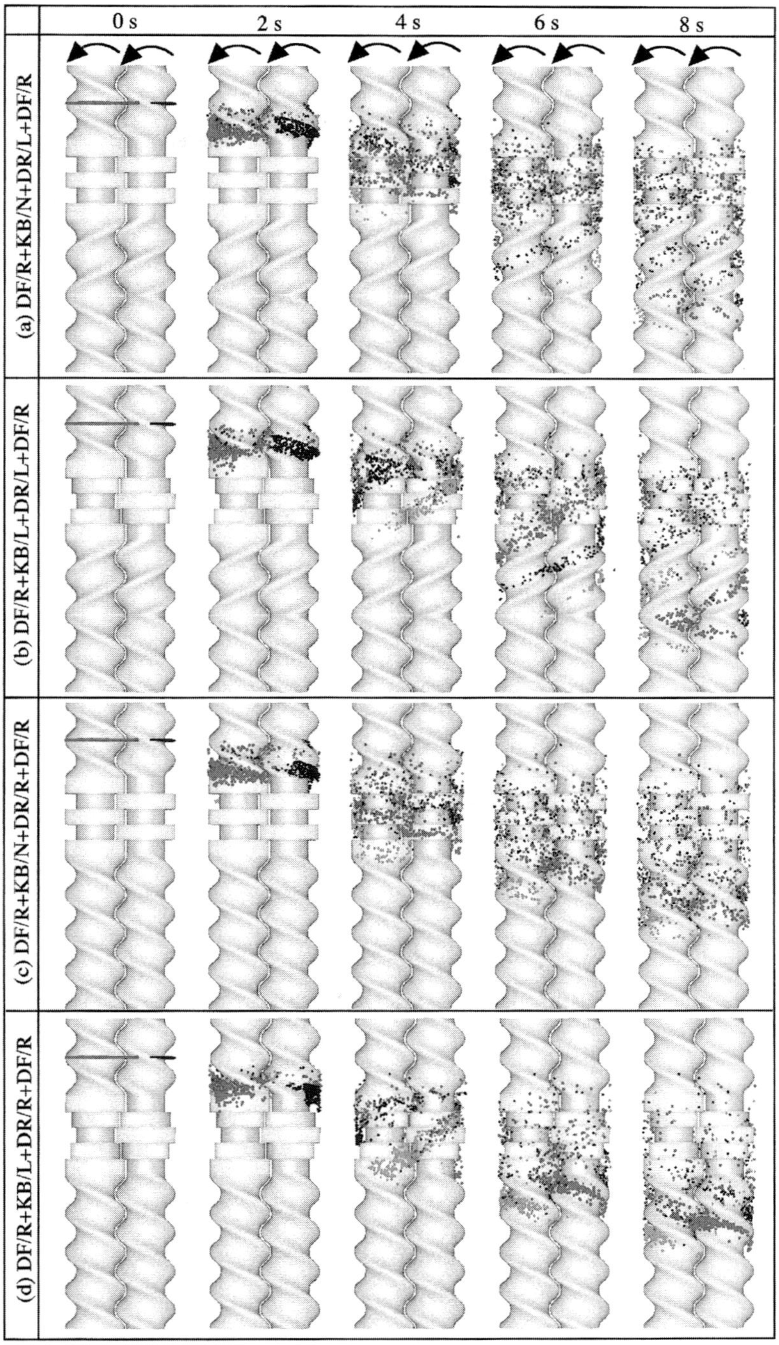

Figure 1
Comparisons of snapshot marker dispersions in four kinds of screw assemblies. The number of markers is 1338. Initial markers are set at X–Y plane of center of first DF/R, with each marker distance being 0.80 mm. Rotational speed is 60 rpm, flow rate is 10 $cm^3\ s^{-1}$. Fluid is pure viscous non-Newtonian. The DF/R and DF/L are right- and left-handed double flighted screws; KB/N and KB/L are both 90° and 135° stagger angles kneading blocks. Barrels are removed (See colour section).

and DF/L are right- and left-handed double flighted screws; KB/N and KB/L are both 90° and 135° stagger angle kneading blocks. These combinations are used in the filled mixing zone to attain effective kneading. Calculation of the four-pitch assembly (DF/R + KB/N + DF/L + DF/R) required 5,984 elements with 58,499 degrees of freedom (about 1.4 GB of memory with the skyline method). Figure 1 shows the comparison of marker tracking results. These snapshots of marker distributions and their animations show

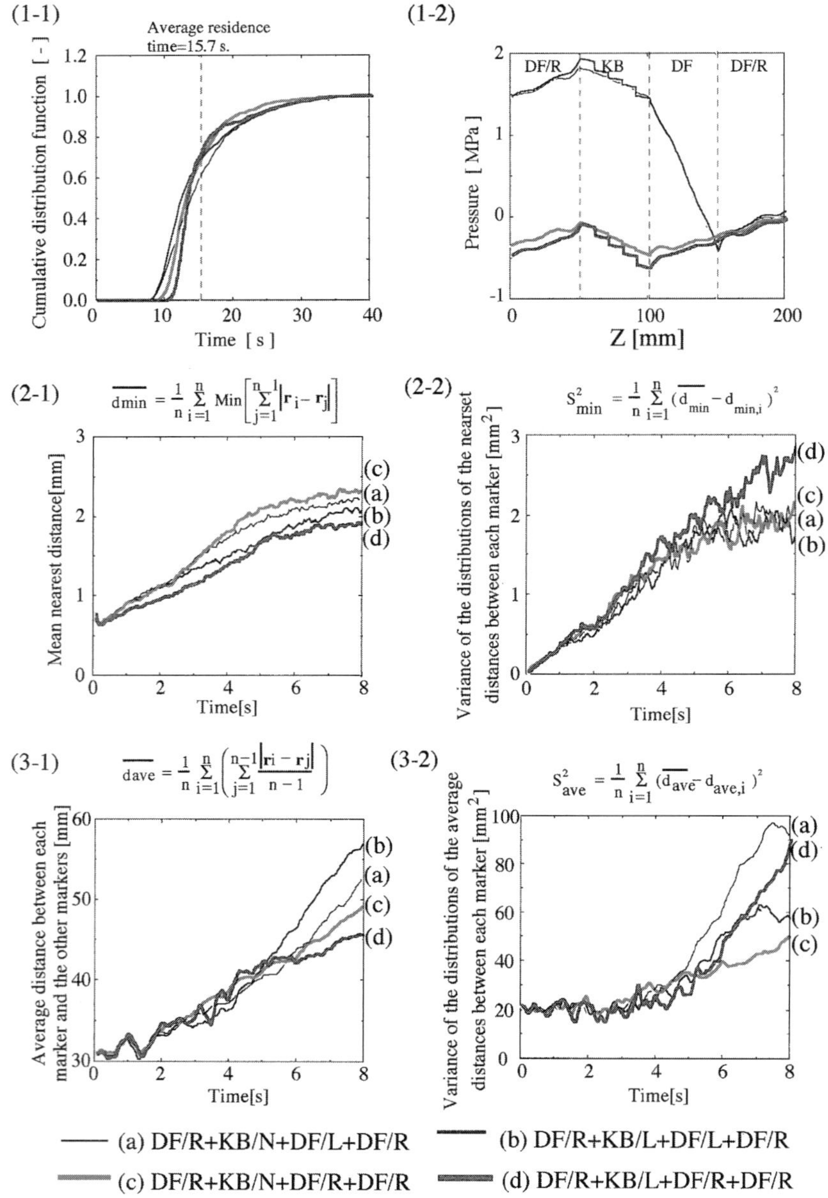

Figure 2

Comparison with (1-1) cumulative residence time distribution, (1-2) cross area average pressure, (2-1) mean nearest distance, (2-2) variance of the mean nearest distance, (3-1) mean value of the average distances and (3-2) variance of the mean value of the average distances for four kinds of screw assemblies. Rotational speed is 60 rpm, flow rate is $10\,cm^3\,s^{-1}$. Fluid is pure viscous non-Newtonian. The DF/R and DF/L are right- and left-handed double flighted screws; KB/N and KB/L are both 90° and 135° stagger angles kneading blocks. (See colour section) detailed comparisons of the different combinations of screw elements, and give new ideas for screw design. From Fig. 1, the KB/N makes markers mix in a circular direction, and KB/L makes markers flow through quasi-channel zones and mix along the transport direction. Figure 2 shows the comparison of distributive mixing performances of 4 kinds of assemblies using equations (5)–(8). Configuration (c) is superior in local mixing, and configuration (b) is superior in dividing fluid elements along the transport direction, considering the results of Fig. 1.

The marker tracking method is also able to analyze the local stresses, strain rate, temperature, and reaction histories, when a non-isothermal reactive simulation is carried out.

See also: Polymer Processing: A Modeling Approach; Polymer Twin Screw Extrusion: 2D Modeling

Bibliography

Cheng H F, Manas-Zloczower I 1998 Distributive mixing in conveying elements of a ZSK-53 co-rotating twin screw extruder. *Polym. Eng. Sci.* **38**, 926–35

Funatsu K, Katsuki S, Takayama T, Kihara S 1999 3-D Numerical analysis on the effects of combination of forward and backward screw elements in twin-screw extruders In: *Conf. Proc., Polymer Processing Society, 15th Annu. Meet.*, 's-Hertogenbosch, The Netherlands. Polymer Processing Society, New York, paper 157 (CD-ROM)

Ishikawa T, Kihara S, Funatsu K 2000 3-D Numerical simulation of non-isothermal flow in co-rotating twin screw extruders. *Polym. Eng. Sci.* **40**, 357–64

Ishikawa T, Kihara S, Funatsu K, Sato T 1999 Numerical simulation of peroxide induced degradation process of polypropylene in twin-screw extruders. In: *Conf. Proc., Polymer Processing Society, 15th Annu. Meet.*, 's-Hertogenbosch, The Netherlands. Polymer Processing Society, New York, paper 227 (CD-ROM)

Yao C H, Manas-Zloczower I 1998 Influence of design on dispersive mixing performance in axial discharge continuous mixer—LCMAX 40. *Polym. Eng. Sci.* **38**, 936–46

Yoshinaga M, Katsuki S, Miyazaki M, Liu L, Kihara S, Funatsu K 2000 Mixing mechanism of three-tip kneading disc in twin screw extruders. *Polym. Eng. Sci.* **40**, 67–77

K. Funatsu and S. Kihara

Porous Ceramic Processing

There are numerous applications in which it is necessary or useful to introduce porosity into a ceramic material. For example, in filtration applications, materials need to be permeable and the processing emphasis is often focused on controlling the size of the pore channels. Applications include the removal of particles or large molecules from liquids (for concentration or purification) and gas separation. Similarly, many sensors, catalytic systems, and gas burners require materials to be permeable as well as providing other functions. Porosity can also be used as a second phase to design materials with tailored properties, and in this case it is necessary to control both porosity volume fraction and its size. For example, porosity can be used to decrease thermal conductivity, which is important in refractory and thermal insulation materials. The presence of porosity can decrease the dielectric constant of a ceramic, which is a property sought in substrates used for microelectronics. The void space in porous ceramics can also be used as a storage medium for the later release of the second phase. For example, this idea has been exploited for medication delivery devices or as diffusion sources in the semiconductor industry. Porous ceramics are also being used as bone implants and the porous structure has been shown to promote bone growth. The decrease of density with the addition of porosity can be used in some structures to produce materials of a given stiffness or strength with minimum weight. The sandwich panel structure, in which a porous (core) material is bonded to strong, stiff faceplates, is an example of this design methodology. The primary function of porous ceramics is, however, not often structural and the presence of porosity usually degrades strength. As many applications require a high degree of mechanical reliability, there is often a need to ensure a compromise between strength and the other properties required for the given application. Porous ceramics are often attractive in applications that involve high temperature or corrosive environments.

The processing techniques required for porous ceramics depend critically on the volume fraction and size of the required porosity. For example, ceramics with porosity levels greater than 70 vol.% can be produced with a (hollow) cellular structure. The processing of cellular ceramics often involves a foaming process and the pore size is typically greater than 100 μm. High-porosity ceramics can also be produced by the bonding or sintering of ceramic fibers. These are termed fibrous ceramics and such materials are commonly used as high-temperature thermal insulation. For example, the thermal protection system of the NASA space shuttle is based on fibrous tiles that use bonded silica or aluminosilicate fibers. An example of the microstructure of this material is given in Fig. 1. Another approach for processing high-porosity materials is to use sol–gel technology and the pore size in these materials is typically less than 1 μm. To produce ceramics with lower levels of porosity, conventional ceramic powder processing is often used but modified to ensure control of the pore volume and size. The main aspects of these various established processing approaches are discussed below. Porous ceramics can also be produced by the oxidation of metals, gas dissolution during solidification, or removal of a (continuous) second phase, e.g., by leaching. An example of the third approach is the production of "thirsty" glass (Vycor), in which one phase is leached from a phase-separated glass. This approach has also been applied to glass ceramics. It is anticipated that porous materials will also be produced by solid freeform processing or by the deposition/etching techniques that are employed in the microelectronics industry.

1. Cellular Ceramics

Cellular materials are generally classified into two broad groups, honeycombs and foams. In honeycombs, the cells form a two-dimensional array whereas foams are comprised of a three-dimensional array of hollow polygons. Ceramic honeycombs have found a substantial market as catalytic converters, heat exchangers, and combustors. Foams are usually divided into two further categories depending on whether or not the individual cells possess solid faces. If the solid is contained only in cell edges, the

Figure 1
Fibrous structure of the high-porosity tiles used in the thermal protection system of the space shuttles.

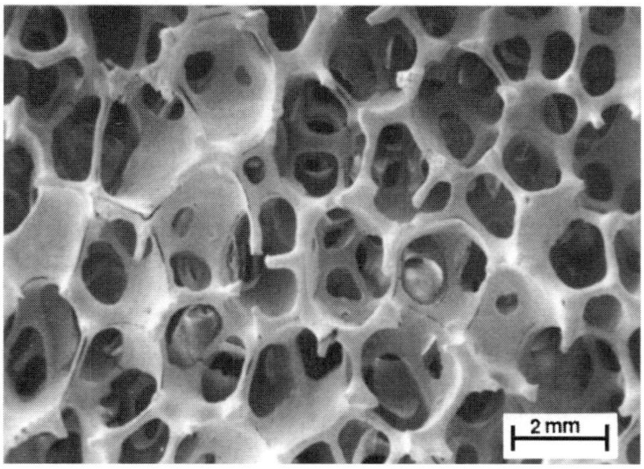

Figure 2
Structure of an open-cell ceramic produced by the sponge replication technique. This material shows the presence of longitudinal strut cracks that probably result from nonuniform coating.

material is termed open cell (or reticulated) and the material is permeable. These materials are often used as filters for molten metals or as kiln furniture. An example of an open-cell ceramic is shown in Fig. 2. If the cell faces are present, the foam is termed closed cell and the individual cells are isolated from each other. Closed-cell glass foams are commercially available as thermal insulation or as abrasives.

Ceramic honeycombs are usually formed by extrusion and this structure then undergoes a conventional binder burn-out and sintering process. The die used in the extrusion process can be designed to produce a wide variety of cell geometric structures, e.g., square, triangular, hexagonal, etc. The extrusion process requires moldable raw materials, the rheological behavior of which can be adjusted by water-soluble additives. In this process, a balance is needed in this rheology to ensure stability of the cellular structure without excessive die wear.

A common processing approach for open-cell ceramics is to coat open-cell polymer foams, usually with a ceramic powder dispersed in a liquid (slurry). This approach approximately replicates the polymer and gives a rather distinct structure in that the cell struts are hollow after the process is complete. The central hole represents the original shape and position of the polymer, which is removed by pyrolysis during the firing process. Various open-cell polymers have been used as substrates for the coating process including polyurethane, poly(vinyl chloride), polystyrene, and latex. The composition of the ceramic slurry used to coat the polymer depends on the final application but has included a wide variety of oxide and nonoxide ceramics. The ceramic slurry often contains various additives to act as binders, rheological control agents, or setting compounds to facilitate the coating process. It is critical that the slurry produces a uniform coating, especially as the polymer foam can often contain sharp corners at the strut edges. The presence of longitudinal strut cracks in Fig. 2 is probably a result of poor coating at the strut corners. In the processing, the polymer "sponge" is immersed in the ceramic slurry. The excess slurry is then removed by mechanical compression or centrifugation. The ceramic foam is dried and fired in a typical manner, except for a pyrolysis step in which the organic additives and polymer foam are removed. During heating, the expansion and gas evolution behavior of the polymer can lead to significant stresses that can damage the ceramic coating if not carefully controlled. For these materials, the pore size in the final product is determined primarily by the cell size of the polymer foam.

An alternative approach for producing open-cell ceramics is to use a physical deposition process on an open-cell substrate. For example, open-cell vitreous carbon foams can be produced by pyrolysis of thermosetting polymer foams. The carbon foam can then be coated with a dense ceramic material, e.g., by chemical vapor deposition. This process has been used to produce a variety of oxide, boride, nitride, and carbide foams. Replication has also been performed on biological materials. For example, a lost wax process has been used to replicate coral and sols have been absorbed into wood to produce a porous structure after the firing process.

In contrast to the above replication methods, there are direct techniques to foam a liquid in which a ceramic powder or a precursor to a ceramic is dispersed. The liquid also contains a constituent which, when treated appropriately, will lead to gas evolution causing bubbles to form in the liquid. The gas evolution process is usually initiated by heat or chemical reaction and a wide variety of processes have been devised. For example, organic compounds can be mixed to produce carbon dioxide or hydrogen in an ambient temperature reaction. Other additives to the liquid are used to stabilize and harden the cellular structure once it is produced. The hardening may occur by a chemical reaction or gelation and is used to "set" the foamed structure. The foam is dried and sintered by conventional means. The final structure of a foamed material often exhibits some morphological anisotropy associated with the "rise" direction of the foam. Glass foams are produced by a high-temperature version of this process in which additives such as hydroxides, carbonates, and sulfates are mixed with glass powder. On heating, the glass becomes fluid and the decomposition of the salts leads to bubble formation. The direct foaming approaches give rise to a closed-cell structure, but if sintering is used the cell faces often perforate and produce an open-cell ceramic. As an alternative to direct foaming, the sintering or bonding of hollow spheres has been used to produce closed-cell ceramics.

2. Fibrous Ceramics

Fibers in a blanket or felt form have been used extensively as thermal insulation but fibers can also be "rigidized" by bonding processes to produce fibrous ceramic monoliths (tiles or boards). These materials are used for a wide variety of thermal insulation applications and are attractive as a result of their low thermal mass, which allows fuel savings. These bodies are usually classified in terms of the refractoriness of the fibers. They may range from low- to high-purity silica glasses or from low- to high-purity alumina fibers, depending on the alumina/silica ratio in the latter fibers. The low-alumina fibers are usually produced by a melt blowing process. One of the processing concerns in blown fibers is the amount of "shot" (solidified particles that escape the fiber-shaping process) that is present in the final product. High-alumina fibers (ICI's Saffil alumina or 3M Corporation's Nextel mullite) are produced by a sol–gel route and can be used at temperatures up to 1600 °C.

Fibrous ceramics are often formed by a uniaxial pressing operation and this can lead to significant anisotropy in the physical properties. In this process, conditioned fibers are mixed with a liquid that contains the appropriate binders and dispersants. This slurry is then mechanically pressed or molded into the required shape and gelation is used to set the structure. The green body is dried and fired by conventional means, although the low thermal conductivity of these materials can lead to longer processing times being required to reach thermal equilibrium.

3. Porous Sol–Gel-derived Ceramics

Sol–gel chemistry can be used to produce porous ceramics either as monoliths or as thin films. The latter are termed ceramic membranes and are used for filtration. The pore size in these sol–gel-derived materials can be controlled from approximately 1 nm to 1 μm. The sol–gel process can be accomplished using a wide variety of techniques. For example, alkoxides such as tetraethyl orthosilicate can be mixed with water to initiate the gelation. Subsequently, hydrolysis, condensation, and polymerization take place to form an alcogel, which is an oxide network that has condensed with liquid (alcohol) filling the porous network. The liquid is then removed by evaporation (xerogels) or by hypercritical drying (aerogels) to produce high-porosity silica. The pore structure can then be further modified by a heat treatment. Alumina, zirconia, and titiania ceramics can be produced by similar sol–gel processes. An alternative sol–gel process is one in which a colloidal suspension of hydoxide or hydrated oxide particles is peptized. Gel formation is induced by increasing particle concentration or by manipulating the zeta potential. The process is controlled by pH, the nature/concentration of the electrolyte, and particle size.

Zeolites are materials in which porosity is available at the molecular level and sol–gel technology has extended this behavior to larger pore sizes. An interesting development of these ideas is the production of porous silica-based material, in which the pores are ordered at the nanoscale. For example, Kresge *et al.* (1992) produced silica-based materials in which the pores are in hexagonal arrays and the pore sizes are in the range 1.6–10 nm. This has proved to be a very active research area and scientists have worked to extend the pore size range and choice of materials. These materials use organic "templates," such as surfactant arrays or emulsion droplets that allow the porous materials to be self-assembled. Latex spheres and block copolymers can also be used to introduce submicrometer pores into sol–gel-derived materials, and it has been demonstrated that three-dimensional ordered structures can be produced by such means.

4. Partially Densified Ceramics

Many ceramics are produced by the (high-temperature) firing of powder compacts. The level of porosity in these materials can be controlled by either firing at a temperature at which densification is incomplete (see Fig. 3) or by controlling the particle size distribution. This latter approach is used extensively in the refractories industry. The microstructure of refractories usually consists of grain and bond phases. The large refractory grains are often derived from the raw materials with a minimum of preprocessing. The bonding phase is usually fine-grained, such as the glassy or silicate bonding derived from clays, carbon

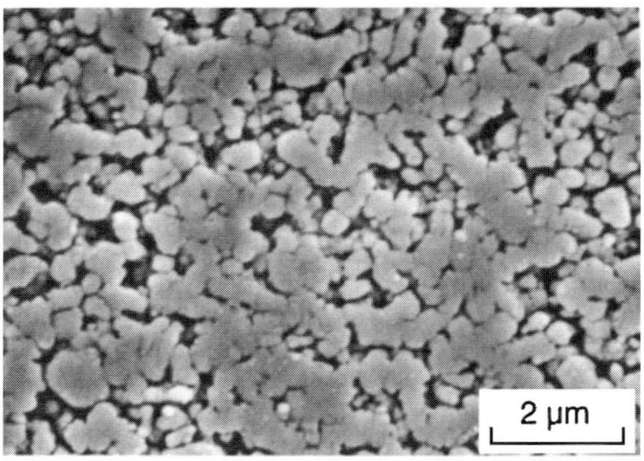

Figure 3
Porous alumina ceramic produced by the incomplete densification of an ultrafine powder (courtesy of Dale Hardy).

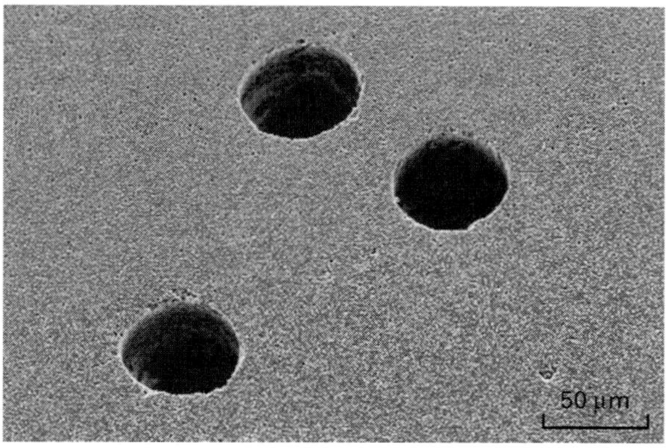

Figure 4
Example of a porous ceramic in which the pores are introduced by the pyrolysis of polymeric fibers. The fibers are aligned to some degree by a tape-casting process (courtesy of Jordan Priest).

bonding derived from resin or pitch, or calcium aluminate cements. During the powder processing, the fractions of the coarse, intermediate, and fine-sized particles are carefully controlled as this defines not only the green density but also the amount of shrinkage that occurs during firing. Unlike other engineering ceramics, refractories may only have limited shrinkage during firing. The firing process is, however, important in controlling the bonding process, which leads to strength development. As refractories can contain several phases, there is the possibility of residual stress development from thermal expansion mismatch and the molar volume changes that arise from the various chemical reactions. Clearly, these stresses must be controlled so that any microcracking does not have an impact on the desired properties in the final product.

Porosity can also be introduced into powder-processed ceramics by adding a fugitive phase to the powder mix. For example, inexpensive raw materials, such as sawdust or starch, are employed. These organic materials are "burnt off" by the firing process leaving the additional void space. Alternatively salts, such as carbonates or hydoxides, which decompose by gas evolution, can be employed. If porosity of well-defined shape is required, polymeric fibers or particles can be used. For example, Fig. 4 shows a porous ceramic in which the pores are produced by the pyrolysis of polymeric fibers. Some ceramic materials are available in the form of hollow spheres, and incorporation of these materials into a powder mix can produce well-defined porosity. In the cement industry, hollow flyash spheres (cenospheres) are sometimes used to produce a lightweight cement or concrete.

Bibliography

Brezny R, Green D J 1994 Mechanical behavior of cellular ceramics. In: *Materials Science and Technology—A Comprehensive Treatment.* VCH, Weinheim, Germany, Vol. 11, Chap. 9

Brinker C J, Scherer G W 1990 *Sol–Gel Science: The Physics and Chemistry of Sol–Gel Processing.* Academic Press, New York

Burggraaf A J, Cot L (eds.) 1996 *Fundamentals of Inorganic Membrane Science and Technology.* Elsevier, New York

Chan K K, Brownstein A M 1991 Ceramic membranes—growth prospects and opportunities. *Bull. Am. Ceram. Soc.* **70**, 703–7

Gibson L J, Ashby M F 1997 *Cellular Solids: Structure and Properties.* Cambridge University Press, Cambridge

Hirschfeld D A, Li T K, Liu D M 1996 Processing of porous oxide ceramics. *Key Eng. Mater.* **115**, 65–79

Holland B T, Blanford C F, Stein A 1998 Synthesis of macroporous minerals with highly ordered three-dimensional arrays of spheroidal voids. *Science* **281**, 538–40

Kresge C T, Leonowicz M E, Roth W J, Vartuli J C, Beck J S 1992 Ordered mesoporous molecular sieves synthesized by a liquid-crystal template mechanism. *Nature* **359**, 710–2

Nettleship I 1996 Applications of porous ceramics. In: Mostaghaci H (ed.) *Applications of Advanced Materials in a High-Tech Society: Part 1.* Trans Tech, Zurich, Switzerland, pp. 305–24

Pierre A C 1998 *Introduction to Sol–Gel Processing.* Kluwer, Boston

Saggio-Woyansky J, Scott C E, Minnear W P 1992 Processing of porous ceramics. *Bull. Am. Ceram. Soc.* **71**, 1674–82

Schaefer D W 1994 Engineered porous materials. *MRS Bull.* **19**, 14–7

Sepulveda P 1997 Gelcasting foams for porous ceramics. *Bull. Am. Ceram. Soc.* **76**, 61–5

Shermann A J, Tuffias R H, Kaplan R B 1991 Refractory ceramic foams—a novel new high-temperature structure. *Bull. Am. Ceram. Soc.* **70**, 1025–9

D. J. Green

Powder Granulation and Compaction

Ceramic components are generally fabricated from inorganic raw materials in their powder form. To improve powder handling and processability, ceramic powders are typically modified (e.g., granulated) prior to processing. The most common method of forming ceramic components is by dry or semi-dry powder compaction, which is also known as dry pressing (Reed and Runk 1976, Reed 1995, pp. 418–49).

Dry pressing is an economical yet versatile technique for high-volume manufacturing of relatively simple geometry powder compacts. It is used readily in the ceramics, powder metallurgy (PM), and pharmaceutical industries. Dry pressing involves compacting a nominally dry powder (i.e., <2 wt.% water) in a mold or die at pressures of 21–690 MPa (3000–100 000 psi). The pressing pressure should be high enough to produce a homogeneous green compact with sufficient strength for subsequent handling/processing, but low enough to avoid excessive wear on the press and tooling (i.e., the forming die/mold). To improve handling and processability, ceramic powders are typically granulated with 1–5 wt.% organic binder for dry pressing.

1. Powder Granulation

Powders used in dry pressing must flow freely to uniformly fill the pressing mold or die, and they must compact uniformly to form a homogeneous microstructure powder compact with reasonable strength. The micrometer-size particles used in ceramic manufacturing exhibit poor flow and compaction behavior in their raw form, and therefore must be granulated. Granulation is the process of intentionally agglomerating fine particles into larger particle clusters that exhibit improved handling and flow characteristics.

1.1 Granulation

Granulated feedstock for ceramic powder pressing is generally produced by either spray granulation or spray drying (Sherrington and Oliver 1981, Lukasiewicz 1991, Reed 1995, pp. 378–93). Both processes rely on the formation of liquid–solid interfaces and on surface tension to produce granules. Additionally, both processes tend to form equiaxial granules (Fig. 1). Relative to spray drying, spray granulation generally produces a granule morphology having a more irregular shape, granules of larger size and size distributions (20–1000 μm, with an average size of 200–300 μm), and granules of higher density. Spray drying produces spherical granules. Spray granulation is a batch process that is well suited to the production of small size batches of powder (i.e., <10 kg), while spray drying is better suited to the production of moderate to large size powder batches.

(*a*) *Spray drying.* Spray drying is the most common method of granulating ceramic powders for dry pressing. It is an indirect granulation process that involves pumping slurry to an atomizer, where small

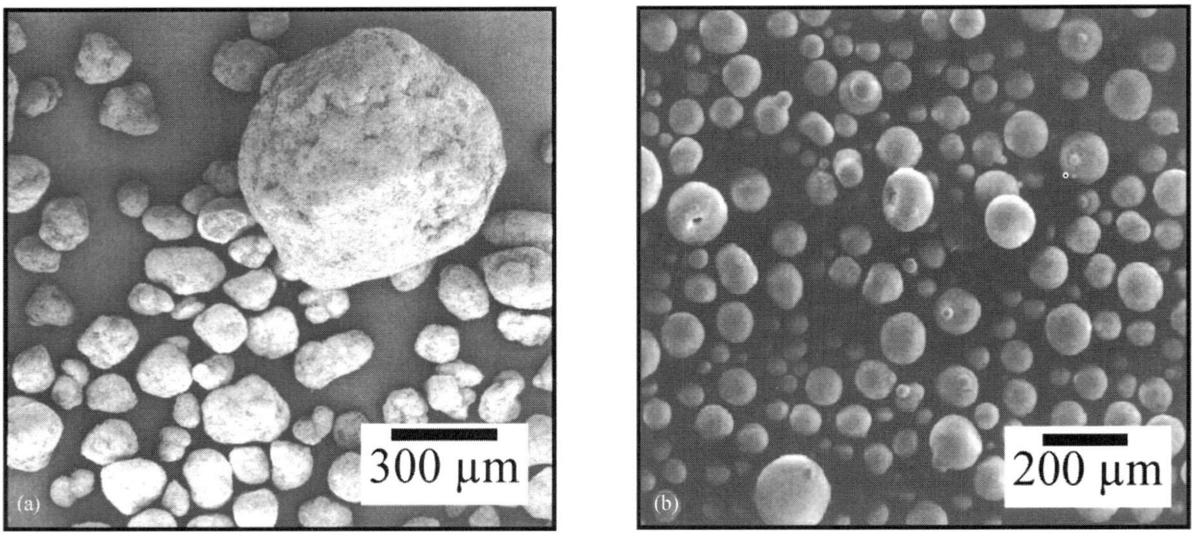

Figure 1
Photomicrographs of granulated alumina powder produced by: (a) containerized batch mixing (i.e., spray granulation); and (b) spray drying.

spherical droplets are sprayed into a large chamber. Heated air flowing in the chamber dries the droplets to form granules that are collected at the base of the drier. Starting with 0.5–10 μm median particle size powders, spray drying typically produces 44–400 μm diameter, spherical granules with an average size of 100–200 μm (Lukasiewicz 1991).

Drying occurs by liquid evaporation from the droplet surface (i.e., the constant-rate drying period) until reaching a critical moisture content, at which time the droplet surface can no longer remain saturated with liquid. A short, constant-rate drying period will limit/avoid binder migration during drying that can produce heterogeneous granules. After achieving the critical moisture content, the evaporation front moves from the surface towards the center of the granule. Concurrently, the rate of evaporation decreases, and the temperature of the granule increases. Care must be taken to ensure that the granule temperature remains below a critical temperature, otherwise the organic binder could be lost or deleteriously altered.

The characteristics of the granules produced by spray drying are largely influenced by the characteristics of the slurry that are sprayed into the heated chamber (Walker *et al.* 2000). The degree of flocculation of the ceramic can have a significant influence on granule morphology, strength, and deformation. Highly dispersed slurries tend to produce irregularly shaped granules, hollow granules, and high density granules (Walker *et al.* 2000). High density and larger size granules also are favored with high solids loadings (i.e., the volume fraction of ceramic particles in the slurry). Partially flocculated slurries favor the production of uniformly dense (i.e., vs. porous) spherical granules that can pack more efficiently, and lower density granules (e.g., 45–55% theoretical density) with lower strength that can deform more readily during pressing.

1.2 Organic Additives

Immediately prior to, during, and/or immediately following granulation, organic additives or processing aids are commonly added to a ceramic powder to enhance processability (Reed 1995, pp. 135–211, Shanefield 1996). Table 1 lists some typical organic additives used in ceramic powder pressing including binders, plasticizers, and lubricants (Reed 1995, pp. 135–211).

Organic binders, like polyvinyl alcohol, are commonly added to powders to aid pressing and to provide strength to the pressed powder compact for subsequent handling and/or machining (Lukasiewicz 1991, Reed 1995, pp. 135–211). The organic binders form bridges between the primary particles in a granule that influence granule strength and deformation under pressure.

Plasticizers are smaller organic molecules that can systematically disrupt the structure of larger molecule binders to make them more flexible by lowering their glass transition temperature. Plasticizers are used in combination with binders to produce more deformable granules with improved compaction behavior.

Lubricants are low molecular weight organic molecules that are added to powders to mitigate friction effects during processing. Lubricants lower interfacial frictional forces between individual particles, and/or between particles and forming die surfaces. Lubricants are added to improve powder compaction, and to reduce the pressure required to eject a pressed compact after pressing.

2. *Press Powder Manufacturing*

In ceramic component manufacturing, granulated ceramic powders are compacted by either isostatic (isopressing) or uniaxial pressing. Isopressing can be used to produce small or large powder compacts with good green density uniformity for sintering. However, isopressing is limited to relatively simple shapes (e.g., spark plug insulators). Furthermore, because of poor dimensional control, green machining is generally required to achieve the desired size and shape compact after isopressing. Appreciably higher production rates of thousands of parts per minute are possible by die pressing and, generally, more complex parts can be formed to near-net-shape by die pressing.

Table 1
Common organic additives used in ceramic powder granulation and compaction.

Binders	Plasticizers	Lubricants
Methyl cellulose	Water	Paraffin wax
Polyvinyl alcohol (PVA)	Ethylene glycol	Aluminum stearate
Polyethylene glycol (PEG)	Triethylene glycol	Magnesium stearate
Acrylic resins	Polyethylene glycol	Stearic acid
Waxes and wax emulsions	Glycerol	Oleic acid
Acrylic–wax emulsions		

2.1 Isostatic Pressing

Isostatic pressing, also known as cold isostatic pressing and isopressing, entails compacting a dry or semi-dry powder in an elastomeric mold submersed in a pressurized liquid (Reed 1995, pp. 418–49). Often rigid tooling (e.g., a steel mandrel) is used in combination with the flexible elastomeric mold to help shape the powder. Typical forming pressures for ceramics are in the range of 21–210 MPa (3000–30 000 psi).

Isopressing can be performed in two ways, wet bag and dry bag. Wet bag isopressing involves compacting a powder in a sealed elastomeric mold that is completely submersed in pressurized fluid. Wet bag isopressing is used for low-volume production of specialty parts, for prototyping, and for research and development.

In dry bag isopressing, the elastomeric mold is an integral part of the isostatic press. As such, the applied pressure is mostly biaxial. A major advantage of dry bag isopressing is that it can be automated for high-volume production, such as in spark plug insulator manufacturing.

2.2 Uniaxial Pressing

Uniaxial pressing, which is also known as dry, die, or dust pressing, involves compacting powder to a prescribed size and shape within a rigid die cavity using one or more rigid plungers (Reed and Runk 1976, Reed 1995, pp. 418–49). The dies and plungers used in uniaxial pressing, also known as the tooling, are typically made of rigid materials that are hard and have good wear resistance. Generally, hardened steel tooling is used, but for greater precision and longer life, more expensive carbide tooling is used. Typical ceramic component production rates for die pressing range from 1 to 300 parts per minute using pressures of 21–140 MPa (3000–20 000 psi).

Die pressing can be accomplished with a single-action, dual-action, or floating die press. A single-action anvil press is a mechanical press with one moving punch. A dual-action press, like a (mechanical) rotary or a hydraulic press, has two or more moving punches. This affords more uniform compaction in larger/thicker parts. The punches and the die body can move independently in the floating die presses most commonly used in ceramic manufacturing. This affords the flexibility to fabricate more complex parts, including parts with tapers, holes, and multiple steps. It also provides the flexibility to balance the forming pressure within different areas of the compact to minimize density gradients in the pressed powder compact.

(*a*) *Net-shape forming.* Economically, a net-shape pressing process is highly desirable in ceramic manufacturing, whereby compact density and densification are well controlled, uniform, and extremely reproducible. Net-shape pressing minimizes waste, and can eliminate expensive finish machining operations (i.e., diamond grinding). Currently, the industry standard for as-sintered tolerance is 1.5%. Many ceramics are routinely sintered to within 1% tolerance, with some specialty cases down to 0.25% (Glass *et al.* 1996). An as-fired tolerance of 1% is adequate for many stand-alone ceramic components; however, more precise fitting components (e.g., engine components) typically require finish machining to size. An as-sintered tolerance of $<0.5\%$ can minimize and possibly even eliminate the expensive finish machining operation, which can account for 90% of the cost of a finished part. The key to achieving a net-shape compaction process is to minimize or eliminate macroscopic density gradients in the particulate assembly during die filling and compaction.

3. *The Powder Pressing Process*

A typical powder pressing operation has three basic steps: (i) filling the mold or die with powder; (ii) compacting the powder to the desired size and shape; and (iii) ejecting the powder compact from the die/mold. Die filling and compaction determine the micro-scale uniformity of the pressed product, while undesirable macroscopic defects can be introduced during ejection.

3.1 Die Filling

The density and packing uniformity of a powder within a die influence the density achieved after compaction and, ultimately, the density, microstructure, and properties of the final sintered product (Ewsuk 1993, Reed 1995, pp. 418–49, Glass *et al.* 1996, Lannutti *et al.* 1997). The packing density and density uniformity achieved during die filling are influenced by granule size/distribution, as well as by the size of the die (German 1989, Dinger and Funk 1997). In general, lower packing densities are obtained with narrow particle/granule size distributions, and with small-dimension die cavities. Additionally, packing density variability increases with increasing aspect ratio (i.e., height/diameter) in narrow diameter dies. Generally speaking, a lower packing density is achieved adjacent to a die wall where packing is disrupted.

Die wall effects can be minimized or eliminated in dies that are sufficiently larger than the granule size. Specifically, smaller size granules should be used to fabricate smaller dimension (e.g., diameter and/or thickness) parts. Observations for one typical, spray-dried alumina powder indicate that the average agglomerate size should be smaller than the compact diameter and thickness by factors of ~250-fold and ~40-fold, respectively.

Free-flowing granulated powders are required to ensure good packing density, packing uniformity, and reproducible die filling. In general, good-flowing powders are spherical or equiaxial, and have a distribution of sizes of 44–400 μm.

(*a*) *Powder flow*. Typically, powder flow is characterized by the time it takes a given mass of powder to flow through a given size of funnel. Powder flow also can be assessed by the angle of repose that a loose powder makes when poured onto a flat surface (i.e., the angle measured between the flat surface and the cone surface of the powder). Granulated ceramic powders with good flow generally have an angle of repose of 40° or less.

Powder bulk density, which is an indirect measure of powder flow, is commonly used in ceramic manufacturing. Powder bulk density is a measure of the ratio of the weight-to-volume of a loose powder (e.g., right out of the hopper or bag). Tap density, which is the ratio of the weight-to-volume of a bulk powder after tapping in a prescribed manner for a set time, is also routinely measured. The ratio of the tap to bulk density, i.e., the Hausner ratio, is also a measure of flow. Better powder flow and die fill reproducibility are achieved with a Hausner ratio that approaches unity.

(*b*) *Fill density*. The bulk density of a powder within the forming die is known as the fill density. Generally, fill densities for granulated ceramic powders are in the range of 25–35% of the true or theoretical density of the powder, based on the inorganic constituents alone. Ideally, high fill densities are desirable to maintain a compaction ratio (i.e., the pressed compact density divided by the fill density) of <2 to avoid pressing defects. Higher fill densities generally indicate better, more uniform packing; however, this can be somewhat misleading because fill density is determined by a combination of granule density and packing.

3.2 *Powder Compaction*

Pressure applied during the powder compaction process promotes granule and particle rearrangement, granule deformation, and consolidation of the particulate assembly. Particle coordination number, green density (i.e., the bulk density of the powder compact prior to sintering), and compact strength increase with increasing compaction pressure (German 1989). In general, a higher pressed compact density is achieved with a higher fill density, and a higher green density compact sinters to a higher end point density (Ewsuk 1993, Glass *et al.* 1996).

(*a*) *Compaction density gradients*. Density gradients are common in pressed powder compacts. While it is not practical to completely eliminate density gradients from a powder compact during pressing, it is necessary to minimize them. As a general rule, density gradients are undesirable because they contribute to differential sintering and shape distortion (Ewsuk 1993). Additionally, density gradients can produce defects that limit the performance and reliability of the finished part.

There are two scales of density gradients that are important in powder compaction: (i) macroscopic density gradients produced by nonuniform die filling and/or pressure gradients during powder compaction (Reed 1995, pp. 418–49, Lannutti *et al.* 1997, Arguello *et al.* 2001); and (ii) microscopic density variations due to packing defects, hollow granules, and/or insufficient granule deformation during compaction (Shinohara *et al.* 1999). To produce a homogeneous green compact microstructure and avoid microscopic density gradients, granules should be strong enough to allow significant rearrangement prior to deformation/fracture (i.e., to pack more efficiently), but weak enough to completely break down, deform, and knit together to produce a uniform microstructure during compaction (Glass *et al.* 1996). In general, granules that are large, irregular-shaped, hollow, and/or that have an excessively high granule density (i.e., relative to the pressed compact green density) are undesirable.

To avoid introducing macroscopic density gradients during pressing, steps must be taken to optimize stress/pressure transmission and stress/pressure uniformity throughout the powder compact. A powder compact is often treated as a continuum, in which the applied pressing stress is transmitted uniformly throughout the continuum. However, powder compacts are comprised of discrete particles, and the stress is transmitted nonuniformly through granule–granule and particle–particle contacts. Models and experiments indicate that a network of nonuniform stresses is developed throughout a compact during pressing. Within this stress network, physical networks of granule/particle bridges form that support stresses higher than the applied stress, and that shield localized regions between the bridges from the applied stress. The result is a complex array of high- and low-stress regions that undoubtedly contribute to the formation of density gradients in a pressed powder compact (Glass *et al.* 1996). Within a given stress network, the smaller granules in a bridge fracture/deform in preference to larger granules. When granule fracture occurs, a bridge collapses and there is a significant drop in pressure, accompanied by a localized increase in density. This complex process makes it difficult to achieve a uniform state of stress and density within a powder compact. Additionally, it may also make it difficult or impossible to achieve the higher network stresses required to deform/fracture the larger granules in some systems, leaving large granule remnants that become the strength-limiting flaws which determine the reliability of the finished component after sintering (Shinohara *et al.* 1999).

In general, lower surface area, coarser median particle size (e.g., 2–5 μm vs. 0.5–1 μm) powders favor more uniform stress transmission during powder pressing. Additionally, the stress uniformity within a compact can be enhanced by minimizing the compaction ratio, by minimizing the compact aspect ratio, and by avoiding single-action uniaxial pressing (Reed 1995, pp. 418–49). Uniform stress transmission becomes particularly problematic in uniaxial pressing with narrow die diameters, and with more complex die/part geometries.

(*b*) *Friction in compaction.* Another important factor that affects the uniformity of stress/pressure transmission during powder compaction is friction. Due to the die wall friction, μ, between a granular material and the forming die, the axial pressure, P_z, at any depth, z, into a cylindrical compact of radius r, decreases exponentially with distance from the applied pressure P_o:

$$P_z = P_o \exp(-2z\mu\alpha/r) \quad (1)$$

where α is an empirical factor related to the internal angle of friction (i.e., interparticle friction) of the powder. A combination of μ and α determines the relationship between the axial pressure, P_z, and radial pressure, P_r, and consequently the uniformity of the pressure throughout the powder compact during forming. Because compaction is pressure dependent, a pressure gradient present during forming produces a density gradient in the powder compact after pressing (Fig. 2(a)).

Die wall friction is influenced by the die material and its surface finish, the powder characteristics and organic additives, and the pressing process. In single-action uniaxial pressing, die wall friction works in opposition to the applied forming pressure, decreasing the pressure transmitted into the powder compact with increasing distance from the pressing punch. For a cylindrical part, the result is a powder compact with the highest density at the outer diameter adjacent to the pressing punch, and the lowest density in the outer diameter adjacent to the stationary punch (Fig. 3). Density decreases axially with distance from the pressing punch, and the density gradient becomes more severe with increasing die wall friction.

To minimize pressing pressure gradients that can create defects in the form of density gradients and/or cracks in a pressed powder compact, die wall friction must be controlled during pressing. Friction can be controlled with internal or external lubricants, or through the design of the die and the pressing process. Die wall friction can be minimized with smooth surface dies and carbide tooling. Frictional effects also are minimized in simple shape, low aspect ratio parts, and by dual-action pressing in preference to single-action pressing (Reed 1995, pp. 418–49). Deleterious die wall frictional effects also can be minimized by isostatic or biaxial pressing in preference to uniaxial die pressing. However, density gradients have been measured even in powder compacts formed by wet-bag isostatic pressing, indicating that interparticle frictional forces also play an important role in the compaction process and in determining the density gradients in a pressed powder compact (Glass *et al.* 1996).

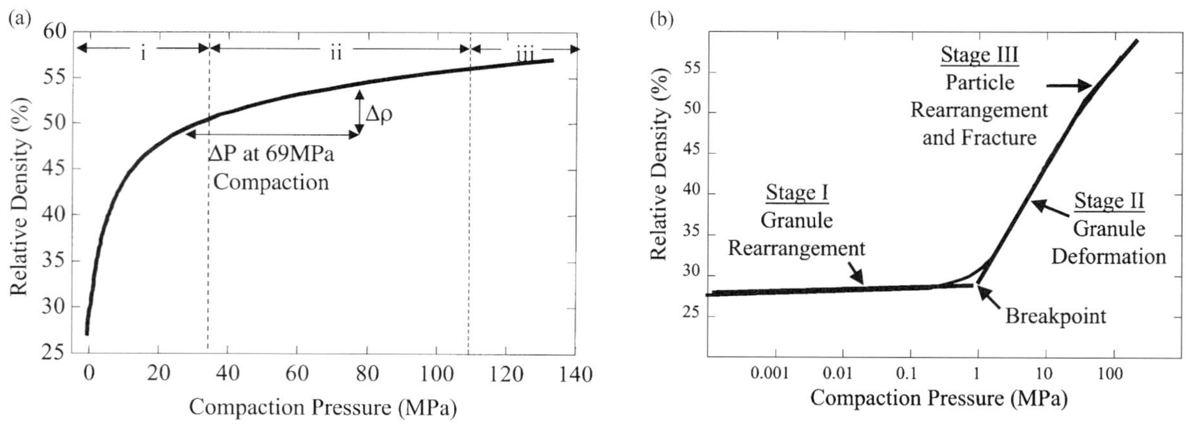

Figure 2
The compaction response of a spray-dried 94 wt.% alumina – 6% glass powder, including: (a) a pressure–density compaction response curve illustrating the density gradient ($\Delta\rho$) that develops as a result of a pressure gradient (ΔP) due to die wall friction during pressing at 69 MPa, and illustrating the preferred pressure range (ii) for powder pressing; and (b) the compaction response curve plotted on a log–pressure scale showing the three stages of compaction and the associated compaction mechanisms, along the granule yield pressure, P_y.

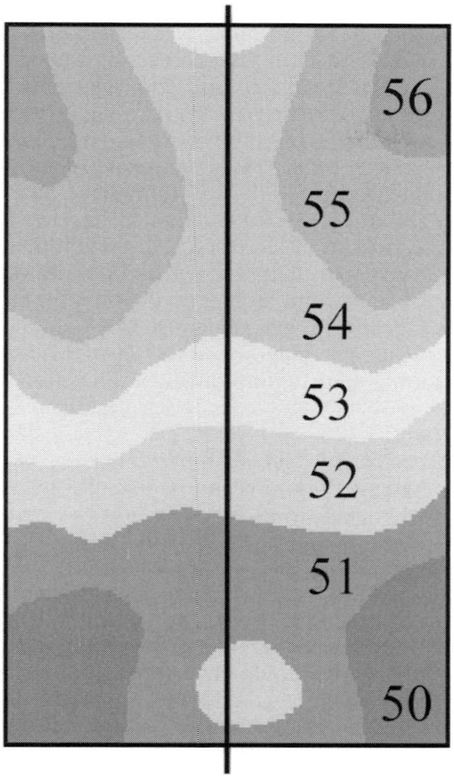

(a)

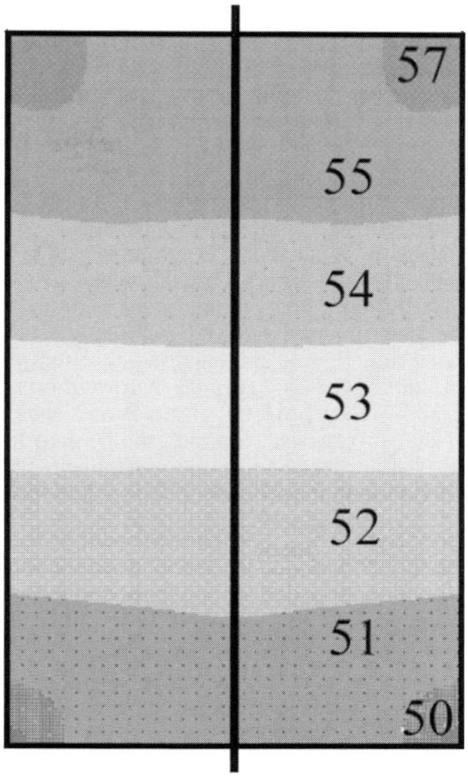

(b)

Figure 3
The density gradients in a cross-section of a cylindrical powder compact of 94% alumina after uniaxial pressing at 69 MPa: (a) as measured by ultrasound velocity measurements completed on a flat and parallel section of a bisque-fired part, and (b) as-predicted by a cap plasticity model within a finite element method framework assuming a die wall friction coefficient of 0.25. Note that the model simulations do not consider the friction effects on the pressing punches, or the initial variability in density after die filling.

(*c*) *Compaction mechanisms.* Particle characteristics, granule characteristics, granule packing, pressing conditions, and discrete particle–particle and particle–die interactions all play an important role in determining compaction behavior. Pressure–density plots, which are also known as compaction response diagrams and pressure compaction diagrams, provide one means of relating compaction behavior to powder characteristics and processing parameters (Fig. 2). Compaction response diagrams show how the density of a powder compact changes as a function of the applied forming pressure (Fig. 2(a)). When the data is plotted on a log–pressure scale, three regimes of nominally linear response become evident that reflect the three stages of powder compaction (Fig. 2(b)).

Mechanistically, compaction is interpreted to occur primarily by granule rearrangement in the low-pressure linear regime designated stage I, and primarily by granule deformation in the higher pressure linear regimes designated stages II and III (Reed and Runk 1976, Reed 1995, pp. 418–49). At the higher pressures, first the large pores between granules are eliminated by granule deformation during stage II. Subsequently, the smaller pores between the particles within a granule shrink by particle fracture and rearrangement, and/or by particle deformation during stage III. The extent of densification due to each mechanism depends on the local pressure, as well as on the properties and characteristics of the granules and primary particles. Typically, granulated ceramic powders are compacted to a point that puts them in the beginning of stage III (Niesz 1996).

(*d*) *Granule characteristics and compaction.* Ideally, during compaction, granules should deform/fracture to eliminate the porosity between granules introduced during die filling, and subsequently knit together to produce a homogeneous microstructure with no remnants of the original granule structure. One means of characterizing granule deformation is by directly measuring the stress–strain behavior of individual granules using mechanical testing equipment. Alternatively, a much more convenient method involves using information contained in a compaction response diagram (Fig. 2(b)). In particular, the granule yield pressure (P_y), also known as the breakpoint, is related to granule strength. The granule yield pressure, which can be determined from the intersection of linear regimes I and II in a compaction response diagram, defines the onset of granule deformation during compaction. Experimental observations have determined that there is very limited granule deformation/fracture during stage I granule rearrangement, while extensive deformation occurs in stage II (Glass *et al.* 1996). The breakpoint scales with the fracture strength, and stronger/harder granules have a higher P_y. Stronger granules with higher breakpoints can undergo more extensive rearrangement at low compaction pressures to increase packing efficiency and density uniformity prior to granule deformation/fracture. Under conditions where die filling is non-uniform on a macroscopic scale, the rearrangement of hard granules prior to deformation/fracture can help homogenize packing and die filling to produce a more uniform green density compact (Glass *et al.* 1996). However, if too hard, the granules will not completely deform to eliminate large intergranular porosity and/or the granule structure after compaction. Excessively hard granules can result in defects such as large residual pores and a cobblestone-like surface finish, as well as a lower compact density after pressing. Soft granules that readily deform during pressing can produce a higher density compact, but if too soft, the granules will not rearrange sufficiently at low pressures to eliminate packing defects present after die filling. For severe packing heterogeneities due to poor die filling, compacting soft granules can result in severe density gradients after pressing, and cracking after sintering. Additionally, excessively soft granules can create problems with powder flow, as well as die sticking that can contribute to surface defects, and that can complicate automated pressing processes. Ideally, the granules used for powder pressing should be hard enough to rearrange during die filling and low-pressure compaction, yet soft enough to deform sufficiently to produce a homogeneous green microstructure under the maximum pressing pressure.

Granule strength, which is also referred to as hardness, is controlled by many factors, including the organic additives used for granulation, moisture content, and granule size (Lannutti *et al.* 1997, Shinohara *et al.* 1999, Walker *et al.* 2000). Generally speaking, weaker granules are produced with softer organic binders that have a lower glass transition temperature, T_g. Plasticizers can be used to lower the T_g of a binder system to make weaker granules, and moisture itself (e.g., humidity) can act as a plasticizer (Shinohara *et al.* 1999, Lannutti *et al.* 1997). In fact, moisture-sensitive binders like PVA can contribute to seasonal pressing variability, with the pressing behavior of the powder being appreciably different during times of low and high humidity. Moisture uniformity within the powder is also important. It has been reported that large spray-dried granules can contain more moisture than the small granules within the same system, and are consequently more deformable. Ideally, a granulated powder should be pressed at a temperature above the T_g to realize optimum granule deformation during compaction.

(e) Characterizing compaction response. Compaction response diagrams can be useful for quality control and to help optimize processes in ceramic powder pressing (Reed 1995, pp. 418–49). For example, to help minimize density gradients in a powder compact during pressing, one can identify a region in the compaction curve where density is relatively insensitive to pressing pressure (Fig. 2(a)). Pressing should be completed in region II, at a pressure high enough to achieve sufficient granule deformation, compact densification, and green strength for handling/machining, but low enough to avoid excessive wear on the press and tooling. Compaction curves also can be used to evaluate binder systems, environmental effects such as temperature and humidity, and effects of aspect ratio on compaction response (Reed 1995, pp. 418–49).

Compaction response diagrams provide general information on compaction, and some information on granule hardness and deformability, but they need to be complemented with other techniques to complete a comprehensive assessment of compaction. In particular, pressed and fracture surfaces of powder compacts formed at successively increasing pressures can be evaluated to characterize microstructure evolution during powder compaction (Niesz 1996). Using this relatively straightforward method, one can observe granule deformation to eliminate the large intergranular packing voids, and the breakdown of the granule structure to produce a homogeneous microstructure (Fig. 4). Pressed surfaces of powder compacts formed at low pressures clearly show the granule structure, intergranular packing voids, and evidence of granule deformation to eliminate those voids. With increasing pressure, the granule structure breaks down.

Concurrently, the strength of the compact increases as the fracture mode changes from intergranular (i.e., around or between granules) to intragranular (i.e., through granules). At even higher forming pressure, the compact microstructure becomes more homogeneous, and the fracture surfaces

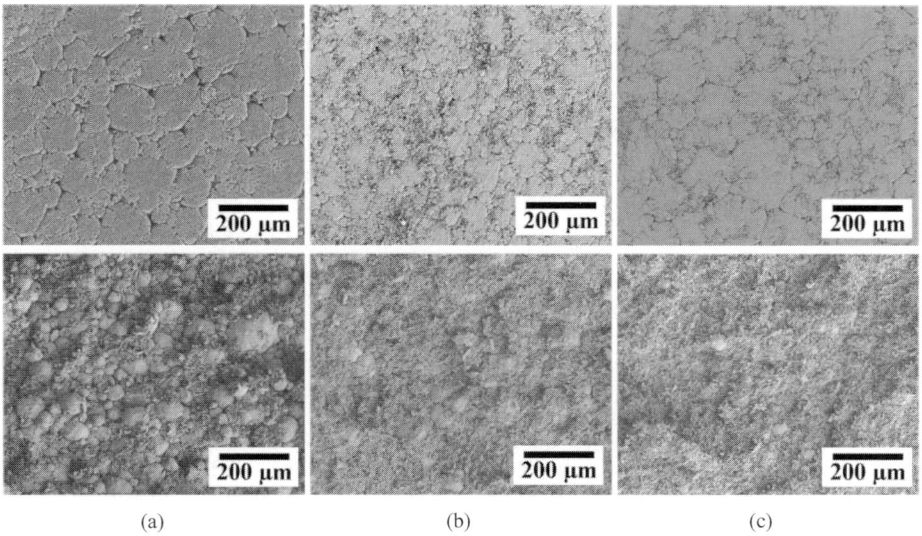

Figure 4
Photomicrographs of a pressed (top) and fracture (bottom) surface of a 94% alumina powder compact after uniaxial die pressing at: (a) 17.2 MPa; (b) 68.9 MPa; and (c) 137.9 MPa. The series shows granule deformation to eliminate the large intergranular packing voids, and the breakdown of the granule structure to produce a homogeneous microstructure with increasing pressing pressure.

of powder compacts show fewer and fewer granule remnants or relics. Ultimately, at a high enough pressure, any evidence of the granule structure will be eliminated altogether. Such flaws can be observed directly under a microscope by immersing thinned sections of powder compacts in specific index of refraction liquids (Shinohara *et al.* 1999). In combination with the compaction response diagram, photomicrographs of compact fracture surfaces, and the liquid immersion technique, the optimum pressing pressure required to manufacture a homogeneously dense powder compact can be determined.

(f) Characterizing powder compacts. Ultrasound velocity measurements, x-ray radiography, x-ray computed tomography (XRCT), and magnetic resonance imaging (MRI) using NMR are used to measure spatial density variations within a powder compact after pressing (Garino *et al.* 1995). These techniques provide quantitative spatial density data over the entire part.

To obtain quantitative data by x-ray radiography and ultrasound, however, carefully prepared samples with flat and parallel surfaces are required. X-ray radiography has slightly lower resolution than ultrasound, but may be preferable when cost is an issue and/or when many samples need to be analyzed (Garino *et. al* 1995). Both are readily adaptable to the ceramic manufacturing environment (e.g., for product quality and assurance testing).

XRCT and NMR are nondestructive techniques that can provide quantitative spatial density information in three dimensions. Advances in XRCT hardware, software, and technology now allow one to quantitatively and efficiently map density gradients and density evolution in powder compacts during and after pressing (Lannutti *et al.* 1997). NMR techniques are less developed and utilized, but in addition to spatial density data, they can provide quantitative microstructure analysis.

(g) Compaction modeling. Experimentally validated compaction models are powerful and promising tools that can be applied to better understand, and ultimately control and improve, particle/granule packing and powder compaction. One approach to modeling has been to develop and use empirical expressions that describe the compaction response curve (German 1989). Although common, empirical pressure–density relations alone lack the physical basis necessary to understand and control compaction behavior, and they cannot predict the formation of density gradients or defects within a powder compact. Additionally, empirical pressure–density models are often system specific, such that they cannot predict compaction response in different powder systems with different characteristics.

In combination with mechanics-based numerical modeling technology such as finite element method (FEM) modeling, a mathematical pressure–density relation or a constitutive model can be used to predict the density gradients in a powder compact after pressing. Realistic predictions of the spatial density variations within a ceramic powder compact have been made using a cap-plasticity constitutive model (i.e., that defines the constitutive behavior of the powder during compaction) within an FEM framework (Arguello *et al.* 2001).

The generic cap-plasticity constitutive model captures the mechanical behavior of granulated ceramic powder during compaction reasonably well (Arguello *et al.* 2001). The model is comprised of a stationary shear failure surface and a strain-hardening cap that define the elastic regime for the compaction of a powder in response to both hydrostatic compression and shear. Hydrostatic and triaxial compression tests have been used successfully to define the cap and the shear failure surface. Additionally, these same tests have been used to quantitatively measure the other bulk properties (e.g., pressure-dependent bulk and shear modulus and Poisson's ratio) required to model ceramic powder compaction (Arguello *et al.* 2001).

In combination with good powder data for input into the model, and a realistic estimate of the die wall friction coefficient, a cap-plasticity model within the FEM framework can predict density gradients that are very consistent with experimental observations (Fig. 3). Currently, FEM compaction model predictions are accurate enough to be useful design tools in ceramic manufacturing. In simpler geometry parts, model predictions can be used to guide die designs and pressing processes to minimize density gradients. Additionally, modeling can provide guidance on the aspect ratios, transition radii, and pressing balance (e.g., the relative displacements of a top and bottom punch in dual action pressing) required to control and minimize density gradients (Figs. 5 and 6). Furthermore, in complex parts in which severe density gradients near forming surfaces cannot be avoided, parts can be designed and formed intentionally oversized, such that regions with severe density gradients can be physically removed by green machining. There is also evidence that severe shear stress gradients predicted by FEM compaction models may be predictors of cracks in green and sintered compacts.

In addition to die design, compaction models also can provide valuable information about ceramic powders and their limitations in pressing. Different powders have different compaction responses, and not all powders can be pressed to all geometries. More pressable powders can be formed into more complex geometry powder compacts, but there will be limitations for harder-to-press powders. Compaction simulations can be used to assess and rank powders in terms of pressability, and to establish use limits for certain powders. Eventually, in combination with readily measurable powder characteristics, it may be possible to employ FEM compaction modeling technology to guide the design and development of more pressable powders.

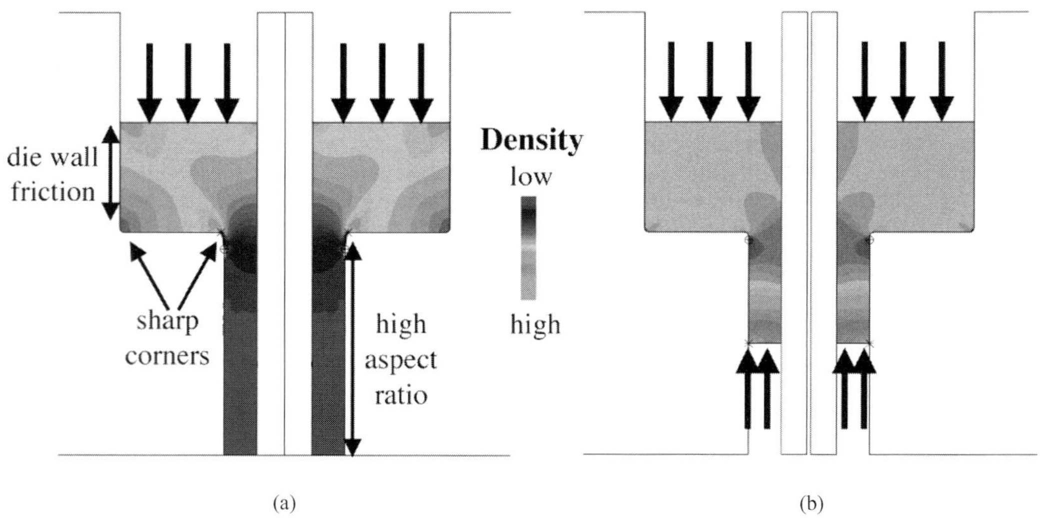

Figure 5
The density gradients in a cross-section of a powder compact (i.e., a bushing) of diatomaceous earth, as predicted by a cap-plasticity model within a FEM framework, for (a) single action; and (b) dual action uniaxial pressing. Some component/die design issues (i.e., high die wall friction, high aspect ratios, and sharp transition radii) are noted, and the benefit of dual action pressing to achieve a more homogeneous density compact is illustrated.

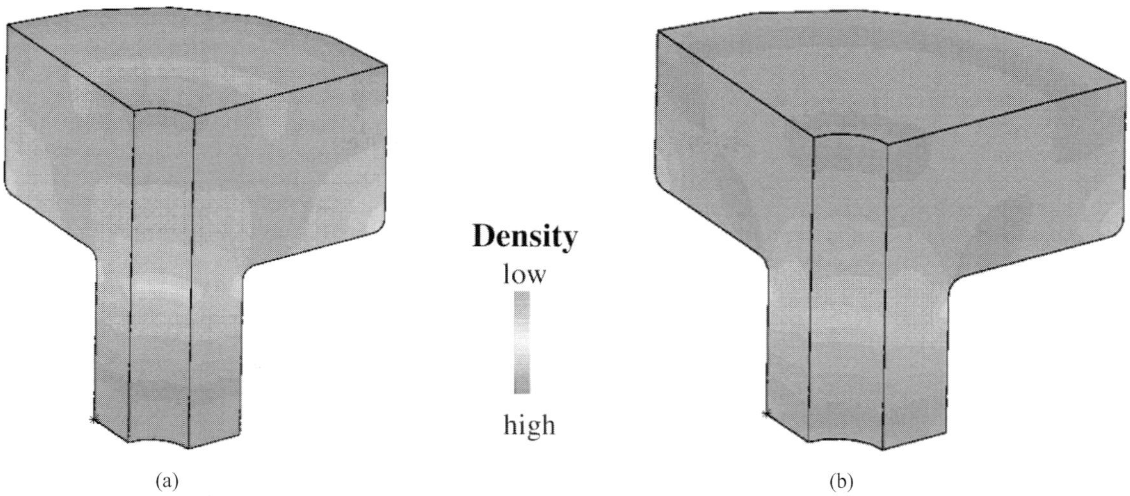

Figure 6
The density gradients in a quarter section of a 94 wt.% alumina compact (i.e., a bushing) pressed uniaxially by dual action pressing. The gradients were predicted using a cap-plasticity model within a FEM framework assuming: (a) unbalanced pressing conditions; and (b) balanced pressing conditions to improve density uniformity.

An appealing aspect of FEM modeling is that realistic and informative simulations of compaction can be completed on relatively complex geometry ceramic compacts, using a desktop or laptop computer, in as little as 30 minutes. A general disadvantage of FEM, and modeling in general, is that reliable and useful simulations are complex, and specialized expertise in computing and mechanics is generally required to obtain meaningful results. Specialized software has been developed that allows a nonexpert to reliably complete compaction simulations on a special class of axisymmetric geometries (Arguello *et al.* 2000); however, it is not yet commercially available.

Another limitation of FEM modeling is that it provides only a continuum analysis. That is, details of the discrete nature of the particle compaction process cannot be ascertained from FEM modeling.

Discrete element method (DEM) modeling has provided evidence that nonuniform packing of granules during die fill contributes to the density gradients observed experimentally in pressed powder compacts (Lannutti *et al.* 1997). Packing defects offer one of the only plausible explanations for the asymmetry in radial density observed in axisymmetric parts (e.g., Fig. 3(a)) in which the radial density variability would otherwise be expected to be axisymmetric (Glass *et al.* 1996, Lannutti *et al.* 1997)

Ideally, a realistic and comprehensive model for powder compaction should be able to account for the discrete nature of the compaction process, and also to predict continuum behavior. Because of the statistical nature of the compaction process, statistical models could ultimately provide the link between DEM and FEM modeling to achieve this end.

3.3 Ejection

The general issues with ejection are green compact strength, ejection pressure (i.e., which is related to die wall friction), and elastic springback. In general, a pressed powder compact must be strong enough to withstand the forces exerted on it during ejection to come out intact. One means of assessing potential problems is by the ratio of the axial crushing strength of a green cylindrical compact relative to the pressure required to eject that compact from the die. A ratio of one or less is an indicator that pressing defects and or compact failure could result on ejection. A diametral compression test is generally a more reliable method of determining the green strength of a cylindrical test pellet. Higher forming pressures produce stronger green compacts, but also increase the pressure required for ejection. An ideal way to avoid ejection defects is with a combination of a high green compact strength and a low ejection pressure. Internal lubricants and die wall lubricants are often used to lower the ejection pressure.

The other critical factor in ejection is springback. Springback is the term that describes the expansion of a powder compact upon ejection from the forming die. The dimensions of a compact can change by as much as 0.5% after ejection. The amount of springback is determined by the elastic energy stored in the sample during the compaction process. Greater springback occurs with higher forming pressures. Additionally, in uniaxial compaction there is differential springback, as the axial springback is greater than the radial. Some springback is required to separate the pressed powder compact from the forming die without sticking; however, excessive springback can produce catastrophic flaws within the compact (Reed 1995, pp. 418–49).

Excessive differential springback (i.e., perpendicular vs. parallel to the pressing direction) is also problematic. Excessively high forming pressures and trapped gasses within a powder compact can exacerbate problems with springback. Problems can be avoided by utilizing the appropriate organic binders to impart good green strength at low forming pressures, and by taking precautions to ensure that air within a bulk powder can escape from the compact during pressing. Problems with differential springback can be minimized/avoided by ejecting the compact under "punch holddown," which maintains axial pressure on the compact during ejection.

(*a*) *Compaction defects.* Many of the defects observed in powder compacts are defects that occur upon ejection. Furthermore, many compaction defects are related to springback (Reed 1995, pp. 418–49). Some typical pressing defects include end-capping, laminations, and vertical cracks (Fig. 7). End-capping describes the cone-shaped separation that forms at the pressing punch face and extends away from the press surface and into the bulk of the compact. It is common in bodies that have poor green strength and high springback. Excessive differential springback, high die wall friction, and high pressing pressure also contribute to end-capping. Ring-capping is an outer ring separation that forms at the outer edge of the pressing punch face, and extends away from the pressed face and out of the bulk of the compact. It is typically caused by poor tolerances between the punch and the die body, and excessive differential springback exacerbates the problem. Laminations are periodic circumferential cracks in a compact that originate on the die surface perpendicular to the pressing direction. These cracks form when green strength is poor, when the pressing pressure is excessively high, and when springback or die wall friction is high. Vertical cracks are elongated cracks that form parallel to the pressing direction when the differential springback and/or the compaction ratio are excessively high. Vertical cracks are common in low green strength bodies that have low gas permeability (e.g., fine particle size powders with fine porosity). High pressing rates also can contribute to the problem.

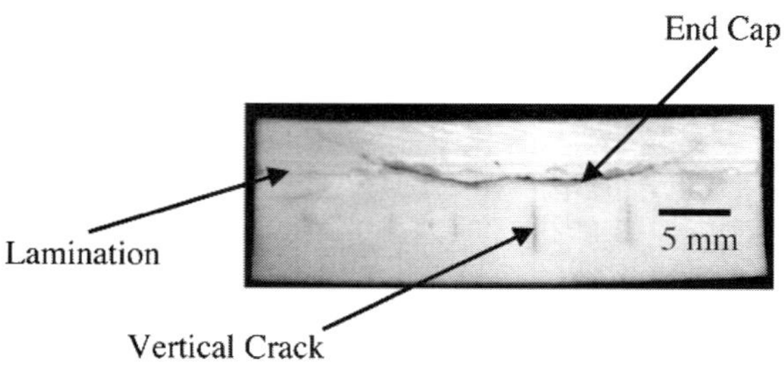

Figure 7
A sintered lead zirconate titanate particulate body, formed by uniaxial compaction, exhibiting typical powder pressing defects including end capping, laminations, and vertical cracks.

4. *Guidelines to Optimize Pressing*

Pressing defects (e.g., laminations or cracks), density gradients, and powder sticking to die surfaces are some problems that are common in dry powder pressing. Experienced press operators generally avoid pressing problems by understanding and controlling parameters such as die fill density, die wall friction, and expansion on ejection. Often, powder pressing problems are related to the powder characteristics/properties, and granule behavior. Understanding and controlling how powder flows into and fills the forming die, how granules deform under the pressing pressure, how uniformly the applied stress is transmitted throughout the compact during forming, and the amount of stored elastic strain energy released upon ejection are all critical to controlled and reproducible dry powder pressing. Optimum pressing performance is achieved with granulated materials that flow freely, and pack and compact to uniformly high relative densities. Ideally, ceramic granules for dry pressing should be spherical in shape, and 44–400 μm in diameter with a median size of 100–200 μm.

Granules should be uniformly dense, and should have a density lower than the pressed compact density in the range of 45–55% of the theoretical density of the powder, based on the inorganic constituents alone. Additionally, the surfaces of granules should be smooth and nonsticky. Granulated powder should have an angle of repose of 40° or lower for good flow, and should pack to a (die) fill density of 25–35% of theoretical density. Granulated powder should easily compact to a uniform density of 55–60% of theoretical density after pressing at 34–103 MPa (5000–15000 psi). And finally, granules should be sufficiently deformable such that large, intergranular packing voids and/or any remnants of the original granule structure are eliminated after compaction (Reed 1995, pp. 418–49).

Acknowledgments

This work was supported by the US Department of Energy (DOE) under Contract DE-AC04-94AL85000. Sandia is a multiprogram laboratory operated by Sandia Corporation, a Lockheed Martin Company, for the US Department of Energy. The author thanks Dr. D. Ellerby of Sandia National Laboratories for his technical review of this article.

Bibliography

Arguello J, Fossum A, Zeuch D, Ewsuk K 2001 Continuum-based FEM modeling of ceramic powder compaction using a cap-plasticity constitutive model. *KONA Powder Particle* **19**, in press

Dinger D R, Funk J E 1997 Particle packing phenomena and its application in materials processing. *MRS Bull.* **22**, 19–23

Ewsuk K G 1993 Consolidation of bulk ceramics. In: Loehmann R E (ed.) *Characterization of Ceramics*. Butterworth-Heinemann, Greenwich, CT, pp. 77–101

Garino T, Readey M J, Mahoney F M, Ewsuk K G, Gieske J, Stoker G, Min S 1995 Characterization techniques to validate models of density variations in pressed powder compacts. In: Martinez R J, Arris H, Emerson J A, Pike G (eds.) *Diversity Into the Next Century. Proc. 27th Int. SAMPE Technical Conf.* SAMPE International, Covina, CA, Vol. 27, pp. 610–21

German R M 1989 *Particle Packing Characteristics*. Metal Powder Industries Federation, Princeton, NJ

Glass S J, Ewsuk K G, Mahoney F M 1996 Ceramic powder compaction. In: Gupta T K, Hiremath B, Nair K M (eds.) *Ceramic Manufacturing Practices and Technology. Ceramic Transactions*. The American Ceramic Society, Westerville, OH, pp. 3–18

Lannutti J J, Deis T A, Kong C M, Phillips D H 1997 Density gradient evolution during dry pressing. *Bull. Am. Ceram. Soc.* **76**, 53–8

Lukasiewicz S J 1991 Granulation and Spray Drying In: *Ceramics and Glasses. ASM Engineered Materials Handbook*. ASM International, Materials Park, OH, Vol. 4, pp. 100–8

Niesz D E 1996 A review of ceramic powder compaction. *KONA Powder Particle* **14**, 43–51

Reed J S 1995 *Introduction to the Principles of Ceramic Processing*, 2nd edn. Wiley, New York, pp. 135–208, 378–45

Reed J S, Runk R B 1976 Dry pressing. In: Wang F F Y (ed.) *Ceramic Fabrication Processes. Treatise on Materials Science and Technology*. Academic Press, New York, Vol. 9, pp. 71–93

Shanefield D J 1996 *Organic Additives and Ceramic Processing, Applications in Powder Metallurgy, Ink, and Paint*, 2nd edn. Kluwer, Boston

Sherrington P J, Oliver R 1981 *Granulation*. In: Monographs in Powder Science and Technology. Heyden, London

Shinohara N, Okumiya M, Hotta T, Nakahira K, Naito M, Uematsu K 1999 Seasonal variation of microstructure and sintered strength of dry-pressed alumina. *J. Am. Ceram. Soc* **82**, 3441–6

Walker Jr W J, Reed J S, Verma S K 2000 Selection of slurry parameters for engineered granule characteristics. In: Puri V M, Adair J H, Knobloch C L (eds.) *1999 Fine Powder Conf. Proc.* Pennsylvania State University. pp. 264–71

K. G. Ewsuk

Primary Recrystallization

Primary recrystallization is the process, driven by the stored energy of plastic deformation, by which the grain structure of the deformed material is transformed to a new structure. It is an important part of processing of metals for two reasons. The first is that material, hardened by deformation processing, is softened to allow further shaping and to produce soft material for final use. The second reason is that it produces a more desirable grain structure. The grain structure is the grain size, shape, and mean orientations (the annealing texture).

1. Stored Energy of Deformation

As in other transformations, it is vital to understand the driving pressure for the reaction. This is usually given as the energy per unit volume ($J\,m^{-3}$), at a pressure ($N\,m^{-2}$). During plastic deformation, energy is expended as the integral of the product of stress, σ, and the increment of strain, $d\varepsilon$. Most of this work ($\approx$ 99%) is released immediately as heat. However, the remaining energy remains stored in the material. This stored energy is found as dislocations at a density, ρ (mm^{-2}). After low temperature deformation, performed below 20% of the absolute melting temperature ($0.2\,T_m$), excess vacancies are also stored. These vacancies, however, contribute little to the stored energy. At higher temperatures, particularly for metals like aluminum and nickel with high values of the normalized stacking fault energy, γ_{sfe}/Gb, almost complete dynamic recovery of the dislocations occurs giving a subgrain structure. (G is the shear modulus and b the Burger's vector of the dislocations.) The structure is then described by the subgrain size, d_s, and boundary energy, γ_s. *Dynamic* recovery is the rearrangement of dislocations that occurs under the combined conditions of strain and temperature. It is distinct from the slower process of *static* recovery that takes place, after straining, under the influence of annealing temperature alone. The stored energies, ΔE_s, in the extreme cases of unrecovered dislocations or strain free subgrains are:

$$\Delta E_s \approx \rho G b^2 (\text{dislocations}) \text{ or } \approx 3\gamma_s/d_s (\text{subgrains}) \quad (1)$$

Typical stored energies are to be found in Martin *et al.* (1997) for low temperature deformation, $T \leqslant 0.3\,T_m$. Much industrial deformation occurs, however, at higher temperatures, where the flow stresses are smaller and stored energy values much less. Little has been reported on actual values of the stored energy after hot working, but the dislocation density, ρ, is much less and the subgrain size, d_s, is much larger.

2. Mechanisms of Recrystallization: Nucleation and Growth

The recrystallization reaction is heterogeneous and is a Gibbs type I process (Doherty *et al.* 1998). In such a process, the reaction is complete within the new grains that are separated by a sharp interface from the deformed material. This type of transformation is called a nucleation and growth process though the name of "nucleation" for the formation of the new grains in recrystallization may be misleading. In other nucleation and growth, reactions, e.g., solidification from the liquid, the new structure *does not pre-exist in the parent structure*, so the new structure has to be formed, atom by atom, to a critical radius, at which the increase of boundary energy is offset by the reduction of volume energy. The usual analysis for nucleation does *not*, however, apply here. There are two reasons for this: (i) even in the most heavily deformed metal, the material retains its crystal structure, so the "new" grains do not have to form; and (ii) the high energy of the high-angle boundaries, combined with the low stored energies, make the formation of a new grain impossible. As a result, grains with totally new orientations, do not "nucleate" in recrystallization. New grains grow from regions present in the deformed state. The problem is to identify these regions and predict how they develop into new grains. The problem is illustrated, for metals with a well-developed subgrain structure, by the size ratio of the initial subgrain diameter, d_s, to the final grain diameter, d_f. The length ratio is of the order of 1×10^{-2}, giving a volume ratio of 1×10^{-6}. The latter ratio gives the low *probability* of a subgrain becoming a recrystallized grain. This low probability arises from two phenomena: (i) the low mobility of the low angle subgrain boundaries, and (ii) the complex details of the deformed microstructure.

3. Grain Boundary Mobility

It is now known that the mobility of the grain boundaries is strongly influenced by two factors; the boundary misorientation and the solute content of the alloy (Humphreys and Hatherly 2004). For recrystallization, the most significant variation in mobility is its dramatic increase from *low-angle* subgrain boundaries, misoriented by 1–3°, to *high-angle* boundaries, misoriented by more than 15–20°. Only new grains with high-angle boundaries can grow. As a result, knowing the density of high local misorientations in the deformed microstructure is critical for the understanding nucleation.

The effect of small traces of solute on grain boundary mobility is also dramatic. Metals at a purity level of about 1 ppm of solute have a low activation energy for atom transfer both *along* and *across* a grain boundary. In the former case, there is boundary diffusion, while in the latter there is boundary migration. In ultrahigh-purity metals, the activation energies are the same and are low, allowing rapid grain boundary movement at temperatures as low as $0.3\,T_m$. This is seen, for example, in very high purity aluminum (with <1 ppm solute) which recrystallizes at room temperature and also in copper that can recrystallize at 370 K or less. Copper is produced commercially at very high purity for conductivity applications, while aluminum almost always contains a trace of iron, which raises the recrystallization temperature to 500 K. The origin of this effect, called solute drag, is the segregation of insoluble elements to the grain boundaries and the necessity of dragging the solute when they move. However, the effect of solute drag appears to saturate at a few hundred parts per million of solute, so changes of solute content above this value have little further influence on boundary mobility.

4. *Microstructure of Deformed Materials*

This topic remains a major challenge for materials science. Ideally the structure of the deformed metal on the basis of starting structure and the applied strain needs to be predicted. Despite a vast number of investigations, both experimental and theoretical, little insight into the details of the deformed microstructure has so far been obtained. Much is known empirically, however, about those aspects of the deformed microstructure that are critical for understanding recrystallization (Humphreys and Hatherly 1995). The most important features are the average stored energy and the development of high local misorientations. The prior grain boundaries, which have high misorientation, increase in a predictable way in area as the sample deforms. Other types of misorientation that develop within grains, which are described below, are much less understood.

4.1 In-grain Misorientations

Large misorientations are produced by low temperature deformation of alloys containing hard particles larger than about 1 μm in diameter. Another source of in-grain misorientation is the development of "transition bands," the interfaces that are formed as grains *break-up* during deformation. This type of grain break-up, an active area of current plasticity research, appears to be most marked in large-grain-size materials. It is known to occur after both room temperature and elevated temperature deformation and is a major source of recrystallization nucleation (e.g., Fig. 1), after moderate deformation.

Additional sources of high local misorientation that promote nucleation are the bands of localized strain, "shear bands," that form most commonly in metals and alloys that show strong strain hardening

Figure 1
Nucleation at a transition band in compressed aluminum. White grains of an orientation similar to that of region A growing into region B and vice versa. Both A and B came from the same initial grain (after Bellier 1971).

either induced by low stacking fault energies or, at least in aluminum by high solute levels. The low-stacking-fault energy f.c.c. brasses and stainless steels are known to form a high density of deformation twins that promote strain hardening and these twins facilitate shear banding. However, the deformation twins also have high-angle misorientations, so for these alloys both the twins and the shear bands are sites for nucleation of new grains.

4.2 Heterogeneity of Stored Energy

For a subgrain to be able to grow, it has to have the mobility provided by the local misorientation and, also, an energy advantage. The dominant source of this energy advantage is the relative subgrain size. The same physics applies here as for grain coarsening, where larger grains grow at the expense of smaller ones. There is increased free energy per unit volume of atoms inside small grains with low radii of curvature. As a result, boundary migration takes place at any curved boundary, by growth

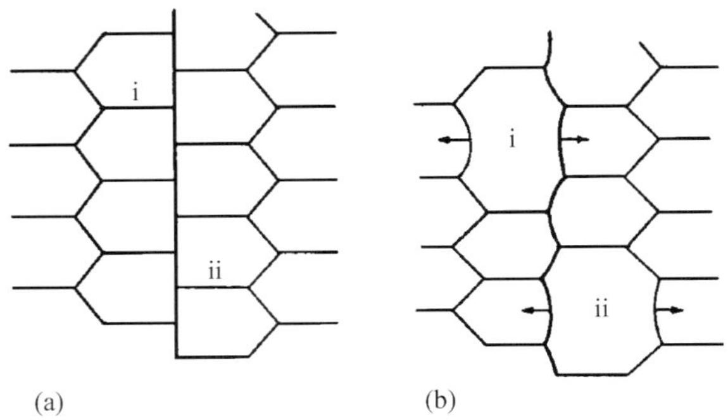

Figure 2
Development of a size advantage by subgrain coalescence at a high angle boundary leading to the formation of new grains, i and ii. The resulting boundary curvatures cause the boundary migration indicated (after Doherty 1978).

towards the center of curvature (Fig. 2). In many cases however, different as-deformed subgrain sizes are found on opposite sides of a grain boundary. The usual origin of subgrain size variation across a grain boundary is the orientation dependence of stored energy. However, it may also arise from orientation-dependent recovery. The former process is the more widely established process, one of the most common examples being in cold rolled iron where grains with {111} planes parallel to the rolling plane, the γ fiber, have a higher stored energy with smaller subgrains and larger misorientations, than in other common rolling orientations such as the α fiber. The latter has $\langle 110 \rangle$ parallel to the rolling direction. In cold-rolled copper, it has been reported that grains with higher Taylor factors have a higher stored energy than those with lower Taylor factors. The Taylor factor is the ratio of slip to the amount of strain, assuming each grain undergoes the same strain.

In aluminum deformed at room temperature, T/T_m of 0.3, little heterogeneity of subgrain size between different orientations is found. However, after extended recovery, regions with a cube orientation {001} parallel to the rolling plane and $\langle 100 \rangle$ parallel to the rolling direction in both aluminum and copper have been reported to form larger subgrains. This development of heterogeneity appears to arise by enhanced recovery. The larger cube subgrains have a major impact on recrystallization. For both copper and aluminum, many nuclei of the expected cube orientation form and give the strong cube annealing texture. For cold-rolled iron, preferred nucleation of γ fiber grains is known to occur—but it is not at present understood why nature gives preferred nucleation of γ grains with *smaller* subgrains.

5. Recrystallization Kinetics and Grain Structure

Textbooks of physical metallurgy list empirical "rules of recrystallization" based on early experiments, mainly on moderately deformed, pure metals. These rules fit well with the current ideas for the processes of nucleation and growth. The major rules are: (i) Recrystallization is thermally activated. (ii) It is accelerated by higher strains which give finer final grain sizes. (iii) There is a critical deformation below which recrystallization does not occur. (iv) Increasing deformation temperature, from "cold" to "warm" working, has the opposite effect, giving lower stored energies, slower kinetics and larger final grain sizes. (v) Finer initial grains accelerate recrystallization and give smaller final grains. (vi) Small traces of solute dramatically raise the required temperature for recrystallization while larger solute contents have limited effect. (vii) Fine second-phase particles inhibit recrystallization, in particular nucleation, giving larger grains. (viii) Large particles stimulate nucleation giving finer grains. All these rules are found to apply, in most cases, to both laboratory investigations and to industrial processing. However, care sometimes needs to be used when the rules are applied to industrial conditions significantly different from those of the laboratory studies from which they were deduced. In particular, industrial processing often involves much higher strains and, during hot rolling, interrupted deformation, for example, during multistage hot rolling. Recrystallization can then occur between passes and there are changes in solute content and precipitate distribution in alloys containing solute. The "controlled rolling" of low-alloy steels, which gives fine ferrite (b.c.c.) grains by careful control of the deformation temperature during rolling, when the steel is in the austenitic (f.c.c.) phase, is now well understood. Here the steel composition is carefully set to allow phases such as NbC, VC, AlN to be initially in solution when hot rolling starts. As the rolling temperature falls, these phases precipitate and inhibit austenite grain growth, before and after recrystallization. Later, when the rolling temperature is even lower, the spacing of the precipitates becomes sufficiently fine to prevent recrystallization. The fine austenite grains are then reduced in thickness by further rolling giving closely spaced austenite boundaries. The transformation to (b.c.c.) ferrite then nucleates on these boundaries, resulting in the desired fine ferrite grain size.

Early observation by optical microscopy of recrystallization led to the analysis of the kinetics of recrystallization. The method used was developed, independently, by Johnson and Mehl (1939), Avrami (1939), and Kolmogorov (1937). This "JMAK" analysis consists of two steps. The first is the prediction, based on the observed nucleation and growth kinetics, of the so-called "extended" volume fraction, $f_v{}^*$, calculated without any correction for the "impingement" of new grains. New grains cannot grow in grains which have already transformed. Correction for this effect, assuming that the sites of nucleation are randomly spaced on a length scale of the recrystallized grain size, gives:

$$df_V = (1 - f_V)df_V^* \quad (2)$$

Integration yields the classic JMAK equation for the true fraction recrystallized, f_V:

$$f_V = 1 - \exp(-f_V^*) \quad (3)$$

For simple power law rates of nucleation and growth ($I_V = dN_V/dt = at^n$, and $G = dr/dt = bt^m$) leads to

$$f_V = 1 - \exp(-kt^p) \quad (4)$$

where $p = n + 3m$. Anderson and Mehl (1945) tested Eqn. (3) using a thin sheet of aluminum with a fine initial grain size, d_o, that had been lightly deformed so that the recrystallized grains grew to a size larger *than the sheet thickness* and gave a final grain size, d_f, where $d_f \gg d_o$. This experiment allowed the grain sizes and mean density of new grains to be measured two-dimensionally, avoiding the difficulties of transforming from two to three dimensions. Also, because nucleation occurred on the prior grain boundaries, at a *length scale much smaller than the final grain size*, the *random* nucleation needed for Eqn. (2) was allowed. The study confirmed the validity of the JMAK analysis. However, it has often been mistakenly assumed that this classic experiment confirmed the validity of the JMAK analysis for recrystallization generally. This is now known not to be true. Doherty *et al.* (1986) discussed this problem and pointed out that in most cases, the assumption that the nucleation sites are spatially distributed at random at the length scale of the recrystallized grain size is rarely valid. Nuclei are often found to be clustered in particular grains and at other particular locations. Examples are the deformed cube bands in rolled f.c.c. metals and the deformed γ bands in rolled iron. As a result of such clustering, the JMAK exponent, p, will be smaller than expected and in general will be neither an integer nor a constant. A further reason for the failure of the simple use of Eqn. (4) to describe recrystallization kinetics is that it is frequently found that the growth rate, G, is not constant ($m < 1$). There are two reasons for this. One is the likelihood of stored energy falling by continued recovery during recrystallization and the second is that the stored energy can vary between deformed grains.

The failure of the JMAK analysis in recrystallization has two results: the difficulty of prediction of the kinetics even when models are developed for nucleation and growth rates, and a more serious difficulty in that the final structure, the grain size, and the texture, cannot be analytically predicted. This is a major problem for an engineering discipline, which needs to be *predictive* for design purposes. At present, the only way to determine how a proposed change in the processing schedule for an industrial alloy may modify the recrystallization kinetics and resulting structure is to study the change experimentally. While that is easy for laboratory studies, for designing a new industrial process, the need for empirical studies, *using industrial plant that has yet to be built*, is a major difficulty. So, although much is now understood about the mechanisms of recrystallization, much more remains to be learnt before this knowledge can be really useful for process design.

See also: Dynamic Recovery and Recrystallization; Primary Recrystallization in Two-phase Alloys

Bibliography

Anderson W A, Mehl R F 1945 Recrystallization of aluminum in terms of the rate of nucleation and growth. *Trans. Metall. Soc. AIME* **161**, 140–67

Avrami M 1939 Kinetics of phase changes. I: general theory. *J. Chem. Phys.* **7**, 1103–12

Bellier S P 1971 Formation of new grains during annealing of cold worked aluminium. D.Phil. thesis, University of Sussex

Doherty R D 1978 In: Haessner F (ed.) *Recrystallization of Metallic Materials*, 2nd edn. Rieder-Verlag GmbH, Stuttgart, Germany

Doherty R D, Hughes D A, Humphreys F J, Jonas J J, Juul-Jensen D, Kassner M E, King W E, McNelley T R, McQueen H J, Rollett A D 1998 Current issues in recrystallization: a review. *Mater. Sci. Eng.* **238A**, 219–74

Doherty R D, Rollett A D, Srolovitz D J 1986 Structural evolution during recrystallization. In: Hansen N, Juul-Jensen D, Leffers T, Ralph B (eds.) *Annealing Processes: Recovery Recrystallization and Grain Growth.* Risø National Laboratory, Denmark, pp. 53–67

Humphreys F J, Hatherly M 1995 *Recrystallization and Related Phenomenon.* Pergamon, Oxford

Johnson W A, Mehl R F 1939 Reaction kinetics in processes of nucleation and growth. *Trans. Metall. Soc. AIME* **135**, 416–42

Kolmogorov A N 1937 *Izv. Skad. Nauk. USSR-Ser-Matemat* **1**(3), 416

Martin J W, Doherty R D, Cantor B 1997 *Stability of Microstructure in Metallic Systems.* Cambridge University Press, Cambridge

R. D. Doherty

Primary Recrystallization in Two-phase Alloys

Recrystallization is a nucleation and growth phenomenon and the presence of a second phase will affect both of these reactions compared to a single-phase situation. The way in which recrystallization of a given alloy is affected depends on whether the second phase is present prior to deformation or is precipitated during the annealing which results in recrystallization. If the second phase is present prior to deformation, recrystallization will be affected in different ways depending on the volume fraction, the size and the spatial distribution of this second phase. This short article will be concerned with conditions where the second phase is present prior to deformation, focusing on the effect of particles on: (a) the deformed state, (b) the nucleation of recrystallization, (c) the kinetics of the reaction, and (d) the resulting grain size and texture. A final section will be devoted to the problems of grain size and texture control in commercial production of sheet metals and extrusions. Humphreys and Hatherly (2004) give an extensive review of this field.

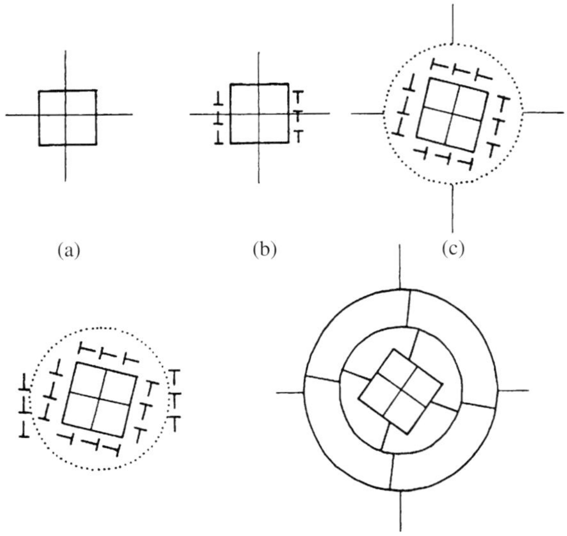

Figure 1
The formation of a deformation zone (after Humphreys and Kalu 1990).

1. *Effect of Particles on the Deformed State*

The presence of hard, nondeformable particles during the deformation of a softer metal matrix will change the dislocation storage pattern compared to that of a particle-free situation. The incompatibility between such particles and the surrounding matrix will cause the generation of geometrically necessary dislocations during deformation, which are accumulated in the particle vicinity and which eventually will form deformation zones around the particles. It follows that such deformation zones will be associated with local misorientation and dislocation density gradients compared to the surrounding more uniformly deformed matrix. The first theoretical treatment which dealt with the deformation behavior of plastically nonhomogeneous systems of this type was the classical work by Ashby (1971), which accounts for the overall effect on the dislocation storage due to a uniform distribution of finely dispersed nondeformable particles. Concerning the effect of particles on recrystallization, however, deformation zones associated with large particles are of particular relevance, because of their role in the nucleation of recrystallization. The current understanding of the deformation zone structure and how it responds to annealing treatments is to a major extent due to the significant contributions by Humphreys and co-workers; for references see Humphreys and Hatherly (2004).

On the basis of the ideas of Ashby (1971), Humphreys and Kalu (1990) proposed the following simple model for the formation of a deformation zone. Consider a particle within a crystal as shown in Fig. 1(a). During deformation, the particle will be surrounded by Orowan loops (Fig. 1(b)). The stress may be relaxed and the loop absorbed into the particle–matrix interface if the particle rotates and if a suitable array of secondary dislocations is generated as shown in Fig. 1(c). These dislocations will make it more difficult for further primary dislocations to penetrate to the particle surface, i.e., they effectively increase the size of the particle as shown in Fig. 1(d). Therefore, relaxation of the primary loops results in further rotation of the volume within the hard shell. This process is repeated as deformation continues (Fig. 1(e)). The result becomes a deformation zone where the misorientation differences between the particle interface and the matrix will exceed more than 20° for micrometer-size particles at strains $\varepsilon > 0.5$. A common observation is that the width of the deformation zone typically compares to the particle diameter. It follows that regions with such microstructural properties become preferential sites for nucleation of recrystallization.

2. *Particle-stimulated Nucleation (PSN) of Recrystallization*

Nucleation of recrystallization during annealing of deformed metals and alloys, single crystals, or polycrystals is heterogeneous in the sense that the reaction invariably appears to be associated with deformation heterogeneities of some sort or another, i.e., transition bands, shear bands, grain boundary regions, and deformation zones around large particles. The reason for this is obvious, since such heterogeneities are regions of high stored energy and steep orientation gradients. To what extent PSN of new grains will affect recrystallization during annealing of a deformed two-phase alloy will depend on how effective particle sites are in competition sites associated with other deformation heterogeneities.

From observations of PSN in single crystals of an aluminum–silicon alloy deformed by rolling, Humphreys (1977) concluded that the critical step is not the nucleation of a new grain within the zone, but growth of such grains beyond the deformation zones. From this it follows that the condition for PSN can the be formulated as a size criterion, i.e., new grains will grow out from zones around particles of size η larger than a critical value η_c given by:

$$\eta_c = \frac{2\gamma_{GB}}{3P_D} \tag{1}$$

where γ_{GB} is the specific grain boundary energy and P_D is the stored deformation energy in the matrix surrounding the deformation zones. Experimental measurements by Humphreys (1977) of the effect of particle size and strain on the occurrence of PSN are shown in Fig. 2. These observations are in qualitative agreement with Eqn. (1). From estimates of the variation in stored energy with strain in aluminum, a reasonable quantitative prediction of the observations in Fig. 2 is also obtained.

In a two-phase alloy which contains a size distribution of particles, $n(\eta)$, and in which deformation has resulted in an average stored energy P_D, the density of PSN during annealing becomes:

$$N_\eta = \int_{\eta_c}^{\infty} n(\eta)\, d\eta \tag{2}$$

How these N_η sites will affect the recrystallization of such a two-phase alloy will depend on the relative strength of the PSN sites compared to the nucleation activities of other deformation heterogeneities. Experimental investigations of the recrystallization behavior of commercial aluminum alloys after cold rolling have confirmed that nucleation of recrystallization is dominated by PSN, and that the effects of altering the particle size distribution and/or the driving pressure on recrystallization kinetics and grain size are well accounted for by Eqns. (1) and (2) (Nes 1980, Nes and Hutchinson 1989, Vatne *et al.* 1996).

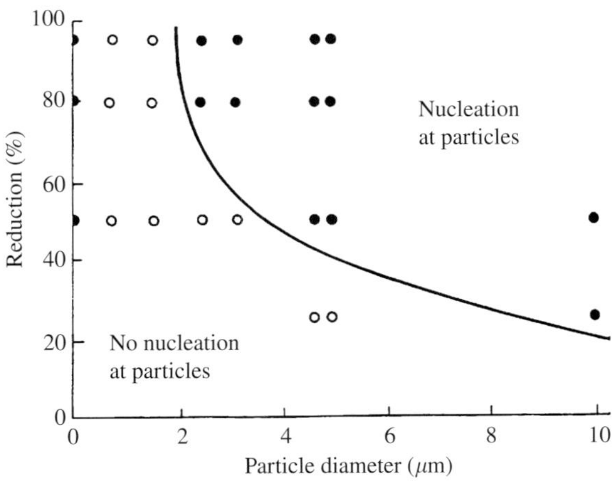

Figure 2
The effect of rolling reduction on the occurrence of particle-stimulated nucleation of recrystallization (after Humphreys 1977).

3. *Effect of Small Particles on Nucleation and Growth*

A fine dispersion of small particles will affect recrystallization through their interaction with the transformation front, i.e., with moving high-angle grain boundaries, a phenomenon referred to as the *Zener drag*. In spite of its name, Zener never published any paper on this effect. His contribution is a personal communication to C. S. Smith, who gave an extremely short presentation of it in a paper published in 1948. Since then a huge literature has emerged on the topic (for a review see). Zener (and Smith 1948) came up with the following approximate solution for restraining pressure, P_Z, due to a volume fraction, f, of randomly distributed spherical particles of radius r:

$$P_Z = \frac{3}{4}\left(\frac{\gamma_{GB} f}{r}\right) \tag{3}$$

Quite a number of more or less ambiguous calculations have been made subsequently; however, all these refinements give results close to Zener's estimate.

The Zener drag will affect both the nucleation of new grains and their subsequent growth and accordingly the kinetics of recrystallization. Nucleation will be suppressed because the particle drag reduces the effective driving pressure for nucleation, i.e., P_D in Eqn. (1) is replaced by $P_D - P_Z$, and it follows that the critical nucleus size increases or the density of sites (Eqn. (2)) decreases. Such a particle effect on nucleation of recrystallization was first reported by Doherty and Martin (1962), who investigated the effect of finely dispersed Al_2Cu particles on the recrystallization of cold-rolled aluminum. Growth of new grains will be affected in a similar way, for the simple reason that the migration rate of the reaction front is determined by the boundary mobility times the effective driving pressure acting on the boundary, i.e., $P_D - P_Z$. Grain growth following the completion of the recrystallization reaction will also be affected, resulting in the well-known *stagnation phenomenon* and associated grain size, D_{gg}, given by:

$$D_{gg} = \alpha(r/f) \quad (4)$$

where α is a constant of order 3 (Humphreys and Hatherly 2004) and Sect. 5 below).

4. *Particle Effects on Recrystallization Texture*

PSN will, as the main effect, contribute to a rather random texture or a diffuse reproduction of the deformation texture components. Under certain conditions PSN may in aluminum sheets also contribute to the P-component, ($\{011\}\langle 122\rangle$), a result interpreted in terms of a growth selection from a random distribution of site orientations (Lücke and Engler 1990). A fine dispersion of small particles may strengthen certain texture components, notably the cube texture in the recrystallization of aluminum sheets, most probably because the reduced Zener pinning effect on the low-energy $\Sigma 7$-boundaries between cube and S-grains (Vatne *et al.* 1996, Higginson *et al.* 1997). For additional details see Sect. 5.

Secondary recrystallization in silicon–iron is a vital stage in the production of the steel sheet having intensely sharp Goss texture for electromagnetic application. The primary structure, which is pinned by fine dispersions of AlN or MnS particles, becomes destabilized as these particles start to dissolve during heating. Goss oriented secondary grains grow faster than grains of other orientations, probably because their boundaries have on average lower energy and so experience a weaker Zener pinning, leading to the marked texture evolution (Hutchinson and Homma 1998).

5. *Grain Size and Texture Control*

Texture and grain size control is synonymous with quality control for a whole range of important engineering materials; from structural steels through a series of sheet metals applications to high-technology materials like modern superplastics. Two of the most successful achievements in the materials area, namely the high-strength, low-alloyed steels and the all-aluminum beer and beverage can, both require stringent structure/texture control in production as well as for final product properties. The control of grain size and texture is intimately connected to the scale of the particle dispersion through PSN and the Zener pinning effect (Nes and Hutchinson 1989).

In cold-worked and annealed metals (or in metals which recrystallize dynamically during processing), Eqn. (4) will determine the grain size provided the as-recrystallized grain size $D_{rec} < D_{gg}$. The recrystallized grain size is given by Eqn. (2), $D_{rec} \propto N^{-1/3}{}_{\eta}$, and it follows that this grain size is determined by PSN through the size distribution of coarse particles and the Zener drag pressure through its effect on the effective driving pressure for recrystallization, $P_D - P_Z$. These dual effects due to the density of large particles, N_0, and the dispersion level of fine particles, f/r, in controlling grain size during processing are schematically outlined in Fig. 3. It follows from this figure that three important cases can be identified:

(i) $f/r < [f/r]_t$. In this case the recrystallized grain size is $D_{rec} < D_{gg}$. This applies to hot rolling of many plain and micro-alloyed steels where abundance of nucleation sites for recrystallization is supplied by the austenite grain boundaries. It also typifies copper and many of its alloys which contain only a small fraction of second phase.

(ii) $[f/r]_t < f/r < [f/r]_{crit}$. In this regime the grain structure and texture is recrystallization controlled. This is generally the case during processing of aluminum alloys. The effect due to the coarse constituent particles is illustrated by the arrow at the top of Fig. 3, showing that an increase in the constituent particle density results in grain refinement. Associated with such a refinement is a texture randomization as illustrated in Fig. 4. Increasing the dispersion level of small particles has a strong effect on grain size, and in addition promotes the formation of the cube recrystallization texture. For a general review

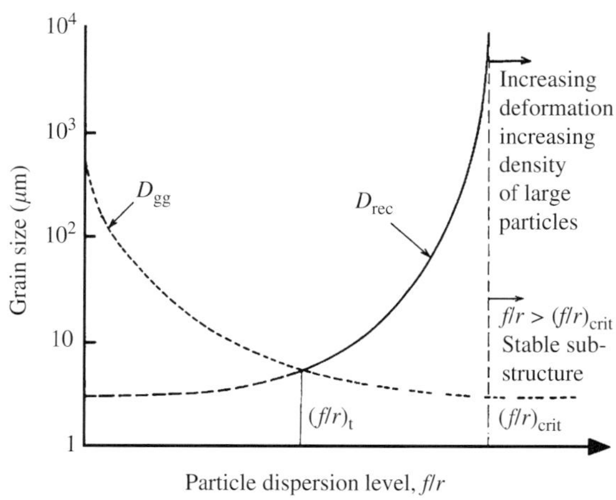

Figure 3
Different ways in which the grain size after annealing of a cold-worked alloy depends on second-phase dispersion level (Zener pressure) (after Nes and Hutchinson 1989).

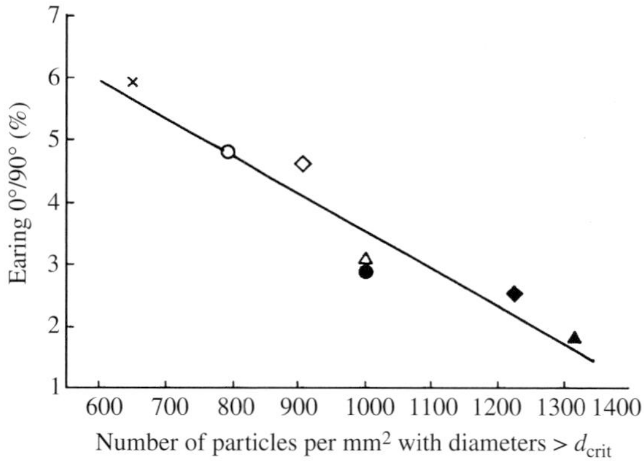

Figure 4
Relationship between earing level and frequency of coarse second-phase particles which nucleate randomly oriented grains (after Hutchinson *et al.* 1989).

see Vatne *et al.* (1996). In the production of aluminum-killed steels precipitation of AlN can be use to optimize the grain size at the completion of recrystallization, and a favorable texture develops at the same time.

(iii) $f/r > [f/r]_{crit}$. The Zener pinning pressure has now reached levels which do not allow recrystallization to occur. Or in other words, the deformation structure has been stabilized by the particle dispersion. Especially in the extrusion of aluminum alloys, many applications require a nonrecrystallized structure. Exceedingly high dispersion levels can be obtained in aluminum alloys by combining small additions of manganese, chromium, and zirconium. The recrystallization temperature can be increased by such additions to well above 500°C, which opens up the possibility for some interesting properties. Plastic deformation of such alloys at temperatures above $T_m/2$ may cause strain-induced continuous recrystallization, a phenomenon which has been used commercially in the development of a new generation of superplastic aluminum alloys.

See also: Primary Recrystallization; Directional Recrystallization

Bibliography

Ashby M F 1971 The deformation of plastically non-homogeneous alloys. In: Kelly A, Nicholson R B (eds.) *Strengthening Methods in Crystals*. Elsevier, Amsterdam, pp. 137–92

Doherty R, Martin J W 1962 The effect of a dispersed second phase on the recrystallization of aluminium–copper alloys. *J. Inst. Met.* **91**, 332

Higginson R L, Aindow M, Bate P S 1997 The effect of finely dispersed particles on primary recrystallization textures in Al–Mn–Si alloys. *Mater. Sci. Eng.* **A225**, 9–21

Humphreys F J 1977 The nucleation of recrystallization at second phase particles in deformed aluminium. *Acta Metall.* **25**, 1323

Humphreys F J, Hatherly M 2004 *Recrystallization and Related Annealing Phenomena*. 2nd ed, Elsevier, UK

Humphreys F J, Kalu P N 1990 The plasticity of particle-containing crystals. *Acta Metall. Mater.* **38**, 917

Hutchinson B, Homma H 1998 Orientation dependence of secondary recrystallisation in silicon–iron—observations and analysis. In: Weiland H, Adams B L, Rollett A D (eds.) *Proc. 3rd Int. Conf. Grain Growth*. The Minerals, Metals and Materials Society, Pittsburgh, PA, pp. 387–96

Hutchinson W B, Oscarsson A, Karlsson Å 1989 Control of microstructure and earing behaviour in aluminium alloy AA3004 canstock. *Mater. Sci. Technol.* **5**, 1118–27

Lücke K, Engler O 1990 Effects of particles on development of microstructure and texture during rolling and recrystallization in fcc alloys. *Mater. Sci. Technol.* **6**, 1113–29

Nes E 1980 Recrystallization in alloys with bimodal particle size distributions. In: Hansen N, Jones A R, Leffers T (eds.) *Proc. 1st Risø Int. Symp. Metallurgy and Materials Science*. Risø National Laboratories, Roskilde, Denmark, pp. 85–95

Nes E, Hutchinson W B 1989 Texture and grain size control during processing of metals. In: Bild-Sørensen J B, Hansen N, Juul Jensen D, Leffers T, Lilholt H, Pedersen O B (eds.) *Proc. 10th Risø Int. Symp. Metallurgy and Materials Science*. Risø National Laboratory, Roskilde, Denmark, pp. 233–49

Smith C S 1948 *Trans. Metall. Soc. AIME* **175**, 15

Vatne H E, Benum S, Daaland O, Nes E 1996 The effect of particles on recrystallization textures and microstructures. *Textures Microstruct.* **26–7**, 385–412

E. Nes

Processing Flexible Polymers to High Performance Fibers

In the 1970s, significant progress was made in exploiting the intrinsic properties of the macromolecular chain, especially in the field of 1D objects such as fibers. Two major routes can be discerned which are completely different in respect to the starting (base) materials, namely rigid as opposed to flexible macromolecules (Lemstra *et al.* 2000). The prime examples of rigid chain polymers are the aromatic polyamides (aramids), notably poly(*p*-phenylene terephthalamide) (PPTA), currently produced under the trade names Kevlar (Du Pont) and Twaron (Akzo Nobel). More recent developments include the PBO fiber from Toyobo, Zylon, and the experimental fiber M-5 from Akzo Nobel based on poly-diimidazo pyridynylene dihydroxy phenylene (Lammers *et al.* 1998). The *primus inter pares* of a high-performance fiber based on flexible macromolecules is polyethylene (PE) which is currently produced by DSM (Dyneema) and its licensee Allied Signal (Spectra). The main difference between rigid and flexible chains is the necessity to force chain extension in the case of flexible polymer molecules in order to exploit the intrinsic possibilities of the chain concerning ultimate mechanical properties, while for rigid rod polymers chain-extension has been built in by synthesis. Figure 1 shows the specific, based on performance by weight, strength versus specific modulus of the various fibers.

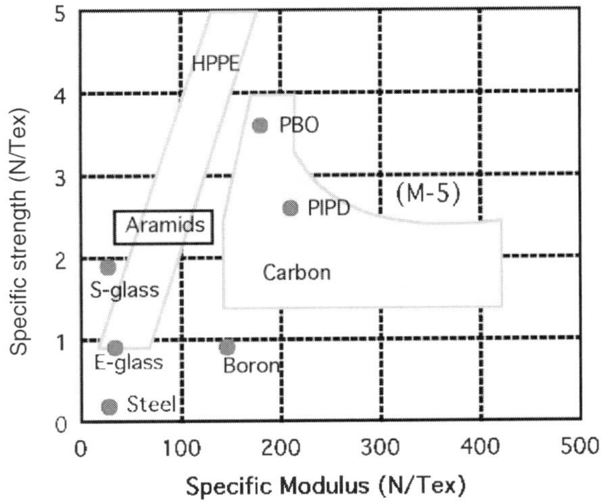

Figure 1
Specific strength versus specific modulus of various fibers. HPPE = high-performance PE fibers.

1. Ultimate Properties

1.1 Tensile Modulus

The ultimate tensile moduli of PE and other polymer systems, can be calculated or obtained experimentally from x ray diffraction measurements on oriented fibers during mechanical loading (Nakamae and Nishino 1992). All the literature data show that the Young's modulus E of PE in the chain direction is extremely high, $\geqslant$200 GPa, due to the small cross-sectional area of the chain, the absence of side groups, and the planar zigzag chain conformation in the orthorhombic crystal lattice. PE is not unique and other linear polymers with similar crystal structures possess similar theoretical moduli, e.g., poly(vinyl alcohol). In contrast, isotactic polypropylene has a lower ultimate modulus (60 GPa) in the chain direction due to its 3_1 helix conformation of the chain in the crystal lattice.

1.2 Ultimate Tensile Strength

In the past, a variety of studies have been devoted to the theoretical tensile strength of oriented and chain extended structures, i.e., the breaking of chains upon loading (He 1986). The theoretical tensile strength of a single, extended polymer chain can be calculated directly from the C–C bond energy. These calculations show that the theoretical tensile strengths are extremely high, in the order of 20–60 GPa (at 0 K). This requires a defect-free, chain-extended structure and *infinite* polymer chains. In practice, however, we are dealing with *finite* chains and a completely different situation is encountered as will be addressed in sect. 2.3.

1.3 Infinite Versus Finite Chains

Upon loading an array of perfectly aligned and extended finite polymer chains, the stress transfer in the system occurs via secondary, intermolecular bonds. Chain overlap is needed in order to be able to transfer the load through the system. Qualitatively, it could be envisage that the bonds in the main chains are only activated when the sum of the small secondary interactions, $\Sigma\varepsilon_i$, approaches the bond energy, E_i, in the main chain. Intuitively, one expects that in order to obtain high-strength structures in the case of PE, a high molar mass is needed, in combination with a high degree of chain-extension, to build up sufficient intermolecular interactions along the chains. Termonia and Smith (1987) used a kinetic model to simulate the fracture behavior of an array of aligned and extended *finite* polymer chains. It was found that the molecular weight (or number of chain ends) has a profound influence both on the fracture mechanism and on the theoretical tensile strength of these hypothetical structures. In order to obtain a strength level of 5 GPa, a molar mass of $>1\times10^5$ Dalton is needed for PE whereas

5×10^3 Dalton is sufficient for PPTA. The conclusion is that polymers possessing strong secondary bonds require a smaller overlap length to obtain a high tenacity.

2. *Melt-spinning/Drawing*

Chain orientation and extension is generated in melt- and solution-spun fibers by: (i) applying a draw-down to the fibers during or immediately after spinning (in the molten state or super-cooled melt) and (ii) drawing of fibers at temperatures close to, but below, the melting or dissolution temperature. Drawing in the (semi-)solid state, i.e., below the melting and/or dissolution temperature is usually much more effective. Relaxation processes are restricted since the chains are trapped into crystals which act as physical network junctions. In the 1970s, Capaccio and Ward (1974) started systematic studies concerning the drawability of linear PEs in the solid state and they developed a technological route. By optimizing the polymer composition and process conditions, PE fibers could be produced possessing tensile (Young) moduli up to 75 GPa and a strength level up to 1.5 GPa. A well-known observation made by Capaccio and Ward (1974) based on numerous *isothermal* drawing experiments, is that with increasing molar mass the maximum draw ratio decreases towards a limiting value of 4–5 at MW values over 1×10^6 Dalton. A limited drawability in the solid state is not unique for PEs. Many other polymers demonstrate a limited drawability, for example polyamides, which are often referred to as the natural draw ratio.

To understand the drawing behavior of PEs in the solid-state, one automatically focuses on the role of crystallites, viz the folded-chain crystals which are organized in more or less well-developed spherulites in melt-crystallized samples. There is, however, no direct correlation between crystal size/crystallinity, or morphology in general, versus the maximum draw ratio. Depending on the crystallization conditions, the drawability can change dramatically for one specific sample. For example, slow cooling from the melt can promote the drawability in the solid state Capaccio and Ward (1974) but also can cause identical PE samples to become brittle in the case of very slow cooling from the melt. In conclusion, melt-spinning/drawing is limited with respect to the molar mass of the PEs. With increasing molar mass, both the spinnability (a strong increase of melt viscosity) and the drawability in the solid-state decrease, which sets an upper limit to melt-spinning of PEs of typically 500 kg mol^{-1}.

3. *Solution-processing and Solution(gel)-spinning*

An obvious route to increase the spinnability of high molar mass PEs is to use solvents to lower the viscosity. In retrospect, the early attempts in the 1960s to employ solution-spinning of high molar mass PEs (Juergenleit 1956), to produce high-strength/high-modulus fibers failed since no proper understanding was available at that time concerning optimized drawing of solution-spun fibers. At the end of the 1970s, solution (gel)-spinning of UHMW-PE was discovered at DSM (DSM/Stamicarbon 1979, Smith and Lemstra 1979). In the solution (gel)-spinning technique, semi-dilute solutions are employed during spinning but the elongation of chains is performed by drawing in the semi-solid state, i.e., below the melting (dissolution) temperature. A solution of UHMW-PE with a low polymer concentration, typically of 1–2%, was spun into water.

Upon cooling a gel-like filament is obtained consisting of a physical network, obtained by thermoreversible gelation, containing a large amount of solvent. The as-spun/quenched filaments are mechanically sufficiently strong (gel-fibers) to be transported into an oven in which drawing is performed. The ultra-drawability of these gel-fibers seems not too surprising in view of the large amount of solvent which could act as a plasticizer during the draw. The remarkable feature, however, is that ultra-drawing is still possible after complete removal of the solvent prior to the drawing process. The solvent is necessary to facilitate processing of the rather intractable polymer UHMW-PE and induces a favorable structure/morphology for ultra-drawing but the solvent is not essential during the drawing process.

4. *The Entanglement Concept*

A very simple model for the enhanced drawability of solution-spun/cast UHMW-PE was put forward by Smith *et al.* (1982) based on a network approach, ignoring completely the morphology and crystal structure. In this model it is assumed that entanglements are trapped in PE upon crystallization and act as semi-permanent cross-links in a physical network upon solid state drawing. The crystals as such do not resist deformation in the temperature domain, close to but below the melting temperature, which is employed for (ultra-) drawing. The maximum draw ratio λ_{max} scales with the ratio between a fully stretched strand, scaling with M_e, between two entanglement points, and the average unperturbed end-to-end distance which scales with the square root of M_e, hence λ_{max} scales with $M_e^{0.5}$. Upon dissolution, the entanglement density decreases and the molar mass between entanglement loci increases, in good solvents about proportional to the inverse of the volume fraction of polymer in solution: M_e (solid) = M_e (melt)/φ. The maximum draw ratio of solution-cast/spun PEs is consequently given by: $\lambda_{max} \propto \varphi^{-0.5}$. For example, as-spun fibers from a 1% solution have an increased drawability with a factor of 10.

The entanglement model is remarkably versatile and can explain various phenomena such as the limited drawability of melt-crystallized UHMW-PE, λ_{max} 4–5, the dependence of the maximum draw ratio of solution-cast/spun UHMW-PE on the initial polymer concentration in solution, $\lambda_{max} \propto \varphi^{-0.5}$, and the increase of the drawability in the solid state upon isothermal crystallization at low supercoolings of UHMW-PE solutions, or from the melt in general, due to the fact that the chains are reeled in, viz pulled out their entanglement network. It is very important to note that the proposed "entanglement network" model is not universally valid. It can be applied to apolar polymers such as PEs and polypropylenes, but not to polymers possessing relatively strong secondary interactions, such as hydrogen bonds. In the case of polyamides, the folded chain crystal resist deformation and unfolding (Postma *et al.* 1990).

5. *Solvent-free Processing of UHMW-PE; Nascent Reactor Powders*

For solution(gel)-spinning of UHMW-PE a lot of solvent is needed for disentangling which has to be fully recovered in view of increased environmental legislation. Consequently, the challenge is to find "solvent-free" routes starting from disentangled precursors. The rationale behind this approach is that once disentangled UHMW-PE structures are obtained via some route, subsequent melt-processing should become feasible. One would at least expect a time-temperature window in which disentangled UHMW-PE should possess a lower initial melt-viscosity in comparison with a standard equilibrium melt. This is in accordance with the long relaxation (reptation) times (1×10^4 seconds per 200°C) of UHMW-PE, viz complete re-entangling is a time consuming process. To prepare (completely) disentangled UHMW-PE structures is feasible and rather straightforward using modern catalysts. Polymerization conditions are known, viz low temperature and rather low catalyst activity, which promote the formation of folded-chain crystals directly on the surface of the (supported) catalyst (Lemstra *et al.* 2000). In the limit of a low concentration of active sites on the catalyst (surface), one could expect that the individual growing chain would form its own *monomolecular* crystal. Despite all efforts made to prepare specific disentangled UHMW-PE precursors for subsequent melt-spinning, the ultimate conclusion at this point in time is that processing disentangled UHMW-PE with the aim to benefit from an initial lower melt-viscosity and to preserve the disentangled state to some extent during processing and prior to drawing, is not feasible at all. The salient feature is that disentangled UHMW-PE, either obtained by collecting precipitated single crystals or via specific low-temperature polymerizations, shows (Engelen and Lemstra 1991) the same high melt viscosity (in shear) upon heating above the melting temperature as standard "equilibrium" UHMW-PE melts. No memory effect from any previous polymerization/crystallization history can be depicted, and moreover, upon re-crystallization from the melt, the favorable drawing characteristics of disentangled UHMW-PE are lost completely and the drawing behavior is indistinguishable from a standard melt-crystallized UHMW-PE sample.

This intriguing problem of no memory effect and immediate loss of drawability upon re-crystallization has been addressed by Sadler and Barham (1990) following the fate of a PE chain using neutron scattering upon melting. They showed that the radius of gyration, R_g, which is rather low in solution-crystallized samples due to chain folding, immediately jumps to its "equilibrium" random coil dimensions upon surpassing the melting temperature (*chain explosion*). The experiments concerning the fast decay in drawability upon re-crystallization, related to the phenomenon of "*coil explosion*," point out that rather local changes, involving chain segments rather than complete chain motion/diffusion (reptation), can dominate the drawing behavior and these results prompted us to modify the "entanglement model" to some extent (Lemstra *et al.* 2000).

Upon crystallization from solution, the molecules fold usually along the {110} plane and the stems of a test chain (heavy dots) are shown in Fig. 2 without indicating the folds. For the sake of simplicity, we assume that adjacent re-entry occurs during crystallization and that the chain is located within one crystal plane. Shearing and unfolding in the direction perpendicular to the chain and along the {110} plane is rather easy in view of the low shear moduli. Upon melting these crystals the chains will immediately adopt a random coil conformation and stems of different molecules will interpenetrate in the "coil explosion" process. Upon re-crystallization, the stems of the test chain become crystallized in a more random order within the crystal and shearing (slip) is more difficult since the chains cannot cross mutually during deformation (Fig. 2).

The instantaneous loss in drawability upon melting and re-crystallization is due to re-arrangement and intermixing of stems involving only local chain motions rather than movement of the complete chains chain as proposed for self-diffusion in polymer melts. The driving force for the mixing of stems and coil-explosion upon melting is the gain in entropy (the entropy of fusion which is absent in the diffusion/reptation processes in quiescent polymer melts.)

6. *Solid-state Processing*

From sect. 5, it may be concluded that all attempts to prepare disentangled UHMW-PE structures by

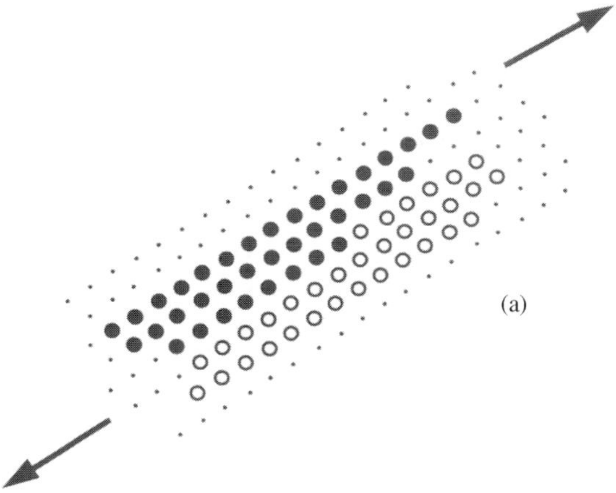

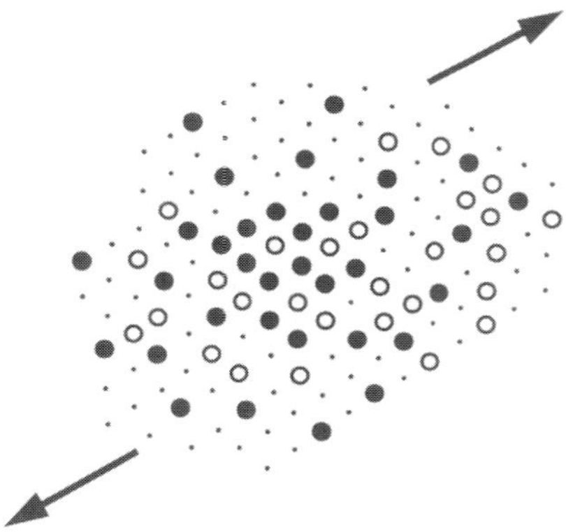

Figure 2
Stem arrangement in PE crystals, adjacent (a) versus re-entry (b).

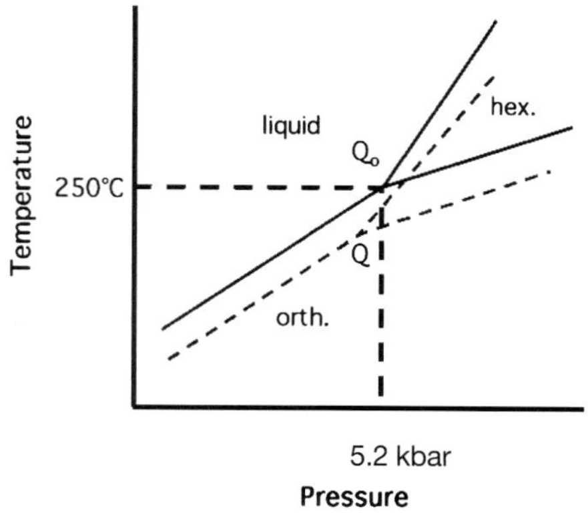

Figure 3
P–T diagram of PE.

specific polymerizations are in vain since no advantage could be obtained in the subsequent melt-processing. This conclusion is, in fact, not true. The only lesson to be learned is that the disentangled UHMW-PE precursors should never be heated above the melting temperature. Below the melting temperature, the disentangled UHMW-PE reactor powders, the so-called nascent or virgin powders, are remarkably ductile and can be processed via calandering or hot-compacting and subsequently drawn into tapes or fibrillated structures. The draw ratio of well-prepared nascent UHMW-PE powders is similar to solution-crystallized UHMW-PE samples.

Processing of UHMW-PE reactor powders has been partly successful for making oriented tapes by sintering/compacting between rollers and subsequent drawing. Kanamoto *et al.* (1987) developed a two-stage drawing process for reactor powders and they obtained Young's moduli of over 100 GPa. Nippon Oil Company developed and patented several solid-state processing routes for making strong UHMW-PE tapes. The only problem with solid-state processing is the handling and limited flow of powders. It is not possible to produce fine denier filaments. To enhance the flowability processing via the hexagonal phase might be a future option.

7. *Processing via the Hexagonal Phase*

The hexagonal phase of (isotropic) PEs is usually observed at elevated pressure and temperature. The (equilibrium) triple point Q is located at 5.2 kbar and 250 °C (see Fig. 3). It was shown by Rastogi *et al.* (1998b) that this triple point is dependent on the crystal size/dimensions and can be lowered to approximately 1 kbar in the case of nascent UHMW-PE reactor powders which possess very small dimensions (< 10 nm). This opens the possibility for processing via the (mobile) hexagonal phase. In the hexagonal phase, the chains are mobile but only in the chain direction via chain-sliding, $\langle c \rangle$, or diffusion. Upon annealing in the hexagonal phase ultimately complete chain extension will be obtained and extended-chain crystals will be obtained. Since there is no motion perpendicular to the chains, the drawability is not lost based on our revised model for drawing (see Fig. 2). In fact, the drawability can even be enhanced in the hexagonal phase as shown by Maxwell *et al.* (1996) since additional "disentangling" will occur as a consequence of extended-chain crystal formation. To date no processing routes via the hexagonal phase for making PE fibers are known. However, processing via the hexagonal phase for making hip joints based on UHMW-PE have been discussed in the literature (Rastogi *et al.* 1998b).

8. *Drawing Behavior of Other Flexible Polymer Systems*

The success of melt-spinning/drawing in general, and of solution(gel)-spinning in the particular case of ultra-high molecular weight PE, to produce oriented PE structures possessing high tensile moduli, stimulated research activities concerning the drawing behavior of other linear polymers, notably polypropylene, polyoxymethylene, and the aliphatic polyamides, nylon 6 and nylon 66. The high melting temperatures, 225°C (nylon 6) to 310°C (nylon 46) compared with PE (~150°C), and the presence of hydrogen bonds, which could reduce the creep make the polyamides attractive candidates. Extensive research effort to produce high modulus and strength fibers based on aliphatic polyamides have failed, however, despite major efforts in industry to perform solution(gel)-spinning based on high molar mass polyamides. The reason for this failure was demonstrated by Postma *et al.* (1990) who showed that the hydrogen bonds in lamellar, solution(gel)-crystallized polyamides, are essentially static up to the melting temperature and act as barriers, thus prohibiting draw. Solution(gel)-spinning promotes the formation of rather well-developed polyamide (single) crystals and, consequently, unfolding into extended-chain structures becomes more difficult.

9. *Properties of PE Fibers*

9.1 *Tensile Strength (1D)*

Currently, UHMW-PE fibers possessing tensile strengths of 3–4 GPa are produced commercially, e.g., Dyneema by DSM and Spectra by its licensee Allied Signal (see Fig. 1). Due to its low density, the specific values for the stiffness and strength of PE fibers are currently superior, at least at *ambient* temperatures.

9.2 *Melting Temperature*

The melting temperature is given by $T_m = \Delta H/\Delta S$ and the chains in the fiber can not melt when the fiber is constrained, because the chains cannot adopt a random coil conformation. Consequently, ΔS is low and the melting temperature increases. In the case of UHMW-PE fibers, this enhanced melting temperature by external constraints is not possible due to the onset of the hexagonal phase at 155°C. Above this temperature, the orthorhombic crystals in the fiber transform into the hexagonal structure. In the hexagonal phase the chain mobility is high and the fiber cannot sustain any load and fails (Lemstra *et al.* 2000).

9.3 *3D Properties*

PE fibers are highly anisotropic structures. The favorable characteristics in the fiber (chain) direction in terms of stiffness and strength are not matched at all in off-axis properties. The low transverse and shear moduli of PE are due to the absence of specific interactions perpendicular to the chain (only weak van der Waals bonding). The absence of strong interactions between adjacent chains causes the strength of the fibers to also be highly anisotropic. This is a major drawback in structural composite applications. It was shown by Peijs (1994) that the low interlaminar shear strength (ILLS) obtained from a three point bending test of unidirectional composites containing UHMW-PE fibers (Table 1) is inherently due to the poor shear and compressive properties of PE fibers. Consequently, UHMW-PE fibers cannot be used in structural composite applications.

9.4 *Long-term Mechanical Properties: Creep*

Soon after the gel-spun fibers were produced, it became apparent that the exceptionally good short-term mechanical properties were not matched by equally good long-term properties. In tensile tests at low strain rates, the UHMW-PE fibers exhibit a drop in Young's modulus and they exhibit a yield stress which is highly dependent on the strain rate. In fact, it is shown that the tensile strength (yield stress) drops by at least an order of magnitude at prolonged time scales. Some representative creep data of UHMW-PE fibers, in this case of a Dyneema fiber (SK66) possessing a tensile modulus of ~100 GPa are shown in Fig. 4. Figure 4 shows that initially the elongation increases proportionally to the logarithm of time.

The creep becomes important at longer loading times. This pronounced, irreversible creep is not too surprising when we take into account that UHMW-PE fibers are made by ultra-drawing and in the drawing process folded-chain crystals are transformed into chain-extended structures. In order to avoid fiber breakage in the drawing operation, the fibers are never drawn to their maximum draw ratio λ_{max} and, since there is no "lock-in" mechanism, the fibers can de drawn further in any subsequent tensile loading experiment, such as a creep

Table 1
Interlaminar shear strength (ILSS) of unidirectional composites; 50/50 fiber/epoxy composites.

Fiber	ILSS (MPa)
UHMW-PE (untreated)	13
UHMW-PE (corona/plasma treated)	20–30
Aramid	45–70
E-glass	75–95
Carbon	80–120

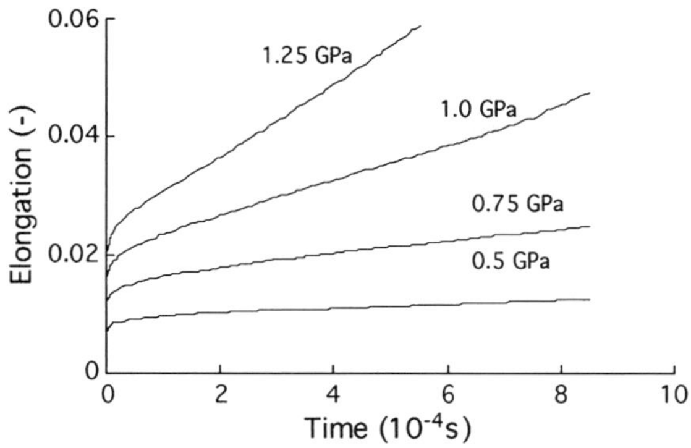

Figure 4
Elongation (creep) of an UHMW-PE fiber at ambient temperature. The Young's modulus (E) of the fiber is ~100 GPa.

test. Consequently, there is no basic difference between the drawing operation and a creep test, especially at elevated temperatures. All known methods for improving the creep resistance such as cross-linking (e.g., electron beam and UV) and chemical modification, require the presence of a certain degree of imperfection in the fiber in order to bring the chains together to form a covalent bond. Creep improvement, therefore, implies a compromise between short-term and long-term mechanical properties (Lemstra *et al.* 2000).

10. Major Applications of PE Fibers

In the previous sections the properties of solution-spun, ultra-drawn UHMW-PE fibers were discussed in terms of their tensile (Young) modulus and tensile strength which show impressive values at short loading times but rather extensive creep at prolonged loading times. These properties are, of course, relevant in a large number of applications such as fishing nets and/or lines, ropes, and composites. Such fibers also possess a number of favorable properties such as a high-impact resistance, high-cutting resistance, low dielectric constant, low dielectric loss, low stretch, high heat conductivity in the fiber direction, and a high sonic modulus. These "secondary" properties are particularly important in applications such as ballistic applications (bulletproof vests), helmets (impact), hybrid composites (impact), gloves (cutting resistance), fishing lines (low stretch), loudspeaker cones (sonic modulus), and radomes/sonar domes (dielectric properties). The properties which motivate the use of highly oriented gel-spun UHMW-PE fibers are high (specific) strength and modulus (all applications), low density, high energy absorption capability (ballistic protection and composites), high sound speed (ballistic protection and speaker cones), flexibility, (ropes and nets), good dielectric properties (radomes), and good chemical resistance (Table 2).

Table 2
Major applications of solution(gel)-spun PE fibers.

Ropes and cables	Ballistic protection	Miscellaneous
Towing lines	Bullet proof vests	Sails
Mooring lines	Insert for vests	Motor helmets
Yacht ropes	Helmets	Cut resistant gloves
Long lines	Car armor panels	Radomes
Trawl nets	Spall liners	Dental floss
Fish farms	Ballistic blankets	Speaker cones
Parapent lines	Containment shields	Cryogenic composites

See also: Polymer Fiber Processing: Modeling; Polymer Fiber Processing: High-performance Fibers; Processing Rigid Polymers to High Performance Fibers; Solid-state Extrusion to High Performance Fibers

Bibliography

Bastiaansen C W M, Meijer H E H, Lemstra P J 1990 Memory effects in polyethylenes–influence of processing and crystallization history. *Polymer* **31**(8), 1435–40

Blades H, White J R 1963 *US Pat.* 3 081 519

Capaccio G, Gibson A G, Ward I M 1979 *Ultra-High Modulus Polymers*. Elsevier Applied Science, London

Capaccio G, Ward I M 1974 *Polymer* **15**, 233

Capaccio G, Ward I M 1975 *Polymer* **16**, 243

DSM/Stamicarbon 1979 *US Pat.* 4 137 394

He T 1986 *Polymer* **27**, 253

Juergenleit W 1956 *US Pat.* 3 048 465

Kanamoto T, Ohama T, Tanaka K, Takeda M, Porter R S 1987 *Polymer* **28**, 1517

Kelly A, Macmillan N H 1986 *Strong Solids*. Clarendon, Oxford, p. 22

Lammers M, Klop E A, Northolt M G, Sikkema D J 1998 Mechanical properties and structural transitions in the new rigid-rod polymer fibre PIPD ('M5') during the manufacturing process. *Polymer* **39**(24), 5999–6005

Lammers M 1998 Ph.D. thesis, PIPD rigid-rod polymer fibres and films; structure formation and mechanical properties. Swiss Federal Institute of Technology (ETH), Zurich

Lemstra P J, Bastiaansen C W M, Rastogi S 2000 In Salem D R Structure Formation in Polymeric Fibers. Hanser, Munich, Germany 185-223. ISBN 3-446-18203-9

Maxwell A S, Unwin A P, Ward I M 1996 The effect of high pressure annealing on the molecular network in polyethylene. *Polymer* **37**(15), 3293–301

Nakamae E K, Nishino T 1992 *Adv. X-Ray Analysis* **35**, 545

Nishino T, Ohkubo H, Nakamae K 1992 Studies on the temperature-dependence of the elastic-modulus of crystalline regions of polymers. 12. Temperature-dependence of the elastic-modulus of crystalline regions of polyethylene with different microstructures–explanation with the kinked-chain model. *Journal of Macromolecular Science-Physics* **31**(2), 191–214

Peijs A A M 1994 Ph.D. thesis Eindhoven University of Technology, High-performance polyethylene fibres in structural composites. ISBN-90-386-0033

Postema A R, Smith P, English A D 1990 Ultra-drawing of polyamides – the hydrogen-bond barrier. *Polymer Communications* **31**(12), 444–7

Rastogi S Koets P, Lemstra P J 1998a *NL Pat. Applic 98/00093*. Method for processing polyethylene having a molecular weight of at least 400.000 by heating at an elevated pressure

Rastogi S, Kurelec L, Lemstra P J 1998b Chain mobility in polymer systems; on the borderline between solid and melt 2. Isothermal phase reversal and sintering of ultra high molecular weight polyethylene. *Macromolecules* **31**, 5022–31

Sadler D M, Barham P J 1990 Structure of drawn fibers. 1. Neutron-scattering studies of necking in melt-crystallized polyethylene (English). *Polymer* **31**(1), 36–42

Smith P, Lemstra P J 1979 *Br. Pat.* 2 051 661

Smith P, Lemstra P J, Booij H C 1982 *J. Polym. Sci. Phys. Ed.* **20**, 2229

Termonia Y, Smith P 1986 *Polymer* **27**, 1845

Termonia Y, Smith P 1987 *Macromolecules* **21**, 835

Tervoortengelen Y M T, Lemstra P J 1991 Morphology of nascent ultra-high-molecular-weight polyethylene reactor powder–chain-extended versus chain-folded crystals. *Polymer Communications* **32**(11), 343–5

Zwick M 1965 *NL Pat. Applic.* 6 501 248

P. J. Lemstra and S. Rastogi

Processing Polymers in Supercritical CO_2

In recent years, supercritical CO_2 (SCF CO_2) has gained increased interest as a solvent for polymer synthesis, for use in a variety of polymer process-related applications, and in the production of specialized materials that cannot be readily produced by other means. Of the many supercritical fluids, there has been a special focus on the use of supercritical CO_2, because it is nontoxic, nonflammable, unreactive under most conditions, and leaves no liquid waste. An important motivation for adoption of SCF CO_2 is its promise to be an environmentally responsible replacement for organic solvents, for other more aggressive or flammable supercritical fluids and water in many applications.

The high compressibility of supercritical CO_2 and the dependence of solubility, transport, and other properties on pressure and temperature near the critical conditions, allow solvent quality to be tuned by simple density control (Fig. 1). Carbon dioxide's easily accessible critical point (31.1°C and 73.8 bar (7.38 MPa)), low cost, and ready availability as a byproduct of many chemical and energy-related processes add to its attractiveness, though specialized pressure cells, high-pressure pumps, and other equipment are necessary for processing. Applications to be presented here capitalize on the special characteristics unique to SCF CO_2 (also including high diffusivities and no surface tension). For the purposes of this article, polymeric behavior in supercritical CO_2, polymer syntheses utilizing CO_2 (as solvent or to achieve unique properties), and applications of polymers utilizing supercritical CO_2 will be discussed.

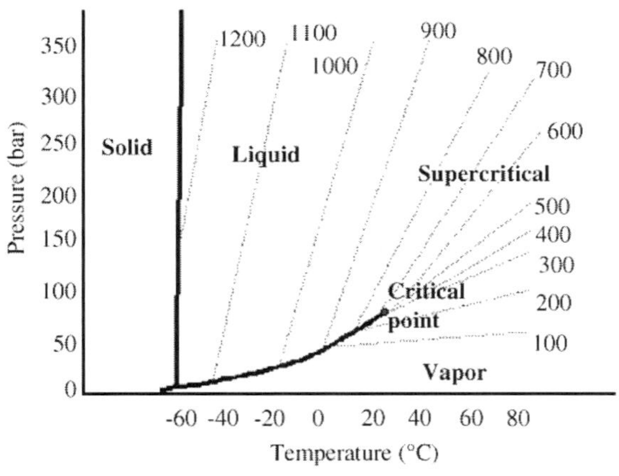

Figure 1
A phase diagram of CO_2 with the critical point at 31.1°C, 73.8 bar (7.38 MPa) is shown. Constant density lines (g l^{-1}) indicate that a large change in density can occur with small variations in pressure.

1. Polymer Behavior in Supercritical CO_2

Most common commercial polymers are unlikely to be soluble in pure supercritical CO_2 under easily attainable conditions. Frequently used conditions are ~35°C and ~175 bar (17.5 MPa), which produce SCF CO_2 with a density in the range of 0.8–1.0 g ml^{-1}, comparable to densities of common organic solvents. At lower densities, the solvent power of CO_2 is usually too low to serve as a polymer solvent, even for CO_2-philic materials. Polymers can be designed to be CO_2-philic by balancing the entropy and enthalpy of solvent–polymer and polymer–polymer interactions (Kirby and McHugh 1999), though rules for solubility are not yet quantitatively defined. High chain flexibility and high free volume of the polymer have been strategically incorporated to enhance solubility by increasing the entropy of mixing since the number of possible configurations available for CO_2 to condense around the polymer are optimized. Since CO_2 has no dipole moment and possesses a low cohesive energy density, a similarly low cohesive energy between polymer segments will also encourage dissolution. Examples of polymers with low cohesive energy densities which are readily soluble in SCF CO_2 include poly(dimethylsiloxane)s, atactic poly(-propylene), and poly(methacrylate)s with long hydrocarbon branches. Lewis base functions, such as carbonyl-containing groups are known to have a favorable specific interaction with CO_2. So do molecules bearing fluorine-containing groups, and these are nearly always required for high solubility. CO_2-phobic polar groups such as hydroxy functions prevent SCF CO_2 solubility, but can be present in a polymer if capped to increase solubility in the nonpolar medium. The quadrupolar nature of CO_2 reduces solubility of completely nonpolar polymers at low temperatures (Kirby and McHugh 1999).

The presence of crystallinity also has an important effect on polymer solubility.

Poly(tetrafluorethylene) (PTFE) is insoluble in SCF CO_2 due to its crystallinity, although it has many of the molecular characteristics associated with CO_2 solubility. The low critical temperature of CO_2 combined with the high melting point of PTFE renders the processing conditions normally used insufficient to melt and then dissolve the polymer. Copolymers of tetrafluor ethene that lack crystallinity are very soluble in CO_2. In short, polymers that have nonpolar character are most readily soluble in SCF CO_2, which has properties most similar to hydrocarbon solvents.

Below the temperatures and pressures that permit a critical CO_2 density for polymer dissolution, polymers will only swell. Although not all polymers dissolve in CO_2, the solvent does have a high permeation rate in virtually all polymers. A number of common polymers that are CO_2 insoluble (e.g., poly(styrene)) will absorb large quantities of CO_2, the presence of which can greatly lower the polymer's effective glass transition (Wang *et al.* 1982). Polymers that are plasticized by CO_2 include poly(ethylene terephthalate), poly(carbonate), poly(ethylene), poly(vinyl chloride), and poly(methacrylate)s. The enhanced mobility caused by CO_2 enables crystallization of poly(carbonate)s that otherwise remain amorphous and the formation of self-assembled monolayers of *n*-alkanethiols. SCF CO_2 plasticization, as will be discussed further, has been used for the formation of various nanostructures by the introduction of additives into polymer hosts to create functional materials.

Polymer solubility is usually determined by a cloud-point test in which a polymer is first dissolved in SCF CO_2, then temperature and pressure are independently reduced to determine under what conditions the polymer precipitates. The cloud-point conditions, defined as a 90% drop in transmitted light through the solution, are determined by optical measurement through a transmitting window in the high-pressure chamber. Solubility limits of low solubility polymers can be inferred. For example, the thickness of poly(styrene) films cast from CO_2 increases as the polymer concentration in the system increases, but only up to the solubility limit. Hybrid instrumentation also enables other techniques for examining polymer behavior *in situ* or on-line, including IR spectroscopy, UV/visible spectroscopy, atomic absorption spectroscopy, fluorescence spectroscopy, thermal lens spectrometry, mass spectrometry, nuclear magnetic resonance, neutron scattering, and dissolution rate monitoring of films (Amador-Hernandez and De Castro 2000).

1.1 Polymeric CO_2-philes as Phase-transfer Agents and Surfactants

The high pressures required for dissolving polar, organometallic, or high molecular weight compounds limit many possible polymer processes conducted in CO_2. CO_2-philes, or CO_2 phase-transfer agents, are compounds with the ability to carry poorly soluble compounds into the supercritical phase (Fig. 2(a)) (Jackson and Fulton 1998, Triolo *et al.* 2000). Surfactant-like molecules (fluorinated or nonfluorinated) have been used to enhance solubility under milder conditions (Sarbu *et al.* 2000) as have functionalized dendrimers such as perfluorinated poly(propylene imine), shown schematically in Fig. 2(b) (Goetheer *et al.* 2000). As discussed, polymer segments are engineered for solubility and specific behavior. It is even possible to design into the polymer architecture the density and temperature at which a transition between free amphiphilic block copolymers and their corresponding micelles (a transition determined by molecular shape, hydrogen bonding, dipole–dipole interactions, electrostatics, temperatures, and solvent quality) will occur.

These polymers have also been applied for stabilization of colloids during SCF CO_2 dispersion polymerization. Polymer surfactants are especially effective for steric stabilization of oligomers during synthesis in solvents (including CO_2) with low dielectric constants. The stabilizer preferentially exists at the polymer–solvent interface to coat the forming particles and impart long-range repulsions between them.

2. *Polymer Synthesis in Supercritical CO_2*

Supercritical CO_2 offers many advantages when used as a reaction medium for polymer synthesis: the inert characteristics of CO_2, the absence of side reactions that could alter polymer structure, the ability to isolate a dry product from the reaction medium by simple depressurization, the abatement or elimination of organic solvents, and polymer plasticization (which enables easy removal of residual monomers). The most commonly used polymerization methods include radical polymerization, cationic polymerization, and transition metal catalyzed reactions. Additional details can be found in polymerization review articles (Kendall *et al.* 1999).

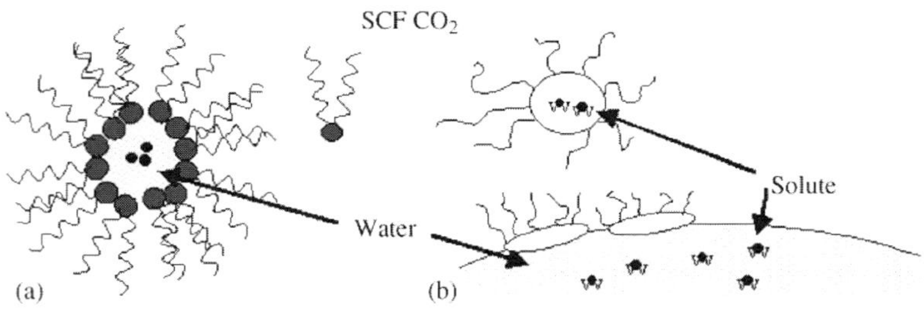

Figure 2
(a) An amphiphilic micelle with polar head and twin polar tails (to aid in spherical micelle formation) is shown. Water molecules and CO_2-phobic solute reside inside the micelle. (b) A dendritic surfactant is able to encapsulate the solute at low concentrations at which an amphiphile would be below critical micelle density.

Homogenous syntheses, in which all reactants are soluble in the supercritical phase, are dictated by solubility requirements and best suited for only highly soluble fluorinated polymers. The free-radical kinetics show higher initiator efficiencies as a result of decreased solvent cage effects in the supercritical phase. Fluorinated methacrylates and copolymers with TFE have been successfully produced and a commercial plant for the polymerization of TFE copolymers has been recently established. In cases where monomers are soluble but polymers are not, precipitation polymerization may be explored (polymers using vinyl monomers and TFE have been produced), the major advantages being lack of chain transfer to solvent and absence of undesirable end groups.

Dispersion polymerization, in which monomers and initiators are soluble but oligomers are not, is considered the best heterogeneous free-radical CO_2 polymerization for insoluble polymers. Traditional solvents used to form particles have posed a severe limitation to commercialization of dispersion polymerization, though with the advent of SCF CO_2, it may finally become commercially useful. Low levels of preformed block copolymers may be used as the surface-active stabilizer necessary to prevent coagulation of growing oligomers, or grafting onto a CO_2-soluble polymer present during the polymerization may form one *in situ*. The resulting product is distinct polymer microspheres with narrow particle size distribution having diameters from 100 nm to 10 μm. Most polymers produced to date by this method are derived from methacrylates and styrenics (Behles and DeSimone 2001).

Other chain-growth polymerizations such as cationic, transition metal-catalyzed, and thermal ring-opening polymerizations have found advantages in this medium. Performing a thermal ring-opening polymerization reaction in a solvent would normally require high boiling compounds that are difficult to remove, but using CO_2 avoids this problem. Likewise, step-growth reactions, including oxidative couplings, melt-phase condensations, and sol–gel polymerizations have been demonstrated in CO_2. Melt-phase condensation reactions in SCF CO_2 result in high molecular weight materials without the usual disadvantage of the high viscosity, since the polymer is plasticized by CO_2.

3. *Processing in Supercritical CO_2*

As a result of the unique combination of features of CO_2 as a solvent, a number of processes can be carried out that are difficult to achieve from either a polymer melt or from solution in more traditional solvents. These include formation of porous structures, fractionation and extraction, formation of blends and nanocomposites, formation of microparticles, polymer impregnation and encapsulation, and preparation of coatings and lithographic polymers. Cooper (2001) discusses some of these processes more extensively in a recent review.

3.1 Porous Structures

Porous silica and poly(silsesquioxane) gels prepared by sol–gel processing are often dried by exchanging the original solvent with SCF CO_2 to prevent shrinking and cracking from drying and the related capillary forces. The resulting aerogels retain more original volume than without CO_2 treatment, and are potentially useful insulating materials due to their low density, high surface area, and low thermal conductivity. Aerogels can also be synthesized in a water-free formulation that eliminates organic solvents altogether (Loy *et al.* 1997).

In addition to direct formation of porous polymers during polymerization, strategic processing of CO_2/polymer mixtures is also used to form foams. Foam formation is generally considered a risk in most CO_2 processing of polymers. The rapid depressurization of CO_2 plasticized or molten polymers leads to foaming of the polymer by pressure-induced phase separation. As CO_2 is removed, T_g increases and the foam structure is trapped in the polymer. The effects of pressure, temperature, and swelling time on the cell size and density, and bulk density of the materials have been researched. Microcellular poly(ethylene tesaphthalate), and poly(carbonate) foams have been produced with pore sizes < 10 μm (Liang and Wang 2000), and poly(methyl methacrylate), poly(styrene) and its copolymers, and poly(propylene) have also been successfully used to produce foamed materials with pores as small as 0.5 μm (Goel and Beckman 1994).

3.2 Fractionation and Extraction

The very specific solubility of SCF CO_2 makes it ideal for fractionation and extraction. It is possible to use CO_2 to separate narrow molecular weight fractions of polymer from polydisperse samples. The change in solvency needed for fractionation is accomplished by isothermal changes in pressure which change solvent density and solvating capability. Polymers as diverse as poly(isobutylene) and novolac resins have been successfully fractionated. Residual monomers or other contaminants have also been extracted from polymers—a feature being used in the microelectronics and pharmaceutical industries where very high purity levels are critical for successful application.

3.3 Polymer Blends and Nanocomposites

It is possible to form polymer blends by taking advantage of the abilities of CO_2 to swell most polymers and dissolve monomers with CO_2. Immiscible polymers experiencing lowered melt viscosities, increased free volume, and reduced interfacial surface tension in the presence of SCF CO_2 can be blended with finer scale mixing than could be achieved in traditional melts (Elkovitch *et al.* 1999). By diffusing monomer and initiator into a plasticized insoluble polymer matrix, polymerization can be carried out to produce blends of polymers that cannot be produced normally. It is also possible to carry in organometallic compounds that serve as initiators for gas phase monomers such as ethylene, to produce blends of poly(ethylene) in a host matrix.

Conductive elastomeric foams have been produced in this manner by *in situ* oxidative polymerization of a conductive polymer with poor mechanical properties, in a more flexible polymeric foam. In one example, a polyurethane foam is impregnated first with ferric triflate oxidant and then exposed to pyrrole vapor to polymerize the pyrrole within the foam pores. Small amounts of cosolvents added to the CO_2 increase the solubility of the oxidant, and simple separation of the waste redox products is achieved by depressurization.

The ability to impregnate a polymer matrix with a variety of molecular and polymer structures using SCF CO_2 as an inert medium for reagent transport offers the possibility to produce a variety of nanocomposites extending beyond blends. Polymer/metal nanocomposites can be produced with metal or semiconductor particles of tens of nanometers small that exhibit size-dependent electrical, chemical, and magnetic properties arising from quantum confinement. Stabilization in the polymer matrix prevents agglomeration and facilitates workability into devices. Organometallic precursors dissolved in SCF CO_2 and infused into solid polymers are left behind after rapid depressurization. A reduction of the precursor to the base metal subsequent to CO_2 removal within the solid matrix results in the desired nanocomposite (Brown and Watkins 2000). Platinum clusters in either poly(4-methyl-1-pentene) or PTFE have been formed using this methodology. The polymer substrate is initially swollen with metal-loaded CO_2. Both the substrate and the reaction product are insoluble in CO_2. Sorption of CO_2 increases diffusion rates in the substrate and enables enhanced penetrant absorption. Manipulating the density of the solvent determines the degree of swelling, diffusion, and partitioning of the organometallic component between SCF CO_2 and the swollen polymer.

Dopants added to polymer matrices may also be colorants or dyes for manufacturing of textiles and accessories, including buttons and zippers (Kazarian *et al.* 1999, Montero *et al.* 2000). Reduction of effluent streams, water usage, and dye losses in processing motivate the incorporation of SCF CO_2 as the carrier for dopants. Although most useful dyes have low solubility in SCF CO_2, successful dyeing occurs when the dye partition coefficient between polymer and solvent is high, indicating the dye's affinity for the polymer. Since transport of dye through the supercritical phase is rapid, diffusion of the dye within the polymer is rate-limiting, though increased by plasticization of the matrix. Supercritical CO_2 dyeing substrates include poly(methyl methacrylate), polyester-based materials, nylons, and cottons, but may expand as research is conducted on other polymers.

3.4 Production of Microparticles

One technique designed to capitalize on the properties of supercritical fluids to create new material structures is Rapid Expansion from Supercritical Solution (RESS) (Fig. 3). A supercritical fluid solution of polymer is expanded across a capillary nozzle, leading to rapid supersaturation and formation of microparticles (Chernyak *et al.* 2001).

Powder coating is one application utilizing RESS. The coatings industry releases a large quantity of volatile organic compounds (VOCs), so process systems with reduced emissions are very desirable. Powder

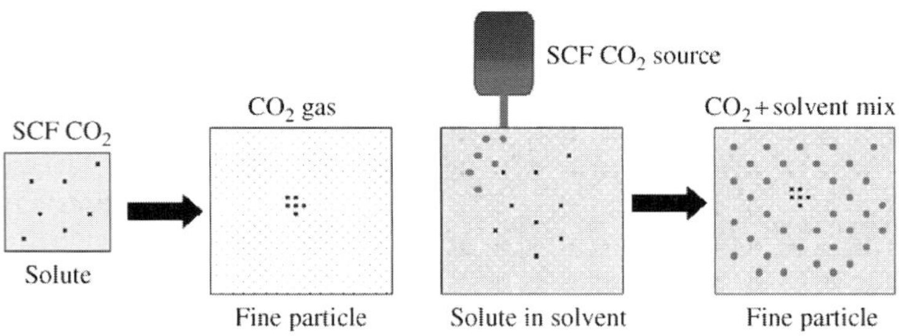

Figure 3
(a) A solute dissolved in supercritical CO_2 is expanded to vapor CO_2 in rapid expansion from a supercritical solvent (RESS). Solubility decreases by a factor of 10^6, so the supersaturated solution precipitates to form uniform particles. (b) A solute dissolved in a carrier solvent is precipitated when SCF CO_2 mixes with the carrier solvent, because SCF CO_2 is a gas antisolvent (GAS) or supercritical nonsolvent (SAS).

coating methods (polymer microparticles several microns in diameter are applied, then baked to form a fused film) may lead to the successful reduction of VOC emissions. Methods that leave residual surfactants in the polymer after melting result in reduced smoothness due to voids caused by the contaminant. Particles produced by RESS for coating have been based on a variety of polymers and include poly (1-butene), poly(ethylene succinate), poly(caprolactone), poly(L-lactic acid), and poly(glycolic acid).

In cases where the polymer is not highly soluble in supercritical CO_2, a co-solvent additive can be used with the CO_2 to improve the RESS process. Ideally, the co-solvent chosen is itself a nonsolvent of the polymer, and the process (RESS-N) will not likely swell the polymer product during expansion or cause agglomeration. For example, the solubility of poly(styrene-*b*-methyl methacrylate) is low in either CO_2 or methanol, but up to 20 wt.% can be dissolved in a mixture of the two used for RESS-N. The particles can be applied to a surface, and then heated to form a continuous film for coating applications (Matsuyama *et al.* 2001).

RESS can also be used to form and coat fibers for fabric applications and coat metal particles as small as 20 μm for solid propellant applications. For coating small particles, a chamber containing polymer dissolved in SCF CO_2 is opened and the contents mixed with a chamber containing the metal powder. The mixture is discharged to the atmosphere and the dissolved polymer precipitates uniformly on all surfaces of the powder.

3.5 *Pharmaceutical and Medical Applications*

Microencapsulation of pharmaceutical compounds in degradable polymers is an important route to new drug-delivery systems (Sihvonen *et al.* 1999). Conventional techniques for particle formation including double-emulsion solvent extraction, liquid antisolvent, and spray drying lead to high residual contents of toxic organic solvent in the final product. When CO_2 is used in preparation, there are no harmful residues remaining in the final product. CO_2 processing of drug-release polymers involves controlling factors such as the particle size, pore structure, drug concentration, and rate of polymer degradation which all affect the rate of drug release. The surface structure and shape of the particle determine the delay time of drug release. Methods including RESS-N and variations have been used to coat or encapsulate proteins and other compounds in particles (Subramaniam *et al.* 1997). Figure 4 illustrates some possible encapsulation schemes that would provide a variety of release kinetics.

In one example, microparticles of poly(lactic acid) loaded with insulin have been formed. A homogenous solution of protein and polymer is first produced, then a supercritical CO_2 antisolvent (SAS) mixture of dichloromethane and dimethyl sulfoxide is added. Unlike RESS particle production under atmospheric conditions, the solution of polymer and drug is atomized through a nozzle into a SCF CO_2 phase, so liquid drops are formed in the SCF bulk. The organic solvent diffuses out of the droplet into the CO_2 until the liquid density of the drop is lowered and precipitation occurs. Supersaturation allows fast precipitation of polymer and drug, trapping drugs in the polymer matrix. Encapsulated proteins may undergo conformational changes due to the high pressure and shearing forces of supercritical processing, but studies indicate that recovery upon redissolution in aqueous media occurs in some cases.

SAS is an example of one technique that may be effective for proteins, peptides, enzymes, hormones, and genetic material. Solution enhanced dispersion (SEDS) by supercritical fluids has also been used to precipitate proteins from aqueous solutions by extracting water with a SCF CO_2 mixture.

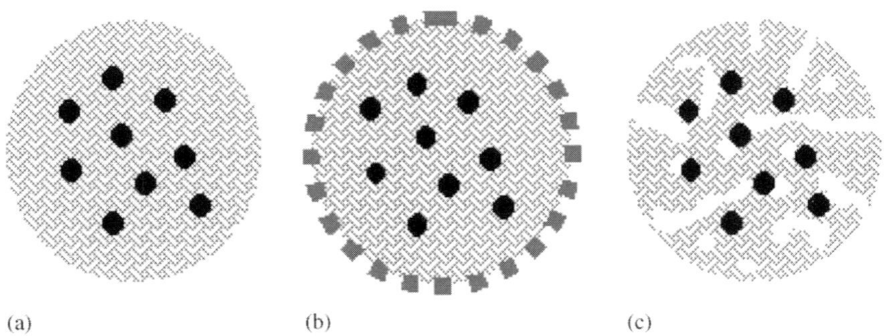

Figure 4
Schematic cross-sections of drugs-in-polymer microspheres to yield varied rates of release. (a) Solid particles with embedded drug released as the polymer degrades. (b) Encapsulated particles for delayed release. (c) Porous particles for higher rates of release. Many aspects of particle structure can be controlled through SCF CO_2 processing conditions.

Using polymer substrates similar to those for drug encapsulation, hydroxyapatite and other calcium salts have been incorporated into polymer matrices as bone replacement materials by using SCF CO_2 as a carrier for impregnation (Mandel and Wang 1999). A biodegradable polymer, e.g., poly(lactide-co-glycolide) (PLGA) was produced with a homogenous distribution of inorganic component and was produced as either powder or monolithic solid. Calcification around the hydroxyapatite particles by living tissue, followed by polymer degradation (into lactic acid and glycolic acid) result in bone construction. Control of porosity through CO_2 processing parameters is important to provide appropriate structure for blood vessel growth.

Another medical application that benefits from SCF CO_2 processing is the sterilization of thermally and hydrolytically sensitive polymeric pharmaceuticals or biomaterials (with or without proteins) that would be damaged by traditional methods. Poly(lactic acid) and other polymers have been treated with SCF CO_2 for 0.5–4.0 h to achieve the successful sterilization of several bacterial microbes and microorganisms including *Staphylococcus aureus* and *Escherichia coli* in the absence of organic solvents or irradiation (Dillow *et al.* 1999). Sterilization might occur as a result of the fluid's special ability to penetrate cell walls and alter the pH by reacting to form carbonic acid. CO_2 sterilization is faster than ethylene oxide or irradiation methods.

3.6 Microelectronics Applications and Lithography

The microelectronics industry is actively investigating the use of supercritical CO_2 for a variety of processes in semiconductor manufacturing. Perhaps the first application for SCF CO_2 processing to be successfully incorporated will be cleaning (McHardy and Sawan 1998). A variety of additives and solvent mixtures have proven successful for cleaning of semiconductor components and removal of lithographic polymer residues, and several companies are now selling SCF CO_2 cleaning equipment for use in the semiconductor FAB (Mullee *et al.* 2001, Rubin *et al.* 1998). Importantly, the CO_2 stripping process significantly reduces chemical usage and does not physically damage underlying metal or novel low-dielectric constant materials, which are usually constructed from nonpolar polymers or inorganic materials and are not compatible with traditional stripping procedures.

The production of fine scale (nanometer) features has been demonstrated using lithographic patterning methods based on photosensitive polymer films that undergo a solubility switch once exposed to radiation. Development (or dissolution) of these imaged photoresists in aqueous solvents creates lateral capillary forces that can damage high-aspect ratio polymer features. Among the useful characteristics of SCF CO_2 are its ability to wet all surfaces, its lack of meniscus formation in drying processes, and its selective solubility of polymers. Drying photoresists by displacing solvents with liquid CO_2 that is transformed to a supercritical state before depressurization prevents coexistence of liquid and vapor phases and prevents damage from pattern collapse (Fig. 5). Patterns smaller than 50 nm which collapse from air-drying can be successfully produced using this solvent exchange method (Namatsu 2000). Most advanced resists are designed to be developed in water-based solutions, so the use of CO_2 in the final drying step often requires mixed solvents or surfactants to remove the developer.

Experimental photoresists have been designed expressly for development directly by CO_2, so that developing and drying take place in a single step. These chemically amplified resists have a protecting group removed by a light-activated photoacid generator to produce a polar, CO_2-insoluble polymer. Both random and block copolymers have been studied and the best CO_2 developable resist to date is poly(-tetrahydropyranyl methacrylate-*b*-perfluorooctyl methacrylate) (Sundararajan *et al.* 2000). Molecular weights of $\sim 15\ 000$ g mol^{-1} and fluorinated block contents of 60 mol.% give the best results allowing minimum feature sizes smaller than 100 nm to be achieved. The inset in Fig. 5 shows a submicron pattern of the polymer mentioned above developed in SCF CO_2. Similar polymers are spin-coatable from liquid CO_2, a process that if coupled to SCF CO_2 development would result in all-dry lithography. To investigate the conditions and time restraints of the development process, a SCF CO_2 dissolution rate monitor was recently constructed. Using interferometric principles, the device monitors the thinning of a resist as it dissolves (Fig. 6). Complete dissolution of polymeric photoresist was shown to proceed in only seconds after introduction of SCF CO_2 in a manner similar to conventional aqueous developers.

Use of SCF CO_2 in microelectronic fabrication also presents opportunities for process simplification. It has been recently demonstrated that CVD deposited fluoropolymers such as PTFE can be exposed to e-beam radiation and developed in SCF CO_2 (Pryce Lewis *et al.* 2001). This process eliminates the need for sacrificial resist coating and permits the development of materials that are otherwise insoluble.

4. *Summary*

Applications and processes both well established and in current development have been reviewed. The attributes of supercritical CO_2 that make it a useful

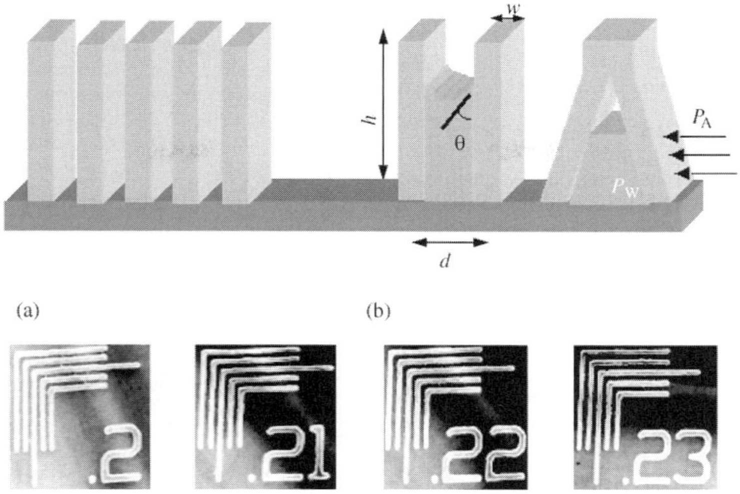

Figure 5
(a) Free-standing high aspect ratio (height : width) polymer features are desired and can be achieved by drying with SCF CO_2 to avoid surface tension. (b) Polymers developed in aqueous developer experience capillary pressure (P_{air}–P_{water}) from the rinse solution arising from the liquid surface tension. The force, a function of surface tension, contact angle, line spacing, line height, and line width, causes polymer features to bend, break, and stick together. Inset shows SCF CO_2 developed resist with features ranging from 200 nm and higher based on poly(tetrahydropyranyl methacrylate-*b*-perfluorooctyl methacrylate).

Figure 6
An *in situ* dissolution rate monitor (DRM) based on laser interferometry or ellipsometry is used to monitor the development of CO_2-soluble polymer photoresists used in microelectronics processing or the deposition of polymer films in coatings processes. The figure shows a commercial SCF CO_2 pump/furnace combination and the inset shows a schematic of interferometric component of DRM.

material for a wide range of polymeric applications will continue to provide impetus for identification of future uses. In the near future increasing use of SCF CO_2 will occur primarily because processing and materials formation can take place that otherwise could not. Its environmental friendliness will be a significant driver in its more widespread adoption as this technology matures.

Bibliography

Amador-Hernandez J, De Castro M D L 2000 On-line detection for supercritical-fluid extraction. *J. Biochem. Biophys. Methods* **43**, 329–43

Behles J A, DeSimone J M 2001 Developments in CO_2 research. *Pure Appl. Chem.* **73**, 1281–5

Brown G D, Watkins J J 2000 Carbon dioxide—dilated block copolymer templates for nanostructured materials. *Mater. Res. Soc. Symp. Proc.* **584**, (Materials Issues and Modeling for Device Nanofabrication) 169–74

Chernyak Y, Henon F, Harris R B, Gould R D, Franklin R K, Edwards J R, DeSimone J M, Carbonell R G 2001 Formation of perfluoropolyether coatings by the rapid expansion of supercritical solutions (RESS) process. Part 1: Experimental results. *Indust. Eng. Chem. Res.* **40**, 6118–26

Cooper A I 2001 Recent developments in materials synthesis and processing using supercritical CO_2. *Adv. Mater.* **13**, 1111-4

Dillow A K, Deghani F, Hrkach J S, Foster N R, Langer R 1999 Bacterial inactivation by using near- and supercritical carbon dioxide. *Proc. Natl. Acad. Sci. USA* **96**, 10344–48

Elkovitch M D, Tomasko D L, Lee L J 1999 Supercritical carbon dioxide assisted blending of polystyrene and poly(methyl methacrylate). *Polym. Eng. Sci.* **39**, 2075–84

Goel S K, Beckman E J 1994 Generation of microcellular polymeric foams using supercritical carbon dioxide: I. *Poly. Eng. Sci.* **34**, 1137–47

Goetheer E L V, Baars M W P L, van den Broeke L J P, Meijer E W, Keurentjes J T F 2000 Functionalized poly(propylene imine) dendrimers as novel phase transfer catalysts in supercritical carbon dioxide. *Ind. Eng. Chem. Res.* **39**, 4634–40

Jackson K, Fulton J L 1998 Surfactants and microemulsions in supercritical fluids. In: McHardy J, Sawan S P (eds.) *Supercritical Fluid Cleaning: Fundamentals, Technology and Applications*. Noyes Publications, Westwood, NJ, pp. 1–26

Kazarian S G, Brantley N H, Eckert C A 1999 Dyeing to be clean: use supercritical carbon dioxide. *CHEMTECH* **29**, 36–41

Kendall J L, Canelas D A, Young J L, DeSimone J M 1999 Polymerizations in supercritical carbon dioxide. *Chem. Rev.* **99**, 543–63

Kirby C F, McHugh M A 1999 Phase behavior of polymers in supercritical fluid solvents. *Chem. Rev.* **99**, 565–602

Liang M, Wang C 2000 Production of engineering plastics foams by supercritical CO_2. *Ind. Eng. Chem. Res.* **39**, 4622–6

Loy D A, Russick E M, Yamanaka S A, Baugher B M, Shea K J 1997 Direct formation of aerogels by sol–gel polymerizations of alkoxysilanes in supercritical carbon dioxide. *Chem. Mater.* **9**, 2264–8

Mandel F S, Wang J D 1999 Manufacturing of speciality materials in supercritical fluid carbon dioxide. *Inorg. Chim. Acta* **294**, 214–23

Matsuyama K, Mishima K, Umemoto H, Yamaguchi S 2001 Environmentally benign formation of polymeric microspheres by rapid expansion of supercritical carbon dioxide solution with a nonsolvent. *Environ. Sci. Technol.* **35**, 4149–55

McHardy J, Sawan S P (eds.) 1998 *Supercritical Fluid Cleaning: Fundamentals, Technology and Applications*. Noyes Publications, Westwood, NJ

Montero G A, Smith C B, Hendrix W A, Butcher D L 2000 Supercritical fluid technology in textile processing: an overview. *Ind. Eng. Chem. Res.* **39**, 4806–12

Mullee W H, Biberger M A, Schilling P E 2001 Removal of photoresist and residue from substrate using supercritical carbon dioxide process. (Supercritical Systems, Inc., USA.) PCT Int. Appl.

Namatsu H 2000 Supercritical drying for water-rinsed resist systems. *J. Vac. Sci. Technol. B* **18**, 3308–12

Pryce Lewis H G, Weibel G L, Ober C K, Gleason K K 2001 E-beam patterning of hot-filament CVD fluorocarbon films using supercritical CO_2 as a developer. *Chem. Vapor Depos* **7**, 195–7

Rubin J B, Davenhall L B, Barton J, Taylor C M V, Tiefert K 1998 A comparison of chilled DI water/ozone and CO_2-based supercritical fluids as replacements for photoresist-stripping solvents In: *Proc. IEEE/CPMT Int. Electronics Manufacturing Technology Symposium*. pp. 308–14

Sarbu T, Styranec T, Beckman E J 2000 Non-fluorous polymers with very high solubility supercritical CO_2 down to low pressures. *Nature* **405**, 165–8

Sihvonen M, Jarvenpaa E, Hietaniemi V, Huopalahti R 1999 Advances in supercritical carbon dioxide technologies. *Trends Food Sci. Tech.* **10**, 217–22

Subramaniam B, Rajewski R A, Snavely K 1997 Pharmaceutical processing with supercritical carbon dioxide. *J. Pharm. Sci.* **86**, 885–90

Sundararajan N, Yang S, Ogino K, Valiyaveettil S, Wang J, Zhou X, Ober C K, Obendorf S K, Allen R D 2000 Supercritical CO_2 processing for submicron imaging of fluoropolymers. *Chem. Mater.* **12**, 41–8

Triolo R, Triolo A, Triolo F, Steytler D C, Lewis C A, Heenan R K, Wignall G, DeSimone J M 2000 Structure of diblock copolymers in supercritical carbon dioxide and critical micellization pressure. *Phys. Rev. E: Stat. Phys., Plasmas, Fluids, Relat. Interdiscip. Top.* **61**(4-B), 4640–3

Wang W, Kramer E J, Sachse W H 1982 Effect of high-pressure carbon dioxide on the glass transition temperature and mechanical properties of polystyrene. *J. Polym. Sci., Polym. Phys. Ed.* **20**, 1371–84

G. Weibel and C. K. Ober

Processing Rigid Polymers to High Performance Fibers

Fiber materials obtained from natural sources have been used through the ages for reinforcement and other structural applications. Well-known examples include the use of straw as reinforcement of clay and plaster, and the use of flax and hemp for cables and ropes. Although such natural fibers are still used extensively, high performance man-made fibers take a steadily increasing share of the market. This is due to the fact that they allow the production of continuous fibers with exceptional mechanical properties, for instance specific tensile strengths that are 5–10 times higher than that of steel. In addition to the purely synthetic materials, also natural sources for man-made fibers are of major importance, e.g., regenerated cellulose. To obtain fiber materials with high performance properties requires processing methods that yield extremely high molecular orientation of the polymer chains along the fiber axis. For the processing of rigid polymers various techniques are possible. With the first method the required high level of molecular orientation is obtained by the processing of rigid polymer chains from liquid-crystalline (LC) nematic solution, using the dry-jet, wet-spinning technique. Alternatively, rigid polymers can be spun from isotropic solutions. In this case a relatively low degree of orientation is induced during the spinning process. The high orientation, which is required for a high tenacity and modulus is imposed in a hot drawing after-treatment. Kevlar (DuPont), Twaron (Acordis), and Zylon (Toyobo) are representatives of the first category, Technora (Teijin), Armos, and Rusar of the second. Because, from an industrial point of view the dry-jet, wet-spinning of nematic solutions is the most important one, this article will concentrate on this process route. Nematic LC spinning solutions are liquids where the polymer chains show spontaneous alignment along a locally defined axis of orientation $\tilde{n}$, the director. Figure 1 shows the molecular arrangement in a nematic polymer solution.

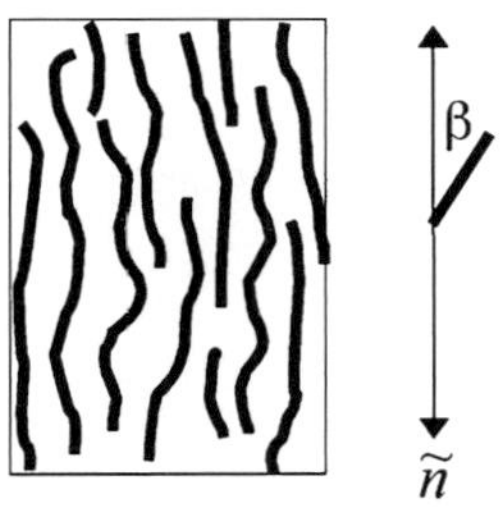

Figure 1
Nematic liquid-crystal structure of a main-chain polymer solution, the angle β indicates the deviation of the molecular chain tangent from the average director $\tilde{n}$.

1. *History*

The first observation of a LC rigid polymer solution was reported by Conmar-Robinson for poly(benzyl-*L*-glutamate) or PBLG in m-cresol, subsequently also p-aramids were found to show such behavior in concentrated sulfuric acid solution. The latter solutions were found to be suitable for the spinning of high performance aramid fibers, presently known as Kevlar (DuPont) and Twaron (Teijin Twaron). Inspired by this success, alternatives to aramid spinning were developed: PBO and PBT fibers, and more recently PIPD (or M5) and a new cellulose yarn. The last example refers to a spinning process that makes use of nematic cellulose solutions and is an excellent example of a high performance man-made fiber from natural sources.

2. *Chemical Structures*

The global requirements for polymer chains that are suitable for fiber spinning from nematic solution are rigid chains, a small molecular cross-section, and strong lateral interactions. Some examples of molecular structures that fit this description are shown in Fig. 2, together with the solvent of choice.

Clearly it is necessary to identify solvents suitable for use in a dry-jet wet-spinning process, where the solvents are preferably cheap and not highly toxic. Contrary to common belief strong acids such as sulfuric acid and phosphoric acid while being corrosive are not particularly toxic or volatile, and, therefore, can be used on an industrial scale without putting the operating personnel at any undue risk. For this reason other known solvents, such as hydrofluoric acid for aramids and trifluoroacetic acid

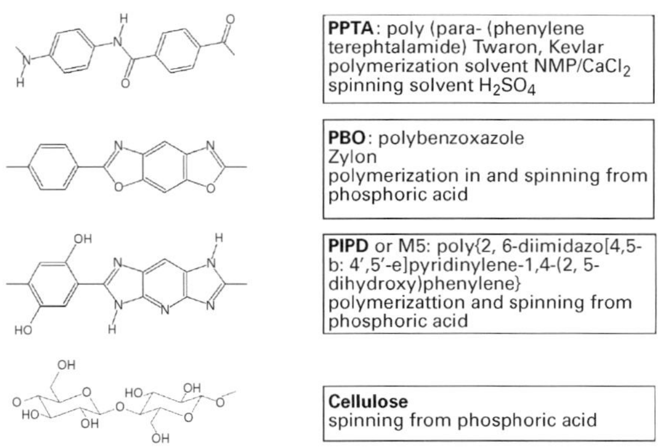

Figure 2
Structures of some rigid main chain LC polymers.

for cellulose-acetate, are not generally considered to be serious options for commercial spinning processes.

3. Liquid-crystalline Order

The remarkable mechanical properties of fibers, made from the above polymers, can be understood from the occurrence of LC order in the spinning solution (see Fig. 1). Due to the stiffness of the polymer chains in these materials there is a tendency for them to pack together in a uniform fashion. This so-called LC order manifests itself in the fact that at a given temperature and for sufficient polymer concentration the polymer solutions show distinctly anisotropic properties. The anisotropy relates to many observable physical properties, such as the optical (birefringence) and the rheological properties, where the collective molecular alignment causes a dependence on the orientation of the sample with respect to the external fields. Many LC phases are known, with varying degrees of orientational and positional order. For reasons of simplicity this article will be confined to the simplest case of the nematic phase. This is a fluid phase where the molecules show collective orientation but without the positional order commonly associated with true crystalline materials. The LC structure shown by these polymer solutions does strongly depend on the polymer concentration and the temperature. For a given polymer concentration, at high temperatures the polymer solution will be isotropic, i.e., without orientation, whereas at low temperatures the molecules will show the collective nematic order described above.

At a well-defined temperature the material will show a phase transition from the nematic to the isotropic phase, which is similar in nature to better known phase transitions like the freezing and boiling point. Figure 3 shows this so-called clearing temperature for three polymer solutions.

In addition to this, the degree of molecular orientation in the nematic phase will depend on temperature, at low temperatures the orientational order will be more perfect than at higher temperatures. This is obvious when the thermal motion of the polymer chains is taken into account. The degree of orientation can be expressed as a number, the orientational order parameter, which by definition is zero for an isotropic phase and is one for a material with perfect molecular alignment. It should be clear that the order parameter will increase with increasing polymer concentration and will decrease with increasing temperature. This can be made more quantitative using a variety of theoretical models such as the Maier-Saupe model and the Onsager model.

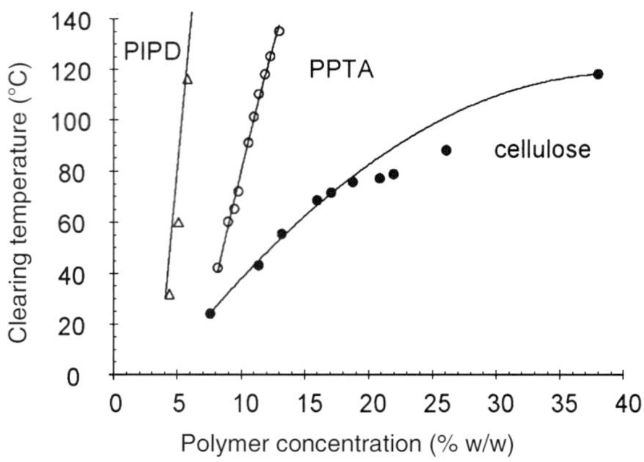

Figure 3
Clearing temperature from the nematic to the isotropic phase of PPTA in sulfuric acid, and PIPD and cellulose in phosphoric acid.

4. Spinning of Liquid-crystalline Polymer Solutions

Poly(p-phenylene terephthalamide) (PPTA), poly-*p*-phenylenebenzobisoxazole (PBO), poly{2,6-diimidazo[4,5-b:4′,5′-e]pyridinylene-1,4-(2,5-dihydroxy)-phenylene} (PIPD), and cellulose are all spun from LC (nematic) solutions. After being filtered and heated, the LC solutions are extruded through gold/platinum spinnerets containing approximately 1000 capillaries with a diameter of 50–200 μm. Figure 4 gives a schematic view of the dry-jet wet-spinning processes.

Rheology of the LC phase plays an important role in processing and structure formation. Shear flow occurs during the melting of the solution, in the pipes of the machine, and in the capillaries of the spinneret. A well-known phenomenon, which was first observed by Kwolek (DuPont) on solutions of aromatic polyamides, is the fact that at low shear rates the shear viscosity exhibits a maximum as a function of the

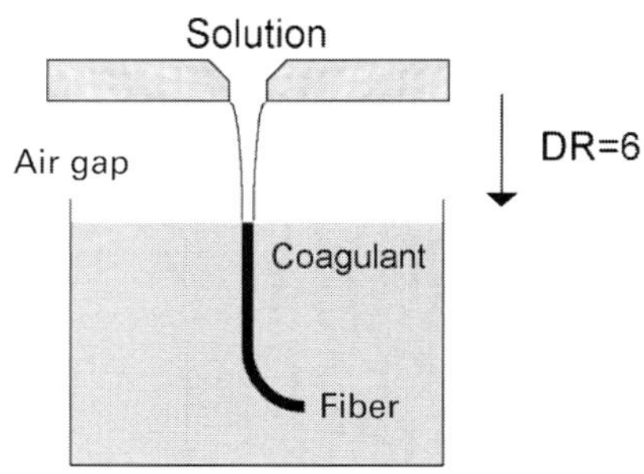

Figure 4
Schematic view of the dry-jet wet-spinning process, the solvent is removed in the coagulation bath containing the coagulant (e.g., H_2O or dilute sulfuric acid for PPTA-spinning).

polymer concentration. This maximum in the viscosity versus concentration corresponds to the onset of the nematic phase as is also explained by the Doi theory. When the shear rate increases, the maximum becomes smaller and can even disappear. This is due to flow-induced orientation of the isotropic phase, which diminishes the difference in flow characteristics between the isotropic and the nematic phase. Main-chain LC phase solutions are strongly shear-thinning, especially in the region of the phase transition. Elongational flow fields occur in the entrance zone of the spinneret capillaries and in the air gap, where a draw ratio of approximately 10 is imposed. The elongational viscosity is much less dependent on the deformation rate than is the case for the shear viscosity. For example, with cellulose solutions in phosphoric acid the velocity profile in the air gap has an exponential shape, indicating a constant elongational viscosity. PPTA solutions show similar behavior. The elongational viscosity of PPTA solutions is in the order of 1 kPa, while for cellulose solutions in phosphoric acid it is about 20 kPa.

For some polymers, e.g., PBO, it can be useful to cool the filaments in the air gap by blowing air onto them. This method is also used in melt spinning of cellulose fibers from NMMO solutions (Tencel, Acordis and Lyocel, Lenzing) and can reduce the occurrence of draw resonance.

Compared to solutions or melts of flexible polymers, the die swell for LC solutions of rigid polymers is rather small. The birefringence in the die swell area is very low, but once the velocity in the air gap passes the extrusion velocity the level of birefringence is close to saturation. Additional further drawing slightly improves the orientation of the nematic phase according to the affine deformation model. In this model the orientation of the director field is calculated by assuming that the strain in the liquid-crystal solution is uniform and equal to the externally applied strain. This allows the calculation of the director orientation in terms of a "director order parameter" that is a function of the externally applied strain only. This director order parameter can be used together with the local orientational order parameter, of the molecules with respect to the director, to calculate the mechanical properties of spun fibers as a function of temperature, concentration, and draw-ratio.

The high level of orientation, which is created during the dry-jet spinning, should be maintained in subsequent washing and drying steps. This is achieved by keeping the yarn under tension in these process steps.

5. *Structure of Main Chain Liquid-crystalline Phase Spun Fibers*

In general the rigid rod polymeric chains of the LC phase fibers bring about a highly crystalline and oriented structure, which is borne out by the results of x ray diffraction, electron diffraction, and in particular by the lattice imaging technique. The chains in cellulose II, PPTA, and PIPD are packed in a 3D order extending in transverse direction over 5–8 nm and in longitudinal direction over more than 40 nm. In the PBO structure the 3D order is disturbed by random translations parallel to the chain axis. In cellulose II and PIPD the chains are laterally bonded by hydrogen bonds in two directions, whereas the crystalline structure of PPTA the hydrogen bonding between the chains extends in a single direction. Different kinds of lateral textures, depending on the coagulation conditions, have been observed in cellulose II and PPTA fibers. In addition, the LC phase fibers can display a pleated sheet morphology with a period of about 250 nm. Typical for the wet-spinning process are the finely distributed and highly elongated voids in the fiber, which give it a fibrillar morphology.

6. *Mechanical Properties*

The mechanical properties of LC phase solution spun fibers can be fully understood on the basis of a single-phase crystalline structure. Numerous experiments show that the continuous chain model provides a good interpretation of the mechanical properties. In this model the fiber properties are determined by the chain modulus (e_c), the shear modulus between adjacent chains (g), the corresponding critical shear stress (τ_c), the orientation distribution of the chain axes ($\langle \sin^2\Theta \rangle$), and the molecular weight distribution (MW). The fiber is modeled as a series arrangement of crystalline domains consisting of perfectly oriented chains. The fiber strain is brought about by elastic chain extension and by elastic shear of the chain axes towards the fiber axis. Both contributions form the equation for the initial modulus of the fiber (E) given by:

$$\frac{1}{E} = \frac{1}{e_c} + \frac{\langle \sin^2\theta \rangle}{2g}$$

where θ is the angle between the chain axis and the fiber axis. For a crack-free fiber the internal shear modulus is about equal to the torsion modulus of the filament.

High draw ratios in the spinning process improve the chain orientation resulting in a higher modulus. Other process conditions, such as hot drawing, can influence the molecular interactions to some degree, e.g., by improving the crystalline order, and hence may increase the shear modulus. Table 1 presents the significant data and the types of interchain bonds. Depending on the degree of chain orientation the tensile curves of polymer fibers below their glass transition temperature are characterized by a more or

Table 1
Summary of the basic elastic constants g and e_c for various organic polymer fibers, together with the highest filament values reached for the modulus (E) and the tensile strength (σ_b). Also shown are the average values for the creep compliance (CC) at 20°C, which is the ratio of creep rate and load stress, and the character of the lateral interchain bonds.

Fiber	e_c (GPa)	g (GPa)	E (GPa)	σ_b (GPa)	CC (%/dec/GPa)	Interchain bond
UHMW PE	250–300	0.7–0.9	150	6.0	0.3	Van der Waals
PET	125	1.0–1.3	38	2.3	0.13	Van der Waals/dipolar
Cellulose II	88	2.0–5.0[a]	55	1.7	0.4[b]	Bi-directional hydrogen
Cellulose I[c]	136	1.5	110	–	–	Unidirectional hydrogen/Van der Waals
PPTA	240	1.5–2.7	140	4.5	0.02–0.04	Unidirectional hydrogen/Van der Waals
PBO	500	1.0–1.5	260	6.1	0.005	Van der Waals/dipolar
PIPD[d]	550	6.0	330	5.0	0.0035	Bi-directional hydrogen

[a]Depending on water content.
[b]Conditioned at 65% RH.
[c]native fiber.
[d]PIPD or M5 is a fiber in a early stage of development so that further improvement of these data is expected.

Table 2
Some applications of LC spun high performance fibers.

Application	Used property	Which polymer
Brake linings	Modulus, T resistance	PPTA
Ballistics	Tenacity, modulus	PPTA, PBO, PIPD
Protective clothing	T resistance	PPTA, PIPD
Mechanical rubber goods	Tenacity, modulus	PPTA, Cellulose, PIPD
(e.g., conveyor belts)	Tenacity, modulus	PPTA, Cellulose, PIPD
Tires, pipes, and hoses	Tenacity, modulus	PPTA/Epoxy composites
Concrete reinforcement	Modulus	PPTA, PBO
Optical cables, fiber reinforced composites	Tenacity, modulus, T resistance	PPTA, PIPD

less pronounced yield at about 0.005–0.025 strain. This yield can be modeled using a critical resolved shear stress, τ_c, at which local loosening of the intermolecular bonds occurs, resulting in a permanent rotation and displacement of adjacent chains and, thus, in a permanent shear deformation of the domains. Hence, above the yield strain the tensile extension of the fiber is accompanied by plastic deformation. It has been shown that the compressive strength of polymer fibers is determined by the same critical shear stress leading to yielding in compression occurring at the same strain as in tensile deformation. Generally, the compressive strength of a crack free polymer fiber is about 0.3 times the shear modulus (this is supported by the yield model and experimental results). The highest experimental value for compressive strength of 1.7 GPa obtained so far is for the PIPD fiber, which is due to the network of hydrogen bonds that extends in two different directions normal to the chain. The hydrogen bond networks of cellulose II and PIPD are rather similar, which causes high shear moduli for both fibers. The intermolecular bonding also to a large extent determines the viscoelastic properties like the creep compliance of the polymer fibers as shown in Table 1.

Strong LC phase-based fibers are obtained for large values of MW, τ_c, and a high degree of chain orientation, i.e., for high values of E and a more uniform structure of the fiber. The highly rigid nature of the polymer chain in PBO and PIPD, and the relatively strong interchain bonding, in particular in PIPD, ensure excellent thermal properties such as a low creep rate and long lifetimes at relatively high temperatures. It is noted that the creep rates for LC phase spun fibers (PPTA, PBO, and PIPD) are much lower than those for UHMW-PE and PET due to the stronger lateral interactions.

7. *Applications*

Due to their impressive mechanical and thermal properties, polymer fibers spun from LC solutions

Table 3
Approximate production volumes of dry-jet wet-spun yarns from rigid polymers.

Brand	Polymer	Producer	Production tons/year
Twaron	PPTA	Teijin Twaron	12000
Kevlar	PPTA	DuPont	18000
Zylon	PBO	Toyobo	400
Cellulose[a]	Cellulose	Acordis	Laboratory scale
M5	PIPD	Magellan	Laboratory scale
Technora	p-aramid	Teijin	2000
Armos			Pilot plant scale
Rusar		Kamensk Chimvolokno	400

[a]New cellulose yarn from phosphoric acid solutions.

are extensively used for various reinforcement and structural applications. In general the advantage of such polymer fibers is that they have a high modulus and strength and are not brittle and have relatively good damage tolerance compared to, for example, carbon fiber and glass fiber. One of the largest applications of PPTA fibers is in the making of brake linings as a replacement of asbestos. Also PPTA aramid fibers are used for ballistics as the main component of bullet proof vests. Some additional applications are listed in Table 2. The various materials each have their own niche in the market and the estimated annual production is shown in Table 3.

See also: Polymer Injection Molding: Multimaterial Molding; Polymer Fiber Processing: High-performance Fibers; Processing Flexible Polymers to High Performance Fibers

Bibliography

Baltussen J J M, Northolt M G 1999 The stress and sonic modulus versus strain curve of polymer fibres with yield. *Polymer* **40**, 6113–24

Collyer A A 1992 *Liquid Crystal Polymers: From Structures to Applications*. Elsevier, London

Doi M, Edwards S F 1986 *The Theory of Polymer Dynamics*. Clarendon, Oxford

Kwolek S L 1972 *US Pat.* 3 671 542

Maier W, Saupe A 1959 Eine einfache molekular-statistiche Theorie der nematischen kristalflüssigen Phase. Teil I. *Z. Naturforsch* **149**, 882–9

Northolt M G, Baltussen J J M 2001 The tensile and compressive deformation of polymer and carbon fibres. *J. Appl. Polym. Sci.* (in press)

Onsager L 1949 The effect of shape on the interaction of colloidal particles. *Ann. N.Y. Acad. Sci.* **51**, 627

Robinson C, Ward J C, Beevers R B 1958 *Disc. Faraday Soc.* **25**, 29–42

S. J. Picken, H. Boerstoel, and M. G. Northolt

R

Rapid Prototyping

Solid freeform fabrication (SFF) techniques involve layerwise fabrication of parts or prototypes from a computer aided design (CAD) representation without the use of any part-specific tooling. There was an explosion in the number of SFF processes over the last decade of the twentieth century. The major application of most of the SFF processes has been in the field of prototyping and hence, many of the SFF techniques are also called "rapid" prototyping techniques. As "rapid" prototyping implies, the advent of the various SFF techniques has resulted in tremendous time saving, especially for complicated models/components. Typically, very complex models that would take several months to fabricate by traditional methods can be made in a few days or less with SFF techniques (Kai and Fai 1997).

There are, in general, two types of prototyping applications. One type of prototyping is used to test the design for "form and fit". In this case the primary concern is with the dimensions of the fabricated prototype. In the other types of prototyping, components (or prototypes) are made to test their functional performance and not just the form or fit. This sort of prototyping is called functional prototyping.

Recently, there has been a thrust, especially in the USA, to develop SFF techniques for prototyping and manufacturing of functional components made of ceramics or metals. Some of the more important SFF techniques that have been developed in this category are stereolithography (SLA) of ceramics (Griffith and Halloran 1997, Griffith *et al.* 1997 p. 403), selective laser sintering (SLS) (Manriquez-Frayre and Bourell 1990, Nutt 1991, Subramanian *et al.* 1993, Lakshminarayan *et al.* 1990, Carter and Jones 1993, Vail and Barlow 1991), computer aided manufacturing of layered engineering materials (CAM-LEM) (Cawley *et al.* 1995, 1998, Liu *et al.* 1999), three-dimensional printing (Sachs *et al.* 1995, 1992, Grau *et al.* 1997) and fused deposition of ceramics and metals (FDC and FDMet) (Danforth *et al.* 1998, Agarwala *et al.* 1996, Bandyopadhyay *et al.* 1997, Wu *et al.* 1999). In the following sections these important SFF processes will be introduced.

1. Stereolithography

Stereolithography (SLA) of polymers is considered to be the first SFF technique and was developed by 3-D systems, Inc. (CA, USA) in 1988 (Kai and Fai 1997). A schematic of the SLA process is shown in Fig. 1. SLA involves UV-assisted photopolymerization of liquid monomers. A UV laser is scanned over a layer of the liquid monomer to cure the monomer in selected areas as dictated by the tool paths. After completion of one layer, another layer of resin is coated on top of the cured layer. This process is called recoating. The process of recoating and curing is repeated until the part is completed.

In 1996, Halloran *et al.* of the University of Michigan modified the SLA process to produce ceramic parts (Griffith and Halloran 1996). In this process, a ceramic particle filled photocurable resin is used in place of the virgin resin. They successfully demonstrated the use of a variety of ceramic materials such as silica, alumina, silicon nitride, and hydroxyapatite in the SLA of ceramics (Brady *et al.* 1996). There are still some problems regarding SLA due to the lower cure depth encountered in ceramic suspensions in comparison with polymer suspensions (Brady *et al.* 1997).

One of the key advantages of the SLA of ceramics is the possibility of fabricating ceramic parts with fine features of about 100 μm. Another advantage of the process is the ability to make parts with excellent surface finish. Surface roughness values as low as 4 μm have been achieved for the top surface and 27.5 μm for the side surfaces (Brady *et al.* 1998).

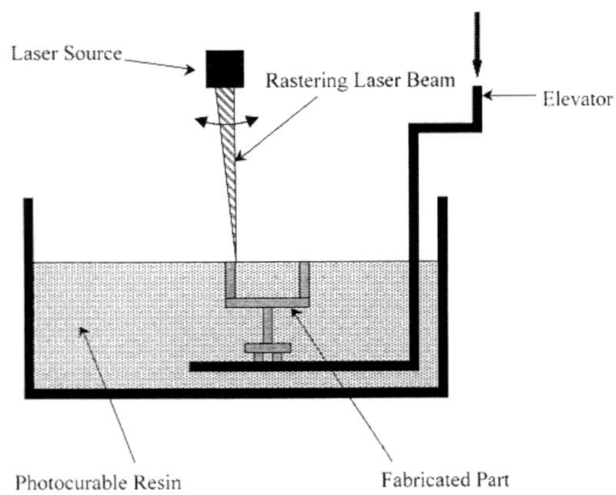

Figure 1
Schematic showing the operation of a typical stereolithography (SLA) apparatus.

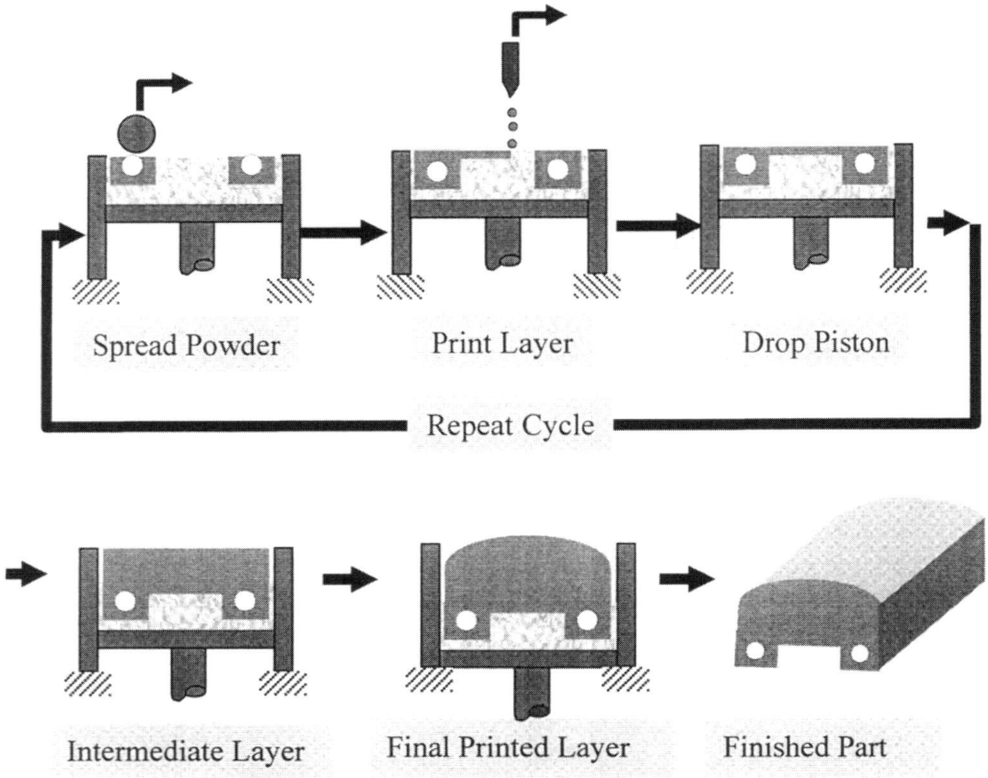

Figure 2
Schematic showing the different steps involved in the fabrication of a component via three-dimensional printing (3DP) technique (Sachs *et al.* 1998).

2. *Three-dimensional Printing*

Three-dimensional printing (3DP) was invented at the Massachusetts Institute of Technology in 1988. The basic idea of 3DP is to build a ceramic or metal green part by printing binder onto a layer of deposited powder. A schematic of the 3DP process is shown in Fig. 2 (Sachs *et al.* 1998b).

The binder is printed using an ink jet printer. Both drop on demand and continuous jet printing techniques were tried with 3DP, but continuous jet was chosen due to its higher rate of binder deposition. The binder has to meet a number of requirements to be useful in 3DP (Sachs *et al.* 1998a). First, the binder solution has to be of sufficiently low viscosity for successful droplet formation. In addition, the binder must dry before the next layer is deposited. In this technique, a continuous stream of droplets emerges from the nozzle. Some of the droplets in the emerging stream may be charged. The uncharged droplets will follow a straight path through the deflection plates to be deposited on the part being built, while charged droplets will be deflected to the side away from the deposition site on the part.

There are essentially two types of 3DP processes depending on the type of powder bed used. Historically, the first type is the dry powder process. In this process, a dry powder is deposited from a roller drum. The powder is fed into the interior of a drum and, as the drum rotates and traverses a substrate, the powder is then dispensed out of the drum through a screen (Sachs *et al.* 1998b). The drum also acts as the compacting device. Upon dispensing the powder, the drum is traversed over the powder bed once more to accomplish compaction. The drum is vibrated for the compaction purposes. The powder has to be selected with the right properties, such as the particle size and size distribution to get good dispensing and packing behavior (Sachs *et al.* 1998b).

In the second method, a wet slurry process, a slurry of the powder is sprayed onto a substrate to form the powder bed. The powder bed is then dried and the binder is printed into the dried powder bed. After part completion, the unprinted regions are removed through redispersion in an ultrasonic bath. Due to the use of a liquid bath to remove the unprinted powder, there are some additional requirements for the binder. The binder should be insoluble in the liquid used for redispersion (essentially water). Also, the binder has to penetrate the powder bed fully and bond with enough strength to withstand ultrasonication. Various binders have been tried for this purpose, including latex, wax emulsions, and homogeneous solution phase binders.

There are various applications for which 3DP is used. Metal parts and tools are fabricated using the dry powder method. A colloidal latex binder solution is used along with stainless steel or tool steel powder for the purpose of fabricating metal parts or tooling (Sachs *et al.* 1998b). The part is usually fired to obtain a sintered skeleton which is then infiltrated with a low melting metal, such as copper or copper alloys to obtain a fully dense part with minimum shrinkage (Sachs *et al.* 1998b). This technology to manufacture metal parts and tools has been licensed by Extrude-Hone (Irwin, PA).

Another major application is the manufacturing of ceramic casting cores or shells for investment casting of metals. Typically, alumina, or silica with amorphous or colloidal silica binder solution is used (Yan and Gu 1996). Soligen Technologies has licensed this MIT technology.

Z Corporation (Somerville, MA) uses a modification of the 3DP dry powder process. In their process, they use starch powder and a proprietary water-based binder to manufacture concept models, which are very simple models for the early stages of prototyping. The models, made of starch, are then impregnated with a resin to give them mechanical strength.

Although the 3DP process is well developed and flexible, there are still some important issues that need to be improved. One of the major areas of research is concerned with improving the dimensional accuracy and control of the parts made by 3DP. At present, the typical shrinkage values of metal parts made by 3DP are about 2% and efforts are underway to reduce that value to about 0.2%. As the absolute value of shrinkage is reduced, it will cause a decrease in the value of uncertainty in shrinkage (Sachs *et al.* 1999). A decrease in the uncertainty of shrinkage corresponds to better dimensional control. In addition to this, there are also problems associated with the removal of loose powder from the fabricated part, especially when the part has a large number of narrow passages.

One of the biggest advantages of 3DP is that it can be used to build almost any geometry including undercuts, overhangs, and through-holes as the unbound powder acts as the support for the rest of the part. Also, 3DP can be used to make very fine features, as small as 100 μm. 3DP is also a very flexible (many materials and geometry) and well-controlled process.

3. *Selective Laser Sintering*

Selective laser sintering (SLS), like SLA, is one of the pioneering SFF techniques. It was commercialized around 1987 by DTM Corporation (Austin, Texas), who purchased the patents and licenses from the

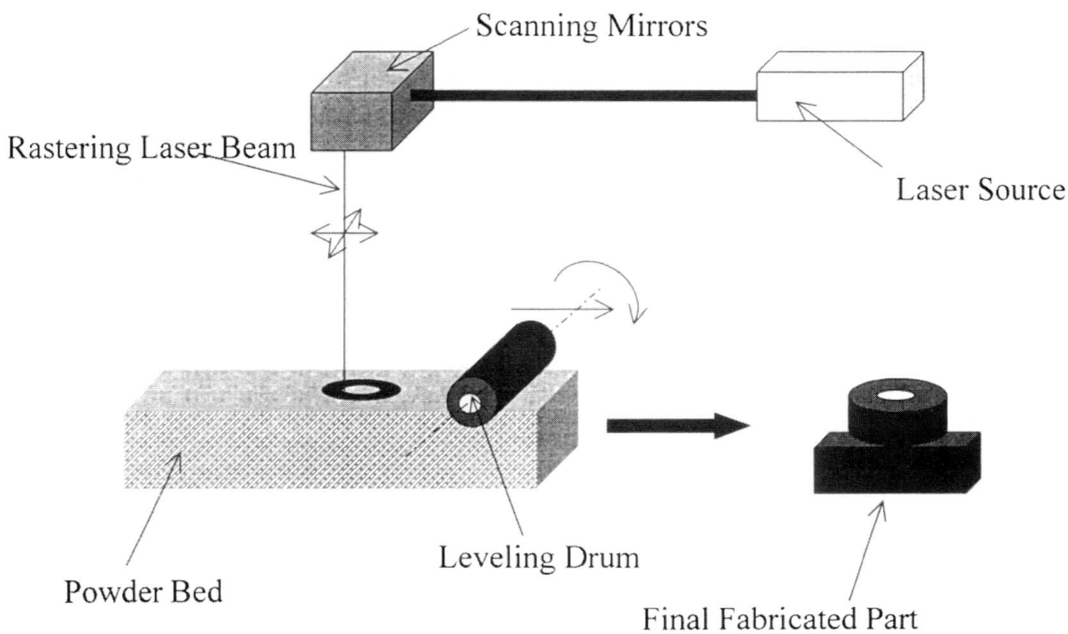

Figure 3
Schematic of the operation of selective laser sintering (SLS) process (Lakshminarayan *et al.* 1990).

University of Texas at Austin (Kai and Fai 1997). The actual process could be thought of as a hybrid of 3DP and SLA, in the sense that it involves deposition of a layer of dry powder, which is then sintered using a laser. The schematic of a typical SLS process is shown in Fig. 3. In SLS, a thin layer (approximately 100–200 µm) of powder is spread on a surface using a cylindrical roller (Lakshminarayan *et al.* 1990). A laser is then scanned over the powder bed, which heats the powder locally and sinter-bonds the adjacent particles to form one layer of the part. The intensity of the laser beam is modulated according to the computer-defined cross-section of the part. Typically, the laser is turned on where the powder needs to be bonded to form a solid section and turned off where no bonding is required. The non-sintered particles act as a support for any hollow section, overhangs or undercuts in the part (like in the case of 3DP). After the formation of the first layer, the next layer of powder is spread over the first layer followed by scanning of the laser. The laser has to have enough intensity to sinter the particles to the adjacent particles and to the previously fabricated layer. Upon completion of a part, it is removed from the chamber, the loose powder removed and the part is post processed, if necessary.

The commercialized SLS process predominantly uses thermoplastic materials such as ABS, polycarbonates, nylon, and wax (Kai *et al.* 1997). There has been extensive research at the University of Texas at Austin to extend this process to include metals and ceramics. The basic problem associated with metals and ceramics is that they have much higher melting points than polymers and, therefore, require much higher energy input for sintering. Two different approaches have been developed to overcome this problem, which can be classified into indirect and direct sintering approaches.

In the case of ceramics, there are two distinct indirect sintering approaches: (i) using an inorganic binder (Lakshminarayan *et al.* 1990, Lakshminarayan and Marcus 1991); and (ii) using an organic binder (Subramanian *et al.* 1993, Vail and Barlow 1991). In the case of the inorganic binder, the laser melts the binder, which then binds the particles together in the green body. In the case of the organic binder, a polymer coated ceramic or metal powder is used to form the powder bed. The laser softens or melts the polymer and bonds the particles in the green body. At the end of build, further post processing is required to obtain a fully dense body. More recently, there has been a thrust to develop a direct sintering approach to sinter ceramic materials. The basic idea behind this approach is to use a high power laser (1 kW, CO_2 laser) and a preheated powder feed to achieve the high temperatures needed for sintering ceramics (Das *et al.* 1991).

One key advantage of the SLS technique is its relatively fast build rates compared with SLA of ceramics or 3DP (a typical 1000 cm^3 ceramic or metal part can be built in about 6–10 hours depending on its complexity). As the SLS part is built from a powder bed, the surrounding loose powder acts as the support material and enables one to build very complex geometries. As mentioned before, SLS can also be used to process a wide variety of materials such as polymers, metals, and ceramics.

4. Laminated Object Manufacturing and CAM-LEM

Laminated object manufacturing (LOM) and computer aided manufacturing of laminated engineering materials (CAM-LEM) are two closely related SFF processing techniques. Both techniques build a part by the lamination of sheet feedstock, which is cut by a laser along contours defined by the CAD file-generated tool path. The sequence in which the cutting and lamination (or stacking) of the sheet feedstock is accomplished differentiates the two techniques.

The general approach of LOM is to stack then cut, as opposed to the cut-then-stack approach of CAM-LEM (Kai and Fai 1997, Klosterman *et al.* 1996, 1997a). In LOM of ceramics, ceramic tapes are used as the sheet feedstock. The tapes are sprayed with a solvent and then stacked (Klosterman *et al.* 1996, 1997a). The solvent acts as an adhesive, helping bond the tapes together. A heated roller is used to further bond the tapes through thermocompression. A laser is then used to cut the tape according to the tool path file. The area of the tape outside the build contour is cut in a crosshatch fashion. At the end of the build, the part is manually removed by cutting the cross-hatched region from the main part by a process called "decubing" (Kai and Fai 1997, Klosterman *et al.* 1997a). After assembly, the layers are laminated by warm isostatic pressing (or other suitable method) to achieve intimate interlayer contact, promoting high-integrity bonding in the subsequent sintering operation. The laminated "green" object is then fired (with an optimized heating schedule) to densify the object and fuse the layers and particles within the layers into a monolithic structure.

Although LOM is attractive, there are some problems. The need to decube makes it labor intensive and it usually involves long processing times (Klosterman *et al.* 1997a). Also, decubing of ceramic parts is difficult and damage to the green part can occur (Klosterman *et al.* 1997a). Research in LOM is being conducted by the University of Dayton Research Institute in collaboration with Helisys, and is focused on trying to overcome the above problems and make LOM of ceramics an automated process (Kai and Fai 1997, Klosterman *et al.* 1997a). Some of the materials so far used in LOM are Si_3N_4, SiC, SiC fiber/SiC composites, and hydroxyapatite (Kai and Fai 1997, Klosterman *et al.* 1996, 1997a, 1997b, Pope *et al.* 1997, Steidle *et al.* 1997). These materials have been used to make parts such as bone implants, turbine blades, and ceramic matrix composites. Ceramic preforms using LOM of Al_2O_3 and silica have also been fabricated (Agarwala *et al.* 1999).

CAM-LEM is a solid freeform fabrication technique capable of fabricating functional components from engineering materials such as ceramics, metals, and engineering plastics. The process was developed as a result of collaboration between Case Western Reserve University and CAM-LEM, Inc., Cleveland, OH (Cawley *et al.* 1995). A schematic of the process is shown in Fig. 4 (Cawley *et al.* 1995). The process uses a powder filled sheet feedstock, which is laser cut using a low powered CO_2 laser along the contours defined by the CAD file-generated tool path. The cut sheets are then robotically stacked to form the three-dimensional object, which undergoes subsequent post-processing to obtain the fully dense functional component. Some of the materials used in CAM-LEM are alumina, PZT (lead zirconate titanate), silicon nitride, and stainless steel (Liu *et al.* 1999).

One important issue in the development of the CAM-LEM technique is feedstock development. The sheet feedstock used in CAM-LEM is fabricated either by tape casting or by compression molding (Cawley *et al.* 1998). Another issue that has been studied extensively is that of tacking, alignment, and lamination. The purpose of tacking is to ensure that the ceramic tapes are bonded prior to (or during) lamination. Several techniques have been used to ensure that the stacked tapes stay in place (Cawley *et al.* 1995, 1998, Liu *et al.* 1999). In one technique, a thin coating of a heat-sensitive adhesive is applied to the surface of the green tapes. Tacking is achieved by application of heat and/or pressure that diffuses the thin coating into the adjacent layers, and produces a bond between them.

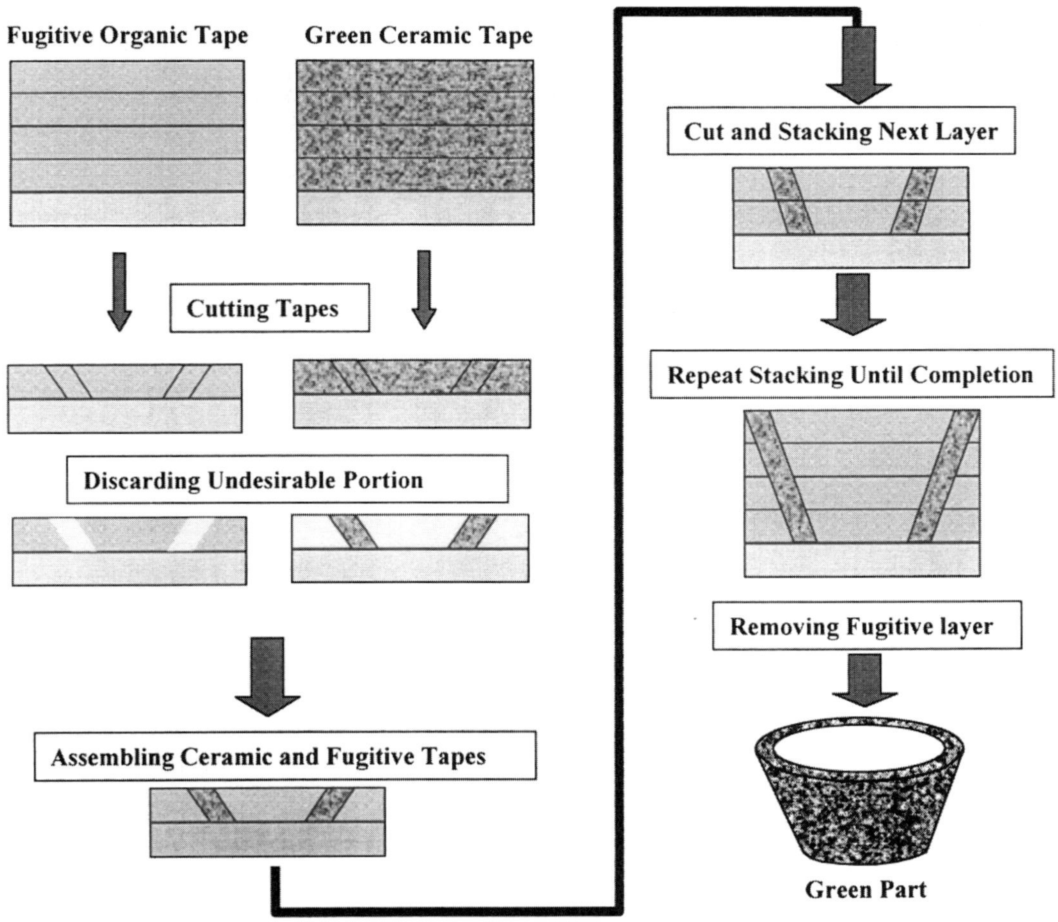

Figure 4
Schematic of the different steps involved in the building of parts by computer-aided manufacturing of laminated engineering materials (CAM-LEM) (Cawley *et al.* 1998).

One of the important advantages of CAM-LEM is the fabrication of multi-material parts both from layer-to-layer and within a layer. During the normal build in CAM-LEM, a fugitive material can also be used in conjunction with the build material. The fugitive sheet is usually cut along contours, which are complementary to the build material contours. The cut sheets of the fugitive material are then stacked along with the build material in each layer similar to a jigsaw puzzle to form a flat layer. This method of fabricating a part with the fugitive material is unique to the CAM-LEM technology. The fugitive material serves as a support during build, and also during post-processing steps. In addition to performing as a support, the fugitive material enables proper pressure transmittance if a thermocompressive lamination procedure is employed (Cawley *et al.* 1995). Some of the typical fugitive materials used are graphite and starch (Cawley *et al.* 1995, 1998). The other significant advantage of the cut-then-stack approach is that a "de-cubing" step is not necessary to remove the excess material as the shape of each layer is cut prior to stacking. The lack of this step in the CAM-LEM process leads to less material loss. This loss may not be important if one uses paper, but in case of some ceramic feedstock, the decrease in material loss during build can lead to a significant reduction in the overall cost of the process.

CAM-LEM has been used to fabricate a variety of products such as fluidic devices, ceramic turbine blades with internal cooling channels, and a dog femur (Liu *et al.* 1996). This technique is a very versatile process with a lot of advantages such as its ability to build multi-material parts.

5. *Fused Deposition Modeling and Fused Deposition of Ceramics*

Fused deposition modeling (FDM), developed by Stratasys, Inc. (Eden Prairie, MN), uses a 1.78 mm diameter thermoplastic filament feedstock. A schematic of the process is shown in Fig. 5. The filament feedstock is drawn from a spool and fed into a heated liquefier via a pair of counter-rotating rollers as shown in Fig. 5. The liquefier is in essence a cylindrical barrel that is resistively heated. The thermoplastic filament melts during its passage through the barrel, and is extruded through a nozzle (i.e., the filament acts as the piston). The liquefier moves in the X–Y plane (Fig. 5) and deposits the extruded material in elliptical cross-section elements, called roads,

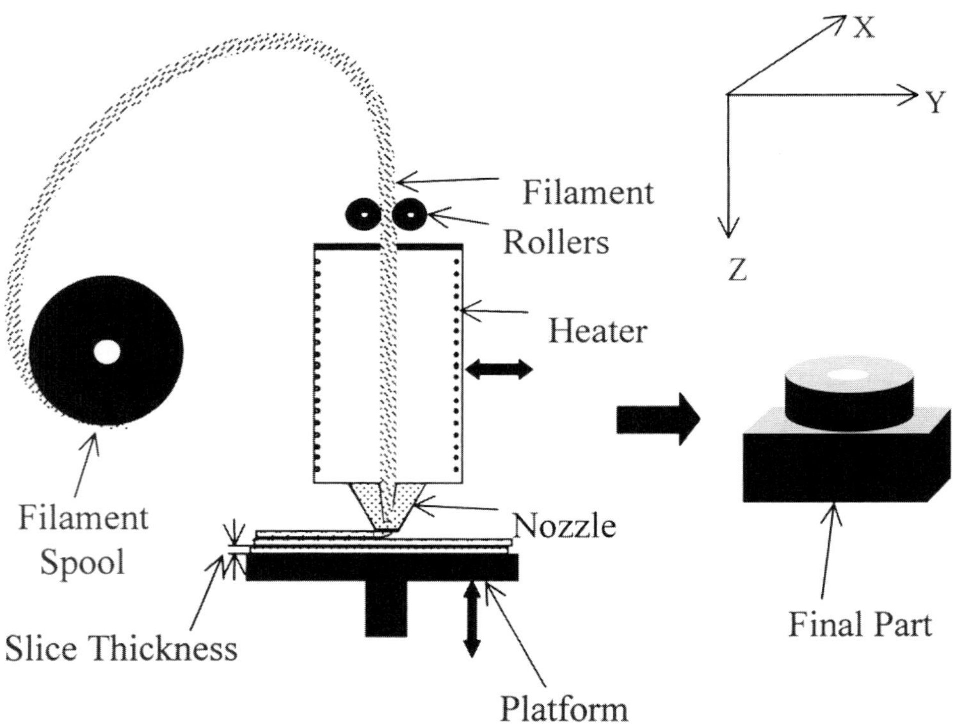

Figure 5
Schematic showing the basic mechanism of operation of a fused deposition machine. The X, Y, Z directions are indicated in the figure to enable the reader to understand the directions of motion of the liquefier assembly and the build platform (Danforth *et al.* 1998).

along paths defined by the CAD file-generated tool path. A single layer is built by repeated deposition of these roads. Upon completion of one layer, the build platform indexes down and the whole process of depositing roads is repeated. The ambient temperature of the build envelope is also controlled above room temperature and below the melting point of the thermoplastic material so that the material can solidify to form a rigid part at the end of the fabrication. The build envelope temperature is also maintained above room temperature (25°C) to allow for sufficient bonding between the deposited roads and layers.

The first commercial FDM machine released was called the three-dimensional modeler. It primarily used crystalline waxes and nylon-based materials (Kai and Fai 1997). Prototypes fabricated with these materials tended to exhibit time-dependent dimensional changes (Yardimci 1999). After a few software updates and hardware changes, the next generation of machines, the FDM 1600 and FDM 1650 was developed. The FDM 1650 has a two-material deposition capability. One material is used as the build material and the other as the support material. Originally, the support structure created using FDM machines needed to be removed manually at the end of a build. Stratasys later introduced a new system called FDM 3000 that uses a water-soluble support material that enables one to remove the support by washing it in a water-based solution (Stratasys Homepage). Another interesting new introduction to the line of FDM products is the FDM Quantum. The FDM Quantum is a machine designed to produce large parts (600 mm × 500 mm × 600 mm) in a short time (Stratasys Homepage). The machine uses patented high speed electromagnetic drives to rapidly move and control the extrusion head (Stratasys Homepage). The use of electromagnetic drives eliminates mechanical parts such as bearings, cables, and gears. This use allows for faster and more accurate movement of the extrusion head compared with the other FDM systems (Stratasys Homepage).

The materials in use include wax, nylon, ABS, MABS (medical grade ABS), and wide variety of elastomers (Kai and Fai 1997). The FDM process is used to make prototypes, patterns for mold making, molds for investment casting and more recently for making medical equipment with the MABS material (Kai and Fai 1997, Stratasys Homepage). Stratasys commercialized a system that uses an alternative to the filament-feed mechanism. The machine is called Genisys and is based on a screw-pump mechanism developed by IBM (Kai and Fai 1997, Stratasys Homepage). The use of a screw-pump mechanism removed some of the restrictions placed on the materials used due to filament fabrication and use. Also, due to the use of a screw pump the magnitude of the extrusion pressure achievable is greater, making the process more robust (Stratasys Homepage). The Genisys system uses three materials (Kai and Fai 1997, Stratasys Homepage): a general purpose polymer for design verification purposes, a nylon-like polymer used to make masters for other processes such as investment casting, and elastomers used, e.g., in door seals.

FDC and FDMet are SFF techniques capable of fabricating functional ceramic and metallic components. The process was invented at Rutgers University in 1994 (Danforth *et al.* 1998) and is based on a commercial FDM technique. The basic idea is to use a particle-loaded thermoplastic filament as the feedstock in place of pure polymer filaments. The use of a wide variety of materials in FDC such as, Si_3N_4, Al_2O_3, PZT (lead zirconate titanate), PMN-PT (lead magnesium niobate–lead titanate), stainless steel, and hydroxyapatite for structural, electronic, and biostructural applications, has been demonstrated (Agarwala *et al.* 1996, Bandyopadhyay *et al.* 1997, Safari *et al.* 1998). A large amount of work was conducted over the last six years of the twentieth century in developing the FDC process. The research spans the various components of the FDC process including development of suitable binders, coating of powders, filament fabrication, FD part building, and post build processes.

Two different binder systems, RU9 and ECG, are in use in FDC (Agarwala *et al.* 1996b). The RU9 and ECG binders consist of four major components: a polymer backbone, a wax component that acts as a viscosity modulator and provides stiffness, a tackifier component that enables the deposited material to bond, and an elastomer component to impart flexibility. A particular binder, termed ECG9, is the more commonly used binder.

The RU9 system has primarily been used with Si_3N_4 and Si_3N_4/SiC. The various powders used with ECG systems are: PZT, PMN–PT, Al_2O_3, hydroxyapatite, and stainless steel. The selection of the binder is only the first in a series of steps necessary to fabricate the filaments.

The quality and uniformity of the extruded filament is another important step that will determine the quality of the FDC process and the fabricated part. A number of processing studies have been conducted on the RU9/Si_3N_4 and ECG9/PZTsystems. These studies have enabled the development of a well-controlled filament fabrication process capable of making hundreds of meters of filaments with uniform properties. Also, there are a variety of parameters related to the software, hardware, and feedstock material that affect the quality of the fabricated part.

In the FDC process in order to build a part, the perimeter or the boundary of a layer (slice) is built first. Then the interior of the part is built using a contour fill, or a vector fill, or a combination fill. There are many parameters associated with building a part: slice thickness, road width, road offset values, angle of vectoring, motion parameters, and flow parameters. Liquefier temperature, build envelope temperature, and nozzle diameter also affect the quality.

One of the important issues involved in FDC is the buckling of filament during building. The buckling of filament is a serious obstacle to the development of a continuous FDC process. An empirical quantity based on the ratio of modulus of filament to the motor torque during FDC has been used to differentiate between filaments that buckle and those that do not buckle. Although, the quantity is useful it does not provide a scientific understanding of the material- and machine-related issues that may affect the buckling of filaments during FDC (Venkataraman 2000, Venkataraman and Rangarajan 1999).

One of the important stages in post processing is the binder removal stage. Detailed studies based on design of experiments have been conducted on the RU9 Si_3N_4 system to establish a standard N_2 heating cycle, flow rate of N_2, the heating cycle in air, and various other procedures involved in the burnout cycle (Bhargava *et al.* 1997).

In the case of the ECG-based system, the binder removal is conducted entirely in static air. The basic process involves slow heating in air to a sufficiently high temperature to remove the polymer. Some part/powder specific modifications to the basic cycle have been developed. These modifications essentially involve different heating rates and hold steps to solve some of the binder removal related problems such as slumping, cracking, and bloating. Upon sintering, the dimensional changes (as-built to final sintered part) are 17% along build plane directions and 20% along the build direction (McIntosh and Danforth 1997). The corresponding shrinkage values for PZT are 16% and 20%, respectively (Lous *et al.* 2000). As can be seen from the shrinkage values for silicon nitride and PZT, there is significant anisotropy in the shrinkage. The shrinkage in the build direction tends to be higher than in the build plane directions (McIntosh and Danforth 1997, Lous *et al.* 2000).

The extrusion freeform fabrication (EFF) process, a method that adapted the basic FDM technique, was developed at Advanced Ceramics Research, Inc. (Ventura and Narang 1998, Prinz and Weiss 1998, Dimos *et al.* 1997). This process uses a piston and plunger-based high-pressure delivery system. A rod-type feedstock (5/8 inch diameter, 6 inches long) is fed into the extruder chamber where it is melted and extruded. A number of high-melting engineering thermoplastics, ceramic (silicon nitride) particle-loaded, and short fiber-loaded thermoplastics are being used with the system (Ventura and Narang 1998, Prinz and Weiss 1998, Dimos *et al.* 1997). The piston plunger mechanism is not a continuous process, whereas the filament-feed mechanism is. This difference has led to restrictions on the part size that can be built.

Researchers at the University of Southern California have developed a process that is based on the FDM approach called contour crafting (Khoshnevis 1999). The process uses a nozzle (usually larger than the ones used in FDM) to lay down the roads corresponding to the outer surface contour of the part. The contour is shaped using typical shaping tools such as trowels to obtain a smooth surface. Once a smooth contour is built the interior of the part can be filled with the material to make the final part. The process is primarily concerned with making parts with a smooth outer surface and in case of hollow parts smooth inner and outer surfaces. As the process uses nozzles much larger than the ones used in FDM, it is typically not useful for making parts with fine details. The process is primarily used to make large conceptual models from clay-based materials.

A novel multimaterial FDC machine is being developed at Rutgers University. The project is funded under a ONR-MURI (Office of Naval Research-Multi University Research Initiative). A picture of the multi-material machine is shown in Fig. 6. One of the objectives of the MURI program at Rutgers University is to develop an intelligent multi-material layered manufacturing system based on the FDC approach. Therefore, one needs to develop the necessary scientific knowledge of the various issues involved in FDC in order to begin developing an intelligent machine.

The MURI machine, also called an intelligent CAD-based system, is equipped with machine vision, which monitors the surface quality of layers during the fabrication. The machine vision is not in a closed loop with the rest of the system, but work is underway on closing the loop so that the defects detected can be removed and the appropriate compensations can be made for the next layer. Such an online process monitoring and control system is essential for the fabrication of high quality parts.

6. Other SFF Techniques for Ceramics and Metals

In addition to the above more mature techniques, there are a few other techniques for fabrication of ceramics and metals being developed in various national laboratories and universities such as laser-engineered net shaping (LENS) (Griffith *et al.* 1996, 1997, Brooks *et al.* 1999, Vasinonta *et al.* 1999), robocasting (Ceasarano *et al.* 1997, King *et al.* 1998, Ceasarano *et al.* 1998), multiphase jet solidification (MJS) (Kai and Fai 1997, Kupp *et al.* 1997, Greulich *et al.* 1995), direct photo shaping of ceramics (DPS) (Greulich *et al.* 1995), and shape deposition manufacturing (SDM) (Prinz and Weiss 1998).

The LENS process and robocasting are two techniques being developed at Sandia National Laboratories. LENS is a laser-based process, where a high

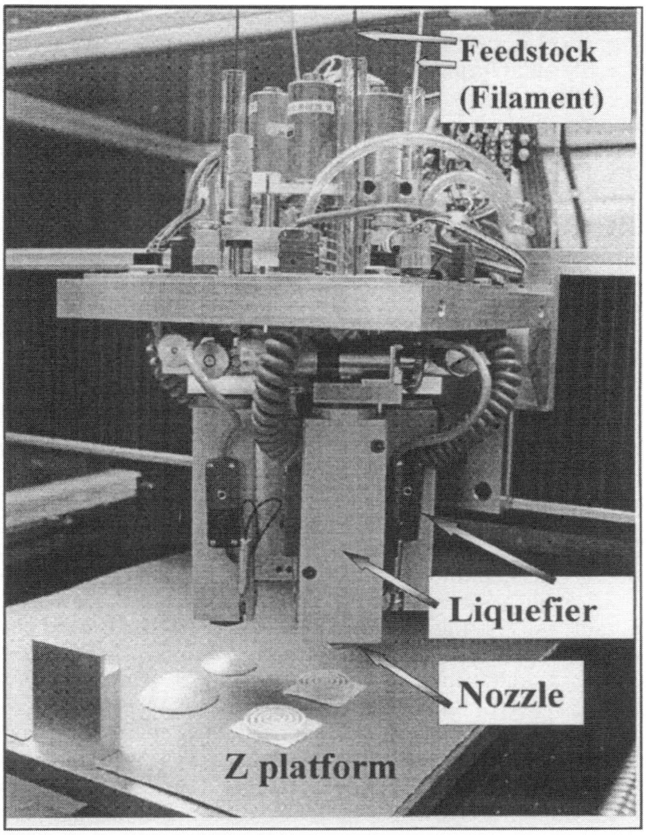

Figure 6
Photograph illustrating the deposition head of the MURI layered manufacturing machine with four liquefiers.

power laser beam (Nd:YAG) is focused on a metal substrate to create a weld pool. At the same time a stream of powder is injected into the weld pool. The laser beam melts the powder locally on top of the substrate. The substrate is then moved under the laser and the part is built line-by-line and layer-by-layer. LENS has been used to fabricate a variety of products ranging from prototypes to actual injection mold tooling using materials such as stainless steel, tool steel, and Inconel alloys (Griffith *et al.* 1997). The typical build rate of the LENS process is about 20 hours for a 100 cm^3 part (Griffith *et al.* 1997). The fabrication of multi-material graded and layered structures using multiple powder feeders has been demonstrated (Griffith *et al.* 1997).

Robocasting is a deposition-based process using a high-solids loading (>50 vol.%) water-based slurry (Ceasarano *et al.* 1997). The slurry is pumped through an orifice onto a moveable substrate to form a continuous bead. The part is subsequently built by the layerwise deposition of the slurry. The typical deposition rates of the slurry in robocasting are about 0.006–0.01 $cm^3 s^{-1}$ (Ceasarano *et al.* 1997). The use of a water-based slurry makes the processing environment friendly. Also, the use of a low volume fraction of organics makes it easier to post-process parts compared with parts made by FDC or other polymer-based forming methods. The process has been used to make parts of varying geometrical complexity using materials such as Al_2O_3, kaolin, PZT, and ZnO (Ceasarano *et al.* 1998). Recently, the ability to deposit multi-materials in a sequential and graded fashion has been demonstrated (Ceasarano *et al.* 1998).

MJS is a SFF technique used to fabricate ceramic and metallic components. It is comparable to the fused deposition process. The MJS process uses either a low melting alloy or a powder-binder mixture in granule or rod form as the feedstock (Kai and Fai 1997, Kupp *et al.* 1997). The feedstock is fed into a patented extrusion apparatus. The low melting alloy or the binder is melted and the low viscosity (<200 Pa s) fluid is deposited on a substrate (Kupp *et al.* 1997). The extruder head is then translated in the X–Y directions to fabricate the part line-by-line and layer-by-layer.

Considerable research is being conducted at the Fraunhofer Institute for Applied Materials Research and Fraunhofer Institute for Manufacturing Engineering and Automation, to optimize the process so as to increase the part complexity and process temperature (Kai and Fai 1997). The list of materials used in MJS includes stainless steel, high speed steels, FeNi alloy, Ti and SiC (Kupp *et al.* 1997, Greulich *et al.* 1995). Efforts to fabricate directly high-melting-point metal alloy parts (without binder) and to include other ceramic materials such as Al_2O_3 and ZrO_2 are in progress (Kupp *et al.* 1997).

Direct photo shaping of ceramics is a novel SFF technique developed at SRI International. The technique involves the use of digitally projected visible light to cure layers of ceramic-loaded photocurable resins (Ventura and Narang 1998). The photocurable resin typically consists of 55 vol.% solids, a dispersant, photoactive monomers, photoinitiator, and a tackifier (to help bond the successive layers). The build process consists of three basic steps. The first step involves the dispensing of the ceramic-loaded photocurable resin. The resin is dispensed from a peristaltic pump and then uniformly applied by a doctor blade (Ventura and Narang 1998). The thickness of the layers is typically about 50 μm. In the curing step, the image of a particular cross-section of the part as defined by the CAD file is digitally projected onto the dispensed layer of the photocurable resin. The portions of the layer exposed to the light cures to form the desired cross-section profile of the part. This sort of a curing is called curing through "flood exposure." The above steps are then repeated to build the part layer-by-layer. The cured resin is then removed and the part sintered to full density in the subsequent post-processing steps. As mentioned earlier, the part is built by curing the resin through flood exposure to light instead of a rastering laser. Therefore, the build time for each layer in this technique is expected to be shorter than for rastering laser techniques such as SLA of ceramics (Ventura and Narang 1998). Typically, a layer of 1 cm × 1 cm area can be cured in about 4 s using the flood exposure technique as opposed to 10 s for SLA of ceramics (Ventura and Narang 1998). The use of visible light with a digital projection system makes the process inexpensive. In addition, the visible light digital projection system leads to high resolution (16 μm). The direct photo shaping method has been shown to be feasible for the fabrication of ceramic (silicon nitride, alumina), metal, and polymer parts (Ventura and Narang 1998). The method is also being applied to the fabrication of micro mechanical and electromechanical devices containing multiple ceramic/polymer/metal interfaces (Ventura and Narang 1998).

SDM differs from all other SFF techniques described before in that it uses a combination of additive and subtractive processes to fabricate parts. The technique was developed at Stanford University (Prinz and Weiss 1998). The objective of the SDM approach is to combine the advantages of the additive techniques with that of the subtractive techniques. In general, the addition of material is accomplished by different techniques depending on the material to be deposited. A computer numerically controlled (CNC) machine is used for material removal. The part is decomposed into slices or "compacts." Then each of the compacts is fabricated to near net shape using a particular deposition technique. During the addition process, a build and a support material are usually deposited. The deposited layer is then machined to net shape. The process is repeated layer-by-layer to build the part. The support material is then removed at the end of the build process to reveal the part. In this technique, the compacts are created in such a way that the compact for the support material has the complement to any undercuts present in the part. This way an undercut in the part is replicated by depositing the build material on the complement of the undercut present in the support region. In the case of ceramic parts, an extruder developed by Advanced Ceramics Research (ACR) is used (Prinz and Weiss 1998). The feedstock used with the extruder is a rod (15 cm long and 1.14 cm diameter) (Prinz and Weiss 1998). The extruder is mounted on a commercially available CNC machine to create an integrated SDM machine. The extruder is used to deposit the support and the build material. This is accomplished by manually switching the extruder tubes preloaded with the materials of choice (Prinz and Weiss 1998). This process has been used to make parts from ceramics such as silicon nitride and alumina. The SDM method has also been used to make metal parts out of copper, nickel, and stainless steel. In case of metal parts, either a laser welder (for steel) or a microcaster (for copper) is used as the deposition device. The laser welder is similar to LENS where a YAG laser is used to create a weld pool into which a stream of powder is injected to build a layer. In a microcaster, a wire of material is fed into a plasma that creates superheated droplets of the metal. The droplets are then deposited onto the substrate to build a part. SDM is also being used to make microelectromechanical systems (MEMS). For this application, sputtering and electroplating are used as deposition methods and micro-CNC and electrical discharge machining (EDM) are used as shaping or removal methods. As mentioned before, SDM combines the advantages of additive and subtractive processes. It has been used with a wide variety of materials (metals, ceramics, and polymers) for a wide variety of applications. The technique can also be used to fabricate multi-material parts. The process has also been demonstrated to fabricate parts with embedded electronics (Prinz and Weiss 1998).

There is yet another interesting SFF technique being developed at Sandia National Laboratories (Dimos *et al.* 1997). This technique is called Direct-Write. The

Direct-Write method is being developed in order to fabricate multi-functional multi-material integrated ceramic components (MMICCs) (Yardimici 1999). The MMICCs usually involve a wide variety of devices such as capacitors, inductors, and conductors embedded in a ceramic substrate. The Direct-Write method is being developed as a rapid, low volume fabrication technique for these components. In the Direct-Write method, ceramic or metal-loaded slurries are printed onto a substrate. The substrate moves along the paths defined by a CAD file to produce one layer of the part. The layer is then dried prior to printing the next layer. In this way, the complete green part (unsintered) is fabricated in a layerwise fashion. Co-firing or co-sintering of the green part produced by the Direct-Write method yields the final integrated component. The Direct-Write method uses a commercial deposition technology called Micropen (Ohmcraft, Inc.). The Micropen system is an automated printing device for ceramic slurries. In Micropen, the ceramic-loaded (or metal particle-loaded) inks are printed through fine orifices (about 25–50 μm diameter). One of the interesting features of the Micropen system is the force feedback control. This feature involves "balancing the upward force exerted on the deposition head (pen) due to the extruding ink and the downward force applied by an electromagnet (placed above the pen)." This sort of control allows one to print inks with reproducible thickness (Dimos *et al.* 1998). It also allows for printing inks with near constant thickness on substrates with variations in height. As mentioned earlier, the Micropen can be used to print inks in lines about 25–50 μm wide and about 25–50 μm high. The typical deposition rate of this technique is about 1×10^{-5} $cm^3 s^{-1}$. Inks of silver (conductor), RuO_2 (resistor), dielectric (Ferro 38N is the commercial name), lead zirconate titanate (dielectric and thick film actuator) have been successfully printed using this technique. Also, components such as integrated RC filters and voltage transformers have been fabricated (Dimos *et al.* 1997). A number of co-firing studies are being conducted to identify possible co-fireable materials (and co-firing conditions) that can be used in Direct-Write method (Dimos *et al.* 1997).

Bibliography

Agarwala M K, Bandyopadhyay A, van Weeren R, Safari A, Danforth S C 1996a FDC, rapid fabrication of structural components. *Am. Ceram. Soc. Bull.* **75**, 60

Agarwala M K, Danforth S C, van Weeren R, Bandyopadhyay A, Safari A, Priedeman W 1996b . In: Bourell D L, Beamen J J, Marcus H L, Crawford R H, Barlow J W (eds.) *Proc. SFF Symp.* Austin, TX, Vol. 7, p. 56

Agarwala M, Klosterman D, Osborne N, Lightman A, Dzugan R, Rhodes G, Nelson C 1999 Hard metal tooling via SFF of ceramics and powder metallurgy. In: Bourell D L, Beamen J J, Marcus H L, Crawford R H, Barlow J W (eds.) *Proc. SFF Symp.* Austin, TX, Vol. 10, p. 759

Bandyopadhyay A, Panda R K, Janas V F, Danforth S C, Safari A 1997 Processing of piezocomposites by fused deposition technique. *J. Am. Ceram. Soc.* **80**, 1366

Bhargava P, Bandyopadhyay A, Rangarajan S, Qi G, Dai C, Wu S, Danforth S C, Safari A 1997 Shrinkage, weight loss, and crack prevention during binder burn out of components produced by fused deposition of ceramics (FDC). In: Bourell D L, Beamen J J, Marcus H L, Crawford R H, Barlow J W (eds.) *Proc. SFF Symp.* Austin, TX, Vol. 8, p. 395

Brady G A, Chu T M, Halloran J W 1996 Curing behavior of ceramic resin for stereolithography. In: Bourell D L, Beamen J J, Marcus H L, Crawford R H, Barlow J W (eds.) *Proc. SFF Symp.* Austin, TX, Vol. 7, p. 403

Brady G A, Chu T, Halloran J W 1997 Stereolithography of ceramic suspensions. *Rapid Prototyping J.* **3**, 61

Brady G A, Halloran J W 1998 *Naval Research Reviews, ONR* Vol. L, Three, p. 39

Brooks J, Robino C, Headley T, Goods S, Griffith M 1999 Microstructure and property optimization of LENS deposited H13 tool steel. In: Bourell D L, Beamen J J, Marcus H L, Crawford R H, Barlow J W (eds.) *Proc. SFF Symp.* Austin, TX, Vol. 10, p. 375

Carter Jr W T, Jones M G 1993 Direct laser sintering of metals. In: Bourell D L, Beamen J J, Marcus H L, Crawford R H, Barlow J W (eds.) *Proc. SFF Symp.* Austin, TX, Vol. 4, p. 51

Cawley J D, Liu Z, Mou J, Heuer A H 1998 Materials issues in laminated object manufacturing of powder-based systems. In: Bourell D L, Beamen J J, Marcus H L, Crawford R H, Barlow J W (eds.) *Proc. SFF Symp.* Austin, TX, Vol. 9, p. 503

Cawley J D, Wei P, Liu Z E, Newman W S, Mathewson B B, Heuer A H 1995 Al_2O_3 Ceramics made by CAM-LEM (computer-aided manufacturing of laminated engineering materials) technology. In: Bourell D L, Beamen J J, Marcus H L, Crawford R H, Barlow J W (eds.) *Proc. SFF Symp.* Austin, TX, Vol. 6, p. 9

Ceasarano III J, Baer T A, Calvert P 1997 Recent developments in freeform fabrication of dense ceramics from slurry deposition. In: Bourell D L, Beamen J J, Marcus H L, Crawford R H, Barlow J W (eds.) *Proc. SFF Symp.* Austin, TX, Vol. 8, p. 25

Ceasarano III J, King B H, Denham H B 1998 Recent development in robocasting of ceramics and multi-material deposition. In: Bourell D L, Beamen J J, Marcus H L, Crawford R H, Barlow J W (eds.) *Proc. SFF Symp.* Austin, TX, Vol. 9, p. 697

Danforth S C, Safari A, Agarwala M K, Bandyopadhyay A, van Weeren R 1998. US Patent No. 5,738,817

Das S, McWilliams J, Wu B, Beaman J J 1991 Design of a high-temperature workstation for the selective laser sintering process. In: Bourell D L, Beamen J J, Marcus H L, Crawford R H, Barlow J W (eds.) *Proc. SFF Symp.* Austin, TX, Vol. 2, p. 164

Dimos D, Yang P, Garino T J, Raymond M V, Rodriguez M A 1997. In: Bourell D L, Beamen J J, Marcus H L, Crawford R H, Barlow J W (eds.) *Proc. SFF Symp.* Austin, TX, Vol. 8, p. 33

Grau J, Moon J, Uhland S, Cima M, Sachs E 1997 High green density ceramic components fabricated by the slurry-based 3DP process. In: Bourell D L, Beamen J J, Marcus H L, Crawford R H, Barlow J W (eds.) *Proc. SFF Symp.* Austin, TX, Vol. 8, p. 371

Greulich M, Greul M, Pintat T 1995 *Rapid Prototyping J.* **1**, 20

Griffith M L, Halloran J W 1996 Freeform fabrication of ceramics via stereolithography. *J. Am. Ceram. Soc.* **79**, 2601

Griffith M L, Halloran J W 1997 Scattering of ultraviolet radiation in turbid suspensions. *J. Appl. Phys.* **81**, 2538

Griffith M L, Harwell L D, Romero J T, Schlienger E, Atwood C L, Smugeresky J E 1997 LENS processing of new materials. In: Bourell D L, Beamen J J, Marcus H L, Crawford R H, Barlow J W (eds.) *Proc. SFF Symp.* Austin, TX, Vol. 8, p. 387

Griffith M L, Keicher D M, Atwood C L, Romero J A, Smugeresky J E, Harwell L D, Greene D L 1996 Freeform fabrication of metallic components using laser-engineered net shaping (LENS). In: Bourell D L, Beamen J J, Marcus H L, Crawford R H, Barlow J W (eds.) *Proc. SFF Symp.* Austin, TX, Vol. 7, p. 125

Kai C C, Fai L K 1997 *Rapid Prototyping*. John Wiley, Singapore

Khoshnevis B 1999 Contour crafting—state of development. In: Bourell D L, Beamen J J, Marcus H L, Crawford R H, Barlow J W (eds.) *Proc. SFF Symp.* Austin, TX, Vol. 10, p. 743

King B H, Morisette S L, Denham H, Ceasarano III J, Dimos D 1998 Influence of rheology on the deposition behavior of ceramic pastes in direct fabrication systems. In: Bourell D L, Beamen J J, Marcus H L, Crawford R H, Barlow J W (eds.) *Proc. SFF Symp.* Austin, TX, Vol. 9, p. 391

Klosterman D, Chartoff R, Osborne N, Graves G 1997a Automated fabrication of monolithic and ceramic matrix composites via laminated object manufacturing (LOM). In: Bourell D L, Beamen J J, Marcus H L, Crawford R H, Barlow J W (eds.) *Proc. SFF Symp.* Austin, TX, Vol. 8, p. 537

Klosterman D, Chartoff R, Osborne N, Graves G, Lightman A 1997b Pre-LOM, LOM, and post-LOM processes for the fabrication of SiC and SiC/SiC components. In: Bourell D L, Beamen J J, Marcus H L, Crawford R H, Barlow J W (eds.) *Proc. SFF Symp.* Austin, TX, Vol. 8, p. 247

Klosterman D, Chartoff R, Priore B, Osborne N, Graves G, Lightman A, Han G, Pak S, Weaver J 1996 Functional ceramic components via laminated object manufacturing. In: Bourell D L, Beamen J J, Marcus H L, Crawford R H, Barlow J W (eds.) *Proc. SFF Symp.* Austin, TX, Vol. 7, p. 356

Kupp D M, Eifert H, Greul M, Kunstner M 1997 Rapid prototyping of functional metal and ceramic components by the multiphase jet solidification (MJS) process. In: Bourell D L, Beamen J J, Marcus H L, Crawford R H, Barlow J W (eds.) *Proc. SFF Symp.* Austin, TX, Vol. 8, p. 203

Lakshminarayan U, Marcus H L 1991 Microstructural and mechanical properties of Al_2O_3/P_2O_5 and Al_2O_3/B_2O_3 composites fabricated by selective laser sintering. In: Bourell D L, Beamen J J, Marcus H L, Crawford R H, Barlow J W (eds.) *Proc. SFF Symp.* Austin, TX, Vol. 2, p. 205

Lakshminarayan U, Ogrydiziak S, Marcus H L 1990 Selective laser sintering of ceramic materials. In: Bourell D L, Beamen J J, Marcus H L, Crawford R H, Barlow J W (eds.) *Proc. SFF Symp.* Austin, TX, Vol. 1, p. 16

Liu Z E, Wei P, Kernan B, Heuer A H, Cawley J D 1996 Metal and ceramic components made by CAM-LEM technology. In: Bourell D L, Beamen J J, Marcus H L, Crawford R H, Barlow J W (eds.) *Proc. SFF Symp.* Austin, TX, Vol. 7, p. 377

Liu Z, Suppakarn N, Cawley J D 1999 Coated feedstock for fabrication of ceramic parts by CAM-LEM. In: Bourell D L, Beamen J J, Marcus H L, Crawford R H, Barlow J W (eds.) *Proc. SFF Symp.* Austin, TX, Vol.10, p. 393

Lous G M, Cornejo I A, McNulty T F, Safari A, Danforth S C 2000 Fabrication of piezoelectric ceramic/polymer composite transducers using fused deposition of ceramics. *J. Am. Ceram Soc* **83**, 124

Manriquez-Frayre J A, Bourell D L 1990 Selective laser sintering of binary metallic powder. In: Bourell D L, Beamen J J, Marcus H L, Crawford R H, Barlow J W (eds.) *Proc. SFF Symp.* Austin, TX, Vol. 1, p. 99

McIntosh J J, Danforth S C Jamalabad V R 1997 Shrinkage and deformation in components manufactured by fused deposition of ceramics. In: Bourell D L, Beamen J J, Marcus H L, Crawford R H, Barlow J W (eds.) *Proc. SFF Symp.* Austin, TX, Vol. 8, p. 159

Nutt K 1991 Selective laser sintering as a rapid prototyping and manufacturing technique. In: Bourell D L, Beamen J J, Marcus H L, Crawford R H, Barlow J W (eds.) *Proc. SFF Symp.* Austin, TX, Vol. 2, p. 131

Pope M J, Patterson C L, Zimbeck W, Fehrenbacher M 1997 Laminated object manufacturing of Si_3N_4 with enhanced properties. . In: Bourell D L, Beamen J J, Marcus H L, Crawford R H, Barlow J W (eds.) *Proc. SFF Symp.* Austin, TX, Vol. 8, p. 529

Prinz F B, Weiss L E 1998. *Naval Research Reviews*, ONR, Vol. L, Three, p. 19

Sachs E *et al.* 1995. US Patent No. 5, 387, 380

Sachs E *et al.* 1998a. US Patent No. 5, 807, 437

Sachs E, Allen S, Cima M, Xu X, Banos J, Serdy J, Brancazio D, Guo H 1998b. *Naval Research Reviews*, ONR, Vol. L, Three, p. 9

Sachs E, Allen S, Hadjiloucas C, Yoo J, Cima M 1999 Low shrinkage metal skeletons by three-dimensional printing. . In: Bourell D L, Beamen J J, Marcus H L, Crawford R H, Barlow J W (eds.) *Proc. SFF Symp.* Austin, TX, Vol. 10, p. 411

Sachs E, Cima M, Williams P, Brancazio D, Cornie J 1992 Three-dimensional printing; rapid tooling and prototypes directly from a CAD model. *J. Eng. Ind.* **114**, 481

Safari A et al. 1998. US Patent No. 5, 818, 149

Steidle C, Klosterman D, Osborne N, Graves G, Chartoff R 1997 Automated fabrication of nonresorbable bone implants using laminated object manufacturing (LOM). In: Bourell D L, Beamen J J, Marcus H L, Crawford R H, Barlow J W (eds.) *Proc. SFF Symp.* Austin, TX, Vol. 8, p. 248

Stratasys, Inc. Homepage, http://www.stratasys.com

Subramanian P K, Zhong G, Marcus H L 1992 Selective laser sintering and reaction sintering of ceramic composites. In: Bourell D L, Beamen J J, Marcus H L, Crawford R H, Barlow J W (eds.) *Proc. SFF Symp.* Austin, TX, Vol. 3, p. 63

Subramanian P K, Zhong G, Vail N K, Barlow J W, Marcus H L 1993 Selective laser sintering of Al_2O_3. In: Bourell D L, Beamen J J, Marcus H L, Crawford R H, Barlow J W (eds.) *Proc. SFF Symp.* Austin, TX, Vol. 4, p. 350

Vail N K, Barlow J W 1991 Effect of polymer coating as intermediate binder on sintering of ceramic parts. In: Bourell D L, Beamen J J, Marcus H L, Crawford R H, Barlow J W (eds.) *Proc. SFF Symp.* Austin, TX, Vol. 2, p. 195

Vasinonta A, Beuth J, Griffith M 1999 Process maps for laser deposition of thin-walled structures. In: Bourell D L, Beamen J J, Marcus H L, Crawford R H, Barlow J W (eds.) *Proc. SFF Symp.* Austin, TX, Vol. 10, p. 383

Venkataraman N 2000 *The Process-property-performance Relationships of Feedstock Materials Used for Fused Deposition of Ceramics (FDC)*. PhD Thesis, Rutgers University

Venkataraman N, Rangarajan S, Matthewson M J, Harper B, Safari A, Danforth S C, Wu G, Langrana N, Yardimci A,

Guceri S 1999 Material property-process relationships in fused deposition of ceramics (FDC) and metals (FDMet). *Proc. Symp. Innovative Processing and Synth. of Ceramics* American Ceramic Society

Ventura S, Narang S 1998. *Naval Research Reviews*, ONR, Vol. L, Three, p. 44

Wu G, Langrana N A, Rangarajan S, McCuiston R, Sadanji R, Danforth S, Safari A 1999 Fabrication of metal components using FDMet: fused deposition of metals. In: Bourell D L, Beamen J J, Marcus H L, Crawford R H, Barlow J W (eds.) *Proc. SFF Symp.* Austin, TX, Vol. 10, p. 775

Yan X, Gu P 1996 *Computer-aided Design*. **28**, 4, Elsevier, Amsterdam, p. 307

Yardimci M A 1999. PhD Thesis, University of Illinois at Chicago

A. Safari, S. C. Danforth, M. Allahverdi, and N. Venkataraman

Rapid Prototyping Processes

Rapid prototyping is a general term for a suite of technologies that employ incremental, additive processing to build objects to net or near-net shape. Most of the commercialized processes produce objects in materials such as wax, plastic or paper (Jacobs). For metals, commercially available technologies are somewhat limited, although a wide range of processing routes are the subject of research. For the purposes of this article, only those processes that result in a metallurgical bond between material increments are considered.

Rapid prototyping processes for metals are generally broken down into two categories; direct processes and indirect processes. Direct processes are those in which a net shape object is produced directly when material increments are deposited onto a substrate or previously deposited material using computer generated instructions. No pattern or mold is employed. These processes are also referred to as solid free-form fabrication (SFF). Indirect processes employ a pattern or mold (often a plastic, wax or paper object produced on a rapid prototyping machine), which is then used to create the metal object desired. To date, indirect processes are the most widely used. This is a result of the limited options available for SFF metal processes, and the need, to date, for substantial finishing to achieve a good surface finish and dimensional accuracy in parts produced by direct metal deposition.

Rapid prototyping in metal is usually closely associated with rapid tooling, a term that refers to the production of molds for prototyping of injection moldings, stampings, die castings, etc., in materials that exactly or closely approximate those used for production. Production tooling for these processes commonly takes four to 12 months to produce and costs in the range of hundreds of thousands of dollars. Very large tools for such processes to produce components such as autobody stampings, can cost more than one million dollars and take up to a year to fabricate. Even prototyping tooling often takes two to three months, and may cost from \$10 000 to $\$3 \times 10^5$ to produce. As a result, there is a high degree of motivation to develop and commercialize faster, lower cost processes for qualifying component designs and production processes. A second benefit of using additive manufacturing processes for producing tooling is the ability to create curved cooling lines, and place them in any location desired. This can increase productivity by 20–70% and may eventually have an even greater economic impact on the tooling and components industry than acceleration of tooling production. Rapid tooling also makes use of indirect and direct object production techniques.

1. Indirect Processes

1.1 Casting

Investment casting is one of the most widely used rapid prototyping techniques for metals, and is the one most rapidly moving towards manufacturing application. Patterns are produced using rapid prototyping processes rather than molded wax. The most commonly used process for producing the patterns is stereolithography. A honeycombed build pattern is often used; this allows the plastic pattern to expand when it is burned out without cracking the mold. Selective laser sintering and laminated object manufacturing are used less often for this purpose due to ash which may remain following pattern burnout. The Sanders model maker, which produces wax parts via an inkjet type of technology with extremely high resolution and accuracy, is also often used. Castings are produced using all castable engineering materials, and the process is being employed to produce very low volume production components when cost savings on pattern tooling may be significant. Figure 1 is an example of an investment casting produced via a stereolithography pattern.

In addition to applying rapid prototyping technology to the fabrication of investment casting patterns, rapid prototyping technologies have been used directly to create casting molds. Soligen uses the three-dimensional printing process developed at MIT to directly print ceramic molds for casting. In this process, known as Direct Shell Production Casting (DSPC), a ceramic powder bed is printed with a binder, which selectively bonds a powder area. Additional powder is added in the vertical axis and the process repeated continually until the shape of the casting mold desired is produced. The powder in the bed that is not printed with binder remains loose, and

Figure 1
Investment casting and stereolithography pattern (from Sandia website).

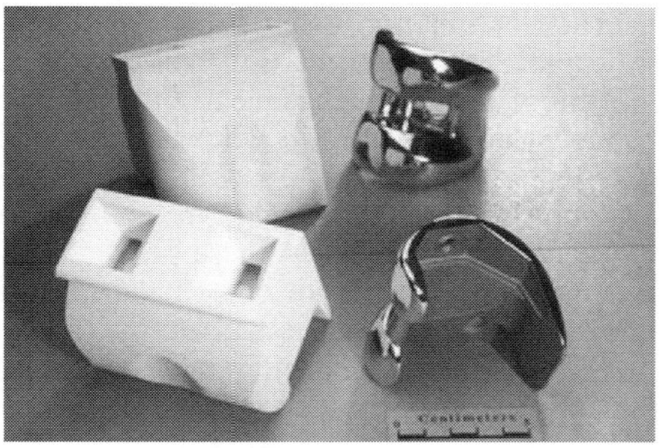

Figure 2
DSPC shells and castings (from Soligen website).

acts as a support structure for overhanging or enclosed features in the shell being printed. The mold is fired in the loose powder, and then the excess powder is removed. Figure 2 shows molds and parts produced via DSPC.

The selective laser sintering process can be used in a similar way to produce sand casting molds. A foundry sand coated with a polymer binder is loaded in a selective sintering system. The laser is rastered across the powder bed, locally melting the binder, and bonding the sand. This process is repeated layer by layer, until a sand mold results. As in the case of DSPC, unbonded sand acts to support the structure, and is removed following completion of the mold. The sand mold is used as in conventional sand casting to produce parts. This material is available commercially for use in selective laser sintering systems manufactured by DTM (Austin, TX, USA) and EOS (Germany).

1.2 Spray Forming

Spray forming is also an indirect process. It is most often applied to production of metal tooling. Because it is a line-of-sight process, spray forming is not suitable for production of components having overhangs or enclosed features. In addition, it is a relatively slow process, with build up rates typically ranging from 2.5 mm h^{-1} to 12.5 mm h^{-1}. However, its relatively low cost and ease of scale-up make it attractive for rapid production of tooling for stamping and injection molding. Spray-formed tools are typically produced by spraying a relatively thin shell and backing it with a pourable material, so lower deposition rates do not present difficulties in this application.

The Tafa spray process was the first spray forming process applied to net shape manufacturing/rapid tooling and is used for very low volume prototype tooling. In the Tafa process, a zinc alloy is sprayed against a pattern that can be produced via stereolithography or other methods. Conventional electric arc thermal spray equipment is used. Because it employs low melting point alloys and standard spraying equipment, TafaSpray is easily installed and operated by small scale shops and is widely used. However, shell thickness is limited, and low temperature materials such as zinc based alloys have limited applications and tool life.

More recently, a steel thermal spraying process for producing thick steel components has been developed. In this process, proprietary techniques (Jordan and Roche 1999) are used to control residual stresses in the growing spray deposit. Cooling through phase transformations involving dimensional changes, as well as the dimensional changes associated with thermal expansion and contraction, often results in the formation of very significant residual stresses in incrementally fabricated parts. These residual stresses can lead to distortion or cracking if the stresses are high enough, and the material sufficiently brittle.

Control of residual stresses in the spray form process allows very thick components to be produced, with high degree of dimensional accuracy in comparison to other processes. Limitations include the line-of-sight nature of the technique, and the poor microstructure produced by thermal spraying processes. High oxide content and the "splat" microstructure of thermally sprayed components leads to poor tensile strength and low ductility. Injection molding and stamping dies are subject primarily to compressive stresses, but this is a consideration where components are to be produced that have a deep draw, and the dies will be subjected to high shear and tensile stresses. The company that developed the process, Spray Forming Developments Ltd. were recently purchased by the Ford Motor Company.

A steel spraying process developed by the Idaho National Engineering and Environmental Laboratory known as Rapid Solidification Processing (RSP) is also being applied to rapid tooling. Like other spray processes, it is confined to line of sight shapes. However, RSP employs a molten metal feed-stock which is aspirated through a de Laval nozzle to produce a molten metal spray with a very small and relatively uniform droplet size. Unlike other spray forming processes, RSP has a relatively high deposition rate, and produces a high quality, small equi-axed grain microstructure, with minimal porosity (McHugh 1998), as seen in Fig. 4. It can be used to produce solid tools rather than shells that require backing.

The method has been used to produce tooling from high strength steel alloys such as H13. RSP results in high-density microstructure with superior properties; as a result it has benefits over other spraying processes for applications where high stresses or thermal fatigue are concerns. In die casting applications, die life in excess of conventionally produced tooling has been reported (McHugh 1999).

Like other solidification-based processes, geometric accuracy at larger sizes has been reported to be

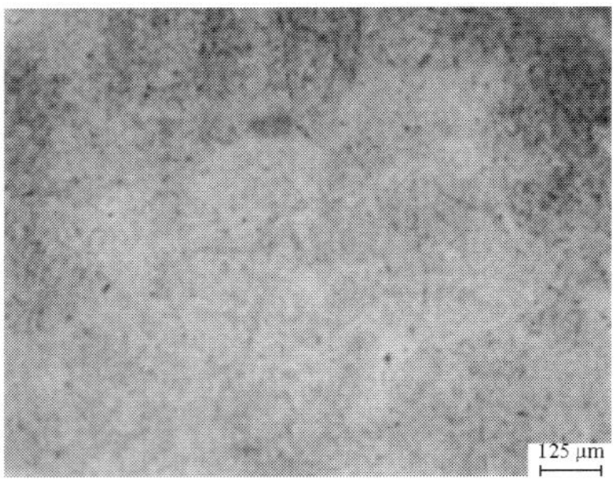

Figure 3
Microstructure of RSP process spray-formed H13 tool steel (from Lockheed Martin/INEEL website).

Figure 4
Laser powder deposition system (from Sandia National Laboratory website).

an issue. However, only finish machining may be required on RSP preforms and superior die life in demanding applications such as die casting tooling, in addition to lower lead time, is a significant benefit.

1.3 KelTool

KelTool is an indirect process that is also primarily directed at the rapid tooling market as evidenced by its name. In this process, a metal powder is combined with a binder and pressed against a pattern, which is usually produced via stereolithography. The green powder compact is treated to remove the binder and then infiltrated with a copper based alloy. This process has been commercialized and is available for licensing from 3D Systems, Inc. Jacobs (2000) has published results that call into question the potential for accuracy of this and other processes that rely on infiltration or any solidification-based process that does not incorporate machining to finish the parts to final dimensions. His results show that the dimensional changes that occur on solidification are not wholly uniform, but incorporate an element of random noise, which means that these processes have an imprecision on the order of 5×10^{-3} to 0.001" per inch. As part size grows, so does the inaccuracy, meaning that finish machining will be required to ensure accuracy as parts increase in size beyond 0.1–0.15 m on a side.

1.4 PHAST

The PHAST process is also a powder and infiltrant-based process. Details of the process, which was developed and patented at Proctor and Gamble, have remained proprietary. However, it involves casting of a powder filled gel against a pattern, burnout of the powder binder, and infiltration with another material. The patent portfolio for the process was recently donated to the Milwaukee School of Engineering.

1.5 Nickel Shell Tooling

Nickel shell tooling is another indirect process for tooling production in which nickel electroforming or chemical vapor deposition (CVD) is used to grow a nickel shell over a mandrel. Since electroforming is a relatively slow process, the nickel shell is then backed with a material such as aluminum-filled epoxy, an engineered concrete, or a ceramic. The mandrels can be produced from any electrically conductive material. Occasionally mandrels are produced via stereolithography and coated with a metal paint to produce surface conductivity, then used as mandrels (Chaneski 1997). Express Tool has commercialized a process employing machined graphite mandrels. This process also incorporates a copper backing layer behind the nickel shell to increase thermal conductivity, which can reduce molding cycle time.

Nickel CVD is a faster process than electroforming, however, the toxicity of the process consumables and cost of facilities has limited the commercialization of the process. Nickel CVD shells are typically grown to a somewhat greater thickness than shells produced via electroforming, but are still backed with a secondary material. This process has been used to produce tooling for household applications such as bath tubs and plastic cabinets.

2. *Direct Processes*

2.1 LENS/DMD

The direct metal process that has probably received the greatest amount of attention is laser metal powder deposition. Several organizations have made significant contributions in this area, including Los Alamos National Laboratory, Sandia National Laboratory, Fraunhofer Institute für Laser Technik (ILT), Stanford University, and the University of Michigan. In essence, these processes, while differing in certain details, are all variations on the theme of laser powder deposition. A powder feeder is used to inject metal powder directly into a laser beam. The metal powder is entrained in the beam and melted. It is then deposited on a substrate. By appropriate substrate motion, the melted metal is deposited in layers to create a component. Figure 4 shows a Laser Engineered Net Shaping (LENS) system, a variant developed by Sandia National Laboratory, in operation.

Laser metal deposition has some important advantages. It can be used with a wide range of engineering materials, although a protective atmosphere may be required for materials that oxidize readily. The small spot size and high cooling rates achievable with laser deposition mean that materials that may be difficult to produce using other routes can be fabricated. For example, nonequilibrium microstructures that demonstrate extended solid solubility can be produced. Powders can also be mixed and fed, allowing alloying in the beam. High cooling rates also result in a fine-grained microstructure. Studies have shown that full density can be achieved, as well as excellent mechanical properties, making the resulting components suitable for production use in demanding applications such as aircraft or propulsion.

In Germany, the Fraunhofer Institutes have been investigating laser powder deposition processes, in particular in combination with milling to improve the accuracy of the finished parts. Figure 5 shows a complex laser-deposited component produced at Fraunhofer ILT. Surface quality and dimensional accuracy of laser DMD parts, produced without finish machining, is limited by both the dimensional variability resulting from solidification shrinkage and the surface ripple produced by the action of gravity on molten metal as it is deposited. Machining, following deposition of one or more layers of material, can eliminate this, although it reduces still further the already low deposition rate characteristic of laser powder deposition.

Laser direct metal deposition is being widely commercialized. Several business strategies are being employed. In North America, Optomec is producing machines for performing laser metal deposition, while Fockele and Schwarz of Germany produce machines for laser deposition combined with machining. Precision Optical Manufacturing (NA) is providing laser metal deposition services to produce parts and repair existing tools and dies. In addition, laser powder deposition is being used on a very large scale to produce commercial preforms for the

Figure 5
Stainless steel (316) rotor with C45 hub created via laser powder deposition (from Fraunhofer ILT website).

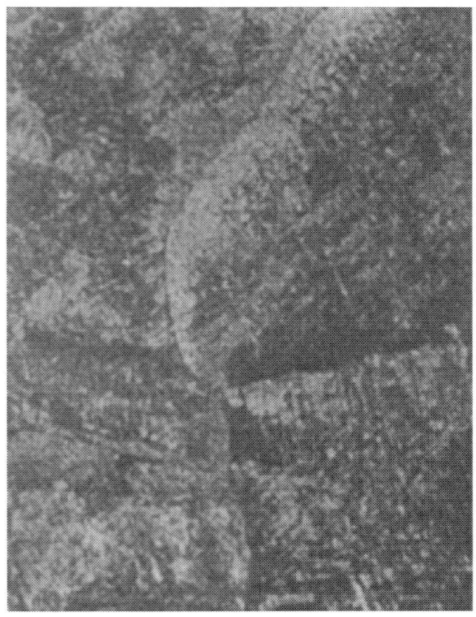

Figure 6
Laser deposited Ti6Al4V 100 x (from Aeromet website).

aerospace industry. Aeromet Inc. is using an 18 kW CO_2 laser to produce components up to $3\,m \times 3\,m \times 3\,m$ in size. Achieving net shape is not an objective of this process. However, by minimizing the amount of finish machining required, economic benefits over competing processes such as forging can be obtained. Figure 6 shows the microstructure of laser deposited Ti6Al4V.

2.2 SDM

Shape Deposition Manufacturing (SDM) developed by Fritz Prinz at Stanford University is a variation on laser powder deposition, in which powders are melted and deposited by laser beam (Link *et al.* 1998). In SDM, unlike previously described processes, a support material is used to ensure that overhangs and complex features can be produced as shown in Fig. 7. For engineering metals, a copper alloy has been used, which must be removed following fabrication using a caustic solution. Subtractive processes are also combined with deposition to achieve accurate net shape production. By milling following deposition, both surface finish and accuracy are greatly improved over simple laser powder deposition. SDM provides excellent mechanical properties and dimensional accuracy. It has been used successfully to produce die casting tooling (Link *et al.* 1998). This process has not been commercialized, but research is continuing on novel applications for structures on the mesoscopic scale.

2.3 Selective Sintering Laser

Selective laser sintering (SLS) is a process widely used to produce components in plastic, wax, sand, rubber, etc. In this process, a laser is rastered over a powder bed, layer by layer, to produce a part. In metals, two approaches are used; one is to cover the metal powder with a thermoplastic binder that creates bonds between the metal powder following application of the SLS process. The binder is burned off and a second material is infiltrated into the green part to achieve full density. The second approach is partially to melt the metal powders via SLS processing. One advantage of this process is that it is self supporting and the powder is easily removed following part fabrication; an important advantage in metal part fabrication. Unfortunately, it is difficult to produce fully dense parts using this process with either of the techniques described above. Secondary processes such as infiltration must almost always be used to produce a fully dense part (Lakshminarayan *et al.* 1995).

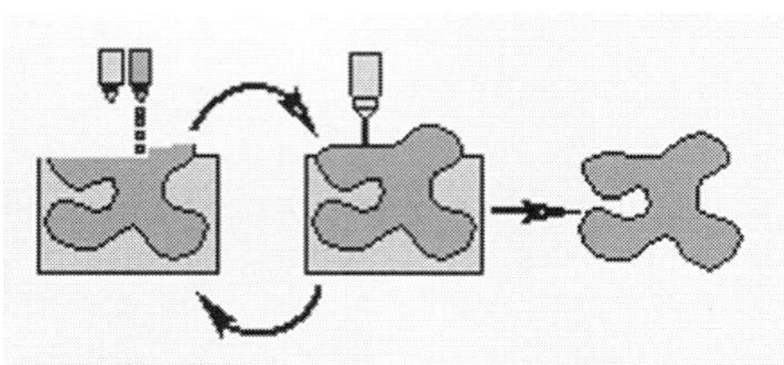

Figure 7
Schematic of SDM process (from Stanford RP laboratory website).

In addition, research is being conducted on the use of selective sintering to produce HIPing cans. The cans are produced to shape using SLS, then filled with powder, and closed. Following the densification process, the can and fill powder are completely fused, and several steps can be eliminated from the conventional HIPing process (Das *et al.* 1999).

2.4 Three-dimensional Printing

Three-dimensional inkjet printing (3DP) was developed at MIT and has been commercialized for a range of engineering and prototyping applications. Extrude Hone Corporation has licensed this technology from MIT and has applied it to metals. The process, known as ProMetal, creates a solid object by printing a sequence of two-dimensional layers. Figure 8 shows a set of injection molding tooling built using 3DP technology.

Each layer is built by spreading a thin layer of powdered material. The powder is then selectively joined together by inkjet printing a binder material. The powder bed is lowered at the completion of each layer and a new layer of powder is spread. Once the complete part has been built by this printing process, the loose powder is removed to leave the green part. The polymeric binder is then burned out, an operation that also lightly sinters the part. The green part is then infiltrated with a copper alloy resulting in a fully dense component. Finishing operations such as machining or polishing are then performed. This process builds the green part relatively rapidly and a fairly wide range of metal powders can be employed. However, the infiltration operation must be performed in a controlled-atmosphere furnace, and the process is subject to the same dimensional accuracy and surface finish problems which plague other infiltration processes. The ProMetal process is

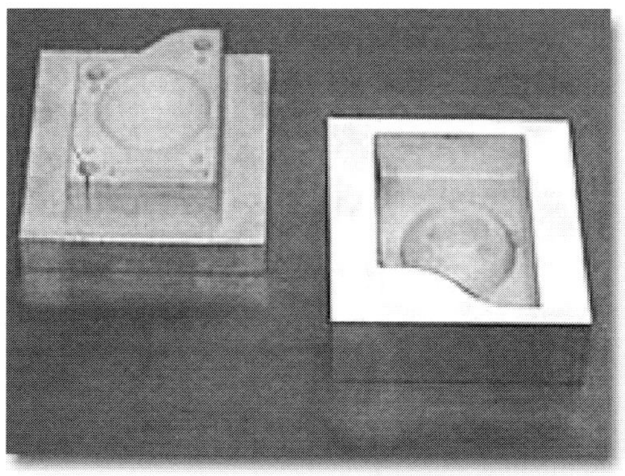

Figure 8
Injection molding tools produced using prometal process (from Extrude Hone website).

available commercially as a complete system and has had limited market acceptance.

2.5 Three-dimensional (3-D) Welding

Development of 3-D welding technology goes back to the 1980s, when it was referred to as shape melting (Bronowski 1986, Edmonds *et al.* 1998) and considered as an alternative manufacturing process for large, difficult-to-form components such as rings and other parts for turbine engines. More recently, other investigators have developed near net shape object technologies based on depositing weld metal sequentially. The process is similar in some respects to laser powder deposition, except that a wire feed stock, melted using a Gas Metal Arc Welding (GMAW) torch rather than a laser, is employed. The weld metal is deposited on a substrate and then onto previous layers to produce a final object. The deposition rate and deposit width are higher in 3-D GMA welding than in laser powder deposition, so achieving dimensional accuracy is more difficult. As a result, techniques for controlling droplet size and deposition rate, and for combining metal deposition with machining to ensure more dimensional accuracy are being developed (Beardsley and Kovacevic 1999, Zhang *et al.* 1998).

Three-dimensional welding has the advantages of offering lower equipment and consumables costs as well as a higher deposition rate than LPD. However, the requirement for a wire feed stock is somewhat limiting, since only materials that can be drawn to the required dimensions can be employed. The use of cored wires could improve this situation, but no research into the potential benefits of this approach appears to have been conducted to date.

2.6 Vapor Deposition Processes

A number of direct and indirect processes based on vapor deposition technologies have been investigated and demonstrated as potential means of producing net shape objects in metal. Many of these processes are being scaled-up from applications in microelectronics fabrication. For indirect applications, a coating is typically deposited over a mandrel of the desired shape and appropriate properties, and allowed to grow to the desired thickness. Processes suitable for direct fabrication, such as laser CVD, may be used without mandrels to produce materials directly using the same types of rastering techniques employed for other processes, such as stereolithography, selective laser sintering, etc.

Physical vapor deposition (PVD), CVD, laser CVD, electron beam PVD, and selective area laser deposition vapor infiltration (SALDVI) are among the technologies that have been applied to rapid prototyping in metals (Jakubenas *et al.* 1997, Duty *et al.* 1999, He *et al.* 2000). Generally speaking, these processes have low build rates and require fairly complex equipment with careful control of the composition of consumables and atmosphere. No commercially available systems have been marketed based on these techniques. It seems likely that the benefits of these processes, including: the potential to fabricate extremely small objects; to tightly control composition and microstructure; and to fabricate functionally gradient materials, may be best realized in the fabrication of mesoscale articles, which are an emerging application for additive manufacturing technologies.

3. Summary

"Additive net shape processing" is a more descriptive and specific term than rapid prototyping for the technical and economic objectives of most of the users of rapid prototyping technology. As described above, a wide range of processes have been applied to additive net shape processing for metals, with varying technical and commercial success. It is clear that innovation in this field is far from over; new processing routes are being researched on an ongoing basis. A simple thought experiment can be used to illustrate the flexibility of the incremental, additive approach to metal object fabrication. Select a materials processing technique and then conceptualize a method to perform it incrementally, with small volumes of material. By combining techniques now in use for rapid prototyping, such as deposition methods, masking techniques, support structures, etc., with the selected process, a route for producing near net shape objects can virtually always be envisioned. Economic, production rate, or accuracy considerations may render development of the technique nonproductive, but most bulk processing approaches can be applied incrementally. Further development of high deposition rate approaches, wider materials selection, and accurate mesoscale device fabrication techniques are among the advances that are certain to occur in this field in the first decade of the twenty-first century.

Bibliography

Agarwala M, Bourell D, Beaman J, Marcus H, Barlow J 1995 Post-processing of selective laser sintered metal parts. *Rapid Prototyping J.* **1**, 36–44

Beardsley H E, Kovacevic R 1999 New rapid prototyping technique based on 3-D welding. *SME Technical Paper (Serials)*, PE99-126-1–PE99-126-6

Bronowski J 1986 *US Pat.* 4 621 762

Chaneski W 1997 Rapid tooling technology: a new competitor. *Modern Machine Shop* **70**, 50, (2)

Das S, Beaman J J, Wohlert M, Bourell D L 1999 Free-form fabrication of high performance titanium components via SLS/HIP. *Proc. Materials Research Society Fall Meeting 1998, Symposium V*. Vol. 542, pp. 45–50

Duty C E, Jean D L, Lackey W J 1999 Design of a laser CVD rapid prototyping system. *Ceramic Engineering and Science Proceedings* **20** (4), 347–54
Edmonds D 1998 US Pat. 4 775 092
He Z, Zhou J G, Tseng A A 2000 Feasibility study of chemical liquid deposition based solid free-form fabrication. *Materials and Design* **21** (2), 83–92
Jacobs P 2000 Effects of random noise shrinkage on rapid tooling accuracy. *Materials and Design* **21** (2), 127–36
Jacobs P F Stereolithography and other RP & M technologies. In: *Rapid Prototyping to Rapid Tooling*. ASME Press, USA
Jakubenas K F, Lee Y L, Shaarawi M S, Marcus H L, Sanchez J M 1997 Selective area laser deposition of titanium oxide. *Rapid Prototyping Journal* **3** (2), 66–70
Jordan R, Roche A 1999 *US Pat*. 5 952 056
Lakshminarayan U, McAlea K, Girouard D, Booth R 1995 Manufacture of iron–copper composite parts using selective laser sintering. *Advances in Powder Metallurgy and Particulate Materials*. Metal Powder Industries Federation, Princeton, NJ, Vol. 3, pp. 13/77–13/85
Link G R, Fessler J, Nickel A, Prinz F 1998 Rapid tooling die case inserts using shape deposition manufacturing. *Materials and Manufacturing Processes* **13** (2), 263–74
McHugh K M 1998 Microstructure transformation of spray-formed H13 tool steel during deposition and heat treatment. In: Marsh S P, Dantzig J A, Trivedi R, Hofmeister W, Chu M G, Lavernia E J, Chun J H (eds.) *Solidification*. The Minerals, Metals, and Materials Society, p. 427
McHugh K M 1999 Fabrication of tooling inserts using RSP tooling technology. *Proceedings of Moldmaking '99*. Communication Technologies, Columbus, OH, p. 383
Zhang Y M, Liguo E, Kovacevic R 1998 Active metal transfer control by monitoring excited droplet oscillation. *Welding Journal* **77** (9), 388s–95s, Conformal cooling reference.

D. White

Rapid Solidification Processing

The advent of rapid solidification of metallic melts at rates of 10^6 K s^{-1} by Pol Duwez (Duwez *et al.* 1960) heralded a new era of materials processing. It was shown that rapidly solidified (RS) materials exhibit increased solid solubility limits, refined microstructural features, and formation of entirely new and totally unexpected non-equilibrium phases such as amorphous phases. This curiosity-driven research of the 1960s witnessed the development of novel processing methods to produce continuous filaments and ribbons of varying dimensions (up to a few centimeters wide) during the 1970s, and maturing into applications for RS products (e.g., transformer core laminations) in the 1980s. The 1980s were also witness to the emergence of quasicrystalline phases as a new set of ordered phases displaying the traditionally forbidden crystal symmetries. During the same period other non-equilibrium processing techniques were also developed to synthesize non-equilibrium or metastable phases, including metallic glasses (Suryanarayana 1999) and aluminum-based amorphous alloys and composites that have tensile strengths approaching 1 GPa. The bulk metallic glasses with a section thickness of a few tens of millimeters, as opposed to a few tens of micrometers in conventional RS alloys, made their appearance in the late 1980s and early 1990s. Thus, while the exploitation of the potential of the rapid solidification processing (RSP) of metallic melts in the 1960s led to publications dealing with a scientific understanding of the extension of solid solubility limits and formation of metastable crystalline and amorphous phases, study of the mechanical, magnetic, and superconducting properties of RSP amorphous alloys were the major preoccupation of the scientists during the 1970s. The 1980s and 1990s saw an explosion of research activity in the field of quasicrystalline phases and bulk metallic glasses.

1. *Techniques for Rapid Solidification Processing*

Since the "gun" technique of quenching metallic melts was introduced in 1960, a number of techniques have been developed to produce RS materials, and these have been summarized by Suryanarayana (1991). The number of techniques (and their variants) is too numerous to list all of them here. But, these have been grouped under different headings by different investigators. Following Jones (1999), these can be listed as (i) droplet, (ii) spinning, and (iii) surface melting methods.

In the droplet methods, the melt is disintegrated into several droplets, which then solidify on a thermally conducting substrate. The process of disintegration of the melt and the method of solidification differ in the different methods. The "gun" technique and the atomization methods fall under this category. The atomization technique has been more popular in recent years for producing large quantities of the rapidly solidified material. However, the solidification rate achieved in the normal atomization processes is low (10^2–10^3 K s^{-1}) (see *Atomization*), and higher rates can be achieved in the "gun" technique (10^6 K s^{-1}) and in some special atomization methods.

The spinning methods involve solidifying a melt stream emerging from an orifice into the form of a continuous strip or ribbon before it can break up into droplets. The free jet impinges on a fast rotating and thermally conducting substrate and solidifies rapidly. The chill block melt spinning, melt-extrusion, and the planar flow casting methods (used to produce wide ribbons of up to about 500 mm width) fall under this category. The solidification rates in this category are typically about 10^5–10^6 K s^{-1}, realizing that these depend on the section thickness, thermal conductivity of the substrate, and the degree of adherence of the liquid to the substrate.

The surface melting methods involve melting a thin surface layer of the material (usually by traversing a laser beam), which then solidifies rapidly due to the rapid extraction of heat into the unmelted bulk (Kobayashi 1999). The solidification rates in this category can reach very high values of 10^{14} K s^{-1}. The reader is advised to refer to Suryanarayana (1991) and Jacobson and McKittrick (1994) for further details of the different techniques. These three groups of rapid solidification methods are shown schematically in Fig. 1.

2. *Non-equilibrium Effects of Rapid Solidification Processing*

The technique of RSP was developed to synthesize materials and alloy phases under non-equilibrium conditions. Accordingly, all the three major constitutional effects of RSP, viz., solid solubility extensions (formation of continuous solid solutions in the Cu–Ag system), formation of metastable intermediate phases (occurrence of the missing h.c.p. Hume–Rothery phase in the Ag–Ge system), and synthesis of amorphous alloys (metallic glasses) (in the Au–Si eutectic alloy) had been reported in 1960, the very first year of the advent of the technique of rapid solidification. Microstructural refinement (of grain size, second-phase particles, and segregation patterns) leading to improved properties was also reported in later years. Improvement in processing techniques led to production of continuous ribbons, thus paving the way for commercial exploitation of rapidly solidified materials. Subsequently, quasicrystalline phases, with the crystallographically forbidden translational symmetries, were discovered in 1984 (Shechtman *et al.* 1984). The exciting developments in this area of RSP have been documented at periodic

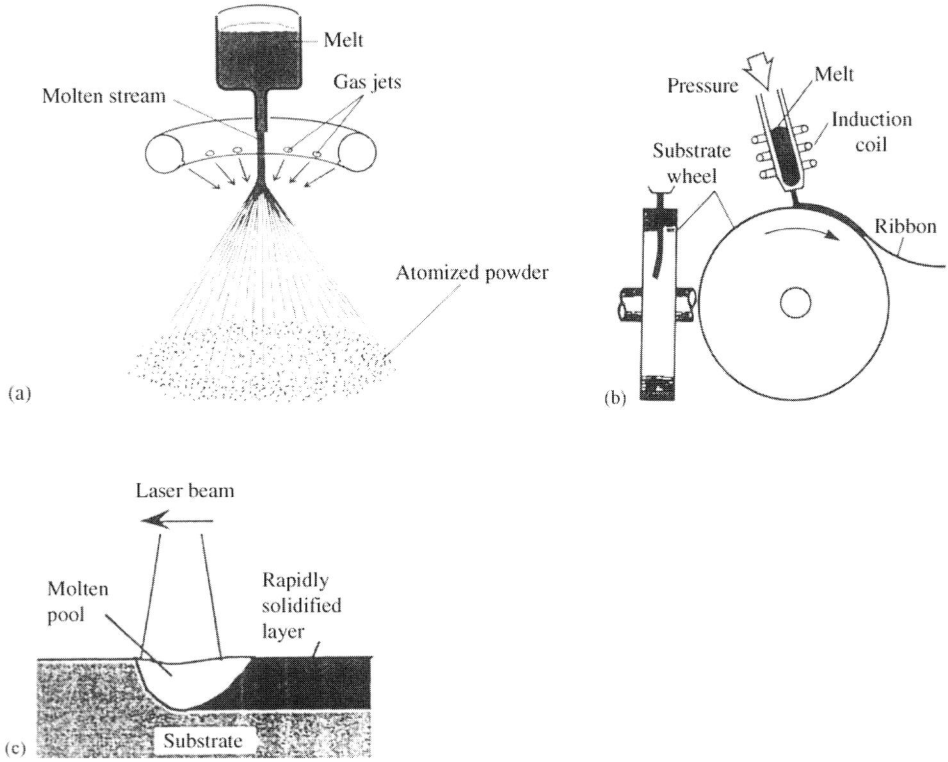

Figure 1
Typical methods to rapidly solidify materials. (a) Droplet method, in which the melt is disintegrated into small droplets which are then allowed to solidify either against a conducting substrate (to achieve high solidification rates, as in the "gun" technique) or by directed gas jets (when the solidification rates are low, as in atomization methods). (b) Spinning method, in which the molten metal is forced through on orifice into the form of a continuous stream and allowed to solidify on a fast rotating thermally conducting substrate. (c) Surface melting method, in which a thin surface layer is melted by a laser beam and is subsequently solidified due to rapid heat extraction by the unmelted bulk.

intervals (Anantharaman and Suryanarayana 1987, Suryanarayana and Jones 1988, Liebermann 1993, Jones 2001).

We will, however, concentrate in this article on the extension of solid solubility limits achieved by rapid solidification processing and its exploitation for producing commercially useful and improved products. We will also discuss the issues related to consolidation of rapidly solidified materials.

3. *Solid Solubility Extensions*

Solid solution alloys show many desirable properties such as increase in strength and modulus, changes in density depending on the density of the solute element, and elimination of undesirable second phases. Uniform precipitation of fine second-phase particles in a metallic matrix has long been known to increase the strength and hardness of precipitation-hardening alloys. Thus, increases in solid solubility limits in alloy systems confer many desirable effects. Solid solubility extension was the first non-equilibrium effect observed in RS alloys. Perplexed by the anomaly that, even though all the four Hume–Rothery rules (similar atomic sizes, close electronegativity values, same valence, and same crystal structure) are satisfied between copper and silver, the Cu–Ag system did not show complete solid solubility of silver in copper and copper in silver, Duwez rapidly solidified the molten Cu–Ag alloys with different solute content by the "gun" technique at solidification rates of about 10^6 K s^{-1}. While the solid solubility of silver in copper is only 4.9 at.% and that of copper in silver 14.1 at.% at eutectic temperature of 1052 K, the rapidly solidified Cu–Ag alloys exhibited complete solid solubility of one element in the other in the entire composition range. Starting with this pioneering investigation, the extension of solid solubility limits has been repeatedly demonstrated in a number of RS alloy systems (Anantharaman and Suryanarayana 1971). Complete solid solution formation is not achieved in all RS alloys; however, significant extensions of solid solubility limits have been reported in several alloy systems. For example, the solid solubility of iron in aluminum has been reported to increase by RSP to 6 at.% from the room temperature equilibrium value of 0.026 at.%. We will refer to this aspect again later in Sect. 7. Even though the frequency of reports dealing exclusively with solid solubility extensions is nowadays rather limited, solid solubility extensions are still being reported. However, the major difference is that these solid solubility extensions are being exploited to improve the product performance.

Table 1 lists the maximum solid solubility extensions achieved in some of the rapidly solidified aluminum- and magnesium-base alloy systems.

3.1 General Features

Some general conclusions can be drawn from the available results. Firstly, the solid solubility extension appears to be a ubiquitous phenomenon, even though the actual magnitude of increase is different in different alloy systems. There are very few alloy systems in which solid solubility extensions have not been reported. Secondly, the solid solubility limits can be increased to higher levels at higher solidification rates. For example, the solid solubility levels in alloys rapidly solidified by the melt spinning technique are higher than those obtained by atomization techniques. Even in the "gun" technique, for example, the thinner foils, which have solidified at higher rates of solidification, show a higher amount of the solute in solid solution than the thicker foils. Having mentioned this, it is only appropriate to point out that systematic studies of the attained solid solubilities in various alloy systems have not been conducted. Thirdly, apart from the actual solid solubility extension achieved, it may be noted that elements, which have a very limited solid solubility under equilibrium conditions, show much extended limits under non-equilibrium conditions. Thus, the extended solid solubility limit, in some cases, can be as much as three orders of magnitude higher than the equilibrium value. Lastly, theoretical predictions of solid solubility limits achieved by RSP methods are woefully inadequate. Most of the investigators report the solid solubility of the alloy after it has cooled down to room temperature, but the theoretical models consider the solid solubility achieved immediately after solidification is completed. This could be an important reason for some of the differences between the reported and estimated solid solubility values.

Table 1
Selected maximum solid solubility extensions reported in rapidly solidified aluminum- and magnesium-base alloys.

		Maximum solid solubility (at.%)	
Solvent	Solute	Equilibrium	Extended
Aluminum	Ag	23.8	40
	Au	0.041	0.35
	Ce	0	1.9
	Co	0.01	5
	Cr	0.38	>6
	Cu	2.5	18
	Fe	0.026	6.0
	Ga	8.8	65.0
	Ge	2.8	33
	Hf	0.183	1.03
	Ir	0	0.29
	Mg	18.9	36.8
	Mn	0.7	>9
	Mo	0.07	1
	Nb	0.067	2.35
	Ni	0.023	7.7
	Pd	0	7.5
	Pt	0	2.2
	Rh	<0.1	0.5
	Ru	0.027	4.5
	Si	1.59	16
	Sn	0.02	0.26
	Ta	0.038	0.94
	V	0.18	2
	W	0.024	1.9
	Zr	0.08	2.2
Magnesium	Al	11.5	22.6
	Ca	0.98	6.5
	Ce	0.09	3.15
	Mn	1.55	2.5
	Y	3.5	9.7
	Zn	2.8	4.8
	Zr	0.2	0.32

3.2 Difficulties in Estimating Solid Solubility Limits

There are generally two methods employed to estimate the solid solubility limits; both utilize the x-ray diffraction technique. The first is by noting that the RS alloy contains only a single-phase solid solution (i.e., the absence of a second phase). The accuracy of this method depends on the detection efficiency of the second phase by the x-ray diffraction technique. Conducting transmission electron microscopy investigations on the RS alloy and ensuring that there is no second phase present in the microstructure of the alloy can overcome this problem. In the second method, the lattice parameter of the solid solution phase is measured as a function of the solute content. The measured lattice parameter is plotted against solute content and the (super)saturation limit is then estimated by noting the solute composition beyond which the lattice parameter does not vary. Alternately, it can be estimated by extrapolating the standard plots of lattice parameter versus solute content available in standard reference books. Thus, the accuracy of the solid solubility values is determined primarily by the slope of the lattice parameter versus solute content plot. This problem is further exacerbated in many cases where the equilibrium solid solubilities are rather limited, and in these cases extrapolations to higher solute contents can be erroneous. Another problem that has been encountered is that the slope of the lattice parameter versus solute content for the rapidly solidified alloys is different from that of the equilibrium alloys.

The maximum solid solubility extension obtained in an alloy is determined by the metastable equilibrium between the extended solid solution and the second phase present in the rapidly solidified alloy, which, depending on the alloy system under consideration, could be (i) an equilibrium intermediate phase, (ii) a non-equilibrium intermediate (crystalline) phase, or (iii) an amorphous phase.

3.3 Models

Equilibrium solid solubility limits have been explained on the basis of the Hume–Rothery rules for solid solution formation. The atomic size factor is considered as the primary responsible factor for this and reasonable success has been achieved using this criterion. The ability to explain the solubility levels is further enhanced by the Darken–Gurry plots, wherein the atomic size (corrected to a coordination number of 12) is plotted against electronegativity. In this case, the more soluble elements cluster around the solvent in the form of an ellipse within the range of $\pm15\%$ of the atomic size and ±0.3 of the electronegativity. Even though this appears to be true in some cases, the Darken–Gurry plot is not able to satisfactorily explain the achieved solid solubility extensions in aluminum alloys (Fig. 2). Note, for example, that copper and magnesium with extended solubilities of 18 at.% and 36.8 at.%, respectively, are outside the ellipse. Better success has been achieved in separating the more soluble and less soluble elements by plotting the semi-empirical parameters to heat of formation and of mixing, or the Wigner–Seitz radius against heat of solution. Only some of these empirical relationships are suited to explain the solubility levels achieved in rapidly solidified alloys, and as such their utility is limited.

Some thermodynamic attempts have been made to predict the maximum solubility values under RSP conditions. These investigations use the concept of T_0, the temperature at which the molar free energies of the liquid and solid phases are equal for the composition of interest. The locus of T_0 over a range of composition constitutes a T_0 curve and represents useful thermodynamic information.

A use of these curves is to determine whether a limit exists for the extension of solid solubility by

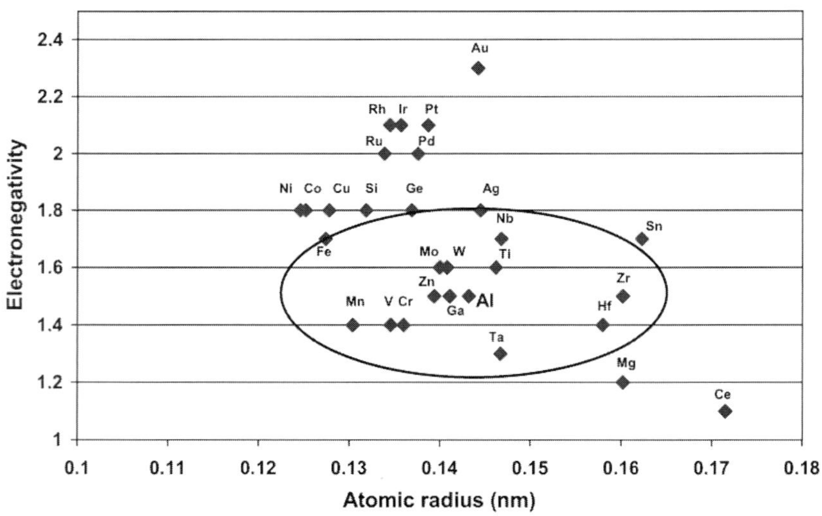

Figure 2
Darken–Gurry (electronegativity vs. atomic size) plot for aluminum alloys. The ellipse represents a deviation of ±0.3 units in electronegativity and $\pm15\%$ of the atomic size of aluminum. Note that elements that show large amounts of solid solubility in aluminum (Table 1) are expected to be inside the ellipse, but they are not.

RSP methods. The T_o curve lies between the liquidus and solidus temperatures and three different possibilities can be visualized as shown for a eutectic phase diagram in Fig. 3. In Fig. 3(a), the T_o curve is continuous from the pure A to the pure B component and the minimum in the T_o curve exists above room temperature. In this situation it is possible to obtain complete solid solubility of A in B (and B in A) by rapidly solidifying the liquid alloy to a temperature below T_o, i.e., massive or partitionless solidification takes place. The Cu–Ag system falls under this category. The T_o curve plunges to very low temperatures in Fig. 3(b). In this situation, the maximum solid solubility extension is limited by the T_o curve, i.e., at compositions to the right of the T_o curve on the A-rich side and to the left of the T_o curve on the B-rich side it will not be possible to obtain massive solidification. The majority of the systems which do not exhibit complete solid solubility under non-equilibrium conditions come under this category. Alloy compositions in the center of these two extremes are good candidate materials for glass formation, provided the alloy is solidified without crystallization to below the glass transition temperature T_g. In alloy systems with retrograde solubility under equilibrium conditions, the T_o curve plunges to absolute zero at a composition corresponding to the liquidus composition at the retrograde temperature. Thus, it is impossible to achieve any extension of solid solubility limit. The Cd–Zn system comes under this category. The third possibility arises when the T_o curves for the A-rich and B-rich alloys intersect in the middle of the phase diagram (Fig. 3(c)), i.e., when the T_o are only slightly depressed below the stable liquidus temperature. In this situation, significant solid solubility extension can occur, but not complete solid solubility of one element in the other.

4. *Metastable Intermediate Phases*

Metastable crystalline intermediate phases synthesized by RSP have been investigated, mostly in the early years of RSP technology. Whereas some of the missing Hume–Rothery phases have been detected in some RS alloy systems, a great number of other crystalline intermediate phases have also been detected. A list of the different crystalline intermediate phases, current to that date, has been included in some of the earlier reviews (see, for example, Anantharaman and Suryanarayana 1971). Since these phases, by themselves, did not appear to have any commercial applications, interest in detailed studies

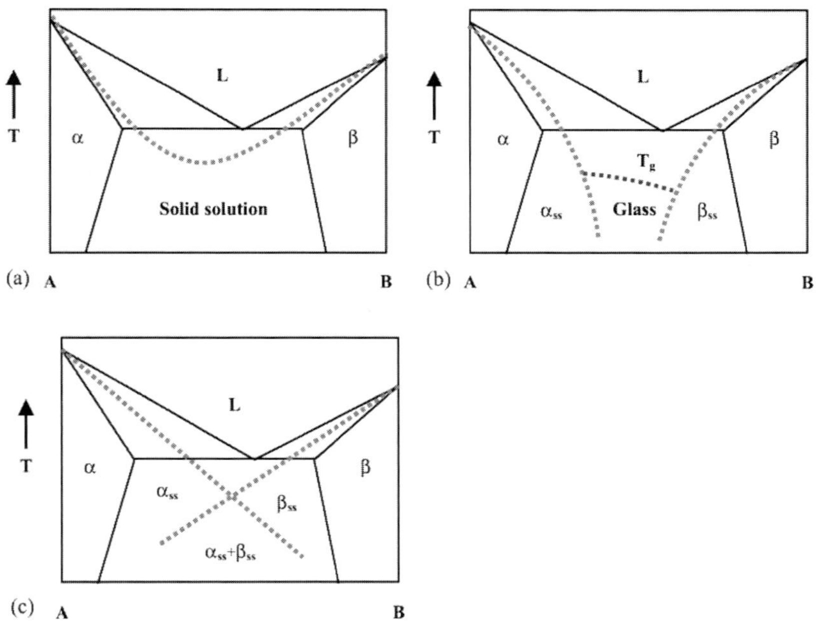

Figure 3
(a)–(c) Schematic representation of the T_o curves (dashed lines) for the liquid-to-solid transformation during RSP of three different types of alloys exhibiting a eutectic reaction. In these diagrams α and β represent the equilibrium solid solutions of B in A and A in B, respectively, α_{ss} and β_{ss} represent the supersaturated solid solutions, and T_g the glass-transition temperature. In (b) and (c), the dashed lines represent the T_o lines for the α/L on the A-rich side and β/L on the B-rich side, respectively.

of these has decreased. However, the detection of a quasicrystalline phase in rapidly solidified Al–Mn alloys in 1984 (Shechtman *et al.* 1984) led to a revival of research activity in this area mainly due to their intriguing crystallographic features: the presence of the crystallographically forbidden 5-fold, 10-fold, and other symmetries (Suryanarayana and Jones 1988, Kelton 1993). There was a plethora of papers for the next ten to fifteen years on different aspects of quasicrystalline phases (formation, structure, decomposition, relation to the corresponding crystalline phases, and synthesis of stable quasicrystals), but since no major applications were found, except as coatings on frying pans, research in this class of materials declined.

5. *Metallic Glasses*

The formation of a noncrystalline phase in a rapidly solidified Au–Si eutectic alloy in 1960 by Duwez and his co-workers was a major unexpected discovery of the potential of RSP. Since then a very large number of noncrystalline phases (referred to as metallic glasses) have been synthesized. It has been noted that iron-based metallic glasses have excellent soft magnetic properties and therefore they have been in commercial use for transformer core laminations. The formation, characteristics, and applications of metallic glasses have been covered in detail in some earlier reviews (Cahn 1991, Cahn and Greer 1996). So far, the metallic glasses produced by RSP are the only materials that have become commercially successful. Since this aspect has also been covered in detail elsewhere in this encyclopedia (see Metallic Glasses), it will not be further discussed here.

Synthesis of metallic glasses requires high critical cooling rates and consequently, the section thickness is limited to a few micrometers. However, a proper choice of the constituent elements and their composition can drastically reduce the critical cooling rate for glassy phase formation to as low as $0.1\,K\,s^{-1}$. These, referred to as bulk metallic glasses, are now produced up to a section thickness of 72 mm (Inoue *et al.* 1997). The bulk metallic glasses exhibit high mechanical strength, high fracture toughness, and good corrosion resistance. A homogeneous dispersion of fine nanoscale particles into a zirconium-base bulk glassy matrix has been shown to further increase the tensile strength without reduction in ductility. Further, the large temperature difference between the glass transition and crystallization temperatures allows these materials to be processed by high-strain-rate superplastic forming methods into complex shapes. The only application of these bulk metallic glasses at present is as face materials in golf clubs. Further details of bulk metallic glasses may be found in Johnson (1999) and Inoue (2000).

6. *Consolidation of Rapidly Solidified Alloys*

Rapidly solidified alloys are often fine-grained materials containing metastable phases such as supersaturated solid solutions, or intermediate crystalline or quasicrystalline phases, and in some cases, the phase is completely amorphous in nature. Thus, consolidation of these alloys is a nontrivial issue if one is interested in maintaining the metastable phases and/or fine microstructures.

Widespread application of RSP materials requires efficient methods of consolidating them (either in the form of powder or ribbons chopped into small flakes) into useful shapes. Conventional consolidation of powders to full density through processes such as cold pressing followed by high-temperature sintering, hot forging, hot extrusion, and hot isostatic pressing normally requires exposure of these materials to high pressures and high-temperatures for extended periods of time. Under these conditions, diffusion takes place rapidly resulting in the transformation of the metastable effects achieved through RSP. Accordingly, precipitation of metastable or stable precipitate phases may occur in supersaturated solid solutions, the metastable crystalline and quasicrystalline phases may transform into equilibrium phases, the nanocrystalline materials may coarsen, and the metallic glasses could crystallize. Consequently the benefits of RSP are lost. Thus, innovative methods of consolidation are required (Groza 1999). The goal of such consolidation methods is to attain essentially 100% theoretical density while preventing or minimizing transformation of the metastable phases.

Rapidly solidified materials are usually consolidated using a multistep approach. The RSP material is cold compacted to 70–80% of theoretical density and placed in a container, which is then vacuum degassed and hot processed. The vacuum-degassing step is necessary to remove the adsorbed gases and water vapor. Internal gas evolution can result in large voids and destruction of the workpiece. Further, RSP aluminum alloys contain a thin hydrous oxide layer formed during atomization, which thickens during subsequent handling. The surface oxide layer can be broken down before or during final consolidation, when the broken oxide particles get uniformly distributed in the form of stringers along the prior particle boundaries. Extrusion appears to be the most effective process to achieve this because the shear forces during extrusion easily break down the oxide layer. Beyond some critical strain level, prior particle boundaries are not observed, and oxide stringers are not continuous.

RSP materials have traditionally been consolidated using hot isostatic pressing methods. However, shock-wave consolidation has been found to be very effective in retaining the metastable phases (Schwarz 1993) and nanocrystallites (Suryanarayana and Korth 1999). Densification and bonding of particles is achieved in this process by passing a shock wave through the powder. As the shock front penetrates the powder the material is compressed to a pressure of several gigapascals, which is much higher than the maximum shear stress that can be developed within the material, and reaches full density in a few microseconds. A pseudo-HIP process called the Ceracon (Ceramic consolidation) process, which develops much higher pressures than in a normal HIP process, has also been investigated, but with mixed results.

7. *Applications of Rapidly Solidified Alloys*

Rapidly solidified alloys have been in existence since the 1960s. A number of practical and potential applications have been envisaged for these materials and the earlier applications for both crystalline and amorphous alloys may be found in Anantharaman and Suryanarayana (1987) and Liebermann (1993). Therefore, the literature subsequent to 1993 only will be considered here.

Apart from the basic scientific interest in exploring the maximum possible limits of solid solubility, the supersaturated solid solutions obtained by RSP have been found to be technologically useful. Major emphasis for these studies has been on producing large quantities of RS alloys using techniques such as atomization and consolidating them into bulk shapes. The RSP route allows a much higher amount of solute to be contained in the solid solution. Such supersaturated solid solutions have been exploited as starting materials to synthesize high-strength alloys containing a large amount of fine dispersoids. An example is the increased solid solubility of lithium in aluminum, which results in a 3% decrease in density and 6% increase in Young's modulus for each percent of lithium present in the solid solution. This, coupled with the microstructural refinement, results in an alloy with much higher stiffness and adequate toughness.

Rapidly solidified aluminum alloys intended for high-temperature applications are based on Al–Fe alloys containing ternary and quaternary additions such as cerium, molybdenum, vanadium, or vanadium and silicon (Davis *et al.* 1994). While the gas-atomized and consolidated Al–Fe–Mo and Al–Fe–V alloys have been found to be stable up to about 590 K (the metastable precipitates responsible for strengthening transform to the equilibrium phases above this temperature), the planar flow cast Al–Fe–V–Si alloys have been found to retain their high-temperature strength up to at least 700 K. The quaternary alloys derive their excellent strength, increased stiffness, and stability from a uniform dispersion of metastable $Al_{13}(Fe,V)_3Si$ particles which remain in equilibrium with the aluminum matrix to temperatures in excess of 773 K. The volume fraction of the silicide dispersoid could be as much as 37 vol.% and the coarsening rate for this dispersoid has been measured to be three to four orders of magnitude slower than for dispersoid phases present in Al–Fe–Ce and Al–Fe–Mo–(V) alloys.

A number of investigations have been directed to a study of RS amorphous Al–Ni–RE (RE = rare-earth) alloys. These have tensile strengths of the order of 1 GPa, which could be further increased by about 50% by allowing fine nanocrystalline α-Al precipitates to form by a suitable heat treatment (Inoue 1998). The Young's modulus of these alloys is also very high (about 90 GPa). Because the RE elements are expensive, substitution by the cheaper misch metal (containing a number of the rare-earth metals) has been found to induce similar properties. These composite materials have high fatigue strength (310 MPa), low coefficient of thermal expansion (20×10^{-6} K^{-1}), good wear resistance, and good thermal stability. The high strength of these alloys seems to be due to a combination of Orowan, Hall–Petch, and solid solution strengthening in the proportion 50:40:10.

Hypereutectic Al–Si alloys (with additional alloying elements like iron, magnesium, manganese, copper, etc.) are good candidate materials for automotive and electronic applications due to their high wear resistance and reduced coefficient of thermal expansion. However, the presence of coarse and hard primary silicon particles reduces the machinability. RSP of Al–Si alloys produces a fine and uniform distribution of the silicon particles in the aluminum matrix and thus overcomes this problem. The wear resistance of these alloys is also higher than that of the conventional alloys.

Inoue *et al.* (2001) have reported development of a high-strength magnesium alloy obtained in a bulk form by warm extrusion of atomized powders of Mg–1 at. % Zn–2 at.% Y alloy at 573 K. This alloy consisted of a solid solution possessing a long periodic (six-layer ABACAB) hexagonal structure and a fine grain size of 100–150 nm in which about 10 vol.% of 7 nm size cubic $Mg_{24}Y_5$ precipitates were dispersed. This composite alloy was reported to have a room-temperature tensile strength of 610 MPa, which is 2.5–5 times higher than those of conventional high-strength magnesium alloys. A reasonably high strength of 400 MPa was maintained at temperatures up to 473 K. This alloy also exhibited superplasticity with elongations up to 700–800% at strain rates of 0.1–0.2 s^{-1}.

Table 2 lists the mechanical properties of some RS alloys and some typical high-strength equilibrium alloys. It may be seen that the increase in strength values is very significant for the lightweight alloys and consequently the specific strength (strength/density) is very high. Thus, it can be seen that the technique of RSP is being utilized commercially to produce high-strength, high-temperature, and

Table 2
Room-temperature mechanical properties of rapidly solidified aluminum- and magnesium-base alloys compared with traditional ingot metallurgy alloys.

Alloy designation and composition	Density ($Mg\ m^{-3}$)	Yield strength (MPa)	Specific yield strength ($MPa\ m^3\ Mg^{-1}$)	Ultimate tensile strength (MPa)	Specific ultimate tensile strength ($MPa\ m^3\ Mg^{-1}$)	Elongation (%)
RS aluminum alloys						
FVS 0611 (5.5 Fe, 0.5 V, 1.0 Si, all wt.%)	2.83	310	110	352	124	16.7
FVS 0812 (8.5 Fe, 1.3 V, 1.7 Si, all wt.%)	2.92	414	142	462	158	12.9
FVS 1212 (11.7 Fe, 1.15 V, 2.4 Si, all wt. %)	3.02	531	176	559	185	7.2
7075[a] (+1Ni+1Fe) extruded	2.80	634	226	717	256	9
RS magnesium alloys						
$Mg_{89.5}Zn_1Al_8Si_{1.5}$ (at.%)	1.84	390	212	429	233	2.8
$Mg_{91}Zn_2Al_5Si_2$ (at.%)	1.80	427	237	455	253	5.7
$Mg_{89}Zn_1Al_8Si_2$ (at.%)	1.88	448	238	468	249	1.7
$Mg_{90}Al_8Ga_2$ (at.%)	1.87	499	267			3.5
$Mg_{70}Ca_{10}Al_{20}$ (at.%)	1.88	600	320			
Traditional ingot metallurgy alloys						
AZ91[b]—T6[c]	1.81	120	66	200	110	3
7075–T6	2.80	500	179	570	204	11
Ti–6wt.%Al–4wt.%V–T6	4.46	1100	247	1170	262	10

[a]7075 is an aluminum-base alloy with 5.1–6.1 Zn, 2.1–2.9 Mg, 1.2–2.0 Cu, 0.5 Fe, 0.4 Si, 0.3 Mn, 0.18–0.28 Cr, and 0.20 Ti (all wt.%).
[b]AZ91 is a magnesium-base alloy with 9.5 Al, 0.5 Zn, 0.3 Mn (all wt.%).
[c]T6 is a heat treatment involving solution treatment and artificial aging.

wear-resistant lightweight alloys via the precipitation and dispersion hardening mechanisms. In addition, the RSP technology is being increasingly exploited to recover value-added products from metal scrap.

The technology of RSP is also being utilized to produce improved tool steels and superalloys, even though for the second group, mechanical alloying appears to be a better technique to achieve the desired properties (Suryanarayana 2001). RSP of tool steels produces homogeneous microstructures by eliminating the segregation patterns and with complete solid solubility of alloying elements. Both these increase the hardness and wear resistance. Subsequent tempering increases the hardness of these materials even further. RSP tool steels do not soften and have a hardness peak at 615 °C, in contrast to commercial tool steels which begin to soften at 300°C and display a secondary hardening peak at 525°C. Another route employed to produce improved tool steels is to synthesize metalloid-lean (5–13 at.%) amorphous iron-base alloys (e.g., as precursors to produce a fine-grained (~0.25 μm) crystalline matrix stabilized by ~25 vol. % of fine borides or carbides. Such materials can find applications as hot-work die tools requiring hot strength and hardness up to 650°C, oxidation resistance up to 700°C, and good impact resistance up to 540°C. RSP tool steels have been shown to have extended cutting tool life and require less maintenance (Suryanarayana 1993).

Nanocrystalline alloys with a grain size of <100 nm have attracted lot of attention in recent years due to an attractive combination of properties such as high strength, increased toughness, formability, enhanced diffusivity, and improved catalytic behavior (Gleiter 2000, Suryanarayana and Koch 2000). One of the methods of producing such materials now is by a low-temperature crystallization of the amorphous melt-spun ribbons, which enables tailoring the microstructure by controlling the crystallization parameters (Lu 1996). Iron-base alloys produced in such a manner, referred to as FINEMET alloys, contain a fine dispersion of α-Fe(Si) particles dispersed in an amorphous matrix and show an excellent combination of soft magnetic characteristics such as low coercivity, high permeability, almost zero magnetostriction, and low core losses (Yoshizawa *et al.* 1988). Such alloys are promising candidates for applications in transformer cores, saturable reactors, high-frequency transformers, and magnetic heads.

8. Summary

Rapid solidification processing started as an academic curiosity in 1960 and has now matured into a fully fledged field of activity in the broad area of materials science and engineering. RSP alloys exhibit increased solid solubility levels, formation of metastable crystalline intermediate phases, and amorphous phases. Recent attributes of RSP include formation of nanocrystalline materials, quasicrystalline phases, and bulk metallic glasses. The field has shifted from the investigation of the fundamental phenomena of metastable phase formation in the early years of RSP to engineering applications in the 1980s and 1990s. A reasonable amount of success has been achieved in this respect. New and fascinating discoveries continue to be made in this field and in recent years the field of RSP has been enriched by inputs from other related non-equilibrium processing techniques. The discovery of bulk metallic glasses has opened up a whole new field to facilitate a basic understanding of the glassy state in temperature regimes that were completely inaccessible in earlier years.

See also: Atomization

Bibliography

Anantharaman T R, Suryanarayana C 1971 A decade of quenching from the melt. *J. Mater. Sci.* **6**, 1111–35
Anantharaman T R, Suryanarayana C 1987 *Rapidly Solidified Materials–A Technological Overview*, Trans Tech. Aedermannsdorf, Switzerland
Cahn R W 1991 Metallic glasses. In: Zarzycki J (ed.) *Materials Science and Technology—A Comprehensive Treatment*, Vol. 9, *Glasses and Amorphous Materials*. VCH, Weinheim, Germany, pp. 493–548
Cahn R W, Greer A L 1996 Metastable states of alloys. In: Cahn R W, Haasen P (eds.) *Physical Metallurgy*, 4th edn. Elsevier, Amsterdam, pp. 1723–830
Davis L A, Das S K, Li J C M, Zedalis M S 1994 Mechanical properties of rapidly solidified amorphous and microcrystalline materials: a review. *Int. J. Rapid Solid.* **8**, 11–73
Duwez P, Willens R H, Klement W Jr 1960 Continuous series of metastable solid solutions in Ag–Cu alloys. *J. Appl. Phys.* **31**, 1136–7
Gleiter H 2000 Nanostructured materials: basic concepts and microstructure. *Acta Mater.* **48**, 1–29
Groza J R 1999 Powder consolidation. In: Suryanarayana C (ed.) *Non-equilibrium Processing of Materials*. Pergamon, Oxford, pp. 347–72
Inoue A 1998 Amorphous, nanoquasicrystalline and nanocrystalline alloys in Al-based systems. *Prog. Mater. Sci.* **43**, 365–520
Inoue A 2000 Stabilization of metallic supercooled liquid and bulk amorphous alloys. *Acta Mater.* **48**, 279–306
Inoue A, Kawamura Y, Matsushita M, Hayashi K, Koike J 2001 Novel hexagonal structure and ultrahigh strength of magnesium solid solution in the Mg–Zn–Y system. *J. Mater. Res.* **16**, 1894–900
Inoue A, Nishiyama N, Kimura H 1997 Preparation and thermal stability of bulk amorphous $Pd_{40}Cu_{30}Ni_{10}P_{20}$ alloy cylinder of 72 mm in diameter. *Mater. Trans, JIM* **38**, 179–83
Jacobson L A, McKittrick J 1994 Rapid solidification processing. *Mater. Sci. Eng.* **R11**, 355–408
Johnson W L 1999 Bulk glass-forming alloys: science and technology. *MRS Bull.* **24** (10), 42–56
Jones H 1999 Rapid solidification. In: Suryanarayana C (ed.) *Non-equilibrium Processing of Materials*. Pergamon, Oxford, pp. 23–45
Jones H 2001 A perspective on the development of rapid solidification and nonequilibrium processing and its future. *Mater. Sci. Eng.* **A304–6**, 11–9
Kelton K F 1993 Quasicrystals: structure and stability. *Int. Mater. Rev.* **38**, 105–37
Kobayashi K F 1999 Laser processing. In: Suryanarayana C (ed.) *Non-equilibrium Processing of Materials*. Pergamon, Oxford, pp. 89–118
Liebermann H H (ed.) 1993 *Rapidly Solidified Alloys*. Dekker, New York
Lu K 1996 Nanocrystalline metals crystallized from amorphous solids: nanocrystallization, structure, and properties. *Mater. Sci. Eng.* **R16**, 161–221
Schwarz R B 1993 Synthesis and consolidation of amorphous metallic powders. In: Liebermann H H (ed.) *Rapidly Solidified Alloys*. Dekker, New York, pp. 157–93
Shechtman D, Blech I, Gratias D, Cahn J W 1984 Metallic phases with long-range orientational order and no translational symmetry. *Phys. Rev. Lett.* **53**, 1951–3
Suryanarayana C 1991 Rapid solidification. In: Cahn R W (ed.) *Materials Science and Technology—A Comprehensive Treatment*, Vol. 15, *Processing of Metals and Alloys*. VCH, Germany, pp. 57–110
Suryanarayana C 1993 Mechanical, chemical, and electrical applications of rapidly solidified alloys. In: Liebermann H H (ed.) *Rapidly Solidified Alloys*. Dekker, New York, pp. 737–54
Suryanarayana C (ed.) 1999 *Non-equilibrium Processing of Materials*. Pergamon, Oxford
Suryanarayana C 2001 Mechanical alloying and milling. *Prog. Mater. Sci.* **46**, 1–184
Suryanarayana C, Jones H 1988 Formation and characteristics of quasicrystalline phases: a review. *Int. J. Rapid Solid.* **3**, 253–93

Suryanarayana C, Koch C C 2000 Nanocrystalline materials—current research and future directions. *Hyperfine Interact.* **130**, 5–44
Suryanarayana C, Korth G E 1999 Consolidation of nanocrystalline powders. *Met. Mater. (Korea)* **5**, 121–8
Yoshizawa Y, Oguma S, Yamauchi K J 1988 New Fe-based soft magnetic alloys composed of ultrafine grain structure. *J. Appl. Phys.* **64**, 6044–6

C. Suryanarayana

Rapid Thermal Annealing

Procedures that entail the heating of semiconductor wafers and other materials to high temperatures for short periods of time are known collectively as rapid thermal processing. The specific term, rapid thermal annealing, for heat treatments usually without melting wafers, and the abbreviation, RTA, are applied to both the annealing equipment as well as the processing itself. RTA typically involves handling one wafer at a time and heating it, for silicon to temperatures usually between 300°C and 1200°C, for times of several minutes or less. Soaking time at peak temperature is often specified in seconds. The transient heating is achieved by either controlling a variable source of IR radiation incident on the wafer, or by momentarily placing the wafer in a preheated chamber or on a hot plate. The heating environment often departs from the isothermal equilibrium of batch processing in a furnace. RTA of a single wafer allows faster heating, shorter dwell time at the soak temperature, and faster cooling back to ambient. The throughput time for individual wafer processing, and thus fabrication cost, is minimized by keeping the total cycle time per wafer as short as possible. For some processes, RTA equipment provides a higher throughput alternative to batch furnaces. Common features among the various approaches are automated execution of preprogrammed recipes, robotic wafer handling, and separate cooling stations to receive wafers after withdrawal from the oven or furnace.

Critical, as distinct from mainly cost-effective, applications of RTA produce results enabled only with short times at high temperatures. Many of these are in the front end of silicon MOS device fabrication. It is in wide use for electrically activating impurity dopant species introduced by ion implantation. The high temperature is required to place dopant atoms on substitutional lattice sites and to anneal out crystal defects associated with the implantation. The short time is required to reduce solid-state diffusion of the dopants within the semiconductor. Other applications, such as the formation and anneal of metal silicides from deposited films, came into general use by way of the ease in purging oxygen from the annealing ambient and controlling diffusion-mediated reactions. Since annealing equipment can also be used for oxidation, there has been extensive study of rapid thermal oxidation in various oxidizing and nitridizing ambients to form thin isolation dielectrics for MOS devices. Single-wafer annealing of low dielectric constant insulators and copper films for back-end fabrication use a lower temperature range, generally under 500°C. Compound semiconductor wafers, e.g., for GaAs VLSI, are encapsulated in a carrier to minimize decomposition.

A diversity of methods for heating and controlling wafer temperature are in use for commercial, developmental, and research purposes. The absence of equilibrium throughout most, if not all, rapid thermal cycles has fostered innovative approaches for assuring sufficient temperature control for microelectronics fabrication. This article describes several widely used, representative examples.

Other related techniques besides annealing are also available. These include scanning high power density lasers and flame plasmas to deliver greater peak power for selectively melting and regrowing surface regions. Rapid resolidification tends to produce more thermal metastability and defects than solid-phase annealing.

1. IR Spiking Process

The challenge of adapting conventional furnaces for a fast and controllable thermal cycle initially led to the alternative idea of an IR spiking process. Here, spiking refers to a sharply peaked function of temperature vs. time. As described in early work (Mammel 1971), a slice of silicon, now called a wafer, is rapidly heated and cooled by exposing it to a transient flux of IR radiation from an array of incandescent lamps. The IR spiking technique was used by R. M. Drexinger and W. K. Mammel at the former Western Electric Company to improve control over dopant diffusion in planar transistor fabrication. Their apparatus, shown in Fig. 1, includes many of the features since adopted for commercial lamp-heated rapid thermal processors. With transfer of heat to the wafer occurring largely by radiation, considerations of oven enclosure design and the wafer emissivity are more critical than they are for equilibrium furnaces. Radiated power depends strongly on temperature as $\varepsilon\sigma T^4$, where ε is an emissivity specific to an object at temperature T, and $\sigma = 5.670 \times 10^{-8}\,\mathrm{W\,m^{-2}\,K^{-4}}$ is the Stefan–Boltzmann constant. At 1000°C a black body ($\varepsilon = 1$) freely radiates $149\,\mathrm{kW\,m^{-2}}$.

Since the early 1970s notable improvements have been seen in heating uniformity, temperature control accuracy, and process stability. Development was driven by the decreasing tolerances for temperature errors with successive generations of integrated circuit technologies. Figure 2 shows the temperature vs. time curves obtained by four IR spiking methods that use either lamps or a fixed-temperature furnace. These examples are the limiting cases where soaking at maximum temperature is reduced to nominally zero time. General methods are described in Sects. 1.1 and 1.2. Section 1.3 discusses the issue of wafer emissivity as it relates to the heating of wafers. Section 2 discusses the various methods of determining wafer temperature and wafer emissivity.

1.1 Lamp Heating Methods

Tungsten lamps controlled by variable electrical power, familiar in numerous industrial applications,

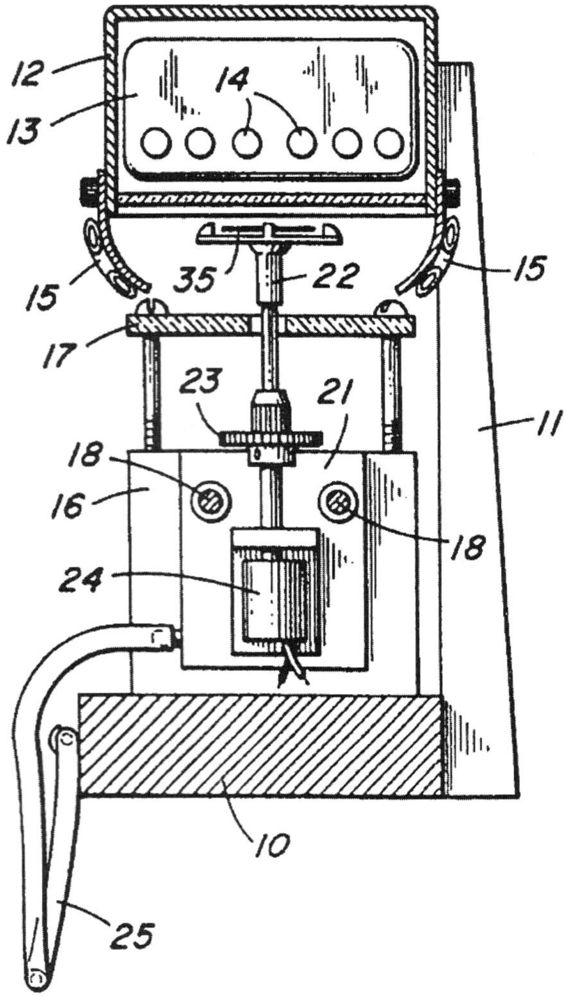

Figure 1
Apparatus for RTA of semiconductor wafers (Mammel 1971). The heater 12 is comprised of reflector 13, quartz lamps 14, water cooling 15, and is fastened to base 10 by supports 11. Wafer 35 on support 22 is levitated by gas from hose 25 and rotated via gear 23 and motor 24 mounted on carriage 21, which moves via guideway 17, rods 18, and bifurcated supports 16.

readily fulfill the special requirements for rapid thermal processing. The tungsten filaments are supported in a protective halogen ambient within quartz envelopes and serve as a conveniently controlled, long-life source of heating radiation from the visible to the near-IR. Pure silicon wafers initially heat up by absorbing radiant energy mainly above the band edge, at wavelengths, λ, shorter than 1 μm. Free carrier excitation, at temperatures above about 500°C, or heavy substrate doping, allows absorption to extend beyond 1 μm.

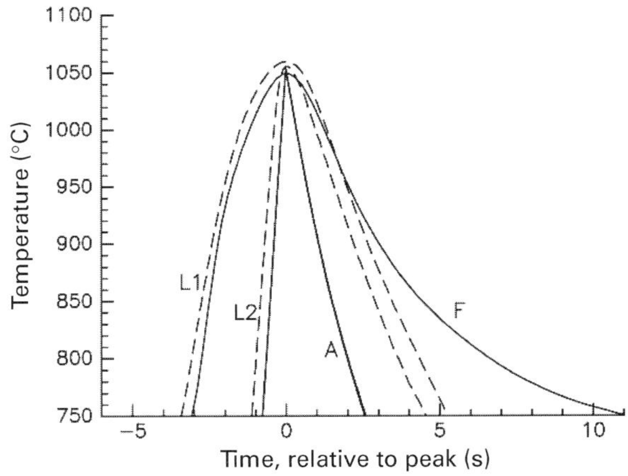

Figure 2
Temperature vs. time for various methods of RTA. Dashed curves are for incandescent lamps, L1 is for single- and L2 for dual-sided heating. Solid curves are F, for bell jar furnace and elevator, and A, for single-sided arc lamp.

There are several basic approaches to the design of a heating chamber with IR lamps. The method shown in Fig. 1 uses heating from one side of the wafer. Space on the other side is used for wafer support and rotation, and access for temperature sensing devices shielded from the lamps. Wafers are supported by quartz pins or by an edge ring. A variant of the method heats the wafer from both sides, boosting the deliverable power and allowing faster heating rates. With incandescent lamps, typically rated from 500 W to 2000 W in a system capable of 30–100 kW power, heating rates up to 400°C s^{-1} are readily achieved. The immediate wafer environment is isolated by a quartz window or a quartz tube surrounding the wafer. Inert annealing applications commonly flow N_2 purge gas at several liters per minute. The chamber is accessed through a door.

Generally, heat is transferred from the lamps to the wafer by radiative coupling aided by reflections in the oven enclosure. The initial rate of increase in wafer temperature scales roughly with radiated lamp power, which is reduced on approaching the desired soak temperature. Power is then regulated to maintain a constant temperature, balancing absorption of lamp radiation against losses by reradiation and gas convection. System heat loss occurs by convection in the ambient gas, air cooling of the lamps and the quartz isolation, and water cooling of the oven enclosure. Parasitic internal heating of the system during annealing is compensated for by gradually reducing the lamp power. Temperature is controlled by measuring the wafer temperature and adjusting the power to the lamps. Fast cooling is achieved by reducing lamp power to zero. For a spike anneal processes, such as that shown in Fig. 2, heater power begins to decrease slightly before reaching the peak temperature.

A reflective oven enclosure, while improving heat retention, also reflects emitted wafer radiation back onto the wafer itself. The center of the wafer tends to receive more radiation from its images than the edges, and radiative emission is greater at the edges. To compensate for this the diameter of the wafer is effectively increased by surrounding the wafer with an annular ring of silicon or silicon carbide material. In some designs this guard ring is detached, while in others it is part of the wafer support structure. The wafer can be floated on gas, as in the apparatus of Fig. 1; however, support at the edge by a heated ring or by pointed quartz pins provides sufficient thermal isolation. Contamination of silicon is avoided by using only clean quartz, silicon, or SiC materials for contacting wafers. Thermophoresis tends to reduce particle contamination on wafers undergoing RTA.

Designs of annealing chambers have employed sophisticated modeling of the heat transfer mechanisms to ensure that wafers are heated uniformly during the ramp-up, soak, and cool-down phases of the cycle, especially when lamp power is dynamically changing. The modeling treats the primary radiant flux from the lamps and reflectors, secondary exchange of radiation between the wafer and quartz shields, heat loss to the chamber, and buoyant gas convention. Convective loss via the ambient gas is more important for low-temperature annealing, such as for metal and dielectric films. Simple systems use a single-point temperature sensor to provide a feedback signal to a proportional integral derivative (PID) closed-loop control system. Power to the various lamps is adjusted to optimize temperature uniformity. More sophisticated model-based temperature control is used in systems with multiple temperature sensors and lamps arranged in multiple banks or zones that are controlled separately.

Incandescent lamps allow control with a resolution in time, about 0.5 s, that depends on the mass and temperatures of the tungsten filaments and the wafer. While the primary heating is by filaments, there is secondary heating by the quartz envelopes of the lamps, quartz isolation, and guard ring. Quartz, absorbing and emitting beyond about 2.5 μm, is heated by the tungsten and the wafer itself. Dashed curves L1 and L2 in Fig. 2 illustrate that the main differences between single- and dual-sided heating are about a factor of two in the heating rate and in the transition time at the peak temperature. The cooling times are about the same and are determined by temperature and radiative loss in the reflective chamber: maximum cooling is typically $80\,^{\circ}\mathrm{C\,s^{-1}}$.

As alternatives to the power limitations of incandescent lamps, annealing systems have been designed around a signal electric arc lamp of power rating sufficient for the desired heating rates. A reflector housing uniformly illuminates the wafer. Advantages of such an approach include unobstructed optics, faster heater response, optimum absorption of dominantly visible radiation, and the ability to capture electronic images of reflectivity and temperature patterns on wafers (Camm and Lojek 1994). Curve A in Fig. 2 is for an arc lamp system with a black chamber having response time of the order of 1 ms. It achieves sharper spiking and faster cooling.

Lamp systems can heat wafers to the melting point of silicon. Top-side heating, in particular, can be used to melt partially the surface of a silicon wafer and to hold the temperature stabile by exploiting the lower emissivity of the melted phase. The silicon melting point has been used in some cases to calibrate radiation pyrometers viewing the bottom side.

1.2 Furnace Heating

Furnace heating methods use approaches that are distinguished by the proximity of the wafer to the heat source. In one case a wafer may be loaded close to a plate heated to near the desired temperature. Wafer placement and thermal equilibration dictate the heating rate. A surrounding heated shield reduces the temperature offset between the heater and wafer. A second method loads a wafer into a vertical furnace with a temperature gradient, as shown schematically in Fig. 3. The hottest region at the top of the furnace is 200–400 °C above the target wafer temperature. Elevator motion raises the wafer close to the top and then lowers it to a sweet zone where a balance of heat flow maintains a nearly constant and uniform wafer temperature. A radiation pyrometer is used to monitor the back side of the wafer. Referring again to Fig. 2, curve F shows the temperature vs. time profile for a spike anneal by this furnace method. The more extended cooling time is the main difference relative to the incandescent lamp method, curve L1, that also uses single-sided heating. As in all methods, final cooling occurs after the wafer, which may still be dull-red hot, is withdrawn from the oven or furnace by a robotic wand.

Variations on the furnace theme include proximity plates embedded with heating and cooling capabilities, and a thin gas layer for thermal coupling to the wafer. A hybrid scheme uses a lamp system containing a passive liner or shielding hot plate to reduce the influence of variable wafer emissivity. A gap between the wafer and the shield, as may be required for wafer pick-up, diminishes the effectiveness of the plate as a thermal susceptor.

1.3 Wafer Emissivity

Usually denoted by ε, the emissivity of a body is the ratio of its spectral radiance to that of a black body at the same temperature. The quantity usually

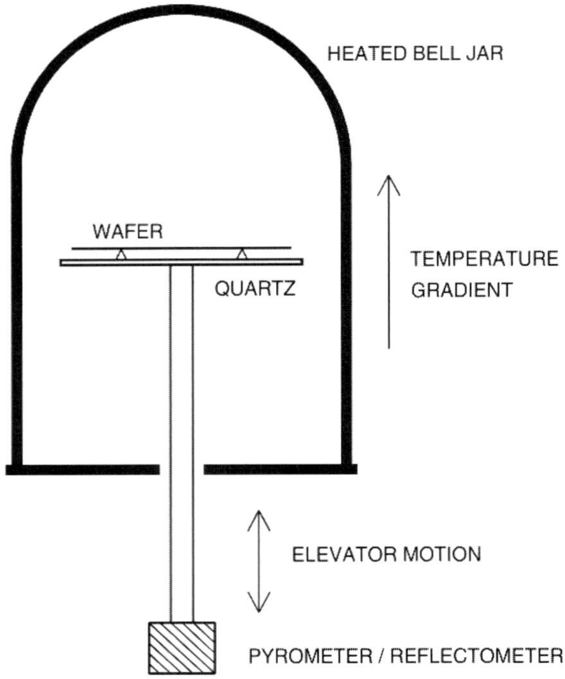

Figure 3
Schematic of furnace method for RTA. The wafer is rapidly elevated into the hot zone at the upper part of the bell jar for heating, and lowered for cooling. A pyrometer and reflectometer for temperature and emissivity measurement is attached at the base of the support column.

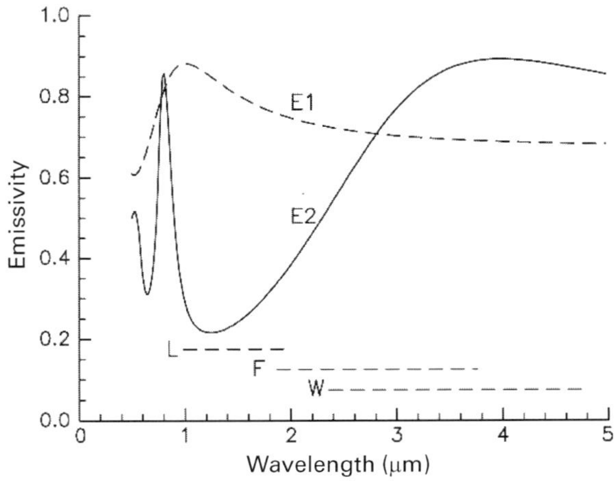

Figure 4
Modeled wavelength dependence of emissivity at 1000°C for silicon wafers coated with thin films: E1 for 200 nm SiO_2 film; E2 for 100 nm silicon film over 200 nm SiO_2 film. Dashed bars denote the central 50% ranges of black body spectra: L for 2700°C (e.g., lamps), F for 1300°C (furnace), and W for 1000°C (wafer).

considered in wafer processing is an average over a solid angle of 2π, which is technically the hemispherical emissivity. Measurement of emissivity is discussed in the context of temperature measurement by pyrometer techniques in Sect. 2.1.

For silicon the emissivity is a function of temperature and wavelength. It is nearly an ideal gray body medium at high temperature; e.g., at 1000°C and 1 μm silicon emissivity is approximately 0.65. Microelectronic fabrication, however, significantly alters the surface and emissivity of silicon wafers, e.g., by film deposition, oxidation, and patterning. Variations in emissivity in the IR lead to differing rates of heat transfer by radiation between the wafer and the heating environment. Device patterns cause lateral variations in emissivity and heat absorption. Temperature may be equilibrated by heat flow through the silicon if the features causing variations in emissivity are sufficiently small, typically under several millimeters. The magnitude of lateral temperature variations is estimated by theoretical treatment of the chamber and wafer emissivity. In practice, emissivity information on process wafers is inadequate for quantitative prediction, even for mean wafer temperature. The practical solution, therefore, is to use wafer temperature sensors. Temperature and emissivity measurement are discussed in Sect. 2.

The major causes of large emissivity variations are film stacks with metals or dielectrics like polycrystalline or amorphous silicon, SiO_2, and Si_3N_4. The lower indices of refraction for the oxide (1.46) and nitride (1.9), compared to silicon (approximately 3.58), cause reflections and interferences at film interfaces and the surface that change the emissivity, owing to changes in reflectivity. Optical interference becomes most pronounced when the effective dielectric film thickness is near an odd multiple of a quarter of the radiation wavelength. While emissivity can be computed if the film stack is known, natural variability in prepared films, such as thickness, may lead to significant uncertainties. Moreover, the index of silicon deposited in an amorphous state is higher and less well known than for bulk single-crystal silicon. Device patterns also cause interference effects that are challenging to calculate *a priori* with any precision. Even in the case of bare silicon, surface roughness, such as may be found on the back sides of wafers, is known to increase the emissivity by several percent relative to a smooth planar surface.

As an illustration of emissivity effects, Fig. 4 shows theoretical curves for the wavelength dependence of the emissivity of a silicon wafer at 1000°C for two types of coatings. Curve E1 is for silicon coated with a 200 nm oxide film and curve E2 is for silicon coated with a 100 nm silicon film on top of a 200 nm oxide film. Three inset dashed bars show the ranges of the central 50% of hypothetical black body spectra at several key temperatures. The peaks of the black body spectra, near the left edges of the bars, are given by the Wein displacement law: $\lambda_{PEAK}T = 2898\,\mu m\,K$. At 1000°C, absorptivity is theoretically equivalent to emissivity. Relative to curve E1, curve E2 in Fig. 4 indicates substantially greater absorption in region L for lamp radiation. Emissivities in region W for wafer radiation are more comparable. Thus, heating by incandescent lamps is expected to be more efficient for the oxide coating (E1), compared to the silicon-on-oxide coating (E2).

2. *Temperature Measurement*

The key metric in RTA is accurate control of temperature. Preprogrammed or open-loop control in the absence of a direct measurement of the wafer temperature has limited value in cases of variable or unknown wafer emissivity, as discussed above. The selection of temperature measurement technique is part of a holistic approach to rapid thermal processor design. The two prominent methods are based on radiation pyrometry and thermocouple junctions. These are discussed below.

2.1 *Radiation Pyrometers*

The favored method for measuring the temperature of wafers in a rapid thermal processor has been IR pyrometry because it does not require wafer contact and is capable of high precision. A pyrometer determines wafer temperature by sampling a portion of the thermally emitted radiation. However, pyrometry requires measurement of wafer emissivity and suppression of, or compensation for, parasitic reflections from the heater and quartz components.

Early pyrometer designs for lamp-heated ovens used an optical port in a quartz isolation tube to view the wafer, while exploiting the opacity of quartz at the pyrometer wavelength to block reflections from the lamps. One implementation of the method operates near 3.5 μm or 4.5 μm, where quartz has absorptivity minima, and uses thinned quartz for the view port. Another method operates at 2.7 μm and uses hydroxyl-doped quartz, except at the view port, to absorb the lamp radiation. A separate detector is used to compensate for radiation from the quartz tube. These methods are suitable for monitoring a passive shielding plate, of stable and known emissivity, next to the wafer. If the wafer were viewed directly, then its emissivity is determined independently.

The logarithmic relationship between temperature and spectral radiant energy gives pyrometry inherently greater precision than more linear alternatives. Semiconductor diode photodetectors and interference filters are used in radiation pyrometers for stable and linear sensivity to the photon flux. Transimpedance amplifiers are used to convert photocurrents to voltage signals. The semiconductor materials in use, and their maximum wavelengths, are silicon (up to 1 μm), InGaAs (up to 2.4 μm) and InAs (up to 3.5 μm). Silicon diodes are used in the 0.85–0.95 μm region to take advantage of the strong IR absorption of the wafers and the superior signal to noise capability of such detectors. Longer wavelengths are better matched to the peak emission wavelengths of the wafers (e.g., region denoted W in Fig. 4). The smaller bandgap diodes for operation beyond 2 μm are usually mounted on thermoelectrically cooled stages to increase the internal shunt resistance and to reduce circuit noise.

Pyrometry applies the Planck radiation law, which for a black body radiator gives the spectral radiance as

$$L_{BB}(\lambda) = c_1\pi^{-1}\lambda^{-5}[\exp(c_2/\lambda T) - 1]^{-1} \qquad (1)$$

where $c_1 = 3.7415 \times 10^{-16}\,W\,m^{-2}$ and $c_2 = 1.4388 \times 10^{-2}\,mK$. The spectral radiance from a heated body such as a wafer depends also on the emissivity, a function of wavelength, temperature, and angle relative to the surface. Emissivity also depends on the absorptivity of the wafer if its thickness is insufficient to fully attenuate incident radiation. For the purposes of IR pyrometry, a potentially intractable general problem is solved with greatly simplified assumptions for the optical properites of wafers. Averages are taken over a 2π solid angle to obtain hemispherical emissivities, $\varepsilon(\lambda)$. Kirkhoff's law of detailed balance is extended to equate hemispherical emissivity with the absorptivity, $\alpha(\lambda)$. Absorptivity in turn is computed from hemispherical reflectivity, $\rho(\lambda)$, and hemispherical transmissivity, $\tau(\lambda)$, by invoking energy conservation and to give the result

$$\varepsilon(\lambda) = \alpha(\lambda) = 1 - \rho(\lambda) - \tau(\lambda) \qquad (2)$$

Thus, the spectral radiation from a wafer is assumed to be given by

$$L(\lambda) = \varepsilon(\lambda)L_{BB}(\lambda) \qquad (3)$$

A pyrometer uses a measurement of $L(\lambda)$ and the freely radiating $\varepsilon(\lambda)$ over its optical pass band to find T by solution of Eqn. (1). Emissivities of both sides of wafers and the absorptivity are measured by using integrating mirrors on both sides to collect the reflected and transmitted rays for a directed incident beam, or the optical inverse under hemispherical illumination. In practice, full hemispherical averaging for $\rho(\lambda)$ and $\tau(\lambda)$ is often replaced by a cone about the specular directions, owing to the near specularity of the surfaces of silicon wafers. The roughened back sides of wafers, where the half-angle of the cone of specularity is typically 15°, are usually closer to being specular, rather than diffuse, scatterers of radiation.

Pyrometry is often further simplified by operating at wavelengths, temperatures, or silicon doping levels where transmission through the wafer can be neglected, i.e., where $\tau(\lambda)=0$. This reduces the problem to finding one unknown parameter, either the reflectivity or the emissivity itself. For wafers in reflective enclosures, the effective emissivity is enhanced, relative to the freely radiating emissivity, owing to reflections from the images of the wafer. A combination of ray tracing or generic modeling with *ex situ* and *in situ* optical measurements are used to deduce the chamber correction and to deduce hemispherical ε from bidirectional measurements. Several of the methods of pyrometry discussed below exploit chamber reflections to advantage.

Suppression of heater interference and *in situ* measurement of reflectivity or emissivity are readily adaptable to processors with single-sided heating. The bell jar furnace method shown in Fig. 3 has a reflectometer at the base of the elevator column. Apart from a quartz support structure, the furnace is a nearly ideal optical absorber. Reflection of modulated IR directed at the back side of the wafer gives a bidirectional reflectivity for estimating emissivity. Reflections of furnace radiation are determined as a function of height before the wafer heats up, and then subtracted from the detector signal during the heating cycle to yield the net thermal radiance of the wafer.

Another approach uses a reflector plate close to the wafer. Figure 5 illustrates the arrangement. Shielding around the perimeter blocks out heater radiation. The wafer and its image in the reflector form an imperfect black body cavity. An optical fiber light pipe, a sapphire rod 1–2 mm in diameter, terminates at the surface of the reflector. The fiber guides a sample of the local radiation to an external photodetector. The effective emissivity to be used in Eqn. (3) is increased relative to the freely radiating emissivity by the internal reflections. The value is less than the ideal of unity, however, because the plate is not a perfect reflector and some radiation escapes at the probe. A simple one-dimensional model gives effective emissivity in the form $\varepsilon[1-\rho_P(1-\varepsilon)]^{-1}$, where ρ_P is an effective reflectivity determined for the plate and probe. Simultaneous measurements by two nearby probes, with differing annular openings for the light pipes, and two corresponding values of ρ_P, are used to extrapolate to the ideal reflector where $\rho_P=1$. An array of probes is used for mapping temperature across a wafer at a spatial resolution that scales with the gap spacing. Accumulation of deposits on the reflector plate, which can cause the ρ_P values to drift over time, are retarded by a flow of purge gas near the probe. Related to this method are schemes placing similar probes in a heater plate below the wafer.

Processing chambers containing lamps surrounding the wafer have already built-in illuminators for determining emissivity. Figure 6 illustrates an application for dual-sided lamp heating with a pyrometer comprised of two light guide sensors. An iris at the wafer sensor restricts its view through the quartz tube to a spot on the wafer. The pointed tip on the lamp sensor, using the light squeezing principle, gives it a broad internal field of view. The lamps are excited with a periodic ripple modulation, e.g., in the range 20–120 Hz, obtained either synchronously with the power line or by programming the lamp supply. The frequency is sufficiently high so that the

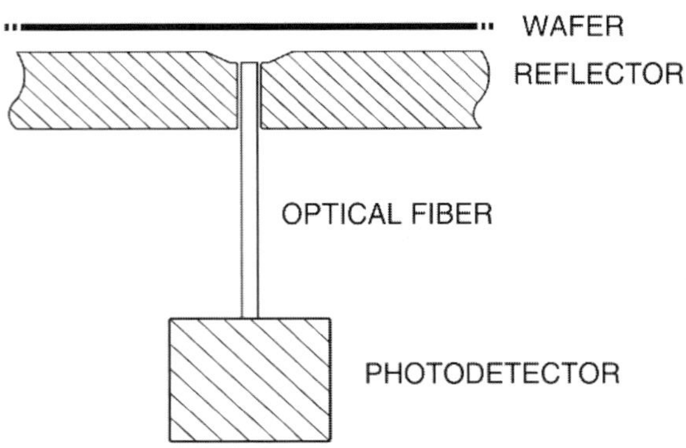

Figure 5
Radiation pyrometer for wafer temperature measurement. Sapphire optical fiber transports a sample of the radiation trapped between hot wafer and reflector plate to a photodetector head. Baffling, not shown, encloses the perimeter.

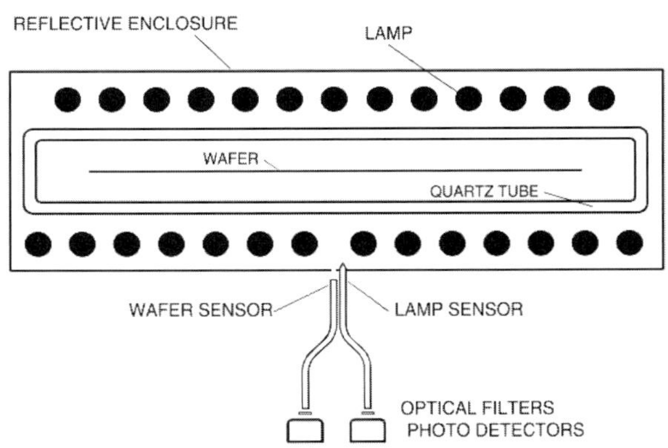

Figure 6
Ripple pyrometer with two sensors for wafer emissivity and temperature measurement. The chamber has lamps on both sides of the wafer and a quartz isolation tube. The wafer sensor receives radiation through a narrow slot that masks a direct view of the lamps. The lamp sensor has a pointed tip for wide-angle reception of reference radiation from lamps and wafer.

wafer temperature does not fluctuate. The modulation signal in the wafer sensor is a direct measure of lamp radiation reflected by the wafer. The corresponding signal in the lamp sensor is used as a reference. Thus the ratio of the fluctuating signals from the two detectors is used to find the effective *in situ* wafer reflectivity. Transmission through the wafer is taken into account by phase-shifting the power to lamps on the other side of the wafer and detecting the out-of-phase signals. The thermal radiance of the wafer is obtained from the wafer sensor signal by subtracting the lamp reflection component, obtained as the product of reflectivity and the lamp sensor reference signal. Figure 7 shows results for an anneal controlled by a pyrometer using ripple modulation. This is a case where the back side of the wafer contains an amorphous polysilicon layer that anneals during the thermal cycle. The emissivity is shown to be dynamically changing during the heating cycle. For comparison, the dashed curve shows the erroneous readings from a second passive pyrometer that uses a fixed emissivity to compute temperature. Emissivity calculation in this method ceases after the lamps are turned off.

2.2 *Thermocouples and Physical Sensors*

Thermocouple junctions play useful roles both as contacting temperature probes for processing wafers and as temperature references for sensor calibration. Sensors based on silicon physical measurements are used in some specialty applications.

The contacting thermocouple probe, while convenient, imperfectly tracks wafer temperature because

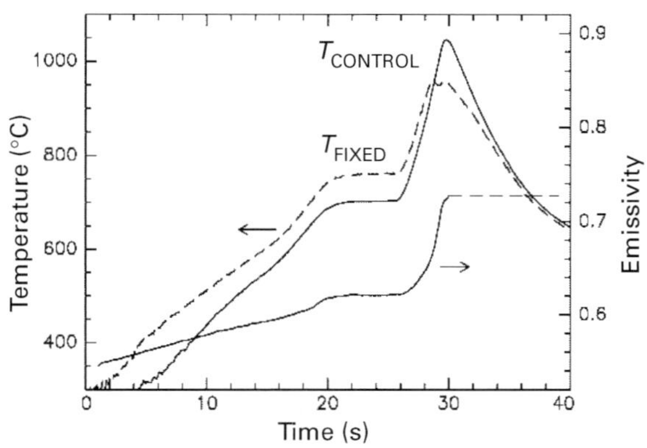

Figure 7
Time dependence of temperature and emissivity during annealing of a wafer with amorphous silicon and silicon dioxide films. Trace $T_{CONTROL}$ is produced by the controlling pyrometer that measures and corrects for emissivity. Dashed trace T_{FIXED} is a false temperature signal from a passive second pyrometer with fixed silicon emissivity setting.

of thermal contact resistance. For high-temperature operation, to avoid reacting with the wafer, metal junctions are encased in a protective coating, such as a silicon carbide sheath. The temperature indicated by the junction is offset relative to that of the wafer because of thermal conduction along the probe and differing radiation from the wafer and heating source. There is also a time delay that depends on the mass at the probe tip. An optimum application of a thermocouple probe would be in combination with a passive hot plate serving as a radiation shield.

Thermocouples bonded to wafers, or in thin-film form deposited over dielectric films on wafers, are used to calibrate pyrometers and other probes. Choice of thermocouple material is governed by maximizing thermal stability at high temperatures and, in the case of wired junctions, minimizing the thermal conductance of the lead wires. Temperature offsets less than 5°C between junction and wafer at temperatures up to about 1100°C are obtainable with type K chromel–alumel. Junctions using 0.13 mm diameter wire are bonded in a cavity with a silica-based cement. Adjacent wire at the junction is also bonded parallel to the silicon surface to reduce thermal gradient errors. Pt–Rh tends to be less accurate for rapid thermal applications, owing to higher thermal conductance of wires and rhodium diffusion. For steady-state calibration purposes, a wafer with either type of embedded junction may be covered with a shielding wafer to reduce heat flow in the wires and the temperature offset between junction and wafer. For thin-film thermocouples, which have superior thermal contact, the Seebeck coefficients differ from bulk materials and are calibrated independently. A wafer with an array of embedded junctions is used to map temperature distributions.

During pyrometer calibration, a wafer with a bonded thermocouple is steadily heated while raw pyrometer signals and corresponding thermocouple temperatures are recorded. Effective wafer emissivity is determined either absolutely or relative to silicon by measurement of the temperature. The emissivity scale is calibrated by profiling the pyrometer signals with wafers of at least two differing emissivities.

Other probe techniques, which have seen more limited or experimental application, use measurements of various temperature-dependent physical properties of the silicon. These include thermal expansion, changes in mid-IR attenuation and index of refraction, the Raman effect, ellipsometry, and the propagation velocity of surface acoustic waves. Some methods, such as the use of interference fringes from surface reflections, combine thermal expansion and index changes.

3. *Process Control*

Temperatures and times for a rapid thermal process are determined from models based on the fundamentals of the physical and chemical transformations and refined by experimentation with device fabrication. For high yields of working circuits, a process window is also determined, expressed as an allowable range of temperature errors, usually taken to be a width of six standard deviations in a statistical distribution. Statistical process control methods specify upper and lower limits to allowable temperature fluctuations. Once the requirements for a process are determined, a capable process controls temperature with the specified limits, or better. The relevant metric is the total fluctuation in temperature within the patterned region that excludes a 3–5 mm border at the edge of the wafer. The components of the metric are the uniformity across the wafer and repeatability of the mean temperature from wafer to wafer.

In addition to the process requirement, the maximum allowable temperature variation across wafers is dictated also by avoiding thermal stress damage. Temperature homogeneity within a 20°C range at temperatures below 600°C are usually adequate to avoid initiating stress fractures and wafer breakage. Temperature uniformity at high temperatures is needed to avoid propagation of slip planes at the edges of the wafer, or at support points, and lateral drift of the device patterns. Maintaining uniformity within 5°C at high temperature is usually sufficient, although the strongly temperature-dependent yield stress of silicon and the wafer stress pattern determine the actual criterion.

Processing is monitored by two basic approaches, involving either device wafers or blanket monitor wafers. Test devices on production wafers that are sensitive to the RTA temperature are used to generate data for statistical analysis. The simplest testers use van der Pauw patterns for measuring sheet resistance. Resistance of silicides formed by metal–silicon reaction, for example, often displays marked temperature dependence over the process window. Temperature fluctuations are inferred by taking the ratio of the width of the distribution of resistance measurements and the separately measured change of resistance with temperature. A similar approach is used at high temperatures, where the monitoring process could be the activation anneal of doped silicon, such as an isolated polysilicon line or a junction depth detected optically. There are some limitations, however, if testing information is delayed until some further processing steps have been completed. Moreover, the correlation of the tester measurement with RTA temperature is compromised by the sensitivity of the tester to fluctuations in other processing steps. For the examples cited these would be, for example, film thickness, surface preparation, and doping.

Special blanket monitor wafers are used to generate immediately available data for periodic monitoring of RTA equipment. A monitor may not necessarily use a process that is actually in the device fabrication flow, but rather is selected because of high sensitivity to temperature and inexpensive and reliable preparation. For low-temperature monitoring, e.g., below 800°C, the process could exploit temperature sensitivity in metal–silicon or metal–N_2 reactions, or amorphous silicon regrowth or crystallization. Reactions that are sensitive at high temperatures include activation anneal of high-dose ion implants and thermal oxidation. For implanted monitors, a screen oxide or an oxidizing annealing ambient is used to inhibit escape of the dopant into the annealing chamber. Wafers for monitoring by oxidation are cleaned as for MOS gate oxidation and may be reused after stripping the oxide in HF.

A convenient definition of monitor sensitivity is the parameter $S(T) = \mathrm{d}\ln M/\mathrm{d}\ln T$, given as the logarithmic dependence of a measured quantity, M, on temperature. For a process governed by thermal activation, and assuming $M \propto \exp(-E_A/k_B T)$, where E_A an effective activation energy and k_B is Boltzmann's constant, one obtains a constant $S = E_A/k_B T$. For example, a monitor using rapid thermal oxidation, where $E_A = 2.5\,\mathrm{eV}$, and taking film thickness for M, one has $S = 23$. Electrical activation of dopants can exhibit higher effective E_A, ranging from 3 eV to 5 eV, and thus greater sensitivity. The monitor wafer method is set up by processing a set of wafers at temperatures in the desired range to obtain an experimental function $M(T)$. The curves in Fig. 8 are an example, showing the temperature dependence of the sheet resistance, here equivalent to M^{-1}, for boron implanted into n-type silicon. Activation of hole carriers by boron diffusion produces strong

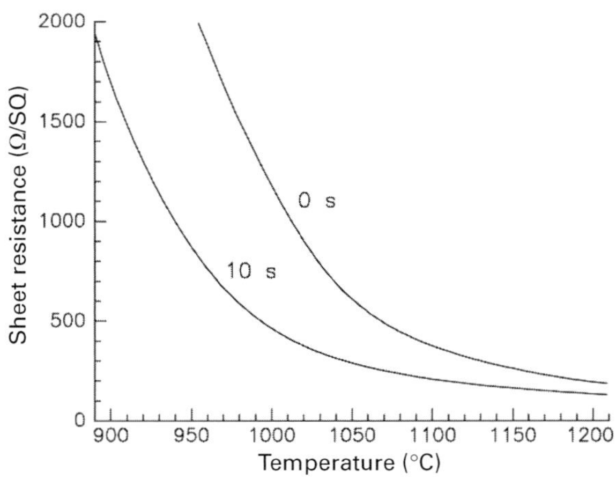

Figure 8
Annealing temperature dependence of the sheet electrical resistance for soak times of 10 s and nominally 0 s at the peak temperature. Silicon is n-type and implanted with $^{11}B^+$ at an energy of 500 eV and nominal dose of $10^{15}\,\mathrm{cm}^{-2}$.

dependence of resistance on annealing temperature and time.

After a monitor wafer receives an anneal or oxidation, M is measured at multiple points across the wafer to generate a contour map. Small measurement deviations, ΔM, are related to temperature deviations, ΔT, through the expression $\Delta T = (T/S)\Delta M/M$. Temperature uniformity is determined from the measurement standard deviation. The measurement mean gives the temperature with reference to a fixed point in the setup function $M(T)$. Methods such as thermocouples are used to trace the temperature reference. Errors in the method arise from fluctuations in the preparation of the monitor wafer, such as surface clean for oxidation, and energy, dose, and screen oxide for ion implants.

Process temperature control accuracy is estimated from the most sensitive critical device parameter. For a MOS transistor, this may be a length, L, like a junction depth or channel length, where the variation, ΔL, must lie within $\pm 5\%$. Assuming square root dependence of L on diffusivity, with thermal activation E_A, the required temperature control accuracy is estimated as $\Delta T = 2(\Delta L/L)k_B T^2/E_A$. For a process at 1000°C governed by 4 eV activation, the RTA control $\Delta T = \pm 3.5$°C.

Bibliography

Camm D A, Lojek B 1994 High power arc-lamp RTP system for high temperature annealing applications. In: Fair R R, Lojek B (eds.) *Proc 2nd Int. Rapid Thermal Processing Conf.*, RTP '94, Round Rock, TX, http://www.rtp-conference.org pp. 259–62

Davis C, Moslehi M M, Bowling A, Luttmer J D 1992 Microelectronics manufacturing science and technology: equipment and sensor technologies. *Texas Instruments Tech. J.* **9**, 20–43

Fair R B (ed.) 1993 *Rapid Thermal Processing: Science and Technology*. Academic Press, New York

Kakoschke R, Bußmann E, Föll H 1991 The appearance of spatially nonuniform temperature distributions during rapid thermal processing. *Appl. Phys. A* **52**, 52–9

Kamgar A 1989 Rapid thermal processing of silicon. In: Watts R K (ed.) *Submicron Integrated Circuits*. Wiley, New York

Mammel W K 1971 Method for heat treatment of workpieces. *US Pat.* 3 627 590

Moslehi M M, Davis C, Bowling A 1992 Microelectronics manufacturing science and technology: single-wafer thermal processing and wafer cleaning. *Texas Instruments Tech. J.* **9**, 44–64

Nguyenphu B, Fiory A T 1999 Wafer temperature measurement in a rapid thermal processor with modulated lamp power. *J. Electron. Mater.* **28**, 1376–84

Peuse B, Miner G, Yam M, Elia C 1998 Advances in RTP temperature measurement and control. *Mater. Res. Soc. Symp. Proc.* **525**, 71–85

Quinn T J 1983 *Temperature*. Academic Press, London

Rogne H, Ahmed H 1998 Emissivity of coated silicon at elevated temperatures. *Mater. Res. Soc. Symp. Proc.* **525**, 27–38

Roozeboom F (ed.) 1996 *Advances in Rapid Thermal and Integrated Processing*. Kluwer, Dordrecht, The Netherlands

Roozeboom F, Parekh N 1990 Rapid thermal processing systems: a review with emphasis on temperature control. *J. Vac. Sci. Technol. B* **8**, 1249–59

Saito T 1967 Spectral emissivity of silicon. *Jpn. J. Appl. Phys.* **6**, 339–47

Schietinger C, Adams B, Yarling C 1991 Ripple technique: a novel non-contact wafer emissivity and temperature method for RTP. *Mater. Res. Soc. Symp. Proc.* **224**, 23–31

Timans P J 1996 The role of thermal radiative properties of semiconductor wafers in rapid thermal processing. *Mater. Res. Soc. Symp. Proc.* **429**, 3–14

Vandenabeele P, Maex K, DeKeersmaecker R 1989 Impact of patterned layers on temperature non-uniformity during rapid thermal processing for VLSI applications. *Mater. Res. Soc. Symp. Proc.* **146**, 149–60

A. T. Fiory

Reaction Forming

The shrinkage upon firing of a porous ceramic-bearing preform can be reduced by increasing the solid volume within the preform prior to or during sintering. This may be accomplished by infiltrating a fluid (liquid or gas) into the porous preform and then converting the fluid into a solid by solidification, condensation, and/or chemical reaction with the preform. Approaches based on solidification or condensation will not be reviewed here. The purpose of the present contribution is to describe recent developments in processes that utilize oxidation reactions between an infiltrant (gas or liquid) and a shaped preform to produce near net-shaped, ceramic-bearing components. Two of these approaches yield all-ceramic bodies by the reactive infiltration of O_2(g) into shaped metal-bearing preforms: the reaction bonded aluminum oxide (RBAO) and volume identical metal oxidation (VIMOX) methods. In two other processes, ceramic/metal (intermetallic) composites are fabricated by the reactive infiltration of molten metals into porous, oxide-bearing preforms: the infiltration alumina aluminide alloy (i-3A) and displacive compensation of porosity (DCP) methods. In these processes the oxygen source is a solid oxide within the solid preform.

1. Near Net-shaped Ceramics from Formable, Metal-bearing Preforms: The RBAO and VIMOX Methods

Oxidation-induced changes in volume are utilized in the patented RBAO (Claussen 1992) and VIMOX (Sandhage 1997) methods to convert shaped metal-bearing precursors into ceramic bodies of similar shape and dimensions. The ratio of the volume of oxide produced to the volume of metal consumed is often referred to as the Pilling–Bedworth ratio (PBR). PBR values for several metals are shown in Table 1.

For most metals, the PBR is in excess of unity. The increase in solid volume associated with the oxidation of such metals within a porous body can be used to offset the contraction associated with postoxidation sintering. Hence, porous shaped preforms comprising intimate mixtures of metal and ceramic can be converted into dense all-ceramic bodies that retain the preform shape and dimensions. For example, the oxidation of a porous mixture of $Al + Al_2O_3 + ZrO_2$ followed by sintering can yield Al_2O_3/ZrO_2 composites with relatively low net shrinkage (Wu *et al.* 1993, Holz *et al.* 1994). This is the basic premise for near net-shape processing by the RBAO method.

Table 1
Pilling–Bedworth ratio (PBR) for several elements.

Metal/oxide	PBR	Metal/oxide	PBR
Al/Al_2O_3	1.28	Mg/MgO	0.804
Si/SiO_2	2.14	Ca/CaO	0.638
Cr/Cr_2O_3	2.01	Sr/SrO	0.611
Ni/NiO	1.66	Ba/BaO	0.670
Ti/TiO_2	1.77	Li/Li_2O	0.567
Zr/ZrO_2	1.51	Na/Na_2O	0.541
Nb/Nb_2O_5	2.69	K/K_2O	0.474

Unlike most metals, the alkali and alkaline earth metals tend to possess PBR values less than unity (see the right half of Table 1). Such oxidation-induced reductions in solid volume can be used to offset volume expansions associated with the oxidation of other elements and/or the formation of ceramic compounds. That is, a preform composed of an appropriate mixture of alkali or alkaline earth elements with other metallic and/or ceramic constituents can be converted by oxidation and further reaction into a near net-shaped monolithic ceramic or ceramic composite. For example, a dense, shaped mixture of $Ba + Sr + Al + Al_2O_3 + SiO_2$ can be oxidized and annealed to produce a dense body of celsian, $(Ba,Sr)Al_2Si_2O_8$, with little or no change in shape and dimensions (Sandhage *et al.* 2000, Viers and Sandhage 1999). This is the basic premise for near net-shape fabrication by the VIMOX process.

For both the RBAO and VIMOX methods, preforms composed of fine, intimate mixtures of metal and ceramic are required in order to obtain desired oxidation and reaction kinetics, and to produce near net-shaped bodies without macrocracking or distortion (Wu *et al.* 1993, Holz *et al.* 1994, Sandhage *et al.* 2000, Viers and Sandhage 1999, Schmutzler *et al.* 1994). High-energy milling can be an effective means of producing such mixtures. However, proper selection of the milling intensity (e.g., rotation speed during attrition milling), milling time, ball-to-charge ratio, and milling liquid are required to produce high yields of fine, intimate metal/oxide mixtures (Wu *et al.* 1993, Holz *et al.* 1994, Sandhage *et al.* 2000, Viers and Sandhage 1999, Schmutzler *et al.* 1994, 1996). Incomplete milling will yield coarse, nonuniform mixtures, whereas excessive milling may result in undesired mechanochemical reactions (e.g., the formation of brittle intermetallic compounds). Excessive mechanochemical reaction and oxidation of the metal-bearing powder mixture must be avoided during milling, so that a sufficient amount of ductile metal is retained to allow for: (i) uniform compaction and forming into a green body of desired shape; and (ii) small net volume changes upon conversion of the green body into the desired ceramic phase(s) (Holz *et al.* 1994, Sandhage *et al.* 2000, Viers and Sandhage 1999). Excessive hydration of the ceramic constituents during milling must also be avoided in order to reduce the risk of bloating during subsequent firing (Holz *et al.* 1994). Finally, owing to the inevitable wear that occurs during high-energy milling, the compositions of the milling media and milling vessel must be properly chosen to avoid degradation of the performance of the final fired components (Wu *et al.* 1993, Holz *et al.* 1994, Sandhage *et al.* 2000, Schmutzler *et al.* 1994).

Due to the presence of appreciable ductile metal in the milled powders (usually >30 vol.%), the green strengths of compacted RBAO and VIMOX mixtures can be appreciably higher than for conventional ceramic green bodies; that is, green strengths of 20–50 MPa are typically achieved in pressed metal-bearing RBAO and VIMOX preforms (Wu *et al.* 1993, Holz *et al.* 1994, Sandhage *et al.* 2000). In the RBAO process, compaction pressures of up to 500 MPa are commonly used to produce Al–Al_2O_3-bearing preforms with a sufficient amount of porosity (usually $\geqslant$25 vol.%) so as to allow for a relatively high oxygen infiltration rate during oxidation and a good match between the oxidation-induced volume increase and the sintering-induced contraction. In the VIMOX process, preforms with densities of up to 93% have been produced using uniaxial compaction stresses as high as 1200 MPa. (Open porosity is not required in VIMOX preforms for near net-shape processing.) After compaction, the robust metal-bearing RBAO and VIMOX preforms can be readily machined into more complicated shapes (Wu *et al.* 1993, Viers and Sandhage 1999). Dense, metal-rich VIMOX preforms have also been formed into wires and tapes by drawing and rolling (Sandhage *et al.* 2000, Schmutzler *et al.* 1994, 1996).

Shaped metal-bearing VIMOX preforms can also be produced by melt infiltration. For example, dense Mg–Al_2O_3-bearing precursors to spinel ($MgAl_2O_4$) have been produced by the pressureless infiltration of molten magnesium into porous Al_2O_3 preforms followed by solidification (Sandhage *et al.* 2000, Kumar and Sandhage 1998). With careful tailoring of the porosity of the Al_2O_3 preform, the desired magnesium content for producing stoichiometric spinel can be achieved. The ductile, interconnected magnesium endows the infiltrated preform with a relatively high green strength (70–100 MPa). As for the milled, powder-derived preforms discussed above, melt-infiltrated preforms can be readily machined into more complex shapes (Kumar and Sandhage 1998).

Shaped, robust VIMOX and RBAO precursors can be converted into dense all-ceramic bodies with the use of appropriate oxidation and postoxidation heat treatments. The porous, nonprotective nature of the oxide scales formed on alkali or alkaline earth metals results in relatively rapid oxidation of VIMOX preforms at $\leqslant$500 °C (Sandhage *et al.* 2000, Schmutzler *et al.* 1994, 1996). Although a dense, slow-growing layer of alumina tends to form on aluminum, a relatively high rate of oxidation of porous RBAO preforms can be achieved if the aluminum is in the form of fine (submicron) particles (Wu *et al.* 1993, Holz *et al.* 1994). (The protective nature of an alumina scale can also be reduced if the scale is allowed to react with other oxides, such as silica or baria, present in a preform (Sandhage *et al.* 2000, Viers and Sandhage 1999, Scheppokat *et al.* 1999)). During the oxidation of RBAO and VIMOX preforms, the following phenomena are of particular concern: (i) the formation of appreciable molten metal due to incomplete oxidation at subsolidus temperatures; and (ii) significant nonuniform thermal excursions (runaway oxidation) resulting from the exothermic nature of the oxidation reaction(s) (Holz *et al.* 1994, Sandhage *et al.* 2000, Schmutzler *et al.* 1994, Gaus *et al.* 1999). The generation of molten metal can lead to appreciable agglomeration and phase coarsening or to the loss of such metal from the preforms due to poor wetting (Holz *et al.* 1994, Sandhage *et al.* 2000, Schmutzler *et al.* 1994, Gaus *et al.* 1999). In order to maximize the extent of oxidation at temperatures where the metallic constituents remain solid, oxidation is conducted with modest rates of heating up to, and/or with isothermal anneals below, subsolidus temperatures. For example, 30–32.5 mm diameter × 35 mm tall Al–Al_2O_3-bearing cylinders have been completely oxidized by heating at 1–2 °C min^{-1} to 1100 °C in air, in some cases with prolonged isothermal anneals at 400–475 °C (Wu *et al.* 1993, Gaus *et al.* 1999), and 10–16 mm lengths of sheathed Ba–Ce-bearing and Ba–Ti-bearing tapes have been completely oxidized (via oxygen migration from the exposed ends inward) within 4–10 h at 300 °C in oxygen (Sandhage *et al.* 2000, Schmutzler *et al.* 1994, 1996). Such modest rates of heating and low-temperature annealing also help to reduce the occurrence of thermal excursions within the preforms that lead to extensive macrocracking (Sandhage *et al.* 2000, Schmutzler *et al.* 1994, Gaus *et al.* 1999). Reductions in the oxygen partial pressure, and the use of gas mixtures containing helium (i.e., to enhance the rate of heat transfer through gas-filled pores), have also been effective means of reducing thermal excursions and associated macrocracking (Viers and Sandhage 1999, Gaus *et al.* 1999).

The oxide grains formed within RBAO and VIMOX preforms upon oxidation at modest subsolidus temperatures tend to be submicron in size (Wu *et al.* 1993, Sandhage *et al.* 2000). Such an intimate mixture of fine-grained oxides can undergo further reaction and sintering under modest annealing conditions (Sandhage *et al.* 2000, Schmutzler *et al.* 1996, Scheppokat *et al* .1999). For example, tapes of $BaTiO_3$ and $Ba(Ce,Nd)O_3$ with densities of $\geqslant$98% have been produced from well-milled and oxidized preforms at peak temperatures of $\leqslant$1085 °C. Intentionally porous bodies of $Ca_{10}(PO_4)_6(OH)_2$ and $(Ba,Pb)TiO_3$ have been produced from well-milled and oxidized VIMOX preforms within times of only

1 h and 10 min, respectively, at peak temperatures of 1120–1150°C. Dense (⩾95%) Al_2O_3/ZrO_2 composites have also been synthesized by firing fully-oxidized RBAO preforms for only 20 min at 1550°C. Some care has to be taken, however, with such fast firing to avoid bloating. Bloating can occur if sintering at a specimen surface precedes complete removal of chemically-bound water (e.g., in the form of hydroxides) located deeper within the specimen. Preforms prepared with high compaction stresses from powders milled for extended times in water-bearing lubricants are particularly susceptible to such bloating (Holz *et al.* 1994).

With proper tailoring of the preform composition, phase content, and microstructure and with appropriate heat treatment conditions, easily-formed RBAO and VIMOX green bodies can be oxidized and converted into near net-shaped ceramics and ceramic composites with attractive properties. For example, dense (pressureless sintered) RBAO-derived Al_2O_3–ZrO_2 composites have been produced with fracture strengths and toughnesses of 700–800 MPa and 5–6 MPa m^{-2}, respectively. By incorporating metal or ceramic additives with larger oxidation-induced volume changes than for aluminum (e.g., Si, SiC, Zr, Cr), nearly zero shrinkage has been achieved by the reaction bonding process (Wu *et al.* 1993, Scheppokat *et al.* 1999). For example, porous preforms of Al– Al_2O_3–SiC–ZrO_2 have been converted into 93–99% dense composites of $Al_6Si_2O_{13}$ (mullite) and ZrO_2 with net linear shrinkages of ⩽1% by slow heating in air (in some cases with several intermediate isothermal anneals) up to 1150–1200°C, holding at this temperature for 15 h, and then sintering at 1550°C (Scheppokat *et al.* 1999). Incomplete oxidation of the SiC has yielded $Al_6Si_2O_{13}$/SiC/ZrO_2-bearing composites with modest net shrinkages (e.g., ⩽7%). The latter composites have exhibited fracture strength and toughness values of up to 610 MPa and 4.9 MPa m^{-2}, respectively.

A variety of functional, near net-shaped ceramics and ceramic composites have been fabricated by the VIMOX process for chemical, electronic, magnetic, gas-sensing, and biomedical applications. Such ceramics include refractory aluminates (e.g., $MgAl_2O_4$, $BaAl_2O_4$, $Ba(Sr)Al_2Si_2O_8$), dielectric or semiconducting titanates (e.g., $BaTiO_3$, $(Ba,Pb)TiO_3$), superconducting cuprates (e.g., $Bi_2Sr_2Ca_1Cu_2O_8$), magnetic ferrites (e.g., $BaFe_{12}O_{19}$), proton-conducting cerates (e.g., Nd_2O_3-doped $BaCeO_3$), water-sensitive chromates (e.g., TiO_2-doped $MgCr_2O_4$), and biocompatible phosphates (e.g., $Ca_{10}(PO_4)_6(OH)_2$). For example, machinable Ba(Sr)–Al–Al_2O_3–SiO_2-bearing preforms have been fully oxidized and converted into $Ba(Sr)Al_2Si_2O_8$ (celsian, a radome ceramic) bodies that retained the preform shapes with dimensional changes well under 1% (Sandhage *et al.* 2000, Viers and Sandhage 1999). Dense, machined Mg–Al_2O_3-bearing preforms have been fully oxidized and converted into dense bodies of spinel ($MgAl_2O_4$, a refractory ceramic) with dimensional changes of ⩽0.6% (Sandhage *et al.* 2000, Kumar and Sandhage 1998). Optical images of green-machined RBAO and VIMOX preforms, and corresponding fully-converted all-ceramic bodies, are shown in Fig. 1.

2. *Net-shaped Ceramic/Metal(Intermetallic) Composites from Porous Oxide-bearing Preforms: The i-3A and DCP Methods*

In the RBAO and VIMOX processes, a fluid oxidant (gaseous oxygen) is allowed to infiltrate into and react with shaped, metal-bearing preforms. A converse approach is to allow a fluid metal to infiltrate into and react with a shaped oxidant-bearing preform. The latter approach is the basis of two more recently developed, reactive infiltration techniques for fabricating dense, near net-shaped ceramic/intermetallic or ceramic/metal composites: the infiltration alumina aluminide alloy (i-3A) process and the displacive compensation of porosity (DCP) process.

A variety of high-melting, alumina–aluminide alloy composites can be fabricated at modest temperatures using the i-3A pressure infiltration process (Wagner *et al.* 1999, Schen *et al.* 1998). In this approach, a shaped, porous preform is first prepared by milling a mixture of a solid oxidant (e.g., TiO_2, Fe_2O_3, Nb_2O_5, ZrO_2) with a quantity of Al_2O_3. The powder mixture is compacted and formed into a shaped green body that is then partially sintered under modest firing conditions (e.g., 30 min at 1250°C). Such sintering produces a rigid (inert) Al_2O_3 network that endows the green body with reasonable mechanical stability during infiltration and reaction. The mechanical stability may be further enhanced by adding water glass or ductile metals to the oxide-rich preform. The partial sintering treatment is also used to adjust the pore fraction of the preform to a predetermined value. By carefully controlling the preform porosity along with the volume fractions of solid oxidant and alumina, it is possible to produce dense, near net-shaped composites composed of tailored amounts of alumina and an intermetallic compound (Wagner *et al.* 1999, Schen *et al.* 1998). Consider the fabrication of dense $Al_2O_3/TiAl_3$ and Al_2O_3/TiAl composites via the following displacement reactions:

$$3TiO_2(s) + 13Al(l) \rightarrow 2Al_2O_3(s) + 3TiAl_3(s) \quad (1)$$

$$3TiO_2(s) + 7Al(l) \rightarrow 2Al_2O_3(s) + 3TiAl(s) \quad (2)$$

The product of Eqn. (1) consists of 30.8 vol.% Al_2O_3 with 69.2 vol.% $TiAl_3$. In order to convert a preform undergoing this reaction into a $TiAl_3$-bearing composite with a higher alumina content, the preform should contain excess (inert) alumina.

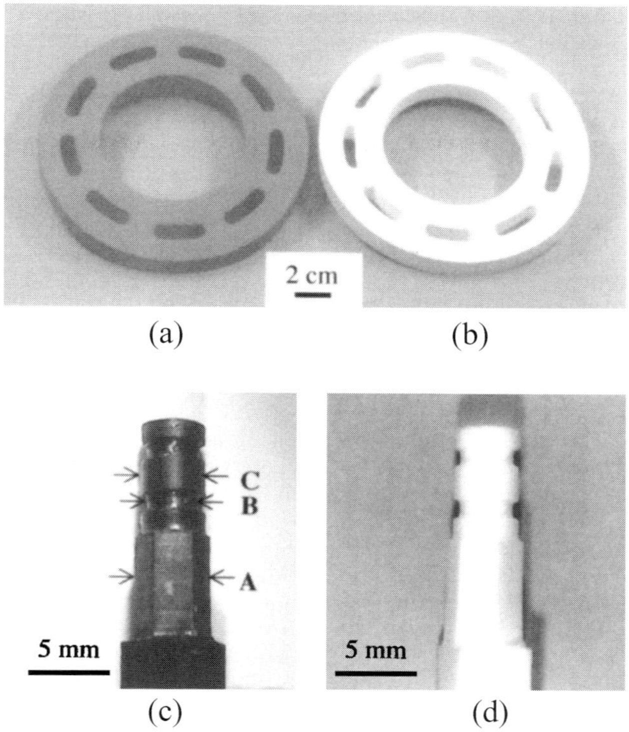

Figure 1
Optical images of: (a) a machined Al–Al_2O_3–ZrO_2-bearing green body; (b) a similar body after full conversion into a low-shrinkage Al_2O_3/ZrO_2 composite (RBAO process); (c) a Mg–Al_2O_3-bearing green body; and (d) the resulting $MgAl_2O_4$ body (VIMOX process (Kumar and Sandhage 1998)). Dimensional changes at the positions A, B, and C in Figs. (c) and (d) were $\leqslant 0.6\%$.

For example, a 60 vol.% Al_2O_3–40 vol.% $TiAl_3$ composite can be produced by the infiltration and reaction of molten aluminum with a 54.2% dense preform comprising 68.3 vol.% Al_2O_3 and 31.7 vol.% TiO_2 (rutile). (This calculation assumes that infiltration is conducted at 850 °C, where the density of Al(l) is 2.332 gcm^{-3}, and that the reaction proceeds along with the infiltration.) If the preform density and alumina phase content are chosen to be 46.8% and 61.3 vol.%, respectively, then Al(l) infiltration and reaction should yield a dense composite of 60 vol.% Al_2O_3–40 vol.% TiAl (via Eqn. (2)). Similar calculations can be conducted for the fabrication of other well-defined 3A composites.

To enhance the rate of Al(l) infiltration, elevated gas pressure and direct squeeze casting have been used in the i-3A process (Wagner *et al.* 1999). With gas pressure infiltration, a porous specimen is immersed under a molten aluminum bath and the pressure of an inert gas (e.g., argon) over the bath is then gradually increased to a modest value for a short time (e.g., 12 MPa for 10–20 min). After infiltration, the preform is raised out of the melt and cooled to room temperature. With squeeze casting, the preform is preheated (e.g., to 600 °C) and then quickly placed into a heated die (e.g., at 300–600 °C). Molten aluminum is then poured into the die and a ram is used to squeeze Al(l) into the porous preform. After solidification of the aluminum, the infiltrated preform and excess solid aluminum are removed from the die. Although squeeze casting results in particularly rapid infiltration (i.e., 6 mm thick preforms can be completely infiltrated within a fraction of a second (Beyer *et al.* 2000)), additional processing must be undertaken to remove the excess solidified aluminum surrounding the specimen and possibly to complete the displacement reaction.

Displacement reactions between molten aluminum and solid oxidants tend to be relatively exothermic (e.g., the standard enthalpy per mole of reaction in Eqn. (1) is 1120 kJ at 850 °C). Such evolved heat can dramatically enhance the rate of reaction at modest applied temperatures. Indeed, dense, fully-reacted Al_2O_3–$TiAl_3$-bearing and Al_2O_3–$NbAl_3$–Nb_2Al-bearing composites have been produced by gas pressure infiltration of Al(l) into Al_2O_3/TiO_2 and $Al_2O_3/Nb_2O_5/NbAlO_4$ preforms, respectively, at peak pressures of 12 MPa for $\leqslant$20 min at $\leqslant$850 °C

(Wagner *et al.* 1999, Schen *et al.* 1998). In the case of squeeze casting, where the residence time of molten aluminum in the preform can be very short ($\ll$1 s), a secondary heat treatment at a modest temperature may be required to complete the reaction (e.g., 640°C for 4 h to produces an Al_2O_3–$TiAl_3$-bearing composite (Wagner *et al.* 1999)). Because the net volume change upon completion of the i-3A displacement reaction is negative (e.g., −19.6% for the reaction in Eqn. (1) above), some internal porosity can develop if such a secondary heat treatment is conducted in the absence of excess aluminum surrounding the preform. Alumina-bearing composites reinforced with a variety of interpenetrating intermetallic compounds (e.g., $TiAl_3$, FeAl, $NbAl_3$, Nb_2Al) have been produced by the i-3A process.

The DCP process also involves the conversion of porous ceramic-bearing preforms into dense composites by the reactive infiltration of molten metals. However, unlike the i-3A method, the ceramic products of DCP-type reactions possess larger volumes than the ceramic reactants (Kumar and Sandhage 1999, Rogers *et al.* 1999). Such reaction-induced increases in ceramic volume can be used to fill the prior pore spaces (i.e., reaction-induced densification of the ceramic preform). The infiltration of shaped, rigid, partially-sintered preforms, followed by complete reaction at internal solid/liquid interfaces, can yield dense ceramic-rich composites that retain the shape and external dimensions of the starting preforms. Examples of two DCP-like reactions are:

$$3Mg(l) + Al_2O_3(s) \rightarrow 3MgO(s) + 2Al(l) \quad (3)$$

$$3Mg(l) + 4Al_2O_3(s) \rightarrow 3MgAl_2O_4(s) + 2Al \quad (4)$$

The reactions in Eqns.(3) and (4) are thermodynamically favored for Al–Mg melts containing more than about 0.3 at.% Mg and about 7.1 at.% Mg, respectively, at 1000°C. Hence, as for the i-3A approach, these DCP reactions can be conducted with Al-rich melts. The volume of 3 moles of MgO is 31.9% larger than the volume of 1 mole of Al_2O_3, whereas the volume of 3 moles of $MgAl_2O_4$ is 16.6% larger than the volume of 4 moles of Al_2O_3. Consequently, a porous Al_2O_3-bearing preform can be converted into a MgO-bearing or $MgAl_2O_4$-bearing composite with a higher ceramic content than in the starting preform. Indeed, by adjusting the porosity and phase content of the DCP preforms along with the melt composition, fully-reacted ceramic/metal composites with a wide range of phase contents (i.e., from discontinuous to continuous metal reinforcements) and compositions can be directly produced. For example, 71%-dense Al_2O_3 preforms have been pressureless-infiltrated and reacted with pure Mg(l) at 1000°C to produce composites of 86 vol.% MgO + 14 vol.% Mg–Al alloy; in this case, the metal phase was discontinuous, as demonstrated by electrical conductivity measurements (Kumar and Sandhage 1999, Rogers *et al.* 1999). Cocontinuous $MgAl_2O_4$/Fe–Ni–Al alloy-bearing composites (with 42–59 vol.% oxide) were fabricated by the pressureless infiltration and reaction of Mg–Al liquid at 900°C with porous preforms comprising mixtures of Fe and $NiAl_2O_4$ (Rogers *et al.* 1999). Such DCP-derived composites were dense and retained the shapes and dimensions of the starting porous preforms to within a few percent (Kumar and Sandhage 1999, Rogers *et al.* 1999).

Near net-shaped oxide/metal and oxide/intermetallic composites with attractive, tailorable properties can be produced by the i-3A and DCP methods using low-temperature, low-cost casting technology. Lightweight $Al_2O_3/TiAl_3$ composites produced by the i-3A method have exhibited average strength and toughness values of 540 MPa and 8.6 MPa m^{-2}, respectively (Wagner *et al.* 1999, Schen *et al.* 1998). $MgAl_2O_4$/Fe–Ni–Al alloy-bearing composites produced by the DCP method have exhibited average strength and toughness values of 380 MPa and 11.3 MPa m^{-2}, respectively (Rogers *et al.* 1999). Composites that are electrically conductive (i.e., with a continuous metal or intermetallic phase) or electrically insulating (i.e., with a discontinuous metal phase) have also been fabricated (Wagner *et al.* 1999, Kumar and Sandhage 1999, Rogers *et al.* 1999).

Bibliography

Beyer P, Janssen B, Claussen N 2000 Synthesis of aluminide-alumina composites by reactive squeeze casting. *Adv. Eng. Mat.* **2**, 734–7

Claussen N 1991 Fabrication and properties of low-shrinkage reaction-bonded mullite. *J. Am. Ceram. Soc.* **74**, 2460–3

Claussen N 1992 *US Pat.* 5158916. Process of producing powder metallurgically produced ceramic formed body

Gaus S P, Sheedy P M, Caram H S, Chan H M, Harmer M P 1999 Controlled firing of reaction-bonded aluminum oxide (RBAO) ceramics: part II, experimental results. *J. Am. Ceram. Soc.* **82**, 909–15

Holz D, Wu S X, Scheppokat S, Claussen N 1994 Effect of processing parameters on phase and microstructure evolution in RBAO ceramics. *J. Am. Ceram. Soc.* **77**, 2509–17

Kumar P, Sandhage K H 1998 The fabrication of near net-shaped spinel bodies by the oxidative transformation of Mg/Al_2O_3 precursors. *J. Mater. Res.* **13**, 3423–35

Kumar P, Sandhage K H 1999 The displacive compensation of porosity (DCP) method for fabricating dense, shaped, high-ceramic-bearing bodies at modest temperatures. *J. Mater. Sci.* **34**, 5757–69

Rogers K A, Kumar P, Citak R, Sandhage K H 1999 Dense, shaped ceramic/metal composites at $\leqslant$1000°C by the displacive compensation of porosity method. *J. Am. Ceram. Soc.* **82**, 757–60

Sandhage K H 1995 *US Pat.* 5447291

Sandhage K H, Allameh S M, Kumar P, Schmutzler H J, Viers D, Zhang X-D 2000 Near net-shaped, alkaline-earth-bearing ceramics for electronic and refractory applications via the oxidation of solid, metal-bearing precursors (the VIMOX process). *Mater. Manuf. Proc.* **15**, 1–28

Schen C, Dehm G, Kaplan W D, Wagner F, Claussen N 1998 Microstructure and phase evolution of niobium-aluminide-alumina composites prepared by melt-infiltration. *Phys. Stat. Sol.* (*a*) **166**, 241–55

Scheppokat S, Janssen R, Claussen N 1999 Phase development and shrinkage of reaction-bonded mullite composites with silicon carbide of different particle sizes. *J. Am. Ceram. Soc.* **82**, 319–24

Schmutzler H J, Antony M M, Sandhage K H 1994 A novel reaction path to $BaTiO_3$ by the oxidation of a solid metallic precursor. *J. Am. Ceram. Soc.* **77**, 721–9

Schmutzler H J, Sandhage K H, Nava J C 1996 The fabrication of dense, shaped barium cerate by the oxidation of solid metal-bearing precursors. *J. Am. Ceram. Soc.* **79**, 1575–84

Viers D S, Sandhage K H 1999 Near net-shaped $(Ba,Sr)Al_2Si_2O_8$ bodies by the oxidation of machinable metal-bearing precursors. *J. Am. Ceram. Soc.* **82**, 249–52

Wagner F, Garcia D E, Krupp A, Claussen N 1999 Interpenetrating Al_2O_3-$TiAl_3$ alloys produced by reactive infiltration. *J. Eur. Ceram. Soc.* **19**, 2449–53

Wu S X, Holz D, Claussen N 1993 Mechanisms and kinetics of reaction-bonded aluminum oxide ceramics. *J. Am. Ceram. Soc.* **76**, 970–80

K. H. Sandhage and N. Claussen

Reactive Polymer Processing: Fundamentals of REX

By *extrusion*, it is meant that a polymer melt is forced to go through a metal die that continuously shapes the melt into a desired form. The heart of a screw extrusion process is one or multi-Archimedean screws rotating in a heated barrel (see e.g., Tadmor and Gogos 1979, White 1990). The polymer, often in the form of particular solids (pellets, powder, etc.), is gravitationally fed to the screw(s) through a hopper or a feeder. The solids are then conveyed forward, plasticated, homogenized, and pressurized along the screw(s). Screw extruders are mainly used:

(i) to manufacture finished or semi-finished products (tubes, pipes, rods, cables, sheets, films, etc.);

(ii) to pelletize polymers;

(iii) to incorporate additives and/or reinforcing agents into polymer resins;

(iv) to purify polymers by eliminating volatile products;

(v) to serve as the last polymerization reactor to further increase the conversion of highly viscous polymerizing systems which cannot be handled readily by classical polymerization reactors.

An extruder can often be used to perform multi-functions in a sequential manner. For example, a screw extruder can be used at the same time as a reactor to house a polymerization reaction, a devolatilizer to remove residual monomer(s) further downstream using a venting port under vacuum, a compounder to incorporate additives such as stabilizing agents, and finally a shaping device to convert the product to a desired form.

1.1. Definition of Reactive Extrusion

In a broad sense, reactive extrusion (REX) refers to the use of screw extruders as chemical reactors to deliberately perform chemical reactions of polymers or polymerizable monomers (Brown and Orlando 1988, Xanthos 1992, Lambla 1993, Hu and Lambla 1997). It is a complex polymer reaction engineering process that combines both the traditionally separated polymer chemistry (polymerization and chemical modification) and extrusion (blending, compounding, structuring, devolatilization, and eventually shaping) into a single process carried out in a screw extruder. Being the most important type of polymer processing equipment, screw extruders have become chemical reactors.

2. Features, Advantages, and Limitations of Reactive Extrusion

The advantages and limitations of reactive extrusion are related to the inherent features of extruders as chemical reactors.

2.1 Features

Extruder reactors have the following nine main inherent features:

(*a*) *Great ability to convey and melt polymer particulate solids, and mix and pump highly viscous media.* This is probably the most important feature of extruder reactors that distinguishes them from other types of reactors.

(*b*) *Longitudinal and (nearly) plug flow reactor.* The geometry of screw extruders is characterized by one or multi-Archimedean screws housed in a heated barrel with a length-to-diameter ratio L/D typically of 15–50. Unless specially designed, there is little back-mixing in standard screw extruders.

(*c*) *Modular screw and barrel design.* This is particularly evident in modern twin-screw extruders. They are constructed based on the building-block principle. Screws and barrels are no longer made in one single block, but are composed of removable screw and barrel elements. A variety of screw and barrel elements is available. Different screw elements have different functions as to conveying, melting, (dispersive or distributive) mixing capacities, devolatilization, etc. The screw and barrel can be altered by simply changing the sequence or the type of screw elements and barrels along the shafts. In this way, almost an infinite number of screw and barrel profiles, or "micro-reactors in series," can be made "at will."

(*d*) *Short residence times.* Residence time in a standard screw extruder is between 0.5 and 5 minutes under typical processing conditions, although shorter or longer residence times are possible under certain circumstances.

(*e*) *Laminar mixing.* Mixing can be defined as a process or an operation to achieve a desired spatial distribution in terms of concentration. The basic mechanism of mixing is to induce relative physical motion of the ingredients involved in the composition. This allows the bringing together of catalysts and reactants for reaction. Because of the high viscosity and consequently very low Reynolds number, laminar flow is the only mechanism for polymer mixing. Laminar mixing can be dispersive (intensive) mixing or distributive (extensive) mixing in character. Dispersive mixing intends to reduce the size of cohesive agglomerates (e.g., carbon black) while distributive mixing is to randomize the component distribution of a non-cohesive mixture (e.g., liquid–liquid). Dispersive mixing requires mostly stress, while distributive mixing needs strain. Orientation of the material elements with respect to the flow direction is also important for distributive mixing. Extruder reactors provide both dispersive and distributive mixing, depending on the types of screw elements.

(*f*) *Limited heat transfer*. For a polymer melt, the thermal conductivity, k is $\sim 0.2\,W°C^{-1}m^{-1}$, the heat capacity, ρC_p is $\sim 2 \times 10^6\,J°C^{-1}m^{-3}$, and the thermal diffusivity, α is $\sim 10^{-7}\,m^2s^{-1}$. Heating polymers in a screw extruder is not a problem, and the specific energy required to bring them from room temperature to a reaction temperature of 150–300°C usually ranges from 3×10^8 to $7 \times 10^8\,J\,m^{-3}$. The energy required to heat up polymers in a screw extruder results from heat conduction from the barrel and viscous dissipation. Pressure-generated heat is often negligible. However, it is difficult to get heat out of polymeric systems because of their low thermal conductivity, as will be discussed further below.

(*g*) *Large viscous heat generation*. The rate of viscous heat generation per unit volume is proportional to the viscosity of the system and the square of the shear rate. Since the viscosity of a polymer melt is typically 10^2–10^4 Pa s, which is 10^5–10^7 times the viscosity of water, viscous heat generation is negligible for low viscosity fluids and very important for highly viscous materials such as polymers.

(*h*) *High pressure and temperature*. Extruder reactors can withstand pressures up to 300 bars and temperatures up to 500°C.

(*i*) *Ease of complete discharge*. Emptying a very viscous product from a conventional reactor can take a long time and leave a substantial "twilight"; an extruder provides continuous and complete discharge which, when passing through a die, forms semi-finished or finished products.

2.2 Advantages

When the above nine inherent features are used to advantage, extruder reactors can be very attractive for many reactive processing applications, and innovative processes/novel products can be developed. Compared with e.g., a stirred tank reactor, an extruder reactor bears the following advantages.

(*a*) *Solventless chemical reactions involving both low and high viscosity reactants*. Carrying out chemical reactions with polymers involved, whose viscosities are typically in the 10 to 10 000 Pa s range, is generally impossible in classical reactors. In a typical solution-process reactor, the amount of solvents/diluents is usually 1–20 times that of the desired polymer product. This means that some big commercial processes of this type use up to more than ten tons of solvents per day. The unique ability of extruders to handle highly viscous polymer fluids without the need for solvents can be used to advantage in reactive extrusion. This can result in dramatic cost reductions in raw materials, solvent storage and recovery, product separation, and drying equipment. Because of the absence of solvents, REX processes are safer, more environmentally friendly and less bulky compared with classical polymer reaction processes. The risk of solvent-induced side reactions is also reduced.

(*b*) *A continuous and flexible process for low-tonnage trials or production*. Making new materials of low-tonnage (say, below 5000 tons per year) usually does not justify the expense for new reactor start-up and associated costs. Screw extruders are just ideal for such processes. Under certain circumstances, specialty compounders can just use their extruders as reactors to make products in a ready-to-use form that meets specific needs, without having to rely on their resin suppliers. Owing to its ability to provide complete melt discharge, an extruder reactor is capable of processing a wide variety of materials, both thermoplastics and elastomers. Also grade changes can be accomplished with minimal heel. In this sense, extruder reactors are excellent multi-utility reactors.

(*c*) *Staging of sequential operations*. As longitudinal and nearly plug flow reactors capable of generating pressure via drag flow, extruder reactors allow for easy staging of multiple operations including feeding, solids conveying, melting, side-stream addition, mixing, reacting, venting, pressurizing, and shaping. Multiple feeding, multiple venting, and multiple reactions can be carried out along the screw length. Alternation of fully filled and partly filled zones along the screw axis can be easily done if required. In this sense, extruder reactors are excellent multi-functional reactors. Sometimes, staging allows the making of products which could not be made in a suitable fashion in classical reactors. The modularity of screw and barrel greatly contributes to staging and flexibility of REX.

(*d*) *Wide process conditions*. Extruder reactors can cope with both low and high temperatures and large temperature gradients. They can also operate under high pressure. Reactions involving catalysts and/or reactants of low boiling points have to be carried out under pressure. Gas-assisted reactions also require a certain pressure. Very high temperature is required for polymers of high melting or glass transition temperatures. However, too high a temperature may be a problem for selectivity control, or for thermally sensitive polymers. Low temperatures are difficult, however, in REX processes, because the temperatures of polymers have to be above their melting temperatures and well above their glass transition temperatures in order for them to gain enough mobility.

(*e*) *Unique processes and/or products.* Extruder reactors are uniquely suited to reactive compatibilization of immiscible polymer blends, dynamic vulcanization of thermoplastic elastomers, and fabrication of nano-materials, in which the control of the morphology and interfaces is facilitated by the absence of solvents.

2.3 *Limitations*

The inherent features of extruders as chemical reactors also bring about limitations or, more exactly, challenges/problems to be solved:

(*a*) *Residence time limitations.* In view of the short residence time, the extruder reactor is restricted to rapid chemical reactions. This reduces considerably the number of reaction systems which can be processed by REX. This is however compensated for by high reacting temperatures, and high concentrations of reacting species because of little or no solvent dilution. Rapid temperature rise as a result of reaction, and viscous dissipation can be used to advantage—but this remains a challenge. For these reasons, reactions that take hours in a solution-process reactor can be processed in an extruder reactor with a residence time of the order of minutes.

(*b*) *Mixing limitations.* In highly viscous media, laminar mixing is the primary mixing mechanism and molecular diffusion can be extremely slow. This is particularly so for reactions between polymer melts. If the polymer melts are immiscible, which is the case for most polymer pairs, reactions are then confined to the interfacial volume elements. Since the total interfacial volume is very small compared with the bulk volume, the extent of reaction between two immiscible polymers is often very low. For this reason, preparing a pure copolymer from two immiscible reactive polymers is likely to be very difficult, unless at least one of them is of low molecular weight. The fact that there is little or no back-mixing in extruder reactors may also be a handicap for certain types of reactions.

(*c*) *Heat removal limitations.* In a screw extruder, heat is removed from the polymer melt through the barrel. The driving force for removing the heat is the temperature difference between the bulk polymer and the coolant:

$$q_c = h_T A (T_P - T_c) \qquad (1)$$

where A is the interior barrel area, h_T is the overall heat transfer coefficient which is a measure of the heat removal capacity of the extruder, and T_p and T_c are the temperatures of the bulk polymer and the coolant, respectively. The overall energy balance for the extruder is as follows:

$$W C_P \Delta T = q_b + E_\eta + q_r - q_c \qquad (2)$$

where W is the mass flow rate, C_p is the specific heat capacity of the polymer, ΔT is the temperature rise of the polymer, and q_b, E_η and q_r are the heat input by conduction through the barrel, the viscous dissipation energy and the exothermic heat of reaction, respectively. If the reaction is endothermic, the term q_r is negative. Limitations on heat removal out of an extruder are the primary concern when reactions releasing a large amount of heat are to be performed. The heat of reaction is usually released in a short time, of the order of the residence time, in a screw extruder and may not be adequately evacuated. This problem is of particular importance for chain polymerization of vinyl monomers. For example, opening the carbon–carbon double bond of vinyl monomers by free radical initiation yields a heat of some 50–90 kJ mol^{-1}. On a weight basis, the amount of the heat of polymerization is more important for small alkylenes like ethylene and propylene. The heat released by the conversion of styrene to polystyrene may reach some 700 kJ kg^{-1}. This amount of heat just cannot be removed fast enough with large-scale extruders, and the polymerization could be of a "run-away" nature. In adiabatic operation, the temperature rise can be several hundred degrees, implying that the system would be burnt out before such a large temperature rise is achieved. Thus, REX is restricted to reacting systems whose heat of reaction can be handled.

(*d*) *Cultural limitations.* Despite its ever-increasing acceptance in the polymer industry, the extruder as a reactor is not a well-known phenomenon in present day process technology, or is barely recognized as such. Chemists and chemical engineers are not used to considering the extruder as a reactor, because they feel much more comfortable with existing processes and stirred tank reactors, for which engineering principles are much better established and databases are much more complete. The lack of instruction on the subject is responsible, at least partly, for this situation. In most engineers' minds, the extruder has been designed and used for compounding and shaping operations, and not as a chemical reactor. Mechanical engineers design their machines for handling polymer solids and polymer melts, but not chemical reactions, and are mostly unaware of the typical requirements for such purposes. However, the design probably has to be adapted considerably in order to optimize the extruder as a reactor. For example, reactions may involve both low- and high-viscosity materials, often with a large disparity in compositions. The extruder is particularly suitable for highly viscous media and does not always work well when low- and high-viscosity materials are present. The extruder has long been designed to handle products whose viscosity decreases in the flow direction, from polymer solids to polymer melts; however, during polymerization, the viscosity of the reacting system increases in the flow direction. It may be difficult to push a viscous material forward with a low-viscosity material, giving rise to process instability and large variation in product quality.

3. *Possibilities of Reactive Extrusion*

Reactive extrusion offers numerous possibilities for making novel polymer products that could not be made in a suitable fashion—at least at reasonable cost—by conventional reactors, as outlined below.

3.1 Bulk Polymerization

This refers to the polymerization of a monomer, or a mixture of monomers, to a high molecular weight (co)polymer directly in extruders, with little or no solvent dilution. In such cases, extruders are used as true polymerization reactors. They are very interesting for polymerization operations that otherwise could not be handled correctly or economically by conventional polymerization reactors. Basically, there are two types of polymerization reactions: polycondensation and polyaddition. Polycondensation reactions are concerned with the covalent connection of monomer molecules, leading to high molecular weight polymers with the release of a small molecule like water, or a low carbon number alcohol. Polyaddition reactions include pseudo-polycondensation, ring-opening polymerization, free radical polymerization, anionic polymerization, and coordination polymerization. Examples of polycondensation include the synthesis of poly(-ether)imides (Schmidt *et al.* 1989), saturated polyesters (Gouinlock *et al.* 1968) and polyamides. Among the examples of polyaddition are the preparation of polyurethanes (Bouilloux *et al.* 1991, Uhland and Wiedman 1994) and polyurethanes-ureas (Quiring *et al.* 1981) (pseudo-polycondensation); polyamide 6 (Illing 1969, Hornsby *et al.* 1994; Key and White 1994); polyamide 12 (Biensan 1976); polyoxymethylene (Seddon and Scarbrough 1968) and epoxides (Heinemeyer and Tatum 1986) (ring-opening polymerization); polystyrene (Kelley 1993); polyacrylates and polymethacrylates (Stuber and Tirrell 1985) (free radical polymerization).

3.2 Chemical Modification of Existing Polymers

Grafting of monomers onto polymer backbones or side chains is probably the most important method of chemical modification of polymers in an extruder reactor (Hu *et al.* 1997). It involves the reaction of a molten polymer with a small molecule or an oligomer capable of forming grafts to the polymer backbone. The properties of the resulting polymer depend very much on the length of the grafts, which in turn is related to the type of molecule employed. If the grafts are long enough, the resulting product is a true graft copolymer, which has physical properties very different from those of the substrate polymer. However, if the grafts are short, the substrate polymer may have almost the same physical properties but markedly different chemical properties. There are four main categories of grafting reactions:

(*a*) *Free radical grafting of vinyl monomers.* This is probably the oldest, most widely practiced, and least expensive one. These advantages are further gained with the use of screw extruders as chemical reactors, which allow the carrying out of free radical grafting without solvents. Free radical grafting has been used to promote functional and/or mechanical properties of various types of polymers, especially hydrocarbon polymers such as polyethylene (PE) and polypropylene (PP). Frequently used vinyl monomers are vinylsilanes (Ultsch and Fritz 1990), maleic anhydride(Gaylord and Mehta 1988, Hu *et al.* 1997), glycidyl methacrylate (Sun *et al.* 1995, Cartier and Hu 1998), 2-(dimethylamino)ethyl methacrylate (Baker and Saleem 1987), 3-isopropenyl-α, α-dimethyl-benzyl isocyanate (or TMI) (Hu and Li 2001).

(*b*) *Ionic grafting.* Plastification of poly(vinyl chloride) (Mijangos *et al.* 1986); sulfonation of unsaturation-bearing polymers (Siadat *et al.* 1980).

(*c*) *Exchange grafting.* Reaction of amines or alcohols with polymer backbones containing anhydride or ester moieties (Lambla *et al.* 1990).

(*d*) *Modification of functional groups attached to polymer backbones.* Examples are conversion of the acetate group in an ethylene-vinyl acetate copolymer to the corresponding alcohol group (Lambla *et al.* 1987); conversion of ester groups in (meth)acrylate-bearing polymers to the corresponding amide or to new ester groups with a low molecular weight amine (Hu *et al.* 1992) or alcohol (Hu and Lambla 1994).

3.3 Molecular Weight Build-up and Controlled Chain Degradation

Molecular weight build-up can be achieved by various methods, including coupling and interchain polymer formation.

(*a*) *Coupling.* Reaction of polymers with polyfunctional coupling or branching agents. For example, bis-2-oxazolines were used to increase the molecular weight of PET (Inata and Matsumua 1986).

(*b*) *Interchain polymer formation.* Reaction between reactive polymers leading to the formation of random, graft, or block copolymers. Various types of reactions can lead to copolymer formation. Very often, the objective is not to obtain pure copolymers but to generate enough copolymers for compatibilizing immiscible blends. For example, the morphology and interfaces of a chemically inert polypropylene and polyamide 6 blend can be controlled by adding a maleic anhydride functionalized polypropylene. This latter can react with the terminal amine group of the polyamide 6, yielding a graft copolymer which then serves as an interfacial agent (Cartier and Hu 1999).

Controlled chain degradation includes the molecular weight degradation of high molecular weight polymer (controlled rheology) and controlled degradation to monomer. The most relevant example is the rheology-controlled polypropylene (Tzoganakis *et al.* 1988). Commercial polypropylene resins produced by Ziegler–Natta catalysts often have poor processability, because of their high polydispersity. Peroxide-promoted degradation allows the narrowing of the molecular weight distribution, thus improving their processability.

3.4 Reactive Compatibilization of Polymer Blends

Obtaining new polymer materials by blending existing polymers has become more and more attractive over the past two decades. Properties of such multiphasic materials depend very much on their morphology and interfaces, which can be controlled by the presence of an appropriate copolymer as a compatibilizer. Such a copolymer can either be prepared separately and then added to the blend (non-reactive compatibilization or *ex situ* compatibilization), or formed *in situ* at the interfaces between the base polymer components directly during the melt blending process (reactive compatibilization, *in situ* compatibilization or reactive blending) (Baker *et al.* 2001). This latter method is very attractive because, in addition to offering the possibility of generating compatibilizers at the interfaces between the base polymer components where they are most needed, without separate synthetic and purification steps which are involved otherwise in non-reactive compatibilization, it eliminates the problem of getting the compatibilizers to the interfaces.

3.5 Dynamic Vulcanization

This refers to the process of crosslinking (or vulcanizing) an elastomer during its melt-mixing with a melt-processable thermoplastic matrix, to give blends which have the processability of thermoplastics and the end-use properties of crosslinked elastomers (Coran and Patel 1997). When mixed in the proper proportions, these blends have a rubbery dispersed phase of crosslinked elastomer and a continuous phase of thermoplastic resin. Crosslinking the rubbery phase stabilizes the rubber particulate morphology, which prevents reagglomeration of rubber particles before the thermoplastic matrix can crystallize or cool to a glassy state, and during any subsequent processing steps.

3.6 Fabrication of Nano-materials (Nano-blends and Nano-composites)

Nano-blends refer to immiscible polymer blends in which the scale of dispersion of one polymer phase in another one is below 100 nm; nano-composites are concerned with polymer-filler systems in which the size of the filler in the polymer matrix is below 100 nm (Hu *et al.* 1999). These types of materials can have superior properties over classical polymer blends or composites.

3.7 Recycling of Plastics

Reactive extrusion can be an interesting process for plastics recycling. For example, used poly(ethylene terephthalate) or PET can be alcoholyzed to a diester and a dialcohol. The latter can be polymerized to PET again. This process can also be applied to polyurethanes. Used poly(methyl methacrylate) (PMMA) can be depolymerized thermally, leading to the formation of the corresponding monomer, methyl methacrylate. Mixtures of used plastics can be reactively compatibilized to give new materials.

4. Practical Aspects of Reactive Extrusion

As described above, an extruder reactor should be seriously considered as a vehicle to carry out reactions when:

(i) at least one component is (or will be) very viscous;

(ii) the reaction time is less than, say, 30 minutes;

(iii) the heat of reaction can be handled.

The success of a REX process depends very much on how well the chemistry fits in the extruder reactor.

4.1 Chemistry

Because there is little or no solvent dilution, reaction conditions in REX are characterized by high viscosity, high temperature, high pressure, high concentration, laminar mixing, little or no solvent effects, and possibly heterogeneity of the reacting media. Before a reaction is processed by REX, the following aspects have to be examined:

(i) Reactions which are well known in solution or under conventional conditions may behave very differently when carried out under REX conditions. When these unique conditions are used to advantage, innovative reactions can be developed. However, the reaction mechanism has to be identified and the main parameters influencing the mechanism and kinetics ought to be quantified. Qualitative or quantitative understanding can be gained using a small-volume batch mixer capable of melting polymer solids and handling viscous media. If the reaction is controlled mainly by reaction time, the main question is whether the kinetics are compatible with the residence time available in the reaction zone of the extruder under defined process conditions. The estimate of reaction time will probably have to be verified in the extruder, because it may differ considerably from batch data. Reaction conditions in an extruder reactor can be very different from those in a batch reactor in terms of temperature, pressure, and mixing. It should be noted that not all reactions are controlled by residence time. For example, the rates of reactions between immiscible polymer melts are controlled primarily by the interfacial area (Hu and Kadri 1998), which is related to the extruder's ability to generate dispersive and distributive mixing. When a reactive gas is to react with a polymer melt, it is the surface of the polymer melt in contact with the gas that controls the rate of reaction, which in turn depends on the degree of fill of the screw channel and the extruder's ability to continuously renew the surface (Kowalski 1992).

(ii) In REX, local hot spots as a result of the heat of reaction or viscous dissipation can largely exceed the set temperature, so that reactions can be in close-to-adiabatic conditions. This is particularly the case in large-scale extruder reactors. Exposure to oxygen may be unavoidable, even when an inert gas stream is operative. Under such conditions, the stability of the reactants and products should be examined.

(iii) The particular reaction conditions in REX also require a reasonable estimate of their effects on possible side reactions, in order to see whether or not they are controllable to within an acceptable limit.

(iv) In addition to catalyst, reactants, and base resins, additives and/or reinforcing agents may be present in the reacting system, or may have to be added to it. Their effects on the reaction need to be evaluated, and solutions have to be found in order to minimize them or to use them to advantage.

(v) The addition sequence of catalyst, reactants, and base resins may influence the product quality. Strategies have to be developed to best take advantage of the staging of REX.

(vi) Some catalysts, reactants and/or products can be corrosive, toxic, and/or explosive. Measures should be taken accordingly.

4.2 Residence Time

The overall mean residence time in a screw extruder, $\bar{t}$, is related to the extruder's volume, V, the degree of fill, f, and the volumetric throughput, Q, by:

$$\bar{t} = \frac{V_{\mathrm{f}}}{Q} = \frac{fV}{Q} \tag{3}$$

where V_{f} is the filled extruder volume, which is the product of fV. The degree of fill, f, is often unity for single-screw extruders, and less than unity for twin-screw extruders. This is because the former are often flood-fed and throughput is closely related to screw speed, whereas the latter are often starve-fed via, e.g., a feeder, and throughput is independent of screw speed. From Eqn. (5), $\bar{t}$ in a twin-screw extruder is dictated mainly by Q and f. The latter often varies along the screw length, and should thus be considered as an average value with respect to the screw length. It is a complicated function of screw and die combination, Q, and screw speed, N.

Local residence times in sequential zones are often very important for process optimization. They can be determined by theoretical and experimental tools (Chen *et al.* 1993; Chen and Hu 1993, Chen *et al.* 1995, Hu *et al.* 1999).

4.3 Heat Transfer

The overall heat transfer coefficient of a screw extruder, h_{T}, is difficult to determine theoretically and experimentally (Domschke and Justus 1993). It is typically in the range between 300 and 900 $\mathrm{W m^{-2} K^{-1}}$, and increases with increasing extruder size. Todd proposed a correlation for a co-rotating twin screw extruder (Todd 1988):

$$Nu = \frac{h_T D}{k} = 0.94 \left(\frac{\rho N D^2}{\eta}\right)^{0.28} \times \left(\frac{C_p \eta}{k}\right)^{0.33} \left(\frac{\eta}{\eta_w}\right)^{0.14} \quad (4)$$

where D is the screw diameter. Cooling polymer melts in screw extruders is difficult, because a maximum heat flux of $3 \times 10^3\,\mathrm{Wm^{-2}}$ is rarely exceeded. This implies that most vinyl monomers cannot be polymerized readily, and especially in large extruders because of reduced surface-to-volume ratio. However, chemical modification of existing polymers, reactive blending, and dynamic vulcanization are good candidates for REX because there is little concern about the heat of reaction. In order to minimize the excessive heat of polymerization to an acceptable level, a number of measures can be taken, however:

(i) *addition of an inert material as a heat sink to the monomer feed.* It can be a previously polymerized product, a different polymer, an inert filler, or a volatile and inert liquid which will be removed downstream.

(ii) *prepolymerization* of part of the monomer in a stirred tank reactor.

(iii) *evaporation of a volatile liquid*, e.g., the monomer, during the heating up of the polymerizing mass.

(iv) *homo- or copolymerization of a macromonomer.*

4.4 Feeding

Twin-screw extruders are usually starve-fed with one or several accurate feeders. Accurate feeding is particularly important for reactions which require a good control of the stoichiometric ratio of reactants, as is the case for polycondensation and pseudo-polycondensation reactions. Major components are usually fed to the main hopper, and minor ones, if not preblended with the major ones, can be fed through a top or a side port. Non-reacting additives (stabilizers, colorants, etc.) may need to be incorporated downstream prior to discharge of the final product, because of incompatibility with the reaction or devolatilization step.

4.5 Staging

Staging in twin-screw extruders is relatively easy because the degree of fill and pressure in screw channels can be controlled, especially with modular co-rotating twin-screw extruders using kneading blocks and left-handed elements. Fully filled channels generate and provide a mixing action that is superior to that in partly filled channels, especially for introducing low-viscosity chemicals like liquid catalyst and reactants (Hu and Cartier 1998). Partly filled channels provide a large melt surface, which is very important for reaction between gas and polymer melt. They facilitate side feeding of low melting point reactants, and elimination of volatile products such as unreacted monomers from the extruder through the application of vacuum—devolatilization. Devolatilization is particularly important for situations where small molecules affect the progress of reactions like polycondensation reactions and reactive compatibilization of immiscible polymer blends by one-step reactive extrusion (Hu *et al.* 1996).

A certain screw length is required to perform a given sequential operation such as reacting and venting. Optimization of the screw length required for each sequential operation is important, but difficult. As a rule of thumb, a screw length of 2-5 L/D is usually required to accomplish a sequential operation. An estimate of the sequence of different operations to be accomplished in a projected REX process helps to identify the zones the extruder must have in order to complete these sequential tasks. A REX process should accomplish at least 5 sequential operations (feeding, solids conveying, melting/plastification, reacting, and pumping), requiring a total screw length of about 10 L/D. 10 sequential operations with possible multiple feeding, multiple reactions, and/or multiple venting are common, requiring a screw length of 20 to 50 L/D.

4.6 Scale-down/Scale-up

Extruders can be as small as 15 mm in diameter and as large as 800 mm. Scale-up is known to be a problem for any engineering process. Actually there is more of a scale-down problem with REX, in that it is difficult to simulate in a small extruder the conditions that prevail in a large production extruder. This is particularly so when heat transfer is the limiting factor. For example, an adiabatic temperature regime, as supposedly will exist in a large extruder, cannot be replicated in a small extruder by imposing a barrel temperature profile, because heating and cooling via the barrel are still operative.

The following five similarities might be used as a starting point for scale-down or scale-up of REX (Meijer and Elmans 1988):

(i) *Product quality similarity.* The product quality, e.g., the conversion of the reaction, is to be kept the same in both extruders. This similarity is the objective of virtually all engineering processes.

(ii) *Residence time similarity.* If the product quality of the REX process is controlled mainly by residence time, special attention should then be paid to this similarity for scale-down/scale-up.

(iii) *Hydrodynamic similarity.* This similarity is likely to exist if the degree of fill, the filled length, and dimensionless flow profiles are the same in both extruders.

(iv) *Thermal similarity*. This requires that the temperature profile in both extruders is the same. Thermal similarity is not easy to achieve, especially for free radical polymerization of vinyl monomers whose heat of reaction is generally very high, or for highly viscous systems whose viscous dissipation energy can be very important. In this latter case, similarity based on specific energy consumption is often used.

(v) *Geometrical similarity*. This similarity exists when the ratio of any two length parameters in the large-scale extruder is the same as the ratio between the corresponding lengths in the small-scale extruder. While this similarity appears to be the least scientific for scale-down/scale-up of REX, it is the easiest to achieve. If the industrial-scale extruder is a co-rotating twin-screw, the laboratory-scale extruder should also be of the same type, preferably with the same L/D ratio.

5. *Future Trends*

REX has been gaining more and more acceptance in the industry as an important post-reactor technology that is complementary to the classical in-reactor technology. Both technologies even compete in certain areas. REX is now reaching its "golden age." Development of new chemistries and new material formulation methods which are applicable to REX, together with new possibilities for on-line process monitoring and control, will surely lead to novel processes/products. Establishment of sound engineering principles for mixing, devolatilization, and scale-down/scale-up will make REX processes more competitive. The future of REX should be bright if long-term commitments continue to develop between the university and industry.

Bibliography

Baker W E, Saleem M 1987 *Polymer* **28**, 2057

Baker W, Scott C, Hu G H 2001 *Reactive Polymer Blending*. Hanser, Munich, Germany

Biensan M 1976 *Fr. Pat. 2 291 231*

Bouilloux A, Macosko C W, Kotnour T 1991 *Ind. Eng. Chem. Res.* **30**, 2431

Brown S B, Orlando C M 1988 In: Kroschwitz J L (ed.) *Encyclopedia of Polymer Science and Engineering*, **14**. Wiley, New York, p. 169

Cartier H, Hu G H 1998 *J. Polym. Sci. A: Chem. Ed.* **36**, 1053

Cartier H, Hu G H 1999 *Polym. Eng. Sci.* **39**, 996

Chen L, Hu G H 1993 *AIChE. J.* **39**, 1558

Chen L, Hu G H, Lindt J T 1995 *Polym. Eng. Sci.* **35**, 598

Chen L, Pan Z, Hu G H 1993 *AIChE. J.* **39**, 1455

Coran A Y, Patel R P 1997 Thermoplastic elastomers based on elastomer/thermoplastic blends dynamically vulcanized. In: Al-Malaika S (ed.) *Reactive Modifiers for Polymers*. Thomson Science and Professional, London, pp. 349–94

Domschke T, Justus C 1993 *Int. Polym. Proc.* **8**, 294

Gaylord N G, Mehta R 1988 *J. Polym. Sci. A: Chem. Ed.* **26**, 1903

Gouinlock E V, Marciniak H W, Shatz M H, Quinn E J, Hindersinn R R 1968 *J. Appl. Polym. Sci.* **12**, 2403

Heinemeyer B W, Tatum S D 1986 *US Pat. 4 612 156*

Hornsby P R, Tung J F, Taverdi K 1994 *J. Appl. Polym. Sci.* **53**, 891

Hu G H, Cartier H 1998 *Int. Polym. Proc.* **13**, 1

Hu G H, Cartier H, Plummer C 1999 *Macromolecules* **32**, 4713

Hu G H, Flat J J, Lambla M 1997 Free-radical grafting of monomers onto polymers by reactive extrusion: principles and applications. In: Al-Malaika S (ed.) *Reactive Modifiers for Polymers*. Thomson Science and Professional, London, pp. 1–80

Hu G H, Holl Y, Lambla M 1992 *J. Polym. Sci.: Chem. Ed.* **30**, 635

Hu G H, Kadri I 1998 *J. Polym. Sci. B: Phys. Ed.* **36**, 2153

Hu G H, Kadri I, Picot C 1999 *Polym. Eng. Sci.* **39**, 930

Hu G H, Lambla M 1994 *Polymer* **35**, 3082

Hu G H, Lambla M 1997 Fundamentals of reactive extrusion: an overview. In: Meijer H E H (ed.) Processing of Polymers, Vol. 18. In: Cahn R W. Haasen P, Kramer E J (eds.) *Encyclopedia of Materials Science and Technology*. Elsevier, Amsterdam, Chap. 6

Hu G H, Li H 2001 Free radical grafting of 3-isopropenyl-α, α-dimethyl-benzyl isocyanate onto polypropylene in the bulk *J. Polym. Sci. A: Chem. Ed.* submitted

Hu G H, Sun Y J, Lambla M 1996 *Polym. Eng. Sci.* **36**, 676

Illing G 1969 *Mod. Plast* **46**(8), 70

Inata H, Matsumura S 1986 *J. Appl. Polym. Sci.* **32**, 5193

Kelley J M 1993 *US Pat. 5 270 029*

Key H, White J L 1994 *J. Appl. Polym. Sci.* **52**, 1249

Kowalski R C 1992 Fit the reactor to the chemistry. In: Xanthos M (ed.) *Reactive Extrusion: Principles and Practice*. Hanser, Munich, Germany, Chap. 1

Lambla M 1993 Reactive processing of thermoplastic polymers. In: Allen G, Bevington J C (eds.) *Comprehensive Polymer Science* (1st suppl.). Pergamon, Oxford, Chap. 21

Lambla M, Druz J, Bouilloux A 1987 *Polym. Eng. Sci.* **27**, 1221

Lambla M, Mazeres F, Druz J 1990 *Plast. Rubber Proc. Appl.* **13**, 75

Meijer H E H, Elmans P H M 1988 Modeling of continuous mixers: the corotating twin-screw extruder. In: Cheremisinoff N P (ed.) *Encyclopedia of Fluid Mechanics, Vol. 9: Polymer Flow Engineering*. Gult Publishing Company, Houston, TX, Chap. 12

Mijangos C, Martinez A, Michel A 1986 *Eur. Polym. J.* **22**, 417

Quiring B, Wenzel W, Niederdellmann G, Wagner H, Goyert W 1981 *Eur. Pat. EP 21 323*

Schmidt L R, Lovgren E M, Meissner P G 1989 *Int. Polym. Proc.* **4**, 270

Seddon R M, Scarbrough L D 1968 *Fr. Pat. 1 508 440*

Siadat B, Lubdberg R D, Lenz R W 1980 *Polym. Eng. Sci.* **20**, 530

Stuber N P, Tirrell M 1985 *Polym. Proc. Eng.* **3**, 71

Sun Y J, Hu G H, Lambla M 1995 *Angew. Makromol. Chem.* **229**, 1–13

Tadmor Z, Gogos C G 1979 *Principles of Polymer Processing*. Wiley, New York
Todd D B 1988 ANTEC Technical Paper 34. p. 54
Tzoganakis C, Vlachopoulos J, Hamielec A E 1988 *Polym. Eng. Sci.* **28**, 170
Uhland E, Wiedman W 1994 *Macromol. Symp.* **83**, 59
Ultsch S, Fritz H G 1990 *Plast. Rubber Proc. Appl.* **13**, 8
White J L 1990 *Twin Screw Extrusion: Technology and Principles*. Hanser, Munich, Germany
Xanthos M 1992 *Reactive Extrusion: Principles and Practice*. Hanser, Munich, Germany

G. H. Hu

Reactive Polymer Processing: Interfacial Aspects

Reactive blending of dissimilar polymers yields block or graft copolymers at polymer–polymer interfaces. The copolymer is expected to reduce the interfacial tension and to prevent coalescence of dispersed particles. Then, the morphology could be tuned up. The presence of copolymers at the interface may improve the adhesion between the two phases and, consequently, material properties can be improved. This seems to be a commonly accepted view of the ambiguous but convenient term "compatibilization." In this article, supplemental evidence for this view is shown for a particular polymer–polymer combination, polysulfone (PSU)/polyamide (PA) as an example, and then other interfacial aspects of the *in situ*-formed block and graft copolymers are presented.

1. *Emulsifier in Polymeric Oil-in-oil Emulsion*

A two-phase polymer melt is essentially a polymeric oil-in-oil (POO) emulsion. Block and graft copolymers play the role of emulsifier ("soap") in the POO-emulsion. When copolymers are formed *in situ* during melt mixing, the particle size is expected to reduce easily.

In Fig. 1(a), PA and PSU were melt-mixed at a weight ratio of 35/65 (PA/PSU) and a small amount of PSU with a phthalic anhydride chain-end (PSU–PhAH; high molecular weight, comparable to those of PSU and PA) was added as a reactive component. The PSU particle size rapidly decreases during mixing. The size decreases by three orders of magnitude, from millimeter scale (pellet size) to micrometer level, in a short time (2 min). A greater amount of PSU–PhAH renders a faster size reduction and finer particle size. The attainable particle size is plotted as a function of PSU–PhAH content in Fig. 1(b), demonstrating clearly the emulsifying effect of the *in situ*-formed block copolymer. Figure 1(c) shows the variations in time of the particle size D during static annealing at 310°C. All specimens here were melt mixed for 1.5 min ($t_r = 1.5$ min). In nonreactive systems (PSU–PhAH = 0 wt.%), D increases with time of annealing. This suggests that coalescence takes place to get bigger droplets during static annealing. In contrast, the particle size D of the reactive system containing a large amount of PSU–PhAH (5, 10 wt.%) remains constant, suggesting that the *in situ*-formed block copolymers prevent the coalescence. This droplet stabilization mechanism is already present in a short mixing (reaction) time.

2. *Pull-out of* In situ-formed *Copolymers*

TEM observation of the block copolymer forming system, PA/PSU–PhAH (without nonreactive PSU), after a long time (13 min) of mixing showed that the PSU particle size was very small (~40 nm) and the size distribution was quite narrow. The size was comparable with the domain size of a premade block copolymer. This suggests a potential for solvent-free preparation of PSU–PA block copolymers. TEM observation of the early stages of reactive blending (~1 min) showed that the *in situ*-formed copolymer chains left from the surface of huge PSU particles (~10 μm) to form micelles in the PA matrix. Such micelle formation was observed only in the melt-mixed blend and not in quiescently annealed blends,

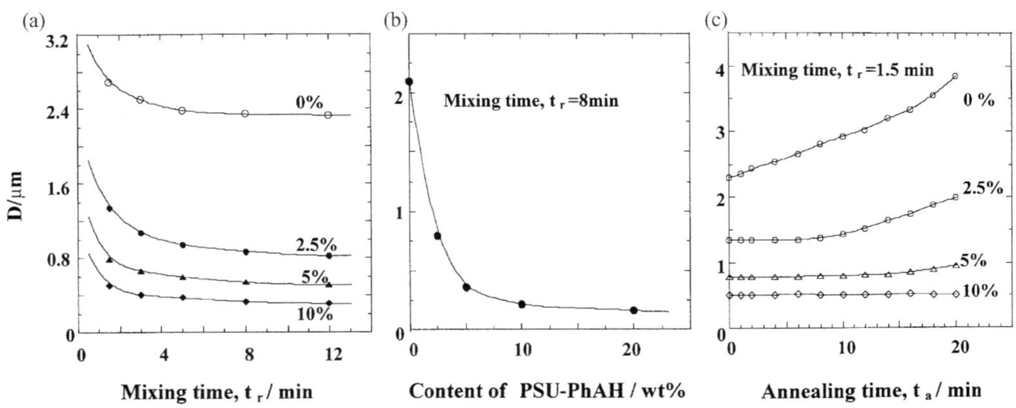

Figure 1
(a) Time variation of the average particle size in 35/65 PA/(PSU + PSU–PhAH) blends during melt mixing at 310°C. PSU–PhAH content varied as shown in the figure. (b) Average particle size (after 8 min mixing) vs. loading amount of reactive PSU. (c) Time variation of the average particle size during static annealing at 310°C (after mixing for 1.5 min).

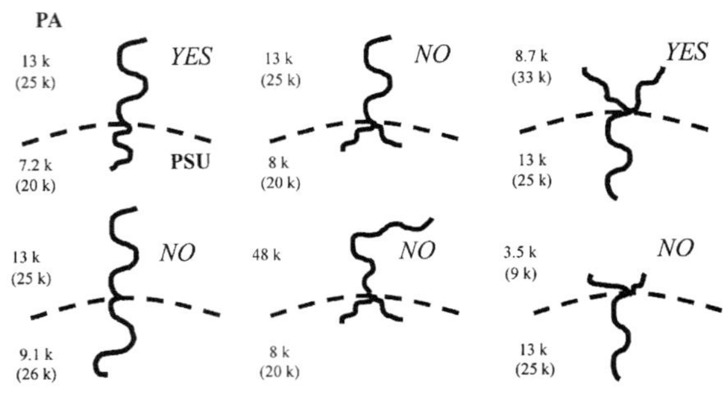

Figure 2
"Pull-out or not" for various copolymers *in situ*-formed in reactive PA/PSU blends. Numbers in and out of parenthesis are Mn and MW of component polymers, respectively.

suggesting that the block copolymers prefer to be located at the interface under static conditions, but that they are easily pulled away from the interface when PA-brushes are subjected to external shear forces during melt-mixing.

When the *in situ*-formed copolymer is pulled out from the interface by external forces, the copolymer does not work as the emulsifier and thus cannot tune up the morphology, e.g., at the sub-micrometer level. On the other hand, the pull-out is an interesting mechanism to create a block copolymer by a "dry" process (solvent-free synthesis). "Pull-out or not" highly depends on the molecular architecture of the *in situ*-formed copolymer, as is shown schematically in Fig. 2. Taking into account the results for other reactive systems, a dense accumulation of *in situ*-formed copolymer chains seems to be a prerequisite for the pull-out mechanism. Note that the results in Fig. 2 are against the thermodynamic theory of polymer–polymer interfaces. The theory claims that symmetric *AB* copolymers (*A* block length $\approx$ *B* block length) prefer to stay at the interface, whereas asymmetric ones (e.g., long *A* block/short *B* block) are unstable at interfaces and tend to leave the interface as micelles into an *A* homopolymer phase. The theory is developed for static conditions, thus it may not be suitable to discuss the stability/instability of the copolymers under shear fields.

3. *Morphology: Particle Size–Asphericity*

During melt mixing of immiscible polymers in, e.g., extruders, the dispersed particles are alternatively subjected to shear fields (in the narrow gap between the barrel surface and the top of screw flight) and left to relax at low shear fields (especially at the bottom of the screw channel). The larger the particle, the more it gets deformed during shearing and the less it recovers to the spherical shape during "rest" time. The relaxation to a spherical shape is driven by the interfacial tension and hindered by the viscosity of the particle and the matrix. Interface effects dominate over bulk effects when the particle size is small enough, and vice versa. Once the blend is quenched, the shape relaxation process is virtually frozen, and all particles remain more or less aspherical; the larger the particle, the greater would be its residual asphericity ζ.

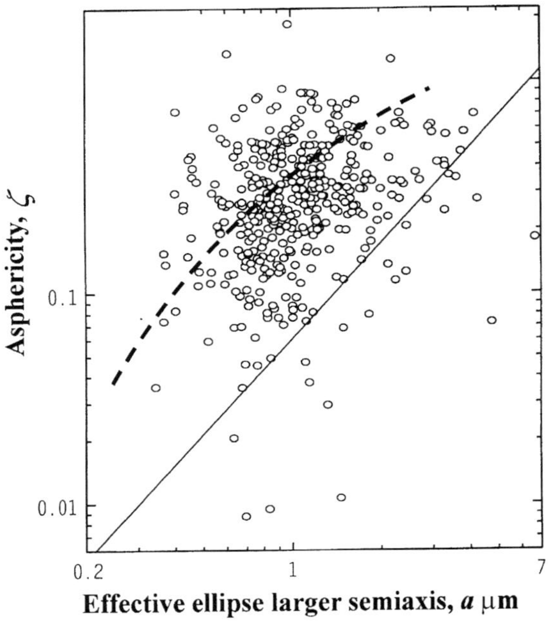

Figure 3
Particle asphericity–size correlation for nonreactive 80/20 PA/PSU blend. Solid line shows the lower boundary above which 95% of data points lie. A lower boundary for reactive system (PA/PSU–MAH) is also shown by the broken line.

Figure 3 shows results from a digital image analysis of melt mixed PA/PSU (nonreactive) and PA/PSU–MAH (reactive) systems (PSU–MAH = maleic anhydride grafted PSU). Data points are for the nonreactive system. As expected from the discussion above, the data points are scattered above a lower boundary of positive slope (the solid line). The upper boundary may be due to breakup of very elongated particles. A perfectly spherical shape may only be reached asymptotically, after infinitely large times. Since the specimens are quenched immediately after sampling, the asphericity parameter should not be less than a certain value corresponding to the lower boundary. In the reactive system, the *in situ*-formed graft copolymer would reduce the interfacial tension and lead to a more aspherical shape of particles. Then, the lower boundary would be shifted upwards, compared with the nonreactive system. This is realized as shown by the broken line in Fig. 3. Such a vertical shift was also observed in ternary systems including nonreactive PSU, PA/PSU/PSU–MAH (small amount); i.e., the ζ–a map is shifted vertically to higher asphericity and smaller particle sizes as the loading amount of reactive PSU increased. It clearly suggests the emulsifying effect of the *in situ*-formed graft copolymer. By contrast, in *in situ* block copolymer-forming (PSU–PhAH) systems, the ζ–a map shifted rather horizontally to smaller particle sizes with increasing loading amount of reactive PSU. The difference in the ζ–a map between the *in situ* graft copolymer- and block copolymer-forming systems may be interpreted by the "pull-out or not" of the *in situ*-formed copolymers from the interfacial region by shear forces during mixing.

4. In situ-*formed Copolymer as a Phase Inversion Aid*

In nonreactive blends of dissimilar polymers, the major component tends to be the continuous phase (matrix), while the minor one forms the dispersed phase. In a symmetric blend (50/50), the lower viscosity component tends to be the matrix. The situation has been empirically described by the following: component A likes to be the matrix when $\varphi_A\eta_B > \varphi_B\eta_A$, where φ and η are the volume fraction and the melt viscosity, respectively. Actually, a 35/65 PA/PSU (nonreactive) blend had a two-phase structure of PA particles dispersed in a PSU matrix. However, when a small amount of PSU–PhAH (Mw = 5.5 k; low molecular weight, compared with PA and PSU) was loaded, phase inversion took place during melt mixing to yield a two-phase structure of PSU particles/PA matrix. In the PSU particles, tiny micelles were occluded, suggesting the pull-out of *in situ*-formed block copolymer. The micelle formation by pull-out was shown to elevate the melt viscosity of the PSU phase and thus to reach a situation of $\varphi_{PA}\eta_{PSU} > \varphi_{PSU}\eta_{PA}$.

5. *Undulation of Polymer–Polymer Interfaces*

A PA/PSU–MAH bilayer specimen was prepared and annealed at 200°C for 30 min. Then, the upper thin PSU layer was removed by immersing the system in a selective solvent for PSU (dichloromethane) and subsequently the exposed interface (surface of PA substrate) was examined by atomic force microscopy (AFM) in a contact mode. The AFM image is shown in Fig. 4. The surface is not flat, but undulated. The average amplitude and wavelength of the undulation were 18 nm and 0.2 μm, respectively. Such undulation was never observed in nonreactive systems. It is caused by the chemical reaction (the *in situ* formation of graft copolymer). The undulation may be driven by an entropic penalty imposed on the *in situ*-formed

Figure 4
AFM image for PA/PSU–MAH system annealed (reacted) at 200°C for 30 min.

graft copolymer chains accumulated densely at the interface.

When the areal density of the copolymer chains formed *in situ* at the interface increases sufficiently, the copolymer chains are forced to stretch perpendicular to the interface leading to a loss in configurational entropy. This entropy loss tends to destabilize the interface by decreasing the interfacial tension. If copolymers at the interface become dense enough, the interfacial tension can even become (transiently) negative, leading to interface instabilities, i.e., undulations, that increase the interfacial area.

6. *Conclusions*

More reaction at polymer–polymer interfaces yields a finer dispersion via a fast size reduction process, suggesting the emulsifying effect of the *in situ*-formed copolymers. The reduction of the interfacial tension by the *in situ*-formation of copolymers leads to a more aspherical shape of particles, in comparison to nonreactive blends. In highly reactive systems, the *in situ*-formed block copolymer chains are pulled out by external forces to form micelles, suggesting a potential for solvent-free synthesis of block copolymers. In contrast, the *in situ*-formed graft copolymers are hardly pulled out. They act as emulsifiers to yield a sub-micrometer dispersion. This may be why graft copolymer-forming systems are often used for commercially important polymer alloys. The *in situ*-formed copolymer can act as a phase inversion aid in asymmetric blends (e.g., 35/65) to provide a morphology of the minor component forming the matrix. The inversion is caused by the increase in viscosity of the minor phase as a result of micelle formation. Undulation occurs at reactive polymer–polymer interfaces and appears to be caused by the entropic penalty imposed on the *in situ*-formed copolymer chains accumulated densely at the interface.

See also: Reactive Polymer Processing: Fundamentals of REX

Bibliography

Charoensirisomboon P, Solomko S I, Inoue T, Weber M 1999 Reactive blending of polysulfone with polyamide: a difference in interfacial behavior between block and graft copolymers. *Polymer* **40**, 6803–10

Inoue T, Marechal P 1997 Reactive processing of polymer blends: interfacial aspects. In: Meijer H E H (ed.) *Materials Science and Technology*. VCH, Weinheim, Germany, Vol. 18, Chap. 8

Paul D R and Bucknall (eds.) 2000 *Polymer Blends*. Wiley, New York

Sigalov G M, Ibuki J, Chiba T, Inoue T 1997 Method of effective ellipses for digital image analysis of size, shape orientation, and interparticle distances in polymer blends. *Macromolecules* **30**, 7759–67

Utracki L A 1998 *Commercial Polymer Blends*. Chapman and Hall, London

T. Inoue and P. Charoensirisomboon

Recent Progress in the Continuous Casting of Steel

Continuous casting of steel by conventional oscillating mold technologies is becoming a very mature technology and current developments are focused, in general, on improvement of productivity and quality. However, a recently completed Japanese National Project on the application of electromotive force to continuous casting has had some very interesting results that could lead to radical changes in the design of future continuous casting molds. In addition, recent developments in thin slab casting have been focused on technology to allow stainless steel grades to be successfully cast. Thus, this update will be focused on: (1) continuous casting technologies that rely on electromotive force and (2) the thin slab casting of stainless steels.

1. The Application of Electromotive Force in Continuous Casting

Electromagnetic stirring both in the mold and in the strand has been applied in billet, bloom, and slab casting. In addition, the electromagnetic brake has also been used in slab casting. These two technologies have been aimed at improved steel quality and/or control of cast structure. Details of developments in stirring and breaking are well covered in the review of Hanley and Kolberg (2003).

It has long been a desire of the steel engineer to develop a process that would mimic the technologies found in aluminum and copper processing, where the Lorentz force is used to control the position of the meniscus in the mold and to prevent meniscus solidification. In this manner, continuous cast surfaces could be developed that are free from oscillation marks, smooth, and potentially defect free. This technology could lead to the elimination of surface grinding and other treatments that are currently routinely used to ensure surface quality. Initial attempts at meniscus control were frustrated by an inability to control bulk fluid flow and joule heating while controlling meniscus position and, although there have been theoretical papers (Roy *et al.* 1995) suggesting that the technology should be possible in steel casting, no industrial prototypes were fully successful.

In 1995 a Japanese National Project was initiated "to improve surface, subsurface and internal quality and to increase the ratio of steel directly sent from the continuous caster to the hot rolling mill" (Ayata *et al.* 2002). This project had three goals: (1) to develop electromagnetic casting (EMC); (2) to develop a superconducting electromagnetic brake (EMBr); and (3) to develop sensors. The focus of the EMC project was to reduce surface roughness and oscillation mark depth to allow elimination of surface conditioning of billets and slabs. The EMBr project was aimed at high speed casting (greater than $3\,\mathrm{m\,min^{-1}}$). The sensor project was focused on the development of noncontact sensors that would allow sub-meniscus flow velocities to be measured and this project would successfully allow sub-meniscus velocities up to $0.4\,\mathrm{m\,min^{-1}}$ to be measured.

EMC technology was proven in plant trials at a commercial billet cater. A schematic of the process is given in Fig. 1 where EMC is shown to alter meniscus shape, reduce solidification along the meniscus, and induce fluid flow near the meniscus. To reduce the induced meniscus fluid flow, which leads to excessive turbulence and the potential for slag emulsification, a pulsative electromagnetic field (Tani *et al.* 2002, Nakata *et al.* 2002, Galpin *et al.* 2002) was employed that successfully reduced the induced meniscus flow and led to a stabilized meniscus region. This development allowed successful implementation of the technology on a conventional billet caster where surface roughness was improved and the oscillation mark depth greatly reduced, allowing elimination of billet surface conditioning. Subsequently, the pulsative technology was trialed on a slab caster with similar positive results.

In slab casting an additional technology was also developed where superimposed a.c. and d.c. magnetic fields were used to allow meniscus control and simultaneous damping of the fluid flow. A schematic of the process is shown in Fig. 2 where the superimposed d.c. field damps the fluid flow caused by the a.c. field. Clearly, the d.c. field must be finely tuned to the a.c. field so that it is not too effective. If sub-meniscus flow is reduced too far, solidification defects of the type that the technology attempts to eliminate will be induced. This technique was successfully trailed at the Kimitsu works of Nippon Steel

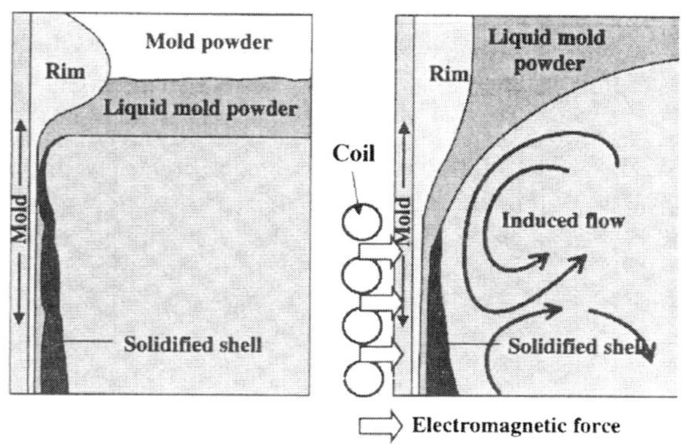

(a) Conventional casting (b) EMC

Figure 1
Schematic comparison of mold solidification in (a) conventional casting and during (b) EMC.

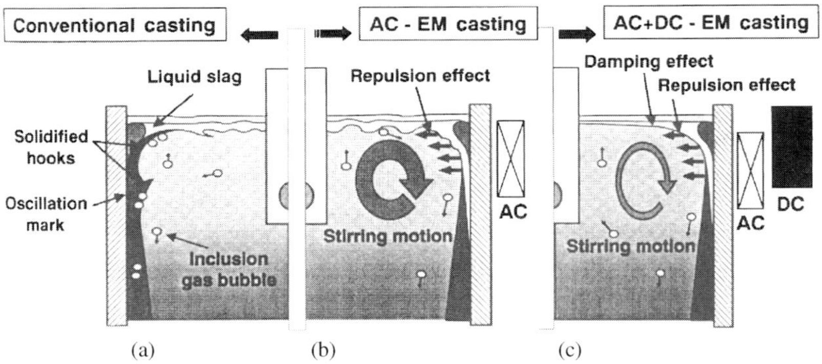

Figure 2
Schematic of: (a) conventional casting, (b) AC EMC, and (c) AC + DC EMC.

Corporation on a 6 ton trial vertical continuous casting machine (Galpin *et al.* 2002). Initial results suggest that this technology can reduce the surface roughness of peritectic carbon steels, a major reason for decreased productivity when casting these grades.

EMBr technology was first developed as a technique to control the natural flow patterns within a mold. Early results suggested that the technology was successful at low casting speeds and that a higher magnetic field strength would be required for the technology to be applied to high throughput rates and higher casting speeds. Thus, a new EMBr technology was developed using superconducting magnets to develop magnetic field densities as high as 1 T. As superconduction occurs at very low temperatures, below that of liquid nitrogen, significant cooling technologies must be applied close to the mold in order for this technology to be successful. Trials using this new technology were successful at casting speeds up to 3 m min^{-1}; however, it remains to be seen if this technology will be industrially applied due to the high refrigeration costs and potential safety concerns.

Of the three developments, EMC has the potential for greatest impact on steel continuous casting as it has the possibility to increase slab surface quality significantly and lead to continuous cast billets, blooms, and slabs that can be rolled directly without surface conditioning. This technology will greatly increase the quality and productivity of conventional continuous casting machines and has the potential to radically improve the current technology.

2. *Thin Slab Casting of Stainless Steels*

Compact strip production (CSP) via the SMS-Demag-developed thin slab casting process has been successfully applied to carbon steel casting and to some stainless grades; however, due to problems in product quality the wide scale application of the technology to a broad range of stainless and alloy steels has lagged behind that of carbon steels. Recently, however, the thin slab casting of a variety of stainless and alloy steels has been proven at a pilot facility in Terni in Italy by a group of developers that included SMS-Demag, Accai Speciali Terni (AST), and Centro Sviluppo Materiali (CSM) (Brascugli *et al.* 2001, Caposti *et al.* 2002). Subsequent to the pilot facility an industrial CSP plant for stainless steels was started in August 1, 2001 and in 2002 was in regular production with a product mix of 60% austenitic stainless steels, 27% ferritic stainless steels, and 13% silicon steels, both grain oriented and non-grain oriented, at a production rate of 30 000 ton per month. This machine is highly instrumented and includes automatic operation, a mold thermal monitoring system, an electromagnetic brake, liquid core reduction, dynamic soft reduction, and dynamic solidification control (Brascugli *et al.* 2001). Currently, strip produced by this machine has been applied to applications such as Bright Annealed AISI 304.

This development allows thin slab casting technology and CSP technology in particular, to be applied to a wide range of stainless and high-alloyed steels. Thus, thin slab technology has now been proven for the majority of steel grades and previous barriers to technology implementation have now been overcome.

3. *Conclusion*

Even though conventional continuous casting operations are becoming mature technologies, there are still significant advances under development that may significantly improve either quality and productivity or the application range of the technology. Both of the above developments will in time lead to significant improvements in continuous casting and will lead to both future and current casting technologies being significantly more technology focused in the past. In addition to these developments, significant effort has been expended on advanced control systems based on multiple sensor input arrays to allow optimized operation with on-line quality prediction. Thus, continuous casting machines have become sophisticated almost-robotic operations, with future trends being towards full automatic control with no manual intervention.

Bibliography

Ayata K, Miyazawa K, Bessho N, Tho T 2002 Development of new continuous casting technologies with electromagnetic force. *4th European Continuous Casting Conference*, Vol. 1, pp. 15–24

Brascugli G, Capotosti R, Cristallini A, Spaccarotella A, Flemming G, Quix F 2001 World's first thin-slab production of stainless steels. *Millennium Steel*, pp. 248–50

Caposti R, Ferranti S, Proietti M, Rossberg W, Cristallini A, Gigli G, Spaccarotella A, Schwellenbach J, Thomanek E, Feldhaus S 2002 Start-up of the first industrial CSP plant for stainless steels at Acciai Speciali Terni Works. *4th European Continuous Casting Conference*, Vol. 2, pp. 680–90

Galpin J M, Gardin P, Lamant J Y, Tani M, Harada H, Toh T 2002 A new step in steel slab quality improvement by superposition of AC and DC magnetic fields in the continuous casting mold. *4th European Continuous Casting Conference*, Vol. 1, pp. 47–55

Hanley P, Kolberg S G 2003 Electromagnetic methods for continuous casting. In: Cramb AW (ed.) *The Making Shaping and Treating of Steel-Vol 3 Continuous Casting*, Chap. 10

Nakata H, Inoue T, Mori H, Ayata K, Murakami T 2002 Improvement of billet surface quality by ultra-high frequency electromagnetic casting. *4th European Continuous Casting Conference*, Vol. 1, pp. 31–8

Roy S, Cramb A W, Hoburg J 1995 Mathematical modelling of meniscus shapes of liquids metals in an electromagnetic field. *Trans. ISS* **16**, 65–72

Tani M, Harada H, Toh T, Yamasaki N, Fujisaki K, Anzai E, Takeuchi E, Matsumiya T, Miyazawa K E 2002 The control of initial solidification by the imposition of pulsative AC electromagnetic field. *4th European Continuous Casting Conference*, Vol. 1, pp. 39–46

A. W. Cramb

S

Sand Casting (Disposable Mold)

The most common casting method is sand casting, which uses silica sand as the mold material. The sand is bonded with clay and water in "green sand" molding, and with resins in "dry sand" molding. Green sand is called "green" because it is not baked. The sand is recovered after casting, and recycled to make new molds.

In sand molding a pattern is used to make the cavity in the sand. The pattern has the same shape as the casting which it makes, but the dimensions are slightly larger to allow for shrinkage as the casting cools to room temperature. Sand is compacted around the pattern. The sand is held in a box, called a flask. The mold is usually split in two parts. Traditionally, the parting plane is horizontal. The top half of the mold is called the cope, and the bottom half is called the drag. The hollow portions of the casting are formed by cores, which are usually made of resin-bonded sand. The cores are supported in the mold by extensions, called core prints. The mold includes a sprue where molten metal enters the mold, and a runner system, consisting of runners and gates that distribute the molten metal to the casting cavity. Risers (also called feeders) are molded into the sand to provide a reservoir of molten metal to compensate for shrinkage as the metal solidifies. The process is illustrated in Fig. 1. Vertical parting lines are also possible.

1. Binders

The most common binding system used in sand casting is the clay-water bond. Sand is mixed with

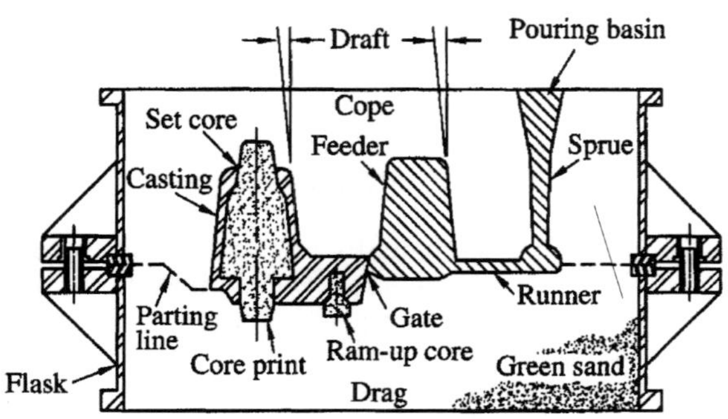

Figure 1
Sand mold with horizontal parting line (after Tlusty 1999).

water and bentonite clay to give the sand strength to hold its shape during casting. Sodium bentonite (also called Western bentonite) imparts toughness and high temperature strength to the mold. Calcium bentonite (also called Southern bentonite) gives the sand flowability and improves removal of the sand from the casting after the casting has solidified. The two types of bentonite are often blended to give a sand with properties tailored to the type of metal and casting that is being produced. Other clays are occasionally used to give the mold special properties.

Resins are used to bond sand for cores and in making precision sand molds for some applications. The resins coat the sand and set as a result of heat and/or chemical reactions during coremaking. Thermosetting resins are used in the hot box and shell processes, in which the sand–resin mixture contacts a heated metal pattern. Chemical reaction binding occurs when two resin components are mixed, as in the no-bake and cold box processes. A catalyst may be introduced to accelerate the reaction. Resins are selected according to the type of metal being poured and the geometry of the casting. Typical resins used are phenol formaldehyde, urea formaldehyde, and furfuryl alcohols. Typical catalysts include phosphoric and sulfuric acid. The resin-coated sand may be injected into a mold, or rammed into a core box (the die that shapes the sand) by hand.

Sodium silicate may also be used as a binder, generally for steel castings, or very large castings. Sand is coated with sodium silicate and poured over the pattern. Passing carbon dioxide gas through the mold accelerates the hardening reaction.

Sand may also be used as a mold material without using a binder. In one process, the sand is held between thin plastic sheets pressed against a pattern, while a vacuum is drawn and held during pouring. The vacuum forces the sand to retain its shape during pouring. When the vacuum is released, the sand falls away from the casting. In the lost foam process, the pattern is made of foamed polystyrene or similar plastic. Loose sand is vibrated around the pattern, and metal poured directly onto the pattern. The pattern vaporizes and is replaced by metal. The mold retains its shape until the casting is solid.

Additives are often added to green sand mixes to give them specific properties. Sea coal (finely divided bituminous coal) is added to improve shake-out (casting removal from the mold) in ferrous castings. When the metal is poured into the mold the coal undergoes a coking reaction, expanding and filling the spaces between the sand grains next to the casting. This prevents molten metal from penetrating between the grains. Cereals may be added to increase strength during mold handling, and cellulose may be added to control the moisture content of the sand.

2. *Sands*

The most commonly used sand is silica, obtained from river banks or lake beds. The sand is washed to remove impurities such as shell residues or other unwanted mineral matter that might interfere with the bonding reaction of the binders. The sand is screened and sorted by size.

The shape of the sand grains is important. Rounded grains have the least surface area, and, therefore, use the least amount of binder. Angular grains use more binder, but tend to interlock, strengthening the mold. Semiangular grains are usually preferred, as they are a compromise between both extremes.

Refractory sands other than silica, may be used. Zircon, chromite, and olivine sands are often used for steel castings because of their superior high temperature properties. Zircon is also used for other metals when a sand having an increased chilling power is desired. Graphite is used for some specialty alloys, such as titanium alloys. The binder used with graphite is a mixture of pitch and other organic compounds, e.g., sugars.

Distribution of sand sizes in a mold is also important. A monosized sand (all of the sand is caught on the same standard sieve) has high permeability, but low strength. Most foundries use sand mixes that are caught on three or four adjacent sieves, so that the strength and permeability of the sand are balanced. To prepare the sand for molding, binder, water, and additives are mixed into the sand in a muller. The muller is a large steel cylinder equipped with a wheel and plows, that mixes and stirs the sand so that all of the sand grains are coated uniformly with the ingredients.

Samples are taken from the mulled sand and tested periodically. The most important test is "compactability," the measure of how well the sand flows and compacts around the pattern. Other tests include moisture and binder content, permeability (required because the heat of the solidifying metal vaporizes moisture), green strength, friability, and hot strength.

Compaction of the sand around the pattern may be done by hand or machines. In hand ramming, a layer of fine sand (called facing sand) is first placed over the cope half of the pattern and tamped down. This fine sand gives a smoother finish to the casting than regular sand. Then the molding sand is added, a few shovels-full at a time and is compacted using a ramming tool. The process is repeated until the flask is full, and all of the sand has been rammed and compacted. The pattern is removed from the sand, and the process is repeated for the drag half. The two halves of the mold are joined together, and clamped, and the mold is ready to pour. Weights are added to the top of the cope to keep the cope from floating on the metal when it is poured into the mold.

Sand compaction is usually mechanized. In the jolt-squeeze process, the cope or drag flask is filled with sand and placed on a table. The table is then raised by air or hydraulic pressure, and allowed to fall back to its original position. This jolts the sand in the flask, compacting it. To be sure that the sand is fully compacted, a plate is then forced into the non-pattern surface of the mold using hydraulic or air pressure to squeeze the remaining sand.

In some molding machines the sand is placed in the flask over the pattern, and compressed air is suddenly released above the mold. The air drives the sand down onto the pattern with great force, producing a mold with excellent surface detail. This is known as impact molding.

Vertical parting line molding machines are also used. In these a swinging plate contains one half of the pattern, and the other half is mounted to a hydraulic ram. The sand is dumped between the swinging plate, which is locked in place, and the ram squeezes the sand to make a mold segment. The swinging plate moves up, and the ram pushes the mold segment out, to join a line of similar segments. In this method, no flask is used.

Mechanized- and automated-molding machines are capable of production rates in excess of four molds per minute. Cores are placed in the molds before they are closed by hand or using robots. Complex cores are often assembled into packages before reaching the molding line, and then the entire package is inserted into the mold before it is closed.

A major advantage of using sand as the molding material is that it is easily reclaimed for reuse. After the castings have been removed from the sand, the sand is screened to remove core butts (core fragments whose resins have not burned out during pouring and cooling), fines, and metal spills. Then it is cooled, and sent back to the muller where water, binder, and other additives are mixed with it. A certain amount of clay binder must be added, as exposure to the high temperatures of molten metal near the casting surface destroys the bonding ability of the clay. This clay is often lost as fines during recycling. The sand can be recycled many times, but eventually it becomes so worn that it must be discarded to a landfill or some other beneficial use.

Sand casting is inherently low cost as silica sand is plentiful, and can be reused. The method can be used to make castings as large as desired (there is no maximum size limit to the process). However, the dimensional accuracy that can be expected from the process is the lowest of all casting processes. For most components, this is not an important problem, as critical surfaces on castings are subsequently machined.

The sand-casting process has been the mainstay of foundry processes for millennia. Although a mature process, improvements continue to be made, assuring that this ancient method of making castings will continue to be used in modern manufacturing technology.

Bibliography

Tlusty J 1999 *Manufacturing Processes and Equipment*. Prentice-Hall, Englewood Cliffs NJ

T. S. Piwonka

Semisolid Processing

Semisolid metal (SSM) processing is a technology that enables the production of near net shape components of superior quality. It was developed from a discovery made at the Massachusetts Institute of Technology in the early 1970s (Flemings 1991) that the stirring of alloys during solidification led to a change in internal structure resulting in a dramatic lowering of apparent viscosity of the semisolid slurry. It was shown by subsequent work (Joly and Mehrabian 1976) that the shear rate and cooling rate, as well as fraction solid had a strong effect on the viscosity (Fig. 1). High shear and low cooling rates can produce slurries of relatively low viscosity (1 Pa s), even up to fraction solids of around 0.6. This is related to the microstructure of the slurry: shearing during solidification changes the solid from the normally dendritic form having well-developed sidearms growing from the seed crystals, to becoming rosette forms and eventually fully spheroidal shapes as illustrated in Fig. 2. The movement of spheroidal solid particles past one another within a liquid matrix would appear to be much easier than the same volume fraction of dendritic solid for hydrodynamic reasons. Furthermore, these spheroidal alloy slurries possess thixotropic properties since the apparent viscosity is time dependent under isothermal conditions (i.e., constant fraction solid). If left undisturbed, the viscosity will increase by orders of magnitude but on renewed stirring it will rapidly fall again (Fig. 3), to become liquid-like. This behavior is quite reversible and is associated with the clustering of the primary solid particles when shearing is reduced, which break up and disperse again when shearing is restarted. The clusters impede the general flow of the slurry producing viscous thickening. The ability of these slurries to change rapidly from solid-like to liquid-like behavior, whereby slugs of the alloy support their own weight initially becoming fluid on sudden shearing, led to the development of a number near net shape forming processes such as thixocasting and thixoforging.

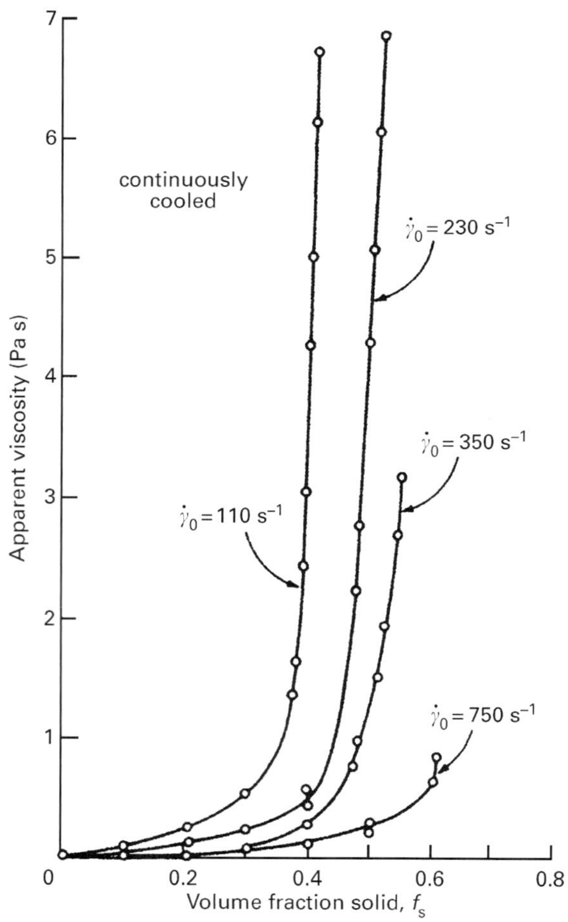

Figure 1
Apparent viscosity vs. fraction solid (f_s) of Sn–15 wt.% Pb sheared continuously and cooled at 0.33 K min^{-1} at different shear rates (after Joly and Mehrabian 1976).

1. *Rheology*

Numerical simulation of SSM forming requires the solution of appropriate rheological equations, which describe how the apparent viscosity of the alloy slurry changes with both shear rate and time. One approach to this problem is to define a structural parameter λ, which takes a value of unity for a fully built-up state and zero when fully broken down, and to write two constitutive equations in terms of the structure. The first deals with flow behavior of

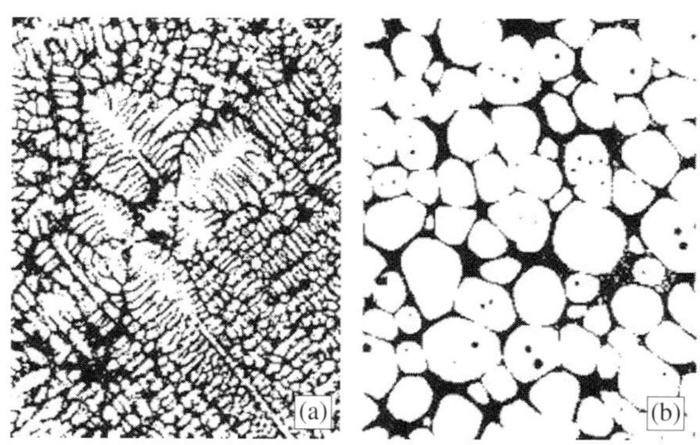

Figure 2
Comparison of (a) dendritic conventionally cast and (b) nondendritic semisolid formed microstructures of aluminum A357 (Al–7Si–0.5 Mg) (both 100 ×) (after Kenney *et al.* 1988).

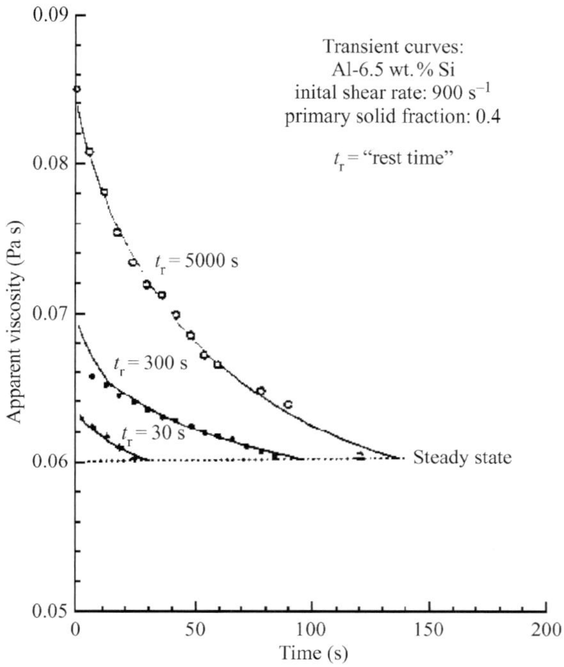

Figure 3
Viscosity of Al–6.5 wt.% Si alloy, at $\dot{\gamma} = 900\,s^{-1}$ vs. time after a "rest time," t_r. All samples were at the steady-state viscosity of 0.06 Pa s at $\dot{\gamma} = 900\,s^{-1}$; shearing was then stopped and restarted at the same shear rate after a time t_r(Flemings 1991).

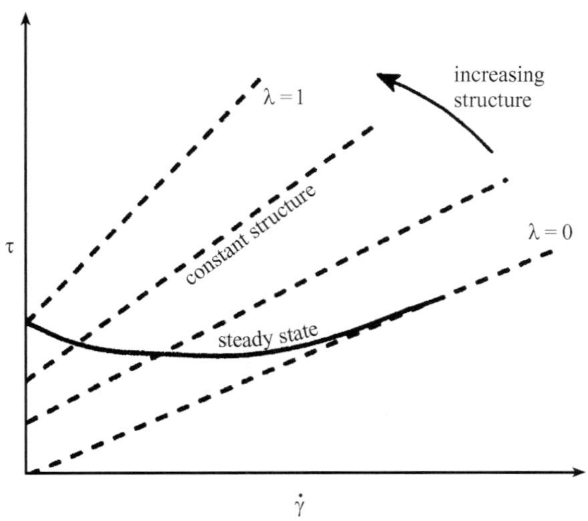

Figure 4
Schematic diagram showing possible rheological behavior of slurries which conform to the isostructural Bingham model (after Kirkwood 1994).

isostructural slurries (i.e., λ constant), and a simple example expressing a yield point is:

$$\tau = \lambda\tau_0 + (\eta_\infty + c\lambda)\dot{\gamma}$$

where τ is the shear stress and τ_0 the yield stress for a fully structured slurry, $\lambda = 1$, η_∞ is the viscosity when the slurry structure has fully broken down, $\lambda = 0$, as $\dot{\gamma} \to \infty$, c is a constant, and γ the shear rate.

The second equation describes the kinetics of structural change:

$$\frac{d\lambda}{dt} = a(1 - \lambda) - b\lambda\dot{\gamma}$$

where a and b are kinetic constants for the build-up and break down of internal structure respectively. At steady-state where $d\lambda/dt = 0$, we have $\lambda = a/(a + b\dot{\gamma})$, which may be substituted into the first equation. Curves showing the relationships between shear stress and strain rate under these simple conditions are given in Fig. 4. More complex forms have been proposed from theoretical considerations of the mechanisms of the clustering of spheroidal particles and their break-up, and from experimental investigations of flow behavior under controlled conditions. It has to be noted, however, that since the forming operation in industrial practice occurs in around 0.1 s, and steady state is achieved in $\sim 10^3$ s, die filling does not occur at the steady-state condition. This is a difficult problem that the modeling must contend with.

2. *Feedstock*

Nondendritic slurries were originally produced by mechanical stirring during solidification of the alloy. This was a batch process and attempts to develop a continuous slurry generator, or rheocaster, by this means failed because of the difficulty of achieving high shear rates and cooling rates needed to generate significant levels of production. Also the use of ceramic moving parts risked the danger of pick up of nonmetallic fragments in the alloy. These problems were overcome by using electromagnetic (magnetohydrodynamic or MHD) stirring deep in the sump of a DC casting unit, thus avoiding gas pick up while providing vigorous stirring of the semisolid alloy, and this has become the standard production route for the manufacture of SSM feedstock.

Several other production routes have been subsequently developed. In spray-forming (the Osprey Process) atomized droplets of liquid metal are sprayed onto a cooled target to produce a fine-grained microstructure; such microstructures can also be produced by hot-working (the SIMA route) or warm-working (the RAP route) followed by reheating an alloy. Yet another way of producing fine equiaxed structure in alloys is through grain refinement additions of certain compounds, e.g., titanium boride in aluminum alloys.

On heating above the solidus temperature, these alloys with fine microstructures will form liquid which penetrates the grain boundaries to generate fine spheroidal particles within a liquid matrix. Semisolid alloy slugs produced by any of these routes have been shown to possess good thixoforming properties.

3. *Thixoforming Processes*

An early proposal was to use the slurry as formed on cooling by directly charging it from the slurry generator into the shot chamber of a die-casting machine and then injecting it into the die. This was termed rheocasting, but because of the difficulty of matching the production rates of slurry manufacture and die casting, it is no longer considered an attractive route.

The alternative approach is to allow a rheocast billet to solidify fully, to cut it up into slugs of appropriate size, which are heated back to the semisolid condition when required at a later date. They regenerate the spheroidal particle structure during this operation and are ready to be formed into complex shapes. This may take place (see Fig. 5) between closed dies (thixoforging), or by loading into the shot chamber of a die-casting machine and injecting into a die (thixocasting). In addition, a process which at present is only used for magnesium alloys (thixomolding), involves loading alloy chips into a heated Archimedes' screw. The screw drives the chips down the barrel under heating and shearing conditions, which generate good semisolid material, and then forces the slurry into a die in an operation similar to plastic injection molding.

4. *Industrial Application*

Effectively all industrial production of SSM components in aluminum alloy is presently carried out using modified die-casting machines available from a number of manufacturers (see Fig. 6). It has been found that somewhat slower injection ram velocities ($\sim 1\,\mathrm{m\,s^{-1}}$) are required than in conventional liquid die-casting, otherwise turbulent flow of the slurry and broken front filling occurs leading to entrapped air and porosity. Also oxide skin can be incorporated within the slurry and all these internal defects will result in deterioration of mechanical properties. Alternatively, if the filling is too slow, premature freezing may occur leading to incomplete filling or the inability to feed shrinkage porosity. These critical requirements emphasize the need for "shot control" of the machine, i.e., control of the ram velocity as a function of distance or time, to achieve thixocast components of consistent reliability and integrity. A high-pressure intensification at the completion of

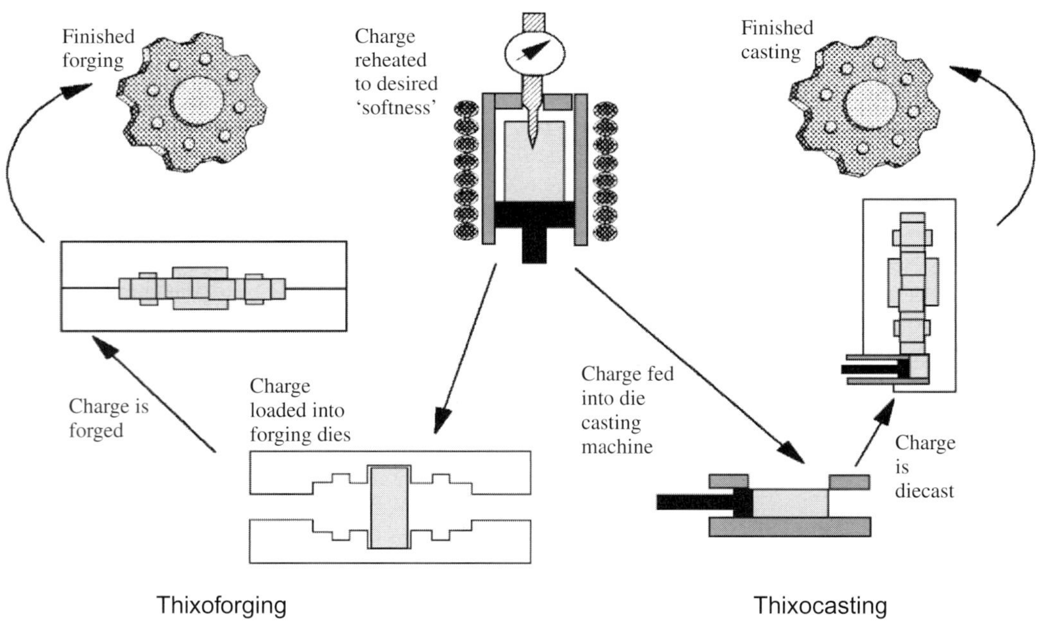

Figure 5
SSM forming processes.

Figure 6
SSM casting cell (courtesy of Buhler).

filling will produce final mass feeding to reduce further any porosity.

Commonly the presized slugs are first rapidly heated by a series of induction coils on a carousel up to the temperature required to produce the semisolid condition, and then homogenized for a period to achieve a uniform spheroidized structure within it. Typically this may take a total of 10 min for each slug. It can be transferred by robot handling to the shot chamber of the die-casting machine to be injected into the die cavity at the rate of around 1 shot per min. On solidification, the die is then opened and the component ejected into a quenching bath. This quench is sufficient to retain enough solute in solution in age-hardenable aluminum alloys, so that they may be low-temperature age hardened at around 170°C (T5 treatment) to produce significant enhancement in mechanical properties (see Table 1) without risk of deformation or dimensional changes in the part.

A range of components are presently being manufactured, mainly for the automotive industry, taking advantage of the lightness and strength in aluminum alloys, together with the complexity of shape, available by this manufacturing route. These include fuel rails, hydraulic suspension cylinders, hydraulic distributors, air conditioner components, engine brackets, welded suspension frames, and airbag canisters. Other high-strength parts have been produced for motorbikes and mountain bikes. Additional spheres of application are to be found in the electronics industry (e.g., computer actuator arms) and in the leisure industry (e.g., bronze soles of golf club heads).

5. *New Alloys*

The vast majority of components produced by semisolid forming are presently from the aluminum

Table 1
Typical mechanical properties in the T5 condition (unless otherwise indicated), measured on thixocast specimen of 19 mm diameter (machined to 13.8 mm diameter) and 200 mm length (Garat *et al.* 1998).

Alloy	Yield strength (MPa)	Ultimate tensile strength (MPa)	Ductility, *E* (%)	Brinell hardness number
Al–Si7–Mg0.3 (A356)	170	260	15	80
Al–Si7–Mg0.6 (A357)	210	275	10	90
Al-Si6–Cu1–Mg (C335)	225	320	7	100
Al-Si6–Cu3–Mg (319)	225	310	3	100
Al–Si17–Cu4–Mg (A390)		250	<0.2	140
[T6]		350	<0.2	165
[Forged]	185	220	1	115

Source: Garat *et al.* (1998)

alloys, A356 and A357. There has been some attempt to develop alloys specifically for thixoforming (such as the Al–6Si–1Mg alloy developed by Pechiney and providing an improvement in strength over A357) but this still remains an area urgently requiring more attention, especially for the high-strength aluminum alloys needed in the aerospace industry. A small amount of commercial activity exists for copper alloys and again this is expected to expand for various hydraulic and marine applications.

In the early development of SSM there was much interest in the machine casting of ferrous alloys, since it was anticipated that the lower temperatures that the expensive dies are subjected to might give them an acceptable life. However, the rapid cooling of the slurry is likely to affect the ability of high-melting point alloys, such as steels, stellites, and nimonic alloys, to fill dies completely and flow into complex cavities, thus limiting the usefulness of the process. However, it may be possible, because of the low-flow stresses, to use cheaper, easily fabricated materials of low thermal conductivity (e.g., graphite, moldable, or machinable ceramics) to overcome this. Again this is an area requiring investigation.

6. *Benefits and Limitations of SSM*

Many benefits arise from the facts that the heat content of a semisolid slurry is much lower, and the viscosity very much higher, than those of the same fully liquid alloy. The low heat content brings about greater thermal efficiency, extended die life, and rapid cooling, allowing a T5 treatment. The high viscosity allows die filling without turbulence and air entrapment. Further benefits are probably to be gained from the uniform grain size and the ability for mass feeding of the slurry in the final compaction. The resulting fine-grain structure, essentially free of porosity, permits heat treatment and the development of the full potential mechanical properties of an alloy (see Table 1).

It has to be admitted that the alloy feedstock is presently more expensive (60–70% above the standard alloy) and that the scrap (from biscuit, overflow, etc.) cannot be immediately recycled. However, scrapped castings are expected to be low owing to control of the process and consequent reduction in defects. Furthermore, some disadvantages (e.g., billet costs) are likely to be temporary and to be reduced as SSM production grows and more suppliers enter the market.

7. *Concluding Remarks*

The immediate future of SSM manufacture must clearly lie in the manufacture of aluminum alloy components of complex shape, high integrity, and medium strength, probably taking advantage of these properties to redesign components by reducing section and producing even lighter structures. Development of new higher strength aluminum alloys should continue this evolution of designing lighter components adequate for the purpose.

Looking further ahead, one might expect to see the use of copper alloys, such as aluminum bronzes, thixoformed in conventional steel dies. Even further into the future, the prospect of die casting complex shapes in stainless steel, tool steels, and cobalt and nickel alloys is an intriguing possibility, which will require development of dies using nonmetallic materials.

Bibliography

Flemings M C 1991 Behavior of metal alloys in the semisolid state. *Metall. Trans.* **22A**, 957–81

Garat M, Blais S, Pluchon C, Loue W R 1998 Aluminum semi-solid processing: from the billet to the finished part. *Proc. 5th Int. Conf. Semi-solid Processing of Alloys and Composites*, pp. xvii–xxxi, Colorado School of Mines, Golden, CO
Joly P A, Mehrabian R 1976 The rheology of a partially solid alloy. *J. Mater. Sci.* **1**, 1393–418
Kenney M P, Courtois J A, Evans R D, Farrior G M, Kyonka C P, Koch A A, Young K P 1988 Semisolid metal casting and forging In: *Metals Handbook*, 9th edn. ASM International, OH, Vol. 15, pp. 3327–38
Kirkwood D H 1994 Semisolid metal processing. *Int. Met. Rev.* **39**, 173–89

D. H. Kirkwood

Sheet Steel: Low Carbon

Low-carbon steel sheet is available for use in both the hot-rolled condition and, after subsequent cold-rolling, annealing, and possibly coating. Modern processing provides a high degree of control of both chemistry and processing to enable a wide range of grades with different combinations of strength and formability to be produced. Close control of gauge, width, shape, and flatness is also achieved. It is the ability of low-carbon steel, however, to be formed economically and satisfactorily into a wide range of complicated shapes without splitting, necking, or wrinkling that is ensuring its continued use as a major engineering material. Other advantages include low cost, high elastic modulus, and good energy absorbing characteristics.

Sheet steels are supplied in two main categories: low-strength or mild steels for which the main requirement is a satisfactory minimum level of formability; and high-strength steels with, as expected, a minimum level of strength. Formability itself has two important aspects—drawability and stretchability. Drawability is usually characterized by the mean Langford value (often known as the r-value) and depends mainly on developing a suitable orientation texture containing a high proportion of (111)-oriented grains and a low proportion of (100)-oriented grains. Stretchability, usually characterized by the elongation value in a tensile test or the work hardening coefficient n, is mainly influenced by the strength of the steel and the strengthening mechanisms used to develop the strength. Some of the relationships between strength and elongation in a tensile test for different types of steel are illustrated in Fig. 1 while equivalent relationships between strength and drawability (mean r-value), are illustrated in Fig. 2. These figures show that there is a general reduction in formability with increasing strength and that steels strengthened by different mechanisms are able to develop various ranges of strength to provide different combinations of strength and formability.

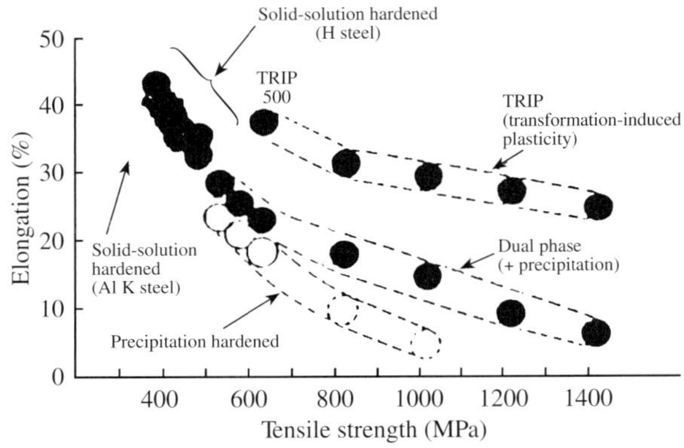

Figure 1
Strength–ductility balance of high-strength steel (after Shimada *et al.* 1996).

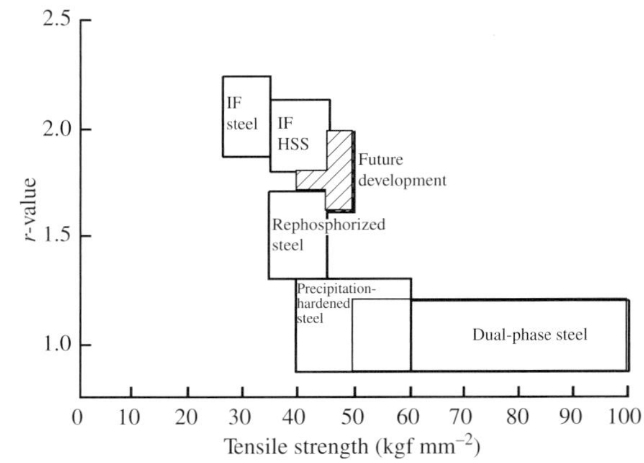

Figure 2
Relationship between mean r-value and tensile strength (after Takechi 1990).

1. *General Processing Considerations*

Almost all sheet steel is now made using a continuous casting process to produce slabs 200–250 mm thick. After cooling these slabs are usually reheated to temperatures up to 1250 °C for hot rolling. The first stage of hot rolling, called roughing, reduces the thickness usually into the range 30–45 mm and the second stage, called finishing, reduces the thickness to the final hot-rolled gauge required, often in the range 1–2 mm up to 5–12 mm. The steel leaves the last stand of the hot mill at a finishing temperature which is determined by metallurgical requirements. Water cooling is then used along a runout table to achieve the required temperature for coiling. Over the last few years thin-slab casting processes have been introduced in a few steel plants which enable intermediate gauges, usually obtained by roughing, to be cast directly.

Cold rolling is used for gauges usually in the range 0.4–3 mm using cold reductions of 50–80%. During cold rolling the steel becomes hard and loses its ductility; therefore an annealing process is used to restore ductility and to develop almost the final properties. In the batch process several tight coils are stacked on a furnace base and heated and cooled under a protective atmosphere, the complete process usually taking several days. In the continuous process each coil is uncoiled and then recoiled after passing through a furnace, the complete process taking several minutes. After annealing, cold-rolled steels may be given a light cold reduction, called temper rolling, primarily to remove the yield point but also to improve flatness and develop the required surface roughness and texture. Cold-rolled and annealed steel would normally have a better formability, gauge control, flatness, and surface than a hot-rolled steel but the former would have a higher cost due to the additional processing.

2. *Mild Steel*

2.1 Hot-rolled Mild Steel

The most formable standard grade of hot-rolled mild steel often contains 0.03–0.06% carbon, 0.003–0.006% nitrogen, 0.2–0.3% manganese, and 0.03–0.06% aluminum and is processed using a moderately high coiling temperature (~650°C). An austenite with a relatively coarse grain size is produced at the slab reheat stage, which, during hot rolling, recrystallizes between each deformation pass to give a smaller grain size. A moderately fine grain size is finally produced after transformation to ferrite depending on the time to the onset of water cooling, the cooling rate (which depends on thickness), and the coiling temperature. The orientation texture is usually fairly random with the result that mean *r*-values are low and close to unity. The drawability is, therefore, moderate. Lower formability and slightly higher-strength grades are produced by relaxing the upper carbon and manganese limits, whereas a superior formability grade may be produced by combining the interstitial nitrogen with a small addition of titanium or boron.

2.2 Cold-rolled Mild Steel

There are two main types of cold-rolled mild steel. The first is the more traditional type described as aluminum-killed (AK) steel, in which the carbon is relatively free in the structure and may form iron carbides. The second, called interstitial free (IF) steel, contains a strong carbide-forming element such as titanium and/or niobium to combine with some or all of the carbon and nitrogen. IF steels of today usually contain ultralow levels of carbon and nitrogen (both <0.003%), whereas the AK mild steels have carbon contents above 0.015% and often higher, depending on the properties required and the annealing method to be used. IF steels often have mean *r*-values above 2.0 and are used for the most difficult forming applications. The highest *r*-values of AK steels are usually a little below 2.0 but grades with mean *r*-values as low as 1.0 may be obtained depending mainly on the composition and annealing method used. These grades are clearly used for applications requiring modest formability.

(*a*) *Interstitial free mild steels.* The IF steels produced in the early 1970s generally contained about 0.01% carbon and needed more than a stoichiometric ratio of titanium to carbon, nitrogen, and sulfur to give the high mean *r*-values. The use of the ultralow carbon contents indicated above has enabled reasonably high *r*-values to be obtained with the addition of less than a stoichiometric amount of titanium, but higher *r*-values can still be obtained using more than a stoichiometric addition. When niobium is used, separately or in combination with titanium, there is evidence that optimum *r*-values and values of elongation are obtained with niobium:carbon atomic ratios of less than unity.

The use of low slab reheat temperatures and high coiling temperatures with IF steels leads to superior forming properties, but a finer hot-band grain size also leads to higher *r*-values. Such a grain size may be obtained by cooling rapidly as soon as possible after the steel emerges from the last stand of the hot mill, commonly referred to as "early cooling."

All IF steels develop the highest *r*-values when high cold reductions, such as 80%, are used. Similar compositions and prior processing may be used for both batch or continuous annealing and similar properties are obtained, but most IF steels are continuously annealed. In general, the annealing requirements of IF steels are not critical and this makes them particularly suitable for processing on hot dip coating lines if a very formable product is required. Properties improve, however, with increasing annealing temperature and temperatures up to 850°C are often used. IF steels do not normally exhibit a yield point. They require, therefore, only the minimum of temper rolling to improve shape and impart the required surface roughness. Many IF steels are also completely non-aging unless specific measures are taken to develop bake hardening (see below).

(*b*) *Aluminum-killed mild steel.* Aluminum-killed mild steels may also be processed by batch or continuous annealing. The metallurgical requirements are, however, completely different for the two processes if optimum formability is to be obtained. With the older batch annealing method, the most important requirement is that aluminum and nitrogen must be held in solution in the hot-rolled coil. Aluminum nitride is then able to cluster or precipitate during slow heating to influence the recrystallization process to give high *r*-values. Retaining aluminum nitride in solution is achieved by the use of a high slab reheat temperature close to 1250°C and a low coiling temperature, usually below 600°C. Optimum properties are usually obtained with 0.003–0.006% nitrogen and 0.025–0.04% soluble aluminum. Carbon contents are usually close to 0.03–0.04%, though higher carbon contents may be used for lower formability grades. With all of these carbon contents, the carbon that is taken into solution at the annealing temperature is able to reprecipitate almost completely during the long slow cool to room temperature. The steel then becomes completely nonaging. Cold reductions close to 70% are often used since higher and lower cold reductions tend to give lower *r*-values. The final temper rolling must be sufficient to remove the yield point. About 0.8% reduction in thickness is often used.

There are two important requirements for achieving optimum properties in AK steel by continuous annealing. First, the nitrogen must be combined as aluminum nitride in the hot-rolled coil to give low interstitial nitrogen contents and second, the carbides must be coarse. The coarse carbides dissolve slowly during the rapid heating. Recrystallization is then able to take place in a substantially interstitial-free matrix (before the carbides have dissolved) to give an adequately coarse grain size and a beneficial texture, to give relatively high *r*-values. The presence of aluminum nitride and a low interstitial nitrogen content in the hot band is promoted by the use of a low slab reheat temperature (down to 1100°C and below) and sometimes a high aluminum content (up to 0.09%). Both of these factors minimize the dissolution during slab reheating of the aluminum nitride formed previously. The most important requirements are, however, that the total nitrogen content is low ($<0.003\%$) and that the coiling temperature is high, well above 700°C. This enables aluminum nitride to form as the steel cools below the coiling temperature. The coarse carbides are also obtained by the use of the high coiling temperature.

The properties of continuously annealed AK steels benefit from high cold reductions and high annealing temperatures like IF steels, but unlike IF steels the final properties depend critically on the cooling part of the annealing cycle. This must be designed to reprecipitate the carbon that goes into solution at the annealing temperature as completely as possible. To achieve this, the steel is first cooled slowly at about $10\,^{\circ}\mathrm{C\,s^{-1}}$ to approximately 675°C to maximize the carbon in solution. This is followed by rapid cooling, ideally at about $100\,^{\circ}\mathrm{C\,s^{-1}}$, to a temperature usually below 450°C. This minimizes the amount of carbon reprecipitated during cooling to maximize the driving force for the nucleation of fine closely-spaced carbides within the grains. These are then able to grow by short-range diffusion. If cooling to about 400°C or above is used the steel is allowed to cool slowly to about 350°C over 2–3 minutes to allow the carbon to precipitate. If temperatures close to 250°C or below are used, the steel is reheated to about 350°C or above and then allowed to cool slowly over approximately two minutes before rapid cooling to room temperature. Cooling to 250°C instead of 400°C leads to the nucleation of more closely spaced carbides and the use of reheating enables them to grow rapidly by taking advantage of a higher diffusion rate. The reprecipitation of carbon then becomes more complete and the steel is less likely to be subject to strain aging. Carbon contents close to 0.01% lead to poor reprecipitation of carbon even with a potentially good cooling sequence. To give the best possible properties, carbon contents tend, therefore, to be in the range 0.015 to 0.025%. The use of relatively low manganese contents such as 0.15% can lead to the formation of relatively more fine manganese sulfide precipitates which may act as easy sites for the nucleation of new carbide precipitates during cooling. A low manganese content is therefore beneficial.

The use of a high coiling temperature can lead to operational problems. An addition of boron may be used to form boron nitride, thereby providing an alternative method of reducing the hot-band interstitial nitrogen content without the need for a high coiling temperature. The benefit to be obtained from coarse carbides is, however, lost with the result that this type of steel develops low *r*-values. Therefore it may only be used for applications requiring relatively low formability. All continuously annealed AK steels exhibit a yield point as in the annealed condition. They all, therefore, require about 1% temper rolling.

3. *High-strength Steels*

3.1 *Solid Solution-strengthened Steels*

Each type of cold-rolled and annealed mild steel may be given a modest increase in strength, to give yield stresses usually in the range 220–300 MPa, by the addition of certain elements that remain in solution in the final product. The element most commonly used is phosphorus because it gives a worthwhile increase in strength for small additions but the strength increase may be supplemented by additions of manganese or silicon. Boron may also be used in titanium IF steels. The phosphorus addition is usually restricted to below 0.1% to avoid problems with welding and other potential problems. The main advantage of these steels is that they retain *r*-values almost as high as those of the mild steels from which they are derived There is, however, the usual progressive drop in formability with increasing strength even though this tends to be less than for other strengthening mechanisms. There is some evidence for a niobium IF steel that the decrease in *r*-value for a given increase in strength is less for phosphorus than for manganese or silicon, whereas for a titanium IF steel the decrease in elongation is greater for phosphorus than for the other two elements.

The base compositions of all these steels are usually similar to those of the mild steels from which they are derived and the general processing requirements are also the same. Thus, for example, a continuously annealed solid solution-strengthened AK steel must be hot-rolled to precipitate aluminum nitride in the hot band if high *r*-values are to be obtained. Similarly, a batch-annealed AK steel must be hot-rolled to retain nitrogen in solution.

Additions of manganese or silicon may be used to give small solid solution-strengthening effects in hot-rolled steels but these additions are more usually used in hot-rolled steels to influence transformation characteristics.

3.2 Bake-hardening Steels

A bake-hardening steel is one that increases its yield stress during a paint stoving operation after forming by means of a strain-aging process. In the as-delivered condition, therefore, the steel has the formability of a relatively low-strength steel but the performance in service of a steel with a higher yield stress. The bake-hardening tendency is usually measured by straining a steel by 2% and then baking at 170°C for 20 minutes. The difference between the flow stress after 2% strain and the lower yield stress after baking is taken as a measure of the bake hardening. The minimum levels of bake hardening that are usually considered to provide a useful benefit are 30–40 MPa and maximum values usually used are 50–60 MPa. Higher values than these would lead to excessive strain aging at room temperature with a consequent loss in formability and the formation of stretcher strain markings on forming, unless just-in-time delivery is organized before forming.

Any steel that exhibits strain-aging will automatically exhibit bake hardening but the commonly used bake-hardening steels are based on solid solution-strengthened steels. Their chemistry or processing is modified, if necessary, to leave sufficient carbon in solution after annealing to give the bake hardening. It is usually considered that 5–20 ppm carbon in solution is adequate to give sufficient bake hardening but different values in this range are preferred by different suppliers depending on other details of the steel. Obtaining a suitable solute carbon content is a natural consequence of the continuous annealing of an AK solid solution-strengthened steel, since even an optimized cooling path is not able to reprecipitate all the carbon completely. A batch-annealed AK bake-hardening steel may be obtained by reducing the carbon content of the steel to about 0.01% or below. The complete reprecipitation of carbon during slow cooling then becomes difficult and sufficient carbon remains in solution to give the bake hardening even after the long slow cool. Batch annealing using hydrogen as the protective atmosphere gives more bake hardening than the older HNX gas annealing due to the faster cooling rate with hydrogen.

Continuously annealed bake-hardening IF steels, containing ultralow carbon contents, may be produced in several ways depending on the *r*-values required but most of these steels are alloyed with niobium and/or titanium. With the lower alloy additions, some of the carbon remains uncombined and this carbon is available to provide the bake hardening. With higher additions, most of the carbon is combined with titanium or niobium prior to annealing. The free carbon needed to provide the bake hardening is then obtained during annealing by the use of a high annealing temperature such as 850°C or more, to take carbon back into solution. The steel is then cooled rapidly to minimize the reformation of precipitates on cooling. As an example, a minimum cooling rate of $30^\circ C\,s^{-1}$ from 850°C is necessary to give adequate bake hardening for a steel with a niobium:carbon atomic ratio of unity. With a lower niobium to carbon atomic ratio, such as 0.3, the very high annealing temperature and the rapid cooling would not be necessary but the resulting *r*-values would be lower.

The production of bake-hardening IF steels by batch annealing is only possible using relatively low alloy additions, since annealing temperatures approaching 850°C are not possible with this annealing method. Clearly obtaining a consistent level of bake hardening becomes difficult when it depends on a very precise control of steel chemistry.

3.3 Microalloyed or High-strength Low-alloy (HSLA) Steels

A useful method of increasing the strength of a hot-rolled steel is to increase the carbon and manganese contents. The potential strength increase is, however, limited if the content of these elements is restricted to retain weldability, unless hard transformation products are obtained. A larger increase in strength may be obtained by making small additions of niobium, titanium, or vanadium combined with a suitable choice of carbon, manganese, and nitrogen content. These steels, called microalloyed or high-strength low-alloy steels (HSLA steels), develop their strength mainly by grain refinement, different amounts of precipitation-strengthening, and a little solid solution-strengthening. They usually have yield stresses in the range of 300–500 MPa or more, but the most commonly used grades have yield stresses a little below the middle of this range.

The grain refinement in these microalloyed steels arises from several mechanisms. At the slab reheat stage, residual precipitates restrict grain growth. At lower temperatures the recrystallization of austenite between each rolling pass is restricted, mainly by strain-induced precipitation but also by solute drag effects. Greater deformation is then accumulated in the austenite before recrystallization can take place and this leads automatically to a finer austenite grain size. Finally, at sufficiently low temperatures, recrystallization is completely inhibited. The final transformation to ferrite then occurs from an unrecrystallized austenite, which again leads to a finer ferrite grain size. It is generally considered that of the three elements mentioned the greatest retardation of recrystallization for a given addition occurs with niobium and the least with vanadium. Niobium tends, therefore, to be the most commonly used element for medium-to-large increases in strength but the effect of niobium may be supplemented by vanadium to provide additional precipitation-strengthening from vanadium nitride if a very high yield stress ($\sim$500 MPa) is required.

Cold-rolled versions of microalloyed steels may be obtained using either batch or continuous annealing but for a given composition higher strengths are usually obtained by continuous annealing. A continuously annealed product also has the potential for giving more consistent properties. In general the strength range possible with a cold-rolled and annealed steel is lower than that for a hot-rolled steel and it is useful to note that all HSLA steels have low r-values. They cannot, therefore, be used for applications requiring high drawability.

3.4 Transformation-strengthened Steels

A further method of producing higher-strength steels is to develop a significant proportion (e.g., up to 20%) of hard transformation products in a matrix of fine ferrite grains. The earliest of these steels to be developed were called dual-phase steels in which the second phase was martensite. These steels may in principle be obtained directly from the hot mill in two ways and by continuous annealing. In the first direct method, a carbon–manganese steel is used containing sufficient manganese and other additions (such as chromium) to suppress the formation of bainite during runout table cooling. The steel develops a suitable structure of ferrite and high-carbon austenite during the early stages of this cooling and the austenite finally transforms to martensite after coiling provided that the coiling temperature is below the M_S (martensite start) temperature ($\sim$250 °C). A difficulty with this method is that not many hot mills are able easily to achieve the required very low coiling temperature. In the second direct method, the steel is more highly alloyed with manganese, silicon, and molybdenum to suppress the formation of bainite even when a coiling temperature well above the M_S temperature is used. A difficulty with this method, however, is that the relatively high silicon level often leads to a poor surface in any hot-rolled product. Neither method is, therefore, in common use.

Dual-phase ferrite–martensite steels are more easily made after cold rolling by continuous annealing. A suitable ferrite–high-carbon austenite microstructure is developed by intercritical annealing and initial slow cooling. The usual rapid cooling available on a continuous annealing line enables the steel to be cooled to below the M_S temperature without the formation of bainite and the austenite then transforms to martensite.

More recently, a hot-rolled product has been developed involving a final structure of ferrite and up to about 20% of bainite. A carbon–manganese steel is hot rolled using a low finishing temperature but still in the austenite region, and a suitable ferrite–high-carbon austenite is developed on the runout table either by continuous cooling or by interrupted cooling part way along the runout table. The use of a coiling temperature usually in the range 400–450 °C enables the austenite that is retained to transform to bainite. This type of product is finding widespread application in structural parts. An equivalent cold-rolled product could be made by continuous annealing.

A further type of steel has been developed in which the hard phase may be a mixture of retained austenite and bainite. These steels may easily have a tensile strength above 600–800 MPa, depending on the precise composition used, and are called multiphase or TRIP steels (transformation-induced plasticity), because the retained austenite transforms to martensite under the action of the forming strains required to make a component. This leads to the good formability already noted from Fig. 1. The steels needed are based on a carbon–manganese composition but with a high level of silicon or possibly aluminum to suppress carbide precipitation. Processing on the hot mill is similar to that for the ferrite–bainite steels mentioned above. In this case, however, the gradual formation of bainite after coiling leads to a further buildup of carbon in the austenite yet to transform. The result is that the M_S temperature of this austenite reduces to below room temperature. The austenite is therefore retained in the structure after final cooling.

A difficulty with this type of hot-rolled steel, however, is that it requires extremely close control of hot-mill processing if acceptably consistent properties are to be obtained, and so it is not yet in commercial production. These types of steel may also in principle be processed by continuous annealing for which both the temperature and time for the partial transformation of austenite to bainite after rapid cooling would be under closer control. In this case certain research has shown that a high holding time is beneficial because the final austenite retained at room temperature transforms gradually under the action of forming strains, thereby imparting the highest elongation. This is because austenite regions with different carbon contents have different stabilities in the presence of deformation.

Ultrahigh-strength steels are steels with a tensile strength above about 1000 MPa, though strengths above 800–900 MPa could also be regarded as ultrahigh strength. Two types are possible, depending on whether the structure consists primarily of martensite or of bainite. However, each may contain a proportion of the other phase together with ferrite or retained austenite and may include precipitation effects. These steels may be processed by continuous annealing involving rapid cooling, using compositions usually based on carbon, manganese, and silicon.

Bibliography

Asfahani R, Tither G (eds.) 1993 *Proc. Int. Symp. Low Carbon Steel for the 90s*. TMS, Pittsburgh, PA

Bleck W (ed.) 1998 *Proc. Int. Symp. Modern LC and ULC Sheet Steels for Cold Forming: Processing and Properties*. Verlag Mainz, Aachen, Germany

Bramfitt B L, Mangonon P L (eds.) 1982 *Proc. Symp. on Metallurgy of Continuous Annealed Sheet Steel, Dallas*. AIME, Warrendale, PA

Collins L E, Baragor D L (eds.) 1991 *Proc. Conf. on Interstitial Free Sheet: Processing, Fabrication and Properties*. CIM/ICM, Ontario

Cuddy L J 1975 *Plastic Deformation of Metals*. Academic Press, New York

Iron and Steel Institute of Japan (ed.) 1994 *Int. Forum Physical Metallurgy of IF Steel 1994*. ISIJ, Tokyo

Korchynsky M (ed.) 1977 *Microalloying '75*. Union Carbide Corporation, New York

Kot R A, Bramfitt B L (eds.) 1981 *Proc. Conf. Fundamentals of Dual Phase Steel*. AIME, Warrendale, PA

Iron and Steel Society 1995 *Microalloying '95, Conf. Proc.* The Iron and Steel Society, Pittsburgh, PA

Pickering F B (ed.) 1992 *Materials Science and Technology*. 1992 VCH, Vol. 7, Chaps. 6 and 7

Pradhan R (ed.) 1984 *Proc. Symp. on Metallurgy of Continuously Annealed Cold Rolled Sheet Steel, Detroit*. AIME, Warrendale, PA

Pradhan R (ed.) 1990 *Proc. Conf. Metallurgy of Vacuum Degassed Steel Products*. The Minerals, Metals and Materials Society, Warrendale, PA

Pradhan R (ed.) 1992 *Proc. Conf. on Developments in the Annealing of Sheet Steels*. The Minerals, Metals and Materials Society, Warrendale, PA

Pradhan R (ed.) 1994 *Proc. Symp. on High-Strength Sheet Steels for the Motor Industry*. The Iron and Steel Society, Baltimore, MD

Pradhan R, Ludkovsky G (eds.) 1988 *Proc. Symp. on Hot and Cold Rolled Sheet Steels*. The Metallurgical Society, Cincinnati, OH

Shimada M, Nashiwa M, Osaki T, Furukawa Y, Shimatani Y, Fujita Y, Aihara H 1996 *Review of Annealing Technology*. Technical Exchange Session, IISI, p. 17

Takechi H 1990 In: Dorgham M A (ed.) *Vehicle Design and Components—Materials Innovation and Automotive Technology, 5th IAVD Conf. Geneva*. Interscience, New York

R. C. Hudd

Sintering

Powders exhibit a fluid-like character that allows their easy shaping into engineering designs. The shaping processes include die compaction, slip casting, tape casting, extrusion, injection molding, isostatic pressing, and rolling (German 1994). After forming the particles have little strength. Consequently, a sintering heat treatment is applied to the powders to strengthen and densify the structure by bonding the particles to each other. This manufacturing route is beneficial in the production of many objects, ranging from automotive connecting rods to pottery. Sintered structures have an advantage since unique, multiple-phase microstructures can be created by mixing different powders.

Sinter bonding of particles occurs at temperatures in excess of approximately one-half of the absolute melting temperature, but can occur at low temperatures for powders in the nanoscale size range. Since engineering materials melt over a wide range of temperatures, there is a corresponding wide range of sintering temperatures, e.g., ice sinters at a low temperature when it forms glacier and clay pottery sinters near 1000°C.

Sintering lowers the surface energy by reducing surface area through the formation of interparticle bonds. The bonds grow by atomic level mechanisms. For most materials, diffusion along the grain boundaries is most important. Alternatively, particle melting occurs during liquid phase sintering, resulting in a solid–liquid mixture that sinters by viscous flow. The liquid phase provides bonding, contributes a capillary force, and enhances mass transport. Other sintering techniques such as hot isostatic pressing apply pressure to enhance particle deformation and bonding. Newer techniques react mixed powders to synthesize intermetallics, borides, carbides, oxides, or nitrides while simultaneously densifying the product.

Sintering is an irreversible process. Once a powder is heated it will not revert back to powder as significant pore structure changes accompany sintering. Additionally, engineering properties, including strength, ductility, conductivity, magnetic permeability, and corrosion resistance, undergo dramatic improvements during sintering. These property changes are the primary justification for industrial sintering. Sintered properties are dependent on the microstructure. Prolonged sintering coarsens the microstructure to a point where the properties peak and then decline with continued heating for longer times or at higher temperatures.

Not only is strength greatly increased, but compact density increases due to shrinkage. The particles attract each other due to capillary forces, leading to self-compression and the elimination of pores. This is common in sintering small particles, especially those $>1\,\mu m$. in contrast, low dimensional change is obtained in components sintered from powders $<100\,\mu m$. In metallic systems, densification is achieved in powder compaction, so there is little need for sintering shrinkage.

1. Brief History

The sintering process has been known for thousands of years (Kingery 1992). Some of the first sintered products were bricks heated in open pit fires to add strength. Still sintering is a primary operation in the production of most ceramics—whitewares, refractories, bricks, abrasives, porcelain, cutting tools, abrasives, and construction materials. The modern era is traced to Coolidge who sintered tungsten powder to develop a durable lamp filament for Edison. Subsequently, spark plugs, cutting tools, bearings, electrical insulators, and electrical contacts were developed in the 1930s. By the 1940s, sintering was used to fabricate nuclear fuel elements and automotive components. In the latter half of the 1900s, attention was directed to high performance materials with exceptional properties not obtainable via other manufacturing routes. In the early twenty-first century, sintering is employed in a diverse range of products that includes dental implants, rocket nozzles, aircraft wing weights, ultrasonic transducers, turbochargers, semiconductor substrates, and golf clubs. Many of the modern applications are justified by factors that include manufacturing economy, improved properties, and novel compositions.

2. Sintering Process

On a microstructural scale, sinter bonding is evident as cohesive necks grow between contacting particles. Figure 1 is a scanning electron micrograph of the neck formation between nickel spheres due to atomic level events that eliminates surface area on the particles. Typically surface energy is assessed by the surface area; smaller particles have high surface areas and more energy per unit volume, which promotes faster sintering. Thus, sintering is faster, or occurs at lower temperatures or in shorter times, for smaller powders.

Not all surface energy is available for sintering. For a crystalline solid, nearly every particle contact forms a grain boundary with an associated energy. The grain boundaries, important to atomic motion, are defective regions with high atomic mobility. For many materials the preferred path for atomic transport during sintering is along the grain boundaries, a process termed grain boundary diffusion. Other mechanisms of mass transport are active, including flow over free surfaces and through the crystal lattice, vapor transport, plastic flow, viscous flow, and dislocation climb. Viscous flow is most important to the sintering of glasses and polymers.

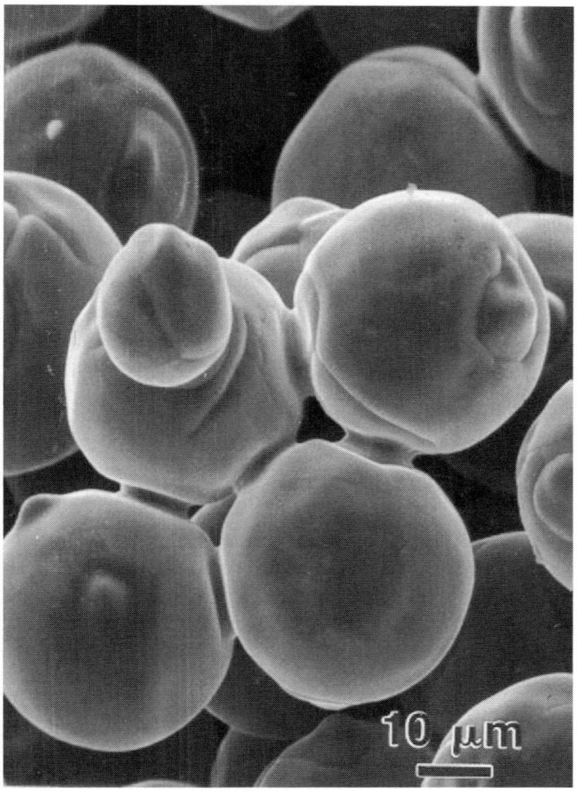

Figure 1
A scanning electron micrograph of sinter bonds formed between spherical particles originally in point contact.

All particles have an inherent driving force for mass flow that is termed the sintering stress which is estimated from the surface energy divided by the particle size (German 1996). Smaller particles provide a larger sintering stress. However, this is only half of the story, since sintering requires mass flow to respond to this stress. When temperature is raised, atoms move faster and the sintering stress works to compress the particle contacts and densify the compact. The sintering stress is larger with smaller particles and the atomic mobility increases with higher temperatures. Thus, as a definition of sintering:

Sintering is a thermal treatment for bonding particles into a coherent, predominantly solid structure via mass transport events that often occur on the atomic scale. The bonding leads to improved strength and a lower system energy.

3. *Sintering Techniques*

Theory is most accurate for the case of single-phase powders sintering by solid-state diffusion (Exner 1979, Shaw 1989). It consists of four stages as detailed in Table 1. Unfortunately, only a small portion of sintering practice relies on solid-state sintering. Most systems consist of multiple phases, possibly forming a liquid, and may be subjected to an external pressure to enhance densification. Pressure-assisted sintering includes hot isostatic pressing and vacuum sintering to a low residual porosity followed by gas pressurization to collapse any remaining pores. Pressurization during sintering is applicable to materials that are unresponsive to traditional sintering cycles, such as composites and intermetallics. At pressures of a few atmospheres densification is controlled by diffusional creep. Alternatively, densification at high pressures (200 MPa) is rapid when the effective stress exceeds the material's yield strength. Pressure is usually hydrostatic (hot isostatic pressing) or uniaxial (forging and hot pressing).

A major distinction among pressureless sintering techniques is between solid-state and liquid phase processes. Single phase, solid-state sintering is applicable to the sintering of pure substances such as nickel, alumina, or copper. Usually, additives are used to promote faster sintering by forming a liquid phase on the grain boundaries. These liquids improve mass transport rates and exert a capillary pull on the particles that is equivalent to a large external pressure. Because of productivity gains, over 70% of the sintered products are formed using a liquid phase, constituting 90% of the product value. Examples include stainless steels, superalloys, sialon (Si_3N_4-based compositions), aluminides, tool steels,

Table 1
Description of sintering stages.

Stage	Process	Surface area loss	Densification	Coarsening
Initial	Neck growth	Significant, up to 50% loss	Small at first	Minimal
Intermediate	Pore rounding and elongation	Near total loss of open porosity	Significant	Increase in grain size and pore size
Final	Pore closure, final densification	Negligible further loss	Slow and little density change	Extensive grain and pore growth

titinates, $AlN–Y_2O_3$, Fe–Nd–B, Co–Sm, TiC–Fe, $ZnO–Bi_2O_3$, W–C–Co, Fe–P, Mo–Cu, W–Ag, Cu–Sn, W–Ni–Fe, and Fe–Cu–C.

There are two forms of liquid phase sintering. Persistent liquid phases exist throughout the high temperature portion of the sintering cycle and can be formed in prealloyed powder (supersolidus sintering) or by inducing melting in a mixture of powders. Alternatively, transient liquid phase sintering forms a liquid that disappears during the sintering cycle due to dissolution into the solid or formation of a new phase, typically a compound. This latter process is reactive liquid phase sintering and is employed in the generation of high temperature compounds, including $MoSi_2$, Ni_3Al, WC, and TiB_2.

Today, sintering is critical to several industries and contributed $23 billion to the US economy in 1999. Sintered products can be tailored for a wide range of engineering properties and applications in automobiles, airplanes, bathroom fixtures, sputtering targets, magnetic recording devices, cutting tools, home appliances, wristwatches, musical instruments, sporting equipment, bearings, filters, heat sinks, batteries, and capacitors. Some of these devices require high surface areas, so it is desirable to obtain sintered strength without densification. In other cases, sintering is performed under conditions where full density is obtained. For full density, sintering requires a high temperature or the application of pressure to ensure pore closure. Such flexibility in processing makes sintering a useful option in many fields, especially since all common engineering materials can be sintered to net-shape.

Bibliography

Exner H E 1979 Principles of single phase sintering. *Rev. Powd. Met. Phys. Ceram.* **1**, 7–251

German R M 1994 *Powder Metallurgy Science*, 2nd edn. Metal Powder Industries Federation, NJ

German R M 1996 *Sintering Theory and Practice*. Wiley, New York

Kingery W D 1992 Sintering from prehistoric times to the present. In: Chaklader A C D, Lund J A (eds.) *Sintering '91*. Trans Tech, VT, pp. 1–10

Shaw N J 1989 Densification and coarsening during solid state sintering of ceramics: a review of the models. *Powd. Metall. Int.* **21**(3), 16–21; **21**(5), 31–3, **21**(6) 25–9

R. M. German

Sintering: Modeling

Sintering was originally used to fire clay pots. Since the 1950s, sintering has moved into the high technology arena. Today, it is used in the fabrication of net-shape components from metals, ceramics, cermets, and composites for diverse applications. Accordingly, much more attention is now devoted to understanding sintering and the control of final dimensions and properties.

During firing the microstructure progression goes from contacting particles to dense, nearly pore-free, bodies. The bonding process interacts with the continuously evolving microstructure via several mass transport processes. Although sintering is described by analytic models, accurate predictions require computer simulations because of complex interactions between the evolving microstructure and mass transport mechanisms. In the time it takes to conduct one laboratory sintering experiment, a contemporary computer simulation of sintering can provide thousands of "experiments." This was not true just a few decades ago. Simulations in the 1960s were 10 times slower than experimental sintering (it took 600 min of computer time to simulate 60 min of sintering for just two particles). As computing speeds improved, simulation progressed and now provide a credible complement to any experimental sintering study. Many aspects of modeling are advancing quickly, especially as numerical tools and reliable material databases become available (German 1993).

Industrial sintering processes involve many variants, including solid-state sintering of single phase-systems, pressure-assisted sintering, liquid-phase sintering, reactive sintering, transient liquid-phase sintering, and activated sintering (German 1996). Besides microstructure simulations, progress is occurring in simulation of final dimensions, furnace heat transfer, component distortion, and atmosphere interaction with the compact. Thus, the range of simulation techniques grows to include continuum mechanics coupled to finite element analysis, discrete element analysis, Monte Carlo simulations, computational fluid dynamics, and simple time-advance simulations. For historical reasons, the latter are most popular, but the important problems in component size, shape, and warpage are being attacked using continuum mechanics coupled with finite element analysis (Olevsky 1998).

Most computer models for sintering start with a solid-state single-phase system. From that base, many of the industrially important variants can be developed. As the sophistication progresses it will soon be possible to simulate production sintering situations. This will be an important transition in sintering technology, where progress comes via simulation as compared to the current situation where experimentation leads progress.

1. Classic Sintering Problem

The classic sintering problem starts with spheres of equal size in point contact. The growth of the sinter bond between these particles is known as the initial stage of sintering. In this stage the neck size is sufficiently small that neighboring necks grow independent of each another (Fig. 1). Mass is transported to the neck by several mechanisms involving diffusion, evaporation, and dislocation motion. The initial stage ends when neighboring necks merge. Intermediate stage sintering occurs as the merging necks shrink the pores to form interconnected pores. Final stage sintering is associated with final pore closure, starting with the pores collapsing into separated spheres that continue shrinking until full density is achieved.

Computer analysis of initial stage sintering is well established (Hillman and German 1992). Early codes assumed sintering took place through the growth of a neck between two equal spheres. Unfortunately, numerical instabilities lead to a loss of volume, making the results of questionable value. With mass conservation, any change in the neck size must have a corresponding relation to the volume of material removed from elsewhere in the microstructure.

Neck growth inherently results from the curvature gradient in the sintering geometry. The curvature gradient drives mass flow, leading to a progressively smoother surface. When mass is moved from the contact between the spheres, neck growth is accompanied by densification. Otherwise, repositioning mass from one point on the surface of a pore to another by evaporation or surface diffusion simply smooths the pores without densification. Any point on the surface curve has an associated curvature. The instantaneous change in the neck profile depends on

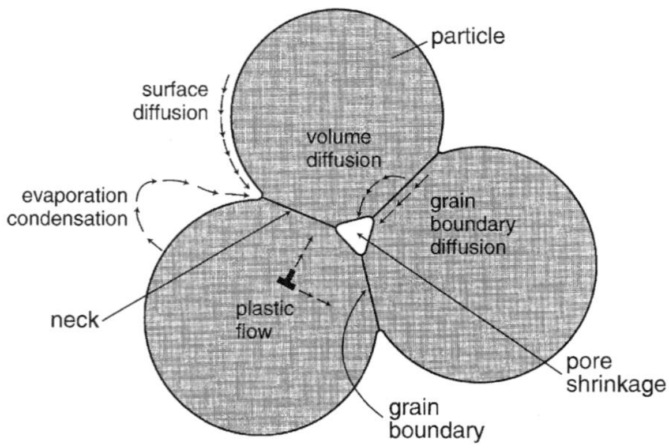

Figure 1
Three particles undergoing sinter bonding. Necks form at the particle contacts and grow during the sintering cycle by various mass transport mechanisms involving diffusion, dislocation flow, and evaporation.

the normal motion at the surface, which depends on the curvature. Sharply curved surfaces exhibit much more sintering. Models for sintering then analyze the point-to-point curvature differences and calculate the motion of mass in proportion to the local curvature. It is the curvature gradient that drives the flux of atoms during initial stage sintering.

$$\frac{dV}{dt} = JA\Omega \tag{1}$$

where V is the neck volume, t is the time, J is the atomic flux, A is the bond area over which the new mass is distributed, and Ω is the volume of a single atom or molecule. The change in curvature with position gives a chemical gradient that directs mass flow into the sinter bond between the particles. The deposited atoms change the neck size and shape. In turn the new curvature gradient influences the flux. An applied pressure during sintering enhances the effective stress (surface energy $\times$ curvature), leading to faster sintering. High temperatures promote faster mass transport and, thereby, contribute to faster neck growth. Likewise, smaller particles decrease the diffusion distance and increase the curvature, both leading to faster sintering. Consequently, accurate calculations of sintering rates require numerical techniques.

Sintering models with simplified forms are available to estimate the neck size from any single transport mechanism. These result in an initial stage sintering model for the neck size ratio X/D as a function of sintering time t under isothermal conditions,

$$\left(\frac{X}{D}\right)^n = \frac{Bt}{D^m} \tag{2}$$

where X is the neck diameter, D is the particle diameter, t is the isothermal sintering time, and B collects material and geometric constants as specified in Table 1. The values of n, m, and B change with time and degree of sintering, but are often assumed constant for each mechanism of mass transport. Many experiments have been analyzed using the exponent n to identify the sintering mechanism. However, in cases where the grain boundary energy is relatively high (dihedral angle is small) neck growth will be paced by grain growth and will not follow Eqn. (2). In these cases there will be a grain boundary groove at the neck.

Many solutions have been proposed for each of the sintering mechanisms and stages, so the form given in Table 1 is only representative of the modeling work and possible solutions. The simplified model represented by Eqn. (2) gives a rough approximation up to a neck size ratio of 0.3. The diffusion coefficient is embedded in the parameter B and follows an Arrhenius temperature dependence. The data for several materials are collected elsewhere (German 1996).

Equation (2) illustrates some key processing factors in sintering. A high sensitivity to the inverse particle size means smaller particles give more rapid sintering. Surface diffusion and grain boundary diffusion have the highest sensitivity to particle size changes, and are favored by smaller particles. In all cases, temperature appears in an exponential term, meaning that small temperature changes can have a large effect. Finally, time has a relatively small effect in comparison to temperature and particle size.

2. *Densification*

Many transport processes decrease the interparticle spacing during neck growth, resulting in compact shrinkage. As the particles move together, new contacts form with delayed onset of neck growth. Densification is often a desirable aspect of sintering, leading to stronger final products. Shrinkage is approximately related to the neck size by a simple model as follows

$$\frac{L}{L_0} = \left(\frac{X}{2D}\right)^2 \tag{3}$$

Table 1
The initial stage sintering equation for spheres $(X/D)^n = Bt/D^m$.

Mechanism	n	m	B
Viscous flow	2	1	$3\gamma/(2\eta)$
Plastic flow	2	1	$9\pi\gamma b D_v/(kT)$
Evaporation–condensation	3	2	$(3P\gamma/\rho^2)(\pi/2)^{1/2}(M/(kT))^{3/2}$
Volume diffusion	5	3	$80D_v\gamma\Omega/(kT)$
Grain boundary diffusion	6	4	$20\delta D_b\gamma\Omega/(kT)$
Surface diffusion	7	4	$56D_s\gamma\Omega^{4/3}/(kT)$

γ, surface energy; D_v, volume diffusivity; η, viscosity; D_s, surface diffusivity; b, Burgers vector; D_b, grain boundary diffusivity; k, Boltzmann constant; P, vapor pressure; T, absolute temperature; M, molecular weight; ρ, theoretical density; Ω, atomic volume; and δ, grain boundary width.

where the linear shrinkage $\Delta L/L_0$ is the compact length change divided by the initial length. Shrinkage is actually a negative value, but the sign is usually ignored. On this basis, shrinkage during initial stage sintering follows a kinetic law similar to Eqn. (2). The parameter B is exponentially dependent on temperature

$$B = B_0 \exp\left(-\frac{Q}{kT}\right) \tag{4}$$

where k is the Boltzmann constant, T is the absolute temperature, and B_0 consists of material parameters (such as surface energy, atomic size, atomic vibration frequency, and system geometry as defined in Table 1).

Sintered strength and other mechanical, thermal, or physical properties are dependent on the combination of density and neck size, e.g., early in sintering strength is dictated by the neck size ratio squared. Accurate calculation of the sintering rate is necessary for predicting sintered properties. Unfortunately, most kinetic models for sintering are complex and cannot be treated by simple approximations. Consequently, computer simulation is enlisted to address the sintering problem. This is an important direction to provide realistic modeling capabilities.

3. *Multiple Transport Mechanisms*

Multiple mechanisms usually act together during sintering. Most simulations recognize that sintering involves the delivery of mass to the interparticle bond by several simultaneous transport mechanisms. A typical assumption is that the instantaneous sintering mass flux is simply the sum of the contributions from each mechanism. At any instant the rate of neck growth is the sum of the individual fluxes,

$$\left.\frac{dX}{dt}\right|_{\text{total}} = \sum \left.\frac{dX}{dt}\right|_i \tag{5}$$

Each individual mechanism, represented by the subscript i, provides an instantaneous contribution to the total neck growth. Accordingly, computer simulations are employed to continuously calculate the contributions from each transport mechanism at a given time. Once the rates of mass flow are determined, the sum is used to enlarge the neck by determining its new volume and shape for the next instant of time. With repositioning of the transported mass, the curvature gradient is recalculated. Time is advanced by a small step and the new gradients are used to recalculate the fluxes and total mass flow. This process goes on with millions of small steps to eventually determine the geometry after a desired sintering time. Only in the past few years has the speed of such calculations exceeded the speed of actual sintering, for just two particles.

In calculating the rate of sintering, the evolution has been from simplistic geometric approximations to sophisticated multiple mechanism computer simulations. The usefulness of such simulations is evident in several of the proprietary codes, where the agreement with experiment is excellent. Usually, surface diffusion dominates initial stage sintering while evaporation-condensation is a minor contributor.

4. *Sintering Diagrams*

A consequence of these computer-based sintering calculations is seen in sintering diagrams such as plotted in Fig. 2 for 6 μm tungsten powder, starting with an initial density near 63% of theoretical. This plot gives the sintered fractional density vs. temperature for two hold times (1 and 10 h). Such diagrams prove useful in condensing and representing sintering behavior (Swinkels and Ashby 1981). These computer simulation products are based on multiple mechanism sintering, with accurate geometric representations and microstructure coarsening. The models continuously calculate the microstructure (grain size, pore size, neck size, and porosity) during

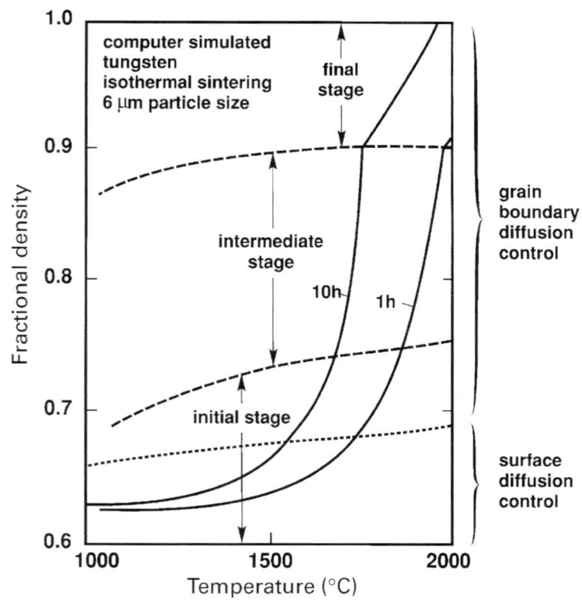

Figure 2
The sintering diagram for 6 μm tungsten powder. This map shows the sintered fractional density vs. sintering temperature for two possible hold times (1 and 10 h), and the stages of sintering are indicated by the dashed lines. Most of the densification is controlled by grain boundary diffusion.

sintering. A sintering diagram can have several forms, but they typically show density or neck size vs. temperature for isothermal sintering at various times. The underlying simulations are sensitive to material properties, but are effective in isolating the roles from the main process variables (time, temperature, pressure, green density, particle size, and grain size). The output might be density, neck size, dimensional change, or final compact shape (includes distortion from green density gradients) as a function of processing conditions. The graphical output is most useful in showing interactions between the key processing variables.

5. Applications

Sintering complexity is recognized in all of the modern models, leading to a reliance on computer simulations. Industrial manufacturing operations that rely on sintering involve complex chemistries, microstructures, and component geometries treated by proprietary models that embraced the practical realities. A major goal is to include the component geometry and to make accurate predictions of final component size as part of net-shape sintering, thereby avoiding postsintering machining. Such simulations require information on the starting component geometry, and knowledge of initial density gradients, and even component placement in the furnace and atmosphere flow conditions. Typical industrial sintering involves a stepwise heating cycle, with possible changes in atmosphere over time. The furnace design will impact on thermal gradients, and gravity is important to component distortion, especially for more massive components.

Computer models are now being employed in the innovation of several products. One example is in the sintering of tantalum capacitors where the process is dominated by just a few transport processes, but the particles are irregular and the heating cycle is nonisothermal (Dubois *et al.* 1996). From this research, new schemes have been isolated for the production of high surface area capacitors. Another target has been the production of precise gears from mixed iron, copper, and graphite powders that alloy and sinter bond during firing (Raman *et al.* 1998). This case has the complexity of starting with mixed composition, highly deformed particles that undergo both solid-state and liquid-phase sintering, with a phase change in the steel. The model includes effects traced to the powder characteristics, heating cycle, and even process atmosphere. Another set of computer simulations is used for heating prealloyed tool steel powder to a temperature where first liquid forms (German 1999). In this case the sintering is by viscous flow of the partially molten prealloyed powder with a rapidly coarsening microstructure. Ranges of time–temperature are defined to help ensure proper cycle definition to attain full density without distortion. Similar progress has been reported on sintering alumina casting refractories and cemented carbide cutting inserts (Favrot *et al.* 1999). These variants add complications of bimodal powder mixtures, shape complexity, and activated sintering.

6. Concluding Remarks

Only recently have sintering models and the associated computer simulations embraced actual production conditions. Early problems traced to a failure to incorporate realistic starting microstructure and density profile. In contrast, production sintering involves a wide range of particle sizes and packing inhomogeneities. One current difficulty traces to use of too few initial particles in the simulations. After sintering, a final grain size might be 10 times larger than the initial particle size. Thus, the initial simulation must start with more than 1000 particles for every final grain in the sintered structure. This is a major computational burden. Further, there are many situations where 1×10^9 particles are consumed to form one sintered grain, which is far beyond current simulation capabilities.

The real payout for sintering models will come when accurate final dimensions can be predicted. Early demonstrations rely on extensive databases, finite element simulations, and consider heat flow during the process (Sanliturk *et al.* 1999, Tsvelikh *et al.* 1995). Models are needed to relate the work of sintering with densification and dimensional change. The long-term goal will be to use these simulations to isolate cycles that improve final product dimensional accuracy. A secondary goal will be to better control microstructure and properties. Major gains in production efficiencies and process control will be enabled by these gains.

Bibliography

Dubois S G, Ganesan R, German R M 1996 Sintering of high surface area tantalum powder. In: Chen E, Crowson A, Lavernia E, Ebihara W, Kumar P (eds.) *Tantalum*. The Minerals, Metals and Materials Society, Warrendale, PA, pp. 319–23

Favrot N, Besson J, Colin C, Delannay F 1999 Cold compaction and solid-state sintering of WC–Co based structures: experiments and modeling. *J. Am. Ceram. Soc.* **82**, 1153–61

German R M 1993 Overview of key directions and problems in computational and numerical techniques in powder metallurgy. In: Madan D S, Anderson I E, Frazier W E, Kumar P, McKimpson M G (eds.) *Computational and Numerical Techniques in Powder Metallurgy*. The Minerals, Metals and Materials Society, Warrendale, PA, pp. 1–15

German R M 1996 *Sintering Theory and Practice*. Wiley, New York

German R M 1999 Computer model for the sintering densification of injection molded M2 tool steel. *Int. J. Powder Metall.* **35**(4), 57–67

Hillman S H, German R M 1992 Constant heating rate analysis of simultaneous sintering mechanisms in alumina. *J. Mater. Sci.* **27**, 2641–8

Olevsky E A 1998 Theory of sintering: from discrete to continuum. *Mater. Sci. Eng.* **R23**, 41–100

Raman R, Griffo A, Zahrah T F, German R M 1998 Computer Simulation of Fe–2Cu–0.8C Sintering. In: (eds.) *Advances in Powder Metallurgy and Particulate Materials.* Metal Powder Industries Federation, Princeton, NJ, pp. 10.89–96

Sanliturk K Y, Aydin I, Briscoe B J 1999 A finite element approach for the shape prediction of ceramic compacts during sintering. *J. Am. Ceram. Soc.* **82**, 1748–56

Swinkels F B, Ashby M F 1981 A second report on sintering diagrams. *Acta Metall.* **29**, 259–81

Tsvelikh A, Thompson W, Easton A, Freshwater I 1995 A geometrical finite element model of the sintering process of advanced ceramics. *Comput. Mater. Sci.* **3**, 457–64

R. M. German

Solid-state Extrusion to High Performance Fibers

Solid-state extrusion and coextrusion (Cappacio *et al.* 1977, Inoue 1985, Porter *et al.* 1988, Zachariades *et al.* 1977), including ram extrusion (Porter *et al.* 1988, Zachariades *et al.* 1977) and hydrostatic extrusion (Cappacio *et al.* 1977, Inoue 1985), have been extensively studied to produce highly oriented polymer morphologies with high mechanical and physical properties. These techniques have been mainly applied to the uniaxial drawing of melt-crystallized morphologies. More recently, it has been shown that solution-crystallized single-crystal aggregates of ultra-high molecular weight polyethylene (UHMW-PE) and polypropylene could be superdrawn by two-stage drawing to draw ratios of 80–300, producing ultra-oriented fibers that have tensile properties approaching their theoretical limits (Kanamoto and Porter 1989). The two-stage draw consists of a first-stage, solid-state coextrusion (or calendering for the industrial process) to a low extrusion draw ratio (EDR), followed by a second-stage, tensile drawing under controlled conditions. This technique has been successfully applied to ultradrawing of nascent reactor powders, including UHMW-PE (Kanamoto *et al.* 1987) and poly(tetrafluoroethylene) (Endo *et al.* 1998). The high ductility of the reactor powders is ascribed to the less-entangled state of the nascent morphologies, depending on the polymerization catalysts and conditions. The advantage of reactor powder draw, compared to the conventional gel spinning/drawing (Lemstra and Kirschbaum 1985), is that this simple process does not involve any solvent treatment that can cause environmental problems related to the toxic solvents. Thus, in this article, the development of high mechanical performance by solid-state extrusion is illustrated by the two-stage drawing of UHMW-PE reactor powders. The industrial application of this technique to produce high performance fibers and films will also be briefly explained.

1. Experimental Procedures

The samples, used "as received," were Montell Hifax 1900 (UHMW-PE) and a series of UHMW-PE prepared by using a high-activity Ziegler catalyst system at the Central Technical Research Laboratory of Nippon Mitsubishi. The powders were initially compression molded into films, always below their T_m values. The compacted powder films, thus obtained, were too brittle to be drawn directly by a tensile force. However, they were ductile after having been solid-state coextruded (or calendered) near their T_m values to a low EDR of 6–20 (first-stage draw). Subsequently, the extrudates were further drawn by a tensile force to achieve a higher draw at controlled temperatures and rates (second-stage draw).

2. Deformation Behavior and Tensile Properties

The nature of the catalyst as well as the polymerization conditions, including the polymerization

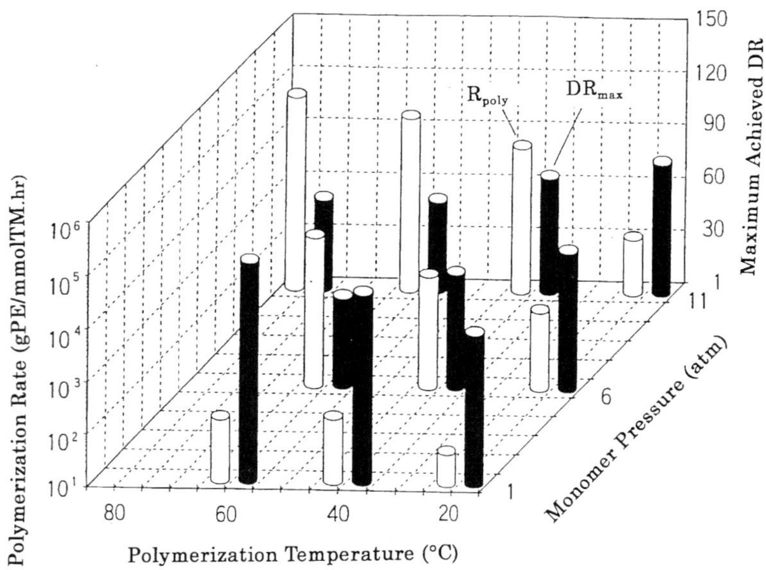

Figure 1
Effects of polymerization temperature (T_{poly}) and pressure (P_{poly}) on the polymerization rate (R_{poly}) and the maximum achieved total draw ratio (DR_{max}) for a series of UHMW-PEs prepared by a high-activity Ziegler catalyst (Sano *et al.* 2001). In order to avoid extensive reorganization, the draw was made at a relatively low temperature of 135°C.

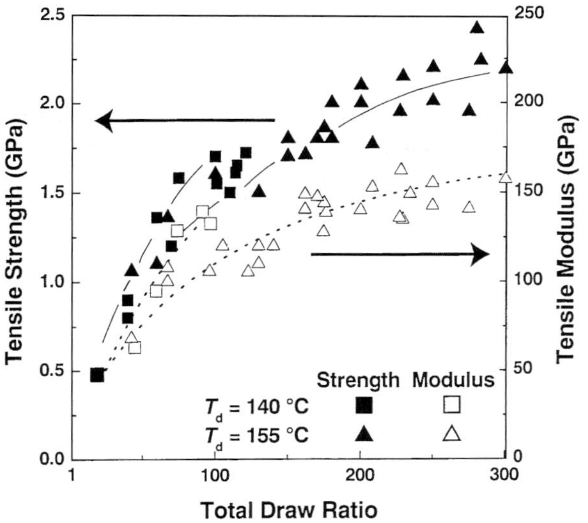

Figure 2
Tensile strength and modulus as a function of total draw ratio for the two-stage drawing of Hifax 1900 reactor powder. The second-stage tensile drawing was made at 140 and 155°C.

temperature (T_{poly}) and monomer pressure (P_{poly}), has an important effect on the ductility of nascent powder (Sano *et al.* 2001, Smith *et al.* 1985). It was shown that polymerization by a low-activity catalyst at low values of T_{poly} gave highly ductile UHMW-PEs (Smith *et al.* 1985). However, even a high-activity catalyst produces UHMW-PEs with fairly high ductility (Sano *et al.* 2001. Figure 1 shows the effects of T_{poly} and P_{poly} on the polymerization rate (R_{poly}) and the maximum achieved DR_t (first-stage EDR × second-stage DR) for a series of UHMW-PEs prepared using a high-activity Ziegler catalyst (Sano *et al.* 2001). These results show that there exist optimal values of T_{poly} and P_{poly}, depending on the characteristics of the catalyst used (Sano *et al.* 2001, Smith *et al.* 1985).

The drawability also depends markedly on the draw techniques and conditions. Some commercial UHMW-PEs show excellent ductility using two-stage draw (Kanamoto *et al.* 1987). The ductility of Hifax 1900 for the second-stage draw increased rapidly with draw temperature (T_d). The highest DR_t, of 200–300 after the second-stage draw, was achieved at T_d = 150–155°C, some 10–15°C above the T_m of the reactor powder. Figure 2 shows tensile modulus and strength as a function of DR_t. The maximum achieved tensile modulus and strength were ~150 and ~2.2 GPa, respectively. This modulus value is comparable to those of commercial UHMW-PE fibers prepared by the gel process. However, the strength is lower than those of commercial fibers.

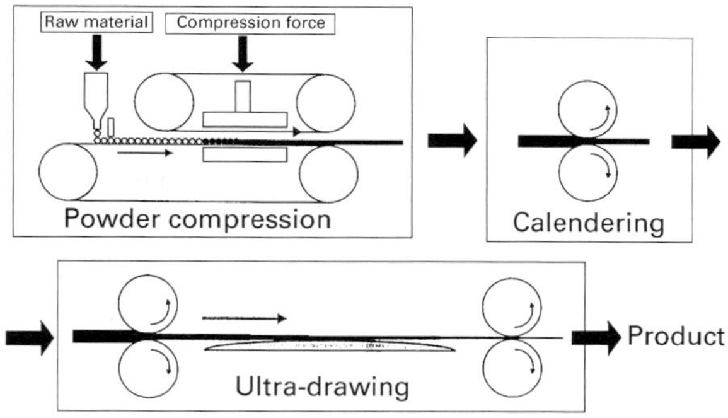

Figure 3
Schematic drawing of an industrial process for the two-stage drawing of reactor powder.

Time-resolved WAXD measurements during the second-stage drawing of a coextrudate with an EDR of 6.5 at 150–160 °C showed that, upon heating the coextrudate at the T_d, with the sample length constant, most of the orthorhombic crystals melted and formed randomly oriented melt, but part of them transformed into the hexagonal crystalline form. Upon the initiation of draw, the hexagonal phase rapidly transformed into the orthorhombic form, and the amorphous chains gradually oriented and crystallized into the oriented orthorhombic form. The coexistence of the three phases indicated that the applied stress was not homogeneously distributed within the deforming sample. These observations revealed that, for the second-stage draw at values of $T_d \leqslant 160$°C, the major deformation proceeded in a highly crystalline state, where the orthorhombic crystal form was predominant. Thus, an efficient draw was achieved at least up to a T_d of 155–160°C, reflecting the characteristics of the UHMW-PE reactor powder.

In the industrial process for producing high performance UHMW-PE tapes and fibers, the first-stage draw is performed by calendering of a powder film previously compression molded below the static melting temperature. The processes used in a pilot plant, built at Nippon Mitsubishi, are schematically shown in Fig. 3. The ultradrawn products show high tensile properties close to those achieved by the laboratory experiments as shown in Fig. 2.

See also: Polymer Fiber Processing: High-performance Fibers; Processing Flexible Polymers to High Performance Fibers; Processing Rigid Polymers to High Performance Fibers

Bibliography

Cappacio G, Gibson A G, Ward I M 1977 *Drawing and Hydrostatic Extrusion of Ultra-High Modulus Polymers*. Applied Science, London

Endo R, Jounai K, Uehara H, Kanamoto T, Porter R S 1998 Uniaxial drawing of polytetrafluoroethylene virgin powder by extrusion plus cold tensile drawing. *J. Polym. Sci., Phys. Ed.* **36**, 2551–62

Inoue N 1985 *Hydrostatic Extrusion of Polymers: Hydrostatic Extrusion, Theory and Application*. Elsevier Applied Science, London

Kanamoto T, Ohama T, Tanaka K, Takeda M, Porter R S 1987 Two-stage drawing of ultrahigh molecular weight polyethylene reactor powder. *Polymer* **28**, 1517–20

Kanamoto T, Porter R S 1989 *Maximum Properties Achieved by Uniaxial Drawing of High Molecular Weight Polyolefins: Integration of Fundamental Polymer Science and Technology-3*. Elsevier Applied Science, London

Lemstra P J, Kirschbaum R 1985 Specialty products based on polymers. *Polymer*. **25**, 1372–84

Porter R S, Chuah H H, Kanamoto T 1988 *Measures for Polymer Chain Extension: High Modulus Polymers, Approaches to Design and Development*. Dekker, New York

Sano A, Iwanami Y, Matsuura K, Yokoyama Y, Kanamoto T 2001 Ultradrawing of ultrahigh molecular weight polyethylene reactor powders prepared by highly active catalyst system. *Polymer*, accepted for publication

Smith P, Chanzy H D, Roztinger B P 1985 Drawing of virgin ultrahigh molecular weight polyethylene: an alternative route to high strength fibers. *Polym. Commun.* **26**, 258–60

Zachariades A E, Mead W T, Porter R S 1977 *Recent Developments in Ultramolecular Orientation of Polyethylene by Solid-State Extrusion: Ultra-High Modulus Polymers*. Applied Science, London

T. Kanamoto

Solid-state Sintering

Solid-state sintering is the bonding and, usually, densification of particles by the application of heat below the melting point of a material. During sintering the surface area of the compact decreases, and this is normally accompanied by an increase in the density, but it is possible for the compact to decrease its surface area and increase its cohesion without a perceptible increase in density. There will be changes in other properties of the compact, such as increased strength, a drop in electrical resistance, or an increase in thermal conductivity. For most applications, an increase in density is the desired result of sintering. As with any physical process, there must be a lowering of the free energy of the system in order for the process to occur. In sintering the reduction in the energy of the compact is due to the reduction in the surface area of the powder particles.

Although sintering is the oldest method of forming materials from powders, an understanding of the processes that occur when powder compacts are heated is only 50 years old. Frenkel (1945), who was the first to propose a model for sintering, assumed that sintering occurred by the deformation of crystalline particles under the influence of surface tension, as would occur for amorphous materials such as glass (see *Viscous Sintering*). However, the predictions of this model did not agree with experimental results using crystalline particles. It was Kuczynski (1949) who demonstrated that sintering occurs by diffusive processes. He proposed several models to describe the early stages of sintering for different transport mechanisms and diffusion paths. Coble (1961) developed models for the later stages of sintering for the case of lattice diffusion or grain boundary diffusion. Although later developments (see reviews by Exner 1979, German 1996, Handwerker *et al.* 1989, 1990, and Rahaman 1999) have refined these basic ideas, the early models serve well to describe sintering processes. As sintering is a result of diffusive processes, two factors determine the rate of sintering: temperature (which controls the diffusion rate) and powder particle size (which controls the length scale for diffusion).

1. Sintering Stages

Sintering is generally divided into three stages: initial, intermediate, and final, expressing transitions in the connectivity of the microstructure. Within each stage the pore shape change is well defined and the shrinkage of pores occurs with the pores maintaining a nearly constant shape. The initial stage refers to the growth of contacts or "necks" between the individual powder particles in the compact (Fig. 1). The radius of the necks grown in the initial stage ranges from zero to ≈ 0.2 of the radius of the particles. In the intermediate stage the pores approximate to continuous channels coincident with the three-particle junctions (Fig. 2). The final stage begins when the pore channels break up into isolated pores located at

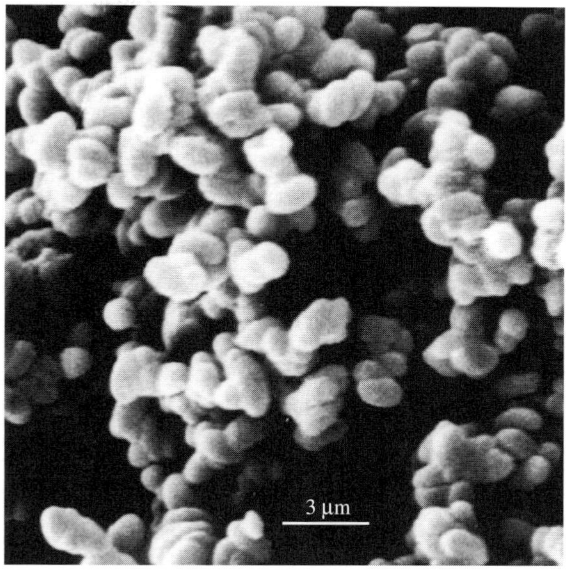

Figure 1
Particles in the initial stage of sintering. Small "necks" can be seen connecting the particles.

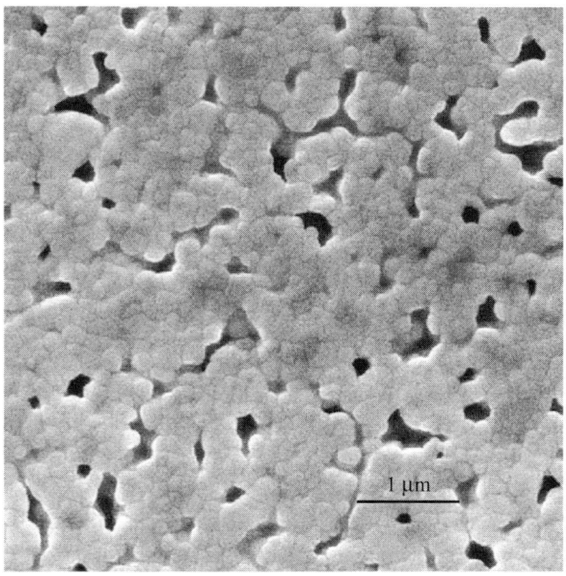

Figure 2
Structure showing the intermediate stage of sintering. Necks between particles have grown, resulting in a structure that contains continuous pore channels through the compact.

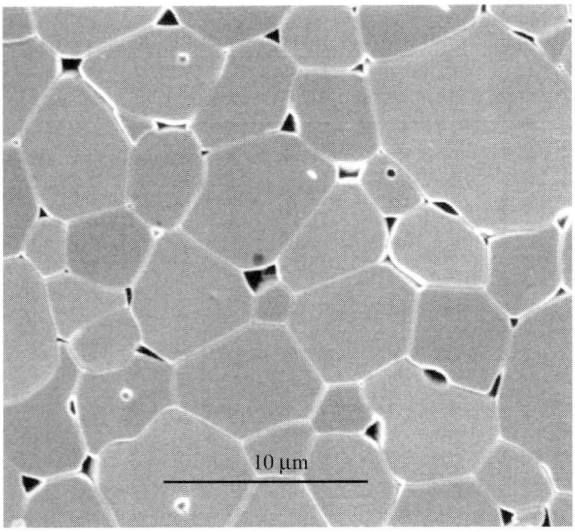

Figure 3
Polished surface of compact showing the final stage of sintering. Pores are isolated and generally located on the grain boundaries. Pore shape differences depending on location can be seen.

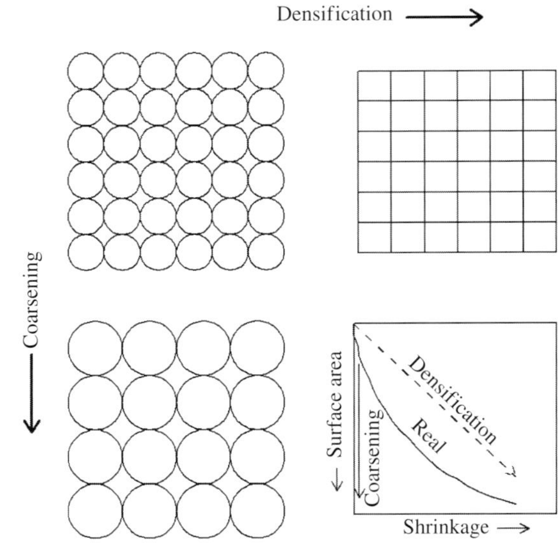

Figure 4
Schematic of the competitive processes that occur during sintering. Densification results in surface area reduction, pore removal, and shrinkage of the compact. Coarsening results only in surface area reduction.

the four-grain junctions (Fig. 3). This usually occurs when the density of the sample has increased to ≈92% of the theoretical density of the material.

2. *Driving Force for Sintering*

During sintering the surface area of the powder compact is reduced. There are two processes that can cause this—coarsening and densification—shown schematically in Fig. 4. For an initial array of n spheres of size R, packed in a cubic arrangement the surface area is $n \cdot 4\pi R^2$. For a coarsening of the spheres to a size $2R$, the number of spheres must decrease (to conserve mass) by a factor of eight and the surface area will be reduced by a factor of two. Thus, the coarsening of the structure will result in a reduction of the total surface energy. Notice that in this schematic there is no growth of the contacts between the particles and no decrease in the porosity of the sample. In contrast, if the spherical particles change their shape to cubes, the centers of the particles can move together and the volume of the sample will decrease. The surface area is completely converted to grain boundaries with an area of $n \cdot 2.15R^2$. The energy of a grain boundary is less than the energy of the two free surfaces that are eliminated when the boundary is formed, and the total energy of the system is reduced. Although both coarsening and densification result in a reduction in the surface area of the compact, coarsening does not result in an increase in density of the sample. To determine which process dominates, it is necessary to determine the kinetic processes that control the rates of the surface area change.

3. *Initial Stage Model*

The simplest model for sintering is the two-sphere model (Fig. 5), which contains most of the important features necessary for understanding sintering (Exner 1979). The basic premise of the model is that matter is transported from regions of high chemical potential to regions of low chemical potential. In this model, the regions of high chemical potential are at the contact region (grain boundary) and along the convex surface of the sphere. The concave neck region is at a lower chemical potential relative to these areas. Matter can be transported to the neck region through the gas phase (by evaporation from the convex surface and condensation on the concave surface), by surface diffusion (from convex surface to concave surface), lattice diffusion (from either the convex surface or the grain boundary), or grain boundary diffusion. When the source of matter is the grain boundary the two particles centers approach (densification) during sintering. When the source of matter is the convex particle surfaces, the two particles change shape without center approach (coarsening). Thus, in the initial stage of sintering, evaporation and condensation and surface diffusion result in coarsening, grain boundary diffusion results in densification, and lattice diffusion can result in

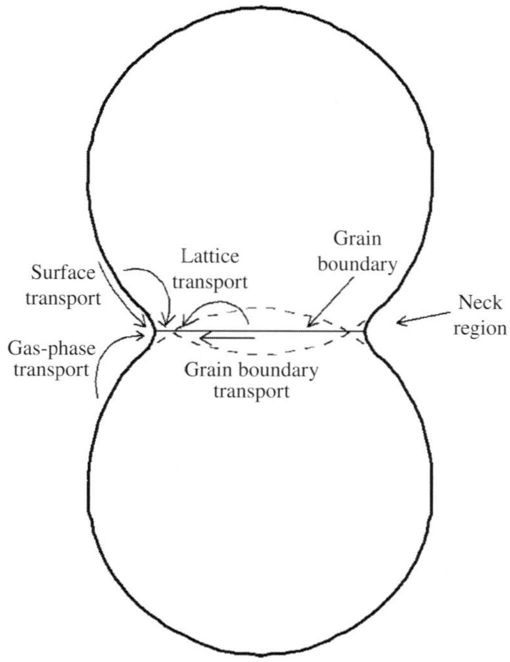

Figure 5
Two-particle model for initial-stage sintering. Transport of matter is from the particle surface or grain boundary to the neck region resulting in neck growth and densification.

either, depending on the source of the matter transported.

The driving force for transport during sintering is the difference in concentration between the source of matter (particle surface or grain boundary) and the sink for matter (neck region). The concentrations are different due to the chemical potential differences that result from the curvature differences (Johnson 1978). Along the particle surface the curvature changes from positive at the particle surface to negative at the neck region. Regions of positive curvature correspond to compressive stresses and will have a higher vapor pressure than a flat surface. Conversely, regions of negative curvature will have a tensile stress and a lower vapor pressure than a flat surface. The vapor pressure difference tends to transport matter through the vapor phase into the neck region. The stress difference tends to transport matter by diffusion to the neck region. Both mechanisms will be active and the total neck growth will be the sum of the two processes. The tensile stress at the neck region is balanced by a compressive stress on the grain boundary. This compressive stress raises the concentration of atoms in the grain boundary region and the tensile stress lowers the concentration of atoms in the neck region. This concentration difference leads to transport of matter from the grain boundary region to the neck region by grain boundary and lattice transport. In this case, there is densification of the compact.

Many different models have been proposed to describe the initial stage of sintering. In general each model first assumes a geometry, determines the chemical potential difference which is driving the transport, calculates the mass flux based on the geometry and the concentrations, and then relates the flux to the neck growth rate or the densification rate. If it is assumed that all the transport processes are independent and additive, the rate of neck growth can be expressed as:

$$\frac{\mathrm{d}x}{\mathrm{d}t} = \frac{C_1 \gamma D_l R}{x^3} + \frac{C_2 \gamma D_b R^2}{x^5} + \frac{C_3 \gamma D_s R^2}{x^5} + \frac{C_4 \gamma p_o R}{x} \quad (1)$$

where x is the radius of the neck, γ is the surface energy, R is the particle radius, C_i is a constant which takes into account the specific diffusion path and the geometry of the model, p_o is the vapor pressure, and D_i is the diffusion coefficient for lattice (l), grain boundary (b), or surface(s) diffusion (Johnson 1978).

4. *Intermediate Stage*

As neck growth proceeds there will be a transition in the pore structure to a system of tubular pores lying along the particle junctions; this is the intermediate stage. In this stage, the pore channels shrink in diameter until the geometry becomes unstable to fluctuations of the surface (Rayleigh instability). While the pore channels can shrink by surface diffusion, the path length for transport becomes very large and thus shrinkage is controlled by lattice or grain boundary diffusion (Rahaman 1999).

5. *Final Stage*

When the tubular pores have shrunk to the point where they are unstable, they will break up into isolated pores located at the grain junctions. This is the final stage of sintering and is the most important stage for most applications. It is during this stage that the properties of the compact reach the desired values. In the final stage several factors must be considered in order to understand the process by which pores are eliminated. The driving force for shrinkage of the pores is the difference in concentration between the atoms on the curved surfaces of the pores and the grain boundaries. As the pores are isolated, surface diffusion can only change the shape of the pores. Shrinkage of the pores is controlled by grain boundary and lattice diffusion. For nearly spherical pores the flux of matter, J, to the pore is approximately:

$$J = K_1 D_i \Delta c \left(\frac{r\mathscr{R}}{\mathscr{R} - r} \right) \quad (2)$$

where K is a geometric constant, D_i is either the lattice or grain boundary diffusion coefficient, Δc is the concentration difference, r is the radius of the pore, and $\mathcal{R}$ is the effective matter source radius.

6. *Particle Size Effect*

One of the paradigms of sintering theory is that reduction in the particle size increases the sintering rate. Countless experiments have all demonstrated the strong dependence of the sintering rate on the particle size. This size dependence is due to increasing both the driving force (through changes in curvature) and the kinetics (through changes in path length). Assuming that one mechanism is dominant and that the degree of sintering is given by the ratio of the neck size, x, to the particle size, R, the sintering time to achieve the same density for two particle sizes, R_1 and $R_2 = \lambda R_1$ is given as:

$$t_2 = \lambda^n t_1 \tag{3}$$

The value of the exponent, n, is related to the specific mechanism that controls densification and is equal to four for grain boundary diffusion and surface diffusion, three for volume diffusion, two for vapor transport, and one for viscous sintering. A twofold reduction in the particle size will result in a 16-fold decrease in the sintering time for a processed controlled by grain boundary or surface diffusion. In addition, for smaller particles it is possible to lower the sintering temperature and still reach the same density in the same time that larger particles would require at higher temperatures. If the relative density, ρ, can be expressed as a function of time and temperature,

$$\rho = t \exp\left(\frac{-Q}{kT}\right) \tag{4}$$

the change in sintering temperature to achieve the same relative density in a given time for a change in particle size can be calculated and is shown schematically in Fig. 6. It is seen that for sintering controlled by grain boundary diffusion a decrease in particle size from 0.6 μm to 0.01 μm lowers the sintering temperature by over 500°C. A decrease in the sintering temperature can be beneficial in preventing decomposition or evaporation of a material, reducing energy use, and inhibiting grain growth. However, changes in sintering temperature may also result in a change in the dominant sintering mechanism leading to a different microstructure with different properties.

7. *Dihedral Angle Effect*

In the discussion of the initial and intermediate stages it was assumed that the dihedral angle of the grain boundary with the pore surface is 180°, implying that the energy of the grain boundary is very much lower than the energy of the surface. As shown in Fig. 3, many pores have dihedral angles of less than 180°. For a pore on a grain boundary, the dihedral angle, ψ, is defined by the ratio of the grain boundary energy, γ_b, to the surface energy, γ_s, such that

$$\frac{\gamma_b}{\gamma_s} = 2\cos\left(\frac{\psi}{2}\right)$$

Figure 7 shows the shapes of pores for different values of the dihedral angle. The curvature of the pore surface increases as the dihedral angle increases, so the driving force for pore shrinkage also increases as the dihedral angle increases. When a pore is located on the junction of three or more grains, the curvature of the pore surface is determined by both the dihedral angle and the number

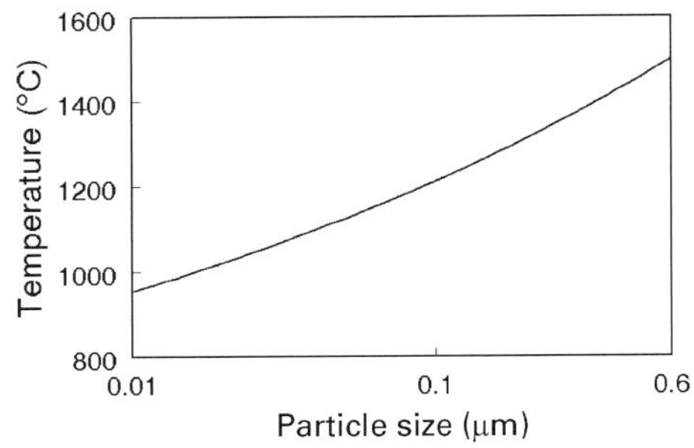

Figure 6
Effect of particle size on the temperature necessary to achieve a given density in a specific time. The line shown is calculated based on sintering being controlled by grain boundary diffusion with an activation energy of 540 kJ mol^{-1}. Courtesy of Prof. Gary L. Messing, Pennsylvania State University.

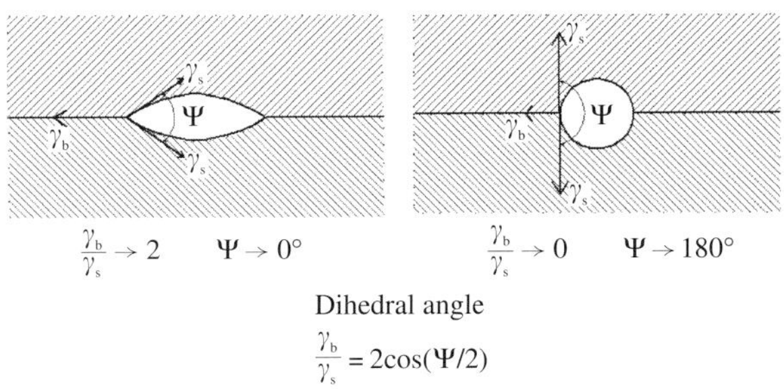

Figure 7
Dihedral angle shows the effect of surface energy to grain boundary energy ratio on the shape of pores. Low dihedral angles result in elongated pores and high dihedral angles result in spherical pores.

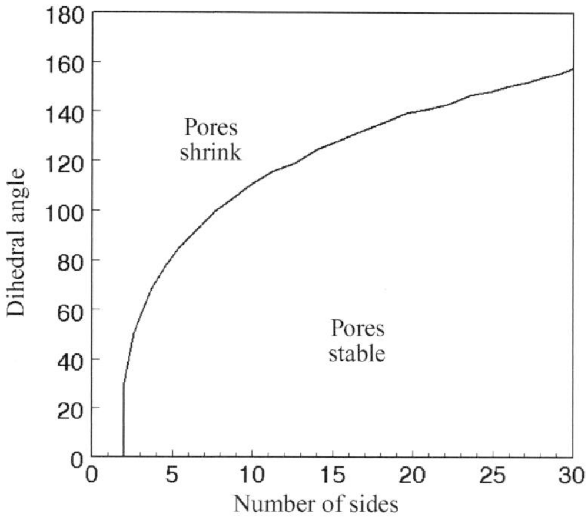

Figure 8
The stability of pores in the final stage of sintering as a function of the number of grains that the pore contacts. Pores with high numbers of sides will tend to be stable and remain in the structure.

of grains the pore contacts. For a three-sided pore when $\psi = 60°$ the pore will have straight sides and there will not be a driving force for shrinkage. Hence, this angle is a critical dihedral angle, ψ_{crit}, for pore shrinkage. For a four-sided pore, $\psi_{crit} = 70.5°$. The critical dihedral angle as a function of the number of surrounding grains is shown in Fig. 8 (Kingery and Francois 1967). Thus, when the dihedral angle is too low, pores will be stable in the structure and there will be a limiting density even though removal of the pores would lower the energy of the system. Ultimately, as grain growth occurs the number of grains intersecting the pore will decrease so that pores will become unstable and shrink. However, as the grain size increases, the diffusion distances increase and the shrinkage rate slows down, making pore removal a relatively slow process.

8. Pore Entrapment

The rate of pore shrinkage given above assumes that the pore is located on a grain boundary. If, during grain growth, the pore is entrapped inside a grain and lattice diffusion is the dominant transport mechanism, the diffusion distance is the pore–boundary separation distance. Such pores cannot be eliminated during sintering. For isolated pores (final stage of sintering) on grain boundaries that contain a gas phase that is insoluble in the lattice, shrinkage will be limited by the rate at which the gas can diffuse to the surface of the sample. This can also lead to a limiting density, but changing the atmosphere during sintering can minimize the effect.

9. Summary

This article is a summary of the basic ideas of solid-state sintering. Many of the features of sintering are well understood. It is clear from the modeling and experiments that particle size is a dominant factor in controlling sintering rates. Modeling has also led to an understanding of the role of the dihedral angle in controlling the microstructure development during sintering. The effects of density and density inhomogeneities in the unsintered (green) powder compact, changes in the heating rate, and the atmosphere are understood. Many other effects, such as crystalline anisotropy, the role of impurities, the interactions of multiple transport paths, and complex geometries are not well understood in a quantitative manner.

Bibliography

Coble R L 1961 Sintering crystalline solids. *J. Appl. Phys.* **32**, 789–92

Exner H E 1979 Principles of single phase sintering In: (ed.) *Reviews on Powder Metallurgy and Physical Ceramics.* Freund Publishing House, Israel, Vol. 1, pp. 1–4

Frenkel J 1945 Viscous flow of crystalline bodies under the action of surface tension. *J. Phys. USSR* **9**, 385

German R M 1996 *Sintering Theory and Practice.* Wiley, New York

Handwerker C A, Blendell J E, Coble R L 1989 Sintering of ceramics. In: Uskokovic D P, Palmour H, Spriggs R M (eds.) *Science of Sintering.* Plenum Press, New York, pp. 3–37

Handwerker C A, Blendell J E, Kaysser W A 1990 *Sintering of Advanced Ceramics.* Ceramic Transactions, American Ceramic Society, Westerville, OH, Vol. 7

Johnson D L 1978 Fundamentals of the sintering of ceramics. In: Palmour H, Davis R F, Hare T M (eds.) *Processing of Crystalline Ceramics.* Plenum Press, New York

Kingery W D, Francois B 1967 Sintering of crystalline oxides. In: Kuczynski G C, Hooton N A, Gibson C F (eds.) *Sintering and Related Phenomena.* Gordon and Breach, New York, p. 471

Kuczynski G C 1949 Self-diffusion in sintering of metal particles. *Trans. AMIE* **185**, 169–78

Rahaman M N 1999 *Ceramic Processing and Sintering.* Marcel Dekker, New York

J. E. Blendell

Solidification of Superalloys

Solidification processing used for the manufacture of superalloys is based on preventing defects which limit the performance of these alloys when used in aircraft turbine engines and other critical applications. The goal is to minimize segregation and microporosity to such an extent that it can be dissipated during subsequent thermo-mechanical processing. For example, at long solidification times, naturally occuring microsegregation creates intermetallic compounds which can only be dissolved by thermomechanical processing at high temperatures. However, other defects of macro scale can occur in various forms ranging from simple nonmetallic inclusions to various levels of severe compositional segregation which result from instabilities in the solidification processing. These macro-scale defects cannot be erased by subsequent processing and the material must be discarded.

The difficulty in solidification processing of superalloys is derived from the nature and complexity of their compositions. Resulting phase relationships are balanced to maximize high temperature mechanical properties, oxidation and corrosion resistance. Three very common compositions of cast and wrought alloys used for rotating turbine hardware are listed in Table 1.

Superalloys are typically nickel-based compositions that have up to 9 wt.% aluminum plus titanium "hardeners" which precipitate as coherent $Ni_3Al(Ti)$. These precipitates are stable at operating temperatures ranging from 500°C to greater than 1100°C and greatly enhance the strength of the matrix. A very common superalloy used in turbine disk applications is nickel–iron based and derives its strength from Ni_3Nb. Alloying additions of refractory elements such as molybdenum and tungsten are also added to strengthen the alloy while chromium is added to help provide protective oxide surface layers. Alloying elements are usually added in balanced amounts and up to levels that avoid regions of deleterious phase instabilities and unacceptably low ductility. As more alloying elements are added, the amount of chemical segregation usually increases, making subsequent metallurgical processing more difficult. High temperature homogenization thermal cycles and forging are used to allow diffusion to equalize the chemical inhomogeneity as much as possible. Insufficient homogenization of the solidification structure results in unacceptable phase distributions, irregular grain growth, inferior mechanical properties and unacceptable final products. Sensitivity to compositional balance is particularly acute during the solidification of large ingots, which are subsequently thermo-mechanically processed into wrought billets then forged into turbine components. The centers of large ingots are particularly prone to segregation because effective heat transfer is reduced. In this region, a small thermal gradient exists at the solid/liquid interface and the solidification rate is slow. Large columnar grains are present, unless convection of the molten metal caused by thermal gradients or electromagnetic fields creates nuclei which promote equiaxed grains. One source of grain nucleation is believed to be dendrite arms segments that become detached from the solidification interface.

During solidification the interdendritic liquid becomes enriched with aluminum and titanium (or niobium) and molybdenum. The dendritic cores become enriched with cobalt and/or nickel and chromium. Carbon and dissolved nitrogen can have a big influence on the solidification structure. Nitrogen in excess of the solubility limits will precipitate out as TiN, which will in turn act as nuclei for carbide precipitation, which is usually TiC(NbC). Depending on the volume of carbide particles that form, the movement of interdendritic liquid can be altered, resulting in porosity, finer grain size and more severe microsegragtion. If these particles agglomerate during solidification processing and pass through to the finished product undetected, they can act as sites for fatigue crack nucleation in the finished product.

Large ingot section sizes ranging from 500 mm in diameter to as large as 900 mm create very demanding solidification conditions. Solidification of these ingots typically occurs in a vacuum during vacuum arc remelting. In this production process, an electrode of the alloy is progressively melted by a metal vapor electric arc and dripped into a water-cooled copper crucible where the molten metal is solidified under very well-controlled conditions. Electrodes weighing 10 000 lb (4550 kg) typically take 24 hours to remelt and resolidify. During this entire time it is imperative that the solidification occurs under steady conditions so that an ingot is produced with a uniform structure, free of severe compositional segregation and minimized clusters of nonmetallic inclusions.

Freckles, or regions of severe solute-rich segregation, are undesirable defects that can occur during

Table 1
Nominal composition (wt.%) of common superalloys used for rotating turbine disks.

	Ni	Fe	Co	Cr	Mo	W	Al	Ti	Nb	C	B
Alloy 718	52	18.5	0	19	3	0	0.5	0.9	5.4	0.03	0.001
Waspaloy	58	0	13.5	19.5	4.3	0	1.3	3	0	0.04	0.001
Alloy 720	58	0	14.8	16	3	1.3	2.5	5	0	0.02	0.01

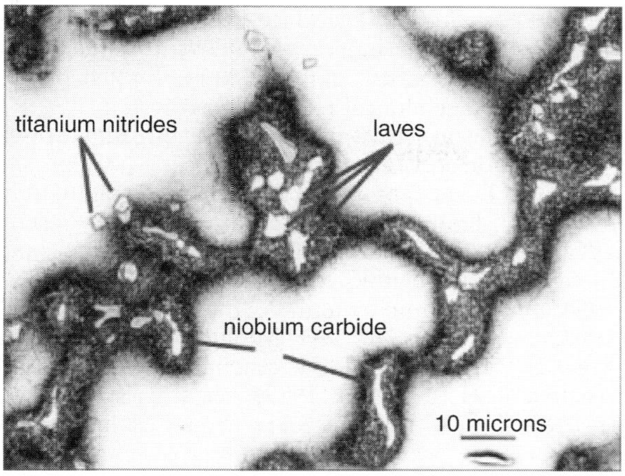

Figure 1
Photomicrograph of Alloy 718 polished and chemically etched to reveal microstructure. Large white areas are dendritic arms. The inter-dendritic regions, rich in niobium, are dark owing to the presence of delta phase and small gamma double-prime precipitates. The larger visible precipitates are niobium carbide, lavés and titanium nitrides.

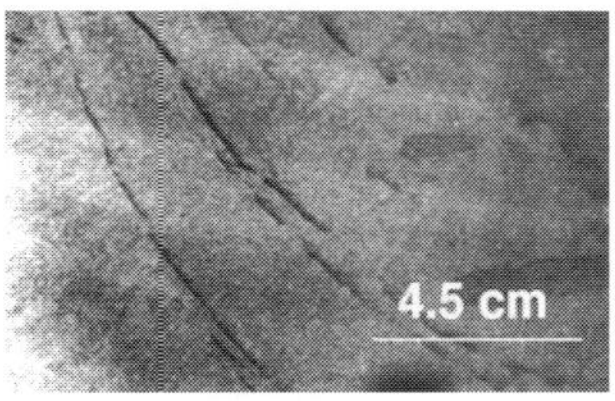

Figure 2
Photomicrograph of dark etching freckles located at mid radius in a longitudinal section cut from a 500 mm diameter Alloy 718 VAR ingot. These channel segregates are nearly aligned with the molten pool surface and grow from left to right.

unsteady ingot solidification. Freckles occur when meltback of the existing solidification interface occurs; this condition is created by arc anomalies.

White spots are another type of defect which occur during superalloy melt processing, but are not a direct result of the solidification process. Instead, white spots occur when solute lean material that has solidified at the water-cooled copper crucible interface detaches and falls into the molten pool and remains unmelted. The material is lean in hardener elements owing to the steep thermal gradient at the crucible wall and, because of its higher density, settles into a solidification mushy zone where there is very little super heat. White spots are so-called because

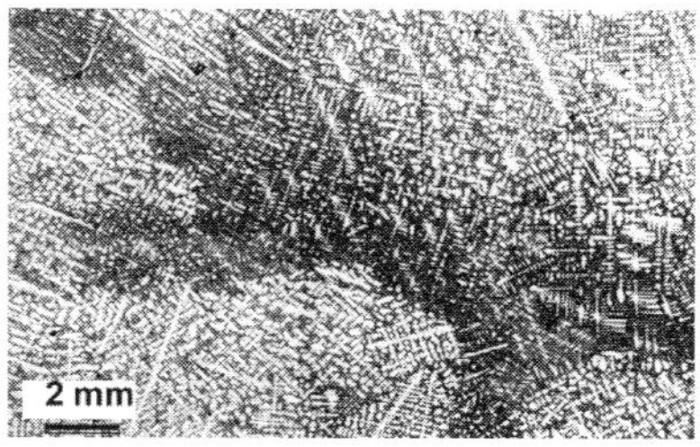

Figure 3
Higher magnification photomicrograph of a selected region of a freckle shown in Fig. 2. The white etching regions are dendrite cores and the dark etching region is the freckle, which is rich in niobium.

they etch lighter than the more solute-rich adjacent matrix. The term "dirty white spot" is coined for white spots that collect oxides and nitrides that have been pushed to the outside surface of the melt pool.

A special type of solute lean defect is called solidification white spot, because it is generated at the solidification interface. This type of white spot is usually small, has a less severe depletion in composition, and is always clean or free of oxide/nitride inclusions. Solidification white spots occur where the depth of the liquid pool is shallow and is susceptible to sudden disturbances in the delicate thermal and compositional conditions at the solidification interface. A potential source for such disturbances may be solid metal particles that occur during plasma arcing and drip shorting of the electrode as it is melted.

In conclusion, the goal of solidification processing of superalloys is directed toward minimizing the distance over which natural microsegregation occurs and the elimination of transient conditions during processing. In general, as the degree of microsegregation increases, the realization of this goal becomes more difficult because of the increased sensitivity of the solidification process.

Bibliography

Cieslak M, Knorovsky G, Headly T, Romig A 1989 The solidification metallurgy of alloy 718 and other Nb-containing superalloys. In: Loria H (ed.) *Superalloys 718, 625, 706, and Various Derivatives*. TMS, Warrendale, PA, pp. 59–68

Flemings M C, Mehrabian R, Nereo G E 1968 Macrosegregation, Part II. *Trans. Metall. Soc. AIME* **242**, 41–9

Helms A, Adasczik C, Jackman L 1996 Extending the size limits of cast/wrought superalloy ingots. In: Kissinger R, *et al.* (eds.) *Proc. Seven Springs Conf.* TMS, Warrendale, PA, pp. 427–33

G. Maurer

Spray Casting: Applications

An efficient means to improve material performance is to introduce reinforcements and/or grain refining phases into metal or intermetallic matrices. In order for this technique to be viable, the *dispersoids* (e.g., oxides, carbides, and nitrides) must be stable thermodynamically, have a relatively small size compared to the grains, and be uniformly distributed. These constraints require the ability to control the dispersoid-producing chemical reactions continuously, as the material is being manufactured. Reactive spray atomization and deposition (RSAD, Fig. 1) is a spray-based processing technique which addresses this challenge by integrating atomization, chemical reaction, and consolidation (Ranganathan *et al.* 1993, Unigame 1989, Unigame *et al.* 1993, Zeng *et al.* 1992, Lawley and Apelian 1994, Dai *et al.* 1998, Delplanque *et al.* 2000). A critical step in the implementation of this technique is the choice of thermodynamically adequate combinations of alloying additives and reactive gas which will produce the target dispersoids. The variety of matrix materials and reinforcing phases that can be processed using RSAD is limited by the kinetics of the chemical reactions involved.

RSAD processes are typically grouped in three categories according to the physical state of the reacting phases: gas–liquid reactions (Ranganathan *et al.* 1993, Unigame 1989, Unigame *et al.* 1993, Zeng *et al.* 1992, Dai *et al.* 1998, Delplanque *et al.* 2000, Lavernia and Wu 1996, Mathur *et al.* 1990, Liu *et al.* 1984); liquid–liquid reactions (Lawley and Apelian 1994, Leatham *et al.* 1993, Majagi *et al.* 1992) (see Fig. 2); and liquid–solid reactions (Lawley and Apelian 1994, Lee *et al.* 1989) (see Fig. 3). Most of the R&D activities of RSAD processes have been

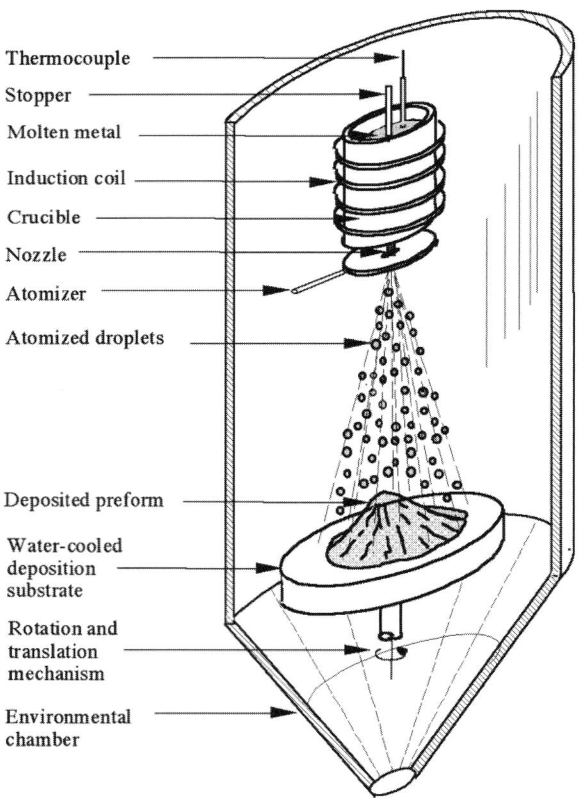

Figure 1
Schematic diagram of the process apparatus utilized for reactive spray deposition experiments (reproduced by permission of The Minerals, Metals and Materials Society from *Metall. Mater. Trans.*, 1994, **29A**, 2597–611).

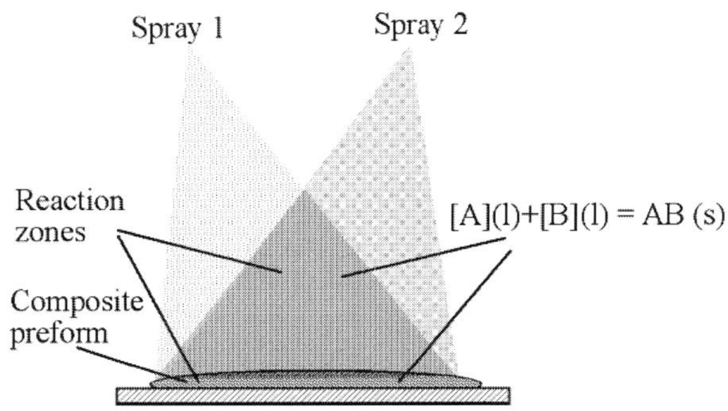

Figure 2
Liquid–liquid RSAD (reproduced by permission of John Wiley and Sons from 'Spray Atomization and Deposition,' 1996).

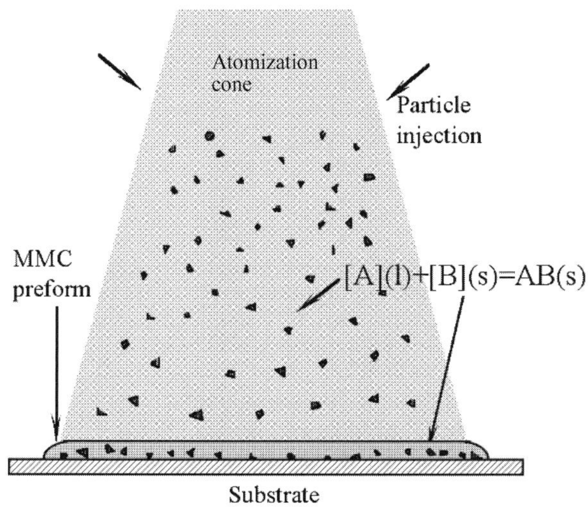

Figure 3
Liquid–solid RSAD (reproduced by permission of John Wiley and Sons from 'Spray Atomization and Deposition,' 1996).

focused on gas–liquid reactions. Consequently, the following overview is limited to that category. Further information regarding RSAD processes using liquid–liquid and liquid–solid reactions can be found in Lavernia and Wu (1996).

1. Gas–Liquid Reactions

This type of RSAD process uses a chemical reaction between the droplet and the atomizing gas to produce the target dispersoids (Ranganathan *et al.* 1993, Unigame 1989, Unigame *et al.* 1993, Zeng *et al.* 1992, Dai *et al.* 1998, Delplanque *et al.* 2000, Lavernia and Wu 1996, Mathur *et al.* 1990, Liu *et al.* 1994). Examples of target reactions include:
For RSAD-processed 5083 Al (Dai *et al.* 1998):

$$2[Al](l) + \tfrac{3}{2}O_2(g) \rightarrow Al_2O_3(s) \tag{1}$$

$$2[Mg](l) + O_2(g) \rightarrow 2MgO(s) \tag{2}$$

$$[Mg](l) + 2[Al](l) + 2O_2(g) \rightarrow MgAl_2O_4(s) \tag{3}$$

For RSAD-processed Cu–1 wt.%Al (Perez and Morris 1994):

$$Cu[Al](l) + O_2(g) \rightarrow Cu + Al_2O_3(s) \tag{4}$$

For RSAD-processed steel alloys (Unigame 1989, Unigame *et al.* 1993):

$$Fe[Al](l) + O_2(g) \rightarrow Fe(l) + Al_2O_3(s) \tag{5}$$

$$Fe[Al](l) + Fe(l) + O_2(g) \rightarrow Fe(l) + FeAl_2O_4(s) \tag{6}$$

For RSAD processed NiAl intermetallics (Zeng *et al.* 1992)

$$NiAl[Al](l) + O_2(g) \rightarrow NiAl(l) + Al_2O_3(s) \tag{7}$$

The feasibility of forming TiC in a steel matrix using CH_4 (methane) as the reactive atomization gas has also been investigated (Mathur *et al.* 1990).

Finally, additives (such as zirconium or yttrium) are sometimes used because of their affinity for oxygen:

$$[Zr](l) + O_2(g) \rightarrow ZrO_2(s) \tag{8}$$

$$NiAl[Y](l) + O_2(g) \rightarrow NiAl(l) + Y_2O_3 \tag{9}$$

2. Process Description

A schematic diagram of a typical reactive spray deposition apparatus is presented in Fig. 1. In order to control the ambient gas thermally and chemically, experiments are performed in an environmental chamber. This chamber is first evacuated down to a vacuum (e.g., 0.1 torr) and then back-filled with an inert gas (e.g., Ar, N_2) to the selected operating pressure (about 1 atm). The alloy (e.g., aluminum alloy, copper–aluminum, steel, or Ni_3Al) is melted and superheated in a crucible and then delivered to a gas-assisted atomizer where the liquid alloy jet is atomized as a result of interactions with high-velocity reactive gas jets (e.g., CH_4, N_2–O_2). The droplets thus produced interact dynamically, thermally, and chemically with the surrounding gas and are carried by the background flow towards a cooled substrate where they impinge, spread, and solidify to form the reactive spray deposited material. As in the case of inert spray deposition, the substrate position is determined so that an "optimum" distribution of liquid, partially solidified, and solid droplets impacts the substrate.

3. Microstructural Characteristics of RSAD Materials

Materials processed using RSAD tend to exhibit an equiaxed grain morphology and a smaller grain size than is typically observed with more conventional processing methods. For instance, as seen in Fig. 4,

(a)

(b)

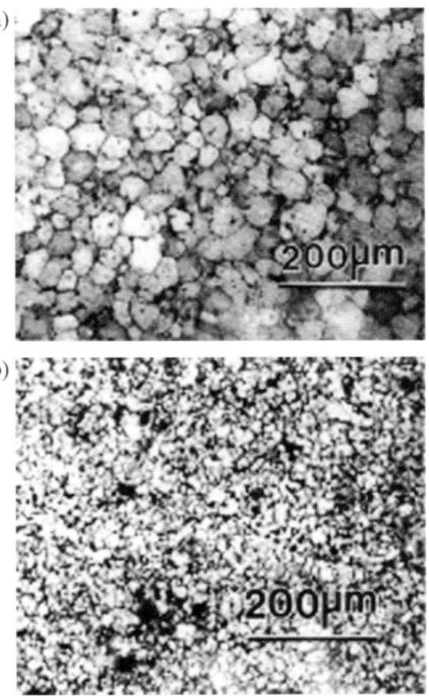

Figure 4
Optical micrographs of as-spray deposited Al alloy (a) 5083 Al alloy processed using N_2 (b) 5083 Al alloy processed using N_2–10% O_2 (reproduced by permission of The Minerals, Metals and Materials Society from *Metall. Mater. Trans.*, 1994, **29A**, 2597–611).

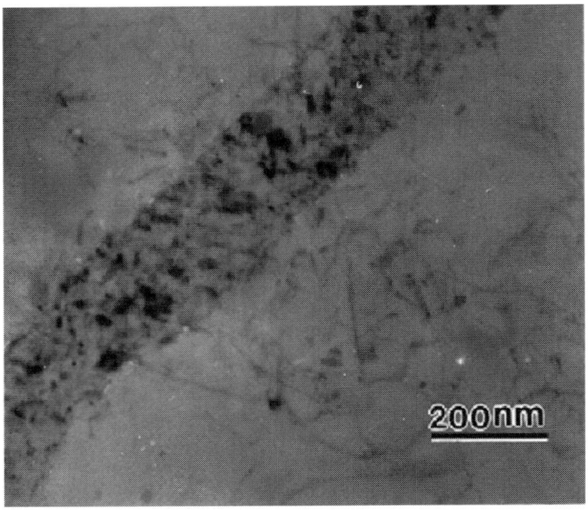

Figure 5
TEM image showing a relatively wide stringer of oxide particles in as-reactive spray deposited 5083 Al alloy processed using N_2–10%O_2 (reproduced by permission of The Minerals, Metals and Materials Society from *Metall. Mater. Trans.*, 1994, **29A**, 2597–611).

adding 10 vol.% O_2 to N_2 atomization gas was found to cause the grain size of 5083 Al to decrease from 42 μm to 18 μm (Dai *et al.* 1998). The grain growth in this alloy was observed to be retarded by stringers (Fig. 5) of oxide (γ-Al_2O_3) particles (0.14 vol.%) distributed along grain boundaries as these pinned the migration of grain boundaries. However, oxide particles were observed only for atomization gas oxygen content of >10%. More generally, the variation of oxide content with oxygen gaseous mixture fraction in RSAD-processed alloys is highly nonlinear. The morphology, size, and nature of the dispersoids are strongly dependent on the system considered and on the processing conditions (Ranganathan *et al.* 1993, Unigame 1989, Unigame *et al.* 1993, Zeng *et al.* 1992, Dai *et al.* 1998, Mathur *et al.* 1990, Perez and Morris 1994).

4. Controlling Mechanisms

Five candidate-controlling mechanisms for dispersoid volume fraction can be identified *a priori* (Zeng *et al.* 1992) (listed here in the order in which they occur spatially from the gas surrounding the droplet inward, for a reactive gaseous component A): (i) transport of A in the boundary layer ("film") surrounding the droplet, (ii) adsorption of A atoms at the droplet surface, (iii) mass transfer of A through the reaction product film, (iv) chemical reaction between A and reactive elements, and (v) mass transport of A in the droplet core. The analysis of Zeng *et al.* (1992) for RSAD of Ni_3Al using N_2–O_2 showed that, for that system, the characteristic time for mass transfer through the boundary layer is shorter than the characteristic times for diffusion within the droplet or through the oxide film. Furthermore, since the diffusivity of oxygen through Al_2O_3 is several orders of magnitude smaller than that through Ni_3Al, mass transfer through the oxide film is, in this case, the limiting step. Zeng *et al.* (1992) also noted that this analysis hinges on the continuity of the oxide film, which is questionable during the initial stages of atomization. If the oxide film is discontinuous, the chemical reaction, rather than diffusion, would be the rate-limiting step. Furthermore, mass transfer of oxygen through the oxide film does not necessarily occur exclusively by diffusion, and this must be taken into account in the evaluation of the characteristic times.

More precise information regarding the controlling mechanisms can be obtained using detailed modeling. Such a model (e.g., see Delplanque *et al.* 2000) must allow the prediction of droplet in-flight thermal and chemical reaction behavior as well as its state at impact as a function of processing parameters such as melt superheat. Droplet position and velocity histories can be obtained from the numerical solution of the one-dimensional equation of motion:

$$\frac{\mathrm{d}V_\mathrm{d}}{\mathrm{d}t} = g\left(1 - \frac{\rho_\mathrm{g}}{\rho_\mathrm{d}}\right) + \frac{3}{4D}\frac{\rho_\mathrm{g}}{\rho_\mathrm{d}}|V_\mathrm{d} - V_\mathrm{g}| \times (V_\mathrm{d} - V_\mathrm{g})C_\mathrm{d}(\mathrm{Re}) \qquad (10)$$

where subscripts d and g refer to the droplet and the ambient gas, D is the droplet diameter and C_d the droplet drag coefficient, which depends on its Reynolds number, Re. In order to predict the spatially resolved temperature field, the energy equation inside the droplet:

$$\frac{\partial T}{\partial t} = \alpha\frac{1}{r^2}\frac{\partial}{\partial r}\left(r^2\frac{\partial T}{\partial r}\right) \qquad (11)$$

can be solved numerically. The nonequilibrium solidification behavior must be considered, including the computation of heterogeneous undercooling and the solid/liquid interface progression rate can be estimated using a power law.

The critical issue in this type of model is the reaction model. In the case of aluminum oxidation, diffusion-based models fail to predict the highly nonlinear dependence of dispersoid volume fraction on oxygen gaseous mass fraction in the atomization gas (Delplanque *et al.* 2000). Recently, a model for oxidation kinetics during reactive spray deposition based

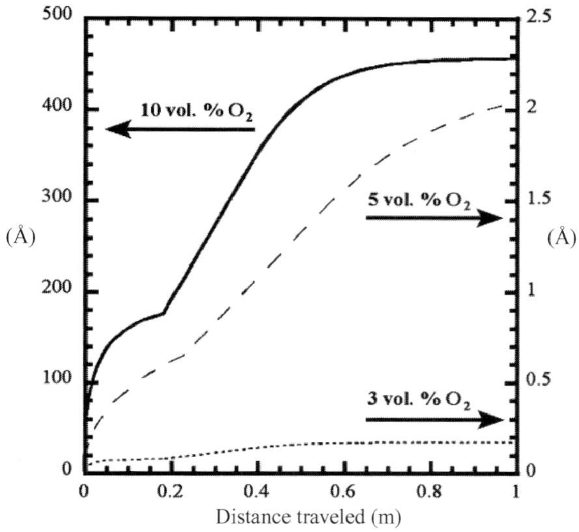

Figure 6
Predicted oxide film thickness for a 50 μm aluminum droplet (140 K superheat) at selected oxygen volume fractions (reproduced by permission of ASME International from *J. Heat Transfer*, 2000).

on the Mott–Cabrera theory of oxidation has been proposed (Delplanque *et al.* 2000). This model assumes that oxidation is limited by ion transport through the oxide layer and yields a growth rate (O denotes the position of the advancing oxide/solid interface):

$$\frac{\mathrm{d}O}{\mathrm{d}t} = -2A_0 \exp\left(-\frac{Q}{\kappa T}\right)\exp\left(K_0\frac{P_{O_2}^{1/2}}{\kappa T}\right) \qquad (12)$$

that depends exponentially on oxygen pressure (see Fig. 6) and, therefore, corroborates the experimentally observed nonlinearity.

Finally, chemical reactions between droplets and atomization gas do not occur only during the droplet flight phase. Indeed, it has been proposed that since the characteristic time for droplet transport (from atomization to impact) is significantly smaller than the characteristic time for deposition (impact, spreading, and solidification), the oxide volume fraction could be more significantly influenced by the deposition phase (Zeng *et al.* 1992). This hypothesis has had some experimental support from Ranganathan *et al.* (1993) who found that the oversprayed Fe–Al powders produced during reactive spray deposition using N_2–O_2 mixtures did not contain oxides. Nevertheless, despite numerous modeling and numerical simulation studies of droplet impact and solidification, no adequately detailed model for oxidation during the deposition phase has been developed yet.

5. Concluding Remarks

The manufacturing flexibility that reactive spray atomization and deposition processes provide is undeniably their main asset. However, despite the significant advances made recently in elucidating some of the mechanisms involved in RSAD, such as grain growth control (as outlined above), the current understanding of these processes is far from that needed to realize their full potential.

One of the most promising fields of application for this technique is probably the manufacturing of engineered/microarchitectured materials. Continuous, in-line actuation of the basic process parameters (e.g., atomization pressure, melt temperature, atomization gas composition, and deposition distance) would allow the production of materials with functionally graded microstructures, composition, and dispersoid distribution and, therefore, exhibiting spatially varying mechanical, thermal, and chemical behavior.

For this type of process to be successful and economically viable, effective process control and optimization strategies must be developed, which, in turn, requires fundamental understanding at the microscale (droplet), mesoscale (spray), and macroscale (deposit) using combined experimental and modeling approaches. Finally, scale-up issues will have to be addressed.

Acknowledgments

The authors would like to thank the US Army Research Office (ARO Grant # DAAG 55-98-1-0088), and ARO AASERT (Grant # DAAG 55-98-1-0141) for financial support on the relevant research conducted at the University of California, Irvine.

Bibliography

Dai S L, Delplanque J P, Lavernia E J 1998 Microstructural characteristics of 5083 Al alloys processed by reactive spray deposition for net-shape manufacturing. *Metall. Mater. Trans.* **29A**, 2597–611

Delplanque J P, Lavernia E J, Rangel R H 2000 Analysis of in-flight oxidation during reactive spray atomization and deposition processing of aluminum. *J. Heat Transfer* **122**, pp. 126–33

Lavernia E J, Wu Y 1996 *Spray Atomization and Deposition.* Wiley, New York

Lawley A, Apelian D 1994 Spray forming of metal matrix composites. *Powd. Metall. Int.* **37**, 123–8

Leatham A G, Ogilvy A, Elias L 1993 The Osprey process—current status and future possibilities. In: Froes F H (ed.) *P/M in Aerospace, Defense and Demanding Applications—1993.* MPIF, Princeton, NJ, pp. 165–75

Lee A K, Sanchez-Caldera L E, Chun J H, Suh N P 1989 Material with novel compositions and fine microstructures produced via the mixalloy process. In: McCandlish L E, Polk D E, Siegel R W, Kear B H (eds.) *Multicomponent Ultrafine*

Microstructures. Materials Research Society, Pittsburgh, PA, pp. 87–92

Liu H, Rangel R H, Lavernia E J 1994 Modeling of reactive atomization and deposition processing of Ni_3Al. *Acta Metall. Mater.* **42**, 3277–89

Majagi S, Ranganathan K, Lawley A, Apelian D 1992 Spray forming of metal matrix composites. In: Lavernia E J, Gungor M N (eds.) *Proc. Symp. Microstructural Design by Solidification Processing*. The Minerals, Metals and Materials Society, Warrendale, PA, pp. 139–49

Mathur P, Kim M H, Lawley A, Apelian D 1990 Spray forming of metal matrix composites In: *Powder Metallurgy: Key to Advanced Materials Technology*. ASM International, Materials Park, OH, p. 55

Perez J F, Morris D G 1994 Copper–Al_2O_3 composites prepared by reactive spray deposition. *Scr. Metall. Mater.* **31**, 231–5

Ranganathan K, Lawley A, Apelian D 1993 *In-situ* spray casting of dispersion strengthened alloys-II: experimental studies. In: Bando Y, Kosuge K (eds.) *Proc. Powder Metallurgy World Congr.* Japan Society of Powder and Powder Metallurgy, Kyoto, Japan, 14E-T6-6

Unigame Y 1989 An analysis of oxide dispersion strengthening of ferrous alloys during spray casting. M.S. thesis, Drexel University, Philadelphia

Unigame Y, Lawley A, Apelian D 1993 *In-situ* spray casting of dispersion strengthened alloys—I: thermodynamics and reaction kinetics. In: Bando Y, Kosuge K (eds.) *Proc. Powder Metallurgy World Congress*. Japan Society of Powder and Powder Metallurgy, Kyoto, Japan, 14E-T6-6

Zeng X, Liu H, Chu M, Lavernia E 1992 An experimental investigation of reactive atomization and deposition processing of Ni_3Al/Y_2O_3 using N_2–O_2 atomization. *Metall. Trans. B Process Metall.* **23**, 3394–9

E. J. Lavernia and J.-P. Delplanque

Spray Casting: Fundamentals

Spray forming is a near net shape casting process carried out by the impact of gas-atomized metal droplets onto a substrate. The metal flow rates are high, of the order of 0.2 to 2 kgs^{-1}, and this is achieved by starting from fully melted liquid as in conventional gas atomization for powder production. Spray forming can be distinguished from other "thermal spray" processes in which solid powder is melted by a source of heat such as an electric arc or a combustion flame. As a result, lower metal flow rates are achieved and the aim of thermal spraying is in general to make coatings on preformed shapes. It has been established, by theoretical models and experimental investigations as described below, that for successful operation the metal droplets must be partially solid at impact on the substrate. If the operating conditions are correctly set, then the spray-formed casting has a predictable shape, it is produced with high sticking efficiencies of over 90% and has a fine grain sized microstructure with only a small fraction (1–2%) of fine gas pores, typically 1–2 μm in diameter. The spray-formed casting has greatly reduced micro and macro-segregation compared to similar products conventionally cast using input material that is 100% liquid and, as a result, spray-formed material has mechanical properties that are closer to wrought alloy properties than to cast. However, only a limited number of shapes have so far been successfully produced. These shapes include cylindrical billets, flat strip or plate, and perhaps most importantly, hollow tubes with a wide range of internal diameters. Other important products include thickly coated bilayers, for example, of stainless steel on carbon steel, and in an important early application, high-carbon chromium steel roll surfaces sprayed onto low carbon substrates to form near net shape metal working rolls. Metal matrix composites have also been produced, usually by injection of the strengthening phase into the spray so that the metal and the reinforcement impact together onto the substrate. Detailed descriptions of the process can be found in various reviews, for example, by Grant (1995), the conference volumes edited by Wood (1990, 1993, 1996, and 1999) and the book by Lavernia and Wu (1996) on the two major steps in spray forming: spray atomization and deposition.

1. Process Description

In the Osprey method of spray forming, which is the dominant method used in current technology (see Wood 1990), a stream of molten metal is "atomized" or broken into a high density spray of molten droplets by impact with a high-velocity inert gas stream. The gas used is typically nitrogen, although for applications where nitrogen contamination in the product is unacceptable, argon gas is used. Large amounts of gas are used, with gas /metal weight ratios of the order of 0.5 to 1. Therefore, for very large-scale production, gas recycling is an important part of the process economics. Unlike modern powder production, it is not usually essential to achieve very fine powder sizes. In spray-forming the powder sizes have a maximum weight fraction at diameters of about 50–100 μm but with a large spread in sizes, typically with a log-normal distribution. The spray forms a cone with a radial variation in mass flux and at present it is necessary to measure the particle size distribution (PSD) and the radial mass flux (RMF) experimentally as a function of the main process variables: gas pressure, metal flow rate, alloy type, super heat, etc. The spray is directed to impact a suitably designed and moving substrate and on this substrate the spray-formed casting builds up. The substrate motion is designed to ensure that, after an initial transient, the point of impact of the spray on the growing casting occurs at a fixed flight distance, typically about 0.2 m, from the point of atomization where the converging gas streams hit the stream of liquid metal. The aim is to have the thermal conditions at the surface of the growing casting, on which the spray is arriving, nearly constant. For billets and strip, the casting remains essentially under the spray at all times. However, for tubes or rolls the rotation of the substrate causes the surface to move under the spray and out again, once on each rotation, and multiple passes under the spray then take place to build up the required casting thickness. In many cases, spray oscillation is used in combination with substrate motion to achieve the required shape. This is clear for strip casting, where with a nonuniform RMF only by oscillating the center of the spray across the width of the strip can an approximately uniform thickness be achieved. Alternative methods for producing strip or plate are to have multiple sprays or to replace the cylindrical initial liquid stream with a slit source of liquid metal. Even for billets and discs it is usual to have the impact point off-center to achieve a controlled nearly cylindrical shape. To achieve the desired near-net shape with a minimum of "over-spray powder" material that was sprayed but did not stick to the substrate, the operating conditions must be controlled to ensure that metal arrives at the required place and that as little as possible of the flux of metal misses the deposit.

2. Thermal Conditions

In any thermal process it is essential to know the temperatures of the material as it passes through the various stages in the process. This is needed both to control the individual steps of the process and also to predict the final microstructure of the product. Atomization takes place with the metal stream at a temperature close to the imposed superheat (temperature above the liquidus). The superheat is set by the conditions in the melt and is typically about 100°C to avoid risk of the metal stream freezing off in the atomizer. After atomization the metal passes through three different stages: (i) the spray, (ii) the deposit surface, and finally (iii) within-sprayed billet strip or tube. Each of these stages needs to be understood. Ideally we would like to be able to measure the conditions in each of these situations but this is difficult and major progress has been made by numerical modeling, for example by Mathur *et al.* (1989), supplemented by experiments to calibrate and validate the models.

2.1 Thermal Conditions and Droplet Microstructure in the Spray

In order to model the thermal conditions in the two-phase, gas–metal, spray, two empirically measured variables must first be known. These variables are the PSD and the gas velocity field. Although the particle velocities and particles sizes can be measured in flight in real time by various diagnostic techniques, such as phase doppler anemometry, it is more common to measure these in separate experiments. The gas flow field can be measured in the absence of metal by standard Pitot tube methods and the PSD and the RMF after collection of powders in cylindrical tube collectors. The collection must be done in the absence of a substrate and with the collectors at a sufficient flight distance to ensure full solidification. It is not correct to sample the over-spray powders for their PSD in a normal spray forming run, since the powders that do not stick to the substrate have been shown not to be representative of the PSD before impact. High-velocity particle collisions with the deposit give a much higher fraction of small particles than was present before impact. That is, droplet fragmentation seems to occur on impact. In addition, the smallest, and thus lightest, droplets are carried away from the casting by the defected gas stream without coming into contact with the deposit.

Current thermal models consider the behavior of individual droplets, of different sizes. Each droplet, after atomization, is accelerated by the gas flow field until it matches the velocity of the gas stream. The gas velocity, however, falls with distance from the atomization point, so the droplet will subsequently be decelerated by the gas. Heat loss from the droplets depends on the local heat transfer condition. The heat transfer is a strong function of the relative velocities of the gas and the droplet. Using these assumptions, the loss of heat from an individual droplet at different flight distances is modeled. This calculation is then repeated for particles of all sizes to predict the average heat content of the spray. Unknowns in this calculation include the under-cooling at which a droplet of a given size is expected to nucleate solid and reheat back towards the liquidus temperature. Solidification growth in under-cooled metallic alloys is very rapid. The under-cooling required for nucleation of small droplets is known to be large. This arises since suitable catalysts for solidification are unlikely to be present in most small droplets. Indeed observation of large under-cooling in small droplets was a classic early result in physical metallurgy that established the validity of the homogeneous nucleation models in structural metallurgy over 50 years ago. Early experiments by Mathur were able to confirm this by intercepting droplets of different sizes at different flight distances onto glass slides passed quickly through the spray. It was shown that there were large under-coolings. However, the average under-cooling fell as the droplet size increased. This data was subsequently used for a validation test for a spray nucleation model as described by Cai *et al.* (1993). In their model it was assumed that the smallest droplets, typically 20 μm or less in diameter, cool very quickly and are unlikely to have any nucleation catalysts. Therefore, the smallest droplets nucleate at the large undercoolings (200–300°C) predicted for homogeneous nucleation. This assumption is confirmed by the previously measured under-coolings in the smallest droplets. The critical new assumption was that, given the large differential velocities of droplets of different sizes, then larger droplets will, once they cooled below the liquidus temperature, be nucleated by impact of smaller, solidified, droplets. Microstructural evidence supporting this idea was clearly confirmed by metallographic studies of over-spray droplets. Using this assumption, Cai *et al.* were able to match successfully the previously measured results for the decrease of under-cooling with increasing droplet size. This match confirms that in the condition of high droplet density of gas atomization, characteristic of spray forming and also of metal powder production, the dominant nucleation process is the impact of small solidified particles onto larger, slower-moving and under-cooled liquid droplets.

These models then predict that just before impact on the substrate, the smallest droplets are fully solidified, the largest ones fully liquid, while the bulk of the droplets are only partially solidified. Since the cooling rates are high and the solidification times short, typically a few milliseconds, the dendrite arm spacing in the droplets is very fine—typically 1 μm or less. This is confirmed by observations of the fine dendritic structure in the over-spray powders and in conventional powder metallurgy.

2.2 Thermal Conditions and Microstructural Development on Impact with the Deposit

The droplets travel at velocities of the order of the gas velocity, 10 to 50 $\mathrm{m\,s^{-1}}$, and so impact onto the deposit at these speeds causes extensive deformation of the partially molten particles. These conditions appear to produce the dendrite fragmentation typically seen during stir casting. Stir casting usually yields grains fragmented to the size of a few dendrite arm lengths. For spray forming, these initial grains, seen in regions of the casting that is rapidly chilled, appear to be only 5–10 μm in size. The details of how fragmentation occurs is not fully agreed. However, the mechanism proposed for stir casting (Doherty *et al.* 1984) seems a reasonable possibility. This involves the following steps: plastic, dendrite arm bending, recrystallization to produce high-angle grain boundaries which are wetted by the liquid metal, producing grain break-up.

These ideas have been experimentally supported for stir casting. A high-speed video of the impact at the surface does indicate considerable liquid movement at the substrate surface. This movement seems typical of liquid on liquid impact, indicating that the solid–liquid mixtures do seem to have the liquid-like viscosities found during high shear rates in stir casting. The surface movement is not the solid-like behavior of solid–liquid mixtures that can readily sinter together at solid fractions greater than about 0.3. Immediately below the surface, when impact-induced movement has died out, there should be solid-like behavior. This must, in fact, occur or the partially solid deposit would break up during the high rotation rates usually imposed in most billet and tube spray-forming processes.

The thermal models of the material under the spray indicate that the material should achieve, in a few milliseconds, a locally uniform temperature determined by the enthalpy content calculated for the spray. That is, the colder small solid particles, the hot fully liquid large droplets and the intermediate partially solid droplets quickly equilibrate in temperature. This is expected by virtue of the high thermal conductivity of metallic alloys and the small heat diffusion distances in the droplets that are flattened by high velocity impact. For a deposit surface under a *stationary* spray, the models indicate that the temperature should, after a short period of metal build-up on the substrate, be that determined by the enthalpy of the spray. For sprays that oscillate and for rotating castings such as tubes that move in and out of the spray, there is additional cooling by the gas.

This additional cooling will reduce the fraction liquid of the deposit significantly below that calculated from the enthalpy content of the metal in the spray just before impact. Reliable thermal models of the complex geometry in, for example, rotating tubes have not yet been fully developed to quantify this effect. Experiments by Doherty *et al.* (1997) indicate that the temperature on the deposit surface and thus the fraction liquid on the deposit are vital factors in successful spray-forming. These experiments also indicate that the deposit surfaces of spray-formed tubes are indeed colder than billets formed under the equivalent sprays. Therefore, reliable thermal models of complex spray-formed billets are clearly needed.

2.3 Thermal Conditions and Microstructural Development within the Deposit

Thermal models for simple shapes such as spray-formed billets are available and these have been validated by temperature measurements. For billets formed continuously under the spray the surface temperatures match the predicted liquid fractions, f_l, predicted for the spray, typically $f_l \sim 0.30$. Under such situations the casting is successful in producing material with a high yield and a low amount of fine gas porosity (Doherty *et al.* 1997). Grain sizes of 20–50 μm are produced with very limited microsegregation with a length scale of the grain size. There is no sign of the usual dendritic structure seen when the segregation is on a scale finer than the grain size. The grain size scales directly with the solidification time, indicating that this is controlled by grain coarsening (Annavarapu and Doherty 1995). A critical result for the development of the fine spherical grain structure appears to be particle or gas pore inhibited grain coarsening. Many results and theoretical models find accelerated coarsening of solid grains in solid–liquid mixtures as the fraction liquid, f_l, decreases towards zero. Coarsening in solidifying spray-formed casting should occur rapidly as f_l falls from 0.3 to zero in times of the order of 10–100 s. However, the grain sizes actually seen after spray forming are much smaller than those predicted from standard coarsening theory. The solution to this important discrepancy between theory and experiment appears to come from grain-boundary pinning by fine particles seen on the boundaries of all fine-grained spray-formed alloys. These particles are sometimes transition metal nitrides in high temperature alloys based on copper, iron or nickel or they are iron impurity particles in aluminum alloys. The role of these boundary-pinning particles has not yet been fully analyzed but it is clear than the fine grain size characteristic of successful spray forming arises from their influence on inhibiting grain coarsening.

3. *Other Aspects of the Structure of Spray-formed Castings*

The limited microsegregation in spray-formed materials directly arises from the unusual solidification condition in spray-forming. The first 50–70% of solidification occurs in a few ms in flight, giving dendrite cells of the size of about 1 μm. These dendrites break up to fine grains on impact and the remaining solidification occurs at high fraction solid, and much more slowly. Thermal equalization for a few seconds near the surface of the deposit allows homogenization of the solid particles giving a solid composition corresponding to the low solidus temperature at a solid fraction, f_s, of ~ 0.7. This gives 70% solid at a composition close to that of the alloy. During the coarsening, as the grains grow in size by five- to 10-fold before solidification is complete, the solid is being continuously remelted and refrozen at temperatures steadily approaching the solidus, resulting in a solid even closer to the alloy composition. This yields the minimal microsegregation characteristic of the process. In addition, with only limited solidification in the deposit (a change of f_s of only 0.3) there is much reduced solidification contraction. This combined with inhibition of liquid flow, owing to the fine liquid channels, greatly reduces the amount of contraction-induced fluid flow. Such liquid flow is responsible for macrosegregation, the chemical inhomogeneity on the length scale of the casting, so this is also greatly reduced in shaped castings made by spray forming.

The fine-scale porosity apparently resulting from trapped atomizing gas is almost always present. The only exceptions are for nitrogen atomization in high-temperature alloys containing nitride-forming elements such as chromium, aluminum, or titanium. The nitride particles can completely remove trapped nitrogen. Atomizing under an inert gas such as argon, however, always leads to trapped porosity, as does aluminum alloy spray forming at low temperatures, where nitrogen reaction appears impossible. Fine gas porosity from insoluble atomizing gases can be reduced by plastic working. This elongates the pores to long tubes, which will break up by the Rayleigh instability (see, for example, Martin *et al.* (1997)) to small spheres. The increased capillarity pressure, ΔP, owing to the solid surface energy, γ, of a pore of radius r, for the smaller spheres, will reduce the volume of the gas pores. For γ of 1 J m^{-2}, and r of 0.1 μm, this gives a pressure of about 100 atmospheres.

$$\Delta P = 2\gamma/r \qquad (1)$$

Under these circumstances porosity can appear to have vanished but it should be remembered that the gas is still present and if the pores coarsen, for example, by pore migration and coalescence during high temperature processing, then the porosity will return. This possible problem in the use of spray-formed components has so far been little studied but its presence should be kept in mind and for critical components be subject to experimental testing.

Bibliography

Annavarapu S, Doherty R D 1995 Inhibited coarsening of solid–liquid microstructures in spray casting at high volume fractions of solid. *Acta Metall. Mater.* **43**, 3207

Cai C, Annavarapu S, Doherty 1993 Modelling based microstructural control in spray casting. In: Wood J V (ed.) *2nd Int. Conf. Spray Forming*. Woodhead, Cambridge, pp. 67–83

Doherty R D, Cai C, Warner-Kohler L K 1997 Modeling and microstructural development in spray forming. *Int. J. Powder Metall.* **33**, 50

Doherty R D, Feest E A, Lee H-I 1984 Microstructure of stir-cast alloys. *Mater. Sci. Eng.* **65**, 181–91

Grant P S 1995 Spray forming. *Prog. Mater. Sci.* **39**, 497

Lavernia E J, Wu Y 1996 *Spray Atomization and Deposition*. Wiley, New York

Mathur P C, Apelian D, Lawley A 1989 Analysis of spray deposition. *Acta Metall.* **37**, 429

Martin J W, Doherty R D, Cantor B 1997 *The Stability of Microstructure in Metallic Systems*. Cambridge University Press, Cambridge, p. 299

Wood J V (ed.) 1990, 1993, 1996, 1999 *Proc. Int. Conf. Spray Forming*. Osprey Metals, Neath, UK

R. D. Doherty

Strip Casting of Steels: Current Developments

Twin roll strip casting of steel, the production of sheets of steel from 1 mm to 4 mm thick directly from liquid steel, has finally transitioned from hot trials to commercial application with three different groups running production scale casting machines (Yamada *et al.* 2002, Hohenbicher *et al.* 2002, Campbell *et al.* 2002). These three groups are: (1) Nippon Steel/Mitsubishi Heavy Industries (MHI) at the Hikari plant of Nippon Steel in Japan, (2) EUROSTRIP®, a consortium of ThyssenKrupp Steel (TKS), Arcelor, and Voest Alpine Industries, at the TKS's Krefeld plant in Germany, and (3) CASTRIP®, a consortium of Nucor Steel Co., Broken Hill Propriety Ltd (BHP) and Ishikawajima-Harima Heavy Industries (IHI). All three plants are capable of producing production scale tonnages and are currently testing the process for commercial applicability.

Twin roll strip casting is the continued development of steel casting processes towards near net shape casting. Traditional casting technologies such as slab casting and thin slab casting produce a slab that is between 50 mm and 250 mm thick. This technology requires that a hot rolling machine must be connected to the continuous caster to convert slabs to a steel sheet that is between 1 mm and 5 mm thick. Thus, one of the goals of twin roll strip casting is to produce directly from the liquid steel, a steel sheet that is between 1 mm and 5 mm thick in order to eliminate the need for a large hot strip mill. The development of strip casting has the potential to simplify the steel production flow sheet by minimizing the physical size of the casting machine and simplifying requirements for hot reduction of the steel strip. Although there was a great interest in the casting of a strip which could be subsequently either used as a product or cold rolled to finished gauge; all current facilities have the ability to complete at least a single pass of hot reduction in order to improve product properties and flatness.

A schematic of the CASTRIP® process is given in Fig. 1. In the figure it can be seen that key technologies include the rotating twin rolls, atmosphere control to avoid scaling, hot reduction and thermal control of the strip after rolling, and a coiler. Other schematics for stainless steel production indicate the presence of strip heating technologies to allow for complete thermal control of the strip after casting (Hohenbicher *et al.* 2002).

The first commercial twin roll production machine was the Hikari machine of Nippon Steel. This machine, which can be viewed as the culmination of the technology first envisioned in 1856 by Sir Henry Bessemer (see Bessemer 1965), was commissioned in October 1997. Of course, no technological breakthrough is without history and this development can be traced back to its initial beginnings in the 1980s when in 1985 Nippon Steel joined with MHI to run an 800 mm wide test caster at the No. 3 R&D Laboratory of Nippon Steel in Yawata.

EUROSTRIP® also has a long and interesting history. Strip casting in Europe can be traced back to the mid-1980s and all three partners had major pilot facilities. For example, TKS and Arcelor had a large project (the Myosotis project at the Ugine Isebergues plant in France), and VA, TKS, and Centro Sviluppo Materiali (CSM) had another project at the AST plant in Terni, Italy. These major forces in European Strip Casting joined together in 1999 (Walther *et al.* 2000) and currently operate two strip casting facilities. The Krefeld caster is focused mainly on stainless steel casting, and a smaller caster at AST in Italy is focused predominantly on carbon steel casting.

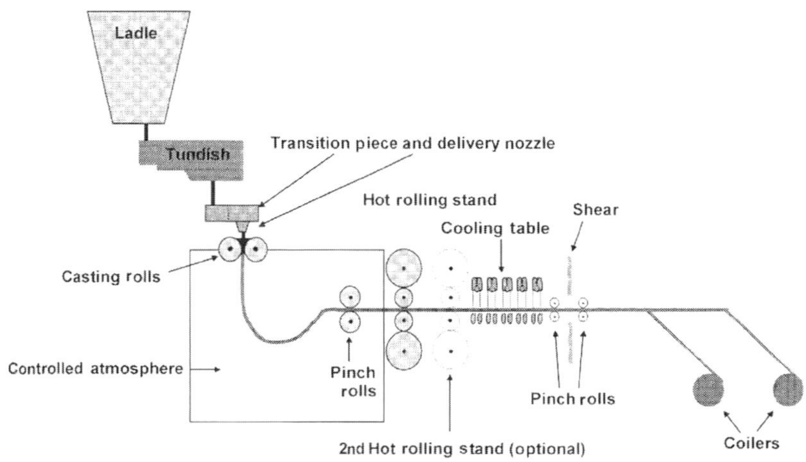

Figure 1
Schematic of the CASTRIP® process (Campbell *et al.* 2002).

Of the three competing twin roll strip casting groups, CASTRIP® was the most secret for many years. This project was initiated in the early 1990s as a joint project between BHP and IHI, under the code name Project "M," and it was not until the end of the 1990s that any significant information was disclosed on the process. Nucor joined this effort in 2000 and the commercial facility was hot-commissioned during May of 2002 in the United States at the Crawfordsville, Indiana plant of Nucor Steel.

The major similarities and differences between the three groups can be seen in Table 1.

The initial focus of both the Hikari and EUROSTRIP® projects was on stainless steel; whereas CASTRIP® was focused primarily on the casting of low carbon steels. This grade differentiation has had a significant impact upon the choice of casting technologies due to the perceived necessity on the part of both Hikari and EUROSTRIP® to produce a thicker cast strip that can then undergo heat treatment and significant deformation in order to develop conventional stainless steel strip properties. Thus, the solution of both Hikari and EUROSTRIP® is based upon a significantly larger roll diameter than CASTRIP®. All three technologies include significant in-line postcasting processing in order to develop conventional properties from the cast strips which tend to be stronger and less ductile in the as-cast condition than expected due to the high cooling rates experienced in the process.

The CASTRIP® process is based upon a small roll diameter philosophy where the strip thickness tends to be lower and the cast speeds higher than that of competing technologies. In addition, the CASTRIP® technology depends upon a high heat transfer roll technology which was developed in Australia (Stresov and Herbertson 1998, Mukuntan *et al.* 2000) by the BHP research team. This technological advantage has focused CASTRIP® towards a strategy of looking for market opportunities based upon ultra-thin hot band products (Mukuntan *et al.* 2000).

Currently, the three major projects have slightly different goals and objectives. The Nippon Steel approach at Hikari is certainly the furthest developed, and the most recent announcements (November 2002, Electric Furnace Conference, ISS) suggest that their goals of productivity (300 ton cast in a five heat sequence) and quality have been met. This is a significant announcement as the major issue of strip casting has been the necessity for economic success.

Strip casting is an interesting process as it is a combination casting/rolling process that must produce a product that is similar to a hot rolled sheet. Thus, the product of strip casting is in many ways defined by the needs of a cold rolling operation. For example, Nippon Steel reports their casting tolerances in a similar manner to a hot strip mill, in that they report parameters such as crown (70 μm), wedge (2 μm), surface roughness (0.6 μm) and undulation (<0.2 μm) for in-line reduced strip, while EUROSTRIP® reports that their material "fulfils DIN standards and the requirements of hot band consumers". CASTRIP® also reports crown (50 μm), surface roughness (0.5 μm), and centerline variation (34 μm) for their in-line reduced strip. All CASTRIP® values are reported to be within the normal tolerances for conventional hot rolled products.

Nippon Steel and EUROSTRIP® both report improved properties for strip cast 304 stainless steels after cold rolling of in-line reduced twin roll cast strip. Nippon Steel has reported workability, as measured by a cylindrical drawing test, plastic anisotropy, bright anneal finish quality, and pitting potential to be improved by strip casting and subsequent processing when compared to the conventional route. CASTRIP® has shown that, in the low carbon steels, a wider variety of properties can be attained with strip casting and in-line processing than is possible from conventional processing.

Given these positive results, the question remains—"What is the future of the strip casting process?" There has been significant interest in the benefits of strip casting. The following were given by EUROSTRIP® (Hohenbicher *et al.* 2002):

(i) reduction of up to 68% of the capital cost of a thin slab plant;

Table 1

Comparison of three strip cast producers (Yamada *et al.* 2002, Hohenbicher *et al.* 2002, Campbell *et al.* 2002).

	Hikari	EUROSTRIP®	CASTRIP®
Roll diameter	1200 mm	1500 mm	500 mm
Coil size	15–20 ton	20–35 ton	25 ton
Grades cast	304 stainless	Austinitic stainless	Low carbon 400 grade stainless
Width	778 mm and 1330 mm	1430 mm	2000 max
Thickness	2–5 mm	1.4–4.5 mm	0.7–2.0 mm
Heat size	60 ton	90 ton	110 ton
Casting speed	90 m min^{-1}max.	15–150 m min^{-1}	80–150 m min^{-1}
Plant output	420 000 ton yr^{-1}	500 000 ton yr^{-1}	500 000 ton yr^{-1}

(ii) space reduction to 40% of a thin slab plant;

(iii) lowered energy consumption, 50% of a thin slab plant; and

(iv) lower atmospheric emissions (70–90% of CO_2, SO_2, and NO_x) when compared to an integrated plant. All of these benefits have great potential. Herbertson (2002) has suggested other potential benefits for strip casting:

(i) economic production of flat products at smaller annual output levels (500000 ton per year);

(ii) streamlined production facilities with improved service of local markets;

(iii) compact plant configurations that will improve delivery performance;

(iv) favorable conditions for ultra-thin "hot band" and the development of a class of products that do not require cold rolling; and

(v) a small "ecological footprint" suitable for a positive contribution to sustainable processing.

However, before anticipating great success, one must also look at operating cost. Is this process economically viable? It is only recently that Nippon Steel has claimed economical viability for the strip casting of 304 stainless steels and this claim was made after five years of operation. This claim has yet to be proven by EUROSTRIP® or CASTRIP®. There are numerous issues in strip casting related to metal delivery, side containment, roll design, and scale reduction that, although technically solved, must also be solved in a manner that allows continuous production at a low operating cost in all grades.

Another issue in strip casting is productivity in the casting of low carbon steels. It is clear that increased productivity will lead to lowered operating cost. Herbertson (2002) has projected that the conversion cost from liquid steel to hot band via strip casting could reach $50 per ton at production levels of 500 ton per year which would make the strip caster competitive on an operating cost basis with a thin slab caster at 1.2 million ton per year. This analysis, however, indicates the major economic issue in strip casting: any technological solutions cannot cost more than $50 per ton or strip casting will be more expensive than current conventional technologies. Thus, all consumables—rolls, nozzles, edge dams, etc.—must be carefully chosen to allow an economical solution. Clearly this also means that long casting times will be necessary to reduce consumable costs.

In addition, to match the productivity of current scrap melting operations, a strip casting machine output level closer to 1 million ton per machine would be best for low carbon steels. Thus, it appears that a combination technology between the lower heat transfer rate, large roll solution of Nippon Steel and EUROSTRIP® and the higher heat transfer rate, and low roll diameter solution of CASTRIP® may be necessary before strip casting of carbon steels could be a stand alone, Greenfield solution for all low carbon steel sheet production. At the moment, the Nucor plant model of twin roll strip casting of carbon steels in a plant that already has a thin slab caster appears to allow the best of both worlds. The potential of strip casting can be developed in a production environment where there is an excess of liquid steel and the strip cast product costs and quality can be readily compared to thin slab cast product.

The future of twin roll strip casting for stainless steels appears to be bright—there is a solution in Japan that fits the economic and quality issues for 304 stainless produced at the Hikari plant of Nippon Steel. The existence of two other economic solutions will be determined by EUROSTRIP® and CASTRIP® by the end of 2004. The advancement of the technology beyond stainless steels will be determined by the detailed nature of the operating cost and quality results obtained by these two operations. In particular, successful strip casting at Nucor will, in a large part, determine the future of the technology for carbon steels. Thus, currently all strip casting operations appear technically successful but the rate of implementation of the technology beyond these three installations will be determined by the detailed nature of the operating costs and product quality.

Bibliography

Bessemer H 1965 *J. Metals* 1189, excerpts from the paper presented in 1891

Campbell P, Gillen G, Blejde W, Mahapatra R 2002 The CASTRIP® process for twin roll casting of steel—start-up experience at Nucor' Crawfordsville Plant. *Proc. 4th European Continuous Casting Conference, October 2002. Birmingham, England, IOM Communications*, Vol. 2. pp. 882–90

Herbertson J 2002 The emergence of strip casting-challenges and impacts of success. *Dr. Manfred Wolf Symposium*, May 10–11, pp. 60–9

Hohenbicher G, Tolve P, Capotosti R, Damasse J M, Albrecht-Fruh U 2002 EUROSTRIP® Direct Strip Casting of Carbon and Stainless Steels? Latest Results from Terni and Krefeld. *Proc. 4th European Continuous Casting Conference, October 2002, Birmingham, England, IOM Communications*, Vol. 2. pp. 869–81

Mukuntan K, Stresov L, Mahapatra R, Blejde W 2000 Evolution of microstrutures and product opportunites in low carbon steel strip casting. *Brimacombe Memorial Symposium Proceedings, ISS*

Stresov L, Herbertson J 1998 *Experimental Studies of Interfacial Heat Transfer and Initial Solidification Pertinent to Strip Casting*. ISIJ, Vol. 38, No. 9

Walther M, Stebner G, Demasse J M, Tolve P, Hohenbicher G 2000 EUROSTRIP®—Initial Operational Results at the Plant at the Krefeld works, CCC 2000, Linz, Austria, June 5–7, 2000

Yamada M, Isogami K, Nakashima H, Okimori M, Sasaki K 2002 World's first commercial Twin Drum Strip Casting Process at Hikari Works of Nippon Steel Corporation. *Proc. 4th European Continuous Casting Conference, October 2002. Birmingham, England, IOM Communications*, Vol. 2. pp. 891–900

A. W. Cramb

T

Tape Casting and Lamination

Tape casting of ceramics was first patented by Howatt in 1952 (Howatt 1952) for the production of thin capacitor sheets. Tape casting is still one of the primary techniques for capacitor manufacture today. Tape casting is also known as the "doctor blading" or "knife coating" process since the doctor blade or knife is the primary instrument involved in the forming process. The goal of tape casting is the creation of a thin, two-dimensional sheet from a liquid suspension of ceramic or metal particles. Both equipment and cap acitor manufacturers report tape thicknesses as low as two microns. Sheets in excess of 3 mm can also be tape-cast but are generally easier to produce by other methods. The bulk of commercial tape production is between 0.025 mm and 1.27 mm in dry thickness.

Single layers of sintered tape, typically 100–1000 μm thick, are used in applications including electronic substrates, porous membranes, monolithic capacitors, and wear-resistant plates. Three-dimensional products such as multilayer ceramic capacitors (MLCC), packages (MLCP) from both high-temperature and low-temperature cofired ceramics, ceramic multichip modules, and functionally graded materials are produced by laminating multiple tape layers. Cast tapes are also used for the production of both simple and complex parts through a process known as laminated object manufacturing.

The tape casting process is similar to the slip casting process in that it is a fabrication technique using a fluid suspension of ceramic or metal particles as the forming medium. There are significant differences, however, between the two processes. Modern tape casting typically utilizes a high-volatility, nonaqueous solvent since the drying process is evaporative rather than absorptive into a plaster mold. It is noteworthy that the Howatt patent reported the use of plaster batts as the tape casting surface and that water was one of the liquid vehicles. In the mid-1950s, Park, of the American Lava Corporation, patented the use of a nonabsorptive carrier in place of the porous casting surface (Park 1961). Nonabsorptive carriers including polymers, metals, and glass are used for modern tape casting.

1. The Doctor Blade Process

The tape casting process, illustrated in Fig. 1, consists of a stationary, adjustable-height doctor blade, a moving carrier surface, and a drying chamber. The "heart" of the tape casting process appears to be mechanically rather simple; however, the design of blades, casting machines, and the slip formulations that make the process work remain subjects of extensive research and development.

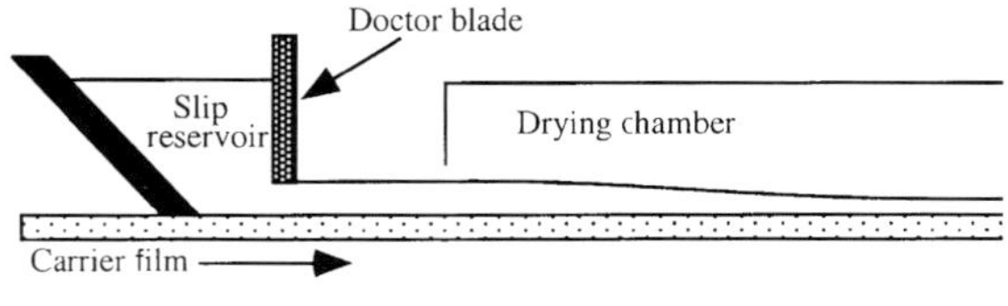

Figure 1
Schematic of the doctor blade tape casting process.

1.1 Materials Selection

Since the material to be cast is predetermined on the basis of required material properties, organic additives are selected depending upon the powder chosen. The most important selection is that of a binder/plasticizer combination that will yield a strong flexible tape. The binder and plasticizer must facilitate formation of the desired part but burn off cleanly during the presintering operation. Most binder systems chosen are long-chain "film-forming" polymers. Many organic systems meet clean burn-off criteria when the sintering is done in an oxidizing atmosphere at relatively high temperatures, i.e., above 1000°C. Many modern ceramic and metallic systems, however, include nitrides, carbides, and reactive braze materials which require the use of organic systems that can be removed in the absence of oxygen and in some cases at temperatures as low as 300°C and in a few seconds rather than in minutes or hours. Acrylic and polycarbonate binders that evaporate or "unzip" upon heating are used in tape casting systems requiring inert or reducing sintering atmospheres. Commonly used binders for oxidizing firing include poly(vinyl acetate), poly(vinyl butyral), poly(ethyl methacrylate), poly(methyl methacrylate), and various copolymers. Other necessary considerations during binder/plasticizer selection include the thickness of the tape to be cast, the surface upon which the cast is to be made, and the solvent(s) desired. Some binders have a greater affinity for and thus adhere more readily to the casting surface chosen. High-volatility solvents such as acetone or methyl ethyl ketone are often used for casting thin tapes

Table 1
Selected organic systems for tape casting.

Nonaqueous solvent systems			
Solvent(s)	Binder	Plasticizer(s)	Deflocculant
MEK/toluene	Poly(propylene carbonate)	Butyl benzyl phthalate	Z-3 Menhaden fish oil
MEK/ethanol	Poly(vinyl butyral)	Dibutyl phthalate/ polyalkylene glycol	Synthetic waxy ester
MEK/ethanol/cyclohexanone	Poly(ethyl methacrylate)	Dibutyl phthalate	Z-3 Menhaden fish oil
Aqueous solvent systems			
Binder	Plasticizer(s)	Cross-linker	Deflocculant
Hydroxypropyl methyl cellulose	Poly(ethylene glycol)/ glycerol	–	Ammonium polyacrylate
Acrylic emulsion	–	Poly(oxyalkylene-diamine)	Ammonium polyacrylate

(below about 0.25 mm) while less volatile solvents are typically used for thicker tapes to avoid the formation of a solvent-depleted "skin" on the tape surface. The type of solvent(s) used limits the binder/plasticizer systems that can be used (or vice versa) since the binder must fully dissolve in the solvent(s). The current trend in many laboratories and production facilities is towards aqueous-based binder systems to address health and safety concerns. Aqueous tape casting tends to employ nonsoluble emulsion binders rather than water-soluble polymers. Extensive lists of solvents and organic additives are available in the literature (see Mistler and Twiname 2000). Table 1 includes some of the more recent systems found in the literature.

Tape casting binders typically require plasticizers to improve room temperature flexibility. It is often beneficial to use dual plasticizers, one to soften the binder and a second to lubricate the tape matrix. The primary plasticizer is a low-volatility solvent for the binder polymer. The second, nonsoluble plasticizer can also act as a carrier release agent. Common plasticizers include various phthalates and glycols. All of the organics should be compatible with each other, if not fully miscible or soluble in the others, to prevent segregation during drying.

The selection of a deflocculant/dispersant is also important since the dispersant prevents reagglomeration of discrete particles after they are broken down during milling. Proper use of a dispersant/deflocculant permits higher solids in the slurry thus accelerating gel formation and drying in the tape cast product. A shorter drying time helps to prevent binder segregation and potential density or chemical gradients in the tape. Organic solvents tend to have low dielectric constants compared with water and thus tend to rely upon the steric hindrance dispersing mechanism to achieve a stable suspension. Aqueous suspensions can make good use of both electrostatic and steric separation forces.

1.2 Batching and Slip Preparation

The preparation of slurries for the tape casting process is similar to standard ceramic processing techniques, i.e., ball milling in a dense container with a charge of grinding media. The major difference encountered is the use of nonaqueous solvents. A powder dispersion step typically precedes binder and plasticizer addition to optimize both tape density and homogeneity. The purposes of a dispersion-milling step are to break up agglomerates or aggregates and prevent reagglomeration prior to binder addition. The solvent(s), dispersant, and powder are processed in a jar mill filled approximately to one third with grinding media. The remainder of the charge fills about one half to three-quarters of the mill, which is rotated for 12 to 24 hours. The slurry usually becomes quite fluid during this step due to the activity of the dispersant on particle surfaces.

Binder(s) and plasticizer(s) are dissolved or mixed into the slurry after dispersion milling. It is recommended that plasticizers are added prior to the binder to promote binder dissolution into the plasticizer. If the binder is added in liquid form this is not as critical. The slurry usually becomes more viscous with binder incorporation since relatively high molecular weight binders are used. There are cases, however, where the addition of a binder makes the slip less viscous since the binder can act as a secondary dispersant and replace the dispersant on particle surfaces during mixing. This type of dispersant/binder interaction must be accounted for in the development of a slip formulation. Tape casting binders are often predissolved in a small amount of solvent to avoid tape defects due to undissolved binder. Binder incorporation is complete within 12 to 24 hours. Tape casting slips tend to have a pseudoplastic rheology after all the components are incorporated.

1.3 Slip Conditioning and Characterization

After thorough mixing of all components, the slip is conditioned in preparation for the tape casting process. The first procedure performed after discharging the slip from the ball mill is a de-airing process to remove entrained air introduced during milling, mixing, and pouring. Slow rolling (without milling media) tends to coalesce small air bubbles and break large bubbles but is a long procedure that can take up to 150 hours to complete at rotation speeds of less than 10 revolutions per minute. Centrifugation of the slip concentrates entrained air on one side of the slip body leading to bubble coalescence and rupture. The most common technique used for nonaqueous slips is vacuum de-airing with continuous agitation or stirring. Vacuum de-airing is done in a vacuum chamber or desiccator under a vacuum of 635–700 torr. The use of higher vacuum pressures is detrimental to the slip since rapid boiling of the slurry will cause excessive solvent loss and the formation of a solvent depleted layer or skin on the slip surface. De-airing time is determined by experimentation and depends upon the slip volume and viscosity as well as other factors. De-airing times range from less than eight minutes to a few hours. A good rule of thumb is to use about 4–8 minutes for a volume of 4 l. A paddle or blade mixer is used to shear the slip during de-airing to lower slip viscosity and to help break bubbles on the slip surface.

The de-aired slip can be used directly in the tape casting process. Manufacturing processes, however, include a filtration process prior to doctor blading. Filtration is essential when processing tapes below 100 μm. Filtration involves the pumping or pressure feeding of slip through a sieve or series of sieves to remove agglomerates or aggregates not broken down during milling and debris introduced during milling, including pieces of mill lining or grinding media that have chipped, and also aids removal of air bubbles not removed during de-airing. Sieve mesh openings are determined by the critical size of the defect-causing objects being removed and the final thickness of the tape being cast. The use of submicron filters is required to prevent critical defects in 5 μm MLCC tapes while 10 μm filtration is adequate for 1000 μm thick film substrates. After filtration the casting slip is typically pseudoplastic with a viscosity between 500 and 6000 MPa s (cP).

1.4 Tape Casting

The heart of the tape casting process is the doctor blade assembly. This assembly consists of an adjustable metering blade with a reservoir to hold the slip. Many variations of doctor blade tape casting exist, all falling into one of two categories: moving blade or stationary blade tape casting. Most moving blade/stationary carrier machines are used in laboratories or very small-scale production facilities whereas the more prevalent, stationary blade/moving carrier systems are used in larger manufacturing processes. The stationary blade configuration permits roll-to-roll continuous production using automated carrier-payout and take-up systems. Roll-to-roll processing can then be used in later forming processes. The doctor blade, whether moving or static, is set at a specific height above the casting surface using micrometers or precision shim stock. For casting very thin tapes, or large quantities of a single tape thickness, a precision-machined, fixed height doctor blade is commonly used.

Blade height, or "blade gap," is one of four main factors that determine the wet film thickness. The other factors governing this thickness, and thus requiring control, are viscosity, casting speed, and reservoir fill depth. The machine operator controls as many of these parameters as possible and then adjusts the blade gap as the last measure of thickness control. The viscosity, being temperature dependent, can be controlled by controlling slip temperature during casting. Control of reservoir height behind the doctor blade is often automated using a sensor to detect slip level in concert with a feed valve to modulate slip flow into the reservoir. Since the ratio of doctor blade height to dry tape thickness can vary from unity to 2.5, trial and error is often the only means to achieve the final desired thickness. The wet thickness of a tape can be monitored using laser reflection or gamma ray backscatter equipment. If thickness monitoring is done just downstream from the doctor blade a feedback loop to the slip level sensor can adjust reservoir depth and thus control wet film thickness very accurately.

Doctor blade assemblies consist of one or more rigid blades mounted in an aluminum or plastic frame. The frame assembly also defines the reservoir that holds the slip. Many different shapes of the metering edge of the blade have been reported ranging from a flat bottom to a rounded bar to a "comma" shape. Angular and parabolic blade bottoms find use mainly in laboratory operations. From a practical standpoint, the use of a flat-bottomed blade has been preferred since it maintains dimensional integrity for a longer time compared with other blade shapes. The thickness of the blade along the casting direction at the casting interface (blade bottom) has also been varied from a knife-edge to over three inches depending, apparently, on individual preferences. Some major considerations influencing blade design include weight, lifetime, and cost of refinishing.

Many tape casters use a double doctor blade—two metering blades in series—in applications where very close tolerances must be maintained (±0.0025 mm over a 200 mm width and long lengths). The first metering blade defines a very precise, if very low, reservoir that aids in the uniformity achievable by the second blade. Coating or "slip lay-down" techniques other than doctor blading include gravure, microgravure, and slot die coating as well as curtain or "waterfall" coating. While doctor blade casting is nearly always performed on a horizontally-oriented machine, gravure or slot die coating can be done on an incline, vertically, or even upside down, depositing slip on the underside of a carrier film.

The doctor blade meters slip on to a moving carrier film that transports the deposited slip into a controlled atmosphere drying chamber. The carrier film itself is chosen based upon several factors including nonreactivity with the slurry, surface quality, surface finish, cost, stiffness, tensile properties under drying conditions, and ease of dry tape release. Common carrier materials include plain, corona treated, fluorinated, or silicone-coated Mylar, plain or silicone-coated polypropylene, Teflon, carbon steel, stainless or valve steel, or polymer- or wax-coated paper. The carrier film is required to withstand downstream drying conditions such as elevated temperatures without degrading the desired properties or characteristics of the cast tape. Very thin thermoplastic carriers can exhibit excessive tensile yield under elevated drying temperatures and thus degrade tape quality. Products such as thin-film electronic substrates that have specific surface finish requirements are cast on very smooth carriers since the tape surface dried against the carrier film displays a lower surface roughness than the air-dried surface.

Casting speed is determined by the length of the drying chamber (or vice versa), the thickness of the tape being cast, and the volatility of the solvent system used. Typical speeds for casting range from $15\,\mathrm{cm\,min^{-1}}$ to over $50\,\mathrm{cm\,min^{-1}}$. Extremely thin tapes dry more rapidly and thus can be cast at speeds of over 10 meters per minute. Some thin capacitor tapes are cast at speeds measured in meters per second.

After a thin, uniform slurry layer is deposited on to the carrier film, the tape is dried to remove the solvent(s). Since the slip is supported by an impermeable carrier, evaporation occurs from only one side of the tape, leading to gradients in solids loading, porosity, and organic content. These gradients can significantly affect densification later in the process. Various heating profiles, slurry solids loading and/or organic compositions, air flow rates, location of heating sources (above vs. underneath), and the use of azeotropic solvent mixtures have been explored to avoid compositional gradients during drying of tapes. While gradients can be minimized by combinations of the above approaches, the one-side drying inherent in the tape casting process makes the formation of gradients through the tape thickness inherent as well.

The use of flowing air speeds the drying process by maintaining low solvent vapor concentrations in the local atmosphere. Air used for drying is often filtered to remove dust particles and preheated to maximize tape drying rate. Airflow is directed countercurrent to the casting direction to form both a solvent concentration gradient and a temperature gradient within the drying chamber. A hot air inlet at the end of the drying chamber results in a maximum temperature at the end of the drying process to remove remnant solvents while allowing a lower temperature at the beginning of the drying process to avoid skinning or "case-hardening." Drying can also be accelerated by the use of underbed heating and overhead radiant heating to heat the wet cast. Underbed heating is very common while overhead radiant heating is more rare.

2. *Shaping*

The green tape, after drying, is a strong, flexible sheet of thermoplastic polymer filled with ceramic, metallic, or other powder. Due to the evaporative drying process, green tapes also include 5–25 vol.% porosity. Green densities of 50–65% of theoretical are routinely achieved for both ceramic and metal powders. This green density can be further increased by reducing the porosity of the green tape through pressing or calendaring. Calendaring reduces tape porosity by compaction between two counter-rotating rollers. It can also be used as a partial solution to the compositional gradients mentioned earlier. Drawbacks to this technique include the additional manufacturing cost as well as degradation of the tape surface finish due to contact with the rolls. Pressing, similar to lamination described below, also reduces porosity and corresponding density gradients.

Long lengths of cast tape are cut or "blanked" into manageable sizes by either cutting the tape or stamping with tool and die. Blanks may be sized to include alignment or registration holes around the perimeter and are often sized to manufacture multiple small parts from a single sheet. The blanking process also avoids use of the tape edges which are tapered and thus unusable. Single layer substrate applications blank desired green substrate sizes and sinter at this point.

Tapes used in applications requiring three-dimensional interconnected circuitry, such as MLCPs, are punched with through-holes at this stage. These holes, or vias, are typically made using punch and die but can also be made by water-jet, laser, or microdrill. The vias are then filled with conductive ink. Circuit patterns are also added to each tape layer by screen-printing conductive ink(s) onto the unfired blanks. Alignment through each of these three stages is paramount and is accomplished by either mechanical (registration holes) or optical (registration marks) means. Some high-performance applications require positioning with better than 5 μm accuracy. Stacking and alignment of the metalized layers forms the three-dimensional structure (Fig. 2). Tape layers are often tacked together at this point to preserve alignment through further processing.

3. Lamination

A low-temperature lamination process fuses multiple tape layers into a single, composite structure. Laminates containing over 300 individual layers are not uncommon for MLCCs. Lamination methods vary depending on both the tape properties and final part requirements. The four most common approaches are uniaxial, isostatic (wet bag), and bladder (dry bag) lamination and tape transfer (Fig. 3). Lamination pressures range from 3 MPa to over 50 MPa while temperatures are generally kept below 100 °C, often between 60 °C and 90 °C. In an ideal case, individual tape layers are no longer discernable in the laminate.

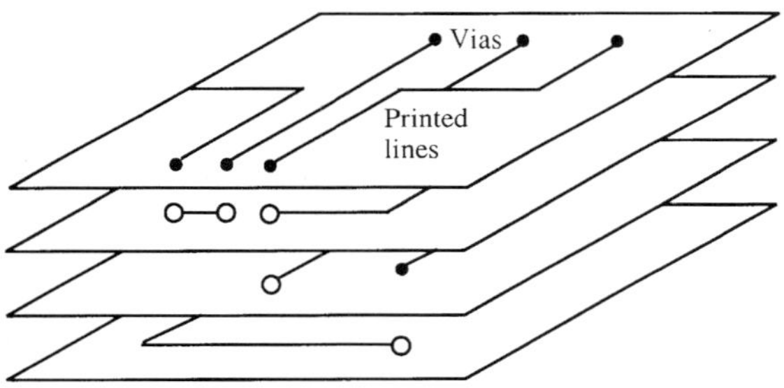

Figure 2
After via punching and metallization, tape layers are stacked to form a three-dimensional, interconnected circuit within the ceramic material.

Uniaxial lamination, the simplest lamination method, fuses single tape layers together by the application of pressure and heat through two opposing heated platens. The edges of a tape stack can be constrained by die walls or left unconstrained during lamination. Isostatic lamination applies uniform temperature and pressure to all surfaces of the tape stack by means of a wet bag isostatic press with a heated fluid bath. Isostatic lamination is more widely used than uniaxial lamination in manufacturing environments due to the uniform application of pressure and the resulting increase in laminate uniformity. Tape layers are protected from the fluid bath within protective envelopes. Vacuum sealing of the envelopes avoids air pockets between tape layers and ensures full contact of one layer to the next. The use of a fluid bath in isostatic lamination also allows the lamination of multiple, separately sealed stacks in a single pressing. Bladder lamination is a combination of the uniaxial and isostatic methods and relates to isostatic lamination in the same manner that dry bag isostatic pressing relates to wet bag isostatic pressing. In bladder lamination a heated fluid-filled bladder is used to press the lamination stack against a heated bottom platen. While the fluid ensures an even application of pressure, vacuum sealing of stacks into protective envelopes is no longer required. The tape-transfer process—sequential stacking and lamination of individual layers prior to carrier removal—is used for very thin or very weak tapes. The tape-transfer process avoids defects introduced during handling of

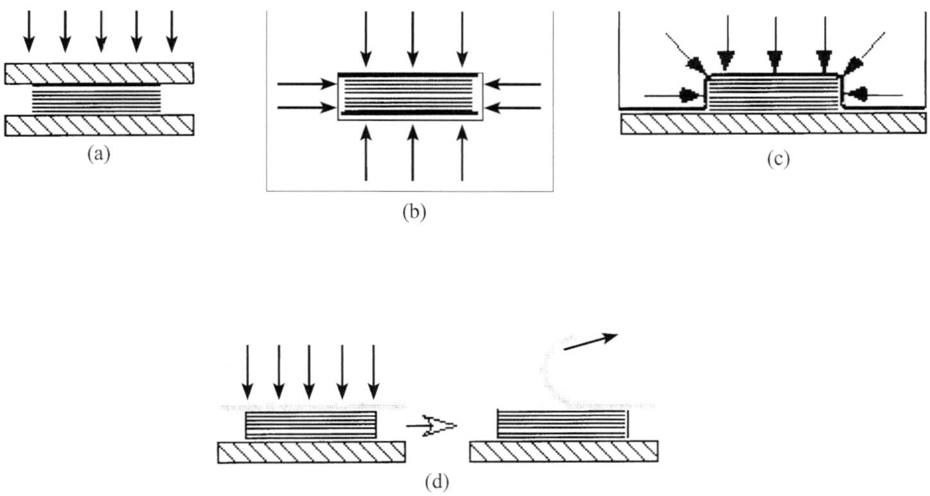

Figure 3
Individual tape layers are formed into a single part by: (a) uniaxial lamination; (b) isostatic lamination in a fluid bath; (c) bladder lamination; or (d) the tape transfer process.

individual layers but can generate density gradients within the laminate by the repeated applications of pressure.

Laminates are considerably more rigid than individual tape layers due to the increased thickness. This increased stiffness allows for a multitude of possible cutting operations. Dicing of multi-omponent laminates, or edge trimming of single component laminates, is accomplished by gang saw, laser cutting, or hot knife cutting. Hot knife cutting, guillotine separation using a heated bed and heated knife, is perhaps the most common method for singulation of green multicomponent arrays due to the desirable balance of speed, cleanliness, and cost. Gang saw and laser cutting can be used for dicing green or sintered structures. Green laminates can also be scored or perforated with rows of vias to create weakened areas between components. The creation of weakened areas in a multicomponent array allows the array to be sintered as a single part and snapped apart after the sintering process.

Bibliography

Howatt G N 1952 Method of producing high-dielectric high-insulation ceramic plates. *US Pat.* 2 582–993

Mistler R E 1992 Tape casting. In: Schneider S J Jr. (ed.) *Engineered Materials Handbook: Ceramics and Glasses*. ASM International, Materials Park, OH, Vol. 4, pp. 161–5

Mistler R E, Twiname E R 2000 *Tape Casting: Theory and Practice*. American Ceramic Society, Westerville, OH

Park J L Jr. 1961 Manufacture of ceramics. *US Pat.* 2 966 719

Shanefield D J 1995 *Organic Additives and Ceramic Processing*. Kluwer Academic, Boston, MA

E. R. Twiname and R. E. Mistler

Titanium Alloys: Thermal Treatment and Thermomechanical Processing

In contrast to many other metallic systems (e.g., steels, aluminum, and magnesium alloys), titanium microstructures depend almost entirely on the size, shape, and distribution of two allotropic phases: the high-temperature b.c.c. β-phase and the low-temperature h.c.p. α-phase. Once the composition (chemistry) of an alloy is defined, thermomechanical processing (TMP)—which is a combination of working (deformation) and thermal (heat) treatment—controls the microstructures and, in turn, the mechanical properties. Therefore, the particular processing route used is selected considering the multitude of required mechanical properties for a specific a408011lication. For example, fracture-critical structures perform most efficiently when fabricated from materials processed to achieve high fracture toughness. Fatigue-limited structures, however, are more sensitive to crack initiation and crack growth rate. Other considerations include the ductility required for applications involving forming or installation, in which case tensile ductility may dictate the processing sequence selection. In general, it is not possible to maximize all these properties simultaneously.

1. Microstructural Development in Titanium Alloys

Most commercially important structural (as opposed to those alloys used in corrosive environments) titanium alloys contain varying amounts of the h.c.p. α- and b.c.c. β-phases. The mechanisms and kinetics of transformation of the β-phase during cooling dominate the development of titanium microstructures. The principal effect of TMP is to perturb or alter these transformation mechanisms providing a powerful technique for controlling or manipulating the microstructure.

Using the simplified schematic phase diagram shown in Fig. 1, and the specific Ti–6Al–4V (wt.%) composition indicated (arrow), microstructure development can be explained. By definition, above the beta transus temperature (upper continuous curved line) only the β-phase is present. The primary breakdown by deformation of the Ti–6Al–4V alloy is carried out at temperatures above the beta transus temperature because of the inherent large (β) grain size in the ingot. This initial processing is concerned with shaping to an intermediate product form (billet/slab, etc.) and refinement of the β-grain size by β-working (i.e., above the beta transus temperature), followed by a β-recrystallization to produce a smaller β-grain size.

Once the β-grain size has been controlled subsequent TMP is aimed at control of the α-phase: size,

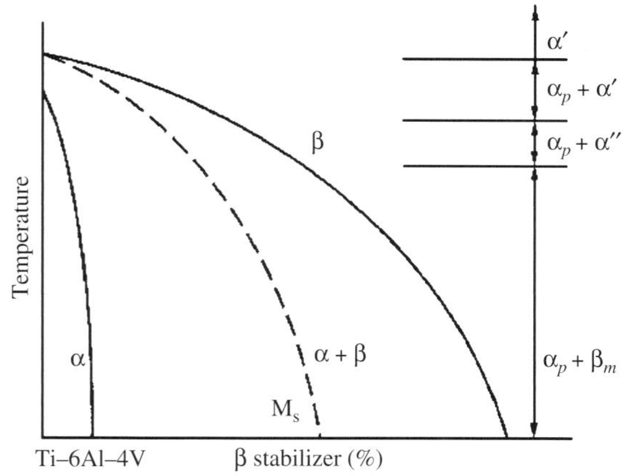

Figure 1
Schematic of Ti–X phase diagram (where X is, for example, molybdenum or vanadium), showing the composition dependence of the martensite start (Ms) temperature, and the constitution that results from quenching from a range of solution treatment temperatures.

shape, and distribution. However, before concentrating on this aspect of TMP, for an alloy such as Ti–6Al–4V it is important to understand the competition between the nucleation and growth of α-phase and the martensitic decomposition of the β-phase which can also be considered with reference to Fig. 1. If an alloy is heated above the beta transus temperature (where it is 100% β-phase) and then cooled, the transformation product that forms depends on the cooling rate. If it is cooled rapidly enough (i.e., avoiding transformation to the α-phase), the β-phase transforms by nucleation and shear to a h.c.p. martensitic product which is designated α'. If it is cooled somewhat more slowly, it transforms by nucleation and growth to a colony microstructure (packets of similarly aligned α-plates) comprised of the α-phase within a β-matrix.

The competition between the mechanisms of nucleation and growth and of martensitic decomposition of the β-phase is demonstrated using a continuous cooling transformation (CCT) diagram, as shown schematically in Fig. 2. The nucleation and growth "C" curve is intersected during slower cooling, whereas it is bypassed during rapid cooling (quenching), in which case martensite is formed. An important additional factor introduced by mechanical working, which is shown schematically in Fig. 2, is the acceleration of the β-decomposition kinetics owing to the continuous introduction of heterogeneous α-phase nucleation sites. This shifts the α-nucleation curve to the left (towards shorter times). Thus, one effect of TMP is to increase the rate at which the β-phase decomposes by nucleation and growth.

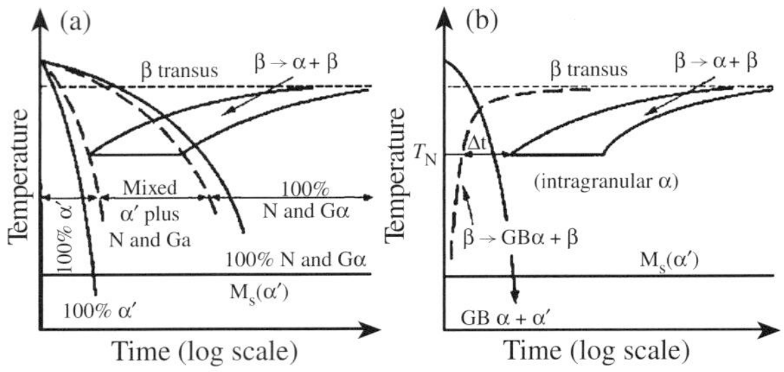

Figure 2
Schematic CCT diagrams showing the cooling rate dependence of the constitution that results from the competition between nucleation and growth and martensitic decomposition of the β-phase: (a) uniform nucleation of α, and (b) heterogeneous nucleation (due to mechanical working).

Unlike steels, the martensite transformation is of little commercial significance in titanium alloys such as Ti–6Al–4V. Hence, the transformation of the β-phase, on cooling below the beta transus temperature (Fig. 1), to the α-phase and the influence that TMP has on the α-phase that forms will be discussed in greater detail. Thermodynamically transformation to the α-phase can only occur below the beta transus temperature, and only when α-phase is present can it be affected by TMP.

2. *Effect of Mechanical Working (TMT) on Microstructure*

For alloys such as Ti–6Al–4V working below the beta transus temperature means that α-phase is present and is being subjected to plastic deformation that can then affect the morphology (shape) of this phase. The extent to which the α-phase morphology is changed depends directly on the amount of plastic deformation to which the material is subjected, which in turn depends on the strain rate and temperature of deformation. The strain rate and temperature at which the plastic working is carried out are critical because there is a dynamic competition between *in situ* recovery and the storage of plastic work which will subsequently drive a change in the shape of the α-phase. Because the rate of recovery decreases with decreasing temperature, smaller amounts of work at lower temperatures can have the same effect as larger amounts at higher temperatures, thus there is not normally a unique combination of deformation amount, temperature, and time to produce a given microstructure.

In practice the morphology of the α-phase is controlled by working in the two-phase α/β-phase field followed by an extended anneal, also in this phase field, to allow the worked α-phase to "relax" to a

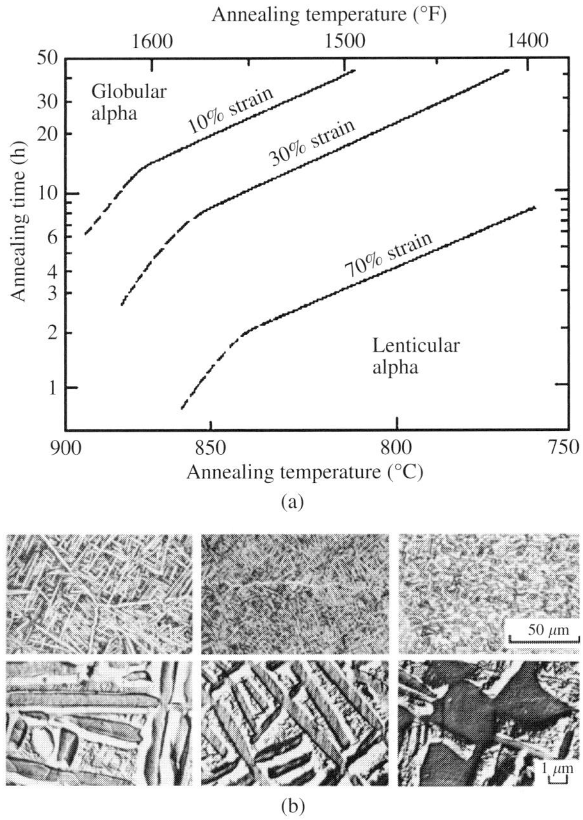

Figure 3
CORONA 5 (Ti–4.5Al–5Mo–1.5Cr): (a) relationship of α-morphology to annealing time and temperature after various amounts of deformation in the α/β-phase field; (b) top line opticals, bottom line surface replicas of microstructure, left and center beta worked and alpha-beta annealed showing lenticular alpha, right alpha-beta worked and alpha-beta annealea showing equiaxed primary alpha due to relaxation of the worked alpha.

more equiaxed morphology. The way in which the amount of working affects the subsequent shape of the α-phase is demonstrated for the CORONA 5 alloy (Ti–4.5Al–5Mo–1.5Cr (wt.%)) in Fig. 3a. This relaxation process is temperature–time dependent, thus a higher temperature α/β-anneal causes the worked α-phase to become equiaxed more quickly than if the anneal is carried out at lower temperature. Fig. 3b demonstrates how beta working does not change the normal lenticular morphology of the alpha phase while alpha-beta working can lead to an equiaxed alpha morphology.

In alloys with higher concentrations of β-stabilizing elements, the kinetics of nucleation and growth are slower. As a result, appreciable competition develops between heterogeneous nucleation of α- at prior β-grain boundaries and in a intragranularmode

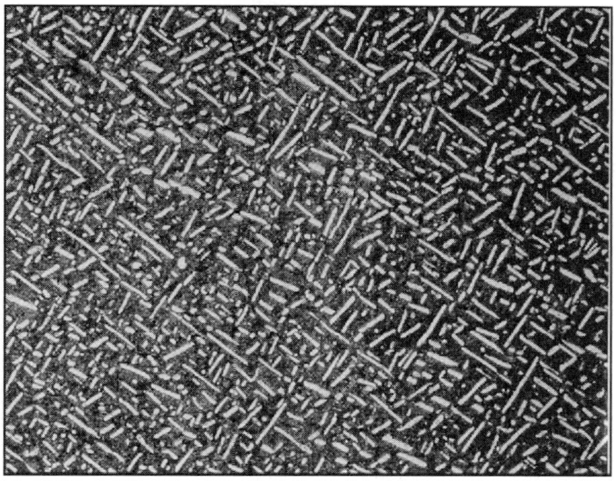

Figure 4
Light photomicrographs showing the absence of continuous grain boundary α and intragranular α with a reduced aspect ratio (length-to-thickness) obtained after 25% α/β-working and annealing in the metastable β-alloy Ti–10V–2Fe–3Al (wt.%). (Courtesy R. R. Boyer, Boeing Commercial Airplane Company.)

(within the grain interior). This competition often results in a continuous or semicontinuous layer of grain boundary α in alloys which have been subjected to thermal treatment alone. The resulting grain boundary α has a deleterious effect on properties, especially tensile ductility. An important benefit of TMP, as previously mentioned, is the acceleration of the nucleation and growth kinetics for intergranular α nucleation. This becomes a useful means of reducing the tendency to form the grain boundary α-phase. Moreover, mechanical working tends to cause the grain boundary α to relax to a more globular "broken-up" morphology making it less continuous and less deleterious to properties.

The working temperatures and amounts of plastic work appropriate to the production of a given microstructure depend on the alloy composition and can be quite critical for some alloys. An example of this is provided by the lean metastable β-alloys such as Ti–10V–2Fe–3Al (wt.%). In these alloys, α/β-working to avoid grain boundary α is essential to obtain adequate tensile ductility levels. An example of an acceptable microstructure in this alloy is shown in Fig. 4.

Another important factor in controlling the final distribution and morphology of the α-phase is the cooling rate from the post-α/β working anneal in the α/β-phase field. During slow cooling, the α-phase grows epitaxially (as dictated by the equilibrium phase diagram), whereas separate nucleation and growth of lenticular α in the remaining metastable β-phase occurs during cooling at higher rates (see Fig. 5).

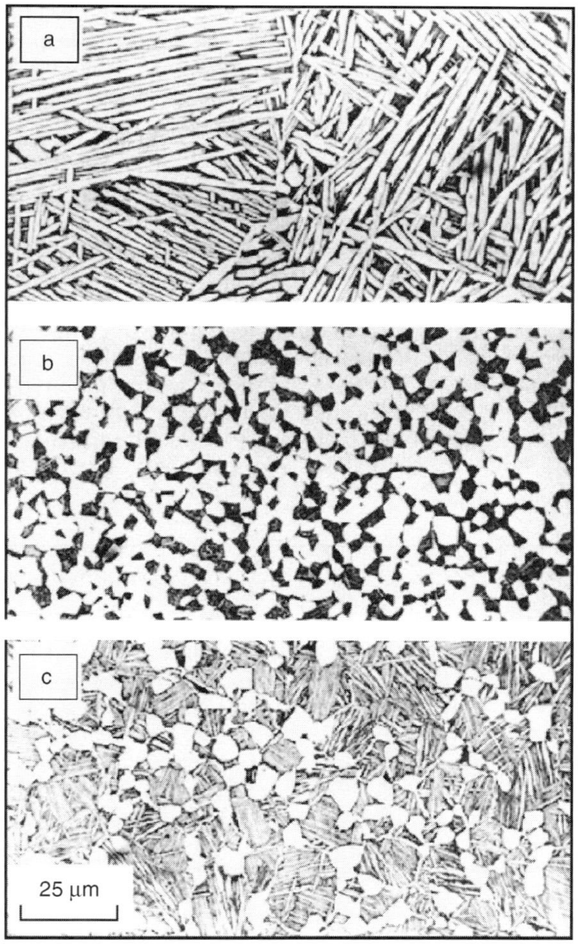

Figure 5
Microstructure of Ti–6Al–2Sn–4Zr–2Mo: (a) β-worked followed by an α/β-anneal to produce a lenticular α-morphology; (b) α/β-worked and α/β-annealed to give predominantly an equiaxed α-shape; (c) α/β-worked followed by a duplex anneal, just below the beta transus temperature (reduced volume fraction of equiaxed α compared to (b)), and significantly below the beta transus temperature (to form the separately nucleated lenticular α between the equiaxed regions).

Thus in applications where a fully equiaxed α-microstructure is required (e.g., for subsequent diffusion bonding) a maximum allowed cooling rate is specified ($90^\circ F\,h^{-1}$ by Rockwell International on the B-1B program) and no separate nucleation of lenticular α-phase occurs. When a balance of mechanical properties is required (e.g., creep and low cycle fatigue in a gas jet engine application) initial cooling is relatively fast to give equiaxed islands of α in the β-matrix. This is then followed by a further lower temperature anneal during which separate nucleation of lenticular α occurs between the equiaxed α-islands (see Fig. 5). Examples of microstructures that can be developed in the Ti–6Al–2Sn–4Zr–2Mo (wt.%) α/β-alloy are shown in Fig. 5. The influence of the α-morphology on mechanical properties is discussed below.

3. *Effects of Microstructure on Mechanical Properties*

The major impact of the development and control of the microstructure is the effect of the final microstructure on the mechanical properties. The schematic shown in Fig. 6 simplifies the situation for fracture toughness and ductility. A lenticular α-morphology favors high values of fracture toughness whereas the globular α-morphology favors an increase in the ductility; thus β-working (leading to a lenticular α-morphology, Fig. 5(a)) raises the former parameter, α/β-working (globular α) favors ductility. This is well illustrated for fracture toughness for the CORONA 5 alloy in Fig. 7. How the fracture toughness–ductility relationship relates to the processing route is summarized for the Ti–6Al–2Sn–4Zr–6Mo (wt.%) alloy in Table 1.

It is beyond the scope of this article to discuss the full details of the relationship between microstructure and mechanical properties. However, a summary of the relationship between the α-morphology and a number of mechanical properties is given in Table 2. Thus the microstructure shown in Fig. 5(c) is a compromise in a gas turbine engine application where both reasonable creep and low cycle fatigue behavior are required.

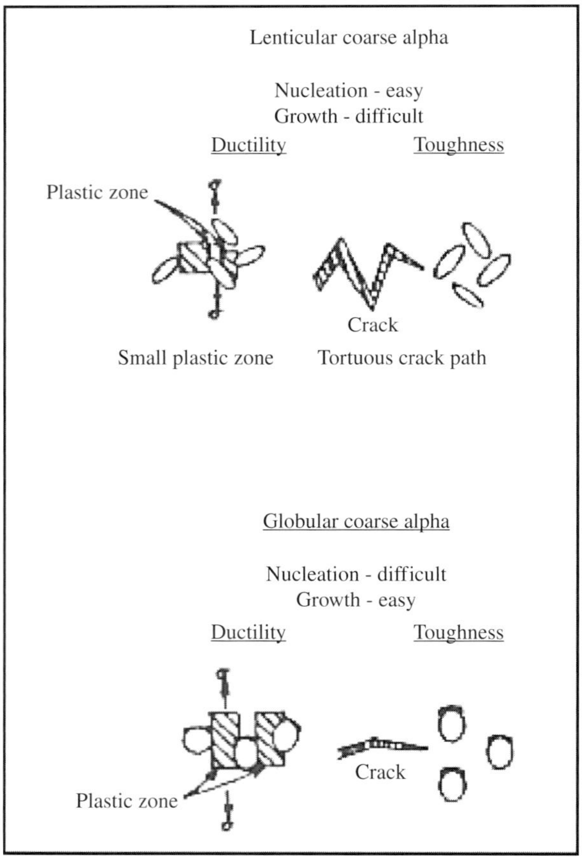

Figure 6
Schematic representation of the effect of the α-morphology on fracture toughness and ductility. The lenticular morphology causes the crack path to be tortuous giving high toughness; the globular shape gives large plastic zones and increased ductility.

4. *Effects of TMT on Preferred Orientation (Texture)*

Because of the hexagonal structure of the α-phase, appreciable anisotropy of properties can be developed in titanium alloys if the α-phase has a significant preferred orientation. This preferred orientation results from working operations and manifests itself as directionality in properties, such as modulus, tensile ductility, and yield stress, as well as many of the fracture-related properties. The detailed effects of TMP temperature, time, and strain on texture are complicated. For example, unidirectional working such as plate rolling produces different textures from those obtained by axisymmetric flow (forging) under the same conditions of time, temperature, and strain. Therefore, the texture to be expected is not only very sensitive to working temperature and amount of strain but is also very sensitive to working methods.

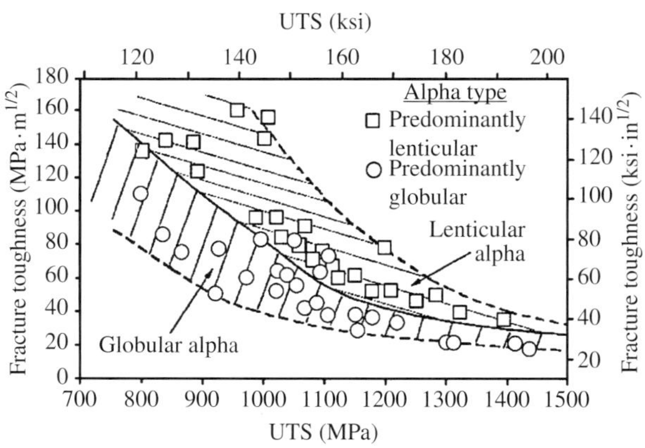

Figure 7
Variation of fracture toughness with strength for CORONA 5 (Ti–4.5Al–5Mo–1.5Cr (wt%)).

Table 1
Typical properties of Ti–6%Al–2%Sn–4%Zr–6%Mo alloy.

Processing[a]	Yield strength (MPa)	Ultimate tensile strength (MPa)	Elongation (%)	Fracture toughness, K_{IC} (MPa $m^{1/2}$)
β	1050	1200	7	1770
α	1120	1210	13	1040

[a]Followed by an α/β-anneal.

Table 2
Effect of α-morphology on various mechanical properties.

Mechanical property[a]	Favored by[b]
Strength	Slightly E
Ductility	E
LCF	E
K_{IC}	L
FCGR	L
Creep resistance	L
DB/SPF	E

[a]LCF, low cycle fatigue; K_{IC}, fracture toughness; FCGR, fatigue crack growth rate; DB/SPF, diffusion bonding/superplastic forming.
[b]E, equiaxed α; L, lenticular α.

Titanium fabricators are aware of the strong influence that texture can have on mechanical properties and control the TMP appropriately. For example, plate is cross-rolled (in two directions at 90°) to minimize the spread in mechanical properties between the rolling direction (longitudinal) and the direction at 90° to this (transverse direction).

Bibliography

Blenkinsop P A, Evans W J, Flowers H M (eds.) 1995 *Titanium '95 Science and Technology*. Institute of Materials, London

Froes F H, Allen P G, Niiomi M (eds.) 1998 *Non-Aerospace Applications of Titanium*. TMS, Warrendale, PA

Froes F H, Caplan I L 1993 *Titanium '92 Science and Technology*. TMS, Warrendale, PA

Froes F H, Eylon D, Bomberger H B (eds.) 1985 *Titanium Technology: Present Status and Future Trends*. Titanium Development Association (now International Titanium Association), Boulder, CO

Hirth J P, Froes F H 1977 Interrelations between fracture toughness and other mechanical properties in titanium alloys. *Metall. Trans* **8A**, 1165

Williams J C, Starke Jr. E A 1984 The role of thermo-mechanical processing in tailoring the properties of aluminum and titanium alloys. In: Krauss G (ed.) *Deformation, Processing, and Structure*. American Society for Metals, Metals Park, OH, p. 279

F. H. Froes

U

Ultrasonic Processing to Produce Nanoparticles

The use of ultrasound radiation for the production of nanoparticles has been a research topic of great interest. This is due to the simplicity of the sonochemical method, the cheap price of the equipment, and that in many cases the as-prepared material is obtained in the amorphous phase. This article describes the various materials that have been prepared by this method. These particles have been used in various fields such as catalysis, substrates for self-assembled coatings on their surfaces, and preparation of metal–polymer composite materials. This review concentrates on sonochemistry and does not deal with the sonoelectrochemical method that has been developed by Reisse and coworkers (Reisse *et al.* 1994), or with ultrasonic spray pyrolysis. In explaining the sonochemical process, we assume the hot-spot mechanism as its origin, although this is not the only explanation for this phenomenon. A second mechanism, the electric discharge, can also explain many of the results (Margulis 1995). We note that an early review on this topic has already been published emphasizing mostly the catalytic activity of these nanoparticles (Suslick *et al.* 1996).

1. Sonochemistry

Sonochemistry is the research area in which molecules undergo chemical reaction due to the application of powerful ultrasound radiation (20 kHz–10 MHz) (Price 1992). The chemical effects of ultrasound do not originate from a direct coupling of the acoustic field to the molecular species. Instead, they come from nonlinear acoustic phenomena, primarily acoustic cavitation. This can be divided into three distinct stages: (i) formation, (ii) growth and (iii) implosive collapse of bubbles (Price 1992). Two types of cavitation have been proposed: (i) stable cavitation, in which a bubble oscillates many times around its equilibrium radius, and (ii) transient cavitation, in which a short-lived bubble undergoes dramatic volume changes in a few acoustic cycles and terminates in a violent collapse (Suslick 1988). Stable and transient cavitation may occur simultaneously in the solution and a bubble undergoing stable cavitation may become a transient cavity. The formation of a bubble in a pure liquid requires a hydrostatic tensile of about 1000 atmospheres. We can, however, follow the formation of a bubble with a pressure wave of just a few atmospheres. This is explained by the presence of a nucleation process where the nuclei can be gas-filled crevices in small particles suspended in solution. The removal of these nuclei (by methods such as microfiltration or vacuum degassing) will increase the cavitation threshold. Once the bubble is formed it can grow by several processes. At low acoustic pressure, rectified diffusion is important. The bubble grows as dissolved gas molecules enter during the expansion phase of the sound wave, and shrinks during the compression phase. Since gas diffusion is dependent on surface area, slightly more gas molecules enter the bubble than exit. Thus, the bubble grows slowly over many acoustic cycles. In strong acoustic fields, the bubble oscillations can become very large through inertial effects. Rapid growth of the bubble during the expansion phase can have sufficient inertia to prevent the bubble from shrinking during the next compression wave. Bubbles growing into resonance size are well coupled to the sound field and can rapidly over-expand. At this point, rapid compression and consequent heating can occur. When the compression is rapid enough, the heating is nearly adiabatic; namely, there is not sufficient time for heat transfer to occur effectively. The ultrasonic irradiation of liquids produces hot spots that can initiate chemical reactions (Suslick 1988). The bubble's collapse, which determines the temperature of the hot spot, is affected by a few parameters that are outlined briefly (Suslick 1988).

First, we emphasize again that the bubble contains dissolved gas and solvent vapor. The first parameter that affects the sonochemical reactivity is $g = C_p/C_v$, which determines the increase of the temperature during compression. Monoatomic gases which have the highest g (1.67) yield the highest temperature. Another factor is the thermal conductivity of the gas. If cavitation were completely adiabatic, helium and argon, when used as pressurizing gases during the sonochemical process, would yield the same final temperature. In fact, the sonochemical yield in argon is much higher than in helium. This is due to the larger thermal conductivity of helium (He, 0.149 $Wm^{-1}K^{-1}$; Ar, 0.0172 $Wm^{-1}K^{-1}$). The second factor is the vapor pressure of the solvent used in the sonochemical process. The vapor of the solvent has a low g and, in addition, may undergo its own sonochemical reaction. Therefore, the preferred solvent is always the one with a low vapor pressure. Solvents such as decalin or long-chain alkanes are the most common ones used in sonochemistry.

Other external factors are the acoustic frequency intensity, ambient temperature, and static pressure. The frequency of the sound waves has relatively little effect on the nature of most sonochemistry. This is due to the fact that there is no direct coupling of the irradiating field with the molecules undergoing sonochemistry. Cavitation ceases above a few megahertz and sonochemistry is not observed. The intensity of the acoustic waves influences the rates of sonochemical reactions. Below a threshold value, the amplitude of the sound field is too small to induce bubble growth. Above the cavitation threshold, the sonochemical rate will increase with increase in the intensity of the sound field. The effect of the temperature of the bulk solution on the sonochemical rate is related to the composition of the bubble before its collapse. In general, as the ambient temperature increases, the sonochemical rate decreases. Contradictory information has been reported as to the effect of the static pressure on the yield of the sonochemical process. It seems that a few factors such as solubility of the gas, the nucleation of the cavity, and the cavity collapse are all dependent on the static pressure. They are expected to behave differently with increase of the static pressure.

2. *Producing Nanophased Amorphous Materials Sonochemically*

(Suslick *et al.* 1991) was the first to demonstrate that when ultrasound radiation passes through $Fe(CO)_5$, either as a neat liquid or through its decalin solution, nanosized amorphous particles of iron are obtained. The amorphicity is demonstrated by XRD, electron diffraction, DSC, and EXAFS measurements. The DSC spectrum detects an exothermic peak, which is attributed to the crystallization temperature, and, indeed, if the amorphous material is heated to this temperature, the XRD of the nanocrystalline

Figure 1
TEM micrographs of the cobalt colloid solutions (a) as-prepared fresh, (b) after a week of aging, (c) after a month of aging.

material is detected. Moreover, if the heated sample is cooled and the spectrum is remeasured, a flat DSC is detected. The reason for obtaining the product in the amorphous phase is related to the high transient temperature, determined to be 5000 K, and pressures of 1000–2000 atmospheres resulting from the cavitation collapse. Since this collapse occurs in less then a nanosecond (Barber and Putterman 1991), cooling rates in excess of $10^{10}\,Ks^{-1}$ are obtained. These rates prevent the crystallization of the sonochemical products. Obtaining amorphous products in sonochemistry is a general phenomenon, which is observed whenever the precursor is a volatile material. In this case the chemical reaction occurs in the gas phase inside the collapsing bubble. However, the sonication of nonvolatile precursor such as salts is also possible. In these cases the reaction takes place at a thin shell surrounding the collapsing bubble. The as-prepared materials under these conditions are usually obtained as nanocrystals, although there are some exceptions where nanoamorphous products are obtained. A second question that is related to all sonochemical products (created inside the collapsing bubble or around the bubble) is why the products are formed as nanoparticles. As a first attempt, we would associate it with the temperature at which the sonication of the metal carbonyls, leading to its dissociation, takes place. At this temperature, $-78\,°C$, the vapor pressure of the carbonyl precursor is low, and calculating the number of molecules in the 100 μm bubble would lead to the creation of nanoparticles. However, it is well known that the dissociation of $Fe(CO)_5$ at room temperature forms nanoparticles of iron as well. It is, therefore, thought that the formation of nanoparticles is a kinetically controlled process related to the formation of small nuclei. These nuclei do not grow further due to the fast collapse of the bubble.

3. *Materials Synthesized*

3.1 *Metals*

In addition to the synthesis of the transition metals produced from the corresponding carbonyls (iron from $Fe(CO)_5$ (Dhas *et al.* 1998), nickel from $Ni(CO)_4$ (Koltypin *et al.* 1996), and cobalt from $Co(CO)_3NO$ (Suslick *et al.* 1994)), other metals have also been synthesized sonochemically, such as nanoscale particles of metallic copper (Suslick *et al.* 1994) by the sonochemical reduction of copper(II) hydrazine carboxylate ($Cu(N_2H_3COO)_2\,2H_2O$) complex in an aqueous solution. When the sonication was carried out under argon, a mixture of Cu_2O and metallic copper is obtained. However, sonicating the precursor solution under a mixture of hydrogen and argon yielded pure copper. The particles were obtained as porous materials having a 50 nm diameter and were smaller than those obtained by the thermal decomposition of the same precursor. A mechanism involving hydrogen radicals as the reducing agent is proposed, where these radicals are obtained from the sonochemical decomposition of water. The reactions are

$$H_2O \rightarrow H^{\bullet} + OH^{\bullet} \tag{1}$$

$$Cu^{2+}_{(aq)} + 2H^{\bullet} \rightarrow Cu^{0}_{(s)} + 2H^{+}_{(aq)} \tag{2}$$

In a separate study, nanoparticles of palladium metallic clusters have been prepared at room temperature by sonochemical reduction of a 1:2 molar mixture of palladium acetate, $Pd(O_2CCH_3)_2$, and myristyltrimethylammonium bromide, $CH_3(CH_2)_{12}N(CH_3)BrNR_4X$, in tetrahydrofuran (THF) or methanol. Apart from its stabilizing effect, NR_4X acts as a reducing agent, probably due to the decomposition that occurs at the liquid-phase region immediately surrounding the collapsing cavity, and provides reducing radicals. A detailed mechanism

Figure 2
TEM micrograph of the elongated copper nanoparticles coated with CTAPTS (cetyl trimethylammonium p-toluene sulfonate).

is proposed explaining the differences between the THF process, where nanoamorphous palladium is obtained, and the methanol process yielding nanocrystalline palladium.

3.2 Metallic Colloids

A comprehensive review on the sonochemical formation of metallic colloids as well as other materials was presented by Grieser (Grieser 1997).

The sonochemical synthesis of noble metal nanoparticles in aqueous solution was carried out. Colloidal silver was prepared by sonicating an aqueous solution of $AgClO_4$ or $AgNO_3$, using sodium dodecylsulfate (SDS), polyoxyethylene-monostearate (PEG40), and polyoxyethylenesorbitan (Tween 20) as surfactants. Nagata and Maeda have summarized in a review paper (Nagata and Maeda 1997) the reduction of different metal ions using sonochemistry. Suslick and co-workers have sonicated an $Fe(CO)_5$ solution of hexadecane in the presence of oleic acid (Suslick *et al.*

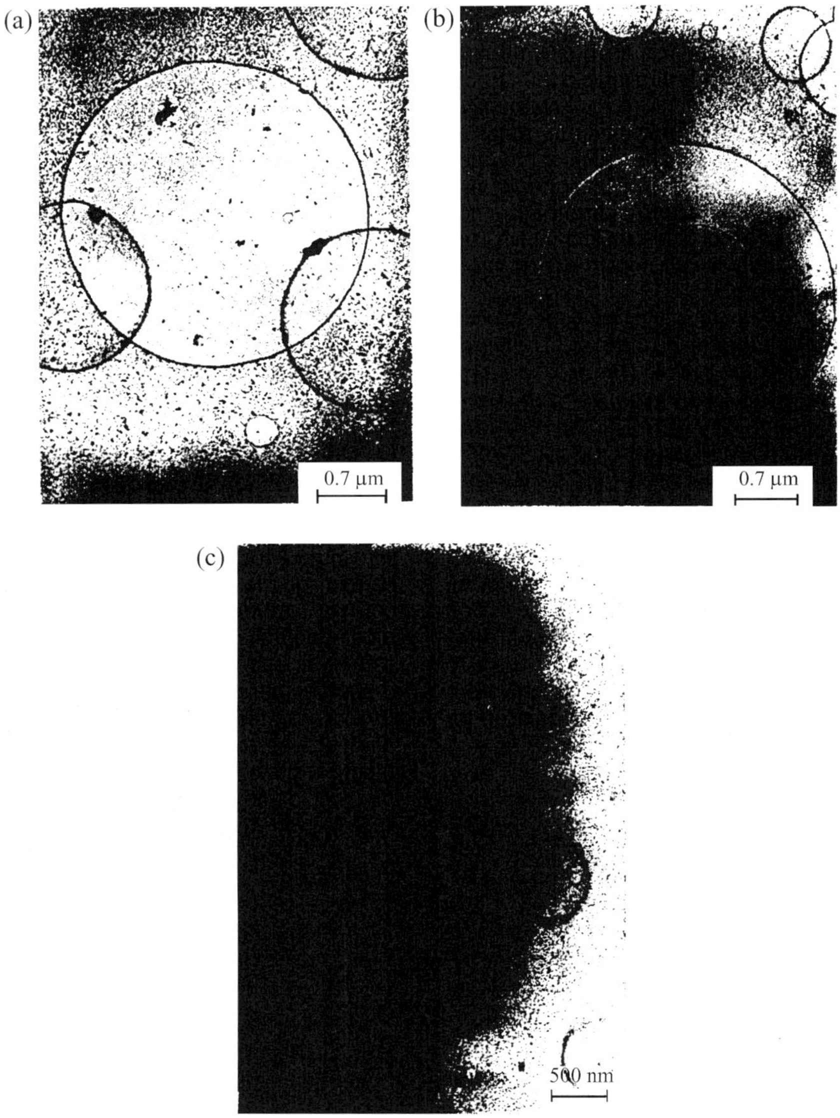

Figure 3
TEM micrograph showing rings of various sizes. The rings seem to intersect each other forming the so-called Olympic Nanorings. The scale bar is 0.7 μm.

1996) and obtained a colloidal solution of metallic iron particles. The iron formed was amorphous and showed high magnetism. The sonochemical preparation of magnetic fluids has also been accomplished (Shafi *et al.* 1998). Three different systems in which stable magnetic fluids were formed have been reported. They represent (i) a cobalt colloidal solution in decalin stabilised by oleic acid, (ii) a colloidal dispersion of amorphous metallic iron in a polymeric matrix, (iii) a Fe_2O_3 colloidal solution in hexadecane stabilized by oleic acid. For the cobalt colloidal solution, it was shown that the particles try to self-organize. When the solution was left to age in air, uniform acicular-shaped particles (Fig. 1) were observed.

Elongated copper nanoparticles were prepared by sonicating copper hydrazine carboxylate in an aqueous solution containing a zwittrionic surfactant, cetyl trimethylammonium p-toluene sulfonate (CTAPTS) (Salkar *et al.* 1999). In Fig. 2, the TEM image of the elongated copper nanoparticles surrounded by surfactant molecules, is presented. The elongated copper nanoparticles were approximately 500 nm in length and 50 nm in width. On the other hand, the copper nanoparticles prepared without using surfactant were spherical in shape, with a diameter of approximately 50 nm. XRD revealed an elongation to be taking place along the [111] and [200] planes of the copper nanocrystals.

The formation of gold sols via the reduction of $AuCl_4^-$ in aqueous solution has been found to be initiated by 20 kHz ultrasound. The presence of aliphatic alcohols was found to enhance significantly the reduction process.

3.3 *Alloys*

Suslick has prepared nanophased amorphous Fe/Co by sonicating a mixture of $Fe(CO)_5$ and $Co(CO)_3NO$ in decalin to obtain the alloy (Suslick *et al.* 1994). A more detailed study, which followed in Suslick's steps, was carried out on the Fe/Ni alloy (Shafi *et al.* 1997). The solid phase composition is very different from the molar ratios of the carbonyl precursors in the solution, due to the ratio of the vapor pressure of the two carbonyls in the gas phase of the collapsing bubble.

3.4 *Oxides*

When the sonication of a decalin solution of $Fe(CO)_5$ is carried out under air, amorphous nanoparticles of Fe_2O_3 are obtained (Cao *et al.* 1997). The magnetization of pure amorphous Fe_2O_3 at room temperature is very low ($< 1 \cdot 5\,emu g^{-1}$) and it crystallizes at 268°C. Ultrafine powders of chromium oxide (Cr_2O_3) and manganese oxide (Mn_2O_3) have also been prepared at room temperature by the sonochemical reduction of an aqueous solution containing ammonium dichromate [$(NH_4)_2Cr_2O_7$] and potassium permanganate ($KMnO_4$), respectively. The powders are nanosized (50–200 nm), and the surface

(a)

(b)

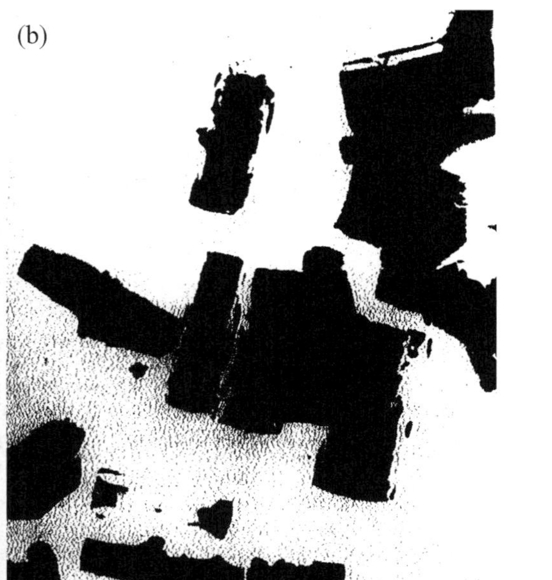

Figure 4
TEM image of the as-prepared material. (a) Magnification 13000, scale: 1.04 cm = 800 nm. (b) Magnification 36000, scale: 1.1 cm = 300 nm.

area varies from 35 to 48 $m^2 g^{-1}$. The amorphous Mn_2O_3 and Cr_2O_3 powders were converted into their crystalline form by annealing them at 600 and 900 K, for 4 h, respectively. Ultrasound irradiation of a slurry of molybdenum hexacarbonyl, $Mo(CO)_6$, in decalin for 3 h under ambient air yields blue-colored $Mo_2O_5 \cdot 2H_2O$. Infrared (FT-IR) analysis reveals that the Mo ions possess molybdenyl bonds (Mo=O) and Mo–O character and also shows the presence of hydrogen-bonded, as well as coordinated, water molecules (Dhas *et al.* 1997).

The mixed transition-metal oxides (ferrites), $NiFe_2O_4$ and $CoFe_2O_4$, were prepared by the sonication of the corresponding carbonyls under air (Shafi *et al.* 1997, 1998). Of special interest is the $BaFe_{12}O_{19}$, which was prepared sonochemically and resulted in the first observation of features such as the Olympic Nanorings (Fig. 3) on transmission electron micrographs (Shafi *et al.* 1999). The barium hexaferrite was formed as a colloidal solution without the use of a surfactant. Observation of rings of smaller dimensions trapped inside the larger ones was also unique.

3.5 *Sulfides*

The sonochemical synthesis of MoS_2 from molybdenum hexacarbonyl and sulfur in isodurene (1, 2, 3, 5-tetramethylbenzene) was reported. The amorphous particles were converted to the crystalline state, hexagonal MoS_2, by heating. The crystalline size was 1.6 nm, while the agglomerated clusters of spherical particles were 15 nm in diameter. The particles were found to have a catalytic activity superior to commercial ReS_2, RuS_2, and MoS_2.

Sonochemistry has also been used for the preparation of cadmium sulfide (CdS) and zinc sulfide (ZnS).

3.6 *Sonohydrolysis of Group 3A Ions*

The sonochemical reaction of an aqueous solution of $GaCl_3$ led to the formation of gallium oxide hydroxide rolled up in a scroll-like layered structure (Fig. 4) to give cylinders of 80–120 nm diameter and 500–600 nm in length (Avivi *et al.* 1999). Small amounts of metallic gallium were incorporated with these tubes.

When an aqueous solution of $InCl_3$ was sonicated, at room temperature or at 0 °C, $In(OH)_3$ was obtained as the sole product. At these temperatures, the non-sonicated hydrolysis does not occur. The proposed mechanism, as in the sonication of $GaCl_3$, is based on the sono-hydrolysis of the indium(III) ions that takes place in the outer ring (the liquid shell) of the collapsing bubble. TEM pictures show bundles of rod-like structures whose width is approximately 10–15 nm and whose length varies from 60 to 150 nm. Heating to 271 °C transferred the $In(OH)_3$ to In_2O_3 with the loss of water.

Bibliography

Avivi S, Mastai Y, Hodes G, Gedanken A 1999 The sonochemical hydrolysis of Ga^{+3} ions: Synthesis of scroll-like cylindrical nanoparticles of gallium oxide hydroxide. *J. Am. Chem. Soc.* **121**, 4196–9

Barber B P, Putterman S J 1991 Observation of synchronous picosecond sonoluminescence. *Nature* **352**, 318–20

Cao X, Prozorov R, Koltypin Yu, Kataby G, Gedanken A 1997 *J. Mater. Research* **12**, 402

Dhas N A, Gedanken A 1997 *Chem. Mater.* **9**, 3144

Dhas N A, Gedanken A 1998 A sonochemical approach to the surface synthesis of cadmium sulfide nanoparticles on submicron silica. *Appl. Phys. Lett.* **72**, 2514–6

Dhas N A, Raj C P, Gedanken A 1998 Synthesis, characterization, and properties of metallic copper nanoparticles. *Chem. Mater.* **10**, 1446–52

Grieser F 1997 Sonochemistry in colloidal systems. *Stud. Surf. Sci. Catal.* **103**, 57–77

Koltypin Y, Katabi G, Cao X, Prozorov R, Gedanken A 1996 Sonochemical preparation of amorphous nickel. *J. Non-Cryst. Solids* **201**, 159–62

Margulis M A 1995 *Sonochemistry and Cavitation*. Gordon and Breach, New York

Nagata Y, Maeda Y 1997 Sonochemical formation of noble metal particles in aqueous solution. *Rec. Res. Dev. Pure Appl. Chem.* **1**, 73–83

Price G J 1992 *Current Trends in Sonochemistry*. Royal Society of Chemistry, Cambridge

Reisse J, Francois H, Vandercammen J, Fabre O, Kirsch-de Mesmaeker A, Maerschalk C, Delplancke J-L 1994 Sonoelectrochemistry in aqueous electrolyte: a new type of sonoelectroreactor. *Electrochim. Acta.* **39**, 37–9

Salkar R A, Jeevanandam P, Anina S T, Koltypin Yu, Palchik O, Gedanken A 2000 *J. Phys. Chem. B.* **104**, 893

Shafi K V P M, Felner I, Mastai Y, Gedanken A 1999 Olympic ring formation from newly prepared barium hexaferrite nanoparticle suspension. *J. Phys. Chem. B* **103**, 3358–60

Shafi K V P M, Gedanken A, Goldfarb R B, Felner I 1997 Sonochemical preparation of nanosized amorphous Fe–Ni alloys. *J. Appl. Phys.* **81**, 6901–5

Shafi K V P M, Gedanken A, Prozorov R 1998 Surfactant-assisted self-organization of cobalt nanoparticles in a magnetic fluid. *Adv. Mater.* **10**, 590–2

Shafi K V P M, Gedanken A, Prozorov R, Balogh J 1998 Sonochemical preparation and size-dependent properties of nanostructured $CoFe_2O_4$ particles. *Chem. Mater.* **10**, 3445–50

Shafi K V P M, Koltypin Y, Gedanken A, Prozorov R, Balogh J, Lendvai J, Felner I 1997 Sonochemical preparation of nanosized amorphous $NiFe_2O_4$ particles. *J. Phys. Chem.* **B 101**, 6409–14

Suslick K S, Choe S-B, Cichowlas A A, Grinstaff M W 1991 Sonochemical synthesis of amorphous iron. *Nature* **353**, 414–6

Suslick K S, Fang M M, Hyeon T 1996 Sonochemical synthesis of iron colloids. *J. Am. Chem. Soc.* **118**, 11960–61

Suslick K S, Fang M, Hyeon T, Cichowlas A A 1994 Nanostructured Fe–Co catalysts generated by ultrasound in molecularily designed nanostructered materials. In: Gonsalves

K E (ed.) *MRS Symp. Proc*, Vol. 351. Materials Research Society, Pittsburgh, PA, pp. 443–8

Suslick K S, Hyeon T, Fang M 1996 Nanostructured materials generated by high-intensity utrasound: sonochemical synthesis and catalytic studies. *Chem Mater.* **8**, 2172–9

Suslick K S, Hyeon T, Fang M, Cichowlas A A 1994 Sonochemical synthesis and catalytic properties of nanostructured molybdenum carbide in molecularily designed nanostructered materials. In: Gonsalves K E (ed.) *MRS Sympos. Proc.*, Vol. 351. Materials Research Society, Pittsburgh, PA, pp. 201–6

Suslick S K 1988 *Ultrasound: Its Chemical, Physical and Biological Effects*. VCH, Weinheim, Germany, Chap. 4

A. Gedanken

V

Vacuum Melting

Technological developments during World War II demonstrated the feasibility for products such as gas turbine engines and nuclear reactors; after the war commercial and military use of these products created a large demand. The combination of high temperature, stress, and oxidative operational environments for these products was much more severe than what was required for other products in the past. Designers and metallurgists used refractory metals (alloys of Ti, Nb, Mo, and W) and nickel base superalloys to function in these hostile environments. In the liquid state these materials are very reactive with oxygen, nitrogen, carbon, and in many cases hydrogen. During solidification, the reaction products precipitate as ceramic and/or intermetallic particles which act as nucleation sites for fatigue cracks. Early work showed that melting these materials in a vacuum or controlled atmosphere produced the improved fracture toughness and resistance to fatigue that is required in the operational environment. The equilibrium thermodynamics of these vacuum melting systems is depicted schematically in an Ellingham diagram (Richardson 1974, vol. 2). Up until the late 1940s vacuum melting processes which included vacuum induction melting (VIM), vacuum arc remelting (VAR), and electron beam melting (EBM) were laboratory scale units used to produce small research quantities of materials. During the 1950s and 1960s the commercial demand for these vacuum-melted materials increased from 10^5 kg to over 2×10^8 kg yr^{-1} (Nisbet 1967). This tremendous growth rate fueled innovations in furnace design, operation, and process control to create the mature vacuum melting industry of today. Vacuum-melt processing is critical to the existence of many of the advanced alloy systems that are used today in the aerospace, industrial, and energy industries. Vacuum melting is what makes possible the design and use of alloys with elements that are highly reactive when molten and exposed to the atmosphere.

1. *Vacuum Induction Melting*

Vacuum induction melting is used to melt and cast superalloys and high-strength steels which require vacuum processing because they contain refractory and reactive elements such as Nb, Ti, Al, etc. Furnaces used in casting operations are primarily "melt and pour" in a vacuum or controlled atmosphere. In this case the charge material chemistry is the same chemistry as that desired for the casting. Casting furnace pour weights range from 1 kg up to 500 kg with the majority of units rated in the 50 kg range. Furnaces used in alloy production are designed with capabilities to manipulate the thermodynamics and chemistry so that desirable elements are retained in the liquid metal at the level required by the recipe and undesirable elements (e.g., O_2, N_2) are removed from the liquid metal. Nitrogen is removed by bulk evaporation and oxygen is removed by carbon in a carbon boil creating CO gas with both gases being removed through the vacuum pumps. Thus, both carbon and oxygen are manipulated in a dependent fashion. The primary strategy during VIM is to get the bath in a "clean state" before the reactive metal additions are made. Many melters also cover the bath with a CaO slag to reduce sulfur.

Pour weights for furnaces used to produce alloy range from 25 kg to 27000 kg with the majority of units rated at $\sim$7000 kg. The first known large commercial furnaces (3600 kg) were installed in Germany in 1923. The vacuum system was only capable of achieving a furnace pressure of $\sim$800 Pa, whereas most modern vacuum induction furnaces are capable of achieving ultimate pressures less than 1 Pa. A 10 kg unit, the first commercially installed unit for casting turbine blades at Special Metals Corp., New Hartford, NY in 1952, is now designated an ASM historical landmark. All VIM furnaces utilize a ceramic crucible to contain the metal, and heating is accomplished by suscepting an a.c. magnetic field generated by a water-cooled coil which surrounds the crucible. The susceptor can either be a hollow graphite cylinder surrounding the melting crucible or the metal charge. In both systems the magnetic field generates eddy currents within the susceptor which create internal resistance heating. The depth of penetration of the eddy currents is inversely proportional to the frequency in the coil. Greater stirring of the molten metal is achieved at larger penetration depths. Thus, the frequency of the power supply must be tuned to the furnace geometry. In addition, the coil must also be carefully insulated to prevent voltage breakdown between coil turns in the vacuum ambient. The types of metal alloys that can be melted in VIM are determined by compatibility of melt and ceramic. Crucibles for superalloys are usually made from mixtures of $MgO–Al_2O_3$. Pour systems such as launders and tundishes are sometimes used to transfer the alloy from the crucible to the molds.

Also ceramic foam filters are sometimes incorporated into the transfer system to reduce turbulance and to remove oxide particles. Filters have become particularly important in investment casting systems (Mancuso *et al.* 1988). Refractory (oxide) crucibles are prone to react with the molten metal. For instance, molten aluminum can reduce the MgO in the crucible lining to form aluminum oxide and magnesium. Crucible reactions are slowed down once a reaction layer is formed at the molten metal/crucible interface (Sutton and Maurer 1979).

The product from alloy production furnaces are usually long cylindrical ingots cast in iron or steel molds. These ingots range in diameter from 0.1 m to 1 m. Since ingots are usually very long (>2 m) compared to the diameter, solidification is almost horizontal from the mold wall into the center. This condition results in entrapped solidification shrinkage or pipe cavity, and for this reason these ingots are primarily used as electrodes for subsequent VAR or electroslag remelting (ESR). A photograph of a 1.4×10^4 kg capacity alloy production furnace is shown in Fig. 1. The primary advantage of VIM is the capability of producing a precision chemistry in a vacuum environment and excellent mixing and the shortfalls include the fact that the melting and processing take place in a ceramic crucible and the difficulty in feeding solidification shrinkage.

A variant of VIM called induction skull melting (ISM) is finding increasing use in Ti and Zr casting furnaces. At present the maximum capacity of these furnaces is ~ 50 kg of Ti. The crucible is composed of many vertical water-cooled Cu segments, each separated by a thin slots (Rishel *et al.* 1999). This arrangement minimizes induced currents in the Cu segments, thus allowing the induction field to pass through the segments and couple to the metal charge. As the name implies a thin skull of the metal being melted forms on the crucible wall and so the melt is contained in metal of the same chemistry. The same skull can be used for many melts as long as the melt chemistry remains the same. A schematic of the ISM crucible is shown in Fig. 2. Advantages of this process include intense induction stirring, the ability to melt scrap and revert, melt in a vacuum or controlled atmosphere, and the absence of ceramics. Disadvantages include the requirement of a much larger power supply than VIM, low melt weight, a limited melt superheat, and skull removal every time the melt chemistry is changed.

2. *Electron Beam Melting*

Up to the 1960s electron beam melting was primarily used to consolidate Ta, Nb, Mo, and W by drip melting into water-cooled Cu crucibles. In some instances the resulting ingots were subsequently vacuum arc remelted. These materials were electron beam melted because they are embrittled by O, N, and in some cases H, they have very high melting points, and there are no ceramics present in the electron beam furnace. Refractory metals were the first candidates for EBM, because the furnace chamber operating pressures were quite low, ~ 0.133 Pa, and the ambient furnace conditions could be tolerated by the available electron gun technology. Electron guns must operate at voltage, power, and pressure levels below the conditions for high voltage breakdown or arc down.

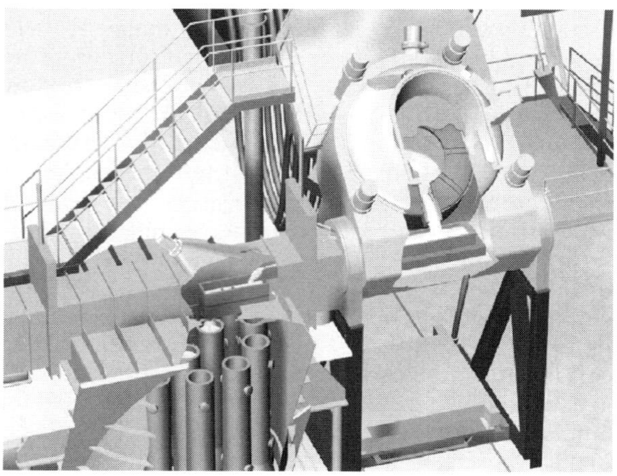

Figure 1
Cutaway of a production 1.4×10^4 kg VIM furnace (reproduced by permission of ALD Vacuum Technologies).

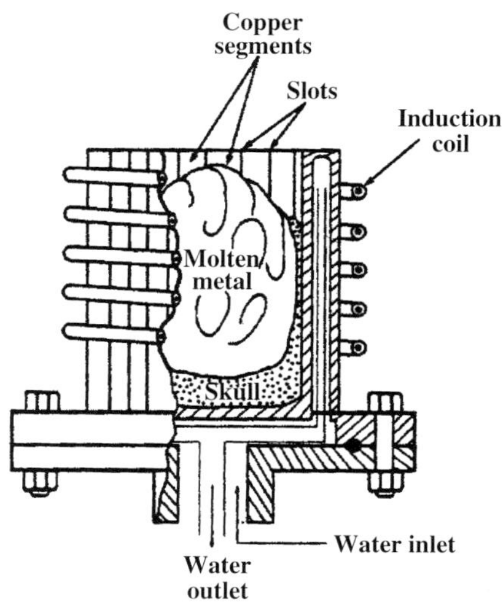

Figure 2
Schematic of an induction skull melting crucible and molten charge (Chronister *et al.* 1986).

From a practical standpoint this means that the pressure in the emitting region of the gun must be as low as possible (10^{-2}–10^{-3} Pa) and decoupled from the ambient conditions in the melt chamber. Arc downs lower productivity because of shortened cathode life, furnace down time, and interrupted solidification. During the 1960s new gun and power supply designs were developed which enabled the gun to tolerate higher-pressure ambient conditions ($\sim$6.7 Pa) that are common during melting of Ni and Ti alloys while in many instances operating at power levels greater than 10^3 kW. Major innovations included rapid gun shutdown and automatic restart, pumping of the gun chamber with separate vacuum pumps, and blocking of ambient gas diffusion from the melt chamber into the gun cavity by bleeding gas into the focusing region of the gun. Electron beams can be easily guided with a magnetic lens system located at the bottom of the gun. More details involving electron gun design and operation can be found in Schiller *et al.* (1982).

Building on these innovations, furnaces to process Ti alloys were designed for continuous introduction of scrap, sponge, and turnings. The molten metal is processed in long, shallow water-cooled copper hearths before casting into round or slab molds by a process similar to continuous casting (Entrekin and Harker 1984). A skull forms on the surface of the water-cooled hearth and becomes the container for the molten metal. An example of such a furnace is illustrated in Fig. 3. Advantages of this processing include the removal of high-density inclusions of WC, originating from machining tools, by sinking to the bottom of the hearth, decomposition and/or capture of hard α-particles (TiN) because of the long dwell time in the liquid, and the ability to recycle Ti scrap. At the present time rotor-grade certification requires that electron beam melted ingots be subsequently vacuum arc remelted before the material can be used in rotating parts.

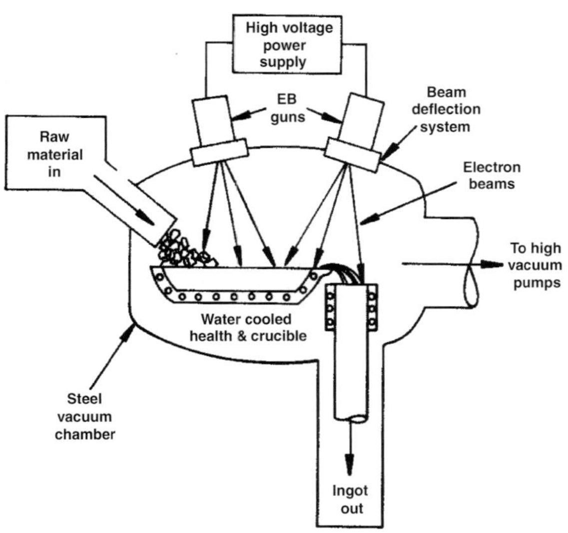

Figure 3
Schematic representation of the electron beam cold hearth refining process (courtesy of Timet Corp.).

One of the biggest, yet unexploited, advantages of electron beam melting is the potential to precisely control solidification. Since the electron beam can be accurately focused and deflected much faster than the thermal diffusion speed in the ingot, it should be possible to control, in space and time, the thermal environment in the solidification region. Online coupling of this control capability with numerical simulation could provide the potential to use EBM as the final melt process and the potential to process superalloys that are subject to severe macrosegration. Disadvantages of EBM include evaporation of high vapor pressure elements, high capital cost, difficulty in controlling melt rate, the fact that the skull must be changed when melt chemistry is changed, and limitation in melt chamber pressure.

3. *Vacuum Arc Remelting*

Vacuum arc remelting is a widely applied vacuum melting process used to control the solidification of segregation sensitive alloys. It is most commonly the final liquid metal processing step before forging. The first furnace, resembling furnaces in operation today, was built by vonBolten in 1903 (Noesen 1967). Ingots produced by air melting, ESR, VIM, and EBM are utilized as electrodes in the VAR process. Materials melted by VAR include high-strength steels, Ni base alloys, Ti base alloys, refractory metals, and U alloys. Premium quality material, called triple melt, processed by VIM/ESR/VAR, is used to produce rotating disks for gas turbine engines. Ingots produced by VAR range in size from 0.05 m to 1 m in diameter and weigh from 1 kg to 1.8×10^4 kg. VAR is carried out in furnaces similar to the illustration in Fig. 4. In this unit the electrode is suspended from a water-cooled Cu ram into a long cylindrical water-cooled Cu crucible. The entire assembly containing the electrode and crucible is evacuated and the electrode and crucible are connected to the negative and positive bus bars, respectively, of a large d.c. power supply. Melting commences after a metal vapor arc is started between the electrode and crucible bottom. Molten metal drips from the electrode into a molten pool contained by the crucible, and at steady state, the average solidification rate is equal to the melt rate. Melting currents range from 0.5 kA to 35 kA, voltages from 18 V to 40 V, and melting rates from 12 g s^{-1} to 250 g s^{-1}. Solidification starts at the crucible wall and proceeds inward with upward curvature so that the columnar dendrites are almost vertical at the ingot centerline. When the thermal

Figure 4
Photograph (a) and cutaway (b) of a production VAR furnace (courtesy of Consarc Corp.).

gradient in the center region is low (caused by a high melt rate creating a deep pool), equiaxed grains will be found in the center region of the ingot. There is a general consensus in the industry that ingots with complete columnar structure yield the highest-quality material. Since molten metal is added to the pool atop the ingot as solidification is proceeding upward, the solidification shrinkage is constantly being fed resulting in a solid ingot.

The arc is the heart of the VAR process. It provides the electrical conduction path which couples the electrode and molten pool, and provides the heat for melting. The arc is sustained by ionized metal vapors evolved by evaporation from the electrode (cathode) and molten pool (anode). Ingot quality is controlled by how uniformly the arc injects current into the molten pool. Control variables melt rate and electrode gap are manipulated to control the arc and provide a steady heat source. Arc behavior is optimized when the arc behaves as a macro-uniform heat source as viewed from within the solidifying region. This optimized arc behavior is coined a diffuse arc. Under some furnace environments, the arc plasma can become constricted and the conditions of macro-uniform heating no longer apply.

From breaker physics work on "vacuum or metal vapor arcs," it is known that this type of arc simultaneously consists of many plasma columns anchored by cathode spots which in effect are electrical parallel circuits (Zanner 1979). The number of columns at any instant is a function of the thermal properties of the cathode. For example, for materials composed of iron or nickel each cathode spot is allocated ~50 A and ~300 A for tungsten. Thus, for a 5 kA Ni or Fe arc one could expect ~100 cathode spots burning simultaneously, or ~100 parallel circuits wherein current is transferred from the cathode to the anode. The cathode spots are highly mobile and travel radially at velocities of $\sim 10\,\mathrm{m\,s^{-1}}$ from the central region of the cathode to the outside edge and up the side wall of the electrode before they are extinguished. The cathode spots in multiple cathode spot arcs appear to have a random motion because of the interaction of magnetic fields between spots. Since the spots move much faster than the thermal diffusion speed in the electrode materials, the arc appears to the electrode face and molten pool as a macro-uniform heat source. Metal transfer (in the form of drips which create shorted columns) from the cathode to the anode must also be considered regarding the uniformity of the arc heat source in a VAR furnace. During the short, current can increase to very high levels depending on the inductance of the system and the lifetime of the short. Normal shorts have lifetimes of from 0.1 ms to 5 ms and frequencies of ~5–20 Hz. It is speculated that these events are damped on the timescales of the fluid motion in the pool, $\sim 5\text{–}10\,\mathrm{cm\,s^{-1}}$. The frequency of drip short formation is inversely proportional to electrode gap length and this relationship is used to dynamically control the gap. Under optimal conditions of a diffuse arc the current is uniformly distributed across the electrode faces creating a constant melt rate and steady flows in the molten pool atop the ingot. The light emitted from the electrode and crucible (annulus region) appears to be of constant flickering (flickering is due to the drip shorts) intensity with the plasma background appearing about the same as the sky on a sunny, but hazy day, as viewed from a monitor in the control room. When the arc is in the diffuse mode, there is no plasma rotation or momentary bright areas in the annulus. Voltage, current, melt rate, and pressure traces appear to be steady over time periods of hours. The background furnace pressure can be less than 1 μm, but the pressure inside the plasma column is speculated to be much higher. Diffuse arcs are stabilized by arc gaps less than 12 mm, low furnace pressure, and a clean molten pool surface (no slag). Diffuse arcs provide optimum material quality and it is very important to maintain a diffuse arc during melting.

A constricted arc is just what the name implies, a constricted plasma column. Constricted arc mode is the most dangerous with respect to ingot quality. When the arc is burning in this mode, current is injected into the molten pool in concentrated fashion by the constricted plasma column. The plasma column usually rotates in a clockwise fashion taking tens of seconds to make a revolution as observed from the control room monitor. The nonuniform current injection disturbs the steady fluid flows within the molten pool and allows the creation of shelf at the ingot/crucible boundary. The easiest way to detect a constriction is by visual appearance on the control room monitor. The light emitted from the annulus region will appear to periodically change intensity going from bright to dark. When the annulus region appears dark, one can sometimes observe a shelf growing in from the crucible wall which will disappear when the region appears bright (the constriction melts away the shelf). In less severe forms of constriction, the arc can exist simultaneously in a diffuse/constricted mode with a portion of the total melting current allocated to the constricted column. At the same time pressure spikes will sometimes be present and the voltage may periodically decrease by 0.5–1.5 V. A constricted arc can cause large variability in the drop short rate and a depression in the melt rate if it occurs over periods of tens of minutes. The primary causes for a constricted arc are slag on the pool surface, the presence of a molecular gas such as CO or N_2, or the evolution of volatile elements from a poorly conditioned electrode surface. The propensity to create an ingot defect with a constricted arc varies with ingot size, the periodicity of the constriction (time on versus time off), the magnitude of current allocated to the column, and whether melt back occurs within the pool. Conditions commonly labeled as "glows" are usually severe arc constrictions. Channel segregates (freckles) and dirty white spots can be caused by constricted arcs. At the present time the melting community is undertaking a large research effort to develop multiple variable control systems based on numerical simulation to control pool shapes and arc behavior.

Advantages of the VAR process include a ceramic free system, control of solidification shrinkage, ability to consistently produce optimal solidification structures in superalloy ingots as large as 1 m in diameter, and melting in a vacuum environment. Disadvantages include the sensitivity of solidification structure to arc behavior and the potential for a hydrogen explosion if a water leak develops while melting Ti.

A variant of the VAR process called skull melting is widely used for casting Ti and other reactive metals. In this case the water-cooled Cu crucible has a tilt capability for pouring castings. The electrode is melted at a very high rate, and after the crucible is filled with metal it is immediately tilted to pour the metal into a mold. Advantages of this process include the absence of ceramics, the ability to melt large amounts of metal ($\sim$1000 kg Ti), and melting in a vacuum ambient. Disadvantages include very low superheat and the need for a fabricated electrode, a large power supply (some units require 80 kA), and the danger of a hydrogen explosion while melting Ti.

4. Summary

Selection of the vacuum melting process used to produce an alloy is based on the reactivity of the alloy, the melting temperature, the size of the ingot required to make the hardware, the cost of the production operation, and the quality requirements needed for the hardware to perform in the operating environment. Vacuum melting and solidification requires tightly controlled processes to produce consistent macrostructures. For example, as mentioned above triple melt is the default process scheme to produce superalloy ingots for gas turbine rotating hardware. In the triple melt process VIM is used to produce the alloy chemistry and microcleanliness, ESR is used to produce a low sulfur consolidated electrode, and VAR is used to produce the desired solidification structure. New technology can have a dramatic effect on process substitution. Improvements in control schemes for EBM could make it possible to meet the triple melt requirements with a single electron beam melt. There is a constant struggle to reduce cost and improve quality, and this produces evolutionary improvements which ripple through the entire process stream from start to finished product creating, for example, a positive effect on the energy and transportation sectors of the world economy.

Bibliography

Bertram L A, Zanner F J 1981 Interaction between computational modeling and experiments for vacuum arc remelting. In: Brody H D, Apelian D (eds.) *Proc. Modeling of Welding and Casting Processes*. AIME, Warrendale, PA, pp. 333–49

Bunshah R F 1962 *The History of Electron Beam Technology*. Wiley, NY

Chronister D J, Scott S W, Stickle D R, Eylon D, Froes F H 1986 Vacuum melting. *JOM* **9**, 51–4

Entrekin C H, Harker H R 1984 Electron beam cast titanium slab. In: Bhat G K, Lherbier L W (eds.) *Proc. Vacuum Metallurgy Conf.* AVS, NY, pp. 45–8

Mancuso S O, Sczerzenie F E, Maurer G E 1988 The investigation of minor element additions on oxide filtering and cleanliness of a nickel based superalloy. In: Duhl D N, Maurer G, Antolovich S, Lund C, Reichman S (eds.) *Proc. Superalloys* 1988. AIME, Warrendale, PA, pp. 377–86

Nisbet J D 1967 Progress of vacuum metallurgy in North America. In: Foster E (ed.) *Trans. Int. Vacuum Metall. Conf.* AVS, NY, p. 17

Noesen S J 1967 Vacuum arc melting from Robert Hare—1839 to the present. In: Foster E (ed.) *Trans. Int. Vacuum Metall. Conf.* AVS, NY, pp. 503–50

Richardson F D 1974 *Physical Chemistry of Melts in Metallurgy*. Academic Press, NY, Vols. 1–2

Rishel L L, Pollock T M, Cramb A W 1999 Induction skull melting of titanium aluminides. In: Mitchell A, Ridgeway L, Baldwin M (eds.) *Proc. 1999 Int. Symp. on Liquid Metal Processing and Casting*. Santa Fe, NM, pp. 287–99

Schiller S, Heisig U, Panzer S 1982 *Electron Beam Technology*. Wiley, NY

Schlatter R 1972 Melting and refining technology of high-temperature steels and superalloys In: *Proc. 2nd Superalloy Conference*. AIME, MCIC-72-10. pp. A1–40

Smith H R 1967 Electron beam processing in vacuum metallurgy. In: Foster E (ed.) *Trans. Int. Vacuum Metallurgy Conference*. AVS, NY, pp. 719–37

Sutton W H, Maurer G E 1979 Influence of VIM crucible composition on the non-metallic content of an advanced hafnium bearing nickel-base superalloy. In: Bhat Slatter G K (eds.) *Proc. 6th International Vacuum Metallurgy Conference on Special Melting*. AVS, SanDiego, CA, pp. 340–56
Winkler O 1960 The theory and practice of vacuum melting. *Met. Rev.* **5**(17), 1–117
Zanner F J 1979 Metal transfer during vacuum arc remelting. *Metall. Trans.* **10B**, 133–42
Zanner F J, Bertram L A 1985 Vacuum arc remelting—an overview In: *Proc. 8th International Conference of Vacuum Metallurgy*. AVS, Linz, Austria, Vol. 1, pp. 512–52

F. J. Zanner

Viscous Sintering

Viscous sintering is a process by which a noncrystalline material reduces its surface area, and thereby reduces its free energy. The process is illustrated in Fig. 1, which shows spherical particles of glass coalescing, just as raindrops on a window merge to reduce their surface area. In contrast to crystalline materials, where sintering involves diffusion of vacancies in the crystal lattice, noncrystalline materials sinter by viscous flow. The rate of viscous sintering is found to be insensitive to the details of the microstructure of the material, so the rate of densification can be estimated given only the viscosity and surface tension of the material and its average pore size. In practice, materials are often constrained during sintering; for example, the densification rate of a film is reduced by its attachment to a substrate that does not shrink as the film does. The sintering rate can be calculated in such cases if the mechanical properties of the porous material are known. An important type of constraint is crystallinity, because crystalline inclusions sinter more slowly than glass, so they retard the overall shrinkage; therefore, it is important to consider how much crystallization can occur during the time required for viscous sintering.

1. Sintering Models

The theoretical basis for viscous sintering was established by Frenkel (1945), who argued that the energy needed to drive viscous flow is provided by reducing the surface area of the body; therefore, the rate of sintering can be found by equating those two energies. To calculate the energy that is dissipated by viscous flow as a certain amount of surface area is eliminated, it is necessary to choose a particular microstructural model. Frenkel analyzed the coalescence of spheres, as in Fig. 1. That is a surprisingly difficult problem, because the region of contact between particles has a complicated shape, so Frenkel had to make severe simplifying assumptions about the pattern of viscous flow. The distance, L, between the centers of a pair of sintering spheres was found to decrease linearly with time, t, according to

$$\frac{L(0)-L(t)}{L(0)}=\frac{3\gamma t}{8\eta a} \tag{1}$$

where γ is the surface tension, η is the viscosity, and a is the initial radius of the spheres. This linear shrinkage rate has been experimentally verified. A detailed numerical simulation of the flow within sintering spheres was performed by Jagota and Dawson (1990). They showed that Frenkel's assumptions are poor, but his result is rather close to the exact solution, evidently as a result of offsetting errors.

The next major step in the analysis of viscous sintering was by Mackenzie and Shuttleworth (MS) (1949), who calculated the rate of shrinkage of a spherical shell. This is an excellent model for the final stages of sintering, when the pores become isolated,

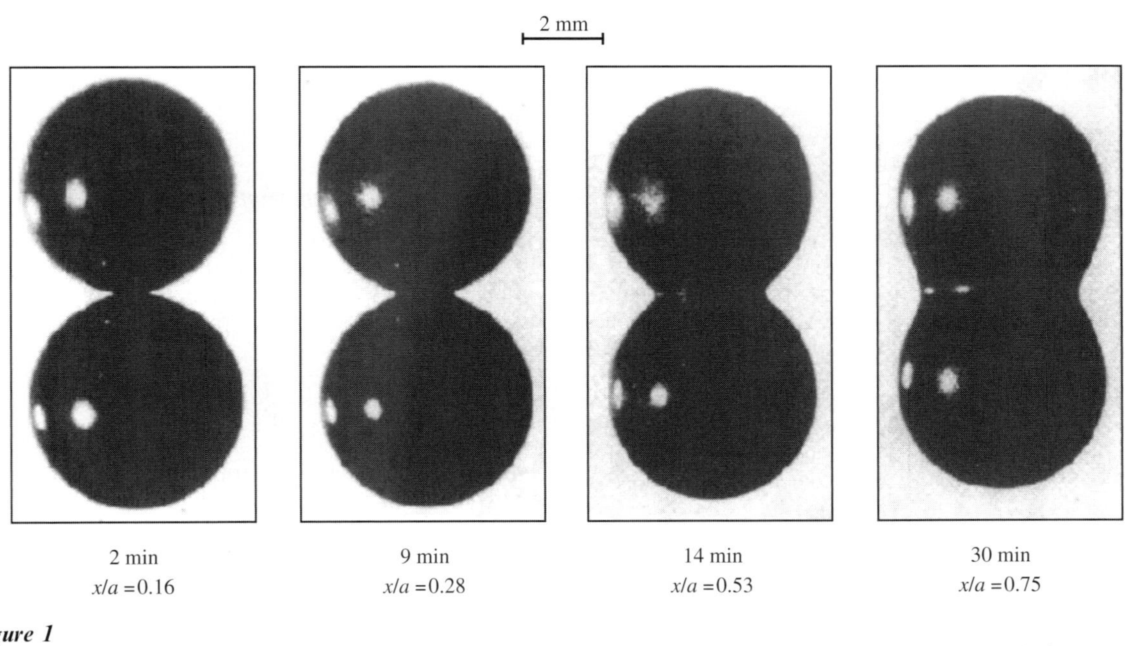

Figure 1
Coalescence of spherical particles of glass (reproduced by permission of Freund Publishing House Ltd. from *Rev. Powder Metall. Phys. Ceram.*, 1979, **1**, 1–4).

and it is relatively easy mathematically because the spherical shape is preserved as the shell contracts. They found that the volume fraction of solids (also called the relative density), ρ, varies with time as

$$\frac{\gamma n^{1/3} t}{\eta} = f(\rho(t)) - f(\rho(0)) \quad (2)$$

where

$$f(\rho) = -\frac{2}{3}\left(\frac{3}{4\pi}\right)^{1/3}\left(\sqrt{3}\tan^{-1}\left(\frac{2(1/\rho-1)^{1/3}-1}{\sqrt{3}}\right) + \frac{3}{2}\log[(1-\rho)^{1/3}+\rho^{1/3}]\right) \quad (3)$$

The dimensionless time axis is $\gamma n^{1/3} t/\eta$, where n is the number of pores per unit volume of solid. In this case, n is 1 divided by the volume of solid in the shell (which is independent of time). The shape of the densification curve is indicated in Fig. 2.

To compare this result with Frenkel's, consider a simple cubic packing of spheres where each unit cell contains one particle (1/8 of each of the 8 particles at the corners of the cell) and has one pore in the center. For that structure, there is one pore per particle, so $n = 1/(4\pi a^3/3)$, and since $L(t)/L(0) = \rho(0)^{1/3}/\rho(t)^{1/3}$ Eqn. (1) can be written as

$$\frac{\gamma n^{1/3} t}{\eta} = \frac{8}{3}\left(\frac{3}{4\pi}\right)^{1/3}\left[1 - \left(\frac{\rho(0)}{\rho(t)}\right)^{1/3}\right] \quad (4)$$

In view of the geometrical assumptions made by Frenkel, Eqn. (4) should only apply to the first 5–10% of linear shrinkage, while the particles are nearly spherical. For the simple cubic packing, $\rho(0) = 0.52$, so the range of applicability would include densities up to ~0.65. As indicated by the dashed curve in Fig. 2, Eqn. (4) agrees well with the MS model, Eqn. (2), for $0.52 \leqslant \rho \leqslant 0.7$, so these models predict almost identical sintering kinetics, even though they are based on completely different geometries.

Frenkel's model is expected to apply only at the beginning of sintering, whereas the MS model only applies near the end of sintering, after the pores have become isolated bubbles. For bodies with low initial densities, such as gels or vapor-deposited optical waveguide preforms, there is a large increase in ρ that is not realistically described by either model. For such materials, Scherer (1977, 1991) introduced a model where the solid phase is represented by cylindrical rods joined at nodes to create a network. The rate of densification for a network of any shape (e.g., cubic, octahedral, or tetrahedral) is found to obey the following equation:

$$\frac{1}{\rho}\frac{d\rho}{dt} = \frac{\gamma S \rho_s}{2\eta} \quad (5)$$

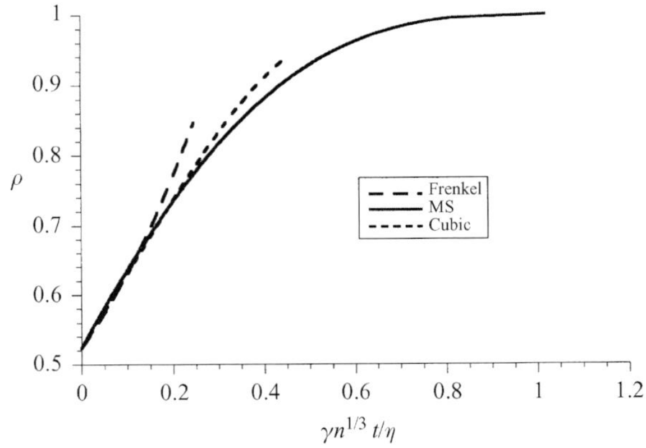

Figure 2
Relative density ρ vs. time for spheres (Frenkel), spherical shell (M–S), and cubic network (Cubic), calculated assuming an initial density of $\rho(0) = \pi/6$ (= 0.52) as for a simple cubic packing of spheres. Frenkel's model is only expected to apply to the early stage of sintering (in this case, up to $\rho \leqslant 0.65$). The curve for the cubic network terminates at $\rho \approx 0.94$, where the pores become isolated; beyond that point the MS model applies.

where S is the specific surface area and ρ_s is the density of the solid phase. The predicted sintering rate for a cubic array of cylinders is indicated by the dotted curve in Fig. 2, and it is seen to agree well with the other models. Thus, the theoretical analyses predict similar sintering kinetics for widely different geometries. Experimental studies (see the review in Brinker and Scherer 1990) verify that the sintering rate of noncrystalline particles follows the curves predicted by the MS or network models.

2. *Factors Affecting Sintering Rate*

If the initial density of the body lies between $0.2 < \rho < 0.8$, which covers the range of practical interest, then Fig. 2 shows that the time needed to complete sintering, t_s, is approximately $\gamma n^{1/3} t_s/\eta \approx 1$, or

$$t_s \approx \frac{\eta}{\gamma n^{1/3}} \approx \frac{1.6\eta a}{\gamma} \quad (6)$$

where the second equality applies for the simple cubic packing of spheres. Equation (6) indicates that smaller particles sinter faster, and Eqn. (5) leads to the same conclusion, since S is inversely proportional to particle size ($S = 2/(a\rho_s)$ for cylinders and $S = 3/(a\rho_s)$ for spheres). For crystalline materials t_s is much more sensitive to particle size, varying as a^3 or a^4. The sintering rate increases (or the t_s decreases) as the surface tension increases, because the surface energy is the driving force for the process. The dependence of γ on temperature and composition is weak for liquids; its value is in the order of $0.1\,\mathrm{J\,m^{-2}}$ for organics and 0.25–$0.4\,\mathrm{J\,m^{-2}}$ for oxide liquids. In contrast, the viscosity is extremely sensitive to both temperature and composition, so the sintering rate can be drastically reduced by raising the temperature (for glass-forming silicate liquids, t_s drops an order of magnitude for an increase of $\sim 100\,^{\circ}\mathrm{C}$). The viscosity of oxides is strongly reduced by hydroxyl (OH) groups, so the sintering rate is greatly accelerated by moisture in the air. For silica the sintering rate is an order of magnitude faster in air than in a dry gas, such as helium (Brinker and Scherer 1990). The solid phase of a gel contains large amounts of hydroxyl that is gradually lost during sintering, resulting in continuously increasing viscosity and decreasing densification rate; consequently, it is advantageous to use a constant heating rate when sintering a gel, so that the effect on η of rising temperature offsets the loss of hydroxyl (Brinker and Scherer 1990).

3. *Sintering Under Constraint*

In practice, sintering often occurs under some sort of constraint. Unless the sintering is done under vacuum, gases will be trapped in the isolated pores and will resist densification; that problem was analyzed by Mackenzie and Shuttleworth (1949). Another sort of internal constraint arises when the body contains a distribution of pore sizes or particle sizes; then, as indicated by Eqn. (5), the sintering rate varies from place to place within the body. Regions that tend to contract more rapidly (i.e., locations where S is large) are subjected to restraint (tensile stress) by the surrounding regions that contract more slowly. A similar situation results when there is a gradient in composition that creates a spatial variation in viscosity. This was found to be a significant factor in the sintering of preforms for optical waveguides (Scherer 1979), where a composition gradient is deliberately created to produce total internal reflection within the fiber; that gradient also causes a strong spatial variation in sintering rate, affecting the course of densification. In a composite there may be filler particles or fibers that impede the contraction of the porous glassy matrix in which they are embedded. In addition to these internal constraints there may be external constraints, as when a sintering film is attached to a rigid substrate, or when a mechanical load is applied during sintering (a process called hot pressing).

3.1 *Constitutive Equation*

To analyze constrained sintering, it is necessary to generalize Frenkel's energy balance, by adding a term for the mechanical work done by the stresses. The result is a constitutive equation for a sintering material with the following form (Scherer 1979):

$$\dot{\varepsilon}_x = \dot{\varepsilon}_f + \frac{1}{F}\left[\sigma_x - N(\sigma_y + \sigma_z)\right] \tag{7}$$

where $\dot{\varepsilon}_x$ is the strain rate in the x direction and σ_x, σ_y, σ_z are the applied stresses in the x, y, and z directions. The free strain rate, $\dot{\varepsilon}_f$, is the linear contraction rate of the sintering body in the absence of applied stresses. Since the linear contraction rate is 1/3 of the volumetric contraction rate,

$$\dot{\varepsilon}_f = \frac{1}{3V}\frac{\mathrm{d}V}{\mathrm{d}t} = -\frac{1}{3\rho}\frac{\mathrm{d}\rho}{\mathrm{d}t} \tag{8}$$

The negative sign reflects the fact that the body contracts as its density increases. Using Eqn. (5), the free strain rate can be written as

$$\dot{\varepsilon}_f = -\frac{\gamma S \rho_s}{6\eta} \tag{9}$$

The quantity $F(\rho)$ in Eqn. (7) is the uniaxial viscosity for the porous viscous material, which is a measure of the additional strain rate resulting from the applied stress. For example, if the material is subjected to a uniaxial stress σ_x, that stress causes the strain rate in the x direction to change by σ_x/F, so that the total strain rate in that direction is

$$\dot{\varepsilon}_x = \dot{\varepsilon}_f + \frac{\sigma_x}{F} \tag{10}$$

If the stress is compressive ($\sigma_x < 0$), then the contraction rate increases (becomes more negative), whereas tensile stress opposes the free strain rate and retards the contraction. For a nonporous liquid, $F(1) = 3\eta$, but for a porous body $F(\rho) < 3\eta$. The quantity $N(\rho)$ in Eqn. (7) is Poisson's ratio for the porous body; for a nonporous liquid, $N(1) = 1/2$, but for a porous body $N(\rho) < 1/2$. A variety of functional forms that have been proposed for F and N is reviewed by Bordia and Scherer (1988) and Olevsky (1998). Most of those models predict that F increases roughly with ρ^2; however, the actual dependence is strongly influenced by the microstructure. For example, if particles are packed into a random dense arrangement, with $\rho \approx 0.63$, the viscosity F may be very small, because the particles initially make point contacts with one another. As sintering causes the contact area to increase (as illustrated in Fig. 1), the resistance to deformation increases and F rises rapidly. This effect is captured in the simulations by Jagota and Dawson (1988), who found that F was approximately proportional to the square root of the contact area between sintering spheres during the initial stage of sintering.

3.2 Practical Applications

Given an expression, such as Eqn. (9), for the free strain rate and theoretical or experimental expressions for $F(\rho)$ and $N(\rho)$, it is possible to use Eqn. (7) to analyze sintering in the presence of internal constraints (such as nondensifying inclusions or gradients in density, viscosity, or pore size) or external constraints (such as applied loads, or attachment of a sintering layer to a substrate). A thorough review of the literature on constitutive equations and numerical analyses of problems in constrained sintering is given by Olevsky (1998). Analytical solutions can be obtained for the rate of densification in some cases of practical importance (Scherer 1992). For a film on a rigid substrate, the rate is

$$\frac{1}{\rho}\frac{d\rho}{dt} = -\left(\frac{1+N}{1-N}\right)\dot{\varepsilon}_f \tag{11}$$

Comparison with Eqn. (8) reveals that the film sinters more slowly than a free body, as long as $(1+N)/(1-N)<3$, which is the case whenever $\rho<1$. For a porous body containing a small volume fraction v of rigid (i.e., nonsintering) inclusions, the viscous sintering rate is

$$\frac{1}{\rho}\frac{d\rho}{dt} = \frac{3(1-v)\dot{\varepsilon}_f}{1+\dfrac{2(1-2N)}{1+N}v} \tag{12}$$

Equation (12) is valid only if the inclusions do not interact, which means that v must be less than ~ 0.1. At higher volume fractions, the inclusions are likely to form a percolating network, so the densification rate is much more strongly retarded than is predicted by Eqn. (12). If a noncrystalline material begins to crystallize while it is sintering, the crystals constitute rigid inclusions. If the fraction crystallized is small, then the effect can be described by Eqn. (12), but if the crystallites form a continuous network then sintering is drastically slowed (Brinker and Scherer 1990).

In hot pressing, the applied load is usually so large that the effect of surface tension on viscous flow is negligible. The MS model then reduces to

$$\frac{\ln[1-\rho(t)]}{\ln[1-\rho(0)]} = \frac{3pt}{4\eta} \tag{13}$$

where p is the applied pressure. Equation (13) indicates that an infinite amount of time is required to reach full density, $\rho=1$, whereas full density is reached in a finite time when no pressure is applied. The discrepancy results from the fact that the capillary pressure in a pore with radius r is equal to $2\gamma/r$, so it becomes infinite as the pore collapses. Therefore, the capillary pressure cannot be neglected with respect to the applied pressure at the end of sintering. That is, Eqn. (13) only applies while the pores are large enough to justify the assumption that $p \gg 2\gamma/r$. Nevertheless, applied pressure can drastically reduce the time required to sinter a body, particularly if the pores are initially large.

4. Future Directions

Compared with crystalline materials, the sintering behavior of noncrystalline materials is quite simple. The rate is relatively insensitive to the details of the microstructure, so the time needed to reach full density can be estimated with reasonable accuracy given only the viscosity and the mean pore size. To predict in detail the evolution of the density and pore size distribution with time, it is necessary to know the initial state of the sample (distribution of pore size, density, and viscosity); however, such calculations are possible with the existing theory. The most challenging problem remaining in the field of viscous sintering is to develop appropriate constitutive equations for the analysis of constrained sintering. The difficulty is that the functions F and N are sensitive to the microstructure (particle shape, contact area, connectivity of voids, and so on). Moreover, the properties may be anisotropic as a result of processing of the unfired body (such as pressing of powder compacts), or may be isotropic at the start of sintering, then become anisotropic as a result of the constraint (e.g., during uniaxial pressing). Considerable progress is being made in this area through careful experimental work and detailed numerical simulations. Much of this work is described in the review by Olevsky (1998).

Bibliography

Bordia R K, Scherer G W 1988 On constrained sintering: I. Constitutive model for a sintering body. *Acta Metall.* **36**, 2393–7; On constrained sintering: II. Comparison of constitutive models. *Acta Metall.* **36**, 2399–409; On constrained sintering: III. Rigid inclusions. *Acta Metall.* **36**, 2411–16

Brinker C J, Scherer G W 1990 *Sol–Gel Science*. Academic Press, Boston, MA, Chap. 11

Exner H E 1979 Principles of single phase sintering. *Rev. Powder Metall. Phys. Ceram.* **1**, 1–251

Frenkel J 1945 Viscous flow of crystalline bodies under the action of surface tension. *J. Phys. (Moscow)* **9**, 385–91

Jagota A, Dawson P R 1988 Micromechanical modeling of powder compacts—I. Unit problems for sintering and traction induced deformation. *Acta Metall.* **36**, 2551–61

Jagota A, Dawson P R 1990 Simulation of the viscous sintering of two particles. *J. Am. Ceram. Soc.* **73**, 173–7

Mackenzie J K, Shuttleworth R 1949 Phenomenological theory of sintering. *Proc. Phys. Soc. London* **62**, 833–52

Olevsky E A 1998 Theory of sintering: from discrete to continuum. *Mater. Sci. Eng.* **R23**, 41–100

Scherer G W 1977 Sintering of low-density glasses: I. Theory. *J. Am. Ceram. Soc.* **60**, 236–9; 239–46
Scherer G W 1979 Sintering inhomogeneous glasses: application to optical waveguides. *J. Non-Cryst. Solids* **34**, 239–56
Scherer G W 1991 Cell models for viscous sintering. *J. Am. Ceram. Soc.* **74**, 1523–31
Scherer G W 1992 Constitutive models for viscous sintering. In: Mehrabadi M M (ed.) *Mechanics of Granular Materials and Powder Systems*. ASME, New York, pp. 1–18

G. W. Scherer

W

Welding

1. Introduction

Welding is the art and science of joining metals. Generally, welding includes the formation of metallurgical bonds in various kinds of fusion—welded, brazed, and soldered joints. New techniques involve frictional heating but welding excludes joints made by mechanical means or adhesives. In the fusion welding process, two metallic parts are joined by heating, causing the parts to melt and flow together (see Fig. 1). The heat source can be in the form of oxyacetylene flame, electric arc, electron beam, laser beam, resistive heating or combinations thereof. The heat source interacts with the two pieces of base material to be welded and heats the material from ambient to above "liquidus temperature," which is the temperature at which the material becomes 100% liquid (e.g., the liquidus temperature of a low-alloy steel is ~1480°C). This liquid mixes and forms a molten pool between the two solid pieces. Depending upon the application, a filler metal may be added to this molten pool. On cooling from the liquidus temperature to ambient temperature, the liquid material solidifies to form the joint between these pieces. In brazing and soldering—which are used extensively in electronic packaging—the bonding takes place by melting and solidification of the brazing or soldering alloy. No melting of the base material takes place. In this review, the recent developments and advances in fusion welding are presented. The reader is referred to handbook for descriptions of brazing, soldering, and other solid-state joining process (Welding Handbook 1991). Fusion welding processes are classified into different subsets, depending upon the heat source, the method of shielding the molten pool, and the modes of filler metal addition. The predominant heat source is an electric arc produced by either direct current (d.c.) or alternating current (a.c.) power sources. Within the arc-welding processes, the classifications are based on shielding and filler metal additions; examples include gas–tungsten arc welding (GTAW), gas–metal arc welding (GMAW), shielded-metal arc welding (SMAW), submerged arc welding (SAW), and flux-cored arc welding (FCAW). In the GTAW process, a nonconsumable tungsten electrode is used for striking an arc with the base material, and the filler metal is added separately to the molten weld pool. In contrast, in the GMAW process, the arc is struck between a consumable wire that is used as an electrode as well as filler metal. In both the GTAW and GMAW processes, the shielding of the molten pool is achieved by inert gases such as helium and argon or by inert gases in combination with carbon dioxide and oxygen for special applications. In the SAW and SMAW process, the arc is struck between a consumable wire and the base metal; however, the shielding is provided by the liquid slag that forms on top of the molten metal. Resistive heating is also used for joining the materials, by localized melting and solidification. In this process, high currents (>5000 A) are passed by two copper electrodes through overlapping materials to be joined. Due to large contact resistance at the faying interface, localized melting occurs, and as soon as the current is terminated, the liquid metal cools and solidifies to form the joint. This resistance welding process takes ~12–15 cycles of alternating current. This process is extensively used in the production of automotive steel bodies. High-energy-density heat sources, such as lasers or electron beams, are also extensively used for welding; heat is generated by the interaction of the high-energy particles (photons or electrons, respectively) with the material to be joined.

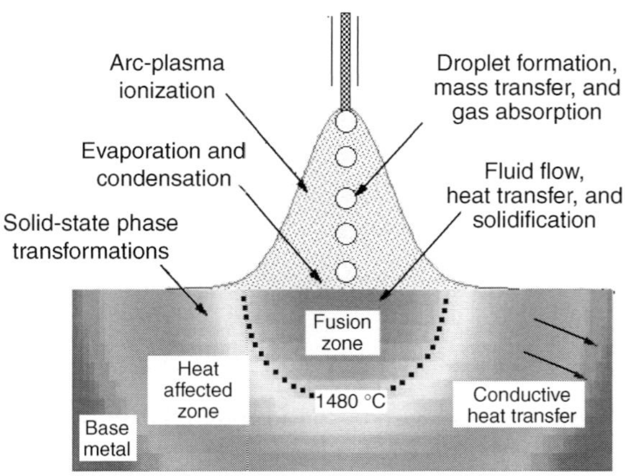

Figure 1
Schematic illustration of the interaction of a heat source with base material and different regions of the weldment, namely, the fusion zone, the heat-affected zone, and the base metal.

In addition to the selection of welding process, various engineering issues, including weld-joint

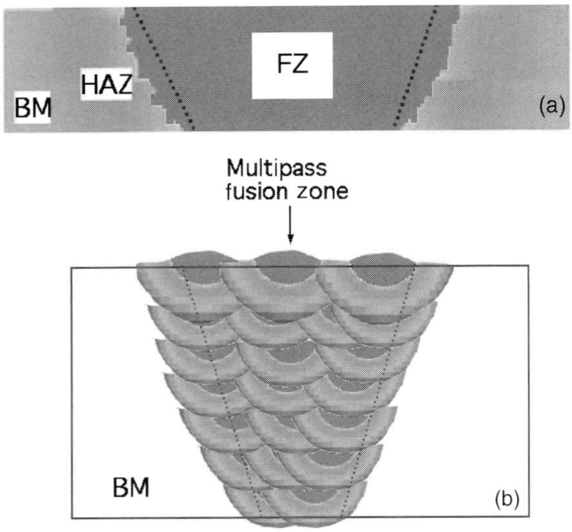

Figure 2
Schematic illustration of single (a) and multipass weld (b) different color contrasts indicate different peak temperatures. Note the presence of As-deposited fusion zone (red color) and thermally affected regions of previously deposited regions.

geometry, weld microstructure (microstructure of material is defined as the spatial arrangement of group of atoms or phases with or without homogeneous distribution of chemical elements. Microstructure of the material is usually characterized by microscopic techniques), weld properties, and welding control and automation are important in determining the quality and reliability of the welds. The weld may be made up of single (Fig. 2(a)) or multiple passes (Fig. 2(b)) of the heat source on the joint with or without the addition of filler metal, depending upon thickness of material. Another important consequence of the weld joint geometry is the development of residual stresses. The residual stresses develop because of large temperature gradients. The residual stresses affect the cracking tendency and final properties of a weld. The cracking tendency and the final properties are controlled by large microstructural gradients resulting from large temperature gradients in the weldment. The region that was melted and allowed to solidify is called the "fusion zone" (FZ). The region that was heated to a temperature between the melting point and the temperature at which noticeable metallurgical phase change occurred is called the "heat-affected zone" (HAZ). The region that is not affected at all due to welding is called the "base metal" (BM). The size of the HAZ and FZ may vary based on joint geometry, the type of welding processes, and process parameters. For example, the welded joints vary in size from small spots (5–10 mm) in automotive and electronic industries to tens of meters long for shipbuilding and aerospace applications. Parts range in thickness from a fraction of a millimeter to tens of centimeters. The time necessary for welding ranges from a fraction of a second to several hours. The weld cooling rates vary from less than 10^2–10^{6}°C s^{-1}.

A major realization in the welding science, engineering and technology is the profound complexity of the welding process and its relationship to quality and productivity in the areas of process–microstructure and microstructure–property relationships. Improving quality and productivity by incorporating cost-effective solutions into the manufacturing processes requires commitment and collaboration among the basic and applied disciplines. In the last few decades major progress has been made in

(i) understanding and modeling the physical processes in welding,

(ii) understanding microstructural evolution and the correlation between microstructure and properties of the welds, and

(iii) intelligent control and automation of welding.

The significant progress made in welding science suggests that welding processes based on scientific principles may be designed to tailor the composition, structure, and properties of a weld in a practical way. However, substantive issues and problems must be addressed and new methodologies must be developed before the dream becomes a reality.

2. *Physical Processes*

During welding, the interaction of the heat source with the material results in melting, vaporization, solidification, subsequent solid-state transformation, stresses and distortions. Since most of the fusion welding processes are fast and are completed within a few minutes, experimental measurement of all of the above processes and their interactions are very difficult. In this regard, computer modeling has been a very valuable method to evaluate the physics, chemistry, and the mechanics of the process. In modeling welding processes, phenomenological models are used to seek insight into various aspects of the processes. Artificial neural network models (ANNs) may be better suited for industrial applications.

2.1 Phenomenological Models

Phenomenological modeling of welding processes has provided unprecedented insight into understanding both the welding process and the welded materials that could not have been obtained by experiment alone. Many of the accomplishments have been greatly facilitated by the spectacular advances in the speed and affordability of computers. While several comprehensive reviews are available (David and DebRoy 1992, 1994, DebRoy and David 1995,

DebRoy 1997), it is worthwhile to mention a few important areas of interest. The simple heat conduction models traditionally used for the prediction of the weld pool geometry, the temperature fields, the cooling rates and the simple features of the solidification structure are being increasingly replaced by accurate models that account for convective heat transfer. In the weld pool, the metal undergoes vigorous recirculatory motion driven by buoyancy, electromagnetic and surface-tension forces. Buoyancy effects originate from the spatial variation of the liquid metal density, mainly because of temperature variations and to a lesser extent, from differences in local composition. Since large variations in temperature are present in the weld pool, the corresponding density gradients produce convective flows.

Electromagnetic effects result from the interaction between the divergent current path in the weld pool and the magnetic field it generates. The effect is important when a large electric current passes through the weld pool. The spatial variation of the surface tension owing to temperature and composition gradients at the weld pool surface often provides the main driving force for the convective flow, known as the "Marangoni flow." Depending on the interplay between the various driving forces, the convective flow can be a simple recirculation or a complex pattern with several convective cells operating (Fig. 3) (Zacharia *et al.* 1986). Fluid flow and heat transfer are important in determining the size and shape of the weld pool and the weld macro- and microstructures (Rosenthal 1941, Szekely 1987, Kou and Lee 1983, Mazumder and Chen 1984, Paul and DebRoy 1988,

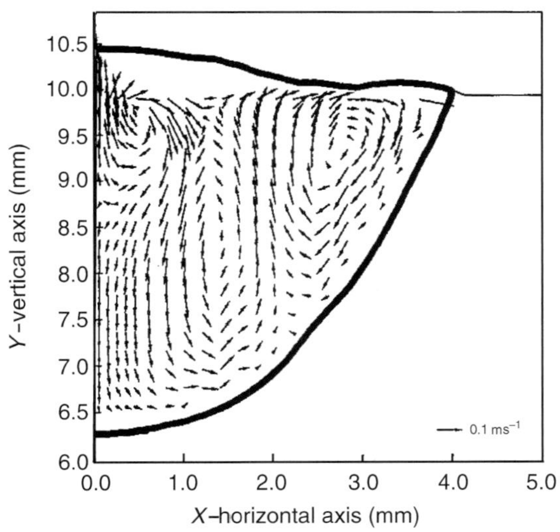

Figure 3
Typical fluid flow patterns predicted using computational mass and heat-transfer models showing convective cells developing across a gas–tungsten arc weld pool cross-section.

Heiple and Roper 1982, Heiple and Burgardt 1985). Also, there is currently a growing recognition that the flow in the weld pool is turbulent in nature, particularly at high heat input (Yang and DebRoy 1999, Hong *et al.* 1996, Choo and Szekely 1994).

The variable depth of penetration during welding of different batches of a commercial material with composition within a given prescribed range has received considerable attention. Previous work has shown that knowledge of the interfacial phenomena in welding is the key to understanding and controlling weld penetration (McNallan 1988, DebRoy and McNallan 1991). Oftentimes the penetration depth is determined by the concentration of the surface-active elements in the commercial alloy (Heiple and Roper 1982, Heiple and Burgardt 1985). These elements can affect the temperature coefficient of surface tension and the resulting direction of fluid flow in the weld pool (Zacharia *et al.* 1989). The convective heat transfer, in turn, controls the weld penetration. Weld pool surface temperatures are normally much higher than the melting points of the weld metals. Consequently, pronounced vaporization of alloying elements takes place, especially when high-energy-density heat sources are used (Block-Bolten and Eagar 1984, Khan and DebRoy 1984, Hettche *et al.* 1981, DebRoy *et al.* 1991). Such losses often change the composition of the weld metal, affect weld properties, and are a serious problem in the welding of many engineering alloys (Collur and DebRoy 1987).

The dissolution of hydrogen, oxygen, and nitrogen in the weld metal affects weldment properties. The gasses may dissolve interstitially in the weld metal, whereupon they escape to form pinholes or bubbles or combine with elements in the alloy to form inclusions. In the welding of steels, hydrogen induces cracking, nitrogen increases the yield strength and the tensile strength but reduces ductility and oxygen promotes inclusion formation. Inclusions form in steel welds because of the reaction between elements including: Al, Ti, Si, Mn, O, N, and S, and they can control weld metal microstructure (Easterling 1993, Bhadeshia and Svensson 1993, Evans and Bailey 1997, Bhadeshia 1992). Various modeling activities have coupled thermodynamic, kinetic and heat transfer and fluid flow models to describe inclusion formation in steel welds (Babu *et al.* 1995, 1996a, 1999, Hong *et al.* 1998, 2000, Kluken and Grong 1989, Koseki *et al.* 1997, Blais *et al.* 1998, Hsieh *et al.* 1996). In addition, there are phenomenological models describing weld solidification and solid-state transformation. They will be described in the next section.

2.2 *Artificial Neural Net Models*

The advances in computer models to describe the physical behavior of the weld pool have characteristically been accompanied by the need for greater

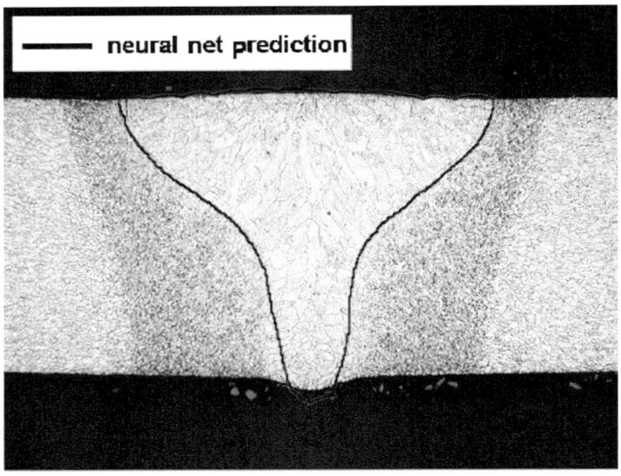

Figure 4
Comparison of weld cross section predicted by artificial neural networks and an actual weld pool of a hybrid laser-assisted arc weld.

computer power. Thus, many of the models require significant computing time or they must be run on sophisticated, high-end supercomputers. Although the insight that is gained is valuable, use and application of the model is not available to many end users. Another methodology for modeling weld behavior is needed. One alternative is the use of ANNs. Such models are empirically based but have the advantage of providing results rapidly, so much so that they can even be used as on-line components in control systems. As an example, ANNs have been used to predict weld pool shapes in hybrid laser-arc steel welds (Vitek *et al.* 2001). Figure 4 shows a comparison of a predicted weld pool cross-section with the actual weld pools; the agreement is quite good. The neural network models can be effectively used for parametric studies to determine the influence of individual variables. In the case of hybrid welds, this demonstrates yet another advantage of the ANNs. Such parametric studies can identify the effect of process variables on the weld pool shape when more sophisticated theory-based models are not yet available. Other researchers have developed ANNs to describe weld metal cracking tendency, microstructure evolution and properties of welds. The reader is referred to a recent review on application of ANNs in material processing for more details (Bhadeshia 1999).

3. *Weldment Structure and Properties*

The integrity of welded joints depends on the weldment microstructure and properties. Rapid heating, cooling and local structural changes cause significant spatial variations in the composition, microstructure, and residual stress in the FZ and HAZ. The various physical processes that dominate the welding process also influence the integrity of the welded structure.

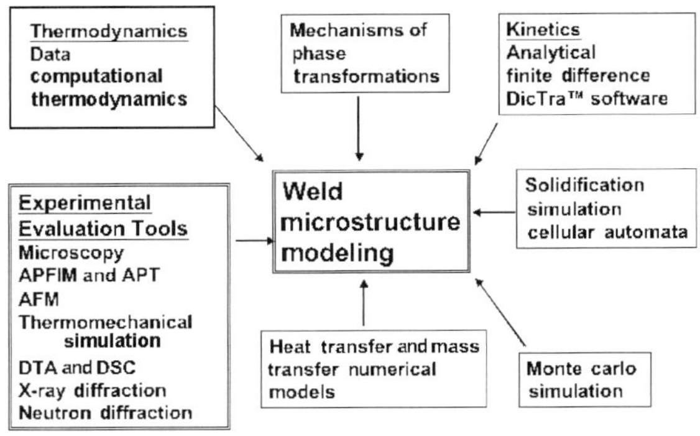

Figure 5
Overview of experimental and modeling tools required for weld microstructure modeling.

Modeling, using both numerical and analytical approaches, is critical to our understanding of the evolution of microstructures in weldments. Figure 5 shows several analytical, experimental and numerical tools that are used in modeling activities. The models for microstructure and properties require various critical assumptions, such as the presence of local thermodynamic equilibrium at liquid–solid or solid–solid interfaces. However, the development of microstructure models still requires a substantial amount of experimental research to understand various physical processes that occur during solidification and solid-state transformation in materials during welding. The following sections describe recent advances in the understanding of microstructure evolution in both the FZ and the HAZ through coupled experimental and theoretical research.

3.1 FZ Microstructure

The FZ undergoes a liquid–to–solid transformation. The size and shape of grains, the distribution of inclusions and the presence of defects such as porosity and hot cracks are controlled by its solidification behavior. Unlike the solidification of castings or ingots, weld pool solidification occurs without a nucleation barrier; therefore, no significant undercooling of the liquid is required for the nucleation of a solid. Solidification occurs spontaneously by epitaxial growth on the partially melted grains. In certain welds, inoculants and other grain-refining techniques are used in much the same way as they are in casting practices. Temperature gradient, growth rate, undercooling and alloy constitution are important factors in determining the FZ microstructure. Depending on the welding process, the weld metal cooling rates may range from 10^2°C s^{-1} to 10^6°C s^{-1}.

Solidification microstructures in welds which are often difficult to interpret, are commonly analyzed with the help of classical theories of nucleation and growth (David and Vitek 1989). The shape of the solid–liquid interface controls the development of microstructural features during growth of the solid in the FZ. Depending on the interface shape, growth of the solid will occur in planar, cellular, or dendritic modes. Sometimes all these distinct microstructural modes of growth can be observed in a weld. The dendritic growth of the solid with its multiple branches has been observed during welding of a single-crystal nickel-base superalloy (Fig. 6). Theories have been developed for interface stability under the conditions of slow cooling, assuming equilibrium at the interface. In recent years, necessary modifications have also been made to these theories to accommodate non equilibrium conditions prevalent during rapid solidification (Mehrabian 1982, Vitek *et al.* 1983, David *et al.* 1987, Katayama and Matsunawa 1984, Elmer *et al.* 1989, Trivedi and Kurz 1986, Boettinger and Coriell 1984, Aziz 1982). Moreover, the rapid solidification theories are also being extended to understand the development of microstructures in rapidly cooled welds (David and Vitek 1989, Flemings 1974, Matsuda *et al.* 1969, Savage 1980, Easterling 1983, Mehrabian 1982, Vitek *et al.* 1983, David *et al.* 1987, Katayama and Matsunawa 1984, Elmer *et al.* 1989). The use of high-energy-density welding processes such as electron-beam and laser-beam welding has made observations of non equilibrium microstructures under rapid cooling conditions common.

The solidification microstructure also affects the weld-cracking tendency. The potential for a weld to crack as the FZ cools, a phenomenon known as "hot cracking," is often attributed to elemental segregation and solidification grain structure during welding. As solidification of the weld metal proceeds spontaneously by epitaxial growth of the partially melted grains in the base metal, the FZ grain structure is mainly determined by the base metal grain structure and the welding conditions (David and Vitek 1989). In certain conditions, it is possible to change the epitaxial columnar growth to equiaxed grain growth through inoculation (Kerr and Villafuerte 1991).

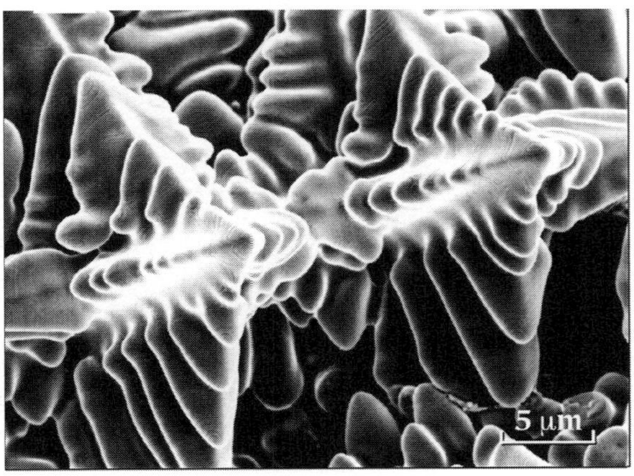

Figure 6
Scanning electron micrograph of a crack surface in a nickel-base superalloy weld showing the presence of primary and side branch dendrites.

Recent work considered the FZ grain structure evolution by using single-phase, single-crystal Fe–Cr–Ni alloy as the BM. In this alloy, the grains grow along a preferred crystallographic direction (⟨100⟩ direction for cubic metals). Conditions for growth are optimum when a preferred direction coincides with the heat flow direction. Therefore, among randomly oriented grains in the polycrystalline cubic base metals, the grains having ⟨100⟩ crystallographic axes most closely aligned with the heat flow direction are favored for growth. However, the development of the FZ grain structure in welds is complicated because of the competing growth of grains with various orientations in the three-dimensional FZ. Recent developments in this area include theoretical and experimental analyses of the dendrite selection process (Rappaz *et al.* 1990, David *et al.* 1990). The above research has proved useful in developing welding repair procedures for single-crystal nickel-base superalloys used in the high-temperature region of the aircraft and land-based turbine engines.

3.2 *HAZ Microstructure*

The HAZ can be heated to temperatures just below the melting point of the material (i.e., there is no melting, and the changes are mainly due to the response of the material to thermal effects in the form of solid-state phase transformations). If the material is a single-phase material, the grains in the HAZ grow and lead to deterioration of properties. Grain growth kinetics as a function of weld thermal cycles can be modeled by using analytical models or by using Monte Carlo simulations to describe the grain growth kinetics under weld thermal cycles (Radhakrishnan and Zacharia 1995).

If the microstructure contains many phases, the changes may involve transformation from one phase to another based on composition and temperature gradients. These changes are closely linked to the thermodynamic stability of these phases and to the diffusivity of various elements in these phases (Grong 1997). These are being modeled with both computational thermodynamic and kinetic models. For example, in low-alloy steels, the transformation of the high-temperature austenite phase (face–centered cubic (f.c.c.) crystal structure) to ferrite phase (body–centered cubic (b.c.c.) crystal structure) is expected to occur with partitioning of both interstitial alloying elements (carbon and nitrogen) and substitutional alloying elements (manganese, silicon, chromium, and nickel) between these phases. This partitioning leads to sluggish transformation of austenite to ferrite. Detailed microstructural characterization using electron microscopy has shown that under rapid cooling conditions, the transformation of austenite phase to ferrite phase occurs with only carbon partitioning while substitutional elements are configurationally frozen. This mode of transformation occurs because of the large difference in the diffusivity of interstitial and substitutional elements and is in agreement with rapid transformation kinetics calculations. This paraequilibrium transformation is also expected in the FZ region. The experimental evidence for such transformations in different steels can be obtained by experimental techniques such as atom-probe field-ion microscopy, transmission electron microscopy (TEM), spatially resolved x-ray diffraction (SRXRD), and time-resolved x-ray diffraction (TRXRD).

An example of using TRXRD to investigate phase changes in the HAZ of a Fe–C–Al–Mn steel is shown in Fig. 7 (Babu *et al.* 2002). The diffraction peaks from f.c.c. (austenite) and b.c.c. (ferrite) phases were measured at a time resolution of 0.05 s during weld heating and cooling from a HAZ location close to the fusion line. During weld heating, after a critical temperature, the f.c.c. diffraction peak was observed. With continued heating, the diffraction peak intensity of f.c.c. increased. As soon as the welding arc was extinguished, the temperature at the measurement location decreased. After this point, the austenite transformed to ferrite. Such experimental verifications of thermodynamic and kinetic models are necessary to control the HAZ properties.

The microstructural changes are linked to peak temperatures and time spent at high temperature. Consequently, the rate of phase transformation is expected to vary from region to region, thereby providing a large microstructural heterogeneity. These microstructural heterogeneities have a large influence on the properties of welded structures. For example, in high-Cr steels used in the energy industry, the presence of a soft zone due to microstructural change in the HAZ leads to premature creep failure of the welded part. Therefore, there is a critical need to describe these changes using thermodynamic and kinetic models and to relate them to weld properties.

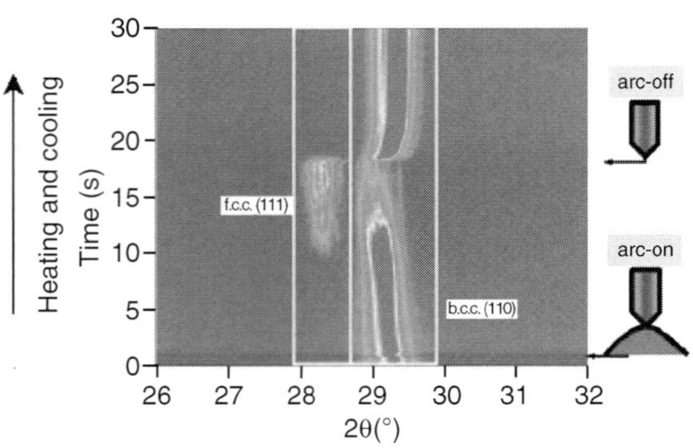

Figure 7
Image representation of time-resolved x-ray diffraction data from the heat-affected zone of a Fe–C–Al–Mn steel during weld heating and cooling. The diffraction peaks from (111) crystallographic planes of the austenite (f.c.c.) phase and diffraction peaks from (110) crystallographic planes of ferrite (b.c.c.) phase are shown.

3.3 Microstructure and Property Characterization

Recent developments in microstructural characterization techniques such as analytical electron microscopy and atom probe field-ion microscopy, have made it possible to characterize microstructures on scales as fine as a few nanometers (Babu *et al.* 1996b, 2000). For example, the stability of FZ microstructure in stainless steel has been characterized using TEM. The TEM analysis showed that the FZ contained both ferrite and austenite phases. However, the same microstructure on aging changes to complex microstructures containing carbides (compound of Cr and C) σ-phase ferrite and austenite (intermetallic compounds) (Fig. 8). Such information

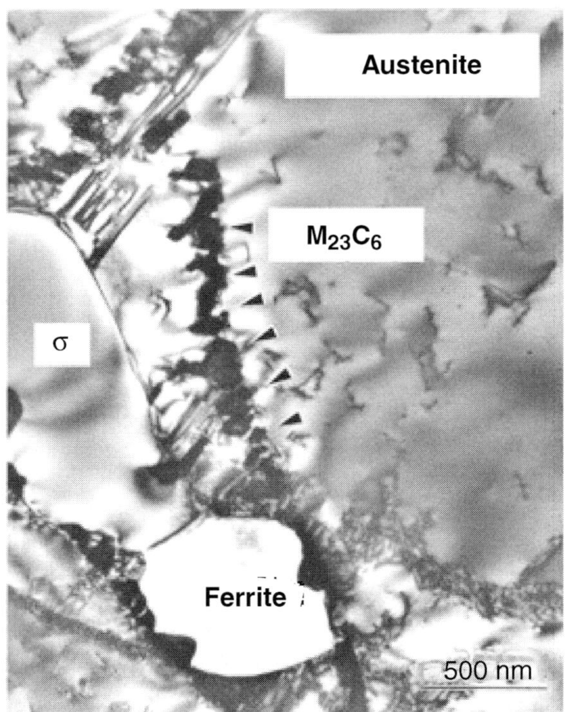

Figure 8
Transmission electron micrograph of a stainless steel fusion zone after being thermally aged at 650°C for 1000 h.

is crucial to understanding high-temperature stability during service at elevated temperatures.

In addition to microscopic techniques, various analytical tools have been developed to track phase changes that occur during weld heating and cooling. These tools include Gleeble® thermomechanical simulation and *in situ* techniques such as differential thermal analysis, SRXRD, TRXRD, and welding of single crystals (Vitek and David 1990, David *et al.* 1979, Cieslak 1991, Elmer *et al.* 1998, 1999, Babu *et al.* 2002).

Because of the spatial variation of weld properties, fine-scale evaluation of properties is also desirable. Recent techniques such as the mechanical property microprobe, indentation creep testing, and miniature impact specimen testing (Vitek and David 1990) have made it possible to test welds either by performing tests on small areas or by miniaturizing test specimens. Also thermal neutrons, which can penetrate 50 mm into steel, have recently been used to characterize residual stresses in weldments on a millimeter scale (Holden *et al.* 1988, Mahin *et al.* 1990, Root *et al.* 1993, Spooner *et al.* 1993).

3.4 Properties and Performance of Welded Structures

Even with the presence of sophisticated models using analytical and numerical approaches for describing heat transfer solidification, solid-state transformation, and residual stress developments (David and DebRoy 1992, 1994, DebRoy and David 1995), an integrated model that comprises all of these physical processes and their interactions has yet to be developed. Such an integrated model must also be capable of describing microstructural degradation during service life. Recently, such an integrated model was developed for describing the performance of low-alloy steel resistance spot welds (Santella *et al.* 1999). The model successfully described the welding process effects, microstructure evolution, and microstructural heterogeneity. The model also evaluated the effect of microstructural heterogeneity on the overall performance of the welded structure. The results showed that, for the spot-welding geometry, a soft zone always forms in the HAZ. This soft zone leads to a concentration of tensile strains during tensile testing of the welded component, which leads to premature failure. Interestingly, this preferential failure in the HAZ is closely linked to weld geometry.

Although the above research showed the feasibility of integrating different sub-process models to describe the overall performance of a welded structure, the integrated model was not developed in a seamless fashion. Each sub-process model (e.g., process and microstructure model) has been developed and executed separately in a disjointed fashion. This still limits the end user application due to the need for expertise and availability of all the sub-process models. Recently, an alternative approach has been developed that is based on a problem-solving environment to integrate sub-process models (Babu *et al.* 2003). This approach uses Internet client–server technology to link with process and microstructure models through one interface that can be readily used by a welding engineer or a scientist. This approach is expected to make a detailed analysis of process and material selection for a specific industrial application possible and available to the end user.

4. Welding Control and Automation

A skilled manual welder uses his sensory perception such as touch, sight, and sound to evaluate the weld process and to take the necessary corrective measures. Success depends on his intuition, experience and skills, and the procedure is labor-intensive and frequently unreliable. As a result, manual welding operations are being progressively replaced by automated systems to achieve increased productivity and quality. Recent progress with automation has been carried out with robots. Defect-free welds could not be made in these operations owing to lack of adaptive controls and appropriate sensors (Cook *et al.* 1990).

Research in welding automation is focused on sensing, modeling, and feedback control (Cook *et al.* 1990, Smartt and Johnson 1991, Jones *et al.* 1995, Richardson *et al.* 1984, Bolstad 1987, Chin *et al.* 1983). Early research in this area dealt with proper placement of the weld bead with respect to the weld joint and control of the weld bead geometry. Current research is focused on extending this base knowledge to achieve stable, robust control methods to control microstructure and mechanical properties. With respect to the latter, modeling efforts are directed toward gaining a more fundamental understanding of the relationship between the input variables and desired properties with the hope that this knowledge will lead to needed innovations in direct and indirect sensing measurement. More stable and robust control requires dynamic process models, real-time system identification and continuous modification of the control laws (i.e., adaptive feedback control to achieve optimum performance over the wide range of operating conditions in which the highly coupled nonlinear welding procedure might be expected to operate).

5. Recent Developments in Welding Processes

The above review shows the complexity involved in fusion welding and attempts to develop science- and engineering-based predictive and control methodologies so that the structural weld will meet the required goals and design life. However, there is an ongoing effort to develop processes for industrial applications that produce high-quality, low-cost welds. Some of the new processes that fit into this category are electromagnetic pulse welding, hybrid laser–assisted arc welding (LAAW), and friction stir welding (FSW) processes. The FSW and LAAW processes are currently being introduced in the industry.

FSW, a relatively new process invented in the 1990s (Thomas 1991, Dawes and Thomas 1995, 1996), is a solid-state joining method that is now well-established for the production of very high quality welds of aluminum alloys of all types. The process is accomplished using a rotating, non-consumable tool that is translated along the length of a joint. The process requires no supplemental power input and no melting of the bulk material occurs; instead, the joint is produced by a deformation process.

The LAAW process uses both an electric arc and a laser beam as heat sources. This process removes the need for the precise joint fit-up requirement of laser-beam welding, and the laser beam increases the productivity of arc welding by increasing the energy density. As a result, the LAAW process is ideally suited for high-productivity sheet-metal welding.

Bibliography

Aziz M J 1982 *J. Apl. Phys.* **53**, 1158

Babu S S, David S A, Vitek J M, Mundra K, DebRoy T 1995 *Mat. Sci. Tech.* **11**, 186–99

Babu S S, David S A, DebRoy T 1996a *Sci. Tech. Weld. Join.* **1**(1), 17–27

Babu S S, David S A, Vitek J M, Miller M K 1996b *J. de Physique IV* **6**, 253

Babu S S, David S A, Vitek J M, Mundra K, DebRoy T 1999 *Sci. Tech. Weld. Join.* **4**, 276–84

Babu S S, David S A, Vitek J M, Miller M K 2000 *Proc. of Advanced Technologies for Superalloy Affordability*. TMS, Nashville, TN

Babu S S, Elmer J W, David S A, Quintana M A 2002 *Proc. R. Soc. Lond.* **A458**, 811

Babu S S, Summers M A, Sarma G, 2003 AWS Technical Convention, (to be published)

Bhadeshia H K D H 1992 *Bainite in Steels*. Institute of Materials, London

Bhadeshia H K D H, Svensson L-E 1993 In: Cerjack H, Easterling K E (eds.) *Mathematical Modeling of Weld Phenomena*. Institute of Materials, London, pp. 109–80

Bhadeshia H K D H 1999 *ISIJ Inter.* **39**, 966

Blais C, Esperance G L, Evans G M 1998 *AWS Convention Abstracts*. Detroit, MI, pp. 244–5

Block-Bolten A, Eagar T W 1984 *Metall. Trans.* **15B**, 461

Boettinger W J, Coriell S R 1984 *Mater. Sci. Eng.* **65**, 27

Bolstad J 1987 *Proc of the Advances in Automation and Precision Arc Welding*. Edison Welding Institute, Columbus, OH

Chin B A, Madsen N H, Gooding J S 1983 *Weld. J.* **62**, 227s

Choo R T C, Szekely J 1994 *Weld. J.* **73**, 25S

Cieslak M J 1991 In: Cieslak M J, Perepezko J H, Kung S, Glicksman M I (eds.) *Metal Science of Joining*. TMS Warrendale, PA, p. 101

Collur M M, DebRoy T 1987 *Metall. Trans.* **18B**, 733

Cook G E, Andersen K, Barrett R J 1990 In: David S A, Vitek J M (eds.) *Recent Trends in Welding Science and Technology*. ASM International, Materials Park, OH, p. 553

David S A, Goodwin G M, Braski D N 1979 *Weld. J.* **58**, 330S

David S A, DebRoy T 1992 *Science* **257**, 497–502

David S A, DebRoy T 1994 *MRS Bull.* **XIX**(1), 29–35

David S A, Vitek J M, Hebble T L 1987 *Weld. J.* **66**, 289S

David S A, Vitek J M 1989 *Int. Mater. Rev.* **34**, 213

David S A, Vitek J M, Rappaz M, Boatner L A 1990 *Metall. Trans.* **21A**, 1753

Dawes C J, Thomas W M 1995 Friction Stir Welding of Aluminum Alloys. *TWI Bull.* **6**, 124

Dawes C J, Thomas W M 1996 *Weld. J.* **75**, 41

DebRoy T 1997 In: Sohn H Y, Evans J W, Apelian D (eds.) *Proceedings of the Julian Szekely Symposium on Materials Processing*. Published by the Minerals, Metals, and Materials Society, Warrendale, PA, pp. 365–79

DebRoy T, Basu S, Mundra K 1991 *J. Appl. Phys.* **70**, 1313

DebRoy T, David S A 1995 *Reviews Modern Phys.* **67**(1), 85–112

DebRoy T, McNallan M J 1991 *Metall. Trans.* **22B**, 557

Easterling K E 1983 *Introduction to Physical Metallurgy of Welding*. Butterworth, London

Easterling K E 1993 In: Cerjack H, Easterling K E (eds.) *Mathematical Modeling of Weld Phenomena*. Institute of Materials, London, pp. 183–200

Elmer J W, Allen S M, Eagar T W 1989 *Metall. Trans.* **20A**, 2117

Elmer J W, Wong J, Ressler T 1998 *Metall. Met. Trans.* **29A**, 2761

Elmer J W, Wong J, Ressler T 1999 In: Vitek J M, *et al.* (eds.) *Trends in Welding Research*. ASM International, Materials Park, OH, p. 107

Evans G M, Bailey N 1997 *Metallurgy of Basic Weld Metal*. Abington Publishing, Cambridge, UK

Flemings M C 1974 *Solidification Processing*. McGraw-Hill, New York

Grong O 1997 *Metallurgical Modeling of Welding*. The Institute of materials, London

Heiple C R, Burgardt P 1985 *Weld. J.* **64**, 159S

Heiple C R, Roper J R 1982 *Weld. J.* **61**, 97S

Hettche L R, Metzbower E A, Ayers J D, Moore P G 1981 *Nav. Res. Rev.* **28**, 4

Holden T M, Root J H, Fidlevis V, Holt R A, Roy G 1988 *Mater. Sci. Forum 27 and 28* 359

Hong K, Weckman D C, Strong A B 1996 In: Smartt H B, Johnson J A, David S A (eds.) *Trends in Welding Research*. ASM International, Materials Park, OH, p. 399

Hong T, DebRoy T, Babu S S, David S A 2000 *Metall. Mat. Trans.* **31B**, 161–9

Hong T, Pitscheneder W, DebRoy T 1998 *Sci. Tech. Weld. Join.* **35**(1), 33–41

Hsieh K C, Babu S S, Vitek J M, David S A 1996 *Mater. Sci. Eng.* **A215**, 84–91

Jones L J, Lang J H, Eagar T W 1995 In: Smartt H B, Johnson J, David S A (eds.) *Int. Trends in Welding Science and Technology*. ASM International, Materials Park, OH, p. 365

Katayama S, Matsunawa A 1984 *Proc. of Int'l Congress on Application of Lasers and Electro Optics*. Laser Institute of America, Boston, Vol. 44, p. 60

Kerr H W, Villafuerte J C 1991 In: Cieslak M J, Pere Pezko J H, Kang S, Glicksman M E (eds.) *Metal Sci. Joining*, TMS Warrendale, PA, p. 11

Khan P A A, DebRoy T 1984 *Metall. Trans.* **15B**, 641

Kluken A, Grong O 1989 *Metall. Trans. A* **20**, 1335–49

Koseki T, Ohkita S, Yurioka N 1997 *Sci. Tech. Weld. Join.* **2**, 65–9

Kou S, Lee Y 1983 *Metall. Trans.* **14A**, 2243

Mahin K W, Winters W, Krafcik J, Holden T, Hosbons R, MacEwen S 1990 In: David S A, Vitek J M (eds.) *Intl Trends in Welding Science and Technology*. ASM International, p. 83
Matsuda J F, Hashimoto T, Senda T 1969 *Trans. Natl. Res. Inst. Met (Jpn)* **11**(1), 83
Mazumder J, Chen M M 1984 *Metall. Trans.* **15A**, 2175
McNallan M J 1988 *Metall. Trans.* **19B**, 483
Mehrabian R 1982 *Intl. Metall. Rev.* **27**, 185
Paul A, DebRoy T 1988 *Metall. Trans.* **1913**, 851
Radhakrishnan B, Zacharia T 1995 *Metall. Mater. Trans. A* **26A**, 2123
Rappaz M, David S A, Vitek J M, Boatner L A 1990 *Metall. Trans.* **21A**, 1767
Richardson R W, Gutow A, Anderson R A, Farson D F 1984 *Weld. J.* **63**, 43s
Root J H, Holden T M, Schroder J, Hubbard C A, Spooner S, Dodson T A, David S A 1993 *Mater. Sci. Tech.* **9**, 754
Rosenthal D 1941 *Weld. J.* **20**, 220S
Santella M L, Babu S S, Riemer B, Feng Z 1999 *Proc. Intl. Conference on Trends in Welding Research*. Pine Mountain, Georgia, 1–5, ASM International, p. 605
Savage W F 1980 *Weld World* **18**, 89
Smartt H B, Johnson J A 1991 In: Dagh C H, Kumara S R T, Shin Y C (eds.) *Proc of the Artificial Neural Networks in Engineering (ANNIE '91)*, ASME, p. 711
Spooner S, Holden J M, Root J H, David S A 1993 In: David S A, Vitek J M (eds.) *Intl' Trends in Welding Science and Technology*. ASM International, Materials Park, OH, p. 139
Szekely J 1987 In: David S A (ed.) *Advances in Welding Science and Technology*. ASM International, Materials Park, OH, p. 3
Thomas W M 1991 Friction Stir Butt Welding. *US Patent No.* 5 460 317
Trivedi R, Kurz W 1986 *Acta. Metall.* **34**, 1663
Vitek J M, DasGupta A, David S A 1983 *Metall. Trans.* **14A**, 1833
Vitek J M, David S A 1990 *Proc. of First US–Japan Symposium on Advances in Welding Metallurgy*, AWS, Japan Welding Society, and Japan Welding Engineering Society, San Francisco, CA, June 7–8, p. 65
Vitek J M, *et al.* 2001 *Sci. Technol. Weld. Join.* **6**, 305
1991 *Welding Handbook*, 8th edn., American Welding Society, Miami, FL
Yang Z, DebRoy T 1999 *Metall. Mater. Trans. B* **30B**, 483
Zacharia T, David S A, Vitek J M, DebRoy T 1989 *Weld. J.* **68**, 499S
Zacharia T, David S A, Vitek J M, Kraus H G 1986 *Metall. Trans.* **17A**, 1865

S. A. David, S. S. Babu, and J. M. Vitek

Wood: Chemical Treatment Processes

The polymeric nature of wood renders it remarkably resistant to degradation, but most wood species will degrade under the proper moisture and temperature regimes (Zabel and Morrell 1992, Eaton and Hale 1993). Wood users have long sought to prevent this damage either through designs that shed water or by treatments that inhibit the agents of decay.

The ancient Greeks and Romans applied various concoctions to wood surfaces (Graham 1973), but most materials were ineffective; however, the Romans did produce the earliest fire retardants when they coated their wooden siege towers with alum. The era of effective wood preservation began in the mid-1830s with the patenting of the Bethell process and the development of creosote. Within a few short years wood users had a chemical system that was highly effective against a variety of agents of deterioration and a method for effectively delivering this material into the wood. Although the types of chemicals used for wood protection continue to be refined, the processes used to deliver them into the wood remain limited by the ability to deliver fluids into a semipermeable material. Methods for delivering them into wood will be discussed in this article; for more information on the chemicals.

1. *Goals of Treatment*

The overall goal of treatment is to extend the useful life of a wood product for periods that vary widely with end use, e.g., utility poles are normally expected to last 30–40 years with a high degree of reliability, whereas wood decking may have a useful life of 15–20 years but must retain its appearance to a far greater extent than a pole. Thus, service life may require protection against biotic and abiotic agents of deterioration. The general purpose of wood treatment is to prevent biological attack; however, treatments can also impart resistance to fire, chemical, mechanical, or UV light damage as well as preventing the risk of water sorption.

2. *Measures of Treatment*

Treatment quality is usually assessed on the basis of the amount of chemical and the depth to which it penetrates into the wood (American Wood Preservers' Association [AWPA] 1998). The amount of chemical or retention is expressed typically on either a mass of chemical/mass of wood basis or a mass per unit volume of wood basis. While mass/mass is more accurate, mass per volume is simpler because it ignores the normal variations in wood density. Retentions are defined normally for a specific zone in the wood. In the US standard, the zone from the surface to 15 mm inward comprises the assay zone for dimension lumber less than 100 mm thick. The retention is set on the basis of the amount needed to provide reliable protection, plus an additional amount to provide a safety factor in the event that individual boards vary in retention. Retentions are determined normally by removing a specified number of increment cores or plugs shortly after treatment. The intensity of sampling varies with the product. Every pole in a charge may be sampled, while only 20 cores are taken from an individual lumber charge that may include 1000 or more pieces. Surface treatments that do not penetrate deeply into the wood are measured normally on a volume or mass per unit area basis (liters of chemical per square meter of wood surface).

Penetration is a second measure of treatment. The depth of penetration is dependent on the wood species and decay risk to which the wood will be exposed (usually 80–100%) and some penetration of heartwood. Penetration is generally measured on cores, but can also be measured on cross-sections cut a minimum of 300 mm from one end. In most standards a specific percentage of cores taken from a unit must meet the treatment requirements. Charges that fail to meet these requirements must be retreated.

3. *Long-term vs. Short-term Treatment*

Preservation tends to be viewed as a relatively long-term practice; however, there are a variety of instances where the period of protection is relatively short (6–12 months). The primary short-term protection is prevention of biological or chemical staining of wood during transport or prior to drying. In this case, the goal is to prevent discoloration until the wood moisture content declines below 20% (the point when most biological activity by agents of decay ceases). Long-term preservation is anything beyond 1 year, but typically means protection in the 10–40 year range.

3.1 *Short-term Treatments*

Short-term protective treatments for wood provide a protective barrier against invasion by the fungi that cause mold and staining of wood (Zabel and Morrell 1992). The protective layer is generally shallow, since the risk of damage is primarily through spores landing on the wood surface. Spores represent a particularly vulnerable stage in the fungus life cycle, but they can also germinate and rapidly penetrate into the wood. Once inside, they are beyond the depth of penetration of antistain treatments. The result will be wood that is largely bright and clear on the surface, but heavily stained within. As a result, application of antistain chemicals within 24 h after sawing is critical for effective stain control. Timely processing coupled with sprinkling remains the best method for limiting discoloration of a log deck.

Most users need to protect their wood from discoloration for a minimum of 6 months, although the protective period may sometimes extend to 12 months. The relatively short period of protection means that the chemicals used for this purpose need not be as broadly toxic as those used for heavy-duty wood protection (Byrne 1998). In fact, because freshly sawn lumber is often handled extensively during processing, there is considerable pressure to use biocides that have little or no toxicity to humans or other nontarget organisms. Finally, most of the chemicals used for short-term protection are water soluble or water miscible.

Antistain chemicals are applied generally by dipping boards individually or as units. Individual boards are usually dipped along the green chain, while units are assembled and dipped in large dip tanks. Dipping individual boards can produce handling problems within a mill since the treated boards are often handled by workers while sorting for grade. Bulk dipping is a more common practice, but even here care must be taken to avoid contaminating the yard with chemical. In an ideal operation, the lumber is dipped, allowed to drain until dripping has ceased, and then stored either under cover or wrapped until shipped. The goal is to minimize potential chemical movement across the plant site.

Developments in antistain treatments are spray systems that coat the board as it passes through an enclosed booth. The spray that runs off the board is collected, filtered, and recycled into the system. Systems apply a concentrated spray to all four surfaces of the lumber as it moves through an enclosed booth. Failure to control sprays adequately will result in incomplete protection and costly claims. Sprayers can also apply lower volumes of more highly concentrated antistain chemicals. These ultra-low volume sprayers have the advantage of reduced dripping, but the risk of poor treatment due to clogged or worn nozzles becomes especially critical. Electrostatically charged particles can also be used to deliver the antistain chemical to the wood, but the technology is still in the proving stages. The process, however, further reduces the volume of material applied, decreasing the potential for dripping or worker exposure. These systems require extensive maintenance to ensure that the required amount of chemical is deposited uniformly on the wood surface.

In general terms, nearly any fungicide will provide reasonable protection against fungal stain and mold at some concentration, provided it is applied uniformly to the wood surface. The selection of chemicals then relates to the wood species, the cost of the chemical, and various user concerns such as nontarget toxicity and stability of solutions. As a result, a market that was once dominated by one chemical has evolved into a much more specialized market with a more diverse array of choices.

3.2 *Long-term Wood Protection*

Many wood applications require protection against various agents of decay for lengths of time that may extend for up to a century. The chemicals and the processes used to deliver them into the wood depend on the wood species, the biodegradation agents, and the environmental exposure conditions. The goal of treatment is to deliver an amount of chemical to a specific zone of the wood that will provide an envelope of protection against wood-degrading organisms. In many instances, 30–40 years is used as the expected service life. Clearly, the amount of chemical and the depth to which it must be delivered for a given species will depend upon the exposure conditions.

Brushing, dipping, or soaking the wood in the treatment chemical are nonpressure methods that depend on capillary uptake and bulk flow of solution into the cell lumens (Hunt and Garratt 1967, Wilkinson 1979). In general, chemical penetration is relatively shallow, ranging from a few mm in difficult-to-treat species such as Douglas fir to 10–15 mm in more receptive species such as southern pine. Nonpressure treatments are used normally where the risk of decay is relatively low, but these surface treatments have also proved useful for providing added protection to naturally durable heartwoods (Newbill and Morrell 1993).

Dipping is the primary method used for treating wood used for windows and door frames in North America. The wood in these exposures is normally protected from extensive wetting through roof overhangs and coatings. The reduced risk of wetting limits the potential for decay development. Other exposures where dipping is used include pallets and crates.

Protection for wood exposed under severe decay conditions is accomplished using combinations of vacuum and pressure to force chemical into the wood. Vacuum treatments involve drawing a short vacuum over the wood, then adding the treatment solution and releasing the vacuum. The vacuum removes air from the cell lumens, creating space for treatment chemicals. The double vacuum process is often used in Europe to treat wood for window and door frames. Vacuum treatment in North America is largely limited to treatment of western red cedar utility poles.

Pressure treatments are performed in a pressure vessel or retort. The first effective use of pressure treatment was the Bethell or full cell process, first patented in 1838 (Hunt and Garratt 1967). The Bethell process begins by placing the wood into the retort and drawing a vacuum to remove air from the wood. The treating solution is added to the retort while under vacuum, and then the pressure is raised and maintained until the desired amount of chemical has been injected into the wood. The pressure is then released and drained from the retort. In some instances, a vacuum is drawn over the treated wood to help relieve any internal pressure and hasten solution recovery. This reduces the risk that chemical will bleed from the wood once in service. There are several measures of chemical uptake during the treatment processes. Gross retention is the amount of chemical absorbed by the wood at the conclusion of the pressure process. Once the pressure is released, air trapped at the center of the wood expands outward carrying with it excessive treatment solution, a process called kickback. The amount of chemical absorbed at the end of the kickback period is the net retention of chemical injected into the wood. After the wood has been treated, samples can be removed and analyzed to determine if the proper amount has been delivered.

The Bethell process delivers the maximum amount of treatment solution into the wood to the maximum depth possible. This process is employed typically for treating wood used in marine environments where the risk of attack is extreme, or for treatments that use water as the solvent, since the solution concentration can be varied to deliver the desired amount of active chemical into the wood.

Alternatives to the Bethell process are the empty cell processes. In the first of these, the Lowry process, wood is placed in the vessel and treating solution is added. The pressure is then raised to the desired level and held until an adequate amount of chemical has been injected. Pressure is released and a vacuum is drawn to hasten solution recovery. In the second, the Rueping process, wood is placed in the vessel and air is injected to increase the pressure to 200–330 kPa. Treatment solution is added, pressure is raised to its final level then released once the desired injection has been achieved. The primary difference between the full and empty cell processes is the presence of air in the wood at the start of the pressure cycle. This air is compressed deeper into the wood as the preservative is injected. Once the pressure is released, the compressed air expands outward, carrying with it excess preservative that reduces retention. The injection of initial air in the Rueping process increases the amount of trapped air, increasing the kickback and further reducing retention. The empty cell processes are normally used for treating wood with oil-based solutions for land-based applications where the amount of chemical required for protection is lower than would be required in more severe marine environments.

The full and empty cell processes represent the primary methods for pressure-treating wood in industrialized countries; however, a number of alternative processes have been developed to address the treatment needs for specific wood species or commodities. For example, modified full cell processes are used increasingly in the southeastern USA because the lower initial vacuum levels reduce preservative uptake and increase kickback. This decreases drying and shipping costs.

In the alternating pressure method (APM) and the oscillating pressure method (OPM) cycles of pressure are used in an attempt to increase preservative uptake. The APM cycles pressure to improve penetration of fluid through the cell pits and is used to treat radiata pine in New Zealand but is not used widely in other countries. The OPM cycles vacuum and pressure in relatively short bursts in an attempt to dislodge aspirated pits and improve fluid flow. This process has been used to treat spruce, a species known for its difficult-to-treat heartwood. A number of studies suggest that pressure changes occur relatively slowly in difficult-to-treat wood species, even when the pressure differentials between the interior and wood surface are quite large. As a result, it is unlikely that the OPM or APM can exert significant influences on treatment.

4. *New Treatment Technologies*

All of the current treatment technologies depend upon the ability of fluids to move into the wood (Hunt and Garratt 1967). While fluids can move relatively easily along the cell lumens of most wood species, movement in the tangential and radial directions is limited by the permeability of the pits; pit membranes can limit the passage of large preservative particles. In addition, encrustations that develop as sapwood is converted to heartwood can completely block fluid flow. Finally, the torus in bordered pits can sometimes aspirate or become blocked. While methods have evolved to "improve" treatment, all have attempted to force conventional liquid systems across pit membranes that are inherently resistant to flow. The result has been incremental improvements in treatment but a continuing inability to treat the heartwood of most species.

In the vapor boron process (Murphy 1995), wood is placed in a vessel and a vacuum is drawn to evacuate the air. The vessel is opened to a second vessel containing trimethyl borate which has a high vapor pressure and volatilizes rapidly. The gaseous borate diffuses into the panels where it reacts with any moisture present to produce methanol and deposits borate in the wood. The process is ideal for panel products because they contain relatively little water. Treating dimension lumber using this same process has been limited because of the need to dry the wood to unrealistically low moisture levels. Failure to do so results in excessive deposition near the surface, and little penetration beyond this zone.

Another approach to improve treatment is to alter the treatment fluid. For example, heating reduces the viscosity so increasing uptake. There have also been attempts to use more volatile liquid solvents such as methylene chloride or liquified petroleum gas in place of conventional heavier oils. While slight improvements were noted, the alternative solvents changed the performance characteristics of the products, which experienced considerable surface decay.

Overcoming resistance to fluid penetration in wood has taken a different turn with the development of supercritical fluid (SCF) treatment systems (Morrell and Levien 1995). Some SCFs have diffusivities similar to those of a gas, allowing them to move through wood like a gas. SCFs can also have solvating properties that approach those of a liquid solvent, creating the potential for identifying a SCF that can move through the wood like a gas while carrying enough preservative to confer protection. SCF treatment technologies are in their infancy, but pilot plant trials suggest that they can be used to impregnate completely a variety of wood species.

5. *Physical Methods for Improving Treatment*

There are a number of methods for physically altering wood to make it more receptive to treatment. Applicability of these methods vary with species and intended end use.

5.1 *Incising*

Incisors drive metal teeth to a specific depth into the wood, increasing the amount of cross-section exposed to preservative flow. Since end-penetration typically is far greater, incising increases preservative penetration to the depth of the incisions. Incising is required in North America for the treatment of many refractory species (AWPA 1998), but it can reduce the material properties of the wood (Winandy *et al.* 1995). Despite the potential for strength losses, incising produces more uniform wood performance.

5.2 *Radial Drilling or Through Boring*

Internal decay is often a major problem in large timbers or round stock, particularly with thin sapwood species. Through boring and radial drilling are two methods for increasing the depth of preservative treatment (Graham 1983) at or near the ground-line. Through boring involves drilling a series of small diameter (11 mm) holes completely through the wood member on one surface. In radial drilling, holes are drilled at regular intervals around a pole or timber to a depth of 75–150 mm, again producing nearly complete penetration of the drilled zone. Both of these processes are used to protect Douglas fir poles and can largely eliminate internal decay at the ground-line.

5.3 *Predrilling and Cutting*

Preservative treatment creates a protective envelope whose effectiveness is compromised quickly when cuts or drill holes are made through this protective barrier. One method for prolonging treated wood service life is to make all cuts and bore holes prior to treatment to produce a more uniform treatment around the holes and cuts.

5.4 *Drying*

For most treatments, chemical uptake will improve substantially when a high percentage of water is removed from wood prior to treatment. The exceptions are boron and fluoride, which require that some free water be present in the cell lumens for substantial chemical diffusion to occur. Conversely, wood of some species becomes less permeable if it is overdried because the pits tend to aspirate. Drying can be accomplished by air and/or kiln drying, steam conditioning, or Boulton seasoning. Each method has specific limitations depending on the wood species, but all are designed to reduce wood moisture content without adversely affecting wood properties. For most treatments, wood moisture contents between 20% and 40% produce the most uniform treatments, although some species can be treated at moisture contents as high as 60–80%.

Bibliography

American Wood Preservers' Association 1998 *Book of Standards*. AWPA, Granbury, TX

Byrne A 1998 Chemical control of biological stain: past, present, and future. In: Morrell J (ed.) *Proc. Biology and Prevention of Sapstain*, Publication 7273. Forest Products Society, Madison, WI, pp. 63–70

Eaton R A, Hale M D C 1993 *Wood: Decay, Pests, and Protection*. Chapman & Hall, London

Graham R D 1973 History of wood preservation. In: Nicholas D D (ed.) *Wood Deterioration and its Prevention by Preservative Treatment*. Syracuse University Press, Syracuse, NY, Vol. I, pp. 1–30

Graham R D 1983 Improving the performance of wood poles. *Proc. American Wood Preservers' Assoc.* **79**, 222–8

Hunt G M, Garratt G A 1967 *Wood Preservation*. McGraw-Hill, New York

Morrell J J, Levien K L 1995 Development of new treatment processes for wood protection. In: *Wood Preservation in the 90s and Beyond*, Publication 7308. Forest Products Society, Madison, WI, pp. 135–41

Murphy R J 1995 Vapor phase treatments for wood products. In: *Wood Preservation in the 90s and Beyond*, Publication 7308. Forest Products Society, Madison, WI, pp. 83–8

Newbill M A, Morrell J J 1993 Ability of over-the-counter preservatives to protect western wood species from fungal attack. *For. Prod. J.* **43**(10), 35–40

Wilkinson J G 1979 *Industrial Timber Preservation*. Associated Business Press, London

Winandy J E, Morrell J J, Lebow S T 1995 Review of the effects of incising on treatability and strength. In: *Wood Preservation in the 90s and Beyond*, Publication 7308. Forest Products Society, Madison, WI, pp. 65–9

Zabel R A, Morrell J J 1992 *Wood Microbiology: Decay and its Prevention*. Academic Press, San Diego, CA

J. J. Morrell

Z

Zinc: Alloying, Thermomechanical Processing, Properties, and Applications

Zinc is the world's fourth most industrially used engineering metal. The use of zinc by humans extends back 5000 years, although it was not always recognized as a separate metal for much of this time. Zinc is known to have been used to produce brasses and bronzes since at least 500 BC by the Romans and Greeks, in Asia and elsewhere. The first known reference to zinc as a metal was made in the late first century BC where it was referred to as "false silver" while describing a region now in the northwestern part of Turkey. Further, a zinc alloy idol has been found at a prehistoric settlement in modern Romania. Zinc was used in China as early as AD 500, and appears to have been known in India about AD 1000. Although first available on a commercial scale in the fourteenth century, its real commercial use in Europe dates roughly from the beginning of the sixteenth century. The birth of the US zinc industry is generally placed in 1860.

Worldwide zinc consumption is now over 7 billion tonnes and growing at an annual rate of $\sim 3\%$. Zinc use is divided into five major categories: (i) coatings for steel including galvanizing, (ii) zinc casting alloys, (iii) alloying element in brass and other alloys, (iv) wrought zinc alloys, and (v) zinc dust and zinc chemicals. Significant amounts of zinc are recycled in the form of zinc, zinc alloys, zinc dust, zinc oxide, and other zinc chemicals. Secondary zinc use is 20–40% of total consumption worldwide.

Zinc and its alloys are used in cast and wrought forms in engineering applications as components and other structural members. Other important engineering applications are based on the properties of zinc and zinc alloy coatings applied to steel for corrosion protection. Zinc and its alloys, designation systems, thermal–mechanical processing (TMP), physical, mechanical, and chemical properties, and areas of application are discussed in this article. Most commercially used zinc grades and alloys have been specified in ASTM standards and these standards are cited in this article. The Unified Numbering System (UNS) has been extensively applied to zinc and zinc alloys, there being nearly 100 UNS designations for zinc. UNS alphanumeric designations start with "Z" followed by five digits. UNS numbers are given for zinc grades and alloys cited in this article.

1. *Basic Physical and Chemical Properties*

Zinc is a bluish-white metal in the fourth period and group 2b of the periodic table. Unalloyed zinc normally is produced by the electrolytic process with a purity of 99.95 wt.% or greater and contains trace impurities from its ores or the refining process. Zinc is generally available in the USA in three grades, which are defined in Table 1, as per ASTM specification B 6. Special High Grade (SHG) (Z13001) is the highest purity and most widely used grade, with 99.990 wt.% minimum zinc content. In Europe grade Z1 (Z12002) has 99.995 wt.% minimum zinc content. High Grade (HG) (Z15001) must contain 99.90 wt.% zinc. Prime Western (PW) (Z19001) grade is normally produced by thermal processing or alloying with lead, has 98.0 wt.% minimum zinc content and is used primarily in batch hot-dip galvanizing.

Some of the more important atomic, structural, chemical, and physical properties of zinc are compiled in Tables 2–7. Unalloyed zinc finds its principal

Table 1
Specification for zinc (ASTM B 6–98).

Grade (UNS)	Lead	Iron (max.)	Cadmium (max.)	Aluminum (max.)	Copper (max.)	Tin (max.)	Total nonzinc (max.)	Zinc (min., by difference)
Special High Grade (Z13001)	0.003 (max.)	0.003	0.003	0.002	0.002	0.001	0.010	99.990
High Grade (Z15001)	0.03 (max.)	0.02	0.02	0.01			0.10	99.90
Prime Western (Z19001)	0.5–1.4	0.05	0.20	0.01	0.20		2.0	98.0

direct uses in corrosion protection. Galvanizing, where a thin layer of zinc is applied to steel is the dominant use along with other zinc coating methods. Sacrificial zinc anodes are attached to steel, structures to provide protection in the presence of an electrolyte such as seawater or moist soils. Numerous other applications result from the basic physical and mechanical properties of zinc.

Table 2
Atomic characteristics of zinc.

Symbol	Zn
Periodic table	Fourth period, group 2b
Atomic number	30
Atomic weight	65.38
Relative abundance of stable isotopes: mass number (%)	64 (48.89), 66 (27.81), 67 (4.00), 68 (8.57), 70 (0.62)
Electron configuration	$1s^22s^25p^63s^23p^63d^{10}4s^2$
Atomic sizes (m):	
Metallic radius (12-coordinated)	1.38×10^{-10}
Ionic (Zn^{2+}) radius	0.74×10^{-10}
Tetrahedral covalent radius	1.31×10^{-10}

2. *Conventional Zinc Die-casting Alloys*

Die-cast components are the dominant form in which zinc alloys are used for structural applications. Zinc castings represent over 20% of zinc consumption in the USA. The conventional die-casting alloys 3, 7, 5, and 2, based on the hypoeutectic Zn–4 wt.% Al composition, have been in widespread use for many years and have by far the greatest commercial use. Higher strengths can be achieved above the eutectic. The ZA alloys based on higher aluminum content are used in both die casting and gravity casting, as are new alloys based on the zinc–copper–aluminum system. These higher alloy content materials as well as other specialty zinc castings alloys are covered in subsequent sections.

The conventional die-casting alloys, which provide the best combination of mechanical properties and casting fluidity, are designed around the Zn–4 wt.% Al composition, with or without copper additions. They have a short freezing range of 5°C with a 387°C melting point. The binary zinc–aluminum eutectic composition at 5 wt.% aluminum provides maximum casting fluidity but lower impact strength. If the aluminum content is reduced below 3.8 wt.%, the tensile strength is reduced and the ease of gravity

Table 3
Structural characteristics of zinc.

Crystal structure	h.c.p.
Lattice parameters (m)	$a = 2.664 \times 10^{-10}$ $c = 4.9469 \times 10^{-10}$ $c/a = 1.857$
Close-packed plane	0001
Close-packed direction	⟨1120⟩
Slip (glide):	
Plane at 20°C	00.1 or 0001
Direction at 20°C	⟨1120⟩
Plane >225°C	1011 or 1012
Direction >225°C	⟨1120⟩
Fracture planes	0001, 1011, 1012, 1010
Twinning plane	1012
Preferred orientations:	
Castings	⟨0001⟩ direction in perpendicular to axis of cast columnar crystals
Wire	0001 plane is parallel to axis of drawn wire
Sheet	0001 plane is parallel to rolling plane and ⟨1120⟩ direction is parallel to rolling direction for sheet rolled at 20°C
Annealed sheet	Texture remains the same as above for sheet rolled at 20°C and annealed
Density ($kg\ m^{-3}$):	
Solid at 25°C	7.133×10^3
Solid at 419.5°C	6.83×10^3
Liquid at 419.5°C	6.62×10^3
Liquid at 800°C	6.25×10^3

Table 4
Thermal properties of zinc.

Melting point (°C)	419.5
Boiling point (°C)	907
Heat of fusion[a] ($J\ kg^{-1}$)	112.95×10^3
Heat of vaporization[b] ($J\ kg^{-1}$)	1755.39×10^3
Molar heat capacity ($J\ mol^{-1}\ K^{-1}$)	$22.36 + 10^{-2}\ T$ (solid[c]), 31.35 (liquid), 20.77 gas)
Specific heat capacity, solid ($J\ kg^{-1}\ K^{-1}$):	
– 200°C	242.4
– 100°C	355.3
0°C	380.4
20°C	385.2
25°C	385.4
100°C	399.6
200°C	413.8
300°C	425.2
400°C	451.9
Specific heat capacity, liquid ($J\ kg^{-1}\ K^{-1}$):	
419.5°C	481
500°C	481
600°C	481
800°C	481
Thermal conductivity, solid ($W\ m^{-1}\ K^{-1}$):	
– 100°C	115.1
0°C	114.9
20°C	113
100°C	110.3
200°C	105.4
300°C	101.1
400°C	96
419.5°C	96
Thermal conductivity, liquid ($W\ m^{-1}\ K^{-1}$):	
419.5°C	49.5
500°C	54.1
600°C	59.9
800°C	(60.17)
Coefficient of linear thermal expansion, polycrystalline solid[d] ($10^6\ K^{-1}$):	
20°C	29
25°C	35
0–100°C	31
20–100°C	31
20–200°C	33
20–300°C	34
Coefficient of linear thermal expansion, single-crystal solid ($10^6\ K^{-1}$):	
0–100°C (*a*-axis)	15
20–100°C (*a*-axis)	14.3
0–100°C (*c*-axis)	61.5
20–100°C (*c*-axis)	60.8
Coefficient of linear thermal expansion, liquid ($10^6\ K^{-1}$):	
500–600°C	60
Coefficient of volumetric thermal expansion, polycrystalline solid, 20–400°C ($10^6\ K^{-1}$)	89
Volumetric contraction due to solidification (%)	4.7

(Continued)

[a]At 419.5°C.
[b] At 907°C.
[c] At $T = 298$–692 K.
[d] $l_t = l_0(1 + 35.4 \times 10^{-3}\ T + 10^{-8}\ T^2)$.

Table 5
Mechanical properties of zinc.

Modulus of elasticity[a] (GPa)	69–138 (no true region of linear stress–strain behavior)
Torsional (rigidity) modulus[a] (GPa)	41.9
Bulk modulus (volume elasticity modulus)[a] (GPa)	69.4
Poisson's ratio[a]	0.249
Surface tension (liquid) ($mN\ m^{-1}$)	$758 \times 0.09(T - 419.5)$, where T is temperature (°C)
Viscosity (liquid) (mPa s)	8.0 at melting point (419.5°C)
Coefficient of friction	0.21 for rolled zinc on rolled zinc

[a]Values for polycrystalline zinc at 20–25°C (modulus of elasticity) or 20°C.

casting in particular decreases. Maximum impact strength is attained at 3.4 wt.% but is reduced above 4.5 wt.% aluminum. The compositions of die-cast alloys 3 (Z33520), 7 (Z33523), 5 (Z35531), and 2 (Z35541) are given in Table 8. Commercially they are known as Zamak (or Mazak in the UK) 3, 7, 5, and 2. Compositions for conventional die-casting alloy and ZA alloy ingot are specified in ASTM B 240 while the compositions for the actual castings are specified in ASTM B 86.

2.1 Zn–4 wt.% Al Alloy System

Conventional Zn–4 wt.% Al zinc die-casting alloys exhibit a fine-grained structure providing optimum tensile strength, hardness, and impact strength. During solidification, primary zinc-rich η-phase crystals with a close-packed hexagonal crystal structure form at 382°C and are embedded in a lamellar matrix of 17.2 wt.% aluminum b.c.c. β phase mixed

Table 6
Electromagnetic properties of zinc.

Electrical resistivity	
Polycrystal at 20°C (Ω m)	5.916×10^{-8}
Polycrystal 0–100°C (Ω m)	$5.46\ (1 + 0.00042T) \times 10^{-8}$
Polycrystal temperature coefficient (Ω m K^{-1})	4.19×10^{-11}
Polycrystal pressure coefficient (Ω m Pa^{-1})	-25×10^{-21}
Single crystals at 20°C (Ω m)	5.89×10^{-8} (*a*-axis), 6.16×10^{-8} (*c*-axis)
Liquid at 423°C (Ω m)	3.7×10^{-5}
Electrical conductivity (% IACS)[b]	28.27
Standard electrode potentials (V)[c]	
vs. standard hydrogen electrode (SHE)	−0.763
vs. standard calomel electrode (SCE)	−1.03
Temperature coefficient (vs. SHE) (V K^{-1})	$+9.1 \times 10^{-5}$
Magnetic susceptibility (diamagnetic):	
Volumetric	-130×10^{6}
Mass	-18.2×10^{-9}
Superconducting characteristics:	
Superconducting transition temperatures in absence of a magnetic field (K)	0.94 ± 0.05
Critical magnetic field to produce super conducting state at absolute zero (A m^{-1})	$(4.3 \pm 0.02) \times 10^{3}$

[a]Between 0.1 MPa and 300 MPa.
[b] % IACS.
[c] Solution potentials for reaction $Zn^{2+} + 2e^{-} \rightarrow Zn$ at 20°C.

Table 7
Optical properties of zinc (normal total emissivity).[a]

State of surface	Temperature (°C)	Emissivity
Zinc, highly polished	"Low"	0.05
Zinc, tarnished	20	0.25
	50	0.21
	280	0.21
Zinc, unspecified surface	100	0.07
Zinc, polished	200–300	0.04–0.05
	325	0.043
Zinc, oxidized at 400°C	400	0.11
Zinc, polished	200–300	0.04–0.05
Zinc, unoxidized	300	0.05
Galvanized sheet iron, bright	28	0.228
Galvanized sheet iron, gray	24	0.276
Galvanized iron	100	0.21

[a]Spectral normal emissivity at room temperature: 0.42–0.58 (at 0.6 μm), 0.50–0.61 (at 1 μm), 0.08 (at 3 μm), 0.05 (at 5 μm), 0.03 (at 1 μm). Coefficient of absorption of solar radiation: zinc, highly polished, 0.55; zinc oxide, 0.15; zinc oxide paint, 0.30.

with the η phase. Aluminum is precipitated from the η phase as the alloy cools to 275°C reducing its aluminum content to 0.6 wt.% while the β phase at this temperature increases to a 22 wt.% aluminum content. A eutectoid decomposition takes place on further cooling into the β phase and a 31.6 wt.% zinc α phase. On cooling to room temperature, the α phase retains excess zinc in solution while aluminum is further precipitated from the η phase. This structure is metastable owing to the rapid cooling rate inherent in pressure die casting, with other structural changes continuing over time. The supercooled β phase decomposes into a eutectoid composed of α and β phase, which creates about 0.06% shrinkage but has little effect on mechanical properties. Supersaturated aluminum precipitates from the η phase as α-phase particles with a 0.06% volume contraction causing a decrease in tensile strength and an increase in elongation over a five-week period. Thermal treatments immediately after casting can stabilize the structure.

2.2 *Zn–4 wt.% Al–Cu Alloys*

Copper-containing zinc–aluminum alloys are comprised of the copper-rich h.c.p. ε phase in addition to the η and α phases at temperatures below the solidus. The η phase decomposes on cooling as in a simple zinc–aluminum binary composition but the copper exerts a restraining effect on the transformation rate. Copper is precipitated as a copper-rich precipitate from the η phase on cooling; however, with rapid cooling this reaction is suppressed creating a metastable microstructure structure. With a ~3 wt.%

Table 8
Composition of zinc die castings (wt.%).[a]

Alloy designation	3	5	7	2
Copper	0.25	0.75–1.25	0.25	2.5–3.0
Aluminum	3.5–4.3	3.5–4.3	3.5–4.3	3.5–4.3
Magnesium	0.020–0.05	0.03–0.08	0.005–0.020	0.020–0.050
Iron	0.100	0.100	0.075	0.100
Lead	0.005	0.005	0.0030	0.005
Cadmium	0.004	0.004	0.0020	0.004
Tin	0.003	0.003	0.0010	0.003
Nickel			0.005–0.020	
Zinc	Remainder	Remainder	Remainder	Remainder

[a]Single figures indicate the maximum amounts permitted.

copper content, as in alloy 2, precipitation can continue over a period of years but thermal treatments can accelerate this rate. Accordingly, the greater stability of alloy 5 with no more than 1.25 wt.% copper is preferred in many countries, although alloy 2 is favored in Germany. Magnesium additions restrain the η phase transformation, as well as providing higher strength and inhibiting intergranular corrosion. Magnesium levels above 0.06 wt.% can lead to hot shortness and reduced impact strength owing to its presence as a lower melting point ternary zinc–aluminum–magnesium eutectic. Lesser levels of magnesium are required with greater purity zinc, leading to the development of alloy 7, which has similar mechanical properties to alloy 3 but superior casting properties allowing higher production rates and better surface finish. Magnesium additions also increase hardness and lower the transition temperature from brittle to ductile behavior during impact loading.

2.3 *Effect of Alloy Additions*

The conventional zinc die-casting alloys are the product of an extensive development program that formulated highly castable, easily finished, economical materials with good mechanical properties. Aluminium additions to zinc have the beneficial effects of increasing strength, reducing grain size, and decreasing the rate at which zinc dissolves in iron and steel, thus making possible the use of higher-speed submerged plunger-type hot chamber die-casting machines. Copper in small quantities increases tensile strength and hardness, but is controlled to maximum limits to prevent an adverse effect on impact strength and dimensional stability. Nickel is specified in alloy 7, in controlled quantities, to neutralize the effect of elements promoting intergranular corrosion. Magnesium is a critical alloying addition to zinc alloys, as it prevents the occurrence of intergranular corrosion.

2.4 *Effect of Impurities*

Early in their development, it was learned that it is absolutely necessary to limit the allowed content of certain impurities and to add certain critical alloying additions to be free of the danger of intergranular corrosion that existed in earlier Zn–4 wt.% Al alloys. Accordingly, SHG zinc of 99.990 wt.% minimum purity is used in die-casting alloy production. Lead, cadmium, and tin contents must be strictly controlled to low limits to prevent any intergranular corrosion. Antimony, arsenic, bismuth, and mercury in amounts above 0.001 wt.% affect the stability of the alloy but are not normally found in SHG zinc or in die-casting operations and accordingly are not specified. Thallium and indium can promote subsurface or intergranular corrosion, but are not found in zinc produced in North America and are not specified in ASTM specifications. However, these elements can be found in zinc produced elsewhere and hence are specified in certain other regional standards. Iron is introduced from contact with cast iron holding pots and tools or as an impurity in aluminum used for alloying. Iron pick-up is minimized owing to the formation of $FeAl_3$. Iron in the allowed concentration does not have a detrimental effect on properties and excess iron combines with aluminum to form a dross that is skimmed off to prevent finishing or machining problems. Limits exist for chromium, silicon, and manganese, although no harmful effects have been observed in the normally specified concentrations and analysis is not required. Manganese, silicon, nickel, and chromium have not been found to promote harmful effects below the specified limits.

2.5 *Conventional Die-casting Alloy Characteristics*

Physical and mechanical properties of these alloys are given in Tables 9 and 10, respectively. Mechanical properties compare favorably with and in some cases exceed those of commonly used aluminum, copper,

Table 9
Physical properties of zinc die castings.

Alloy designation	3 (7)[a]	5	2
Density at 21°C ($g\,cm^{-3}$)	6.6	6.7	6.6
Solidification temperature range (°C)	387–381	386–380	390–379
Specific heat capacity at 20–100°C ($J\,kg^{-1}\,K^{-1}$)	418.7	418.7	418.7
Solidification shrinkage (%)	1.17	1.17	1.25
Coefficient of thermal expansion ($10^{-6}\,K^{-1}$)	27.4	27.4	27.7
Thermal conductivity at 70–140°C ($W\,m^{-1}K^{-1}$)	113.0	108.9	104.7
Electrical conductivity (1% IACS)	27	26	25
Electrical resistivity at 20°C ($10^{-8}\,\Omega m$)	6.4	6.5	6.9
Diemakers shrinkage $mm\,mm^{-1}$	0.007	0.007	0.007

[a]Alloys 3 and 7 have nearly identical physical properties. Values given here are for alloy 3.

Table 10
Mechanical properties of zinc die castings at room temperature.

Alloy designation	3	5	7	2
Ultimate tensile strength (MPa)	283	328	283	359
Yield stress, 0.2% offset (MPa)	221	228	221	
Elongation in 50 mm (%)	10	7	13	7
Hardness, Brinell 500 kg	82	91	80	100
Shear strength (MPa)	214	262.0	214	317
Compressive strength, 0.1% offset (MPa)	414	600	414	641
Impact strength[a] (J)	58	65	58	47
Fatigue strength, 5×10^8 cycles (MPa)	47.6	56.5	46.9	58.6
Young's modulus (MPa)	85.5	85.5	85.5	85.5

[a]6.35% 6.35 men bar unnotched.

and iron casting alloys, particularly impact strength and toughness. The yield strength of zinc die-casting alloys is actually greater than the yield strength of common aluminum die-casting alloys. Hence zinc alloys are used for a wide variety of functional and light structural as well as decorative components. Zinc die castings are in a metastable condition as cast because of the very rapid cooling on contact with the steel die surface during manufacture. Over time, the metastable microstructure moves toward equilibrium, which can result in small dimensional and property changes. Although not normally employed, stabilization heat treatments are available to accelerate this aging effect.

Alloy 3 exhibits excellent mechanical and physical properties, has good long-term dimensional stability, is easy to finish, and has good castability. Some 80–90% of all zinc castings are made from versatile alloy 3. Alloy 7 is a higher purity form of alloy 3 that has better castability, and exhibits a better surface finish making it popular for decorative applications. Greater ductility, however, also requires greater care while casting and trimming alloy 7 to avoid problems with excessive flash. Alloy 2 is the strongest of the 4 wt.% aluminum die-casting alloys and retains strength and stability better after aging but has reduced hardness and impact strength. Alloy 5 has a similar metallurgy to alloy 3 but contains 1 wt.% copper that provides higher tensile strength and hardness along with decreased elongation. Creep resistance is better than the other 4 wt.% aluminum alloys and room temperature stability is better than that of alloy 2.

Special consideration must be given to applications involving sustained loads at elevated temperatures, because of low resistance to creep under certain conditions. Successful use of zinc alloys under such conditions can be achieved by proper attention to design stress considerations. However, other zinc die-casting alloys offer better creep resistance. ZA-27, the Zn–27 wt.% Al casting alloy, has greatly improved creep resistance at room and moderately elevated temperatures (up to 150°C). The ZA alloys have largely replaced alloy 16, a zinc–copper–aluminum–chromium–titanium alloy, developed for better creep properties. Zn–5 wt.% Cu–Al and Zn–10 wt.% Cu–Al die-casting alloys have been developed with superior creep properties. Secondary creep rates for these zinc alloys at various use temperatures have been determined and are available in the literature.

2.6 Zinc Die-casting Process

The ability of the die-casting process to fabricate large numbers of complex parts rapidly, in intricate detail, to tight dimensional tolerances and with excellent surface finish reaches its zenith with zinc. Accordingly, zinc die casting has been called the fastest way from molten metal to finished part. The low melting temperature and low attack rate of molten zinc alloys on iron and steel allows the faster and more economical hot-chamber die-casting method to be used, compared to aluminum and other higher melting point alloys. The development of thin-wall zinc die-casting technology since the 1970s has further extended the ability of die casting to fabricate weight-saving, material-efficient components using only the metal required by the part function. Wall thicknesses over large areas of 1 mm or even 0.5 mm are possible, which are much thinner than normally possible with aluminum. This technological advance has significantly extended the product capability and range of the die-casting process. Die lives can be $5 \times 10^5 - 2 \times 10^6$ shots.

Zinc alloys exhibit excellent machinability at high speeds with minimal tool wear and can be fastened by a wide range of joining techniques including welding and soldering. The surface of a zinc die casting is highly receptive to many commercial organic or inorganic finishes used for cosmetic appeal, improved corrosion protection, and greater wear resistance. Zinc castings are painted, dyed, electroplated, anodized, and chromate-treated. Both decorative bright chrome and wear-resistant hard chrome finishes are commonly applied, as are many other electroplated coatings including cadmium, nickel, copper, and brass. Zinc anodizing produces a mossy green finish, which is hard and highly resistant to corrosion and wear. Chemical conversion treatments are used to provide corrosion protection resulting from handling and moisture and also serve as an excellent paint base.

Because of this combination of good material properties, thin-wall casting capability, finishing characteristics, and economy of manufacture, zinc die castings have become widely used, with major applications in automotive components, builders' hardware, electrical fittings, appliances, commercial machinery, sporting goods, computer housings, and scientific and electronic equipment.

3. ZA Casting Alloys

A separate family of high-performance engineering alloys is growing in commercial use. These zinc–aluminum alloys, commonly referred to as ZA alloys, are based on higher aluminum contents than the conventional die-casting alloys. Three alloys, ZA-8, ZA-12, and ZA-27, have been developed; the numerical suffix indicates the nominal aluminum content (Table 11). Copper additions improve the strength, hardness, creep resistance, and corrosion resistance and are balanced to prevent excessive property changes and to maintain dimensional stability on aging. Magnesium additions increase strength and hardness and are also effective in preventing intergranular corrosion. As with the conventional die-casting alloys, lead, cadmium, and tin levels are strictly controlled to prevent the occurrence of intergranular corrosion. These hypereutectic zinc–aluminum alloys were designed to take advantage of solid-solution hardening, precipitation hardening, and changes in the amounts and nature of the aluminum- and zinc-rich phases formed, compared with the conventional hypoeutectic zinc–aluminum die-casting alloys. The microstructure of ZA alloys consists of aluminum-rich α-phase dendrites, an α- and η-phase eutectic, with a small amount of copper-rich ε phase.

The ZA alloys exhibit excellent mechanical properties, with strength and hardness significantly greater than those of commonly used gravity- and die-cast zinc, aluminum, copper, and iron alloys. Bearing characteristics and wear resistance also

Table 11
Composition of ZA alloy castings (wt.%).[a]

	ZA-8 (UNS Z25630)	ZA-12 (UNS Z35630)	ZA-27 (UNS Z35840)
Aluminum	8.0–8.8	10.5–11.5	25.0–28.0
Copper	0.8–1.3	0.5–1.25	2.0–2.5
Magnesium	0.015–0.030	0.015–0.030	0.010–0.020
Iron	0.075	0.075	0.075
Lead	0.006	0.006	0.006
Cadmium	0.006	0.006	0.006
Tin	0.003	0.003	0.003

[a]Single figures indicate the maximum amounts; zinc forms the remainder in all cases.

Table 12
Physical properties of the ZA alloy castings.

	ZA-8[a]	ZA-12[b]	ZA-27[c]
Density ($g\,cm^{-3}$)	6.3	6.03	5.0
Melting range (°C)	375–404	377–432	375–487
Solidification shrinkage (%)	1.1	1.3	1.3
Specific heat capacity ($J\,kg^{-1}\,K^{-1}$)	435	450	525
Coefficient of thermal expansion ($10^{-6}\,K^{-1}$)	23.3	24.2	26.0
Thermal expansion ($10^{-6}\,K^{-1}$)	23.3	24.2	26.0
Thermal conductivity ($W\,m^{-1}\,K^{-1}$)	115	116	126
Electrical conductivity (% IACS)	27.7	28.3	29.7
Electrical resistivity RCM at 20°C	6.2	6.1	5.8
Pattern/mold/die shrinkage ($mm\,m^{-1}$)[d]	10.4	13.0	13.0

[a]Sand cast or permanent mold cast.
[b]Sand cast, permanent mold cast, or die cast.
[c]Sand cast HT (heat treated at 320°C for 3 h and furnace cooled) or die cast.

compare favorably with those of widely used bearing bronzes. Physical and mechanical properties of these alloys are given in Tables 12 and 13, respectively.

ZA alloys and zinc alloys in general are used in virtually every casting process allowing the economical fabrication of parts in quantities from millions per year down to a single component. ZA-8, with 8 wt.% aluminum, is die cast by the faster, more economical hot-chamber process, while ZA-12 and ZA-27, with 11 wt.% and 27 wt.% aluminum, respectively, are cold-chamber die cast. The ZA alloys are cast by sand, permanent mold, graphite permanent mold, shell mold, and high-pressure die-casting methods. The graphite permanent mold process was specifically developed on the basis of the castability and low casting temperatures of the ZA alloys. Plaster mold casting in ZA-12 is used for prototyping all zinc alloys and processes to prove new designs. Investment, low-pressure permanent mold, centrifugal, continuous, and less traditional rubber, bronze, and aluminum mold casting processes are also used. ZA alloy applications take advantage of the unique engineering properties and fabrication capability of these alloys including bearing and wear-resistant parts including components fabricated without the need for pressed-in bushings or cast-in wear inserts. ZA alloys have proved to be successful replacements for components previously produced in bronze, brass, aluminum, and cast iron or steel, and provide a new cost-effective alternative in materials selection. Automotive applications, in particular, take advantage of the higher strength to weight ratio of ZA castings.

4. Other Casting Alloys

4.1 Zinc–Copper–Aluminum Alloys

Alloys based on the zinc–copper–aluminum system provide greater strength and improved creep resistance. The Zn–5 wt.% Cu–3wt.% Al alloy was developed initially with higher strength, higher temperature automotive applications in mind. Ingot and casting specifications are found in ASTM B 892 and B 894, respectively. The microstructure shows ε-phase dendrites surrounded by an α–η–ε ternary eutectic. The copper-rich ε phase constitutes ~30% of the microstructure and is harder and stronger than the other two aluminum-rich phases and acts to reinforce and strengthen the alloy matrix. The alloy is either hot-chamber die cast or gravity cast. A separate but related 10 wt.% copper alloy has been developed for cold-chamber die casting, which exhibits higher strength based on a greater volume fraction of ε phase in the ternary eutectic microstructure. Lower secondary creep rates and higher strength than other zinc casting alloys characterize these zinc–copper–aluminum alloys. A low friction coefficient and high wear characteristics allows for their use in bearing applications. Other applications include automotive components, snow mobile shock absorbers, tractor parts, window security system gearbox components, electrical connectors, and video recorder items.

4.2 Slush-casting Alloys

Zinc slush-casting alloys are used primarily for casting of decorative items including small statues, lighting fixtures, lamp bases, and casket hardware. ASTM B 792 specifies slush casting alloy A (Z34510) with 4.75 wt.% aluminum and 0.25 wt.% copper, and alloy B (Z30500) containing 5.5 wt.% aluminum. These alloys have tensile strengths of 175–195 MPa and unnotched Charpy impact strengths of 1.4–4.2 N m at room temperature.

4.3 Forming Die/Injecting Molding Alloys

A separate family of alloys exhibiting high impact and compressive strengths is used in the preparation of

Table 13
Mechanical properties of ZA castings.

	ZA-8			ZA-12			ZA-27		
	Sand cast	Permanent mold cast	Die cast	Sand cast	Permanent mold cast	Die cast	Sand cast	Sand cast HT[a]	Die cast
Ultimate tensile strength (MPa)	263	221–225	374	276–317	310–345	404	400–440	310–325	414–427
Tensile yield strength, 0.2% offset (MPa)	198	208	290	211	268	320	371	257	376
Compressive yield strength, 0.1% offset (MPa)	199	210	252	230	235	269	330	257	385
Elongation (% in 50 mm)	1–2	1–2	6–10	1–3	1–3	4–7	3–6	8–11	1–3
Hardness, Brinell	85	87	103	94	89	100	113	94	119
Shear strength (MPa)		241	275	253	241	296	292	225	325
Impact strength unnotched Charpy (J)	20		42	25	29		48	58	13
Fatigue strength[b] (MPa)		51.7	103	103	103	117	172	103	145
Young's modulus (GPa)	85.5	85.5	85.5	83	83	83	78	78	78

[a]Heat treated at 320 °C for 3 h and furnace cooled.
[b]Rotary bend at 5×10^8 cycles.

forming dies used in the pressing and stamping of metal sheet, and for injection molds for plastics. These forming die alloys, commonly known as Kirksite, typically contain 4 wt.% aluminum and 3 wt.% copper with or without a small amount of magnesium. Fabrication is by plaster or sand casting. ASTM B 793 specifies alloy A (Z35543) with 4.0 wt.% aluminum, 3.0 wt.% copper, and 0.06 wt.% magnesium, and alloy B (Z35542) containing 4.1 wt.% aluminum, 2.7 wt.% copper, and 0.04 wt.% magnesium. Tensile strengths are 210–280 MPa and unnotched Charpy impact strengths ~21–28 N m at room temperature.

5. *Wrought Zinc Alloys*

Rolled zinc sheet and strip is the major wrought zinc product form, but wire, extrusions, and forgings are also in important commercial use.

5.1 Rolled Zinc Sheet and Strip

The overwhelming majority of wrought zinc alloys and applications are for rolled zinc sheet and strip. The conventional rolled zinc alloys may be conveniently classified according to major alloy groups: (i) unalloyed zinc (Z13004 and Z15006), (ii) zinc–low copper (Z40101), (iii) zinc–high copper (Z40301), (iv) zinc–low copper–titanium (Z41121), (v) zinc–high copper–titanium (Z41321), (vi) zinc–lead (Z20301), (vii) zinc–lead–cadmium (Z21721), (viii) zinc–lead–manganese (Z24311), and (ix) a range of zinc–aluminum alloys. The nominal compositions or composition ranges for commercial alloys in each of these major groups are summarized in Table 14. These are all relatively low alloy content metals, with the exception of some zinc–aluminum alloys. Rolled zinc alloys are specified in ASTM B 69.

Because of their h.c.p. crystal structure, rolled zinc alloy properties are anisotropic and they exhibit rolling and annealing textures. Unalloyed rolled zinc creeps at room temperature, with additions of copper or copper and titanium employed to improve its creep resistance. The zinc–copper–titanium alloys, in particular, are widely used in a variety of thin-sheet architectural applications such as roofing and rain ware, which are popular in Europe. The creep resistance of these alloys, however, is sensitive to mechanical and thermal processing and is directional within the sheet. Proper heat treatment after rolling minimizes the amount of directional creep behavior. The room temperature mechanical properties of rolled zinc alloys are summarized in Table 15, as measured in the longitudinal direction. It should be noted that the properties of rolled zinc alloys are to an unusually high degree a result of a particular combination of their composition and thermal–mechanical processing history. Commercially, very

Table 14
Chemical composition of rolled zinc alloys.[a]

Alloy	Cu	Pb	Fe	Cd	Ti	Al	Sn	Mn	Mg
Rolled special high-grade zinc (Z13004)	0.003 max.	0.003 max.	0.003 max.	0.003 max.		0.002 max.	0.001 max.		
Commercially pure rolled zinc (Z15006)	0.08 max.	0.03 max.	0.02 max.	0.01 max.	0.02 max.	0.01 max.	0.003 max.		
Zinc–low copper rolled zinc alloy (Z40101)	0.08–0.040	0.01 max.	0.01 max.	0.005 max.	0.02 max.	0.01 max.	0.003 max.		
Zinc–high copper–titanium rolled zinc alloy (Z41121)	0.50–1.0	0.01 max.	0.01 max.	0.005 max.	0.04 max.	0.01 max.	0.003 max.		
Zinc–low copper–titanium rolled zinc alloy (Z41121)	0.08–0.49	0.01 max.	0.01 max.	0.005 max.	0.05–0.18	0.01 max.	0.003 max.		
Zinc–high copper–titanium rolled zinc alloy (Z41321)	0.50–1.00	0.01 max.	0.01 max.	0.005 max.	0.08–0.18	0.01 max.	0.003 max.		
Zinc–lead rolled zinc alloy (ZZ20301)	0.005 max.	0.10 max.	0.01 max.	0.01 max.	0.02 max.	0.002 max.			
Zinc–lead–cadmium rolled zinc alloy (X21721)	0.005 max.	1.0 max.	0.01 max.	0.07 max.	0.02 max.	0.002 max.			
Zinc–lead–manganese rolled zinc alloy (Z24311)	0.005 max.	0.03–0.08	0.01 max.	0.005 max.	0.02 max.	0.002 max.		0.015 max.	0.0015 max.
Zinc–aluminum rolled zinc alloy	5.0 max.	0.05 max.	0.1 max.	0.15 max.	0.02 max.	1.4–34.0	0.003 max.		0.10 max.

[a]Zinc: balance by difference.

similar properties are obtained with different starting chemistries because of different subsequent thermal–mechanical processing schemes. Because of this characteristic, zinc sheet and strip is frequently specified based on required performance properties rather than primarily by chemistry. Rolled zinc can be readily formed by a number of mechanical working techniques and is easily joined by soldering or resistance welding.

A major application for rolled zinc is in coinage. The US penny has been produced since 1982 from the Zn–0.8 wt.% Cu alloy, with or without a small amount of titanium (under 0.02 wt.%). The penny looks like the previous 95 wt.% Cu–5 wt.% Zn alloy penny because it is copper plated but is less costly and achieves a 19% weight saving. Worldwide, significant usage remains for dry cell batteries where callots punched from zinc sheet containing small additions of lead and cadmium are impact extruded into battery cans for the traditional zinc–carbon cell. However, in the USA and Western Europe, alkaline batteries using zinc powder have largely replaced the zinc–carbon cell. Other applications for rolled zinc are found in architecture and building materials, automotive bright trim, photoengraving, electrical contacts, superplastic alloys, and miscellaneous products similar to those for which brass is employed.

The zinc–aluminum rolled alloys include a group of "superplastic" metals based on the composition 78 wt.% Zn–22 wt.% Al with small additions of copper and/or magnesium. These unique materials can be uniformly and plastically deformed in much the same manner as plastic and molten glass. They are superplastic at moderately elevated temperatures and low strain rates and exhibit good mechanical properties at room temperature. The great advantage of these alloys is that they can be easily fabricated into the desired shape and then heat treated so that they exhibit good mechanical properties. However, commercial application has been limited. Superplastic zinc exhibits fascinating metallurgical characteristics. The room temperature mechanical properties of three zinc–aluminum superplastic alloys are given in Table 16. The effects of adding copper and magnesium to the 78 wt.% Zn–22 wt.% Al base composition include increased tensile, yield, and creep strengths, and hardness, accompanied by decreased ductility. Impact strength is only moderately affected. Annealing at 350°C followed by air cooling also produces significant improvements in tensile, yield, and creep strengths, and hardness, although the degree of improvement is less in the more highly alloyed materials. This heat treatment decreases ductility and impact strength except in the case of the zinc–aluminum–copper–magnesium alloy, where impact strength increases slightly. The superplastic zinc–aluminum alloys are normally fabricated by techniques such as thermal forming, compression molding, deep drawing, stretch forming, or hot pressing of powders.

Table 15
Mechanical properties of rolled zinc alloys (tensile tested in longitudinal direction).

Alloy (UNS)	Tensile strength (ksi)	Tensile strength (MPa)	Elongation (%)	Hardness (HR15T)
Rolled special high-grade zinc (Z13004)	9–18	62–124	15–45	24–40
Commercially pure rolled zinc (Z15006)	10–18	69–124	15–60	28–45
Zinc–low copper rolled zinc alloy (Z40101)	12–27	83–186	28–60	34–64
Zinc–high copper rolled zinc alloy (Z40301)	20–34	138–234	30–90	54–72
Zinc–low copper–titanium rolled zinc alloy (Z41121)	14–28	96–193	30–70	35–67
Zinc–high copper–titanium rolled zinc alloy (Z41321)	20–34	138–234	30–90	54–72
Zinc–lead rolled zinc alloy (Z20301)	18–28	124–193	40–65	50–68
Zinc–lead–cadmium rolled zinc alloy (Z21721)	14–30	96–207	30–50	52–70
Zinc–lead–manganese rolled zinc alloy (Z24311)	14	96–193	45–80	40–70
Zinc–aluminum rolled zinc alloy	21–80	145–550	3–90	46–84

5.2 *Zinc Wire*

Drawn zinc wire products comprise the second major category of wrought zinc alloys. Zinc wire is produced by continuous casting, rolling, and finally drawing. Wire is normally made from 1 mm to 6 mm in diameter. The largest use of zinc wire is in thermal spraying (metallizing) where it is used primarily for corrosion protection of steel structures such as bridges, but also for electromagnetic shielding applications on electronic housings. ASTM B 833 specifies two commercially pure zinc grades with 99.99 wt.% (Z13005) and 99.9 wt.% (Z15005) minimum zinc contents and the 85 wt.% zinc–15 wt.% aluminum alloy (Z30700) that offers greater corrosion resistance in many environments and retains sufficient flexibility for easy use. Another significant use of zinc alloy wire is for solders employed to solder aluminum and to repair damaged galvanized coatings. These solder alloys include zinc–aluminum, zinc–aluminum–copper, zinc–tin, zinc–tin–copper, zinc–cadmium–tin, and zinc–cadmium alloys, as specified in ASTM B 907. Zn–0.7 wt.% Cu and Zn–2-4 wt.% Al solders with small additions of nickel, chromium, and titanium are of interest for soldering copper–brass automotive radiators for a combination of vibration and creep resistance along with resistance to corrosion. Other zinc wire applications include nails, screws, slate hooks, and wire gauze.

5.3 *Forged and Extruded Zinc*

Wrought zinc alloys are also fabricated into forged and extruded products, but have found limited application. These alloys are of two systems, zinc–aluminum–copper and zinc–copper–titanium. ZA-27 is used for extrusion and forging, as well as for high-strength gravity and die castings. Extrusion of ZA-27 improves its tensile and yield strengths and imparts greater ductility compared to as-cast properties. The Zn–14.5 wt.% Al–0.7 wt.% Cu–0.02 wt.% Mn alloy was developed for specialty forging applications requiring good impact strength at low temperatures, and also exhibits high strength and ductility. Zinc–copper–titanium and zinc–copper–titanium–manganese forging alloys are also in use. Typical compositions and room temperature mechanical properties of the forged and extruded zinc alloys are given in Table 17. Zinc alloy forgings and extrusions can be readily machined, joined by soldering, welding, or adhesives, and finished with paints, polymers, or electroplated coatings. Secondary forming operations, such as bending, swaging, flaring, stamping, and coining can be easily carried out.

6. *Zinc as an Alloying Element*

A significant use of zinc is as an alloying element in several other alloy systems. The predominant use is with copper to form brasses and bronzes where zinc frequently constitutes 30–40 wt.% of the alloy composition. Zinc additions to cast and wrought aluminum are another important use. Magnesium- and tin-based alloys contain zinc as do a number of complex, low-melting brazing and soldering alloys which may contain, in addition to zinc, elements such as cadmium, silver, lead, tin, aluminum, copper, antimony, iron, nickel, bismuth, manganese, phosphorus, silicon, or gold.

7. *Zinc Coatings for Corrosion Protection*

Zinc coatings are universally recognized as an important commercial method for protecting iron and steel from rust and represent more then 50 wt.% of all zinc use. Because of its position relative to iron in the electromotive series, zinc corrodes preferentially in a galvanic cell of the two metals. However, the protective zinc coating dissolves at only a moderate rate because its standard electrode potential is only

Table 16
Room temperature mechanical properties of superplastic zinc–aluminum alloys.

Alloy	Heat treatment	Tensile strength (MPa)	0.2% yield strength (MPa)	Elongation (%)	Elastic modulus (GPa)	Hardness, Rockwell (T-15)	Creep strength[a] (MPa)	Impact strength (J)
Zn–22% Al	As rolled	193.0	158.6	98	42.7	46	12.0	32.5
	Annealed 350°C, air cooled	351.6	310.3	13	63.4	79	114.0	24.4
Zn–22% Al–0.5% Cu	As rolled	310.0	255.1	27	80.0	70	206.8	32.5
	Annealed 350°C, air cooled	400.0	351.6	11	67.6	84	413.6–482.6	24.4
Zn–22% Al–0.5% Cu–0.02% Mg	As rolled	379.2	296.5	25	93.1	79	248.2	28.5
	Annealed 350°C, air cooled	441.2	386.1	9	68.3	85	551.6–689.5	29.8

[a]Stress to produce a creep rate of 0.01% per 1000 h (1% in 11.4 years).

slightly more electronegative than the values for various iron reactions. Furthermore, in the pH range 5–10, it has been found that a dense, protective film forms that slows down the rate of zinc dissolution. This complex film consists primarily of zinc carbonate and zinc oxide. Thus, zinc may be used to protect iron and steel from corrosion for long periods. Galvanic protection of iron or steel by zinc can take place by two methods: coating the ferrous metal with a thin layer of zinc, or attaching a sacrificial zinc anode to the steel structure in the presence of an electrolyte such as seawater or moist soils.

Historically, only commercial zinc grades were used for coatings. However, zinc alloys have become important because of greater corrosion protection, superior engineering properties such as formability, and their ability to successfully coat difficult steel compositions. Alloy coating microstructures have a significant effect on engineering properties. Several methods of applying zinc coatings are in general use: hot-dip galvanizing after fabrication (batch galvanizing), continuous hot-dip galvanizing, continuous electrogalvanizing, electroplating, mechanical plating, spraying, and painting with zinc-rich paints.

7.1 Hot-dip Galvanizing after Fabrication (Batch Galvanizing)

Hot-dip galvanizing has been employed to protect iron and steel since 1741. In hot-dip galvanizing after fabrication or batch galvanizing, the iron or steel article is completely immersed in a bath of molten zinc at ~450°C. The process is applied primarily to finished articles and semifinished shapes such as large steel beams, castings, or forgings. Galvanizing forms a durable, protective zinc coating that completely seals edges, rivets, seams and welds so that there are no unprotected areas from which rusting can start. In the hot-dip galvanizing process, a series of intermetallic iron–zinc compound layers are formed. The intermetallic phases are formed by a diffusion reaction during immersion with increasing zinc content moving out from the ferrous surface of: 75 wt.% (γ), 90 wt.% (δ), 94 wt.% (ζ), and 100 wt.% (η), respectively. The free zinc outer η phase forms during withdrawal of the steel article from the molten zinc bath. Hot-dip galvanized coatings are thoroughly bonded metallurgically to the steel substrate. Historically, in actual practice Prime Western zinc has been used which contains ~1 wt.% lead to improve zinc fluidity and drainage. Because a 1 wt.% lead content is above the solubility level of lead in zinc, a molten lead layer forms on the bottom of the galvanizing kettle. This practice allows easier removal of the zinc–iron bottom dross that floats on

Table 17
Typical compositions (wt.%)[a] and room temperature mechanical properties of forged or extruded zinc alloys.

Alloy	Al	Mg	Cu	Ti	Mn	Tensile strength (MPa)	0.2% yield strength (MPa)	Elongation (%)	Brinell hardness	Shear strength (MPa)	Creep strength[b] (MPa)
Zn–11 Al–Cu	11.0	0.02	0.75			344–496	310–414	18–13	100–120	248	21
Zn–14.5 Al–Cu	14.5	0.02	0.70			345–517	310–448	20–14	100–125	248	21
Zn–Cu—Ti			1.0	0.1		241	159	30	65	172	69
Zn–Cu—Ti–Mn			1.0	0.1	1.0	331	241	20	90	228	103

[a]Balance zinc.
[b]Stress to produce creep rate of 1% in 10 years.

the lead but under the zinc. Lead also promotes the formation of larger and more pronounced spangles or surface grains. Some galvanizers have started operating with Special High Grade or High Grade zinc but these are in the minority. The use of a 0.1 wt.% bismuth concentration in the zinc bath has been found to provide most of the benefits of lead with reduced environmental concerns. Aluminum at a low level of ~0.005 wt.% is used to brighten the appearance of the galvanized coating and is normally added using master alloys. In some cases, tin is now substituted for aluminum producing a bright appearance having a deeper sheen, without the harmful interactions with the flux employed in galvanizing.

The thickness of the hot-dip galvanized coating depends on the time of immersion in the molten zinc bath, the bath temperature, the composition of the steel being coated, and the thickness of the steel being galvanized. Traditional steels rapidly approach a saturation coating thickness, which is self-regulating and not increased by additional immersion time in the bath. However, reactive steels such as those produced in continuous casting, which contain silicon and phosphorous in particular concentrations, exhibit an excessively thick dull gray galvanized coating, which in some instances is so thick and brittle it flakes off. The reason for this is that the structure is dominated by long needles of the hard and brittle ζ phase that extend to the surface, consuming the pure zinc η layer after withdrawal from the bath, and resulting in a brittle coating that can be up to three times thicker than normal. The use of a Zn–0.05 wt.% Ni concentration in the bath solves the problem with additions in the form of a Zn–0.5 wt.% Ni master alloy or the introduction of nickel powder. In other cases, strict process control via lower bath temperatures, shorter immersion times, rapid cooling, and smoother steel surfaces allows satisfactory galvanizing.

7.2 Continuous Hot-dip Galvanizing

The continuous hot-dip galvanizing process is applied to sheet, strip, wire, tube, and pipe products manufactured in continuous operations. The advantages of this technique are high production rates and a more ductile galvanized product, which is readily formed. The overwhelming use of zinc in galvanizing is for sheet galvanized products and this is where major technological improvements and alloy coating development have taken place. A small addition of aluminum to the galvanizing bath and a limited immersion time in the bath produce a thin zinc coating with only a thin intermetallic layer, in contrast to the thick multiphase iron–zinc coatings characteristic of batch hot-dip galvanizing. The use of Continuous Galvanizing Grade (CGG) zinc alloys containing 0.25–1.0 wt.% zinc has become commonplace. The trend is for minimum lead and/or antimony additions, producing a smaller size spangle that is more ductile and allows a superior painted surface. Standardized CGG grades are recognized in ASTM B 852. Zinc–aluminum and zinc–antimony master alloys are specified in ASTM B 860. The process is flexible enough to allow application of varying coating thicknesses. Applications are in steel framing, roofing, painted building panels, construction, automotive appliances, and a multitude of other end-use areas.

Alloy coatings have been developed that offer enhanced engineering performance. In these instances, the enhanced performance of the zinc alloy coating has removed prior limitations and allowed the potential of the underlying steel sheet to be more fully used. Galvanneal is a zinc–iron coating produced in much the same way as conventional galvanized. However, the bath has a reduced aluminum level to promote greater intermetallic phase formation and the surface is heated after removal from the bath during which the zinc η surface layer is converted by diffusion to iron–zinc intermetallic phases. Galvanneal is widely used in the automotive industry because it has better weldability and paintability than conventional galvanized.

Zinc–aluminum coatings are produced by passing the steel through a bath containing 1–55 wt.% aluminum. Two coatings offering enhanced corrosion resistance and other performance features, Galfan and Galvalume, have become widely used worldwide. Galvalume with ~55 wt.% aluminum, 1.5 wt.% silicon, and the balance zinc provides the better bold corrosion resistance of aluminum with many of the features of galvanized. The microstructure consists of the outer coating and a thin intermetallic layer, which bonds to the steel. The silicon addition restricts the formation of the intermetallic layer that consists of two intermetallic compounds, with the inner sublayer being a quaternary Al–Fe–Si–Zn compound and the outer layer a ternary Al–Si–Fe compound. The multiphase outer coating contains cored aluminum-rich dendrites and zinc-rich interdendritic areas, with occasional silicon particles interspersed. Because ~80% of the volume is aluminum, the coating performs more like an aluminum coating. Galvalume uses include roofing and siding, appliance uses based on its high reflectivity and heat oxidation resistance, plus applications based on its general utility.

Galfan contains 95 wt.% zinc, 5 wt.% aluminum, and a small amount of mischmetal. Superzinc is a similar product using a small magnesium addition instead of mischmetal. The multiphase microstructure of Galfan is characteristic of its eutectic composition, exhibiting a lamellar structure of alternating zinc- and aluminum-rich phases. The fineness of the structure increases with higher cooling rate and is completely eutectic when fast cooled. With slower cooling rates, primary zinc- or aluminum-rich crystals appear, depending on composition. Commercially, Galfan is normally fast cooled to generate the desirable fine eutectic microstructure. In addition to two to three times the corrosion life of conventional galvanized steel, Galfan provides superior sacrificial cut edge protection. A unique feature of Galfan is the virtual absence of an intermetallic phase between the steel substrate and the Zn–5 wt.% Al coating. This condition is the result of the mischmetal addition, consisting primarily of lanthanum and cerium, which allows complete wetting of the steel by the alloy coating and suppresses intermetallic phase formation. The absence of an intermetallic phase renders Galfan highly formable as crack nucleation at the intermetallic is avoided and the ductile fine eutectic microstructure itself is highly resistant to crack propagation. The surface chemistry of Galfan is highly receptive for painting. Thus the intrinsic corrosion resistance and sacrificial protection of Galfan can be used in fabricated and painted products. Painted building applications, formed automotive components, bare and painted appliance parts and cabinets, deep-drawn motor housings, and fencing are among the applications of Galfan.

7.3 Electrogalvanizing

Electrogalvanizing is employed to obtain a thin, readily formed coating on steel sheet and wire. A spangle-free, uniform finish characterizes electrogalvanized products with excellent coating adherence during forming. They are readily prepared for painting by phosphatizing, a chemical conversion coating process carried out by dipping in a mild phosphoric acid solution, and have excellent paint adherence. Electrogalvanizing allows close dimensional tolerances to be achieved. Electrogalvanized sheet can be produced with a zinc or zinc alloy coating on one or both sides or with a heavier coating thickness on one side. A thin zinc coating (flash coat) has traditionally found use for limited corrosion exposure where a smooth finish is desired. Heavier coating weight products have been widely used in the automotive industry for corrosion-resistant sheet applications where a smooth surface finish is paramount, either as pure zinc or zinc–iron coatings. Other alloy coatings, particularly zinc–nickel, have become popular, especially in Japan where their greater corrosion resistance allows the use of thinner coatings and a lower electrical cost during plating onto the sheet. However, advances in continuous hot–dip sheet technology have allowed parallel results to be achieved at reduced cost which has caused a reduction in electogalvanized use.

7.4 Electroplating

Zinc electroplating is used to apply a zinc or zinc alloy coating to steel parts by electrodeposition and is closely related to electrogalvanizing. It is a controlled zinc coating process and is ideal for parts where close dimensional tolerances are required. Electroplating is often applied to large numbers of small parts coated in a rack or barrel plating operation, because it produces a thin, uniform coating on all parts. Another advantage of these processes is that the base material is never exposed as all surfaces are coated. Zinc electroplating is ideal to apply a corrosion-resistant or decorative coating to fasteners, hardware, and other small parts. Various zinc alloy coatings are also electroplated.

7.5 Mechanical Plating

Mechanical plating relies on mechanical forces to bond a thin zinc layer to the iron or steel part. There are two zinc mechanical plating processes. The first is a cold welding process commonly known as mechanical plating or mechanical galvanizing. It is accomplished by first tumbling the part to be coated in a mixture of glass beads, water, an inhibited acidic powder cleaner, and an acid copper salt at pH 1.8. This process cleans the parts and produces an immersion coating of copper. A catalyst and zinc powder, sized 8–20 μm, is then added and the parts tumbled again. The result is a thin, uniform zinc coating of closely controlled thickness. Zinc–cadmium and other alloy coatings may also be applied by this method. The second process, called sherardizing, is much less widely used, and primarily in Europe. It produces a thin, uniform zinc coating by heating the part with zinc dust and sand in a slowly rotating drum. Both mechanical plating processes are advantageous for coating large numbers of small parts where close control of coating thickness and coating uniformity is required. The mechanical galvanizing system has the additional advantage that the parts are not exposed to elevated temperatures.

7.6 Thermal Spraying (Metallizing)

Coating by thermal spraying, also called metallizing, is accomplished by projecting droplets of molten zinc onto the surface of the item to be coated. The atomized droplets can be produced from powder, wire, or molten zinc; a stream of compressed air or inert gas impels them to the target. Most thermally sprayed zinc is produced using zinc or 85 wt.% Zn–15 wt.% Al wire, as specified in ASTM B 833. The sprayed zinc coating is slightly rough and porous compared with the coatings produced by other methods. However, these characteristics do not detract from the production of a fully protective coating. Moreover, their roughness and porosity make zinc-sprayed coatings ideal for subsequent painting. Zinc spraying has the additional advantage that it can be applied to almost any size or shape of fabricated assembly and can achieve virtually complete coverage. It is the only method by which unusually heavy zinc coatings can be applied. It is generally not the process of choice for small parts or for parts where very thin, very uniform, and closely controlled coating thicknesses are required. Hence, most applications are for large structural applications such as bridges and exposed steelwork.

7.7 Zinc-rich Paints

A protective zinc coating can be applied to articles by painting, provided the volume fraction of the zinc dust or powder in the dry paint film is high enough to allow electrical contact from particle to particle and to the steel substrate. The paint thus functions both as a barrier coating and as a sacrificial coating. In addition, there are several zinc chemicals, such as zinc oxide and zinc chromate, which perform other functions in corrosion-resistant paints. Only the galvanically protective or "zinc-rich" paints are discussed here. The exact nature of the protective mechanism is uncertain and it may be somewhat different than traditional galvanic protection.

To achieve the particle–particle and particle–substrate electrical contact, the zinc dust content generally must be 80–95 vol.% of the total solids in the paint. The actual weight percent depends on the nature of the binder and other additives. The vehicles used to bind the pigment into the paint film may be organic or inorganic, and there are numerous zinc-rich paint formulations designed to obtain different paint characteristics. Although zinc-rich paints are generally used as primers for subsequent topcoats, they have been used alone or as coatings for galvanized products to obtain even greater corrosion protection.

Applications for zinc-rich paints include pipelines, oil tankers, steel structures such as bridges, electrical power transmission towers, automotive body panels and repair of defects in galvanized coatings.

8. Sacrificial Zinc Anodes

Sacrificial zinc anodes provide cathodic protection of metal structures immersed in an electrolyte. This is a different approach than that of zinc coatings, as a separate external zinc anode is physically and electrically attached to the article to be protected. As is the case with zinc coatings, corrosion protection is realized because zinc is sacrificed in the galvanic cell formed by the steel structure, the zinc anode, and the current-carrying electrolyte. Seawater and moist soils with sufficient contents of positive and negative ions are examples of solutions capable of carrying currents. Zinc anodes generally conform to ASTM B 418 specification for cast and wrought galvanic zinc anodes for use in saline electrolytes and have the following compositions (wt.%, balance zinc). Type 1: 0.1–0.4 Al, 0.03–0.10 Cd, 0.005 Fe; Type 2: 0.005 Al, 0.003 Cd, 0.0014 Fe. Ship hulls, pier pilings immersed in seawater, spillways and lock gates, and water storage tanks are all examples of underwater applications where zinc anodes may be used for corrosion protection. Underground uses include transmission pipelines, gas distribution systems, transmission tower footings and guy anchors, and domestic oil storage tanks. Zinc anodes can also be used for ground rods and grounding cells.

Bibliography

American Zinc Association 1998 *Zinc Recycled Forever*. Washington, DC

ASM 1979 *Metals Handbook: Vol. 2. Properties and Selection: Non-Ferrous Alloys and Pure Metals*, 9th edn. ASM International, Materials Park, OH

Fink D G (ed.) 1968 *Standard Handbook of Electrical Engineers*, 10th edn. McGraw-Hill, New York

Hampel C A, Hawley G O (eds.) 1973 *The Encyclopedia of Chemistry*, 3rd edn. Reinhold, New York

International Lead Zinc Research Organization 1988 *Engineering Properties of Zinc Alloys*, 3rd edn. Research Triangle Park, NC

Jolly J H 1997 *US Zinc Industry, A History, Statistics, and Glossary*. American Literary Press, Baltimore, MD

Lynch R F 1987 Hot-dip galvanizing alloys. *J. Metals* **39**(8), 39–41

Machler M 1996 Zinc alloy enhances strength and creep resistance. *Adv. Mater. Process.*

Marder A R 1990 Microstructural characteristics of zinc coatings In: *Conf. on Zinc-Based Steel Coating Systems:*

Metallurgy and Performance. Minerals, Metals & Materials Society, Warrendale, PA

Mathewson C H 1964 *Zinc, The Science and Technology of the Metal, Its Alloys and Compounds*. Rheinhold, New York

Morgan S W K 1985 *Zinc and Its Alloys and Compounds*. Wiley, New York

Porter F 1991 *Zinc Handbook, Properties, Processing and Use in Design*. Marcel Dekker, New York

Raznjevic K 1976 *Handbook of Thermodynamic Tables and Charts*, Hemisphere. Washington, DC

Slunder C J, Boyd W K 1979 *Zinc: Its Corrosion Resistance*. Zinc Institute, New York

Smithells C J 1976 *Metals Reference Book*, 5th edn. Metals Society, London

Weast R C (ed.) 1979 *CRC Handbook of Chemistry and Physics*. CRC, Boca Raton, FL

R. F. Lynch

LIST OF CONTRIBUTORS

Contributors are listed in alphabetical order together with their addresses. Titles of articles that they have authored follow in alphabetical order. Where articles are coauthored, this has been indicated by an asterisk preceding the title.

Agassant, J. F.
UMR Ecole des Mines de Paris
Sophia Antipolis
France
Polymer Processing: A Modeling Approach

Allahverdi, M.
Rutgers University
Piscataway, New Jersey
USA
Electroceramics: Rapid Prototyping
* *Rapid Prototyping*

Apelian, D.
Worcester Polytechnic Institute
Massachusetts
USA
Particulate Processing (Powder Metallurgy)

Askeland, D. R.
University of Missouri-Rolla
Missouri
USA
Lost Foam Casting (Disposable Mold)

Atkins, A. G.
University of Reading
UK
Metal Forming: Formability

Avitzur, B.
Metalforming Incorporated
Allentown, Pennsylvania
USA
Metal Working: Drawing of Bars and Wires

Baaijens, F. P. T.
Eindhoven University of Technology
The Netherlands
Polymer Injection Molding: Modeling for Properties
Polymer Rheology for Melt Processing: Complex Flows

Babu, S. S.
Oak Ridge National Laboratory
TN
USA
Welding

Barnett, M. R.
Deakin University
Geelong, Victoria
Switzerland
Metal Forming (Hot): Microstructural Aspects

Belforte, D. A.
Belforte Associates
Sturbridge, Massachusetts
USA
Laser Cutting

Blendell, J. E.
National Institute of Standards and Technology, Gaithersburg
Maryland
USA
Solid-state Sintering

Boerstoel, H.
Teijin Twaron
Arnhem
The Netherlands
Processing Rigid Polymers to High Performance Fibers

Bourban, P.-E.
Ecole Polytechnique Fédérale de Lausanne
Switzerland
Polymer Matrix Composites: Matrices and Processing

Brauns, J.
Modification of Wood: Mechanical Properties and Application

Brown, C. A.
Worcester Polytechnic Institute
Worcester, Massachusetts
USA
Machining of Metals: Fundamentals

Butler, W. A.
Bloomington, Indiana
USA
Die Casting (Permanent Mold)

Cadden, C. H.
Brazing

Cahn, R. W.[†]
University of Cambridge
UK
Mechanical Milling

Calvert, P.
University of Arizona
Tucson, Arizona
USA
Biomimetic Materials: Properties and Processing
Freeform Fabrication

Caron, R. N.
Olin Corporation
New Haven, Connecticut
USA
Copper Alloys: Thermal and Thermomechanical Processing

Cascone, P.
Gold Casting Alloy, New, for Dental Applications

Charoensirisomboon, P.
BASF
Ludwigshafen
Germany
Reactive Polymer Processing: Interfacial Aspects

Cho, H.
Seoul National University
North Korea
Grinding

Claussen, N.
Technische Universitat Hamburg–Harburg
Hamburg
Germany
Reaction Forming

Coates, P. D.
University of Bradford
UK
Polymer Processing: On-line Monitoring

Cramb, A. W.
Carnegie Mellon University
Pittsburgh, PA
USA
Recent Progress in the Continuous Casting of Steel
Strip Casting of Steels: Current Developments

Crook, P.
Haynes International Inc.
Kokomo, Indiana
USA
Cobalt Alloys: Alloying and Thermomechanical Processing

[†]deceased.

Dahle, A. K.
University of Queensland
Brisbane
Australia
Aluminum Alloys, Heat Treatment of

Danforth, S. C.
Rutgers University
Piscataway, New Jersey
USA
Rapid Prototyping

Dautzenberg, J. H.
Eindhoven University of Technology
Netherlands
Metal Forming at High Energy and High Velocity

David, S. A.
Oak Ridge National Laboratory
TN
USA
Welding

Delplanque, J.-P.
Colorado School of Mines
Golden, Colorado
USA
Spray Casting: Applications

Dilthey, U.
RWTH Aachen
Germany
Joining of Metals

Doege, E.
University of Hanover
Germany
Metal Forming: Close Die Forging
Metal Working: Stretching of Sheets

Doherty, R. D.
Drexel University
Philadelphia, Pennsylvania
USA
Primary Recrystallization
Spray Casting: Fundamentals

Dörr, T.
Technical University of Brandenburg at Cottbus
Cottbus
Germany
Mechanical Surface Treatments for Decorative Purposes

Dreyer, G.
University of Hanover
Germany
Metal Forming: Close Die Forging

Easler, T. E.
COI Ceramics Inc.
San Diego, California
USA
Green Machining

Edelstein, A. S.
Naval Research Laboratory
Washington, DC
USA
Nanomaterials

Edirisinghe, M. J.
Queen Mary, University of London
UK
Injection Molding

Eisenblätter, G.
RWTH Aachen
Germany
Machining: Wear of Tools

Elbing, F.
Technical University Berlin
Germany
Machining and Nonmechanical Shaping: A Comparison

Elemans, P. H. M.
DSM Research
Geleen
The Netherlands
Polymer Twin Screw Extrusion: 2D Modeling

Ewsuk, K. G.
Sandia National Laboratories
Albuquerque, New Mexico
USA
Powder Granulation and Compaction

Fangmeier, R.
R. Fangmeier Engineering
Solingen
Germany
Metal Working: Straightening

Fiory, A. T.
Agere Systems
Murray Hill, New Jersey
USA
Rapid Thermal Annealing

Froes, F. H.
University of Idaho
Moscow, Idaho
USA
Titanium Alloys: Thermal Treatment and Thermomechanical Processing

Frost, T.
Institut für Werkzeugmaschinen und Fabrikbetrieb, Technische Universität Berlin
Germany
Cutting and Drilling of Metals and Other Materials: A Comparison

Funatsu, K.
Kyushu University
Fukuoka Japan
Polymer Twin Screw Extrusion: 3D Modeling

Ganguly, C.
Nuclear Fuel Complex
Hyderabad
India
Nuclear Reactor Fuel Fabrication (Including Quality Control)

Gedanken, A.
Bar-Ilan University
Ramat-Gan
Israel
Ultrasonic Processing to Produce Nanoparticles

Geiger, M.
Lehrstuhl für Fertigungstechnologie, Erlangen
Germany
Laser Forming of Metals

German, R. M.
Pennsylvania State University
University Park, Pennsylvania
USA
Liquid Phase Sintering: Metals
Sintering
Sintering: Modeling

Gottstein, G.
RWTH Aachen
Germany
Metal Forming (Cold): Dislocation Mechanisms and Microstructural Changes

Green, D. J.
Pennsylvania State University
University Park, Pennsylvania
USA
Ceramics and Glasses, Tempering of: Residual Stresses
Porous Ceramic Processing

Großmann, K.
Technische Universität, Dresden
Germany
Machining and Precision

Grünschloss, E.
Universität Stuttgart
Stuttgart
Germany
Polymer Single Screw Extrusion: Modeling

Haagh, G. A. A. V.
Eindhoven University of Technology
Netherlands
Polymer Injection Molding: Gas-assisted Molding

Haferkamp, H.
Universität Hannover
Germany
Deformation Processing and Primary Shaping of Ceramics and Polymers: Comparison with Metals

Hallfeld, T.
Hannover
Germany
Metal Working: Stretching of Sheets

Hauck, J.
Technische Universität Aachen
Germany
Polymer Processing, Process Control of

Hennige, T.
Lehrstuhl für Fertigungstechnologie
Erlangen Germany
Laser Forming of Metals

Hodgson, P. D.
Deakin University
Geelong, Victoria
Switzerland
Metal Forming (Hot): Microstructural Aspects

Hogg, R.
Pennsylvania State University
University Park, Pennsylvania
USA
Grinding

Hu, G. H.
Institut National Polytechnique de Lorraine
Nancy
France
Reactive Polymer Processing: Fundamentals of REX

Hudd, R. C.
Porthcawl
UK
Sheet Steel: Low Carbon

Inoue, T.
Yamagata University
Yonezawa
Japan and
Toyota Central R&D Laboratories
Nagakute, Aichi
Japan
Polymer Injection Molding: 3D Modeling
Reactive Polymer Processing: Interfacial Aspects

Jones, A. R.
University of Liverpool
UK
Mechanical Alloying

Jorstad, J. L.
J. L. J. Technologies, Inc.
Richmond, Virginia
USA
Casting of Semisolid Metals: Engineering Applications

Kanamoto, T.
Science University of Tokyo
Japan
Solid-state Extrusion to High Performance Fibers

Kennedy, P.
Moldflow Pty. Ltd.
Victoria
Australia
Polymer Injection Molding: Moldflow

Kihara, S.
Kyushu University
Fukuoka
Japan
Polymer Twin Screw Extrusion: 3D Modeling

Kikutani, T.
Tokyo Institute of Technology
Japan
Polymer Fiber Processing: Modeling
Polymer Fibers: Formation and Structure

Kirkwood, D. H.
University of Sheffield
UK
Semisolid Processing

Kitagawa, T.
Toyobo Research Center
Shiga
Japan
Polymer Fiber Processing: High-performance Fibers

Klarstrom, D.
Haynes International Inc.
Kokomo, Indiana
USA
Cobalt Alloys: Alloying and Thermomechanical Processing

Klocke, F.
Rheinisch-Westfälische Technische Hochschule
Aachen
Germany
Machining and Machinability
Machining: Wear of Tools

Koehler, W.
DaimlerChrysler Aerospace Dornier
Friedrichshafen
Germany
Metal Forming: Specialized Procedures for the Aircraft Industry

Komanduri, R.
Mechanical & Aerospace Engineering
Oklahoma State University
USA
Cutting-tool Materials

Kopp, J.
Institute of Plastics Processing
Aachen
Germany
Forming Procedures in Polymer Component Manufacturing

Kopp, R.
Institut für Bildsame Formgebung
Aachen
Germany
Metal Forming: General and Economic Aspects

Krauss, G.
Colorado School of Mines
Golden, Colorado
USA
Ferrous Metals, Heat Treatment of

Krieg, T.
RWTH Aachen
Germany
Machining: Wear of Tools

Kuzman, K.
University of Ljubljana
Slovenia
Metal Working: Cold Extrusion

Kwon, O.-H.
Saint-Gobain Ceramics and Plastics, Inc.
Northborough, Massachusetts
USA
Liquid Phase Sintering: Ceramics

Lavernia, E. J.
University of California
Irvine, California
USA
Spray Casting: Applications

Lawley, A.
Drexel University
Philadelphia, Pennsylvania
USA
Atomization

Lemstra, P. J.
Dutch Polymer Institute/Eindhoven University of Technology, Eindhoven
The Netherlands
Processing Flexible Polymers to High Performance Fibers

Lynch, R. F.
Lynch & Associates Inc.
Wyckoff, New Jersey
USA
Zinc: Alloying, Thermomechanical Processing, Properties, and Applications

Månson, J. -A. E.
Ecole Polytechnique Fédérale de Lausanne
Switzerland
Polymer Matrix Composites: Matrices and Processing

Marano, R. S.
Ford Motor Co.
Dearborn, Michigan
USA
Cutting Fluids for Machining

Marder, J. M.
Isotech LLC
Euclid, Ohio
USA
Beryllium: Alloying, Thermomechanical Processing, Properties and Applications

Martin, J. W.
University of Oxford
UK
Directional Recrystallization

Matsuoka, T.
Toyota Central R&D Laboratories
Nagakute, Aichi
Japan
Polymer Injection Molding: 3D Modeling

Matthews, F. L.
Imperial College of Science
Technology and Medicine
London
UK
Composites, Joining of

Maurer, G.
Special Metals Corporation
New Hartford, New York
USA
Solidification of Superalloys

McLeish, T.
University of Leeds
UK
Polymer Rheology for Melt Processing: Molecular Modeling

McQuay, P. A.
Howmet Research Corporation
Whitehall, Michigan
USA
Casting of Intermetallics

McQueen, H. J.
Concordia University
Montreal, Quebec
Canada
Dynamic Recovery and Recrystallization

Michaeli, E. W.
Rheinisch Westfalische Technische Hochschule
Aachen
Germany
Forming Procedures in Polymer Component Manufacturing
Polymer Processing, Process Control of

Misiolek, W. Z.
Lehigh University
Bethlehem, Pennsylvania
USA
Metal Working: Drawing of Bars and Wires

Mistler, R. E.
Morrisville
Pennsylvania
USA
Tape Casting and Lamination

Mitsoulis, E.
National Technical University of Athens
Greece
Polymer Film Casting: Modeling

Montmitonnet, P.
Ecole Nationale Supérieure des Mines de Paris
Sophia-Antipolis
France
Metal Working: Cold Rolling

Morrell, J. J.
Oregon State University
Corvallis, Oregon
USA
Wood: Chemical Treatment Processes

Müller, B.
Lehrstuhl für Fertigungstechnologie
Erlangen
Germany
Laser Forming of Metals

Müller, G.
Universität Erlangen-Nürnburg
Erlangen
Germany
Crystal Growth from the Melt

Müller, K.
Technische Universität Berlin
Germany
Metal Working: Extrusion

Naslain, R. R.
Université Bordeaux *1*
Pessac
France
Ceramic Matrix Composites: Matrices and Processing

Nes, E.
Norwegian University of Science and Technology
Trondheim
Norway
Primary Recrystallization in Two-phase Alloys

Northolt, M. G.
AKZO Nobel, Arnhem
The Netherlands
Processing Rigid Polymers to High Performance Fibers

Ober, C. K.
Cornell University
Ithaca
New York
USA
Processing Polymers in Supercritical CO_2

Padmanabhan, K. A.
Indian Institute of Technology
Kanpur
India
Metal Forming at Very Low Strain Rates

Pape, J.
Universität-Paderborn
Germany
Polymer Single Screw Extrusion: Software Packages

Paton, N. E.
Howmet Research Corporation
Whitehall, Michigan
USA
Investment Casting (Disposable Mold)

Peters, G. W. M.
Eindhoven University of Technology
The Netherlands
Polymer Rheology for Melt Processing: Complex Flows

Phulé, P. P.
University of Pittsburgh
Pennsylvania
USA
Electroceramics, Colloidal Stability and Processing of

Picken, S. J.
Delft University of Technology
The Netherlands
Processing Rigid Polymers to High Performance Fibers

Piwonka, T. S.
University of Alabama
Tuscaloosa, Alabama
USA
Sand Casting (Disposable Mold)

Plege, B.
DaimlerChrysler Aerospace Dornier
Friedrichshafen
Germany
Metal Forming: Specialized Procedures for the Aircraft Industry

Plummer, C. J. G.
Ecole Polytechnique Fédérale de Lausanne
Switzerland
Polymer Matrix Composites: Matrices and Processing

Pollinger, J. P.
AlliedSignal Ceramic Components
Torrance, California
USA
Gel Casting

Potente, H.
Universität-Paderborn
Germany
Polymer Single Screw Extrusion: Software Packages

Rabinkin, A.
Honeywell Int.
New Jersey
USA
Brazing Filler Metals

Rastogi, S.
Dutch Polymer Institute/Eindhoven University of Technology, Eindhoven
The Netherlands
Processing Flexible Polymers to High Performance Fibers

Rocens, K.
Modification of Wood: Mechanical Properties and Application

Rudolph, P.
Institut für Kristallzüchtung
Germany
Crystal Growth from the Melt

Ruesch, S.
University of Hanover
Germany
Metal Forming: Close Die Forging

Safari, A.
Rutgers University
Piscataway, New Jersey
USA
Electroceramics: Rapid Prototyping
Rapid Prototyping

Sahm, K. F.
DaimlerChrysler Aerospace Dornier
Friedrichshafen
Germany
Metal Forming: Specialized Procedures for the Aircraft Industry

Samarasekera, I. V.
Centre for Metallurgical Process Engineering and University of British Columbia
Vancouver
Canada
Hot Rolling

Sandhage, K. H.
Ohio State University
Columbus, Ohio
USA
Reaction Forming

Scherer, G. W.
Princeton University
New Jersey
USA
Viscous Sintering

Schilling, C. H.
Ames Laboratory and Iowa State University,
Ames, Iowa
USA
Colloid Casting
Plastic Forming

Schmoeckel, D.
Stuttgart
Germany
Metal Forming (Warm): Comparison with Hot and Cold Forming

Schneider, M. H.
Polymer Modified Wood

Scholtes, B.
Universität Gh Kassel
Germany
Mechanical Surface Treatment

Sikka, V. K.
Oak Ridge National Laboratories
Oak Ridge, Tennessee
USA
Casting of Intermetallics

Sisson, R. D. Jr.
Worcester Polytechnic Institute
Massachusetts
USA
Heat Treatment

Smith, P.
Eidgenössische Technische Hochschule Zurich
Switzerland
Conducting Polymers, Solution- and Gel-processing of

Smolenski, D. J.
General Motors Corporation
Detroit, Michigan
USA
Cutting Fluids for Machining

Spur, G.
Technical University Berlin
Germany
Machining and Nonmechanical Shaping: A Comparison

Starke,, E. A. Jr.
University of Virginia
Charlottesville, Virginia
USA
Aluminum Alloys: Thermomechanical Processing

Suryanarayana, C.
University of Central Florida
Orlando, Florida
USA
Rapid Solidification Processing

Tönshoff, H. K.
Institut für Fertigungstechnik und Spanende Werkzeugmaschinen
Hannover
Germany
Machining: Generalities and Economics

Tandon, R.
Phillips Metal Injection Molding
Menomonie, Wisconsin
USA
Metal Injection Molding

Tatartchenko, V. A.
CRISMATEC, Nemours
France
Growth of Shaped Crystals from the Melt

Thomas, B. G.
University of Illinois
Urbana, Illinois
USA
Continuous Casting
Continuous Casting: Complex Models

Turng, L.-S.
C-MOLD
Ithaca, New York
USA
Polymer Injection Molding: C-Mold

Twiname, E. R.
Morrisville, Pennsylvania
USA
Tape Casting and Lamination

Uhlmann, E.
Institut für Werkzeugmaschinen und Fabrikbetrieb, Technische Universität Berlin
Germany
Cutting and Drilling of Metals and Other Materials: A Comparison

Venkataraman, N.
Rutgers University
Piscataway, New Jersey
USA
Rapid Prototyping

Verbeeten, W. M. H.
Eindhoven University of Technology
The Netherlands
Polymer Rheology for Melt Processing: Complex Flows

Vitek, J. M.
Oak Ridge National Laboratory
TN
USA
Welding

Voehringer, O.
Universität Karlsruhe (TH)
Germany
Mechanical Surface Treatment

Vollertsen, F.
Universität Paderborn
Germany
Metal Forming: Microparts

Wagner, L.
Technical University of Brandenburg at Cottbus
Cottbus
Germany
Mechanical Surface Treatments for Decorative Purposes

Wang, K. K.
Cornell University
Ithaca, New York
USA
Polymer Injection Molding: C-Mold

Weber, J. H.
Special Metals Corporation
Huntington, West Virginia
USA
Nickel Alloys: Thermal Treatment and Thermomechanical Processing
Nickel-based Superalloys: Alloying Methods and Thermomechanical Processing

Weder, C.
Case Western Reserve University
Cleveland, Ohio
USA
Conducting Polymers, Solution- and Gel-processing of

Weibel, G.
Cornell University
Ithaca, New York
USA
Processing Polymers in Supercritical CO_2

Westengen, H.
Norsk-Hydro ISI
Porsvrunn
Norway
Magnesium Alloys: Thermal Treatment and Thermomechanical Processing

White, D.
Ford Motor Corporation
Dearborn, Michigan
USA
Rapid Prototyping Processes

Wise, R. J.
TWI, Cambridge
UK
Polymer Glasses, Welding of: Processes and Mechanisms

Zanner, F. J.
Sandia Park
NM
USA
Vacuum Melting

Zheng, R.
Moldflow Pty. Ltd.
Victoria
Australia
Polymer Injection Molding: Moldflow

Zoetelief, W. F.
DSM Research
Geleen
Netherlands
Polymer Injection Molding: Multimaterial Molding

SUBJECT INDEX

The Subject Index has been compiled to assist the reader in locating all references to a particular topic in the Encyclopedia. Index entries are filed in letter-by-letter sequence, and arranged in set-out style with up to three levels of heading. Major coverage of a subject is indicated by ***bold italic*** page numbers. Page numbers followed by *T* (or *F*) refer to Tables (or Figures). *See* references point the user to preferred terms; *see also* references indicate terms of related interest. Every effort has been made to make the index as comprehensive as possible and to standardize the terms used.

A

B

C

D

E

F

G

H

I

J

K

L

M

N

O

P

Q

R

S

U

V

W

X

Y

Z